INTERMEDIATE ALGEBRA

LAURA J. BRACKEN
LEWIS-CLARK STATE COLLEGE

EDWARD S. MILLER
LEWIS-CLARK STATE COLLEGE

BROOKS/COLE
CENGAGE Learning

Australia • Brazil • Japan • Korea • Mexico • Singapore • Spain • United Kingdom • United States

Intermediate Algebra
Laura J. Bracken, Edward S. Miller

Senior Publisher: Charlie Van Wagner

Senior Developmental Editors: Erin Brown,
　Katherine Greig, Rita Lombard

Assistant Editor: Lauren Crosby

Senior Editorial Assistant: Jennifer Cordoba

Managing Media Editor: Heleny Wong

Associate Media Editor: Guanglei Zhang

Senior Brand Manager: Gordon Lee

Market Development Manager: Danae April

Senior Content Project Manager: Cheryll
　Linthicum

Senior Art Director: Vernon Boes

Senior Manufacturing Planner: Becky Cross

Rights Acquisitions Specialist: Roberta Broyer

Production Service: Martha Emry

Photo Researcher: Terri Wright

Text Researcher: Terri Wright

Copy Editor: Barbara Willette

Art Editor: Leslie Lahr

Illustrator: Chris Ufer, Graphic World, Inc.

Text Designer: Diane Beasley

Cover Designer: Morris Design

Cover Image: © Jiang Hongyan/Shutterstock

Compositor: Lachina Publishing Services

For product information and technology assistance, contact us at
Cengage Learning Customer & Sales Support, 1-800-354-9706.

For permission to use material from this text or product,
submit all requests online at **cengage.com/permissions.**
Further permissions questions can be emailed to
permissionrequest@cengage.com.

Library of Congress Control Number: 2012946937
ISBN-13: 978-0-618-94668-6
ISBN-10: 0-6-189-4668-3

Brooks/Cole
20 Davis Drive
Belmont, CA 94002-3098
USA

Cengage Learning is a leading provider of customized learning solutions with office locations around the globe, including Singapore, the United Kingdom, Australia, Mexico, Brazil, and Japan. Locate your local office at **www.cengage.com/global.**

Cengage Learning products are represented in Canada by Nelson Education, Ltd.

To learn more about Brooks/Cole, visit **www.cengage.com/brookscole.**

Purchase any of our products at your local college store or at our preferred online store **www.cengagebrain.com.**

Printed in the United States of America
1 2 3 4 5 6 7 16 15 14 13 12

To Dr. Micheal Vernon—mathematician,
teacher, colleague, mentor,
and friend

About the Authors

LAURA BRACKEN, co-author of *Investigating Prealgebra and Investigating College Mathematics*, teaches developmental mathematics at Lewis-Clark State College. As developmental math coordinator, Laura led the process of developing objectives, standardizing assessments, and enforcing placement including a mastery skill quiz program. She has worked collaboratively with science faculty to make connections between developmental math and introductory science courses. Laura has presented at numerous national and regional conferences and currently serves as the regional representative for the AMATYC Placement and Assessment Committee. Her blog, *Dev Math Diary*, is at http://devmathdiary.wordpress.com/

ED MILLER is a professor of mathematics at Lewis-Clark State College. He earned his PhD in general topology at Ohio University in 1989. He teaches a wide range of courses, including elementary and intermediate algebra. His terms as chair of the General Education Committee, the Curriculum Committee, and the Division of Natural Sciences and Mathematics have spurred him to look at courses as part of an integrated whole rather than discrete units. A regular presenter at national and regional meetings, Ed is exploring the use of multiple choice questions as a teaching tool as well as an assessment tool.

About the Cover

The cover image symbolizes the importance of self-empowerment or "stepping up" in order to achieve your goals. Success requires independence, persistence, responsibility, and a willingness to put your "best foot" forward—all key skills this developmental mathematics series seeks to foster.

Dear Colleagues,

Several years ago, we began writing materials to supplement our developmental mathematics textbooks and improve student success in our classes. Although we work at a four-year state college, it essentially offers open admission, serving as a community college and technical school for a large region. About 60% of our students must take developmental mathematics; many are enrolled in developmental English courses and are first-generation college students.

At our school and many others, Intermediate Algebra is the prerequisite for College Algebra or Pre-calculus as well as other courses that are application intensive. To be successful in upcoming courses, Intermediate Algebra students must develop both conceptual understandings and procedural skills in algebra, functions, and problem solving. They also need to learn about the culture of college. In support of these goals, we developed features and content for our textbook that include:

- An early introduction of function and function concepts. To improve students' understanding of functions, we introduce the concepts of domain, range, function notation, graphing, maximum, minimum, real zero, and translation at a basic level in Chapter 3. We then revisit these characteristics and concepts in successive chapters as other function families are introduced.
- A five-step problem-solving plan based on the work of George Polya to organize student work on applications. Rather than memorize solutions to individual problems, the five steps help students develop strategies both for solving problems and for reflecting about the reasonability of their answers.
- Short readings called Success in College Mathematics along with follow-up exercises that address topics of relevance to students who are making the transition to college-level courses and help instructors identify issues outside of mathematics that may affect student success and retention.
- Examples with detailed, step-by-step explanation. Since students often rely heavily on examples when working on assignments and studying, each step of the detailed worked examples is accompanied by an annotation that details the mathematical operation performed. Color-coding and bold type are also used to help enhance the explanation of the work performed in each step of the solution.
- End-of-chapter Study Plan and Review Exercises. These help guide students through test or exam preparation, step by step. The Study Plan includes study tables for each section in the chapter along with "Can I . . ." study questions that are linked to Review Exercises. References to Examples and Practice Problems from within the sections are also included for additional review, if needed.
- Algorithmic online homework capabilities through Enhanced WebAssign® that include unique question types from the textbook as well as an interactive and customizable eBook (YouBook).
- Tools for instructors, such as Teaching Notes in the Annotated Instructor's Edition, along with Classroom Examples that mirror the examples provided in the student textbook. PowerLecture is also available to instructors, and it contains ExamView® algorithmic computerized testing, Solution Builder (link to the complete textbook solutions), PowerPoint slides including selected worked examples from the textbook along with the corresponding Classroom Examples, and other resources for reference and instructional support.

We welcome your feedback as you use these materials. If you have any questions or comments please contact us at **bracken@lcsc.edu** and **edmiller@lcsc.edu**. We also invite you to follow our blog at **http://devmathdiary.wordpress.com/**

All the best to you and to your students.

Laura Bracken and Ed Miller

What Instructors and Students Are Saying About Bracken and Miller's *Intermediate Algebra*

"I really like [Bracken and Miller's *Intermediate Algebra*]. . . . The book delivers on what it promises in every aspect."

—Shahrokh Parvini, *San Diego Mesa College*

"[*Intermediate Algebra*] deserves high marks for clarity, explicitness, and readability."

—Haile Haile, *Minneapolis Community and Technical College*

"I like the revisiting of topics [early-and-often functions] because students think they understand until they are asked the same question two days later. It will help to have the consistent review. I have always believed in lots of review."

—Barbara Biggs, *Utah Valley University*

"The fact that the applications include excerpts from current research and reports provides the real-life context that the authors desire as well as increasing student interest, and the number of problems with extraneous information will force students to read carefully for important information. These problems do not appear periodically; they are intentional and consistent."

—Shelly Hansen, *Colorado Mesa University*

"I love these [Success in College Mathematics]. I think that many students take developmental math (and need to repeat it so much) because they're lacking in study skills and/or experiencing a transition-to-college culture shock. This feature is very useful in reminding the student that the burden of learning is on them, and it also helps the students be more aware of their own learning."

—Daniel Kleinfelter, *College of the Desert*

"The five-step problem-solving plan is outstanding. The real-world applications are easier solved if this five-step strategy is used. This is one of the strongest parts of the textbook."

—Nicoleta Bila, *Fayetteville State University*

"Problem-solving plans help me organize data to complete problems."
"Very helpful. Gives great examples and explains problems very well."
"This book is easy to understand. It's very helpful and gives clear examples."

—Class Test Students, *North Seattle Community College*

Brief Contents

Contents

CHAPTER 4	**Systems of Linear Equations**	**297**

SUCCESS IN COLLEGE MATHEMATICS
Finding and Learning from Mistakes   297

CHAPTER 5	**Rational Expressions, Equations, and Functions**	**401**

SUCCESS IN COLLEGE MATHEMATICS
Making Connections   401

CHAPTER 9

Conic Sections and Systems of Nonlinear Equations

781

CHAPTER 10

Sequences, Series, and the Binomial Theorem

849

APPENDIX

Index of Applications

Page numbers in black represent where the given entries appear in end-of-section exercises.
Page numbers in **boldface** represent where the given entries appear in Examples.
Page numbers in blue represent where the given entries appear in end-of-chapter exercises.
Page numbers in red represent where the given entries appear in Practice Problems.

Page numbers in black represent where the given entries appear in end-of-section exercises.
Page numbers in **boldface** represent where the given entries appear in Examples.
Page numbers in blue represent where the given entries appear in end-of-chapter exercises.
Page numbers in red represent where the given entries appear in Practice Problems.

Preface

Our goal in writing *Intermediate Algebra* is to share strategies and materials that we have developed to prepare our students for success in college-level mathematics courses. We believe that successful students must develop both conceptual understanding and procedural fluency. They must improve their ability and confidence in solving application problems. They frequently need to develop new habits of self-sufficiency and personal responsibility.

Success in college-level mathematics courses often depends on an understanding of function. Experience in our classrooms showed us that students build a much more robust understanding of function if domain, range, maximum, minimum, and real zeros are introduced early in the course, at a very basic level, and then are revisited with increasing detail throughout the course. The use of multiple representations helps students build both algebraic skills and conceptual understanding of these concepts. Although it may not be traditional to introduce real zeros and maximum/minimum this early in Intermediate Algebra, our experience has shown that it makes a significant difference in developing a conceptual rather than just a procedural understanding of functions. More opportunities to practice and make connections make a difference.

To be successful in a college-level math course, students must also be able to decode and solve a wide variety of application problems. Since many of our students struggled with even starting the process, we developed a five-step problem-solving plan based on the work of George Polya. We have included many opportunities for students to practice problem solving throughout the text and use contexts that are familiar from daily life, pertain to various careers, or are relevant to their other academic courses.

Being successful in college requires more than learning academic content. Integrating student success information in courses like Intermediate Algebra can help students feel part of a community, learn coping skills, and result in increased persistence and retention. We wrote the Success in College Mathematics readings and corresponding exercises to help students learn about the culture of college and to consider strategies for improving their performance.

When students pass an intermediate algebra course with the skills, knowledge, and attitudes prerequisite for a college-level math class, they are empowered to move forward to a degree or certificate. Our experience class testing this textbook tells us that Intermediate Algebra can be an important part of preparing students for that future.

Step Up to Success

Success in College Mathematics

To help students transition to college-level mathematics, Success in College Mathematics appear at the beginning of each chapter and address topics such as personal responsibility, study skills, and time management. Follow-up exercises appear at the end of each section to help students reflect on their own attitudes and habits and how they can improve their performance.

Fundamentals of Algebra   1

SUCCESS IN COLLEGE MATHEMATICS
Personal Responsibility

College is a place where freedom and personal responsibility can lead to great accomplishments. Your long-term goal may be a degree, a certificate, or some other credential. A shorter-term goal is to pass your classes this term. An even shorter-term goal is to do well on your assignments and to pass the first test.
... many people at your college who want to help you achieve these goals.
... is a new world that offers great opportunities. To be successful, you need
... how college works. You also need to make choices and develop habits that
... ess. With freedom and choices comes personal responsibility. The only
... can do the work and learn the math is you.
... duction to each chapter in this book focuses on strategies for being
... e Intermediate Algebra. Each exercise set includes follow-up questions
... these strategies. In this first chapter, these questions assess your
... f your college and your math class.

1.1 Sets and Numerical Expressions
1.2 Algebraic Expressions
1.3 Equations and Inequalities in One Variable
1.4 Scientific Notation and Unit Analysis
1.5 Applications and Problem Solving
1.6 Slope and Linear Equations in Two Variables
1.7 Writing the Equation of a Line
1.8 Linear Inequalities in Two Variables

SUCCESS IN COLLEGE MATHEMATICS

77. Describe your instructor's policy about late assignments.

78. Some instructors include attendance, class participation, or participation in an on-line discussion board as part of the student's grade. Explain how, if at all, attendance or participation affects your grade.

Examples with Step-by-Step Explanation

Each step of the worked examples is accompanied by an annotation that explains how the solution progresses, from the first line to the final answer. Color-coding and **bold type** are also used to help students easily identify the operation that occurs in each step, enabling the text to act as tutor when students are not in the classroom.

EXAMPLE 15   (a) Solve: $-3x - 21 < 9$

SOLUTION ▶

$$-3x - 21 < 9$$
$$\underline{+21 \;+21}$$ Addition property of inequality.
$$-3x + \mathbf{0} < \mathbf{30}$$ Simplify.
$$\frac{\mathbf{-3}x}{-3} > \frac{30}{-3}$$ Division property of inequality; reverse the inequality sign.
$$x > -10$$ Simplify.

(b) Use a number line graph to represent the solution.

▶   ←—+—+—(—+—+—)—+—+—+—+—→ x
      −12 −10 −8 −6

(c) Use interval notation to represent the solution.

▶ $(-10, \infty)$

Side-by-Side Examples

To help students apply their knowledge of arithmetic to algebra, some of the worked examples are set up side by side with shared annotations. In this format, students can more readily see how certain principles and procedures are used repeatedly in both arithmetic and algebra.

EXAMPLE 1   Simplify: $\dfrac{15}{24}$ and $\dfrac{x^2 + 5x + 6}{x^2 - 9}$

SOLUTION ▶

$$\frac{15}{24} \qquad\qquad \frac{x^2 + 5x + 6}{x^2 - 9}$$

$$= \frac{5 \cdot 3}{8 \cdot 3} \qquad \text{Factor the numerator and denominator.} \qquad = \frac{(x + 2)(x + 3)}{(x + 3)(x - 3)}$$

$$= \frac{5 \cdot \cancel{3}}{8 \cdot \cancel{3}} \qquad \text{Find common factors.} \qquad = \frac{(x + 2)(x + \cancel{3})}{(x + \cancel{3})(x - 3)}$$

$$= \frac{5}{8} \cdot 1 \qquad \text{Simplify fractions that are equal to 1.} \qquad = \frac{x + 2}{x - 3} \cdot 1$$

$$= \frac{5}{8} \qquad \text{Simplify.} \qquad = \frac{x + 2}{x - 3}$$

Practice Problems

Following a set of worked examples in each section are a set of Practice Problems that mirror the examples. Students can use the Practice Problems to check their understanding of the concepts or skills presented.

Practice Problems

For problems 1–4, simplify.

1. $\dfrac{x^2 - 9}{x^2 - 7x + 12}$ 2. $\dfrac{9 - w^2}{w^2 - 7w + 12}$

3. $\dfrac{2x^2 + 5x + 3}{2x^2 - 7x - 15}$ 4. $\dfrac{54p^5qr^9}{72p^8q^2r}$

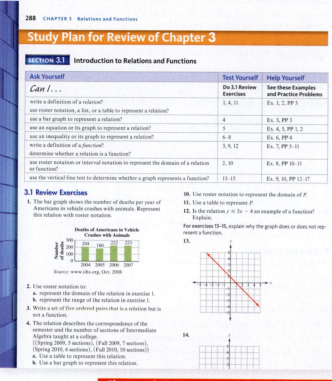

End-of-Chapter Study Plan and Review Exercises

This tool offers students an effective and efficient way to prepare for quizzes or exams. The Study Plan, appearing at the end of each chapter, includes study tables for each section that will help students get organized as they prepare for a test or quiz. Each table contains a set of *Can I . . .* questions, based on the section objectives, for self-reflection. *Can I . . .* questions are tied to one or more Review Exercises that students can use to test themselves. Examples and/or Practice Problems are also linked to each question should students need reference for review.

End-of-Chapter Test

Each chapter concludes with a sample test that students can take in preparation for an upcoming test or exam as a means to assess their understanding of the chapter.

End-of-Chapter Cumulative Review

To provide review during the course, Cumulative Reviews appear at the end of Chapters 3, 6, and 9. To assess readiness and provide practice for a final exam, a Cumulative Review for Chapters 1–10 appears after Chapter 10.

Problem Solving, Critical Thinking, and Reasoning

Five Steps for Problem Solving

Based on the work of George Polya, the Five Steps provide a framework for jump-starting and organizing student problem solving. Step-by-step worked examples throughout the textbook support students as they learn to solve a wide range of applications.

THE FIVE STEPS

Step 1 Understand the problem.

Step 2 Make a plan.

Step 3 Carry out the plan.

Step 4 Look back.

Step 5 Report the solution.

Checking for Accuracy and Reasonableness

To be successful in mathematics and other disciplines, an important skill for students to develop is checking their work for reasonability and accuracy. The explanations of reasonability in examples of the Five Steps and the checking of solutions in many of the worked examples provide a model for students learning to reflect on their own work. Being able to justify an answer helps students develop confidence and self-sufficiency.

Real Sources

Among the application problems appearing in this textbook are specially developed exercises and examples that reference information taken directly from news articles, research studies, and other fact-based sources. Many of these applications are set up in two parts with a problem statement given first followed by an excerpt. Since the question appears first and then the information needed to answer the question follows, students experience problem solving in a more true-to-life way. Though similar in length to texts or tweets, the authentic excerpts often contain more information than is needed to solve the applications. Students will need to think critically to select the relevant information, and in doing so practice the skills needed to solve problems outside of the classroom.

SOLUTION ▶ **Step 1 Understand the problem.**

What is unknown? The number of people with Twitter accounts

Assign the variable. N = number of people with Twitter accounts

Needed information. The survey polled 3018 people; 24% said they had Twitter accounts. In decimal form, 24% = 0.24.

Extraneous information. The information about Facebook and MySpace, the previous percent of people with Twitter accounts

Step 2 Make a plan.

Since number of parts = (percent)(total parts), a word equation is

number with Twitter accounts = (percent with Twitter)(total number)

Step 3 Carry out the plan.

number with Twitter accounts = (percent with Twitter)(total number)   Word equation.

$N = (0.24)(3018 \text{ people})$ Algebraic equation.

$N = \textbf{724.32} \text{ people}$ Simplify.

$N \approx \textbf{724} \text{ people}$ Round.

Step 4 Look back.

Is the answer reasonable? Working backwards, $\frac{724}{3018} \cdot 100$ equals 23.98. . .%, which rounds to 24%. Since this is equal to the percent in the problem, the answer seems reasonable.

What can we learn from this problem? When multiplying a number by a percent, rewrite the percent as a decimal number.

Step 5 Report the solution.

About 724 people said they had Twitter accounts.

EXAMPLE 4   (a) Use the zero product property to solve $4z^3 - 36z = 0$.

SOLUTION ▶
$$4z^3 - 36z = 0 \qquad \text{The greatest common factor is } 4z.$$
$$4z(z^2 - 9) = 0 \qquad \text{Factor out the greatest common factor.}$$
$$4z(z + 3)(z - 3) = 0 \qquad \text{Difference of two squares pattern.}$$

$4z = 0$ or $z + 3 = 0$ or $z - 3 = 0$ Zero product property.

$\dfrac{4z}{4} = \dfrac{0}{4}$ $\dfrac{-3\ -3}{z + 0 = -3}$ $\dfrac{+3\ +3}{z + 0 = 3}$ Properties of equality. Simplify.

$z = 0$ or $z = -3$ or $z = 3$ This equation has three solutions.

(b) Check.

▶ Check: $z = 0$ Check: $z = -3$ Check: $z = 3$

$4z^3 - 36z = 0$ $4z^3 - 36z = 0$ $4z^3 - 36z = 0$

$4(0)^3 - 36(0) = 0$ $4(-3)^3 - 36(-3) = 0$ $4(3)^3 - 36(3) = 0$

$\mathbf{0} - \mathbf{0} = 0$ $4(-27) + \mathbf{108} = 0$ $4(27) - \mathbf{108} = 0$

$\mathbf{0} = \mathbf{0}$ True. $\mathbf{0} = \mathbf{0}$ True. $\mathbf{0} = \mathbf{0}$ True.

EXAMPLE 2   In spring 2011, the 19th Biannual Youth Survey on Politics and Public Service polled 3018 U.S. citizens between the ages of 18 and 29. Find the number of the people surveyed who had Twitter accounts. Round to the nearest percent.

Over the past year, Millennial Facebook adoption has grown significantly from 64% to 80% (90% adoption among four-year college students), while MySpace has shed six percentage points over the same period. Although Twitter is clearly a less relevant tool for young adults than Facebook, Twitter accounts among young adults also rose over the past year from 15% to 24%. (*Source:* www.iop.harvard.edu, March 31, 2011)

Problem Solving: Practice and Review

Beginning in Chapter 2, each section-ending exercise set includes a short set of applications-based exercises called *Problem Solving: Practice and Review*. Because these problems do not involve the concepts or skills taught in the section, students need to think critically about the information and relationships in the problem. Even in sections that have few or no applications, students can continue to practice their problem-solving skills by completing one or more of these exercises.

Problem Solving: Practice and Review

Follow your instructor's guidelines for using the five steps as outlined in Section 1.5, p. 51.

97. Find the percent increase in athletic fees approved by the president. Round to the nearest percent.

In April 2008, in a general fee referendum, students at Fresno State voted, 777–412, against raising athletic fees to $50 per semester, from $7. Based on a recommendation from a campus fee advisory committee, John D. Welty, the Fresno State president, overrode the vote and approved an increase to $32 per semester for the 2008–9 academic year. (*Source:* www.nytimes.com, May 30, 2009)

98. Miami Dade College has eight campuses and over 300 programs of study. In fall 2008, Miami Dade College had 61,288 students taking classes for credit. This was a 4.9%

Find the Mistake

The section-ending exercise sets include *Find the Mistake* exercises. Each exercise is a problem and a step-by-step solution that includes one mistake. Students are asked to identify the error and then rework the problem correctly. These exercises help students learn how to find errors in their own work, which improves student persistence and self-sufficiency.

---

104. Problem: Find the slope of the line that passes through $(6, -8)$ and $(4, 7)$.

Incorrect Answer: $m = \dfrac{y_2 - y_1}{x_2 - x_1}$

$$m = \frac{7 - 8}{4 - 6}$$

$$m = \frac{-1}{-2}$$

$$m = \frac{1}{2}$$

Concept Development

Learning the Language of Math

Vocabulary matching exercises that appear before the section-ending exercises help students improve their knowledge of vocabulary and notation.

5.1 VOCABULARY PRACTICE

Match the term with its description.

1. The result of division
2. We can write each number in this set as a fraction in which the numerator and denominator are integers.
3. All numbers in this set are nonrepeating, nonterminating decimal numbers.
4. A polynomial with one term
5. When the numerator and denominator of an expression have no common factors
6. The quotient of two polynomials
7. This set of numbers includes the set of whole numbers and their opposites.
8. When we multiply these together, the result is a product.
9. This set of numbers is the union of the rational numbers and the irrational numbers.
10. Division by zero

A. factors
B. integers
C. irrational numbers
D. lowest terms
E. monomial
F. quotient
G. rational expression
H. rational numbers
I. real numbers
J. undefined

Writing Exercises

Integrated within the Practice Problems, end-of-section Exercises, and end-of-chapter Reviews, Tests, and Cumulative Reviews are exercises that require a written response. Students must reflect on concepts presented within the section in order to form a written response in their own words.

45. Explain why we cannot use the zero product property to solve *all* polynomial equations that are equal to 0.
46. Many quadratic equations in one variable have two solutions. The equation $x^2 + 8x + 16 = 0$ has only one solution. Explain why.

Multiple Representations

Algebraic concepts are often presented in conjunction with a graph and/or a table. By presenting relationships in multiple ways, students are more likely to build connections between their concrete understandings and the abstract language of algebra.

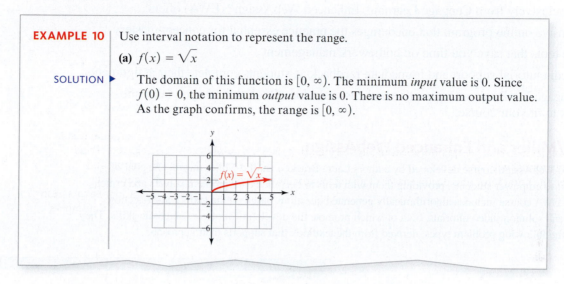

EXAMPLE 10 Use interval notation to represent the range.

(a) $f(x) = \sqrt{x}$

SOLUTION ▶ The domain of this function is $[0, \infty)$. The minimum *input* value is 0. Since $f(0) = 0$, the minimum *output* value is 0. There is no maximum output value. As the graph confirms, the range is $[0, \infty)$.

Using Technology

For instructors who wish to integrate calculator technology, optional scientific and graphing calculator instruction is provided at the end of selected sections, where appropriate. The instruction that appears within the Using Technology boxes includes examples with keystrokes, screen shots, and Practice Problems. Follow-up *Technology* exercises appear at the end of section-ending exercises for continued practice.

Using Technology: The Minimum or Maximum Output Value of a Quadratic Function

The greatest or least output of a function is its **maximum** or **minimum output**.

EXAMPLE 11 Find the minimum output of $y = 2x^2 - 5x - 4$. Use interval notation to represent the range of this function. Round to the nearest thousandth.

Begin with a standard window. Press [Y=]. Type the function. Press [GRAPH]. Go to the CALC menu. Choose 3. Press [ENTER].

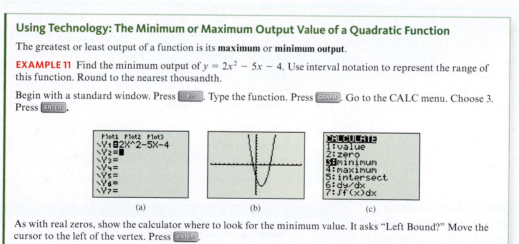

(a)   (b)   (c)

As with real zeros, show the calculator where to look for the minimum value. It asks "Left Bound?" Move the cursor to the left of the vertex. Press [ENTER].

Enhanced WebAssign®

The perfect homework management tool to reinforce the material in this textbook.

Available exclusively from Cengage Learning, Enhanced WebAssign® (EWA) offers:

- An extensive online program that encourages the practice essential for concept mastery
- Intuitive tools that save you time on homework management
- Multimedia tutorial support and immediate feedback as students complete their assignments
- The Cengage YouBook, an interactive and customizable eBook that lets you tailor the textbook to fit your course.

Bracken/Miller and Enhanced WebAssign

The Enhanced WebAssign course developed by authors Laura Bracken and Ed Miller mirrors the goal of their textbook: to empower students, providing them with tools to become better problem-solvers and critical thinkers. The EWA course includes algorithmically generated questions based on the text's end-of-section exercise sets and solution video tutorials, both of which promote the development of core algebraic skills. They also provide the following problem types, derived from the textbook that supports their approach.

1. The *Five Steps* for problem solving, based on the work of George Polya, is a framework the authors use to demonstrate how to solve a wide variety of application problems. By working through applications that involve the use of the five steps, students will learn how to follow an organized plan for solving problems, no matter their type.

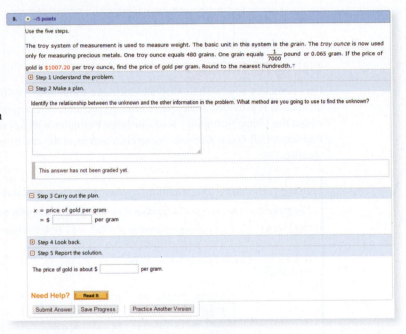

2. The *Find the Mistake* problems encourage students to think critically about their work and correct it as they analyze common errors and pitfalls. Drawn from section-ending exercises in the textbook, students are asked to identify the error and then rework the problem correctly. These exercises cultivate analytical skills and self-sufficiency.

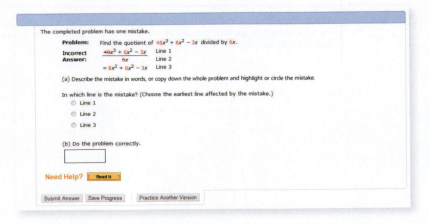

3. The *Success in College Mathematics* problems give students the chance to take charge of their learning. Based on the chapter opener narratives and corresponding end-of-section exercises from the textbook, these questions help students better understand college culture and address such topics as personal responsibility, study skills, and time management.

4. The *Solving Equations Exercises with Checks* encourage students to go through the analytical process of checking their solutions. These problems help promote self-sufficiency, an important skill students need at the college level.

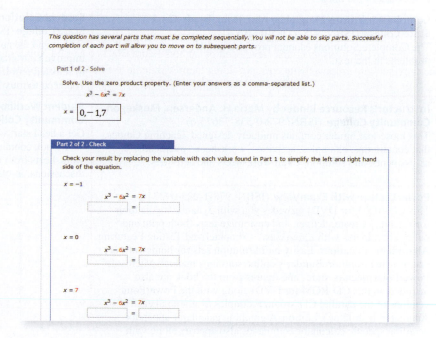

Other Enhanced WebAssign Resources

Also available for this course are interactive tools that address various learning styles helping students to study and improve their success.

Cengage YouBook

Engage your students with a customizable and interactive eBook that lets you embed web links, modify the textbook narrative as needed, reorder entire sections and chapters, and hide any content you don't teach to create an eBook that perfectly matches your syllabus.

Concept Mastery Videos

Extensive video resources—from Rena Petrello and Dana Mosely—reinforce concepts and help students prepare for exams. These are accessible through the Resources Tab in every Enhanced WebAssign course. Select videos are also embedded in the Cengage YouBook.

Resources

FOR THE INSTRUCTOR	FOR THE STUDENT
Annotated Instructor's Edition (ISBN: 978-0-618-94670-9) The Annotated Instructor's Edition features several teaching tools. Classroom Examples that are parallel to the worked examples in the student edition appear in the margin. These may be used to supplement lectures and also appear in PowerPoint form on the PowerLecture CD-ROM. Also included in the margin are Teaching Notes that offer instructors tips and suggestions for presenting the content. Answers to most exercises appear on page in close proximity. Answers not appearing on page are located in an appendix in the back of the book.	
Complete Solutions Manual by Alan Hain, Lewis-Clark State College (ISBN: 978-1-285-09216-4) The Complete Solutions Manual provides detailed solutions to all problems in the text.	**Student Solutions Manual by Alan Hain, Lewis-Clark State College** (ISBN: 978-1-285-09217-1) Go beyond the answers—see what it takes to get there and improve your grade! This manual provides step-by-step solutions to selected problems in the series, giving you the information you need to truly understand how these problems are solved.
Instructor's Resource Binder by Maria H. Andersen, Muskegon Community College (ISBN: 978-0-538-73675-6) This loose-leaf binder contains uniquely designed Teaching Guides that contain tips, examples, activities, worksheets, overheads, assessments, and solutions to all worksheets and activities.	**Student Workbook by Maria H. Andersen, Muskegon Community College** (ISBN: 978-1-285-09222-5) Get a head start with this hands-on resource! The Student Workbook contains all of the assessments, activities, and worksheets from the Instructor's Resource Binder for classroom discussions, in-class activities, and group work.
PowerLecture with ExamView (ISBN: 978-1-285-09224-9) This CD-ROM (or DVD) provides you with dynamic media tools for teaching. Create, deliver, and customize tests (both print and online) in minutes with ExamView® Computerized Testing Featuring Algorithmic Equations. Easily build solution sets for homework or exams using Solution Builder's online solutions manual. Microsoft® PowerPoint® lecture slides and figures from the book are also included on this CD-ROM (or DVD) along with the PowerPoint versions of the parallel Classroom Examples available in each Annotated Instructor's Edition. A graphing calculator appendix detailing specific calculator commands also appears along with additional support materials for use specifically with the textbook.	
Solution Builder This online instructor database offers complete worked solutions to all exercises in the text, allowing you to create customized, secure solutions printouts (in PDF format) matched exactly to the problems you assign in class. For more information, visit www.cengage.com/solutionbuilder.	
Enhanced WebAssign (ISBN: 978-0-538-73810-1) Exclusively from Cengage Learning, Enhanced WebAssign® combines the exceptional Mathematics content that you know and love with the most powerful online homework solution, WebAssign. Enhanced WebAssign engages students with immediate feedback, rich tutorial content and interactive, fully customizable eBooks (YouBook) helping students to develop a deeper conceptual understanding of their subject matter. Online assignments can be built by selecting from thousands of text-specific problems or supplemented with problems from any Cengage Learning textbook. Consistent with the authors' approach, unique features and problem types from the text have been integrated into Enhanced WebAssign including the 5-Steps, Success in College Mathematics, Find the Mistake, and Solving Equation Exercises with Checks.	**Enhanced WebAssign** (ISBN: 978-0-538-73810-1) Enhanced WebAssign (assigned by the instructor) provides you with instant feedback on homework assignments. This online homework system is easy to use and includes helpful links to textbook sections, video examples, and problem-specific tutorials.
Enhanced WebAssign: Start Smart Guide for Students (ISBN: 978-0-495-38479-3) Author: Brooks/Cole The Enhanced WebAssign: Start Smart Guide for Students helps students get up and running quickly with Enhanced WebAssign so that they can study smarter and improve their performance in class.	**Enhanced WebAssign: Start Smart Guide for Students** (ISBN: 978-0-495-38479-3) Author: Brooks/Cole If your instructor has chosen to package Enhanced WebAssign with your text, this manual will help you get up and running quickly with the Enhanced WebAssign system so that you can study smarter and improve your performance in class.

Ensuring Quality and Accuracy

Over the course of the development of this textbook, numerous quality assurance checks were applied. We gathered feedback by sharing each draft of the manuscript with instructors of Intermediate Algebra and collecting their written responses to extensive questionnaires. The feedback received was carefully read by the authors and publisher and then analyzed. The authors then revised the content accordingly. During the review and revision process, more than 100 instructors commented on various aspects of the manuscript, including the overall approach, features, organization, depth of topic coverage, clarity of writing, and the quality, variety, and quantity of examples and exercises. Additionally, special advisors commented on specific issues related to content, art, and design throughout the process. Feedback was also collected at a number of focus groups that were held at various stages of manuscript development and at a number of locations across the country. Our focus group participants offered insightful comments and guidance on an array of topics that helped shape many of the elements that appear in the finished textbook.

The authors' colleagues at Lewis-Clark State College also played a vital role in the development of *Intermediate Algebra*. They class-tested several iterations of the manuscript, with the authors continuously updating the manuscript based on comments from instructors as well as students.

The start of the production process ushered another level of quality assurance. A group of highly experienced art and text editors reviewed the manuscript in preparation for copyediting, proofreading, and art rendering. Once in composition, the text and art underwent multiple rounds of examination by a proofreader, copyeditor, and production editor as well as a mathematical accuracy checker and the authors. The network of quality assurance personnel then expanded to include authors of supplements for electronic as well as print products to help ensure product compatibility and accuracy across the whole of the program.

Our ultimate goal in delivering this product is to ensure that it meets the highest standards for quality and accuracy. At every step of the way, we considered how to best meet the needs of students and instructors. We consider our immediate and extended family of educational, editorial, art, and production professionals to be among the best, committed to producing materials of solid worth and lasting value.

Acknowledgments

A textbook is the result of the efforts of a very large group of contributors. In writing these books, we incorporated the suggestions and insight of colleagues and students across the country. As classroom instructors, we recognize and are grateful for the time, commitment, and feedback of the following faculty and students.

Advisory Board Members

Kirby Bunas, *Santa Rosa Junior College*
Michael Kinter, *Cuesta College*
Marie St. James, *St. Clair Community College*

Reviewers and Class Testers

Dr. Laura Adkins, *Missouri Southern State University*
Darla Aguilar, *Pima Community College*
Kathleen D. Allen, *Hinds Community College–Rankin*
Jerry Allison, *Trident Technical College–Berkeley*
Jacob Amidon, *Finger Lakes Community College*
Sheila Anderson, *Housatonic Community College*
Jan Archibald, *Ventura College*
Dimos Arsenidis, *California State University–Long Beach*

Benjamin Aschenbrenner, *North Seattle Community College*
Michele Bach, *Kansas City Community College*
Brian Balman, *Johnson County Community College*
Leann Beaven, *University of Southern Indiana*
Alison Becker–Moses, *Mercer County Community College*
Rosanne Benn, *Prince George's Community College*
Rebecca Berthiaume, *Edison State College*
Armando Bezies-Kindling, *Pima Community College–Downtown*

Barbara Biggs, *Utah Valley University*
Dr. Nicoleta Bila, *Fayetteville State University*
Ina Kaye Black, *Bluegrass Community and Technical College*
Kathleen Boehler, *Central Community College–Grand Island*
Michael Bowen, *Ventura College*
Gail Brewer, *Amarillo College*
Shane Brewer, *College of Eastern Utah–San Juan*
Susan Caldiero, *Cosumnes River College*
Nancy Carpenter, *Johnson County Community College*
Jeremy Carr, *Pensacola Junior College*
Edie Carter, *Amarillo College*
Gerald Chrisman, *Gateway Technical College–Racine*
Suzanne Christian-Miller, *Diablo Valley College*
Dianna Cichocki, *Erie Community College–South*
John Close, *Salt Lake Community College*
Stacy Corle, *Pennsylvania State University–Altoona*
Kyle Costello, *Salt Lake Community College*
Debra Coventry, *Henderson State University*
Pam Cox, *East Mississippi Community College*
Edith Cranor-Buck, *Western State College of Colorado*
William D. Cross, *Palomar College*
Ken Culp, *Northern Michigan University*
Steven I. Davidson, *San Jacinto College–Central*
Susan Dimick, *Spokane Community College*
Dr. Julien Doucet, *Louisiana State University–Alexandria*
Karen Edwards, *Diablo Valley College*
Patricia Elko, *SUNY Morrisville*
Mike Everett, *Santa Ana College*
Rob Farinelli, *Community College of Allegheny County*
Stuart Farm, *University of North Dakota*
Julie Fisher, *Austin Community College*
Dr. Dorothy French, *Community College of Philadelphia*
Gada Dharmesh J., *Jackson Community College*
Scott Gentile II, *City College of San Francisco*
Jane Golden, *Hillsborough Community College*
Barbara Goldner, *North Seattle Community College*
John Greene, *Henderson State University*
Kathy Gross, *Cayuga Community College–Fulton*
Susan Hahn, *Kean University*
Shawna Haider, *Salt Lake Community College*
Dr. Haile Haile, *Minneapolis Community and Technical College*
Shelly Hansen, *Colorado Mesa University*
John T. Harris
Dr. Jennifer Hegeman, *Missouri Western State University*
Elaine Hodz, *Florida Community College at Jacksonville–Downtown*
Laura Hoye, *Trident Technical College*
Kimberly Johnson, *Mesa Community College*
Dr. Tina Johnson, *Midwestern State University*
Todd Kandarian, *Reedley College–Madera Center*
Dr. Fred Katiraie, *Montgomery College–Rockville*
Dr. John Kawai, *Los Angeles Valley College*
Dr. David Keller, *Kirkwood Community College*
Catherine Carroll Kiaie, *Cardinal Stritch University*
Barbara Kistler, *Lehigh Carbon Community College*

Daniel Kleinfelter, *College of the Desert*
Kandace Kling, *Portland Community College*
Alexander Kolesnik, *Ventura College*
Randa Kress, *Idaho State University*
Thang Le, *College of the Desert*
Kevin Leith, *Central New Mexico Community College*
Edith Lester, *Volunteer State Community College*
Mark Marino, *Erie Community College–North*
Amy Marolt, *Northeast Mississippi Community College*
William Martin, *Pima Community College*
Dr. Derek Martinez, *Central New Mexico Community College*
Carlea McAvoy, *South Puget Sound Community College*
Caroyln McCallum, *Yakima Valley Community College*
Carrie A. McCammon, *Ivy Tech State College*
Margaret Michener, *The University of Nebraska at Kearney*
Debbie Miner, *Utah Valley University*
Mary Mizell, *Northwest Florida State College*
Ben Moulton, *Utah Valley University*
Bethany R. Mueller, *Pensacola State College*
Dr. Ki-Bong Nam, *University of Wisconsin–Whitewater*
Sandi Nieto, *Santa Rosa Junior College–Santa Rosa*
Dr. Sam Obeid, *Richland College*
Jon Odell, *Richland Community College*
Michael Orr, *Grossmont College*
Becky L. Parrish, *Ohio University–Lancaster*
Dr. Shahrokh Parvini, *San Diego Mesa College*
John Pflughoeft, *Northwestern Michigan College*
Tom Pomykalski, *Madison Area Technical College*
Carol Ann Poore, *Hinds Community College–Rankin*
Michael Potter, *The University of Virginia's College at Wise*
Beth Powell, *MiraCosta College*
Brooke Quinlan, *Hillsborough Community College*
Genele Rhoads, *Solano Community College*
Tanya Rivers, *Western State College of Colorado*
Vicki Schell, *Pensacola State College*
Randy Scott, *Santiago Canyon College*
Jane Serbousek, *Northern Virginia Community College*
Sandra Silverberg, *Bergen Community College*
M. Terry Simon, *The University of Toledo*
Roy Simpson, *Cosumnes River College*
Zeph Smith, *Salt Lake Community College*
Donald Solomon, *University of Wisconsin–Milwaukee*
J. Sriskandarajah, *Madison College–Madison, WI*
Dr. Panyada Sullivan, *Yakima Valley Community College*
Mary Ann Teel, *University of North Texas*
Rosalie Tepper, *Shoreline Community College*
Rose Toering, *Kilian Community College*
Dr. Suzanne Tourville, *Columbia College*
Calandra Walker, *Tallahassee Community College*
Dr. Bingwu Wang, *Eastern Michigan University*
Dr. Jane West, *Trident Technical College*
Darren Wiberg, *Utah Valley University*
Dale Width, *Central Washington University*
Peter Willett, *Diablo Valley College*
Dr. Douglas Windham, *Tallahassee Community College*
Dr. Tzu-Yi Alan Yang, *Columbus State Community College*

Developmental Focus Group Participants

Dimos Aresenidis, *California State University–Long Beach*
Angelica Ascencio, student, *Cerritos College*
Mohammad Aslam, *Georgia Perimeter College–Clarkston*
Patricia J. Blus, *National Louis University*
Dona V. Boccio, *Queensborough Community College*
Dmitri Budharin, *Cerritos College*
Kirby Bunas, *Santa Rosa Junior College*
Ashraful Chowdhury, *Georgia Perimeter College–Clarkston*
Elisa Chung, *Riverside Community College*
Mariana Coanda, *Broward College*
Joseph S. de Guzman, M.S., *Norco College*
Cheryl Eichenseer, *St. Charles Community College*
Dr. Susan Fife, *Houston Community College*
Kathryn S. Fritz, *North Central Texas College*
Adrianne Guzman, student, *Cerritos College*
Dr. Kim Tsai Granger, *St. Louis Community College–Wildwood*
Edna G. Greenwood, *Tarrant County College–Northwest*
Larry D. Hardy, *Georgia Perimeter College–Newton*
Mahshid Hassani, *Hillsborough Community College*
Richard Hobbs, *Mission College*
Linda Hoppe, *Jefferson College*
Laurel Howard, *Utah Valley University*
Elizabeth Howell, *North Central Texas College*
James Johnson, *Modesto Junior College*
Janhavi Joshi, *De Anza College*
Susan Keith, *Georgia Perimeter College–Newton*
Michael Kinter, *Cuesta College*
Alex Kolesnik, *Ventura College*
Edith Lester, *Volunteer State Community College*
Bob Martin, *Tarrant County College*

Arda Melkonian, *Victor Valley College*
Ashod Minasian, *El Camino College*
Lydia Morales, *Ventura College*
Christina Morian, *Lincoln University*
Joyce Nemeth, *Broward College*
Marla Owens, *North Central Texas College*
Michael Papin, *Yuba College*
Svetlana Podkolzina, *Solano Community College*
Linda Retterath, *West Valley Mission College*
Marcus H. Rhymes, *Georgia Perimeter College–Clarkston*
Weldon Ritchie, *Le Cordon Bleu*
Kheck Segemeny, *Solano Community College*
Charlene Snow, *Solano Community College*
Pamela R. Sheehan, *Solano Community College and Los Medanos Community College*
Cara Smyczynski, *Trident Technical College*
Marie St. James, *St. Clair County Community College*
Deborah Strance, *Allan Hancock College*
Francesco Strazzullo, *Reinhardt University*
Chetra Talwinder, *Woodland Community College*
Jennie Thompson, *Leeward Community College*
J.B. Thoo, *Yuba College*
Binh Truong, *American River College*
Sally Vandenberg, *Barstow College*
Barbara Villatoro, *Diablo Valley College*
Dahlia N. Vu, *Santa Ana College*
Carol M. Walker, *Hinds Community College*
Tracy Welch, *North Central Missouri College*
Jason Wetzel, *Midlands Technical College*
Emily C. Whaley, *Georgia Perimeter College–Clarkston*
Rebecca Wong, *West Valley College*

We would also like to extend special thanks to other people who have helped us write. We are indebted to Charlie Van Wagner, our publisher, for his support, and to Lynn Cox, Richard Stratton, Mary Finch, Bill Hoffman, Angus McDonald, Maria Morelli, Peter Galuardi, and Jay Campbell for their encouragement. The insight and experience of Erin Brown was our mainstay during writing. We cannot thank Cheryll Linthicum and Martha Emry enough for being in our corner during production. Rita Lombard, Katherine Greig, Gordon Lee, and Danae April helped us clarify the story behind this book and have developed a powerful marketing plan. We also are grateful for the efforts of Leslie Lahr, Barbara Willette, Marian Selig, Heleny Wong, Guanglei Zhang, Carrie Jones, Lauren Crosby, Jennifer Cordoba, Carrie Green, Roger Lipsett, Scott Barnett, Tammy Morgan, Alex Kolesnik, and Sally Kalpakoff.

We have been very fortunate to have the support of colleagues and staff as we have class-tested this manuscript at Lewis-Clark State College. Our thanks to Alan Hain, Misti Dawn Henry, Burma Hutchinson, Matt Johnston, Masoud Kazemi, Jean Sawyer, Karen Schmidt, the staff in the print shop, and the people in the physical plant who delivered the endless boxes of copies.

Above all, we thank the students who gave their forthright feedback as they used preliminary and class test editions. Their efforts were invaluable as we wrote and accuracy checked this first edition.

Laura Bracken
Ed Miller

Dear Student,

In both our campus and on-line classes, we observe students developing the attitudes and habits that lead to success. Some of those observations are outlined below. As you begin the course, we encourage you to consider the tips provided. They may help you improve your performance as you work through the course.

TIP #1 — *Ask questions—discover what method works best for you.*

Students who are most likely to succeed expect that some things will come more easily than others. When they are confused or discouraged, they seek help. Some students who are enrolled in a course that meets in a classroom may feel awkward about asking questions in front of peers or the instructor. These students often develop strategies for getting the help they need in a more private way, such as sending an email or text message to an instructor or campus tutor. Some students may be able to contact an instructor or tutor by phone. These students often ask questions about the current assignment—they don't let too many days go by before they ask for help. By doing the work and asking for help to complete the work when needed and as the work is assigned, these students seem to have a more rewarding experience.

> **Textbook Hint**
>
> Throughout the textbook, you will find many worked examples, and **Practice Problems** that correspond to the worked examples, that you can try on your own. **Answers** to most of the Practice Problems are found in the back of the textbook so that you can check to see how well you did. Answers to the odd-numbered section-ending exercises are also found in the back of the book so that you can check your work.

TIP #2 — *Make choices that promote long-term goals.*

College is an environment where students are regarded as adults, responsible for their own learning. Tracking assignments and due dates for multiple classes, preparing for exams, and getting to know the policies of different instructors can overwhelm some students—especially those entering college for the first time. Students most likely to succeed often develop strategies for managing the requirements of their courses. They may create schedules to track assignment due dates or important exam dates. They may turn down opportunities to attend social events in order to allow themselves enough time to complete a project or to study. Successful students understand that attending college sometimes means making hard choices. A short-term sacrifice may be what it takes to achieve a more meaningful, long-term goal.

© lightpoet/Shutterstock.com

> **Textbook Hint**
>
> **Success in College Mathematics**, found at the beginning of each chapter, offers practical tips on making the transition to college-level courses and managing your time effectively so that you stay on top of your assignments. Refer to the **Study Plan and Review** at the end of each chapter, along with the **Chapter Tests** and **Cumulative Reviews**, to help with exam preparation.

© wavebreakmedia/Shutterstock.com

TIP #3   *Learn to be self-sufficient.*

Students most likely to succeed in math are students who have learned to be self-sufficient by checking their work for accuracy and reasonability. They often show steps of a solution, line by line, and look for any errors in arithmetic or algebra. These students also may have learned to determine whether the final answers they provide make sense—for example, they might verify that values reported as final answers are not too big or small in the context of the problems they are solving. Becoming a self-sufficient learner takes practice, but is a big asset to students in helping them improve their grades and achieve their academic goals.

> ### Textbook Hint
>
> Throughout the text, the worked examples will show you how to check your work for accuracy and reasonability and the section exercises will provide practice with these skills. You will also practice identifying errors in the **Find the Mistake** exercises.

TIP #4   *Memorize less, understand more.*

Students who are successful in math seem to rely on their understanding of basic principles and concepts rather than on memorizing how to do a specific problem. They look for connections and patterns and learn how to apply basic principles to new problems.

> ### Textbook Hint
>
> Section 1.5 outlines a series of **five steps** that can be applied to solve a wide variety of problems. You will also find that as you read the explanations next to each step of worked examples certain rules and concepts are used repeatedly—but to solve very different problems. Important properties, rules, and patterns are listed in boxes throughout the book. Be sure to complete the exercises in each section and in the **Study Plan** that ask you to explain your understanding of different concepts.

Fundamentals of Algebra

SUCCESS IN COLLEGE MATHEMATICS

Personal Responsibility

College is a place where freedom and personal responsibility can lead to great accomplishments. Your long-term goal may be a degree, a certificate, or some other credential. A shorter-term goal is to pass your classes this term. An even shorter-term goal is to do well on your assignments and to pass the first test. There are many people at your college who want to help you achieve these goals.

College is a new world that offers great opportunities. To be successful, you need to know how college works. You also need to make choices and develop habits that lead to success. With freedom and choices comes personal responsibility. *The only person who can do the work and learn the math is you.*

The introduction to each chapter in this book focuses on strategies for being successful in Intermediate Algebra. Each exercise set includes follow-up questions about using these strategies. In this first chapter, these questions assess your knowledge of your college and your math class.

© Turtleman/Shutterstock

A group of horses is a herd. A group of sheep is a flock. A list of the days of the week that begin with T includes Tuesday and Thursday. We can also describe these groups and list as a set of horses, a set of sheep, or a set of the days of the week that begin with T. In this section, we begin our work with sets and expressions.

Sets and Numerical Expressions

After reading the text, working the practice problems, and completing assigned exercises, you should be able to:

1. Represent a set using roster notation, words, a number line graph, and/or interval notation.
2. Determine whether a number is prime, composite, or neither.
3. Determine whether two sets are disjoint and whether a set is a subset of another set.
4. Determine whether a number belongs to the set of real numbers, rational numbers, irrational numbers, integers, and/or whole numbers.
5. Evaluate an exponential expression.
6. Evaluate a square root or cube root.
7. Evaluate an expression following the order of operations.

Sets and Subsets

A **set** is a group or collection. The things in the set are its **elements**. We use **roster notation** to represent some sets. In roster notation, commas separate the elements of the set. The elements are inside **set brackets** { }.

EXAMPLE 1 Use roster notation to represent the set of the days of the week that begin with T.

SOLUTION ▶ {Tuesday, Thursday}

Some sets follow a pattern. An ellipsis, . . . , shows that a pattern in a set continues.

EXAMPLE 2 Use roster notation to represent the set of whole numbers.

SOLUTION ▶ {0, 1, 2, 3, . . .}

A graph on a **number line** can represent some sets. Each point graphed on the number line represents a number in the set. The numbers below the number line are the **scale**. The vertical lines on the number line are **tick marks**. To prepare for later work in graphing, we will label number line graphs with an x.

EXAMPLE 3 **(a)** Use a number line graph to represent the set $\{0, 2, 4, 5\}$.

SOLUTION ▶

(b) Write the set that represents the number line graph.

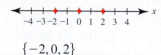

$\{-2, 0, 2\}$

When we multiply two or more numbers together, the result is a **product**. The numbers that we multiply together are **factors**. A whole number greater than 1 is either a **prime number** or a **composite number**. The only factors of a prime number are itself and 1. For example, since the only factors of 17 are itself and 1, 17 is a prime number. Since $6 = (1)(6)$ and $6 = (2)(3)$, 6 is a composite number.

> **Prime Number** A whole number greater than 1 that has exactly two distinct factors: itself and 1.
>
> **Composite Number** A whole number greater than 1 that is not prime. It has at least one factor that is not 1 or itself.

EXAMPLE 4 Use roster notation to represent the set of the first six composite numbers greater than 1.

SOLUTION ▶ $\{4, 6, 8, 9, 10, 12\}$

A number and its **opposite** are the same distance from 0 on the number line. The sum of a number and its opposite is 0. For example, the opposite of 2 is -2. The opposite of -45 is 45. The set of **integers** includes all of the whole numbers and their opposites.

EXAMPLE 5 Use roster notation to represent the set of integers.

SOLUTION ▶ $\{\ldots, -3, -2, -1, 0, 1, 2, 3, 4, \ldots\}$

If set A is a **subset** of set B, every element in A is an element of B. For example, since every element in the set of whole numbers, $\{0, 1, 2, 3, \ldots\}$ is also an element of the set of integers, $\{\ldots, -2, -1, 0, 1, 2, 3, \ldots\}$, the set of whole numbers is a subset of the set of integers.

EXAMPLE 6 **(a)** Set M is {cup, plate, spoon}. Set P is {cup, spoon}. Is set P a subset of set M?

SOLUTION ▶   Since all of the elements in set P are also elements in set M, set P is a subset of set M.

(b) If set B is $\{-2, 2, 4, 6, 8\}$, is set B a subset of the set of whole numbers?

▶ Since -2 is an element of set B but is not an element of the set of whole numbers, $\{0, 1, 2, 3, \ldots\}$, set B is not a subset of the set of whole numbers.

If two sets do not have any elements in common, they are **disjoint** sets. If set M is {cup, plate, spoon} and set Q is {knife, fork}, set M and set Q are disjoint sets.

EXAMPLE 7 Are the set of integers and the set of whole numbers disjoint sets? Explain.

SOLUTION ▶ The set of integers is $\{\ldots, -2, -1, 0, 1, 2, 3, \ldots\}$ and the set of whole numbers is $\{0, 1, 2, 3, \ldots\}$. Since these two sets both include the elements $0, 1, 2, 3, \ldots$, they are not disjoint sets.

Answers to all Practice Problems may be found in the back of the book.

Practice Problems

1. Use a number line graph to represent each set.
   a. $\{-2, 0, 3, 6\}$ b. $\{10, 20, 30, 40\}$

2. Use roster notation to represent the number line graph.

3. Identify the prime numbers.
   a. 13   b. 21   c. 0   d. -12 e. 63   f. 37

4. Set W is {bear, lion, tiger}. Set Z is {bear, tiger}.
   a. Is W a subset of Z?   b. Is Z a subset of W?

5. Set K is all of the integers greater than -4. Set M is all of the integers less than 0. Are these disjoint sets? Explain.

Sets and Interval Notation

If we can write a number as a fraction in which the numerator and denominator are integers, it is a **rational number**. The denominator of a fraction cannot equal 0. As we will see in Chapter 4, a repeating decimal number can be written as a fraction in which the numerator and denominator are integers. Repeating decimals such as $2.1563563563\ldots$ and $0.\overline{78}$ are rational numbers. The bar on $0.\overline{78}$ shows the numbers that repeat. For example, we can rewrite $2.1563563563\ldots$ as $2.1\overline{563}$.

EXAMPLE 8   (a) Explain why $\dfrac{3}{4}$ is a rational number.

SOLUTION ▶   Since 3 and 4 are integers, $\dfrac{3}{4}$ is a rational number.

(b) Explain why 0.23 is a rational number.

▶   We can rewrite 0.23 as a fraction, $\dfrac{23}{100}$. Since 23 and 100 are integers, 0.23 is a rational number.

(c) Explain why $\dfrac{3}{7}$ is a rational number.

▶   Since 3 and 7 are integers, $\dfrac{3}{7}$ is a rational number. The quotient of $3 \div 7$ equals $0.\overline{428571}$, a repeating decimal.

(d) Explain why -6 is a rational number.

▶   We can rewrite -6 as a fraction, $\dfrac{-6}{1}$. Since -6 and 1 are integers, -6 is a rational number.

An **irrational number** is a nonrepeating, nonterminating decimal number. We cannot write an irrational number as a fraction in which the numerator and denominator are integers. The number π is irrational. Although we often use a rounded value of $\pi \approx 3.14$, it does not terminate. When writing an irrational number, use an ellipsis to show that it does not terminate. For example, $\pi = 3.1415926535\ldots$, a number that neither repeats nor ends.

The elements of the set of **real numbers** are the rational numbers and the irrational numbers.

Sets of Real Numbers

Whole Numbers	$\{0, 1, 2, 3, \ldots\}$
Integers	$\{\ldots, -3, -2, -1, 0, 1, 2, \ldots\}$
Rational Numbers	Numbers that can be written as a fraction in which the numerator and denominator are integers.
	If a and b are integers and $b \neq 0$, $\frac{a}{b}$ is a rational number.
Irrational Numbers	Numbers that cannot be written as a fraction in which the numerator and denominator are integers; nonrepeating, nonterminating decimal numbers.
Real Numbers	The rational and irrational numbers.

Figure 1

We can use a number line graph to represent the set of real numbers (Figure 1). The red line with arrows shows that every position on this number line represents a real number. The arrowed line extends on the right toward **positive infinity**. The symbol for positive infinity is ∞. The arrowed line extends on the left toward **negative infinity**. The symbol for negative infinity is −∞. Infinity is not a real number. There is no "largest" or "smallest" real number.

EXAMPLE 9

(a) Use a number line graph to represent the set of real numbers that are greater than or equal to 1.

SOLUTION ▶

The number 1 is a **boundary value** of this set. Draw a bracket, [, on the number line at 1 to show that 1 is an element of the set. (In the past, you may have used a solid dot instead of a bracket.) The arrowed line shows that all real numbers greater than 1 are included in the set.

(b) Use a number line graph to represent the set of real numbers that are less than 0.

▶

The number 0 is a **boundary value** of this set. Draw a parenthesis,), on the number line at 0 to show that 0 is *not* an element of the set. (In the past, you may have used an open circle instead of a parenthesis.) The arrowed line shows that all real numbers less than 0 are elements of this set.

To show the connection between a number line graph and interval notation, we use brackets and parentheses in number line graphs instead of solid dots and open circles. **Interval notation** is another way to represent some sets of numbers. The interval begins and ends with a bracket or parenthesis. A comma separates the boundary values. If a set extends to infinity or negative infinity, the interval begins or ends with a parenthesis.

EXAMPLE 10

Use a number line graph and interval notation to represent the set.

(a) The set of real numbers greater than or equal to –5

SOLUTION ▶

Number line graph: ◄―┼―┼―[―┼―┼―►―┼―┼―► x
−7 −6 −5 −4 −3 −2 −1 0

Interval notation: $[-5, \infty)$ The boundary value –5 is included in the set.

(b) The set of real numbers less than 3

▶ Number line graph:

Interval notation: $(-\infty, 3)$ The boundary value 3 is not included in the set.

(c) The set of real numbers greater than or equal to 2 and less than or equal to 7

▶ Number line graph:

Interval notation: $[2, 7]$ Both boundary values are included in the set.

(d) The set with one element, the number 2

▶ Number line graph:

Interval notation: $[2, 2]$

Interval Notation

An interval represents a set of numbers.

An interval begins and ends with a parenthesis or a bracket.

- If the interval begins or ends with negative or positive infinity, use a parenthesis.
- If the interval begins or ends with a boundary value that is not included in the interval, use a parenthesis.
- If the interval begins or ends with a boundary value that is included in the interval, use a bracket.

Set	Number line graph	Interval notation
The set of real numbers less then a		$(-\infty, a)$
The set of real numbers less than or equal to a		$(-\infty, a]$
A set with one real number, a		$[a, a]$
The set of real numbers greater than or equal to a		$[a, \infty)$
The set of real numbers greater than a		(a, ∞)

Practice Problems

For problems 6–13, match the number to the set(s) of which it is an element.

6. 18   **7.** $\dfrac{5}{6}$ **8.** -9 **9.** -0.3 **A.** Real numbers

10. π **11.** 0.25 **12.** $0.141414\ldots$ **B.** Rational numbers

13. $0.\overline{6}$ **C.** Irrational numbers

D. Integers

E. Whole numbers

For problems 14–16,
(a) use a number line graph to represent the set.
(b) use interval notation to represent the set.

14. The real numbers less than or equal to 5
15. The real numbers greater than 2
16. The real numbers greater than or equal to 4 and less than 10

© Brian K./Shutterstock

Expressions, Exponents, and Roots

We can represent six cups of coffee with the number 6. This number is an **expression**. We can also represent these cups as an expression that is a sum: $4 + 2$. Because they represent the same amount, 6 and $4 + 2$ are **equivalent expressions**.

Expressions may include exponents. An exponent is a notation used to represent repeated multiplications: $3^4 = (3)(3)(3)(3)$. The **base** in this expression is 3. The **exponent** is 4. An exponent of 1 tells us that the expression is equal to the base: $7^1 = 7$.

To **evaluate** an expression, we rewrite it as an equivalent expression.

EXAMPLE 11 | Evaluate.

(a) 4^1 **(b)** 4^2 **(c)** $(-4)^2$ **(d)** $(-4)^3$

SOLUTION ▶
$$= 4 \qquad = (4)(4) \qquad = (-4)(-4) \qquad = (-4)(-4)(-4)$$
$$\qquad\qquad = 16 \qquad\qquad = 16 \qquad\qquad = -64$$

To *square* a number, we multiply it by itself: $4^2 = 16$. To find the **square root** of a number, we undo the process of squaring. Since $(4)(4) = 16$ and $(-4)(-4) = 16$, the square roots of 16 are 4 and -4. The positive square root of a number is its **principal square root**. The notation for the principal square root is $\sqrt{}$. So $\sqrt{16} = 4$. The number 16 in $\sqrt{16}$ is the **radicand**.

The square root of a real number may be an irrational number. We can use a calculator to find an approximate value for the square root and round it to a given place value. For example, $\sqrt{5} = 2.23606\ldots$. Rounded to the nearest hundredth, $\sqrt{5} \approx 2.24$. The $\approx$ symbol means "approximately equal."

The product of a positive real number and itself is a positive real number. The product of a negative real number and itself is a positive real number. So the product of a real number (except 0) and itself must be a positive real number. This means that the *square root of a negative real number is not a real number.*

EXAMPLE 12 | Evaluate. Round any irrational numbers to the nearest hundredth.

(a) $\sqrt{25}$ **(b)** $\sqrt{10}$ **(c)** $\sqrt{-3}$

SOLUTION ▶
$$= 5 \qquad\quad = 3.162\ldots \qquad \text{Not a real number}$$
$$\qquad\qquad \approx 3.16$$

To *cube* a number, we multiply it three times: $4^3 = (4)(4)(4) = 64$. To evaluate the **cube root** of a number, we undo the process of cubing. Since $(2)(2)(2) = 8$, the cube root of 8 is 2. The cube root of a negative number is a real number. For example, since $(-4)(-4)(-4) = -64$, the cube root of -64 is -4.

The notation for a cube root is $\sqrt[3]{}$. The number 3 is the **index** of the root. So $\sqrt[3]{64} = 4$ and $\sqrt[3]{8} = 2$. Cube roots can be irrational. We often use a calculator to find the approximate value of these roots. However, we will not evaluate irrational cube roots until we study rational exponents.

EXAMPLE 13   Evaluate.

(a) $\sqrt[3]{27}$ **(b)** $\sqrt[3]{-27}$ **(c)** $\sqrt[3]{1000}$

SOLUTION ▶ $= 3$ $= -3$ $= 10$

Practice Problems

For problems 17–25, evaluate. Round irrational numbers to the nearest tenth.

17. 7^2 **18.** $(-7)^2$ **19.** 7^3 **20.** $(-7)^3$ **21.** $\sqrt[3]{343}$

22. $\sqrt{49}$ **23.** $\sqrt{-49}$ **24.** $\sqrt{15}$ **25.** $\sqrt[3]{-64}$

Expressions and the Order of Operations

The order of operations describes the rules that we use to evaluate expressions.

The Order of Operations

1. Working from left to right, evaluate expressions in parentheses or in other grouping symbols (whichever comes first).
2. Working from left to right, evaluate exponential expressions or roots (whichever comes first).
3. Working from left to right, do multiplication or division (whichever comes first).
4. Working from left to right, do addition or subtraction (whichever comes first).

In the next example, we follow the order of operations to evaluate an expression. Since each expression is equal to the one above it, each new line begins with an = sign.

EXAMPLE 14   Evaluate: $-12 - 15 \div (-3) + 6(4 - 9)$

SOLUTION ▶

$-12 - 15 \div (-3) + 6(4 - 9)$

$= -12 - 15 \div (-3) + 6(\mathbf{-5})$ Evaluate inside of parentheses.

$= -12 \mathbf{+ 5} + 6(-5)$ Multiply or <u>divide</u> from left to right.

$= -12 + 5 \mathbf{- 30}$ <u>Multiply</u> or divide from left to right.

$= \mathbf{-7} - 30$ <u>Add</u> or subtract from left to right.

$= -37$ Add or <u>subtract</u> from left to right.

To remember the order of operations, you may have learned the mnemonic PEMDAS: Parentheses, Exponents, Multiplication, Division, Addition, Subtraction. Be careful. This does not include the very important directions to do these operations from **left to right**, whichever comes first.

EXAMPLE 15 | Evaluate: $-10^2 - 40 \div 4 \cdot 2 - 4$

SOLUTION ▶

$$-10^2 - 40 \div 4 \cdot 2 - 4$$
$$= \mathbf{-100} - 40 \div 4 \cdot 2 - 4 \qquad \text{Evaluate the exponential expression: } -10^2 = -100$$
$$= -100 - \mathbf{10} \cdot 2 - 4 \qquad \text{Multiply or \underline{divide} from left to right.}$$
$$= -100 - \mathbf{20} - 4 \qquad \text{\underline{Multiply} or divide from left to right.}$$
$$= \mathbf{-120} - 4 \qquad \text{Add or \underline{subtract} from left to right.}$$
$$= -124 \qquad \text{Add or \underline{subtract} from left to right.}$$

Practice Problems

For problems 26–28, evaluate.

26. $-13 - 7^2 - 45 \div (-9) - 3(2)$ **27.** $-6^2 - \sqrt{225} \div 3 - (4 - 9)^2$

28. $\left(\dfrac{5 + 3^2}{9 - 7} \right)^3$

1.1 VOCABULARY PRACTICE

Match the term with its description or representation.

1. Two sets that have no elements in common
2. A whole number greater than 1 whose only factors are itself and 1
3. A number that we can write as a fraction in which the numerator and denominator are integers
4. In the expression 5^3, 5 is an example of this.
5. $\{0, 1, 2, 3, \ldots\}$
6. A set is a collection of these.
7. The number 3 in $\sqrt[3]{8}$ is an example of this.
8. These numbers are the same distance from 0 on the number line.
9. The elements of this set are the rational numbers and the irrational numbers.
10. $\{\ldots, -3, -2, -1, 0, 1, 2, 3, \ldots\}$

A. base
B. disjoint sets
C. elements
D. index
E. integers
F. opposites
G. prime number
H. rational number
I. real numbers
J. whole numbers

1.1 Exercises

These exercises are opportunities for practice. A college basketball player spends hours shooting free throws and practicing layups. Successful math students practice in the same way. They practice until a skill is automatic and they can do it almost always without mistakes.

Follow your instructor's guidelines for showing your work. Read the directions carefully for each group of exercises. You can find the answers for many of the odd-numbered exercises in the back of the book. Check your work so that you do not practice doing something the wrong way. Get help if you need it.

For exercises 1–4, use a number line graph to represent the set.

1. $\{-5, -1, 2, 3, 4\}$ 2. $\{-4, 0, 1, 2, 4\}$

3. $\{5, 6, 7, 8\}$ 4. $\{4, 5, 6, 7\}$

For exercises 5–8, use roster notation to represent the number line graph.

5.

6.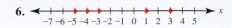

7.

8.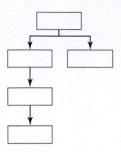

9. Use roster notation to represent the set of integers.

10. Use roster notation to represent the set of whole numbers.

For exercises 11–14, identify the number as prime or composite.

11. 101

12. 103

13. 147

14. 177

For exercises 15–16, set *F* is the set of whole numbers. Set *G* is {0, 1, 2, 3, 4, 5}.

15. Is set *F* a subset of set *G*? Explain.

16. Is set *G* a subset of set *F*? Explain.

For exercises 17–18, set *D* is {yellow, green}, and set *E* is {blue, yellow, red, green}.

17. Is set *E* a subset of set *D*? Explain.

18. Is set *D* a subset of set *E*? Explain.

19. Is the set of integers a subset of the set of rational numbers? Explain.

20. Is the set of irrational numbers a subset of the real numbers? Explain.

21. Are the set of irrational numbers and the set of rational numbers disjoint sets? Explain.

22. Are the set of integers and the set of rational numbers disjoint sets? Explain.

23. To show that it is a rational number, rewrite 0.399 as a fraction with an integer in the numerator and denominator.

24. To show that it is a rational number, rewrite 0.2341 as a fraction with an integer in the numerator and denominator.

25. Each box in the diagram represents a set. A set below another set must be a subset of the set above. Copy the diagram. Fill in the boxes, using the following sets: integers, irrational numbers, rational numbers, real numbers, whole numbers.

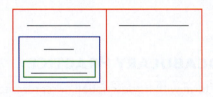

26. The largest red rectangle in the diagram represents the set of real numbers. Each rectangle inside it represents a subset of the real numbers. Copy the diagram. Fill in each blank with the name of one of these sets: integers, irrational numbers, rational numbers, whole numbers.

For exercises 27–40, match the number to the set(s) of which it is an element.

27. −8

28. $\frac{2}{3}$

29. −0.37

30. $\sqrt{11}$

31. π

32. $\sqrt{81}$

33. $\sqrt[3]{27}$

34. $\sqrt{-16}$

35. $\frac{3}{11}$

36. $\sqrt[3]{-64}$

37. $\sqrt{-25}$

38. −0.2

39. $\sqrt{5}$

40. −54

A. real numbers
B. rational numbers
C. irrational numbers
D. integers
E. whole numbers
F. prime numbers
G. composite numbers
H. none of the above

41. Copy the table below. Put a checkmark or X in the box if the number is an element of the set at the top of the column.

Table for exercise 41

	Real numbers	Rational numbers	Irrational numbers	Integers	Whole numbers
$\frac{3}{5}$					
π					
−8					
$\sqrt{13}$					
$\sqrt{100}$					
$-0.\overline{43}$					

42. Copy the table below. Put a checkmark or X in the box if the number is an element of the set at the top of the column.

For exercises 43–50, use interval notation to represent the number line graph.

43.
$\xleftarrow{\quad}\overset{(\quad\quad)}{\underset{2\ 3\ 4\ 5\ 6\ 7\ 8\ 9}{\,|\,|\,|\,|\,|\,|\,|\,|\,}}\xrightarrow{\quad} x$

44. $\xleftarrow{\quad}\underset{3\ 4\ 5\ 6\ 7\ 8\ 9\ 10}{\,|\,|\,|\,|\,|\,|\,|\,|\,}\xrightarrow{\quad} x$

45. $\xleftarrow{\quad}\underset{-1\ 0\ 1\ 2\ 3\ 4\ 5\ 6}{\,|\,|\,|\,|\,|\,|\,|\,|\,}\xrightarrow{\quad} x$

46. $\xleftarrow{\quad}\underset{0\ 1\ 2\ 3\ 4\ 5\ 6\ 7}{\,|\,|\,|\,|\,|\,|\,|\,|\,}\xrightarrow{\quad} x$

47. $\xleftarrow{\quad}\underset{2\ 3\ 4\ 5\ 6\ 7\ 8\ 9}{\,|\,|\,|\,|\,|\,|\,|\,|\,}\xrightarrow{\quad} x$

48. $\xleftarrow{\quad}\underset{3\ 4\ 5\ 6\ 7\ 8\ 9\ 10}{\,|\,|\,|\,|\,|\,|\,|\,|\,}\xrightarrow{\quad} x$

49. $\xleftarrow{\quad}\underset{-1\ 0\ 1\ 2\ 3\ 4\ 5\ 6}{\,|\,|\,|\,|\,|\,|\,|\,|\,}\xrightarrow{\quad} x$

50. $\xleftarrow{\quad}\underset{0\ 1\ 2\ 3\ 4\ 5\ 6\ 7}{\,|\,|\,|\,|\,|\,|\,|\,|\,}\xrightarrow{\quad} x$

For exercises 51–60,
(a) use a number line graph to represent the set.
(b) use interval notation to represent the set.

51. The set of real numbers less than 7

52. The set of real numbers less than 8

53. The set of real numbers greater than or equal to -4

54. The set of real numbers greater than or equal to -6

55. The set of real numbers greater than 2 and less than 10

56. The set of real numbers greater than 1 and less than 8

57. The set of real numbers less than or equal to -15

58. The set of real numbers less than or equal to -2

59. The set of real numbers greater than or equal to 25 and less than 100

60. The set of real numbers greater than or equal to 30 and less than 80

For exercises 61–82, evaluate.

61. 9^2 **62.** 8^2

63. $(-9)^2$ **64.** $(-8)^2$

65. 2^3 **66.** 5^3

67. $(-2)^3$ **68.** $(-5)^3$

69. 9^1 **70.** 8^1

71. $\sqrt{49}$ **72.** $\sqrt{100}$

73. $\sqrt{1}$ **74.** $\sqrt{0}$

75. $\sqrt{36}$ **76.** $\sqrt{121}$

77. $\sqrt{-144}$ **78.** $\sqrt{-9}$

79. $\sqrt[3]{1}$ **80.** $\sqrt[3]{125}$

81. $\sqrt[3]{-125}$ **82.** $\sqrt[3]{-1}$

83. The square roots of 36 are 6 and -6. However, $\sqrt{36}$ equals only 6. Explain why.

84. The square root of -100 is not a real number. Explain why.

85. A prime number is a whole number greater than 1 whose only factors are itself and 1. List the first 15 prime numbers.

86. A composite number is a whole number greater than 1 that is not prime. List the first 15 composite numbers.

For exercises 87–96, evaluate. Use Examples 14 and 15 in this section as models of how to show your work.

87. $12 + 6 \div (-3) - 8 \cdot 3$

88. $-18 + 12 \div (-4) - 6 \cdot 2$

89. $\left(\dfrac{12 - 24}{-5 - 1}\right)^3$ **90.** $\left(\dfrac{-35 - 20}{-4 - 7}\right)^3$

91. $8 \div 2 \cdot 4 - (-3) - 9^2$

92. $6 \div 2 \cdot 10 - (-8) - 4^2$

93. $-2(18 \div 9) + 4(8 - 15)$

94. $3(28 \div 7) + 6(9 - 17)$

95. $14 + 72 \div 3^2 - (11 - 15)$

96. $11 + 36 \div 3^2 - (2 - 21)$

Technology

For exercises 97–100, use a calculator to evaluate. If the number is irrational, round to the nearest hundredth.

97. $\sqrt{257}$ **98.** $\sqrt{327}$

99. -21^2 **100.** $(-21)^2$

Table for exercise 42

	Real numbers	Rational numbers	Irrational numbers	Integers	Whole numbers
$\sqrt{36}$					
$0.\overline{67}$					
π					
-2					
$\dfrac{1}{7}$					
$\sqrt{3}$					

Find the Mistake

For exercises 101–104, the completed problem has one mistake.
(a) Describe the mistake in words.
(b) Do the problem correctly.

101. Problem: Use interval notation to represent the set of real numbers greater than or equal to −5.

   Incorrect Answer: $(-5, \infty)$

102. Problem: Use interval notation to represent the set of real numbers less than 6.

   Incorrect Answer: $(6, -\infty)$

103. Problem: Evaluate: $\sqrt{-16}$

   Incorrect Answer: -4

104. Problem: List the first five prime numbers.

   Incorrect Answer: 1, 2, 3, 5, 7

105. In college, you may need to communicate with your instructors in person, by e-mail, by text message, or by voice mail. Your instructors may vary in how they wish to be addressed. Some may prefer that you use their titles and last names. Others may prefer that you use only their first names. Describe how the instructor of your math class prefers to be addressed.

106. Your math class may be in a computer lab, a regular classroom, a large lecture hall, on-line, or some combination of these locations. You may use a printed textbook or an e-book. Your instructor may require that you bring other materials to class. Describe the materials that you should have when you go to class or start working on-line.

© Mark R./Shutterstock

An **axiom** of a number system is a basic principle. Taken together, the axioms of the real numbers act as a constitution for evaluating and simplifying expressions. In this section, we will use axioms of the real numbers and other rules to simplify expressions.

SECTION 1.2

Algebraic Expressions

After reading the text, working the practice problems, and completing assigned exercises, you should be able to:

1. Use the commutative properties, associative properties, distributive property, the rules of exponents, and/or the order of operations to simplify an algebraic expression.

2. Evaluate or simplify expressions that include operations with 0.

3. Evaluate or simplify expressions that include operations with 1.

4. Simplify exponential expressions.

Expressions and Axioms

The **commutative properties**, the **associative properties**, and the **distributive property** are axioms of the real numbers. To state the axioms, we use words and/or variables. A **variable** is a symbol that represents an unknown number. In the descriptions of the properties, the variables a, b, and c represent real numbers.

The **commutative properties** state that we can change the *order* of addition or multiplication of real numbers without changing the result.

If a and b are real numbers,

$a + b = b + a$ **Commutative property of addition**

$ab = ba$ **Commutative property of multiplication**

The **associative properties** state that we can change the *grouping* of addition or multiplication without changing the result.

If a, b, and c are real numbers,

$a + (b + c) = (a + b) + c$ **Associative property of addition**

$a(bc) = (ab)c$ **Associative property of multiplication**

If a, b, and c are real numbers,

$a(b + c) = ab + ac$ **Distributive property of multiplication over addition**

$ab + ac = a(b + c)$

EXAMPLE 1   Identify the property that is best illustrated by the equation.

(a) $3 \cdot k = k \cdot 3$

SOLUTION ▶   The *order* changed: commutative property of multiplication.

(b) $(5a)b = 5(ab)$

▶   The *grouping* changed: associative property of multiplication.

(c) $2(3 + x) = 6 + 2x$

▶   The distributive property: $a(b + c) = ab + ac$

(d) $6 + (3 + n) = 6 + (n + 3)$

▶   The order changed: commutative property of addition.

(e) $6x + 6y = 6(x + y)$

▶   The distributive property: $ab + ac = a(b + c)$

We use the axioms and other rules to evaluate and simplify expressions. If an expression only includes numbers, it is a **numerical expression** and we **evaluate** it. If an expression includes one or more variables, it is an **algebraic expression** and we **simplify** it.

In the expression $3p + 5$, $3p$ and 5 are **terms**. Terms are separated by addition or subtraction. The number 3 in the term $3p$ is a **coefficient**. The number 5 is a **constant**. **Like terms** have the same variable(s) with the same exponent on each variable. For example, $5x$ and $2x$ are like terms, as are $12cd$ and $9cd$. To simplify an expression, combine like terms.

EXAMPLE 2   Simplify: $5x + 9 + 3x + 2y + 11$

SOLUTION ▶ $5x + 9 + 3x + 2y + 11$ Identify like terms.

       $= 8x + 2y + 20$ Combine like terms.

Use the distributive property to simplify expressions that include unlike terms in parentheses such as $-2(4a - 8)$. Although we cannot combine $4a$ and -8, we can use the distributive property to rewrite this expression as $-8a + 16$.

EXAMPLE 3 Simplify: $12cd - 2(4a - 8) + 9cd$

SOLUTION ▶

$12cd - 2(4a - 8) + 9cd$ Like terms can have two variables.

$= 12cd - \mathbf{2}(4a) - \mathbf{2}(-8) + 9cd$ Distributive property.

$= 12cd - 8a + 16 + 9cd$ Simplify; identify like terms.

$= -8a + 21cd + 16$ Combine like terms.

Since a decimal number is a real number, the distributive property also applies to expressions that include decimal numbers.

EXAMPLE 4 Simplify: $0.2(3x - 6) - 0.1(8x + 7)$

SOLUTION ▶

$0.2(3x - 6) - 0.1(8x + 7)$

$= \mathbf{0.2}(3x) + \mathbf{0.2}(-6) - \mathbf{0.1}(8x) - \mathbf{0.1}(7)$ Distributive property.

$= 0.6x - 1.2 - 0.8x - 0.7$ Simplify.

$= -0.2x - 1.9$ Combine like terms.

Practice Problems

1. Identify the property that is best illustrated by the equation.
   a. $3(y + 2) = 3y + 3(2)$
   b. $x \cdot 8 = 8 \cdot x$
   c. $2w + 2z = 2(w + z)$
   d. $(9c)d = 9(cd)$
   e. $a + (b + 7) = a + (7 + b)$

For problems 2–5, simplify.

2. $6(3x - 2y) + 5(4x - 7y)$ 3. $6(3x - 2y) - 5(4x - 7y)$
4. $9z - 4z^2 - 15 - 3z^2 + 21z$ 5. $5xy - 0.7x + 6xy - 0.11x - xy$

Expressions and Zero

Zero is the **additive identity** of the real numbers. When 0 is added to any real number, the number does not change. The sum of a real number and its **opposite** is 0. For example, $6 + (-6) = 0$.

The opposite of a real number is also called its **additive inverse**. The product of any real number and 0 is 0. For example, $5(0) = 0$.

In the real number system, division undoes multiplication and multiplication undoes division.

$15 \div 3 = 5$ Divide 15 by 3. The result is 5.

$(5)(3) = 15$ Multiply 5 by 3. The result is 15.

This relationship of multiplication and division is true for all real numbers *except for zero*. If $5 \div 0 = 0$, we cannot undo this result by multiplying because $(0)(0) \neq 5$. So $5 \div 0$ is undefined. *Division of a real number by 0 is undefined.*

Additive Identity If n is a real number, then $n + 0 = n$. The additive identity of the real numbers is 0.

Additive Inverse (Opposite) If n is a real number, then $n + (-n) = 0$. The additive inverse of n is $-n$.

> **Zero Product Property**   If a and b are real numbers and $a \cdot b = 0$, then $a = 0$ or $b = 0$.
>
> **Division of a real number by 0 is undefined.**   If a is a real number, $a \div 0$ is undefined and $\dfrac{a}{0}$ is undefined.

EXAMPLE 5   Evaluate.

(a) $8 + 0$ (b) $(8)(0)$ (c) $\dfrac{0}{8}$ (d) $\dfrac{8}{0}$ (e) $8 \div 0$

SOLUTION ▶ $= 8$ $= 0$ $= 0$ undefined   undefined

Practice Problems

For problems 6–11, evaluate.

6. $0 \div 3$ **7.** $0 + 3$ **8.** $\dfrac{3}{0}$ **9.** $\dfrac{0}{3}$ **10.** $0 - 3$ **11.** $3 \div 0$

Expressions and One

In the real number system, the **multiplicative identity** is 1. When we multiply a real number by 1, the number does not change. We use the multiplicative identity to rewrite fractions with different denominators as equivalent fractions with the same (common) denominator. Multiplying by 1 does not change the value of the fraction.

EXAMPLE 6   Simplify: $\dfrac{3}{5}w + \dfrac{2}{9}w$

SOLUTION ▶

$\dfrac{3}{5}w + \dfrac{2}{9}w$ The least common denominator is 45.

$= \left(\dfrac{3}{5}w\right)\left(\dfrac{9}{9}\right) + \left(\dfrac{2}{9}w\right)\left(\dfrac{5}{5}\right)$ Multiply each term by a fraction that is equal to 1.

$= \dfrac{27}{45}w + \dfrac{10}{45}w$ Simplify.

$= \dfrac{37}{45}w$ Combine like terms.

In the next example, we rewrite $-(2x^2 - 6)$ as $-1(2x^2 - 6)$. Multiplying by 1 does not change the value of the expression, but it may make it easier to use the distributive property.

EXAMPLE 7   Simplify: $5(x^2 + 9) - (2x^2 - 6)$

SOLUTION ▶

$5(x^2 + 9) - (2x^2 - 6)$

$= 5(x^2 + 9) - \mathbf{1}(2x^2 - 6)$ Multiplying by 1 does not change the value.

$= \mathbf{5}x^2 + \mathbf{5}(9) - \mathbf{1}(2x^2) - \mathbf{1}(-6)$ Distributive property.

$= 5x^2 + 45 - 2x^2 + 6$ Simplify; identify like terms.

$= 3x^2 + 51$ Combine like terms.

The product of a real number (except 0) and its **reciprocal** is 1. For example, $(6)\left(\dfrac{1}{6}\right) = 1$. The reciprocal of a real number is its **multiplicative inverse**.

EXAMPLE 8 | Simplify: $\dfrac{3}{7}\left(\dfrac{7}{3}h + 7\right) - 15$

SOLUTION ▶

$$\dfrac{3}{7}\left(\dfrac{7}{3}h + 7\right) - 15$$

$$= \dfrac{3}{7}\left(\dfrac{7}{3}h\right) + \dfrac{3}{7}(7) - 15 \qquad \text{Distributive property.}$$

$$= 1h + 3 - 15 \qquad\qquad \text{Simplify; } \dfrac{3}{7} \cdot \dfrac{7}{3} = 1; \dfrac{3}{7} \cdot 7 = 3$$

$$= h - 12 \qquad\qquad\qquad \text{Simplify.}$$

Multiplicative Identity   If n is a real number, then $n \cdot 1 = n$. The multiplicative identity of the real numbers is 1.

Multiplicative Inverse (Reciprocal)   If n is a real number and $n \neq 0$, then $n \cdot \dfrac{1}{n} = 1$. The multiplicative inverse (reciprocal) of n is $\dfrac{1}{n}$.

Practice Problems

For problems 12–14, simplify.

12. $\dfrac{2}{5}\left(\dfrac{2}{9}x + \dfrac{1}{9}\right) + \dfrac{4}{15}$ **13.** $\dfrac{1}{4}(c - 12) - \left(\dfrac{9}{2}c - 8\right) + \dfrac{1}{3}c$

14. $\dfrac{3}{2}\left(\dfrac{2}{3}h - 8\right) - \left(\dfrac{17}{9}h + 6\right)$

Simplifying Exponential Expressions

We can use repeated multiplications to simplify exponential expressions, or we can use the **exponent rules**.

The **addition and subtraction rule of exponents** restates what we know about combining like terms. We cannot simplify the expression $9x^2 - 3y^2$ because the *variables* in each term are different. We cannot simplify $4x^2 + 2x^3$ because the *exponents* in each term are different.

Addition and Subtraction Rule of Exponents

We can only add or subtract terms with exactly the same variable(s) with the same exponent(s).

EXAMPLE 9 | Simplify: $7x^3 + 5x^2 - 9x + 6 - 12x^2 - 4x + 15$

SOLUTION ▶

$$7x^3 + 5x^2 - 9x + 6 - 12x^2 - 4x + 15 \qquad \text{Identify like terms.}$$

$$= 7x^3 - 7x^2 - 13x + 21 \qquad\qquad\qquad \text{Addition and subtraction rule of exponents.}$$

We can multiply exponential expressions by first rewriting each expression as repeated multiplications, or we can use the **product rule of exponents**.

Product Rule of Exponents

$$x^m \cdot x^n = x^{m+n}$$

When multiplying exponential expressions with the same base, add the exponents and keep the same base; x, y are real numbers; m, n are integers; x and $m + n$ cannot both be zero.

We can show that the product rule of exponents is true by simplifying $(a^3)(a^5)$ first by using repeated multiplications and then by using the rule. The result is the same, a^8.

Using repeated multiplications	Using the product rule of exponents
$(a^3)(a^5)$	$(a^3)(a^5)$ The bases are the same.
$= (a \cdot a \cdot a)(a \cdot a \cdot a \cdot a \cdot a)$	$= a^{(3+5)}$ Add the exponents.
$= a \cdot a \cdot a \cdot a \cdot a \cdot a \cdot a \cdot a$	$= a^8$ Simplify.
$= a^8$	

EXAMPLE 10 Simplify: $(z^3)(z^4)$

SOLUTION ▶ $(z^3)(z^4)$ The bases are the same.

$= z^{(3+4)}$ Product rule of exponents.

$= z^7$ Simplify.

In the next example, we multiply the coefficients and use the product rule of exponents.

EXAMPLE 11 Simplify: $(5p^4)(-9p^5)$

SOLUTION ▶ $(5p^4)(-9p^5)$

$= (5)(-9)p^{(4+5)}$ Multiply coefficients; product rule of exponents.

$= -45p^9$ Simplify.

To simplify a quotient of exponential expressions, we can first rewrite the exponential expression using repeated multiplications, or we can use the **quotient rule of exponents**.

Quotient Rule of Exponents

$$\frac{x^m}{x^n} = x^{m-n} \qquad x \neq 0$$

When dividing exponential expressions with the same base, subtract the exponents and keep the same base; x, y are real numbers; m, n are integers; x and $m - n$ cannot both be zero.

We can show that the quotient rule of exponents is true by simplifying $\dfrac{b^7}{b^2}$ first by using repeated multiplications and then by using the rule. The result is the same, b^5.

Using repeated multiplications	**Using the quotient rule of exponents**
$\dfrac{b^7}{b^2}$	$\dfrac{b^7}{b^2}$ The bases are the same.
$= \dfrac{b \cdot b \cdot b \cdot b \cdot b \cdot b \cdot b}{b \cdot b}$	$= b^{(7-2)}$ Subtract the exponents.
$= \dfrac{b}{b} \cdot \dfrac{b}{b} \cdot \dfrac{b \cdot b \cdot b \cdot b \cdot b}{1}$	$= b^5$ Simplify.
$= 1 \cdot 1 \cdot b \cdot b \cdot b \cdot b \cdot b$	
$= b^5$	

EXAMPLE 12   Simplify: $\dfrac{36x^9}{8x^2}$

SOLUTION ▶

$$\dfrac{36x^9}{8x^2}$$ The exponential expressions have the same base.

$$= \dfrac{36}{8} x^{(9-2)}$$ Quotient rule of exponents.

$$= \dfrac{9}{2} x^7$$ Simplify. Do not change the fraction to a decimal.

To simplify an exponential expression that is raised to a power, we can first rewrite the exponential expression using repeated multiplications, or we can use the **power rule of exponents**.

Power Rule of Exponents

$$(x^m)^n = x^{mn}$$

When raising an exponential expression to a power, multiply the exponents and keep the same base; x, y are real numbers; m, n are integers; x and mn cannot both be zero.

We can show that the power rule of exponents is true by simplifying $(z^3)^2$ first by using repeated multiplications and then by using the rule. The result is the same, z^6.

Using repeated multiplications	**Using the power rule of exponents**
$(z^3)^2$	$(z^3)^2$
$= (z^3)(z^3)$	$= z^{(3)(2)}$ Multiply the exponents.
$= (z \cdot z \cdot z)(z \cdot z \cdot z)$	$= z^6$ Simplify.
$= z \cdot z \cdot z \cdot z \cdot z \cdot z$	
$= z^6$	

The base of an exponential expression can be a product of two or more numbers or variables. We can simplify this expression using repeated multiplications, or we can use the **many bases rule for a product**.

Many Bases Rule for a Product

$$(xy)^m = x^m y^m$$

When raising the product of two or more bases to a power, raise each base to the power; x, y are real numbers; m is an integer; x or y and m cannot both be zero.

We can show that the many bases rule for a product is true by simplifying $(ac)^5$ first by using repeated multiplications and then by using the rule. The result is the same, a^5c^5.

Using repeated multiplications	Using the many bases rule for a product
$(ac)^5$	$(ac)^5$
$= (ac)(ac)(ac)(ac)(ac)$	$= a^5c^5$
$= a \cdot c \cdot a \cdot c \cdot a \cdot c \cdot a \cdot c \cdot a \cdot c$	
$= a \cdot a \cdot a \cdot a \cdot a \cdot c \cdot c \cdot c \cdot c \cdot c$	
$= a^5c^5$	

EXAMPLE 13   Simplify: $(2x^4y)^3$

SOLUTION ▶

$(2x^4y)^3$

$= (2^3)(x^4)^3y^3$ Many bases rule for a product.

$= 8x^{(4)(3)}y^3$ Simplify; power rule of exponents.

$= 8x^{12}y^3$ Simplify.

The base of an exponential expression can be a quotient of two or more numbers or variables. We can simplify this expression using repeated multiplications, or we can use the **many bases rule for a quotient**.

Many Bases Rule for a Quotient

$$\left(\frac{x}{y}\right)^m = \frac{x^m}{y^m} \qquad y \neq 0$$

When raising the quotient of two or more bases to a power, raise each base to the power; x, y are real numbers; m is an integer; x and m cannot both be zero.

We can show that the many bases rule for a quotient is true by simplifying $\left(\dfrac{a}{b}\right)^3$ first by using repeated multiplications and then by using the rule. The result is the same, $\dfrac{a^3}{b^3}$.

Using repeated multiplications	Using the many bases rule for a quotient
$\left(\dfrac{a}{b}\right)^3$	$\left(\dfrac{a}{b}\right)^3$
$= \left(\dfrac{a}{b}\right)\left(\dfrac{a}{b}\right)\left(\dfrac{a}{b}\right)$	$= \dfrac{a^3}{b^3}$
$= \dfrac{a \cdot a \cdot a}{b \cdot b \cdot b}$	
$= \dfrac{a^3}{b^3}$	

EXAMPLE 14 | Simplify: $\left(\dfrac{7x}{8y}\right)^2$

SOLUTION ▶ $\left(\dfrac{7x}{8y}\right)^2$

$= \dfrac{(7x)^2}{(8y)^2}$ Many bases rule for a quotient.

$= \dfrac{7^2 x^2}{8^2 y^2}$ Many bases rule for a product.

$= \dfrac{49x^2}{64y^2}$ Simplify.

In the next example, notice that the numbers and variables in the expressions are the same. However, the positions of the parentheses are different. The simplified expressions are *not* the same.

EXAMPLE 15 | **(a)** Simplify: $(-4x^3)(-5x^2)$

SOLUTION ▶ $(-4x^3)(-5x^2)$

$= (-4)(-5)x^{(3+2)}$ The base of each exponential expression is the same, x.

$= 20x^5$ Multiply coefficients; product rule of exponents.

Simplify.

(b) Simplify: $(-4x)^3(-5x)^2$

▶ $(-4x)^3(-5x)^2$

$= (-4)^3(x)^3(-5)^2(x)^2$ Many bases rule for a product.

$= (-64)(25)x^{(3+2)}$ Simplify; product rule of exponents.

$= -1600x^5$ Simplify.

Practice Problems

For problems 15–20, simplify.

15. $7x^2 + 9x^2 + 3x$ **16.** $\left(\dfrac{10x^{12}}{2x^7}\right)^2$ **17.** $(2y^5)^3$

18. $(6a^2)(3a)(2a^7)$ **19.** $p^4z^3p^2z$ **20.** $4z(3z^2 + 1)$

Exponents That Are Negative or Equal to 0

We can use repeated multiplications and the quotient rule of exponents to show that $\dfrac{1}{x^3}$ and x^{-3} are equivalent expressions.

Using repeated multiplications	**Using the quotient rule of exponents**
$\dfrac{x^2}{x^5}$	$\dfrac{x^2}{x^5}$
$= \dfrac{x \cdot x}{x \cdot x \cdot x \cdot x \cdot x}$	$= x^{(2-5)}$
$= \dfrac{x}{x} \cdot \dfrac{x}{x} \cdot \dfrac{1}{x \cdot x \cdot x}$	$= x^{-3}$
$= 1 \cdot 1 \cdot \dfrac{1}{x^3}$	
$= \dfrac{1}{x^3}$	

This example illustrates the meaning of a negative exponent: $\dfrac{1}{x^3} = x^{-3}$.

Negative Exponents

If x and m are real numbers and x is not equal to 0, $x^{-m} = \dfrac{1}{x^m}$ and $\dfrac{1}{x^{-m}} = x^m$.

By tradition, simplified expressions do not include negative exponents. In the next example, we follow this tradition by rewriting x^{-4} as $\dfrac{1}{x^4}$.

EXAMPLE 16 Simplify: $\dfrac{x^5 y^7}{x^9}$

SOLUTION ▶

$$\dfrac{x^5 y^7}{x^9}$$

$$= x^{(5-9)} y^7 \qquad \text{Quotient rule of exponents.}$$

$$= x^{-4} y^7 \qquad \text{Simplify.}$$

$$= \dfrac{1}{x^4} \cdot y^7 \qquad x^{-4} = \dfrac{1}{x^4}$$

$$= \dfrac{y^7}{x^4} \qquad \text{All of the exponents are positive.}$$

We can use repeated multiplications and the quotient rule of exponents to show that x^0 and 1 are equivalent expressions.

Using repeated multiplications	Using the quotient rule of exponents
$\dfrac{x^3}{x^3}$	$\dfrac{x^3}{x^3}$
$= \dfrac{x \cdot x \cdot x}{x \cdot x \cdot x}$	$= x^{(3-3)}$
$= 1$	$= x^0$

This example illustrates the meaning of an exponent that is equal to 0: $x^0 = 1$.

Zero as an Exponent

If x is a real number and x is not equal to 0, $x^0 = 1$.

EXAMPLE 17 Simplify: $\left(\dfrac{a^5}{b^3}\right)^0$

SOLUTION ▶

$$\left(\dfrac{a^5}{b^3}\right)^0 \qquad \text{The exponent is 0.}$$

$$= 1 \qquad \text{Simplify.}$$

Practice Problems

For problems 21–27, simplify. All exponents must be positive.

21. $\dfrac{x^7}{x^{12}}$ **22.** $(y^{-5})^3$ **23.** $a^{-2}a^{-7}a^9$ **24.** $p^{-4}z^3p^{-2}z$

25. $\dfrac{f^{-3}}{f}$ **26.** $\left(\dfrac{m^2}{n^5}\right)^0$ **27.** $(3ab)^0$

Using Technology: Order of Operations on a Calculator

Some calculators include the order of operations as part of their programming. Others do not. Some calculators have parentheses that we can use. Others do not. *Find out what kind of calculator you can use, if any, on homework and tests.*

Scientific calculators have one or more entry lines.

EXAMPLE 18 Use a scientific calculator with one entry line to evaluate $64 - 3 \cdot 5 + 4^2$.

This calculator follows the order of operations for arithmetic. Enter the expression from left to right. As the keys are pressed, the display changes. Use the y^x key to enter the exponent. *Do not press = until the end.*

Press `6` `4` `–` `3` `×` `5` `+` `4` `yˣ` `2` `=`   The answer is 65.

EXAMPLE 19 Use a scientific calculator with two entry lines to evaluate $64 - 3 \cdot 5 + 4^2$.

The entire expression is visible in the entry line. Type in the expression from left to right. Use the `^` key to enter the exponent. *Do not press = until the end.*

Press `6` `4` `–` `3` `×` `5` `+` `4` `^` `2` `=`   The answer is again 65.

Most scientific calculators include keys for parentheses.

EXAMPLE 20 Evaluate $64 - 3 \cdot (5 + 4^2)$ on a scientific calculator with two entry lines.

Press `6` `4` `–` `3` `×` `(` `5` `+` `4` `^` `2` `)` `=`   The answer is 1.

Graphing calculators have more than two entry lines. Instead of pressing `=`, press `ENTER`.

Always take the time to think about your answer. If you think "that can't be right," it is time to redo the problem. You may have made an error as you entered numbers and operations into the calculator.

Practice Problems For problems 28–30, use a calculator to evaluate.

28. $60 - 54 \div 3$ **29.** $92 \div 2 \cdot 3 + 3^2 - 4$

30. $-6(-9 - 2) - 4(3 + 1)$

1.2 VOCABULARY PRACTICE

Match the term with its description.

1. When multiplying exponential expressions with the same base, add the exponents and keep the same base.
2. Rules for evaluating expressions
3. Terms that include exactly the same variables with the same exponents
4. A basic unproved principle of a number system
5. Changing the order of multiplication or the order of addition does not change the result.
6. Changing the grouping of multiplication or addition does not change the result.
7. In our number system, the result of adding a number and this is always 0.
8. Another name for a multiplicative inverse
9. The number that tells us how many times to multiply a base
10. When dividing two exponential expressions with the same base, subtract the exponents and keep the same base.

A. additive inverse
B. associative property
C. axiom
D. commutative property
E. exponent
F. like terms
G. the order of operations
H. product rule of exponents
I. quotient rule of exponents
J. reciprocal

1.2 Exercises

Follow your instructor's guidelines for showing your work.

For exercises 1–30, simplify. Use the examples in this section as models of how to show your work.

1. $3w - 11y + 18w - 20y$
2. $5h - 14m + 47h - 11m$
3. $2(3x + y) - 14x + y$
4. $7(8y - x) - 21y + x$
5. $\frac{3}{4}(12x - 6) - \frac{2}{3}(15x + 2)$
6. $\frac{2}{5}(20x - 3) - \frac{4}{7}(14x + 5)$
7. $24\left(\frac{2}{3}x^2 + \frac{5}{8}x - \frac{1}{6}\right)$
8. $30\left(\frac{5}{6}y^2 - \frac{7}{15}y + \frac{1}{2}\right)$
9. $-\frac{2}{3}\left(\frac{3}{2}a - 1\right) + \frac{5}{6}$
10. $-\frac{4}{5}\left(\frac{5}{4}a - 1\right) + \frac{3}{10}$
11. $7(3x - 4)$
12. $6(4x - 9)$
13. $7x(3x - 4)$
14. $6x(4x - 9)$
15. $7x^2(3x - 4)$
16. $6x^2(4x - 9)$

17. $3(p - 1) - (p - 1)$
18. $5(a - 1) - (a - 1)$
19. $15(2f) - 3(-8f)$
20. $21(3d) - 4(-9d)$
21. $16 - (2n - 4) - 8$
22. $28 - (7m - 11) - 3$
23. $-16 - 3(2n - 4) - 8$
24. $-28 - 5(7m - 11) - 3$
25. $-\frac{5}{9} - \left(\frac{2}{15} + 4g\right)$
26. $-\frac{7}{10} - \left(\frac{3}{8} + 2k\right)$
27. $0.25(4b - 8) - 0.3(b - 10)$
28. $0.75(12z - 24) - 0.5(z - 10)$
29. $-3.5(2n - 3p + 20z)$
30. $-4.5(3a - 2b + 30c)$
31. Addition is commutative. Subtraction is not commutative. Write an example using only numbers (no variables) that shows that subtraction is not commutative.
32. Multiplication is associative. Division is not associative. Write an example using only numbers (no variables) that shows that division is not associative.

For exercises 33–42, identify the property that is best illustrated by the example. A property or term may be used more than once.

33. $3 \cdot 7 = 7 \cdot 3$

34. $5(2x - 7) = 10x - 35$

35. $5 + 0 = 5$

36. $9 \cdot 1 = 9$

37. $2(x \cdot 9) = (2x) \cdot 9$

38. $4 + x = x + 4$

39. $a \cdot 0 = 0$

40. $4 + (x + 3) = 4 + (3 + x)$

41. $p \cdot \dfrac{1}{p} = 1$

42. $6 + (-6) = 0$

A. Additive identity
B. Additive inverse
C. Associative property of addition
D. Associative property of multiplication
E. Commutative property of addition
F. Commutative property of multiplication
G. Distributive property
H. Multiplicative identity
I. Multiplicative inverse
J. Zero product property

43. Is the additive identity equal to the sum of a real number and its additive inverse? Explain.

44. Is the multiplicative identity equal to the product of a real number and its multiplicative inverse? Explain.

45. Write an example that shows that addition undoes subtraction.

46. Write an example that shows that subtraction undoes addition.

47. Explain why $3 \div 0$ cannot equal 0.

48. A pie can be a model of a fraction. For example, a model of the fraction $\dfrac{4}{5}$ is a pie cut into five pieces. One of the pieces is gone. Four pieces remain. Use the model of a pie to explain why $\dfrac{4}{0}$ does not make sense.

For exercises 49–72, simplify. *Note:* A simplified expression includes only positive exponents.

49. $20h^2 - 7h^2 + 5h^3$

50. $13k^3 - 5k^2 + 9k^2$

51. $x^8 y^9 x^3 y$

52. $c^4 d^9 c^{17} d$

53. $\dfrac{u^9 x^3}{u^5}$

54. $\dfrac{c^6 d^3}{c^2}$

55. $(x^3)^4$

56. $(a^5)^3$

57. $(a^4 b)^6$

58. $(x^5 y)^4$

59. $(2x^5 y^7)^3$

60. $(2b^4 c^9)^3$

61. $(5j^2)(3jk)^2$

62. $(2p^3)(5vz)^2$

63. $\left(\dfrac{x}{2}\right)^3$

64. $\left(\dfrac{y}{4}\right)^3$

65. $\dfrac{h^4}{h^9}$

66. $\dfrac{p^6}{p^{10}}$

67. $3x^0$

68. $5y^0$

69. $\dfrac{z^2}{z^2}$

70. $\dfrac{p^6}{p^6}$

71. $\left(\dfrac{d^5 v^3}{d^5 v^{11}}\right)^0$

72. $\left(\dfrac{x^7 y^8}{x^7 y^{20}}\right)^0$

For exercises 73–90, simplify. *Note:* A simplified expression includes only positive exponents.

73. $6x(3x^2 - 4x) - 9(2x^3 + 5x^2 + 1)$

74. $10z(2z^2 - 9z) - 3(5z^3 + 2z^2 - 1)$

75. $\dfrac{w^7 z^{20}}{wz^{13}}$

76. $\dfrac{x^4 y^{15}}{xy^8}$

77. $\dfrac{12a}{5} - \dfrac{16a}{3}$

78. $\dfrac{14u}{3} - \dfrac{9u}{7}$

79. $\dfrac{12k^5}{15k^2} + \dfrac{9k^8}{10k^5}$

80. $\dfrac{20m^8}{30m^6} + \dfrac{35m^{12}}{50m^{10}}$

81. $\dfrac{12a^3 b^9}{4a^7 b^3}$

82. $\dfrac{14x^9 y^2}{30x^6 y^{13}}$

83. $(3x^6 y^{11})(4x^{-19} y^4)$

84. $(2p^3 k^{21})(7p^{-18} k^2)$

85. $\dfrac{d^5 v^3}{d^5 v^{11}}$

86. $\dfrac{x^7 y^8}{x^7 y^{20}}$

87. $\left(\dfrac{d^5 v^3}{d^5 v^{11}}\right)^4$

88. $\left(\dfrac{x^7 y^8}{x^7 y^{20}}\right)^5$

89. $\left(-\dfrac{3x^2 y}{5}\right)^2 \left(\dfrac{4x^3 y^6}{7}\right)$

90. $\left(-\dfrac{2a^4 b}{5}\right)^2 \left(\dfrac{2a^3 b^5}{7}\right)$

For exercises 91–92, simplify the expression
(a) using repeated multiplications and simplifying.
(b) using a rule of exponents.

91. $\dfrac{z^8}{z^3}$

92. $(z^4)(z^5)$

Technology

For exercises 93–96, use a scientific calculator to evaluate the expression. Use the parentheses on the calculator.

93. $-20 - 6^2 \div (2 + 1)$

94. $-18 - 9^2 \div (2 + 1)$

95. $-4^2 + 8 - 4(9 - 1)^2$

96. $-3^2 + 13 - 6(8 - 1)^2$

Find the Mistake

For exercises 97–100, the completed problem has one mistake.
(a) Describe the mistake in words, or copy down the whole problem and highlight or circle the mistake.
(b) Do the problem correctly.

97. Problem: Simplify: $(x^7)^3$

   Incorrect Answer: $(x^7)^3$

   $$= x^{10}$$

98. Problem: Simplify: $\dfrac{x^7}{x^{10}}$

   Incorrect Answer: $\dfrac{x^7}{x^{10}}$

   $$= x^3$$

99. Problem: Simplify: $3(h - 5) - (h - 11)$

Incorrect Answer: $3(h - 5) - (h - 11)$
$$= 3h - 15 - h - 11$$
$$= 2h - 26$$

100. Problem: Simplify: $x^{-7}y^2x^{-9}y^{-13}$

Incorrect Answer: $x^{-7}y^2x^{-9}y^{-13}$
$$= x^{-16}y^{-11}$$

101. In the first week of a term, you may find that you need to change your schedule and add or drop a class. Explain how to add or drop a class at your college.

102. If you rely on financial aid to buy textbooks and your financial aid has not yet arrived, what can you do to keep up with the reading and assignments until you receive your financial aid?

© DOE

A student at the City College of New York has a $4600 Pell Grant that she uses to pay her tuition bill of $4000. The amount of money remaining in her grant is an *unknown number*. We can represent this unknown number with a variable, m, and write an equation that describes the relationship between the other information and this unknown number: $m = \$4600 - \$4000$. In this section, we will write and solve equations and inequalities in one variable.

SECTION 1.3

Equations and Inequalities in One Variable

After reading the text, working the practice problems, and completing assigned exercises, you should be able to:

1. Solve a linear equation in one variable and check the solution(s).

2. Use a number line graph or interval notation to represent the solution of a linear equation in one variable.

3. Solve a linear inequality in one variable and check the solutions.

4. Use a number line graph or interval notation to represent the solution of a linear inequality in one variable.

Solutions and Equations

An equation always includes an equals sign, $=$. Equations such as $3 + 5 = 8$ do not have variables. Other equations such as $m = \$4600 - \$4000$ have one variable. The **solution** of an equation with one variable is the "unknown number." If we replace the variable in an equation with a solution of the equation and evaluate, the equation is true.

We can solve some equations "in our heads." This is **solving an equation by inspection**. By inspection, the solution of $m = \$4600 - \$4000$ is $m = \$600$. We can also use the **properties of equality** to solve an equation.

> ### The Properties of Equality
>
> **The Addition Property of Equality**   We can add any real number to both sides of an equation without changing its solution. If a, b, and c are real numbers and $a = b$, then $a + c = b + c$.
>
> **The Subtraction Property of Equality**   We can subtract any real number from both sides of an equation without changing its solution. If a, b, and c are real numbers and $a = b$, then $a - c = b - c$.
>
> **The Multiplication Property of Equality**   We can multiply both sides of an equation by any real number except 0 without changing its solution. If a, b, and c are real numbers, $c \neq 0$, and $a = b$, then $ac = bc$.
>
> **The Division Property of Equality**   We can divide both sides of an equation by any real number except 0 without changing its solution. If a, b, and c are real numbers, $c \neq 0$, and $a = b$, then $\dfrac{a}{c} = \dfrac{b}{c}$.

Takeiteasy Art/Shutterstock

The properties of equality are like a balanced double-pan scale. If exactly the same amount is added to or taken from both pans, the scale stays balanced. If we multiply or divide the amount in each pan by the same number, the scale stays balanced.

In the next example, we use the addition property of equality to isolate the variable. When the variable is isolated, we know the number that the variable represents. This number is the **solution of the equation**.

EXAMPLE 1   Solve: $a - 20 = 4$

SOLUTION ▶

$a - 20 = 4$	The variable, a, is not isolated.
$a - 20 + 20 = 4 + 20$	Addition property of equality; add 20 to each side.
$a + \mathbf{0} = \mathbf{24}$	Simplify.
$\boldsymbol{a} = \mathbf{24}$	Simplify. The variable is isolated.

We can also add or subtract from both sides vertically.

▶

$$a - 20 = 4$$
$$\underline{+20 \quad +20}$$ Addition property of equality; add 20 to each side.
$$a + \mathbf{0} = \mathbf{24}$$ Simplify.
$$\boldsymbol{a} = \mathbf{24}$$ Simplify. The variable is isolated.

In the next example, we use two properties of equality to isolate the variable. Find out the work and steps that your instructor wants you to show as you solve equations.

EXAMPLE 2   Solve: $3x + 2 = 14$

SOLUTION ▶

$$3x + 2 = 14$$
$$\underline{-2 \quad -2}$$ Subtraction property of equality; subtract 2 from each side.
$$3x + \mathbf{0} = \mathbf{12}$$ Simplify.
$$\dfrac{3x}{3} = \dfrac{12}{3}$$ Division property of equality; divide both sides by 3.
$$\mathbf{1}x = \mathbf{4}$$ Simplify; $\dfrac{3}{3} = 1$.
$$\boldsymbol{x} = \mathbf{4}$$ Simplify. The solution of this equation is $x = 4$.

If an expression in an equation can be simplified, do this before using the properties of equality. In the next example, we use the distributive property and combine like terms before using the addition property of equality. An expression is simplified when it has at most one term with a variable and one term that is a constant.

EXAMPLE 3 Solve: $5(x - 3) - 2x = 20$

SOLUTION ▶

$$5(x - 3) - 2x = 20$$

$$\mathbf{5x - 15} - 2x = 20 \qquad \text{Distributive property.}$$

$$\mathbf{3x} - 15 = 20 \qquad \text{Combine like terms.}$$

$$\underline{+15 \quad +15} \qquad \text{Addition property of equality; add 15 to both sides.}$$

$$3x + \mathbf{0} = \mathbf{35} \qquad \text{Simplify.}$$

$$\frac{3x}{3} = \frac{35}{3} \qquad \text{Division property of equality; divide both sides by 3.}$$

$$x = \frac{35}{3} \qquad \text{Simplify; } \frac{3}{3} = 1. \text{ Do not rewrite } \frac{35}{3} \text{ as a decimal.}$$

To **check** a solution, replace each variable in the original equation with the solution. Then evaluate each side of the equation. If the solution is correct, the left side of the equation will equal the right side of the equation.

EXAMPLE 4 The solution of $5(x - 3) - 2x = 20$ is $x = \dfrac{35}{3}$. Check this solution.

SOLUTION ▶

$$5\left(\frac{35}{3} - 3\right) - 2\left(\frac{35}{3}\right) = 20 \qquad \text{Replace the variable with the solution.}$$

$$5\left(\frac{35}{3} - \frac{9}{3}\right) - 2\left(\frac{35}{3}\right) = 20 \qquad \text{Rewrite 3 as a fraction: } \frac{3}{1} \cdot \frac{3}{3} = \frac{9}{3}$$

$$5\left(\frac{26}{3}\right) - 2\left(\frac{35}{3}\right) = 20 \qquad \text{Subtract inside the parentheses.}$$

$$\frac{130}{3} - \frac{70}{3} = 20 \qquad \text{Multiply.}$$

$$\frac{60}{3} = 20 \qquad \text{Subtract.}$$

$$20 = 20 \qquad \text{True; the two sides are equal; the solution is correct.}$$

When solving an equation that includes fractions, do not rewrite the numbers in the equation or the solution as a decimal number. Do not rewrite improper fractions (the numerator is greater than the denominator) as mixed numbers.

EXAMPLE 5 (a) Solve: $\dfrac{1}{2}x + \dfrac{4}{5} = \dfrac{53}{60}$

SOLUTION ▶

$$\frac{1}{2}x + \frac{4}{5} = \frac{53}{60} \qquad \text{The variable, } x, \text{ is not isolated.}$$

$$\underline{-\frac{4}{5} \quad -\frac{4}{5}} \qquad \text{Subtraction property of equality.}$$

$$\frac{1}{2}x + \mathbf{0} = \frac{53}{60} - \frac{4}{5} \qquad \text{The least common denominator of 60 and 5 is 60.}$$

$$\frac{1}{2}x = \frac{53}{60} - \frac{4}{5} \cdot \frac{12}{12} \qquad \text{Multiply by a fraction that is equal to 1.}$$

$$\frac{1}{2}x = \frac{53}{60} - \frac{48}{60} \qquad \text{Simplify.}$$

$$\frac{1}{2}x = \frac{5}{60} \qquad \text{Subtract numerators; denominator does not change.}$$

$$\frac{1}{2}x = \frac{5 \cdot 1}{5 \cdot 12} \qquad \text{Find common factors in numerator and denominator.}$$

$$\frac{1}{2}x = 1 \cdot \frac{1}{12} \qquad \text{Simplify; } \frac{5}{5} = 1$$

$$\frac{1}{2}x = \frac{1}{12} \qquad \text{Simplify.}$$

$$\frac{2}{1}\left(\frac{1}{2}x\right) = \frac{2}{1} \cdot \frac{1}{12} \qquad \text{Multiplication property of equality.}$$

$$1x = \frac{2}{1} \cdot \frac{1}{2 \cdot 6} \qquad \text{Simplify; find common factors.}$$

$$x = \frac{1}{6} \qquad \text{Simplify; } \frac{2}{2} = 1; \frac{1}{6} \text{ is in lowest terms.}$$

(b) Check.

$$\frac{1}{2}x + \frac{4}{5} = \frac{53}{60} \qquad \text{Use the original equation.}$$

$$\frac{1}{2}\left(\frac{1}{6}\right) + \frac{4}{5} = \frac{53}{60} \qquad \text{Replace the variable, } x, \text{ with } \frac{1}{6}.$$

$$\frac{1}{12} \cdot 1 + \frac{4}{5} \cdot 1 = \frac{53}{60} \qquad \text{The product of 1 and a number is the number.}$$

$$\frac{1}{12} \cdot \frac{5}{5} + \frac{4}{5} \cdot \frac{12}{12} = \frac{53}{60} \qquad \text{The least common denominator of 12 and 5 is 60.}$$

$$\frac{5}{60} + \frac{48}{60} = \frac{53}{60} \qquad \text{Multiply the fractions.}$$

$$\frac{53}{60} = \frac{53}{60} \qquad \text{True; the two sides are equal; the solution is correct.}$$

Before solving an equation that includes fractions, we can **clear the fractions**. To clear fractions from an equation, multiply all of the terms in the equation by the least common denominator of the fractions. After simplifying, the coefficients and constants in the equation are integers.

EXAMPLE 6 Clear the fractions and solve: $\frac{4}{5}(3u - 1) = \frac{1}{6}u + 7$

SOLUTION ▸ The least common denominator of $\frac{4}{5}$ and $\frac{1}{6}$ is 30. To clear the fractions, multiply both sides by 30.

$$\frac{4}{5}(3u - 1) = \frac{1}{6}u + 7$$

$$30\left(\frac{4}{5}(3u - 1)\right) = 30\left(\frac{1}{6}u + 7\right) \qquad \text{Multiplication property of equality.}$$

$$30\left(\frac{4}{5}\right)(3u - 1) = 30\left(\frac{1}{6}u\right) + 30(7) \qquad \text{Distributive property.}$$

$$\left(\frac{120}{5}\right)(3u - 1) = \frac{30}{6}u + 210 \qquad \text{Simplify.}$$

$$(24)(3u - 1) = 5u + 210 \qquad \text{Simplify; the fractions are cleared.}$$

$$(24)(3u) + (24)(-1) = 5u + 210 \qquad \text{Distributive property.}$$

$$72u - 24 = 5u + 210 \qquad \text{Simplify.}$$

$$\underline{-5u \qquad\qquad -5u} \qquad\qquad \text{Subtraction property of equality.}$$

$$67u - 24 = 0 + 210 \qquad \text{Simplify.}$$

$$67u - 24 = 210 \qquad\qquad \text{Simplify.}$$

$$\underline{+24 \qquad +24} \qquad\qquad \text{Addition property of equality.}$$

$$67u + 0 = 234 \qquad\qquad \text{Simplify.}$$

$$\frac{67u}{67} = \frac{234}{67} \qquad\qquad \text{Division property of equality.}$$

$$u = \frac{234}{67} \qquad\qquad \text{Simplify.}$$

Instead of first subtracting $5u$ from both sides, we can instead first subtract 210 from both sides or add 24 to both sides. The solution is the same.

We can check this solution by replacing the variable with the exact value for the solution, $\frac{234}{67}$, or we can check an approximate solution. To use an approximation, rewrite the solution and the fractions in the equation as decimal numbers rounded to the nearest hundredth. Replace the variable(s) with the approximate solution. If the solution is correct, the left side and the right side of the equation will be very close to equal. Because of rounding, they may not be exactly equal. In the next example, we do an approximate check for the solution of the equation in Example 6.

EXAMPLE 7   Use an approximation to check the solution $u = \dfrac{234}{67}$.

SOLUTION ▶

$$\frac{4}{5}(3u - 1) = \frac{1}{6}u + 7 \qquad\qquad \text{Use the original equation.}$$

$$(0.8)(3u - 1) \approx 0.17u + 7 \qquad\qquad \frac{4}{5} = 0.8; \frac{1}{6} \approx 0.17; u \approx 3.49$$

$$0.8(3(3.49) - 1) \approx (0.17)(3.49) + 7 \qquad \text{Replace the variable, } u, \text{ with 3.49.}$$

$$(0.8)(10.47 - 1) \approx 0.5933 + 7 \qquad\qquad \text{Follow the order of operations.}$$

$$(0.8)(9.47) \approx 7.5933 \qquad\qquad\qquad \text{Follow the order of operations.}$$

$$7.576 \approx 7.5933 \qquad\qquad\qquad\quad \text{Follow the order of operations.}$$

Since the left side of the equation is approximately equal to the right side, the solution is probably correct.

Some instructors prefer that their students do only exact checks. Ask your instructor how you should check the solution(s) of an equation.

Clearing the Fractions from an Equation in One Variable

1. Find the least common denominator of the fractions in the equation.
2. Multiply both sides of the equation by this denominator (multiplication property of equality).
3. Simplify. The coefficients and constants in the simplified equation are integers.

In the next example, the equation includes coefficients that are decimal numbers.

EXAMPLE 8   **(a)** Solve: $2.65x + 15,000 = 3.85x$

SOLUTION ▶

$$2.65x + 15,000 = 3.85x$$

$$\underline{-2.65x \qquad\qquad -2.65x} \qquad \text{Subtraction property of equality.}$$

$$0 + 15,000 = \mathbf{1.20}x \qquad \text{Simplify.}$$

$$\frac{\mathbf{15,000}}{1.20} = \frac{1.20x}{1.20} \qquad \text{Division property of equality.}$$

$$12,500 = x \qquad \text{Simplify.}$$

We can *clear the decimals* from an equation like $2.65x + 15,000 = 3.85x$ by multiplying both sides of the equation by a multiple of 10. Multiplying both sides of $2.65x + 15,000 = 3.85x$ by 100 results in an equation without decimal numbers: $265x + 1,500,000 = 385x$. The solution of this equation is the same, $x = 12,500$.

(b) Clear the decimals and solve $2.65x + 15,000 = 3.85x$.

▶

$$2.65x + 15,000 = 3.85x$$

$$100(2.65x + 15,000) = 100(3.85x) \qquad \text{Multiplication property of equality.}$$

$$265x + 1,500,000 = 385x \qquad \text{Simplify.}$$

$$\underline{-265x \qquad\qquad -265x} \qquad \text{Subtraction property of equality.}$$

$$0 + 1,500,000 = \mathbf{120}x \qquad \text{Simplify.}$$

$$\frac{\mathbf{1,500,000}}{120} = \frac{120x}{120} \qquad \text{Division property of equality.}$$

$$12,500 = x \qquad \text{Simplify.}$$

Practice Problems

For problems 1–3,
(a) clear any fractions and solve.
(b) check.

1. $5(7x + 3) - 9 = 76$ **2.** $\frac{4}{5}y - \frac{2}{3} = \frac{7}{15}$ **3.** $4(3w - 8) + 12 = 2(5w - 34)$

For problems 4–6,
(a) clear the decimals and solve.
(b) check.

4. $1.35n + 2500 = 4.55n$ **5.** $0.6x + 8.4 = 0.9x - 12.6$ **6.** $1.5(4c + 2) = 18$

Identities and Contradictions

If an equation in one variable has one solution, it is a **conditional equation**. Since the solution of $3x + 2 = 14$ is $x = 4$, this is a conditional equation. An equation with no solution is a **contradiction**. For example, there is no real number that is a solution of $x - 4 = x + 3$.

EXAMPLE 9   Solve: $2(5x - 3) = 10x + 7$

SOLUTION ▶

$$2(5x - 3) = 10x + 7$$

$$\mathbf{10x - 6} = 10x + 7 \qquad \text{Distributive property.}$$

$$\underline{-10x \qquad -10x} \qquad \text{Subtraction property of equality.}$$

$$\mathbf{0} - 6 = \mathbf{0} + 7 \qquad \text{Simplify. There are no terms with variables.}$$

$$-6 = 7 \qquad \text{False.}$$

The final equation has no variables and is false: $-6 \neq 7$. The equation $2(5x - 3) = 10x + 7$ has no solution.

For some equations, called **identities**, every real number is a solution. Since there are infinitely many numbers in the set of real numbers, these equations have infinitely many solutions.

EXAMPLE 10   Solve: $12p - 24 = 3(4p - 8)$

SOLUTION ▶

$$12p - 24 = 3(4p - 8)$$

$12p - 24 = \mathbf{12p - 24}$	Distributive property.
$\underline{-12p \qquad -12p}$	Subtraction property of equality.
$\mathbf{0} - 24 = \mathbf{0} - 24$	Simplify. There are no terms with variables.
$-24 = -24$	True.

The final equation has no variables and is true: $-24 = -24$. The solution of $12p - 24 = 3(4p - 8)$ is the set of real numbers.

If a linear equation in one variable has two or more solutions, it is an identity. So to confirm that an equation is an identity, check two real numbers.

EXAMPLE 11   The solution of $12p - 24 = 3(4p - 8)$ is the set of real numbers. Check this solution.

SOLUTION ▶ Choose any two numbers from the set of real numbers: $p = 0$ and $p = 1$.

$$12p - 24 = 3(4p - 8) \qquad\qquad 12p - 24 = 3(4p - 8)$$
$$12(\mathbf{0}) - 24 = 3(4(\mathbf{0}) - 8) \qquad\qquad 12(\mathbf{1}) - 24 = 3(4(\mathbf{1}) - 8)$$
$$\mathbf{0} - 24 = 3(\mathbf{0} - 8) \qquad\qquad \mathbf{12} - 24 = 3(\mathbf{4} - 8)$$
$$\mathbf{-24} = 3(\mathbf{-8}) \qquad\qquad \mathbf{-12} = 3(\mathbf{-4})$$
$$-24 = \mathbf{-24} \ \text{ True.} \qquad\qquad -12 = \mathbf{-12} \ \text{ True.}$$

Both numbers are solutions. This equation is an identity.

Solutions of Linear Equations in One Variable

Conditional Equation   One solution

Contradiction   No solution

Identity   Infinitely many solutions (the set of real numbers)

Practice Problems

For problems 7–10,
(a) solve.
(b) check.

7. $9x - 22 = 4(x + 2) + 5(x - 6)$ **8.** $9x - 1 = 4(x + 2) + 5(x - 6)$

9. $5x - (6x + 7) = -x + 2$ **10.** $\dfrac{3}{4}(4x - 12) = 3x - 9$

Number Line Graphs and Interval Notation

We can use a number line graph or interval notation to represent the solution(s) of an equation in one variable.

EXAMPLE 12 | The solution of $x - 1 = 2$ is $x = 3$.

(a) Use a number line graph to represent this solution.

SOLUTION ▶

$$-1\ 0\ 1\ 2\ 3\ 4\ 5\ 6\ 7$$

(b) Use interval notation to represent this solution.

▶ $[3, 3]$

We can also use a number line graph or interval notation to represent the set of real numbers.

EXAMPLE 13 | The solution of $x + 2 = x + 6 - 4$ is the set of real numbers.

(a) Use a number line graph to represent this solution.

SOLUTION ▶

$$0\ 1\ 2\ 3\ 4\ 5\ 6\ 7$$

(b) Use interval notation to represent this solution.

▶ $(-\infty, \infty)$

Practice Problems

For problems 11–13,
(a) solve.
(b) use a number line graph to represent the solution.
(c) use interval notation to represent the solution.

11. $\dfrac{3}{4}x - \dfrac{1}{12} = \dfrac{1}{24}$ **12.** $7x = 2x$ **13.** $2(3x - 7) = 6x - 14$

Inequalities in One Variable

The symbols used in writing inequalities include $>$ (greater than), $<$ (less than), $\geq$ (greater than or equal to), and $\leq$ (less than or equal to).

To solve an equation in one variable, we use the properties of equality. An **inequality** is not an equation. To solve an inequality in one variable, we use the **properties of inequality**.

The Properties of Inequality

The Addition and Subtraction Property of Inequality   Adding a real number to or subtracting a real number from both sides of an inequality does not change the solution of the inequality.

- If a, b, and c are real numbers and $a > b$, then $a + c > b + c$.
- If a, b, and c are real numbers and $a > b$, then $a - c > b - c$.

The Multiplication and Division Property of Inequality   We can multiply both sides or divide both sides of an inequality by a real number other than 0.

- If the number is *greater than* 0, the solution is not changed.
- If the number is *less than* 0, reverse the direction of the inequality sign.
- If a, b, and c are real numbers, $c > 0$, and $a > b$, then $ac > bc$.

- If a, b, and c are real numbers, $c < 0$, and $a > b$, then $ac < bc$.

- If a, b, and c are real numbers, $c > 0$, and $a > b$, then $\dfrac{a}{c} > \dfrac{b}{c}$.

- If a, b, and c are real numbers, $c < 0$, and $a > b$, then $\dfrac{a}{c} < \dfrac{b}{c}$.

EXAMPLE 14   **(a)** Solve: $3x - 6 < 2x + 11$

SOLUTION ▶

$$3x - 6 < 2x + 11$$

$\underline{-2x \qquad -2x}$	Subtraction property of inequality.
$x - 6 < \mathbf{0} + 11$	Simplify.
$\underline{+6 \qquad +6}$	Addition property of inequality.
$x + \mathbf{0} < 0 + \mathbf{17}$	Simplify.
$x < 17$	The solution is the set of real numbers less than 17.

(b) Check.

$3x - 6 < 2x + 11$	Original inequality.
$3(\mathbf{16}) - 6 < 2(\mathbf{16}) + 11$	Replace x with a number that is less than 17.
$\mathbf{48} - 6 < \mathbf{32} + 11$	Evaluate.
$42 < 43$	True.

(c) Use a number line graph to represent the solution.

(d) Use interval notation to represent the solution.

▶ $(-\infty, 17)$

In the next example, we will divide both sides of the inequality by a negative number and reverse the direction of the inequality sign. Why do we reverse the direction of the inequality sign?

Think about the inequality $4 < 12$.	$4 < 12$	$4 < 12$
If we multiply or divide both sides of the inequality by a positive number, the inequality is still true.	$2 \cdot 4 < 2 \cdot 12$ $8 < 24$ True.	$\dfrac{4}{2} < \dfrac{12}{2}$ $2 < 6$ True.
If we multiply or divide both sides of the inequality by a negative number, the inequality is false.	$4 < 12$ $-2 \cdot 4 < -2 \cdot 12$ $-8 < -24$ False.	$4 < 12$ $\dfrac{4}{-2} < \dfrac{12}{-2}$ $-2 < -6$ False.
If we reverse the direction of the inequality sign when we multiply or divide both sides of the inequality by a negative number, the inequality is true.	$4 < 12$ $-2 \cdot 4 > -2 \cdot 12$ $-8 > -24$ True.	$4 < 12$ $\dfrac{4}{-2} > \dfrac{12}{-2}$ $-2 > -6$ True.

So to find the correct solution of an inequality, reverse the sign when multiplying or dividing by a negative number.

EXAMPLE 15 (a) Solve: $-3x - 21 < 9$

SOLUTION ▶

$$-3x - 21 < 9$$
$$\underline{+21 \ +21}\qquad\text{Addition property of inequality.}$$
$$-3x + \mathbf{0} < \mathbf{30}\qquad\text{Simplify.}$$
$$\frac{-\mathbf{3}x}{-3} > \frac{30}{-3}\qquad\text{Division property of inequality; }\textcolor{red}{\text{reverse the inequality sign.}}$$
$$x > -10\qquad\text{Simplify.}$$

(b) Use a number line graph to represent the solution.

▶
$$-12 \quad -10 \quad -8 \quad -6$$

(c) Use interval notation to represent the solution.

▶ $(-10, \infty)$

An inequality with $>$ or $<$ is a **strict inequality**. An inequality with $\geq$ or $\leq$ is not strict. In the next example, the solution of the inequality is $x \geq 10$. The number 10 is a boundary value of this inequality.

EXAMPLE 16 (a) Solve: $-\dfrac{1}{2}(x - 6) \leq -2$

SOLUTION ▶

$$-\frac{1}{2}(x - 6) \leq -2$$

$$-\frac{\mathbf{1}}{\mathbf{2}}x - \frac{\mathbf{1}}{\mathbf{2}}(-6) \leq -2\qquad\text{Distributive property.}$$

$$-\frac{1}{2}x + \mathbf{3} \leq -2\qquad\text{Simplify.}$$

$$\underline{\phantom{-\tfrac{1}{2}x}-3\qquad -3}\qquad\text{Subtraction property of inequality.}$$

$$-\frac{1}{2}x + \mathbf{0} \leq \mathbf{-5}\qquad\text{Simplify.}$$

$$(-2)\left(-\frac{\mathbf{1}}{\mathbf{2}}x\right) \geq (-2)(-5)\qquad\text{Multiplication property of inequality; }\textcolor{red}{\text{reverse the sign.}}$$

$$x \geq 10\qquad\text{Simplify.}$$

(b) Check.

▶   The solution of this inequality is the set of real numbers *greater than or equal to* 10. If we check 10, we cannot tell whether the inequality sign is pointing in the right direction. So we choose to check 11, a number that is greater than 10.

$$-\frac{1}{2}(x - 6) \leq -2\qquad\text{Use the original inequality.}$$

$$-\frac{1}{2}(\mathbf{11} - 6) \leq -2\qquad\text{Check a real number greater than 10.}$$

$$-\frac{1}{2}(\mathbf{5}) \leq -2\qquad\text{Follow the order of operations.}$$

$$-\frac{\mathbf{5}}{\mathbf{2}} \leq -2\qquad\text{True; } -2 = \frac{-4}{2}; -\frac{5}{2} \leq -\frac{4}{2}$$

(c) Use a number line graph to represent the solution.

▶
$$8 \quad 9 \quad 10 \ 11 \ 12 \ 13 \ 14 \ 15$$

(d) Use interval notation to represent the solution.

▶ $[10, \infty)$

In the next example, we rewrite the inequality $-\dfrac{13}{18} \geq p$ as $p \leq -\dfrac{13}{18}$. This makes it easier to use a number line graph or interval notation to represent the inequality.

EXAMPLE 17 (a) Solve: $-\dfrac{1}{2}p - \dfrac{1}{3} \geq \dfrac{3}{4} + p$

SOLUTION ▶

$$-\frac{1}{2}p - \frac{1}{3} \geq \frac{3}{4} + p \qquad \text{The least common denominator is 12.}$$

$$12\left(-\frac{1}{2}p - \frac{1}{3}\right) \geq 12\left(\frac{3}{4} + p\right) \qquad \begin{array}{l}\text{Clear fractions; multiplication}\\\text{property of equality.}\end{array}$$

$$12\left(-\frac{1}{2}p\right) + 12\left(-\frac{1}{3}\right) \geq 12\left(\frac{3}{4}\right) + 12p \qquad \text{Distributive property.}$$

$$-\frac{12}{2}p - \frac{12}{3} \geq \frac{36}{4} + 12p \qquad \text{Simplify.}$$

$$-6p - 4 \geq 9 + 12p \qquad \text{Simplify.}$$

$$\underline{+6p \qquad\qquad +6p} \qquad \text{Addition property of equality.}$$

$$0 - 4 \geq 9 + 18p \qquad \text{Simplify.}$$

$$\underline{-9 \quad -9} \qquad \text{Subtraction property of equality.}$$

$$-13 \geq 0 + 18p \qquad \text{Simplify.}$$

$$\frac{-13}{18} \geq \frac{18p}{18} \qquad \text{Division property of equality.}$$

$$-\frac{13}{18} \geq p \qquad \text{Simplify; this is equivalent to } p \leq -\frac{13}{18}.$$

(b) Check.

▶ The solution of this inequality is the set of real numbers less than or equal to $-\dfrac{13}{18}$. To make the arithmetic easier, we will check the first integer less than $-\dfrac{13}{18}$, which is -1.

$$-\frac{1}{2}p - \frac{1}{3} \geq \frac{3}{4} + p \qquad \text{Check in the original inequality.}$$

$$-\frac{1}{2}(-1) - \frac{1}{3} \geq \frac{3}{4} + (-1) \qquad \text{Replace the variable with } -1.$$

$$\frac{1}{2} - \frac{1}{3} \geq \frac{3}{4} - 1 \qquad \text{Follow the order of operations.}$$

$$\frac{1}{2}\left(\frac{6}{6}\right) - \frac{1}{3}\left(\frac{4}{4}\right) \geq \frac{3}{4}\left(\frac{3}{3}\right) - 1\left(\frac{12}{12}\right) \qquad \text{The least common denominator is 12.}$$

$$\frac{6}{12} - \frac{4}{12} \geq \frac{9}{12} - \frac{12}{12} \qquad \text{Evaluate.}$$

$$\frac{2}{12} \geq -\frac{3}{12} \qquad \text{True.}$$

(c) Use a number line graph to represent the solution.

▶ Use an approximate value of $-\dfrac{13}{18} \approx -0.72$ to estimate the location of the bracket on the number line.

(d) Use interval notation to represent the solution.

▶ $$\left(-\infty, -\frac{13}{18}\right]$$

Practice Problems

For problems 14–21,
(a) solve.
(b) check.
(c) use a number line graph to represent the solution.
(d) use interval notation to represent the solution.

14. $3x + 18 > 12$ **15.** $-3x + 8 > 12$ **16.** $13x + 8 \geq -12$

17. $3(5c - 1) \leq 45 + 9c$ **18.** $-\frac{1}{2}a - 9 < \frac{1}{6}a$ **19.** $\frac{2}{3}x + \frac{1}{5} \leq -2x + 9$

20. $\frac{3}{7}(14p - 8) \geq -21$ **21.** $-\frac{3}{4}w + \frac{1}{2} > \frac{5}{8}$

1.3 VOCABULARY PRACTICE

Match the term with its description or symbol.

1. A symbol that represents an unknown number
2. We can subtract any real number from both sides of an equation without changing its solution.
3. $\geq$
4. $\leq$
5. A number or numbers that can replace the variable in an equation and result in a true statement
6. We can add any real number to both sides of an equation without changing its solution.
7. An inequality that includes a $>$ or $<$ sign
8. $\approx$
9. An equation with no solution
10. An equation with infinitely many solutions

A. addition property of equality
B. approximately equal to
C. contradiction
D. greater than or equal to
E. identity
F. less than or equal to
G. solution
H. strict inequality
I. subtraction property of equality
J. variable

1.3 Exercises

Follow your instructor's guidelines for showing your work.

For exercises 1–16,
(a) use a number line graph to represent the solution.
(b) use interval notation to represent the solution.

1. $x = 5$
2. $x = 3$
3. $x > 5$
4. $x > 3$
5. $x < 5$
6. $x < 3$
7. $x \geq 5$
8. $x \geq 3$
9. $x \leq 5$
10. $x \leq 3$
11. $x = \frac{3}{7}$
12. $x = \frac{4}{9}$
13. $x > 100$
14. $x > -150$
15. x is a real number less than 5
16. x is a real number less than 2

For exercises 17–78,
(a) solve.
(b) check.
17. $p - 8 = 21$
18. $w - 12 = 34$
19. $-31 = n + 40$
20. $-24 = m + 13$
21. $8b + 2 = -14$
22. $6c + 8 = -34$

23. $2(x - 1) = 15$

24. $3(x - 1) = 17$

25. $2(12 - z) = 4(3 - z)$

26. $8(10 - y) = 6(14 - y)$

27. $7(2x + 4) - 9 = 15x - 3$

28. $3(2x - 9) + 8x = 15x - 11$

29. $c - \dfrac{3}{4} = \dfrac{17}{4}$

30. $d - \dfrac{5}{8} = \dfrac{27}{8}$

31. $\dfrac{3}{4}h = 15$

32. $\dfrac{4}{5}k = 60$

33. $0.8 = w + 1.9$

34. $0.29 = y + 0.81$

35. $\dfrac{a}{0.3} = 15$

36. $\dfrac{m}{0.7} = 12$

37. $\dfrac{2}{3}x = \dfrac{8}{9}$

38. $\dfrac{3}{4}x = \dfrac{9}{16}$

39. $k - 0.35k = 78$

40. $z - 0.48z = 208$

41. $\dfrac{1}{8}a + \dfrac{3}{8}a = \dfrac{5}{6}$

42. $\dfrac{1}{10}d + \dfrac{7}{10}d = 48$

43. $49 = \dfrac{1}{4}k + \dfrac{1}{3}k$

44. $62 = \dfrac{2}{5}n + \dfrac{3}{8}n$

45. $-c - c = 21$

46. $-p - p = 13$

47. $q - 0.9q = 45$

48. $a - 0.8a = 54$

49. $2v - 9 = 2(v + 4)$

50. $3u - 17 = 3(u + 9)$

51. $7c - 18 = 2(4c - 9) - c$

52. $9w - 20 = 5(2w - 4) - w$

53. $7(2x + 4) - 9 = 15x - (x + 3)$

54. $3(2x - 9) + 8x = 15x - (x + 16)$

55. $11n - (n + 5) = 3(2n + 1) + 4(n - 2)$

56. $9u - (u + 11) = 3(2u + 1) + 2(u - 7)$

57. $2z = 9z$

58. $12y = 2y$

59. $\dfrac{3}{4}(12p - 8) = 16$

60. $\dfrac{2}{3}(15k - 4) = 18$

61. $\dfrac{h}{12} = 9$

62. $\dfrac{k}{15} = 6$

63. $\dfrac{M}{75} = \dfrac{8}{15}$

64. $\dfrac{P}{128} = \dfrac{11}{16}$

65. $\dfrac{8}{15} = \dfrac{n}{20}$

66. $\dfrac{9}{20} = \dfrac{a}{30}$

67. $\dfrac{3}{4}x + \dfrac{5}{6}x - 12 = 2x - 12 - \dfrac{5}{12}x$

68. $\dfrac{4}{5}x + \dfrac{2}{7}x - 18 = 2x - 18 - \dfrac{32}{35}x$

69. $\dfrac{3}{4}a + \dfrac{5}{6} = \dfrac{5}{8}$

70. $\dfrac{5}{6}c + \dfrac{7}{8} = \dfrac{3}{4}$

71. $-\dfrac{1}{2}v = 8$

72. $-\dfrac{1}{4}q = 6$

73. $5x - (6 - 4x) = 84$

74. $4x - (10 - 5x) = 44$

75. $2 - (3z + 1) = 8z$

76. $4 - (5p + 2) = 12p$

77. $\dfrac{3}{7}x = x$

78. $\dfrac{4}{9}z = z$

79. A student solved $7(3x - 9) + 2 = 14x + 6 + 7x$.

$$7(3x - 9) + 2 = 14x + 6 + 7x$$
$$21x - 63 + 2 = 14x + 6 + 7x$$
$$21x - 61 = 21x + 6$$

After this step, the student knew that the equation has no solution. Explain how the student could know this before she finished solving it.

80. A student solved $7(3x - 9) + 2 = 14x - 61 + 7x$.

$$7(3x - 9) + 2 = 14x - 61 + 7x$$
$$21x - 63 + 2 = 14x - 61 + 7x$$
$$21x - 61 = 21x - 61$$

After this step, the student knew that the solution is the set of real numbers. Explain how the student could know this before he finished solving it.

81. Write your own example of an equation in one variable that has no solution.

82. Write your own example of an equation in one variable that has infinitely many solutions.

83. You are teaching someone to solve an equation in one variable. Write directions that tell this person what to do.

84. You are teaching someone to clear the fractions from an equation in one variable. Write directions that tell this person what to do.

85. Explain how to use interval notation to represent $x > 6$.

86. Explain how to use interval notation to represent $x \leq 2$.

87. Explain how to use a number line graph to represent $x \leq 2$.

88. Explain how to use a number line graph to represent $x > 6$.

For exercises 89–144,
(a) solve.
(b) check.
(c) use interval notation to represent the solution.

89. $3x - 9 \leq 15$

90. $4x - 12 \leq 16$

91. $-3x - 9 < 15$

92. $-4x - 12 < 16$

93. $-3x - 9 > -15$

94. $-4x - 12 > -16$

95. $3x - 9 \geq -15$

96. $4x - 12 \geq -16$

97. $2(d + 1) - (d + 8) < -15$

98. $2(r + 1) - (r + 10) < -35$

99. $4(n - 50) > 3(2n + 6)$

100. $8(w - 15) > 2(5w + 14)$

101. $-\dfrac{3}{4}x + 8 < 15 - \dfrac{2}{3}x$

102. $-\dfrac{3}{8}x - 18 > 34 + \dfrac{5}{4}x$

103. $-\dfrac{1}{2}x > 36$

104. $-\dfrac{1}{3}x < 52$

105. $-12y \geq -2y$

106. $-24h \geq -2h$

107. $\dfrac{x}{2} + \dfrac{5}{9} \leq 0$

108. $\dfrac{z}{3} + \dfrac{5}{8} \leq 0$

109. $-f > 3$

110. $-p > 4$

111. $0 < -2k$

112. $0 \leq -8c$

113. $8x - 9 \geq 6x - 6$

114. $12x - 3 \geq 10x - 4$

115. $7x - 12 > 6x - 12$

116. $5x - 34 > 4x - 34$

117. $4(2x - 6) \leq 5(x - 9)$

118. $3(4x - 1) \leq 9(x - 15)$

119. $\dfrac{x}{340} \geq \dfrac{7}{20}$

120. $\dfrac{x}{945} \geq \dfrac{7}{15}$

121. $\dfrac{15}{16} < \dfrac{x}{336}$

122. $\dfrac{11}{12} < \dfrac{x}{276}$

123. $3(x + 6) - x < 5x + 1$

124. $2(x + 4) - x < 7x + 1$

125. $8p < 3p$

126. $7k < 3k$

127. $-8p < 3p$

128. $-7k < 3k$

129. $8p < -3p$

130. $7k < -3k$

131. $-8p < -3p$

132. $-7k < -3k$

133. $\dfrac{3}{4}m + \dfrac{1}{6} > 1$

134. $\dfrac{5}{6}n + \dfrac{1}{4} > 1$

135. $\dfrac{1}{3}k - 8 \leq \dfrac{3}{5}k + 2$

136. $\dfrac{1}{2}h - 6 \leq \dfrac{5}{4}h + 2$

137. $\dfrac{2}{3}w - 5 < -\dfrac{4}{9}$

138. $\dfrac{4}{5}m - 8 < -\dfrac{3}{10}$

139. $5h + \dfrac{1}{2} > -2h + \dfrac{3}{4}$

140. $9c + \dfrac{2}{3} > -5c + \dfrac{5}{6}$

141. $\dfrac{3}{4}(2u + 8) < \dfrac{5}{6}(3u - 12)$

142. $\dfrac{1}{2}(3v - 4) < \dfrac{4}{5}(2v - 10)$

143. $\dfrac{1}{2}(x + 9) - \dfrac{6}{7}(x + 2) \leq 0$

144. $\dfrac{2}{5}(x + 6) - \dfrac{7}{8}(x + 5) \leq 0$

Find the Mistake

For exercises 145–148, the completed problem has one mistake.
(a) Describe the mistake in words, or copy down the whole problem and highlight or circle the mistake.
(b) Do the problem correctly.

145. Problem: Solve: $-7x + 2 < 30$

  Incorrect Answer: $-7x + 2 < 30$
  $$-7x + 2 - 2 < 30 - 2$$
  $$-7x < 28$$
  $$\frac{-7x}{-7} < \frac{28}{-7}$$
  $$x < -4$$

146. Problem: Solve: $3x + 8 = 13$

  Incorrect Answer: $3x + 8 = 13$
  $$\frac{+8 \quad +8}{3x = 21}$$
  $$\frac{3x}{3} = \frac{21}{3}$$
  $$x = 7$$

147. Problem: Use interval notation to represent $x < 5$.

  Incorrect Answer: $(\infty, 5)$

148. Problem: Use a number line graph to represent $x < -2$.

  Incorrect Answer:

  $\longleftarrow \!\!|\!\!-\!\!|\!\!-\!\!|\!\!\blacktriangleleft\!\!|\!\!-\!\!|\!\!-\!\!|\!\!-\!\!|\!\!-\!\!|\!\!\longrightarrow x$
  $\quad -7\ -6\ -5\ -4\ -3\ -2\ -1\ \ 0$

SUCCESS IN COLLEGE MATHEMATICS

149. Describe the purpose of a syllabus.

150. What grade do you need to earn in this class to be able to move on to the next class?

© wavebreakmedia ltd/Shutterstock

Many measurements include very large or very small numbers. Scientists often write these measurements in **scientific notation**. Measurements used in scientific formulas and in problem solving must be in the correct units. Scientists use **unit analysis** to change measurements into the correct units. In this section, we will study scientific notation and unit analysis.

SECTION 1.4

Scientific Notation and Unit Analysis

After reading the text, working the practice problems, and completing assigned exercises, you should be able to:

1. Write a measurement in scientific notation.
2. Do arithmetic in scientific notation.
3. Use the meaning of an SI prefix to rewrite a measurement using its base unit.
4. Use a conversion factor to show the relationship of two units of measurement.
5. Use unit analysis to convert a measurement into another unit of measurement.
6. Simplify expressions that include units of measurement.

Powers of Ten and Scientific Notation

The number 432 represents 4 hundreds +3 tens +2 ones. The number 432 is in **place value notation**. As we move to the left, each position has ten times more value.

A **power of ten** is an exponential expression with a base of **ten** raised to a **power**, the exponent.

When the exponent is a whole number, the value of the power of ten is greater than or equal to 1.	When the exponent is an integer that is less than 0, the value of the power of ten is less than 1.
$1 = 10^0$	$10^{-1} = \dfrac{1}{10^1} = 0.1$
$10 = 10^1$	
$100 = 10^2$	$10^{-2} = \dfrac{1}{10^2} = 0.01$
$1000 = 10^3$	
$10{,}000 = 10^4$	$10^{-3} = \dfrac{1}{10^3} = 0.001$

We describe some very large numbers using special names or using a power of ten. For example, one billion is 1,000,000,000 or 10^9. One trillion is 1,000,000,000,000 or 10^{12}.

A number in **scientific notation** is the product of a number between 1 and 10 (including 1 but not including 10) and a power of ten. We use $\times$ as the notation for multiplication. In 4×10^8, the number 4 is the **mantissa**, and 10^8 is the power of ten.

> **Scientific Notation** A number $a \times 10^b$ is in scientific notation if b is an integer, a is a real number, and $1 \le |a| < 10$. The mantissa is a. The power of ten is 10^b.

EXAMPLE 1 The speed of sound in helium gas at 0°C is $927 \dfrac{\text{meter}}{\text{second}}$. Write this measurement in scientific notation.

SOLUTION ▶

$$927 \frac{\text{meter}}{\text{second}}$$ The original measurement is in place value notation.

$$= \mathbf{9.27 \times 100} \; \frac{\text{meter}}{\text{second}}$$ Rewrite 927 (the mantissa) as a number between 1 and 10.

$$= 9.27 \times 10^2 \; \frac{\text{meter}}{\text{second}}$$ Rewrite 100 as a power of ten.

To quickly write a number that is *greater than* 1 in scientific notation, move the decimal point until the mantissa is between 1 and 10. The number of places the decimal point moves is equal to the exponent on the power of ten.

$$927. \frac{\text{meter}}{\text{second}}$$ The decimal point is to the right of 7.

$$= 9.27 \times 10^2 \; \frac{\text{meter}}{\text{second}}$$ Move the decimal point two place values to the left.

A simplified exponential expression has only positive exponents. However, a number that is in *scientific notation* can include a negative exponent.

EXAMPLE 2 The diameter of a bacterium is 0.000007 meter. Write this measurement in scientific notation.

SOLUTION ▶

$$0.000007 \text{ meter}$$

$$= \mathbf{7 \times 0.000001} \text{ meter}$$ The mantissa, 7, is a number between 1 and 10.

$$= \mathbf{7 \times 10^{-6}} \text{ meter}$$ Rewrite 0.000001 as 10^{-6}.

To quickly write a number that is *less than* 1 in scientific notation, move the decimal point until the mantissa is between 1 and 10. The exponent will be a negative number.

0.000007 meter

$= 7 \times 10^{-6}$ meter   Move the decimal point six places to the right.

Practice Problems

For problems 1–4, write the measurement in scientific notation. Include the units.

1. 0.000560 meter   **2.** 299,000,000 $\dfrac{\text{meter}}{\text{second}}$ **3.** 6200 kilograms

4. 0.041 liter

Arithmetic in Scientific Notation

To multiply in scientific notation, use the product rule of exponents, $x^m \cdot x^n = x^{m+n}$, to multiply the powers of ten. To multiply units that are the same, use exponential notation. For example, (meter)(meter) = meter^2. If the units are different, just write them as two separate words or connect them with a dot. For example, (acre)(foot) = acre foot or acre · foot.

EXAMPLE 3   Evaluate: $(7 \times 10^2 \text{ meter})(8 \times 10^4 \text{ meter})$

SOLUTION ▶ $(7 \times 10^2 \text{ meter})(8 \times 10^4 \text{ meter})$

$= (7 \times 8)(10^2 \times 10^4)(\text{meter} \times \text{meter})$ Regroup the factors.

$= 56 \times 10^{(2+4)} \text{ meter}^2$ Simplify; product rule of exponents.

$= 56 \times 10^6 \text{ meter}^2$ Simplify; the mantissa is not between 1 and 10.

$= \mathbf{5.6 \times 10^7} \text{ meter}^2$ The mantissa is now between 1 and 10.

When we rewrite 56 as 5.6, the *mantissa decreases* by one power of ten. So the *exponent* on the power of ten *increases* by 1 from 6 to 7.

To divide in scientific notation, use the quotient rule of exponents, $\dfrac{x^m}{x^n} = x^{m-n}$.

EXAMPLE 4   Evaluate: $\dfrac{1.56 \times 10^6}{2.5 \times 10^9}$

SOLUTION ▶ $\dfrac{1.56 \times 10^6}{2.5 \times 10^9}$

$= \dfrac{1.56}{2.5} \times \dfrac{10^6}{10^9}$ Divide the mantissas; divide the powers of ten.

$= \dfrac{1.56}{2.5} \times 10^{(6-9)}$ Quotient rule of exponents.

$= \mathbf{0.624 \times 10^{-3}}$ Simplify; the mantissa is not between 1 and 10.

$= 6.24 \times 10^{-4}$ The mantissa is now between 1 and 10.

When we rewrite 0.624 as 6.24, the *mantissa increases* by one power of ten. So the *exponent* on the power of ten *decreases* by 1 from -3 to -4.

When raising a measurement in scientific notation to a power, use the many bases rule for a product, $(xy)^m = x^m y^m$. If there is a unit in the measurement, it is also raised to the power.

EXAMPLE 5 | Evaluate: $(7 \times 10^{-6} \text{ meter})^2$

SOLUTION ▶ $\quad (7 \times 10^{-6} \text{ meter})^2$

$\quad = 7^2 \times (10^{-6})^2 \text{ meter}^2 \qquad$ Many bases rule for a product.

$\quad = \mathbf{49} \times \mathbf{10^{-12}} \text{ meter}^2 \qquad$ Simplify.

$\quad = \mathbf{4.9} \times \mathbf{10^{-11}} \text{ meter}^2 \qquad$ The mantissa is now between 1 and 10.

When adding or subtracting exponential expressions, we can combine only like terms. The base and the exponent in the expressions must be the same. Similarly, *we can add or subtract numbers in scientific notation only if the powers of ten are the same and the units are the same.* If the powers of ten are different, we must rewrite one of the measurements before adding.

EXAMPLE 6 | The mass of the earth is 5.97×10^{24} kilograms. The mass of Mars is 6.42×10^{23} kilograms. Find the total mass of the earth and Mars.

SOLUTION ▶ The powers of ten in the measurements are not the same. We can rewrite either of the measurements. Choosing Mars, we decrease the mantissa by one power of ten, from 6.42 to 0.642, and increase the power of ten from 10^{23} to 10^{24}.

mass of earth + mass of Mars

$\quad = 5.97 \times 10^{24} \text{ kilograms} + 6.42 \times 10^{23} \text{ kilograms} \qquad$ Powers of ten are different.

$\quad = 5.97 \times 10^{24} \text{ kilograms} + \mathbf{0.642 \times 10^{24}} \text{ kilograms} \qquad$ Powers of ten are the same.

$\quad = \mathbf{6.612 \times 10^{24}} \text{ kilograms} \qquad$ Power of ten does not change.

Practice Problems

For problems 5–8, evaluate. The final answer must be in scientific notation.

5. $(6.2 \times 10^8 \text{ kilometer})(4.5 \times 10^7 \text{ kilometer})$

6. $\dfrac{6.45 \times 10^{-3} \text{ gram}}{7.5 \times 10^2 \text{ milliliter}}$

7. $(2.5 \times 10^{-4} \text{ kilogram})\left(9.8 \dfrac{\text{meter}}{\text{second}^2}\right)$

8. $5.1 \times 10^4 \text{ liters} + 6.7 \times 10^3 \text{ liters}$

Systems of Measurement

Scientists use units from the **International System of Measurement (SI)**. Each type of measurement has a **base unit**. Table 1 lists some of the base units.

Table 1 Base Units in the SI System of Measurement

Base unit	What the base unit measures	Abbreviation
meter	distance	m
gram	mass	g
second	time	s or sec

A few other units act like base units. These include the liter (unit of volume), the bel (unit of sound intensity), and the newton (unit of force).

To create smaller or larger units, write a prefix in front of the base unit. The commonly used prefixes are listed in Table 2.

Table 2 SI Prefixes

Prefix	Abbreviation	Multiplier	Name for multiplier
giga-	G	10^9	1 billion
mega-	M	10^6	1 million
kilo-	k	10^3	1 thousand
hecto-	h	10^2	1 hundred
deka-	da	10^1	1 ten
Base unit		1	1
deci-	d	10^{-1}	1 tenth
centi-	c	10^{-2}	1 hundredth
milli-	m	10^{-3}	1 thousandth
micro-	μ	10^{-6}	1 millionth
nano-	n	10^{-9}	1 billionth

EXAMPLE 7 Use place value notation, and rewrite the measurement in its base unit.

(a) 6 kilometers   The base unit is meters. The prefix is kilo-.

SOLUTION ▶ $= 6 \times 10^3$ **meters**   The multiplier for the prefix kilo- is 10^3.

$= $ **6000** meters   Place value notation.

(b) 9 deciliters   The base unit is liter. The prefix is deci-.

▶ $= 9 \times 10^{-1}$ **liter**   The multiplier for the prefix deci- is 10^{-1}.

$= $ **0.9** liter   Place value notation.

In many daily activities in the United States we use the U.S. customary system, which includes units such as inch, foot, mile, gallon, and acre. The units in this system come from tradition. We cannot use prefixes to make units in this system smaller or larger.

Practice Problems

For problems 9–14, use place value notation and rewrite the measurement in its base unit.

9. 7 milliliters   **10.** 9 kilograms   **11.** 9 micrometers

12. 8×10^4 nanometers   **13.** 2×10^1 kilonewtons   **14.** 3×10^2 megatons

Relationship of Units

When working with units of measurement, we often use abbreviations, as listed in Table 3.

Table 3 Abbreviations of Units

Units of length	meter	kilometer	centimeter	millimeter	mile	yard	foot	inch
Abbreviation	m	km	cm	mm	mile	yd	ft	in.

Units of volume	liter	milliliter	cubic centimeter	gallon	quart	pint	cup	liquid ounce	table-spoon	tea-spoon
Abbreviation	L	mL	cc or cm^3	gal	qt	pt	c	oz	T	t

Table 3 Abbreviations of Units (*continued*)

Units of mass/weight	gram	kilogram	milligram	microgram
Abbreviation	g	kg	mg	μg

Units of time	hour	minute	second
Abbreviation	hr	min	s

Units of force	newton	pound	ounce
Abbreviation	N	lb	oz

The relationships of some units are listed in Table 4. The "approximately equal" sign, $\approx$, shows that the value of a measurement is rounded.

Table 4 Relationship of Units

SI to U.S. customary	U.S. customary to U.S. customary	Other
2.54 cm = 1 in.	1 ft = 12 in.	1 astronomical unit (AU) $\approx$ 1.496 $\times$ 10^8 km
1.6 km $\approx$ 1 mi	1 mi = 5280 ft	
1 L $\approx$ 1.06 qt	1 gal = 4 qt	1 hr = 60 min
5 mL $\approx$ 1 t	1 qt = 2 pt	1 min = 60 s
4047 m^2 $\approx$ 1 acre	1 pt = 2 c	1 day = 24 hr
1 kg $\approx$ 2.2 lb*	16 T = 1 c	1 year $\approx$ 365 days
	1 pt = 16 fluid oz	\$1.00 $\approx$ 0.69 euro†
	1 T = 3 t	\$1.00 $\approx$ 10.687 Mexican pesos†
	1 lb = 16 oz	\$1.00 $\approx$ 1.23 Canadian dollars†
	1 ton = 2000 lb	1 mL = 1 cm^3
	1 acre $\approx$ 43,560 ft^2	1 cc = 1 cm^3

*At the surface of the earth.
†Currency values change.

We can write an equation that shows the relationship of two units of measurement.

EXAMPLE 8 Write an equation that shows the relationship of the units.

(a) centimeter and meter

SOLUTION ▶ The prefix *centi-* means 10^{-2} or 0.01.

1 cm = 0.01 m

(b) megawatt and watt

▶ The prefix *mega-* means 10^6 or 1,000,000.

1 megawatt = 1,000,000 watts

(c) foot and inch

▶ There are 12 in. in 1 ft.

12 in. = 1 ft

Unit Analysis

A fraction with the same numerator and denominator is equal to 1. A fraction with an *equivalent* numerator and denominator is also equal to 1. For example, $\dfrac{1\text{ qt}}{4\text{ c}} = 1$. The fraction $\dfrac{1\text{ qt}}{4\text{ c}}$ is a **conversion factor**.

In unit analysis, we change the units of a measurement by multiplying by conversion factors. Multiplying by 1 does not change the value of the measurement.

EXAMPLE 9   Use unit analysis to convert $65\ \dfrac{\text{mi}}{\text{hr}}$ into $\dfrac{\text{ft}}{\text{s}}$. Round to the nearest whole number.

SOLUTION ▶ From Table 4: 1 hr = 60 min; 1 min = 60 s; 1 mi = 5280 ft.

Since we do not have the relationship for hours and seconds, first change hours into minutes and then change minutes into seconds. To change hours into minutes, multiply by a conversion factor that shows the relationship of minutes and hours. We can use either $\dfrac{1\text{ hr}}{60\text{ min}}$ or $\dfrac{60\text{ min}}{1\text{ hr}}$. Choose the conversion factor $\dfrac{1\text{ hr}}{60\text{ min}}$ so that the unit being changed (hr) is in the numerator and the denominator. This creates the fraction $\dfrac{\text{hr}}{\text{hr}}$, that is equal to 1. Simplify fractions that are equal to 1 by putting a line through the units in the numerator and denominator.

$$\frac{65\text{ mi}}{1\text{ hr}}$$

$$= \left(\frac{65\text{ mi}}{1\ \cancel{\text{hr}}}\right)\left(\frac{1\ \cancel{\text{hr}}}{60\text{ min}}\right)$$

$$65\,\frac{\text{mi}}{\text{hr}} = \frac{65\text{ mi}}{1\text{ hr}}$$

$$\frac{1\text{ hr}}{60\text{ min}} = 1$$

Now multiply again by a factor equal to 1 to change minutes into seconds.

$$= \left(\frac{65\text{ mi}}{1\ \cancel{\text{hr}}}\right)\left(\frac{1\ \cancel{\text{hr}}}{60\ \cancel{\text{min}}}\right)\left(\frac{1\ \cancel{\text{min}}}{60\text{ s}}\right)$$

$$\frac{1\text{ min}}{60\text{ s}} = 1$$

Next, multiply again by a conversion factor equal to 1 to change miles into feet.

$$= \left(\frac{65\ \cancel{\text{mi}}}{1\ \cancel{\text{hr}}}\right)\left(\frac{1\ \cancel{\text{hr}}}{60\ \cancel{\text{min}}}\right)\left(\frac{1\ \cancel{\text{min}}}{60\text{ s}}\right)\left(\frac{5280\text{ ft}}{1\ \cancel{\text{mi}}}\right)$$

$$\frac{5280\text{ ft}}{1\text{ mi}} = 1$$

$$= \frac{343{,}200\text{ ft}}{3600\text{ s}}$$

Simplify.

$$\approx 95\,\frac{\text{ft}}{\text{s}}$$

Simplify; round.

To make this process more efficient, multiply by all three conversion factors in one step.

$$\frac{65 \text{ mi}}{1 \text{ hr}}$$

$$= \left(\frac{65 \text{ mi}}{1 \text{ hr}}\right)\left(\frac{1 \text{ hr}}{60 \text{ min}}\right)\left(\frac{1 \text{ min}}{60 \text{ s}}\right)\left(\frac{5280 \text{ ft}}{1 \text{ mi}}\right) \qquad \text{Multiply by a factor equal to 1.}$$

$$= \frac{343{,}200 \text{ ft}}{3600 \text{ s}} \qquad \text{Simplify the numerator; simplify the denominator.}$$

$$\approx 95 \frac{\text{ft}}{\text{s}} \qquad \text{Divide the numerator by the denominator; round.}$$

In Example 9, the original measurement included two different units: miles and hours. In the next example, the original measurement has only one unit: gallons.

EXAMPLE 10 Use unit analysis to convert 4.5 gal into cups.

SOLUTION ▶ From Table 4: 1 gal = 4 qt; 1 qt = 2 pt; 1 pt = 2 c.

$$4.5 \text{ gal}$$

$$= \left(\frac{4.5 \text{ gal}}{1}\right)\left(\frac{4 \text{ qt}}{1 \text{ gal}}\right)\left(\frac{2 \text{ pt}}{1 \text{ qt}}\right)\left(\frac{2 \text{ c}}{1 \text{ pt}}\right) \qquad \text{Multiply by a factor equal to 1.}$$

$$= 72 \text{ c} \qquad \text{Simplify.}$$

When converting U.S. dollars into another currency, use the word *dollar* rather than the $ symbol.

EXAMPLE 11 If the price of gasoline in France is 1.14 euros per liter, use the currency values from Table 4 to find the price of gasoline in U.S. dollars per gallon. Round to the nearest hundredth.

SOLUTION ▶ From Table 4: 1.06 qt = 1 L; 1 gal = 4 qt; 1 dollar = 0.69 euro.

$$1.14 \frac{\text{euro}}{\text{liter}}$$

$$= \left(\frac{1.14 \text{ euro}}{1 \text{ L}}\right)\left(\frac{1 \text{ dollar}}{0.69 \text{ euro}}\right)\left(\frac{1 \text{ L}}{1.06 \text{ qt}}\right)\left(\frac{4 \text{ qt}}{1 \text{ gal}}\right) \qquad \text{Multiply by a factor equal to 1.}$$

$$= \frac{4.56 \text{ dollar}}{0.7314 \text{ gal}} \qquad \text{Simplify.}$$

$$\approx 6.23 \frac{\text{dollar}}{\text{gal}} \qquad \text{Round; \$6.23 per gallon.}$$

Health care providers use unit analysis in their work with medications. In the next example, we rewrite the relationship 50 mg Demerol = 5 mL of syrup as a conversion factor that is equal to 1, $\frac{5 \text{ mL}}{50 \text{ mg}}$.

EXAMPLE 12 Demerol syrup contains 50 mg of Demerol in 5 mL of syrup. A patient needs 65 mg of Demerol. The units on the syringe are cubic centimeters (cc). Use unit analysis to find the number of cubic centimeters of syrup to give the patient.

SOLUTION ▶ Relationships: 50 mg Demerol = 5 mL syrup; 1 cc = 1 mL.

$$65 \text{ mg} \qquad \text{Amount of Demerol needed by patient.}$$

$$= \left(\frac{65 \text{ mg}}{1}\right)\left(\frac{5 \text{ mL}}{50 \text{ mg}}\right)\left(\frac{1 \text{ cc}}{1 \text{ mL}}\right) \qquad \text{Multiply by conversion factors.}$$

$$= 6.5 \text{ cc} \qquad \text{Simplify.}$$

Practice Problems

For problems 18–21, use unit analysis and the information from Table 4 to convert the measurement into the given unit(s). If necessary, round to the nearest hundredth.

18. $15 \dfrac{\text{mi}}{\text{hr}}$ to $\dfrac{\text{ft}}{\text{s}}$ **19.** $\dfrac{\$2.60}{1 \text{ gal}}$ to $\dfrac{\text{euro}}{\text{L}}$ **20.** $8 \dfrac{\text{g}}{\text{dL}}$ to $\dfrac{\text{mg}}{\text{mL}}$ **21.** 13 qt to T

Simplifying Expressions and Equations That Include Units

Scientists usually include the units of measurement in all steps of problem solving.

EXAMPLE 13   On January 27, 2010, the distance between Mars and the earth was about 9.9×10^{10} m. The speed of light in a vacuum is about $3.0 \times 10^{8} \dfrac{\text{m}}{\text{s}}$. Since $\text{time} = \dfrac{\text{distance}}{\text{rate}}$, the time for light to travel from Mars to the earth on that day was about $\dfrac{9.9 \times 10^{10} \text{ m}}{3.0 \times 10^{8} \dfrac{\text{m}}{\text{s}}}$. Simplify this expression to find the time. Write the answer in scientific notation. (*Source:* www.nasa.gov)

SOLUTION ▶

$$\dfrac{9.9 \times 10^{10} \text{ m}}{3.0 \times 10^{8} \dfrac{\text{m}}{\text{s}}}$$

$$= \dfrac{9.9}{3.0} \times \dfrac{10^{10}}{10^{8}} \times \dfrac{\text{m}}{\dfrac{\text{m}}{\text{s}}} \qquad \text{Divide mantissas, exponential expressions, and units.}$$

$$= 3.3 \times 10^{(10-8)} \times \text{m} \div \dfrac{\text{m}}{\text{s}} \qquad \text{Quotient rule; rewrite fraction as division.}$$

$$= 3.3 \times 10^{2} \times \cancel{\text{m}} \times \dfrac{\text{s}}{\cancel{\text{m}}} \qquad \text{Rewrite division as multiplication by reciprocal.}$$

$$= 3.3 \times 10^{2} \text{ s} \qquad \text{Simplify; time for light to travel from Mars to the earth.}$$

Practice Problems

For problems 22–24, simplify each expression. These expressions are similar to expressions used in introductory college science classes.

22. $(60.0 \text{ kg})\left(88 \dfrac{\text{m}}{\text{s}}\right)$ **23.** $(0.5)(60 \text{ kg})\left(88 \dfrac{\text{m}}{\text{s}}\right)^{2}$ **24.** $\dfrac{8.92 \dfrac{\text{g}}{\text{cm}^{3}}}{100 \text{ g}}$

25. Simplify $\dfrac{3.85 \times 10^{10} \text{ m}}{3.0 \times 10^{8} \dfrac{\text{m}}{\text{s}}}$. Write in scientific notation. Round the mantissa to the nearest hundredth.

Using Technology: Scientific Notation

Many science instructors in introductory courses recommend that students use a scientific calculator.

For arithmetic in place value notation, a scientific calculator is in Normal or Floating Point mode. When working with numbers in scientific notation, change the mode to Scientific Notation. On some calculators, look for the letters SCI printed above a key. Press ⌨2nd⌨ ; press this key. On other calculators, press the ⌨MODE⌨ key. The screen will show different modes. Move the cursor to select **Sci**. On other calculators, when the ⌨MODE⌨ key is pressed, a list of different modes paired with numbers appears. Press the number paired with SCI.

Most scientific calculators allow the entry of numbers in scientific notation using an ⌨EE⌨ , ⌨E⌨ , or ⌨EXP⌨ key. If the EE symbol is in small letters above the key, it may be necessary to first press a shift or 2nd key. *Do not use the exponent keys,* ⌨y^x⌨ *or* ⌨^⌨ *to enter numbers in scientific notation.*

EXAMPLE 14 Enter 6.02×10^{23} on a scientific calculator.

Change the mode of the calculator to scientific notation.

Press these keys: ⌨6⌨ ⌨·⌨ ⌨0⌨ ⌨2⌨ ⌨EE⌨ ⌨2⌨ ⌨3⌨

Practice Problems For problems 26–29, use a scientific calculator in scientific notation mode to simplify each expression. Round to the nearest tenth.

26. $(6.2 \times 10^8 \text{ km})(4.5 \times 10^7 \text{ km})$ **27.** $\dfrac{6.45 \times 10^{-3} \text{ g}}{7.5 \times 10^2 \text{ mL}}$

28. $(2.5 \times 10^{-4} \text{ kg})\left(9.8 \dfrac{\text{m}}{\text{s}^2}\right)(3 \times 10^2 \text{ m})$ **29.** $5.1 \times 10^4 \text{ L} + 6.7 \times 10^3 \text{ L}$

1.4 VOCABULARY PRACTICE

Match the term with its description.

1. A notation in which we write a number as a product of a number between 1 and 10 and a power of ten
2. The SI base unit of length
3. $(x^m)^n = x^{mn}$
4. A fraction that is equal to 1
5. $x^m \cdot x^n = x^{m+n}$
6. The SI base unit of time
7. The SI prefix that represents one-thousandth
8. The SI prefix that represents one-thousand
9. The SI prefix that represents one-hundredth
10. The SI prefix that represents one million

A. centi-
B. conversion factor
C. kilo-
D. mega-
E. meter
F. milli-
G. power rule of exponents
H. product rule of exponents
I. scientific notation
J. second

1.4 Exercises

Follow your instructor's guidelines for showing your work.

The relationships in Table 4 were used to create these exercises. Using information from other sources may result in slightly different answers.

For exercises 1–10, write the **bold** number or measurement in scientific notation. Include any units of measurement.

1. The diameter of a red blood cell is about **0.000007 meter**.

2. The SARS virus is **0.0000001 meter** in diameter.

3. In 2004 the atmosphere contained about **2,700,000,000,000 metric tons** of carbon dioxide.

4. In 1995 an asteroid passed within **7,500,000 kilometers** of the earth.

5. A planetary nebula is **10,400 light-years** from the earth.

6. The area of Lake Powell is **65,800 square hectometers**.

7. Researchers from the Yale University School of Medicine studied the itch sensation produced by injecting **0.0001 microgram** of histamine into human subjects.

8. To remove giardia from drinking water, the Centers for Disease Control and Prevention recommends the use of a water filter with a maximum absolute pore size of **0.000001 meter**.

9. The strength rating on a carabiner for rope climbing is **25 kilonewtons**.

10. Researchers in Rochester, New York, were working to reduce lead in the blood of urban children. They hoped for a decrease of **0.000003** $\dfrac{\textbf{gram}}{\textbf{deciliter}}$.

For exercises 11–40, evaluate. The final answer must be in scientific notation.

11. $(6.2 \times 10^8)(4.5 \times 10^5)$

12. $(7.5 \times 10^4)(6.2 \times 10^9)$

13. $(6.2 \times 10^{-8})(4.5 \times 10^5)$

14. $(7.5 \times 10^4)(6.2 \times 10^{-9})$

15. $(6.2 \times 10^8)(4.5 \times 10^{-5})$

16. $(7.5 \times 10^{-4})(6.2 \times 10^9)$

17. $(6 \times 10^5)^2$

18. $(8 \times 10^9)^2$

19. $(6 \times 10^{-5})^2$

20. $(8 \times 10^{-9})^2$

21. $(3 \times 10^4)(5 \times 10^3)^2$

22. $(6 \times 10^3)(5 \times 10^4)^2$

23. $(4 \times 10^{-4})^3$

24. $(5 \times 10^{-6})^3$

25. $(2 \times 10^7)^3$

26. $(3 \times 10^5)^2$

27. $\dfrac{3 \times 10^5}{6 \times 10^2}$

28. $\dfrac{4 \times 10^7}{8 \times 10^3}$

29. $\dfrac{3 \times 10^5}{6 \times 10^{-2}}$

30. $\dfrac{4 \times 10^7}{8 \times 10^{-3}}$

31. $\dfrac{3 \times 10^{-5}}{6 \times 10^2}$

32. $\dfrac{4 \times 10^{-7}}{8 \times 10^3}$

33. $3.2 \times 10^4 + 9.4 \times 10^4$

34. $5.9 \times 10^6 + 8.3 \times 10^6$

35. $3.2 \times 10^{-4} + 9.4 \times 10^{-4}$

36. $5.9 \times 10^{-6} + 8.3 \times 10^{-6}$

37. $3.2 \times 10^4 + 9.4 \times 10^5$

38 $5.9 \times 10^6 + 8.3 \times 10^7$

39. $3.2 \times 10^6 + 9.4 \times 10^4$

40. $5.9 \times 10^8 + 8.3 \times 10^6$

For exercises 41–50, use place value notation and rewrite each measurement in its base unit.

41. 3 centigrams

42. 5 centimeters

43. 6 kilometers

44. 9 kilograms

45. 8 decibels

46. 4 deciliters

47. 2 megatons

48. 6 meganewtons

49. 7 micrograms

50. 3 microliters

For exercises 51–60, use the information in Table 2 and Table 4 to write an equation that shows the relationship of the units.

51. km and m

52. kg and g

53. c and pt

54. T and c

55. min and s

56. hr and min

57. qt and L

58. mi and km

59. U.S. dollar and Mexican peso

60. U.S. dollar and euro

For exercises 61–70, use unit analysis and the relationships in Table 4 to convert the measurement into the given unit(s). Use Examples 9–12 as models for showing your work.

61. $35 \dfrac{\text{mi}}{\text{hr}}; \dfrac{\text{ft}}{\text{min}}$

62. $45 \dfrac{\text{mi}}{\text{hr}}; \dfrac{\text{ft}}{\text{min}}$

63. $25 \dfrac{\text{mi}}{\text{hr}}; \dfrac{\text{km}}{\text{hr}}$

64. $35 \dfrac{\text{mi}}{\text{hr}}; \dfrac{\text{km}}{\text{hr}}$

65. $60 \dfrac{\text{km}}{\text{hr}}; \dfrac{\text{ft}}{\text{s}}$

66. $30 \dfrac{\text{km}}{\text{hr}}; \dfrac{\text{ft}}{\text{s}}$

67. $2.5 \dfrac{\text{g}}{\text{mL}}; \dfrac{\text{kg}}{\text{L}}$

68. $1.9 \dfrac{\text{g}}{\text{mL}}; \dfrac{\text{kg}}{\text{L}}$

69. $2.5 \dfrac{\text{g}}{\text{cm}^3}; \dfrac{\text{kg}}{\text{L}}$

70. $1.9 \dfrac{\text{g}}{\text{cm}^3}; \dfrac{\text{kg}}{\text{L}}$

71. Change $3.0 \times 10^8 \dfrac{\text{m}}{\text{s}}$ into $\dfrac{\text{AU}}{\text{s}}$. Round the mantissa to the nearest tenth.

72. Change $3.0 \times 10^{10} \dfrac{\text{cm}}{\text{s}}$ into $\dfrac{\text{AU}}{\text{s}}$. Round the mantissa to the nearest tenth.

For exercises 73–78, use unit analysis to solve the problem.

73. One capsule of ampicillin contains 250 mg. A patient needs 0.75 g of ampicillin. Find the number of capsules the patient needs.

74. A patient needs 0.25 g of Tegetol. A bottle contains 100 mg of Tegetol in 5 mL of solution. Find the amount of solution in milliliters that the patient needs.

75. At a gas station in Toronto, the price of gasoline was 0.849 Canadian dollars per liter. Using the currency values from Table 4, find the price of this gasoline in U.S. dollars per gallon. Round to the nearest hundredth.

76. At a gas station in Mexico City, the price of gasoline was 6.5 Mexican pesos per liter. Using the currency values from Table 4, find the price of this gasoline in U.S. dollars per gallon. Round to the nearest hundredth.

77. An *acre·foot* of water is the volume of water in a space that has a surface area of 1 acre and a depth of 1 foot. The storage capacity behind Grand Coulee Dam is 125,000,000 acre·feet. Change this measurement into gallons (1 cubic ft = 7.481 gal). Write the answer in scientific notation. Round the mantissa to the nearest hundredth.

© aricyhmeister/Shutterstock

78. In the United States, the unit of barometric pressure is usually inches of mercury. In the SI system, the unit of barometric pressure is a *pascal* (101.325 kilopascals = 760 mm of mercury). A meteorologist in the United States reports that the barometric pressure is 30.04 in. of mercury. Change this measurement into kilopascals. Round to the nearest whole number.

Exercises 79–80 are problems from an introductory college astronomy or physics class.

79. The time for light to travel from the sun to the earth on a certain day is $\dfrac{1.5 \times 10^{11}\ \text{m}}{3.0 \times 10^{8}\ \dfrac{\text{m}}{\text{s}}}$. Simplify. Write the answer in scientific notation.

80. The time for light to travel from the moon to the earth on a certain day is $\dfrac{4.06 \times 10^{8}\ \text{m}}{3.0 \times 10^{8}\ \dfrac{\text{m}}{\text{s}}}$. Simplify. Round to the nearest hundredth.

Exercises 81–82 are problems from an introductory college chemistry class.

81. Gasoline includes octane, C_8H_{18}. The expression

$$\left(\frac{50\ \text{g}\ C_8H_{18}}{1}\right)\left(\frac{1\ \text{mol}\ C_8H_{18}}{114\ \text{g}\ C_8H_{18}}\right)\left(\frac{25\ \text{mol}\ O_2}{2\ \text{mol}\ C_8H_{18}}\right)\left(\frac{32\ \text{g}\ O_2}{1\ \text{mol}\ O_2}\right)$$

equals the amount of oxygen, O_2, needed to burn 50 g of octane. Simplify the expression. Round to the nearest whole number.

82. Silicon dioxide (SiO_2) is one of the ingredients used to make silicon carbide (SiC). Sandpaper includes silicon carbide. The expression

$$\left(\frac{500\ \text{g}\ SiO_2}{1}\right)\left(\frac{1\ \text{mol}\ SiO_2}{60.1\ \text{g}\ SiO_2}\right)\left(\frac{1\ \text{mol}\ SiC}{1\ \text{mol}\ SiO_2}\right)\left(\frac{40.1\ \text{g}\ SiC}{1\ \text{mol}\ SiC}\right)$$

equals the amount of silicon carbide that can be made from 500 grams of silicon dioxide. Simplify the expression. Round to the nearest whole number.

Exercises 83–84 are problems from an introductory college biology class. Simplify each expression. If necessary, round the mantissa to the nearest whole number.

83. The time for water to diffuse 50 micrometers in a corn leaf is $\dfrac{(5 \times 10^{-5}\ \text{m})^2}{\left(\dfrac{1.1 \times 10^{-9}\ \text{m}^2}{1\ \text{s}}\right)}$.

84. The time for water to diffuse 5 centimeters in a corn leaf is $\dfrac{(5 \times 10^{-2}\ \text{m})^2}{\left(\dfrac{1.1 \times 10^{-9}\ \text{m}^2}{1\ \text{s}}\right)}$.

85. In May 2009, the U.S. national debt limit was \$12.104 trillion. Write the word *trillion* as a power of ten.

86. The Milky Way galaxy contains over 400 billion stars. Write the word *billion* as a power of ten.

87. Explain why you think banks do not use scientific notation.

88. Explain why you think pediatricians do not use scientific notation to record the length of a newborn baby.

Technology

For exercises 89–92, use a scientific calculator in scientific notation mode to evaluate.

89. $(3.2 \times 10^8)(4.1 \times 10^6)$

90. $\dfrac{1.8 \times 10^{15}}{2 \times 10^{-9}}$

91. $(3.2 \times 10^{-8})(4.1 \times 10^{-6})$

92. $\dfrac{1.8 \times 10^{-15}}{2 \times 10^{-9}}$

Find the Mistake

For exercises 93–96, the completed problem has one mistake.
(a) Describe the mistake in words, or copy down the whole problem and highlight or circle the mistake.
(b) Do the problem correctly.

93. **Problem:** Convert $35\ \dfrac{\text{mi}}{\text{hr}}$ into $\dfrac{\text{ft}}{\text{s}}$. Round to the nearest whole number.

   Incorrect Answer: $\left(\dfrac{35\ \text{mi}}{1\ \text{hr}}\right)\left(\dfrac{5280\ \text{ft}}{1\ \text{mi}}\right)\left(\dfrac{1\ \text{hr}}{60\ \text{s}}\right)$

   $= 3080\ \dfrac{\text{ft}}{\text{s}}$

94. **Problem:** The distance formula is distance = (speed)(time). Find the distance that a car with a speed of $60\ \dfrac{\text{mi}}{\text{hr}}$ travels in 10 minutes.

   Incorrect Answer: distance $= \left(60\ \dfrac{\text{mi}}{\text{hr}}\right)(0.1\ \text{hr})$

   distance = 6 mi

95. **Problem:** Simplify: $2 \times 10^3 + 6 \times 10^4$

   Incorrect Answer: $2 \times 10^3 + 6 \times 10^4$

   $= 8 \times 10^7$

96. Problem: Simplify: $(6 \times 10^{-5})^3$

   Incorrect Answer: $(6 \times 10^{-5})^3$

$$= 6^3 \times (10^{-5})^3$$
$$= 18 \times 10^{-15}$$
$$= 1.8 \times 10^{-14}$$

SUCCESS IN COLLEGE MATHEMATICS

97. Some instructors in college do not admit students who arrive late to class. Others allow late students to enter if they do so quietly. Describe your instructor's policy about arriving on time to class.

98. Some college students own a portable computer, MP3 player, tablet, smartphone, cell phone, and/or other electronic device. Describe your instructor's policy about the use of these electronic devices in class.

© fstockfoto/Shutterstock

In this section, we solve many application problems. Think about the tennis player who practices serving for hours. In the same way, improving your ability to solve application problems takes a lot of practice.

SECTION 1.5

Applications and Problem Solving

After reading the text, working the practice problems, and completing assigned exercises, you should be able to:

Use the five steps to solve an application problem.

Introduction to Problem Solving

Why solve application problems? People who can solve problems have an advantage in real life and in the workplace. Employers want employees who are confident problem solvers. Solving problems can also help us learn more mathematics.

There is no magic formula for becoming a good problem solver. But it does seem that successful problem solvers do a lot of problems; *they practice*. They don't give up when their first attempt doesn't work; *they persevere*. They don't always expect immediate success; *they are patient*.

In this book, we use **five steps** to solve many different kinds of problems in many different situations. The steps are based on the work of George Polya (1887–1985). Polya believed that students learn to solve problems the way they learn a skill such as swimming, with practice and coaching. To improve your problem-solving ability, you need to do many problems. As Dr. Polya wrote, "Mathematics is not a spectator sport."

THE FIVE STEPS

Step 1 Understand the problem.
Step 2 Make a plan.
Step 3 Carry out the plan.
Step 4 Look back.
Step 5 Report the solution.

Step 1 Understand the problem.

- Read the problem, perhaps more than once. Look up any unfamiliar words.
- Identify what you are trying to find, the *unknown*. Assign a variable that represents the unknown.
- Identify the information needed to solve the problem. Identify information that is not needed. This is **extraneous information**.

Step 2 Make a plan.

- Identify the relationship between the unknown and the other information in the problem. For example, the relationship could be a difference, a sum, or a percent, or it could be described by a formula.
- If this problem is similar to a problem that you have done before, think about how you solved the original problem. The examples in the book and examples completed by your instructor can also help you make a plan to solve a new problem.
- Decide how to solve the problem.

Step 3 Carry out the plan.

- Write and solve the equation or formula. It is often helpful to write a word equation first and then write a matching algebra equation.

Step 4 Look back.

- Think about whether your answer solves the problem. Does it make sense? Is it reasonable? If it does not make sense, make and carry out a new plan.
- Check your algebra or arithmetic, line by line, to make sure it is correct.
- Think about whether you have learned something new from this problem, something that you may use to solve similar problems in the future.

Step 5 Report the solution.

- How you report the solution depends on the audience. In a college mathematics class, writing the solution to the problem in a complete sentence is often what the audience (your instructor) wants.

THE FIVE STEPS

Step 1 Understand the problem.
Read the problem. Identify the unknown. Assign a variable. Identify extraneous information.

Step 2 Make a plan.
Identify the relationship between the unknown and the other information in the problem. Decide how to find the unknown: write and solve an equation or inequality, use a formula, draw and use a graph, evaluate a function, solve a system of equations.

Step 3 Carry out the plan.

Step 4 Look back.
Does the answer make sense? Is it reasonable?
Check for errors in arithmetic or algebra.
Think about what you have learned from solving this problem.

Step 5 Report the solution.
Write the answer to the problem in a complete sentence.

How much of this understanding, planning, and looking back should you write down? Ask your instructor for guidance.

EXAMPLE 1 A survey of 1160 American adults born between 1946 and 1964 asked about life as an older employee. Find the number of the people who said they were able to stay abreast of developments in their field and keep up with technology. Round to the nearest whole number.

> But 61 percent of the baby boomers surveyed said their age is not an issue at work, while 25 percent called it an asset. . . . About two-thirds of poll respondents said they were able to stay abreast of developments in their field and keep up with technology. (*Source:* www.washingtonpost.com, Apr. 26, 2011)

SOLUTION ▶ **Step 1 Understand the problem.**

What is unknown? The number of people able to keep up

Assign the variable. N = number of people able to keep up

Needed information. The survey polled 1160 people; two-thirds said they were able to keep up.

Extraneous information. The percents given in the article are not needed.

Step 2 Make a plan.

To find a fraction of a whole, multiply the fraction and the whole. A word equation is

number of people able to keep up = (fraction able to keep up)(total people)

Step 3 Carry out the plan.

number of people able to keep up = (fraction able to keep up)(total people)   Word equation.

$$N = \left(\frac{2}{3}\right)\left(\frac{1160 \text{ people}}{1}\right)$$ Algebraic equation.

$$N = \frac{2320 \text{ people}}{3}$$ Multiply.

$$N = 773.3 \dots \text{ people}$$ Divide.

$$N \approx 773 \text{ people}$$ Round.

Step 4 Look back.

Is the answer reasonable? Thinking about this problem another way, we can divide the total into three groups of about 387 people. Two of these groups, a total of 774 people, say that they can keep up. Since this is very close to the rounded answer of 773 people, the answer seems reasonable.

What can we learn from this problem? When multiplying a number by a fraction, we usually leave the answer as a fraction or mixed number. In an application problem, we often do not write the answer as a fraction but instead write it as a rounded decimal number.

Step 5 Report the solution.

To make sure that you are answering the problem, read both the problem and your solution to yourself or aloud. In this example, the problem is "Find the number of the people who said they were able to stay abreast of developments in their field and keep up with technology." The answer is "About 773 people said they were able to stay abreast of developments in their field and keep up with technology."

A **percent** is the number of parts out of 100 total parts. Since a percent can be written as a fraction in which the numerator and denominator are integers, it is a rational number.

Percent

$$\text{percent} = \left(\frac{\text{number of parts}}{\text{total parts}}\right)(100\%)$$

$$\text{number of parts} = (\text{percent})(\text{total parts})$$

EXAMPLE 2 In spring 2011, the 19th Biannual Youth Survey on Politics and Public Service polled 3018 U.S. citizens between the ages of 18 and 29. Find the number of the people surveyed who had Twitter accounts. Round to the nearest percent.

Over the past year, Millennial Facebook adoption has grown significantly from 64% to 80% (90% adoption among four-year college students), while MySpace has shed six percentage points over the same period. Although Twitter is clearly a less relevant tool for young adults than Facebook, Twitter accounts among young adults also rose over the past year from 15% to 24%. (*Source:* www.iop.harvard.edu, March 31, 2011)

SOLUTION ▶ **Step 1 Understand the problem.**

What is unknown? The number of people with Twitter accounts

Assign the variable. N = number of people with Twitter accounts

Needed information. The survey polled 3018 people; 24% said they had Twitter accounts. In decimal form, 24% = 0.24.

Extraneous information. The information about Facebook and MySpace, the previous percent of people with Twitter accounts

Step 2 Make a plan.

Since number of parts = (percent)(total parts), a word equation is

number with Twitter accounts = (percent with Twitter)(total number)

Step 3 Carry out the plan.

number with Twitter accounts = (percent with Twitter)(total number)   Word equation.

$$N = (0.24)(3018 \text{ people})$$ Algebraic equation.
$$N = \mathbf{724.32} \text{ people}$$ Simplify.
$$N \approx \mathbf{724} \text{ people}$$ Round.

Step 4 Look back.

Is the answer reasonable? Working backwards, $\frac{724}{3018} \cdot 100$ equals 23.98. . .%, which rounds to 24%. Since this is equal to the percent in the problem, the answer seems reasonable.

What can we learn from this problem? When multiplying a number by a percent, rewrite the percent as a decimal number.

Step 5 Report the solution.

About 724 people said they had Twitter accounts.

When finding a percent increase or decrease, the "number of parts" is the amount of increase or decrease. The "total parts" is the original amount.

EXAMPLE 3 Find the percent decrease in the number of reports of larceny/theft on college campuses in Tennessee from 2009 to 2010. Round to the nearest tenth of a percent. (*Source:* www.tbi.state.tn.us, *Crime on Campus* 2010)

Type of crime	2008	2009	2010
Assault	578	741	629
Destruction/damage/vandalism	852	809	735
Larceny/theft	2624	2903	2799

SOLUTION ▶ **Step 1 Understand the problem.**

What is unknown? The percent decrease in the number of reports of larceny/theft from 2009 to 2010

Assign the variable. P = percent decrease in larceny/theft

Needed information. The number of reports of larceny/theft in 2009, the number of reports of larceny/theft in 2010

Extraneous information. The other information in the table

Step 2 Make a plan.

Since percent $= \left(\dfrac{\text{number of parts}}{\text{total parts}}\right)(100\%)$, percent $= \left(\dfrac{\text{decrease}}{\text{original amount}}\right)(100\%)$.

A word equation for this problem is

$$\text{percent} = \left(\dfrac{\text{amount in 2010} - \text{amount in 2009}}{\text{amount in 2009}}\right)(100\%)$$

Step 3 Carry out the plan.

$$\text{percent} = \left(\dfrac{\text{amount in 2010} - \text{amount in 2009}}{\text{amount in 2009}}\right)(100\%) \quad \text{Word equation.}$$

$$P = \left(\dfrac{2799 \text{ reports} - 2903 \text{ reports}}{2903 \text{ reports}}\right)(100\%) \quad \text{Algebraic equation.}$$

$$P = \left(\dfrac{-104 \text{ reports}}{2903 \text{ reports}}\right)(100\%) \quad \text{Simplify.}$$

$$P \approx -3.6\% \quad \text{Round.}$$

Step 4 Look back.

Is the answer reasonable? Working backwards, 3.6% of 2903 reports equals about 105 reports. Since this is close to the decrease of 104 reports, the answer seems reasonable.

What can we learn from this problem? A percent decrease can be represented by a negative percent.

Step 5 Report the solution.

The number of reports of larceny/theft decreased about 3.6% from 2009 to 2010.

Taxes are often based on a percent of a price or income. In calculating the total cost of an item including sales tax, the tax is added to the price.

EXAMPLE 4 A resident of Dade County paid $26,700 for a 2010 Nissan Altima Hybrid with a curb weight of 3470 pounds. Additional costs included a 7% sales tax on the price of the car and $256 for plate and license. This car qualified for a $2350 federal income tax credit. Including the credit, find the real cost of this car.

SOLUTION ▶ **Step 1 Understand the problem.**

What is unknown? The real cost of the car

Assign the variable. C = real cost of the car

Needed information. Price of car, sales tax rate, cost of plate/license, amount of tax credit

Extraneous information. Curb weight of the car

Step 2 Make a plan.

Since number of parts equals (percent)(total parts), amount of tax equals (tax rate)(price of car). A word equation for the real cost of the car is

$$\text{cost} = \text{price} + (\text{tax rate})(\text{price}) + (\text{plate/license}) - \text{credit}$$

Step 3 Carry out the plan.

$$\text{cost} = \text{price} + (\text{tax rate})(\text{price}) + (\text{plate/license}) - \text{credit} \quad \text{Word equation.}$$

$$C = \$26,700 + (0.07)(\$26,700) + \$256 - \$2350 \quad \text{Algebraic equation.}$$

$$C = \$26,700 + \$1869 + \$256 - \$2350 \quad \text{Simplify.}$$

$$C = \$26,475 \quad \text{Simplify.}$$

Step 4 Look back.

Is the answer reasonable? Estimating, the tax and plate/license cost total about $2100. Since the rebate is $2350, we expect the real cost of the car to be reduced by the difference of these two amounts, or about $250. Since the difference of the original price ($26,700) and the real cost ($26,475) is $225, the answer seems reasonable.

What can we learn from this problem? When we see the word *credit,* we often think of addition. In this problem, however, it reduces the real cost of the car, so we need to subtract the credit.

Step 5 Report the solution.

The real cost of the car is $26,475.

When working with percent, we often use the relationship number of parts equals (percent)(whole). However, if we know *the sale price and the percent discount, we cannot find the original price by multiplying the percent discount and the sale price and adding this to the sale price.* Why not? The discount in dollars comes from multiplying the original price by the percent. The sale price is less than the original price. If we multiply the smaller sale price by the percent, the discount is too small.

EXAMPLE 5 The sale price of a camera is $129. The percent discount is 20%. Find the original price.

SOLUTION ▸ **Step 1 Understand the problem.**

What is unknown? The original price of the camera

Assign the variable. p = original price of the camera

Needed information. The sale price of the camera; the percent discount

Extraneous information. None

Step 2 Make a plan.

Since number of parts equals (percent)(whole), the discount equals (percent discount)(original price). To find the sale price, subtract. A word equation is

original price − discount = sale price

Step 3 Carry out the plan.

$$p - 0.20p = \$129 \qquad \text{original price − discount = sale price}$$
$$\mathbf{0.80p} = \$129 \qquad \text{Simplify; } 1p - 0.20p = 0.80p$$
$$\frac{0.80p}{0.80} = \frac{\$129}{0.80} \qquad \text{Division property of equality.}$$
$$p = \$161.25 \qquad \text{Simplify.}$$

Step 4 Look back.

Working backwards, since $(0.20)(\$161.25) = \$32.25$ and $\$161.25 - \$32.25$ equals the sale price of $129, the answer seems reasonable.

Step 5 Report the solution.

The original price of the camera was $161.25.

In Example 5, we knew the sale price and the percent discount, and we used this information to find the original price. In the next example, we know the amount that the airlines spent on fuel in 2011 and the percent increase, and we use this information to find the amount that the airlines spent on fuel in 2010.

EXAMPLE 6 Find the amount that the airlines spent on fuel during the first three months of 2010. Round to the nearest tenth of a billion.

During the first three months of 2011, the airlines spent $8.7 billion on fuel, 31 percent more than last year. (*Source:* www.washingtonpost.com, May 30, 2011)

▶ **Step 1 Understand the problem.**

What is unknown? The amount spent on fuel during the first three months of 2010

Assign the variable. A = the amount spent in 2010

Needed information. The amount spent in 2011; the percent increase

Extraneous information. None

Step 2 Make a plan.

The amount spent in 2010 plus the percent increase equals the known amount spent in 2011. Since number of parts equals (percent)(whole), the increase is (percent increase)(amount spent in 2010). A word equation is

amount spent in 2010 + (percent increase)(amount spent in 2010)
= amount spent in 2011

Step 3 Carry out the plan.

$$A + 0.31A = \$8.7 \text{ billion}$$ amount in 2010 + increase = amount spent in 2011

$$\mathbf{1.31}A = \$8.7 \text{ billion}$$ $1A + 0.31A = 1.31A$

$$\frac{1.31A}{1.31} = \frac{\$8.7 \text{ billion}}{1.31}$$ Division property of equality.

$$A = \$6.64\ldots \text{ billion}$$ Simplify.

$$A \approx \mathbf{\$6.6} \text{ billion}$$ Round.

Step 4 Look back.

Estimating, 31% is about one-third, and one-third of \$6.6 billion is about \$2.2 billion. The sum of the answer of \$6.6 billion, and the estimated increase of \$2.2 billion is \$8.8 billion. Since this is very close to the actual amount in 2011 of \$8.7 billion, the answer seems reasonable.

Step 5 Report the solution.

The airlines spent about \$6.6 billion on fuel in the first three months of 2010.

A survey asked 80 students whether they owned a pet. Of these students, 51 owned a dog. We can write this as a rate: $\dfrac{51 \text{ students own a dog}}{80 \text{ total students}}$. A **proportion** is an equation with two equivalent rates. In the next example, the problem includes a *known rate* from a survey. To solve the problem, we write and solve a proportion: *known rate = unknown rate.*

EXAMPLE 7   In 2011, there were about 15,507,300 licensed drivers in the state of Florida. Use a proportion to predict the number of these drivers who fell asleep at the wheel.

Two out of every five drivers . . . admit to having fallen asleep at the wheel at some point. (*Sources:* www.flhsmv.gov, 2011; www.aaafoundation.org, 2010)

SOLUTION ▶ **Step 1 Understand the problem.**

What is unknown? The number of drivers who fell asleep

Assign the variable. N = number of drivers who fell asleep

Needed information. Total number of drivers, known ratio of drivers who fell asleep to total number of drivers

Extraneous information. None

Step 2 Make a plan.

In a proportion, known rate = unknown rate. The rate is

$$\frac{\text{number of drivers who fell asleep}}{\text{total number of drivers}}$$

Step 3 Carry out the plan.

known rate $=$ unknown rate		Word equation.
$\dfrac{\text{2 drivers who fell asleep}}{\text{5 drivers}} = \dfrac{N}{15{,}507{,}300 \text{ drivers}}$		Algebraic equation.
$\dfrac{2}{5} = \dfrac{N}{15{,}507{,}300}$		To solve, remove the units.
$\dfrac{15{,}507{,}300}{1} \cdot \dfrac{2}{5} = \dfrac{N}{15{,}507{,}300} \cdot \dfrac{15{,}507{,}300}{1}$		Multiplication property of equality.
$\mathbf{6{,}202{,}920 = 1}N$		Simplify.
$6{,}202{,}920 = N$		Simplify.

Step 4 Look back.

Is the answer reasonable? The known rate should equal the unknown rate. Since $\dfrac{2}{5} = 0.4$ and $\dfrac{6{,}202{,}920}{15{,}507{,}300} = 0.4$, the answer seems reasonable.

What can we learn from this problem? Before removing units to use the property of equality, be sure that the units simplify and make sense. Replace the units in the final answer.

Step 5 Report the solution.

Of 15,507,300 drivers, 6,202,920 drivers would admit to falling asleep at the wheel.

In government documents, formulas are often used to calculate payments or fees. When evaluating a formula, follow the order of operations.

EXAMPLE 8   Find the monthly base child support owed by a parent when the total monthly base support is $434.58, the monthly net family income is $3000, the monthly income level is $2415, the marginal percentage is 16.66%, and the parent's percentage share of family income is 40%. The cost of daycare for these children is about $800 per month. Round to the nearest whole number.

In Michigan, the base child support payment is found using information in the General Care Support Table and the General Care equation, $G = \{A + [B \times (C - D)]\} \times E$. In this equation, G is the base child support owed by a parent, A is the monthly base support . . . , B is the marginal percentage . . . , C is the monthly net family income, D is the monthly income level . . . , and E is the parent's percentage share of family income. (*Source:* 2008 Michigan Child Support Formula Manual and Supplement)

SOLUTION ▸ **Step 1 Understand the problem.**

What is unknown? The base child support

Assign the variable. Use the variable in the formula; $G =$ amount of base child support

Needed information. Values for A ($434.58), B (0.1666), C ($3000), D ($2415), and E (0.40)

Extraneous information. The cost of daycare per month

Step 2 Make a plan.

Use the formula, replacing the variables with the given amounts.

Step 3 Carry out the plan.

$G = \{A + [B \times (C - D)]\} \times E$		Formula.
$G = \{\$434.58 + [0.1666 \times (\$3000 - \$2415)]\} \times 0.40$		Replace the variables.
$G = \{\$434.58 + [0.1666(\mathbf{\$585})]\} \times 0.40$		Work from the inside out.

$$G = \{\$434.58 + \$97.461\} \times 0.40 \qquad \text{Follow the order of operations.}$$
$$G = \$532.041 \times 0.40 \qquad \text{Follow the order of operations.}$$
$$G = \$212.816 \qquad \text{Follow the order of operations.}$$
$$G \approx \$213 \qquad \text{Round.}$$

Step 4 Look back.

Is the answer reasonable? Rounding, $G = \{\$435 + [0.17 \times (\$3000 - \$2400)]\} \times 0.40 = \$214.80$. Since this is close to $213, the answer seems reasonable.

What can we learn from this problem? When using a formula, follow the order of operations. Some formulas use the symbol $\times$ for multiplication.

Step 5 Report the solution.

The base child support owed by a parent is about $213 per month.

In the next example, a picture is included in the "understand the problem" step. Pictures often make it easier to understand a problem situation.

EXAMPLE 9   Concrete is a mixture of cement, sand, and gravel. For a given volume of concrete, one part is cement, two parts are sand, and three parts are gravel. Find the amount of sand in cubic feet needed to make a sidewalk that is 30 ft long, 4 ft wide, and 4 in. deep. Round to the nearest tenth.

SOLUTION ▶ **Step 1 Understand the problem.**

What is unknown? The amount of sand

Assign the variable. A = amount of sand

Needed information. The volume of concrete and the parts of concrete that are sand

Extraneous information. None

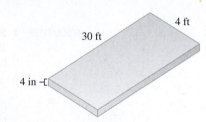

Step 2 Make a plan.

The fraction of the concrete that is sand is $\dfrac{\text{number of parts of sand}}{\text{total parts of concrete}}$, which equals $\dfrac{2 \text{ parts}}{6 \text{ parts}}$ or $\dfrac{1}{3}$. So amount of sand $= \dfrac{1}{3}(\text{volume of concrete})$.

The sidewalk is a *rectangular solid*. The formula for the volume of a rectangular solid is $V = LWH$. Since the volume is in cubic feet, the depth of the sidewalk must be changed from inches to feet.

Step 3 Carry out the plan.

$$\text{amount of sand} = (\text{fraction of sand})(\text{volume of concrete}) \qquad \text{Word equation.}$$

$$A = \frac{1}{3}(LWH) \qquad \text{Algebraic equation.}$$

$$A = \frac{1}{3}(30 \text{ ft})(4 \text{ ft})(4 \text{ in.}) \qquad \text{Replace variables.}$$

$$A = \frac{1}{3}\left(\frac{30 \text{ ft}}{1}\right)\left(\frac{4 \text{ ft}}{1}\right)\left(\frac{4 \text{ in.}}{1}\right)\left(\frac{1 \text{ ft}}{12 \text{ in.}}\right) \qquad \begin{array}{l}\text{Multiply by a conversion}\\ \text{factor equal to 1.}\end{array}$$

$$A = 13.33\dots \text{ ft}^3 \text{ sand} \qquad \text{Simplify.}$$

$$A \approx 13.3 \text{ ft}^3 \text{ sand} \qquad \text{Round.}$$

Step 4 Look back.
Is the answer reasonable? Working backwards, the amount of *one* part of sand is the *total amount of sand,* 13.3 ft^3, divided by 2 or about 6.7 ft^3. Six parts should equal the total amount of concrete. Since $(6)(6.7 \text{ ft}^3)$ equals 40.2 ft^3 and the total amount of concrete is about 40 ft^3, the answer seems reasonable.

What can we learn from this problem? When we use a formula, we can change the units by multiplying by a conversion factor such as $\dfrac{1 \text{ ft}}{12 \text{ in.}}$ that is equal to 1.

Step 5 Report the solution.
About 13.3 ft^3 of sand is needed to make the concrete.

We can use an inequality to describe a situation that includes the phrase *more than* or *less than.*

EXAMPLE 10   In 2009, about 1,400,000 children were born in the United States. Find the number of fathers of these children who viewed themselves as stay-at-home fathers. Round to the nearest thousand. (*Source:* www.cdc.gov)

> Many fathers who had primary child-care responsibility at home while working part-time or pursuing a degree viewed themselves as stay-at-home fathers. When those factors are included as well as unmarried and single dads, the share of fathers who stay at home to raise children jumps . . . to more than 6 percent. (*Source:* www.chicago tribune.com, April 26, 2011)

SOLUTION ▶ **Step 1 Understand the problem.**
What is unknown? The number of fathers who viewed themselves as stay-at-home fathers

Assign the variable. N = number of stay-at-home fathers

Needed information. The total number of fathers, the minimum percent who view themselves as stay-at-home fathers

Extraneous information. None

Step 2 Make a plan.
Since number of parts equals (percent)(whole), the minimum number of fathers who view themselves as stay-at-home fathers is the product of the percent and the total number of fathers. A word inequality is

  number of stay-at-home fathers $>$ (percent of stay-at-home fathers) (total fathers)

Step 3 Carry out the plan.
number of stay-at-home fathers $>$ (percent of stay-at-home fathers) (total fathers)
$$N > (0.06)(1{,}400{,}000 \text{ fathers})$$
$$N > \mathbf{84{,}000} \text{ fathers}$$

Step 4 Look back.
Is the answer reasonable? Since $\left(\dfrac{84{,}000 \text{ fathers}}{1{,}400{,}000 \text{ fathers}}\right)(100\%)$ equals 6%, the answer seems reasonable.

What can we learn from this problem? The words *more than* indicate a minimum number. The unknown is greater than this minimum number.

Step 5 Report the solution.
There were more than 84,000 fathers who viewed themselves as stay-at-home fathers.

Practice Problems

For problems 1–9, use the five steps.

1. A headline on zdnet.com said that "two thirds of small business owners use Facebook for marketing." This headline was based on the responses of 1132 small business owners. Find the number of these small business owners who use Facebook for marketing. Round to the nearest whole number. (*Source:* www.zdnet.com, May 26, 2011)

2. Find the percent of the monitoring locations that received very good to excellent water quality marks. Round to the nearest percent.

> Overall water quality during the summer dry (AB411) time period in California this past year was very good. . . . Of the 445 ocean water quality monitoring locations throughout California, 400 . . . . received very good to excellent water quality marks (A or B grades) from April through October 2010. (*Source:* brc.healthebay.org)

3. Find the percent increase in the Dow Jones Industrial Average between Sept 15, 2010, and May 26, 2011. Round to the nearest percent.

Date	Dow Jones Industrial Average
July 2, 2010	9686.48
Sept. 15, 2010	10,572.73
March 16, 2011	11,613.30
May 26, 2011	12,876.00

4. A meal at a restaurant cost $28.90. The person paying the bill is going to leave a tip of 18% on the amount before tax. The sales tax on the amount paid for the meal (before tip) is 8.5%. Find the total cost of the meal, tip, and tax. Round to the nearest hundredth.

5. Use a proportion to find the number of 28,000 uninsured people with asthma who could not afford their prescription medicines.

> About 2 in 5 (40%) uninsured people with asthma could not afford their prescription medicines and about 1 in 9 (11%) insured people with asthma could not afford their prescription medicines. (*Source:* www.cdc.gov, May 2011)

6. The formula for calculating the earned run average of a pitcher (ERA) is

$$\text{ERA} = 9\left(\frac{\text{number of earned runs}}{\text{number of innings pitched}}\right)$$

On May 29, 2010, Roy Halladay of the Philadelphia Phillies threw a perfect game, striking out 11 of 27 batters. For the 2010 season, Halladay pitched 250.2 innings and gave up 68 earned runs. Find his ERA for the 2010 season. Round to the nearest hundredth. (*Source:* www.mlb.com)

7. Find the amount of rice that was destroyed by flooding. Write the answer in place value notation (do not use the word million in the answer).

> Two rounds of flooding in Sri Lanka . . . have destroyed at least 35 percent of the staple rice crop. . . . Total expected rice production this season had been 2.7 million metric tonnes from 739,000 hectares. (*Source:* uk.reuters.com, Feb. 8, 2011)

8. The formula $A = 2HL + 2HW + 2\left(\frac{1}{2}GW\right)$ finds the area of siding, A, needed for a house with length L, width W, height up to the eaves H, and height of the gable G. Find the area of siding needed for a house with a length of 46 ft, width of 22 ft, height up to the eaves of 9 ft, and height of the gable of 8 ft.

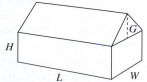

9. The number of passengers that arrived at or departed from McCarran International Airport in Las Vegas increased 2.3 percent in April 2011, compared to April 2010. In April 2011, 3,462,134 passengers arrived at or departed from the airport. Find the number of passengers that arrived or departed in 2010. Round to the nearest whole number. (*Source:* www.vegasinc.com, May 23, 2011)

1.5 VOCABULARY PRACTICE

The solution to a problem is either a *reasonable answer* or *not a reasonable answer*. Use each example to practice thinking about the meaning of the term *reasonable answer*. Choose the term that best describes the measurement printed in **bold**.

1. A nurse calculates that an injection of **2 L** of lidocaine is needed to numb a finger for stitches.

2. A savings account pays 3% annual interest. $200 is invested. The annual interest is **$600 per year**.

3. A trucker averaged a speed of 60 mi per hour. She drove 7 hours and traveled **420 mi**.

4. It is 200 mi on Interstate 15 from Butte, Montana, to Idaho Falls, Idaho. The speed of a car is 75 mi per hour. The trip will take **5 hr**.

5. The sales tax rate in a city in New York State is 8.5%. A car costs $26,000. The sales tax is **$2210**.

6. A student has test scores of 75%, 80%, 40%, 60%, and 75%. The student wants to finish the class with a 70% average. The student needs to score **at least 90%** on the last test.

7. The sale price of a television is $100. It is marked down 20% from its original price. The original price is **$120**.

8. The price of a 12-can case of soda is $5.99. The price per can is **$0.25**.

9. The adult dose of a medicine is 40 mg every 4 hr. A child's dose should be **400 mg** every 4 hr.

10. The formula for finding the area of a circle is $A = \pi r^2$. The diameter of the bottom of a tart pan is 10 in. The area of the bottom of the tart pan is about **300 in.^{2}**.

A. reasonable answer
B. not a reasonable answer

1.5 Exercises

Follow your instructor's guidelines for showing your work.

For exercises 1–70, use the five steps.

1. A student at the City College of New York has a $4600 Pell Grant. Her tuition bill for two semesters is $4000. She uses her Pell Grant to pay this tuition. Find the amount remaining in her grant after paying tuition.

2. The monthly charge in a parking garage increased from $158 a month to $175 a month. Find the increase in the monthly charge.

3. At the Iceberg Drive-In, a double cheeseburger costs $5.25, and a vanilla shake costs $3.80. Find the cost of 14 double cheeseburgers.

4. The cost of regular grade gasoline, which is 10% ethanol, is $3.73 per gallon. The cost of premium grade gasoline is $3.93 per gallon. Find the cost of 12 gallons of premium grade gasoline.

5. A paving contractor said that the cost to pave a parking lot was about $3 per square foot. Find the cost to pave a rectangular parking lot that is 900 ft wide and 1200 ft long.

6. A living room carpet costs $32 per square yard. Find the cost to carpet a room that is 13 ft wide and 20 ft long.

7. For students taking at least six credits, a college daycare charges $350 a month for preschoolers, $400 a month for toddlers, and $450 a month for infants. Find the cost of daycare for 9 months for a family with two preschoolers and one toddler.

8. For students taking at least six credits, a college daycare center charges $350 a month for preschoolers, $400 a month for toddlers, and $450 a month for infants. Find the cost of daycare for 9 months for one preschooler and two infants.

9. A student works 12 hr a week at a job that pays $7.25 an hour. His bills are about $225 per week. If he can find a job that pays $8.50 an hour after taxes, find the number of additional hours that he needs to work to pay his bills. Round *up* to the nearest whole number.

10. A student works 18 hr a week at a job that pays $7.50 an hour. Her bills are about $250 per week. If she can find a job that pays $9.25 an hour after taxes, find the number of additional hours that she needs to work to pay her bills. Round *up* to the nearest whole number.

11. A Basic wireless phone plan costs $39.99 a month. Each text message costs $0.20. A Select plan costs $59.99 a month and has unlimited text messages. Find the number of text messages at which the monthly cost of the plans are equal.

12. A Select wireless phone plan costs $59.99 a month. Each megabyte of data costs $1.99. A Connect wire-

less phone plan costs $69.99 a month. Data transfer is free. Find the number of megabytes of data at which the monthly costs of the plans are equal. Round to the nearest whole number.

13. Find the decrease in the number of new citizenship applications from 2007 to 2008.

In the Southern California district, for instance, applications plunged to 58,433 last year (2008) from 253,666 the previous year (2007), U.S. immigration statistics show. Most experts say that a 69% increase in application fees to $675 was one reason for the steep decline. (*Source:* www.latimes.com, May 11, 2009).

14. Find the number of jobs lost in the last year in Phoenix that were not in construction.

Median home prices for resold homes peaked at $268,000 in June 2006. Now the median price is $120,000.... In the last year, Phoenix lost 41,000 construction jobs and 136,000 overall, accounting for 7% of its workforce. (*Source:* www.latimes.com, May 18, 2009)

15. Find the number of gallons of water per year that a "typical" home uses in San Diego. Write the answer in place value notation.

San Diego's top single-family residential water customer used 5.5 million gallons during a recent 12-month period—enough to serve about 44 typical homes. (*Source:* www.signonsandiego.com, May 16, 2009)

16. Find the average cost per mile for the new toll roads. Write the answer in place value notation. Round to the nearest million.

Construction of the 65 miles of new toll roads in Central Texas will cost approximately $2.9 billion. (*Source:* www.centraltexasturnpike.org)

17. The tuition and fees for an out-of-state student for one semester at Lewis-Clark State College are $6393. A student registers for 15 credits. This represents 15 hours of class time per week. The semester is 16 weeks long. There are two semesters in each school year. Find the price per hour of class time. Round to the nearest whole number.

18. The tuition and fees for a state resident for one semester at Lewis-Clark State College is $2298. A student registers for 15 credits. This represents 15 hours of class time per week. The semester is 16 weeks long. Seven of the credits are for courses in the student's major. Find the price per hour of class time. Round to the nearest whole number.

19. According to the U.S. Census Bureau, the population of Kansas in 2008 was 2,802,134 people. Find the number of dentists in Kansas. Round to the nearest whole number.

The majority of Kansas counties (86 out of 105), along with the cities of Topeka and Wichita, are designated as "Health Professions Shortage Areas" for dentistry by the federal government's Health Resources and Services Administration. Kansas has 36.8 dentists per 100,000 residents, 33 percent below national targets. (*Source:* www.wichita.edu, Oct 28, 2008)

20. According to the U.S. Census Bureau, the population of Arizona in 2006 was 6,166,318 people. Predict how many

of these people contracted valley fever. Round to the nearest whole number.

Incidence [of valley fever] appears to have increased in Arizona over the past decade. Incidence in 2006 was 91 cases per 100,000 population. (*Source:* www.cdc.gov)

21. In the 2008–2009 school year, the public schools in Prince William County, Virginia, had an enrollment of 73,657 students. Find the number of students who now rely on schools for two meals a day. Round to the nearest hundred.

In many communities, growing numbers of students rely on schools for two meals a day.... In Prince William County, the number rose ... to 33 percent and in Montgomery County ... to 28 percent. (*Source:* www.washingtonpost.com, May 23, 2009)

22. Find the maximum number of people in the study who became nauseated after chemotherapy. Round to the nearest ten.

Up to 70% of patients become nauseated after chemo, according to a study of 644 people released Thursday, in advance of the annual meeting of the American Society of Clinical Oncology. (*Source:* www.usatoday.com, May 14, 2009)

23. Find the percent of the full-time positions that were cut.

Effective last week, the University of Utah museum cut three of its 15 full-time positions and reduced the working hours of remaining full-time staff by 20 percent. (*Source:* www.sltrib.com/news, May 26, 2009)

24. Find the percent of the pay boxes that had problems. Round to the nearest percent.

Many of the newly-installed pay-and-display boxes in the downtown [Chicago] area were not working properly today.... The company has sent out 15 crews to fix problems affecting roughly 125 of 556 pay boxes in the city. (*Source:* newsblogs.chicagotribune.com, May 27, 2009)

For exercises 25–26, use the information in the table.

Cost to employee	Traditional plan	PPO plan
Cost per month	$29.50	$23
Deductible	$350	$250
Percent of major medical bills (after deductible) paid by the employee (upper limit $4300)	20%	15%
Cost per doctor visit (not applied towards deductible)	$0	$20

25. Healthcare costs per year for a state employee include the cost per month, the deductible, her share of major medical bills, and any charges for doctor visits. A state employee in good health expects to visit her doctor twice a year. She estimates that her major medical expenses per year will be $600. Find the difference in cost per year between the traditional plan and the PPO plan for this employee.

26. Health care costs per year for a state employee include the cost per month, the deductible, his share of major medical bills, and any charges for doctor visits. An older state employee visits the doctor once a month. He estimates that his major medical expenses for the year will be $3000. Find the difference in cost per year between the traditional plan and the PPO plan for this employee.

27. Find the Hispanic population in Wisconsin in 2000. Round to the nearest thousand.

Since 2000, the Hispanic population in Wisconsin has increased by 48.2% to 285,827 people, or 5.1% of the state's population. (*Source:* www.JSonline.com, May 14, 2009)

28. Find the number of student athletes who competed in the 1981–1982 school year. Round to the nearest thousand.

According to NCAA figures, a record 17,682 college teams competed in the 2007–8 academic year, 60 percent more than in 1981–82. During that time, the number of student-athletes grew 78 percent, to a record 412,768. (*Source:* www.nytimes.com, May 4, 2009)

29. Find the percent decrease in traffic deaths from 2007 to 2008. Round to the nearest percent.

Massachusetts had 346 traffic deaths in 2008, according to preliminary state data, down from 434 in 2007. (*Source:* www.nytimes.com, April 7, 2009)

30. Find the percent increase in the number of patients infected with MRSA. Round to the nearest percent.

Over the past decade, the number of Washington hospital patients infected with a frightening, antibiotic-resistant germ called MRSA has skyrocketed from 141 a year to 4723. (*Source:* seattletimes.nwsource.com, Nov. 16, 2008)

For exercises 31–32, the formula $I = PRT$ can be used to estimate the interest charge on the balance on a credit card. The monthly interest is I, the monthly balance is P, the monthly interest rate in decimal form is R, and the time in months is T.

31. A student has a good credit history. She predicts that her average unpaid balance will be $150. She has $2500 in federal student loans. Find the difference in the cost of using each card per *year*. Round to the nearest hundredth.

Credit card	AT&T Universal Card	Discover
Annual fee	$0	$0
Annual interest rate	15.9%	15.99%
Monthly interest rate	1.325%	1.3325%

32. A student has a poor credit history. He predicts that his average unpaid balance will be $900. He has $3500 in student loans. Find the difference in the cost of using each card per *year*. Include the annual fee in this cost. Round to the nearest hundredth.

Credit card	Wells Fargo	Bank of America
Annual fee	$0	$29
Annual interest rate	19.49%	16.99%
Monthly interest rate	1.6242%	1.4158%

33. Cotton muslin fabric is 44 in. wide. The supplies for a quilting class include ten square "sandwiches." They are 14 in. long and 14 in. wide. Each sandwich is one piece of $\frac{1}{4}$-in.-thick polyester batting placed between the two square pieces of cotton muslin fabric. Find the minimum number of yards of cotton muslin fabric needed to make the sandwiches. Round *up* to the nearest tenth of a yard.

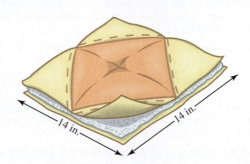

34. Roofing shingles are sold by the "square." A square can cover an area of roof that is 10 ft long and 10 ft wide. A home improvement store sells shingles by the square or by the bundle. The area of a roof is 2160 ft². Find the minimum number of squares needed to shingle this roof. Round *up* to the nearest square.

35. Find the number of American children in 2007. Round to the nearest tenth of a million.

In 1970, about 3% of all children under 18 lived in households headed by a grandparent. By 2007, 4.7 million kids—or 6.5% of American children—were living in households headed by a grandparent, according to U.S. Census Bureau data. (*Source:* online.wsj.com, April 4, 2009)

36. Find the number of single track downloads in 2007. Write the answer in millions. Round to the nearest ten million.

The United States is the world leader in digital music sales. . . . Single track downloads crossed the billion mark for the first time in 2008, totalling 1.1 billion, up 27 percent from 2007. (*Source:* www.ifpi.org, Digital Music Report, 2009)

37. To maintain eligibility, an athlete needs a test average in math class of 75%. There are six tests in the class. On the first five tests, the athlete scored 85%, 72%, 61%, 79%, and 70%. Find the score needed on the last test to have a test average of 75%.

38. A final semester score depends on tests (55%), homework (20%), final exam (20%), and participation (5%). A student has a test score average of 85, a homework score average of 79, a score on the final exam of 91, and a class participation score average of 2. Find the final semester score for this student. Round to the nearest whole number.

39. The trash bags sold at a local grocery store are available in 30-gal, 33-gal, and 39-gal sizes. Each box contains 30 trash bags. The inner top diameter of a circular plastic trashcan is 20 in. The height of the can is 28 in. Find the minimum size of trash bag for this trashcan (1 gal = 231 in.³; the volume of a cylinder is $V = \pi r^2 h$; $\pi \approx 3.14$).

20 in.

28 in.

40. Mark Parker remodeled his boat to include a marine toilet with a fiberglass holding tank located under the V-berth. The inside of the finished tank is a rectangular trapezoid. It is about 29.5 in. wide at the base, 17.5 in. wide at the top, 18.5 in. long, and 17.5 in. high.

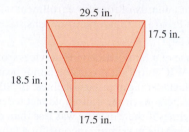

29.5 in.

17.5 in.

18.5 in.

17.5 in.

Find the volume of the holding tank in gallons. Round to the nearest tenth of a gallon (1 gal = 231 in.³). The formula for the area of a trapezoid is $A = \dfrac{1}{2}(b_1 + b_2)(h)$.

b_1

h

b_2

(*Source:* www.goodoldboat.com, Nov./Dec. 1999)

For exercises 41–42, a newsprint recycling company recycles 750,000 metric tons of old newsprint a year. Pulpers turn the newspaper into a liquid. The liquid is 4.5% fiber and 95.5% water. It is screened many times and washed to remove ink. It is then treated to control pH and brightness. Machines turn it into finished newsprint.

Machine	Rate of newsprint production	Width of newsprint
No. 1	$4450\ \dfrac{\text{ft}}{\text{min}}$	308 in.
No. 2	$5800\ \dfrac{\text{ft}}{\text{min}}$	336 in.

41. If the machines both work 8 hr without stopping, find the difference in the area of newsprint in square feet made by the two machines.

42. Find the difference in the production rate of the machines in *miles* of newsprint produced *per hour*. Round to the nearest whole number.

43. An employee earns $2300 in gross wages each month. For medical and dental insurance, the employee pays $34 per month and the employer pays $594 per month. *Total compensation* is the amount of gross wages plus the amount that the employer pays for insurance. Find the total compensation per year for the employee.

44. A report on the costs of growing broccoli in California estimates that the *employer's real cost* for hand labor is $9.25 per hour. Of this, $6.75 per hour is wages paid to the employee. The other costs include employer payments for Social Security, unemployment insurance, workman's compensation, transportation costs, and cost of supervision. For a 40-hr work week, find the difference between the employer's real costs and the amount of wages paid to the employee.

45. An employee in the produce department at Walmart in El Paso earns $11.18 an hour. This employee works 40 hours per week for 50 weeks per year. The total amount he earns per year before taxes or other deductions is his *gross pay*. He pays $5824 per year for daycare for his toddler. It is a 10-min drive from the daycare center to work. Find the percent of this employee's gross pay that he spends on daycare per year. Round to the nearest percent.

46. A certified nursing assistant in Florida earns $10.61 an hour. This employee works 40 hr per week for 50 weeks per year. The total amount he earns per year before taxes or other deductions is his *gross pay*. He pays $5835 per year for daycare for his toddler. His car gets 20 mi per gallon. Find the percent of his gross pay that he spends on daycare per year. Round to the nearest percent.

47. A traveler has 1500 U.S. dollars. In Amsterdam, the traveler changes this money into euros ($1 ≈ 0.69 euro) and spends 750 euros. In London, he changes his remaining euros into British pounds (1 euro ≈ 0.736 pound) and spends 50 pounds. He finally changes his remaining pounds into dollars (1 pound ≈ 1.996 U.S. dollars). Ignoring any currency change fees, find the final number of U.S. dollars. Round to the nearest whole number.

48. A traveler asks a bank to change $375 into Mexican pesos ($1 ≈ 10.687 pesos) and $375 into Costa Rican colons ($1 ≈ 480.20 colons). Find the number of pesos and the number of colons. Round to the nearest whole number.

49. Find the annual reduction in sales tax for an Arkansas family that spends an average of $250 per week on food (1 year = 52 weeks).

State sales tax on food and food ingredients will be reduced from 3% to 2% effective July 1, 2009. This act does not apply to local tax rates. (*Source:* www.dfa.arkansas.gov, April 2009)

50. If the sales tax were removed, find the annual reduction in the total cost of food and sales tax for an Alabama family that spends an average of $275 per week on food (1 year = 52 weeks).

An effort to remove the state's 4-percent sales tax from groceries failed. . . . Alabama currently levies a 4 percent state tax on groceries. (*Source: www.clantonadvertiser.com,* March 25, 2009)

For exercises 51–52, the FDA approved the use of a new drug (ranibizumab) for the treatment of macular degeneration. One dose of the drug is injected each month into the eye. The wholesale cost of one dose of the drug is $1950. (*Source: New England Journal of Medicine,* Oct. 5, 2006)

51. Find the wholesale cost for 5 years of this treatment.

52. In the United States, patients who have Plan B Medicare insurance coverage are responsible for 20% of the cost of treatment with ranibizumab. Find the cost to the patient for 10 years of treatment for a patient with this insurance.

53. Find the total number of positions in the nation's school systems. Round to the nearest thousand.

In the economic stimulus bill passed in February 2009, Congress appropriated about $100 billion in emergency education financing. States spent much of that in the current fiscal year, saving more than 342,000 school jobs, about 5.5 percent of all the positions in the nation's 15,000 school systems. (*Source: www.nytimes.com,* April 20, 2010)

54. A total of 453 administrators responded to the survey. Find the number who expect to lay off school workers for the fall. Round to the nearest whole number. (*Source:* American Association of School Administrators, April 2010)

A survey by the American Association of School Administrators found that 9 of 10 superintendents expected to lay off school workers for the fall, up from two of three superintendents last year. The survey also found that the percentage considering a four-day school week had jumped to 13 percent, from 2 percent a year ago. (*Source: www.nytimes.com,* April 20, 2010)

For exercises 55–56, a formula for a high school graduation rate is

$$\text{graduation rate} = \frac{G_4}{G_4 + D_1 + D_2 + D_3 + D_4} \cdot 100\%$$

where G_4 is the number of graduates at the end of the fourth year of high school, D_1 is the number of dropouts in the first year (grade 9), D_2 is the number of dropouts in the second year, D_3 is the number of dropouts in the third year, and D_4 is the number of dropouts in the fourth year.

55. Find the graduation rate in 2007–2008 for the Oakland Unified School District. Round to the nearest percent. (*Source:* California Department of Education)

Oakland Unified School District, 2007–2008		
Grade	Dropouts	Graduates
9	126	—
10	198	—
11	282	—
12	287	1992

56. Find the graduation rate in 2007–2008 for the Los Angeles Unified School District. Round to the nearest percent. (*Source:* California Department of Education)

Los Angeles Unified School District, 2007–2008		
Grade	Dropouts	Graduates
9	1460	—
10	2031	—
11	2375	—
12	3324	31,165

57. The income of a family receiving services from the Texas Nutrition Program for Women, Infants, and Children (WIC) cannot be more than 185% of the federal poverty guidelines. In 2011, a family with four persons in the household met federal poverty guidelines if their income was less than or equal to $22,350. Find the maximum income for a family of four for enrollment in the Texas Nutrition Program. (*Sources: www.dshs.state.tx.us; www.aspe.hhs.gov*)

58. The income of a family receiving services from the North Carolina WIC nutrition program cannot be more than 185% of the federal poverty guidelines. In 2011, a family with three persons in the household met federal poverty guidelines if their income was less than or equal to $18,530. Find the maximum income for a family of three (mother, infant, and 18-month old child) for enrollment in the North Carolina WIC program. (*Sources: www.nutritionnc.com; www.aspe.hhs.gov*)

For exercises 59–60, $G = \dfrac{6.67 \times 10^{-11} \text{ m}^3}{1 \text{ s}^2 \cdot \text{kg}}$ and $M = 5.98 \times 10^{24}$ kg. Write the answer in scientific notation; round the mantissa to the nearest hundredth.

59. The formula $V = \sqrt{\dfrac{GM}{r}}$ represents the relationship of the velocity, V, of an object in orbit around the earth, the gravitational constant, G, the mass of the earth, M, and the radius of the orbit r. If $r = 3.84 \times 10^8$ m, find V.

60. The formula $V = \sqrt{\dfrac{2GM}{r}}$ represents the relationship of the escape velocity, V, from the surface of the earth, the gravitational constant, G, the mass of the earth, M, and the radius of the earth, r. If $r = 6.38 \times 10^6$ m, find V.

61. An IV uses 15 drops to deliver 1 mL of solution $\left(\dfrac{15 \text{ drops}}{1 \text{ mL}}\right)$. A patient needs to receive a solution at a rate of $\dfrac{75 \text{ mL}}{1 \text{ hr}}$. Find the needed speed of the IV in drops per minute. Round to the nearest whole number.

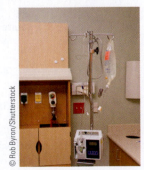
© Rob Byron/Shutterstock

62. An IV uses 10 drops to deliver 1 mL of solution $\left(\dfrac{10 \text{ drops}}{1 \text{ mL}}\right)$. An IV is set to deliver solution at a rate of $\dfrac{20 \text{ drops}}{1 \text{ min}}$. An IV bag contains 400 mL of solution.
Find the time in minutes it will take to empty the bag.

63. Find the maximum amount of the economic stimulus package that had been paid out. Round to the nearest tenth of a billion.

Nearly three months after President Obama approved a $787 billion economic stimulus package, intended to create or save jobs, the federal government has paid out less than 6 percent of the money, largely in the form of social service payments to states. (*Source:* www.nytimes.com, May 13, 2009)

64. Find the maximum number of U.S. residents in 2008. Round to the nearest million.

The largest and fastest-growing minority group was Hispanics, who reached 46.9 million in 2008, up by 3.2 percent from 2007. In 2008, nearly one in six U.S. residents was Hispanic. (*Source:* www.census.gov, May 14, 2009)

65. Find the maximum number of hogs and pigs in the nation.

The chance that the hot dogs and pork sausages consumed on the Fourth of July originated in Iowa [is more than 1 in 4]. The Hawkeye State [Iowa] was home to 19.3 million hogs and pigs on March 1, 2009. This represents more than one-fourth of the nation's total. North Carolina (9.4 million) and Minnesota (7.3 million) were the runners-up. (*Source:* www.census.gov, May 4, 2009)

66. The Pensacola Beach Pier in Florida is 1471 ft long. Find the minimum length of railings on one side of the pier that will have to be no more than 34 in. high to meet the new regulations. Round to the nearest foot.

The 215,000-word proposal includes these new requirements. . . . At least 25 percent of the railings at fishing piers would have to be no more than 34 inches high, so that a person in a wheelchair could fish over the railing. (*Source: San Diego Union Tribune*, June 16, 2008)

67. A potential resident of Boulevard Gardens has an annual income of $65,000. Find the maximum amount of this income that can be used for mortgage and maintenance fees at this co-op building.

At the northwestern end, closest to Astoria, is Boulevard Gardens, the co-op where Ms. Gallagher lives, in a complex of 10 buildings with 6 stories each, and a total of 960 units. . . . In the wake of the mortgage crisis, though, some would-be buyers have had difficulty meeting the co-op's rule that no more than 30 percent of a resident's income can go toward mortgage and maintenance fees. (*Source:* www.nytimes.com, March 16, 2008)

68. Legislative Bill 500 was approved by the Governor of Nebraska on May 26, 2009. A community has a cemetery with a perpetual fund balance of $75,500. Find the maximum amount of this principal that can be used for the cemetery in any fiscal year.

The principal of the perpetual fund may be used for the general care, management, maintenance, improvement, beautifying, and welfare of the cemetery as long as no more than twenty percent of the principal is so used in any fiscal year and no more than forty percent of the principal is so used in any period of ten consecutive fiscal years. (*Source:* uniweb.legislature.ne.gov, May 2009)

69. In June 2010, a sinkhole opened under a clothing factory, destroying it. A local cement factory suggested that ash from the Pacaya Volcano could be combined with cement and used to fill the hole. Assuming that the sinkhole is cylindrical in shape, find the amount of cement in cubic feet needed to fill the sinkhole (volume of a cylinder $= \pi r^2 h$; $\pi \approx 3.14$). Round to the nearest thousand.

Some 66 feet (20 meters) across, nearly 100 feet (30 meters) deep and almost a perfect circle at its gaping mouth, the sinkhole opened up suddenly May 30 in Ciudad Nueva [Guatemala]. (*Source:* Associated Press, June 5, 2010)

70. Assume that the destruction of the tornado was a rectangular area. Find the area of destruction in square feet (1 yd = 3 ft; 1 mi = 5280 ft).

Storms collapsed a movie-theater roof in Illinois and ripped siding off a building at a Michigan nuclear plant, forcing a shutdown. But the worst destruction was reserved for a 100-yard-wide, 7-mile-long strip southeast of Toledo left littered Sunday with wrecked vehicles, splintered wood, and family possessions. (*Source:* CBS News/AP, June 7, 2010)

71. In 1996, there were three breeding pairs of wolves in Idaho. By 2004, there were 27 breeding pairs in Idaho. What was the percent increase in breeding pairs?
   a. Choose the best estimate: 11%, 88%, or 800%.
   b. Explain your reasoning in choosing this estimate.

72. A savings account pays 5% annual interest, compounded daily. When interest is compounded daily, the interest owed is calculated every day and added to the account. The initial investment is $1000. The money is kept in the savings account for two years.
   a. Choose the best estimate for the value of the investment after two years: $1010, $1100, $2000, or $6000.
   b. Explain your reasoning in choosing this estimate.

Find the Mistake

For exercises 73–76, the completed problem has one mistake.
(a) Describe the mistake in words, or copy down the whole problem and highlight or circle the mistake.
(b) Redo only the step in the problem solving plan that is shown.

73. **Problem:** The sale price of a television is $100. It is marked down 20% from its original price. Find the original price.

   Incorrect Answer: Carry out the plan.

P = original price	Assign the variable.
$P = (0.20)(\$100) + \$100$	Write an equation.
$P = \$120$	Solve the equation.

74. Problem: The diameter of the bottom of a tart pan is 10 in. Find the area of the bottom of the tart pan. The formula for the area of a circle is $A = \pi r^2$.

© Aleksandra Duda/Shutterstock

Incorrect Answer: Carry out the plan.

$$A = \pi r^2$$
$$A = (3.14)(10 \text{ in.})^2$$
$$A = 3.14(100 \text{ in.}^2)$$
$$A = 314 \text{ in.}^2$$

75. Problem: The energy of the sun is produced by hydrogen fusion. When hydrogen atoms fuse together to make helium, some mass is changed to energy. The equation $E = mc^2$ calculates the amount of energy E released by fusion of a mass m. The speed of light in a vacuum is c, $3 \times 10^8 \frac{m}{s}$. When four hydrogen nuclei combine to make one helium nucleus, 4.8×10^{-29} kg of mass changes into energy. Find the amount of energy produced by this fusion.

Incorrect Answer: Carry out the plan.

$$E = mc^2$$
$$E = (4.8 \times 10^{-29} \text{ kg})\left(3 \times 10^8 \frac{m}{s}\right)^2$$
$$E = (4.8 \times 3) \times (10^{-29} \times 10^{16}) \times \left(\text{kg} \cdot \frac{m^2}{s^2}\right)$$
$$E = 14.4 \times 10^{-13} \frac{\text{kg} \cdot m^2}{s^2}$$
$$E = 1.44 \times 10^{-12} \frac{\text{kg} \cdot m^2}{s^2}$$

76. Problem: A package of dried plums weighs 12 oz. The nutrition information on the label uses a serving size of $1\frac{1}{2}$ oz. How many servings are in each package?

Incorrect Answer: Carry out the plan.

$$S = \text{servings}$$
$$S = \left(1\frac{1}{2} \text{ oz}\right)\left(\frac{12 \text{ oz}}{1 \text{ package}}\right)$$
$$S = \left(\frac{3}{2} \text{ oz}\right)\left(\frac{12 \text{ oz}}{1 \text{ package}}\right)$$
$$S = 18 \text{ servings}$$

SUCCESS IN COLLEGE MATHEMATICS

77. Describe your instructor's policy about late assignments.

78. Some instructors include attendance, class participation, or participation in an on-line discussion board as part of the student's grade. Explain how, if at all, attendance or participation affects your grade.

© Rob Byron/Shutterstock

An average homeowner in Massachusetts buys 8000 kilowatt · hours of electrical power per year. One hundred square feet of solar panels can generate 1200 kilowatt · hours of electrical power. A **linear equation** in two variables can describe the relationship between the area of solar panels used by the homeowner and the amount of electricity the homeowner still needs to buy. In this section, we will study linear equations.
(*Source*: Massachusetts Technology Collaborative Renewable Energy Trust, 2008)

SECTION 1.6 Slope and Linear Equations in Two Variables

After reading the text, working the practice problems, and completing assigned exercises, you should be able to:

1. Rewrite a linear equation in standard form.

2. Given the equation of a line, find its intercepts and graph the line.

3. Given two points on a line, the equation of a line, or the graph of a line, calculate its slope.

4. Identify the slope of a line that is parallel or perpendicular to a given line.

5. Rewrite a linear equation in slope-intercept form.

6. Use slope-intercept graphing to graph a linear equation.

7. Given the equation of a horizontal or a vertical line, graph the line.

Linear Equations in Standard Form

The **standard form** of a linear equation is $ax + by = c$, where a, b, and c are real numbers. A **form** of an equation is a useful pattern.

Form of a Linear Equation in Two Variables

Standard Form $ax + by = c$ a, b, and c are real numbers

EXAMPLE 1 Rewrite $y - 7 = -\dfrac{1}{2}(4x - 18)$ in standard form.

SOLUTION ▶

$$y - 7 = -\frac{1}{2}(4x - 18)$$

$$y - 7 = \left(-\frac{1}{2}\right)(4x) - \frac{1}{2}(-18) \qquad \text{Distributive property.}$$

$$y - 7 = -2x + 9 \qquad \text{Simplify.}$$

$$\underline{+7 \qquad\qquad +7} \qquad \text{Addition property of equality.}$$

$$y + 0 = -2x + 16 \qquad \text{Simplify.}$$

$$\underline{+2x \quad +2x} \qquad\qquad \text{Addition property of equality.}$$

$$2x + y = 16 \qquad \text{Simplify; standard form.}$$

Practice Problems

For problems 1–2, rewrite the equation in standard form.

1. $y = -7x + 9$ **2.** $y - 6 = -\dfrac{3}{4}(8x + 12)$

Graphing Linear Equations

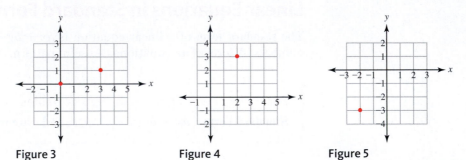

Tick mark

Scale

Figure 1

We often use a number line graph (Figure 1) to represent a real number.

To graph an **ordered pair** (x, y), we use a **rectangular coordinate system** (Figure 2). A coordinate system includes a horizontal number line and a vertical number line. Each number line is an **axis**. The horizontal number line is the **x-axis**. The vertical number line is the **y-axis**.

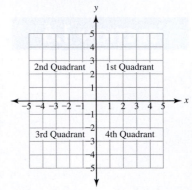

Figure 2

Each axis is labeled with its variable and a **scale**. The scales do not have to be the same on both axes. (The plural of *axis* is *axes*.) The axes divide the coordinate system into four regions. Each region is a **quadrant** (*quad* means "four").

To represent an ordered pair, we graph a **point**. The **origin** is the point that represents the ordered pair $(0, 0)$. A point that is 3 units to the right of the origin and 1 unit up from the origin (Figure 3) represents the ordered pair $(3, 1)$. The point that is 2 units to the right of the origin and 3 units up from the origin (Figure 4) represents the ordered pair $(2, 3)$. The point on the graph in Figure 5 represents the ordered pair $(-2, -3)$. The x-coordinate of this ordered pair is -2. The y-coordinate is -3. The point is 2 units to the left and 3 units down from the origin.

Figure 3   **Figure 4**   **Figure 5**

To graph a linear equation in standard form, we can use its **intercepts** and one other point. The **x-intercept** is the point where a line crosses the x-axis. The y-coordinate of an x-intercept is 0. The **y-intercept** is the point where a line crosses the y-axis. The x-coordinate of a y-intercept is 0.

Finding an x-intercept

To find an x-intercept $(x, 0)$:

1. Replace y in the equation with 0.

2. Solve the equation to find the value of x in the ordered pair.

> **Finding a y-intercept**
> To find a y-intercept $(0, y)$:
> **1.** Replace x in the equation with 0.
> **2.** Solve the equation to find the value of y in the ordered pair.

EXAMPLE 2   The equation of a line is $-2x + 3y = 12$.

(a) Find the x-intercept.

SOLUTION ▶

$$-2x + 3y = 12$$ Use the original equation.

$$-2x + 3(\mathbf{0}) = 12$$ Replace y with 0.

$$-2x + \mathbf{0} = 12$$ Simplify.

$$\frac{\mathbf{-2x}}{-2} = \frac{12}{-2}$$ Division property of equality.

$$x = -6$$ Simplify. The x-intercept is $(-6, 0)$.

(b) Find the y-intercept.

▶

$$-2x + 3y = 12$$ Use the original equation.

$$-2(\mathbf{0}) + 3y = 12$$ Replace x with 0.

$$\mathbf{0} + 3y = 12$$ Simplify.

$$\frac{\mathbf{3y}}{3} = \frac{12}{3}$$ Division property of equality.

$$y = 4$$ Simplify. The y-intercept is $(0, 4)$.

(c) Find one more ordered pair that is a solution of this equation.

▶   Replace either x or y with a real number other than 0.

$$-2x + 3y = 12$$ Use the original equation.

$$-2(\mathbf{2}) + 3y = 12$$ Replace x with a real number; here we use 2.

$$\mathbf{-4} + 3y = 12$$ Simplify.

$$\underline{+4 \qquad\qquad +4}$$ Addition property of equality.

$$\frac{\mathbf{3y}}{3} = \frac{16}{3}$$ Division property of equality.

$$y = \frac{16}{3}$$ Simplify. The ordered pair is $\left(2, \dfrac{16}{3}\right)$.

(d) Organize the ordered pairs in a table; graph the equation.

▶   Graph each point, and draw a line through the points. Each point on this line represents an ordered pair that is a solution of $-2x + 3y = 12$.

x	y
-6	0
0	4
2	$\dfrac{16}{3}$

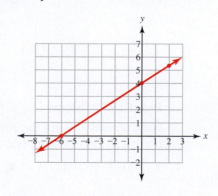

A graph of an equation can help us visualize the relationship between two variables.

EXAMPLE 3   A landscape architect is creating a plan that includes 5500 ft² to be planted in either shrubs or grass.

(a) Assign a variable to represent each area.

SOLUTION ▶ $x =$ shrub area $y =$ grass area

(b) Write a linear equation that shows the relationship of the area with shrubs and the area with grass.

▶ $x + y = 5500$ ft²

(c) Find the x-intercept and the y-intercept.

▶

$x + y = 65,000$ ft²

x-intercept		**y-intercept**
$x + y = 5500$ ft²	Use the original equation.	$x + y = 5500$ ft²
$x + \mathbf{0}$ **ft²** $= 5500$ ft²	Replace the variable.	$\mathbf{0}$ **ft²** $+ y = 5500$ ft²
$x = 5500$ ft²	Simplify.	$y = 5500$ ft²

The x-intercept is $(5500 \text{ ft}^2, 0 \text{ ft}^2)$; the y-intercept is $(0 \text{ ft}^2, 5500 \text{ ft}^2)$.

(d) Find one more ordered pair that is a solution of this equation.

▶   Replace either x or y with a measurement other than 0.

$x + y = 5500$ ft²	Use the original equation.
$\mathbf{4000}$ **ft²** $+ y = 5500$ ft²	Replace x with 4000 ft².
$\underline{-4000 \text{ ft}^2 \qquad -4000 \text{ ft}^2}$	Subtraction property of equality.
$\mathbf{0} + y = \mathbf{1500}$ **ft²**	Simplify.
$\mathbf{y} = 1500$ ft²	Simplify. The ordered pair is $(4000 \text{ ft}^2, 1500 \text{ ft}^2)$.

(e) Organize the ordered pairs in a table; graph the equation.

▶   Since the area must be greater than or equal to 0, this line stops at the intercepts. Each point on this line is a possible combination of the area with shrubs and the area with grass. Each axis is labeled with the units.

x (ft²)	y (ft²)
0	5500
5500	0
4000	1500

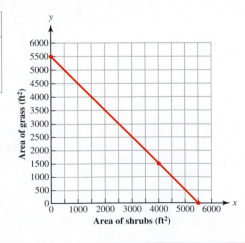

Practice Problems

For problems 3–5, use the intercepts and one other point to graph the linear equation on an xy-coordinate system.

3. $-4x + 5y = 16$ **4.** $4x - y = 12$ **5.** $x + y = \$350$

Slope

To describe the appearance of a friend, we may use hair color or height. To describe a line, we often use its **slope**.

For any two points on a line, the vertical distance between the points is the **rise**. The horizontal distance between the points is the **run**. The slope of the line is the **ratio of the rise over the run**: slope $= \dfrac{\text{rise}}{\text{run}}$. To develop a formula for finding slope, we name two points on the line: (x_1, y_1) and (x_2, y_2). The rise is the difference of the y-coordinates. The run is the difference of the x-coordinates (Figure 6).

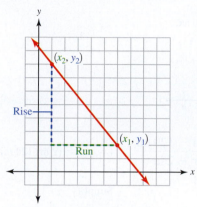

Figure 6

Slope Formula

slope $= \dfrac{\text{rise}}{\text{run}}$ The slope of the line is m.

$m = \dfrac{y_2 - y_1}{x_2 - x_1}$ Two points on the line are (x_1, y_1) and (x_2, y_2).

EXAMPLE 4   Find the slope of the line that passes through $(3, -2)$ and $(1, 6)$.

SOLUTION ▶

$m = \dfrac{y_2 - y_1}{x_2 - x_1}$ The slope formula.

$m = \dfrac{6 - (-2)}{1 - 3}$ $(x_1, y_1) = (3, -2); (x_2, y_2) = (1, 6)$

$m = \dfrac{8}{-2}$ Simplify.

$m = -\dfrac{4}{1}$ We can write the slope as $-\dfrac{4}{1}$ or as -4.

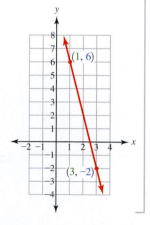

A slope can include units of measurement. We often say the slope $\dfrac{60 \text{ mi}}{1 \text{ hr}}$ as "60 miles per hour." This slope is an example of an **average rate of change**, which is an important topic in Section 3.4.

EXAMPLE 5 Find the slope of the line that passes through (2 hr, 120 mi) and (5 hr, 300 mi).

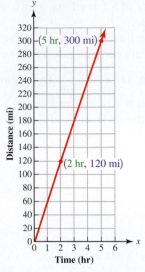

SOLUTION ▶ $m = \dfrac{300 \text{ mi} - 120 \text{ mi}}{5 \text{ hr} - 2 \text{ hr}}$ $(x_1, y_1) = (2 \text{ hr}, 120 \text{ mi}); (x_2, y_2) = (5 \text{ hr}, 300 \text{ mi})$

$m = \dfrac{180 \text{ mi}}{3 \text{ hr}}$ Simplify.

$m = \dfrac{60 \text{ mi}}{1 \text{ hr}}$ We can write the slope as $\dfrac{60 \text{ mi}}{1 \text{ hr}}$ or $60 \dfrac{\text{mi}}{\text{hr}}$.

We can use the slope formula to find the slope of horizontal lines and vertical lines.

Since the y-coordinate of each point on a horizontal line is the same, the rise is 0. So **the slope of a horizontal line is 0**. When the slope of a line is 0, do not say it has "no slope." The meaning of "no slope" is not precise enough to use in mathematics.

EXAMPLE 6 Find the slope of the line that passes through $(-3, -5)$ and $(4, -5)$.

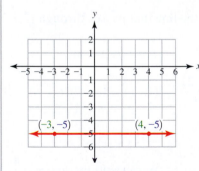

SOLUTION ▶ $m = \dfrac{y_2 - y_1}{x_2 - x_1}$ The slope formula.

$m = \dfrac{-5 - (-5)}{4 - (-3)}$ $(x_1, y_1) = (-3, -5); (x_2, y_2) = (4, -5)$

$m = \dfrac{0}{7}$ Simplify. The slope is 0.

Each point on a vertical line has the same x-coordinate. So the run, $x_2 - x_1$, is 0. Since division by zero is undefined, *the slope of a vertical line is undefined*.

EXAMPLE 7 Find the slope of the line that passes through (2, 6) and (2, 1).

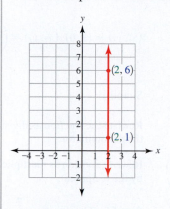

SOLUTION ▶

$$m = \frac{y_2 - y_1}{x_2 - x_1} \qquad \text{The slope formula.}$$

$$m = \frac{1 - 6}{2 - 2} \qquad (x_1, y_1) = (2, 6); (x_2, y_2) = (2, 1)$$

$$m = \frac{-5}{0} \qquad \text{Simplify. The slope is undefined.}$$

To use the slope formula to find the slope of a line, we need to know two points on the line. In the next example, we use the x-intercept and y-intercept.

EXAMPLE 8 The equation of a line is $7x + 2y = 21$. Find the slope of the line.

(a) Find the x-intercept.

SOLUTION ▶

$$\begin{aligned} 7x + 2y &= 21 \qquad \text{The equation of the line.} \\ 7x + 2(\mathbf{0}) &= 21 \qquad \text{Replace } y \text{ with 0.} \\ 7x + \mathbf{0} &= 21 \qquad \text{Simplify.} \\ \frac{\mathbf{7x}}{7} &= \frac{21}{7} \qquad \text{Division property of equality.} \\ x &= 3 \qquad \text{Simplify.} \end{aligned}$$

The x-intercept is $(3, 0)$.

(b) Find the y-intercept.

▶

$$\begin{aligned} 7x + 2y &= 21 \qquad \text{The equation of the line.} \\ 7(\mathbf{0}) + 2y &= 21 \qquad \text{Replace } x \text{ with 0.} \\ \mathbf{0} + 2y &= 21 \qquad \text{Simplify.} \\ \frac{\mathbf{2y}}{2} &= \frac{21}{2} \qquad \text{Division property of equality.} \\ y &= \frac{21}{2} \qquad \text{Simplify.} \end{aligned}$$

The y-intercept is $\left(0, \dfrac{21}{2}\right)$.

(c) Use the x-intercept and y-intercept to find the slope of the line.

$$m = \frac{y_2 - y_1}{x_2 - x_1}$$ The slope formula.

$$m = \frac{\frac{21}{2} - 0}{0 - 3}$$ $(x_1, y_1) = (3, 0); (x_2, y_2) = \left(0, \frac{21}{2}\right)$

$$m = \frac{\frac{21}{2}}{-3}$$ Simplify.

$$m = \frac{21}{2} \div (-3)$$ Rewrite the fraction as division by the denominator.

$$m = \left(\frac{21}{2}\right)\left(-\frac{1}{3}\right)$$ Rewrite division as multiplication by the reciprocal.

$$m = -\frac{7}{2}$$ Simplify. The slope is $-\frac{7}{2}$.

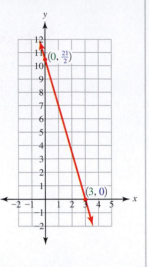

To find the slope of a graphed line, identify two points on the line. "Count" the rise and the run between the points or use the slope formula.

EXAMPLE 9 | Find the slope of the line in the graph at the right.

SOLUTION ▸ Identify two points on the line. Moving from left to right, the rise between these points is -3. The run is 4.

Since slope is $\dfrac{\text{rise}}{\text{run}}$, the slope of this line is $\dfrac{-3}{4}$ or $-\dfrac{3}{4}$.

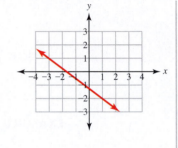

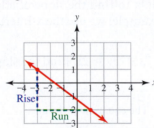

Practice Problems

For problems 6–10, find the slope of the line that passes through the given points.

6. $(5, 9); (2, 15)$ **7.** $(-2, 5); (-3, -7)$ **8.** $(4 \text{ hr}, 1900 \text{ mi}); (7 \text{ hr}, 3325 \text{ mi})$

9. $(-4, 3); (9, 3)$ **10.** $(2, 4); (2, -5)$

For problems 11–12, use the intercepts to find the slope of the line.

11. $3x - 8y = 24$ **12.** $4x + 5y = -16$

For problems 13–14, find the slope of the graphed line.

13.

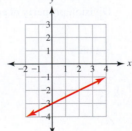

14.

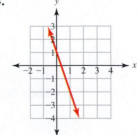

Parallel and Perpendicular Lines

Parallel lines do not intersect, and their slopes are equal.

EXAMPLE 10   Find the slope of the graphed parallel lines.

SOLUTION ▶ Identify two points on one of the lines. Moving from left to right, the rise between these points is 2. The run is 3. Since slope is $\dfrac{\text{rise}}{\text{run}}$, the slope of this line is $\dfrac{2}{3}$. Since the other line is parallel to this line, its slope is also $\dfrac{2}{3}$.

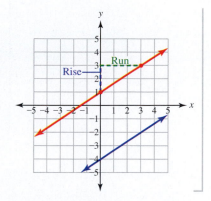

When **perpendicular** lines intersect, they create a 90-degree (right) angle. If neither line is vertical, the slopes of perpendicular lines are **opposite reciprocals**. The product of their slopes equals -1.

EXAMPLE 11   The slope of line A is 2. Identify the slope of a line that is perpendicular to line A.

SOLUTION ▶ The slope of line A is 2 or $\dfrac{2}{1}$. The opposite reciprocal of $\dfrac{2}{1}$ is $-\dfrac{1}{2}$. So the slope of a line that is perpendicular to line A is $-\dfrac{1}{2}$.

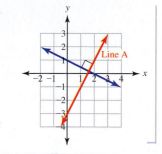

Practice Problems

For problems 15–16, find the slope of the graphed parallel lines.

15.

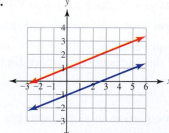

16.

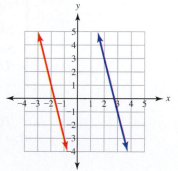

17. The slope of line A is $-\dfrac{5}{6}$. Identify the slope of a line that is perpendicular to line A.

Slope-Intercept Form

In the next example, we use the properties of equality to isolate y on one side of the equation. This is called **solving the equation for y**.

EXAMPLE 12 | Solve $5x - 4y = 8$ for y.

SOLUTION ▶

$$5x - 4y = 8 \qquad \text{y is not isolated.}$$

$$\underline{-5x \qquad\qquad -5x} \qquad \text{Subtraction property of equality.}$$

$$0 - 4y = 8 - \mathbf{5x} \qquad \text{Simplify; 8 and $-5x$ are not like terms.}$$

$$\frac{-4y}{-4} = \frac{8}{-4} - \frac{5x}{-4} \qquad \text{Division property of equality. Divide each term.}$$

$$y = -2 + \frac{5}{4}x \qquad \text{Simplify.}$$

$$y = \frac{5}{4}x - 2 \qquad \text{Use the commutative property; write the variable term first.}$$

A linear equation in two variables solved for y is in **slope-intercept form**, $y = mx + b$. The value of m is the slope of the line represented by this equation. The value of b is the y-coordinate of the y-intercept.

EXAMPLE 13 | A linear equation is $y - 2 = -3(5x + 8)$.

(a) Rewrite this equation in slope-intercept form.

SOLUTION ▶

$$y - 2 = -3(5x + 8)$$

$$y - 2 = \mathbf{-3}(5x) - \mathbf{3}(8) \qquad \text{Distributive property.}$$

$$y - 2 = \mathbf{-15x - 24} \qquad \text{Simplify.}$$

$$\underline{+2 \qquad\qquad +2} \qquad \text{Addition property of equality.}$$

$$y + \mathbf{0} = -15x - \mathbf{22} \qquad \text{Simplify.}$$

$$\mathbf{y} = -15x - 22 \qquad \text{Simplify; slope-intercept form.}$$

(b) Identify the slope and y-intercept.

▶ In slope-intercept form, $y = mx + b$, m is the slope and b is the y-coordinate of the y-intercept. The slope of $y = -15x - 22$ is -15, and the y-intercept is $(0, -22)$.

Forms of a Linear Equation in Two Variables

Standard Form $\quad ax + by = c \qquad a$, b, and c are real numbers

Slope-Intercept Form $\quad y = mx + b \qquad m$ and b are real numbers
The slope of the line is m. The y-coordinate of the y-intercept is b.

To graph an equation in slope-intercept form, we can find three ordered pairs that are solutions of the equation.

EXAMPLE 14 | Find three ordered pairs that are solutions of $y = \frac{2}{3}x - 1$ and graph this equation.

(a) When x is 0, what is y? (This is the y-intercept.)

x	y
0	?

SOLUTION ▶

$$y = \frac{2}{3}x - 1$$

$$y = \frac{2}{3}(\mathbf{0}) - 1 \qquad \text{Replace x with 0.}$$

$$y = \mathbf{-1} \qquad \text{One solution of this equation is $(0, -1)$.}$$

Since y is isolated, it is easier to continue to choose values for x rather than choosing $y = 0$ and finding the x-intercept.

(b) When x is 3, what is y?

$$y = \frac{2}{3}x - 1$$

$$y = \frac{2}{3}(\mathbf{3}) - 1 \qquad \text{Replace } x \text{ with 3.}$$

$$y = \mathbf{2} - 1 \qquad \text{Simplify.}$$

$$y = \mathbf{1} \qquad \text{A second solution is } (3, 1).$$

x	y
0	−1
3	?
▪	▪

Choosing an x-value that is a multiple of the denominator, 3, results in easier arithmetic.

(c) When x is −3, what is y?

$$y = \frac{2}{3}x - 1$$

$$y = \frac{2}{3}(\mathbf{-3}) - 1 \qquad \text{Replace } x \text{ with –3.}$$

$$y = \mathbf{-2} - 1 \qquad \text{Simplify.}$$

$$y = \mathbf{-3} \qquad \text{A third solution is } (-3, -3).$$

x	y
0	−1
3	1
−3	?

(d) Graph the ordered pairs. Draw a line through the points.

Notice that the values for x spread the points out across the graph. If the points are too close together, it is difficult to draw a line that is a good representation of the equation.

x	y
0	−1
3	1
−3	−3

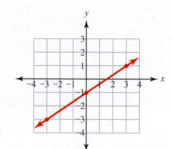

Slope-intercept graphing is another way to graph an equation that is in slope-intercept form. Graph the y-intercept and use the slope to find the location of another point on the line.

EXAMPLE 15 Use slope-intercept graphing to graph $y = -\frac{3}{2}x + 5$.

SOLUTION ▶ Graph the y-intercept, $(0, 5)$. Rewrite the slope as $\frac{-3}{2}$.

Beginning at the y-intercept, move *down 3 units* and to the *right 2 units*. This is the location of another point on this line. Graph a point here. Draw a line through the points.

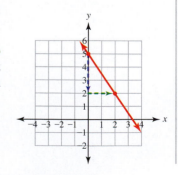

Or, rewrite the slope as $\dfrac{3}{-2}$. From the y-intercept, move *up 3 units* and to the *left 2 units*. This is the location of another point on the line. Draw a line through these points.

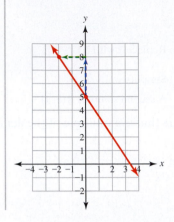

Slope-Intercept Graphing

1. Graph the y-intercept.
2. Use the slope, $\dfrac{\text{rise}}{\text{run}}$, to locate another point on the line. Graph this point.
3. Draw a line through the two points.

A slope is a fraction, $\dfrac{\text{rise}}{\text{run}}$. We can write the slope of the line $y = -2x - 5$ as $-\dfrac{2}{1}, \dfrac{-2}{1}$, or $\dfrac{2}{-1}$.

EXAMPLE 16   Use slope-intercept graphing to graph $y = -2x - 5$.

SOLUTION ▶ Graph the y-intercept $(0, -5)$. Use the slope $\dfrac{-2}{1}$ to graph the point $(1, -7)$. Draw the line x through the points.

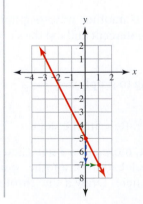

The slope of a horizontal line is 0. The slope of the line $y = 0x + 4$ is 0, and the y-intercept is $(0, 4)$. Since $0x$ equals 0, the equation of this line is usually simplified to $y = 4$.

EXAMPLE 17 | Use slope-intercept graphing to graph $y = 4$.

SOLUTION ▶ The equation $y = 4$ can be rewritten as $y = 0x + 4$. The slope is 0, and the y-intercept is $(0, 4)$. The rise between any two points is 0 so the line is horizontal. To graph the line, draw a horizontal line through the y-intercept, $(0, 4)$. Since the line is horizontal, it will include any point with a y-coordinate of 4.

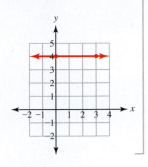

The equation of a vertical line that passes through the point $(2, 0)$ is $x = 2$. The x-coordinate of every point on this vertical line is 2. Since the slope of a vertical line is undefined, the equation of a vertical line cannot be written in slope-intercept form. In standard form, this equation is $1x + 0y = 2$.

EXAMPLE 18 | Graph: $x = 6$

SOLUTION ▶ The x-coordinate of every point on this line is 6. It is a vertical line with an x-intercept of $(6, 0)$. In standard form, this equation is $x + 0y = 6$.

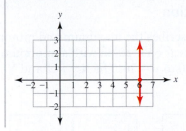

In the next example, the scales on the axes of the graphs are not the same. A unit of electric power is kilowatt · hour $(\text{kW} \cdot \text{hr})$.

EXAMPLE 19 | An average homeowner in Massachusetts buys 8000 kW · hr of electric power per year. A solar panel with an area of 100 ft² can generate 1200 kW · hr of electrical power. The equation $y = \left(\dfrac{-1200 \text{ kW} \cdot \text{hr}}{1 \text{ panel}} \right) x + 8000 \text{ kW} \cdot \text{hr}$ describes the relationship between the number of panels used by the homeowner, x, and the number of kilowatt · hours a homeowner still needs to buy, y. Use slope-intercept graphing to graph this equation.

SOLUTION ▶ Graph the y-intercept, $(0, 8000 \text{ kW} \cdot \text{hr})$. From the y-intercept, move down 1200 kW · hr and to the right 1 panel. Pay close attention to the scale on each axis.

This is the location of another point on this line. Graph this point. Draw a line through the points. Since the number of panels or kilowatt · hours is not a negative number, the line stops at the intercepts.

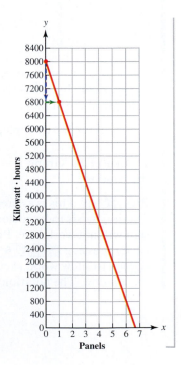

Practice Problems

18. An equation is $3x + 5y = 15$.
    a. Rewrite this equation in slope-intercept form.
    b. Identify the slope.
    c. Identify the y-intercept.
    d. Find three ordered pairs that are solutions of the equation and graph this equation.
    e. Use slope-intercept graphing to graph this equation.

For problems 19–22, use slope-intercept graphing to graph the equation.

19. $y = 4x - 5$ 20. $y = -\dfrac{2}{5}x + 6$ 21. $y = -2$ 22. $y = 0$

23. Graph: $x = 1$ 24. Graph: $x = 0$

25. Use slope-intercept graphing to graph $y = \left(\dfrac{-\$500}{1\ \text{year}}\right)x + \$12{,}000$.

Using Technology: Graphing a Linear Equation

The **window** of a graph on a graphing calculator describes the maximum and minimum values on each axis. An equation graphed in a **standard window** will appear on a graph with a minimum x-value of -10 and a maximum x-value of 10. The y-axis extends from $y = -10$ to $y = 10$. The $X_{scl} = 1$ and $Y_{scl} = 1$ mean that the distance between tick marks on each axis is 1.

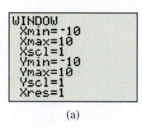
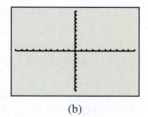
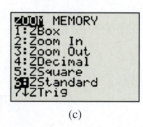

(a)   (b)   (c)

To change a window, type in new values for X_{min}, X_{max}, Y_{min}, and Y_{max} in the WINDOW screen. Or, press [ZOOM]. **Zoom In** makes the window smaller. **Zoom Out** makes the window larger. **ZStandard** resets the window to the Standard settings.

Before doing the example, press [WINDOW]. If the window is not standard, press [ZOOM]. Choose 6. Press [ENTER].

EXAMPLE 20 Graph $y = -\dfrac{3}{4}x + 6$. Choose a window that shows both intercepts. Sketch the graph. Describe the window.

Press [Y=]. This is a screen for entering equations. If there are equations already entered, press [CLEAR]. The fraction needs to be enclosed in parentheses. Press [(]. Press [(-)] for the negative sign; the subtraction sign will not work here. Press [3], press [÷], press [4], press [)]. For x, press the [X,T,θ,n] key. Finish by pressing [+] [6]. Press [ENTER]. The equation appears. To see the graph in a standard window, press [GRAPH]. To show work, sketch the graph. If there are no scales on the sketched graph, it is important to describe the window. The notation $[-10, 10, 1, -10, 10, 1]$ represents $[X_{min}, X_{max}, X_{scl}, Y_{min}, Y_{max}, Y_{scl}]$.

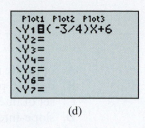

(d)

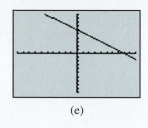

(e)

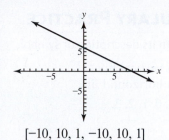

$[-10, 10, 1, -10, 10, 1]$

In the next example, the intercepts of the line are not visible in a standard window. Use **Zoom Out** to change the window.

EXAMPLE 21 Graph $y = -2x + 30$. Choose a window that shows both intercepts.

Type the equation on the Y= screen; press GRAPH. No line appears. The line is outside of the standard window.

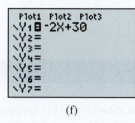

(f)

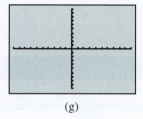

(g)

Press ZOOM. Choose 3. Press ENTER. The screen now shows the axes with X = 0 and Y = 0. We can change the center of the graph when we zoom out. But since we do not want to change this, just press ENTER. The line and both intercepts are visible. Notice that the axes look thick. To fix this, change X_{scl} and Y_{scl}.

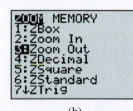

(h)

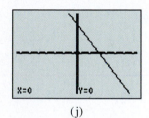

(i)

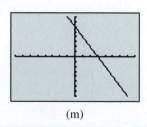

(j)

Press WINDOW. Change X_{scl} to 5 and Y_{scl} to 5. Press ENTER. Press GRAPH.

```
WINDOW
 Xmin=-40
 Xmax=40
 Xscl=1
 Ymin=-40
 Ymax=40
 Yscl=1
 Xres=1
```

(k)

```
WINDOW
 Xmin=-40
 Xmax=40
 Xscl=5
 Ymin=-40
 Ymax=40
 Yscl=5
 Xres=1
```

(l)

(m)

Practice Problems For problems 26–29,
(a) graph each equation. Choose a window that shows both intercepts.
(b) sketch the graph; describe the window. On one of the problems, it will be necessary to zoom out twice.

26. $y = 5x + 2$ **27.** $y = \dfrac{3}{5}x - 12$

28. $y = 3$ **29.** $y = -10x + 60$

1.6 VOCABULARY PRACTICE

Match the term with its description or symbol.

1. The slope of a vertical line
2. The slope of a horizontal line
3. $\{\ldots, -2, -1, 0, 1, 2, 3, \ldots\}$
4. $m = \dfrac{y_2 - y_1}{x_2 - x_1}$
5. (x_1, y_1)
6. A point that always has a y-coordinate of 0
7. The point where a line crosses the y-axis
8. $y = mx + b$
9. $ax + by = c$
10. $\dfrac{a}{b}$ and $-\dfrac{b}{a}$ are this.

A. opposite reciprocals
B. ordered pair
C. set of integers
D. slope-intercept form
E. slope formula
F. standard form
G. undefined
H. x-intercept
I. y-intercept
J. zero

1.6 Exercises

Follow your instructor's guidelines for showing your work. For graphs, clearly label the axes and include a scale on each axis.

For exercises 1–6, rewrite the equation in standard form.

1. $y = -12x + 3$
2. $y = -11x + 8$
3. $y - 3 = -4(x - 9)$
4. $y - 12 = -6(x - 11)$
5. $y + 1 = -\dfrac{5}{6}(18x + 54)$
6. $y + 1 = -\dfrac{3}{8}(24x + 72)$

For exercises 7–22,
(a) find the x-intercept, the y-intercept, and one other ordered pair that is a solution of the equation. Organize these ordered pairs in a table.
(b) graph the equation.

7. $2x + y = 6$
8. $3x + y = 6$
9. $3x - 4y = 24$
10. $5x - 7y = 35$
11. $x - 5y = 15$
12. $x - 6y = 18$
13. $-6x + 5y = -30$
14. $-2x + 7y = -28$
15. $2x + 5y = 12$
16. $4x + 5y = 32$
17. $y = 2x + 3$
18. $y = 3x + 1$
19. $y = \dfrac{3}{4}x - 7$
20. $y = \dfrac{5}{6}x - 9$

21. The equation $x + y = 8$ cups describes the relationship of cups of wheat flour, x, and cups of rye flour, y, in a recipe for bread.

22. The equation $x + y = 3$ pounds describes the relationship of pounds of beef, x, and pounds of pork, y, in a recipe for chili.

For exercises 23–42, find the slope of the line that passes through the given points.

23. $(15, 4); (18, 40)$
24. $(9, 17); (11, 31)$
25. $(-6, 5); (-9, 2)$
26. $(-7, 12); (-3, 6)$
27. $(-7, 8); (3, -4)$
28. $(-11, 2); (5, -6)$
29. $\left(\dfrac{1}{3}, \dfrac{5}{7}\right); \left(\dfrac{5}{3}, \dfrac{4}{7}\right)$
30. $\left(\dfrac{1}{4}, \dfrac{5}{9}\right); \left(\dfrac{7}{4}, \dfrac{7}{9}\right)$
31. $(4, 2); (4, -7)$
32. $(-7, 2); (-7, -3)$
33. $(-6, 8); (2, 8)$
34. $(4, -3); (6, -3)$
35. $(0, 4); (-3, 0)$
36. $(0, 9); (-2, 0)$
37. $(-2, -6); (0, -7)$
38. $(3, -8); (0, -4)$
39. $\left(7, \dfrac{3}{8}\right); \left(9, \dfrac{1}{6}\right)$
40. $\left(9, \dfrac{5}{6}\right); \left(11, \dfrac{1}{8}\right)$
41. $\left(0, \dfrac{1}{2}\right); \left(-\dfrac{5}{8}, 0\right)$
42. $\left(0, \dfrac{3}{8}\right); \left(-\dfrac{1}{2}, 0\right)$

For exercises 43–46, find the slope of the line that passes through the given points. Include the units of measurement.

43. $(3 \text{ years}, \$50{,}500); (7 \text{ years}, \$50{,}500)$
44. $(4 \text{ min}, 6 \text{ L}); (20 \text{ min}, 6 \text{ L})$
45. $(5.75 \text{ s}, 20 \text{ m}); (8.75 \text{ s}, 170 \text{ m})$
46. $(2.5 \text{ s}, 30 \text{ km}); (6.5 \text{ s}, 62 \text{ km})$

For exercises 47–58,
(a) find two ordered pairs that are solutions of the equation.
(b) use these ordered pairs and the slope formula to find the slope.

47. $4x + 9y = 72$
48. $3x + 10y = 60$
49. $x - 3y = 27$
50. $x - 4y = 48$

51. $5x - y = 32$ **52.** $8x - y = 43$

53. $2x + 5y = 15$ **54.** $3x + 2y = 9$

55. $x - y = 8$ **56.** $x - y = 7$

57. $y = -x$ **58.** $y = x$

For exercises 59–68,
(a) find two points on the line.
(b) find the slope of the line.

59.

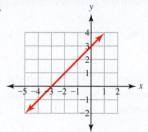

60.

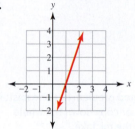

61.

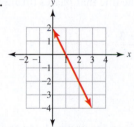

62.

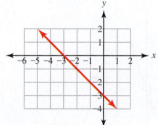

63.

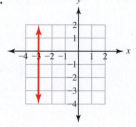

64.

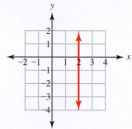

65.

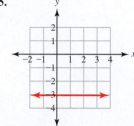

66.

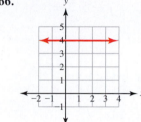

67.

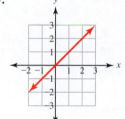

68.

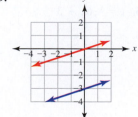

For exercises 69–72, find the slope of the graphed parallel lines.

69.

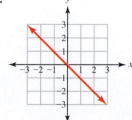

70.

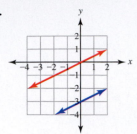

71.

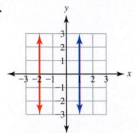

72.

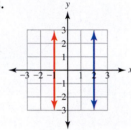

For exercises 73–78, the slope of line *A* is given. Identify the slope of a line that is perpendicular to line *A*.

73. 13

74. 6

75. $-\dfrac{7}{8}$

76. $-\dfrac{9}{10}$

77. 0

78. undefined

For exercises 79–82,
(a) rewrite the equation in slope-intercept form.
(b) identify the slope.
(c) identify the *y*-intercept

79. $2x - 3y = 9$

80. $4x - 5y = 15$

81. $y - 2 = 3(x - 1)$

82. $y - 4 = 3(x - 8)$

For exercises 83–90,
(a) identify the slope of the line.
(b) identify the *y*-intercept of the line.
(c) use slope-intercept graphing to graph the equation.

83. $y = \dfrac{5}{8}x - 6$

84. $y = \dfrac{4}{9}x - 10$

85. $y = -2x + 5$

86. $y = -6x + 3$

87. $y = -\dfrac{1}{2}x$

88. $y = -\dfrac{3}{4}x$

89. $y = \left(-\dfrac{\$4000}{1 \text{ year}}\right)x + \$16,000$

90. $y = \left(-\dfrac{\$600}{1 \text{ year}}\right)x + \$3000$

For exercises 91–94, graph the equation.

91. $y = -2$ **92.** $y = -3$ **93.** $x = 3$ **94.** $x = 4$

For exercises 95–96, if a triangle is a *right* triangle, two of the sides of the triangle must be perpendicular and form a 90-degree angle. The product of the slopes of these two sides equals –1. Show that the triangle with the given vertices is a right triangle.

95. $(4, 1); \ (6, 5); \ (8, -1)$ **96.** $(6, 1); \ (8, 7); \ (12, -1)$

Technology

For exercises 97–100, use a graphing calculator to graph the linear equation. Choose a window that shows both intercepts. Sketch the graph; describe the window.

97. $y = 2x - 9$

98. $y = \dfrac{3}{4}x - 8$

99. $y = -10x + 50$

100. $y = -\dfrac{5}{6}x + 18$

Find the Mistake

For exercises 101–104, the completed problem has one mistake.
(a) Describe the mistake in words, or copy down the whole problem and highlight or circle the mistake.
(b) Do the problem correctly.

101. **Problem:** Find the slope of the line that passes through the points (5, 3) and (5, 12).

Incorrect Answer: $m = \dfrac{12 - 3}{5 - 5}$

$$m = \dfrac{9}{0}$$

$$m = 0$$

102. **Problem:** Identify two points on the graphed line and use the slope formula to find the slope of the line.

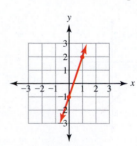

Incorrect Answer: Two points are $(-1, 0)$ and $(2, 1)$.

$$m = \dfrac{y_2 - y_1}{x_2 - x_1}$$

$$m = \dfrac{1 - 0}{2 - (-1)}$$

$$m = \dfrac{1}{3}$$

103. Problem: The slope of line A is $-\dfrac{3}{8}$. Identify the slope of a line that is perpendicular to line A.

   Incorrect Answer: The slope of a line that is perpendicular to line A is $-\dfrac{8}{3}$.

104. Problem: Find the slope of the line that passes through $(6, -8)$ and $(4, 7)$.

   Incorrect Answer:
   $$m = \frac{y_2 - y_1}{x_2 - x_1}$$
   $$m = \frac{7 - 8}{4 - 6}$$
   $$m = \frac{-1}{-2}$$
   $$m = \frac{1}{2}$$

SUCCESS IN COLLEGE MATHEMATICS

105. If you communicate with an instructor using e-mail, many instructors prefer that you present the information more formally than you would if you were text messaging or using social media such as Twitter. Write an e-mail to your instructor that explains that you are too ill to come to class for a test. Ask what you should do about missing the test. If your college does not have guidelines for writing e-mail, use the format shown below.

   To: (Instructor e-mail address)
   Subject Line: (include the name of your class and section number or class time)
   Salutation: (use the title and name preferred by your instructor)
   Body: (write in complete standard English sentences; check your spelling. Do not use smiley faces, other emoticons, or inappropriate language.)
   Signature: (use your full name)

© topseller/Shutterstock

Because of a power outage, the temperature inside a house is 40 degrees (the y-coordinate of the y-intercept). When the power comes on and the furnace begins to work, the temperature in the house increases at an average rate of 5 degrees per hour (the slope). Using the slope and the y-intercept, we can write a linear equation that describes this situation. In this section, we will write linear equations.

SECTION 1.7   Writing the Equation of a Line

After reading the text, working the practice problems, and completing assigned exercises, you should be able to:

1. Given the slope and y-intercept, write the equation of a line.

2. Given two points, write the equation of a line.

3. Write the equation of a horizontal or vertical line that passes through a point.

4. Given the equation of a line, write the equation of a parallel or perpendicular line that passes through a given point.

Slope-Intercept Form

Given the slope of a line and the y-intercept, we can write its equation in slope-intercept form, $y = mx + b$.

EXAMPLE 1 The slope of a line is $\dfrac{3}{4}$. The y-intercept is $(0, -6)$. Write the equation of this line in slope-intercept form.

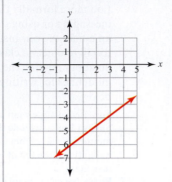

SOLUTION ▶

$$y = mx + b \qquad \text{Slope-intercept form.}$$

$$y = \frac{3}{4}x + (-6) \qquad m = \frac{3}{4}; b = -6$$

$$y = \frac{3}{4}x - 6 \qquad \text{Simplify.}$$

The average change in the temperature of a house in a given amount of time is a slope, $\dfrac{\text{change in temperature}}{\text{change in time}}$.

EXAMPLE 2 Because of a power outage, the temperature inside a house is 40 degrees. When the power comes on and the furnace begins to work, the temperature in the house increases at an average rate of $\dfrac{5 \text{ degrees}}{1 \text{ hr}}$. Write a linear equation in slope-intercept form that describes the relationship between the time that the furnace has worked, x, and the temperature in the house, y.

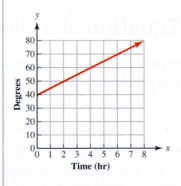

SOLUTION ▶

$$y = mx + b \qquad\qquad \text{Slope-intercept form.}$$

$$y = \left(\frac{5 \text{ degrees}}{1 \text{ hr}}\right)x + 40 \text{ degrees} \qquad m = \frac{5 \text{ degrees}}{1 \text{ hr}}; b = 40 \text{ degrees}$$

If we can identify the slope and y-intercept of a graphed line, we can write its equation in slope-intercept form.

EXAMPLE 3 Write the equation of the graphed line in slope-intercept form.

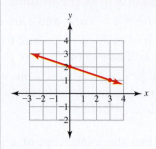

SOLUTION ▶ The y-intercept is $(0, 2)$. The slope between two points on the line, $(0, 2)$ and $(3, 1)$, is $-\dfrac{1}{3}$. (From left to right, move down 1 unit and to the right 3 units). The equation of the line in slope-intercept form, $y = mx + b$, is $y = -\dfrac{1}{3}x + 2$.

Practice Problems

1. Write the equation in slope-intercept form of a line that has a slope of -8 and that passes through $(0, -1)$.

2. A business purchased a copy machine for $47,000. For tax purposes, the value of the copy machine decreases at an average rate of $-\dfrac{\$4700}{1 \text{ year}}$. Write a linear equation in slope-intercept form that describes the relationship between the time that the business has owned the machine, x, and the value of the machine, y.

3. Write the equation of the graphed line in slope-intercept form.

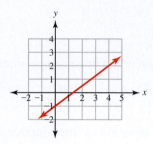

Point-Slope Form

If we change the subscripts of the slope definition, $m = \dfrac{y_2 - y_1}{x_2 - x_1}$, to $m = \dfrac{y - y_1}{x - x_1}$, we can use the properties of equality to create the **point-slope form** of a linear equation.

$$m = \frac{y - y_1}{x - x_1}$$ Slope formula.

$$(x - x_1)m = \frac{\cancel{(x - x_1)}}{1} \cdot \frac{y - y_1}{\cancel{x - x_1}}$$ Clear fractions; multiplication property of equality.

$$(x - x_1)m = y - y_1$$ Simplify.

$$y - y_1 = m(x - x_1)$$ Switch sides; change order; this is point-slope form.

> ### Forms of a Linear Equation in Two Variables
>
> **Standard Form** $ax + by = c$ a, b, and c are real numbers
>
> **Slope-Intercept Form** $y = mx + b$ m and b are real numbers
> The slope of the line is m. The y-coordinate of the y-intercept is b.
>
> **Point-Slope Form** $y - y_1 = m(x - x_1)$ m, x_1, and y_1 are real numbers
> The slope of the line is m. A point on the line is (x_1, y_1).

If we know the slope and the y-intercept of a line, we can use slope-intercept form to write its equation. If we know the slope and a point on the line *but we do not know the y-intercept*, we cannot use slope-intercept form. However, we can begin by writing the equation in point-slope form and then rewrite the equation in slope-intercept form.

EXAMPLE 4 A line passes through the point $(5, -1)$ and has a slope of 2.

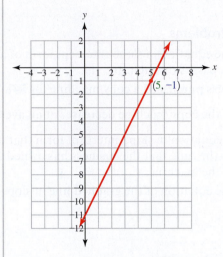

(a) Write the equation of the line in slope-intercept form.

SOLUTION ▶
$$y - y_1 = m(x - x_1) \qquad \text{Begin with point-slope form.}$$
$$y - (-1) = 2(x - 5) \qquad (x_1, y_1) = (5, -1)$$

Now rewrite the equation in slope-intercept form, $y = mx + b$.

$$y + 1 = 2x + 2(-5) \qquad \text{Simplify; distributive property.}$$
$$y + 1 = 2x - 10 \qquad \text{Simplify.}$$
$$\underline{ -1 \qquad\qquad -1} \qquad \text{Subtraction property of equality.}$$
$$y + 0 = 2x - 11 \qquad \text{Simplify.}$$
$$y = 2x - 11 \qquad \text{Simplify; the equation is in slope-intercept form.}$$

(b) Identify the y-intercept.

Since $b = -11$, the y-intercept is $(0, -11)$.

> ### Writing the Equation of a Line in Slope-Intercept Form Given the Slope and One Point on the Line
>
> **1.** Replace x_1, y_1, and m in point-slope form, $y - y_1 = m(x - x_1)$.
>
> **2.** Use the distributive property and the properties of equality to rewrite the equation in slope-intercept form.

To write the equation of a line in slope-intercept form given two points on the line, first use the slope formula to find the slope. Then use point-slope form, the slope, and *one of the points* to write the equation of the line. Finally, rewrite the equation in slope-intercept form.

Writing the Equation of a Line in Slope-Intercept Form Given Two Points on the Line

1. Use the slope formula to find the slope of the line.

2. Use point-slope form and either one of the points to write the equation of the line.

3. Rewrite the equation in slope-intercept form.

EXAMPLE 5 | A line passes through the points $(5, 3)$ and $(7, -9)$.

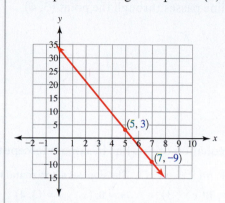

(a) Find the slope.

SOLUTION ▶

$$m = \frac{y_2 - y_1}{x_2 - x_1} \qquad \text{Slope formula.}$$

$$m = \frac{-9 - 3}{7 - 5} \qquad (x_1, y_1) = (5, 3); (x_2, y_2) = (7, -9)$$

$$m = \frac{-12}{2} \qquad \text{Simplify.}$$

$$m = -6 \qquad \text{The slope of the line.}$$

(b) Write the equation of the line in point-slope form. Rewrite in slope-intercept form.

$$y - y_1 = m(x - x_1) \qquad \text{Point-slope form}$$

$$y - 3 = -6(x - 5) \qquad m = -6; (x_1, y_1) = (5, 3)$$

$$y - 3 = -6x + 30 \qquad \text{Distributive property.}$$

$$\underline{+3 \qquad\qquad +3} \qquad \text{Addition property of equality.}$$

$$y + 0 = -6x + 33 \qquad \text{Simplify.}$$

$$y = -6x + 33 \qquad \text{The equation is in slope-intercept form.}$$

(c) Identify the y-intercept of this line.

▶ Since $b = 33$, the y-intercept is $(0, 33)$.

> ### Practice Problems
>
> For problems 4–6,
> **(a)** write the equation of the line in slope-intercept form.
> **(b)** find the y-intercept.
>
> **4.** The slope of the line is -9, and the line passes through the point $(-4, 2)$.
>
> **5.** The slope of the line is $\dfrac{3}{8}$, and the line passes through the point $(-2, -15)$.
>
> **6.** The line passes through the points $(-1, 8)$ and $(-6, 43)$.

Equations of Horizontal and Vertical Lines

Since the rise between any two points on a horizontal line is 0, the slope of a horizontal line is 0.

EXAMPLE 6 | A horizontal line passes through the point $(3, 4)$.

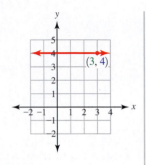

(a) Write the equation of this line in slope-intercept form.

SOLUTION ▶

$$y - y_1 = m(x - x_1)$$ The slope of a horizontal line is 0.
$$y - 4 = \mathbf{0}(x - 3)$$ $m = 0; (x_1, y_1) = (3, 4)$
$$y - 4 = \mathbf{0}$$ Simplify.
$$\underline{+4 \ +4}$$ Addition property of equality.
$$y + \mathbf{0} = \mathbf{4}$$ Simplify.
$$y = 4$$ Simplify.

In slope-intercept form, this equation is $y = 0x + 4$ or $y = 4$.

(b) Write the equation of this line in standard form.

▶   The y-coordinates of all the points on a horizontal line are the same. In this example, the y-coordinate of each point on the line is 4. In standard form, this equation is $0x + 1y = 4$ or $y = 4$.

Since the run between any two points on a vertical line is 0 and division by 0 is undefined, the slope of a vertical line is undefined. Since we cannot replace m with a real number, we cannot write the equation of a vertical line in slope-intercept form or point-slope form. However, we can write its equation in standard form.

EXAMPLE 7 | A vertical line passes through the point $(3, 4)$. Write the equation of this line in standard form.

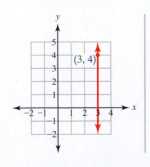

SOLUTION ▶ The x-coordinates of all the points on a vertical line are the same. In this example, the x-coordinate of each point on the line is 3. In standard form, this equation is $1x + 0y = 3$ or $x = 3$.

Horizontal and Vertical Lines

Horizontal Lines

The slope of a horizontal line is 0. The y-intercept is $(0, b)$.
In slope-intercept form, its equation is $y = b$.
In standard form, its equation is $0x + y = b$ or $y = b$.

Vertical Lines

The slope of a vertical line is undefined. Unless the line is $x = 0$, it has no y-intercept.
The equation of a vertical line cannot be written in slope-intercept form.
In standard form, its equation is $x + 0y = c$ or $x = c$.

Given a point, we can write the equation of the horizontal or vertical line that passes through the point.

EXAMPLE 8 **(a)** Write the equation of the horizontal line that passes through $(6, -2)$.

SOLUTION ▶ $\qquad y = -2$

(b) Write the equation of the vertical line that passes through $(0, 3)$.

▶ $\qquad x = 0$

Practice Problems

For problems 7–8, write the equation of the line that passes through the given points.

7. $(4, 9); (2, 9)$ **8.** $(2, 7); (2, -3)$
9. Write the equation of the horizontal line that passes through $(7, -1)$.
10. Write the equation of the vertical line that passes through $(7, -1)$.

Parallel and Perpendicular Lines

Parallel lines never intersect. Since parallel lines have the same rise and run, their slopes are equal.

EXAMPLE 9 The equation of a line is $y = -2x + 6$. Write the equation in slope-intercept form of a parallel line that passes through $(-3, 2)$.

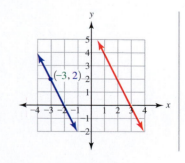

SOLUTION ▶ Since these lines are parallel, their slopes are equal, $m = -2$.

$y - y_1 = m(x - x_1)$	Point-slope form.
$y - 2 = -2(x - (-3))$	$m = -2; (x_1, y_1) = (-3, 2)$
$y - 2 = -2(x + 3)$	Simplify.
$y - 2 = -2x - 2(3)$	Distributive property.
$y - 2 = -2x - 6$	Simplify.
$\underline{\quad +2 \qquad \quad +2\quad}$	Addition property of equality.
$y + 0 = -2x - 4$	Simplify.
$y = -2x - 4$	The equation of the parallel line in slope-intercept form.

When perpendicular lines intersect, they create a 90-degree (right) angle. If neither line is vertical, the slopes of perpendicular lines are **opposite reciprocals**. The product of their slopes equals -1.

EXAMPLE 10   In the graph, line A and line B appear to be perpendicular.

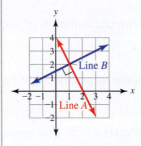

(a) Find the slope of line A.

SOLUTION ▶   Identify two points on line A: $(1, 2)$ and $(2, 0)$

$$m_A = \frac{y_2 - y_1}{x_2 - x_1} \qquad \text{The slope of line } A \text{ is } m_A.$$

$$m_A = \frac{0 - 2}{2 - 1} \qquad (x_1, y_1) = (1, 2); (x_2, y_2) = (2, 0)$$

$$m_A = \frac{-2}{1} \qquad \text{Simplify.}$$

(b) Find the slope of line B.

▶   Identify two points on line B: $(1, 2)$ and $(3, 3)$.

$$m_B = \frac{y_2 - y_1}{x_2 - x_1} \qquad \text{The slope of line } B \text{ is } m_B.$$

$$m_B = \frac{3 - 2}{3 - 1} \qquad (x_1, y_1) = (1, 2); (x_2, y_2) = (3, 3)$$

$$m_B = \frac{1}{2} \qquad \text{Simplify.}$$

(c) Use the slopes to show that line A and line B are perpendicular.

▶ $m_A \cdot m_B$ The product of the slopes of the lines.

$$= \frac{-2}{1} \cdot \frac{1}{2} \qquad \text{Replace the variables with the slopes.}$$

$$= -1 \qquad \text{The product is } -1; \text{ the lines are perpendicular.}$$

To write the equation of a line that is perpendicular to a given line and passes through a given point, identify the slope of the perpendicular line and use point-slope form.

EXAMPLE 11   The equation of line A is $y = -\dfrac{5}{6}x + 3$. Line B is perpendicular to line A.

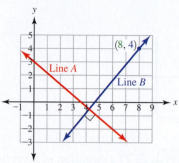

(a) Identify the slope of line B.

SOLUTION ▶   The equation $y = -\dfrac{5}{6}x + 3$ is in slope-intercept form. The slope of line A is $-\dfrac{5}{6}$. The slope of line B, $\dfrac{6}{5}$, is the opposite reciprocal of the slope of line A.

(b) A point on line B is $(8, 4)$. Write the equation in slope-intercept form of line B.

$$y - y_1 = m(x - x_1)$$ Begin with point-slope form.

$$y - 4 = \frac{6}{5}(x - 8)$$ $m = \dfrac{6}{5}; (x_1, y_1) = (8, 4)$

$$y - 4 = \frac{6}{5}x + \frac{6}{5}(-8)$$ Distributive property.

$$y - 4 = \frac{6}{5}x - \frac{48}{5}$$ Simplify.

$$\underline{\quad +4 \qquad\qquad +4 \quad}$$ Addition property of equality.

$$y + 0 = \frac{6}{5}x - \frac{48}{5} + \frac{4}{1} \cdot \frac{5}{5}$$ Simplify; multiply by a fraction equal to 1.

$$y = \frac{6}{5}x - \frac{48}{5} + \frac{20}{5}$$ Simplify.

$$y = \frac{6}{5}x - \frac{28}{5}$$ Simplify; this is the equation of line B.

(c) Find the y-intercept of line B.

▶   Since $b = -\dfrac{28}{5}$, the y-intercept of line B is $\left(0, -\dfrac{28}{5}\right)$.

Slopes of Parallel and Perpendicular Lines

Parallel Lines   Slopes are equal.

Perpendicular Lines   Slopes are opposite reciprocals. (Neither line can be vertical.)

We can also write the equation of a line that is parallel or perpendicular to a horizontal or vertical line.

EXAMPLE 12   (a) Write the equation of the line that is parallel to $x = -2$ and passes through the point $(4, 1)$.

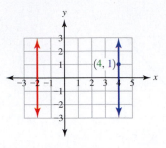

SOLUTION ▶   Both lines are vertical. The equation of a vertical line that passes through the point $(4, 1)$ is $x = 4$.

(b) Write the equation of the line that is perpendicular to $x = 3$ and passes through the point $(2, 1)$.

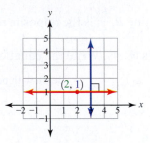

▶   The line $x = 3$ is vertical; a perpendicular line is horizontal. The equation of a horizontal line that passes through the point $(2, 1)$ is $y = 1$.

Practice Problems

For problems 11–12, the equation of a line is $y = -3x + 4$.

11. Write the equation in slope-intercept form of the parallel line that passes through the point $(6, 8)$.

12. Write the equation in slope-intercept form of the perpendicular line that passes through the point $(-2, 7)$.

13. Write the equation of the line that is parallel to $x = 1$ and that passes through the point $(6, -4)$.

14. Write the equation of the line that is parallel to $y = -3$ and that passes through the point $(6, -4)$.

15. Write the equation of the line that is perpendicular to $x = -4$ and that passes through the point $(3, 8)$.

16. Write the equation of the line that is perpendicular to $y = 5$ and that passes through the point $(3, 8)$.

1.7 VOCABULARY PRACTICE

Match the term with its description or symbol.

1. $ax + by = c$
2. $y - y_1 = m(x - x_1)$
3. $y = mx + b$
4. $(0, b)$
5. m
6. $m = \dfrac{y_2 - y_1}{x_2 - x_1}$
7. The slope of a horizontal line
8. The slope of a vertical line
9. $y = b$
10. $x = c$

A. equation of a horizontal line
B. equation of a vertical line
C. point-slope form
D. slope
E. slope formula
F. slope-intercept form
G. standard form
H. undefined
I. y-intercept
J. zero

1.7 Exercises

Follow your instructor's guidelines for showing your work.

For exercises 1–10, write the equation in slope-intercept form of the line with the given slope and y-intercept.

1. $m = \dfrac{5}{8}$, $(0, 3)$

2. $m = \dfrac{3}{4}$, $(0, 9)$

3. $m = -6$, $(0, -4)$

4. $m = -8$, $(0, -2)$

5. $m = 0$, $(0, 15)$

6. $m = 0$, $(0, 29)$

7. $m = -7$, $(0, 0)$

8. $m = -10$, $(0, 0)$

9. $m = \dfrac{\$10.25}{1 \text{ hr}}$, $(0, \$80)$

10. $m = \dfrac{\$8.85}{1 \text{ hr}}$, $(0, \$20)$

For exercises 11–18,
(a) use the graph to identify the y-intercept of the line.
(b) use the graph to identify the slope of the line.
(c) write the equation of the line in slope-intercept form.

11.

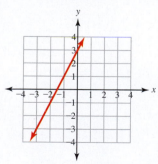

12.

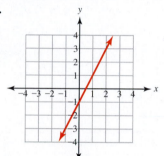

13.

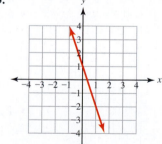

14.

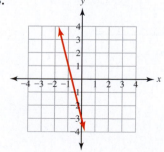

15.

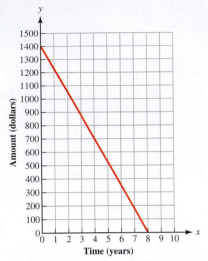

Time (years)

16.

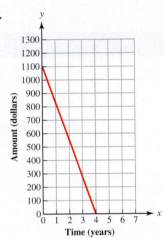

Time (years)

17.

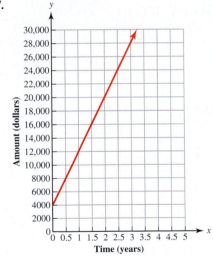

Time (years)

18.

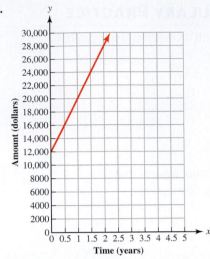

Time (years)

For exercises 19–28, a line with the given slope passes through the given point.
(a) Write the equation of the line in slope-intercept form.
(b) Identify the *y*-intercept.

19. slope = 8; point $(2, -9)$

20. slope = 6; point $(3, -5)$

21. slope = -8; point $(2, 9)$

22. slope = -6; point $(3, 5)$

23. slope = $\dfrac{2}{3}$; point $(5, 8)$

24. slope = $\dfrac{5}{8}$; point $(4, 7)$

25. slope = 0; point $(6, -3)$

26. slope = 0; point $(8, -1)$

27. slope = 0.2; point $(1, 6.4)$

28. slope = 0.4; point $(2, 1.6)$

For exercises 29–38, a line passes through the given points.
(a) Find the slope.
(b) Write its equation in slope-intercept form.
(c) Identify the *y*-intercept.

29. $(6, 1); (8, 11)$

30. $(5, 9); (7, 17)$

31. $(9, 12); (3, 36)$

32. $(10, 15); (6, 27)$

33. $(1, -3); (4, -18)$

34. $(2, -5); (-3, 20)$

35. $(-4, 15); (4, 21)$

36. $(-8, 30); (-16, 24)$

37. $(5, -9); (16, -9)$

38. $(3, -11); (20, -11)$

For exercises 39–40, write the equation of the vertical line that passes through the point.

39. $(-2, 5)$ **40.** $(5, 6)$

For exercises 41–42, write the equation of the horizontal line that passes through the point.

41. $(-2, 5)$ **42.** $(5, 6)$

For exercises 43–46, write the equation of the graphed line.

43.

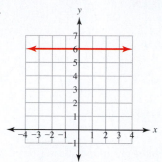

44.

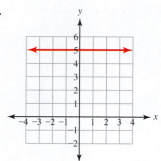

45.

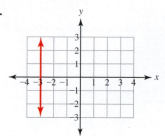

46.

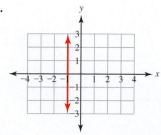

For exercises 47–50, identify the slope of a line that is:

47. Parallel to the line $y = 8x - 11$

48. Parallel to the line $y = 3x - 14$

49. Perpendicular to the line $y = 8x - 11$

50. Perpendicular to the line $y = 3x - 14$

51. What is the product of two numbers that are opposite reciprocals?

52. What is the product of two numbers that are reciprocals?

For exercises 53–60, write the equation in slope-intercept form of a line that is:

53. Parallel to $y = 9x + 2$ and passes through the point $(5, -13)$

54. Parallel to $y = 5x - 16$ and passes through the point $(7, -15)$

55. Parallel to $y = \dfrac{3}{4}x + 8$ and passes through the point $(-2, -1)$

56. Parallel to $y = \dfrac{3}{8}x + 2$ and passes through the point $(-6, -1)$

57. Perpendicular to $y = \dfrac{5}{8}x + 10$ and passes through the point $(3, -11)$

58. Perpendicular to $y = \dfrac{3}{4}x + 2$ and passes through the point $(4, -15)$

59. Perpendicular to $y = -\dfrac{2}{3}x + \dfrac{1}{2}$ and passes through the point $(-4, 6)$

60. Perpendicular to $y = -\dfrac{6}{5}x + \dfrac{3}{8}$ and passes through the point $(-6, 2)$

61. A line passes through the points $(1, 8)$ and $(3, 18)$. Write the equation in slope-intercept form of the perpendicular line that passes through the point $(4, -6)$.

62. A line passes through the points $(3, 10)$ and $(6, 34)$. Write the equation in slope-intercept form of the perpendicular line that passes through the point $(-1, 5)$.

63. A line passes through the points $(1, 8)$ and $(3, 18)$. Write the equation in slope-intercept form of the parallel line that passes through the point $(4, -6)$.

64. A line passes through the points $(3, 10)$ and $(6, 34)$. Write the equation in slope-intercept form of the parallel line that passes through the point $(-1, 5)$.

65. A line passes through the points $(2, 15)$ and $(10, 19)$. Write the equation in slope-intercept form of the perpendicular line that passes through the point $(-4, 7)$.

66. A line passes through the points $(14, 9)$ and $(18, 11)$. Write the equation in slope-intercept form of the perpendicular line that passes through the point $(-3, 2)$.

67. A line passes through the points $(21, 15)$ and $(24, 17)$. Write the equation in slope-intercept form of the parallel line that passes through the point $(-7, 3)$.

68. A line passes through the points $(38, 2)$ and $(44, 7)$. Write the equation in slope-intercept form of the parallel line that passes through the point $(-3, 9)$.

For exercises 69–76, write the equation of the line that is:

69. Parallel to $x = 6$ and passes through the point $(10, 11)$

70. Parallel to $x = 9$ and passes through the point $(7, 5)$

71. Parallel to $y = -8$ and passes through the point $(10, 11)$

72. Parallel to $y = -12$ and passes through the point $(7, 5)$

73. Perpendicular to $x = -4$ and passes through the point $(8, 1)$

74. Perpendicular to $x = -7$ and passes through the point $(4, 1)$

75. Perpendicular to $y = 20$ and passes through the point $(6, -3)$

76. Perpendicular to $y = 18$ and passes through the point $(2, -9)$

For exercises 77–84,
(a) write the equation of the graphed line.
(b) write the equation in slope-intercept form of the line that is parallel to this line and passes through the graphed point.

77.

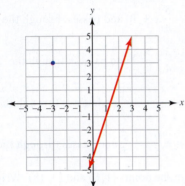

78.

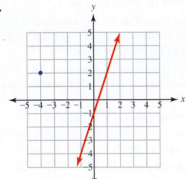

79.

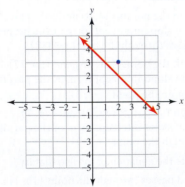

80.

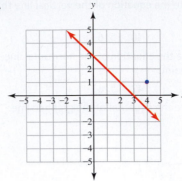

81.

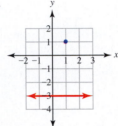

82.

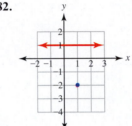

83.

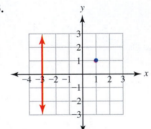

84.

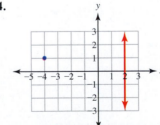

For exercises 85–92,
(a) write the equation of the line with the given slope and y-intercept.
(b) graph the equation.

85. $m = \dfrac{2}{3}$; $(0, 2)$

86. $m = \dfrac{3}{4}$; $(0, 1)$

87. $m = -1$; $(0, 0)$

88. $m = 1$; $(0, 0)$

89. $m = -3$; $(0, 5)$

90. $m = -4$; $(0, 6)$

91. $m = 0$; $(0, 3)$

92. $m = 0$; $(0, 2)$

Find the Mistake

For exercises 93–96, the completed problem has one mistake.
(a) Describe the mistake in words, or copy down the whole problem and highlight or circle the mistake.
(b) Do the problem correctly.

93. Problem: Write the equation in slope-intercept form of the line that passes through the points $(5, 2)$ and $(3, 12)$.

Incorrect Answer: $m = \dfrac{y_2 - y_1}{x_2 - x_1}$ $(x_1, y_1) = (5, 2)$; $(x_2, y_2) = (3, 12)$

$$m = \dfrac{12 - 2}{3 - 5}$$

$$m = \dfrac{10}{-2}$$

$$m = -5$$

Equation: $y = -5x + 12$

94. Problem: Write the equation of the vertical line that passes through the point $(9, 4)$.

Incorrect Answer: $y = 4$

95. Problem: The equation of a line is $x = 6$. Write the equation of a line that is perpendicular to this line and passes through the point $(2, -1)$.

Incorrect Answer: $x = 2$

96. Problem: Write the equation in slope-intercept form of the line with a slope of 7 that passes through the point $(-2, 8)$.

Incorrect Answer: $y - 8 = 7(x - 2)$

$$y - 8 = 7x - 14$$

$$\underline{+8 \qquad\qquad +8}$$

$$y + 0 = 7x - 6$$

$$y = 7x - 6$$

SUCCESS IN COLLEGE MATHEMATICS

97. If you are having trouble logging onto the campus network, how do you get help?

98. If you don't own a computer of your own or don't have Internet access, where are computers located either on or off campus that you can use?

© Sergei Bachlakov/Shutterstock

At the Beijing Olympics, the top competitors on the balance beam had a combined A score and B score of at least 14 points. We can model this situation with an inequality in two variables: $A + B \geq 14$ points. In this section, we will write and graph inequalities in two variables.

SECTION 1.8

Linear Inequalities in Two Variables

After reading the text, working the practice problems, and completing assigned exercises, you should be able to:

1. Graph a linear inequality.

2. Use a linear inequality to represent a constraint.

Linear Inequalities

The linear equation $x + y = 3$ represents "the sum of two numbers is *equal to* 3." The graph of this equation is a line (see Figure 1 on the next page); each point on the line represents a solution of the equation.

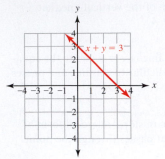

Figure 1

The **linear inequality** $x + y \leq 3$ represents the sentence "The sum of two numbers is *less than or equal to* 3." The graph of this equation (Figure 2) includes a **boundary line** and a **solution region**. Each point on the boundary line and each point in the solution region represent a solution of the inequality. For example, $(0, 0)$ is a point in the solution region. It is a solution of $x + y \leq 3$: $0 + 0 \leq 3$ is true.

A **strict inequality** such as $x + y < 3$ includes a > or < sign. The points on the boundary line are not solutions of the inequality. To show that the points on the line are not included in the graph of the inequality, we draw a dashed rather than a solid line (Figure 3). For example, $(2, 1)$ is a point on the dashed line. Since $2 + 1 < 3$ is not true, $(2, 1)$ is not a solution of $x + y < 3$.

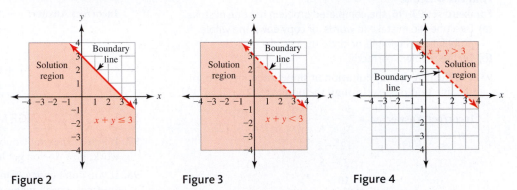

Figure 2   **Figure 3**   **Figure 4**

If we switch the direction of the inequality sign and graph $x + y > 3$ (Figure 4), the location of the solution region changes to the other side of the boundary line. If we replace the variables in $x + y > 3$ with the coordinates of any point in the solution region, the inequality is true. For example, the point $(2, 4)$ is in the solution region, and $2 + 4 > 3$ is true.

Graphs of Linear Inequalities in Two Variables

If the inequality is strict ($>$ or $<$), the boundary line is dashed. The points on the boundary line are not solutions of the inequality.

If the inequality is not strict ($\geq$ or $\leq$), the boundary line is solid. The points on the boundary line are solutions of the inequality.

All of the points in the shaded area are solutions of the inequality. The shaded area is the solution region.

EXAMPLE 1 Graph: $x + y \leq 5$

(a) Use the intercepts and one other point to graph the solid line $x + y = 5$.

SOLUTION ▶ $x + y = 5$ is the boundary line of the solution region. Each of the points on this line represents a solution of $x + y \leq 5$.

x	y
0	5
5	0
2	3

The boundary line divides the coordinate system into two regions. All the points in one of these regions are also solutions of $x + y < 5$.

(b) Use a test point to identify the solution region.

Any point that is not on the boundary line can be a test point. Replace the variables in the inequality with the coordinates of the test point. If the inequality is true, the test point is a solution of the inequality. All of the rest of the points in this region are also solutions.

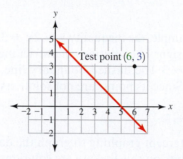

In this example, we test $(6, 3)$ to find out whether this ordered pair is a solution of the inequality. However, any point that is not on the boundary line can be tested.

$x + y < 5$	Use the inequality $x + y < 5$.
$6 + 3 < 5$	Replace the variables with the coordinates of the test point $(6, 3)$.
$9 < 5$	Evaluate. This inequality is false.

When we replace the variables with the coordinates of the test point, the inequality is false. Neither $(6, 3)$ nor any of the other points on this side of the boundary line are solutions of the inequality. The solutions are in the region on *the other side of the boundary line*. We can check this by testing a point in the region on the other side of the boundary line: $(0, 0)$.

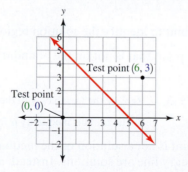

$x + y < 5$	Use the inequality $x + y < 5$.
$0 + 0 < 5$	Replace the variables with the coordinates of the test point, $(0, 0)$.
$0 < 5$	Evaluate. This inequality is true.

Since the inequality is true, $(0, 0)$ is a solution of $x + y < 5$. Every point in the region on this side of the boundary line is a solution.

(c) Shade the solution region.

To shade the region that includes $(0, 0)$, use a highlighter or some other method. Ask your instructor how you should show shading.

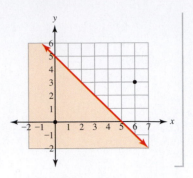

In the next example, the inequality $y > 2x + 3$ is a **strict inequality**. A strict inequality has a $<$ or $>$ sign. The solutions of a strict inequality do *not* include the points on the line $y = 2x + 3$ The dashed boundary line, $y = 2x + 3$, shows the edge of the solution region. Since its points are not solutions of the inequality, it is a dashed line.

EXAMPLE 2   Graph: $y > 2x + 3$

SOLUTION ▶ **(a)** Use slope-intercept graphing to graph the dashed boundary line $y = 2x + 3$. Since the points on this line are *not* solutions of the inequality $y > 2x + 3$, the boundary line is dashed. Graph the y-intercept, $(0, 3)$. Since the slope is $\dfrac{2}{1}$, move up 2 units and to the right 1 unit to the location of another point on the line.

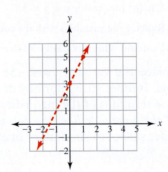

(b) Use a test point to identify the solution region.

▶   Choose any point that is not on the boundary line. We will test $(4, 1)$.

$y > 2x + 3$

$1 > 2(4) + 3$ Replace the variables with the coordinates of the test point, $(4, 1)$.

$1 > \mathbf{11}$ Evaluate. This inequality is false.

This test point is not a solution of the inequality. None of the points on this side of the boundary line are solutions. Instead, all of the points on the other side of the line from the test point are solutions.

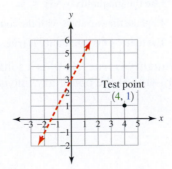

(c) Shade the solution region.

▶ Shade the region on the opposite side of the boundary line. The graph of this inequality includes all of the points in the solution region but does not include the points on the dashed boundary line.

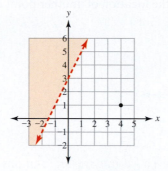

Graphing a Linear Inequality

1. Determine whether the boundary line is solid or dashed. If the inequality is not strict, it includes a $\geq$ or $\leq$ sign, and the boundary line is a solid line. If the inequality is strict, it includes a $>$ or $<$ sign, and the boundary line is dashed.

2. Graph the boundary line.

3. Select a test point that is *not* on the boundary line. Replace the variables in the inequality with the coordinates of the test point. The resulting inequality will be true or false.

4. If the inequality is true, the test point is in the solution region. Shade the area on this side of the boundary line. If the inequality is not true, the test point is *not* in the solution region. Shade the area on the *other* side of the boundary line.

If an inequality is in the form $ax + by \leq c$, we can rewrite it in slope-intercept form and then use slope-intercept graphing to graph the boundary line. If we multiply or divide both sides of the inequality by a negative number, we must reverse the direction of the inequality sign.

EXAMPLE 3   Graph: $4x - 3y \leq 9$

(a) Rewrite the inequality in slope-intercept form.

SOLUTION ▶

$$4x - 3y \leq 9 \qquad \text{This equation is in the form } ax + by \leq c.$$

$$\underline{-4x \qquad\qquad -4x} \qquad \text{Subtraction property of inequality.}$$

$$0 - 3y \leq -4x + 9 \qquad \text{Simplify}$$

$$\frac{-3y}{-3} \geq \frac{-4x}{-3} + \frac{9}{-3} \qquad \text{Division property of inequality; reverse the sign.}$$

$$y \geq \frac{4}{3}x - 3 \qquad \text{Slope-intercept form.}$$

(b) Use slope-intercept graphing to graph the solid boundary line $y = \frac{4}{3}x - 3$.

▶ Graph the y-intercept, $(0, -3)$. Since the slope is $\frac{4}{3}$, move up 4 units and to the right 3 units to the location of another point on the line.

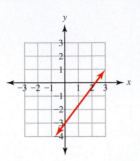

(c) Use a test point to identify the solution region.

▶ Choose any point that is not on the boundary line. Since the arithmetic is easy, we will test $(0, 0)$.

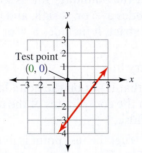

$$y \geq \frac{4}{3}x - 3$$

$0 \geq \frac{4}{3}(0) - 3$ Replace the variables with the coordinates of the test point.

$0 \geq \mathbf{0} - 3$ Evaluate.

$0 \geq \mathbf{-3}$ Evaluate. This inequality is true.

Since the test point is a solution of the inequality, all of the points on this side of the boundary line are solutions of the inequality.

(d) Shade the solution region.

▶ Shade the region that includes $(0, 0)$.

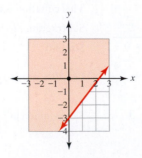

In the previous examples, the inequalities had two variables. We can also graph inequalities in one variable on an xy-coordinate system.

EXAMPLE 4 | Graph: $x < 3$

(a) Graph the dashed boundary line $x = 3$.

SOLUTION ▶   The inequality is strict. Since the points on this line are *not* solutions of the inequality $x < 3$, the vertical boundary line is dashed.

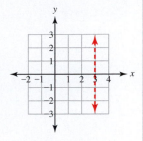

(b) Use a test point to identify the solution region.

▶   Choose any point that is not on the boundary line. We will test $(2, 1)$. Since the inequality includes only the variable x, we only use the x-coordinate of the test point.

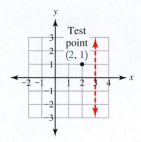

$x < 3$

$2 < 3$ Replace the variable with a coordinate of the test point, (2, 1). Evaluate. This inequality is true.

This test point is a solution of the inequality. All of the points on this side of the boundary line are solutions.

(c) Shade the solution region.

▶   Shade the region on the same side of the boundary line as the test point. The graph of this inequality includes all of the points in the solution region but does not include the points on the dashed boundary line.

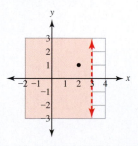

In the next example, the inequality is not strict. The horizontal boundary line is solid.

EXAMPLE 5 | Graph: $y \geq 150$

(a) Graph the solid boundary line $y = 150$.

SOLUTION ▶   The inequality is not strict. Since the points on this line are solutions of the inequality $y \geq 150$, the horizontal boundary line is solid.

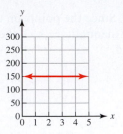

(b) Use a test point to identify the solution region.

▶   Choose any point that is not on the boundary line. We will test $(3, 200)$. Since the inequality includes only the variable y, we only use the y-coordinate of the test point.

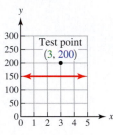

$$y \geq 150$$

$$200 \geq 150 \qquad \text{Replace the variable with a coordinate of the test point, } (3, 200).$$
$$\text{Evaluate. This inequality is true.}$$

This test point is a solution of the inequality. All of the points on this side of the boundary line are solutions.

(c) Shade the solution region.

▶   Shade the region on the same side of the boundary line as the test point. The graph of this inequality includes all of the points in the solution region and the points on the boundary line.

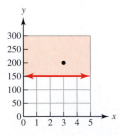

Practice Problems

For problems 1–6, graph each inequality. Show your work with the test point.

1. $7x + 6y < 42$ **2.** $y \geq -3x + 5$ **3.** $5x - 4y > -20$

4. $y < -2$ **5.** $y \geq \dfrac{3}{5}x - 8$ **6.** $x < -3$

Writing a Linear Inequality

In Examples 1–5, we graphed linear inequalities in one or two variables. In the next example, we write the inequality represented by a graph.

Writing the Inequality Represented by a Graph

1. Find the equation of the boundary line.
2. Determine the type of inequality sign: strict ($>$, $<$) or not strict ($\geq$, $\leq$).
3. Find the direction of the inequality sign.

    Guess what the inequality sign should be.

    Pick a test point in the solution region.

    Determine whether this test point is a solution of the inequality. If the test point is a solution, the sign is correct. If the test point is not a solution, reverse the direction of the sign.

EXAMPLE 6 | Write the inequality represented by the graph.

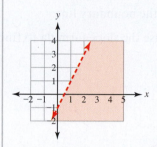

(a) Find the equation of the boundary line.

SOLUTION ▶ The y-intercept of the boundary line is $(0, -1)$. The slope is $\dfrac{2}{1}$. So the boundary line is $y = 2x - 1$.

(b) Identify the type of inequality sign.

▶ Since the boundary line is dashed, this is a strict inequality. The sign will be $<$ or $>$.

(c) Find the direction of the inequality sign.

▶ We can guess that the inequality is $y > 2x - 1$. To test the direction of the sign, use a test point that is in the shaded region on the graph, $(4, 3)$.

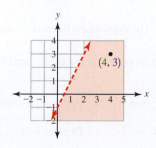

$$y > 2x - 1$$
$$3 > 2(4) - 1 \qquad \text{Replace the variables with the coordinates of the test point.}$$
$$3 > 8 - 1 \qquad \text{Evaluate.}$$
$$3 > 7 \qquad \text{The inequality is false; we chose the wrong sign.}$$

Since the test point is not a solution, the sign in $y > 2x - 1$ is not correct. Change the sign. This graph represents the inequality $y < 2x - 1$.

In the next example, since the *y*-intercept is difficult to estimate, we use point-slope form to write the equation of the boundary line.

EXAMPLE 7 | Write the inequality represented by the graph.

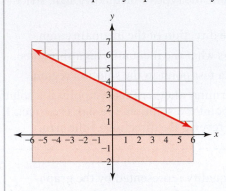

(a) Find the equation of the boundary line.

SOLUTION ▶   Identify two points; use the slope formula to find the slope.

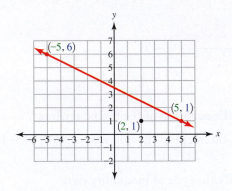

$$m = \frac{y_2 - y_1}{x_2 - x_1}$$ Slope formula.

$$m = \frac{6 - 1}{-5 - 5}$$ $(x_1, y_1) = (5, 1); (x_2, y_2) = (-5, 6)$

$$m = \frac{5}{-10}$$ Simplify.

$$m = -\frac{1}{2}$$ The slope of the boundary line.

Write the equation in point-slope form, and then rewrite the equation in slope-intercept form.

$$y - y_1 = m(x - x_1)$$ Point-slope form.

$$y - 1 = -\frac{1}{2}(x - 5)$$ Replace m, x_1, and y_1.

$$y - 1 = -\frac{1}{2}x + \frac{5}{2}$$ Distributive property.

$$\underline{ +1 \phantom{= -\frac{1}{2}x} +1}$$ Subtraction property of equality.

$$y + 0 = -\frac{1}{2}x + \frac{5}{2} + \frac{2}{2}$$ Common denominator is 2; $1 = \frac{2}{2}$

$$y = -\frac{1}{2}x + \frac{7}{2}$$ Slope-intercept form.

(b) Identify the type of inequality sign.

▶   Since the boundary line is solid, the sign is ≥ or ≤.

(c) Find the direction of the inequality sign.

▶   **Guess:** $y \leq -\dfrac{1}{2}x + \dfrac{7}{2}$ Guess: the correct sign is ≤.

Test point: $(2, 1)$ Pick a test point in the shaded region.

$1 \leq -\dfrac{1}{2}(2) + \dfrac{7}{2}$ Replace the variables with coordinates of the test point.

$1 \leq -\dfrac{2}{2} + \dfrac{7}{2}$ Simplify.

$1 \leq \dfrac{5}{2}$ Evaluate. This inequality is true.

Since the test point is a solution, the sign in $y \leq -\dfrac{1}{2}x + \dfrac{7}{2}$ is correct. This

graph represents the inequality $y \leq -\dfrac{1}{2}x + \dfrac{7}{2}$.

Practice Problems

For problems 7–9, write the inequality represented by the graph.

7.

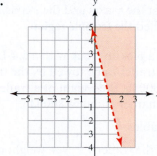

8.

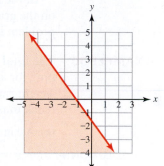

9.

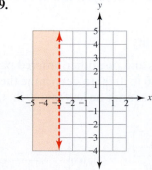

Constraints

When an owner puts a dog in a travel carrier, the dog is *constrained*. This constraint limits the movement of the dog. An inequality is also a constraint. It limits the allowed values of the variable(s).

EXAMPLE 8 | A student can spend *no more than* $500 on books and supplies.

(a) Assign two variables.

SOLUTION ▶ x = amount in dollars spent on books

y = amount in dollars spent on supplies

(b) Write an inequality that represents this constraint.

▶ The total amount must be less than or equal to $500.

$$x + y \leq \$500$$

In the next example, only the amount of paper that can be shredded per hour is constrained. The inequality that describes the constraint has only one variable.

EXAMPLE 9 | The newspaper clipping describes the constraint on the amount of paper that can be shredded per hour.

> Westchester County spent $217,000 for the shredder truck. . . . The larger truck can exert 2,000 pounds of pressure on a steel blade that can shred up to 9,000 pounds of paper an hour. (*Source:* www.nytimes.com, Sept. 19, 2008)

(a) Assign one variable.

SOLUTION ▶ x = amount of paper that can be shredded per hour

(b) Write an inequality that represents the constraint.

▶ $x \leq 9000$ lb

In the next example, the minimum value for x and y is 0 pounds. So the shading on the graph stops at the x-axis and the y-axis. The solution region in an application problem is often called the **feasible region**.

EXAMPLE 10 | The total weight of a pickup truck with a 5.4-L engine is 5000 lb. A family is shopping for a camping trailer to be towed by the truck. Since the gross combined weight rating is 13,000 lb, the weight of a trailer plus the weight of the load in the truck must be less than or equal to 8000 lb.

(a) Assign two variables.

SOLUTION ▶ x = weight of the trailer y = weight of the load in the truck

(b) Write an inequality that represents the constraint.

▶ $x + y \leq 8000$ lb

(c) Graph the inequality.

▶ Use the intercepts and one other ordered pair to graph the solid boundary line. Use a test point, (0 lb, 0 lb) to determine the shading. Since the weight of either the trailer or load cannot be less than 0 lb, the shading stops at the y-axis and at the x-axis.

x (lb)	y (lb)
0	8000
8000	0
2000	6000

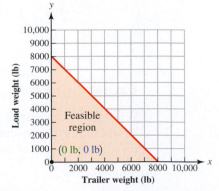

(d) Describe what the shaded area represents.

▶   The points in the shaded area represent the allowed combinations of the weight from a trailer and weight of a load in the truck.

Practice Problems

10. A large egg contains about 6 g of protein. One cup of milk contains about 8 g of protein. A diet restricts the amount of protein from eggs and milk to no more than 48 g per day.
 a. Assign two variables.
 b. Write an inequality that represents this constraint.

11. A student will spend at least $50 per month on prescription drugs.
 a. Assign one variable.
 b. Write an inequality that represents this constraint.

12. Graph $3x + y \le 2000$ lb. Show your work with the test point.

Using Technology: Graphing an Inequality in Two Variables

A graphing calculator can graph an inequality in two variables. Type the boundary line equation on the Y= screen. After checking a test point, use the "graph style icon" to shade the solution region.

EXAMPLE 11   Graph $y < -4x + 9$. Sketch the graph. Describe the window.

Enter the equation.   Graph.

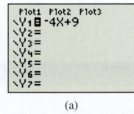

(a)

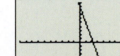

(b)

Choose a test point: $(0, 0)$

$y < -4x + 9$

$0 < -4(0) + 9$

$0 < 9$ True.

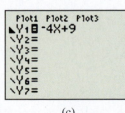

(c)

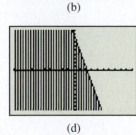

(d)

Move the cursor to the left of the equals sign. Press [ENTER] repeatedly until the "shaded below" icon appears. Press [GRAPH].

The "graph style icon" can draw the line as dashed or solid, or it can shade one side of the line. However, it cannot do both at the same time.

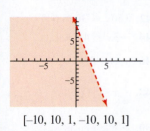

Sketch the graph. Draw the boundary line as a dashed line or as a solid line. Describe the window.

[–10, 10, 1, –10, 10, 1]

For Texas Instruments graphing calculators, there is an application for graphing inequalities. This application can be downloaded from www.education.ti.com.

Practice Problems For problems 13–14, graph the inequality. Choose a window that includes the *x*-intercept and the *y*-intercept of the boundary line. Use the shading option. Sketch the graph, drawing the boundary line as dashed or solid; describe the window.

13. $y \geq 2x - 7$ 14. $y < -\dfrac{3}{4}x + 8$

1.8 VOCABULARY PRACTICE

Match the term with its description.

1. $<$
2. $>$
3. $\leq$
4. $\geq$
5. We use this as we determine the shaded area on the graph of a linear inequality.
6. This line marks the edge of the solution region on the graph of a linear inequality.
7. On the graph of a linear inequality, this area includes all the points that represent solutions of the inequality.
8. On the graph of a constraint, this area includes all the points that represent the ordered pairs that are within the limits of the constraint.
9. A boundary line used to represent a strict inequality
10. A boundary line used to represent an inequality that is not strict

A. boundary line
B. dashed line
C. feasible region
D. greater than
E. greater than or equal to
F. less than
G. less than or equal to
H. solid line
I. solution region
J. test point

1.8 Exercises

Follow your instructor's guidelines for showing your work.

For exercises 1–50, graph the inequality. Show your work with the test point.

1. $y > 5x + 3$
2. $y > 3x + 5$
3. $y \leq 2x - 5$
4. $y \leq 3x - 9$
5. $y < -4$
6. $y < 3$
7. $x \geq 1$
8. $x \geq -4$
9. $3x + 2y \leq 18$
10. $5x + 3y \leq 15$
11. $9x - 2y \leq 18$
12. $2x - 7y \leq 14$
13. $y > x$
14. $y < x$
15. $y \leq -x$
16. $y \geq -x$
17. $y < -\dfrac{1}{3}x + 7$
18. $y > -\dfrac{1}{2}x + 6$
19. $x - y \geq 5$
20. $x - y \geq 4$
21. $5x + 2y < -15$
22. $3x + 2y > -9$
23. $y \geq -5x$
24. $y \geq -3x$
25. $x > 0$
26. $x < 0$

27. $y \leq 0$
28. $y \geq 0$
29. $x < y - 4$
30. $x < y - 5$
31. $3 < x$
32. $4 > x$
33. $5x - y > 30$
34. $6x - y > 54$
35. $-3x - 4y \leq 15$
36. $-4x - 5y \leq 24$
37. $\dfrac{1}{3}x + \dfrac{1}{6}y \geq 6$
38. $\dfrac{1}{2}x + \dfrac{1}{4}y \geq 6$
39. $\dfrac{1}{3}x - \dfrac{2}{3}y < 4$
40. $\dfrac{2}{3}x - \dfrac{4}{3}y < 4$
41. $150x + 2y > 600$
42. $150x + 3y > 600$
43. $-4x - 3y \leq 120$
44. $-3x - 4y \leq 120$
45. $y < -5x + \dfrac{1}{2}$
46. $y < -3x + \dfrac{1}{2}$
47. $\dfrac{1}{3}x < -y$
48. $\dfrac{1}{4}x < -y$
49. $\dfrac{2}{5}x \geq -\dfrac{3}{4}y$
50. $\dfrac{2}{3}x \geq -\dfrac{4}{5}y$

For exercises 51–62, write an inequality that represents the graph.

51.

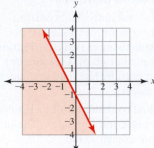

52.

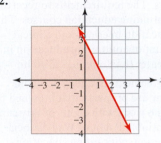

53.

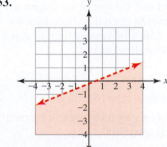

54.

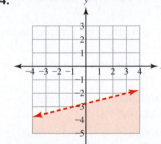

55.

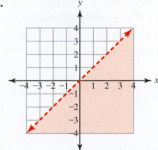

56.

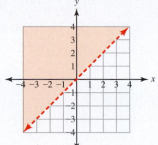

57.

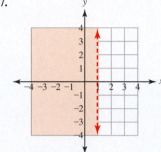

58.

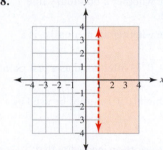

59.

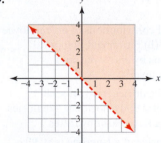

60.

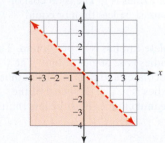

61.

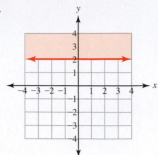

62.

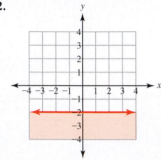

63. Explain how to determine whether a boundary line should be solid or dashed.

64. Explain how to determine whether an inequality is a strict inequality.

65. Explain why you think $(0, 0)$ is often used as a test point in graphing an inequality.

66. Describe a situation in which $(0, 0)$ cannot be used as a test point.

For exercises 67–70,

(a) use the assigned variables to write an inequality that represents the constraint.

(b) graph the constraint.

67. The maximum combined contributions of an employee and an employer to a 403(b) retirement plan total $44,000: $x =$ contributions by employee, and $y =$ contributions by employer.

68. The maximum combined calories from fat and carbohydrates in a day total 900 calories: $x =$ calories from fat, and $y =$ calories from carbohydrates.

69. At the iTunes Store, the price of a ringtone is $0.99. The price of an iPod game is $4.99. The number of ringtones is x. The number of games is y. The total cost of the ringtones and games must be less than or equal to $75.

70. The price of some chocolate candies is $3 per pound. The price of some peanut candies is $2.50 per pound. The number of pounds of chocolate candies is x. The number of pounds of peanut candies is y. The total cost of the chocolate candies and peanut candies must be less than or equal to $30.

For exercises 71–74,

(a) assign two variables.

(b) write an inequality that represents the constraint.

71. Federal election law limits total individual contributions to any national party committee to $28,500 per year. A business owner plans to make two contributions. The amount in her first contribution plus the amount in her second contribution must be less than or equal to $28,500. (*Source:* www.fec.gov)

72. Federal law limits individual contributions to a presidential candidate in a primary election campaign. The sum of two contributions cannot exceed $2300.

73. In Texas, an advanced addiction counselor must be recertified every two years. The requirements to recertify include completion of continuing education courses. The amount of continuing education in ethics and clinical supervision plus the amount of continuing education in other subjects must be greater than or equal to 40 hr.

74. A *right of way* is a legal agreement that gives someone the right to travel across or use property that is owned by another person. People who work with rights of way can be certified by the International Right of Way Association. To maintain certification, continuing education units (CEU) must be completed. The amount of continuing education units earned as a student or instructor at IRWA courses plus the amount of continuing education units earned in other ways must be greater than or equal to 75 hr. (*Source:* www.irwaonline.org)

For exercises 75–82,

(a) assign one variable.

(b) write an inequality that represents the constraint.

75. Members of the board of a charity must make an annual personal contribution of at least $1000.

76. Directors on the board of a corporation must acquire and maintain ownership of at least 1500 shares of the corporation's stock.

77. A homeowner can park no more than six vehicles on his property in a subdivision.

78. The maximum number of housing units in a planned unit development is 350.

79. The minimum initial investment in the Vanguard Admiral Treasury Money Market Fund is $50,000.

80. The minimum initial investment in the Fidelity Blue Chip Growth Fund is $2500.

81. The minimum amount of folate that a woman of child-bearing age should ingest per day is 400 micrograms.

82. The maximum amount of preformed vitamin A that a pregnant woman should ingest each day is 10,000 international units (I.U.).

For exercises 83–88,
(a) assign two variables.
(b) write an inequality that represents the constraint.

83. A state allows teachers to retire early with a reduced benefit package if they meet the rule of 70: The sum of the teacher's age and the years of service must be at least 70 years.

84. For Missouri employees in the Local Government Employees Retirement System, the rule of 80 is an optional early retirement provision in which an employee may retire with an unreduced benefit if the sum of the employee's age and years of service is at least 80 years.

85. The maximum amount of credits that can be paid for with financial aid is 125% of the credits required for a degree or certificate program.

86. The amount of a guaranteed loan used for agricultural production can be no more than 50% of the total loan amount.

87. The annual amount of mortgage payments can be no more than 40% of the annual take-home pay of the loan applicant.

88. The annual amount of car payments can be no more than 15% of the annual take-home pay of the loan applicant.

Technology

For exercises 89–92, use a graphing calculator to graph the inequality. Shade the solution region. Choose a window that shows the y-intercept of the boundary line. Sketch the graph, drawing the boundary line as solid or dashed; describe the window.

89. $y > 3x - 8$

90. $y < -2x + 16$

91. $y \le 1500x + 9000$

92. $y \ge 0.5x + 6$

Find the Mistake

For exercises 93–96, the completed problem has one mistake.
(a) Describe the mistake in words.
(b) Do the problem correctly.

93. **Problem:** Graph: $y > \dfrac{1}{2}x - 3$

Incorrect Answer:

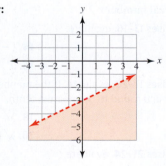

94. **Problem:** Graph: $x + y \le -3$

Incorrect Answer:

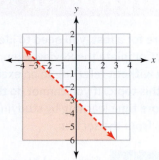

95. **Problem:** Graph: $x - y \ge 0$

Incorrect Answer:

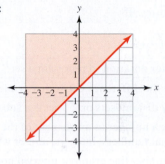

96. **Problem:** Graph: $y > 2$

Incorrect Answer:

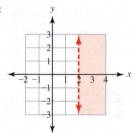

SUCCESS IN COLLEGE MATHEMATICS

97. Many colleges have a student code of conduct that describes the consequences of academic dishonesty (cheating). Some instructors include this information in the class syllabus. What is the consequence in your math class for academic dishonesty?

Study Plan for Review of Chapter 1

To prepare for a test, you need to identify what you know and where you need more study and practice. Complete the Review Exercises for each question. If you can do each Review Exercise correctly without referring to an example or your notes, you probably do not need to spend more time on this topic. If you cannot do the Review Exercises correctly without looking back, you need to spend more time learning by studying the examples or your notes or by redoing Practice Problems or Exercises from the section.

SECTION 1.1   Sets and Numerical Expressions

Ask Yourself	Test Yourself	Help Yourself
Can I . . .	**Do 1.1 Review Exercises**	**See these Examples and Practice Problems**
use roster notation to represent a set?	1, 2	Ex. 1, 2, 4, 5, PP 2
explain the difference between a prime number and a composite number?	3	Ex. 4, PP 3
identify disjoint sets or if a set is a subset of another set?	4–6	Ex. 6, 7, PP 4, 5
identify a number as an element of the set of real numbers, rational numbers, irrational numbers, integers, and/or whole numbers?	7, 8	Ex. 8, PP 6–13
use a number line graph or interval notation to represent a set of numbers?	9, 10–12	Ex. 3, 9, 10, PP 1, 14–16
evaluate an exponential expression?	13–16	Ex. 11, PP 17–20
evaluate a square root or a cube root?	17–20	Ex. 12, 13, PP 21–25
describe the order of operations? evaluate an expression by following the order of operations?	21–26	Ex. 14, 15, PP 26–28

1.1 Review Exercises

1. Use roster notation to represent the number line graph.

$$-6\ -5\ -4\ -3\ -2\ -1\ \ 0\ \ 1\ \ 2\ \ 3\ \ 4$$

2. Use roster notation to represent the set of integers.

3. Copy the table. Put a checkmark or X in the box if the number is an element of the set.

	Prime	Composite	Neither prime nor composite
5			
0			
−3			
12			
20			
11			

4. **a.** Is every integer an element of the set of rational numbers?
   b. Explain.

5. Set A is $\{0, 5, 6, 9\}$. Set B is $\{2, 3, 4, 9\}$. Are these sets disjoint? Explain.

6. Set A is $\{-3, -2, -1\}$. Set B is the set of integers.
   a. Is set A a subset of set B?
   b. Explain.

7. Describe the difference between a rational number and an irrational number.

8. Copy the table at the top of the next page. Put a checkmark or X in the box if the number is an element of the set.

9. Use a number line graph to represent $\{-2, 0, 4, 5\}$.

For exercises 10–12,
(a) use a number line graph to represent the set.
(b) use interval notation to represent the set.

10. The real numbers greater than 8

11. The real numbers less than or equal to 2

12. The real numbers

For exercises 13–16, evaluate.

13. 7^2 14. $(-7)^2$

15. 7^3 16. 7^1

For exercises 17–20, evaluate. If the number is irrational, use a calculator and round to the nearest tenth.

17. $\sqrt{9}$ 18. $\sqrt[3]{-64}$

19. $\sqrt{5}$ 20. $\sqrt{12}$

For exercises 21–24, evaluate.

21. $-24 \div 6 \cdot 2 - 9$

22. $-3^2 \cdot 12 \div 6 - 15$

23. $\dfrac{4^2 + 24 \div 6 \cdot 2}{2 - 4}$

24. $6 - (2 - 7)^2 - (-3)$

Table for exercise 8

	Real numbers	Rational numbers	Irrational numbers	Integers	Whole numbers
5					
−1.1					
0					
π					
$-\dfrac{2}{9}$					
−14					
$\dfrac{3}{4}$					

25. An expression is $-80 + 32 \div 4^2 \cdot 2$. What should be done first?

26. An expression is $-80 + 32 \div 4 \cdot 2$. What should be done first?

SECTION 1.2 **Algebraic Expressions**

Ask Yourself	Test Yourself	Help Yourself
Can I . . .	Do 1.2 Review Exercises	See these Examples and Practice Problems
identify examples of the commutative property, associative property, or distributive property?	27–32	Ex. 1, PP 1
simplify an expression by combining like terms and/or using the distributive property?	35, 40, 41, 43–45, 47	Ex 2–4, 6–9, PP 2–5, 12–14
evaluate an expression that includes operations with 0? identify the additive and multiplicative identities of the real numbers? describe the difference between the opposite and the reciprocal of a number?	28, 30, 49–54	Ex. 5, PP 6–11
find the common denominator of two fractions and rewrite the fractions with this denominator? identify when I need to find the common denominator of two fractions?	45, 46	Ex. 6, PP 12–14
write the product rule of exponents, the quotient rule of exponents, and the power rule of exponents? use the exponent rules to simplify an expression?	33, 34, 36–38, 40–42, 46	Ex. 10–16, PP 15–20
rewrite an expression with a negative exponent as an equivalent expression with a positive exponent?	37, 38, 48	Ex. 16, PP 21–25
simplify an exponential expression in which the exponent is 0 or is 1?	36, 39, 54	Ex. 17, PP 26, 27

1.2 Review Exercises

For exercises 27–32, choose the property or term that is best illustrated by the equation.

27. $4 \cdot p = p \cdot 4$

28. $c + 0 = c$

29. $c \cdot 0 = 0$

30. $c \cdot 1 = c$

31. $2(x + 3) = 2x + 6$

32. $(c + 3) + 4 = c + (3 + 4)$

    A. additive identity
    B. associative property
    C. commutative property
    D. distributive property
    E. multiplicative identity
    F. zero product property

For exercises 33–46, simplify. Exponents in the final expression should be positive.

33. $\dfrac{12x^9}{30x^2}$

34. $(3x^4)(5x^7)$

35. $3(x^2 - 5x^2)$

36. $\dfrac{p^7}{p^7}$

37. $(2m^{-4}p)(9m^{-12}p^6)$

38. $(3h^{-5})^2$

39. $5d^0$

40. $3x^2(8x - 4)$

41. $5x^3(6x^2 + 4x)$

42. $\left(\dfrac{2h^3}{5k^2}\right)^2$

43. $2(3x - 9) - 5(4x + 8)$

44. $6p - (-4p + 3)$

45. $\dfrac{3}{4}\left(\dfrac{8}{3}a - \dfrac{1}{2}\right) - \left(\dfrac{2}{7}a - \dfrac{4}{5}\right)$

46. $\left(\dfrac{5}{6}w\right)\left(\dfrac{3}{10}w^4\right)$

47. Explain why it is necessary to use the distributive property instead of following the order of operations to simplify $8(3x - 9)$.

48. One of the expressions is not simplified. Identify and simplify this expression.

a. $8x^3 + 9x^2 - x$

b. $\dfrac{12x^4}{13y^5}$

c. $2 - 4w^5$

d. $7x^4y^{-5}$

For exercises 49–53, evaluate.

49. $\dfrac{7}{0}$

50. $0 \div 7$

51. $(0)(7)$

52. $0 - 7$

53. $\dfrac{0}{7}$

54. Identify the expressions that are undefined.

a. $\dfrac{0}{2}$

b. $(0)(2)$

c. $2 \div 0$

d. 2^0

e. $0 \div 2$

f. $\dfrac{2}{0}$

SECTION 1.3   Equations and Inequalities in One Variable

Ask Yourself	Test Yourself	Help Yourself
Can I...	**Do 1.3 Review Exercises**	**See these Examples and Practice Problems**
solve a linear equation in one variable and check the solution?	55–59, 62	Ex. 1–13, PP 1–13
clear the fractions from a linear equation?	59	Ex. 6, PP 2
identify a linear equation in one variable that has infinitely many solutions or that has no solution?	57, 59	Ex. 9–11, PP 7–10
use a number line graph or interval notation to represent the solution of a linear equation in one variable?	60, 61	Ex. 12, 13, PP 11–13
solve a linear inequality in one variable and check the solution?	65–67	Ex. 14–17, PP 14–21
use a number line graph or interval notation to represent the solution of a linear inequality in one variable?	63–67	Ex. 14–17, PP 14–21

1.3 Review Exercises

For exercises 55–58,
(a) solve.
(b) check.

55. $3(7x - 2) + 9 = 4(5x + 8) - 13$

56. $\dfrac{1}{8}x - 15 = -12$

57. $4(3x + 1) = 12x - 9$

58. $3x = 9x$

59. $\dfrac{3}{4}x + 9 = -\dfrac{11}{8}x + 15 + \dfrac{17}{8}x - 6$

   a. Clear the fractions from the equation and solve.
   b. Check.

60. The solution of $x + 6 = 10$ is $x = 4$.
   a. Use a number line graph to represent this solution.
   b. Use interval notation to represent this solution.

61. The solution of $3(x - 1) = 3x - 3$ is the set of real numbers.
   a. Use a number line graph to represent this solution.
   b. Use interval notation to represent this solution.

62. The solution of one of the equations is $x = 6$. Choose this equation.

   a. $6 - 3x = 12$

   b. $-\dfrac{x}{3} - 2x = -10$

   c. $\dfrac{x}{3} - 2x = -10$

   d. $-5x + 4 = -34$

63. Choose the interval that represents this number line graph.

   a. $(\infty, 2)$

   b. $(-\infty, 2)$

   c. $(-\infty, 2]$

   d. $(\infty, 2]$

64. Choose the number line graph that represents the interval $[-3, \infty)$.

   a.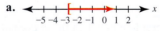

   b.

   c.

For exercises 65–67,
(a) solve.
(b) check.
(c) use a number line graph to represent the solution.
(d) use interval notation to represent the solution.

65. $3x + 9 \le -21$

66. $-3x + 9 < -21$

67. $-2(4x + 7) < x - 9$

SECTION 1.4 **Scientific Notation and Unit Analysis**

Ask Yourself	Test Yourself	Help Yourself
Can I . . .	**Do 1.4 Review Exercises**	**See these Examples and Practice Problems**
write a number or measurement in scientific notation?	68–71	Ex. 1, 2, PP 1–4
multiply and divide numbers in scientific notation?	72–74	Ex. 3–5, PP 5–7
add and subtract numbers in scientific notation that have different powers of ten?	75	Ex. 6, PP 8
use place value notation and write an SI unit in its base unit?	76–77	Ex. 7, PP 9–14
use an equation or a conversion factor to show the relationship of two units?	78–80	Ex. 8, PP 15–17
use unit analysis to convert a measurement into another unit?	81–83	Ex. 9–12, PP 18–21
simplify expressions that include units of measurement?	84	Ex. 13, PP 22–25

1.4 Review Exercises

For exercises 68–69, write the measurement in scientific notation. Include the units of measurement.

68. In 1 mole of carbon, there are about 602,000,000,000,000,000,000,000 atoms.

69. The gravitational constant is $0.00000000006673 \dfrac{m^3}{kg \cdot s^2}$.

70. a. Explain why 35×10^8 is not in scientific notation.
b. Rewrite this number in scientific notation.

71. a. Explain why 0.35×10^{-6} is not in scientific notation.
b. Rewrite this number in scientific notation.

For exercises 72–75, evaluate. Write the answer in scientific notation.

72. $(6.0 \times 10^{-7})^2$

73. $(3.5 \times 10^4)(2.6 \times 10^8)$

74. $\dfrac{2 \times 10^8}{4 \times 10^{-3}}$

75. $6.2 \times 10^3 + 5.4 \times 10^2$

For exercises 76–77, use place value notation and rewrite the measurement in its base unit.

76. 8 kilometers

77. 4 microliters

For exercises 78–79, write an equation that shows the relationship of the units.

78. mile and feet

79. milliliters and liter

80. The relationship of kilometers and miles is 1.6 km ≈ 1 mi. Identify the conversion factor needed to convert $\dfrac{75\ mi}{1\ hr}$ into $\dfrac{km}{hr}$.

81. To convert $\dfrac{18\ L}{1\ hr}$ into $\dfrac{qt}{s}$, a student wrote this expression:
$\dfrac{18\ L}{1\ hr} \cdot \dfrac{1.06\ qt}{1\ L} \cdot \dfrac{1\ hr}{60\ min}$. Identify the missing conversion factor.

For exercises 82–83, use unit analysis and the relationships in Table 4 to convert the measurement into the given unit(s).

82. $75\ \dfrac{cm^3}{s}$ into $\dfrac{L}{hr}$

83. 6 qt into tablespoons

84. The expression $\left(\dfrac{4.58 \times 10^9\ km}{3.00 \times 10^5\ \frac{km}{s}} \right)\left(\dfrac{1\ min}{60\ s} \right)\left(\dfrac{1\ hr}{60\ min} \right)$ represents the time for a radio signal to travel from Pluto to the earth. Simplify this expression. Round the mantissa to the nearest hundredth.

SECTION 1.5 **Applications and Problem Solving**

Ask Yourself	Test Yourself	Help Yourself
Can I . . .	**Do 1.5 Review Exercises**	**See these Examples and Practice Problems**
use the five steps to solve an application problem?	85–90	Ex 1–10, PP 1–9
identify extraneous information in an application problem?	85–87, 89, 90	Ex. 1–4, 8, PP 3, 5, 6, 7

1.5 Review Exercises

For problems 85–90, use the five steps.

85. Find the number of victims of stalking who accrued out-of-pocket costs. Write the answer in place value notation. Round to the nearest tenth of a million.

During a 12-month period, an estimated 3.4 million persons age 18 or older were victims of stalking. . . . About 3 in 10 of stalking victims accrued out-of-pocket costs for things such as attorney fees, damage to property, child care costs, moving expenses, or changing phone numbers. (*Source:* www .justice.gov, Jan. 2009)

86. A nursing assistant worked 55 hr in one week. He earned $12 per hour for the first 40 hr of his shifts and $18 for each additional hour. Because he worked overtime, his children were in daycare 15 additional hours. Find his earnings for this week.

87. In June 2010, Clearwater Paper announced that Shelby, North Carolina, will be the site of a new $260 million factory. The factory will have 250 full-time employees with an average wage of $38,000 per year and will make 70,000 tons of tissue products per year. The average wage in this North Carolina county is $31,200. Find the percent increase in the average wage of the new jobs compared to the current average wage in the county. Round to the nearest percent. (*Source:* www.lmtribune.com, June 11, 2010)

88. The Deepwater Horizon oil platform exploded on April 20, 2010. On June 10, 2010, government scientists estimated that at least 40,000 gal per day of oil had been flowing into the Gulf of Mexico from the well. Find the minimum number of barrels of oil that had gone into the Gulf of Mexico from April 21 up to and including June 10 (April has 30 days; May has 31 days; 1 barrel = 32 gal). (*Source:* www.nytimes.com, June 11, 2010)

89. Find the number of full-time undergraduates not attending community college in 2008. Round to the nearest tenth of a million.

Community college students represent about a quarter (24%) of all full-time undergraduates. Nationwide, 2.2 million students attend community college full time, a quarter of a million more than attend private colleges full time. . . . Forty percent of community college students have such low incomes that they have no resources to pay for a college education. (*Source:* www.ticas.org, May 2009)

90. The amount of paint required to paint the exterior of a house is given by the formula $G = \dfrac{Ac}{C}$, where G is the number of gallons of paint, A is the area to be painted, C is the *coverage* of the paint on the surface, and c is the number of coats. For wood siding, the coverage is $\dfrac{470\ \text{ft}^2}{1\ \text{gal}}$. For medium-texture stucco, the coverage is $\dfrac{128\ \text{ft}^2}{1\ \text{gal}}$. The area of wood siding on a house is 1750 ft². Find the number of 1-gal cans of paint needed to apply two coats. (*Source:* R. Scharff, *Workshop Math*, Sterling Publishing Co., 1989)

SECTION 1.6   **Slope and Linear Equations in Two Variables**

Ask Yourself	Test Yourself	Help Yourself
Can I...	**Do 1.6 Review Exercises**	**See these Examples and Practice Problems**
write the standard form of a linear equation?	91	Ex. 1, PP 1, 2
rewrite a linear equation in standard form?		
use the equation of a line to find the intercepts and one other point and graph the line?	92, 93	Ex. 2, 3, 14, PP 3–5
find the slope of a line, given two points on the line?	94–96	Ex. 4–7, PP 6–10
find the slope of a line, given its equation?	97, 103	Ex. 8, PP 11–12
find the slope of a line, given its graph?	98	Ex. 9, PP 13–16
find the slope of a line that is parallel or perpendicular to a given line?	99, 100	Ex. 10, 11, PP 15–17
write the slope-intercept form of a linear equation?	101, 103	Ex. 12, 13, PP 18
rewrite the equation of a line in slope-intercept form and identify its slope and y-intercept?		
use slope-intercept graphing to graph a linear equation?	101, 102	Ex. 15–17, PP 19–22, 25
graph a horizontal or vertical line, given its equation?	104, 105	Ex. 17, 18, PP 21–24

1.6 Review Exercises

91. Rewrite $y - 5 = 3(x + 2)$ in standard form.

For exercises 92–93,
(a) find the intercepts and one other ordered pair that is a solution of the equation.
(b) graph the equation.

92. $3x - 8y = 24$

93. $y = -3x + 5$

For exercises 94–96, find the slope of the line that passes through the given points.

94. $(5, 2)$; $(5, -8)$

95. $(-7, 4)$; $(2, -3)$

96. $(6, 2)$; $(-1, 2)$

97. Use the intercepts to find the slope of $2x - 9y = 27$.

98. Identify the slope of the graphed line.

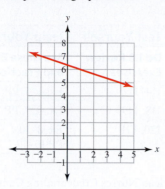

99. The slope of line A is $\dfrac{1}{8}$.

    a. Identify the slope of a line that is parallel to line A.

    b. Identify the slope of a line that is perpendicular to line A.

100. Choose the equation that has a graph that is perpendicular to the graph of $y = -\dfrac{2}{3}x + 5$.

    a. $y = -\dfrac{3}{2}x + 6$ **b.** $3x + 2y = 7$

    c. $3x - 2y = 5$ **d.** $y = \dfrac{2}{3}x + 1$

101. The equation of a line is $6x - 5y = 20$.

    a. Rewrite this equation in slope-intercept form.

    b. Identify the slope.

    c. Identify the y-intercept.

    d. Use slope-intercept graphing to graph the equation.

102. The equation of a line is $y = -\dfrac{3}{5}x + 7$.

    a. Identify the slope.

    b. Identify the y-intercept.

    c. Use slope-intercept graphing to graph the equation.

103. Copy and complete the table.

Equation	Slope of the line	y-intercept of the line
$y = -3x + 1$		
$x = 2$		
$y = 9x$		
$y = 8$		
$3x - 5y = 24$		

For exercises 104–105, graph the equation.

104. $x = 5$ **105.** $y = 2$

SECTION 1.7 **Writing the Equation of a Line**

Ask Yourself	Test Yourself	Help Yourself
Can I . . .	**Do 1.7 Review Exercises**	**See these Examples and Practice Problems**
given the slope and y-intercept or a graph, write the equation of a line in slope-intercept form?	106	Ex. 1–3, PP 1–3
given the slope and a point, write the equation of a line in slope-intercept form?	107	Ex. 4, PP 4, 5
write the point-slope form of a linear equation?	108	Ex. 5, PP 6
given two points, write the equation of a line in slope-intercept form?		
write the equation of the horizontal or vertical line that passes through a given point?	109, 110	Ex. 6–8, PP 7–10
write the equation of a parallel or perpendicular line that passes through a given point?	111, 112	Ex. 9–12, PP 11–16

1.7 Review Exercises

106. The slope of a line is -11, and its y-intercept is $(0, 7)$. Write its equation in slope-intercept form.

107. The slope of a line is $\dfrac{1}{2}$, and the line passes through the point $(4, 10)$. Write its equation in slope-intercept form.

108. A line passes through $(3, 11)$ and $(5, 9)$. Write its equation in slope-intercept form.

109. Write the equation of the vertical line that passes through the point $(9, 6)$.

110. Write the equation of the horizontal line that passes through the point $(9, 6)$.

111. The equation of a line is $y = 3x - 9$. Write the equation in slope-intercept form of a parallel line that passes through the point $(-4, 6)$.

112. The equation of a line is $y = \dfrac{2}{3}x + 6$. Write the equation in slope-intercept form of a perpendicular line that passes through the point $(3, 4)$.

113. An equation is $6x + \underline{}\, y = 18$. Fill in the blank with a coefficient so that the y-intercept is $(0, 2)$.

Ask Yourself	Test Yourself	Help Yourself
Can I . . .	**Do 1.8 Review Exercises**	**See these Examples and Practice Problems**
graph a linear inequality?	114–117	Ex. 1–5, PP 1–6
write a linear inequality, given its graph?	118, 119	Ex. 6, 7, PP 7–9
write a linear inequality in one or two variables that represents a constraint and graph the constraint?	120, 121	Ex. 8–10, PP 10–12

1.8 Review Exercises

For exercises 114–117, graph the inequality. Show your work with the test point.

114. $8x - 5y < 40$

115. $x \le 2$

116. $x + 7y \ge 14$

117. $y > 4$

118. a. The inequality represented by the graph is $-3x + 2y$ __ 6. Should the blank in the inequality be $\ge$ or $\le$?

b. Explain.

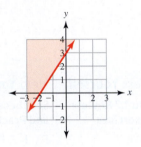

119. Write the inequality represented by the graph. Show your work with the test point.

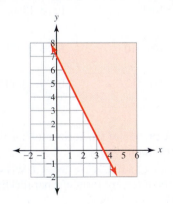

120. The American Nurses Credentialing Center requires at least 75 continuing education hours for certificate renewal. A nurse will earn some of these hours in formally approved seminars, x, and some of these hours in completing study modules, y. (*Source:* www.nursecredentialing.org)

a. Write an inequality that represents this constraint.

b. Graph this inequality.

121. A press release from the Centers for Disease Control and Prevention describes the number of adults killed by heart disease, stroke, and other cardiovascular diseases.

Heart disease, stroke, and other cardiovascular (blood vessel) diseases are among the leading cause of death and now kill more than 800,000 adults in the U.S. each year. (*Source:* www.cdc.gov, Feb. 1, 2011)

a. Assign one variable.

b. Write an inequality that represents this constraint.

Chapter 1 Test

Show your work.

1. Evaluate: $-5^2 - 12 \div 3 \cdot 2 - (-6 + 1)$

For problems 2–11, evaluate.

2. $\sqrt{64}$

3. 4^3

4. $\sqrt{-64}$

5. $\dfrac{8}{0}$

6. $0 \div 3$

7. 2^0

8. 6^{-2}

9. -4^2

10. $(-4)^2$

11. $\dfrac{0}{4}$

For problems 12–14, simplify.

12. $(7x^4)^2(3x^5y)$

13. $\dfrac{18xy^5}{40x^3y}$

14. $(5k^3)^3$

15. Identify the property that allows us to rewrite $a(b + c)$ as $(b + c)a$.

For problems 16–18,
(a) solve.
(b) check.

16. $\dfrac{5}{8}z - \dfrac{1}{2} = \dfrac{13}{4}$

17. $7(3x - 5) + 9 = 4(5x + 2) - 8 + x$

18. $12 - p = p$

19. The solution of an equation is the set of real numbers.
    a. Use a number line graph to represent this solution.
    b. Use interval notation to represent this solution.

For problems 20–21,
(a) solve.
(b) check.

20. $9x - 27 > -108$

21. $-9x - 27 > -108$

22. The solution of an inequality is $x \le 4$.
    a. Use a number line graph to represent this solution.
    b. Use interval notation to represent this solution.

23. **a.** Find three solutions of $-4x + 5y = 32$. Organize the solutions in a table.
    b. Graph the equation.

24. Find the slope of the line $2x - 9y = 15$.

25. Explain why the slope of a vertical line is undefined.

26. Write the equation in slope-intercept form of the line that passes through the points $(1, 9)$ and $(3, -13)$.

27. An equation of a line is $y = -\dfrac{5}{6}x + 7$.
    a. Identify the slope of this line.
    b. Identify the y-intercept.
    c. Use slope-intercept graphing to graph the line.

28. Graph: $x = 2$

29. Write the equation of the horizontal line that passes through the point $(-3, 9)$.

For problems 30–31, evaluate. Write the answer in scientific notation.

30. $\left(3.2 \times 10^4 \dfrac{\text{ft}}{\text{s}}\right)(4.8 \times 10^2 \text{ s})$

31. $3.5 \times 10^{-4} + 6.1 \times 10^{-3}$

32. Use unit analysis to convert $852 \dfrac{\text{m}}{\text{s}}$ into $\dfrac{\text{km}}{\text{hr}}$ $(1000 \text{ m } = 1 \text{ km})$.

For problems 33–34, use the five steps from Section 1.5.

33. In designing a college dormitory, the plumbing code requires that one shower be provided for every eight people up to 150 people. If more than 150 people live in the dorm, one shower must be provided for every additional 20 people. Find the minimum number of showers required in a dorm that will house 250 people.

34. Find the price per credit before the increase. Round to the nearest whole number.

    Portland Community College is raising its fall tuition by 5.7 percent to $74 per credit, or $1110 for a full-time student taking 15 credits. (*Source:* www.oregonlive.com, May 27, 2009)

35. Graph: $3x - 8y > -24$

36. Write the equation of the graphed line.

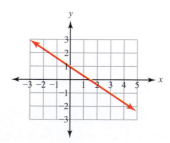

Polynomials and Absolute Value

2

Studying for Tests

The concepts and skills of Intermediate Algebra are built on the foundation of what you learned in a previous algebra course. *As you learn new concepts and skills in class or on-line, you need to study and practice them until they are also part of your algebra foundation.* This does not happen the night before a test. Instead, you need to manage your time so that you can read your math book before class, read and revise your notes as soon as possible after class, and complete your homework with time to get any help you might need. You should keep a separate list in your notes of important terms to be memorized, and you should also have a place to write down anything that your instructor emphasizes as very important and likely to be on the test.

Before the test, evaluate whether you are ready to take it. Completing the Study Plan for Review assignments can help you do this. You should also redo the exercises in your homework that you found most difficult.

Although all of this work takes time, it is worth it. As well as being prepared to take a test, you will more likely remember what you have learned long term. You are more likely to receive the grade you want in this course, bringing you one step closer to meeting your academic and professional goals.

© Craig Hanson/Shutterstock

In Greek, the prefix *poly* means "many." *Poly*unsaturated fats have many unsaturated chemical bonds. *Poly*ester is a large molecule created from many ester molecules. *Poly*phonic means "many voices." In this section, we will study *poly*nomials. Although *nomial* does not directly translate as "term," a polynomial can be one, two, or *many* terms.

SECTION 2.1

Adding, Subtracting, and Multiplying Polynomial Expressions

After reading the text, working the practice problems, and completing assigned exercises, you should be able to:

1. Determine whether an expression is a polynomial.
2. Identify the degree of a polynomial in one variable.
3. Identify a monomial, binomial, or trinomial.
4. Simplify a polynomial expression by combining like terms.
5. Use the distributive property to multiply polynomial expressions.
6. Use a pattern to multiply polynomial expressions.
7. Write a polynomial expression that represents the perimeter or area of a rectangle.

Polynomial Vocabulary

A **variable** is a letter or other symbol that represents an unknown number. A **constant** is a number. When we multiply a number and a variable, the number is a **coefficient**. A **term** is a variable, a constant, or the product of a number and a variable(s). A **polynomial expression** is a single term or a sum or difference of terms. In a polynomial expression, a variable can never be an exponent, in a denominator, or under a radical sign such as a square root. If a variable is raised to a power, this power is always an integer greater than or equal to 0.

EXAMPLE 1 | Identify the number of terms, the variable, the coefficient, and the constant in the expression $3x + 7$.

SOLUTION ▶ This expression has two terms, $3x$ and 7. The variable is x. The coefficient is 3. The constant is 7.

EXAMPLE 2 | Is the expression a polynomial? If it is not, explain.

(a) $7p$

SOLUTION ▶ Yes.

(b) $6x - \dfrac{5}{x}$

▶ No, there is a variable in a denominator.

(c) $2^x + 3^x$

▶ No, a variable is an exponent.

(d) $7a^3 - 2a + 1$

▶ Yes.

(e) 9

▶ Yes. A constant is a polynomial.

(f) $x^2 + xy + y^2$

▶ Yes.

We usually write the terms in a polynomial in one variable in **descending order**. From left to right, the exponents on the variables decrease. The terms in $2x^4 + 7x^2 - 9x + 8$ are in descending order. In descending order, the coefficient of the first term is the **lead coefficient**.

The **degree** of a polynomial expression in one variable is the value of the largest exponent on the variable. For example, the degree of $x^2 - 5x + 6$ is 2. The polynomial expression 3 has no variable. However, since $x^0 = 1$, we can rewrite this expression as $3x^0$, and its degree is 0.

EXAMPLE 3 | Identify the degree of each polynomial.

(a) $7x^3 - 2x + 1$

SOLUTION ▶ The degree is 3.

(b) $-6p^2 - 9p$

▶ The degree is 2.

(c) $2w + 17$

▶ The degree is 1.

(d) 8

▶ The degree is 0.

Polynomial Expression

A polynomial expression is a single term or a sum or difference of terms. A polynomial expression in one variable can be written in the form

$$a_n x^n + a_{n-1} x^{n-1} + \cdots + a_2 x^2 + a_1 x + a_0$$

where x is a variable, n is an integer ≥ 0, and $a_0, a_1, a_2, \ldots, a_n$ are real numbers. The **degree** of this expression is n. The **lead coefficient** is a_n; a_n does not equal 0.

Polynomial expressions with three or fewer terms have special names. A polynomial with one term is a **monomial**, with two terms is a **binomial**, and with three terms is a **trinomial**.

Practice Problems

For problems 1–5,
(a) is the expression a polynomial?
(b) If it is, identify its degree.

1. $7y^3 - 4y^2 + 2y + 15$ **2.** $\dfrac{5a - 9}{2a + 1}$ **3.** 5^x

4. $7x^9 + 2$ **5.** 4

Simplifying Polynomial Expressions

Like terms have exactly the same variables, and the exponent on each variable is the same. To simplify a polynomial expression, combine like terms.

EXAMPLE 4 Simplify: $5w^3 + 7w^2 - 9w + 8 + 10w^3 - 11$

SOLUTION ▶ $5w^3 + 7w^2 - 9w + 8 + 10w^3 - 11$ Identify like terms.

$= 15w^3 + 7w^2 - 9w - 3$ Combine like terms.

When adding or subtracting polynomials, we can also arrange them vertically. In Chapter 5, we will use long division to divide polynomials. As part of this process, we will do vertical subtraction of polynomials.

EXAMPLE 5 Simplify: $(9c^3 - 6c^2 + 4c - 8) + (2c^3 + 5c^2 + c - 1)$

SOLUTION ▶ $(9c^3 - 6c^2 + 4c - 8) + (2c^3 + 5c^2 + c - 1)$ **Vertical Addition**

$= 11c^3 - 1c^2 + 5c - 9$

$$9c^3 - 6c^2 + 4c - 8$$
$$\underline{+2c^3 + 5c^2 + c - 1}$$
$$11c^3 - 1c^2 + 5c - 9$$

When we subtract two polynomials, the second polynomial is inside parentheses. We use the distributive property to simplify.

EXAMPLE 6 Simplify: $(9x^3 - 6x^2 + 4x - 8) - (2x^3 + 5x^2 + x - 1)$

SOLUTION ▶ $(9x^3 - 6x^2 + 4x - 8) - (2x^3 + 5x^2 + x - 1)$ **Vertical Subtraction**

$= (9x^3 - 6x^2 + 4x - 8) - \mathbf{1}(2x^3 + 5x^2 + x - 1)$

$= 9x^3 - 6x^2 + 4x - 8 - 2x^3 - 5x^2 - x + 1$

$= 7x^3 - 11x^2 + 3x - 7$

$$9x^3 - 6x^2 + 4x - 8$$
$$\underline{-(2x^3 + 5x^2 + x - 1)}$$
$$7x^3 - 11x^2 + 3x - 7$$

In the next example, one of the polynomials does not have a y-term. When doing vertical addition or subtraction, use $0y$ as a **placeholder**.

EXAMPLE 7 Simplify: $(6y^2 - 4y + 2) - (9y^2 - 8)$

SOLUTION ▶ $(6y^2 - 4y + 2) - (9y^2 - 8)$ **Vertical Subtraction**

$= (6y^2 - 4y + 2) - \mathbf{1}(9y^2 - 8)$

$= 6y^2 - 4y + 2 - 9y^2 + 8$

$= -3y^2 - 4y + 10$

$$6y^2 - 4y + 2$$
$$\underline{-(9y^2 + 0y - 8)}$$
$$-3y^2 - 4y + 10$$

We can also add and subtract polynomials with more than one variable. For two terms to be like terms, each variable and its exponent must be the same.

EXAMPLE 8 Simplify: $(7x^2 + 4xy + 5y^2) - (x^2 - xy + y^2)$

SOLUTION ▶ $(7x^2 + 4xy + 5y^2) - (x^2 - xy + y^2)$ **Vertical Subtraction**

$= (7x^2 + 4xy + 5y^2) - \mathbf{1}(x^2 - xy + y^2)$

$= 7x^2 + 4xy + 5y^2 - 1x^2 + 1xy - 1y^2$

$= 6x^2 + 5xy + 4y^2$

$$7x^2 + 4xy + 5y^2$$
$$\underline{-(x^2 - xy + y^2)}$$
$$6x^2 + 5xy + 4y^2$$

Practice Problems

For problems 6–9, simplify.

6. $(x^2 + 8x - 9) + (7x^2 + 2x - 3)$ 7. $(z^2 + 8z - 9) - (7z^2 + 2z - 3)$

8. $(-2x^3 + 5x + 1) - (7x^2 - x + 8)$

9. $(h^3 + 2h^2k - 3hk^2 - 9) - (4h^3 - hk^2 - 1)$

Multiplying Polynomial Expressions

To find the product of two monomials, multiply the coefficients and use the product rule of exponents, $x^m \cdot x^n = x^{m+n}$, to multiply the exponential expressions. In doing this, we are *simplifying* the expression.

EXAMPLE 9   Simplify: $(3x^5)(2x^2)$

SOLUTION ▶ $(3x^5)(2x^2)$ The exponential expressions have the same base.

$= 6x^{(5+2)}$ Multiply coefficients; product rule of exponents.

$= 6x^7$ Simplify.

To multiply a monomial and a binomial, use the distributive property, $a(b + c) = ab + ac$.

EXAMPLE 10   Simplify: $(3a^5)(2a^2 + 5)$

SOLUTION ▶ $(3a^5)(2a^2 + 5)$

$= 3a^5(2a^2) + 3a^5(5)$ Distributive property.

$= 6a^{(5+2)} + 15a^5$ Multiply coefficients; product rule of exponents.

$= 6a^7 + 15a^5$ These are not like terms and cannot be combined.

We can use a rectangle to visualize the multiplication of two binomials. The area of a rectangle equals the product of its length and its width. One of the binomials is the length. The other binomial is the width. The area of the rectangle is equal to the product of the binomials.

EXAMPLE 11   Use a rectangle to simplify $(x + 3)(x + 2)$.

SOLUTION ▶ The area of the entire rectangle equals the product of the length and the width.

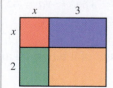

The length of the rectangle is $x + 3$.

The width is $x + 2$.

The area of the entire rectangle is A.

Area of the entire rectangle = (length)(width)

$$A = (x + 3)(x + 2)$$

The area of the entire rectangle also equals the sum of the four different areas.

area = x^2 area = $3x$ area = $2x$ area = 6

Area of the entire rectangle = sum of the four areas shown in the diagram.

$$A = x^2 + 3x + 2x + 6$$
$$A = x^2 + 5x + 6$$

Since $A = (x + 3)(x + 2)$ and $A = x^2 + 5x + 6$, $(x + 3)(x + 2)$ is equal to $x^2 + 5x + 6$.

Instead of using a rectangle to multiply two binomials, we can use the distributive property.

EXAMPLE 12 Simplify: $(x + 3)(x + 2)$

SOLUTION ▶

$(x + 3)(x + 2)$

$= \mathbf{x}(x + 2) + \mathbf{3}(x + 2)$ Distributive property.

$= \mathbf{x}(x) + \mathbf{x}(2) + \mathbf{3}(x) + \mathbf{3}(2)$ Distributive property.

$= x^2 + 2x + 3x + 6$ Simplify.

$= x^2 + \mathbf{5}x + 6$ Combine like terms.

In the next example, each binomial is a difference of two terms.

EXAMPLE 13 Simplify: $(7x - 2)(4x - 8)$

SOLUTION ▶

$(7x - 2)(4x - 8)$

$= \mathbf{7x}(4x - 8) - \mathbf{2}(4x - 8)$ Distributive property.

$= \mathbf{7x}(4x) + \mathbf{7x}(-8) - \mathbf{2}(4x) - \mathbf{2}(-8)$ Distributive property.

$= 28x^2 - 56x - 8x + 16$ Simplify.

$= 28x^2 - \mathbf{64}x + 16$ Combine like terms.

Instead of thinking about the distributive property, some students use the phrase "**Firsts, Outers, Inners, Lasts** (**FOIL**)" to remember how to multiply binomials. However, this phrase is limited to binomials. It cannot be used to multiply a binomial and a trinomial. The distributive property is not limited in this way.

EXAMPLE 14 Simplify: $(3a^5 - 9)(2a^2 + a + 5)$

SOLUTION ▶

$(3a^5 - 9)(2a^2 + a + 5)$

$= \mathbf{3a^5}(2a^2 + a + 5) - \mathbf{9}(2a^2 + a + 5)$ Distributive property.

$= \mathbf{3a^5}(2a^2) + \mathbf{3a^5}(a) + \mathbf{3a^5}(5) - \mathbf{9}(2a^2) - \mathbf{9}(a) - \mathbf{9}(5)$ Distributive property.

$= 6a^{(5+2)} + 3a^{(5+1)} + 15a^5 - 18a^2 - 9a - 45$ Product rule of exponents.

$= 6a^7 + 3a^6 + 15a^5 - 18a^2 - 9a - 45$ Simplify.

In the next example, the product of a binomial and a trinomial is a binomial.

EXAMPLE 15 Simplify: $(y - 2)(y^2 + 2y + 4)$

SOLUTION ▶

$(y - 2)(y^2 + 2y + 4)$

$= \mathbf{y}(y^2 + 2y + 4) - \mathbf{2}(y^2 + 2y + 4)$ Distributive property.

$= \mathbf{y}(y^2) + \mathbf{y}(2y) + \mathbf{y}(4) - \mathbf{2}(y^2) - \mathbf{2}(2y) - \mathbf{2}(4)$ Distributive property.

$= y^3 + 2y^2 + 4y - 2y^2 - 4y - 8$ Simplify.

$= y^3 - 8$ Combine like terms.

Notice that the first term in the binomial, y^3, is a perfect cube and that the last term, 8, is also a perfect cube (2^3). This binomial is a *difference of cubes*.

Practice Problems

For problems 10–14, simplify.

10. $3k(7k^5)$ **11.** $(2y)(-3y + 8)$ **12.** $(2y + 7)(-3y + 8)$

13. $5x(x^2 + 8x - 9)$ **14.** $(6m + 2)(3m^2 + 9m - 4)$

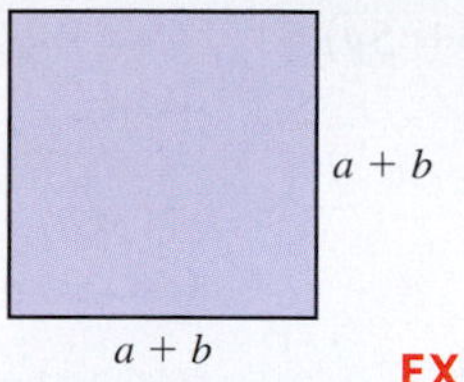

$a + b$

$a + b$

Figure 1

Patterns

When we multiply a binomial by itself, we are **squaring** it (Figure 1). The product is a **perfect square trinomial**.

EXAMPLE 16 | Simplify: $(a + b)^2$

SOLUTION ▶

$$(a + b)^2$$

$= (a + b)(a + b)$	Multiply a binomial by itself.
$= \mathbf{a}(a + b) + \mathbf{b}(a + b)$	Distributive property.
$= a^2 + ab + ba + b^2$	Distributive property.
$= a^2 + ab + \mathbf{ab} + b^2$	$ba = ab$; commutative property.
$= a^2 + \mathbf{2ab} + b^2$	Combine like terms; $1ab + 1ab = 2ab$

The result of Example 16 is a **perfect square trinomial pattern**: $(a + b)^2 = a^2 + 2ab + b^2$. To use a pattern to multiply a binomial by itself, first identify what a and b represent. For example, to use the pattern $(a + b)^2$ to simplify $(2x + 3)^2$, $a = 2x$ and $b = 3$.

EXAMPLE 17 | **(a)** Use the pattern $(a + b)^2 = a^2 + 2ab + b^2$ to simplify $(2x + 3)^2$.

SOLUTION ▶

$(2x + 3)^2$	$a = 2x, b = 3$
$= (2x)^2 + 2(2x)(3) + (3)^2$	$(a + b)^2 = a^2 + 2ab + b^2$
$= 4x^2 + 12x + 9$	Simplify.

(b) Use the distributive property to simplify $(2x + 3)^2$.

▶

$(2x + 3)^2$	
$= (2x + 3)(2x + 3)$	Rewrite the square as a product.
$= \mathbf{2x}(2x + 3) + \mathbf{3}(2x + 3)$	Distributive property.
$= 4x^2 + 6x + 6x + 9$	Distributive property.
$= 4x^2 + \mathbf{12x} + 9$	Combine like terms.

Whether we use a pattern or the distributive property, the product is the same.

Why do we use patterns to multiply polynomials? In some situations, it is quicker to use a pattern than to use the distributive property. Later in this chapter, we will reverse the patterns and use them in factoring polynomials. Multiplying polynomials with a pattern also provides more practice in recognizing a pattern and applying it. Standard form, slope-intercept form, and point-slope form of a linear equation are also examples of patterns. In Chapter 3, we will use patterns as we begin our work with functions.

EXAMPLE 18 | Use the distributive property to simplify $(a + b)(a - b)$.

SOLUTION ▶

$(a + b)(a - b)$	
$= \mathbf{a}(a - b) + \mathbf{b}(a - b)$	Distributive property.
$= a^2 - ab + ba - b^2$	Distributive property.
$= a^2 - ab + \mathbf{ab} - b^2$	$ba = ab$; commutative property.
$= a^2 - b^2$	Combine like terms.

Since $a^2 - b^2$ is a difference of squares, $(a + b)(a - b) = a^2 - b^2$ is the **difference of squares pattern**.

EXAMPLE 19 | Use the difference of squares pattern to simplify $(3w - 5y)(3w + 5y)$.

SOLUTION ▶

$$
\begin{aligned}
(3w - 5y)(3w + 5y) \qquad & a = 3w; b = 5y \\
= (3w)^2 - (5y)^2 \qquad & (a + b)(a - b) = a^2 - b^2 \\
= 9w^2 - 25y^2 \qquad & \text{Simplify.}
\end{aligned}
$$

In the next example, the product of a binomial and a trinomial is the **difference of cubes pattern**, $(a - b)(a^2 + ab + b^2) = a^3 - b^3$.

EXAMPLE 20 | Simplify: $(a - b)(a^2 + ab + b^2)$

SOLUTION ▶

$$
\begin{aligned}
(a - b)(a^2 + ab + b^2) & \\
= \boldsymbol{a}(a^2 + ab + b^2) - \boldsymbol{b}(a^2 + ab + b^2) \qquad & \text{Distributive property.} \\
= \boldsymbol{a}(a^2) + \boldsymbol{a}(ab) + \boldsymbol{a}(b^2) - \boldsymbol{b}(a^2) - \boldsymbol{b}(ab) - \boldsymbol{b}(b^2) \qquad & \text{Distributive property.} \\
= a^3 + a^2b + ab^2 - a^2b - ab^2 - b^3 \qquad & \text{Simplify.} \\
= a^3 - b^3 \qquad & \text{Simplify.}
\end{aligned}
$$

When polynomial factors match a pattern, we can either use a pattern or use the distributive property to simplify. The product is the same.

EXAMPLE 21 | **(a)** Use the difference of cubes pattern to simplify $(2y - 3z)(4y^2 + 6yz + 9z^2)$.

SOLUTION ▶

$$
\begin{aligned}
(2y - 3z)(4y^2 + 6yz + 9z^2) \qquad & a = 2y; b = 3z \\
= (2y - 3z)((2y)^2 + (2y)(3z) + (3z)^2) \qquad & (a - b)(a^2 + ab + b^2) = a^3 - b^3 \\
= (2y)^3 - (3z)^3 \qquad & \text{Follow the pattern: } a^3 - b^3. \\
= 2^3y^3 - 3^3z^3 \qquad & \text{Many bases rule for a product.} \\
= 8y^3 - 27z^3 \qquad & \text{Simplify.}
\end{aligned}
$$

(b) Use the distributive property to simplify $(2y - 3z)(4y^2 + 6yz + 9z^2)$.

▶

$$
\begin{aligned}
(2y - 3z)(4y^2 + 6yz + 9z^2) & \\
= \boldsymbol{2y}(4y^2 + 6yz + 9z^2) - \boldsymbol{3z}(4y^2 + 6yz + 9z^2) \qquad & \text{Distributive property.} \\
= \boldsymbol{2y}(4y^2) + \boldsymbol{2y}(6yz) + \boldsymbol{2y}(9z^2) - \boldsymbol{3z}(4y^2) - \boldsymbol{3z}(6yz) - \boldsymbol{3z}(9z^2) & \\
= 8y^3 + 12y^2z + 18yz^2 - 12y^2z - 18yz^2 - 27z^3 \qquad & \text{Simplify.} \\
= 8y^3 - 27z^3 \qquad & \text{Simplify.}
\end{aligned}
$$

Patterns for Multiplying Polynomials

$(a + b)(a + b) = a^2 + 2ab + b^2$	Perfect square trinomial pattern
$(a - b)(a - b) = a^2 - 2ab + b^2$	Perfect square trinomial pattern
$(a + b)(a - b) = a^2 - b^2$	Difference of squares pattern
$(a - b)(a^2 + ab + b^2) = a^3 - b^3$	Difference of cubes pattern
$(a + b)(a^2 - ab + b^2) = a^3 + b^3$	Sum of cubes pattern

Unlike a *difference* of squares, a *sum* of squares like $a^2 + b^2$ has no factors other than itself and 1. It is a prime polynomial.

Practice Problems

For problems 15–20,
(a) use a pattern to simplify.
(b) use the distributive property to simplify.

15. $(x + 8)(x + 8)$ **16.** $(x + 8)(x - 8)$ **17.** $(x - 8)(x - 8)$
18. $(3p + 5)(3p + 5)$ **19.** $(z - 5)(z^2 + 5z + 25)$
20. $(z + 5)(z^2 - 5z + 25)$

Writing Polynomial Expressions

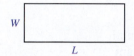

W

L

Figure 2

If W = width and L = length (Figure 2), the formula for the perimeter of a rectangle is $P = 2W + 2L$. The formula for the area of a rectangle is $A = LW$.

In the next example, we write polynomial expressions that represent the perimeter and area of a rectangle. When an expression includes a variable, we often say it is "in" that variable. So $W + 3$ is a polynomial expression "in W."

EXAMPLE 22   The length, L, of a rectangle is 3 ft more than its width, W.

(a) Write a polynomial expression in W that represents the length.

SOLUTION ▶ $L = W + 3$ The width plus 3.

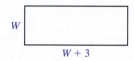

W

$W + 3$

(b) Write a polynomial expression in W that represents the perimeter.

▶
$P = 2W + 2L$ Formula for perimeter.
$P = 2W + 2(W + 3)$ Replace L with $W + 3$.
$P = 2W + 2(W) + 2(3)$ Distributive property.
$P = 4W + 6$ Combine like terms.

(c) Write a polynomial expression in W that represents the area.

▶
$A = LW$ Formula for area.
$A = (W + 3)W$ The area equals (length)(width).
$A = W^2 + 3W$ Distributive property.

Practice Problems

21. The width of a rectangle is W. The length, L, is 6 ft more than the width.
 a. Write a polynomial expression in W that represents the length of this rectangle.
 b. Write a polynomial expression in W that represents the perimeter of this rectangle.
 c. Draw a diagram of the rectangle.
 d. Write a polynomial expression in W that represents the area of this rectangle.

22. The length of a rectangle is L. The width, W, is 25 ft less than twice the length.
   a. Write a polynomial expression in L that represents the width of this rectangle.
   b. Write a polynomial expression in L that represents the perimeter of this rectangle.
   c. Draw a diagram of the rectangle.
   d. Write a polynomial expression in L that represents the area of this rectangle.

2.1 VOCABULARY PRACTICE

Match the term with its description.

1. $(a + b)(a - b) = a^2 - b^2$
2. $(a - b)(a^2 + ab + b^2) = a^3 - b^3$
3. $(a + b)(a^2 - ab + b^2) = a^3 + b^3$
4. $(a + b)(a + b) = a^2 + 2ab + b^2$
5. $3xy$ and $7xy$
6. A polynomial with two terms
7. The exponents in the terms decrease from left to right.
8. A polynomial with three terms
9. $a(b + c) = ab + ac$
10. In the expression $5x + 9$, 9 is an example of this.

A. binomial
B. constant
C. descending order
D. difference of cubes pattern
E. difference of squares pattern
F. distributive property
G. like terms
H. perfect square trinomial pattern
I. sum of cubes pattern
J. trinomial

2.1 Exercises

Follow your instructor's guidelines for showing your work.

For exercises 1–8,
(a) identify the degree of the polynomial expression.
(b) identify the type of polynomial (monomial, binomial, or trinomial).

1. $4x - 3$
2. $5x + 11$
3. $3x^3 + 4x^2 - 7$
4. $7x^3 - 6x + 2$
5. $-5x$
6. $9x$
7. -7
8. 3

9. The prefix *tri* means "three." Write an English word besides *trinomial* that includes the prefix *tri*. Explain the meaning of the word.

10. The prefix *mono* means "one." Write an English word besides *monomial* that includes the prefix *mono*. Explain the meaning of the word.

11. For the expression $a(b + c)$, assume that a, b, and c are real numbers. To evaluate this expression using the order of operations, what is done first? Explain why.

12. A beginning algebra student simplified $3x + 7y$ as $10xy$. Explain why this is not correct.

For exercises 13–20,
(a) is the expression a polynomial?
(b) If it is not, explain.

13. $\frac{3}{4}x^2 + \frac{x}{9}$
14. $\frac{2}{3}x^2 + \frac{x}{5}$
15. $\frac{5}{m + 2}$
16. $\frac{3}{7 - k}$
17. $x^4 + 0.3x - 5$
18. $0.5x^2 - 12.31$
19. $x^2 + x^{\frac{3}{4}}$
20. $x + x^{\frac{1}{2}}$

For exercises 21–64, simplify. Do not use a pattern.

21. $(5x + 15) + (7x - 2)$
22. $(8x + 13) + (9x - 6)$
23. $(5x + 15) - (7x - 2)$
24. $(8x + 13) - (9x - 6)$
25. $(5x + 15)(7x + 2)$

26. $(8x + 13)(9x + 6)$

27. $(5x + 15)(7x - 2)$

28. $(8x + 13)(9x - 6)$

29. $(q^2 - 5q + 15) + (7q - 2)$

30. $(p^2 - 8p + 13) + (9p - 6)$

31. $(q^2 - 5q + 15) - (7q - 2)$

32. $(p^2 - 8p + 13) - (9p - 6)$

33. $(3h^2 + 6h - 9) - (41h^2 - 5h + 20)$

34. $(7k^2 + 8k - 21) - (18k^2 - 11k + 60)$

35. $(3h^3 + 6h - 9) + (41h^2 - 5h + 20)$

36. $(7k^3 + 8k - 21) + (18k^2 - 11k + 60)$

37. $(8x^2 + 5xy + 2y^2) + (3x^2 - 7xy + 6y^2)$

38. $(5x^2 + 3xy + 4y^2) + (9x^2 - 8xy + 5y^2)$

39. $(2a^2 + 3ab - 8b^2) - (15a^2 - 10ab + 7b^2)$

40. $(3c^2 + 4cd - 9d^2) - (14c^2 - 7cd + 2d^2)$

41. $(a^2 - ab + b^2) - (a^2 + ab - b^2)$

42. $(c^2 - cd + d^2) - (c^2 + cd - d^2)$

43. $(x + 3)(x^2 - 4x + 9)$

44. $(x + 5)(x^2 - 7x + 4)$

45. $(x - 3)(x^2 + 4x + 9)$

46. $(x - 5)(x^2 + 7x + 4)$

47. $(x + 3)(x^2 + 4x + 9)$

48. $(x + 5)(x^2 + 7x + 4)$

49. $(c + 9)^2$

50. $(d + 7)^2$

51. $(c - 9)^2$

52. $(d - 7)^2$

53. $(9 - c)^2$

54. $(7 - d)^2$

55. $(c - 9)(c + 9)$

56. $(d - 7)(d + 7)$

57. $(5y + 3)(5y - 3)$

58. $(4w + 5)(4w - 5)$

59. $(x + y)(y - x)$

60. $(h + k)(k - h)$

61. $(x + y) - (x - y)$

62. $(h + k) - (h - k)$

63. $(x + y) - (y - x)$

64. $(h + k) - (k - h)$

For exercises 65–69, the equation is a pattern for multiplying polynomials. To show that this pattern is correct, simplify the left side of the equation.

65. $(a + b)(a + b) = a^2 + 2ab + b^2$

66. $(a - b)(a - b) = a^2 - 2ab + b^2$

67. $(a + b)(a - b) = a^2 - b^2$

68. $(a - b)(a^2 + ab + b^2) = a^3 - b^3$

69. $(a + b)(a^2 - ab + b^2) = a^3 + b^3$

70. Explain why *difference of cubes* is a good name for the pattern $(a - b)(a^2 + ab + b^2) = a^3 - b^3$.

For exercises 71–80, use a pattern to find the product. Use Examples 17 and 21 as models of how to show your work.

71. $(3x + y)^2$

72. $(5x + y)^2$

73. $(3x - y)^2$

74. $(5x - y)^2$

75. $(x - 6y)(x^2 + 6xy + 36y^2)$

76. $(x - 3y)(x^2 + 3xy + 9y^2)$

77. $(4x - 5y)^2$

78. $(2x - 9y)^2$

79. $(2h + 3k)(4h^2 - 6hk + 9k^2)$

80. $(5c + 2d)(25c^2 - 10cd + 4d^2)$

81. The width of a rectangle is W. The length is 8 ft more than the width.
   a. Write a polynomial expression in W that represents the length of this rectangle.
   b. Draw a diagram of the rectangle.
   c. Write a polynomial expression in W that represents the perimeter of this rectangle.
   d. Write a polynomial expression in W that represents the area of this rectangle.

82. The length of a rectangle is L. The width is 3 m less than the length.
   a. Write a polynomial expression in L that represents the width of this rectangle.
   b. Draw a diagram of the rectangle.
   c. Write a polynomial expression in L that represents the perimeter of this rectangle.
   d. Write a polynomial expression in L that represents the area of this rectangle.

83. The width of a rectangle is W. The length of the rectangle is equal to twice the width plus 7 m.
   a. Write a polynomial expression in W that represents the length of this rectangle.
   b. Draw a diagram of the rectangle.
   c. Write a polynomial expression in W that represents the perimeter of this rectangle.
   d. Write a polynomial expression in W that represents the area of this rectangle.

84. The width of a rectangle is W. The length of the rectangle is equal to twice the width minus 2 cm.
   a. Write a polynomial expression in W that represents the length of this rectangle.
   b. Draw a diagram of the rectangle.
   c. Write a polynomial expression in W that represents the perimeter of this rectangle.
   d. Write a polynomial expression in W that represents the area of this rectangle.

85. A photographer making enlargements of a rectangular photograph wants to keep the relative width and length of the photograph the same in each enlargement. The width of the photograph is W. The length of the photograph is always five-fourths of the width.
  a. Write a polynomial expression in W that represents the length of each photograph.
  b. Draw a diagram of the photograph.
  c. Write a polynomial expression in W that represents the perimeter of this photograph.
  d. Write a polynomial expression in W that represents the area of this photograph.

86. A quilter is making enlargements of a machine-quilting pattern for rectangular areas. She wants to keep the relative width and length of the pattern the same in each enlargement. The length of the quilt pattern is L. The width of the pattern is one-fourth of the length.
  a. Write a polynomial expression in L that represents the width of the quilt pattern.
  b. Draw a diagram of the quilt pattern.
  c. Write a polynomial expression in L that represents the perimeter of this pattern.
  d. Write a polynomial expression that represents the area of this pattern.

L

Problem Solving: Practice and Review

Follow your instructor's guidelines for using the five steps as outlined in Section 1.5, p. 51.

87. Find Dell's sales to large enterprises a year ago. Round to the nearest tenth of a billion.

Dell's sales to large enterprises totaled $3.4 billion for the quarter, down 31 percent from a year ago. . . . Consumer sales totaled $2.8 billion, down 16 percent. (*Source:* www.statesman.com, May 29, 2009)

88. Nearly 5 million people a year in the United States receive a blood transfusion. Jim Murray, age 69, of Alaska has donated 21 gal of blood. The volume of a blood donation is usually 1 pt. Find the number of 1-pt donations made by Jim Murray. (1 gal = 4 qt; 1 qt = 2 pt.) (*Sources:* www.nlm.nih.gov; www.usatoday.com, January 15, 2008)

89. The diagram shows a design for a livestock manure lagoon. The volume of this lagoon is equal to the product of the maximum depth, the average width, and the average length. The maximum depth of a manure lagoon is 12 ft. The surface is a square. The length of the surface is 192 ft, and its width is 192 ft. The bottom is also a square. The length of the bottom is 144 ft, and its width is 144 ft. Find the volume of the lagoon. Round to the nearest thousand. (*Source:* Hermanson, Ronald E., 1991.

"Livestock Manure Lagoons Protect Water Quality," Washington State University)

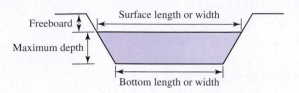

90. Find the number of Florida foster children. Round to the nearest hundred.

In all, 2669 children—or 13 percent of Florida foster children—are being given powerful psychiatric drugs, said the study. (*Source:* www.miamiherald.com, May 29, 2009)

Find the Mistake

For exercises 91–94, the completed problem has one mistake.
(a) Describe the mistake in words, or copy down the whole problem and highlight or circle the mistake.
(b) Do the problem correctly.

91. Problem: Simplify: $-4x(5x^3 - 2x^2)$
  Incorrect Answer: $-4x(5x^3 - 2x^2)$
$$= -4x(3x)$$
$$= -12x^2$$

92. Problem: Simplify: $(6x^2 + 4x - 1) - (2x^2 - 9x + 8)$
  Incorrect Answer: $(6x^2 + 4x - 1) - (2x^2 - 9x + 8)$
$$= 6x^2 + 4x - 1 - 2x^2 - 9x - 8$$
$$= 4x^2 - 5x - 9$$

93. Problem: Simplify: $(5x)(8x)$
  Incorrect Answer: $(5x)(8x)$
$$= 13x^2$$

94. Problem: Simplify: $(3x - 4)^2$
  Incorrect Answer: $(3x - 4)^2$
$$= (3x - 4)(3x - 4)$$
$$= 3x(3x) - 4(-4)$$
$$= 9x^2 + 16$$

Review

95. Use roster notation to write the set of whole numbers.

96. Use roster notation to write the set of integers.

97. Explain the difference between a rational number and an irrational number.

98. a. Are the rational numbers a subset of the irrational numbers?
  b. Explain.

SUCCESS IN COLLEGE MATHEMATICS

99. Copy the schedule on the next page and identify when you plan to be in class (C), studying (S), working (W), commuting (T), sleeping (R), and personal/family activities (P). In each box, either write the indicated letter or use a different color of highlighter for each activity.

100. Now circle the times when you plan to be in math class, working on-line for your math class, or studying for math. How many hours do you plan to spend on math per week?

Table for exercises 99–100

Time	Sun.	Mon.	Tues.	Wed.	Thurs.	Fri.	Sat.
12 midnight–1 a.m.							
1 a.m.–2 a.m.							
2 a.m.–3 a.m.							
3 a.m.–4 a.m.							
4 a.m.–5 a.m.							
5 a.m.–6 a.m.							
6 a.m.–7 a.m.							
7 a.m.–8 a.m.							
8 a.m.–9 a.m.							
9 a.m.–10 a.m.							
10 a.m.–11 a.m.							
11 a.m.–12 noon							
12 noon–1 p.m.							
1 p.m.–2 p.m.							
2 p.m.–3 p.m.							
3 p.m.–4 p.m.							
4 p.m.–5 p.m.							
5 p.m.–6 p.m.							
6 p.m.–7 p.m.							
7 p.m.–8 p.m.							
8 p.m.–9 p.m.							
9 p.m.–10 p.m.							
10 p.m.–11 p.m.							
11 p.m.–12 midnight							

© Dmitriy Shironosov/Shutterstock

The "undo" command in a word processing program lets us reverse something we have done. The document returns to its original appearance. In Section 2.1, we multiplied factors together. The product was a polynomial. In this section, we will "undo" this multiplication and rewrite the polynomial as a product of factors. We call this process **factoring**.

SECTION 2.2

Factoring Polynomials: Greatest Common Factor and Grouping

After reading the text, working the practice problems, and completing assigned exercises, you should be able to:

1. Find the greatest common factor of two or more terms.

2. Use the greatest common factor to factor a polynomial.

3. Use grouping to factor a polynomial.

4. Use a pattern to factor a polynomial.

5. Use more than one method of factoring to factor a polynomial completely.

Greatest Common Factor

A **prime number** is a whole number greater than 1 whose only factors are itself and 1. The **prime factorization** of a number is a product of prime factors. Use exponential notation for repeated factors.

EXAMPLE 1   **(a)** Write the prime factorization of 12.

SOLUTION ▶
$$12$$
$$= 2 \cdot 6 \qquad \text{Choose two factors of 12.}$$
$$= 2 \cdot 2 \cdot 3 \qquad \text{Factor again; all factors are prime.}$$
$$= \mathbf{2^2} \cdot 3 \qquad \text{Rewrite } 2 \cdot 2 \text{ as an exponential expression.}$$

(b) Write the prime factorization of 60.

▶
$$60$$
$$= 6 \cdot 10 \qquad \text{Choose two factors of 60.}$$
$$= 2 \cdot 3 \cdot 2 \cdot 5 \qquad \text{Factor again; all factors are prime.}$$
$$= \mathbf{2^2} \cdot 3 \cdot 5 \qquad \text{Rewrite } 2 \cdot 2 \text{ as an exponential expression.}$$

We can also write a prime factorization for a term that includes variables.

EXAMPLE 2   Write the prime factorization of $24xy^2$.

SOLUTION ▶
$$24xy^2$$
$$= 4 \cdot 6 \cdot x \cdot y^2 \qquad \text{Choose two factors of 24.}$$
$$= 2 \cdot 2 \cdot 2 \cdot 3 \cdot x \cdot y^2 \qquad \text{Factor again; all factors are prime.}$$
$$= \mathbf{2^3} \cdot 3 \cdot x \cdot y^2 \qquad \text{Rewrite } 2 \cdot 2 \cdot 2 \text{ as an exponential expression.}$$

To factor a polynomial, we may need to find the **greatest common factor** (GCF) of each term in the polynomial. The greatest common factor of a polynomial is the monomial of highest degree and largest coefficient that is a factor of each term.

One way to find the greatest common factor is to complete a prime factorization of each term in the polynomial. The greatest common factor is the product of the shared factors and variables in the factorizations. The exponent on each factor is the *smallest* exponent on this factor in all of the factorizations.

EXAMPLE 3   Use prime factorization to find the greatest common factor of $24xy^2z$ and $60x^3y^2$.

SOLUTION ▶ Prime factorization of $24xy^2z$

$$= 2^3 \cdot 3^1 \cdot x^1 \cdot y^2 \cdot z$$

Prime factorization of $60x^3y^2$

$$= 2^2 \cdot 3^1 \cdot 5^1 \cdot x^3 \cdot y^2$$

Greatest common factor of $24xy^2z$ and $60x^3y^2$

$$= 2^2 \cdot 3^1 \cdot x^1 \cdot y^2 \qquad \text{Shared factors with } smallest \text{ exponents.}$$
$$= 12xy^2 \qquad \text{Simplify. The greatest common factor is } 12xy^2.$$

> **Using Prime Factorization to Find the Greatest Common Factor of a Polynomial Expression**
> 1. Write the prime factorization of each term.
> 2. Identify shared prime factors and variables and the smallest exponent on these factors or variables.
> 3. The greatest common factor is the product of the shared prime factors and variables with these exponents.

We can also use factor lists to find the greatest common factor of a polynomial.

EXAMPLE 4 | Use factor lists to find the greatest common factor of $24xy^2$ and $60x^3y^2$.

SOLUTION ▶ Factors of 24: 1, 2, 3, 4, 6, 8, 12, 24

Factors of 60: 1, 2, 3, 4, 5, 6, 10, 12, 15, 20, 30, 60

What is the largest number that is a factor of 24 and 60? 12

What is the smallest exponent on x in the two terms? 1

What is the smallest exponent on y in the two terms? 2

The greatest common factor is $12xy^2$.

> **Practice Problems**
>
> For problems 1–6, use prime factorization or factor lists to find the greatest common factor.
>
> **1.** $28xy^2$; $98x^4y$ **2.** $28x^2y^2$; $98x^3y^2$ **3.** $28xz^2$; $98x^4y^2$
>
> **4.** $90a^2c^3$; $108d^2f$ **5.** $30x^2y$; $12x^2yz$; $75xy^2z$ **6.** $17de$; $31ab$; $11xy$

Factoring and the Greatest Common Factor

When we rewrite a polynomial as a product of factors, we are "undoing" a multiplication. For example, we can rewrite some polynomials as the product of the greatest common factor and a new polynomial. If the only factors of a polynomial are 1 and itself, the polynomial is **prime**. The polynomial $3x - 5z$ is prime.

EXAMPLE 5 | **(a)** Factor: $6x^2y + 15xy^2$

SOLUTION ▶   Find the greatest common factor of $6x^2y$ and $15xy^2$.

$6x^2y$ $15xy^2$
$= 2 \cdot 3 \cdot x^2 \cdot y^1$ Prime factorizations. $= 3 \cdot 5 \cdot x^1 \cdot y^2$

The greatest common factor is $3xy$.

$6x^2y + 15xy^2$

$= \mathbf{3xy}(\underline{} + \underline{})$ Rewrite as a product of the **GCF** and a binomial.

$= 3xy(\mathbf{2x} + \underline{})$ $(3xy)(2x) = 6x^2y$

$= 3xy(2x + \mathbf{5y})$ $(3xy)(5y) = 15xy^2$

(b) Check.

▶ $3xy(2x + 5y)$ Check by multiplying the factors.

$= \mathbf{3xy}(2x) + \mathbf{3xy}(5y)$ Distributive property.

$= 6x^2y + 15xy^2$ Simplify.

Since the product of the factors is the original polynomial, the factoring is correct.

We combine some of these steps in the next example.

EXAMPLE 6 | Factor: $36x + 28y - 8z$

SOLUTION ▶

$36x + 28y - 8z$ The greatest common factor is 4.

$= 4(\underline{\quad} + \underline{\quad} - \underline{\quad})$ Factor out the greatest common factor, 4.

$= 4(9x + 7y - 2z)$ $4(9x) = 36x$; $4(7y) = 28y$; $4(-2z) = -8z$

> **Factoring Out the Greatest Common Factor**
>
> 1. Identify the greatest common factor.
> 2. Rewrite the polynomial as the product of the greatest common factor and a new polynomial. This is "factoring out the greatest common factor."

In Examples 5 and 6, the greatest common factor is positive. As we will see in Example 9, it is sometimes helpful to instead choose a greatest common factor that is negative.

EXAMPLE 7 | **(a)** Factor $-3ac - bc$. Choose a positive greatest common factor.

SOLUTION ▶

$-3ac - bc$

$= c(-3a - b)$ Factor out a positive greatest common factor, c.

(b) Factor $-3ac - bc$. Choose a negative greatest common factor.

▶

$-3ac - bc$

$= -c(3a + b)$ Factor out a negative greatest common factor, $-c$.

> **Practice Problems**
>
> For problems 7–10,
> **(a)** factor.
> **(b)** check.
>
> **7.** $12ab^4 + 18b^3$ **8.** $6a^2c^5 + 18ac$
> **9.** $14m^2p^3 + 3mp^4 + 11m^3p^2$ **10.** $20xy + 21z$

Factor by Grouping

The terms in $3x + 3z + yx + yz$ do not have a common factor. However, the terms $3x$ and $3z$ have a common factor, 3, and the terms yx and yz have a common factor, y. If we use parentheses to group these terms and factor out the common factor from each group, another common factor, a binomial, appears. To complete the factoring, we factor out this binomial. This is called **factoring by grouping**.

EXAMPLE 8 | Factor: $3x + 3z + yx + yz$

SOLUTION ▶

$3x + 3z + yx + yz$ These four terms have no common factor.

$= (3x + 3z) + (yx + yz)$ Group terms that have a common factor.

$= 3(\underline{\quad} + \underline{\quad}) + y(\underline{\quad} + \underline{\quad})$ Identify a common factor of each group.

$= 3(x + z) + y(x + z)$ Another common factor appears: $x + z$.

$= (x + z)(\underline{\quad} + \underline{\quad})$ Factor out the common factor: $x + z$.

$= (x + z)(3 + y)$ Complete the factoring.

We cannot group terms that follow subtraction. Rewrite the subtraction as addition of a negative number: $a + b - c + d$ is equal to $a + b + -c + d$.

EXAMPLE 9 Factor: $6a^2 + 2ab - 3ac - bc$

SOLUTION ▶

$6a^2 + 2ab - 3ac - bc$	These four terms have no common factor.
$= 6a^2 + 2ab + -3ac - bc$	Rewrite: $2ab - 3ac = 2ab + -3ac$
$= (6a^2 + 2ab) + (-3ac - bc)$	Group terms that have a common factor.
$= 2a(\underline{} + \underline{}) + (-c)(\underline{} + \underline{})$	Identify a common factor of each group.
$= 2a(3a + b) + (-c)(3a + b)$	Another common factor appears: $3a + b$.
$= (3a + b)(\underline{} + \underline{})$	Factor out the common factor: $3a + b$.
$= (3a + b)(2a + (-c))$	Complete the factoring.
$= (3a + b)(2a - c)$	Rewrite: $2a + (-c) = 2a - c$

Factoring by Grouping

1. If the polynomial includes subtraction between the middle terms, rewrite subtraction as addition of a negative number.

2. Group the first two terms and the second two terms.

3. Factor the greatest common factor from each group. This factor may be 1.

4. If the groups now have a common binomial factor, factor again.

5. If the groups do not have a common binomial factor, change the order of the terms and repeat the process.

6. To check, use the distributive property to multiply the factors. If the factoring is correct, the product is the original polynomial.

When factoring by grouping, the terms in one of the groups may have only a common factor of 1. Factor out the 1 from this group, and factor out the common factor from the other group.

EXAMPLE 10 Factor: $7p - 6h + 14px^2 - 12hx^2$

SOLUTION ▶

$7p - 6h + 14px^2 - 12hx^2$	These terms have no common factor.
$= (7p - 6h) + (14px^2 - 12hx^2)$	Group terms that have a common factor.
$= 1(\underline{} + \underline{}) + 2x^2(\underline{} + \underline{})$	Identify a common factor of each group.
$= 1(7p - 6h) + 2x^2(7p - 6h)$	Another common factor appears: $7p - 6h$.
$= (7p - 6h)(\underline{} + \underline{})$	Factor out the common factor: $7p - 6h$.
$= (7p - 6h)(1 + 2x^2)$	Complete the factoring.

In the next example, we rewrite $-a$ as $-1a$ and choose to factor out -1 from the second group of terms.

EXAMPLE 11 Factor: $2ac - 2bc - a + b$

SOLUTION ▶

$2ac - 2bc - a + b$	These four terms have no common factor.
$= 2ac - 2bc + -1a + b$	Rewrite as addition of a negative number; $-a = -1a$
$= (2ac - 2bc) + (-1a + b)$	Group terms that have a common factor.
$= 2c(\_\_ - \_\_) + (-1)(\_\_ - \_\_)$	Identify a common factor of each group.
$= 2c(a - b) + (-1)(a - b)$	Another common factor appears: $a - b$.
$= (a - b)(2c + (-1))$	Factor out $a - b$; complete the factoring.
$= (a - b)(2c - 1)$	Rewrite: $2c + (-1) = 2c - 1$

If we had instead chosen to factor 1 from the second group of terms, the expression would have been $2c(a - b) + 1(-a + b)$. Since $a - b$ and $-a + b$ are not the same, we could not have completed the factoring.

In the next example, changing the sign of the common factors results in a different answer. However, the answers are equivalent, and both are correct.

EXAMPLE 12 Factor: $16wz - 40kz - 2w + 5k$

SOLUTION ▶

$16wz - 40kz - 2w + 5k$	These four terms have no common factor.
$= 16wz - 40kz + -2w + 5k$	Rewrite: $-40kz - 2w = -40kz + -2w$
$= (16wz - 40kz) + (-2w + 5k)$	Group terms that have a common factor.
$= 8z(\_\_ - \_\_) + (-1)(\_\_ - \_\_)$	Identify a common factor of each group.
$= 8z(2w - 5k) + (-1)(2w - 5k)$	Another common factor appears: $2w - 5k$.
$= (2w - 5k)(8z + (-1))$	Factor out $2w - 5k$; complete the factoring.
$= (2w - 5k)(8z - 1)$	Rewrite: $8z + (-1) = 8z - 1$

We factored $8z$ from the first group of terms and -1 from the second group of terms. What happens if we instead factor $-8z$ from the first group and 1 from the second group?

▶

$16wz - 40kz - 2w + 5k$	
$= (16wz - 40kz) + (-2w + 5k)$	Group terms that have a common factor.
$= -8z(\_\_ + \_\_) + 1(\_\_ + \_\_)$	Identify a common factor of each group.
$= -8z(-2w + 5k) + 1(-2w + 5k)$	Another common factor appears: $-2w + 5k$.
$= (-2w + 5k)(-8z + 1)$	Factor out $-2w + 5k$; complete the factoring.

The factors of the polynomial are $(2w - 5k)(8z - 1)$ or $(-2w + 5k)(-8z + 1)$. The signs on the factors are opposites. When we multiply these factors, the products are the same. Both answers are correct.

When checking your factoring with the answers provided in the back of the book, remember that the signs on the terms in your factoring may be opposite to the signs in the given answer.

Practice Problems

For problems 11–14, factor.

11. $15x^2 - 27x + 5xy - 9y$ **12.** $12hk - 10h - 6k + 5$

13. $7ac^2 + 7bc^2 - 2ad - 2bd$ **14.** $2fw - f + 2w - 1$

Patterns

In Section 2.1, we used patterns to find the product of polynomials. If we reverse the order of these patterns, we can use them to factor some polynomials.

Patterns for Factoring Polynomials

$a^2 + 2ab + b^2 = (a + b)(a + b)$	Perfect square trinomial pattern
$a^2 - 2ab + b^2 = (a - b)(a - b)$	Perfect square trinomial pattern
$a^2 - b^2 = (a + b)(a - b)$	Difference of squares pattern
$a^3 - b^3 = (a - b)(a^2 + ab + b^2)$	Difference of cubes pattern
$a^3 + b^3 = (a + b)(a^2 - ab + b^2)$	Sum of cubes pattern

In the next example, the polynomial is a difference of squares, $a^2 - b^2$. To factor, rewrite the polynomial to match the pattern. Then use the pattern to find the factors.

EXAMPLE 13 Use a pattern to factor $x^2 - 16y^2$.

SOLUTION ▶

$$x^2 - 16y^2 \qquad \text{A difference of squares.}$$
$$= (x)^2 - (4y)^2 \qquad \text{Match the pattern; } a = x; b = 4y$$
$$= (x + 4y)(x - 4y) \qquad \text{Write the factors: } (a + b)(a - b).$$

When two factors are the same, we can use exponential notation. For example, $(4h - 3k)(4h - 3k) = (4h - 3k)^2$.

EXAMPLE 14 Use a pattern to factor $16h^2 - 24hk + 9k^2$.

SOLUTION ▶

$$16h^2 - 24hk + 9k^2 \qquad \text{The first and last terms are perfect squares.}$$
$$= (4h)^2 - 2(4h)(3k) + (3k)^2 \qquad \text{Match the pattern; } a = 4h; b = 3k$$
$$= (4h - 3k)(4h - 3k) \qquad \text{Write the factors: } (a - b)(a - b).$$
$$= (4h - 3k)^2 \qquad \text{Use exponential notation.}$$

When factoring with a pattern, the polynomial must match the pattern exactly.

EXAMPLE 15 **(a)** Use a pattern to factor $16x^2 - 3y^2$.

SOLUTION ▶

Although this polynomial is a difference, it is not a difference of squares. The first term, $16x^2$, is a perfect square because $4x \cdot 4x = 16x^2$. However, $3y^2$ is not a perfect square. This polynomial has no common factors; it is prime.

(b) Use a pattern to factor $25n^2 + 49p^2$.

▶

Although the terms are both perfect squares, this polynomial is a *sum* of squares, not a *difference* of squares. It does not match the pattern. This polynomial has no common factors; it is prime.

Using a Pattern to Factor

1. Choose a pattern that is similar to the polynomial.
2. Rewrite the polynomial to match the pattern.
3. Use the pattern to identify the factors.

To use the difference or sum of perfect cubes pattern, write a list of perfect cubes such as $2^3 = 8$, $3^3 = 27$, $4^3 = 64$, $5^3 = 125$, and $6^3 = 216$.

EXAMPLE 16 **(a)** Use a pattern to factor $27p^3 + 8w^3$.

SOLUTION ▶

$27p^3 + 8w^3$	Perfect cubes: $3^3 = 27$ and $2^3 = 8$
$= (3p)^3 + (2w)^3$	Match the pattern; $a = 3p$; $b = 2w$
$= (3p + 2w)((3p)^2 - (3p)(2w) + (2w)^2)$	Write the factors: $(a + b)(a^2 - ab + b^2)$.
$= (3p + 2w)(9p^2 - 6pw + 4w^2)$	Simplify.

(b) Check.

$(3p + 2w)(9p^2 - 6pw + 4w^2)$	
$= 3p(9p^2 - 6pw + 4w^2) + 2w(9p^2 - 6pw + 4w^2)$	Distributive property.
$= 27p^3 - 18p^2w + 12pw^2 + 18p^2w - 12pw^2 + 8w^3$	Distributive property.
$= 27p^3 + 8w^3$	Simplify.

Practice Problems

For problems 15–19, use a pattern to factor.

15. $x^2 - y^2$　　　**16.** $25x^2 - 49y^2$　　　**17.** $64x^2 - 16xy + y^2$

18. $8x^3 - y^3$　　　**19.** $36a^2 + 12a + 1$

Factor Completely

To factor a polynomial as much as possible, we may need to use more than one factoring strategy. When each factor is prime, the polynomial is **factored completely**.

Methods for Factoring Polynomial Expressions

1. Factor out the greatest common factor

2. Factor by grouping

3. Use patterns to factor

When factoring completely, first look for a greatest common factor of all the terms. Then use another method to try to factor the remaining polynomial. Since the remaining polynomial in the next example is a difference of two squares, we use the difference of two squares pattern.

EXAMPLE 17 Factor $6x^2p - 6py^2$ completely.

SOLUTION ▶

$6x^2p - 6py^2$	The greatest common factor is $6p$.
$= 6p(x^2 - y^2)$	Factor out the greatest common factor, $6p$.
$= 6p(x^2 - y^2)$	Difference of two squares. Match the pattern; $a = x$, $b = y$
$= 6p(x + y)(x - y)$	Write the factors: $(a + b)(a - b)$.

A variable term that is a perfect square has an exponent that is divisible by 2. For example, y^{10} is a perfect square; $10 \div 2 = 5$. Using the power rule of exponents, we can rewrite y^{10} as $(y^5)^2$.

EXAMPLE 18 | Factor $75k^2x - 48xy^{10}$ completely.

SOLUTION ▶

$$75k^2x - 48xy^{10}$$ The greatest common factor is $3x$.

$$= \mathbf{3x}(25k^2 - 16y^{10})$$ Factor out the greatest common factor, $3x$.

$$= 3x((5k)^2 - (4y^5)^2)$$ Difference of two squares. Match the pattern; $a = 5k$; $b = 4y^5$

$$= 3x(5k + 4y^5)(5k - 4y^5)$$ Write the factors: $(a + b)(a - b)$.

If the remaining polynomial has four terms, factor by grouping.

EXAMPLE 19 | Factor $90a^2 - 18a + 45ah - 9h$ completely.

SOLUTION ▶

$$90a^2 - 18a + 45ah - 9h$$ The greatest common factor is 9.

$$= \mathbf{9}(10a^2 - 2a + 5ah - h)$$ Factor out the greatest common factor, 9.

$$= 9[(10a^2 - 2a) + (5ah - h)]$$ Group; use brackets and parentheses.

$$= 9[2a(5a - 1) + h(5a - 1)]$$ Factor each group.

$$= 9[(5a - 1)(2a + h)]$$ Factor again; the common factor is $5a - 1$.

$$= 9(5a - 1)(2a + h)$$ Remove the brackets to complete the factoring.

Do not expect to always use two or more factoring strategies. In the next example, the remaining polynomial is prime.

EXAMPLE 20 | Factor $50x^2 + 72y^2$ completely.

SOLUTION ▶

$$50x^2 + 72y^2$$ The greatest common factor is 2.

$$= \mathbf{2}(25x^2 + 36y^2)$$ Factor out the greatest common factor, 2.

Although $25x^2$ and $36y^2$ are perfect squares, $25x^2 + 36y^2$ is a *sum* of squares, not a *difference* of squares. The polynomial $25x^2 + 36y^2$ is prime.

Practice Problems

For problems 20–23, factor completely. If the only factors of a polynomial are itself and 1, identify it as prime.

20. $6x^2 + 60xy + 150y^2$ **21.** $16m^2 + 25n^2$

22. $2x^3 - 128y^3$ **23.** $2x^2 + 12y + 18z$

2.2 VOCABULARY PRACTICE

Match the term with its description.

1. The result of multiplication
2. The only factors of this polynomial are itself and 1.
3. The monomial of highest degree and largest coefficient that is a factor of each term
4. A polynomial with three terms
5. This operation can be rewritten as addition of a negative number.
6. $a^3 - b^3 = (a - b)(a^2 + ab + b^2)$
7. $x^2 + 2xy + y^2 = (x + y)(x + y)$
8. $p^3 + q^3 = (p + q)(p^2 - pq + q^2)$
9. $m^2 - n^2 = (m + n)(m - n)$
10. $a \cdot b = b \cdot a$

A. commutative property
B. difference of cubes pattern
C. difference of squares pattern
D. greatest common factor of two or more terms
E. perfect square trinomial pattern
F. prime polynomial
G. product
H. subtraction
I. sum of cubes pattern
J. trinomial

2.2 Exercises

Follow your instructor's guidelines for showing your work.

For exercises 1–4, write the prime factorization of the term.

1. $18x^2y$

2. $12xy^2$

3. $294ab^3$

4. $150a^3b$

For exercises 5–16, use prime factorization or factor lists to find the greatest common factor of the terms.

5. $120x^2y$; $180x^2$

6. $252e^2f$; $268e^2$

7. $42abc$; $70ade$

8. $30wxy$; $105wz^2$

9. $1078ab^2c$; $1365ac$

10. $550ab^2c$; $525a^2bc$

11. $90x^2yz$; $50xy^2z$; $40x^2y^3z^2$

12. $280h^2kp^3$; $105hk^2p$; $315hk^3p^2$

13. $34k^2z$; $33k^3z^4$

14. $22x^2y$; $39x^3y^5$

15. $72xy$; -72

16. $36bc$; -36

For exercises 17–28, factor out the greatest common factor.

17. $3x^2 + 15x - 27$

18. $2y^2 + 12y - 16$

19. $3x^3 + 15x^2 - 27x$

20. $2y^3 + 12y^2 - 16y$

21. $10a^2b^2 - 45a^2c$

22. $14d^3f^2 - 77d^3g$

23. $12fg + 4fgh$

24. $20ab + 5abc$

25. $7x^4 - 21x^2 + 14x^5 - x$

26. $5z^3 - 35z^2 + 60z^5 - z$

27. $-7x^4 + 21x^2 - 14x^5 + x$

28. $-5z^3 + 35z^2 - 60z^5 + z$

For exercises 29–44, factor by grouping. If the only factors of a polynomial are itself and 1, identify the polynomial as prime.

29. $6ax + 10x + 3ay + 5y$

30. $6cw + 21w + 2cz + 7z$

31. $6ax + 3ay + 10x + 5y$

32. $6cw + 2cz + 21w + 7z$

33. $6ax - 10x + 3ay - 5y$

34. $6cw - 21w + 2cz - 7z$

35. $27xy - 3x + 45y - 5$

36. $42xy - 7x + 12y - 2$

37. $110pq - 20p - 99q + 18$

38. $99ab - 55a - 90b + 50$

39. $3x^2 + 3xz + xy + yz$

40. $5d^2 + 5dw + dx + wx$

41. $3x^2 + 3xz - xy - yz$

42. $5d^2 + 5dw - dx - wx$

43. $-21a^2 + 7aq - 3ap + pq$

44. $-63af + 9fh - 7ag + gh$

For exercises 45–66, use a pattern to factor.

45. $b^2 - c^2$

46. $z^2 - a^2$

47. $25y^2 - 81$

48. $16w^2 - 49$

49. $25y^2 - 81z^{12}$

50. $16w^2 - 49b^{10}$

51. $z^2 + 2wz + w^2$

52. $f^2 + 2fg + g^2$

53. $9z^2 + 30z + 25$

54. $36f^2 + 84f + 49$

55. $125x^3 + y^3$

56. $216c^3 - d^3$

57. $a^3 - 27$

58. $x^3 + 8$

59. $8a^3 - 27$

60. $27x^3 + 8$

61. $8a^3 + 27$

62. $27x^3 - 8$

63. $64p^3 + w^3$

64. $125u^3 - z^3$

65. $216b^3 - 125c^3$

66. $64c^3 + 343d^3$

For exercises 67–82, more than one of the factoring methods may need to be used. Factor completely.

67. $3x^2 - 12y^2$

68. $5a^2 - 500b^2$

69. $18hx - 30h + 9wx - 15w$

70. $40xy - 70x + 20yz - 35z$

71. $5x^2 + 10xy + 5y^2$

72. $7h^2 + 14hk + 7k^2$

73. $10x^3 - 10y^3$

74. $18a^3 + 18b^3$

75. $4a^2 + 32b^2$

76. $5b^2 + 30c^2$

77. $3xy + 12bx - 3fx + 9$

78. $9de^2 + 27de - 81d^2e^3 + 18$

79. $28x^2 + 28xy + 7y^2$

80. $36x^2 + 24xy + 4y^2$

81. $36a^2 - 4$

82. $20a^2 - 5$

83. a. Factor $x^3 + 2x + 3x^2 + 6$ by grouping.

   b. We can change the order of this polynomial to $x^3 + 6 + 2x + 3x^2$. Show that the polynomial in this arrangement cannot be factored by grouping.

   c. Think about the answers to parts a and b. Can the order of terms in a polynomial have an effect on whether the polynomial can be factored by grouping?

84. Show that $a^2 + b^2 = (a + b)(a + b)$ is not true.

85. The radius of the small inner circle in the diagram is n. The radius of the large circle is N. To find the area of a circle, use the formula area $= \pi r^2$, where r is the radius of the circle.

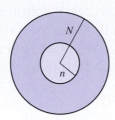

   a. Write an expression in n for the area of the inner circle.

   b. Write an expression in N for the area of the outer circle.

   c. Write an expression for the area of the doughnut-shaped region.

   d. Factor the expression in part c completely.

86. The smaller inner rectangle has a width of y and a length of $2y$. The large rectangle has a width of $2x$ and a length of $4x$.

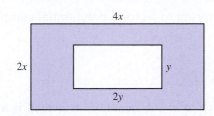

   a. Write an expression in y for the area of the inner rectangle.

   b. Write an expression in x for the area of the outer rectangle.

   c. Write an expression for the area of the shaded region.

   d. Factor the expression in part c completely.

Problem Solving: Practice and Review

Follow your instructor's guidelines for using the five steps as outlined in Section 1.5, p. 51.

87. Find the average cost per mile for the pipeline. Write the answer in place value notation. Round to the nearest ten thousand.

And then there is the case of Shell Lake in Washburn County, where chronic flooding for property owners came to a head in 2003. Officials spent $1.5 million to build a 4 1/2-mile pipeline to the Yellow River in 2004 to serve as a release valve for chronic flooding. (*Source:* www.jsonline .com, May 23, 2009)

88. Find the number of consumers who fell victim to a phishing attack in 2007. Round to the nearest tenth of a million.

Five million consumers in the United States fell victim [to phishing attacks] during 2008, an increase of 40 percent over 2007. (*Source:* www.scmagazineus.com, April 15, 2009)

89. Find the percent that the annual cost of rent is of the average annual income of residents. Round to the nearest percent.

There are more than 400,000 residents in the New York City Housing Authority's 2611 buildings at any given time. Judge Sotomayor, President Obama's nominee for the United States Supreme Court, is just one of more than 100 marquee names on a city list of alumni. . . . Today, the average [annual] income of residents is $22,728, the average [monthly] rent $324. (*Source:* www.nytimes.com, May 31, 2009)

90. Harris Interactive conducted an online study of 4435 U.S. workers for CareerBuilder.com. Find the minimum number of these workers that had not gone on or were not planning to take a vacation in 2009. Round to the nearest ten.

More than a third (35 percent) of workers say they haven't gone on or aren't planning to take a vacation in 2009. (*Source:* www.careerbuilder.com, May 18, 2009)

Find the Mistake

For exercises 91–94, the completed problem has one mistake.

(a) Describe the mistake in words, or copy down the whole problem and highlight or circle the mistake.

(b) Do the problem correctly.

91. Problem: Factor $9x^2 - 81y^2$ completely.

   Incorrect Answer: $9x^2 - 81y^2$
$$= ((3x)^2 - (9y)^2)$$
$$= (3x - 9y)(3x + 9y)$$

92. Problem: Factor $2fx - 2gx + 6fy - 6gy$ completely.

   Incorrect Answer: $2fx - 2gx + 6fy - 6gy$
$$= (2fx - 2gx) + (6fy - 6gy)$$
$$= 2x(f - g) + 6y(f - g)$$
$$= (f - g)(2x + 6y)$$

93. Problem: Factor $x^2 + 4x + x + 4$ completely.

   Incorrect Answer: $x^2 + 4x + x + 4$
$$= (x^2 + 4x) + (x + 4)$$
$$= x(x + 4) + (x + 4)$$
$$= x(x + 4)$$

94. Problem: Factor $x^2y + 3x^2 - 7y - 21$ completely.

   Incorrect Answer: $x^2y + 3x^2 - 7y - 21$
$$= (x^2y + 3x^2) - (7y - 21)$$
$$= x^2(y + 3) - 7(y - 3)$$

This polynomial is prime.

Review

95. Find two whole numbers that have a sum of 10 and a product of 21.

96. Find two whole numbers that have a sum of 10 and a product of 24.

97. Find two whole numbers that have a sum of 10 and a product of 25.

98. Find two integers that have a sum of 10 and a product of –24.

SUCCESS IN COLLEGE MATHEMATICS

99. Before you go to class, you should read the textbook section(s) that will be discussed and identify the main topic of the section and the topics of the subsections. Looking ahead to Section 2.3, what is the main topic of the section and the topics of the subsections?

100. What do the objectives at the beginning of Section 2.3 tell you?

© Lim Yong Hian/Shutterstock

In a completed Sudoku puzzle, every row, every column, and every 3×3 box contains the digits 1 through 9. To factor a trinomial, we also complete a mathematical puzzle. We must find two whole numbers with a specific product and sum. In this section, we will use two strategies to factor trinomials.

SECTION 2.3

Factoring Polynomials: Trinomials and a Strategy for Factoring Completely

After reading the text, working the practice problems, and completing assigned exercises, you should be able to:

1. Use the *ac* method to factor a trinomial.

2. Use guess and check to factor a trinomial.

3. Use the discriminant to determine whether a quadratic trinomial is prime.

4. Use more than one method of factoring to factor a polynomial completely.

The *ac* Method

A **trinomial** is a polynomial with three terms. A **quadratic trinomial** is a polynomial with three terms that can be written in the form $ax^2 + bx + c$. In Section 2.2, we factored polynomials by grouping. To factor a trinomial by grouping, we first rewrite the trinomial as an equivalent polynomial with four terms.

EXAMPLE 1   Use grouping to factor $10x^2 + 17x + 3$.

SOLUTION ▶

$$10x^2 + 17x + 3$$
$$= 10x^2 + 2x + 15x + 3 \qquad \text{Rewrite with four terms: } 17x = 2x + 15x$$
$$= (10x^2 + 2x) + (15x + 3) \qquad \text{Group terms that have a common factor.}$$
$$= 2x(\underline{\quad} + \underline{\quad}) + 3(\underline{\quad} + \underline{\quad}) \qquad \text{Identify a common factor of each group.}$$
$$= 2x(5x + 1) + 3(5x + 1) \qquad \text{Another common factor appears: } 5x + 1.$$
$$= (5x + 1)(\underline{\quad} + \underline{\quad}) \qquad \text{Factor out the common factor, } 5x + 1.$$
$$= (5x + 1)(2x + 3) \qquad \text{Complete the factoring.}$$

Notice that the middle term, $17x$, is rewritten as $2x + 15x$. Other pairs of terms, such as $8x + 9x$ or $20x - 3x$, also equal $17x$. However, factoring by grouping will be successful only if we choose $2x + 15x$.

$$10x^2 + 17x + 3$$
$$= 10x^2 + 8x + 9x + 3 \qquad \text{What if we change from } 2x + 15x \text{ to } 8x + 9x?$$
$$= (10x^2 + 8x) + (9x + 3) \qquad \text{Group terms that have a common factor.}$$
$$= 2x(\underline{\quad} + \underline{\quad}) + 3(\underline{\quad} + \underline{\quad}) \qquad \text{Identify a common factor of each group.}$$
$$= 2x(5x + 4) + 3(3x + 1) \qquad \text{We cannot complete the factoring.}$$

Since we do not want to have to try factoring each different way of rewriting the middle term of a trinomial, we need a guide for rewriting it as a sum or difference. This guide is the **ac method** for factoring trinomials in the form $ax^2 + bx + c$. The coefficient of the first term, the **lead coefficient**, is a. The coefficient of the middle term is b. The constant is c.

---

ac Method for Factoring Trinomials in the Form $ax^2 + bx + c$

1. Identify a, b, and c.

2. Rewrite the middle term of the trinomial as a sum or difference of two terms. The product of the coefficients of these terms must equal ac. The sum of the coefficients must equal b.

3. Factor this equivalent polynomial by grouping.

EXAMPLE 2   Use the *ac* method to factor $2p^2 + 13p + 6$.

SOLUTION ▶   For $2p^2 + 13p + 6$, $a = 2$, $b = 13$, $c = 6$, and $ac = 12$.

Find two numbers with a product of 12 (ac) and a sum of 13 (b).

ac = 12 Factors of 12	Sum of factors	b = 13 Does sum equal 13?
(2)(6)	8	No
(3)(4)	7	No
(1)(12)	**13**	**Yes**

Since $(1)(12) = 12$ and $1 + 12 = 13$, rewrite the middle term, $13p$, as $12p + 1p$.

$$2p^2 + 13p + 6 \qquad a = 2; b = 13; c = 6; ac = 12$$
$$= 2p^2 + 12p + 1p + 6 \qquad \text{Rewrite } 13p \text{ as } 12p + 1p.$$
$$= (2p^2 + 12p) + (1p + 6) \qquad \text{Group terms that have a common factor.}$$
$$= 2p(\underline{\quad} + \underline{\quad}) + 1(\underline{\quad} + \underline{\quad}) \qquad \text{Identify a common factor of each group.}$$
$$= 2p(p + 6) + 1(p + 6) \qquad \text{Another common factor appears: } p + 6.$$
$$= (p + 6)(2p + 1) \qquad \text{Factor out } p + 6; \text{ complete the factoring.}$$

When ac is a negative number, one of the coefficients of the new middle terms is a positive number, and the other coefficient is a negative number.

EXAMPLE 3   (a) Use the *ac* method to factor $6x^2 - x - 12$.

SOLUTION ▶   For $6x^2 - x - 12$, $a = 6$, $b = -1$, $c = -12$, and $ac = -72$.

Find two numbers with a product of -72 and a sum of -1.

$ac = -72$ Factors of –72	Sum of factors	$b = -1$ Does sum equal –1?
$(1)(-72)$ or $(-1)(72)$	-71 or 71	No
$(2)(-36)$ or $(-2)(36)$	-34 or 34	No
$(3)(-24)$ or $(-3)(24)$	-21 or 21	No
$(4)(-18)$ or $(-4)(18)$	-14 or 14	No
$(6)(-12)$ or $(-6)(12)$	-6 or 6	No
$\mathbf{(8)(-9)}$ or $(-8)(9)$	-1 or 1	**Yes**

Since $(8)(-9) = -72$ and $8 + (-9) = -1$, rewrite the middle term, $-x$, as $8x + -9x$.

$$
\begin{aligned}
6x^2 - x - 12 & & -x = -1x \\
= 6x^2 - \mathbf{1}x - 12 & & a = 6;\, b = -1;\, c = -12;\, ac = -72 \\
= 6x^2 + \mathbf{8x} + \mathbf{-9x} - 12 & & \text{Rewrite } -x \text{ as } 8x + -9x. \\
= (6x^2 + 8x) + (-9x - 12) & & \text{Group terms with a common factor.} \\
= 2x(\underline{} + \underline{}) + -3(\underline{} + \underline{}) & & \text{Identify a common factor of each group.} \\
= 2x(\mathbf{3x + 4}) + -3(\mathbf{3x + 4}) & & \text{Another common factor appears, } 3x + 4. \\
= (\mathbf{3x + 4})(2x + -3) & & \text{Complete the factoring.} \\
= (3x + 4)(2x - 3) & & \text{Rewrite } 2x + -3 \text{ as } 2x - 3.
\end{aligned}
$$

(b) Check the factoring.

▶   When we factor a polynomial, we are rewriting it as a product of factors. To check, we multiply the factors. The product should be the original polynomial.

$$
\begin{aligned}
(3x + 4)(2x - 3) & & \\
= \mathbf{3x}(2x - 3) + \mathbf{4}(2x - 3) & & \text{Distributive property.} \\
= 6x^2 - 9x + 8x - 12 & & \text{Distributive property.} \\
= 6x^2 - \mathbf{x} - 12 & & \text{Simplify.}
\end{aligned}
$$

The product of the factors is the original polynomial; the factoring is correct.

To use the *ac* method to factor $6x^2 - x - 12$, we rewrote $-x$ as $8x + -9x$. We could have instead rewritten the middle term as $-9x + 8x$. The factors are the same.

In the next example, *b* is a negative number and *ac* is a positive number. Since the sum of two negative numbers is a negative number and the product of two negative numbers is a positive number, the new coefficients must both be negative.

EXAMPLE 4   Use the *ac* method to factor $3z^2 - 32z + 45$.

SOLUTION ▶   For $3z^2 - 32z + 45$, $a = 3$, $b = -32$, $c = 45$, and $ac = 135$.

Find two numbers with a product of 135 and a sum of -32.

$ac = 135$ Factors of 135	Sum of factors	$b = -32$ Does sum equal –32?
$(-1)(-135)$	-136	No
$(-3)(-45)$	-48	No
$(-5)(-27)$	-32	**Yes**

Since $(-5)(-27) = 135$ and $-5 + (-27) = -32$, rewrite the middle term, $-32z$, as $-5z + -27z$.

$3z^2 - 32z + 45$	$a = 3; b = -32; c = 45; ac = 135$
$= 3z^2 - \mathbf{5z} + \mathbf{-27z} + 45$	Rewrite: $-32z = -5z + -27z$
$= (3z^2 - 5z) + (-27z + 45)$	Group terms that have a common factor.
$= z(\underline{\ \ } + \underline{\ \ }) + (-9)(\underline{\ \ } + \underline{\ \ })$	Identify a common factor of each group.
$= z(\mathbf{3z - 5}) + -9(\mathbf{3z - 5})$	Another common factor appears: $3z - 5$.
$= (\mathbf{3z - 5})(z + -9)$	Factor out $3z - 5$; complete the factoring.
$= (\mathbf{3z - 5})(z - 9)$	Rewrite: $z + -9 = z - 9$

If we choose to factor 9 out of the second group instead of -9, the result is $z(3z - 5) + 9(-3z + 5)$. Since there is no common factor in these two groups, this choice does not allow us to finish the factoring.

Practice Problems

For problems 1–4, use the ac method to factor. Identify any prime polynomials.

1. $2x^2 + 17x + 30$ **2.** $2x^2 - 17x + 30$ **3.** $2x^2 + x - 30$
4. $12x^2 - 16x + 5$

Using Guess and Check to Factor Trinomials

In the ac method, we rewrite the middle term of a trinomial as the sum of two terms. The product of the coefficients of these new terms equals ac. The sum of the coefficients equals b. If the lead coefficient, a, is 1, then $ac = c$. In this situation, it may be faster to use **guess and check** to factor rather than the ac method.

To **guess**, write two sets of empty parentheses and guess the terms that are needed in each factor. The product of the numbers in the factors equals c; the sum of the numbers is b. To **check**, multiply the factors. The product should be the original polynomial.

EXAMPLE 5 **(a)** Use guess and check to factor $x^2 + 5x + 6$.

SOLUTION ▶

$x^2 + 5x + 6$	$a = 1; b = 5; c = 6$
$= (x + \underline{\ \ })(x + \underline{\ \ })$	The first term is x^2, so both factors begin with x.
$= (x + \mathbf{2})(x + \mathbf{3})$	Identify two numbers whose sum is 5 and whose product is 6.

(b) Check.

$(x + 2)(x + 3)$	The product should equal the original polynomial.
$= \mathbf{x}(x + 3) + \mathbf{2}(x + 3)$	Distributive property.
$= x^2 + 3x + 2x + 6$	Distributive property.
$= x^2 + \mathbf{5x} + 6$	Combine like terms.

Since the product of the factors is the original polynomial, $x^2 + 5x + 6$, the factoring is correct.

In the next example, we need to choose two numbers whose sum is negative and whose product is negative. One of the numbers will be negative, and one of the numbers will be positive.

EXAMPLE 6   (a) Use guess and check to factor $x^2 - 5x - 14$.

SOLUTION ▶

$$x^2 - 5x - 14 \qquad\qquad a = 1; b = -5; c = -14$$
$$= (x - \underline{})(x + \underline{}) \qquad \text{The first term is } x^2, \text{ so both factors begin with } x.$$
$$= (x - \mathbf{7})(x + \mathbf{2}) \qquad \text{Identify two numbers whose sum is } -5 \text{ and whose product is } -14.$$

(b) Check.

▶

$$(x - 7)(x + 2) \qquad\qquad \text{The product should equal the original polynomial.}$$
$$= \mathbf{x}(x + 2) - \mathbf{7}(x + 2) \qquad \text{Distributive property.}$$
$$= x^2 + 2x - 7x - 14 \qquad \text{Distributive property.}$$
$$= x^2 - \mathbf{5x} - 14 \qquad\qquad \text{Combine like terms.}$$

Since the product of the factors is the same as the original polynomial, $x^2 - 5x - 14$, the factoring is correct.

In the next example, notice that a is used in two different ways. It is a variable in the polynomial $a^2 - 11a + 28$, and it is the lead coefficient in the pattern $ax^2 + bx + c$.

EXAMPLE 7   (a) Use guess and check to factor $a^2 - 11a + 28$.

SOLUTION ▶

$$a^2 - 11a + 28 \qquad\qquad a = 1 \ (a \text{ is the lead coefficient}); b = -11; c = 28$$
$$= (a - \underline{})(a - \underline{}) \qquad \text{The first term is } a^2, \text{ so both factors begin with the variable } a.$$
$$= (a - \mathbf{7})(a - \mathbf{4}) \qquad \text{Identify two numbers whose sum is } -11 \text{ and whose product is 28.}$$

(b) Check.

▶

$$(a - 7)(a - 4) \qquad\qquad \text{The product should equal the original polynomial.}$$
$$= \mathbf{a}(a - 4) - \mathbf{7}(a - 4) \qquad \text{Distributive property.}$$
$$= a^2 - 4a - 7a + 28 \qquad \text{Distributive property.}$$
$$= a^2 - \mathbf{11a} + 28 \qquad\qquad \text{Simplify.}$$

Since the product of the factors is the same as the original polynomial, $a^2 - 11a + 28$, the factoring is correct.

Using Guess and Check to Factor Trinomials in the Form $ax^2 + bx + c$

1. Write down two empty sets of parentheses.
2. Guess the first term in each factor.
3. Guess the second term in each factor.

   If $b > 0$ and $c > 0$, both of the second terms will be positive.

   If $b > 0$ and $c < 0$, one of the second terms will be positive and one will be negative.

   If $b < 0$ and $c < 0$, one of the second terms will be positive and one will be negative.

   If $b < 0$ and $c > 0$, both of the second terms will be negative.

4. Check.

Both the ac method and guess and check are effective methods for factoring trinomials. When $a = 1$, guess and check is usually faster. When a does not equal 1, it is usually quicker to use the ac method. It is important to show your work and check the result.

Practice Problems

For problems 5–8, use guess and check to factor.

5. $x^2 + 14x + 40$ **6.** $x^2 - 8x + 7$ **7.** $x^2 - 9x - 22$
8. $x^2 + 7x + 6$

The Discriminant and Trinomials

A trinomial in the form $ax^2 + bx + c$ is a **quadratic trinomial in standard form**. The **discriminant** of a quadratic trinomial is the number equal to $b^2 - 4ac$. We can use the square root of the discriminant to determine whether the trinomial is prime. If a, b, and c have no common factor and the principal square root of the discriminant is an irrational number or is not a real number, the trinomial is prime. The only factors of a prime trinomial are itself and 1.

Standard Form of a Quadratic Trinomial

$ax^2 + bx + c$

where a, b, and c are real numbers; a is the lead coefficient; and c is a constant.

Discriminant of a Quadratic Trinomial

$b^2 - 4ac$

If the principal square root of the discriminant is a whole number, the trinomial is not prime. If a, b, and c have no common factor and the principal square root of the discriminant is an irrational number or is not a real number, the trinomial is prime.

EXAMPLE 8   **(a)** Find the discriminant for $2p^2 + 13p + 6$.

SOLUTION ▶

$b^2 - 4ac$	The discriminant.
$= (\mathbf{13})^2 - 4 \cdot \mathbf{2} \cdot \mathbf{6}$	$a = 2; b = 13; c = 6$
$= 169 - 48$	Evaluate.
$= 121$	Evaluate.

(b) Use the discriminant to determine whether $2p^2 + 13p + 6$ is prime.

▶ Since $\sqrt{121} = 11$ and 11 is a whole number, this trinomial can be factored. It is not prime. We factored this trinomial in Example 2 in this section.

When we use the *ac* method or guess and check, we look for factors of b. If a factor pair is missed, we may incorrectly think that the trinomial is prime. The discriminant is a quick way to check whether a trinomial is prime.

EXAMPLE 9   **(a)** Find the discriminant for $2x^2 - x + 6$.

SOLUTION ▶

$b^2 - 4ac$	The discriminant.
$= (\mathbf{-1})^2 - 4 \cdot \mathbf{2} \cdot \mathbf{6}$	$a = 2; b = -1; c = 6$
$= 1 - 48$	Evaluate.
$= -47$	Evaluate.

(b) Use the discriminant to determine whether $2x^2 - x + 6$ is prime.

▶ Since $\sqrt{-47}$ is not a real number, this polynomial is prime.

The discriminant is part of the quadratic formula. In Chapter 7, we will study quadratic equations and the quadratic formula. We will then learn why the discriminant can be used to identify prime quadratic trinomials.

Practice Problems

For problems 9–12,
(a) find the discriminant.
(b) use the discriminant to determine whether the trinomial is prime. Do not factor.

9. $x^2 + 15x + 40$ **10.** $x^2 - 3x + 7$

11. $x^2 - 8x + 15$ **12.** $2x^2 - 27$

Factoring Trinomials That Are Quadratic in Form

A **quadratic trinomial** can be written in the form $ax^2 + bx + c$. The exponent in the lead term is 2. We can use the ac method or guess and check to factor some trinomials that are not quadratic. The degree of these trinomials is not 2, but they can be rewritten to match the pattern $ax^2 + bx + c$. These trinomials are **quadratic in form**.

EXAMPLE 10 **(a)** Use guess and check to factor $n^4 + 16n^2 + 63$.

SOLUTION ▶

$n^4 + 16n^2 + 63$ $a = 1; b = 16; c = 63$

$= (n^2 + \underline{\quad})(n^2 + \underline{\quad})$ The factors begin with n^2, since $(n^2)(n^2) = n^4$.

$= (n^2 + 7)(n^2 + 9)$ Identify two numbers whose sum is 16 and whose product is 63.

(b) Check.

▶

$(n^2 + 7)(n^2 + 9)$

$= \mathbf{n^2}(n^2 + 9) + \mathbf{7}(n^2 + 9)$ Distributive property.

$= n^4 + 9n^2 + 7n^2 + 63$ Distributive property.

$= n^4 + \mathbf{16n^2} + 63$ Simplify. The factoring is correct.

The trinomial in Example 10, $n^4 + 16n^2 + 63$, is quadratic in form. We can rewrite it as $(n^2)^2 + 16n^2 + 63$; this matches the pattern $ax^2 + bx + c$. The exponent in the lead term of a trinomial that is quadratic in form equals two times the exponent in the middle term.

The trinomial $6x^8 - 7x^4 - 3$ is also quadratic in form. We can rewrite it as $6(x^4)^2 - 7x^4 - 3$. The exponent in the lead term, 8, is equal to two times the exponent in the middle term, 4.

EXAMPLE 11 Factor: $6x^8 - 7x^4 - 3$

SOLUTION ▶

$6x^8 - 7x^4 - 3$ $a = 6; b = -7; c = -3; ac = -18;$ use the ac method.

$= 6x^8 - \mathbf{9x^4} + \mathbf{2x^4} - 3$ Rewrite $-7x^4$ as $-9x^4 + 2x^4$.

$= (6x^8 - 9x^4) + (2x^4 - 3)$ Group terms that have a common factor.

$= 3x^4(\underline{\quad} + \underline{\quad}) + 1(\underline{\quad} + \underline{\quad})$ Identify a common factor of each group.

$= 3x^4(\mathbf{2x^4 - 3}) + 1(\mathbf{2x^4 - 3})$ Another common factor appears: $2x^4 - 3$.

$= (\mathbf{2x^4 - 3})(3x^4 + 1)$ Factor out $2x^4 - 3$; complete the factoring.

Practice Problems

For problems 13–15, factor.

13. $x^4 + 8x^2 + 15$ **14.** $x^8 - 5x^4 + 6$ **15.** $p^6 + 4p^3 - 45$

Factor Completely

To **factor a polynomial completely**, we may need to use more than one factoring method.

Methods for Factoring Polynomial Expressions

- Greatest common factor
- Grouping
- Pattern
- *ac* method (trinomials)
- Guess and check (trinomials)

To begin factoring a polynomial completely, look for a common factor of all the terms. Then use another method to factor the remaining polynomial. The polynomial is not factored completely until each factor is prime.

EXAMPLE 12 | Factor $36k^3 + 30k^2 - 24k$ completely.

SOLUTION ▶

$$36k^3 + 30k^2 - 24k$$ The greatest common factor is $6k$.

$$= \mathbf{6k}(6k^2 + 5k - 4)$$ Factor out the greatest common factor, $6k$.

$$= 6k[6k^2 + \mathbf{8k} + \mathbf{-3k} - 4]$$ *ac* method; $a = 6$; $b = 5$; $c = -4$; $ac = -24$

$$= 6k[(6k^2 + 8k) + (-3k - 4)]$$ Group. Use brackets and parentheses.

$$= 6k[2k(\mathbf{3k + 4}) + -1(\mathbf{3k + 4})]$$ Factor the common factor from each group.

$$= 6k[(\mathbf{3k + 4})(2k + -1)]$$ Factor again; the common factor is $3k + 4$.

$$= 6k(3k + 4)(2k - 1)$$ Simplify; remove the brackets.

Since $6k$ is prime, $3k + 4$ is prime, and $2k - 1$ is prime, this polynomial is factored completely.

In the next example, we first factor out the greatest common factor. We then use the discriminant to show that the remaining quadratic trinomial is prime.

EXAMPLE 13 | Factor $3x^5 + 6x^4 - 21x^3$ completely.

SOLUTION ▶

$$3x^5 + 6x^4 - 21x^3$$ The greatest common factor is $3x^3$.

$$= \mathbf{3x^3}(x^2 + 2x - 7)$$ Factor out the greatest common factor, $3x^3$.

If $x^2 + 2x - 7$ is prime, we have completely factored the original polynomial. If it is not prime, we need to continue factoring.

$$b^2 - 4ac$$ The discriminant.

$$= \mathbf{2}^2 - 4(\mathbf{1})(\mathbf{-7})$$ $a = 1$; $b = 2$; $c = -7$

$$= 32$$ $\sqrt{32} = 5.65\ldots$, which is an irrational number.

Since the principal square root of the discriminant is an irrational number, $x^2 + 2x - 7$ is prime. Since the other factor, $3x^3$, is also prime, the polynomial is factored completely.

In the next example, we factor out a greatest common factor, use the *ac* method, and then use a pattern to factor completely. This polynomial has four prime factors.

EXAMPLE 14 | Factor $12p^4 - 3p^2 - 54$ completely.

SOLUTION ▶

$$12p^4 - 3p^2 - 54$$ The greatest common factor is 3.

$$= \mathbf{3}(4p^4 - p^2 - 18)$$ Factor out the greatest common factor, 3.

$$= 3[4p^4 - \mathbf{9}p^2 + \mathbf{8}p^2 - 18]$$ *ac* method; $a = 4$; $b = -1$; $c = -18$; $ac = -72$

$$= 3[(4p^4 - 9p^2) + (8p^2 - 18)]$$ Group. Use brackets and parentheses.

$$= 3[p^2(\mathbf{4p^2 - 9}) + 2(\mathbf{4p^2 - 9})]$$ Factor the common factor from each group.

$$= 3[(\mathbf{4p^2 - 9})(p^2 + 2)]$$ Factor again; the common factor is $4p^2 - 9$.

$$= 3[((2p)^2 - 3^2)(p^2 + 2)]$$ Difference of squares. Match the pattern; $a = 2p$ and $b = 3$

$$= 3[(2p + 3)(2p - 3)(p^2 + 2)]$$ $a^2 - b^2 = (a + b)(a - b)$

$$= 3(2p + 3)(2p - 3)(p^2 + 2)$$ Remove the brackets; $p^2 + 2$ is prime.

Since 3 is prime, $2p + 3$ is prime, $2p - 3$ is prime, and $p^2 + 2$ is prime, this polynomial is factored completely.

The first step in factoring completely should be to factor out any greatest common factor. In the next example, we see what happens if this is not done.

EXAMPLE 15 | Factor $3x^2 + 15x - 42$ completely.

SOLUTION ▶ The greatest common factor is 3. What happens if we do not first factor out the greatest common factor?

$$3x^2 + 15x - 42$$ $a = 3$; $b = 15$; $c = -42$; $ac = -126$

$$= 3x^2 - \mathbf{6}x + \mathbf{21}x - 42$$ Rewrite: $15x = -6x + 21x$

$$= (3x^2 - 6x) + (21x - 42)$$ Group terms that have a common factor.

$$= 3x(\mathbf{x - 2}) + 21(\mathbf{x - 2})$$ Factor the common factor from each group.

$$= (\mathbf{x - 2})(3x + 21)$$ Factor again; the common factor is $x - 2$.

$$= (x - 2)(\mathbf{3})(\mathbf{x + 7})$$ $3x + 21$ is not prime; factor out 3.

$$= 3(x - 2)(x + 7)$$ Change order.

Since 3 is prime, $x - 2$ is prime, and $x - 7$ is prime, the polynomial is factored completely. If we factor out the greatest common factor, 3, in the first step, the lead coefficient of the remaining trinomial is 1, and the remaining factoring is easier.

$$3x^2 + 15x - 42$$ The greatest common factor is 3.

$$= \mathbf{3}(x^2 + 5x - 14)$$ Factor out the greatest common factor, 3.

$$= 3(x + \underline{})(x - \underline{})$$ Trinomial; use guess and check.

$$= 3(x + \mathbf{7})(x - \mathbf{2})$$ Identify two numbers whose sum is 5 and whose product is -14.

The factors are the same.

For many polynomials, we can use a visual image for factoring completely. To completely factor a polynomial, first factor out any greatest common factor. Then try to factor the remaining polynomial. The method that is used to factor the remaining polynomial depends on the number of terms. Remember that the remaining polynomial may be prime. Factoring out the greatest common factor may be all that can be done.

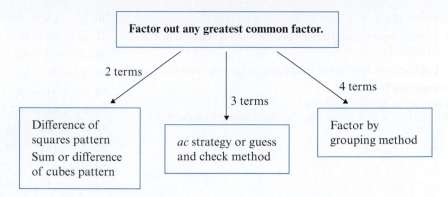

Factoring is not always obvious. Try a method. If it does not work, try a different method. You will not always find the correct factors on your first try.

Practice Problems

For problems 16–27, factor the polynomial completely.

16. $8x^2 + 50x - 42$ **17.** $15w^3 + 6w^2 - 3w$ **18.** $5x^3 - 80x$

19. $7x^2 + 28$ **20.** $12x^2 - 17x - 7$ **21.** $4ac + 2ad + 20c + 10d$

22. $4p^7 - 144p^5$ **23.** $2n^3 + 10n^2 - 18n - 90$

24. $w^4 + 15w^3 - 56w^2$ **25.** $3x^2 - 42x - 147$

26. $u^3 + 8u^2 + 16u$ **27.** $b^4 - 2b^2 - 63$

2.3 VOCABULARY PRACTICE

Match the term with its description.

1. $ax^2 + bx + c$
2. The result of multiplication
3. A nonrepeating, nonterminating decimal number
4. A polynomial with three terms
5. The coefficient of the first term in a trinomial in standard form
6. The only factors of this polynomial are itself and 1.
7. The monomial of highest degree and largest coefficient that is a factor of each term
8. $b^2 - 4ac$
9. $a^2 - b^2 = (a - b)(a + b)$
10. $a(b + c) = ab + ac$

A. difference of squares pattern
B. discriminant
C. distributive property
D. greatest common factor of two or more terms
E. irrational number
F. lead coefficient
G. prime polynomial
H. product
I. quadratic trinomial
J. trinomial

2.3 Exercises

Follow your instructor's guidelines for showing your work.

For exercises 1–8,
(a) find the discriminant.
(b) use the discriminant to determine whether the trinomial is prime.

1. $2x^2 + 11x + 5$
2. $2x^2 + 13x + 6$
3. $3x^2 + 5x + 6$
4. $3x^2 + 10x + 9$
5. $3x^2 + 5x + 2$
6. $3x^2 + 10x + 7$
7. $4x^2 - 5x + 2$

8. $4x^2 - 10x + 7$

9. Explain how to determine whether $\sqrt{15}$ is rational or irrational.

10. Explain how to check the factoring of a polynomial.

For exercises 11–34,
(a) use the ac method to factor.
(b) if the trinomial is prime, use the discriminant to show that it is prime.

11. $2x^2 + 11x + 5$

12. $2x^2 + 13x + 6$

13. $2x^2 - 9x - 5$

14. $2x^2 - 11x - 6$

15. $2x^2 + 9x - 5$

16. $2x^2 + 11x - 6$

17. $2x^2 + 5x + 6$

18. $2x^2 + 4x + 7$

19. $2x^2 - 11x + 5$

20. $2x^2 - 13x + 6$

21. $2x^2 - 7x - 15$

22. $2x^2 + 3x - 14$

23. $3x^2 + 10x + 8$

24. $3x^2 + 10x + 7$

25. $12x^2 + 19x + 5$

26. $12x^2 + 13x + 3$

27. $3x^2 - 8x + 4$

28. $3x^2 - 8x + 5$

29. $z^2 + 10z + 25$

30. $a^2 + 12a + 36$

31. $w^2 - 10w + 25$

32. $b^2 - 12b + 36$

33. $a^2 + 5a + 6$

34. $d^2 + 8d + 15$

For exercises 35–50,
(a) use guess and check to factor.
(b) if the polynomial is prime, use the discriminant to show that it is prime.

35. $f^2 + 11f + 18$

36. $h^2 + 10h + 16$

37. $b^2 + 12b + 11$

38. $f^2 + 16f + 15$

39. $y^2 - 9y - 36$

40. $g^2 - 12g - 28$

41. $c^2 - 6c + 9$

42. $h^2 - 8h + 16$

43. $p^2 - 15p + 56$

44. $z^2 - 15z + 54$

45. $y^2 - 8y - 15$

46. $k^2 - 13k - 36$

47. $q^2 - 9q + 18$

48. $j^2 - 16j + 63$

49. $x^2 + 2x - 99$

50. $x^2 + 3x - 54$

For exercises 51–62, factor. These trinomials are quadratic in form.

51. $x^4 + 14x^2 + 45$

52. $x^4 + 15x^2 + 54$

53. $4x^4 + 25x^2 - 21$

54. $3x^4 + 19x^2 - 14$

55. $9n^4 - 3n^2 - 2$

56. $9m^4 - 9m^2 - 4$

57. $2u^{10} - 11u^5 - 40$

58. $2a^{10} - 3a^5 - 14$

59. $9a^4 - 24a^2 + 16$

60. $4p^6 - 20p^3 + 25$

61. $35w^8 + 2w^4 - 1$

62. $52w^6 + 9w^3 - 1$

For exercises 63–92, use the methods from Sections 2.2 and 2.3 to factor completely. Identify any prime polynomials.

63. $8h^2 - 44h - 84$

64. $6m^2 + 28m - 10$

65. $x^3 + 3x^2 - 4x - 12$

66. $x^3 + 5x^2 - 9x - 45$

67. $7x^4 - 7xy^3$

68. $11p^5 - 11p^2w^3$

69. $6c^3 + c^2 - 35c$

70. $10a^3 - 13a^2 - 3a$

71. $2x^3 + 14x^2 + 30x$

72. $9x^2 + 81x + 153$

73. $28p^2 + 84p + 63$

74. $18a^2 + 12a + 2$

75. $25x^2 - 16y^3$

76. $49z^4 - 81w^5$

77. $21p^2 - 75p + 36$

78. $36h^2 - 38h + 10$

79. $-4x^5 - 16x^7$

80. $-9g^8 - 27g^{11}$

81. $y^3 - 2y^2 - 9y + 18$

82. $z^3 - 7z^2 - 4z + 28$

83. $20ab - 40ac + 4b - 4c$

84. $12np - 12w + 4p - 4w$

85. $3w^3 + 30w^2 + 75w$

86. $2y^3 + 24y^2 + 72y$

87. $5x^2 + 65x - 180$

88. $6a^2 + 66a - 144$

89. $c^3 + 3c^2 + 16c + 48$

90. $m^3 + 4m^2 + 9m + 36$

91. $25a^3 - 10a^2 + a$

92. $36u^3 - 12u^2 + u$

For exercises 93–94, we can use a pattern to factor a difference of squares polynomial, or we can rewrite it as a trinomial and use the ac method to factor. Finish the factoring.

93. $x^2 - 9$

$= x^2 + 0x - 9$

$= x^2 - 3x + 3x - 9$

$= (x^2 - 3x) + (3x - 9)$

94. $16a^2 - 25b^2$

$= 16a^2 + 0ab - 25b^2$

$= 16a^2 - 20ab + 20ab - 25b^2$

$= (16a^2 - 20ab) + (20ab - 25b^2)$

95. The binomial $x^2 + 9$ can be rewritten as a quadratic trinomial: $x^2 + 0x + 9$.

  a. What is b in this trinomial?

  b. Use the discriminant to find out whether $x^2 + 9$ is prime.

96. The binomial $2x^2 + 16$ can be rewritten as a quadratic trinomial: $2x^2 + 0x + 16$.

  a. What is b in this trinomial?

  b. Use the discriminant to find out whether $2x^2 + 16$ is prime.

Problem Solving: Practice and Review

Follow your instructor's guidelines for using the five steps as outlined in Section 1.5, p. 51.

97. Find the percent increase in athletic fees approved by the president. Round to the nearest percent.

  In April 2008, in a general fee referendum, students at Fresno State voted, 777–412, against raising athletic fees to $50 per semester, from $7. Based on a recommendation from a campus fee advisory committee, John D. Welty, the Fresno State president, overrode the vote and approved an increase to $32 per semester for the 2008–9 academic year. (*Source:* www.nytimes.com, May 30, 2009)

98. Miami Dade College has eight campuses and over 300 programs of study. In fall 2008, Miami Dade College had 61,288 students taking classes for credit. This was a 4.9% increase from fall 2007. Find the number of students taking classes for credit in fall 2007. Round to the nearest hundred. (*Source:* www.mdc.edu)

99. Find the new cost of a 4.5-mile ambulance ride that includes advanced life support.

  Fenty . . . raised the cost of advanced life support from $471 to $832 . . . [and] a fee of $6.06 per mile traveled. (*Source:* www.washingtonexaminer.com, March 31, 2008)

100. The low bid to build a 5.5-mile guardrail along Airline Drive in St. Charles Parish, Louisiana, was $1.56 million. The population of St. Charles Parish in 2009 was 51,611 people. Find the cost per *foot* to build the guardrail. Round to the nearest whole number. (*Sources:* www.nola.com, May 27, 2009; quickfacts.census.gov)

Find the Mistake

For exercises 101–104, the completed problem has one mistake. **(a)** Describe the mistake in words, or copy down the whole problem and highlight or circle the mistake. **(b)** Do the problem correctly.

101. **Problem:** Factor: $x^2 + 5x - 6$

  Incorrect Answer: $x^2 + 5x - 6$

  $= x^2 + 3x + 2x - 6$

  $= x(x + 3) - 2(x + 3)$

  $= (x + 3)(x - 2)$

102. **Problem:** Factor: $x^2 - 9x - 14$

  Incorrect Answer: $x^2 - 9x - 14$

  $= x^2 - 7x - 2x - 14$

  $= x(x - 7) - 2(x - 7)$

  $= (x - 7)(x - 2)$

103. **Problem:** Factor: $x^2 + 25$

  Incorrect Answer: $x^2 + 25$

  $= (x - 5)(x + 5)$

104. **Problem:** Factor: $x^2 - 15x - 56$

  Incorrect Answer: $x^2 - 15x - 56$

  $= x^2 - 7x - 8x - 56$

  $= x(x - 7) + 8(x - 7)$

  $= (x - 7)(x + 8)$

Review

105. Solve: $3x - 2 = 0$

106. Solve: $-3x - 2 = 0$

107. What is the product of any number and zero?

108. What is the sum of any number and zero?

SUCCESS IN COLLEGE MATHEMATICS

109. When reading the textbook before class, it is important to read the examples slowly, step by step. Some students find it helpful to work each example as they read. What is the purpose of the text that is written to the right of each line of the example? (These are called annotations.)

110. The practice problems at the end of each subsection are similar to the examples in the subsection. The answers to the practice problems are in the answer section in the back of the book. For Section 2.4, read each subsection and do one of the practice problems for each subsection.

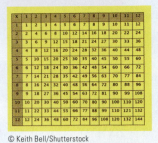

© Keith Bell/Shutterstock

Multiplication tables usually include the whole numbers from 1 to 12 and their multiples. Zero is not included because the product of any number and zero is zero. In this section, we will use this multiplication fact to develop the zero product property and to solve polynomial equations.

Polynomial Equations

After reading the text, working the practice problems, and completing assigned exercises, you should be able to:

1. State the zero product property.

2. Use the zero product property to solve a polynomial equation in one variable and check the solution(s).

3. Solve an application problem by solving a polynomial equation in one variable.

The Zero Product Property

If the product of two different factors is 0, then one of the factors must be 0.

EXAMPLE 1   Solve: $5b = 0$

SOLUTION ▶

$5b = 0$ The factors are 5 and b.

$\dfrac{5b}{5} = \dfrac{0}{5}$ Division property of equality.

$b = 0$ One of the factors is 0.

If the product of 5 and an unknown factor is 0, then the unknown factor must be 0. This relationship is an example of the **zero product property**.

> **Zero Product Property**
>
> If a and b are real numbers and $a \cdot b = 0$, then $a = 0$ or $b = 0$.

If a polynomial equation can be factored and if the product of the factors is 0, then at least one of the factors must equal 0. Mathematicians use the word *or* in a precise way. The statement "$a = 0$ or $b = 0$" tells us that $a = 0$ is true, $b = 0$ is true, or both are true. (Mathematicians use the same variables to represent very different things. Previously, we have seen a and b used to represent the coefficients in the standard form of a linear equation, $ax + by = c$, and the coefficients of a quadratic trinomial, $ax^2 + bx + c$. Here, a and b represent factors.)

EXAMPLE 2 Use the zero product property to solve $x(x - 5) = 0$.

SOLUTION ▶

$$x(x - 5) = 0 \qquad a = x; b = x - 5$$

$$x = 0 \quad \text{or} \quad x - 5 = 0 \qquad \text{Zero product property: if } ab = 0, \text{ then } a = 0 \text{ or } b = 0.$$

$$\underline{\qquad +5 \quad +5} \qquad \text{Addition property of equality.}$$

$$x + 0 = 5 \qquad \text{Simplify.}$$

$$x = 5 \qquad \text{Simplify.}$$

$$x = 0 \quad \text{or} \quad x = 5$$

In Section 2.3, we factored **quadratic** *expressions* in the form $ax^2 + bx + c$. Now we will use the zero product property to solve a **quadratic** *equation* in **standard form**, $ax^2 + bx + c = 0$.

EXAMPLE 3 **(a)** Use the zero product property to solve $10x^2 + 11x - 8 = 0$.

SOLUTION ▶

$$10x^2 + 11x - 8 = 0 \qquad a = 10; b = 11; c = -8; ac = -80$$

$$10x^2 + 16x + -5x - 8 = 0 \qquad 16 + -5 = 11 \text{ and } (16)(-5) = -80$$

$$(10x^2 + 16x) + (-5x - 8) = 0 \qquad \text{Group terms with a common factor.}$$

$$2x(\underline{\quad} + \underline{\quad}) + -1(\underline{\quad} + \underline{\quad}) = 0 \qquad \text{Identify common factors.}$$

$$2x(5x + 8) + -1(5x + 8) = 0 \qquad \text{Another common factor appears: } 5x + 8.$$

$$(5x + 8)(2x - 1) = 0 \qquad \text{Factor out } 5x + 8; \text{ complete the factoring.}$$

$$5x + 8 = 0 \quad \text{or} \quad 2x - 1 = 0 \qquad \text{Zero product property.}$$

$$\underline{-8 \ -8} \qquad\qquad \underline{+1 \ +1} \qquad \text{Properties of equality.}$$

$$5x + 0 = -8 \qquad 2x + 0 = 1$$

$$\frac{5x}{5} = \frac{-8}{5} \qquad\qquad \frac{2x}{2} = \frac{1}{2} \qquad \text{Division property of equality.}$$

$$x = -\frac{8}{5} \quad \text{or} \quad x = \frac{1}{2} \qquad \text{Simplify.}$$

(b) Check.

▶ Check: $x = -\dfrac{8}{5}$

$$10x^2 + 11x - 8 = 0$$

$$10\left(-\frac{8}{5}\right)^2 + 11\left(-\frac{8}{5}\right) - 8 = 0$$

$$10\left(\frac{64}{25}\right) - \frac{88}{5} - 8 = 0$$

$$(2 \cdot 5)\left(\frac{64}{5 \cdot 5}\right) - \frac{88}{5} - 8 = 0$$

$$\frac{128}{5} - \frac{88}{5} - 8 = 0$$

$$\frac{128}{5} - \frac{88}{5} - \frac{40}{5} = 0$$

$$0 = 0 \quad \text{True.}$$

Check: $x = \dfrac{1}{2}$

$$10x^2 + 11x - 8 = 0$$

$$10\left(\frac{1}{2}\right)^2 + 11\left(\frac{1}{2}\right) - 8 = 0$$

$$10\left(\frac{1}{4}\right) + \frac{11}{2} - 8 = 0$$

$$(2 \cdot 5)\left(\frac{1}{2 \cdot 2}\right) + \frac{11}{2} - 8 = 0$$

$$\frac{5}{2} + \frac{11}{2} - 8 = 0$$

$$\frac{5}{2} + \frac{11}{2} - \frac{16}{2} = 0$$

$$0 = 0 \quad \text{True.}$$

We can extend the zero product property to solve an equation with more than two factors whose product is 0. For example, if a, b, and c are real numbers and $abc = 0$, then $a = 0$ or $b = 0$ or $c = 0$.

EXAMPLE 4 | **(a)** Use the zero product property to solve $4z^3 - 36z = 0$.

SOLUTION ▶

$$4z^3 - 36z = 0 \qquad \text{The greatest common factor is } 4z.$$
$$4z(z^2 - 9) = 0 \qquad \text{Factor out the greatest common factor.}$$
$$4z(z + 3)(z - 3) = 0 \qquad \text{Difference of two squares pattern.}$$

$$
\begin{array}{lll}
4z = 0 \quad \text{or} \quad z + 3 = 0 \quad \text{or} \quad z - 3 = 0 & \text{Zero product property.} \\
\dfrac{4z}{4} = \dfrac{0}{4} \qquad\qquad \dfrac{-3 \ -3}{z + 0 = -3} \qquad \dfrac{+3 \ +3}{z + 0 = 3} & \begin{array}{l}\text{Properties of equality.} \\ \text{Simplify.}\end{array} \\
z = 0 \quad \text{or} \qquad z = -3 \quad \text{or} \qquad z = 3 & \text{This equation has three solutions.}
\end{array}
$$

(b) Check.

▶ Check: $z = 0$ Check: $z = -3$ Check: $z = 3$

$$
\begin{array}{lll}
4z^3 - 36z = 0 & 4z^3 - 36z = 0 & 4z^3 - 36z = 0 \\
4(0)^3 - 36(0) = 0 & 4(-3)^3 - 36(-3) = 0 & 4(3)^3 - 36(3) = 0 \\
0 - 0 = 0 & 4(-27) + 108 = 0 & 4(27) - 108 = 0 \\
0 = 0 \quad \text{True.} & 0 = 0 \quad \text{True.} & 0 = 0 \quad \text{True.}
\end{array}
$$

If a factor does not include a variable, it will not produce a solution.

EXAMPLE 5 | Use the zero product property to solve $5x + 10 = 0$.

SOLUTION ▶

$$
\begin{array}{ll}
5x + 10 = 0 & \\
5(x + 2) = 0 & \text{Factor out the greatest common factor, 5.} \\
5 = 0 \quad \text{or} \quad x + 2 = 0 & \text{Zero product property.} \\
\qquad\qquad\quad \dfrac{-2 \ -2}{x + 0 = -2} & \text{Subtraction property of equality.} \\
\qquad\qquad\quad x + 0 = -2 & \text{Simplify.} \\
\qquad\qquad\quad x = -2 & \text{Simplify.}
\end{array}
$$

The equation $5 = 0$ is false. The only solution of this equation is $x = -2$.

The zero product property applies only when the product of two factors equals 0. In the next example, we rewrite the equation so that it equals 0 before factoring and using the zero product property.

EXAMPLE 6 | **(a)** Use the zero product property to solve $(x + 4)(x - 3) = 8$.

SOLUTION ▶ This equation does not equal 0. Before factoring and using the zero product property, simplify and rewrite the equation so that it does equal 0.

$$
\begin{array}{ll}
(x + 4)(x - 3) = 8 & \text{The product of two factors is not 0.} \\
x(x - 3) + 4(x - 3) = 8 & \text{Distributive property.} \\
x^2 - 3x + 4x - 12 = 8 & \text{Distributive property.} \\
x^2 + x - 12 = 8 & \text{Combine like terms.} \\
\qquad\qquad\qquad \dfrac{-8 \ -8}{x^2 + x - 20 = 0} & \text{Subtraction property of equality.} \\
x^2 + x - 20 = 0 & \text{Simplify. The equation equals 0.} \\
\\
(x + 5)(x - 4) = 0 & \text{Factor; } ac \text{ method or guess and check.} \\
x + 5 = 0 \quad \text{or} \quad x - 4 = 0 & \text{Zero product property.} \\
\dfrac{-5 \ \ -5}{x + 0 = -5} \qquad \dfrac{+4 \ +4}{x + 0 = 4} & \text{Property of equality.} \\
x + 0 = -5 \qquad\quad x + 0 = 4 & \text{Simplify.} \\
x = -5 \quad \text{or} \qquad x = 4 & \text{Simplify.}
\end{array}
$$

(b) Check.

Check: $x = -5$	Check: $x = 4$
$(x + 4)(x - 3) = 8$	$(x + 4)(x - 3) = 8$
$(-5 + 4)(-5 - 3) = 8$	$(4 + 4)(4 - 3) = 8$
$(-1)(-8) = 8$	$(8)(1) = 8$
$8 = 8$ True.	$8 = 8$ True.

Using the Zero Product Property to Solve a Quadratic Equation

1. If the quadratic equation is not in standard form, $ax^2 + bx + c = 0$, use the properties of equality to rewrite it in standard form.

2. Factor.

3. Use the zero product property.

4. Solve each equation.

5. Check.

We use the zero product property to solve quadratic equations in standard form that can be factored. In Chapter 7, we will learn other methods for solving quadratic equations that cannot be factored.

Practice Problems

For problems 1–6, use the zero product property to solve.

1. $x^2 + 7x + 12 = 0$	**2.** $5x^3 + 35x^2 + 60x = 0$	**3.** $x^2 - 9x = 0$
4. $4x^3 - 100x = 0$	**5.** $(x - 7)(x + 1) = 9$	**6.** $(x - 1)^2 = 25$

Applications

A polynomial equation in one variable can represent the relationship of the area and width of a rectangle. In the next example, the length of a rectangle is three feet shorter than twice its width and the area is 20 ft². If W = width, then the length equals $2W - 3$. Since the area of a rectangle equals the product of the length and width, $W(2W - 3) = 20$. To find W, rewrite this equation in standard form, factor, and use the zero product property.

EXAMPLE 7 The length of a rectangle is 3 ft shorter than twice its width. The area of the rectangle is 20 ft². Find its length and width.

SOLUTION ▶ **Step 1 Understand the problem.**

Although there are two unknowns, the length and the width, we need to use a polynomial equation in one variable to find these unknowns. The information about the length depends on the width, so the unknown is the width of the rectangle.

W = width

Width ▢

Length

Step 2 Make a plan.

Since the length is 3 ft less than twice the width, the length equals $2W - 3$ ft. Since we know the area of the rectangle, we can use the formula $A = LW$.

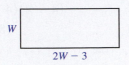

W

$2W - 3$

Step 3 Carry out the plan.

$A = LW$	Formula for area of a rectangle.
$20 = (2W - 3)W$	Replace variables; remove the units.
$20 = 2W^2 - 3W$	Distributive property.
$\underline{-20 \qquad\qquad -20}$	Subtraction property of equality.
$0 = 2W^2 - 3W - 20$	Simplify; $a = 2; b = -3; c = -20; ac = -40$
$0 = 2W^2 - 8W + 5W - 20$	$(-8)(5) = -40; -8 + (5) = -3$
$0 = (2W^2 - 8W) + (5W - 20)$	Group terms with common factors.
$0 = 2W(W - 4) + 5(W - 4)$	Factor the greatest common factor from each group.
$0 = (W - 4)(2W + 5)$	Factor again; the common factor is $W - 4$.
$0 = W - 4 \quad$ or $\quad 0 = 2W + 5$	Zero product property.
$\underline{+4 \qquad +4} \qquad \underline{-5 \qquad\quad -5}$	Properties of equality.
$4 = W + 0 \quad$ or $\quad -5 = 2W + 0$	Simplify.
$4 = W \qquad$ or $\qquad \dfrac{-5}{2} = \dfrac{2W}{2}$	Simplify.
$-\dfrac{5}{2} = W$	Simplify.

Although $W = -\dfrac{5}{2}$ is a solution of the equation $20 = 2W^2 - 3W$, distances are positive measurements, and $-\dfrac{5}{2}$ ft cannot be an answer to the problem. The only solution of the problem is $W = 4$ ft.

$L = 2W - 3$ ft	The length is 3 ft less than twice the width.
$L = 2(\mathbf{4\ ft}) - 3$ ft	Use only the positive value for W.
$L = \mathbf{8\ ft} - 3$ ft	Simplify.
$L = \mathbf{5\ ft}$	Simplify.

Step 4 Look back.

Since $2(4\text{ ft}) - 3$ ft equals the length of 5 ft and $(4\text{ ft})(5\text{ ft})$ equals the given area of 20 ft^2, the answer seems reasonable. We can learn from this problem that in some situations, only one of the solutions of an equation can be the answer to a problem.

Step 5 Report the solution.

The length of the rectangle is 5 ft, and the width of the rectangle is 4 ft.

Practice Problems

For problems 7–8, use the five steps from Section 1.5.

7. The length of a rectangle is 2 ft more than its width. The area of the rectangle is 15 ft². Find the length and the width of the rectangle.

8. The length of a rectangle is 5 in. less than three times the width. The area of the rectangle is 112 in.². Find the length and the width of the rectangle.

2.4 VOCABULARY PRACTICE

Match the term with its description.

1. A polynomial whose only factors are itself and 1
2. The product of any number and zero is always this.
3. The product of the length and width of a rectangle
4. A nonrepeating, nonterminating decimal number
5. A polynomial with three terms
6. The degree of $ax^2 + bx + c$
7. A unit of area
8. $b^2 - 4ac$
9. $a(b + c) = ab + ac$
10. If $ab = 0$, then $a = 0$ or $b = 0$.

A. area
B. discriminant
C. distributive property
D. irrational number
E. prime polynomial
F. square feet
G. trinomial
H. two
I. zero
J. zero product property

2.4 Exercises

Follow your instructor's guidelines for showing your work.

For exercises 1–12, use the zero product property to solve.

1. $(z + 8)(z - 2) = 0$

2. $(x + 5)(x - 7) = 0$

3. $x(x + 1)(2x - 3) = 0$

4. $p(p + 3)(4p - 5) = 0$

5. $3d(2d - 7)(5d + 1) = 0$

6. $4y(6y - 7)(3y + 1) = 0$

7. $-2x(x + 8) = 0$

8. $-4x(x + 6) = 0$

9. $5(c + 3) = 0$

10. $6(d - 9) = 0$

11. $7(p + 3)(p - 3) = 0$

12. $8(b - 1)(b + 1) = 0$

13. A student using the zero product property said that the solutions of the equation $(x + 2)(x - 6) = 15$ are $x = -2$ and $x = 6$. Explain what is wrong with this thinking.

14. A student using the zero product property said that the solutions of the equation $3(x + 8) = 0$ are $x = 3$ and $x = -8$. Explain what is wrong with this thinking.

For exercises 15–44,
(a) use the zero product property to solve.
(b) check the solution(s).

15. $x^2 - 4x - 45 = 0$

16. $x^2 - 4x - 21 = 0$

17. $j^2 - 9 = 0$

18. $m^2 - 36 = 0$

19. $d^3 - 5d^2 + 4d = 0$

20. $k^3 - 5k^2 + 6k = 0$

21. $4x^2 + 20x + 25 = 0$

22. $4x^2 + 28x + 49 = 0$

23. $4a^2 - 11a - 3 = 0$

24. $5y^2 - 9y - 2 = 0$

25. $4x^2 - 25 = 0$

26. $9x^2 - 4 = 0$

27. $6p^2 + 41p + 63 = 0$

28. $6h^2 + 43h + 55 = 0$

29. $15x^2 + x - 2 = 0$

30. $28x^2 + x - 2 = 0$

31. $c^2 - 7c = 0$

32. $b^2 - 3b = 0$

33. $c^2 + c = 0$

34. $b^2 + b = 0$

35. $6w^2 - 13w - 15 = 0$

36. $5u^2 - 2u - 16 = 0$

37. $12w^2 - 26w - 30 = 0$

38. $15u^2 - 6u - 48 = 0$

39. $4x - 32 = 0$

40. $8x - 32 = 0$

41. $3m^2 - 48 = 0$

42. $5k^2 - 125 = 0$

43. $6x^2 + 11x - 7 = 0$

44. $6x^2 - 5x - 25 = 0$

45. Explain why we cannot use the zero product property to solve *all* polynomial equations that are equal to 0.

46. Many quadratic equations in one variable have two solutions. The equation $x^2 + 8x + 16 = 0$ has only one solution. Explain why.

For exercises 47–72,
(a) use the zero product property to solve.
(b) check.

47. $x^2 + 13x = -42$

48. $x^2 + 11x = -18$

49. $f^2 - f + 10 = 16$

50. $m^2 + m + 12 = 32$

51. $p^2 = 3p$

52. $j^2 = 8j$

53. $p^2 = -3p$

54. $j^2 = -8j$

55. $-p^2 = 3p$

56. $-j^2 = 8j$

57. $v^2 = 16$

58. $w^2 = 25$

59. $9v^2 = 16$

60. $49w^2 = 25$

61. $12x^2 = -11x - 2$

62. $10x^2 = 23x + 5$

63. $6x^2 - 11x = 35$

64. $6x^2 + x = 35$

65. $3f^2 - 21f = 54$

66. $2y^2 - 2y = 84$

67. $x^3 - 3x^2 = 4x$

68. $x^3 - 3x^2 = 10x$

69. $(x - 6)(x + 4) = 24$

70. $(x - 8)(x + 5) = 14$

71. $(a + 3)^2 = 1$

72. $(x + 5)^2 = 9$

For exercises 73–84, use a polynomial equation in one variable and the five steps to find the length and width of the rectangle.

73. The length of a rectangle is 3 ft shorter than twice its width. Its area is 54 ft².

74. The length of a rectangle is 3 ft shorter than twice its width. Its area is 35 ft².

75. A rectangle is 2 m longer than it is wide. Its area is 80 m².

76. A rectangle is 3 in. longer than it is wide. Its area is 70 in.².

77. The length of a rectangle is 1 ft more than twice its width. Its area is 105 ft².

78. The length of a rectangle is 1 ft more than three times its width. Its area is 52 ft².

79. The width of a rectangle is 7 in. less than its length. Its area is 294 in.².

80. The width of a rectangle is 12 in. less than its length. Its area is 325 in.².

81. The length of a rectangle is 2 cm less than three times its width. Its area is 176 cm².

82. The length of a rectangle is 4 cm less than twice its width. Its area is 126 cm².

83. The width of a rectangle is one-half of its length. Its area is 968 m².

84. The width of a rectangle is one-fourth of its length. Its area is 1156 m².

For exercises 85–88, the formula for the area of a triangle is area $= \frac{1}{2}$(base)(height). Use a polynomial equation in one variable to find the base and height of the triangle.

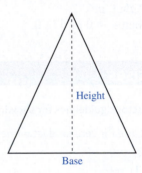

85. The height of a triangle is 2 ft more than its base. Its area is 60 ft².

86. The height of a triangle is 4 ft more than its base. Its area is 70 ft².

87. The base of a triangle is three times its height. Its area is 54 in.².

88. The base of a triangle is five times its height. Its area is 160 in.².

Problem Solving: Practice and Review

Follow your instructor's guidelines for using the five steps as outlined in Section 1.5, p. 51.

89. In November 2007, 138 million people each watched an average of 3 hr, 15 min per month of on-line video. This was about three-quarters of the Internet users in the United States in 2007. Find the number of Internet users in the United States in November 2007. (*Source:* www .nytimes.com, Jan. 17, 2008)

90. The formula for the future value A of an investment earning simple interest is $A = P + PRT$. The principal (original amount invested) is P, the annual simple interest rate in decimal form is R, and T is the number of years of the investment. Find the annual interest rate required for an investment of \$7500 to have a future value of \$9900 in 4 years.

91. A student is beginning an exercise program. Each week, she wants to walk an average of 2.5 mi per day. In the first six days of a week, she walked 1.5 mi, 1.75 mi, 5 mi, 1 mi, 3 mi, and 2 mi. Find the distance she needs to walk on the seventh day for her average distance to be 2.5 mi per day.

92. A semi truck is hauling a container. The volume of the container is given in cubic meters and cubic feet. Use these measurements to find the number of cubic feet in a cubic meter. Round to the nearest tenth.

Courtesy of the Author

Find the Mistake

For exercises 93–96, the completed problem has one mistake.
(a) Describe the mistake in words, or copy down the whole problem and highlight or circle the mistake.
(b) Do the problem correctly.

93. **Problem:** Solve: $x^2 + 5x = -6$

Incorrect Answer: $x^2 + 5x = -6$
$$x(x + 5) = -6$$
$$x = -6 \quad \text{or} \quad x + 5 = -6$$
$$\frac{-5 \quad -5}{x = -11}$$
$$x = -6 \quad \text{or} \quad x = -11$$

94. **Problem:** Solve: $2x^2 + 4x - 126 = 0$

Incorrect Answer: $2x^2 + 4x - 126 = 0$
$$2(x^2 + 2x - 63) = 0$$
$$2[x^2 + 9x - 7x - 63] = 0$$
$$2[x(x + 9) - 7(x + 9)] = 0$$
$$2(x + 9)(x - 7) = 0$$
$$2 = 0 \quad \text{or} \quad x + 9 = 0 \quad \text{or} \quad x - 7 = 0$$
$$\frac{-9 \ -9}{x = -9} \qquad \frac{+7 \ +7}{x = 7}$$
$$x = 2 \quad \text{or} \quad x = -9 \quad \text{or} \quad x = 7$$

95. **Problem:** Solve: $2x^2 + 11x - 21 = 0$

Incorrect Answer: $2x^2 + 11x - 21 = 0$
$$2x^2 - 14x + 3x - 21 = 0$$
$$(2x^2 - 14x) + (3x - 21) = 0$$
$$2x(x - 7) + 3(x - 7) = 0$$
$$(x - 7)(2x + 3) = 0$$
$$x - 7 = 0 \quad \text{or} \quad 2x + 3 = 0$$
$$\frac{+7 \ +7}{x + 0 = 7} \qquad \frac{-3 \quad -3}{2x + 0 = -3}$$
$$x = 7 \qquad \qquad \frac{2x}{2} = \frac{-3}{2}$$
$$x = -\frac{3}{2}$$
$$x = 7 \quad \text{or} \quad x = -\frac{3}{2}$$

96. **Problem:** Use the zero product property to solve $x^2 + 9x + 16 = -4$.

Incorrect Answer: Since $x^2 + 9x + 16$ is prime, it cannot be factored. We cannot use the zero product property to solve the equation.

Review

97. a. Graph 6 on a number line graph.
    b. What is the distance between 0 and 6?

98. a. Graph –6 on a number line graph.
    b. What is the distance between 0 and –6?

99. Solve: $4x - 12 = 32$

100. Solve: $-(4x - 12) = 32$

SUCCESS IN COLLEGE MATHEMATICS

101. Attach a copy of your notes for today's class.

102. Rewriting notes as soon as possible after class is often recommended to improve your memory and understanding. To rewrite notes, rewrite anything that you cannot read well, and fill in words or steps that you did not have time to write. At the bottom of the notes, write summary sentences such as "The main topic of today's class was . . . ," "It is important to know that . . . ," and "The most common mistake made in these kinds of problems is . . . ." Rewrite the notes from today's class. If you did not take notes, explain why.

© John Kwan/Shutterstock

Many issues in American politics are complex. Sometimes we have reasons to be in favor of a policy. At the same time, we have other reasons to oppose it. Our responses to a survey about this policy may be *ambivalent*. A formula for measuring ambivalence includes absolute value. In this section, we will study the definition of absolute value, evaluate formulas that include absolute value, and solve absolute value equations.

<table>
<tr><td>

SECTION 2.5

</td><td>

Absolute Value Equations

</td></tr>
</table>

After reading the text, working the practice problems, and completing assigned exercises, you should be able to:

1. Evaluate a formula that includes an absolute value.

2. Solve an absolute value equation in one variable and check the solution(s).

Absolute Value

$|-3| = 3$ and $|3| = 3$

3 units 3 units

−3   0   3

Figure 1

The **absolute value of a number** is equal to its distance from 0 on the real number line (Figure 1). The notation for absolute value is $|\ |$. Because we measure distance with numbers that are greater than or equal to 0, the absolute value of a real number is also greater than or equal to 0. For example, $|-3| = 3$ and $|17| = 17$.

To find the vertical distance between two points, (x_1, y_1) and (x_2, y_2) (Figure 2), we can use the formula $d = |y_2 - y_1|$. For the horizontal distance between two points, the formula is $d = |x_2 - x_1|$. The absolute value in these formulas ensures that the distance will be greater than or equal to 0.

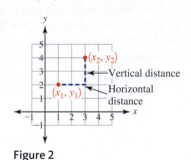

Figure 2

EXAMPLE 1   Find the vertical distance between $(-7, 4)$ and $(5, -1)$.

SOLUTION ►

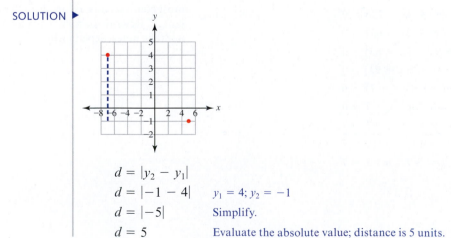

$d = |y_2 - y_1|$

$d = |-1 - 4|$ $y_1 = 4; y_2 = -1$

$d = |-5|$ Simplify.

$d = 5$ Evaluate the absolute value; distance is 5 units.

In choosing investments, stockbrokers study changes in stock prices. In a given day, a stock will have a high price, a low price, and a close price, which is the price at the end of trading.

The *true range* value of a stock is the greatest of three differences:

- The absolute value of the high price of the stock in a given day minus the low price of the stock on the same day, |high − low|
- The absolute value of the high price minus the previous day close price, |high − previous day close|
- The absolute value of the low price minus the previous day close price, |low − previous day close|

EXAMPLE 2 The table shows the opening, high, low, and previous day closing for the stock of Apple, Inc. on June 25, 2010. Find the true range value for this stock on this date.

SOLUTION ▶

Date	Open	High	Low	Previous day close
June 25	$270.06	$270.27	$265.81	$269

Source: http://Finance.yahoo.com

The greatest of |high − low|, |high − previous day close|, and |low − previous day close| is the true range value.

$$|\text{high} - \text{low}|$$
$$= |\$270.27 - \$265.81|$$
$$= \$4.46$$

$$|\text{high} - \text{previous day close}|$$
$$= |\$270.27 - \$269|$$
$$= \$1.27$$

$$|\text{low} - \text{previous day close}|$$
$$= |\$265.81 - \$269|$$
$$= \$3.19$$

The true range value of Apple, Inc. stock for June 25, 2010, is $4.46.

Practice Problems

1. Find the horizontal distance between $(-3, 10)$ and $(1, -17)$.

2. On a political survey, "ambivalent" people have both positive and negative responses about a policy or candidate. The Griffin formula for ambivalence A is $A = \dfrac{P + N}{2} - |P - N|$. The number of positive responses is P, and the number of negative responses is N. Find the value for ambivalence if a person responds positively to six questions and responds negatively to four questions about a policy.
(*Source:* Steenbergen, M.R., et al., 2004. "The Not so Ambivalent Public: Policy Attitudes in the Political Culture of Ambivalence." In Saris, et al., Eds., 2004 *Studies in Public Opinion, Attitudes, Non-attitudes. Measurement Error and Change,* Princeton, NJ: Princeton University Press)

Absolute Value Equations

The absolute value of a number is its distance from 0 on the real number line. We can also write a definition for absolute value using variables, mathematical symbols, and a minimum of words.

> **Absolute Value**
>
> For any real number x,
>
> if $x \geq 0$, then $|x| = x$.
>
> if $x < 0$, then $|x| = -x$.

The definition tells us that if a number is greater than or equal to 0, then it is equal to its absolute value. For example, $|6| = 6$. If a number is less than 0, then its absolute value is equal to its opposite. For example, $|-6| = 6$ and $-(-6) = 6$. When we write "if $x < 0$, then $|x| = -x$," we are not saying that the absolute value of x is a negative number. We are saying that when x is a negative number, its absolute value is the opposite of this negative number and is therefore a positive number.

To solve absolute value equations in the form $|x| = a$, we need to explore two possible cases. In the first case, the value of the expression inside the absolute value sign, x, is greater than or equal to 0. From the definition, we know that $x = a$. In the second case, the value of the expression inside the absolute value sign, x, is less than 0. From the definition, we know that $-x = a$. So, the solutions of $|x| = a$ are $x = a$ or $-x = a$.

In the next example, we solve $|p - 2| = 5$. We need to solve two equations, $p - 2 = 5$ and $-(p - 2) = 5$.

EXAMPLE 3 | (a) Solve: $|p - 2| = 5$

SOLUTION ▶ If $p - 2$ is greater than or equal to 0, $p - 2 = 5$.

If $p - 2$ is less than 0, $-(p - 2) = 5$.

$p - 2 = 5$	First case: $p - 2 \geq 0$
$\underline{+2 \ \ +2}$	Addition property of equality.
$p + 0 = 7$	Simplify.
$p = 7$	One solution is 7.

$-(p - 2) = 5$	Second case: $p - 2 < 0$
$-p + 2 = 5$	Distributive property.
$\underline{-2 \ \ -2}$	Subtraction property of equality.
$-p + 0 = 3$	Simplify.
$\dfrac{-p}{-1} = \dfrac{3}{-1}$	Division property of equality.
$p = -3$	The other solution is -3.

(b) Check.

To check, replace the variable in the original equation with each solution.

Check: $p = 7$ Check: $p = -3$

$|p - 2| = 5$ $|p - 2| = 5$

$|7 - 2| = 5$ $|-3 - 2| = 5$

$|5| = 5$ $|-5| = 5$

$5 = 5$ True. $5 = 5$ True.

We use the same method to solve the next equation.

EXAMPLE 4   (a) Solve: $|2x - 8| - 10 = -4$

SOLUTION ▶   If $2x - 8$ is greater than or equal to 0, $2x - 8 - 10 = -4$.

If $2x - 8$ is less than 0, $-(2x - 8) - 10 = -4$.

$$2x - 8 - 10 = -4 \qquad \text{First case: } 2x - 8 \geq 0$$
$$2x - 18 = -4 \qquad \text{Simplify.}$$
$$\underline{ +18 \quad +18} \qquad \text{Addition property of equality.}$$
$$2x + 0 = 14 \qquad \text{Simplify.}$$
$$\frac{2x}{2} = \frac{14}{2} \qquad \text{Division property of equality.}$$
$$x = 7 \qquad \text{One solution is 7.}$$

$$-(2x - 8) - 10 = -4 \qquad \text{Second case: } 2x - 8 < 0$$
$$-2x + 8 - 10 = -4 \qquad \text{Distributive property.}$$
$$-2x - 2 = -4 \qquad \text{Simplify.}$$
$$\underline{ +2 \quad +2} \qquad \text{Addition property of equality.}$$
$$-2x + 0 = -2 \qquad \text{Simplify.}$$
$$\frac{-2x}{-2} = \frac{-2}{-2} \qquad \text{Division property of equality.}$$
$$x = 1 \qquad \text{The other solution is 1.}$$

(b) Check.

▶
$$\text{Check: } x = 7 \qquad\qquad \text{Check: } x = 1$$
$$|2x - 8| - 10 = -4 \qquad\qquad |2x - 8| - 10 = -4$$
$$|2(7) - 8| - 10 = -4 \qquad\qquad |2(1) - 8| - 10 = -4$$
$$|14 - 8| - 10 = -4 \qquad\qquad |2 - 8| - 10 = -4$$
$$|6| - 10 = -4 \qquad\qquad |-6| - 10 = -4$$
$$6 - 10 = -4 \qquad\qquad 6 - 10 = -4$$
$$-4 = -4 \quad \text{True.} \qquad\qquad -4 = -4 \quad \text{True.}$$

In the next example, the absolute value follows a subtraction sign. The equation $12 - -(3x + 15) = -21$ is equivalent to $12 + (3x + 15) = -21$.

EXAMPLE 5   (a) Solve: $12 - |3x + 15| = -21$

SOLUTION ▶   If $3x + 15$ is greater than or equal to 0, $12 - (3x + 15) = -21$.

If $3x + 15$ is less than 0, $12 - -(3x + 15) = -21$.

Simplifying, we have $12 + (3x + 15) = -21$.

$$12 - (3x + 15) = -21 \qquad \text{First case: } 3x + 15 \geq 0$$
$$12 - 3x - 15 = -21 \qquad \text{Distributive property.}$$
$$-3x - 3 = -21 \qquad \text{Combine like terms.}$$
$$\underline{ +3 \quad\quad +3} \qquad \text{Addition property of equality}$$
$$-3x + 0 = -18 \qquad \text{Simplify.}$$
$$\frac{-3x}{-3} = \frac{-18}{-3} \qquad \text{Division property of equality.}$$
$$x = 6 \qquad \text{One solution is 6.}$$

$$12 + (3x + 15) = -21 \qquad \text{Second case: } 3x + 15 < 0$$
$$3x + 27 = -21 \qquad \text{Simplify.}$$
$$\underline{-27 \quad -27} \qquad \text{Subtraction property of equality.}$$
$$3x + 0 = -48 \qquad \text{Simplify.}$$
$$\frac{3x}{3} = \frac{-48}{3} \qquad \text{Division property of equality.}$$
$$x = -16 \qquad \text{The other solution is } -16.$$

(b) Check.

▶

Check: $x = 6$	Check: $x = -16$
$12 - |3x + 15| = -21$	$12 - |3x + 15| = -21$
$12 - |3(6) + 15| = -21$	$12 - |3(-16) + 15| = -21$
$12 - |18 + 15| = -21$	$12 - |-48 + 15| = -21$
$12 - |33| = -21$	$12 - |-33| = -21$
$12 - 33 = -21$	$12 - 33 = -21$
$-21 = -21$ True.	$-21 = -21$ True.

Some absolute value equations in one variable have no solution.

EXAMPLE 6   **(a)** Solve: $|a + 4| = -9$

SOLUTION ▶   Since an absolute value cannot equal a negative number, this equation has no solution. If we follow the process of solving an absolute value equation and replace the solutions in the original equation, the equation is not true. These are **extraneous solutions**.

$$a + 4 = -9 \qquad \text{First case: } a + 4 \geq 0$$
$$\underline{-4 \quad -4} \qquad \text{Subtraction property of equality.}$$
$$a + 0 = -13 \qquad \text{Simplify.}$$
$$\cancel{a = -13} \qquad \text{Simplify.}$$

$$-(a + 4) = -9 \qquad \text{Second case: } a + 4 < 0$$
$$-a - 4 = -9 \qquad \text{Distributive property.}$$
$$\underline{+4 \quad +4} \qquad \text{Addition property of equality.}$$
$$-a + 0 = -5 \qquad \text{Simplify.}$$
$$\frac{-a}{-1} = \frac{-5}{-1} \qquad \text{Division property of equality.}$$
$$\cancel{a = 5} \qquad \text{Simplify.}$$

(b) Check.

▶

Check: $a = -13$	Check: $a = 5$
$|a + 4| = -9$	$|a + 4| = -9$
$|-13 + 4| = -9$	$|5 + 4| = -9$
$|-9| = -9$	$|9| = -9$
$9 = -9$ False.	$9 = -9$ False.

Since $a = -13$ and $a = 5$ are extraneous solutions, this equation has no solution. Ask your instructor how to mark extraneous solutions.

Solving an Absolute Value Equation in One Variable

$$|x - h| + k = C \qquad h, k, \text{ and } C \text{ are real numbers}$$

1. Use the definition of absolute value to rewrite the equation as two equations. For $|x - h| + k = C$, these equations are $(x - h) + k = C$ and $-(x - h) + k = C$.

2. Solve each equation.

3. Check the solution(s).

Practice Problems

For problems 3–8,
(a) solve.
(b) check.

3. $|3x| - 27 = 12$ 4. $|3x - 27| = 12$ 5. $|3x + 6| - 27 = -12$
6. $8 - |4a - 20| = -56$ 7. $|2x - 14| = 0$ 8. $|x + 9| + 21 = 3$

Equations with Two Absolute Values

In the next example, the equation includes two absolute values. The expression in each absolute value can be greater than or equal to 0, or it can be less than 0. When we use the absolute value definition, this results in four different equations to solve. However, after simplifying, we find that there are only two unique equations to solve. For example, we can show that $c + 6 = 2c - 4$ and $-(c + 6) = -(2c - 4)$ are equivalent.

$-(c + 6) = -(2c - 4)$	Show that this equation is equivalent to $c + 6 = 2c - 4$.
$\mathbf{-1}(c + 6) = \mathbf{-1}(2c - 4)$	Multiply by 1.
$\dfrac{-1(c + 6)}{-1} = \dfrac{-1(2c - 4)}{-1}$	Division property of equality.
$c + 6 = 2c - 4$	Simplify. The equations are equivalent.

EXAMPLE 7

(a) Solve: $|c + 6| = |2c - 4|$

SOLUTION ▶

If $c + 6 \geq 0$ and $2c - 4 \geq 0$, then $c + 6 = 2c - 4$.

If $c + 6 < 0$ and $2c - 4 \geq 0$, then $-(c + 6) = 2c - 4$.

If $c + 6 \geq 0$ and $2c - 4 < 0$, then $c + 6 = -(2c - 4)$.

If $c + 6 < 0$ and $2c - 4 < 0$, then $-(c + 6) = -(2c - 4)$.

Since $c + 6 = 2c - 4$ and $-(c + 6) = -(2c - 4)$ are equivalent and $-(c + 6) = 2c - 4$ and $c + 6 = -(2c - 4)$ are equivalent, we need to solve two equations.

Solve: $c + 6 = 2c - 4$	First equation.
$\underline{\quad -6 \qquad\qquad -6\quad}$	Subtraction property of equality.
$c + 0 = 2c - \mathbf{10}$	Simplify.
$\underline{-2c \qquad\qquad -2c\quad}$	Subtraction property of equality.
$-c = \mathbf{0} - 10$	Simplify.
$\dfrac{-c}{-1} = \dfrac{\mathbf{-10}}{-1}$	Division property of equality.
$c = 10$	Simplify. One solution is 10.

Solve: $-(c + 6) = 2c - 4$ Second equation.

$-\mathbf{1}(c + 6) = 2c - 4$ Multiply by 1.

$-\mathbf{c} - \mathbf{6} = 2c - 4$ Distributive property.

$\underline{+6 \qquad +6}$ Addition property of equality.

$-c + \mathbf{0} = 2c + \mathbf{2}$ Simplify.

$\underline{-2c \qquad\quad -2c}$ Subtraction property of equality.

$-\mathbf{3}c + \mathbf{0} = \mathbf{0} + 2$ Simplify.

$-3c = 2$ Simplify.

$\dfrac{-3c}{-3} = \dfrac{2}{-3}$ Division property of equality.

$c = -\dfrac{2}{3}$ Simplify. The other solution is $-\dfrac{2}{3}$.

(b) Check.

▶ Check: $c = 10$

$|c + 6| = |2c - 4|$

$|\mathbf{10} + 6| = |2(\mathbf{10}) - 4|$

$|16| = |16|$

$16 = 16$ True.

Check: $c = -\dfrac{2}{3}$

$|c + 6| = |2c - 4|$

$\left|-\dfrac{2}{3} + 6\right| = \left|2\left(-\dfrac{2}{3}\right) - 4\right|$

$\left|-\dfrac{2}{3} + \dfrac{18}{3}\right| = \left|-\dfrac{4}{3} - \dfrac{12}{3}\right|$

$\left|\dfrac{16}{3}\right| = \left|-\dfrac{16}{3}\right|$

$\dfrac{16}{3} = \dfrac{16}{3}$ True.

Absolute value equations can include two absolute values and a constant, and absolute value equations can also have infinitely many solutions. Solving these absolute value equations is often a topic in college algebra or precalculus classes.

Practice Problems

For problems 9–11,
(a) solve.
(b) check.

9. $|2x + 9| = |x - 3|$ **10.** $|2a + 1| = |6a - 3|$

11. $|3x| = |2x + 10|$

2.5 VOCABULARY PRACTICE

Match the term with its description. Terms may be used more than once.

1. The distance a number is from 0 on the real number line
2. If a real number is greater than 0, its absolute value is this.
3. If a real number is less than 0, its absolute value is this.
4. If a real number is equal to 0, its absolute value is this.
5. $\geq$
6. $\leq$
7. $|x_2 - x_1|$
8. $|y_2 - y_1|$
9. $a(b + c) = ab + ac$
10. We can subtract any real number from both sides of an equation without changing its solution.

A. absolute value
B. distributive property
C. greater than or equal to
D. greater than 0
E. horizontal distance between two points (x_1, y_1) and (x_2, y_2)
F. less than or equal to
G. subtraction property of equality
H. vertical distance between two points (x_1, y_1) and (x_2, y_2)
I. zero

2.5 Exercises

Follow your instructor's guidelines for showing your work.

For exercises 1–4, use the formulas in this section.

1. Find the vertical distance between $(-5, -3)$ and $(1, -6)$.

2. Find the vertical distance between $(-9, -6)$ and $(-3, -15)$.

3. Find the horizontal distance between $(-5, -3)$ and $(1, -6)$.

4. Find the horizontal distance between $(-9, -6)$ and $(3, -15)$.

For exercises 5–6, the Lavine Test,

$$A = \frac{P_1 + P_2 + N_1 + N_2}{4} - (|P_1 - P_2| + |N_1 - N_2|)$$

is a measure of ambivalence for the choice between two political candidates. The number of positive answers about Candidate 1 is P_1, the number of positive answers about Candidate 2 is P_2, the number of negative answers about Candidate 1 is N_1, and the number of negative answers about Candidate 2 is N_2. (*Source:* G. Glasgow, "Reconsidering Tests for Ambivalence in Political Choice Survey Data," unpublished paper, 2004)

5. Find the value for ambivalence if a person responded positively to 3 questions about Candidate 1, positively to 6 questions about Candidate 2, negatively to 7 questions about Candidate 1, and negatively to 4 questions about Candidate 2.

6. Find the value for ambivalence if a person responded positively to 1 question about Candidate 1, positively to 8 questions about Candidate 2, negatively to 9 questions about Candidate 1, and negatively to 2 questions about Candidate 2.

For exercises 7–8, use the definition of true range value and the formulas in Example 2 in this section.

Date	Open	High	Low	Previous day close
Nov. 28	44.2400	44.3125	40.0625	40.6250
Nov. 29	40.8125	41.2188	37.6250	39.8750
Nov. 30	38.1562	39.3750	36.5000	38.0312

7. Find the true range value for Nov. 29.

8. Find the true range value for Nov. 30.

For exercises 9–10, a *forecast* is a prediction of the number of parts needed to meet customer demand for a product. A measure of the error in a forecast is *forecast error percent*:

$$\text{forecast error percent} = \left(\frac{|\text{actual demand} - \text{forecast demand}|}{\text{actual demand}} \right) \cdot 100$$

9. A forecast predicted that 76,000 parts would be needed for production in May. The actual number of parts needed was 58,000. Find the forecast error percent. Round to the nearest percent.

10. A forecast predicted that 2500 parts would be needed for production in June. The actual number of parts needed was 2350. Find the forecast error percent. Round to the nearest percent.

For exercises 11–50,
(a) solve.
(b) check.

11. $|p| = 2$

12. $|a| = 9$

13. $|x| + 9 = 15$

14. $|x| + 8 = 21$

15. $|x| - 9 = 15$

16. $|x| - 8 = 21$

17. $|x + 9| = 2$

18. $|x + 8| = 3$

19. $|x + 9| = -2$

20. $|x + 8| = -3$

21. $|w - 3| - 6 = 2$

22. $|d - 5| - 9 = 3$

23. $|z - 3| - 6 = -2$

24. $|z - 5| - 9 = -3$

25. $8 - |7 - v| = 15$

26. $9 - |5 - u| = 23$

27. $|2x| - 6 = -6$

28. $|3x| - 6 = -6$

29. $|4x + 8| = 24$

30. $|6x + 18| = 48$

31. $|3z + 21| - 15 = 24$

32. $|3h + 36| - 18 = 54$

33. $12 - |2x - 8| = -36$

34. $20 - |2x + 10| = -60$

35. $-6 + |-3x + 9| = 27$

36. $-8 + |-4x + 12| = 60$

37. $-40 = |5w - 20| - 80$

38. $-45 = |5k - 20| - 85$

39. $-20 = |-5w - 30| - 50$

40. $-18 = |-9p - 27| - 45$

41. $-2 = -|-w - 4| - 8$

42. $-3 = -|-n - 6| - 9$

43. $|3x + 7| = 19$

44. $|4x + 9| = 21$

45. $|5x + 6| - 3 = 24$

46. $|5x + 9| - 21 = 42$

47. $|2x + 1| = 0$

48. $|3x + 1| = 0$

49. $-|2x + 1| = 0$

50. $-|3x + 1| = 0$

51. A linear equation such as $x + 2 = 7$ has one solution. An absolute value equation such as $|x + 2| = 7$ has two solutions. Explain why an absolute value equation often has two solutions.

52. Many absolute value equations have two solutions. The absolute value equation $|x + 5| = 0$ has only one solution. Explain why.

53. **a.** Create an absolute value equation with one solution.
    b. Solve the equation.
    c. Check.

54. Create an absolute value equation with no solution.

For exercises 55–62,
(a) solve.
(b) check.

55. $|2c - 6| = |c + 15|$

56. $|2b - 3| = |b + 9|$

57. $|2a + 8| = |4a + 12|$

58. $|3w + 12| = |6w + 18|$

59. $|z + 3| = |z - 5|$

60. $|h + 4| = |h - 8|$

61. $|x + 6| = -|2x + 5|$

62. $|x + 11| = -|2x + 1|$

For exercises 63–80, the equations are a mixture of polynomial equations (Section 2.4) and absolute value equations (Section 2.5).
(a) Solve.
(b) Check.

63. $3x + 12 = 54$

64. $4x - 28 = 56$

65. $|2z + 5| + 9 = 40$

66. $|2p - 15| + 7 = 60$

67. $a^2 + 14a + 49 = 0$

68. $f^2 + 18f + 81 = 0$

69. $6x^2 + 31x = -40$

70. $6x^2 + 23x = -7$

71. $-8 = |2x + 4| - 20$

72. $-12 = |2x + 6| - 60$

73. $3(x - 9) + x = 2(2x + 9)$

74. $6(x - 2) + 3x = 9(x - 5)$

75. $\frac{1}{2}(6x + 8) = 3x + 4$

76. $\frac{1}{4}(12x + 20) = 3x + 5$

77. $p^2 - 36 = 0$

78. $c^2 - 144 = 0$

79. $3z^2 + 12z - 20 = 3z^2 + 10z + 42$

80. $4k^2 + 15k + 21 = 4k^2 + 18k - 54$

Problem Solving: Practice and Review

Follow your instructor's guidelines for using the five steps as outlined in Section 1.5, p. 51.

81. Find the amount of the total freight carried by air into or out of the United States on cargo planes.

    Every day, 20 million pounds of cargo, or 16 percent of the total freight carried by air into or out of the United States, are transported by passenger planes, according to the International Air Cargo Association. The vast majority, or 84 percent, is carried on cargo planes. (*Source:* www.nytimes.com, Nov. 1, 2010)

82. An individual counted her paces as she walked 200 feet several times. The first time, she walked 80 paces. The second time, she walked 77 paces. The third time she walked 71 paces. Find her pace factor. Round to the nearest tenth.

When measuring distance by pacing, an individual must know his or her pace factor (PF). The pace factor is determined by pacing a known distance, 100 to 300 feet, several times and dividing the distance by the average number of paces to determine the feet per pace:

$$PF = \frac{\text{Distance}}{\text{Average Number of Paces}}$$

(*Source:* Field, H. *Landscape Surveying.* Delmar Cengage Learning, 2003)

83. Find the percent increase from 1999 to 2008 in students majoring in hospitality management at the University of Central Florida. Round to the nearest percent.

San Diego State's School of Hospitality & Tourism Management started with 13 students in 2001 and now has 500. At the University of Central Florida, the number of students majoring in hospitality management rose from about 85 in 1999 to about 2000 today [2008]. And at Florida Atlantic University, the number of hospitality students has grown from three in 2004 to 200 today [2008]. (*Source:* www.usatoday.com, Jan. 8, 2008)

84. According to the U.S. Census Bureau, the population of the United States in 2010 was about 308,746,000 people. Predict the number of these people who got sick from a foodborne disease. Round to the nearest thousand.

CDC estimates that each year roughly 1 in 6 Americans ... gets sick, 128,000 are hospitalized, and 3,000 die of food-borne diseases. (*Source:* www.cdc.gov, Apr 15, 2011)

Find the Mistake

For exercises 85–88, the completed problem has one mistake.
(a) Describe the mistake in words, or copy down the whole problem and highlight or circle the mistake.
(b) Do the problem correctly.

85. Problem: Solve: $|x - 6| + 15 = 2$

Incorrect Answer: $|x - 6| + 15 = 2$

$$
\begin{array}{lll}
x - 6 + 15 = 2 & \text{or} & -(x - 6) + 15 = 2 \\
x + 9 = 2 & & -x + 6 + 15 = 2 \\
\underline{-9 \quad -9} & & -x + 21 = 2 \\
x + 0 = -7 & & \underline{-21 \quad -21} \\
x = -7 & & -x + 0 = -19 \\
& & \dfrac{-x}{-1} = \dfrac{-19}{-1} \\
& & x = 19
\end{array}
$$

$$x = -7 \quad \text{or} \quad x = 19$$

86. Problem: Solve: $|x - 3| - 9 = -4$

Incorrect Answer: This equation has no solution. An absolute value cannot equal a negative number.

87. Problem: Solve: $|-3x + 6| = 9$

Incorrect Answer: $|-3x + 6| = 9$

$$
\begin{array}{lll}
-3x + 6 = 9 & \text{or} & -(-3x + 6) = 9 \\
\underline{-6 \quad -6} & & 3x + 6 = 9 \\
-3x + 0 = 3 & & \underline{-6 \quad -6} \\
\dfrac{-3x}{-3} = \dfrac{3}{-3} & & 3x + 0 = 3 \\
x = -1 & & \dfrac{3x}{3} = \dfrac{3}{3} \\
& & x = 1
\end{array}
$$

$$x = -1 \quad \text{or} \quad x = 1$$

88. Problem: Solve: $|x - 6| + 13 = 4$

Incorrect Answer: $|x - 6| + 13 = 4$

$$
\begin{array}{lll}
x - 6 + 13 = 4 & \text{or} & -(x - 6) + 13 = 4 \\
x + 7 = 4 & & -x + 6 + 13 = 4 \\
\underline{-7 \quad -7} & & -x + 19 = 4 \\
x = -3 & & \underline{-19 \quad -19} \\
& & -x + 0 = -15 \\
& & -x = -15 \\
& & \dfrac{}{-1} \quad \dfrac{}{-1} \\
& & x = 15
\end{array}
$$

$$x = -3 \quad \text{or} \quad x = 15$$

Review

For exercises 89–92, graph the equation on an *xy*-coordinate system.

89. $x = 2$

90. $3x - 7y = -21$

91. $y = -\dfrac{3}{7}x + 9$

92. $y = 1$

SUCCESS IN COLLEGE MATHEMATICS

93. After completing part of a chapter, you can use the Study Plan for Review at the end of the chapter to assess how well you understand what you have learned. Read the Study Plan questions for Sections 2.1–2.5. Are there any Study Plan questions that your instructor does not expect you to be able to do?

94. Answer the Study Plan questions with a "Yes" or "No." How many of these questions did you answer with "No"?

95. Do you think you are prepared to take a test on the material from Sections 2.1–2.5?

© Thomas M. Perkins/Shutterstock

A parent may tell a child, "You must make your bed *and* put your laundry away before you can go outside." The child knows that he or she must do *both* things before being allowed to go outside. This is an example of a **conjunction**. But if the parent says, "You must make your bed *or* put your laundry away before you can go outside," the child can choose to do either chore (or both). This is a **disjunction**. In this section, we will study conjunctions, disjunctions, and absolute value inequalities.

SECTION 2.6

Conjunctions, Disjunctions, and Absolute Value Inequalities

After reading the text, working the practice problems, and completing assigned exercises, you should be able to:

1. Find the intersection or union of two sets.

2. Identify a conjunction or disjunction.

3. Graph a conjunction or disjunction of sets.

4. Solve a compound inequality.

5. Solve an absolute value inequality.

Intersection, Union, Conjunction, and Disjunction

A **set** is a collection of elements. The **intersection** of two sets is a new set that includes only the elements that are common to both sets. The symbol for intersection is ∩. The **union** of two sets is a new set that includes each of the elements found in the sets. The symbol for union is ∪.

EXAMPLE 1 Set A is {paper, book, pencil, eraser}. Set B is {book, pencil, calculator}.

(a) Describe $A \cap B$.

SOLUTION ▶ $A \cap B$ is {book, pencil} because only "book" and "pencil" are elements of both sets A and B.

(b) Describe $A \cup B$.

▶ $A \cup B$ is {paper, book, pencil, eraser, calculator} because each element is found in either set A or set B or in both sets A and B.

The solution of an inequality in one variable is a set. For example, the solution of the inequality $x > 0$ is the set of real numbers greater than 0. The number 0 is a **boundary value**. We can represent this solution with an inequality, a number line graph, or interval notation.

EXAMPLE 2   (a) Use a number line graph and interval notation to represent the inequality $x > 0$.

SOLUTION ▶   To show that 0 is a boundary value but is not included in the set, use a parenthesis on the graph and in the interval.

$-2\ -1\quad 0\quad 1\quad 2\quad 3\quad 4\quad 5$

$(0, \infty)$

(b) Use a number line graph and interval notation to represent the inequality $x \leq -3$.

▶   To show that -3 is a boundary value and is included in the set, use a bracket on the graph and in the interval.

$-8\ -7\ -6\ -5\ -4\ -3\ -2\ -1$

$(-\infty, -3]$

The **intersection** of sets is a **conjunction**. A conjunction is itself a set. We can use a **compound inequality** to represent some conjunctions. For example, $3 < x < 7$ is a compound inequality. It is the set of real numbers that are greater than 3 and less than 7.

EXAMPLE 3   Set A is the set of the solutions of the inequality $x > 3$. Set B is the set of solutions of the inequality $x < 7$.

(a) Describe $A \cap B$.

SOLUTION ▶ $A \cap B$ is the set of real numbers that are greater than 3 and less than 7.

(b) Use a compound inequality to represent $A \cap B$.

▶ $3 < x < 7$

(c) Use a number line graph to represent $A \cap B$.

▶ $1\quad 2\quad 3\quad 4\quad 5\quad 6\quad 7\quad 8\quad 9$

(d) Use interval notation to represent $A \cap B$.

▶ $(3, 7)$

(e) Use the two inequalities and the word *and* to represent $A \cap B$.

▶ $x > 3$ and $x < 7$

In the next example, the weight of an animal must be between two given amounts. We can represent the set of allowed weights as a compound inequality or as two inequalities connected with the word *and*.

EXAMPLE 4   To compete in a fair, a hog must weigh at least 220 lb and no more than 260 lb. The weight of the hog is w.

(a) Use two inequalities with the word *and* to represent this conjunction.

SOLUTION ▶ $w \geq 220$ lb and $w \leq 260$ lb

(b) Use a compound inequality to represent this conjunction.

▶ 220 lb $\leq w \leq 260$ lb

Only the hogs that weigh greater than or equal to 220 lb and less than or equal to 260 lb can compete in the fair. These hogs belong to both sets in the conjunction.

A **union** of two sets is a **disjunction**. Two separate inequalities connected by the word *or* are a disjunction. To represent a disjunction with interval notation, use the union symbol, ∪. For example, the union of the sets $x < 4$ and $x > 6$ is $(-\infty, 4) \cup (6, \infty)$.

EXAMPLE 5 Set A is the set of the solutions of the inequality $x > 6$. Set B is the set of the solutions of the inequality $x < 4$.

(a) Describe $A \cup B$.

SOLUTION ► $A \cup B$ is the set of real numbers that are greater than 6 or that are less than 4.

(b) Use a number line graph to represent $A \cup B$.

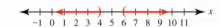

(c) Use interval notation to represent $A \cup B$.

► $(-\infty, 4) \cup (6, \infty)$

(d) Use the word *or* and two inequalities to represent $A \cup B$.

► $x > 6$ or $x < 4$

(e) Use a compound inequality to represent $A \cup B$.

►   We cannot write a disjunction as a single compound inequality.

In the next example, the length of a fish must be greater than one amount or less than another amount, an example of what fishery managers call a "slot limit." We use a disjunction to describe the set of allowed lengths.

EXAMPLE 6 At Battle Lake, anglers may keep walleyes that are less than 17 in. long or greater than 26 in. long. If $L = $ length of fish, use a disjunction to describe the allowed lengths.

SOLUTION ► $L < 17$ in.   or $L > 26$ in.

A fish that can be kept belongs to *only one set* in the disjunction.

Conjunction   The intersection of sets is a conjunction. These sets can be the solutions of two inequalities. A number line graph, a compound inequality, and interval notation can represent the conjunction of two inequalities. The word *and* is used to describe a conjunction.

Disjunction   The union of sets is a disjunction. These sets can be the solutions of two inequalities. A number line graph and interval notation can represent the disjunction of two inequalities. The word *or* is used to describe a disjunction.

Practice Problems

For problems 1–2, set A is {3, 4, 5, 6, 7}, and set B is {3, 5, 7, 9, 11}.

1. Describe $A \cap B$.   2. Describe $A \cup B$.

3. At a local video store, a parent rents only movies with a G rating that are also available on DVD.
   a. Is this an example of a conjunction or a disjunction?
   b. Explain.

4. Parents may bring either fruit juice boxes or bottles of water for snacks after soccer games.
   a. Is this an example of a conjunction or a disjunction?
   b. Explain.

5. Set A is the set of real numbers greater than 0. Set B is the set of real numbers less than 3.
   a. Use a number line graph to represent $A \cap B$.
   b. Use interval notation to represent $A \cap B$.
   c. Use a compound inequality to represent $A \cap B$.

6. Set A is the set of real numbers less than 0. Set B is the real numbers greater than 3.
   a. Use a number line graph to represent $A \cup B$.
   b. Use interval notation to represent $A \cup B$.

Solving Conjunctions

A conjunction can be the intersection of two inequalities. To visualize this intersection, we can represent each inequality with a number line graph and with interval notation.

EXAMPLE 7   **(a)** Use a number line graph and interval notation to represent $x \geq 2$.

SOLUTION ▶

$[2, \infty)$

(b) Use a number line graph and interval notation to represent $x \leq 7$.

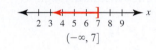

$(-\infty, 7]$

(c) Use a number line graph and interval notation to represent the intersection of $x \geq 2$ and $x \leq 7$.

$[2, 7]$

(d) Use a compound inequality to represent the intersection of $x \geq 2$ and $x \leq 7$.

$2 \leq x \leq 7$

In Section 1.3, we used the properties of inequality to isolate the variable and solve an inequality in one variable.

The Addition and Subtraction Property of Inequality   Adding a real number to or subtracting a real number from both sides of an inequality does not change the solution of the inequality.

The Multiplication and Division Property of Inequality   We can multiply both sides or divide both sides of an inequality by a real number other than 0. If the number is greater than 0, the solution is not changed. If the number is less than 0, reverse the direction of the inequality sign.

To solve $x + 3 > 8$, we use the subtraction property of inequality and subtract 3 from *both sides* of the inequality. With compound inequalities (conjunctions), we also use the properties of inequality to isolate the variable. However, since there are three expressions in the inequality, we must apply the property of inequality to each expression. For example, to solve $-12 \le 2x + 4 \le 18$, first subtract 4 from *each expression* in the compound inequality.

EXAMPLE 8   **(a)** Solve: $-12 \le 2x + 4 \le 18$

SOLUTION ▶

$$-12 \le 2x + 4 \le 18$$

$$\underline{\quad -4 \qquad\qquad -4 \quad -4 \quad}$$ Subtraction property of inequality.

$$-16 \le 2x + 0 \le 14$$ Simplify.

$$\frac{-16}{2} \le \frac{2x}{2} \le \frac{14}{2}$$ Division property of inequality.

$$-8 \le x \le 7$$ Simplify. The variable is isolated.

(b) Use a number line graph and interval notation to represent the solution.

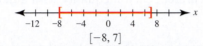

$$[-8, 7]$$

(c) Describe the solution.

▶   The solution of $-12 \le 2x + 4 \le 18$ is the set of real numbers that are greater than or equal to -8 and less than or equal to 7.

If each expression in the compound inequality is divided by a negative number, reverse the direction of *each* inequality sign.

EXAMPLE 9   **(a)** Solve: $-6 \le -4x - 3 < 12$

SOLUTION ▶

$$-6 \le -4x - 3 < 12$$

$$\underline{+3 \qquad\qquad +3 \quad +3 \quad}$$ Addition property of inequality.

$$-3 \le -4x + 0 < 15$$ Simplify.

$$\frac{-3}{-4} \ge \frac{-4x}{-4} > \frac{15}{-4}$$ Division property of inequality; reverse the signs.

$$\frac{3}{4} \ge x > -\frac{15}{4}$$ Simplify. The variable is isolated.

(b) Use a number line graph and interval notation to represent the solution.

▶   The numbers in an interval on the real number line increase from left to right. We need to rewrite this solution with the least number on the left. The direction of the inequality signs reverse. The compound inequality $\frac{3}{4} \ge x > -\frac{15}{4}$ is equivalent to $-\frac{15}{4} < x \le \frac{3}{4}$.

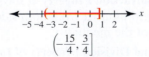

$$\left(-\frac{15}{4}, \frac{3}{4}\right]$$

Practice Problems

For problems 7–9,
(a) use a number line graph to represent each inequality.
(b) use interval notation to represent each inequality.

7. $-6 \le x \le 2$ **8.** $0 < x < 4$ **9.** $-8 \le x < -1$

For problems 10–11,
(a) solve.
(b) use interval notation to represent the solution.

10. $12 < 4x + 8 \leq 28$ **11.** $-20 < -4x + 8 \leq 36$

Solving Disjunctions

A disjunction can be the union of two inequalities. To visualize this union, we can represent each inequality with a number line graph and with interval notation.

EXAMPLE 10   **(a)** Use a number line graph and interval notation to represent $x \geq 4$.

SOLUTION ▶

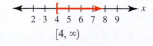

$[4, \infty)$

(b) Use a number line graph and interval notation to represent $x < 1$.

▶

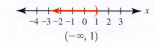

$(-\infty, 1)$

(c) Use a number line graph and interval notation to represent the union $x \geq 4$ or $x < 1$.

▶

$(-\infty, 1) \cup [4, \infty)$

In the next example, the sets in the disjunction share some elements.

EXAMPLE 11   **(a)** Use a number line graph and interval notation to represent $x > 1$.

SOLUTION ▶

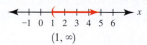

$(1, \infty)$

(b) Use a number line graph and interval notation to represent $x > 3$.

▶

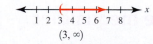

$(3, \infty)$

(c) Use a number line graph and interval notation to represent the union $x > 1$ or $x > 3$.

▶

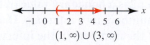

$(1, \infty) \cup (3, \infty)$

To solve a disjunction, use the properties of inequality to solve each inequality.

EXAMPLE 12 (a) Solve: $3x + 9 \geq 12$ or $3x + 9 < -3$

SOLUTION ▶

$3x + 9 \geq 12$	or	$3x + 9 < -3$	
$\underline{-9 -9}$		$\underline{-9 -9}$	Subtraction property of inequality.
$3x + 0 \geq 3$		$3x + 0 < -12$	Simplify.
$\dfrac{3x}{3} \geq \dfrac{3}{3}$		$\dfrac{3x}{3} < \dfrac{-12}{3}$	Division property of inequality.
$x \geq 1$	or	$x < -4$	Simplify.

(b) Use a number line graph and interval notation to represent the solution.

▶
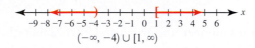

$$(-\infty, -4) \cup [1, \infty)$$

(c) Describe the solution.

▶ The solution of $3x + 9 \geq 12$ or $3x + 9 < -3$ is the union of the set of real numbers that are less than -4 with the set of real numbers that are greater than or equal to 1.

A disjunction cannot be represented as a compound inequality. For example, the disjunction $x \geq 1$ or $x < -4$ is *not* equivalent to $-4 < x \geq 1$.

Practice Problems

For problems 12–14, represent each disjunction with
(a) a number line graph.
(b) interval notation.

12. $x \leq -6$ or $x \geq 1$ **13.** $x < -4$ or $x > -1$
14. $x < -3$ or $x < 1$

For problems 15–16,
(a) solve.
(b) use a number line graph to represent the solution.
(c) use interval notation to represent the solution.

15. $0 \leq 6x - 12$ or $4x \geq 0$
16. $3x - 9 \geq 24$ or $3x - 9 < -72$

Absolute Value Inequalities

The solution of the linear equation $4x = 8$ is $x = 2$. If we replace x in the equation with any other number, the equation is not true. In contrast, the linear inequality $4x > 8$ has infinitely many solutions. All real numbers greater than 2 are solutions of this inequality. The set that includes all of the solutions of the inequality is its **solution set**. The number 2 is a **boundary value** of this inequality. The boundary value divides the set of real numbers into intervals. If a number in an interval is a solution of the inequality, all of the numbers in the interval are solutions.

In Chapter 1, we used the properties of inequality to solve linear inequalities. We can instead find the boundary values of the inequalities, use the boundary values to divide the set of real numbers into intervals, and choose a number in each interval to check if the interval includes solutions of the inequality.

EXAMPLE 13   Solve the linear inequality $-3x + 6 < 18$.

(a) Find any real boundary values.

SOLUTION ▶   To find the boundary values, solve the linear equation $-3x + 6 = 18$.

$$-3x + 6 = 18$$

$\underline{\quad -6 \quad -6}$	Subtraction property of equality.
$-3x + 0 = 12$	Simplify.
$\dfrac{-3x}{-3} = \dfrac{12}{-3}$	Division property of equality.
$x = -4$	Simplify; -4 is the boundary value.

(b) Use the boundary value to solve the inequality.

▶   The boundary value divides the set of real numbers into intervals. Choose a number in each interval, and check whether it is a solution of the inequality.

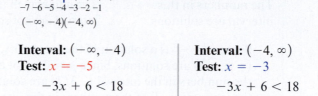

$(-\infty, -4)(-4, \infty)$

Interval: $(-\infty, -4)$	**Interval:** $(-4, \infty)$
Test: $x = -5$	**Test:** $x = -3$
$-3x + 6 < 18$	$-3x + 6 < 18$
$-3(-5) + 6 < 18$	$-3(-3) + 6 < 18$
$21 < 18$ False.	$15 < 18$ True.

Since $x = -5$ is not a solution of the inequality, none of the numbers in the interval $(-\infty, -4)$ are solutions. Since $x = -3$ is a solution of the inequality, all of the numbers in the interval $(-4, \infty)$ are solutions.

(c) Use a number line graph and interval notation to represent the solution.

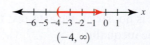

$(-4, \infty)$

To solve an absolute value inequality, we follow a similar process. To find the boundary values for a strict inequality $(>$ or $<)$, rewrite the absolute value inequality as an equation and solve it.

EXAMPLE 14   Solve the absolute value inequality $|p - 2| > 5$.

(a) Find any real boundary values for $|p - 2| > 5$.

SOLUTION ▶   To find any real boundary values, solve $|p - 2| = 5$.

$p - 2 = 5$	Case 1: If $p - 2 \geq 0$, then $p - 2 = 5$.
$\underline{\quad +2 \quad +2}$	Addition property of inequality.
$p + 0 = 7$	Simplify.
$p = 7$	Simplify; 7 is one boundary value.
$-(p - 2) = 5$	Case 2: If $p - 2 < 0$, then $-(p - 2) = 5$.
$-p + 2 = 5$	Distributive property.
$\underline{\quad -2 \quad -2}$	Subtraction property of equality.
$-p + 0 = 3$	Simplify.
$\dfrac{-p}{-1} = \dfrac{3}{-1}$	Division property of inequality.
$p = -3$	Simplify; -3 is another boundary value.

(b) Use the boundary values to solve the inequality.

▶ The boundary values $p = 7$ and $p = -3$ divide the set of real numbers into three intervals, $(-\infty, -3)$, $(-3, 7)$, and $(7, \infty)$. Choose a number in each interval and check whether it is a solution of the inequality.

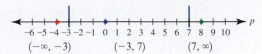

$$(-\infty, -3) \qquad (-3, 7) \qquad (7, \infty)$$

Interval: $(-\infty, -3)$	**Interval:** $(-3, 7)$	**Interval:** $(7, \infty)$
Test: $p = -4$	**Test:** $p = 0$	**Test:** $p = 8$
$\lvert p - 2 \rvert > 5$	$\lvert p - 2 \rvert > 5$	$\lvert p - 2 \rvert > 5$
$\lvert -4 - 2 \rvert > 5$	$\lvert 0 - 2 \rvert > 5$	$\lvert 8 - 2 \rvert > 5$
$\lvert -6 \rvert > 5$	$\lvert -2 \rvert > 5$	$\lvert 6 \rvert > 5$
$6 > 5$ True.	$2 > 5$ False.	$6 > 5$ True.
The numbers in this interval are solutions.	The numbers in this interval are not solutions.	The numbers in this interval are solutions.

Since $p = -4$ is a solution of the inequality, all of the numbers in the interval $(-\infty, -3)$ are solutions. Since $p = 0$ is not a solution of the inequality, none of the numbers in the interval $(-3, 7)$ are solutions. Since $p = 8$ is a solution of the inequality, all of the numbers in the interval $(7, \infty)$ are solutions. The solution of the inequality is the union of the set of real numbers less than -3 with the set of real numbers greater than 7. This is a disjunction.

(c) Use a number line graph and interval notation to represent the solution.

▶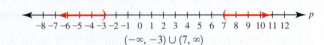

$$(-\infty, -3) \cup (7, \infty)$$

When an inequality is not strict, we also test the boundary values to see whether they are solutions of the inequality.

EXAMPLE 15 Solve: $\lvert 2a + 14 \rvert - 3 \le -1$.

(a) Find any real boundary values for $\lvert 2a + 14 \rvert - 3 \le -1$.

SOLUTION ▶ To find any real boundary values, solve $\lvert 2a + 14 \rvert - 3 = -1$.

$(2a + 14) - 3 = -1$	Case 1: If $2a + 14 \ge 0$, then $(2a + 14) - 3 = -1$.
$2a + 11 = -1$	Combine like terms.
$\underline{-11 \quad -11}$	Subtraction property of inequality.
$2a + 0 = -12$	Simplify.
$\dfrac{2a}{2} = \dfrac{-12}{2}$	Division property of inequality.
$a = -6$	Simplify; -6 is one boundary value.
$-(2a + 14) - 3 = -1$	Case 2: If $2a + 14 < 0$, then $-(2a + 14) - 3 = -1$.
$-1(2a + 14) - 3 = -1$	Multiply by 1.
$-2a - 14 - 3 = -1$	Distributive property.
$-2a - 17 = -1$	Combine like terms.
$\underline{+17 \quad +17}$	Addition property of equality.
$-2a + 0 = 16$	Simplify.

$$\frac{-2a}{-2} = \frac{16}{-2} \quad \text{Division property of inequality.}$$

$$a = -8 \quad \text{Simplify; } -8 \text{ is another boundary value.}$$

(b) Use the boundary values to solve the inequality.

The boundary values $a = -6$ and $a = -8$ divide the set of real numbers into intervals. Choose a number in each interval, and check whether it is a solution of the inequality.

$$(-\infty, -8) \quad (-8, -6) \quad (-6, \infty)$$

Interval: $(-\infty, -8)$	**Interval:** $(-8, -6)$	**Interval:** $(-6, \infty)$
Test: $a = -9$	**Test:** $a = -7$	**Test:** $a = -5$
$|2a + 14| - 3 \le -1$	$|2a + 14| - 3 \le -1$	$|2a + 14| - 3 \le -1$
$|2(-9) + 14| - 3 \le -1$	$|2(-7) + 14| - 3 \le -1$	$|2(-5) + 14| - 3 \le -1$
$|-4| - 3 \le -1$	$|0| - 3 \le -1$	$|4| - 3 \le -1$
$1 \le -1$	$-3 \le -1$	$1 \le -1$
False.	True.	False.
The numbers in this interval are not solutions.	The numbers in this interval are solutions.	The numbers in this interval are not solutions.

Since $a = -9$ is not a solution of the inequality, none of the numbers in the interval $(-\infty, -8)$ are solutions. Since $a = -7$ is a solution of the inequality, all of the numbers in the interval $(-8, -6)$ are solutions. Since $a = -5$ is not a solution of the inequality, none of the numbers in the interval $(-6, \infty)$ are solutions.

Since the inequality $|2a + 14| - 3 \le -1$ is not strict, we need to also test the boundary values to see whether they are solutions.

Test: $a = -8$	**Test:** $a = -6$
$|2a + 14| - 3 \le -1$	$|2a + 14| - 3 \le -1$
$|2(-8) + 14| - 3 \le -1$	$|2(-6) + 14| - 3 \le -1$
$|-2| - 3 \le -1$	$|2| - 3 \le -1$
$-1 \le -1$ True.	$-1 \le -1$ True.

Both $a = -8$ and $a = -6$ are solutions of the inequality.

The solution of the inequality is the intersection of the set of real numbers greater than *or equal to* -8 with the set of real numbers less than *or equal to* -6. This is a conjunction.

(c) Use a number line graph and interval notation to represent the solution of $|2a + 14| - 3 \le -1$.

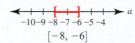

$$[-8, -6]$$

In the next example, the inequality has no boundary values. An absolute value inequality with no boundary values may have no solution, or its solution may be the set of real numbers. To determine whether an absolute value inequality with no boundary values has a solution, check any real number. If the number is a solution of the inequality, then any real number is a solution. If the number is not a solution, then the inequality has no solution.

EXAMPLE 16 Solve: $|x - 12| + 5 < 2$

(a) Find any real boundary values for $|x - 12| + 5 < 2$.

SOLUTION ▶ To find any real boundary values, solve $|x - 12| + 5 = 2$.

$$|x - 12| + 5 = 2$$

$$\underline{\;\; -5 \;\; -5} \qquad \text{Subtraction property of equality.}$$

$$|x - 12| + \mathbf{0} = \mathbf{-3} \qquad \text{Simplify.}$$

$$|x - 12| = -3 \qquad \text{Simplify.}$$

Since the absolute value of an expression cannot equal a negative number, this equation has no solution, and the inequality has no boundary values.

(b) To find whether the inequality has a solution, check any real number.

▶ Since arithmetic with 0 is relatively easy, check $x = 0$.

Test: $x = 0$

$$|x - 12| + 5 < 2$$

$$|0 - 12| + 5 < 2$$

$$12 + 5 < 2$$

$$17 < 2 \quad \text{False.}$$

Since $17 < 2$ is false, $x = 0$ is not a solution, and no real number is a solution. The inequality $|x - 12| + 5 < 2$ has no solution.

Solving an Absolute Value Inequality

1. Find the boundary values of the inequality.

2. Use the boundary values to divide the set of real numbers into intervals.

3. Choose a number from each interval. If this number is a solution of the inequality, every number in the interval is a solution.

4. If the inequality is not strict, test the boundary value(s) to see whether it is a solution.

5. If the inequality has no real boundary values, either it has no solution or its solution is the set of real numbers. Choose a real number. If this number is a solution of the inequality, then every real number is a solution. If this number is not a solution, then the inequality has no solution.

Practice Problems

For problems 17–19,
(a) find any real boundary values.
(b) solve the inequality.
(c) use interval notation to represent the solution.

17. $|x + 3| < 12$ **18.** $|2x - 8| \geq 24$ **19.** $|x - 4| + 6 > 1$

2.6 VOCABULARY PRACTICE

Match the term with its description.

1. The union of two sets
2. The intersection of two sets
3. ∩
4. ∪
5. | |
6. We can divide both sides of an inequality by a term (that is not equal to 0). If the divisor is less than 0, we reverse the direction of the inequality sign.
7. ≥
8. ≤
9. >
10. <

A. absolute value
B. conjunction
C. disjunction
D. division property of inequality
E. greater than
F. greater than or equal to
G. intersection
H. less than
I. less than or equal to
J. union

2.6 Exercises

Follow your instructor's guidelines for showing your work.

1. Set A is {red, black, blue, yellow, purple}. Set B is {black, yellow, purple, green}.
   a. Describe $A \cap B$.
   b. Describe $A \cup B$.

2. Set Q is {bed, chair, couch, lamp}. Set R is {bed, couch, rug, lamp}.
   a. Describe $Q \cap R$.
   b. Describe $Q \cup R$.

3. Set A is the set of real numbers greater than 4. Set B is the set of real numbers less than 9.
   a. Describe $A \cap B$.
   b. Describe $A \cup B$.

4. Set Q is the set of integers less than 7. Set R is the set of integers greater than 3.
   a. Describe $Q \cap R$.
   b. Describe $Q \cup R$.

5. Describe two sets for which the intersection of the sets contains no elements.

6. Describe two sets for which the intersection of the sets contains only one element.

7. At a restaurant, a customer can order any item that appears on the breakfast menu or that appears on the lunch menu. Is this an example of a conjunction or a disjunction?

8. The following ad is from a British website. Is the restriction on height in this ad an example of a conjunction or a disjunction?

   Extreme Dodgems Stock Car Racing: Customers buying this activity will receive a voucher valid for ten months. Extreme Dodgems operates throughout the year on selected dates. Drivers must be reasonably fit and aged over 16 years. The minimum height is 4 ft 10 and the maximum height is 6 ft 6. The maximum weight is 20 [stone]. (*Source:* www .experiencemad.co.uk)

© Chris Hellyar/Shutterstock

For exercises 9–18, graph each set on a real number line.

9. a. Graph: $x \leq 3$
   b. Graph: $-2 \leq x$
   c. Graph the set that is the intersection of the sets in parts a and b.
   d. Write the compound inequality that represents this intersection.

10. a. Graph: $x \leq 4$
    b. Graph: $-3 \leq x$
    c. Graph the set that is the intersection of the sets in parts a and b.
    d. Write the compound inequality that represents this intersection.

11. a. Graph: $x \leq -2$
    b. Graph: $-6 \leq x$
    c. Graph the set that is the intersection of the sets in parts a and b.
    d. Write the compound inequality that represents this intersection.

12. a. Graph: $x \leq -1$
    b. Graph: $-7 \leq x$
    c. Graph the set that is the intersection of the sets in parts a and b.
    d. Write the compound inequality that represents this intersection.

13. **a.** Graph: $x \geq 2$
    b. Graph: $x \geq 6$
    c. Graph the set that is the union of the sets in parts a and b.

14. **a.** Graph: $x \geq 3$
    b. Graph: $x \geq 7$
    c. Graph the set that is the union of the sets in parts a and b.

15. **a.** Graph: $x \geq 2$
    b. Graph: $x \leq 6$
    c. Graph the set that is the union of the sets in parts a and b.

16. **a.** Graph: $x \geq 3$
    b. Graph: $x \leq 7$
    c. Graph the set that is the union of the sets in parts a and b.

17. **a.** Graph: $x > -3$
    b. Graph: $x < 0$
    c. Graph the set that is the union of the sets in parts a and b.

18. **a.** Graph: $x > -4$
    b. Graph: $x < 0$
    c. Graph the set that is the union of the sets in parts a and b.

For exercises 19–26,
(a) use a number line graph to represent each conjunction.
(b) use interval notation to represent each conjunction.

19. $2 < x < 8$

20. $-3 < x < 7$

21. $-4 < x \leq 0$

22. $-5 < x \leq 0$

23. $-10 \leq p \leq -3$

24. $-12 \leq w \leq 2$

25. $0 \leq x < 5$

26. $0 \leq x < 4$

For exercises 27–44,
(a) solve the conjunction.
(b) use interval notation to represent the solution.

27. $-5 < x + 7 \leq 10$

28. $-3 < x + 8 \leq 17$

29. $-5 < x - 7 \leq 10$

30. $-3 < x - 8 \leq 17$

31. $4 < x + 9 < 15$

32. $3 < x + 7 < 16$

33. $3 < x - 2 < 10$

34. $6 < x - 1 < 13$

35. $-10 \leq x + 2 \leq -3$

36. $-15 \leq x + 4 \leq -2$

37. $2 \leq 4x + 14 < 34$

38. $6 \leq 3x + 15 < 27$

39. $2 \leq -4x + 14 < 34$

40. $6 \leq -3x + 15 < 27$

41. $6 \leq 3x + 12 < 27$

42. $12 \leq 3x + 6 < 30$

43. $5 \leq 5x + 30 \leq 75$

44. $10 \leq 5x + 35 \leq 90$

45. Use two inequalities and the word *and* to represent $-2 < x < 9$.

46. Use two inequalities and the word *and* to represent $-7 < x < 15$.

For exercises 47–52,
(a) solve the disjunction.
(b) use interval notation to represent the solution.

47. $x + 12 \geq 15$ or $x - 4 \leq -11$

48. $x - 10 \geq 16$ or $x + 8 \leq -14$

49. $5x + 8 \geq 43$ or $5x + 8 \leq 28$

50. $6x + 10 \geq 52$ or $6x - 10 \leq 26$

51. $-2x - 5 > -11$ or $-2x - 5 < -21$

52. $-4x - 12 > 20$ or $-4x - 9 > -33$

For exercises 53–72,
(a) find the boundary values.
(b) solve the inequality.
(c) use a number line graph to represent the solution.
(d) use interval notation to represent the solution.

53. $|x + 8| > 2$

54. $|x + 6| > 2$

55. $|x + 8| < 2$

56. $|x + 6| < 2$

57. $|x - 3| < 9$

58. $|x - 4| < 10$

59. $|a + 2| + 5 \geq 11$

60. $|p + 1| + 8 \geq 20$

61. $|x - 7| - 4 \leq 6$

62. $|c - 3| - 9 \leq 12$

63. $|x + 4| + 9 > 1$

64. $|x + 7| + 8 > 3$

65. $|w| + 4 > 12$

66. $|m| + 5 > 14$

67. $|2x - 6| \geq 20$

68. $|2x - 8| \geq 30$

69. $|3x| + 12 < 21$

70. $|3x| + 30 < 45$

71. $|-3x| + 12 < 21$

72. $|-3x| + 30 < 45$

Problem Solving: Practice and Review

Follow your instructor's guidelines for using the five steps as outlined in Section 1.5, p. 51.

73. According to the U.S. Census, the population in the United States in 2009 was about 307,006,600 people. Find the number of these people who had asthma. Round to the nearest hundred.

    The number of people with asthma continues to grow. One in 12 people . . . had asthma in 2009, compared with 1 in 14 . . . in 2001. (*Source:* www.cdc.gov, May, 2011)

74. Find the percent increase in the number of identity theft incidents from 2008 to 2010. Round to the nearest percent.

In 2010, the IRS identified almost 245,000 identity theft incidents, up from 169,087 in 2009 and just 51,702 in 2008. (*Source:* wsj.com, June 1, 2011)

75. U.S. newspaper advertising revenue in the first three months of 2011 fell 7% compared to the revenue in the first three months of 2010. The total revenue from January through March 2011 was $5.6 billion. This was a dramatic change from the last three months of 2005, when the revenue was $14.3 billion. Find the amount of newspaper advertising revenue in the first three months of 2010. Round to the nearest tenth of a billion. (*Source:* www.miamiherald.com, June 1, 2011)

76. According to the 2010 Census, the population of adults in the United States was about 234.5 million people. Predict how many of these people were overweight or obese. Round to the nearest tenth of a million. (*Source:* www.census.gov, Age and Sex Composition, 2010)

Because more than one-third of children and more than two-thirds of adults in the United States are overweight or obese, the 7th edition of Dietary Guidelines for Americans places stronger emphasis on reducing calorie consumption and increasing physical activity. (*Source:* www.cnpp.usda.gov, Jan. 31, 2011)

Find the Mistake

For exercises 77–80, the completed problem has one mistake.
(a) Describe the mistake in words, or copy down the whole problem and highlight or circle the mistake.
(b) Do the problem correctly.

77. Problem: Solve: $-12 \le -2x + 6 < 10$

Incorrect Answer: $-12 \le -2x + 6 < 10$

$$\underline{ -6 -6 -6}$$
$$-18 \le -2x + 0 < 4$$
$$\frac{-18}{-2} \le \frac{-2x}{-2} < \frac{4}{-2}$$
$$9 \le x < -2$$

78. Problem: Use a number line graph to represent $x \le 6$ or $x \ge 1$.

Incorrect Answer:

$$[1, 6]$$

79. Problem: Solve $|x + 3| > 12$. Use interval notation to describe the solution.

Incorrect Answer: $|x + 3| > 12$

$$x + 3 > 12$$
$$\underline{ -3 -3}$$
$$x > 9 \quad (9, \infty)$$

80. Problem: Use interval notation to represent $-3 \le x < 5$.

Incorrect Answer: $[-3, \infty) \cup (-\infty, 5)$

Review

For exercises 81–84, graph the equation on an *xy*-coordinate system.

81. $2x - 5y = 15$

82. $y = \dfrac{3}{5}x - 9$

83. $y = -x$

84. $x = 4$

SUCCESS IN COLLEGE MATHEMATICS

85. Does your college have a place such as a tutoring center where you can get help?

86. If so, what are the hours that this tutoring center is open?

Study Plan for Review of Chapter 2

SECTION 2.1   Adding, Subtracting, and Multiplying Polynomial Expressions

Ask Yourself	Test Yourself	Help Yourself
Can I ...	Do 2.1 Review Exercises	See these Examples and Practice Problems
identify an expression that is a polynomial?	1–4	Ex. 2, PP 1, 4, 5
identify the degree of a polynomial in one variable?	6–8, 9–12	Ex. 3, PP 1, 4, 5
describe the difference between a monomial, a binomial, and a trinomial?		
identify like terms?	14, 15	Ex. 4–8, PP 6–9
add or subtract polynomial expressions?		
multiply polynomial expressions?	17–18	Ex. 9–17, PP 10–20
use a pattern to multiply polynomial expressions?	20–21	Ex. 17, 19, 21, PP 15–20

Ask Yourself	Test Yourself	Help Yourself
Can I . . .	Do 2.1 Review Exercises	See these Examples and Practice Problems
write a polynomial expression in L or in W that represents the relationship of the length and width of a rectangle?	23, 24	Ex. 22, PP 21, 22
write a polynomial expression that represents the perimeter or area of a rectangle?	25, 26	Ex. 22, PP 21, 22

2.1 Review Exercises

1. Explain why $6^x - 9$ is not a polynomial expression.

For exercises 2–4, is the expression a polynomial?

2. $3x^5 - \dfrac{2}{x}$

3. $7x^2 - 6x$

4. 8

5. Explain how to determine the degree of a polynomial in one variable.

For exercises 6–8, identify the degree of the polynomial.

6. $5x^7 - 9x^4 + 8$

7. $x^3 - 2x^2 + x - 1$

8. $x + 8$

9. Describe the difference between a monomial and a trinomial.

For exercises 10–12, identify each polynomial as a monomial, binomial, or trinomial.

10. $x - 9$

11. $5x^2$

12. $6x^2 - 5x + 2$

13. Explain how to identify terms that are like terms.

For exercises 14–15, simplify.

14. $(x^2 + 5x - 8) + (6x^2 + 2x - 11)$

15. $(x^2 + 5x - 8) - (6x^2 + 2x - 11)$

16. Write the distributive property.

For exercises 17–18, use the distributive property to simplify.

17. $(2x - 9)(5x^2 + 7x - 3)$

18. $6x^2(7x^3 + 9x^2 - 10x + 25)$

19. What is the difference of squares pattern?

For exercises 20–21, use a pattern to simplify.

20. $(3x - 5)(3x + 5)$

21. $(x - 15)(x - 15)$

22. a. Write the formula for finding the area of a rectangle.
b. Write the formula for finding the perimeter of a rectangle.

23. The width of a rectangle is 6 in. less than three times its length, L. Write an expression in L that represents the width.

For exercises 24–26, the length of a rectangle is 3 ft more than five times its width, W. Write an expression or equation in W that represents its

24. length.

25. perimeter.

26. area.

SECTION 2.2   **Factoring Polynomials: Greatest Common Factor and Grouping**

Ask Yourself	Test Yourself	Help Yourself
Can I . . .	Do. 2.2 Review Exercises	See these Examples and Practice Problems
write a prime factorization? identify the greatest common factor of two or more terms?	27–29	Ex. 1–4, PP 1–6
factor out a greatest common factor from a polynomial?	32, 33, 35, 37	Ex. 5–7, PP 7–10
factor a polynomial by grouping?	31, 34, 36	Ex. 8–12, PP 11–14
use a pattern to factor a polynomial?	35, 37–39	Ex. 13–16, PP 15–19
describe what it means if a polynomial is prime? use more than one method of factoring to factor a polynomial completely?	30, 35, 37	Ex. 17–20, PP 20–23

2.2 Review Exercises

27. Explain how to find the greatest common factor of two or more terms.

For exercises 28–29, find the greatest common factor of the terms.

28. $9x^4$; $21x^2$; $15x^3$

29. $105d^2f^3$; $108df^2$

30. To factor $18x^2 + 162x + 360$ completely, explain why factoring out the greatest common factor is the first method to use.

31. Describe a polynomial that can be factored by grouping.

For exercises 32–39, factor completely. Identify prime polynomials.

32. $9a^3b^5 + 27ab$

33. $75p^2x^2 + 49q^3y^2$

34. $7kx - 7ky - 3x + 3y$

35. $36x^2 - 64$

36. $8xy - 2kx + 20y - 5k$

37. $10x^2 - 90$

38. $z^3 + 8$

39. $27z^3 + 125$

SECTION 2.3   ## Factoring Polynomials: Trinomials and a Strategy for Factoring Completely

Ask Yourself	Test Yourself	Help Yourself
Can I . . .	**Do 2.3 Review Exercises**	**See these Examples and Practice Problems**
identify a quadratic trinomial?	41–45, 55	Ex. 2–4, PP 1–4
use the *ac* method to factor a quadratic trinomial?		
use guess and check to factor a quadratic trinomial in which $a = 1$?	40, 46, 47	Ex. 5–7, PP 5–8
find the discriminant of a quadratic trinomial?	57–59	Ex. 8–9, PP 9–12
use the discriminant to determine whether a quadratic trinomial is prime?	58–59	Ex. 8–9, PP 9–12
factor a trinomial that is quadratic in form?	54, 56	Ex. 10, 11, PP 13–15
use more than one method of factoring to factor a polynomial completely?	42, 48, 49, 51–53	Ex. 12–14, PP 16–27

2.3 Review Exercises

For exercises 40–56, factor completely.

40. $x^2 + 9x + 20$

41. $81p^2 + 36p + 4$

42. $24u^2 + 72u + 54$

43. $6w^2 - 19w - 7$

44. $4x^2 + 5x - 6$

45. $2x^2 + 11x - 90$

46. $h^2 - 7h - 30$

47. $k^2 - 13k + 42$

48. $10x^2y - 700xy^3 + 70xy$

49. $-10hm + 15h + 10km - 15k$

50. $12x^2 + 18y^2$

51. $24ac - 6a + 12bc - 3b$

52. $16a^3 - 54b^3$

53. $32p^2 - 50w^2$

54. $a^4 + 15a^2 + 56$

55. $2x^2 - 13x - 7$

56. $2w^6 - 14w^3 - 36$

57. Explain how to find the discriminant of a quadratic trinomial.

For exercises 58–59,
(a) find the discriminant.
(b) use the discriminant to determine whether the quadratic trinomial is prime.

58. $14x^2 + 61x - 9$

59. $x^2 + 3x + 11$

SECTION 2.4   ## Polynomial Equations

Ask Yourself	Test Yourself	Help Yourself
Can I . . .	**Do 2.4 Review Exercises**	**See these Examples and Practice Problems**
state the zero product property?	60–64	Ex. 2–6, PP 1–6
use the zero product property to solve a polynomial equation in one variable and check the solution(s)?		
use a polynomial equation to find the length and the width of a rectangle?	66, 67	Ex. 7, PP 7, 8

2.4 Review Exercises

60. A factored polynomial equation is $x(x + 2)(x - 5) = 0$. Using a, b, and c to represent the factors, write the zero product property that can be used to solve this equation.

For exercises 61–64,
(a) use the zero product property to solve.
(b) check.

61. $x^2 + 10x - 24 = 0$

62. $4z^3 - 4z^2 - 3z = 0$

63. $x^2 - 9x = 0$

64. $16p^2 - 1 = 0$

65. The length of a rectangle is 5 m less than three times its width. If W = width, write an expression in W that represents the length.

For exercises 66–67, use the five steps from Section 1.5.

66. A rectangle is 3 in. longer than twice its width. The area of the rectangle is 44 in.^{2}. Find the length and width of the rectangle.

67. The length of a rectangle is 2 cm less than three times its width. The area of the rectangle is 96 cm^2. Find the length and width of the rectangle.

SECTION 2.5 Absolute Value Equations

Ask Yourself	Test Yourself	Help Yourself
Can I . . .	**Do 2.5 Review Exercises**	**See these Examples and Practice Problems**
evaluate a formula that includes an absolute value?	68	Ex. 1, 2, PP 1, 2
solve an absolute value equation in one variable and check the solution(s)?	71–74	Ex. 3–5, 7, PP 3–7, 9–11
identify an absolute value equation that has no solution?	69	Ex. 6, PP 8

2.5 Review Exercises

68. *Price elasticity of demand* is a measure of how the demand for an item changes if the price changes. For some items, a price increase means that the demand goes down. This is *elastic demand*. For other items, a price increase does not result in a change of demand. People will buy the item at any price. This is *inelastic demand*. A formula for price elasticity of demand is

$$E = \left| \frac{P_2 - P_1}{P_1} \div \frac{Q_2 - Q_1}{Q_1} \right|$$

The price elasticity of demand is E, the current price is P_1, the new price is P_2, the demand at the current price is Q_1, and the demand at the new price is Q_2. A photographer has a digital website. At a price of $80, he sold 20 prints a week.

When he changed the price to $64, he sold 28 prints a week. Find the price elasticity of demand.

69. Explain why the equation $|x - 5| = -7$ has no solution.

70. Many absolute value equations have two solutions. Explain why $|x - 5| = 0$ only has one solution.

For exercises 71–74,
(a) solve.
(b) check.

71. $|x - 3| + 9 = 17$

72. $|x - 3| + 9 = 2$

73. $15 = |2x + 11|$

74. $|3x + 1| = |x - 8|$

SECTION 2.6 Conjunctions, Disjunctions, and Absolute Value Inequalities

Ask Yourself	Test Yourself	Help Yourself
Can I . . .	**Do 2.6 Review Exercises**	**See these Examples and Practice Problems**
describe the intersection or the union of two sets? use a number line graph or interval notation to represent an inequality?	75	Ex. 1, 2, PP 1, 2
describe the difference between a conjunction and a disjunction?	76	PP 3, 4
use a compound inequality, number line graph, interval notation, or words to represent the intersection of two inequalities?	77, 79	Ex. 3, 4, 7, PP 5, 7–11
use a compound inequality, number line graph, interval notation, or words to represent the union of two inequalities?	78, 80	Ex. 5, 6, PP 6, 12–16
solve a compound inequality?	79	Ex. 8, 9, PP 7–11
solve a disjunction?	80	Ex. 10, 11, PP 12–16

Ask Yourself		Test Yourself	Help Yourself
Can I . . .		Do 2.6 Review Exercises	See these Examples and Practice Problems
identify the boundary values for an absolute value inequality?		81–83	Ex. 14, 15, PP 17–19
solve an absolute value inequality?		81, 83	Ex. 14, 15, PP 17–18
identify an absolute value inequality that has no solution?		82	Ex. 16, PP 19

2.6 Review Exercises

75. Set A is $\{1, 3, 5, 7, 9, 11\}$. Set B is $\{3, 5, 8, 11, 14\}$.
  a. Describe $A \cap B$.
  b. Describe $A \cup B$.

76. The solution of an absolute value inequality is $x > 2$ or $x < -1$. Is this a conjunction or a disjunction? Explain.

77. $-3 < x \le 10$ is a conjunction.
  a. Use a number line graph to represent this conjunction.
  b. Use interval notation to represent this conjunction.

78. $x > 4$ or $x \le 0$ is a disjunction.
  a. Use a number line graph to represent this conjunction.
  b. Use interval notation to represent this conjunction.

For exercises 79–80,
(a) solve.
(b) use interval notation to represent the solution.

79. $-8 \le 2x + 24 < 40$

80. $2x + 12 > 18$ or $3x - 6 < -15$

For exercises 81–83,
(a) find the boundary values.
(b) solve.
(c) use interval notation to represent the solution, if any.

81. $|x - 4| < 15$

82. $|p - 6| + 9 < 1$

83. $|2x + 8| - 15 \ge 25$

Chapter 2 Test

For problems 1–2, simplify.

1. $(5x - 9)(2x^2 - 7x - 4)$

2. $(x^3 - 6x^2 + 5x - 9) - (4x^3 - x^2 - 15)$

3. Write an example of a fourth-degree trinomial in one variable.

4. The expression $6x^9 - \dfrac{1}{x}$ is not a polynomial. Explain why.

For problems 5–14, factor completely. Identify prime polynomials.

5. $24kx + 6hx - 12ky - 3hy$

6. $6k^3 - k^2 - 15k$

7. $64b^2 - 25c^2$

8. $2x^2 + 2x + 24$

9. $p^3 - 27$

10. $a^2 + 9a + 12$

11. $h^2 - 14h + 49$

12. $6z^2 + 13z - 63$

13. $x^2 + 16$

14. $63x^2 - 28$

15. Write the zero product property.

For problems 16–17,
(a) use the zero product property to solve.
(b) check.

16. $2x^2 - 5x - 60 = 3$

17. $w^3 - 100w = 0$

For problem 18, use the five steps.

18. A rectangle is 6 in. longer than it is wide. The area of the rectangle is 91 in.^{2}. Find the length and width of the rectangle.

For problems 19–21,
(a) solve.
(b) check.

19. $|c + 8| + 20 = 9$

20. $|6x + 42| - 9 = 15$

21. $8 - |2x + 1| = -15$

22. a. Solve: $-21 < 3x + 12 < 54$
  b. Use a number line graph to represent the solution.
  c. Use interval notation to represent the solution.

23. a. Solve: $|x + 9| - 11 \le 48$
  b. Use a number line graph to represent the solution.
  c. Use interval notation to represent the solution.

Relations and Functions

SUCCESS IN COLLEGE MATHEMATICS

Problem Solving

THE FIVE STEPS
Step 1 Understand the problem.
Step 2 Make a plan.
Step 3 Carry out the plan.
Step 4 Look back.
Step 5 Report the solution.

In this textbook, the five steps organize the work needed to solve application problems. The steps are based on the work of mathematician George Polya (1887–1985). Polya believed that students develop their problem-solving skills by solving many different kinds of problems.

In each set of exercises, there are four problems under the heading "Problem Solving: Practice and Review" that do not match examples from the section. However, they are similar to problems you have solved before. For example, the many kinds of percent problems are all based on the definition

of percent: $percent = \left(\dfrac{number\ of\ parts}{total\ parts} \right) (100) \%$

The goal of these Problem Solving: Practice and Review applications is to improve your ability to read, understand, and solve a variety of problems without depending on finding a specific matching example.

A missed call screen on a cell phone shows a list of names and the date of each missed call. This list is an example of a **correspondence**. We can arrange this correspondence in pairs: (name, date). Since the name is always first, these are **ordered pairs**. A collection of these ordered pairs is a **relation**. In this section, we will study multiple ways to represent a relation.

Introduction to Relations and Functions

After reading the text, working the practice problems, and completing assigned exercises, you should be able to:

1. Use a list, roster notation, a table, and/or a bar graph to represent a relation.
2. Use an equation, an inequality, and/or their graphs to represent a relation.
3. Represent the domain and range of a relation or function.
4. Determine whether a set is a function.

Correspondence and Relation

A **relation** is a set of ordered pairs. The first value in each ordered pair is the **input value**. The second value is the **output value**. We can represent some relations as a table, as a list, or in roster notation. A table has columns with headings. The headings describe the input and output values but are not part of the relation. A list may be written in any format, but the correspondence of the input and output is clear. A missed call screen on a cell phone is often written as a list. Roster notation is a way of writing a set of ordered pairs in which each ordered pair is in parentheses and all of the ordered pairs are in set brackets.

EXAMPLE 1 The list on the missed call screen represents a relation. The input values are the contact names; the output values are the dates of the missed calls.

Missed Calls	
Barb	May 4
Masoud	May 4
Galvez	May 5
Charles	May 6

(a) Use roster notation to represent the relation.

SOLUTION ▶ {(Barb, May 4), (Masoud, May 4), (Galvez, May 5), (Charles, May 6)}

(b) Use a table to represent the relation.

Name	Day of call
Barb	May 4
Masoud	May 4
Galvez	May 5
Charles	May 6

In the next example, we choose only one correspondence from the table: the correspondence between the step in the salary schedule and the salary for a regular certified teacher with a bachelor's degree.

EXAMPLE 2   Use roster notation to represent the correspondence between the step in the salary schedule and the salary for a regular certified teacher with a bachelor's degree.

Salary Schedule for Regular Certified Teachers in the Philadelphia School District		
	Bachelor's degree	**Master's degree**	**Master's degree + 30 credits**
Step 1	$44,039	$45,334	$48,169
Step 2	$45,901	$47,520	$50,676
Step 3	$49,625	$51,729	$55,372
Step 4	$52,781	$54,884	$58,692

Source: webgui.phila.k12.pa.us, 2010

SOLUTION ▶   {(Step 1, $44,039), (Step 2, $45,901), (Step 3, $49,625), (Step 4, $52,781)}

In presenting a relation to an audience, a bar graph can be a strong visual image. Bar graphs can be created by hand or with a spreadsheet software program.

EXAMPLE 3   Use a bar graph to represent the correspondence between the date and the amount of snowfall.

Jackson Hole Mountain Resort Snowfall 2010–2011						
Date	Nov.	Dec.	Jan.	Feb.	Mar.	Apr.
Snowfall (in.)	112	94	73	54	117	89

Source: bestsnow.net

SOLUTION ▶ The input values for this relation are on the horizontal axis. The output values are on the vertical axis.

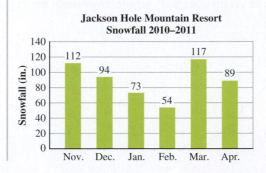

In algebra, we often use an equation to represent a relation.

EXAMPLE 4   Some friends have $250 to spend on two days of vacation. The amount of money they spend on the first day plus the amount of money they spend on the second day equals $250. Write an equation that represents this relation.

SOLUTION ▶ If x represents the money spent on the first day and y represents the money spent on the second day, each ordered pair in the relation is (x, y). Since the total amount is $250, x + y = $250.

A graph on a rectangular coordinate system can also represent a relation.

EXAMPLE 5   Use a graph to represent the relation $x + y = $250.

SOLUTION ▶ Graph three points, and draw a straight line through the points. Each ordered pair in the relation is represented by a point on this line. Since the values for x and y must be greater than or equal to $0, the line ends at the x-axis and the y-axis.

x	y
$125	$125
$175	$75
$100	$150

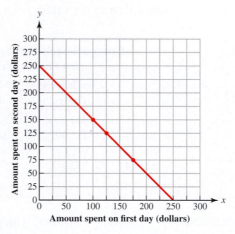

In Example 4, the friends spent $250 on two days of vacation. What if the friends could also spend *less than* $250 in two days? We can still represent this situation with a collection of ordered pairs: (amount spent on first day, amount spent on second day). The set of these ordered pairs is a relation. However, we use an *inequality* rather than an equation to represent it.

EXAMPLE 6   Some friends will spend *no more than* $250 on two days of vacation. The amount of money they spend on the first day plus the amount of money they spend on the second day will be *less than or equal to* $250. Use an inequality and a graph to represent this relation.

SOLUTION ▶ If the amount of money spent on the first day is x and the amount of money spent on the second day is y, each ordered pair in the relation is (x, y). Since the total amount is $250, x + y \leq $250. Since the total money must be greater than or equal to $0, the graph does not include any points with coordinates that are less than $0.

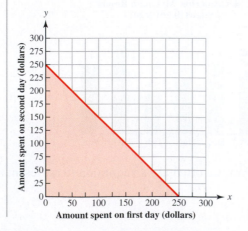

> ### Relation
>
> A **relation** is a set of ordered pairs. The first value in each ordered pair is the **input value**. The second value in each ordered pair is the **output value**.
>
> Representations of a relation include roster notation, a table, a list, a bar graph, an equation, an inequality, and a graph on a rectangular coordinate system.

Practice Problems

1. The amount of rain that fell on Saturday and the amount of rain that fell on Sunday totaled 3 in. Represent this relation with:
   a. an equation.   **b.** a graph.
2. A contractor has 80 employee-hours to divide between two job sites. Represent this relation with:
   a. an equation.   **b.** a graph.
3. In 2008, the U.S. orange production was 10.1 million tons. In 2009, the U.S. orange production was 9.1 million tons. In 2010, the U.S. orange production was 8.2 million tons. Represent this relation with:
   a. a table.   **b.** a bar graph.
4. Together, the area of vegetables and the area of flowers in a garden are less than or equal to 200 ft^2. Represent this relation with:
   a. an inequality.   **b.** a graph.

Functions

A **function** is a special kind of relation in which each input value corresponds to exactly one output value. No two ordered pairs have the same input value.

EXAMPLE 7   Is the set a function?

(a) $\{(0, 1), (2, 3), (5, 3)\}$

SOLUTION ▶   Yes, the inputs are 0, 2, and 5. No ordered pairs have the same input value.

(b) $\{(0, 1), (2, 3), (0, 6)\}$

▶   No, the same input value corresponds to different output values. Two ordered pairs have the same input value, 0.

(c) The correspondence of the set of the shoe size of every student in a class and the height of every student in the same class

▶   Probably not. If at least two students wear the same size shoe, then at least two ordered pairs have the same input value.

(d) The correspondence of the student ID number of every student in a class and the height of each student

▶   If every student at a school has a unique ID number, this is a function.

(e) The result of pushing down a turn signal lever in a car

▶   Yes, this set is a function with one ordered pair. The input is pushing down a turn signal lever. The output is the blinking of the left turn signal lights.

(f) The result of stepping on a floor switch for a light

▶   No, the input is stepping on the switch. The output is "light on" or "light off." This relation has two ordered pairs: (step on switch, light on) and (step on switch, light off). The input repeats; the set is not a function.

(g) The set of solutions of the equation $y = 2x - 3$. The input value is x. The output value is y.

▶ The solutions of this equation are ordered pairs. If we change the input in this equation, the output also changes. It is impossible for an input to have two different outputs. This is a function.

(h) The set of solutions of the inequality $y \geq 2x - 3$. The input value is x. The output value is y.

▶ The solutions of this inequality are ordered pairs. If we change the input in this equation, *the output may not change.* The same input may correspond to different outputs. For example, the ordered pairs $(0, 1)$ and $(0, 2)$ are both solutions of this inequality.

The turn signal is an example of one reason why functions are so important. A function is *predictable*. When the turn signal is pressed down, the result is always the same. If the input of pressing down could result in more than one output, the turn signal would not be reliable.

In some functions, there is a numerical relationship between the input value and the output value. For example, a function can be a set of ordered pairs in which the output value, y, is equal to the sum of the input value, x, and 1: $y = x + 1$ (Figure 1). The expression $x + 1$ is the **rule** of this function.

We can visualize using a rule and a function machine to create a set of ordered pairs. We input a number, x. The function machine adds 1 to this number. The result is the output value, y.

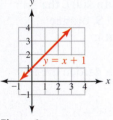

Figure 1

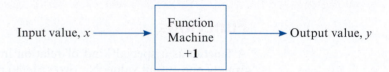

For each input value in a function, there is a corresponding output value. We often write the corresponding values as ordered pairs: (input value, output value). When we graph a function, each point (x, y) on the graph represents one of the ordered pairs.

Function

A **function** is a relation. It is a set of ordered pairs in which each input corresponds to exactly one output. No two ordered pairs have the same input value.

Representations of a function include roster notation, a table, a list, a bar graph, an equation, and a graph on a rectangular coordinate system.

Practice Problems

For problems 5–9, determine whether the set is a relation, a function, both a relation and a function, or neither a relation nor a function.

5. $\{(3, 4), (4, 5), (5, 6)\}$ **6.** $y = 8x$ **7.** $\{0, 1, 2, 3, \ldots\}$

8. The correspondence between the ID number of each student at a college and the birth date of the student

9. The correspondence between the birth date of each student at a college and the ID number of the student

Domain and Range

A relation is a set of ordered pairs. We can describe each ordered pair as (input value, output value). The **domain** of a relation is the set of its input values. The **range** of a relation is the set of its output values. Since a function is a relation, the domain of a function is also the set of its input values and the range of a function is the set of its output values.

EXAMPLE 8

(a) Represent the relation in the bar graph as a set of ordered pairs.

Monthly Payment on a Five Year
Loan for $10,000 by Interest Rate

Source: Laura Bracken and Edward Miller

SOLUTION ▶

$\{(5\%, \$189), (6\%, \$194), (7\%, \$198), (8\%, \$203), (9\%, \$208)\}$

(b) Is this relation a function?

▶ Yes, each input value corresponds to exactly one output value. No input values repeat.

(c) Use roster notation to represent the domain.

▶ $\{5\%, 6\%, 7\%, 8\%, 9\%\}$ Set of input values.

(d) Use roster notation to represent the range.

▶ $\{\$189, \$194, \$198, \$203, \$208\}$ Set of output values.

Domain and Range

The **domain** of a relation is the set of its input values. The **range** of a relation is the set of its output values.

Since a function is a relation, the domain of a function is the set of its input values and the range of a function is the set of its output values.

Practice Problems

10. a. Is the relation in Example 1 of this section a function?
   b. Use roster notation to represent the domain.
   c. Use roster notation to represent the range.

11. a. Is the relation in Example 3 of this section a function?
   b. Use roster notation to represent the domain.
   c. Use roster notation to represent the range.

Vertical Line Test

Figure 2

A vertical line represents a relation. Each ordered pair in the relation is represented by a point on the line. However, since every ordered pair on a vertical line has the same input value, it is not a function.

If every vertical line crosses a graph in at most one point, the graph represents a function. No ordered pairs on the graph have the same input. If a vertical line crosses a graph in more than one point (Figure 2), at least two ordered pairs on the graph have the same input. So the graph does not represent a function. This is the **vertical line test**. *Given a graph, if every vertical line crosses the graph in at most one point, the graph represents a function.*

EXAMPLE 9 Use the vertical line test to determine whether the graph represents a function.

(a)

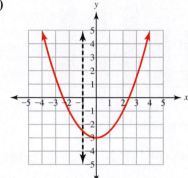

(b)

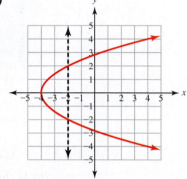

SOLUTION ▶ Every vertical line crosses the graph in at most one point. This is a function.

At least one vertical line crosses the graph in more than one point. This is not a function.

(c)

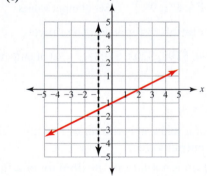

(d)

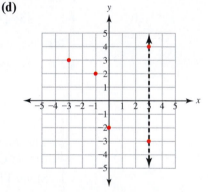

▶ Every vertical line crosses the graph in at most one point. This is a function.

At least one vertical line crosses the graph in more than one point. This is not a function.

Since the solution of a linear inequality such as $y < \frac{1}{2}x - 5$ is a set of ordered pairs, (x, y), it is a relation. However, the graph of a linear inequality fails the vertical line test. Although the boundary line may be a function, more than one point in the solution region has the same input value. A linear inequality is not a function.

EXAMPLE 10 | Use the vertical line test to determine whether the graph represents a function.

(a)

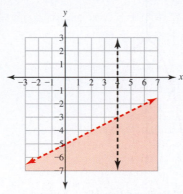

SOLUTION ▶ At least one vertical line crosses the graph in more than one point. This is not a function.

(b)

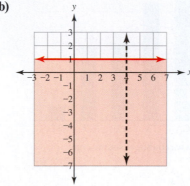

▶ At least one vertical line crosses in more than one point. This is not a function.

Practice Problems

For problems 12–17, determine whether the graph represents a function.

12.

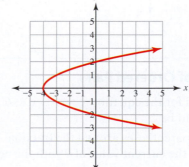

13.

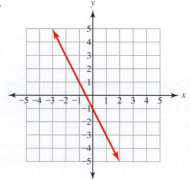

14.

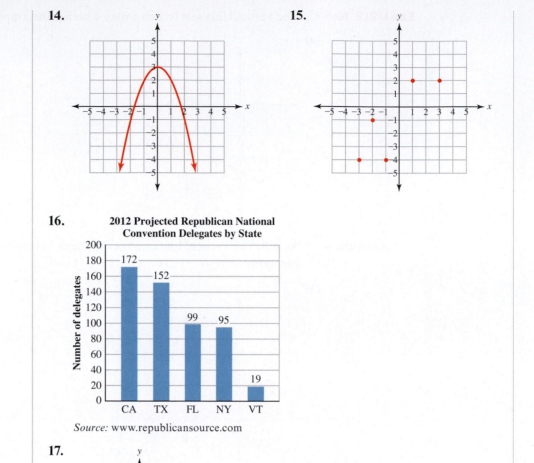

15.

16.

2012 Projected Republican National Convention Delegates by State

Source: www.republicansource.com

17.

Using Technology: Building a Table of Ordered Pairs

EXAMPLE 11 Build a table of ordered pairs for the equation $y = -\dfrac{3}{4}x + 6$. Use a starting x-value of 10. The interval between x-values in the table should be 2.

Press ⬛Y= . Type the equation, including parentheses around the fraction. Go to the TBLSET menu (press ⬛2nd ⬛WINDOW). In the TblStart line, replace the current value of 0 with the new starting value of 10.

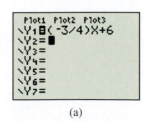

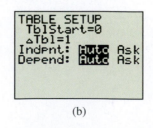

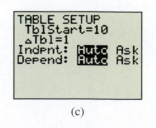

(a)   (b)   (c)

Use the down arrow button to move the cursor to the line that begins with ΔTbl. Replace the 1 with the new interval value, 2. To see the table of ordered pairs, go to the TABLE screen (press [2nd] [GRAPH]). To see other ordered pairs, move the cursor up and down the table.

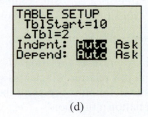
(d)

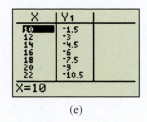
(e)

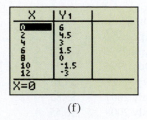
(f)

Practice Problems For problems 18–20, use a graphing calculator to build a table of ordered pairs for $y = 10x - 40$ with the given beginning x-value and interval between x-values. Report the answer as a table with the first five ordered pairs.

18. $x = 5$; interval $= 2$ **19.** $x = 0$; interval $= 20$ **20.** $x = -20$; interval $= 5$

3.1 VOCABULARY PRACTICE

Match the term with its description or symbol.

1. (x, y)
2. $\geq$
3. $\leq$
4. A set of ordered pairs
5. A relation in which each input value corresponds to exactly one output value
6. For the function $y = x + 1$, $x + 1$ is called this.
7. The set of inputs of a relation or function
8. The set of outputs of a relation or function
9. In (5%, \$189), 5% is this.
10. In (5%, \$189), \$189 is this.

A. domain
B. function
C. greater than or equal to
D. input value
E. less than or equal to
F. ordered pair
G. output value
H. range
I. relation
J. rule

3.1 Exercises

Follow your instructor's guidelines for showing your work.

In the exercises that ask you to create a graph, clearly label the scale and axes. Units of measurement should be included in the labels on the axes.

1. A list of music tracks and their play time is a correspondence. Use roster notation to represent this correspondence: (name, time).

1	Time of the Preacher	2:26
2	I Couldn't Believe It Was True	1:32
3	Time of the Preacher Theme	1:13
4	Medley: Blue Rock Montana/Red . . .	1:36
5	Blue Eyes Crying in the Rain	2:21
6	Red Headed Stranger	4:00
7	Time of the Preacher Theme	0:27
8	Just As I Am	1:48
9	Denver	0:53

2. Use a table to represent the correspondence in Exercise 1.

3. A list of the ranking of the top four movies and the name of the movie is a correspondence. Use a table to represent this correspondence: (rank, name).

Top 100
1. Crash
2. The Departed
3. Mr. and Mrs. Smith
4. Walk the Line

4. Use roster notation to represent the correspondence in Exercise 3.

5. The Fair Labor Standards Act sets the federal minimum wage. The table shows the correspondence between the date and the federal minimum wage. Draw a bar graph that represents this relation.

Jan. 1981	$3.35
Apr. 1990	$3.80
Apr. 1991	$4.25
Oct. 1996	$4.75
Sept. 1997	$5.15
July 2007	$5.85
July 2008	$6.55
July 2009	$7.25

Source: www.bls.gov

6. The minimum annual salary for teachers in Oklahoma is set by the state. As shown in the table, it depends on the years of experience and degree earned. Draw a bar graph that represents the correspondence between the number of years of experience and the minimum salary for a teacher with a bachelor's degree.

Years of experience	Bachelor's degree	Master's degree	Doctoral degree
0	$31,600	$32,800	$34,000
1	$31,975	$33,175	$34,375
2	$32,350	$33,550	$34,750
3	$32,725	$33,925	$35,125
4	$33,100	$34,300	$35,500

Source: www.sde.state.ok.us, 2011

For exercises 7–8, use the bar graph.

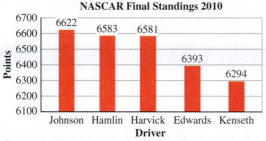

NASCAR Final Standings 2010

Source: www.nascar.com

7. Use roster notation to represent this relation.

8. Use a table to represent this relation.

9. Is the set of whole numbers a relation? Explain.

10. Is the set of integers a relation? Explain.

For exercises 11–12, $W = \{(0, 4), (1, 7), (2, 10), (3, 13)\}$.

11. Use roster notation to represent the domain of W.

12. Use roster notation to represent the range of W.

13. a. The table at the top of the next column represents a relation. Use roster notation to represent this relation. Name this set N.
    b. Use roster notation to represent the range of N.
    c. Is this relation a function?

Year	Net income of Exxon Mobil Corporation
2008	$45.2 billion
2009	$19.3 billion
2010	$30.5 billion

Source: www.exxonmobil.com

14. a. The table represents a relation. Use roster notation to represent this relation. Name this set M.
    b. Use roster notation to represent the domain of M.
    c. Is this relation a function?

2011 Hybrid car	EPA city (miles per gallon)
Toyota Prius	51
Ford Fusion	41
Honda Civic	40
Honda Insight	40

Source: www.fueleconomy.gov, 2011

For exercises 15–22, graph the relation and use the vertical line test to show that it is a function. To review graphing linear equations, see Section 1.6.

15. $6x + 7y = 42$

16. $5x + 9y = 45$

17. $x - y = -100$

18. $x - y = -200$

19. $y = \left(\dfrac{\$5000}{1 \text{ year}}\right)x + \$10,000$

20. $y = \left(\dfrac{\$2000}{1 \text{ year}}\right)x + \$8000$

21. $y = \left(-\dfrac{\$5000}{1 \text{ year}}\right)x + \$10,000$

22. $y = \left(-\dfrac{\$2000}{1 \text{ year}}\right)x + \$8000$

For exercises 23–26, graph the relation.

23. A meatloaf recipe requires 4 lb of ground meat in any combination of ground beef and ground pork. If x = number of pounds of ground beef and y = number of pounds of ground pork, this relation is represented by the equation $x + y = 4$ lb.

24. A punch recipe requires 5 gal of fruit juice in any combination of orange juice and pineapple juice. If x = number of gallons of orange juice and y = number of gallons of pineapple juice, this relation is represented by the equation $x + y = 5$ gal.

25. Workers decorating a ballroom will use 150 rolls of streamers that are either blue or silver. If x = number of blue rolls and y = number of silver rolls, this relation is represented by the equation $x + y = 150$ rolls.

26. Workers at a tree farm will plant 7500 tree seedlings that are either Scotch Pines or Grand Firs. If x = number of Scotch Pine seedlings and y = number of Grand Fir seedlings, this relation is represented by the equation $x + y = 7500$ seedlings.

For exercises 27–38, graph the relation and use the vertical line test to show that the relation is not a function. To review graphing linear inequalities, see Section 1.8.

27. $x - y > 6$

28. $x - y > 8$

29. $2x - 5y \leq 30$

30. $3x - 5y \leq 30$

31. $y > 2x$

32. $y > 3x$

33. $y < \dfrac{5}{9}x - 12$

34. $y < \dfrac{2}{7}x - 15$

35. $x \geq 0$

36. $x > 0$

37. $y < 0$

38. $y \leq 0$

39. Describe the difference between a relation and a function.

40. Explain why the graph of a linear inequality does not represent a function.

41. Create a set with six ordered pairs that is a function.

42. Create a set with six ordered pairs that is a relation but is not a function.

43. The table shows the correspondence of some Chicago ZIP codes with the number of real estate loan foreclosures during December 2007.
   a. Does this table represent a function?
   b. Explain.

ZIP code	Number of foreclosures
60620	67
60621	60
60623	38
60628	109
60629	68
60632	22

Source: www.ilfls.com

44. The table shows the correspondence of the year and the winner of the Stanley Cup.
   a. Does this table represent a function?
   b. Explain.

Year	Winner of the Stanley Cup
2010	Chicago Blackhawks
2009	Pittsburgh Penguins
2008	Detroit Red Wings
2007	Anaheim Ducks
2006	Carolina Hurricanes
2004	Tampa Bay Lightning

Source: proicehockey.about.com

For exercises 45–48,
(a) is the relation a function?
(b) explain why it is or is not a function.

45. {(coffee, Kenya), (tea, Sri Lanka), (coffee, Kona), (tea, India)}

46. {(heart, 10), (diamond, 9), (diamond, king), (spade, queen)}

47. {(queen, heart),(king, diamond),(jack, diamond)}

48. {(Sri Lanka, tea), (China, tea), (Hawaii, coffee), (India, tea)}

49. a. Does the bar graph represent a function?
   b. Explain.

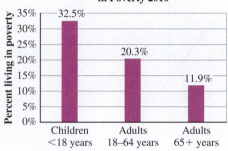

Source: www.census. gov, 2011

50. a. Does the bar graph represent a function?
   b. Explain.

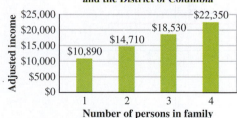

Source: www. aspe.hhs.gov.

51. Explain why you think the word *relation* is used to describe a set of ordered pairs.

52. Explain why you think the word *correspondence* is used to describe a relationship between the inputs and outputs of a function.

For exercises 53–54,
(a) use roster notation to represent the domain of the function.
(b) use roster notation to represent the range of the function.

53. {(small, $1.29), (medium, $1.69), (large, $2.29)}

54. {(70%, C−), (71%, C−), (84%, B), (96%, A)}

For exercises 55–68, does the graph represent a function?

55.

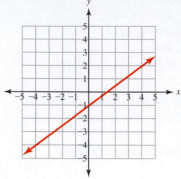

56.

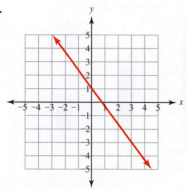

57.

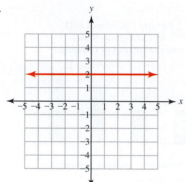

58.

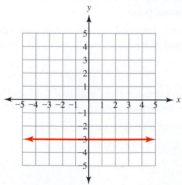

59.

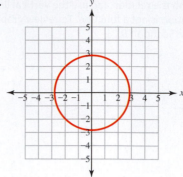

60.

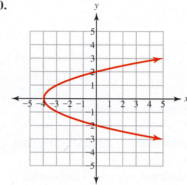

61.

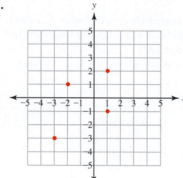

62.

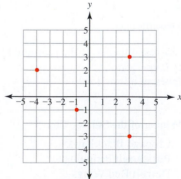

63.

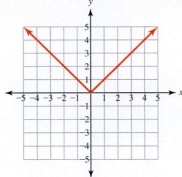

64.

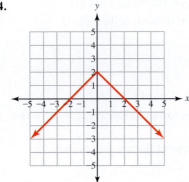

65.

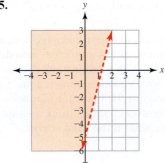

66.

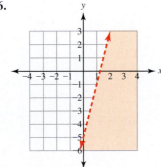

67.

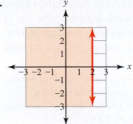

68.

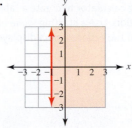

69. Explain why we can use a vertical line to test whether a graph represents a function.

70. Explain why the graph of a vertical line does not represent a function.

71. Create a set of four ordered pairs that is a function.

72. Create a set of four ordered pairs that is a relation but is not a function.

Problem Solving: Practice and Review

Follow your instructor's guidelines for using the five steps as outlined in Section 1.5, p. 51.

73. Find the percent decrease in allowable walleye harvest by the Lake Erie Commission. Round to the nearest percent.

The LEC cut the lakewide allowable walleye catch for 2009 to 2.45 million fish, down from 3.6 million walleye in 2008. (*Source:* www.cleveland.com, March 24, 2009)

74. In 2008, there were 320 domestic violence cases in the Bannock County, Idaho, court. Find the number of victims who didn't think anyone would believe their stories.

Six of every 10 victims didn't think anyone would believe their stories. (*Source:* www.lmtribune.com, June 1, 2009)

75. Assume that the population of the United States in 2018 will be about 328,395,000 people. Find the amount of money predicted to be spent in the economy in 2018. Round to the nearest tenth of a trillion.

Government statisticians estimated that health costs will reach $13,100 per person in 2018, accounting for $1 out of every $5 spent in the economy. (*Source:* www.msnbc.msn.com, Feb. 23, 2009)

76. Glencoe Beach is on Lake Michigan. An adult is going to go to the beach two days a week before 6 p.m. for 11 weeks. Find the difference in cost for a resident between paying admission for each visit and paying for an individual pass.

    The Glencoe Beach prices for daily admission before 6 p.m. are $6 for adult residents and $9 for non-residents. An individual pass is $65 for residents and $85 for non-residents. (*Source:* www.triblocal.com, May 29, 2009)

Technology

For exercises 77–80, use a graphing calculator to create a table of ordered pairs for the linear equation. Use a starting value of 0 and an interval of 2. Report five of the ordered pairs displayed on the calculator screen.

77. $y = 2x - 9$

78. $y = \dfrac{3}{4}x - 8$

79. $2x + 9y = 40$

80. $3x + 5y = 40$

Find the Mistake

For exercises 81–84, the completed problem has one mistake.
(a) Describe the mistake in words, or copy down the whole problem and highlight or circle the mistake.
(b) Do the problem correctly.

81. **Problem:** A relation is $\{(1, 9), (6, 4), (2, 3)\}$. Use roster notation to represent its domain.

    Incorrect Answer: The domain is 1, 2, 6.

82. **Problem:** Graph: $y > -x$

    Incorrect Answer:

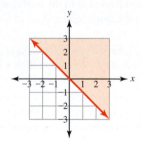

83. **Problem:** Graph: $x \geq 3$

    Incorrect Answer:

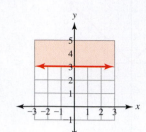

84. **Problem:** Use roster notation to represent the domain of the function $\{(5, 2), (6, 0), (8, 1), (-2, 1)\}$.

    Incorrect Answer: The domain is $\{0, 1, 2\}$.

Review

85. Solve: $0 = \dfrac{3}{4}x - \dfrac{5}{6}$

86. Evaluate $4x^2 - 5x - 7$ when $x = -6$.

87. Write the point-slope form of a linear equation in two variables.

88. Identify the y-intercept of the line represented by $y = -4x + 9$.

SUCCESS IN COLLEGE MATHEMATICS

89. List the five steps used in this book to organize problem solving.

90. In exercise 74, the information in the problem includes a ratio: "six of every 10 victims." When a problem includes a ratio, it often can be solved with a proportion. A proportion is an equation in which a *known ratio* is equal to an *unknown ratio*. Read exercise 74. If x is the number of victims in Bannock County who didn't think anyone would believe their stories, what is the *unknown ratio*?

© Jim Lopes/Shutterstock

The correspondence between the price per pound of hamburger and the cost of a certain number of pounds of hamburger is a function. Since we can describe this function with an equation in the form $y = mx + b$, it is a linear function. In this section, we will investigate linear and constant functions.

SECTION 3.2   Linear and Constant Functions

After reading the text, working the practice problems, and completing assigned exercises, you should be able to:

1. Evaluate a function.

2. Graph a linear or constant function.

3. Use interval notation to represent the domain and range of a linear function.

4. Use interval notation to represent the domain and range of a constant function.

5. Use an algebraic method to find the exact value of the real zero of a linear or constant function.

6. Use a graphical method to estimate the real zero of a linear or constant function.

Function Notation

A function is a set of ordered pairs in which each input corresponds to exactly one output. To **evaluate a function**, find the output value that corresponds to a given input value. To evaluate a function represented with roster notation, find the output value that is in the same ordered pair as the given input value.

EXAMPLE 1   A function is $R = \{(\text{Chavez, catcher}),(\text{Camacho, shortstop}),(\text{Dulz, left field})\}$. Evaluate $R(\text{Camacho})$.

SOLUTION ▸   When the input value is Camacho, the output value is shortstop.

$$R(\text{Camacho}) = \text{shortstop}$$

In Section 3.1, we described a function in which the output value, y, is equal to the input value plus 1. An equation that represents this function is $y = x + 1$. The input value, x, is the **independent variable**. Since the value of the output value, y, depends on x, y is the **dependent variable**. When we find the value of y that corresponds to a given value of x, we are **evaluating** the function.

EXAMPLE 2   A function is $y = 2x + 3$. Find y when $x = 1$.

SOLUTION ▸
$$y = 2x + 3$$
$$y = 2(\mathbf{1}) + 3 \qquad \text{Replace } x \text{ with 1.}$$
$$y = \mathbf{5} \qquad\qquad \text{Simplify.}$$

Instead of writing $y = 2x + 3$, we can use **function notation**: $f(x) = 2x + 3$. The notation $f(x)$ means "the output value of the function f when the input value is x."

The output value can be represented by y or by $f(x)$; $y = f(x)$. The expression $2x + 3$ is the **function rule** for the function f.

"The output value of the function f when the input value is x" is a long sentence. We often instead just say "f of x." It is important to recognize that the f in $f(x)$ is *not* a variable. We are not multiplying f and x; we cannot divide both sides by f or x.

To **evaluate** $f(x) = 2x + 3$ for a given input value, replace the input variable in the rule with the given input value and evaluate the expression on the right side. Do not change the left side.

EXAMPLE 3 | A function is $f(x) = 2x + 3$. Evaluate $f(1)$.

SOLUTION ▶

$$f(x) = 2x + 3$$
$$f(\mathbf{1}) = 2(\mathbf{1}) + 3 \qquad \text{Replace } x \text{ with 1.}$$
$$f(1) = \mathbf{2} + 3 \qquad \text{Simplify.}$$
$$f(1) = \mathbf{5} \qquad \text{Simplify.}$$

When the input is 1, the output is 5. This function is a set of ordered pairs that includes $(1, 5)$.

Some functions have only one output value.

EXAMPLE 4 | A function is $f(x) = 9$. Evaluate $f(500)$.

SOLUTION ▶ The only output of this function is 9. For every real number input value, the output value is 9. So $f(500) = 9$.

To use a graph to evaluate a function, we estimate the coordinates of an ordered pair.

EXAMPLE 5 | Use the graph of the function f to evaluate $f(3)$.

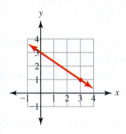

SOLUTION ▶ Find the ordered pair with an input value of 3: $(3, 1)$. Since $y = f(x)$ and the y-coordinate of $(3, 1)$ is 1, $f(3) = 1$.

In function notation, the name of the function is often a letter such as f. Other letters or symbols can also be names of a function. For example, the name of the function $h(x) = 8x - 1$ is h. We use letters to name functions because they are easy to write and say. We could name a function "smiley face," $\odot(x) = 8x - 1$, but this is not as convenient as a letter name.

When the input value of a function is a negative number, it is often necessary to use parentheses.

EXAMPLE 6 | A function is $g(x) = x^2 - x + 7$. Evaluate $g(-4)$.

SOLUTION ▶

$$g(x) = x^2 - x + 7$$
$$g(\mathbf{-4}) = (\mathbf{-4})^2 - (\mathbf{-4}) + 7 \qquad \text{Replace } x \text{ with } -4.$$
$$g(-4) = \mathbf{16} + \mathbf{4} + 7 \qquad \text{Simplify the right side; follow the order of operations.}$$
$$g(-4) = \mathbf{27} \qquad \text{Simplify the right side.}$$

Practice Problems

1. A function is $M = \{(\text{Mon., salad}), (\text{Tues., pizza}), (\text{Wed., stew})\}$. Evaluate $M(\text{Tues.})$.

2. A function is $y = \dfrac{3}{4}x + 1$. Find y when $x = 8$.

3. A function is $g(x) = -6$. Evaluate $g(4)$.

4. A function is $f(x) = 4x^2 - 5x + 2$. Evaluate $f(3)$.

5. A function is $f(x) = 4x^2 - 5x + 2$. Evaluate $f(-3)$.

6. Use the graph of the function f to evaluate $f(3)$.

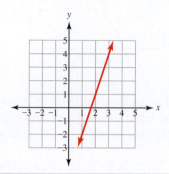

Linear Functions

We use three forms of a linear equation in two variables: slope-intercept form, point-slope form, and standard form. These linear equations describe the relationship of an input value, x, and an output value, y. If the solutions of these equations are sets of ordered pairs in which each input corresponds to exactly one output, these equations are functions. If the slope is defined and is not equal to 0, these equations represent **linear functions**.

Forms of Linear Functions

Standard Form $ax + by = c$

Slope-Intercept Form $y = mx + b$ or $f(x) = mx + b$

Point-Slope Form $y - y_1 = m(x - x_1)$

The graph of a linear function (Figure 1 and Figure 2) is a straight line that passes the vertical line test. Since the slope cannot be 0, horizontal lines do not represent linear functions. (As we will see later in the section, horizontal lines represent **constant functions**.)

The graph of a vertical line represents a relation but this relation is not a function. Since all of the points on a vertical line have the same input value, the graph fails the vertical line test.

To graph the function $y = 2x + 3$, we can build a table of ordered pairs, graph the points, and draw a line through them. Or we can graph the y-intercept, use the slope to find the next point, and draw a line through the points (slope-intercept graphing). Since $f(x) = y$, we can use the same methods to graph the linear function $f(x) = 2x + 3$.

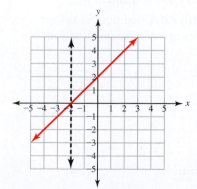

Figure 1 Any line with a positive slope passes the vertical line test.

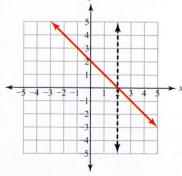

Figure 2 Any line with a negative slope passes the vertical line test.

EXAMPLE 7 | Graph: $f(x) = 2x + 3$

SOLUTION ▶ To graph $f(x) = 2x + 3$, graph $y = 2x + 3$. Build a table of three ordered pairs, graph these points, and draw a line through the points. Or use slope-intercept graphing. The y-intercept is $(0, 3)$, and the slope is $\dfrac{2}{1}$.

x	y
-1	1
0	3
1	5

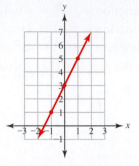

The domain of a function is the set of its input values. For a linear function such as $y = 2x + 3$, the input value can be any real number. So *the domain of any linear function is the set of real numbers.* In interval notation, the domain of a linear function is $(-\infty, \infty)$.

The range of a function is the set of its output values. If the input value of a linear function such as $y = 2x + 3$ is a real number, the output value will also be a real number. Each real number is an output of some ordered pair in this function. *For a linear function* (with a defined slope that is not 0), *the range is the set of real numbers.*

EXAMPLE 8 | Use interval notation to represent:

(a) the domain of $y = x + 1$.

SOLUTION ▶   This function is a linear function in the form $y = mx + b$. The domain of a linear function is the set of real numbers, $(-\infty, \infty)$.

(b) the range of $y = x + 1$.

▶   The range of this linear function is the set of real numbers, $(-\infty, \infty)$.

Linear Function

The forms of a linear function include $y = mx + b$, $f(x) = mx + b$, and $y - y_1 = m(x - x_1)$, where b, m, x_1, and y_1 are real numbers and $m \neq 0$.

The graph of a linear function is a straight line with a defined nonzero slope that passes the vertical line test. The domain and range of a linear function are both $(-\infty, \infty)$.

Practice Problems

7. A function is $f(x) = -2x + 5$.
   a. Graph this function.
   b. Use interval notation to represent the domain.
   c. Use interval notation to represent the range.
   d. Identify the slope of the graph.
   e. Identify the y-intercept of the graph.
   f. Evaluate: $f(-6)$

Constant Functions

A **constant function** in slope-intercept form is $f(x) = C$ or $y = C$, where C is a real number. The graph of a constant function is a horizontal line that passes the vertical line test (Figure 3). Each ordered pair has the same output value. Since any real number can be an input value, the domain of a constant function is the set of real numbers. The range includes only one output value, C. So when we evaluate a constant function, the result is always C. For example, if $f(x) = 4$, then $f(1) = 4$ and $f(-9) = 4$.

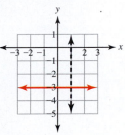

Figure 3

EXAMPLE 9 | A function is $g(x) = -1$.

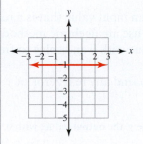

(a) Use interval notation to represent the domain.

SOLUTION ▶ This function is a constant function in the form $f(x) = C$. The domain of a constant function is the set of real numbers, $(-\infty, \infty)$.

(b) Use interval notation to represent the range.

▶ The only output value of this constant function is -1. To use interval notation to represent a set with only one real number, begin and end the interval with the real number. The range is $[-1, -1]$.

(c) Use the graph to evaluate $g(2)$.

▶ Find the ordered pair with an input value of 2: $(2, -1)$. Since the y-coordinate (the output value) of $(2, -1)$ is -1, $g(2) = -1$. Every point on the line has a y-coordinate of -1.

(d) Evaluate: $g(100)$

▶ $g(x) = -1$

$g(100) = -1$ The only output of a constant function is C.

Constant Function

The forms of a constant function include $y = C$ and $f(x) = C$, where C is a real number.

The graph of a constant function is a horizontal line with a slope of 0 that passes the vertical line test.

The domain of a constant function is $(-\infty, \infty)$, and the range of a constant function is $[C, C]$.

Practice Problems

8. A function is $h(x) = 3$.
   a. Graph this function.
   b. Use interval notation to represent the domain.
   c. Use interval notation to represent the range.
   d. Identify the slope of the graph.
   e. Identify the y-intercept of the graph.
   f. Use the graph to evaluate $h(-1)$.
   g. Evaluate: $h(-350)$

A Real Zero of a Function

A **real zero** of a function is an input value that is a real number and that corresponds to an output value of 0. To use an algebraic method to find the exact value of a real zero, replace the output value with 0 and solve to find the input value.

EXAMPLE 10 Use an algebraic method to find any real zeros of $y = 2x + 3$.

SOLUTION ▶

$$y = 2x + 3$$

$$0 = 2x + 3 \qquad \text{Replace } y, \text{ the output value, with } 0.$$

$$\underline{-3 \qquad\qquad -3} \qquad \text{Subtraction property of equality.}$$

$$-3 = 2x + 0 \qquad \text{Simplify.}$$

$$\frac{-3}{2} = \frac{2x}{2} \qquad \text{Division property of equality.}$$

$$-\frac{3}{2} = x \qquad \text{Simplify.}$$

The input value that corresponds to an output value of 0 is $-\frac{3}{2}$. The real zero of this function is $x = -\frac{3}{2}$.

In the next example, the function is written in function notation. To find the real zero of the function, replace the output value, $f(x)$, with 0 and solve for the input value, x.

EXAMPLE 11 Use an algebraic method to find any real zeros of $f(x) = -\frac{3}{4}x - 1$.

SOLUTION ▶

$$f(x) = -\frac{3}{4}x - 1$$

$$0 = -\frac{3}{4}x - 1 \qquad \text{Replace } f(x), \text{ the output value, with } 0.$$

$$\underline{+1 \qquad\qquad\qquad +1} \qquad \text{Addition property of equality.}$$

$$1 = -\frac{3}{4}x + 0 \qquad \text{Simplify.}$$

$$\left(-\frac{4}{3}\right)(1) = \left(-\frac{4}{3}\right)\left(-\frac{3}{4}x\right) \qquad \text{Multiplication property of equality.}$$

$$-\frac{4}{3} = x \qquad \text{Simplify.}$$

The input value that corresponds to an output value of 0 is $-\frac{4}{3}$. The real zero of this function is $x = -\frac{4}{3}$.

In the next example, the constant function does not have a real zero. There is no real number input that corresponds to an output value of 0.

EXAMPLE 12 Use an algebraic method to find any real zeros of the constant function $y = 6$.

SOLUTION ▶ $y = 6$

$\mathbf{0 = 6}$ False.   Replace y, the output value, with 0.

Since $0 = 6$ is false, there is no input value that corresponds to an output value of 0. This function has no real zero.

The only constant function that has a real zero is $f(x) = 0$. For this function, every input is a real zero of the function because every output is 0.

In Examples 10 and 11, we used an algebraic method to find the real zero of a function. In the next example, we use a graphical method to identify the real zero of a function.

EXAMPLE 13 Use the graph of $y = 2x + 4$ to identify any real zeros.

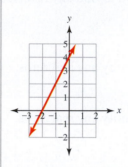

SOLUTION ▶ On its graph, a real zero of a function is the x-coordinate of an x-intercept. On this graph, the x-intercept appears to be $(-2, 0)$. The real zero of this function is $x = -2$.

If a function has no real zero, its graph will not intersect the x-axis. It has no x-intercept.

EXAMPLE 14 Use the graph of $y = 2$ to identify any real zeros.

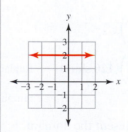

SOLUTION ▶ On a graph, a real zero of a function is the x-coordinate of an x-intercept. On this graph, there is no x-intercept. This function has no real zero. There is no input value that corresponds to an output value of 0.

> ### A Real Zero of a Function
>
> A real zero of a function is an input value that is a real number and that corresponds to an output value of 0.
>
> To use an algebraic method to find any real zeros, replace the output value with 0 and solve for the input value.
>
> To use a graphical method to find any real zeros, identify the x-coordinate of any x-intercepts. If the graph of a function has no x-intercept, the function has no real zeros.

Practice Problems

9. A function is $y = -x + 5$.
   a. Use an algebraic method to find any real zeros.
   b. Use interval notation to represent the domain.
   c. Use interval notation to represent the range.
10. Use an algebraic method to find any real zeros of $f(x) = -4$.
11. A function is $f(x) = \dfrac{1}{4}x - 8$.

   a. Graph this function.
   b. Use the graph to identify any real zeros.
12. A function is $y = -1$.
   a. Graph this function.
   b. Use the graph to identify any real zeros.

Applications

In the next example, the price of a shirt depends on the number of shirts that are ordered.

EXAMPLE 15   The cost for an order of 100 to 199 custom T-shirts is \$4.39 per T-shirt. The function $C(x) = \left(\dfrac{\$4.39}{1 \text{ shirt}}\right)x$ represents the relationship of the number of shirts ordered, x, and the cost of an order, $C(x)$.

(a) Evaluate: $C(120 \text{ shirts})$

SOLUTION ▶
$$C(x) = \left(\frac{\$4.39}{1 \text{ shirt}}\right)x$$

$$C(\mathbf{120\ shirts}) = \left(\frac{\$4.39}{1 \text{ shirt}}\right)(\mathbf{120\ shirts}) \qquad \text{Replace } x \text{ with 120 shirts.}$$

$$C(120 \text{ shirts}) = \mathbf{\$526.80} \qquad \text{Simplify; the units are dollars.}$$

(b) Use roster notation to represent the domain.

▶ This function applies only for orders of 100 to 199 T-shirts. The domain is {100 shirts, 101 shirts, 102 shirts, . . . , 199 shirts}. Since we cannot buy a fraction of a shirt, the domain includes only whole numbers of shirts.

(c) Use roster notation to represent the range.

▶ Since the domain is {100 shirts, 101 shirts, 102 shirts, . . . , 199 shirts}, the range is the output values that correspond to these input values: {\$439, \$443.39, \$447.78, . . . , \$873.61}.

In Example 15, since we cannot order part of a T-shirt, the domain includes only whole numbers of shirts. We use roster notation to represent the domain. In the next example, since we can order part of a pound, we use interval notation to represent the domain.

EXAMPLE 16 A store is running a special on hamburger for $2.29 per pound with a maximum purchase of 10 lb. The linear function $C(x) = \left(\dfrac{\$2.29}{1\ \text{lb}}\right)x$ represents the relationship of the number of pounds of hamburger, x, and the cost of the hamburger, $C(x)$.

(a) Evaluate: $C(9\ \text{lb})$

SOLUTION ▶

$$C(x) = \left(\frac{\$2.29}{1\ \text{lb}}\right)x$$

$$C(\textbf{9 lb}) = \left(\frac{\$2.29}{1\ \text{lb}}\right)(\textbf{9 lb}) \qquad \text{Replace } x \text{ with 9 lb.}$$

$$C(9\ \text{lb}) = \textbf{\$20.61} \qquad \text{Simplify; the units are dollars.}$$

(b) Use interval notation to represent the domain.

▶ A customer can order any amount from 0 to 10 lb, including 0 lb and 10 lb. The domain is $[0\ \text{lb}, 10\ \text{lb}]$.

(c) Use interval notation to represent the range.

▶ Since the domain is $[0\ \text{lb}, 10\ \text{lb}]$, the range is the output values that correspond to these input values: $[\$0, \$22.90]$.

In the next example, the linear function is $C(x) = \left(\dfrac{\$2}{1\ \text{mi}}\right)x + \$2.50$. The y-intercept is $(0, \$2.50)$. The cost of a taxi, $C(x)$, is the sum of the beginning cost of $2.50 plus an additional cost that depends on the distance the taxi travels.

EXAMPLE 17 The cost to take a taxi, $C(x)$ depends on the distance traveled, x:

$$C(x) = \left(\frac{\$2}{1\ \text{mi}}\right)x + \$2.50$$

Find the cost to travel 2 mi.

SOLUTION ▶

$$C(x) = \left(\frac{\$2}{1\ \text{mi}}\right)x + \$2.50$$

$$C(\textbf{2 mi}) = \left(\frac{\$2}{1\ \text{mi}}\right)(\textbf{2 mi}) + \$2.50 \qquad \text{Replace } x \text{ with 2 mi.}$$

$$C(2\ \text{mi}) = \textbf{\$4} + \$2.50 \qquad \text{Simplify.}$$

$$C(2\ \text{mi}) = \textbf{\$6.50} \qquad \text{Simplify; the units are dollars.}$$

The real zero of a function is the input value that corresponds to an output value of 0. In the next example, the real zero is the input value, time, when the output value, amount of water, is 0 gallons.

EXAMPLE 18 A homeowner collects water from the roof in a 60-gal rain barrel. The barrel is full. She uses 2.5 gal of water each day to water pots of tomato plants. Assuming that no more rain falls, the function $V(x) = \left(\dfrac{-2.5\ \text{gal}}{1\ \text{day}}\right)x + 60\ \text{gal}$ describes the relationship of the number of days she waters, x, and the volume of water in gallons in the barrel, $V(x)$.

(a) Find any real zeros.

SOLUTION ▶

$$V(x) = \left(\frac{-2.5 \text{ gal}}{1 \text{ day}}\right)x + 60 \text{ gal}$$

$$0 \text{ gal} = \left(\frac{-2.5 \text{ gal}}{1 \text{ day}}\right)x + 60 \text{ gal} \qquad \text{Replace } V(x) \text{ with } 0.$$

$$\underline{-60 \text{ gal} \qquad\qquad\qquad\qquad -60 \text{ gal}} \qquad \text{Subtraction property of equality.}$$

$$-60 \text{ gal} = \left(\frac{-2.5 \text{ gal}}{1 \text{ day}}\right)x + 0 \qquad \text{Simplify.}$$

$$\left(\frac{1 \text{ day}}{-2.5 \text{ gal}}\right)\left(\frac{-60 \text{ gal}}{1}\right) = \left(\frac{1 \text{ day}}{-2.5 \text{ gal}}\right)\left(\frac{-2.5 \text{ gal}}{1 \text{ day}}\right)x \qquad \text{Multiplication property of equality.}$$

$$24 \text{ days} = x \qquad \text{Simplify.}$$

(b) Describe what the real zero represents in this situation.

▶ When the output value is 0 gal, the input value is 24 days. The homeowner can water for 24 days from this barrel. The barrel will then be empty.

Practice Problems

13. The cost for an order of 15–24 picture frames is $17.85 per frame. The function $C(x) = \left(\dfrac{\$17.85}{1 \text{ frame}}\right)x$ describes the relationship of the number of frames ordered, x, and the cost of the order, C.
  a. Evaluate: $C(18 \text{ frames})$
  b. Use roster notation to represent the domain.
  c. Use roster notation to represent the range.

14. In the City of San Diego, the water department measures the water used in hundred cubic feet (HCF). The charge for the first 14 HCF used is $3.612 per HCF. The function $C(x) = \left(\dfrac{\$3.612}{1 \text{ HCF}}\right)x$ describes the relationship of the number of HCF used, x, and the cost for this water, $C(x)$. Round to the nearest hundredth. (*Source:* www.sandiego.gov, March 2011)
  a. Evaluate: $C(11 \text{ HCF})$
  b. Use interval notation to represent the domain.
  c. Use interval notation to represent the range.

15. During April, EKO Compost sold compost at a bulk rate of $20 per full-size pickup load. A full-size pickup can haul about 2 yd³. The function $C(x) = \left(\dfrac{\$20}{1 \text{ load}}\right)x$ describes the relationship of the number of loads of compost, x, and the cost of the compost. Find the cost of 12 yd³ of compost. (*Source:* www.ekocompost.com)

Courtesy of the Author

16. A homeowner uses about 0.6 lb of charcoal briquets each time he grills a meal. The charcoal comes in an 18-lb bag. The function $W(x) = \left(\dfrac{-0.6 \text{ lb}}{1 \text{ meal}}\right)x + 18 \text{ lb}$ describes the relationship of the number of meals grilled, x, and the weight of charcoal in the bag.
  a. Find any real zeros.
  b. Describe what a real zero represents in this situation.

Using Technology: Estimate the Real Zero of a Linear Function

EXAMPLE 19   Find any real zeros of $y = 3x + 20$. Round to the nearest tenth.

Press Y=. Type the function. Press GRAPH. The x-intercept of the line is visible in a standard window. The x-coordinate of this point is the real zero of the function.

Open the CALC menu. (The CALC command is printed above the TRACE key. Press 2nd, press TRACE.)

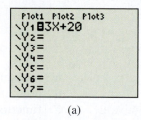

(a)

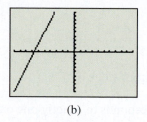

(b)

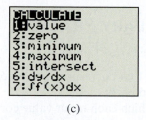

(c)

Choose 2: zero. Press ENTER. The calculator asks, "Left Bound?" We need to mark a point with an x-value that is less than the zero. But where is the cursor? The values for x and y at the bottom of the screen show us the location of the cursor. It is out of the window, at $(0, 20)$.

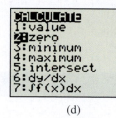

(d)

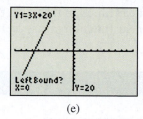

(e)

Press the left arrow key until the cursor appears. Keep pressing the left arrow key until the cursor is to the left of the x-intercept. Press ENTER. The calculator asks for the "Right Bound?" Use the right arrow key to move the cursor to the right of the x-intercept.

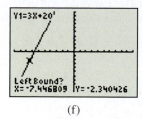

(f)

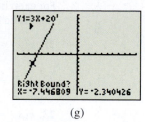

(g)

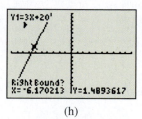

(h)

Press ENTER. The calculator asks "Guess?" Move the cursor right on top of the x-intercept. Press ENTER.

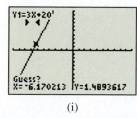

(i)

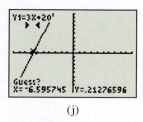

(j)

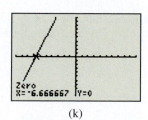

(k)

The real zero is about -6.666667. Rounding to the nearest tenth, $x \approx -6.7$. This is the estimated input value that corresponds to an output value of 0.

Practice Problems For problems 17–19,

(a) graph each function. Choose a window that shows the x-intercept. Sketch the graph; describe the window.

(b) find any real zeros. If necessary, round to the nearest hundredth.

17. $y = 4x - 12$ **18.** $y = -6x + 20$ **19.** $f(x) = 5$

3.2 VOCABULARY PRACTICE

Match the term with its description.

1. An input value that corresponds to an output value of 0
2. A set of ordered pairs
3. A relation in which each input value corresponds to exactly one output value
4. $y = mx + b$
5. The set of the inputs of a function
6. The set of the outputs of a function
7. $y - y_1 = m(x - x_1)$
8. $f(x) = C$; C is a real number.
9. The point on the graph of a linear function that is on the y-axis
10. The point on the graph of a linear function that is on the x-axis

A. constant function
B. domain
C. function
D. point-slope form
E. range
F. relation
G. slope-intercept form
H. y-intercept
I. x-intercept
J. real zero of a function

3.2 Exercises

Follow your instructor's guidelines for showing your work.

1. Is every relation also a function?

2. Is every function also a relation?

3. Is $y = 3$ a function?

4. Is $y = 2$ a function?

5. Is $x = 1$ a function?

6. Is $x = 7$ a function?

For exercises 7–8,
(a) graph the function.
(b) identify any real zeros.

7. $f(x) = 7x$

8. $f(x) = 2x$

For exercises 9–20,
(a) graph the relation.
(b) does the graph represent a function?
(c) explain why the graph does or does not represent a function.

9. $y = -8x + 15$

10. $y = -5x + 20$

11. $y < -2x + 5$

12. $y < -3x + 4$

13. $f(x) = -\dfrac{3}{4}x + 1$

14. $f(x) = -\dfrac{2}{3}x + 5$

15. $g(x) = 5$

16. $h(x) = 2$

17. $\{(4, 9), (2, 9), (7, 9), (9, 0)\}$

18. $\{(7, 3), (4, 3), (2, 3), (3, 0)\}$

19. $7x - 2y = 21$

20. $8x - 3y = 16$

For exercises 21–32, evaluate the function for the given input value. Use Examples 2 and 3 as a model for showing your work.

21. $f(x) = -\dfrac{2}{3}x + 11$; $x = 9$

22. $f(x) = -\dfrac{5}{6}x - 14$; $x = 12$

23. $g(x) = \dfrac{x}{6} + \dfrac{2}{5}$; $x = -3$

24. $h(x) = \dfrac{x}{6} + \dfrac{2}{5}$; $x = -4$

25. $y = \dfrac{3}{4}x - 8$; $x = 12$

26. $y = \dfrac{2}{3}x - 5$; $x = 12$

27. $f(x) = 0.4x + 5.9$; $f(8)$

28. $f(x) = 0.6x + 2.3$; $f(13)$

29. $g(x) = 5$; $g(-6)$

30. $h(x) = 11$; $h(-3)$

31. $k(x) = -x$; $k(-7)$

32. $k(x) = -x$; $k(-2)$

For exercises 33–34, a function is $y = x + 1$.

33. When the input value of this function is 3, what is the output value?

34. When the input value of this function is 4, what is the output value?

For exercises 35–36, a function is $y = 3x - 5$.

35. When the input value of this function is -4, what is the output value?

36. When the input value of this function is -9, what is the output value?

37. A function is $y = x + 1$. How many ordered pairs are in this set?

38. A function is $y = 3x - 5$. How many ordered pairs are in this set?

39. The sales tax rate in Texas is 6.25%. The city of Lubbock has an additional 2.343% community sales tax. The function $T(x) = 0.0625x + 0.02343x$ describes the total sales tax in dollars, $T(x)$, when x is the number of dollars spent in Lubbock. Find the sales tax on a $16.50 ticket to the Joyland Amusement Park in Lubbock. Round to the nearest hundredth.

40. The sales tax rate in Washington is 6.5%. The city of Seattle has a sales tax rate of 2.6%. The Regional Transit Authority has a 0.4% tax rate. The function $T(x) = 0.065x + 0.026x + 0.004x$ describes the total sales tax in dollars, $T(x)$, when x is the number of dollars spent in Seattle. Find the sales tax on a $19.95 ticket to the Experience Music Project in Seattle. Round to the nearest hundredth.

41. a. Use the graph of the function f to evaluate $f(-3)$.
   b. If $f(x) = 5$, what is x?

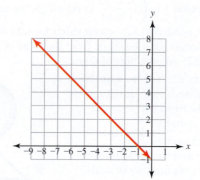

42. Refer to the graph in exercise 41.
   a. Use the graph of the function f to evaluate $f(-2)$.
   b. If $f(x) = 7$, what is x?

43. a. Use the graph of the function f to evaluate $f(1)$.
   b. If $f(x) = 0$, what is x?

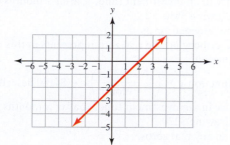

44. Refer to the graph in exercise 43.
   a. Use the graph of the function f to evaluate $f(-2)$.
   b. If $f(x) = -1$, what is x?

45. Use the graph of the function f to evaluate $f(-3)$.

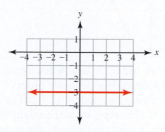

46. Refer to the graph in exercise 45. Use the graph of the function f to evaluate $f(1)$.

For exercises 47–52, to review finding the slope of a line and writing the equation of a line, see Section 1.6 and Section 1.7.

47. What is the slope of the graph of $f(x) = 5$?

48. What is the slope of the graph of $f(x) = 1$?

For exercises 49–52,
(a) find the slope of the line that passes through the given points.
(b) write the equation of the line in slope-interept form.
(c) identify the y-intercept of the line.
(d) explain why this line represents a function.

49. $(5, 1); (3, 7)$

50. $(6, 2); (4, 8)$

51. $(4, 3); (6, 3)$

52. $(2, 1); (3, 1)$

For exercises 53–56, using $f(x)$ notation, write the function that the graph represents. To review writing the equation of a line, see Section 1.6.

53.

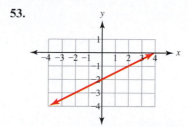

54.

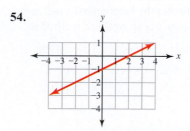

55.

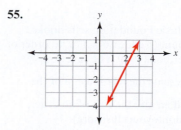

56.

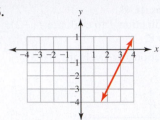

57. The price of an order changes when a customer orders more than 12 items. If a customer orders 13 to 23 items, the regular price is lowered by 12.5%. The function $P(x) = x - 0.125x$ represents the relationship of the original price, x, and the quantity discount price, $P(x)$. The regular price of an order with 20 items is $175. Find the quantity discount price. Round to the nearest hundredth.

58. The price of an order of batteries is lowered by 15% when a customer orders 26 or more batteries. The function $P(x) = x - 0.15x$ represents the relationship of the number of the regular price, x, and the quantity discount price, $P(x)$. The regular price for an order of 50 photo lithium batteries is $575. Find the quantity discount price.

59. If $C = \{$(ham, 290 cal), (roast beef, 290 cal), (chicken teriyaki, 370 cal), (Veggie Delite®, 230 cal)$\}$, evaluate $C($ham$)$. (*Source:* www.subway.com)

60. If $C = \{$(dbl qtr pounder with cheese, 740 cal), (Big Mac®, 450 cal), (Filet-O-Fish®, 380 cal), (Egg McMuffin®, 300 cal)$\}$, evaluate $C($Big Mac®$)$. (*Source:* www .mcdonalds.com)

For exercises 61–66, use interval notation to:
(a) represent the domain.
(b) represent the range.
61. $y = 4x + 1$
62. $y = -2x + 11$
63. $f(x) = 6$
64. $f(x) = -4$
65. $g(x) = -2x$
66. $g(x) = 3x$

For exercises 67–72,
(a) graph this function.
(b) does this function have a real zero?
(c) explain how you know whether the function has a real zero.
(d) if the function has a real zero, use an algebraic method to find it.
67. $f(x) = 3x - 15$
68. $f(x) = 4x - 12$
69. $y = 8$
70. $y = 13$
71. $g(x) = 2x$
72. $g(x) = 3x$
73. The Hold On To Your Music Foundation sells books, offering a quantity discount to educators. For an order of 20–49 books, the cost per book is $9. The function $C(x) = \left(\dfrac{\$9}{1 \text{ book}}\right)x$ represents the relationship of the number of books ordered, x, and the cost of an order, $C(x)$. (*Source:* www.holdontoyourmusic.org)
a. Evaluate: $C(25$ books$)$

b. Use roster notation to represent the domain.
c. Use roster notation to represent the range.
74. A company sells cases of 12 cans of marking paint for surveyors. For purchases of 12–49 cases, the price per case is $34.00. The function $C(x) = \left(\dfrac{\$34}{1 \text{ case}}\right)x$ shows the relationship of the number of cases ordered, x, and the cost of an order, $C(x)$. (*Source:* www.measureandlevel .com)
a. Evaluate: $C(39$ cases$)$
b. Use roster notation to represent the domain.
c. Use roster notation to represent the range.

For exercises 75–78, use roster notation to represent the
(a) domain.
(b) range.
75. Plastic landscape edging is sold in rolls that are 20 ft long and cost $12.93. The function $y = \left(\dfrac{\$12.93}{1 \text{ roll}}\right)x$ describes the cost in dollars of edging a lawn, y, where x is the number of rolls of edging. The function is used only for lawns that have a maximum perimeter of 1000 ft.
76. Landscapers use pine needles for ground covering. Each bale of pine needles covers 60 ft². There is no delivery charge for an order of at least 700 bales. The function $y = \left(\dfrac{\$3.25}{1 \text{ bale}}\right)x$ describes the cost in dollars, y, for x bales. The company can deliver a maximum of 5000 bales. This function is only used when there is no delivery charge.
77. The function $y = \left(\dfrac{1.25 \text{ yd}}{1 \text{ shirt}}\right)x$ estimates the amount of fabric, y, needed to sew x knit shirts for school uniforms. The uniforms must be made from the same bolt of fabric; there are 75 yd on a bolt of fabric.
78. A 6-penny nail is 2 in. long. Each pound of nails includes about 165 nails. The function $y = \left(\dfrac{1 \text{ pound}}{165 \text{ nails}}\right)x$ describes the total number of pounds, y, when x is the number of nails. Assume that the nails are sold at a home improvement store using a scale that can weigh no more than 20 lb. Round to the nearest thousandth.

© Miguel Angel Salinas Salinas/Shutterstock

79. The function $y = \left(-\dfrac{\$550}{1 \text{ month}}\right)x + \$9350$ describes the balance in an investment fund, y, after x months of withdrawing $550.
a. Find any real zeros.
b. Describe what the real zero represents in this situation.
80. The function $y = \left(-\dfrac{\$220}{1 \text{ month}}\right)x + \$4180$ describes the balance in an investment fund, y, after x months of withdrawing $220.
a. Find any real zeros.
b. Describe what the real zero represents in this situation.

81. The function $y = \left(\dfrac{-12 \text{ doughnuts}}{1 \text{ customer}}\right)x + 600 \text{ doughnuts}$
describes the number of doughnuts in a bakery, y, after x customers each buy a dozen (12) doughnuts.
a. Find any real zeros.
b. Describe what the real zero represents in this situation.

82. The function $y = \left(\dfrac{-1 \text{ bag}}{1 \text{ customer}}\right)x + 100 \text{ bags}$ describes the number of bags of coffee beans for sale at a coffee shop, y, after x customers buy 1 bag of beans.
a. Find any real zeros.
b. Describe what the real zero represents in this situation.

Problem Solving: Practice and Review

Follow your instructor's guidelines for using the five steps as outlined in Section 1.5, p. 51.

83. The American Conservatory Theater (A.C.T.) in San Francisco rents costumes for commercial purposes such as photo shoots. A customer takes 45 costumes with a rental price of $140 each per week and returns 25 of them before opening them. Find the total cost of renting these costumes for 1 week.

We will charge you only for the items you use, if the unused items are returned before opening. . . . If you use less than half of the items that were taken out on a trial rental, then we will charge 10% of the total of the original invoice, or a $2040 pull fee, whichever is greater. (*Source:* www.act-sf.org, 2011)

84. Find the percent increase in the annual net price for low-income students to attend a 2-year public college. Round to the nearest percent.

Over the last two decades, the annual net price of public college has increased for low- and moderate-income students. Between 1992–1993 and the most recent year for which national data are available, 2007–2008, total grant aid from all sources failed to keep pace with increases in the price of public colleges. At 4-year public colleges, net prices rose from $7,570 to $10,620 for low-income students and from $8,790 to $14,650 for moderate-income students. At 2-year public colleges, net prices rose from $6,260 to $8,017 for low-income students and from $7,020 to $10,830 for moderate-income students. (*Source:* Advisory Committee on Student Financial Assistance, 2010)

85. Find the number of the interviewed adults who have lifetime alcohol dependence where N = number of adults interviewed. Round to the nearest whole number.

Design, Setting, and Participants Face-to-face interviews with a representative US adult sample ($N = 43,093$). **Results** Prevalence of lifetime and 12-month alcohol abuse was 17.8% and 4.7%; prevalence of lifetime and 12-month alcohol dependence was 12.5% and 3.8%. (*Source:* Hasin, et al., "Prevalence, Correlates, Disability, and Comorbidity of *DSM-IV* Alcohol Abuse and Dependence in the United States," *Archives of General Psychiatry*, 2007; 64(7):830–842)

86. Zoysia is a semitropical warm season grass. In Mississippi, the direct cost of growing an acre of Zoysia sod is $3537. The market price per square yard of Zoysia sod is $2. There are 4840 yd^2 per acre. The profit to the grower is the difference between the cost of growing the sod and the revenue from selling the sod. Find the profit to the grower for

a sale of 40 acres of Zoysia sod. (*Source:* Mississippi Agricultural and Forestry Experiment Station, 2001)

Technology

For exercises 87–90,
(a) use a graphing calculator to graph the function. Choose a window that shows the x-intercept. Sketch the graph; describe the window.
(b) use a graphing calculator to find any real zeros. If necessary, round to the nearest hundredth.

87. $y = 2x - 10$

88. $f(x) = 3x - 9$

89. $y = -3x + 11$

90. $y = -2x + 9$

Find the Mistake

For exercises 91–94, the completed problem has one mistake.
(a) Describe the mistake in words, or copy down the whole problem and highlight or circle the mistake.
(b) Do the problem correctly.

91. Problem: Use an algebraic method to find any real zeros.

Incorrect Answer: $f(x) = 3x + 15$
$$f(0) = 3(0) + 15$$
$$f(0) = 15$$
The real zero of the function is 15.

92. Problem: The price of personalized calendars for fundraising depends on the quantity. For orders of 51–99 calendars, the price is $6.25 plus a $40 set-up fee. The function $C(x) = \left(\dfrac{\$6.25}{1 \text{ calendar}}\right)x + \$40$ describes the cost, $C(x)$, of ordering x calendars. What is the domain of this function?

Incorrect Answer: The domain is [51 calendars, 99 calendars].

93. Problem: A function is $f(x) = -x - 21$. Evaluate $f(-3)$.

Incorrect Answer: $f(x) = -x - 21$
$$f(-3) = -3 - 21$$
$$f(-3) = -24$$

94. Problem: The graph represents the relation $y = -2$. Does this graph represent a function?

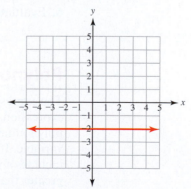

Incorrect Answer: No, it does not represent a function because it has repeating output values.

Review

95. Use the zero product property to solve $x^2 - 4 = 0$.

96. Use the zero product property to solve
$x^2 - 15x + 56 = 0$.

97. Use interval notation to represent the set of real numbers greater than or equal to 4.

98. Use interval notation to represent the set of real numbers less than or equal to -3.

SUCCESS IN COLLEGE MATHEMATICS

99. In exercise 84, you need to find the percent increase. The basic relationship of percent is

$$\text{percent} = \left(\frac{\text{number of parts}}{\text{total parts}}\right)(100)\%.$$ In this exercise, what is the *part* and what is the *whole*?

100. In exercise 85, you are given the total number of interviewed adults and the percent of these adults who have lifetime alcohol dependence. You need to find the number of these adults who have lifetime alcohol dependence. The basic relationship of percent is

$$\text{percent} = \left(\frac{\text{number of parts}}{\text{total parts}}\right)(100)\%.$$ In this exercise, what is the *unknown* in this relationship?

© emin kuliyev/Shutterstock

The function $N(x) = -0.0213x^2 + 0.3674x + 60.869$ is a model of the relationship between the number of years since 1975, *x,* and the number of Sunday and daily newspapers sold per year in millions, $N(x)$. Since the function rule is a degree 2 polynomial expression, this is a **quadratic function**. In this section, we will study quadratic and other polynomial functions. (*Source:* U.S. Statistical Abstract, 2011)

SECTION 3.3

Quadratic and Cubic Functions

After reading the text, working the practice problems, and completing assigned exercises, you should be able to:

1. Identify the degree of a polynomial function.

2. Use the graph of a quadratic function to estimate its vertex, its axis of symmetry, and/or its maximum or minimum output value.

3. Use a table of ordered pairs to graph a quadratic function.

4. Use interval notation to represent the domain and range of a quadratic or cubic function.

5. Use an algebraic or graphical method to find any real zeros of a quadratic function.

6. Evaluate a polynomial function.

Polynomial Functions

The expression $12x^2 + 5x + 6$ is a **polynomial expression**. If a term in a polynomial expression includes an exponent, the exponent is never a variable. If a term in a polynomial expression includes a denominator, the denominator does not include a variable. If a term in a polynomial expression includes a radical, the radicand never includes a variable. If a variable is raised to a power, this power is always an integer greater than or equal to 0.

The rule of a **polynomial function** is a polynomial expression. The **degree** of a polynomial function in one variable is the value of the largest exponent on a variable in its rule.

EXAMPLE 1   Identify the degree and type of each polynomial function.

(a) $f(x) = 3x^3 + 2x^2 + 5x + 9$

SOLUTION ▶   Degree: 3   Cubic

(b) $g(x) = x^2 + 7$

▶   Degree: 2   Quadratic

(c) $y = 4x + 6$

▶   Degree: 1   Linear

(d) $R(x) = 5$

▶   Degree: 0   Constant

Since any real number can be an input value, the domain of a polynomial function is the set of real numbers, $(-\infty, \infty)$. For example, the domain of the polynomial function $f(x) = 3x^3 + 2x^2 + 5x + 9$ is $(-\infty, \infty)$. Any real number can replace x in this function. In Chapter 5 and beyond, we will study functions in which the domain is not the set of real numbers.

Polynomial Function

A polynomial function can be written in the form

$$f(x) = a_n x^n + a_{n-1} x^{n-1} + \cdots + a_2 x^2 + a_1 x + a_0$$

where x is a variable, n is an integer ≥ 0, and $a_0, a_1, a_2, \ldots, a_n$ are real numbers and $a_n \neq 0$. The **degree** of this function is n. The **lead coefficient** is a_n. The **domain** of a polynomial function is $(-\infty, \infty)$.

Constant function	Degree 0	$f(x) = C$	C is a real number
Linear function	Degree 1	$f(x) = mx + b$	m, b are real numbers, $m \neq 0$
Quadratic function	Degree 2	$f(x) = ax^2 + bx + c$	a, b, c are real numbers, $a \neq 0$
Cubic function	Degree 3	$f(x) = ax^3 + bx^2 + cx + d$	a, b, c, d are real numbers, $a \neq 0$

Practice Problems

For each function,
(a) identify its degree.
(b) identify whether it is a constant, linear, quadratic, or cubic function.

1. $y = x^2 - 2$ 2. $f(x) = 9x$ 3. $g(x) = x^3 + 5x - 1$
4. $h(x) = 4$

Quadratic Functions

The **standard form** of a **quadratic function** is $y = ax^2 + bx + c$. The **domain** of any quadratic function is the set of real numbers, $(-\infty, \infty)$. The graph of a quadratic function (see Figure 1 on the next page) is a **parabola** that opens up or opens down.

If we draw a vertical line down the middle of the graph, each "arm" of the parabola is a mirror image of the other. This vertical line is the **axis of symmetry**.

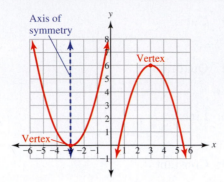

Figure 1

The "bottom" point or "top" point is the **vertex**. A quadratic function always has a **minimum or maximum output value**. It is the y-coordinate of the vertex.

To use a table of ordered pairs to graph a quadratic function, we need to identify enough points to sketch both arms of the parabola.

EXAMPLE 2   A quadratic function is $f(x) = x^2 + 3$.

(a) Identify a, b, and c.

SOLUTION ▶ $f(x) = 1x^2 + 0x + 3$

$a = 1, b = 0, c = 3$

(b) Graph the function.

▶   We need to graph enough points to show the parabola. Unlike graphing a line, this means that we need to graph more than three points.

$f(0) = 0^2 + 3$	
$f(0) = 3$	
$f(1) = 1^2 + 3$	
$f(1) = 4$	
$f(2) = 2^2 + 3$	
$f(2) = 7$	

x	y
0	3
1	4
2	7

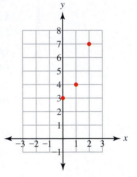

This appears to be the "right arm" of the parabola. We need to find more points to the left of the y-axis.

$f(-1) = (-1)^2 + 3$
$f(-1) = 4$
$f(-2) = (-2)^2 + 3$
$f(-2) = 7$

x	y
0	3
1	4
2	7
-1	4
-2	7

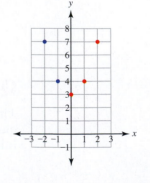

Sketch a parabola through the points. The axis of symmetry is a vertical line passing through the point $(0, 3)$.

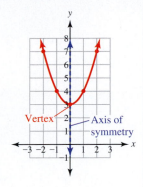

(c) Identify the vertex.

▶   The vertex is $(0, 3)$.

(d) Identify whether the parabola opens up or opens down.

▶   The parabola opens up.

(e) Identify the maximum or minimum output value.

▶   Since the parabola opens up, there is a minimum output value. The minimum output value is 3.

(f) Write the equation of the axis of symmetry.

▶   The axis of symmetry is a vertical line passing through $(0, 3)$. The equation of this line is $x = 0$.

(g) Use interval notation to represent the domain of this function.

▶   A quadratic function is a polynomial function. The domain of a polynomial function is the set of real numbers, $(-\infty, \infty)$. Any real number can be an input for this function.

(h) Use interval notation to represent the range of this function.

▶   The minimum output value is 3. All of the other output values are greater than 3. The range is $[3, \infty)$.

In Example 2, the graph of $f(x) = x^2 + 3$ is a parabola that **opens up**. Since the lead coefficient, a, is 1, $a > 0$. In the next example, $a < 0$. The graph **opens down**.

EXAMPLE 3   A quadratic function is $f(x) = -x^2 + 6x - 5$.

(a) Identify a, b, and c.

SOLUTION ▶ $f(x) = -1x^2 + 6x - 5$

$a = -1, b = 6, c = -5$

$f(0) = 0^2 + 6(0) - 5$
$f(0) = -5$

$f(1) = -(1^2) + 6(1) - 5$
$f(1) = 0$

(b) Graph the function.

x	y
0	−5
1	0
2	3
3	4
4	3
5	0

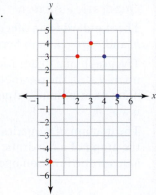

$f(2) = -(2^2) + 6(2) - 5$
$f(2) = 3$

$f(3) = -(3^2) + 6(3) - 5$
$f(3) = 4$

$f(4) = -(4^2) + 6(4) - 5$
$f(4) = 3$

$f(5) = -(5^2) + 6(5) - 5$
$f(5) = 0$

When the input value is greater than 3, the output values begin to decrease. The point (4, 3) is on the right arm of the parabola. We can finish the graph by finding more ordered pairs. Or we can draw the right arm of the parabola as a mirror image of the already graphed left arm. Sketch a parabola through the points.

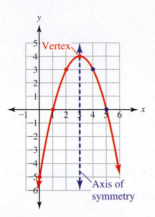

(c) Identify the vertex.

▶ The vertex is $(3, 4)$.

(d) Identify whether the parabola opens up or opens down.

▶ The parabola opens down (the lead coefficient, -1, is less than 0).

(e) Identify the maximum or minimum output value.

▶ Since the parabola opens down, there is a maximum output value. The maximum output value is 4.

(f) Write the equation of the axis of symmetry.

▶ The axis of symmetry is a vertical line passing through $(3, 4)$. The equation of this line is $x = 3$.

(g) Use interval notation to represent the domain of this function.

▶ A quadratic function is a polynomial function. The domain of a polynomial function is the set of real numbers, $(-\infty, \infty)$. Any real number can be an input for this function.

(h) Use interval notation to represent the range of this function.

▶ Since the maximum output value is 4 and all of the other output values are less than 4, the range is $(-\infty, 4]$.

Quadratic Function

The standard form of a quadratic function is $f(x) = ax^2 + bx + c$, where a, b, and c are real numbers and $a \neq 0$.

The graph of a quadratic function is a parabola. If $a > 0$, the parabola opens up. If $a < 0$, the parabola opens down. If the vertex is (h, k), then the maximum or minimum output value is k and the equation of the axis of symmetry is $x = h$.

The domain of a quadratic function is the set of real numbers, $(-\infty, \infty)$. If the graph of a quadratic function opens up, the range is $[k, \infty)$. If the graph of a quadratic function opens down, the range is $(-\infty, k]$.

EXAMPLE 4   (a) The graph of a quadratic function is a parabola with a vertex of $(5, -6)$. The parabola opens up. Use interval notation to represent the domain and the range.

SOLUTION ▶   The domain of a quadratic function is the set of real numbers, $(-\infty, \infty)$. Since the parabola opens up, the y-coordinate of the vertex, -6, is the minimum output value. The range is the set of real numbers greater than or equal to -6, $[-6, \infty)$.

(b) The graph of a quadratic function is a parabola with a vertex of $(-3, 8)$. The parabola opens down. Use interval notation to represent the domain and the range of this function.

▶   The domain of a quadratic function is the set of real numbers, $(-\infty, \infty)$. Since the parabola opens down, the y-coordinate of the vertex, 8, is the maximum output value. The range is the set of real numbers less than or equal to 8, $(-\infty, 8]$.

When evaluating a quadratic function, use parentheses around the input value.

EXAMPLE 5   A quadratic function is $f(x) = x^2 - x - 6$. Evaluate $f(-3)$.

SOLUTION ▶
$$f(x) = x^2 - x - 6$$
$$f(-3) = (-3)^2 - (-3) - 6 \qquad \text{Replace } x \text{ with } -3.$$
$$f(-3) = 9 + 3 - 6 \qquad \text{Simplify.}$$
$$f(-3) = 6 \qquad \text{Simplify.}$$

In the next example, we evaluate a quadratic function using an input value that is the number of years since 1975. The output value is the number of newspapers sold in millions. Since the function rule does not include units, we need to use the information in the problem to determine the units in the answer.

EXAMPLE 6   The function $N(x) = -0.0213x^2 + 0.3674x + 60.869$ is a model of the relationship of the number of years since 1975, x, and the number of Sunday and daily newspapers sold per year in millions, $N(x)$. Evaluate $N(40)$ to predict the number of Sunday and daily newspapers that will be sold in 2015. Round to the nearest tenth of a million. (*Source:* U.S. Statistical Abstract, 2012)

SOLUTION ▶
$$N(x) = -0.0213x^2 + 0.3674x + 60.869$$
$$N(40) = -0.0213(40)^2 + 0.3674(40) + 60.869 \qquad \text{Replace } x \text{ with } 40.$$
$$N(40) = -0.0213(1600) + 0.3674(40) + 60.869 \qquad \text{Simplify.}$$
$$N(40) = -34.08 + 14.696 + 60.869 \qquad \text{Simplify.}$$
$$N(40) = 41.485 \qquad \text{Simplify.}$$
$$N(40) \approx 41.5 \text{ million newspapers} \qquad \text{Round; write the units.}$$

In 30 years, the number of Sunday and daily newspapers sold will be about 41.5 million.

Practice Problems

For problems 5–7,
(a) identify a, b, and c.
(b) use a table of ordered pairs to graph the function.
(c) identify the vertex.
(d) identify the direction that the parabola opens.
(e) identify the maximum or minimum output value.
(f) write the equation of the axis of symmetry.
(g) use interval notation to represent the domain.
(h) use interval notation to represent the range.

5. $y = x^2 - 8x + 17$ **6.** $f(x) = x^2 - 2x - 2$
7. $g(x) = -x^2 + 2x + 2$

For problems 8 and 9, the vertex and the direction that the parabola opens is given. Use interval notation to represent:
(a) the domain.
(b) the range.

8. Opens up; vertex $(2, 4)$ **9.** Opens down; vertex $(2, 4)$

For problems 10–12, evaluate the function for the given input value.

10. $f(x) = x^2 - 12x + 27$; $f(6)$ **11.** $f(x) = x^2 - 12x + 27$; $f(-6)$
12. The function $V(x) = 0.7329x^2 + 5.8044x + 19.687$ represents the relationship of the value of U.S. e-commerce in billions of dollars, y, and the number of years since 1999, x. Evaluate $V(20 \text{ years})$. Round to the nearest tenth of a billion. (*Source:* www.census.gov, May, 2011)

Cubic Functions

A cubic function is a degree 3 polynomial function. As with all polynomial functions, the domain of a cubic function is $(-\infty, \infty)$. Like linear functions, a cubic function has no minimum or maximum output value. The range of a cubic function is the set of real numbers, $(-\infty, \infty)$.

EXAMPLE 7 Use the vertical line test to determine whether the graph represents a function.

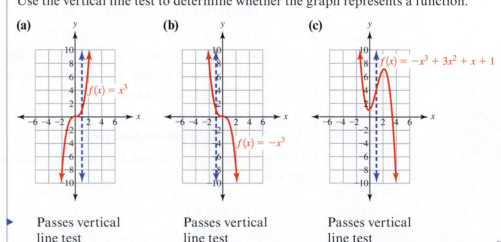

(a) $f(x) = x^3$ (b) $f(x) = -x^3$ (c) $f(x) = -x^3 + 3x^2 + x + 1$

SOLUTION ▶   Passes vertical line test   Passes vertical line test   Passes vertical line test

When evaluating a cubic function, use parentheses around the input value.

EXAMPLE 8 | A function is $g(x) = 6x^3 - x^2 + 7x - 9$. Evaluate $g(-1)$.

SOLUTION ▶

$$g(x) = 6x^3 - x^2 + 7x - 9$$

$g(-1) = 6(-1)^3 - (-1)^2 + 7(-1) - 9$ Replace x with -1.

$g(-1) = 6(-1) - 1 - 7 - 9$ Simplify.

$g(-1) = -23$ Simplify.

Practice Problems

13. A cubic function is $g(x) = x^3 - 4x^2 + 2x - 1$.

    a. Evaluate: $g(3)$

    b. Use interval notation to represent the domain of the function.

    c. Use interval notation to represent the range of the function.

Real Zeros

A real zero of a function is an input value that corresponds to an output value of 0. A linear function has one real zero. A quadratic function can have one, two, or no real zeros. To find the real zeros of $f(x) = x^2 - x - 6$, replace $f(x)$ with 0 and use factoring and the zero product property to solve the quadratic equation.

EXAMPLE 9 | Find any real zeros of $f(x) = x^2 - x - 6$.

SOLUTION ▶

$f(x) = x^2 - x - 6$ A quadratic function.

$0 = x^2 - x - 6$ Replace $f(x)$ with 0.

$0 = (x - 3)(x + 2)$ Factor.

$x - 3 = 0$ or $x + 2 = 0$ Zero product property.

$\underline{+3 \ +3} \qquad\qquad \underline{-2 \ -2}$ Properties of equality.

$x + 0 = 3$ or $x + 0 = -2$ Simplify.

$x = 3$ or $x = -2$ Simplify.

Real zeros: $x = 3,$ $x = -2$ This function has two real zeros.

In the next example, the function is a degree 3 cubic function. After replacing the output value with 0, we factor out the greatest common factor and then factor the remaining trinomial.

EXAMPLE 10 | Find any real zeros of $f(x) = x^3 + x^2 - 90x$.

SOLUTION ▶

$f(x) = x^3 + x^2 - 90x$

$0 = x^3 + x^2 - 90x$ Replace $f(x)$ with 0.

$0 = x(x^2 + x - 90)$ Factor out the greatest common factor, x.

$0 = x(x + 10)(x - 9)$ Factor $x^2 + x - 90$.

$x = 0$ or $x + 10 = 0$ or $x - 9 = 0$ Zero product property.

$\underline{-10 \ \ -10} \qquad\qquad \underline{+9 \ +9}$ Properties of equality.

$x + 0 = -10$ or $x + 0 = 9$ Simplify.

$x = -10$ or $x = 9$ Simplify.

Real zeros: $x = 0,$ $x = -10,$ $x = 9$ This function has three real zeros.

On the graph of a function, the x-coordinate of an x-intercept is a real zero. Check by replacing the input variable in the function with the real zero.

EXAMPLE 11 **(a)** Use the graph of $f(x) = x^2 - x - 6$ to identify any real zeros.

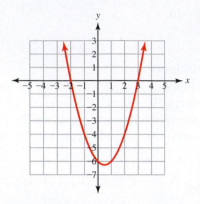

SOLUTION ▶ The x-intercepts are $(-2, 0)$ and $(3, 0)$. The real zeros are $x = -2$ and $x = 3$.

(b) Show that $x = -2$ is a real zero of $f(x) = x^2 - x - 6$.

$$f(x) = x^2 - x - 6$$
$$f(-2) = (-2)^2 - (-2) - 6 \qquad \text{Evaluate } f(-2).$$
$$f(-2) = 4 + 2 - 6 \qquad \text{Simplify.}$$
$$f(-2) = 0 \qquad \text{Simplify.}$$

(c) Show that $x = 3$ is a real zero of $f(x) = x^2 - x - 6$.

$$f(x) = x^2 - x - 6$$
$$f(3) = (3)^2 - 3 - 6 \qquad \text{Evaluate } f(3).$$
$$f(3) = 9 - 3 - 6 \qquad \text{Simplify.}$$
$$f(3) = 0 \qquad \text{Simplify.}$$

Since $x = -2$ and $x = 3$ correspond to output values of 0, they are real zeros of the function.

If the vertex of a quadratic function is an x-intercept, the function only has one real zero. If the graph has no x-intercepts, the function does not have a real zero.

EXAMPLE 12 **(a)** Use the graph to identify any real zeros of the function.

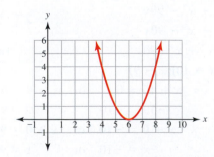

SOLUTION ▶ Since the x-intercept is $(6, 0)$, the real zero of the function is $x = 6$.

(b) Use the graph to identify any real zeros of the function.

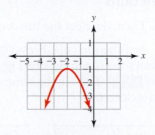

▶ Since the graph of the function has no x-intercept, the function has no real zeros.

Practice Problems

For problems 14–16, use an algebraic method to find any real zeros of the function.

14. $f(x) = x^2 - 12x + 27$ **15.** $y = 4x - 30$

16. $g(x) = x^2 - 6x + 9$

For problems 17–22, use the graph to identify any real zeros of the function.

17.

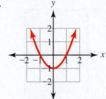

18.

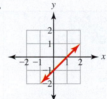

19.

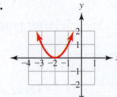

20.

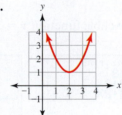

21.

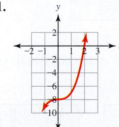

22.

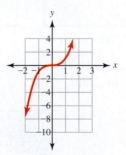

Using Technology: Graphs of Polynomial Functions

EXAMPLE 13 Graph the linear function $y = 2x + 8$. Then, deselect the function.

Begin with a standard window. Press [Y=]. Type in the function rule. Press [GRAPH]. The graph is a straight line. To keep the function without showing its graph, deselect the function. Press [Y=]. Move the cursor up until it is on top of the equals sign to the left of $2x + 8$. Press [ENTER]. The highlighting on the equals sign disappears. The graph of this function will not appear until it is reselected.

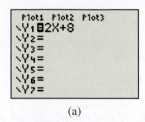
(a)

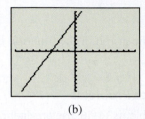
(b)

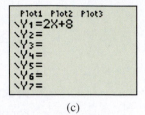
(c)

EXAMPLE 14 Graph the quadratic function $f(x) = x^2 + 7x + 3$. After graphing, deselect the function.

Press [Y=]. The first equation is still on the screen. On the next line, type in the new function rule. Press [GRAPH]. The graph is a parabola. Use the same procedure to deselect the function. The equals sign for this function is no longer highlighted.

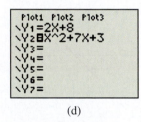
(d)

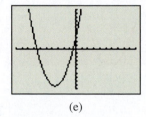
(e)

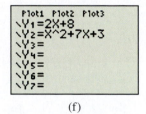
(f)

EXAMPLE 15 Graph the cubic function $y = x^3 + 2x^2 - 5x - 7$. After graphing, deselect the function.

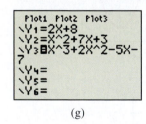
(g)

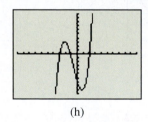
(h)

There is no single standard shape for the graph of a cubic function.

To reselect the first two functions, move the cursor back onto each equals sign. Press [ENTER]. The screen shows the graphs of all three functions.

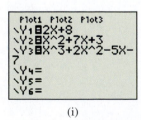
(i)

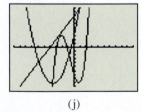
(j)

Practice Problems For problems 23–28, graph each equation. Choose a window that shows any *x*-intercepts. Sketch the graph; describe the window.

23. $y = 0.5x^2 - 6x + 2$　　**24.** $y = -0.5x^2 - 6x + 2$　　**25.** $y = x^3$

26. $y = x^3 - 3x^2$　　**27.** $y = -x^3 - 2x^2 + 5x$　　**28.** $y = x^4 - 3x^2 - 6$

3.3 VOCABULARY PRACTICE

Match the term with its description.

1. An input value that corresponds to an output value of 0
2. The set of outputs of a function
3. The set of inputs of a function
4. A set of ordered pairs
5. A relation in which each input corresponds to exactly one output value
6. A test that determines whether a graph represents a function
7. A vertical line that passes through the vertex of a parabola
8. The greatest output value of a function
9. The least output value of a function
10. A point on a graph with a y-coordinate of 0

A. axis of symmetry
B. domain
C. function
D. maximum
E. minimum
F. range
G. relation
H. vertical line test
I. x-intercept
J. real zero of a function

3.3 Exercises

Follow your instructor's guidelines for showing your work.

For exercises 1–6,
(a) identify the degree of the function.
(b) describe the function as linear, quadratic, or cubic.

1. $f(x) = 3x + 1$
2. $f(x) = x^2 + 7x + 12$
3. $g(x) = x^2 - 6x + 8$
4. $g(x) = x^3 - 9$
5. $y = 8x^3 + x^2 - 7x + 5$
6. $y = 17x - 3$

For problems 7–10, explain why the graph does or does not represent a function.

7.

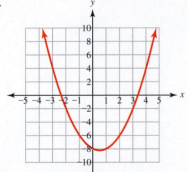

8.

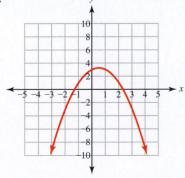

9.

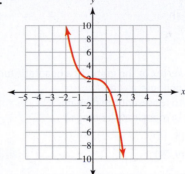

10.

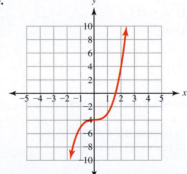

11. Use interval notation to represent the domain of a cubic function.

12. Use interval notation to represent the domain of a quadratic function.

For exercises 13–24,
(a) identify a, b, and c.
(b) use a table of ordered pairs to graph the function.
(c) identify the vertex.
(d) identify the direction that the parabola opens.
(e) identify the maximum or minimum output value.
(f) write the equation of the axis of symmetry.

(g) use interval notation to represent the domain.
(h) use interval notation to represent the range.

13. $f(x) = x^2 + 1$

14. $f(x) = x^2 - 4$

15. $f(x) = -x^2 + 1$

16. $f(x) = -x^2 - 4$

17. $y = x^2 - 6$

18. $y = x^2 - 9$

19. $f(x) = x^2 - 2x + 1$

20. $f(x) = x^2 + 2x + 1$

21. $g(x) = x^2 + 4x + 1$

22. $h(x) = x^2 + 6x + 7$

23. $f(x) = -x^2 + 8x - 13$

24. $f(x) = -x^2 + 4x + 1$

For exercises 25–28,
(a) does the graph of the function open up or down?
(b) explain how you know whether it opens up or down.

25. $f(x) = x^2 - 9x + 8$

26. $f(x) = x^2 - 11x + 10$

27. $f(x) = -x^2 + 4$

28. $f(x) = -x^2 + 8$

For exercises 29–36, the vertex of the graph of a quadratic function and the direction it opens are given.
(a) Use interval notation to represent the domain.
(b) Use interval notation to represent the range.

29. opens down; vertex $(8, 15)$

30. opens down; vertex $(9, 21)$

31. opens up; vertex $(3, -2)$

32. opens up; vertex $(5, -8)$

33. opens down; vertex $(-4, -11)$

34. opens down; vertex $(-12, -1)$

35. opens up; vertex $(0, -9)$

36. opens up; vertex $(0, 13)$

For exercises 37–42,
(a) use the graph of the quadratic function to estimate its vertex.
(b) use interval notation to represent the domain.
(c) use interval notation to represent the range.

37.

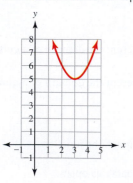

38.

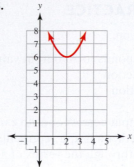

39.

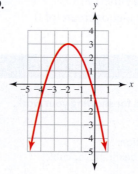

40.

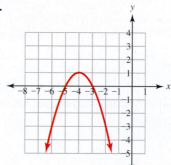

41.

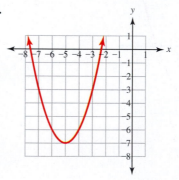

42.

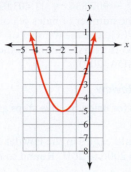

62.

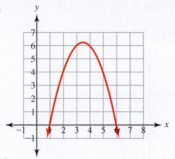

For exercises 43–46, use interval notation to:
(a) represent the domain.
(b) represent the range.

43. $f(x) = -2x + 9$

44. $f(x) = 2x + 9$

45. $f(x) = 2x^3 + 5x^2 - 3x + 1$

46. $f(x) = 8x^3 + x^2 + x + 4$

63.

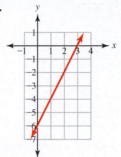

For exercises 47–60, use an algebraic method to find any real zeros. If the function does not have any real zeros, state this.

47. $f(x) = 6x - 30$

48. $g(x) = 7x - 28$

49. $h(x) = x^2 + 5x - 24$

50. $k(x) = x^2 + 3x - 28$

51. $y = 2x^2 + 7x - 15$

52. $y = 2x^2 + x - 15$

53. $f(x) = x^2 - 7x$

54. $g(x) = x^2 + 9x$

55. $y = x^3 - 6x^2 - 27x$

56. $y = x^3 + 3x^2 - 70x$

57. $y = x$

58. $y = -x$

59. $y = 5$

60. $y = 6$

64.

65.

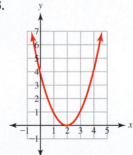

For exercises 61–66, use the graph to identify any real zeros.

61.

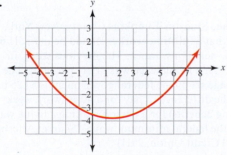

66.

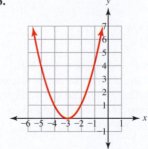

67. Use the graph of the function f
   a. to evaluate $f(-2)$.
   b. If $f(x) = -4$, identify x.

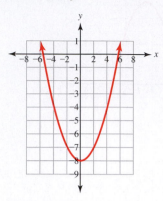

68. Use the graph of the function f
   a. to evaluate $f(2)$.
   b. If $f(x) = -2$, identify x.

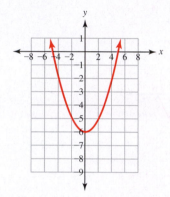

For exercises 69–80, evaluate the function for the given input value.

69. $f(x) = x^2 + 12x - 9$; $f(-2)$

70. $h(x) = x^2 + 15x - 7$; $h(-3)$

71. $C(x) = 2x^3 - 8x^2 - x$; $C(10)$

72. $P(x) = 8x^3 - 2x^2 - x$; $P(10)$

73. $y = -x^2 - 9$; $x = -2$

74. $y = -x^2 + 11$; $x = -8$

75. $k(x) = x^2 + 3x + \dfrac{2}{5}$; $k\left(\dfrac{1}{3}\right)$

76. $k(x) = x^2 + 4x + \dfrac{2}{3}$; $k\left(\dfrac{1}{4}\right)$

77. $g(x) = x^2 + 3x + 2$; $x = a + h$

78. $h(x) = x^2 + 2x + 3$; $x = a + h$

79. The function $N(x) = 109.05x^2 + 4185.5x + 187{,}040$ describes the relationship of the number of active doctors of medicine in the United States, $N(x)$, and the number of years since 1949, x. Evaluate $N(70 \text{ years})$. Round to the nearest thousand. (*Source:* www.cdc.gov, 2010)

80. The function $N(x) = 21{,}478x^2 - 907{,}834x + 11{,}403{,}029$ describes the relationship of the number of bales of cotton processed by U.S. mills, $N(x)$, and the number of years since 1997, x. Evaluate $N(15 \text{ years})$. Round to the nearest thousand. (*Source:* USDA, 2010)

Problem Solving: Practice and Review

Follow your instructor's guidelines for using the five steps as outlined in Section 1.5, p. 51.

81. Find the percent increase in bald eagles in the Powder River Basin in Wyoming from 2006 to 2011. Round to the nearest percent.

On the morning of January 8, 44 volunteers searched for bald and golden eagles across the Powder River Basin. Volunteers counted 200 bald eagles . . . on established survey routes along 1,340 miles of public roads. . . . The midwinter bald eagle survey has been conducted in the Powder River Basin since 2006, with 119 eagles counted in that year. The 2007 through 2010 surveys found 300, 162, 269, and 288 eagles, respectively. These survey totals vary due to the number of routes covered each year, but these totals are also influenced by weather and the availability of food sources including carrion, prairie dogs and rabbits. (*Source:* www.blm.gov, Feb. 18, 2011)

82. According to the California Department of Finance, the population of California on January 1, 2009, was about 38,293,000 people. Predict how many of these people will get Wilson disease. Round to the nearest ten.

Wilson disease is a genetic disorder that prevents the body from getting rid of extra copper. . . . About one in 40,000 people get Wilson disease, which affects men and women equally. Symptoms usually appear between ages 5 to 35, but new cases have been reported in people between ages 2 to 72. (*Source:* www.nih.gov, June 1, 2009)

83. Bowling alley owners proposed placing electronic slot machines at 300 bowling centers. Fifty percent of the profits from the machines would be turned over to the state of Ohio. Find the average total profit per slot machine. Write the answer in place value notation. Round to the nearest thousand.

They estimate placing 10 of the electronic slot machines at each bowling center could generate $130 million in revenue for the state each year. (*Source:* news.cincinnati.com, June 1, 2009)

84. A Living Wage Study in Tompkins County, New York, determined that a full-time worker should be paid $11.67 per hour plus health insurance. The previous study, in 2009, found that a living wage was $11.11 per hour. The minimum wage set by the federal government is $7.25 per hour. Find the difference in pay for a 40-hr work week (not including benefits) for a person making a living wage rather than the federal minimum wage. (*Source:* Alternatives Federal Credit Union, 2011)

Technology

For exercises 85–88,
(a) use a graphing calculator to graph each function. Choose a window that shows any x-intercepts. Sketch the graph; describe the window.
(b) identify the function as linear, quadratic, or cubic.

85. $f(x) = x^2 - 10x + 22$

86. $g(x) = 2x^3 - 4x^2$

87. $y = x^3 - 2x^2 - 3$

88. $y = -\dfrac{3}{5}x + 6$

Find The Mistake

For exercises 89–92, the completed problem has one mistake.
(a) Describe the mistake in words, or copy down the whole problem and highlight or circle the mistake.
(b) Do the problem correctly.

89. Problem: Use the graph of the cubic function to identify its range.

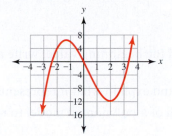

Incorrect Answer: The range of this function is $[-12, \infty)$.

90. Problem: Use an algebraic method to find any real zeros of $f(x) = x^2 + 8x + 12$.

Incorrect Answer: $f(0) = (0)^2 + 8(0) + 12$

$f(0) = 0 + 12$

$f(0) = 12$

The real zero of this function is $x = 12$.

91. Problem: A function is $f(x) = -x^2 - 4x + 9$. Evaluate $f(8)$.

Incorrect Answer: $f(8) = -8^2 - 4(8) + 9$

$f(8) = 64 - 32 + 9$

$f(8) = 41$

92. Problem: Use interval notation to represent the range of a quadratic function that opens down and has a vertex of $(9, 2)$.

Incorrect Answer: The minimum output value is 2. The range is $[2, \infty)$.

Review

For exercises 93–96, write the equation of the line with the given slope and y-intercept.

93. slope $= 4$; y-intercept is $(0, 6)$

94. slope $= 8$; y-intercept is $(0, 15)$

95. slope $= \dfrac{60 \text{ mi}}{1 \text{ hr}}$; y-intercept is $(0, 200 \text{ mi})$

96. slope $= \dfrac{-88 \text{ ft}}{1 \text{ s}}$; y-intercept is $(0, 500 \text{ ft})$

SUCCESS IN COLLEGE MATHEMATICS

97. In exercise 83, the unknown is the average total profit per slot machine. The problem gives the number of slot machines to be placed at each bowling center, the number of bowling centers, the percent of the profits from the slot machines that would be turned over to the state of Ohio, and the total amount of revenue expected from the slot machines for Ohio. Using words, explain how to solve this problem and find the average total profit per slot machine. Do not write an equation.

98. In solving applications, making a plan can include writing a *word equation*. A word equation can include operation symbols such as $+$, $-$, $\cdot$, and $\div$ but it does not include any numbers. In exercise 84, the problem gives the living wage pay per hour, the federal minimum wage per hour, and number of hours worked per week. The unknown is the difference in pay for a 40-hr work week between someone who earns the living wage and someone who earns the federal minimum wage. Write a word equation that describes this problem situation.

A function represents the relationship of two variables. We can use polynomial functions to model the relationship of two variables such as the time in years and the demand for registered nurses. In this section, we will use polynomial functions to model and solve application problems.

SECTION 3.4

Polynomial Models

After reading the text, working the practice problems, and completing assigned exercises, you should be able to:

1. Identify the average rate of change and the "beginning output value" in a linear model.

2. Find an average rate of change.

3. Given the average rate of change and the beginning value, write a linear model.

4. Find the real zero of a linear model and explain what it represents.

5. Use a polynomial model or the graph of a polynomial model to make predictions.

Linear Models

A linear function that represents the relationship of two variables in a real-life situation is a **linear model**. The slope in the model is the **average rate of change**. An average rate of change such as $\dfrac{65 \text{ mi}}{1 \text{ hr}}$ or $\dfrac{\$6}{1 \text{ lb}}$ compares two measurements with different units.

EXAMPLE 1 The linear function $y = \left(\dfrac{65 \text{ mi}}{1 \text{ hr}}\right)x$ represents the relationship of the time spent traveling, x, and the distance traveled by a car, y. Describe what the slope represents.

SOLUTION ▶ This function is in slope-intercept form, $y = mx + b$. The slope, $\dfrac{65 \text{ mi}}{1 \text{ hr}}$, is the change in distance divided by the change in time. This is the speed of the car, an average rate of change.

The y-coordinate of the y-intercept in a linear function is the output value when the input value is 0. In a linear model in which the input value is time, the y-intercept is the "*beginning* output value."

EXAMPLE 2 A student exercises at her top effort on an elliptical machine until the machine shows that she has used 200 calories (cal). She then lowers her effort to an average

rate of 15 cal per minute. The linear function
$$y = \left(\frac{15 \text{ cal}}{1 \text{ min}}\right)x + 200 \text{ cal represents the relationship}$$
of the time in minutes that she exercises at the lower
rate, x, and the total calories used in her workout, y.

© fckng/Shutterstock

(a) Describe what the slope represents.

SOLUTION ▶ The slope is $\dfrac{15 \text{ cal}}{1 \text{ min}}$. This is the average rate of change in the second part of her
workout.

(b) Describe what the y-coordinate of the y-intercept represents.

▶ The y-intercept is 200 cal. It represents the calories she has burned in her workout before she lowers her effort. When $x = 0$ min, $y = 200$ cal.

An average rate of change is a slope. For example, the average rate of change
$\dfrac{-11 \text{ gal}}{1 \text{ min}}$ describes the average change in volume for a change in time. To find an
average rate of change, identify two ordered pairs and use the slope
formula, $m = \dfrac{y_2 - y_1}{x_2 - x_1}$.

EXAMPLE 3 A full swimming pool holds 3400 gal of water. It takes 5 hr to drain the pool. Find
the average rate of change in gallons per minute. Round to the nearest integer.

Change the units of time from hours to minutes.

SOLUTION ▶ $5 \text{ hr} \cdot \dfrac{60 \text{ min}}{1 \text{ hr}}$ Multiply by a fraction equal to 1.

$= 300 \text{ min}$ Simplify.

Write the information as two ordered pairs.

▶ $(x_1, y_1) = (0 \text{ min}, 3400 \text{ gal})$ $(x_2, y_2) = (300 \text{ min}, 0 \text{ gal})$

Use the slope formula to find the average rate of change, m.

▶ $m = \dfrac{y_2 - y_1}{x_2 - x_1}$ m is the average rate of change.

$m = \dfrac{0 \text{ gal} - 3400 \text{ gal}}{300 \text{ min} - 0 \text{ min}}$ $(x_1, y_1) = (0 \text{ min}, 3400 \text{ gal}); (x_2, y_2) = (300 \text{ min}, 0 \text{ gal})$

$m = \dfrac{-3400 \text{ gal}}{300 \text{ min}}$ Simplify.

$m = \dfrac{-11.3 \ldots \text{ gal}}{1 \text{ min}}$ The drain rate is negative; the volume is decreasing.

$m \approx \dfrac{-11 \text{ gal}}{1 \text{ min}}$ Round.

Average Rate of Change

$$\text{average rate of change} \ = \ \frac{\text{change in the output value}}{\text{change in the input value}}$$

The average rate of change is a slope. For two ordered pairs (x_1, y_1) and (x_2, y_2),

$$\text{average rate of change} \ = \ \frac{y_2 - y_1}{x_2 - x_1}$$

Practice Problems

1. The linear function $y = \left(\dfrac{\$200}{1 \text{ month}}\right)x + \$850$ describes the relationship between the time in months, x, and the total amount in a savings account, y.
   a. Describe what the slope represents.
   b. Describe what the y-coordinate of the y-intercept represents.
2. On the first day of the month, an electric meter is at 30,478 kilowatt·hours. Thirty days later, the meter is at 31,150 kilowatt·hours. Find the average rate of change in kilowatt·hours per day. Round to the nearest whole number.

Writing Linear Models

We can use a linear model to make predictions about the future. These predictions are accurate only if the average rate of change remains the same.

EXAMPLE 4   The linear model $A(x) = \left(\dfrac{0.4686 \text{ lb}}{1 \text{ year}}\right)x + 21.643$ lb represents the relationship of the time in years since 1984, x, and the average consumption per person of cheese, $A(x)$, in the United States. The model is based on data from 1984 to 2009.

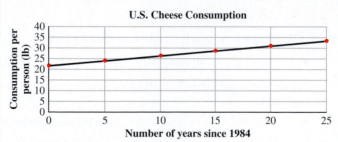

U.S. Cheese Consumption

Source: www.eatwisconsincheese.com

(a) Describe what the slope of the linear model represents.

SOLUTION ▶   The slope, $\dfrac{0.4686 \text{ lb}}{1 \text{ year}}$, is the average rate of change of the amount of cheese consumed per person per year.

(b) Describe what the y-coordinate of the y-intercept of the linear model represents.

▶   The y-coordinate of the y-intercept, 21.643 lb, is the amount of cheese consumed per person in 1984.

(c) Evaluate this function to predict the amount of cheese consumed per person in 2014. Round to the nearest tenth.

Since $2014 - 1984 = 30$ years, the input value is not in the given data. When evaluating this function for an input value of 30 years, we are assuming that the average rate of change does not change after 2009.

$$A(x) = \left(\frac{0.4686 \text{ lb}}{1 \text{ year}} \right) x + 21.643 \text{ lb}$$

$$A(\mathbf{30 \text{ years}}) = \left(\frac{0.4686 \text{ lb}}{1 \text{ year}} \right)(\mathbf{30 \text{ years}}) + 21.643 \text{ lb} \qquad \text{Replace } x \text{ with 30 years.}$$

$$A(30 \text{ years}) = \mathbf{14.058 \text{ lb}} + 21.643 \text{ lb} \qquad \text{Simplify.}$$

$$A(30 \text{ years}) = \mathbf{35.701 \text{ lb}} \qquad \text{Simplify.}$$

$$A(30 \text{ years}) \approx \mathbf{35.7 \text{ lb}} \qquad \text{Round.}$$

To write a linear model in slope-intercept form, we need to know the average rate of change (the slope) and the beginning output value when the input value is 0 (the y-coordinate of the y-intercept).

EXAMPLE 5 A programmer is developing the shopping basket for an on-line company. The total shipping cost equals the base cost of $14.95 for the first item plus $5 per additional item. A linear model describes the relationship of the total shipping cost, y, and the number of additional items, x.

(a) Describe what the slope of the linear model represents.

SOLUTION The slope, $\dfrac{\$5}{1 \text{ additional item}}$, is the average rate of change in the total shipping cost per additional item.

(b) Describe what the y-coordinate of the y-intercept of the linear model represents.

The y-coordinate of the y-intercept is the total shipping cost, $14.95, when x is 0 additional items.

(c) Write a linear function in slope-intercept form that predicts the total shipping cost, y, for x additional items.

Using slope-intercept form, $y = mx + b$, the function is

$$y = \left(\frac{\$5}{1 \text{ additional item}} \right) x + \$14.95$$

(d) Evaluate this function to find the total shipping cost for ordering eight items.

The first item is included in the base cost; there are seven additional items.

$$y = \left(\frac{\$5}{1 \text{ additional item}} \right) x + \$14.95$$

$$y = \left(\frac{\$5}{1 \text{ additional item}} \right)(\mathbf{7 \text{ additional items}}) + \$14.95 \qquad \text{Replace } x.$$

$$y = \mathbf{\$35} + \$14.95 \qquad \text{Simplify.}$$

$$y = \mathbf{\$49.95} \qquad \text{Simplify.}$$

(e) Graph this function for input values from 0 to 20 additional items.

x (additional items)	y (dollars)
0	$14.95
10	$64.95
20	$114.95

Slope-Intercept Form of a Linear Model

$$y = (\text{average rate of change})x + \text{the output value when the input value is } 0$$
$$y = mx + b$$

In Example 5, the slope is positive. As the input value (number of items) increases, the output value (total shipping cost) also increases. In the next example, the slope is negative. As the input value (time) increases, the output value (empty storage volume) decreases.

EXAMPLE 6 A manure lagoon is a pond used to store manure from animals such as cows or pigs. Twice a year, liquid is pumped from the lagoon to be used as fertilizer. However, solids (sludge) remain after pumping out the liquid, decreasing the volume of manure that the lagoon can hold. A dairy farm with 250 cows has a new lagoon with a storage volume of 2,000,000 gal. A linear model describes the relationship of the remaining storage volume, y, after x years of putting manure in the lagoon, pumping out the liquid, and sludge remaining after pumping.

> A major disadvantage of lagoons for cattle manure is the problem of sludge removal. Solids will accumulate in dairy-cow lagoons at an annual rate of about 260 cubic feet per head. (*Source:* Ronald E. Hermanson, "Livestock Manure Lagoons Protect Water Quality," Washington State University, 1991).

(a) Describe the beginning output value, identify the y-intercept, and identify b.

SOLUTION The beginning output value is the volume of the empty lagoon, 2,000,000 gal. The y-intercept is (0 years, 2,000,000 gal); $b = 2,000,000$ gal.

(b) Find the average rate of change, m. (1 ft$^3 \approx 7.48$ gal.)

The average rate of change is the amount of storage volume lost to sludge buildup per year. Since the unit of storage volume is gallons, we need to change the storage volume used by 1 cow per year, $\dfrac{260 \text{ ft}^3}{1 \text{ year}}$, into gallons and multiply by 250 cows.

$$m = \left(\frac{-260 \text{ ft}^3}{1 \text{ cow} \cdot 1 \text{ year}}\right)\left(\frac{7.48 \text{ gal}}{1 \text{ ft}^3}\right)(250 \text{ cows})$$

$$m = \frac{-486,200 \text{ gal}}{1 \text{ year}}$$

(c) Write the linear model in slope-intercept form, $y = mx + b$.

$$y = \left(\frac{-486{,}200 \text{ gal}}{1 \text{ year}}\right)x + 2{,}000{,}000 \text{ gal}$$

(d) Evaluate this function to predict the remaining storage volume after 3 years.

$$y = \left(\frac{-486{,}200 \text{ gal}}{1 \text{ year}}\right)x + 2{,}000{,}000 \text{ gal}$$

$$y = \left(\frac{-486{,}200 \text{ gal}}{1 \text{ year}}\right)(3 \text{ years}) + 2{,}000{,}000 \text{ gal} \qquad \text{Replace } x \text{ with 3 years.}$$

$$y = \textbf{541{,}400 gal} \qquad\qquad\qquad\qquad\qquad \text{Simplify.}$$

After 3 years, the buildup of sludge has decreased the storage volume in the lagoon to about 541,400 gal.

Practice Problems

3. A student has \$5500 in his account to pay rent during the school year. The monthly rent payment is \$550. A linear model represents the relationship of the amount in his account, y, after x months of paying rent.
   a. Describe what the slope represents.
   b. Describe what the y-coordinate of the y-intercept represents.
   c. Write the linear model in slope-intercept form.
   d. Evaluate this function to find the balance in his account after 6 months.
   e. Graph this function for input values from 0 to 10 months.

4. A young bamboo plant is 150 cm tall. Its average rate of growth is $\frac{50 \text{ cm}}{1 \text{ day}}$.

   A linear model represents the relationship of the height of the bamboo in centimeters, $H(x)$, after x more days of growth.
   a. Describe what the slope in the model represents.
   b. Describe what the y-coordinate of the y-intercept in the model represents.
   c. Write the linear model in slope-intercept form.
   d. Evaluate this function to predict the height of the plant after 30 days.
   e. Graph this function for input values from 0 to 10 days.

Models and Zeros

In Example 6, we wrote a linear model that predicts the remaining storage volume of a manure lagoon. The real zero of this function is the number of years it takes for the storage volume to decrease to 0 gal.

EXAMPLE 7 **(a)** Find the real zero of $y = \left(\dfrac{-486{,}200 \text{ gal}}{1 \text{ year}}\right)x + 2{,}000{,}000 \text{ gal}$. Round to the nearest whole number.

SOLUTION

$$y = \left(\frac{-486{,}200 \text{ gal}}{1 \text{ year}}\right)x + 2{,}000{,}000 \text{ gal}$$

$$\textbf{0 gal} = \left(\frac{-486{,}200 \text{ gal}}{1 \text{ year}}\right)x + 2{,}000{,}000 \text{ gal} \qquad \text{Replace the output value with 0.}$$

$$0 = -486{,}200x + 2{,}000{,}000 \qquad\qquad\qquad \text{Remove the units.}$$

$$\underline{-2{,}000{,}000 \qquad\qquad\qquad -2{,}000{,}000} \qquad \text{Subtraction property of equality.}$$

$$-\textbf{2{,}000{,}000} = -486{,}200x + 0 \qquad\qquad\quad \text{Simplify.}$$

$$\frac{-2,000,000}{-486,200} = \frac{-486,200x}{-486,200}$$ Division property of equality.

$$4.11\ldots = x$$ Simplify.

$$\mathbf{4\ years} \approx x$$ Round.

(b) Describe the situation when the output value is 0 gal.

▶ When the output value is 0 gal, the lagoon is filled with sludge, and there is no remaining storage volume. Since the value is rounded down to the nearest whole number, the actual output value is not exactly 0 gal. When rounding a real zero, keep in mind the consequences of rounding. In this situation, it is better to round down when the lagoon is not quite full than to round up to when the lagoon is overflowing.

Why solve problems about manure lagoons? If we want to drink milk, there must be efficient dairy cow operations and manure disposal. Designers of all kinds of lagoons and storage tanks need to think about the average fill rate. Owners of oil tanks, barbeque pits, septic tanks, and cement mixers need to know how frequently to schedule cleanouts of sludge and grease.

Practice Problems

5. A faucet is leaking at an average rate of $\dfrac{0.5\ \text{in.}^3}{1\ \text{hr}}$ into a sink with a closed drain. The volume of the sink is $1700\ \text{in.}^3$. A model of the relationship of the empty volume of the sink, $V(x)$, and the time that the sink is leaking in hours, x, is $V(x) = \left(-\dfrac{0.5\ \text{in.}^3}{1\ \text{hr}}\right)x + 1700\ \text{in.}^3$.

a. Find any real zeros.
b. Describe the situation when the output value is $0\ \text{in.}^3$.

The Line of Best Fit

In real-life situations, the relationship between inputs and outputs may not be perfectly linear. The **line of best fit** is the straight line that best represents the relationship. This line may pass through some, none, or all of the points. Mathematicians use regression methods to find the line of best fit. We will not learn regression here, but we will use information from lines of best fit to write linear models.

In the next example, a spreadsheet program created the graph. The program finds the line of best fit for the graphed points. It reports the slope and y-intercept of the line.

EXAMPLE 8 The graph of the line of best fit represents the relationship of the revenue from Internet publishing and broadcasting, y, and the number of years since 2004, x.

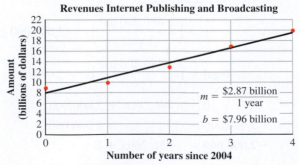

Source: www.census.gov, 2011

(a) Use the information about the line of best fit to write a linear model that represents this relationship.

SOLUTION ▶ In slope-intercept form, $y = mx + b$, the function is

$$y = \left(\frac{\$2.87 \text{ billion}}{1 \text{ year}}\right)x + \$7.96 \text{ billion}$$

(b) Evaluate this function to predict the revenue from Internet publishing and broadcasting in 2020.

▶ Since $2020 - 2004 = 16$ years, $x = 16$ years.

$$y = \left(\frac{\$2.87 \text{ billion}}{1 \text{ year}}\right)x + \$7.96 \text{ billion}$$

$$y = \left(\frac{\$2.87 \text{ billion}}{1 \text{ year}}\right)(\textbf{16 years}) + \$7.96 \text{ billion} \qquad \text{Replace } x.$$

$$y = \textbf{\$45.92 billion} + \$7.96 \text{ billion} \qquad \text{Simplify.}$$

$$y = \textbf{\$53.88 billion} \qquad \text{Simplify.}$$

Practice Problems

6. a. Use the information about the line of best fit to write a linear model that represents the relationship between the amount of retail sales at health and personal care stores, y, and the number of years since 1995, x.
  b. Evaluate this function to predict the value of retail sales at health and personal care stores in 2015. Round to the nearest tenth of a billion.

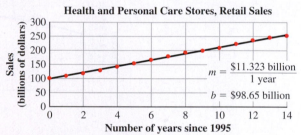

Health and Personal Care Stores, Retail Sales

$$m = \frac{\$11.323 \text{ billion}}{1 \text{ year}}$$

$$b = \$98.65 \text{ billion}$$

Source: www.census.gov

Other Polynomial Models

When the input value is a measure of time, the input variable is often t. In the next example, the model is a degree 2 quadratic function. When a model or formula like this one does not include units, use the content of the problem to identify the units of the answer.

EXAMPLE 9 A model of the demand for registered nurses, $N(t)$, is

$$N(t) = 603.6t^2 + 28{,}208x + 2{,}003{,}606$$

where t is the number of years since 2000. Predict the demand for registered nurses in 2015. Round to the nearest thousand.

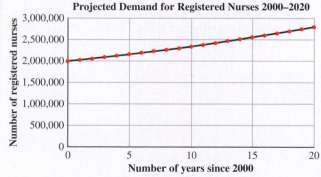

Projected Demand for Registered Nurses 2000–2020

Source: www.hrsa.gov, 2002

SOLUTION ▶ Since $2015 - 2000 = 15$, the input value is 15 years.

$$N(t) = 603.6t^2 + 28{,}208x + 2{,}003{,}606$$

$$N(t) = 603.6(\mathbf{15})^2 + 28{,}208(\mathbf{15}) + 2{,}003{,}606 \qquad \text{Replace } x.$$

$$N(t) = \mathbf{135{,}810} + \mathbf{423{,}120} + 2{,}003{,}606 \qquad \text{Simplify.}$$

$$N(t) = \mathbf{2{,}562{,}536} \qquad \text{Simplify.}$$

$$N(t) \approx \mathbf{2{,}563{,}000} \qquad \text{Round.}$$

The model predicts that in 2015, about 2,563,000 registered nurses will be needed.

In Example 9, as the input values increase, the output values also increase. In the next example, as the input values increase, the output values instead decrease.

EXAMPLE 10   A model of the percent of high school students who smoke, y, is

$$y = -0.32x^2 + 3.45x + 26.46$$

where x is the number of years since 1991. Predict the percent of high school students who smoke in 2011. Round to the nearest percent. (*Source:* www.cdc.gov)

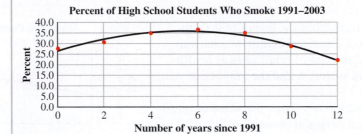

SOLUTION ▶ Since $2011 - 1991 = 20$, the input value is 20 years.

$$y = -0.32x^2 + 3.45x + 26.46$$

$$y = -0.32(\mathbf{20})^2 + 3.45(\mathbf{20}) + 26.46 \qquad \text{Replace } x.$$

$$y = \mathbf{-128} + \mathbf{69} + 26.46 \qquad \text{Simplify.}$$

$$y = \mathbf{-32.54} \qquad \text{Simplify.}$$

$$y \approx \mathbf{-33\%} \qquad \text{Round.}$$

The model predicts that in 2011, -33% of high school students will smoke. However, the output value in real life cannot be less than 0%. This output value is not reasonable; the model is *not* an accurate predictor in 2011.

Practice Problems

7. A model of the total fall enrollment in degree-granting higher education institutions, $f(t)$, is $f(t) = 43.5t^3 - 3021t^2 + 292{,}252t + 8{,}805{,}906$. The number of years since 1970 is t. Predict the enrollment in 2014. Round to the nearest thousand. (*Source:* NCES, 2008)

Using Technology: The Minimum or Maximum Output Value of a Quadratic Function

The greatest or least output of a function is its **maximum** or **minimum output**.

EXAMPLE 11 Find the minimum output of $y = 2x^2 - 5x - 4$. Use interval notation to represent the range of this function. Round to the nearest thousandth.

Begin with a standard window. Press [Y=]. Type the function. Press [GRAPH]. Go to the CALC menu. Choose 3. Press [ENTER].

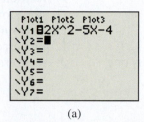
(a)

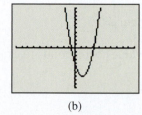
(b)

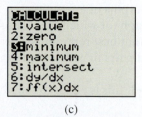
(c)

As with real zeros, show the calculator where to look for the minimum value. It asks "Left Bound?" Move the cursor to the left of the vertex. Press [ENTER].

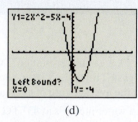
(d)

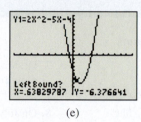
(e)

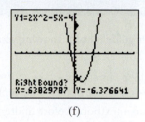
(f)

The calculator asks "Right Bound?" Move the cursor to the right of the vertex. Press [ENTER]. The calculator asks "Guess?" Move the cursor right on top of the vertex. Press [ENTER]. The calculator reports estimated values for the vertex. We need only the y-coordinate, since this is the minimum output: $y \approx -7.124$. The estimated range is $[-7.124, \infty)$.

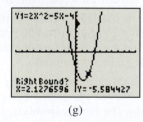
(g)

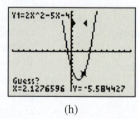
(h)

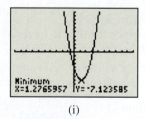
(i)

Practice Problems For problems 8–10,
  (a) graph each function. Choose a window that shows the x-intercepts and the vertex. Sketch the graph; describe the window.
  (b) find the minimum or maximum output value. If necessary, round the output value to the nearest hundredth.
  (c) use interval notation to represent the range.

  8. $f(x) = x^2 + 2x - 15$ **9.** $g(x) = -x^2 + 7x - 10$
  10. $f(x) = -x^2 + 10x - 23$

3.4 VOCABULARY PRACTICE

Match the term with its description.

1. An input value that corresponds to an output value of 0
2. $m = \dfrac{y_2 - y_1}{x_2 - x_1}$
3. A set of ordered pairs
4. The set of the inputs of a function
5. $y = mx + b$
6. The set of the outputs of a function
7. A point on a graph for which the x-coordinate is 0
8. A straight line that best represents given data
9. A set of ordered pairs in which each input value corresponds to exactly one output value
10. In a linear model, this is equal to the slope.

A. average rate of change
B. domain
C. function
D. line of best fit
E. range
F. relation
G. slope formula
H. slope-intercept form
I. y-intercept
J. real zero of a function

3.4 Exercises

Follow your instructor's guidelines for showing your work.

For exercises 1–4,
(a) describe what the slope represents.
(b) describe what the y-coordinate of the y-intercept represents.

1. A student must read a textbook before midterm. The function $P(t) = \left(\dfrac{-15 \text{ pages}}{1 \text{ day}}\right)t + 400$ pages represents the relationship between the time in days, t, and the pages left to read in the textbook, $P(t)$.

2. A doctor puts a patient on a weight-loss program. The function $W(t) = \left(\dfrac{-2 \text{ lb}}{1 \text{ week}}\right)t + 95$ lb represents the relationship between the time in weeks, t, and the remaining pounds that must be lost, $W(t)$.

3. The total cost to make a product is the sum of the over-head cost and the cost of materials to make each product. The function $C(x) = \left(\dfrac{\$4}{1 \text{ product}}\right)x + \$1500$ represents the relationship between the number of products, x, and the total cost to make the products, $C(x)$.

4. The monthly earnings of a salesperson are the sum of the guaranteed salary and a commission of 4% on the total value of her sales. The function $M(x) = 0.04x + \$900$ represents the relationship between the value of her sales per month in dollars, x, and her total monthly earnings, $M(x)$.

5. A fence is 30 ft long. Five days later, the fence is 120 ft long. Find the average rate of change in the length of fence per day.

6. A trail crew has cleared 3 mi of trail. Nine days later, they have cleared 21 mi of trail. Find the average rate of change in the length of cleared trail per day.

7. On March 20, a gas meter is at 6673 CCF (1 CCF = 100 ft^3 of natural gas). Thirty days later, the gas meter is at 6694 CCF. Find the average rate of change of the meter per day.

8. On July 1, a water meter is at 437 TG (1 TG = 1000 gal). Sixty days later, the water meter is at 461 TG. Find the average rate of change in the meter per day.

9. The area of a grass lawn is 10,000 ft^2. After 30 min of mowing, 4575 ft^2 remains to be mowed. Find the average rate of change in the area of unmowed lawn per minute. Round to the nearest whole number.

10. An old house has 750 ft of baseboard that needs to be sanded. After 20 hr of work, 600 ft still needs to be sanded. Find the average rate of change in the length of unsanded baseboard per hour.

11. A student is saving money for spring break. On January 1, she has $150 saved. She then decides to start saving the $3.75 per day she usually spends on lattes.
    a. Write a linear function that predicts the total amount of money saved, y, after x days of saving her latte money.
    b. Evaluate the function when $x = 20$ days.
    c. Graph this function for x values from 0 days to 50 days. The scales on the x-axis and the y-axis might not be the same.

12. At midterm, a student has played a total of 80 hr of computer solitaire. She decides to play only 30 min per day for the rest of the term.
    a. Write a linear function that predicts the total amount of time *in hours* during the semester spent playing solitaire, y, after x days past midterm.
    b. Evaluate the function when $x = 28$ days.
    c. Graph this function for x values from 0 to 60 days. The scales on the x-axis and y-axis might not be the same.

13. A preschool has a 1-gal container of milk. Each preschooler receives a 4-oz glass of milk at snack time.
    a. Write a linear function that predicts the amount *in ounces* of milk left in the container, *y*, after *x* glasses of milk have been poured (1 gal = 128 oz).
    b. Use an algebraic method to find the real zero of this function. Round *down* to the nearest whole number.
    c. Describe the situation when the output value is 0 gal.

14. A homeowner collects rainwater from a roof during the winter using a 55-gal barrel. In the summer, he uses about 3.5 gal of the water per day for his garden.
    a. Assume that the barrel is full. Write a linear function that predicts the amount of water left in the barrel, *y*, after *x* days of watering, assuming that it does not rain during that time.
    b. Use an algebraic method to find the real zero of this function. Round *down* to the nearest day.
    c. Describe the situation when the output value is 0 gal.

15. The growth rate of head hair is $\frac{0.02 \text{ in.}}{1 \text{ day}}$. At the beginning of January, the bangs on a student's forehead were 2 in. long.
    a. Write a linear function that is a model of the length of her bangs, $L(x)$, after *x* days of growth.
    b. The student decided to grow her bangs out for a new hairstyle. Use the function to find the time when the output value is 6 in.
    c. Is it reasonable to find an input value for this function for an output value of 120 in.? Explain.

16. Malawi is a nation in southern Africa. A team of researchers discovered that the growth rate of a native fish is about $\frac{0.22 \text{ g}}{1 \text{ day}}$. At the beginning of the study, each fish weighed about 5 g.
    a. Write a linear function that is a model of the weight of the fish, $W(x)$, after *x* days of the study.
    b. Evaluate this function after 90 days of feeding. Round to the nearest whole number.
    c. Is it reasonable to evaluate this function after 20,000 days of feeding? Explain.

17. From 1970 to 2007, the number of families that were married couples without children under age 18 increased at a linear rate of about 345,189 families per year. In 1970, there were about 2,001,000 of these families. (*Source:* www.nces.ed.gov, 2008)
    a. Write a function that is a model of the total number of these families, $N(x)$, for *x* years since 1970.
    b. Evaluate this function to predict the number of these families in 2012. Round to the nearest thousand.

18. From 1974 to 2006, the number of degree-granting two-year postsecondary institutions (including community colleges) increased at a linear rate of about $\frac{43.9 \text{ institutions}}{1 \text{ year}}$. In 1974, there were about 3004 of these institutions. (*Source:* www.nces.ed.gov, 2009)
    a. Write a linear function that is a model of the number of these institutions, $N(x)$, for *x* years since 1974.

b. Evaluate this function to predict the number of these institutions in 2015. Round to the nearest whole number.

19. In large-scale production of broiler chickens, the chickens drink water from troughs, bell drinkers, or nipples.

Type	Recommended amount
Troughs	2.5 cm per bird
Bell drinkers	1 bell per 120 birds
Nipples	5 to 20 birds per nipple

Source: www.agbio.ca

    a. Write a linear function that predicts the length of trough, *y*, needed for *x* birds.
    b. Evaluate this function to find the length of trough needed for a poultry house with 18,800 birds.

20. In Wheeling, Illinois, parking regulations require that retail establishments with a floor area greater than 90,000 ft² provide 3.5 parking spaces per 1000 ft². (*Source:* www.wheeling.il.us)
    a. Write a linear function that predicts the number of parking spaces, *y*, needed for *x* ft² of floor area.
    b. Evaluate this function to find the number of parking spaces for a "big box" store with a floor area of 140,000 ft².

21. From 1967 to 2006, the percent of women ages 18–24 enrolled in postsecondary education grew at an approximately linear rate. The graph shows the line of best fit model of these data.

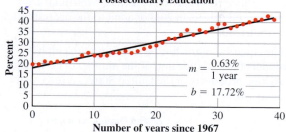

Percent of Women Ages 18–24 Enrolled in Postsecondary Education

Source: www.nces.ed.gov, 2007

    a. Write a linear function that is a model of the percent of women ages 18–24 enrolled in postsecondary education, *y*, in *x* years since 1967.
    b. Evaluate this function to predict the percent of women ages 18–24 enrolled in postsecondary education in 2013.

22. The graph on the next page shows the number of registered autos in the United States. From 1947 to 1992, the line of best fit has a slope of $\frac{2,499,410 \text{ autos}}{1 \text{ year}}$ and a beginning value of 29,457,398 autos.
    a. Write a linear function that is a model of the number of auto registrations, *y*, in *x* years since 1947.

b. Why do you think there was a decrease in registrations from early in the 1940s until about 1945?

U.S. Auto Registrations 1935–2006

Source: www.fhwa.dot.gov

23. From 1949 to 1995, the number of licensed drivers grew at an approximately linear rate. The graph shows the line of best fit model of this data.

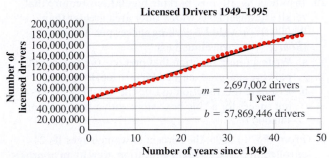

Licensed Drivers 1949–1995

$$m = \frac{2{,}697{,}002 \text{ drivers}}{1 \text{ year}}$$

$$b = 57{,}869{,}446 \text{ drivers}$$

Source: www.fhwa.dot.gov

a. Write a linear function that is a model of the number of licensed drivers, y, in x years since 1949.

b. Evaluate this function to estimate the number of licensed drivers in 2013. Round to the nearest hundred thousand.

24. The equation of the line of best fit on the graph is a model of the relationship of the number of certified organic producers in the United States, y, for x years since 2000.

U.S. Certified Organic Producers

$$y = 472.84x + 6420.3$$

Source: www.usda.gov

a. Rewrite this function including the units of measurement.

b. Evaluate this function to predict the number of certified organic producers in 2015. Round to the nearest hundred.

25. From 1980 to 2005, the number of days of hospital care per year for people from age 65 to 74 years declined at an approximately linear rate. The equation of the line of best fit on the graph is a model of the relationship of the number of days of hospital care, y, in x years since 1980.

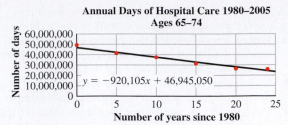

Annual Days of Hospital Care 1980–2005
Ages 65–74

$$y = -920{,}105x + 46{,}945{,}050$$

Source: www.cdc.gov, 2004

a. Rewrite this function to include the units of measurement.

b. Find the real zero. Round to the nearest whole number.

c. Do you think that this model will be an effective predictor in 2040? Explain.

26. From 2004 to 2007, the number of households that had landlines but had no wireless phone declined at an approximately linear rate. The line of best fit on the graph is a model of the relationship of the number of households with landlines but no wireless phone, y, in x years since 2004.

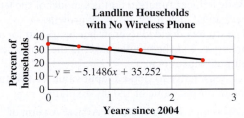

Landline Households
with No Wireless Phone

$$y = -5.1486x + 35.252$$

Source: www.cdc gov, 2007

a. Rewrite this function to include the units of measurement.

b. Find the real zero. Round to the nearest tenth.

c. Do you think that this model will be an effective predictor in 2015? Explain.

27. Between 2005 and 2007, the number of Internet-related fraud complaints such as identify theft increased at an approximately linear rate of $\dfrac{12{,}071 \text{ complaints}}{1 \text{ year}}$. The y-coordinate of the y-intercept of the line of best fit of a model of the relationship of the number of fraud complaints filed, y, for x years since 2005 is 195,790 complaints. Write a linear model of this relationship. Include the units of measurement. (*Source:* www.ftc.gov, Feb. 2008)

28. Since 2004, the percent of state and local law enforcement agencies in the southeastern United States reporting gang activity has increased at a linear rate of about $\dfrac{5\%}{1 \text{ year}}$. The y-coordinate of the y-intercept of the line of best fit of the relationship of the percent reporting gang activity, y, for x years since 2004 is 49.2%. Write a linear model of this relationship. Include the units of measurement. (*Source:* www.fbi.gov, 2009)

29. A bottle of prescription drugs holds 180 pills. A patient takes two pills a day.
   a. Write a function that is a model of the number of pills remaining in the bottle, $f(x)$, after x days.
   b. Find the real zero.
   c. Describe the situation when the output value is 0 pills.

30. A 1-qt container of Roundup® concentrate weed killer contains 32 oz. A homeowner mixes 1.5 oz of the concentrate with water to make 1 gal of ready-to-use weed killer.
   a. Write a function that is a model of the number of ounces of Roundup concentrate remaining in the container, $f(x)$, after mixing x gal of ready-to-use weed killer.
   b. Find the real zero. Round down to the nearest whole number.
   c. Describe the situation when the output value is 0 oz.

31. A model of the relationship of the percent of American adults over age 20 who are obese, $P(x)$, and the years since 1988, x, is $P(x) = \left(\dfrac{0.64\%}{1 \text{ year}}\right)x + 22.89\%$. Evaluate this function to predict the percent of American adults who will be obese in 2015. Round to the nearest tenth of a percent. (*Source:* www.cdc.gov, Dec. 2008)

32. A model of the relationship of the number of years since 1999, t, and the average hourly earnings in dollars for construction workers, $E(t)$, is $E(t) = \left(\dfrac{\$0.49}{1 \text{ year}}\right)t + \$16.65$. Evaluate this function to predict the average hourly earnings in dollars for these workers in 2015. Round to the nearest hundredth. (*Source:* www.bls.gov, 2009)

33. A model of the relationship of the number of years since 2000, t, and the percent of eighth graders who said that alcohol is fairly easy or very easy to get, y is
$y = \left(\dfrac{-1.37\%}{1 \text{ year}}\right)x + 69.8\%$. Evaluate this function to predict the percent of eighth graders who will say that alcohol is fairly easy or very easy to get in 2012. Round to the nearest tenth of a percent. (*Source:* www.monitoringthefuture.org, 2008).

34. A model of the relationship of the number of years since 1991, t, and the rate of injury and illness per 100 full-time workers per year in private industry, $R(t)$, is
$R(t) = \left(\dfrac{-0.367 \text{ cases}}{1 \text{ year}}\right)t + 8.973$ cases. Evaluate this function to predict the rate of injury and illness per 100 full-time workers in 2009. Round to the nearest tenth. (*Source:* www.bls.gov)

35. A model of the relationship of the number of twists in Arselon thread, n, and the maximum force in centinewtons that can be applied to the thread before it breaks, $F(n)$, is $F(n) = (3 \times 10^{-8})n^4 - (4 \times 10^{-5})n^3 + 0.0148n + 2482.9$. Evaluate this function to find the maximum force for a thread with 200 twists. Round to the nearest hundred. (*Source:* Mazurk et al., "Development of Rational Technology for Manufacture of Arselon Sewing Thread," *Fibre Chemistry*, Vol. 37, No. 2, 2005)

36. A model of the relationship of the number of years since 1981, x, and the number of AIDS cases in Illinois, Y, is $Y = -41.2857 + 61.6666x - 23.25x^2 + 4.0833x^3$. Evaluate this function to predict the number of AIDS cases in Illinois in 2000. Round to the nearest hundred. (*Source:* Cove et al., "The Present and the Future of AIDS and Tuberculosis in Illinois," *American Journal of Public Health*, Vol. 8, No. 8, 1990)

37. A model of the relationship of the number of years since 1970, t, and annual national health expenditures in billions of dollars, $E(t)$, is $E(t) = 1.451t^2 + 1.698t + 62.22$. Evaluate this function to predict the amount of national health expenditures in 2009. Round to the nearest billion. (*Source:* www.cms.gov)

38. A model of the relationship of the number of years since 1988, x, and the number of violent crimes, $N(x)$, is $N(x) = 212.6x^3 - 704.3x^2 - 73,148x + 1,977,401$. Evaluate this function to predict the number of violent crimes in 2012. Round to the nearest thousand. (*Source:* www2.fbi.gov, Sept. 2008).

For exercises 39–40, VO2 max is the maximum amount of oxygen in milliliters used in 1 min by 1 kg of body weight.

39. In a study of endurance athletes, VO2 max declined linearly from $\dfrac{66 \text{ mL}}{1 \text{ kg} \cdot \text{min}}$ at an altitude of 300 m to $\dfrac{55 \text{ mL}}{1 \text{ kg} \cdot \text{min}}$ at an altitude of 2800 m.
   (*Source:* Wehrlin et al., "Linear Decrease in VO2 max and Performance with Increasing Altitude in Endurance Athletes," *European Journal of Applied Physiology*, Vol. 96, No. 4, 2006)
   a. Write two ordered pairs that describe this data: (altitude, VO2 max).
   b. Find the slope of a line that passes through these points.
   c. Use point-slope form to write a linear model of the relationship of VO2 max and altitude. Rewrite the model in slope-intercept form.
   d. Evaluate this function to predict the VO2 max at 1500 m. Round to the nearest whole number.

40. Researchers found that a strenuous program of exercises resulted in a linear increase in aerobic power. The average change in VO2 max was $\dfrac{0.12 \text{ mL}}{1 \text{ kg} \cdot \text{min}}$ per week, which simplifies to $\dfrac{0.12 \text{ mL}}{1 \text{ kg} \cdot \text{min} \cdot \text{week}}$. (*Source:* Hickson et al., "Linear Increase in Aerobic Power Induced by a Strenuous Program of Endurance Exercises," *Journal of Applied Physiology*, Vol. 42, No. 3, 1977)
   a. An athlete starting a strenuous exercise program has a VO2 max of $\dfrac{40 \text{ mL}}{1 \text{ kg} \cdot \text{min}}$. Write a function in slope-intercept form that is a model of the relationship of the number of weeks of the exercise program, x, and the VO2 max, y.
   b. Evaluate this function to find the athlete's VO2 max after 5 weeks.

Problem Solving: Practice and Review

Follow your instructor's guidelines for using the five steps as outlined in Section 1.5, p. 51.

41. Find the original cost of the summer school offerings.

   The Capistrano Unified School District in Orange County cut its summer school offerings by roughly three-quarters, saving about $600,000. "In every area, we have waiting lists," said Ron Lebs, a deputy superintendent of the 51,000-student district. (*Source:* www.latimes.com, May 29, 2009)

42. The Fahrenheit roller coaster at Hersheypark in Pennsylvania goes up 121 ft before plummeting down a 97-degree drop. Each ride lasts 85 s. There are 12 riders on a train. The ride has 3 trains. The theoretical capacity of the ride is 850 riders per hour. To reach this theoretical capacity, find the number of trips each of the 3 trains must complete per hour. Round to the nearest whole number. (*Source:* www.hersheypark.com)

43. Batman™ The Ride is a roller coaster at Six Flags® Great Adventure in New Jersey. It has a top speed of 50 mi per hour. Its length is 2693 ft, and its maximum height is 10.5 stories. Each ride lasts 1 min 40 s. Find the ride's average speed in miles per hour (1 mi = 5280 ft). Round to the nearest whole number. (*Source:* www.sixflags.com/GreatAdventure)

44. In Henderson, Nevada, about 2 million yd³ of contaminated dirt will be moved from a hazardous waste site to a landfill 3 mi away. Assume that the dirt will be moved in belly dump trucks with a capacity of 17 yd³. The average fuel economy for these trucks is $\dfrac{5 \text{ mi}}{1 \text{ gal diesel fuel}}$. Find the amount of diesel fuel needed to move the dirt. Round to the nearest thousand. (*Source:* www.ngem.com)

Technology

For exercises 45–48,
(a) use a graphing calculator to graph the function. Choose a window that shows the vertex and the *x*-intercepts. Sketch the graph; describe the window.
(b) estimate the minimum or maximum output value. Round to the nearest hundredth.
(c) use interval notation to represent the range.

45. $f(x) = 2x^2 - 12x + 22$

46. $g(x) = -x^2 - x + 15$

47. $y = -x^2 + 5x + 1$

48. $y = x^2 - 5x - 30$

Find the Mistake

For exercises 49–52, the completed problem has one mistake.
(a) Describe the mistake in words, or copy down the whole problem and highlight or circle the mistake.
(b) Do the problem correctly.

49. **Problem:** A Realtor® has budgeted $500 per month for advertising in a local newspaper. A reasonable one-day ad to sell a house costs $4.60. Write a linear function that is a model of the amount of money remaining in the budget, *y*, after running *x* one-day ads.

   Incorrect Answer: $y = \left(\dfrac{\$4.60}{1 \text{ ad}}\right)x + \$500$

50. **Problem:** An ad read, "Dog Owners! Complete dog waste removal service. Cleans yards and pens. $10 wk 1 dog; $2.50 additional dog. Satisfaction guaranteed. Call Dooty Free." Write a linear function that represents the total cost per week, *y*, of having the service remove the waste of *x* dogs. (*Source:* www.moneysaver.com)

© Billie Tribitt

   Incorrect Answer: $y = \left(\dfrac{\$2.50}{1 \text{ dog}}\right)x + \$10$

51. **Problem:** The line of best fit for a graph that represents the relationship of the number of years since 1961, *x*, and the number of licensed drivers in the United States, *y*, is

   $$y = \left(\dfrac{2.68 \text{ million drivers}}{1 \text{ year}}\right)x + 87.2 \text{ million drivers.}$$

   Predict the number of licensed drivers in the United States in 2008. Round to the nearest million. (*Source:* www.fhwa.dot.gov)

   Incorrect Answer:

   $$y = \left(\dfrac{2.68 \text{ million drivers}}{1 \text{ year}}\right)x + 87.2 \text{ million drivers}$$

   $$y = \left(\dfrac{2.68 \text{ million drivers}}{1 \text{ year}}\right)(2008) + 87.2 \text{ million drivers}$$

   $$y = 5469 \text{ million drivers}$$

52. **Problem:** A model of the relationship of the number of years since 1982, *t*, and the annual direct expenditures by the judicial branch in billions of dollars, $E(t)$, is $E(t) = 0.037t^2 + 0.831t + 7.80$. Evaluate this function to find the direct expenditures in 2008. Round to the nearest tenth of a billion.

   Incorrect Answer:

   $$E(t) = 0.037t^2 + 0.831t + 7.80$$
   $$E(2008) = 0.037(2008)^2 + 0.831(2008) + 7.80$$
   $$E(2008) = \$150,862.8 \text{ billion}$$

Review

For exercises 53–56, simplify.

53. $(3x - 6)^2$

54. $-9(3x - 6) + 11$

55. $(3x - 6)^2 + 2(3x - 6) + 7$

56. $(x^2 + 5x + 4)^2$

SUCCESS IN COLLEGE MATHEMATICS

57. In some application problems, a formula is used to solve the problem. In exercise 43, the length of the roller coaster and the time for one ride are given. The unknown is the speed of the roller coaster. What formula describes the relationship of length, time, and speed?

58. Extraneous information is information that is not needed to solve a problem. In exercise 43, what information in the problem is extraneous?

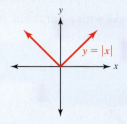

A function is a set of ordered pairs in which each input value corresponds to exactly one output value. We have studied polynomial functions. In this section, we will study another "family" of functions, the absolute value functions.

SECTION 3.5

Absolute Value Functions

After reading the text, working the practice problems, and completing assigned exercises, you should be able to:

1. Use a table of ordered pairs to graph an absolute value function.

2. Given an absolute value function in the form $y = a|x - h| + k$, identify the vertex.

3. Use the graph of an absolute value function to estimate its vertex, write the equation of the axis of symmetry, and identify the maximum or minimum output value.

4. Use interval notation to represent the domain and range of an absolute value function.

5. Use an algebraic or graphical method to find the real zeros of an absolute value function.

Graphs of Absolute Value Functions

The **absolute value** of a real number is its distance from 0 on the real number line. Since a distance is greater than or equal to 0, an absolute value is also greater than or equal to 0. The graph of an **absolute value function** in the form $f(x) = a|x - h| + k$, where a, h, and k are real numbers, is V-shaped (Figure 1) and can open up or down. The vertex of the graph is (h, k), the equation of the axis of symmetry is $x = h$, and the minimum or maximum output value is k.

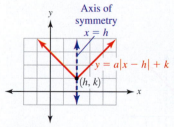

Figure 1

EXAMPLE 1 An absolute value function is $y = |x|$.

(a) Identify a, h, and k.

SOLUTION ▶
$$y = a|x - h| + k$$
$$y = 1|x - 0| + 0 \qquad a = 1, \ h = 0, \ k = 0$$

(b) Use a table of ordered pairs to graph the function.

▶ We need to graph enough points to show the V shape. As with a parabola, we need to graph more than three points.

$$f(0) = |0|$$
$$f(0) = 0$$
$$f(1) = |1|$$
$$f(1) = 1$$
$$f(2) = |2|$$
$$f(2) = 2$$

x	y
0	0
1	1
2	2

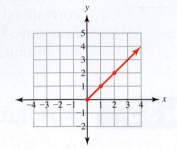

This appears to be the right side of the graph. We need to find more points to the left of the y-axis.

$$f(-1) = |-1|$$
$$f(-1) = 1$$
$$f(-2) = |-2|$$
$$f(-2) = 2$$

x	y
0	0
1	1
2	2
-1	1
-2	2

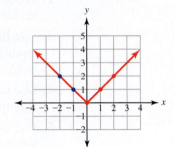

(c) Identify the vertex.

▶ On the graph, the vertex appears to be (0, 0). This makes sense because the output value of the function cannot be less than 0.

(d) Identify the minimum output value.

▶ The minimum output value is the y-coordinate of the vertex, 0.

(e) Use the vertical line test to show that the graph represents a function.

▶ A vertical line crosses the graph in at most one point. This graph represents a function.

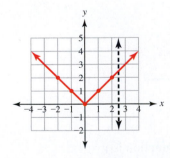

If $a > 0$, the graph of an absolute value function in the form $y = a|x - h| + k$ opens up. In the next example, we graph $y = -|x| + 4$. Since we can rewrite this function as $y = -1|x - 0| + 4$, $a = -1$. The graph opens down.

EXAMPLE 2   An absolute value function is $y = -|x| + 4$.

(a) Identify a, h, and k.

SOLUTION ▶ $y = a|x - h| + k$

$y = \mathbf{-1}|x - \mathbf{0}| + 4$ $a = -1$, $h = 0$, $k = 4$

(b) Use a table of ordered pairs to graph the function.

x	y
−2	2
−1	3
0	4
1	3
2	2

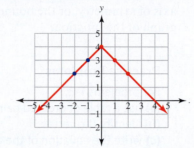

(c) Identify the vertex.

▶   The vertex is (0, 4).

(d) Identify the maximum output value.

▶   The maximum output value is the y-coordinate of the vertex, 4.

The graph of an absolute value function is symmetric about a vertical line, the **axis of symmetry**. This line includes the vertex of the graph of the function.

EXAMPLE 3   An absolute value function is $f(x) = |x + 4| - 1$.

(a) Use a table of ordered pairs to graph the function.

SOLUTION ▶   Since $a = 1$, the graph opens up. These points look like they are on the "right arm" of the graph. We need more negative input values to find the "left arm" of the graph.

x	y
−2	1
−1	2
0	3
1	4

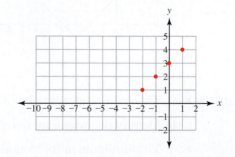

As we graph more points, the symmetric left arm of the graph appears. Sketch the graph through the points.

x	y
−3	0
−4	−1
−5	0
−6	1
−7	2

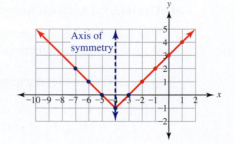

(b) Identify the vertex.

▶   The vertex appears to be $(-4, -1)$.

(c) Write the equation of the axis of symmetry.

▶   The axis of symmetry passes through the vertex. Its equation is $x = -4$.

Dashed lines on graphs are used for vertical line tests, horizontal line tests, an axis of symmetry, or the boundary of an inequality. Dashed lines are *not* part of the graph of the function.

Practice Problems

For problems 1–3,
(a) use a table of ordered pairs to graph the function.
(b) identify the vertex.
(c) write the equation of the axis of symmetry.

1. $y = |x| + 2$ **2.** $f(x) = -|x| - 1$ **3.** $g(x) = |x - 3| - 2$

Domain and Range

Any real number can be an input for the absolute value function $y = a|x - h| + k$. So the **domain** of an absolute value function is the set of real numbers. The **range** of an absolute value function depends on the minimum or maximum output value. If the graph opens up, the function has a **minimum** output value.

EXAMPLE 4   **(a)** Use the graph to identify the vertex.

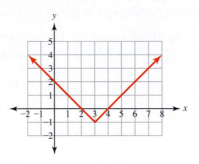

SOLUTION ▶   The vertex appears to be $(3, -1)$.

(b) Identify the minimum output value.

▶   The minimum output value is the y-coordinate of the vertex, -1.

(c) Use interval notation to represent the domain.

▶   The domain of an absolute value function is the set of real numbers, $(-\infty, \infty)$.

(d) Use interval notation to represent the range.

▶   The output values are greater than or equal to the minimum output. The range is the set of real numbers greater than or equal to -1, $[-1, \infty)$.

In the next example, the graph opens down. The function has a **maximum** output value.

EXAMPLE 5   (a) Use the graph to identify the vertex.

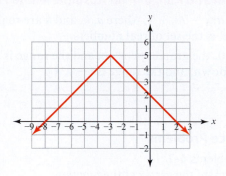

SOLUTION ▶   The vertex is $(-3, 5)$.

(b) Identify the maximum output value.

▶   The maximum output value is the y-coordinate of the vertex, 5.

(c) Use interval notation to represent the range.

▶   The range is the set of real numbers less than or equal to 5, $(-\infty, 5]$.

If we know the vertex and the direction that the graph of an absolute value function opens, we can identify its range.

EXAMPLE 6   The vertex of an absolute value function is $(3, 1)$.

(a) Use interval notation to represent the range when the graph opens up.

SOLUTION ▶   Since the graph opens up, the output value of the vertex, $y = 1$, is a minimum. The range is $[1, \infty)$.

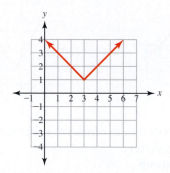

(b) Use interval notation to represent the range when the graph opens down.

▶   Since the graph opens down, the output value of the vertex, $y = 1$, is a maximum. The range is $(-\infty, 1]$.

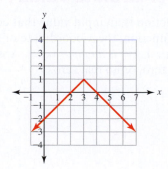

> **Domain and Range of an Absolute Value Function**
>
> If $y = a|x - h| + k$, where a, k, and h are real numbers and $a \neq 0$, the domain is the set of real numbers.
>
> If $a > 0$, the graph opens up, and the range is $[k, \infty)$. If $a < 0$, the graph opens down, and the range is $(-\infty, k]$.

Practice Problems

For problems 4–7,
(a) estimate the vertex of the function.
(b) identify the minimum or maximum output value.
(c) use interval notation to represent the domain.
(d) use interval notation to represent the range.

4.

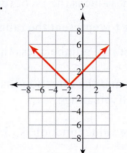

5.

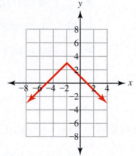

6.

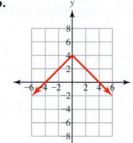

7.

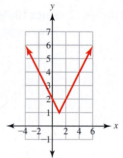

8. Use interval notation to represent the range of the absolute value function with the given vertex.
   a. $(6, 9)$; opens up
   b. $(6, 9)$; opens down

Real Zeros of Absolute Value Functions

A real zero of a function is an input value that corresponds to an output value of 0. On the graph of a function, the x-coordinate of an x-intercept is a real zero. *A zero is not an ordered pair; a zero is an input value.* Absolute value functions can have two real zeros, one real zero, or no real zero.

EXAMPLE 7   Use the graph to identify any real zeros.

(a) $y = |x - 2|$ **(b)** $y = -|x + 2| + 3$ **(c)** $y = |x - 1| + 1$

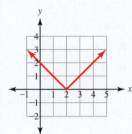   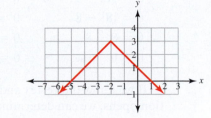   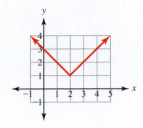

SOLUTION ▶   x-intercept: $(2, 0)$ x-intercepts: $(-5, 0)$, $(1, 0)$ No x-intercepts

Real zero: $x = 2$ Real zeros: $x = -5$, $x = 1$ No real zeros

To find any real zeros of an absolute value function, replace the output value with 0 and solve for the input value. We are solving an absolute value equation in one variable. (To review, see Section 2.5.)

Solving an Absolute Value Equation

To solve $|x - h| + k = C$, where h, k, and C are real numbers,

1. Use the definition of absolute value to rewrite as two equations,
   $(x - h) + k = C$ and $-(x - h) + k = C$.

2. Solve each equation.

3. Check. If both solutions are extraneous, the equation has no solution.

EXAMPLE 8   **(a)** Use an algebraic method to find any real zeros of $y = |x - 2| - 8$.

SOLUTION ▶

$\mathbf{0 = |x - 2| - 8}$	Replace the output value, y, with 0.
$0 = (x - 2) - 8$	First case: $x - 2 \geq 0$
$0 = x - 10$	Simplify.
$\underline{+10 \qquad +10}$	Addition property of equality.
$\mathbf{10 = x + 0}$	Simplify.
$10 = x$	One solution is 10.
$0 = -(x - 2) - 8$	Second case: $x - 2 < 0$
$0 = \mathbf{-x + 2} - 8$	Distributive property.
$0 = \mathbf{-x - 6}$	Simplify.
$\underline{+6 \qquad\quad +6}$	Addition property of equality.
$\mathbf{6 = -x + 0}$	Simplify.
$\dfrac{6}{-1} = \dfrac{\mathbf{-x}}{-1}$	Division property of equality.
$-6 = x$	Simplify. The other solution is -6.

Real zeros: $x = 10$, $x = -6$

(b) Check.

$0 = \lvert x - 2 \rvert - 8$	$0 = \lvert x - 2 \rvert - 8$
$0 = \lvert \mathbf{10} - 2 \rvert - 8$	$0 = \lvert \mathbf{-6} - 2 \rvert - 8$
$0 = \lvert \mathbf{8} \rvert - 8$	$0 = \lvert \mathbf{-8} \rvert - 8$
$0 = \mathbf{8} - 8$	$0 = \mathbf{8} - 8$
$0 = \mathbf{0}$ True.	$0 = \mathbf{0}$ True.

If we know the vertex and the direction that the graph of an absolute value function opens, we can determine whether it has any real zeros.

EXAMPLE 9 | The vertex of an absolute value function is $(3, 1)$.

(a) If the graph opens up, does the function have any real zeros?

SOLUTION ▶

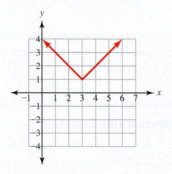

The vertex is above the *x*-axis. If the graph opens up, the graph does not intersect the *x*-axis. Since there are no *x*-intercepts, the function has no real zeros.

(b) If the graph opens down, does the function have any real zeros?

▶

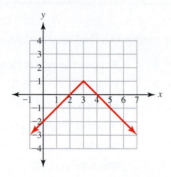

The vertex is above the *x*-axis. If the graph opens down, the graph intersects the *x*-axis in two points. Since there are two *x*-intercepts, there are two real zeros.

Practice Problems

For problems 9–12, use the graph of the function to identify any real zeros.

9.

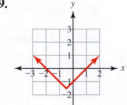

10.

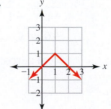

11.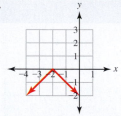

12.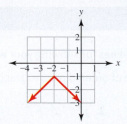

13. Use an algebraic method to find any real zeros of $y = |x - 5| - 1$.

14. The vertex of an absolute value function is $(2, -3)$. Its graph opens down.
  a. Does this function have any real zeros?
  b. Explain.

Using Technology: Graphing an Absolute Value Function

EXAMPLE 10 Graph: $y = |x|$

Press [Y=]. To insert the absolute value, press [MATH]. Move the cursor to the right and highlight NUM. Choose 1. Press [ENTER]. This inserts the three letters and a parenthesis: abs (. Type the expression inside the absolute value. In this example, it is just x. Type a parenthesis,). Press [GRAPH]. We see the graph of the function.

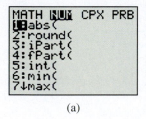
(a)

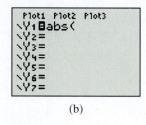
(b)

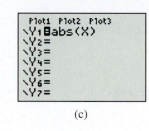
(c)

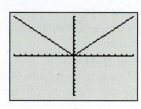
(d)
This is the graph of $y = |x|$.

Practice Problems For problems 15–18,
  (a) graph each function. Choose a window that shows the vertex. Sketch the graph; describe the window.
  (b) find the minimum or maximum output value.
  (c) use interval notation to represent the range.

15. $y = |x| - 6$ **16.** $y = |x + 9| - 12$ **17.** $y = 7 - |x|$ **18.** $y = |x - 12|$

3.5 VOCABULARY PRACTICE

Match the term with its description.

1. The set of the input values in a function

2. The set of the output values in a function

3. A vertical line that passes through the vertex of the graph of an absolute value function

4. The first coordinate of an ordered pair in a function

5. The second coordinate of an ordered pair in a function

6. A set of ordered pairs in which each input value corresponds to exactly one output value

7. The distance a number is from 0 on the real number line

8. A test used to determine whether a graph represents a function

9. On the graph of an absolute value function, the ordered pair that includes the minimum or maximum output value

10. If $a > 0$, the output value in the vertex is this.

A. absolute value
B. axis of symmetry
C. domain
D. function
E. input value
F. minimum
G. output value
H. range
I. vertex
J. vertical line test

3.5 Exercises

Follow your instructor's guidelines for showing your work.

For exercises 1–24, use a table of ordered pairs to graph the function.

1. $y = |x| + 3$
2. $y = |x| + 2$
3. $y = |x| - 3$
4. $y = |x| - 2$
5. $y = |x + 3|$
6. $y = |x + 2|$
7. $y = |x - 3|$
8. $y = |x - 2|$
9. $y = |x - 3| - 6$
10. $y = |x - 2| - 4$
11. $y = |x - 3| + 6$
12. $y = |x - 2| + 4$
13. $y = -|x| + 3$
14. $y = -|x| + 2$
15. $y = -|x| - 3$
16. $y = -|x| - 2$
17. $y = 3|x|$
18. $y = 2|x|$
19. $y = -3|x|$
20. $y = -2|x|$
21. $y = |3x|$
22. $y = |2x|$
23. $y = |-3x|$
24. $y = |-2x|$

25. Draw the graph of an absolute value function with only one x-intercept.

26. Draw the graph of an absolute value function with no x-intercept.

For exercises 27–32,
(a) use the graph to identify the vertex.
(b) use interval notation to represent the range.

27.

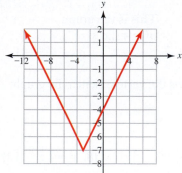

28.

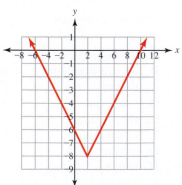

29.

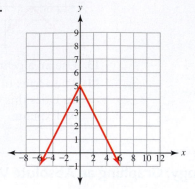

30.

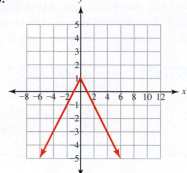

31.

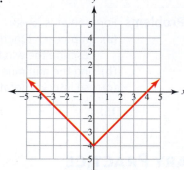

32.

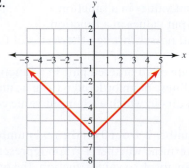

For exercises 33–42, use interval notation to represent the range of the absolute value function with the given vertex.

33. $(5, 2)$; opens up

34. $(2, 8)$; opens up

35. $(-1, 2)$; opens down

36. $(-3, 7)$; opens up

37. $(5, 2)$; opens down

38. $(2, 8)$; opens down

39. $(4, -3)$; opens down

40. $(9, -6)$; opens down

41. $(0, 0)$; opens up

42. $(0, 0)$; opens down

For exercises 43–48, use the graph to estimate any real zeros.

43.

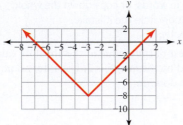

44.

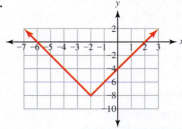

45.

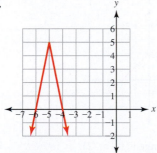

46.

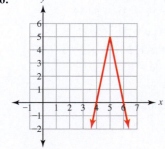

47.

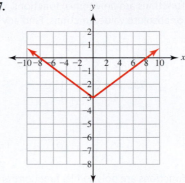

48.

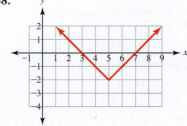

For exercises 49–60, use an algebraic method to find any real zeros.

49. $f(x) = |x| - 2$

50. $f(x) = |x| - 9$

51. $g(x) = |x + 10| - 2$

52. $g(x) = |x + 16| - 9$

53. $y = |x + 10|$

54. $y = |x + 16|$

55. $f(x) = |3x| + 12$

56. $f(x) = |2x| + 14$

57. $y = |3x + 9| - 12$

58. $y = |2x + 20| - 14$

59. $y = |-3x + 9| - 12$

60. $y = |-2x + 20| - 14$

For exercises 61–64,
(a) does an absolute value function with the given vertex and that opens in the given direction have any real zeros?
(b) explain.

61. $(8, 9)$; opens down

62. $(-2, 11)$; opens down

63. $(-3, -4)$; opens down

64. $(4, -15)$; opens down

For exercises 65–68, use interval notation to:
(a) represent the domain.
(b) represent the range.

65. $y = |x + 3| - 2$; vertex $(-3, -2)$

66. $y = |x + 4| - 5$; vertex $(-4, -5)$

67. $y = x^2 + 6x + 7$; vertex $(-3, -2)$

68. $y = x^2 + 8x + 11$; vertex $(-4, -5)$

For exercises 69–76, the functions are polynomial functions (see Sections 3.2 and 3.3) or absolute value functions. Find any real zeros.

69. $y = \dfrac{3}{4}x + 6$

70. $y = \dfrac{3}{5}x + 6$

71. $y = 2x^2 - 13x - 7$

72. $y = 2x^2 - 11x - 6$

73. $f(x) = x^3 - 16x$

74. $f(x) = x^3 - 25x$

75. $y = |2x + 6| - 18$

76. $y = |2x + 4| - 16$

For exercises 77–82, the functions are polynomial functions or absolute value functions. If the function has a minimum or maximum output value, the vertex is given.
(a) Use interval notation to represent the domain.
(b) Use interval notation to represent the range.

77. $f(x) = x - 9$

78. $f(x) = x - 10$

79. $y = 2x^2 - 12x + 22$; vertex $(3, 4)$

80. $y = 2x^2 - 16x + 35$; vertex $(4, 3)$

81. $g(x) = -2|x + 3| - 4$; vertex $(-3, -4)$

82. $g(x) = -2|x + 5| - 9$; vertex $(-5, -9)$

Problem Solving: Practice and Review

Follow your instructor's guidelines for using the five steps as outlined in Section 1.5, p. 51.

83. A brochure for the Scafco Corporation describes the dimensions of a water tank. The diameter is 18 ft. The height up to the eaves is 10 ft 9 in. The reported capacity is 19,400 gal. Find the difference in the capacity of a cylinder with these measurements and the reported capacity of the water tank. Round to the nearest hundred. (Volume of a cylinder $= \pi r^2 h$; 1 ft^3 $\approx$ 7.481 gal; 1 ft^3 = 1728 in.^{3}; $\pi \approx 3.14$.) (*Source:* www.scafco.com)

84. Find the number of tomatoes produced per acre in the previous generation of greenhouses. Round to the nearest whole number.

Designed by Kubo Greenhouse Projects, a Dutch company, the temperature- and humidity-controlled glass-sheeted farm is expected to produce 482 tons of tomatoes per acre, 15% more than Houweling's previous generation of greenhouses. The plants live far longer than field crops and are replaced every six months. (*Source:* www.latimes.com, May 14, 2009)

85. Find the amount that the bail bondsman collects in non-refundable fees during the 35-month period.

Wayne Spath is a bail bondsman. . . . If Mr. Spath considers a potential client a good risk, he will post bail in exchange for a nonrefundable 10 percent fee. In a 35-month period ending in November, his records show Mr. Spath posted about \$37 million in bonds—7934 of them. (*Source:* www.nytimes.com, Jan. 29, 2008)

86. Find the percent decrease in the number of students that will be helped by the State Work Study program. Round to the nearest percent.

Several other student-aid programs took a hit: The State Work Study program, which this year helped about 7,600 undergraduate and graduate students work their way through school in jobs related to their field of study. There's only enough money to help about 2,500 students next year. (*Source:* www.seattletimes.nwsource.com, June 6, 2011)

Technology

For exercises 87–90,
(a) graph each function on a graphing calculator. Choose a window that shows the vertex. Sketch the graph; describe the window.
(b) use interval notation to represent the domain.
(c) identify the minimum or maximum output value.
(d) use interval notation to represent the range.

87. $f(x) = |x + 3| - 4$

88. $f(x) = |x + 2| - 3$

89. $y = -|x - 2| + 15$

90. $y = -|x - 4| + 18$

Find the Mistake

For exercises 91–94, the completed problem has one mistake.
(a) Describe the mistake in words, or copy down the whole problem and highlight or circle the mistake.
(b) Do the problem correctly.

91. Problem: The smallest output value of the absolute value function represented by the graph is 2. Use interval notation to represent the domain.

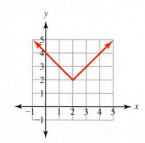

Incorrect Answer: The domain is $[2, \infty)$.

92. Problem: The graph represents $f(x) = -|x| + 2$. Use interval notation to represent the range.

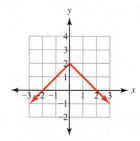

Incorrect Answer: The range is $(-\infty, 2)$.

93. Problem: Use an algebraic method to find any real zeros of $f(x) = |x + 3| - 5$.

Incorrect Answer: $f(x) = |x + 3| - 5$
$$0 = |x + 3| - 5$$
$$0 = x + 3 - 5$$
$$0 = x - 2$$
$$2 = x$$

The real zero of this function is $x = 2$.

94. Problem: Use the graph of the function to identify any real zeros.

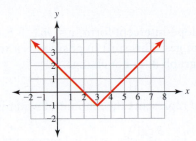

Incorrect Answer: The zeros are $(2, 0)$ and $(4, 0)$.

Review

95. Write the slope-intercept form of a linear equation in two variables.

96. Write the point-slope form of a linear equation in two variables.

97. Write the standard form of a quadratic equation in two variables.

98. The equation of a line is $y - 4 = \frac{2}{3}(x - 7)$.
   a. Identify the slope.
   b. Identify the y-intercept.

SUCCESS IN COLLEGE MATHEMATICS

99. In exercise 83, you need to find the difference in the volume of a tank and a cylinder. Describe how to change the units of measurement in solving this problem.

100. In exercise 84, you know the percent increase and the future amount of tomatoes that will be produced. You need to find the original amount of tomatoes. A word equation that describes this relationship is original amount + amount of increase = new amount. If $x =$ the original amount of tomatoes, write an algebraic equation that describes this relationship.

For a **60-gal** rainwater barrel, the model $y = \left(\dfrac{-5 \text{ gal}}{1 \text{ day}}\right)x + 60 \text{ gal}$ describes the amount of water in the barrel, y, after x days. For an **80-gal** rainwater barrel, the model $y = \left(\dfrac{-5 \text{ gal}}{1 \text{ day}}\right)x + 80 \text{ gal}$ describes the amount of water in a rainwater barrel, y, after x days. If we graph these functions, the lines have the same slope but different y-intercepts. In this section, we will study how changing a function can cause horizontal or vertical translations in its graph.

| SECTION 3.6 | Translation of Polynomial Functions |

After reading the text, working the practice problems, and completing assigned exercises, you should be able to:

1. Change a linear function in slope-intercept form to cause a vertical translation of its graph.

2. Change a linear function in point-slope form to cause a horizontal or vertical translation of its graph.

3. Change a quadratic function in standard form to cause a vertical translation of its graph.

4. Change an absolute value function to cause a horizontal or vertical translation of its graph.

5. Given a graph, use translation to graph a function.

Translation of Linear Functions in Slope-Intercept Form

The y-intercept of a linear function in slope-intercept form, $y = mx + b$, is $(0, b)$. If b changes, the position of the graph changes. This shift in position is a **vertical shift** or **vertical translation**. The slope of the graph does not change.

EXAMPLE 1 The function $y = \left(-\dfrac{5 \text{ gal}}{1 \text{ day}}\right)x + 60 \text{ gal}$ represents the amount of water in a rain barrel, y, after x days. The function $y = \left(-\dfrac{5 \text{ gal}}{1 \text{ day}}\right)x + 80 \text{ gal}$ represents the amount of water in a different rain barrel, y, after x days.

(a) Graph both functions.

SOLUTION ▶ Use slope-intercept graphing.

The y-intercept of $y = \left(-\dfrac{5 \text{ gal}}{1 \text{ day}}\right)x + 60 \text{ gal}$ is $(0 \text{ days}, 60 \text{ gal})$.

The slope is $-\dfrac{5 \text{ gal}}{1 \text{ day}}$.

The y-intercept of $y = \left(-\dfrac{5 \text{ gal}}{1 \text{ day}}\right)x + 80 \text{ gal}$ is $(0 \text{ days}, 80 \text{ gal})$.

The slope is $-\dfrac{5 \text{ gal}}{1 \text{ day}}$.

The graphs of the functions are parallel lines with the same slope and different y-intercepts.

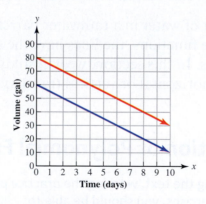

▶ **(b)** Describe the difference (the shift) in the graphs of these functions.

The graph of $y = \left(-\dfrac{5 \text{ gal}}{1 \text{ day}}\right)x + 80 \text{ gal}$ is *shifted vertically up* 20 gal from the graph of $y = \left(-\dfrac{5 \text{ gal}}{1 \text{ day}}\right)x + 60 \text{ gal}$. The difference in the values of b is equal to the vertical shift in the position of the graphs: $80 \text{ gal} - 60 \text{ gal} = 20 \text{ gal}$.

In the next example, subtracting 5 from b shifts the graph vertically down 5 units. To **graph by translation**, we graph the shifted y-intercept and draw the original line in this new position. Although the y-intercepts of the two lines are different, the slopes are the same.

EXAMPLE 2 The graph represents $f(x) = -\dfrac{5}{4}x + 4$. The y-intercept is $(0, 4)$.

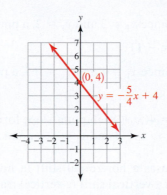

(a) Change this function so that its graph is shifted vertically down 5 units.

SOLUTION ▶ The y-coordinate of the y-intercept is 4. To shift the y-intercept vertically down, subtract: $4 - 5 = -1$. The y-intercept of the shifted function is $(0, -1)$, and $b = -1$.

$$f(x) = -\frac{5}{4}x - 1 \qquad y = mx + b$$

(b) Graph the new function by translation.

▶ Copy the original line in its new position with the new shifted y-intercept. The slope of the lines is the same.

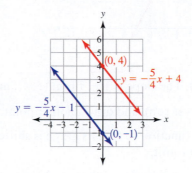

Practice Problems

1. A function is $f(x) = 2x + 9$.
  a. Graph the function.
  b. Change the function so that its graph is shifted vertically down 3 units.
  c. Graph the new function by translation.
2. A function is $g(x) = -6x + 5$.
  a. Graph the function.
  b. Change the function so that its graph is shifted vertically up 7 units.
  c. Graph the new function by translation.

Translation of Linear Functions in Point-Slope Form

To shift the graph of a linear function in point-slope form, $y - y_1 = m(x - x_1)$, we change (x_1, y_1).

EXAMPLE 3 For each function, identify the slope, m, and a point on the line, (x_1, y_1).

(a) $y - 2 = 4(x - 3)$

SOLUTION ▶ The slope is 4. Since $x_1 = 3$ and $y_1 = 2$, a point on the line is $(3, 2)$.

(b) $y - (-5) = 4(x - 1)$

▶ The slope is 4. Since $x_1 = 1$ and $y_1 = -5$, a point on the line is $(1, -5)$.

(c) $y - 7 = 4(x + 8)$

▶ Rewrite $y - 7 = 4(x + 8)$ in point-slope form, $y - 7 = 4(x - (-8))$. The slope is 4. Since $x_1 = -8$ and $y_1 = 7$, a point on the line is $(-8, 7)$.

Changing x_1 causes a horizontal shift (a horizontal translation) of the graph. Changing y_1 causes a vertical shift (a vertical translation) of the graph.

EXAMPLE 4 A function is $y - 4 = -2(x - 5)$.

(a) Identify: (x_1, y_1)

SOLUTION ▶ Since this function is in point-slope form, $y - y_1 = m(x - x_1)$, (x_1, y_1) is $(5, 4)$.

(b) Graph this function.

▶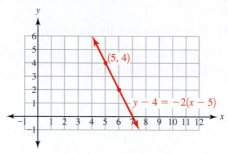

Graph $(5, 4)$. The slope is $-\dfrac{2}{1}$. From $(5, 4)$, move down 2 units and to the right 1 unit (rise over run), and draw another point. Draw a line through the points.

(c) Change this function so that its graph is shifted horizontally right 3 units and vertically up 1 unit.

▶ To shift the graph horizontally right 3 units, add 3 to x_1: $5 + 3 = 8$. To shift the graph vertically up 1 unit, add 1 to y_1: $4 + 1 = 5$. The shifted line contains the point $(8, 5)$. Rewrite the equation in point-slope form with the original slope and the new point, $y - 5 = -2(x - 8)$.

(d) Graph the new function by translation.

▶ Graph $(8, 5)$, and draw a line through it that is parallel to the original line.

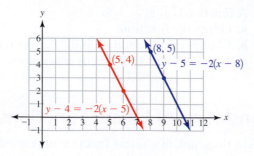

> **Translation of Linear Functions**
>
> If a linear function is in slope-intercept form, $y = mx + b$, changing b causes a vertical shift in the position of its graph.
>
> If a linear function is in point-slope form, $y - y_1 = m(x - x_1)$, changing x_1 causes a horizontal shift in its graph. Changing y_1 causes a vertical shift in its graph.

In the next example, we first rewrite the function $y - 3 = 2(x + 4)$ as $y - 3 = 2(x - (-4))$ so that it exactly matches point-slope form, $y - y_1 = m(x - x_1)$.

EXAMPLE 5 A function is $y - 3 = 2(x + 4)$.

(a) Identify: (x_1, y_1)

SOLUTION ▶ Rewrite $y - 3 = 2(x + 4)$ in point-slope form as $y - 3 = 2(x - (-4))$. Then (x_1, y_1) is $(-4, 3)$.

(b) Graph this function.

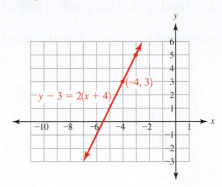

Graph $(-4, 3)$. The slope is $\dfrac{2}{1}$. From $(-4, 3)$, move up 2 units and to the right 1 unit and draw another point. Draw a line through the points.

(c) Change the function so that its graph is shifted horizontally left 6 units and vertically down 5 units.

To shift the graph horizontally left 6 units, subtract 6 from x_1: $-4 - 6 = -10$. To shift the graph vertically down 5 units, subtract 5 from y_1: $3 - 5 = -2$. The shifted line contains the point $(-10, -2)$. In point slope form, the function is $y - (-2) = 2(x - (-10))$. Simplifying, we get the function $y + 2 = 2(x + 10)$.

(d) Graph the new function by translation.

Graph $(-10, -2)$, and draw a line through it that is parallel to the original line.

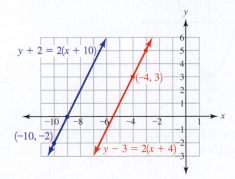

Practice Problems

3. A function is $y - 8 = 3(x - 2)$.
   a. Identify (x_1, y_1).
   b. Graph this function.
   c. Change the function so that its graph is shifted horizontally right 2 units and vertically up 1 unit.
   d. Graph the new function by translation.

4. A function is $y + 4 = 2(x - 5)$.
   a. Identify (x_1, y_1).
   b. Use (x_1, y_1) and the slope to graph this function.
   c. Change this function to shift its graph horizontally right 2 units and vertically up 1 unit.
   d. Graph the new function by translation.

Translation of Quadratic Functions

The standard form of a quadratic function is $y = ax^2 + bx + c$. Changing the value of c causes a vertical shift in the graph of a quadratic function.

EXAMPLE 6 The graph represents $y = x^2 + 3$. Use this graph and translation to graph $y = x^2 - 4$.

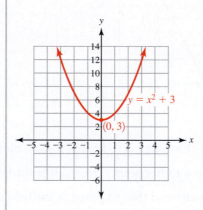

SOLUTION ▶ For $y = x^2 + 3$, $c = 3$. For $y = x^2 - 4$, $c = -4$. Since -4 is 7 less than 3, shift the graph of $y = x^2 + 3$ vertically down 7 units.

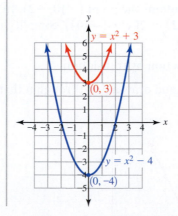

In the next example, the vertex is not on the y-axis. A change in c still causes a vertical shift in the graph.

EXAMPLE 7 The graph represents $f(x) = -x^2 - 8x - 12$. Use this graph and translation to graph $f(x) = -x^2 - 8x - 4$.

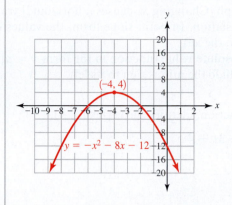

SOLUTION ▶ For $f(x) = -x^2 - 8x - 12$, $c = -12$. For $f(x) = -x^2 - 8x - 4$, $c = -4$. Since -4 is 8 more than -12, shift the graph of $f(x) = -x^2 - 8x - 12$ vertically up 8 units. When you copy the original parabola by hand, the copy might not be perfect. The goal is to clearly show the shift in the vertex.

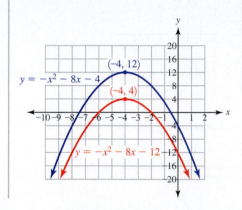

In Chapter 7, we will learn how to write a quadratic equation in vertex form rather than standard form. We can then do horizontal as well as vertical translations.

Practice Problems

5. Use the graph and translation to graph $y = x^2 - 8x + 14$.

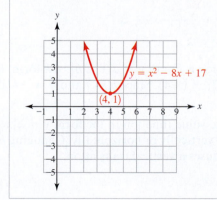

6. Use the graph and translation to graph $g(x) = -x^2 + 2x + 5$.

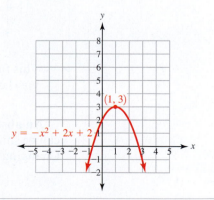

Translation of Absolute Value Functions

For a linear function in point-slope form, $y - y_1 = m(x - x_1)$, changing (x_1, y_1) shifts its graph. Changing x_1 causes a horizontal translation. Changing y_1 causes a vertical translation. In point-slope form, the values of x_1 and y_1 do not include the minus sign to the left.

For an absolute value function in the form $y = a|x - h| + k$, the value of h also does not include the minus sign to its left.

EXAMPLE 8   For each function, identify the vertex (h, k).

(a) $y = |x - 3| + 4$

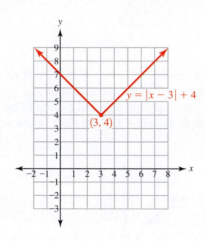

SOLUTION ▶   Since $h = 3$ and $k = 4$, the vertex is $(3, 4)$.

(b) $y = |x - 7| - 4$

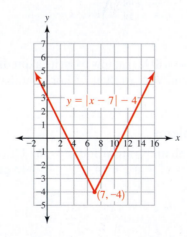

▶   This function is not in $y = a|x - h| + k$ form. Rewrite it as $y = |x - 7| + (-4)$. Since $h = 7$ and $k = -4$, the vertex is $(7, -4)$.

An absolute value function in the form $y = a|x - h| + k$ with a vertex of (h, k) can be shifted vertically or horizontally. Changing h causes a horizontal translation. Changing k causes a vertical translation.

> **Translation of an Absolute Value Function**
>
> For an absolute value function $y = a|x - h| + k$, where a, h, and k are real numbers, $a \neq 0$, and (h, k) is the vertex, changing h results in a horizontal translation and changing k results in a vertical translation.

To graph an absolute value function by translation, graph the vertex and copy the graph of the original function in its new position. The slope of each arm of the V-shaped graph does not change.

EXAMPLE 9 | The graph represents $y = |x - 4| + 2$. The vertex is $(4, 2)$.

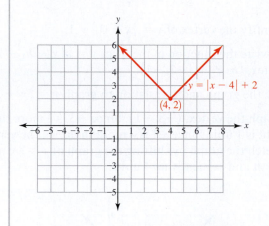

(a) Change the function so that its graph is shifted horizontally left 6 units and vertically down 5 units.

SOLUTION ▶ To shift the graph horizontally left 6 units, subtract 6 from h: $4 - 6 = -2$.

To shift the graph vertically down 5 units, subtract 5 from k: $2 - 5 = -3$.

The vertex of the shifted graph, (h, k), is $(-2, -3)$. The new function is $y = |x - (-2)| - 3$, which simplifies to $y = |x + 2| - 3$.

(b) Graph the new function by translation.

▶ Graph the vertex and sketch the graph by shifting the graph of $y = |x - 4| + 2$ horizontally left 6 units and vertically down 5 units.

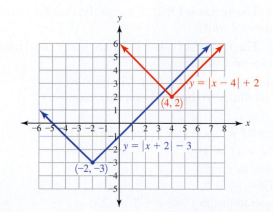

In the next example, to identify the vertex of $y = |x + 3| - 1$, we rewrite it in the form $y = a|x - h| + k$.

EXAMPLE 10 | The graph represents $y = |x + 4| - 5$. The vertex is $(-4, -5)$.

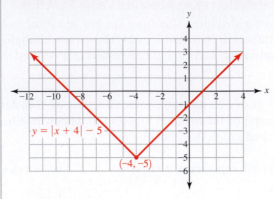

(a) Identify the vertex of $y = |x + 3| - 1$.

SOLUTION ▶   Rewrite the function in the form $y = a|x - h| + k$ as $y = |x - (-3)| - 1$. The vertex, (h, k), is $(-3, -1)$.

(b) Use the graph and translation to graph $y = |x + 3| - 1$.

▶   Graph the vertex, $(-3, -1)$. Since -3 is 1 more than -4, the horizontal shift right is 1 unit. Since -1 is 4 more than -5, the vertical shift up is 4 units. Sketch the graph by shifting the graph of $y = |x + 4| - 5$ horizontally right 1 unit and vertically up 4 units.

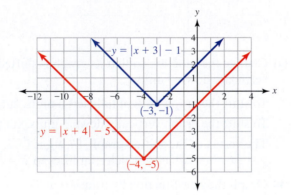

Practice Problems

7. A function is $y = |x - 3| + 2$. The vertex is $(3, 2)$. Change the function to shift its graph horizontally to the left 7 units and vertically down 1 unit.

8. A function is $y = |x + 4| - 1$. The vertex is $(-4, -1)$. Change the function to shift its graph horizontally to the right 2 units and vertically up 5 units.

9. A function is $y = |x + 2| + 6$. The vertex is $(-2, 6)$. Change the function to shift its graph horizontally to the left 3 units and vertically down 10 units.

For problems 10–13, the graph represents $y = |x - 1| + 2$. The vertex is (1, 2).

10. a. Identify the vertex of $y = |x - 1| + 4$.
   b. Use the graph and translation to graph $y = |x - 1| + 4$.

11. a. Identify the vertex of $y = |x - 3| + 2$.
   b. Use the graph and translation to graph $y = |x - 3| + 2$.

12. a. Identify the vertex of $y = |x - 1| - 5$.
   b. Use the graph and translation to graph $y = |x - 1| - 5$.

13. a. Identify the vertex of $y = |x + 3| + 2$.
   b. Use the graph and translation to graph $y = |x + 3| + 2$.

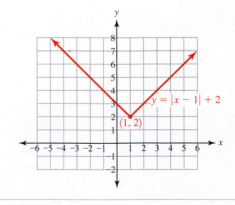

3.6 Vocabulary Practice

Match the term with its description.

1. $f(x) = mx + b$
2. $f(x) = ax^2 + bx + c$
3. $f(x) = ax^3 + bx^2 + cx + d$
4. $f(x) = C$
5. $f(x) = a|x - h| + k$
6. The horizontal or vertical shift in a graph caused by changing a function
7. The graph of a quadratic function
8. The graph of a constant function
9. In a quadratic function in standard form, if $a > 0$, then the function opens in this direction.
10. In an absolute value function in standard form, if $a < 0$, then the function opens in this direction.

A. absolute value function
B. constant function
C. cubic function
D. down
E. horizontal line
F. linear function
G. parabola
H. quadratic function
I. translation
J. up

3.6 Exercises

Follow your instructor's guidelines for showing your work.

For exercises 1–4,
(a) identify the slope.
(b) identify the y-intercept.

1. $f(x) = \dfrac{3}{4}x - 5$

2. $f(x) = \dfrac{2}{3}x - 6$

3. $f(x) = -6x + 1$

4. $f(x) = -3x + 1$

5. To cause a vertical translation in the function $y = 6x + 3$, what number is changed?

6. To cause a vertical translation in the function $y = 8x + 5$, what number is changed?

For exercises 7–12,
(a) graph the function.
(b) change the function so that its graph is shifted the given amount.
(c) graph the new function by translation.

7. $f(x) = \dfrac{3}{4}x - 5$; shift up 2 units

8. $f(x) = \dfrac{2}{3}x - 6$; shift up 3 units

9. $f(x) = -6x + 1$; shift down 9 units
10. $f(x) = -3x + 1$; shift down 7 units

11. $y = \left(\dfrac{-3 \text{ lb}}{1 \text{ week}}\right)x + 30 \text{ lb}$; shift down 15 pounds

12. $y = \left(\dfrac{-3 \text{ lb}}{1 \text{ week}}\right)x + 30 \text{ lb}$; shift up 15 pounds

For exercises 13–18,
(a) identify the slope.
(b) identify (x_1, y_1).

13. $y - 7 = 3(x - 4)$
14. $y - 5 = 3(x - 4)$
15. $y + 7 = 3(x - 4)$
16. $y + 5 = 3(x - 4)$
17. $y - 7 = 3(x + 4)$
18. $y - 5 = 3(x + 4)$

19. To cause a vertical translation in $y - 7 = 3(x - 4)$, what number is changed?

20. To cause a vertical translation in $y - 5 = 3(x - 4)$, what number is changed?

21. To cause a horizontal translation in $y - 7 = 3(x - 4)$, what number is changed?

22. To cause a horizontal translation in $y - 5 = 3(x - 4)$, what number is changed?

For exercises 23–26,

(a) identify (x_1, y_1).

(b) graph the function.

(c) change the function so that the graph is shifted the given amounts.

(d) graph the new function by translation.

23. $y - 7 = 3(x - 12)$; shift horizontally right 6 units and vertically up 4 units

24. $y - 5 = 3(x - 4)$; shift horizontally right 2 units and vertically up 6 units

25. $y - 8 = \frac{1}{2}(x - 2)$; shift horizontally left 9 units and vertically down 5 units

26. $y - 6 = \frac{1}{3}(x - 1)$; shift horizontally left 6 units and vertically down 7 units

For exercises 27–30, change the function so that the graph is shifted horizontally right 2 units and vertically up 4 units.

27. $y - 1 = 3(x - 2)$

28. $y - 1 = 2(x - 3)$

29. $y - 5 = -3(x + 1)$

30. $y - 6 = -3(x + 5)$

For exercises 31–34, change the function so that the graph is shifted horizontally left 3 units and vertically up 6 units.

31. $y + 8 = 5(x + 1)$

32. $y + 7 = 6(x + 1)$

33. $y - 6 = -2(x - 4)$

34. $y - 9 = -4(x - 5)$

For exercises 35–36,

(a) write the function in slope-intercept form that represents the graphed line.

(b) change the function so that the graph is shifted vertically down 2 units.

35.

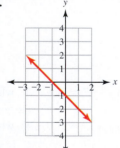

36.

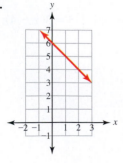

For exercises 37–38,

(a) identify two points on the line.

(b) identify the slope of the line.

(c) write the function in point-slope form that represents the graphed line.

(d) change the function so that the graph is shifted horizontally right 1 unit and vertically up 3 units.

37.

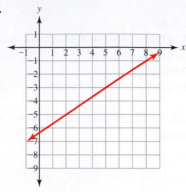

38.

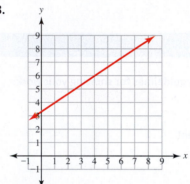

39. To cause a vertical translation in $y = 5x^2 + 7x + 2$, what number is changed?

40. To cause a vertical translation in $y = 8x^2 + 5x + 4$, what number is changed?

For exercises 41–48, use translation to graph the function.

41. $g(x) = 2x^2 + 4$

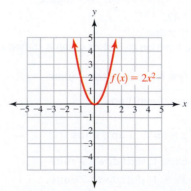

42. $g(x) = 3x^2 + 2$

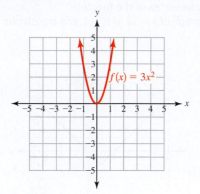

$f(x) = 3x^2 -$

43. $y = \dfrac{1}{2}x^2 - 3$

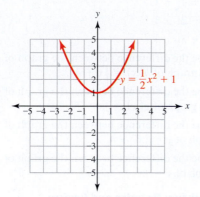

$y = \dfrac{1}{2}x^2 + 1$

44. $y = \dfrac{1}{2}x^2 - 1$

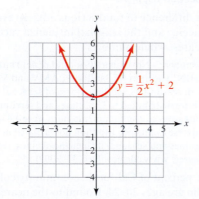

$y = \dfrac{1}{2}x^2 + 2$

45. $y = x^2 + 6x + 7$

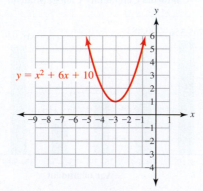

$y = x^2 + 6x + 10$

46. $y = x^2 + 4x + 2$

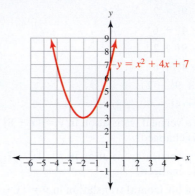

$y = x^2 + 4x + 7$

47. $f(x) = -x^2 + 6x - 6$

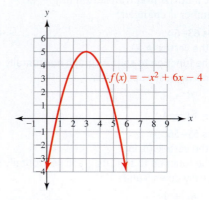

$f(x) = -x^2 + 6x - 4$

48. $f(x) = -x^2 + 8x - 15$

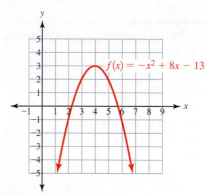

$f(x) = -x^2 + 8x - 13$

For exercises 49–50, change the function so that its graph is shifted vertically up 6 units.

49. $y = 3x^2 + 5x - 8$

50. $y = 2x^2 + 7x - 9$

For exercises 51–52, change the function so that its graph is shifted vertically down 6 units.

51. $y = 3x^2 + 5x - 8$

52. $y = 2x^2 + 7x - 9$

For exercises 53–58, identify the vertex (h, k).

53. $y = |x - 9| + 2$

54. $y = |x - 4| + 6$

55. $y = |x + 9| + 2$

56. $y = |x + 4| + 6$

57. $y = 5|x + 9| - 2$

58. $y = 3|x + 4| - 6$

59. To cause a vertical translation in $y = |x - 8| + 3$, what number is changed?

60. To cause a vertical translation in $y = |x - 7| + 15$, what number is changed?

61. To cause a horizontal translation in $y = |x - 8| + 3$, what number is changed?

62. To cause a horizontal translation in $y = |x - 7| + 15$, what number is changed?

For exercises 63–64,
(a) identify the vertex (h, k).
(b) change the function to shift its graph horizontally right 3 units and vertically up 2 units.

63. $y = |x - 7| + 8$

64. $y = |x - 8| + 4$

For exercises 65–66,
(a) identify the vertex (h, k).
(b) change the function to shift its graph horizontally left 2 units and vertically down 5 units.

65. $y = |x + 6| + 1$

66. $y = |x + 9| + 2$

For exercises 67–68,
(a) identify the vertex (h, k).
(b) change the function to shift its graph horizontally left 4 units and vertically up 6 units.

67. $y = |x + 10| - 2$

68. $y = |x + 12| - 5$

For exercises 69–72,
(a) identify the vertex of the function.
(b) use the graph of $y = |x + 2| + 1$ and translation to graph the function.

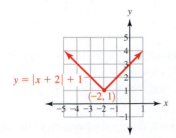

69. $y = |x + 2| + 4$

70. $y = |x + 2| + 2$

71. $y = |x + 2| - 5$

72. $y = |x + 2| - 4$

For exercises 73–76,
(a) identify the vertex of the function.
(b) use the graph of $y = |x + 1| - 1$ and translation to graph the function.

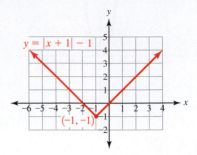

73. $y = |x + 3| - 1$

74. $y = |x + 4| - 1$

75. $y = |x - 2| - 1$

76. $y = |x - 3| - 1$

77. Describe the difference between the graph of $h(x) = |x|$ and the graph of $h(x) = |x + 4|$.

78. Describe the difference between the graph of $g(x) = |x|$ and the graph of $g(x) = |x - 5|$.

79. Describe the difference between the graph of $y = |x|$ and the graph of $y = |x| + 4$.

80. Describe the difference between the graph of $y = |x|$ and the graph of $y = |x| - 5$.

Problem Solving: Practice and Review

Follow your instructor's guidelines for using the five steps as outlined in Section 1.5, p. 51.

81. Find the difference in total serious adverse events for the H1N1 vaccine and the seasonal influenza vaccine. Round to the nearest whole number.

During October 5–November 20, a total of 46.2 million doses of H1N1 vaccines (11.3 million LAMV and 34.9 million MIV doses) and 98.9 million doses of seasonal influenza vaccines were distributed to U.S. states and territories. . . . The serious adverse event reporting rates were 4.4 and 2.9 serious adverse events per 1 million doses distributed for H1N1 vaccines and seasonal influenza vaccines, respectively. (*Source:* www.cdc.gov, Dec. 11, 2009)

82. Find the percent of California community college students who are ages 20–24. Round to the nearest percent. (*Source:* www.ccleague.org)

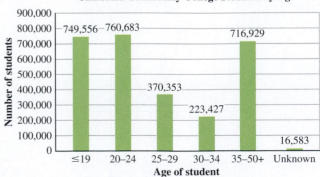

California Community College Students by Age

83. Predict how many of 650 Austin construction workers did not get rest breaks from their employer.

Four in 10 Austin construction workers reported that their employer did not give them rest breaks, and 27 percent said they were not provided with drinking water. (*Source:* www.statesman.com, June 23, 2010)

84. In 2009, Trident Technical College students received $28,598,659 in federal student loans. This is a 113% increase over the amount of federal student loans received in 2005. Find the amount of federal student loans received in 2005. Round to the nearest ten thousand. (*Source:* www.tridenttech.edu, 2010 Factbook)

Technology

For exercises 85–88,

(a) graph both functions on a graphing calculator. Choose a window that shows each vertex. Sketch the graph; describe the window.

(b) identify any horizontal shift in the graph.

(c) identify any vertical shift in the graph.

85. $y = 3x - 5; y = 3x - 8$

86. $y = x^2 - 2x + 4; y = x^2 - 2x - 7$

87. $y = |x + 1| - 5; y = |x + 3| - 2$

88. $y = -|x - 3| + 4; y = -|x + 1| + 1$

Find the Mistake

For exercises 89–92, the completed problem has one mistake.

(a) Describe the mistake in words, or copy down the whole problem and highlight or circle the mistake.

(b) Do the problem correctly.

89. Problem: Use the slope and (x_1, y_1) to graph $y + 2 = -3(x - 7)$.

Incorrect Answer: The slope is -3, and (x_1, y_1) is $(7, 2)$.

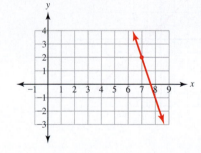

90. Problem: Identify the vertex of the graph of $y = |x + 5| - 9$.

Incorrect Answer: The vertex is $(5, -9)$.

91. Problem: The graph represents $f(x) = |x| + 2$. Compare the appearance of this graph and the graph that represents $f(x) = |x + 5| + 2$.

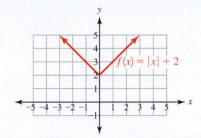

Incorrect Answer: The graph of $f(x) = |x + 5| + 2$ will be shifted horizontally right 5 units.

92. Problem: Change $y + 5 = 3(x - 2)$ to shift its graph vertically down 7 units.

Incorrect Answer: The vertex of the given function is $(-5, 2)$. The vertex of the changed function is $(-5, -5)$. The changed function is $y + 5 = 3(x + 5)$.

Review

For exercises 93–96, simplify the expression.

93. $(4x^2)(3x^3)$

94. $(4x)^2(3x^3)$

95. $\dfrac{12x^5y^3}{50xy^9}$

96. $\left(\dfrac{2x}{3y^2}\right)^3$

SUCCESS IN COLLEGE MATHEMATICS

97. Your plan for solving an application may be the same as you have used before to solve a similar problem. Describe a problem that you have solved that is similar to exercise 83.

98. Describe your plan for solving exercise 83.

Study Plan for Review of Chapter 3

SECTION 3.1   Introduction to Relations and Functions

Ask Yourself	Test Yourself	Help Yourself
Can I . . .	**Do 3.1 Review Exercises**	**See these Examples and Practice Problems**
write a definition of a *relation*?	1, 4, 11	Ex. 1, 2, PP 3
use roster notation, a list, or a table to represent a relation?		
use a bar graph to represent a relation?	4	Ex. 3, PP 3
use an equation or its graph to represent a relation?	5	Ex. 4, 5, PP 1, 2
use an inequality or its graph to represent a relation?	6–8	Ex. 6, PP 4
write a definition of a *function*?	3, 9, 12	Ex. 7, PP 5–11
determine whether a relation is a function?		
use roster notation or interval notation to represent the domain of a relation or function?	2, 10	Ex. 8, PP 10–11
use the vertical line test to determine whether a graph represents a function?	13–15	Ex. 9, 10, PP 12–17

3.1 Review Exercises

1. The bar graph shows the number of deaths per year of Americans in vehicle crashes with animals. Represent this relation with roster notation.

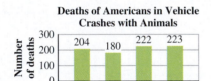

Deaths of Americans in Vehicle Crashes with Animals

Source: www.iihs.org, Oct. 2008

2. Use roster notation to:
   a. represent the domain of the relation in exercise 1.
   b. represent the range of the relation in exercise 1.

3. Write a set of five ordered pairs that is a relation but is not a function.

4. The relation describes the correspondence of the semester and the number of sections of Intermediate Algebra taught at a college.
   {(Spring 2009, 5 sections), (Fall 2009, 7 sections), (Spring 2010, 6 sections), (Fall 2010, 10 sections)}
   a. Use a table to represent this relation.
   b. Use a bar graph to represent this relation.

For exercises 5–8, graph the equation or inequality on an *xy*-coordinate system.

5. $8x - 9y = 64$ 6. $x < 2$

7. $x + 7y \geq 14$ 8. $y > 4$

For exercises 9–11,
$P = \{(\text{Washington, 1st}), (\text{Adams, 2nd}), (\text{Reagan, 40th}), (\text{Obama, 44th})\}$

9. Is this relation a function? Explain.

10. Use roster notation to represent the domain of *P*.

11. Use a table to represent *P*.

12. Is the relation $y \leq 3x - 4$ an example of a function? Explain.

For exercises 13–15, explain why the graph does or does not represent a function.

13.

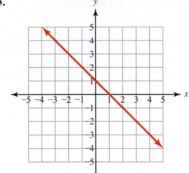

14.

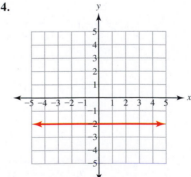

15.

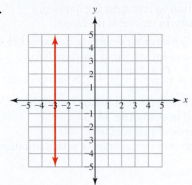

| SECTION 3.2 | Linear and Constant Functions |

Ask Yourself	Test Yourself	Help Yourself
Can I . . .	**Do 3.2 Review Exercises**	**See these Examples and Practice Problems**
evaluate a function that is represented by roster notation?	23	Ex. 1, PP 1
evaluate a function that is represented by an equation?	18	Ex. 2, PP 2
evaluate a function that is written in function notation?	16, 17	Ex. 3, 4, 6, 9, 15–17, PP 3–5, 7, 8, 13–15
evaluate a function that is represented by a graph?	19	Ex. 5, PP 6
write slope-intercept form, standard form, and point-slope form of a linear function? graph a linear function?	22	Ex. 7, PP 7, 8
use interval notation to represent the domain and range of a linear function?	20	Ex. 8, PP 7
use interval notation to represent the domain and range of a constant function?	20	Ex. 9, PP 8, 9
use an algebraic method to find the real zero(s) of a linear or constant function?	21	Ex. 10–12, PP 9, 10
use a graph to estimate the real zero(s) of a linear or constant function?	22	Ex. 13, 14, PP 11, 12
use roster notation or interval notation to represent the domain and range of a function used in an application?	24	Ex. 15, 16, PP 13, 14
find the real zero of a function used in an application problem and describe what the zero represents?	25	Ex. 18, PP 16

3.2 Review Exercises

16. A function is $f(x) = \dfrac{3}{4}x - 9$. Evaluate $f(-8)$.

17. In Henry County, Georgia, the Board of Commissioners charges a flat fee of $300 plus $25 per acre for a conditional exception to a zoning regulation. A function that represents the relationship of the total cost for a conditional exception, $C(x)$, and the number of acres affected by the exception, x, is $C(x) = \left(\dfrac{\$25}{1 \text{ acre}}\right)x + \$300$.

Evaluate $C(7.5 \text{ acres})$.
(*Source:* www.co.henry.ga.us, Dec.2011)

18. A function is $y = \dfrac{3}{4}x - \dfrac{5}{6}$. Find y when $x = 2$.

19. Use the graph to find $f(2)$.

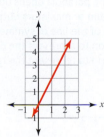

20. Use interval notation to:
  a. represent the domain of $f(x) = x + 5$.
  b. represent the range of $y = 3x - 7$.
  c. represent the range of $g(x) = 4$.

21. Use an algebraic method to find any real zeros of $g(x) = -6x + 20$.

22. a. Graph: $y = -4x + 10$
    b. Use the graph to identify any real zeros of this function.

23. A function is
    $C = \{(ID, Boise),(TN, Nashville),(MI, Lansing)\}$.
    Find $C(MI)$.

24. At a grocery store a cereal is on sale for $1.99 per box, with a limit of 8 boxes per customer. The linear function $C(x) = \left(\dfrac{\$1.99}{1\ \text{box}}\right)x$ represents the relationship of the number of boxes of cereal, x, and the total cost, $C(x)$.
    a. Use roster notation to represent the domain.
    b. Use roster notation to represent the range.

25. A bottle of liquid dishwashing detergent contains 125 oz. Washing a dishwasher full of dishes requires 0.5 oz. The function $V(x) = \left(\dfrac{-0.5\ \text{oz}}{1\ \text{load}}\right)x + 125\ \text{oz}$ represents the relationship of the number of loads, x, and the volume of detergent in the bottle, $V(x)$.
    a. Find any real zeros of this function.
    b. Describe what a real zero represents in this situation.

SECTION 3.3 Quadratic and Cubic Functions

Ask Yourself	Test Yourself	Help Yourself
Can I . . .	**Do 3.3 Review Exercises**	**See these Examples and Practice Problems**
identify the degree of a polynomial function in one variable?	26	Ex. 1, PP 1–4
identify whether a polynomial function is a constant, linear, quadratic or cubic function?	26	Ex. 1, PP 1–4
use a table of ordered pairs to graph a quadratic function?	29, 30	Ex. 2, 3, PP 5–7
use the graph of a quadratic function to estimate its vertex, axis of symmetry, and/or minimum or maximum output value?	27, 28	Ex. 2, 3, PP 5–7
use interval notation to represent the domain and range of a quadratic function?	27, 28	Ex. 2–4, PP 5–9
evaluate a quadratic or cubic function?	31	Ex. 5, 6, 8, PP 10–12, 13
use interval notation to represent the domain and range of a cubic function?		
use an algebraic method to find any real zeros of a polynomial function?	32, 33	Ex. 9, 10, PP 14–16
use a graphical method to find any real zeros of a polynomial function?	32, 34	Ex. 11, 12, PP 17–22

3.3 Review Exercises

26. a. Write an example of a quadratic function.
    b. Write an example of a function that is degree 0.

For exercises 27–28, use the graph to:
(a) identify the vertex.
(b) identify whether the parabola opens up or down.
(c) identify the minimum or maximum output value.
(d) write the equation of the line of symmetry.
(e) use interval notation to represent the domain.
(f) use interval notation to represent the range.

27.

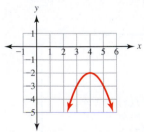

28.

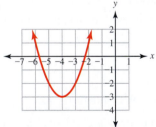

For exercises 29–30, use a table of ordered pairs to graph the function.

29. $y = x^2 + 4x + 3$

30. $f(x) = -x^2 + 6x - 5$

31. A cubic function is $f(x) = x^3 + 6$.
    a. Use interval notation to represent the domain.
    b. Use interval notation to represent the range.
    c. Evaluate: $f(-4)$
    d. Identify the degree of this function.

32. The graph represents the function $f(x) = x^2 - 2x - 8$.

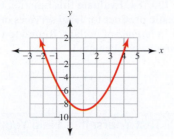

a. Use the graph to identify any real zeros.
b. Use an algebraic method to find any real zeros.

33. Use an algebraic method to find any real zeros of $g(x) = x^3 - 15x^2 + 54x$.

34. Does every quadratic function have at least one real zero? Explain.

SECTION 3.4 **Polynomial Models**

Ask Yourself	Test Yourself	Help Yourself
Can I . . .	**Do 3.4 Review Exercises**	**See these Examples and Practice Problems**
describe what the slope and the y-coordinate of the y-intercept of a linear model represent?	37, 38, 39	Ex. 1, 2, 4–6, PP 1, 3, 4
find an average rate of change?	35	Ex 3, 6, PP 2
evaluate a model for a given input value?	36, 37, 40, 41	Ex. 4–6, 8–10, PP 3, 4, 6, 7
write a linear model, given the average rate of change and the beginning output value?	36, 38, 39, 40	Ex. 5, 6, 8, PP 3, 4
graph a linear model?	37	Ex. 5, PP 3, 4
find the zero of a polynomial model and explain what it represents?	39	Ex. 7, PP 5

3.4 Review Exercises

35. A bicyclist is at mile marker 200. After 30 min, she is at mile marker 206. Find her average rate of change in miles per *hour*.

36. A student has completed 24 credits for his bachelor's degree in nursing, which requires a total of 128 credits. He is successfully completing 12 credits per semester.
  a. Write a linear function that predicts the total amount of successfully completed credits, y, after x more semesters of classes.
  b. Evaluate the function when $x = 3$ semesters.

37. The model $R(x) = \left(\dfrac{\$50}{1 \text{ year}}\right)x + \$575$ describes the relationship of the number of years since 2009, x, and the rent of an apartment, $R(x)$.
  a. Describe what the slope represents.
  b. Describe what the y-coordinate of the y-intercept represents.
  c. Evaluate this function to predict the rent in 2015.
  d. Graph this function for values from 0 year to 10 years.

38. If a sale of surface management maps from the Bureau of Land Management (BLM) involves 1 through 49 maps, the cost per map is $4. (*Source:* www.blm.gov/co/st, Dec. 2011)
  a. Write a linear model that describes the relationship of the total cost of the sale, y, for x maps.
  b. Describe what the slope represents.
  c. Identify the y-intercept.
  d. Describe what the y-coordinate of the y-intercept represents.

39. The volume of a diesel storage tank is 500 gal. In the spring, a farmer has the tank filled. Every 10-hr work day a tractor uses 44 gal of diesel.
  a. Write a linear model that represents the relationship of the volume of diesel in the storage tank, y, after x work days.
  b. Describe what the slope represents.
  c. Describe what the y-coordinate of the y-intercept represents.
  d. Find any real zeros. Round to the nearest whole number.
  e. Describe the situation when the output value is 0.

40. From 1998 to 2010, the wages and salaries paid by all levels of government grew at an approximately linear rate. The graph shows the line of best fit model of this data.

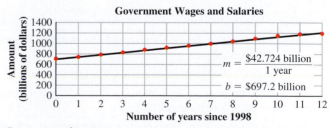

Government Wages and Salaries

$m = \dfrac{\$42.724 \text{ billion}}{1 \text{ year}}$

$b = \$697.2 \text{ billion}$

Source: www.bea.gov

  a. Write a linear model that represents the relationship of government wages and salaries, y, and the number of years since 1998, x.

b. Evaluate this function to find the wages and salaries paid by all levels of government in 2013. Round to the nearest billion.

41. A model of the gross domestic product from legal services in *millions* of dollars, *y*, and the number of years since 1998, *x*, is $y = 211.32x^3 - 1735.8x^2 + 10{,}563x + 120{,}184$. Evaluate this function to predict the gross domestic product for legal services in 2010. Write the answer in *billions* of dollars. Round to the nearest tenth. (*Source:* www.bea.gov)

SECTION 3.5 Absolute Value Functions

Ask Yourself	Test Yourself	Help Yourself
Can I . . .	**Do 3.5 Review Exercises**	**See these Examples and Practice Problems**
write a definition of *absolute value*?	42	Ex. 1–3 , PP 1–3
use a table of ordered pairs to graph an absolute value function?		
given the graph of an absolute value function, identify the vertex, write the equation of the axis of symmetry, and identify the maximum or minimum output value?	43	Ex. 1–4, PP 1–7
use interval notation to represent the domain and range of an absolute value function?	43	Ex. 4–6, PP 4–8
use an algebraic or graphical method to identify any real zeros of an absolute value function?	43, 44	Ex. 7–9, PP 9–14

3.5 Review Exercises

42. Use a table of ordered pairs to graph $y = |x| - 1$.

43. The graph represents an absolute value function.

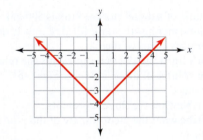

a. Identify the vertex.
b. Write the equation of the axis of symmetry.
c. Identify the minimum output value.
d. Use interval notation to represent the domain.
e. Use interval notation to represent the range.
f. Identify any real zeros.

44. Use an algebraic method to find any real zeros of $f(x) = |x - 4| - 10$.

SECTION 3.6 Translation of Polynomial Functions

Ask Yourself	Test Yourself	Help Yourself
Can I . . .	**Do 3.6 Review Exercises**	**See these Examples and Practice Problems**
change a linear function in slope-intercept form to cause a vertical translation?	45	Ex. 1, 2, PP 1, 2
change a linear function in point-slope form to cause a vertical and/or horizontal translation?	46–48	Ex. 3–5, PP 3, 4
change a quadratic function to cause a vertical translation?	49	Ex. 6, 7, PP 5, 6
identify the vertex of an absolute value function, given its equation in the form $y = a|x - h| + k$ or its graph?	50	Ex. 8–10, PP 10–13
change an absolute value function to cause a vertical or horizontal translation?	51	Ex. 8–10, PP 7–9
use translation to graph an absolute value function?	52	Ex. 9, 10, PP 10–13

3.6 Review Exercises

45. A function is $h(x) = 3x - 6$.
   a. Graph the function.
   b. Change the function so that its graph is shifted vertically up 8 units.
   c. Graph the new function by translation.

46. A function is $y - 6 = 9(x + 1)$.
   a. Graph the function.
   b. Change the function so that its graph is shifted horizontally right 7 units and vertically up 3 units.
   c. Graph the new function by translation.

47. A function is $y - 4 = 3(x - 2)$. If this function is changed to $y - 1 = 3(x + 5)$, describe the shift in its graph.

48. The graph represents the functions $y - 3 = 2(x - 4)$ with $(x_1, y_1) = (4, 3)$ and $y - 8 = 2(x - 5)$ with $(x_1, y_1) = (5, 8)$. Compared to the original function, the graph of $y - 8 = 2(x - 5)$ is shifted 1 unit to the right and 5 units up. A student looked at the graph and said, "But the graph of the second function is to the *left* of the graph of the first function. How can you say that there has been a horizontal shift to the *right* of 1?" Explain to this student why there has been a horizontal shift of 1 to the right.

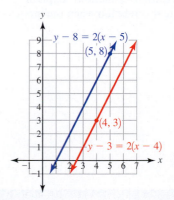

49. A function is $y = x^2 + 5x + 9$. Change this function to shift its graph vertically down 2 units.

50. Identify the vertex of the graph that represents the function.
   a. $y = |x - 3| + 1$
   b. $y = |x + 4| - 2$

51. A function is $f(x) = |x - 2| - 9$. Change this function to shift the graph horizontally left 3 units and vertically up 5 units.

52. The graph represents the function $f(x) = |x - 3| + 4$. Use this graph and translation to graph
   a. $g(x) = |x - 3| + 7$
   b. $h(x) = |x - 1| + 4$

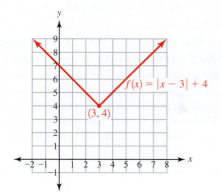

Chapter 3 Test

1. Describe the difference between a relation and a function.

2. A function f is $\{(5, 2), (7, 4), (-2, -5), (0, 4)\}$.
   a. Use roster notation to represent the domain.
   b. Use roster notation to represent the range.
   c. Evaluate: $f(7)$

3. The cost per package of pizza flour for orders of 3 packages to 10 packages is $5.25. Shipping is an additional $6 per order. The function $y = \left(\dfrac{\$5.25}{1\ package}\right)x + \$6$ represents the relationship of the number of packages ordered, x, and the total cost of the order, y. Use roster notation to represent the domain of this function.

For problems 4–9, $f(x) = 3x + 9$.

4. Graph the function.

5. Use the graph to identify any real zero(s) of the function.

6. Use interval notation to represent the domain and range.

7. Evaluate: $f(-8)$

8. Identify the degree of this function.

9. Change the function so that its graph is shifted vertically up 8 units.

10. A farm combine that is 20 ft wide can harvest an average of 6.4 acres of soybeans per hour. The combine has already harvested 15 acres of soybeans.
   a. Write a linear function in slope-intercept form that represents the relationship of the number of total

acres harvested by the combine, y, and the number of additional hours spent harvesting, x.

b. The combine continues to harvest for 12 more hours. Evaluate the function to find the total number of harvested acres.

For problems 11–18, use $f(x) = -x^2 - 2x + 3$.

11. Build a table of ordered pairs, and graph this function.

12. Use interval notation to represent the domain.

13. Identify the vertex of the graph.

14. Write the equation of the axis of symmetry.

15. Identify the maximum or minimum output value.

16. Use interval notation to represent the range.

17. Identify the degree of this function.

18. Evaluate: $f(-10)$

19. A function is $f(x) = x^2 - 2x - 63$. Use an algebraic method to find any real zeros of this function.

20. A model of the relationship of projected annual expenditures on hospital care in billions of dollars, y, and the number of years since 2005, x, is $y = 7.1452x^2 + 61.314x + 2087.8$. Find the projected annual expenditures in 2020. Round to the nearest billion. (*Source:* www.cms.gov, 2009)

For problems 21–24, use the graph of $f(x) = |x - 5| - 2$.

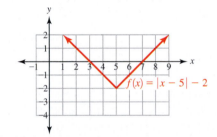

21. Use interval notation to represent the domain.

22. Identify the vertex.

23. Use interval notation to represent the range.

24. Identify any real zeros of this function.

25. Use an algebraic method to find any real zeros of the function $y = |x + 2| - 6$.

26. Explain why the function $y = |x| + 3$ has no real zeros.

27. The bar graph represents a relation.

Source: www.dishnetwork.com, June 3, 2009

a. Represent this relation as a set of ordered pairs.
b. Use roster notation to represent its range.
c. Is this relation a function? Explain.
d. Evaluate this relation when the input value is Silver.

Cumulative Review Chapters 1–3

Follow your instructor's directions for showing your work.

For exercises 1–8, evaluate.

1. $(1 - 5)^2 - 36 \div 2 \cdot 6(-1 + 3)$

2. $\dfrac{3}{4} \div \dfrac{1}{2} - \dfrac{9}{10}$

3. $\dfrac{0}{9}$

4. $\dfrac{9}{0}$

5. 2^0

6. $\dfrac{6 - 3(2 - 9)}{-3^2}$

7. $(6 \times 10^8)(3 \times 10^{-1})$

8. $\dfrac{2 \times 10^{-7}}{5 \times 10^{-4}}$

For exercises 9–14, simplify.

9. $3(3x - 4) - (9x - 2)$

10. $\dfrac{30c^3df^2}{16cd^3f^8}$

11. $(7n^5)(-4n^9)^2$

12. $(3c - 8)(4c^2 + 6c - 7)$

13. $(4w - 3)(4w + 3)$

14. $(8x^3 - 6x^2 + 9) + (2x^3 - x - 14)$

For exercises 15–21, factor completely. Identify prime polynomials.

15. $10ax^3 + 5a^2x^2 - 15ax^2$

16. $49p^2 - 36w^2$

17. $3x^2 - 7x - 20$

18. $2x^2 - 26x + 72$

19. $x^2 + 9x + 20$

20. $6px + 14p - 3wx - 7w$

21. $8x^3 - 27$

For exercises 22–30,
(a) solve.
(b) check.

22. $9x - 15 = 6$

23. $\dfrac{2}{3}p + \dfrac{1}{6} = \dfrac{2}{9}$

24. $0.85 = x - 0.42$

25. $6(5 - 8k) = -2(25k - 4)$

26. $5d - 15 = 3(2d - 4) - d$

27. $x^2 - 6x - 27 = 0$

28. $5 = 6u^2 + 13u$

29. $|x + 9| - 3 = 15$

30. $12 = |2x - 8| + 16$

For exercises 31–32,
(a) solve.
(b) check.
(c) use interval notation to represent the solution.

31. $-8a - 15 < 9$

32. $2(6x - 3) \geq 3(8x + 4)$

33. a. Solve: $-8 < x + 2 < 19$
   b. Use interval notation to represent the solution.

34. Write the equation in slope-intercept form of the line that passes through $(7, -9)$ and $(-2, 1)$.

35. Write the equation of the vertical line that passes through $(4, -11)$.

36. Write the equation of a line that is perpendicular to the line $7x - 2y = 14$ and passes through $(1, 3)$.

37. Explain why the slope of a vertical line is undefined.

For exercises 38–41, graph on a rectangular coordinate system. Label the axes, and include a scale on each axis.

38. $y = \dfrac{3}{5}x - 9$

39. $7x + 3y = -14$

40. $x < -4$

41. $3x - y \geq 15$

42. Use a table of ordered pairs to graph $f(x) = x^2 - 3$.

43. The solution of an inequality is $x \leq -2$.
   a. Use a number line graph to represent this solution.
   b. Use interval notation to represent this solution.

44. The domain of the function $y = 8x - 1$ is the set of real numbers. Use interval notation to represent this domain.

45. A conjunction is $-8 \leq x < -2$.
   a. Use a number line graph to represent this conjunction.
   b. Use interval notation to represent this conjunction.

46. Explain why the set of ordered pairs $\{(0, 3), (2, 4), (7, 9), (0, 8)\}$ is not a function.

47. A function is {(gelato, chocolate), (ice cream, strawberry), (sorbet, lemon)}.
   a. Use roster notation to describe the domain.
   b. Use roster notation to describe the range.

48. A model of the relationship of the life expectancy at birth of American women in years, y, and the number of years since 1970, x, is

$$y = \left(\frac{0.2773 \text{ year life expectancy}}{1 \text{ year}}\right)x + 67.356 \text{ years}.$$

life expectancy. Predict the life expectancy at birth of women in 2015. Round to the nearest tenth. (*Source:* www.cdc.gov, 2008)

49. Evaluate $f(x) = x^2 - 6x + 2$ when $x = -4$.

For exercises 50–55, a function is $f(x) = -2x + 6$.

50. Graph this function.

51. Use the graph to identify any real zeros.

52. Use an algebraic method to find any real zeros.

53. a. Use interval notation to describe the domain.
   b. Use interval notation to describe the range.

54. Identify the degree of this function.

55. Change this function so that its graph is shifted vertically down 10 units.

For exercises 56–59, refer to the graph of $y = -|x - 2| + 4$.

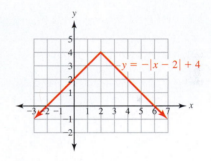

56. Identify the minimum or maximum output value.

57. Use interval notation to describe the domain.

58. Use interval notation to describe the range.

59. Use an algebraic method to find any real zeros.

60. A student has saved $1200. Her goal is to save $6000 to buy a used truck. She is saving $150 a month.
   a. Write a linear function that predicts the total amount in savings, y, after x more months of saving.
   b. Evaluate the function when $x = 9$ months.

61. In 2011, the enrollment at a state college was 3000 students. In 2007, the enrollment was 2600 students. Find the average rate of change in enrollment per year.

62. A rectangle is 3 in. longer than two times the width. The area of the rectangle is 44 in.². Use a quadratic equation to find the length and width.

For exercises 63–66, use the five steps.

63. Find the amount of trash (refuse) that New York City produced each day in 1985.

The Fountain Avenue Landfill opened in 1961, filling up with residential trash, construction debris, asbestos incinerator ash and, notoriously, the bodies of mob victims. In its last year of operation, 1985, an average of 8,200 tons of trash arrived there each day—some 40 percent of the city's [New York's] refuse. (*Source:* www.nytimes.com, Sept. 7, 2009)

64. Find the number of superintendents surveyed who said they had cut librarians, nurses, cooks, and bus drivers.

About half of the 160 school superintendents from 37 states surveyed by the American Association of School Administrators said that despite receiving stimulus money, they were forced to cut teachers in core subjects. Eight out of 10 said they had cut librarians, nurses, cooks, and bus drivers. (*Source:* www.nytimes.com, Sept. 8, 2009)

65. Al Fischer of Massapequa, New York, donated his 320th pint of blood in September 2009. The 75-year-old man has been donating blood for 58 years. Find the number of gallons of blood he has donated (2pt = 1qt; 4qt = 1gal). (*Source:* www.usatoday.com, Sept. 7, 2009)

66. The sale price of a computer is $1108.80. It is marked down 12%. Find the original price.

Systems of Linear Equations

SUCCESS IN COLLEGE MATHEMATICS

Finding and Learning from Mistakes

We all occasionally make minor arithmetic or algebra mistakes such as writing down the wrong product for a multiplication fact, making a sign error, or misplacing a decimal point. As algebra problems get more complicated, such mistakes can lead to major difficulties. An arithmetic mistake can result in a prime rather than a factorable polynomial. A sign error in the process of solving an equation means that the check of the solution also will not work.

In the past, when you discovered that your answer was wrong, you may have erased and started over. Now that the solutions are longer and more complicated, this method is not efficient and does not help you avoid making the mistake again on a future problem. If you start over, you must redo the entire problem, not just the part after the mistake. If you make this same mistake frequently, erasing does not allow you to discover what you have been doing wrong.

In this textbook, the Find the Mistake exercises provide an opportunity to look at incorrect answers and identify the mistakes. The mistakes in these incorrect answers occur frequently in student work. If you learn where mistakes often occur, you may avoid making them yourself when doing similar problems.

The Sage Bakery is a small business in Lewiston, Idaho. After being open for several years, the owner carefully determined the time it took to make each item and the cost of the ingredients. He found that the cost of making some items was more than their price in the bakery. He was not "breaking even." In this section, we will learn how to use a system of equations to find the break-even point.

SECTION 4.1   Systems of Linear Equations

After reading the text, working the practice problems, and completing assigned exercises, you should be able to:

1. Determine whether an ordered pair is a solution of a system of linear equations.
2. Find the solution of a system of linear equations by graphing.
3. Use a system of equations to solve an application.

Graphing a System of Linear Equations

The **standard form** of a linear equation in two variables is $ax + by = c$. A **solution** of this linear equation is an ordered pair. Each solution can be represented as a point on a rectangular coordinate system. The graph of the equation is the line formed by these points.

A **system of linear equations** is two or more linear equations. The graph of each equation in the system is a line. A **solution of the system** is an ordered pair(s) that is a solution of all of the equations. The solution of the system is the intersection point of the graphs of the equations. In the first example, the graphs of the equations in the system are two lines with different slopes.

EXAMPLE 1 | The graph represents the system of equations $\begin{aligned} y &= 3x - 7 \\ y &= -2x - 2 \end{aligned}$. Solve by graphing.

SOLUTION ▶ The intersection point is $(1, -4)$. This system of equations has one solution.

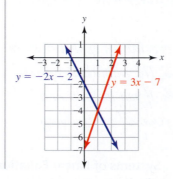

In the next example, the lines are parallel and do not intersect. Their slopes are equal.

EXAMPLE 2 | The graph represents the system of equations $\begin{array}{l} y = -2x + 4 \\ y = -2x + 1 \end{array}$. Solve by graphing.

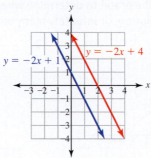

SOLUTION ▶ Since the lines do not intersect, there is no ordered pair that is a solution of both equations. This system of equations has **no solution**.

A system may include two equations that look different. However, when we rewrite these equations in slope-intercept form, the equations are the same, and their graphs are the same. These are **coinciding lines**. Since the lines share every point, this system has infinitely many solutions. In **set-builder notation** the solution is $\{(x, y) \mid y = mx + b\}$, where the vertical line represents the words "such that." This notation tells us that the solution of this system of equations is the set of infinitely many ordered pairs (x, y) *such that* each ordered pair is a solution of the equation $y = mx + b$.

EXAMPLE 3 | The graph represents the system of equations $\begin{array}{l} y = 3x + 2 \\ -3x + y = 2 \end{array}$. Solve by graphing.

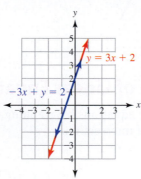

SOLUTION ▶ If we rewrite $-3x + y = 2$ in slope-intercept form, we see that it is the same equation as $y = 3x + 2$.

$$-3x + y = 2$$

$+3x \qquad\qquad +3x$		Addition property of equality.
$0 + y = 2 + 3x$		Simplify.
$y = 3x + 2$		Slope-intercept form; $y = mx + b$.

The lines share every point. This system has infinitely many solutions. In set-builder notation, the solution of this system of equations is $\{(x, y) \mid y = 3x + 2\}$.

A System of Two Linear Equations

One solution   Lines intersect in one point.

No solution   Lines are parallel.

Infinitely many solutions   Lines are coinciding.

Practice Problems

For problems 1–4, use the graph to determine whether the system of equations has one solution, no solution, or infinitely many solutions.

1.

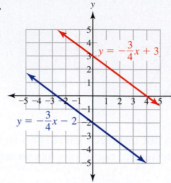

2.

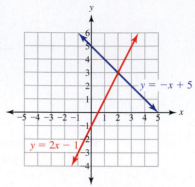

3.

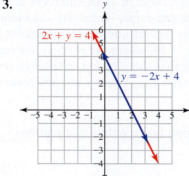

4.

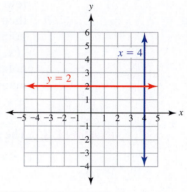

Solving a System of Linear Equations by Graphing

To solve a system of linear equations by graphing, graph each line. The solution of the system is the point(s) of intersection.

EXAMPLE 4   (a) Solve $\begin{aligned} y &= \dfrac{3}{5}x - 2 \\ 2x - 7y &= 3 \end{aligned}$ by graphing.

SOLUTION ▸   Since $y = \dfrac{3}{5}x - 2$ is in slope-intercept form, use slope-intercept graphing.

Graph the y-intercept, $(0, -2)$. Use the slope, $\dfrac{3}{5}$, to find the location of another point. Draw a line through the points.

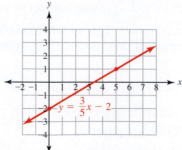

Since the equation $2x - 7y = 3$ is in standard form, find the intercepts and one other point to graph. The lines intersect at $(5, 1)$.

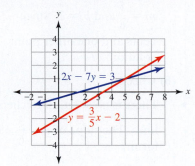

x	y
0	$-\dfrac{3}{7}$
$\dfrac{3}{2}$	0
3	$\dfrac{3}{7}$

(b) Check.

▶ The solution of the system must be a solution of both equations.

$$y = \frac{3}{5}x - 2 \qquad \text{Original equation.} \qquad 2x - 7y = 3$$

$$1 = \frac{3}{5}(\mathbf{5}) - 2 \qquad \text{Replace variables.} \qquad 2(\mathbf{5}) - 7(\mathbf{1}) = 3$$

$$1 = 1 \quad \text{True.} \qquad \text{Evaluate.} \qquad \qquad 3 = 3 \quad \text{True.}$$

Solving a system of equations by graphing provides an estimate. For example, the solution of the graphed system in Figure 1 appears to be $(5, 1)$. However, its actual solution is $\left(\dfrac{111}{22}, \dfrac{111}{110}\right)$. On this graph, our eyes cannot tell the difference between $(5, 1)$ and $\left(\dfrac{111}{22}, \dfrac{111}{110}\right)$. In Section 4.2, we will learn algebraic methods that find the exact solution(s) of a system of equations.

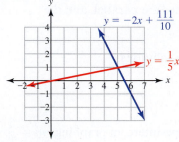

Figure 1

Solving a System of Two Linear Equations by Graphing

1. Graph each line. Use slope-intercept graphing or build a table of ordered pairs.
2. Identify the point of intersection. This is the solution of the system.
3. To check, replace the variables in each equation with the solution.

EXAMPLE 5 **(a)** Solve $\begin{aligned} -5x + 6y &= 54 \\ y &= \dfrac{5}{6}x + 2 \end{aligned}$ by graphing.

SOLUTION ▶

x	y
0	9
$-\dfrac{54}{5}$	0
-4	$\dfrac{17}{3}$

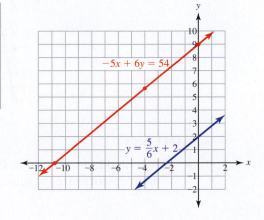

Since $y = \dfrac{5}{6}x + 2$ is in slope-intercept form, use slope-intercept graphing:

$m = \dfrac{5}{6}$ and $b = 2$. Since $-5x + 6y = 54$ is in standard form, graph the intercepts and one other point. The lines appear to be parallel. The system has no solution.

(b) Check.

There is no solution to check. However, we can compare the slopes of the two lines. The slope of $y = \dfrac{5}{6}x + 2$ is $\dfrac{5}{6}$. Rewrite $-5x + 6y = 54$ in slope-intercept form.

$$
\begin{array}{ll}
-5x + 6y = 54 & \text{Rewrite this equation in slope-intercept form.} \\
\underline{+5x \qquad\quad +5x} & \text{Addition property of equality.} \\
0 + 6y = 5x + 54 & \text{Simplify.} \\
\dfrac{6y}{6} = \dfrac{5x}{6} + \dfrac{54}{6} & \text{Division property of equality.} \\
y = \dfrac{5}{6}x + 9 & \text{Slope intercept form; } y = mx + b.
\end{array}
$$

Since the slopes of the lines are the same, $\dfrac{5}{6}$, the lines are parallel, and the system has no solution.

In the next example, the system has infinitely many solutions.

EXAMPLE 6 **(a)** Solve $\begin{array}{l} y = 6x - 7 \\ 12x - 2y = 14 \end{array}$ by graphing.

SOLUTION ▶ Since $y = 6x - 7$ is in slope-intercept form, use slope-intercept graphing: $m = \dfrac{6}{1}$ and $b = -7$.

Since $12x - 2y = 14$ is in standard form, graph the intercepts and one other point. These lines appear to be coinciding. The solution of the system is the set that includes all of the points on this line. In set-builder notation, the solution is $\{(x, y) \mid y = 6x - 7\}$.

x	y
0	-7
$\dfrac{7}{6}$	0
2	5

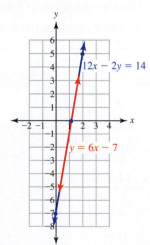

(b) Check.

Rewrite $12x - 2y = 14$ in slope-intercept form. If the equations in slope-intercept form are the same, then the lines are coinciding.

$$12x - 2y = 14$$ Rewrite this equation in slope-intercept form.

$$\underline{-12x \qquad\qquad -12x}$$ Subtraction property of equality.

$$0 + -2y = 14 - 12x$$ Simplify.

$$\frac{-2y}{-2} = \frac{-12x}{-2} + \frac{14}{-2}$$ Commutative property; division property of equality.

$$y = 6x - 7$$ Slope-intercept form; $y = mx + b$.

In slope-intercept form, the equations in the system are the same. These lines are coinciding. The system has infinitely many solutions.

A system of equations may include a vertical or horizontal line.

EXAMPLE 7 **(a)** Solve $\begin{array}{l} y = \dfrac{1}{2}x - 4 \\ x = 6 \end{array}$ by graphing.

SOLUTION ▶ Using slope-intercept graphing for $y = \dfrac{1}{2}x - 4$: $m = \dfrac{1}{2}$ and $b = -4$. The graph of $x = 6$ is a vertical line. The x-coordinate of every point on the line is 6. The lines intersect at $(6, -1)$.

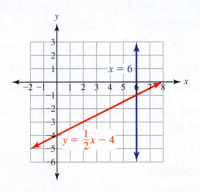

(b) Check.

$$y = \frac{1}{2}x - 4 \qquad \text{Original equations.} \qquad x + 0y = 6$$

$$-1 = \frac{1}{2}(6) - 4 \qquad \text{Replace variables.} \qquad 6 + 0(-1) = 6$$

$$-1 = -1 \quad \text{True.} \qquad \text{Evaluate.} \qquad\qquad 6 = 6 \quad \text{True.}$$

Practice Problems

For problems 5–8,
(a) solve by graphing.
(b) if the solution is an ordered pair, check.

5. $y = 4x - 5$
$\quad 3x + 2y = 12$

6. $x = 4$
$\quad y = 2x - 7$

7. $3x - 4y = 4$
$\quad -6x + 8y = -40$

8. $-2x + 10y = 25$
$\quad y = \dfrac{1}{5}x + \dfrac{5}{2}$

Solutions of a System of Linear Equations

To check the solution of a system of equations using an algebraic method, replace the variables in each equation in the system. If the ordered pair is a solution of each equation, it is the solution of the system.

EXAMPLE 8   Use an algebraic method to determine whether $(1, 4)$ is the solution of the system of equations represented by the graph.

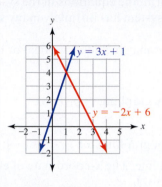

SOLUTION ▶ The equation of each line is given in the graph. Replace the variables in each equation with the coordinates of the given point, $(1, 4)$.

$y = -2x + 6$	Original equations.	$y = 3x + 1$
$4 = -2(1) + 6$	Replace variables.	$4 = 3(1) + 1$
$4 = 4$ True.	Evaluate.	$4 = 4$ True.

Since the ordered pair is a solution of each equation in the system, it is the solution of the system of equations.

Practice Problems

9. Use an algebraic method to determine whether $(2, 1)$ is the solution of the system of equations represented by the graph.

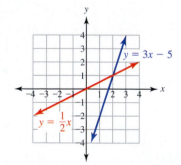

10. Use an algebraic method to determine whether $\left(\dfrac{3}{2}, \dfrac{5}{2}\right)$ is the solution of the system of equations represented by the graph.

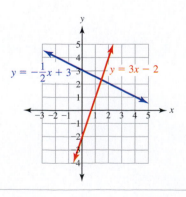

Applications

We can use a system of equations and graphing to estimate the **break-even point** for a product. A business "breaks even" when the costs of making a product are equal to the revenue from selling the product. If the revenue is greater than the costs, the business will make a **profit**. If the revenue is less than the costs, the business has a **loss**.

A graph can have a powerful visual impact. It may be the best way to present a solution such as a break-even point to an audience.

EXAMPLE 9   The cost to make a product is $5. The fixed overhead costs to make the product each month are $6500. A model of the total costs, y, to make x products is $y = \left(\dfrac{\$5}{1 \text{ product}}\right)x + \$6500$. The price of each product is $10. A model of the total revenue, y, from the sale of x products is $y = \left(\dfrac{\$10}{1 \text{ product}}\right)x$. Assume that the company will sell all the products. Use the graph of this system of equations to find the break-even point, and then describe the costs and the revenue at the break-even point.

SOLUTION ▶ **Step 1 Understand the problem.**
The unknowns are the number of products at the break-even point and the costs/revenue at the break-even point.

$x =$ number of products at the break-even point

$y =$ costs or revenue at the break-even point

Step 2 Make a plan.
We are given the linear model for the total costs to make x products and a linear model for the total revenue for selling x products. The intersection point of the graphs of these models is the break-even point. At the break-even point, the costs and revenue are equal.

Step 3 Carry out the plan.
Since the y-intercept is large ($6500) and the rise from the slope is small ($5), it is difficult to use slope-intercept graphing. To graph, build a table of ordered pairs.

$$y = \left(\frac{\$5}{1 \text{ product}}\right)x + \$6500$$

x (products)	y (dollars)
0	6500
500	9000
1000	11,500

$$y = \left(\frac{\$10}{1 \text{ product}}\right)x$$

x (products)	y (dollars)
0	0
500	5000
1000	10,000

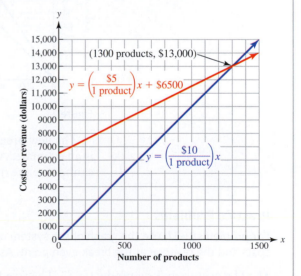

The break-even point appears to be (1300 products, $13,000). When 1300 products are made and sold, the costs equal the revenue. The costs to make 1300 products are $13,000. The revenue from selling 1300 products is $13,000.

Step 4 Look back.

The costs to make 1300 products are $\left(\dfrac{\$5}{1\ \text{product}}\right)(1300\ \text{products}) + \$6500$,

which equals \$13,000. The revenue from selling 1300 products is

$\left(\dfrac{\$10}{1\ \text{product}}\right)(1300\ \text{products})$, which also equals \$13,000. Since the costs

equal the revenue, the answer seems reasonable.

Step 5 Report the solution.

To break even, the company should make and sell 1300 products. The costs and the revenue at the break-even point are equal, \$13,000.

If the price of a product increases and the costs remain the same, fewer products need to be sold to break even. For example, if the price of the product in Example 9

increases to \$15, the model of the total revenues becomes $y = \left(\dfrac{\$15}{1\ \text{product}}\right)x$.

$$y = \left(\dfrac{\$5}{1\ \text{product}}\right)x + \$6500$$

$$y = \left(\dfrac{\$10}{1\ \text{product}}\right)x$$

$$y = \left(\dfrac{\$15}{1\ \text{product}}\right)x$$

x (products)	y (dollars)
0	0
500	7500
1000	15,000

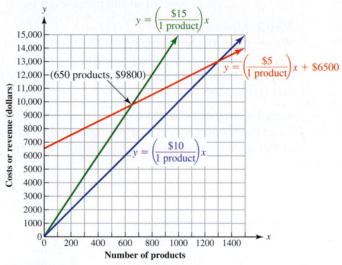

Figure 2

The new break-even point (Figure 2) appears to be (650 products, \$9800). This is an estimate. Although the actual intersection point is (650 products, \$9750), we cannot see the difference between \$9750 and \$9800 on this graph.

Practice Problems

For problems 11–12, use the graph of the system of equations to find the break-even point, and then describe the costs and the revenue at the break-even point. Assume that the company will sell all the products.

11. The cost to make a product is \$3. The fixed overhead costs each month to make the product are \$4500. A

model of the total costs, y, to make x products is $y = \left(\dfrac{\$3}{1\ \text{product}}\right)x + \$4500$. The price of each product

is \$18. A model of the total revenue, y, from the sale of x products is $y = \left(\dfrac{\$18}{1\ \text{product}}\right)x$.

12. The cost to make a product is $3. The fixed overhead costs to make the product each month are $4500. A model of the total costs, y, to make x products is $y = \left(\dfrac{\$3}{1 \text{ product}}\right)x + \$4500$. The price of each product is $6. A model of the total revenue, y, from the sale of x products is $y = \left(\dfrac{\$6}{1 \text{ product}}\right)x$.

Using Technology: Solving a System of Linear Equations

To estimate the solution of a system of linear equations, graph the lines. Use the **intersect** command in the CALC menu to estimate the intersection point.

EXAMPLE 10 Solve $\begin{aligned} y &= \dfrac{3}{5}x - 2 \\ 2x - 7y &= 3 \end{aligned}$ by graphing.

To graph $2x - 7y = 3$, rewrite it in slope-intercept form, $y = \dfrac{2}{7}x - \dfrac{3}{7}$. The graphing window should be standard. Type both equations on the [Y=] screen. Press [GRAPH]. The intersection point is visible. In the CALC menu, choose 5: intersect.

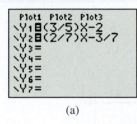

(a)

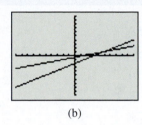

(b)

(c)

The cursor will appear on one of the lines. This is the "first curve." Press [ENTER]. The cursor moves to the other line. This is the "second curve." Press [ENTER]. The calculator asks "Guess?" Move the cursor to the intersection point. Press [ENTER].

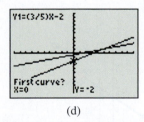

(d)

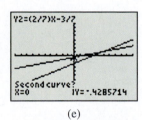

(e)

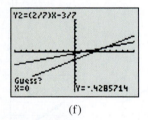

(f)

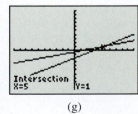

(g)

The calculator displays the estimated point of intersection: $x = 5$ and $y = 1$. The solution of the system of equations is the ordered pair (5, 1).

Practice Problems For problems 13–15,
(a) use a graphing calculator to graph the system of equations. Sketch the graph; describe the window.
(b) identify the solution.

13. $y = -4x + 7$
$3x + 5y = 1$

14. $y = -4x + 7$
$2x + y = -9$

15. $y = \dfrac{2}{3}x + 8$
$2x - 3y = 5$

4.1 VOCABULARY PRACTICE

Match the term with its description. A term may be used more than once.

1. $ax + by = c$
2. $y = mx + b$
3. Lines with the same slope but different y-intercepts
4. The point at which two lines cross
5. The equations of these lines may look different, but their graphs are the same.
6. Lines that never intersect
7. Lines that share every point
8. When the revenue received from selling a product equals the costs of making the product
9. The solution of a system of equations represented by two coinciding lines
10. The solution of a system of equations represented by two parallel lines

A. break-even point
B. coinciding lines
C. infinite set of ordered pairs
D. intersection point
E. no solution
F. parallel lines
G. slope-intercept form of a linear equation
H. standard form of a linear equation

4.1 Exercises

Follow your instructor's guidelines for showing your work.

For exercises 1–12, the graph represents a system of equations. Solve by graphing.

1.
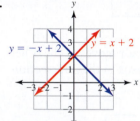
$y = -x + 2$
$y = x + 2$

2.
$y = \frac{1}{4}x + 4$ $y = -\frac{1}{4}x + 4$

3.
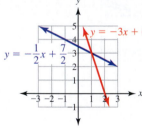
$y = -3x + 6$
$y = -\frac{1}{2}x + \frac{7}{2}$

4.
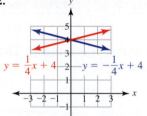
$y = -\frac{1}{3}x - \frac{13}{3}$
$y = -\frac{2}{3}x - \frac{14}{3}$

5.
$y = 2x + 3$
$y = 2x - 4$

6.
$y = -2x + 3$
$y = -2x - 3$

7.
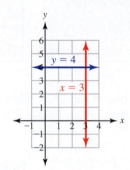
$y = 4$
$x = 3$

8.
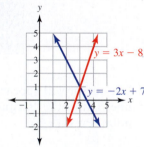
$y = 3x - 8$
$y = -2x + 7$

9.
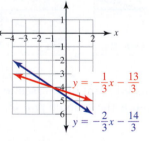
$y = \frac{2}{3}x + 2$
$-2x + 3y = 6$

10.
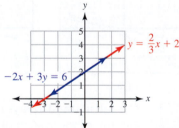
$2x + 3y = 9$
$y = -\frac{2}{3}x + 3$

11.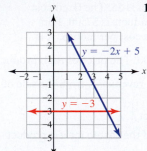

12.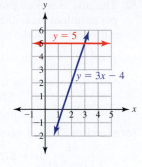

For exercises 13–14,
(a) complete the table of ordered pairs for each equation.
(b) use the tables to find the solution of the system of equations.

13. $x + 2y = 9$ $\qquad$ $y = 2x - 8$

x	y
-1	
0	
1	
2	
3	
4	
5	
6	

x	y
-1	
0	
1	
2	
3	
4	
5	
6	

14. $x + 6y = -3$ $\qquad$ $y = 2x - 7$

x	y
-1	
0	
1	
2	
3	
4	
5	
6	

x	y
-1	
0	
1	
2	
3	
4	
5	
6	

15. When we use a graph to solve a system of equations, the solution is an estimate. Explain why.

16. We seldom solve a system of equations by comparing tables of values. Describe a disadvantage of using this method.

For exercises 17–46, solve by graphing.

17. $y = 5x + 3$
$\quad y = -6x + 3$

18. $y = 4x + 5$
$\quad y = -7x + 5$

19. $2x - y = 6$
$\quad y = -x + 9$

20. $2x + y = 14$
$\quad y = 3x - 11$

21. $y = \dfrac{2}{3}x - 4$
$\quad y = 2x$

22. $y = \dfrac{3}{4}x + 2$
$\quad y = x + 3$

23. $y = \dfrac{1}{2}x - 6$
$\quad x - 2y = 10$

24. $y = \dfrac{1}{4}x - 9$
$\quad x - 4y = -12$

25. $3x + 5y = -9$
$\quad 9x + 2y = 12$

26. $4x + 3y = -12$
$\quad 5x + 6y = -33$

27. $y = \dfrac{2}{3}x + 8$
$\quad -2x + 3y = 24$

28. $y = \dfrac{3}{5}x + 9$
$\quad -3x + 5y = 45$

29. $x + 2y = 10$
$\quad y = \dfrac{1}{2}x - 2$

30. $x + 2y = 8$
$\quad y = \dfrac{1}{2}x - 3$

31. $x = -4$
$\quad y = 2$

32. $x = -3$
$\quad y = 1$

33. $3x + 4y = -4$
$\quad y = 2$

34. $2x + 5y = 2$
$\quad y = 4$

35. $x = 6$
$\quad x = 2$

36. $x = 8$
$\quad x = 3$

37. $y = \dfrac{3}{4}x - 2$
$\quad y = -x + 5$

38. $y = \dfrac{3}{5}x - 1$
$\quad y = -x + 7$

39. $2x + 5y = 3$
$\quad -x + y = 9$

40. $3x + 10y = -4$
$\quad -x + y = 10$

41. $7y = 21$
$\quad -4y + 20 = 8$

42. $9y = 36$
$\quad -5y + 38 = 18$

43. $y = 3x$
$\quad -x + y = 0$

44. $y = -2x$
$\quad -x + y = 0$

45. $\quad 5x + 2y = 20$
$\quad -3x - 4y = -12$

46. $\quad 4x + 3y = 12$
$\quad -5x + 2y = -15$

47. Use an algebraic method to determine whether $(2, 3)$ is the solution of the system of equations represented by the graph.

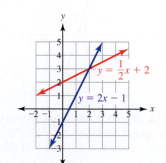

48. Use an algebraic method to determine whether $(3, 2)$ is the solution of the system of equations represented by the graph.

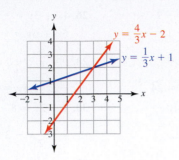

49. Use an algebraic method to determine whether $\left(\dfrac{1}{2}, \dfrac{1}{2}\right)$ is the solution of the system of equations represented by the graph.

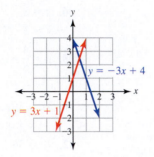

50. Use an algebraic method to determine whether $\left(\dfrac{3}{2}, \dfrac{1}{2}\right)$ is the solution of the system of equations represented by the graph.

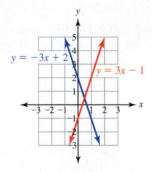

For exercises 51–52, use the graph of the system of equations to find the break-even point, and then describe the costs and the revenue at the break-even point. Assume that the company will sell all the products.

51. The cost to make a product is $5. The fixed cost per month to make the products is $7500. For x products, a model of the cost per month, y, is

$y = \left(\dfrac{\$5}{1\ \text{product}}\right)x + \$7500$. The price of each product is $55. A model of the total revenue, y, from the sale of x

products is $y = \left(\dfrac{\$55}{1\ \text{product}}\right)x$.

52. The cost to make a product is $5. The fixed cost per month to make the products is $7500. For x products, a model of the cost per month, y, is

$y = \left(\dfrac{\$5}{1\ \text{product}}\right)x + \$7500$. The price of each product is $105. A model of the total revenue, y, from the sale of x

products is $y = \left(\dfrac{\$105}{1\ \text{product}}\right)x$.

For exercises 53–54, a model of the cost to make a product is $y = \left(\dfrac{\$5}{1\ \text{product}}\right)x + \$7500$. The graph of this model is a straight line.

53. What does the slope of this line represent?

54. What does the y-intercept of this line represent?

For exercises 55–56, use the graph of the system of equations to find the break-even point, and then describe the costs and the revenue at the break-even point. Assume that the company will sell all the products.

55. The cost to make a product is $100. The fixed cost per month to make the products is $80,000. For x products, a model of the cost per month, y, is

$y = \left(\dfrac{\$100}{1\ \text{product}}\right)x + \$80,000$. The price of each product is $125. A model of the total revenue, y, from the sale of x

products is $y = \left(\dfrac{\$125}{1\ \text{product}}\right)x$.

56. The cost to make a product is $100. The fixed cost per month to make the products is $80,000. For x products, a model of the cost per month, y, is

$y = \left(\dfrac{\$100}{1\ \text{product}}\right)x + \$80,000$. The price of each product is $150. A model of the total revenue, y, from the

sale of x products is $y = \left(\dfrac{\$150}{1\ \text{product}}\right)x$.

For exercises 57–58, a model of the cost to make a product is $y = \left(\dfrac{\$100}{1\ \text{product}}\right)x + \$80,000$. The graph of this model is a straight line.

57. What does the y-intercept of this line represent?

58. What does the slope of this line represent?

For exercises 59–66, use a graph of the system of equations and the five steps.

59. An investor has $36,000. He wants to put twice as much in short-term investments as he puts in long-term investments. The amount in short-term investments is x. The amount in long-term investments is y. A model of this situation is the system of equations $x + y = \$36,000$ and $y = \dfrac{1}{2}x$. Find the amount to put in each investment.

60. An investor has $50,000. She wants to put three times as much in real estate as she puts in stocks. The amount invested in real estate is x. The amount invested in stocks is y. A model of this situation is the system of equations $x + y = \$50{,}000$ and $y = \dfrac{1}{3}x$. Find the amount to put in each investment.

61. The cost of 50 chicken wraps and 30 ham sandwiches is $300. The cost of 30 chicken wraps and 50 ham sandwiches is $340. The price of a wrap is x. The price of a sandwich is y. A model of this situation is the system of equations $50x + 30y = \$300$ and $30x + 50y = \$340$. Find the price of a chicken wrap and the price of a ham sandwich.

62. The cost of 10 scones and 20 doughnuts is $40. The cost of 20 scones and 10 doughnuts is $50. Let x represent the price of a scone. Let y represent the price of a doughnut. A model of this situation is the system of equations $10x + 20y = \$40$ and $20x + 10y = \$50$. Find the price of a scone and the price of a doughnut.

63. A cell phone customer is comparing the cost of text message plans. In both plans, 200 text messages per month are free. Plan A costs $5 per month. Each additional text message costs $0.10. A model of the monthly cost, y, for x additional messages is $y = \left(\dfrac{\$0.10}{1 \text{ text message}} \right) x + \$5$.

In Plan B, the charge per month is $10. Each additional message costs $0.05. A model of the monthly cost, y, for x additional messages is $y = \left(\dfrac{\$0.05}{1 \text{ text message}} \right) x + \$10$.

Find the number of additional text messages at which the cost of the two plans are equal.

64. A band needs to rent a vehicle to go on tour. A rental company in Miami rents 15-passenger vans and 15-passenger minibuses. The cost to rent a van is $1350 plus any additional mileage costs. For every mile over 1000 miles, the cost per mile is $0.49. A model of the total cost, y, per x total miles is

$y = \left(\dfrac{\$0.49}{1 \text{ mi}} \right)(x - 1000) + \$1350$. The cost to rent a

minibus is $1425. For every mile over 1000 miles, the cost per mile is $0.19. A model of the total cost, y, per x total miles is $y = \left(\dfrac{\$0.19}{1 \text{ mi}} \right)(x - 1000) + \$1425$. Find the number of miles at which the costs of the two vehicles are equal.

65. A food stand at a baseball game sold a total of 3000 hotdogs and Polish sausages. The price of a hotdog was $1. The price of a Polish sausage was $3. The total receipts for hotdogs and sausages were $4200. Let x represent the number of hotdogs sold. Let y represent the number of Polish sausages sold. A model of this situation is the system of equations $x + y = 3000$ and $\$1x + \$3y = \$4200$. Find the number of hotdogs and the number of Polish sausages sold.

66. At a warehouse store, the price of a mixed bouquet with many kinds of flowers is $14. The price of a grower's bunch with a single kind of flower is $8. On Mother's Day weekend, the store sold 2000 bouquets. The total receipts were $19,000. Let x represent the number of mixed bouquets sold. Let y represent the number of grower's bunches sold. A model of this situation is the system of equations $x + y = 2000$ and $\$14x + \$8y = \$19{,}000$. Find the number of each kind of bouquet sold.

© Luti/Shutterstock

Problem Solving: Practice and Review

Follow your instructor's guidelines for using the five steps as outlined in Section 1.5, p. 51.

67. In fiscal year 2010, the IRS found 11,840,441 math errors on individual U.S. income tax returns. This was a 4.93% decrease from the number of errors found in fiscal year 2009. Find the number of errors found in 2009. Round to the nearest thousand. (*Source:* www.irs.gov)

68. Find the company's first quarter revenue a year earlier, in 2010. Round to the nearest hundredth of a billion.

For the company's [Amazon] first quarter, which ended March 31, revenue was $9.857 billion, above the average analyst estimate of $9.57 billion and 38.2 percent above a year earlier. (*Source:* www.reuters.com, Apr. 26, 2011)

69. Researchers from Penn State studied the effect of sewage on insect populations in Fourmile Creek. They collected insects upstream and downstream from where the sewage entered the creek. The discharge rate of the sewage was $\dfrac{19{,}300 \text{ dekaliters}}{1 \text{ day}}$. Change this measurement into liters per hour (1 dekaliter = 10 liters). Round to the nearest whole number. (*Source:* Masteller, Wagner, "The impact of sewage influence on the occurrence of psychodiae in a stream," *Freshwater Invertebrate Biology*, 1984).

70. The Karvonen heart rate formula finds the target heart rate zone for aerobic exercise. Find the target heart rate zone for a 24-year-old person with a resting heart rate of 65 beats per minute. Round to the nearest whole number. (*Source:* www.acefitness.org)

Low end of the zone = 0.65(220 − age − resting heart rate) + resting heart rate

High end of the zone = 0.85(220 − age − resting heart rate) + resting heart rate

Technology

For exercises 71–74,
(a) use a graphing calculator to graph and solve each system of equations. Sketch the graph, and describe the window.
(b) state the solution as an ordered pair.

71. $5x + 2y = 16$
$3x - 4y = 7$

72. $3x + 4y = 23$
$7x - 8y = 45$

73. $2x + 9y = 261$
$8x + 5y = 393$

74. $6x + 5y = -99$
$x + 13y = 166$

Find the Mistake

For exercises 75–78, the completed problem has one mistake.
(a) Describe the mistake in words, or copy down the whole problem and highlight or circle the mistake.
(b) Do the problem correctly.

75. Problem: Solve $\begin{aligned} y &= -\dfrac{3}{2}x + 2 \\ 3x + 2y &= 4 \end{aligned}$ by graphing.

Incorrect Answer: The solution is the set of real numbers.

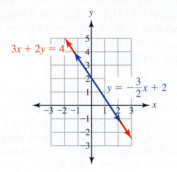

76. Problem: Solve $\begin{aligned} y &= -3x + 5 \\ y &= 2 \end{aligned}$ by graphing.

Incorrect Answer: The solution is (2, 1).

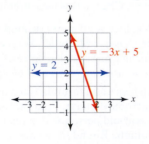

77. Problem: Solve $\begin{aligned} y &= 3x + 8 \\ y &= 6x - 4 \end{aligned}$ by graphing.

Incorrect Answer: The lines do not intersect. This system has no solution.

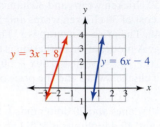

78. Problem: Use an algebraic method to check whether $(2, -5)$ is a solution of $\begin{aligned} 3x + 8y &= -34 \\ x - y &= -3 \end{aligned}$.

Incorrect Answer:

$3x + 8y = -34$	$x - y = -3$
$3(2) + 8(-5) = -34$	$2 - 5 = -3$
$-34 = -34$ True.	$-3 = -3$ True.

The ordered pair $(2, -5)$ is a solution of this system.

Review

For exercises 79–82,
(a) solve each equation.
(b) check the solution.

79. $3x + 2(6x - 3) = 9$

80. $4x - 5\left(\dfrac{3}{2}x + 5\right) = -32$

81. $\dfrac{3}{2}x - \dfrac{11}{2} = \dfrac{1}{8}x$

82. $9\left(3y - \dfrac{17}{3}\right) - 7y = -11$

SUCCESS IN COLLEGE MATHEMATICS

83. A common mistake in graphing was made in this incorrect graph of $y = 3x + 2$. What is not correct on this graph?

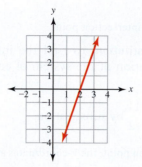

84. a. In a y-intercept, is the x-coordinate or the y-coordinate 0?

   b. In an x-intercept, is the x-coordinate or the y-coordinate 0?

To *estimate* the solution of a system of linear equations, we graph each line. If there is a single intersection point, this is the solution of the system. To find the *exact* solution, we use an algebraic method. In this section, we will use the **substitution method** and the **elimination method** to find the exact solution of a system of linear equations in two variables.

SECTION 4.2 Algebraic Methods

After reading the text, working the practice problems, and completing assigned exercises, you should be able to:

1. Use substitution to solve a system of linear equations in two variables.

2. Use elimination to solve a system of linear equations in two variables.

The Substitution Method

The lines on the graph (Figure 1) represent the equations $y = 3x - 7$ and $y = -2x + 8$. These lines intersect at $(3, 2)$. This is the solution of this system of equations.

At this intersection point, the y-coordinates of both lines are the same: $y = y$. So $3x - 7 = -2x + 8$. We can solve this equation for x.

Figure 1

$$3x - 7 = -2x + 8 \qquad \text{At the intersection point, the } y\text{-coordinates are equal.}$$

$$\underline{+7 \qquad\qquad +7} \qquad \text{Addition property of equality.}$$

$$3x + 0 = -2x + 15 \qquad \text{Simplify.}$$

$$\underline{+2x \qquad\qquad +2x} \qquad \text{Addition property of equality.}$$

$$\mathbf{5x} + 0 = \mathbf{0} + 15 \qquad \text{Simplify.}$$

$$\frac{5x}{5} = \frac{15}{5} \qquad \text{Division property of equality.}$$

$$x = 3 \qquad \text{Simplify; the } x\text{-coordinate of the intersection point.}$$

Without using the graph, we found the x-coordinate of the intersection point. Now we can use this value to find the y-coordinate of the intersection point.

$y = 3x - 7$	To find the y-coordinate, choose either one of the equations.
$y = 3(3) - 7$	Replace x with the x-coordinate of the intersection point, 3.
$y = 9 - 7$	Simplify.
$y = 2$	The y-coordinate of the intersection point.

We just used an **algebraic method**, **substitution**, to find the intersection point of these lines, $(3, 2)$. This is the exact solution of the system of equations, not an estimate.

EXAMPLE 1 **(a)** Solve the system of equations $\begin{array}{l} y = 2x + 11 \\ y = -x + 2 \end{array}$ by substitution.

SOLUTION ▶

$2x + 11 = -x + 2$	At the intersection point, the y-coordinates are equal: $y = y$
$\underline{\quad -11 \qquad\quad -11\quad}$	Subtraction property of equality.
$2x + 0 = -x - 9$	Simplify.
$\underline{+x \qquad\qquad +x\quad}$	Addition property of equality.
$3x = 0 - 9$	Simplify.
$\dfrac{3x}{3} = \dfrac{-9}{3}$	Division property of equality.
$x = -3$	The x-coordinate of the intersection point.
$y = 2x + 11$	To find the y-coordinate, choose either one of the equations.
$y = 2(-3) + 11$	Replace x with the x-coordinate of the intersection point.
$y = -6 + 11$	Simplify.
$y = 5$	The y-coordinate of the intersection point.

The solution of this system of equations is $(-3, 5)$.

(b) Check $(-3, 5)$ by replacing the variables in both equations.

▶

$y = 2x + 11$	Check both equations.	$y = -x + 2$
$5 = 2(-3) + 11$	Replace variables.	$5 = -(-3) + 2$
$5 = -6 + 11$	Evaluate.	$5 = 3 + 2$
$5 = 5$ True.	The solution is correct.	$5 = 5$ True.

Since the ordered pair $(-3, 5)$ is a solution of both of the equations, it is the solution of the system.

Solving a System of Two Linear Equations by Substitution

1. Solve one of the equations for a variable. The expression on the other side of the solved equation is the "substitution expression."

2. Replace the variable in the other equation with the substitution expression.

3. Solve for the remaining variable.

4. Replace the variable in one of the original equations with the value found in step 3. Solve for the other variable.

5. To check, replace the variables in each equation with the solution.

In the Example 1, both equations in the system are solved for y. In the next example, only one of the equations is solved for y. We can use the same algebraic method to find the solution.

EXAMPLE 2 (a) Solve $\begin{array}{l} y = 6x - 9 \\ 5x + 7y = 31 \end{array}$ by substitution.

SOLUTION ▶

$y = 6x - 9$	The substitution expression is $6x - 9$.
$5x + 7(6x - 9) = 31$	Replace y in $5x + 7y = 31$ with $6x - 9$.
$5x + 42x - 63 = 31$	Distributive property.
$47x - 63 = 31$	Combine like terms.
$\underline{\;+63\;\;+63}$	Addition property of equality.
$47x + 0 = 94$	Simplify.
$\dfrac{47x}{47} = \dfrac{94}{47}$	Division property of equality.
$x = 2$	The solution of the system is $(2, y)$.

Use this value of x to find y.

$y = 6(2) - 9$	To find y, replace x in either equation with 2.
$y = 3$	Simplify.

The solution of the system is $(2, 3)$.

(b) Check. The solution of the system must be a solution of both equations.

▶

$y = 6x - 9$	Check both equations.	$5x + 7y = 31$	
$3 = 6(2) - 9$	Replace variables.	$5(2) + 7(3) = 31$	
$3 = 12 - 9$	Evaluate.	$10 + 21 = 31$	
$3 = 3$ True.	The solution is correct.	$31 = 31$ True.	

The graph of a system of equations with no solution is two parallel lines. In the next example, we use the substitution method to try to solve a system with no solution. We replace a variable with the substitution expression. As we solve this equation, the variables disappear, and the simplified equation is false.

EXAMPLE 3 Solve $\begin{array}{l} y = -3x + \dfrac{4}{9} \\ 6x + 2y = 1 \end{array}$ by substitution.

SOLUTION ▶

$y = -3x + \dfrac{4}{9}$	The substitution expression is $-3x + \dfrac{4}{9}$.
$6x + 2\left(-3x + \dfrac{4}{9}\right) = 1$	Replace y with the substitution expression.
$6x - 6x + \dfrac{8}{9} = 1$	Distributive property.
$0 + \dfrac{8}{9} = 1$	The variables disappear; this equation is false.

This system has no solution.

A system of equations represented by coinciding lines has infinitely many solutions. In the next example, we use the substitution method to solve a system with infinitely many solutions. We replace a variable with the substitution expression. As we solve this equation, the variables disappear, and the simplified equation is true. We can use set-builder notation to describe the solution.

EXAMPLE 4 **(a)** Solve $\begin{array}{l} y = 2x + 5 \\ -\dfrac{2}{3}x + \dfrac{1}{3}y = \dfrac{5}{3} \end{array}$ by substitution.

SOLUTION ▶

$y = 2x + 5$	The substitution expression is $2x + 5$.
$-\dfrac{2}{3}x + \dfrac{1}{3}(2x + 5) = \dfrac{5}{3}$	Replace y with the substitution expression.
$-\dfrac{2}{3}x + \dfrac{2}{3}x + \dfrac{5}{3} = \dfrac{5}{3}$	Distributive property.
$\mathbf{0} + \dfrac{5}{3} = \dfrac{5}{3}$	The variables disappear; this equation is true.

This system has infinitely many solutions: $\{(x, y) \mid y = 2x + 5\}$.

(b) Check. Choose two ordered pairs that are solutions of $y = 2x + 5$, such as $(0, 5)$ and $(1, 7)$. Replace the variables in the other equation with these values. Any ordered pair that is a solution of one equation should also be a solution of the other equation.

$-\dfrac{2}{3}x + \dfrac{1}{3}y = \dfrac{5}{3}$	Use the other equation.	$-\dfrac{2}{3}x + \dfrac{1}{3}y = \dfrac{5}{3}$
$-\dfrac{2}{3}(\mathbf{0}) + \dfrac{1}{3}(\mathbf{5}) = \dfrac{5}{3}$	Replace variables.	$-\dfrac{2}{3}(\mathbf{1})x + \dfrac{1}{3}(\mathbf{7}) = \dfrac{5}{3}$
$\mathbf{0} + \dfrac{1}{3}(\mathbf{5}) = \dfrac{5}{3}$	Evaluate.	$-\dfrac{2}{3} + \dfrac{7}{3} = \dfrac{5}{3}$
$\dfrac{5}{3} = \dfrac{5}{3}$ True.	The solution is correct.	$\dfrac{5}{3} = \dfrac{5}{3}$ True.

In the next example, both equations are solved for x. Either $-6y + \dfrac{11}{4}$ or $3y - \dfrac{1}{4}$ can be the substitution expression.

EXAMPLE 5 Solve $\begin{array}{l} x = -6y + \dfrac{11}{4} \\ x = 3y - \dfrac{1}{4} \end{array}$ by substitution.

SOLUTION ▶

$x = -6y + \dfrac{11}{4}$	The substitution expression is $-6y + \dfrac{11}{4}$.
$-6y + \dfrac{11}{4} = 3y - \dfrac{1}{4}$	Replace x in the other equation with $-6y + \dfrac{11}{4}$.
$4\left(-6y + \dfrac{11}{4}\right) = 4\left(3y - \dfrac{1}{4}\right)$	Clear fractions; multiplication property of equality.
$4(-6y) + 4\left(\dfrac{11}{4}\right) = 4(3y) + 4\left(-\dfrac{1}{4}\right)$	Distributive property.
$-24y + 11 = 12y - 1$	Simplify.
$\underline{-12y \qquad\quad -12y}$	Subtraction property of equality.
$-36y + 11 = 0 - 1$	Simplify.
$\underline{\qquad -11 \qquad -11}$	Subtraction property of equality.
$-36y + 0 = -12$	Simplify.

$$\frac{-36y}{-36} = \frac{-12}{-36}$$ Division property of equality.

$$y = \frac{1}{3}$$ Simplify; the solution of the system is $\left(x, \frac{1}{3}\right)$.

Use this value of y to find x.

$$x = 3\left(\frac{1}{3}\right) - \frac{1}{4}$$ To find x, replace y in either equation with $\frac{1}{3}$.

$$x = 1 - \frac{1}{4}$$ Simplify.

$$x = \frac{3}{4}$$ Simplify.

The solution is $\left(\frac{3}{4}, \frac{1}{3}\right)$.

To solve a system by substitution when both equations are in standard form, first solve one of the equations for a variable. In the next example, we choose to solve $2x + 3y = -4$ for x.

EXAMPLE 6 Solve $\begin{matrix} 6x - 5y = -40 \\ 2x + 3y = -4 \end{matrix}$ by substitution.

SOLUTION ▶

$$2x + 3y = -4$$ Solve either of the equations for a variable.

$$\underline{\quad -3y \qquad\qquad -3y\quad}$$ Subtraction property of equality.

$$2x + 0 = -4 - 3y$$ Simplify.

$$\frac{2x}{2} = \frac{-4}{2} - \frac{3y}{2}$$ Division property of equality.

$$x = -2 - \frac{3y}{2}$$ The substitution expression is $-2 - \frac{3y}{2}$.

$$6x - 5y = -40$$ The other equation in the system.

$$6\left(-2 - \frac{3y}{2}\right) - 5y = -40$$ Replace x with the substitution expression.

$$6(-2) + 6\left(-\frac{3y}{2}\right) - 5y = -40$$ Distributive property.

$$-12 - 9y - 5y = -40$$ Simplify.

$$-12 - 14y = -40$$ Simplify.

$$\underline{+12 \qquad\qquad +12\quad}$$ Addition property of equality.

$$0 - 14y = -28$$ Simplify.

$$\frac{-14y}{-14} = \frac{-28}{-14}$$ Division property of equality.

$$y = 2$$ Simplify; the solution of the system is $(x, 2)$.

$$2x + 3y = -4$$ Use either of the original equations to find x.

$$2x + 3(2) = -4$$ Replace y with 2.

$$2x + 6 = -4$$ Simplify.

$$\underline{\quad -6 \quad -6\quad}$$ Subtraction property of equality.

$$2x + 0 = -10$$ Simplify.

$$\frac{2x}{2} = \frac{-10}{2} \qquad \text{Division property of equality.}$$

$$x = -5 \qquad \text{Simplify.}$$

The solution of the system is $(-5, 2)$.

Practice Problems

For problems 1–4,
(a) solve the system of equations by substitution.
(b) check.

1. $y = -2x + 5$
   $3x + 7y = -53$

2. $y = \frac{3}{2}x + 9$
   $3x - 2y = 8$

3. $-4x - 5y = -10$
   $y = -\frac{4}{5}x + 2$

4. $x = y - \frac{5}{2}$

   $x = -2y + \frac{13}{2}$

Figure 2

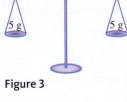

Figure 3

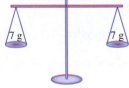

Figure 4

The Elimination Method

The scale in Figure 2 is balanced.

The scale in Figure 3 is also balanced. The weight in the left pan equals the weight in the right pan.

Now imagine that we combine the weights from the first two scales on another scale (Figure 4). The weight in the left pan equals the weight in the right pan. The scale is balanced.

This idea also applies to linear equations. *If we add both sides of two linear equations together, the sum is a new equation. The left side of this new equation equals the right side.* We are using the addition property of equality to add equivalent expressions to both sides of an equation. This does not change the solution(s) of the equation. When solving a system of linear equations by **elimination**, we add the equations in the system together.

In the next example, the coefficients of the y-terms are opposites. When the equations are added, the y-term is eliminated.

EXAMPLE 7   **(a)** Solve $\begin{array}{c} 2x + 3y = 22 \\ 7x - 3y = 23 \end{array}$ by elimination.

SOLUTION ▶   Add the equations vertically to eliminate the y-terms. Solve for x.

$$2x + 3y = 22 \qquad \text{Add the equations vertically; the } y\text{-terms are opposites.}$$
$$\underline{+\ 7x - 3y = 23} \qquad \text{Addition property of equality.}$$
$$9x + 0 = 45 \qquad \text{Simplify.}$$
$$9x = 45 \qquad \text{Simplify; the term with } y \text{ is eliminated.}$$
$$\frac{9x}{9} = \frac{45}{9} \qquad \text{Division property of equality.}$$
$$x = 5 \qquad \text{Simplify; the solution of the system is } (5, y).$$

Replace x in one of the equations and solve for y.

$$2x + 3y = 22 \qquad \text{To find } y, \text{ choose either one of the equations in the system.}$$
$$2(5) + 3y = 22 \qquad \text{Replace } x \text{ with 5.}$$

$$10 + 3y = 22 \qquad \text{Simplify.}$$

$$\underline{-10 \qquad\quad -10} \qquad \text{Subtraction property of equality.}$$

$$0 + 3y = 12 \qquad \text{Simplify.}$$

$$\frac{3y}{3} = \frac{12}{3} \qquad \text{Division property of equality.}$$

$$y = 4 \qquad \text{Simplify.}$$

The solution of the system of equations is $(5, 4)$.

(b) Check.

$2x + 3y = 22$	Check both equations.	$7x - 3y = 23$
$2(5) + 3(4) = 22$	Replace variables with the solution.	$7(5) - 3(4) = 23$
$10 + 12 = 22$	Evaluate.	$35 - 12 = 23$
$22 = 22$ True.	The solution is correct.	$23 = 23$ True.

In the elimination method, we add two equations. In Example 7, the coefficients of $3y$ and $-3y$ are opposites, and their sum is 0. When we add the equations together, the y-terms are eliminated. The system in the next example is $\begin{array}{l} 2x + 3y = -13 \\ 5x + 4y = -15 \end{array}$. If we add these equations together, no variable is eliminated. So we need to first multiply each of the equations by a different number to create terms that are opposites.

We choose to eliminate the x-terms. Since the coefficient of $5x$ is 5, we multiply both sides of $2x + 3y = -13$ by 5. The x-term in this equation is now $10x$. Now we need to multiply $5x + 4y = -15$ by a constant that results in an x-term of $-10x$, the opposite of $10x$. Since $(-2)(5) = -10$, we multiply both sides of $5x + 4y = -15$ by -2.

EXAMPLE 8   Solve $\begin{array}{l} 2x + 3y = -13 \\ 5x + 4y = -15 \end{array}$ by elimination.

SOLUTION ▶ Choose either x or y to eliminate. In this example, we will eliminate x.

Create x-terms that are opposites.
Multiply $2x + 3y = -13$ by 5.

$$5(2x + 3y) = 5(-13) \qquad \text{Multiplication property of equality.}$$

$$10x + 15y = -65 \qquad \text{Simplify; the } x\text{-term is now } 10x.$$

Multiply $5x + 4y = -15$ by -2.

$$-2(5x + 4y) = -2(-15) \qquad \text{Multiplication property of equality.}$$

$$-10x - 8y = 30 \qquad \text{Simplify; the } x\text{-term is now } -10x.$$

Add the equations vertically to eliminate the x-terms. Solve for y.

$$10x + 15y = -65 \qquad \text{Add the equations vertically; the } x\text{-terms are opposites.}$$

$$\underline{+\ -10x - 8y = 30} \qquad \text{Addition property of equality.}$$

$$0 + 7y = -35 \qquad \text{Simplify; the term with } x \text{ is eliminated.}$$

$$7y = -35 \qquad \text{Simplify.}$$

$$\frac{7y}{7} = \frac{-35}{7} \qquad \text{Division property of equality.}$$

$$y = -5 \qquad \text{Simplify; the solution of the system is } (x, -5).$$

Replace y in either one of the equations and solve for x.

$$2x + 3y = -13 \qquad \text{To find } x, \text{ choose either of the original equations in the system.}$$
$$2x + 3(-5) = -13 \qquad \text{Replace } y \text{ with } -5.$$
$$2x - 15 = -13 \qquad \text{Simplify.}$$
$$\underline{ +15 \quad +15} \qquad \text{Subtraction property of equality.}$$
$$2x + 0 = 2 \qquad \text{Simplify.}$$
$$\frac{2x}{2} = \frac{2}{2} \qquad \text{Division property of equality.}$$
$$x = 1 \qquad \text{Simplify; the solution of the system is } (1, -5).$$

The solution of the system of equations is $(1, -5)$.

It does not matter whether we choose x or we choose y to eliminate. The solution of the system is the same.

EXAMPLE 9   Solve $\begin{array}{l} 3x + 4y = 11 \\ -5x + 6y = -12 \end{array}$ by elimination.

SOLUTION ▶ Choose either x or y to eliminate. In this example, we will eliminate y.

Create y-terms that are opposites.
Multiply both sides of $3x + 4y = 11$ by 6.

$$6(3x + 4y) = 6(11) \qquad \text{Multiplication property of equality.}$$
$$18x + 24y = 66 \qquad \text{Simplify; the } y\text{-term is now } 24y.$$

Multiply both sides of $-5x + 6y = -12$ by -4.

$$-4(-5x + 6y) = -4(-12) \qquad \text{Multiplication property of equality.}$$
$$20x - 24y = 48 \qquad \text{Simplify; the } y\text{-term is now } -24y.$$

Add the equations vertically to eliminate the y-terms. Solve for x.

$$18x + 24y = 66 \qquad \text{Add the equations vertically; the } y\text{-terms are opposites.}$$
$$\underline{+\;20x - 24y = 48} \qquad \text{Addition property of equality.}$$
$$38x + 0 = 114 \qquad \text{Simplify. The term with } y \text{ is eliminated.}$$
$$38x = 114 \qquad \text{Simplify.}$$
$$\frac{38x}{38} = \frac{114}{38} \qquad \text{Division property of equality.}$$
$$x = 3 \qquad \text{Simplify; the solution of the system is } (3, y).$$

Replace x in either one of the equations and solve for y.

$$3x + 4y = 11 \qquad \text{To find } y, \text{ choose either of the original equations in the system.}$$
$$3(3) + 4y = 11 \qquad \text{Replace } x \text{ with } 3.$$
$$9 + 4y = 11 \qquad \text{Simplify.}$$
$$\underline{-9 -9} \qquad \text{Subtraction property of equality.}$$
$$0 + 4y = 2 \qquad \text{Simplify.}$$
$$\frac{4y}{4} = \frac{2}{4} \qquad \text{Division property of equality.}$$
$$y = \frac{1}{2} \qquad \text{Simplify; the solution of the system is } \left(3, \frac{1}{2}\right).$$

The solution is $\left(3, \dfrac{1}{2}\right)$.

Solving a System of Two Linear Equations by Elimination

1. Clear fractions and write each equation in standard form, $ax + by = c$. Choose the variable to eliminate.

2. Multiply both sides of either one or both of the equations by a constant. Choose this constant(s) so that addition of the equations eliminates the chosen variable.

3. Add the equations. At least one of the variables is eliminated.

4. Solve this new equation for the remaining variable.

5. Replace the variable in one of the original equations with the value found in step 4. Solve for the other variable.

6. To check, replace the variables in each equation with the solution.

In the next example, we use the elimination method to solve a system of equations with infinitely many solutions. When we add the equations together, all of the terms with variables are eliminated. The sum of the equations is a true equation without variables. To create like terms that are opposites, we need to multiply only one of the equations by a constant.

EXAMPLE 10 | Solve $\begin{aligned} -a - 3b &= -4 \\ 3a + 9b &= 12 \end{aligned}$ by elimination.

SOLUTION ▶ Choose either a or b to eliminate. In this example, we will eliminate a.

Create a-terms that are opposites.
Multiply $-a - 3b = -4$ by 3.

$$3(-a - 3b) = 3(-4) \qquad \text{Multiplication property of equality.}$$
$$-3a - 9b = -12 \qquad \text{Simplify; } -3a \text{ is the opposite of } 3a.$$

We do not need to multiply the other equation by a constant.

$$3a + 9b = 12 \qquad 3a \text{ is the opposite of } -3a.$$

Add the equations vertically to eliminate the a-terms.

$$\begin{array}{rl} -3a - 9b = -12 & \text{Add the equations vertically; the } b\text{-terms are opposites.} \\ +\quad 3a + 9b = 12 & \text{Addition property of equality.} \\ \hline 0 + 0 = 0 & \text{Simplify.} \\ \mathbf{0 = 0} & \text{The variables disappear; the equation is true.} \end{array}$$

Both variables are eliminated; the equation is true. The solution of this system of equations is a set with infinitely many ordered pairs. Use one of the equations and set-builder notation to describe the solution: $\{(a, b) \mid -a - 3b = -4\}$ or $\{(a, b) \mid 3a + 9b = 12\}$.

In the next example, the equations include coefficients that are fractions. Because integers are often easier to add than fractions, clear the fractions from both equations. One of the equations is not in standard form, $ax + by = c$. So that the like terms in the equation are in the same position, rewrite this equation in standard form.

EXAMPLE 11 (a) Solve $\begin{aligned} \frac{2}{3}x + \frac{4}{5}y &= \frac{25}{6} \\ 4x &= 3 - \frac{2}{5}y \end{aligned}$ by elimination.

SOLUTION ▶ Choose either x or y to eliminate. In this example, we will eliminate x.

Clear the fractions from $\frac{2}{3}x + \frac{4}{5}y = \frac{25}{6}$.

$$30\left(\frac{2}{3}x + \frac{4}{5}y\right) = 30\left(\frac{25}{6}\right) \qquad \text{Multiplication property of equality.}$$

$$30 \cdot \frac{2}{3}x + 30 \cdot \frac{4}{5}y = 125 \qquad \text{Distributive property; simplify.}$$

$$20x + 24y = 125 \qquad \text{Simplify; the coefficients are integers.}$$

Clear the fractions from $4x = 3 - \frac{2}{5}y$. Rewrite in standard form.

$$5(4x) = 5\left(3 - \frac{2}{5}y\right) \qquad \text{Multiplication property of equality.}$$

$$20x = 5 \cdot 3 + 5\left(-\frac{2}{5}y\right) \qquad \text{Simplify; distributive property.}$$

$$20x = 15 - 2y \qquad \text{Simplify.}$$

$$\underline{+2y \qquad\qquad +2y} \qquad \text{Addition property of equality.}$$

$$20x + 2y = 15 \qquad \text{The equation is now in standard form.}$$

Create x-terms that are opposites.
Multiply $20x + 24y = 125$ by -1.

$$(-1)(20x + 24y) = (-1)(125) \qquad \text{Multiplication property of equality.}$$

$$-20x - 24y = -125 \qquad \text{Simplify; the x-term is now $-20x$.}$$

We do not need to multiply the other equation by a constant.

$$20x + 2y = 15 \qquad \text{$20x$ is the opposite of $-20x$.}$$

Add the equations vertically to eliminate the x-terms. Solve for y.

$$\begin{array}{ll} -20x - 24y = -125 & \text{Add equations vertically; the x-terms are opposites.} \\ \underline{+20x + 2y = 15} & \text{Addition property of equality.} \\ 0x - 22y = -110 & \text{Simplify.} \\ -22y = -110 & \text{Simplify.} \end{array}$$

$$\frac{-22y}{-22} = \frac{-110}{-22} \qquad \text{Division property of equality.}$$

$$y = 5 \qquad \text{Simplify; the solution of the system is $(x, 5)$.}$$

Replace y in either one of the original equations and solve for x.

$$4x = 3 - \frac{2}{5}y \qquad \text{To find x, choose either of the original equations.}$$

$$4x = 3 - \frac{2}{5}(5) \qquad \text{Replace y with 5.}$$

$$4x = 3 - 2 \qquad \text{Simplify.}$$

$$4x = 1 \qquad \text{Simplify.}$$

$$\frac{4x}{4} = \frac{1}{4}$$ Division property of equality.

$$x = \frac{1}{4}$$ Simplify; the solution of the system is $\left(\frac{1}{4}, 5\right)$.

The solution is $\left(\frac{1}{4}, 5\right)$.

(b) Check.

$\frac{2}{3}x + \frac{4}{5}y = \frac{25}{6}$	Check in both equations.	$4x = 3 - \frac{2}{5}y$
$\frac{2}{3}\left(\frac{1}{4}\right) + \frac{4}{5}(5) = \frac{25}{6}$	Replace variables.	$4\left(\frac{1}{4}\right) = 3 - \frac{2}{5}(5)$
$\frac{1}{6} + 4 = \frac{25}{6}$	Evaluate.	$1 = 3 - 2$
$\frac{1}{6} + \frac{24}{6} = \frac{25}{6}$	Common denominator	$1 = 1$ True.
$\frac{25}{6} = \frac{25}{6}$ True.	The solution is correct.	

Practice Problems

For problems 5–8,
(a) solve the system of equations by elimination.
(b) check.

5. $7x + 2y = 62$
  $3x - 5y = 9$

6. $12a + 3b = 29$
  $-6a + 5b = 44$

7. $3x - 4y = -2$
  $2y = \frac{3}{2}x - \frac{5}{3}$

8. $2x - y = -1$
  $3y - 6x = 3$

4.2 VOCABULARY PRACTICE

Match the term with its description.

1. $ax + by = c$
2. $y = mx + b$
3. An algebraic method for solving a system of equations in which the equations are added together
4. An algebraic method for solving a system of equations in which at least one equation is solved for a variable
5. In the term $3x$, x is the variable and 3 is an example of this.
6. In the expression $4x + 7$, 7 is an example of this.
7. The sum of a number and its opposite
8. $\{\ldots, -3, -2, -1, 0, 1, 2, \ldots\}$
9. The solution of a system of equations represented by two coinciding lines
10. The solution of a system of equations represented by two parallel lines

A. coefficient
B. constant
C. elimination method
D. an infinite set of ordered pairs
E. integers
F. no solution
G. slope-intercept form of a linear equation
H. standard form of a linear equation
I. substitution method
J. zero

4.2 Exercises

Follow your instructor's guidelines for showing your work.

For exercises 1–24,
(a) solve by substitution.
(b) if the solution is an ordered pair, check.

1. $y = 5x + 9$
   $2x - 3y = -14$

2. $y = 3x - 8$
   $3x - 2y = 13$

3. $y = 2x + 5$
   $8x - 3y = 5$

4. $y = 2x + 5$
   $6x - 5y = -85$

5. $y = \frac{4}{5}x + 8$
   $-8x + 10y = 4$

6. $y = \frac{2}{7}x + 3$
   $-2x + 7y = 49$

7. $y = \frac{2}{9}x - 8$
   $y = \frac{1}{3}x - 9$

8. $y = \frac{3}{10}x - 7$
   $y = \frac{2}{5}x - 8$

9. $y = \frac{6}{7}x + 11$
   $6x - 7y = -77$

10. $y = \frac{8}{9}x + 4$
    $8x - 9y = -36$

11. $5x + 2y = 2$
    $x = y + \frac{11}{10}$

12. $7x + 3y = 7$
    $x = y - \frac{19}{21}$

13. $y = -x + 8$
    $x = \frac{1}{8}y - 1$

14. $y = -x + 11$
    $x = \frac{1}{11}y - 1$

15. $y = -x + 8$
    $y = \frac{1}{8}x - 1$

16. $y = -x + 11$
    $y = \frac{1}{11}x - 1$

17. $-2x + 9y = 3$
    $y = \frac{2}{9}x - \frac{1}{3}$

18. $-5x + 12y = 3$
    $y = \frac{5}{12}x - \frac{1}{4}$

19. $d = -\frac{2}{13}a + \frac{1}{13}$
    $5a + 40d = 10$

20. $h = -\frac{3}{4}k + \frac{1}{4}$
    $6h + 11k = 47$

21. $c = 500n + 7500$
    $c = 2000n$

22. $c = 250n + 2500$
    $c = 500n$

23. $c = 2.50m + 5$
    $c = 0.75m + 36.5$

24. $p = 0.35n + 8$
    $p = 0.15n + 12$

For exercises 25–36,
(a) solve one of the equations for a variable.
(b) solve by substitution.
(c) if the solution is an ordered pair, check.

25. $2x + 3y = 29$
    $x - y = -23$

26. $3x + 2y = 27$
    $x - y = -26$

27. $\frac{1}{2}x + 3y = -22$
    $5x - 6y = 32$

28. $\frac{1}{3}x + 5y = -43$
    $7x + 8y = -30$

29. $3a + 4b = 6$
    $-8a + 7b = -16$

30. $5d + 6f = 15$
    $-7d + 8f = -21$

31. $5x + 6y = -12$
    $-10x - 12y = 24$

32. $3x + 7y = -14$
    $-6x - 14y = 28$

33. $4x - 7y = -13$
$-12x + 21y = 15$

34. $9x - 12y = -5$
$-18x + 24y = 11$

35. $3x - 5y = -17$
$-9x + 15y = 11$

36. $6x - 13y = -2$
$-12x + 26y = 5$

For exercises 37–58,
(a) solve by elimination.
(b) if the solution is an ordered pair, check.

37. $3x + 2y = 28$
$4x - y = 19$

38. $4x + 5y = 52$
$7x - 5y = -19$

39. $7x + 3y = -2$
$-7x - 2y = 6$

40. $4x + 11y = -6$
$-4x - 9y = 10$

41. $2x + 9y = -69$
$-11x + 4y = 5$

42. $6x + 13y = -90$
$-8x + 3y = -2$

43. $8x - y = 9$
$24x - 3y = 15$

44. $5x - y = 14$
$20x - 4y = -21$

45. $3x + 5y = 4$
$7x - 3y = 2$

46. $7x + 5y = 3$
$15x - 7y = 2$

47. $27x + 5y = 8$
$53x + 19y = 18$

48. $9x + 7y = 8$
$13x + 15y = 14$

49. $-6x + 5y = -17$
$12x - 10y = 34$

50. $-11x + 2y = 8$
$33x - 6y = -24$

51. $7x + 3y = 18$
$9x - 2y = -12$

52. $17x + 4y = 48$
$8x + 3y = 36$

53. $14x + 25y = 0$
$-21x + 10y = 0$

54. $-9x + 13y = 0$
$19x - 22y = 0$

55. $2a + 5c = -15$
$4a + 15c = -31$

56. $3c + 6d = -25$
$5c + 12d = -40$

57. $p - w = -2$
$p + w = -14$

58. $h - k = -5$
$h + k = -9$

For exercises 59–66,
(a) rewrite the equations in standard form, if necessary. Clear any fractions.
(b) solve by elimination.

59. $y = 3x + 1$
$5x + 8y = 66$

60. $y = 4x + 1$
$2x + 7y = 97$

61. $\dfrac{3}{4}x + \dfrac{2}{5}y = \dfrac{7}{8}$
$\dfrac{5}{8}x + \dfrac{1}{2}y = \dfrac{11}{16}$

62. $\dfrac{1}{2}x + \dfrac{5}{8}y = \dfrac{17}{16}$
$\dfrac{3}{4}x + \dfrac{1}{2}y = \dfrac{15}{16}$

63. $y = -\dfrac{3}{4}x + 10$
$y = \dfrac{1}{2}x + 15$

64. $y = -\dfrac{5}{6}x + 15$
$y = \dfrac{1}{2}x + 7$

65. $x + y = 180$
$0.08x + 0.14y = 21.6$

66. $x + y = 420$
$0.06x + 0.15y = 32.4$

For exercises 67–76,
(a) solve the system of equations by graphing.
(b) solve the system of equations using either the substitution method or the elimination method.

67. $3x + 2y = 18$
$-x - y = -7$

68. $2x + 3y = 24$
$-x - y = -10$

69. $y = 3x + 2$
$5x + 2y = 15$

70. $y = 2x + 6$
$4x + 5y = 44$

71. $y = -4x + 3$
$y = 5x + 3$

72. $y = -6x + 2$

$y = 7x + 2$

73. $y = \dfrac{3}{4}x + 2$

$-6x + 8y = 16$

74. $y = \dfrac{2}{3}x + 4$

$-6x + 9y = 36$

75. $y = \dfrac{5}{6}x - 9$

$y = \dfrac{5}{6}x + 3$

76. $y = \dfrac{3}{5}x - 8$

$y = \dfrac{3}{5}x + 2$

77. Explain how to decide whether to use substitution or elimination to solve a system of linear equations.

78. Explain why you think we often clear fractions from the equations before solving a system of equations by elimination.

79. Write a system of two linear equations in standard form in which the solution of the system is $(3, 8)$.

80. Write a system of two linear equations in standard form in which the solution of the system is $(4, 5)$.

Problem Solving: Practice and Review

Follow your instructor's guidelines for using the five steps as outlined in Section 1.5, p. 51.

81. The rule of 72 is

$$\text{years to double} = \frac{72}{\text{interest rate in percent form}}$$

This rule finds a rough estimate of the time it will take an investment to double in value. An investor wants the doubling time of his investment to be about 5 years. What interest rate does his investment need to earn?

82. A San Francisco resident rides the cable car five days a week from her apartment to her job and home again. A one-way ticket is $6. A Fast Pass® for unlimited rides for 1 month costs $74. For four weeks of travel to work, find the difference in cost between the Fast Pass and one-way tickets.

83. Find the total energy used in 2009 by all sectors of the economy in Oregon.

Energy savings for 2009 from recycling—approximately 27 trillion BTU—the equivalent of 216,000,000 gallons of gasoline, or roughly 2.4 percent of total energy used [2009] by all sectors of the economy in Oregon. (*Source:* www.deq.state.or.us, Sept. 2010)

84. Find the amount of methamphetamine in grams in an individual dose (1 metric ton = 1000 kg; 1 kg = 1000 g). Round to the nearest hundredth.

Sailors found 49,640 liters (13,000 gallons) of ephedrine, a chemical used to make methamphetamine . . . enough to produce 40.2 metric tons . . . about 309 million individual doses. (*Source:* www.miamiherald.com, June 17, 2009)

Technology

For exercises 85–88,

(a) use either substitution or elimination to solve each system of equations.

(b) check the solution by graphing the system on a graphing calculator. Sketch the graph; describe the window.

(c) identify any point(s) of intersection.

85. $y = -3x + 6$

$3x + 7y = -30$

86. $y = -2x + 7$

$4x + 5y = 5$

87. $6x - 11y = 43$

$5x - 4y = 41$

88. $3x - 7y = 8$

$5x - 8y = 17$

Find the Mistake

For exercises 89–92, the completed problem has one mistake.

(a) Describe the mistake in words, or copy down the whole problem and highlight or circle the mistake.

(b) Do the problem correctly.

89. Problem: Solve $\begin{array}{l} 2x + y = 11 \\ 3x + y = 14 \end{array}$ by elimination.

Incorrect Answer: $2x + y = 11$

$$\dfrac{+ \ 3x + y = 14}{5x + 0 = 25}$$

$$5x = 25$$

$$x = 5$$

$$2x + y = 11$$

$$2(5) + y = 11$$

$$y = 1$$

The solution is $(5, 1)$.

90. Problem: Solve $\begin{array}{l} y = -x + 4 \\ 3x - 8y = -87 \end{array}$ by substitution.

Incorrect Answer: $3x - 8(-x + 4) = -87$

$$3x - 8x - 32 = -87$$

$$-5x - 32 = -87$$

$$-5x = -55$$

$$x = 11$$

$$y = -x + 4$$

$$y = -11 + 4$$

$$y = -7$$

The solution of the system is $(11, -7)$.

91. Problem: Solve $\begin{array}{l}6x + 5y = 16\\3x + 2y = 7\end{array}$ by elimination.

Incorrect Answer:
$$6x + 5y = 16$$
$$+\ -6x - 4y = 7$$
$$\overline{0 + y = 23}$$
$$y = 23$$

$$3x + 2y = 7$$
$$3x + 2(23) = 7$$
$$3x + 46 = 7$$
$$3x = -39$$
$$x = -13$$

The solution is $(-13, 23)$.

92. Problem: Solve $\begin{array}{l}x - y = 1\\3x - 2y = 7\end{array}$ by elimination.

Incorrect Answer:
$$-3x - 3y = 3$$
$$+\ \ \ 3x - 2y = 7$$
$$\overline{0 - 5y = 10}$$
$$-5y = 10$$
$$\frac{-5y}{-5} = \frac{10}{-5}$$
$$y = -2$$

$$x - y = 1$$
$$x - (-2) = 1$$
$$x + 2 = 1$$
$$x = -1$$

The solution is $(-1, -2)$.

Review

93. Rewrite 5.8% in decimal notation.

94. Rewrite 0.5% in decimal notation.

95. Write an equation that shows the relationship of distance, rate, and time using the variables d, r, and t.

96. Three liters of pure orange juice and 8 L of water are mixed together to make 11 L of a juice drink. What percent of the juice drink is pure orange juice? Round to the nearest tenth of a percent.

SUCCESS IN COLLEGE MATHEMATICS

97. When solving a system of equations by substitution, a student incorrectly simplified $3x + 2(7x - 6) = 5$ as $3x + 14x - 6 = 5$. Describe the mistake that the student made.

98. What property is used to simplify this equation correctly?

In an episode of the television show *House M.D.*, two doctors are traveling by plane from Singapore to New York. The flight path is over the polar ice cap. During the flight, one of the passengers becomes ill. He has a strange rash. The doctors suspect meningococcus, a highly contagious disease. They must advise the pilot whether to turn back to Singapore before the plane reaches the *point of no return*. In this section, we use a system of equations to find a point of no return as well as to solve other application problems.

© Shutter Lover/Shutterstock

SECTION 4.3 Applications

After reading the text, working the practice problems, and completing assigned exercises, you should be able to:

Use a system of two linear equations to solve an application problem.

Applications with Two Unknowns

The relationships in some application problems can be described with two variables and two equations. We may be able to model and solve these problems with a system of equations.

At the break-even point, the costs to make a product are equal to the revenue from selling the product. To solve a break-even problem, write one equation that is a model of the costs and another equation that is a model of the revenue. The solution of this system of equations is the break-even point. As with many applications in this section, it is often easier to first write a word equation that describes the relationships in the problem and then replace the variables.

A System of Equations for Finding the Break-Even Point

$$\text{costs} = \left(\frac{\text{cost}}{1 \text{ product}}\right)(\text{number of products}) + \text{fixed overhead costs}$$

$$\text{revenue} = \left(\frac{\text{revenue}}{1 \text{ product}}\right)(\text{number of products})$$

EXAMPLE 1 The cost to make a product is $6500. The fixed overhead costs to make the product are $79,000 per month. The company sells each product for $8000. Find the number of products to make and sell each month to break even. Find the revenue at the break-even point. Round up to the nearest whole number.

SOLUTION ▶ **Step 1 Understand the problem.**
The unknowns are the number of products at the break-even point and the costs/revenue at the break-even point.

x = number of products at the break-even point

y = costs or revenue at the break-even point

Step 2 Make a plan.
Write and solve a system of equations in which one equation represents the costs to make the products and the other equation represents the revenue from selling the products. A linear model in slope intercept form, $y = mx + b$, of the costs is

$$\text{costs} = \left(\frac{\text{cost}}{1 \text{ product}}\right)(\text{number of products}) + \text{ fixed overhead costs}$$

or $\qquad y = \left(\dfrac{\$6500}{1 \text{ product}}\right)x + \$79,000$

A linear model of the revenue is

$$\text{revenue} = \left(\frac{\text{revenue}}{1 \text{ product}}\right)(\text{number of products}) \qquad \text{or} \qquad y = \left(\frac{\$8000}{1 \text{ product}}\right)x$$

At the break-even point, the costs and revenue are equal, $y = y$. So the solution of this system of equations is the break-even point.

Step 3 Carry out the plan.
Remove the units: $y = 6500x + 79{,}000$ and $y = 8000x$. Since both equations are solved for a variable, solve by substitution.

$y = 6500x + 79{,}000$	Choose one of the equations.
$8000x = 6500x + 79{,}000$	Replace y with the substitution expression, $8000x$.
$\underline{-6500x \quad -6500x}$	Subtraction property of equality.
$1500x = 0 + 79{,}000$	Simplify.
$\dfrac{1500x}{1500} = \dfrac{79{,}000}{1500}$	Division property of equality.
$x = 52.66\ldots$	Simplify; the solution of the system is $(52.66\ldots, y)$.
$x \approx \mathbf{53}$ products	Round up to the nearest whole number.

To find the revenue at the break-even point, replace x in the original revenue equation with 53 products.

$$y = \left(\frac{\$8000}{1 \text{ product}}\right)x \qquad \text{To find } y, \text{ choose one of the equations.}$$

$$y = \left(\frac{\$8000}{1 \text{ product}}\right)(53 \text{ products}) \qquad \text{Replace } x \text{ with 53 products.}$$

$$y = \mathbf{\$424{,}000} \qquad \text{Simplify.}$$

The solution of the system is the break-even point, about (53 products, $424,000).

Step 4 Look back.
The cost to make 53 products, $423,500, and the revenue from selling 53 products, $424,000, are nearly equal. The difference is due to rounding of the number of products. The ordered pair (53 products, $424,000) is a reasonable break-even point.

In this problem, we see that rounding the number of products up may result in a small profit: The revenue is greater than the costs. Rounding the number of products down may result in a small loss: The revenue is less than the costs.

Step 5 Report the solution.
To break even, the company should make and sell 53 products. The costs at the break-even point are $423,500, and the revenue is $424,000.

In the next example, we know that "x is 4 times greater than y." So we can write an *inequality* with x and y: $x > y$. To write an *equation* with x and y, we need to multiply one of the variables by 4. Since x is greater than y, multiply y by 4 to make an expression that is equal to x: $x = 4y$.

EXAMPLE 2 A psychology instructor is writing a test with 150 points with two kinds of questions: short answer and multiple choice. She wants the total points from short answer questions to be four times more than the total points from multiple choice questions. How many total points should she assign to each kind of question?

SOLUTION ▶ **Step 1 Understand the problem.**
The unknowns are the total number of points for short answer questions and the total number of points for multiple choice questions.

x = total number of points for short answer questions

y = total number of points for multiple choice questions

Step 2 Make a plan.
Write and solve a system of equations. One equation represents the relationship of the points from each kind of question and the total points. The other equation represents the points from short answer questions compared to the points from multiple choice questions. Word equations representing this system are

short answer points + multiple choice points = total points

short answer points = 4(multiple choice points)

Step 3 Carry out the plan.

$x + y = 150$ points	short answer points + multiple choice points = total points
$x = 4y$	short answer points = 4(multiple choice points)

Since one of the equations is solved for x, solve by substitution.

$x + y = 150$	Remove the units.
$4y + y = 150$	Replace x with the substitution expression, $4y$.
$5y = 150$	Combine like terms.
$\dfrac{5y}{5} = \dfrac{150}{5}$	Division property of equality.
$y = 30$	Simplify; the solution is $(x, 30 \text{ points})$.

$$x + y = 150 \qquad \text{To find } x, \text{ choose one of the original equations.}$$
$$x + 30 = 150 \qquad \text{Replace } y \text{ with 30.}$$
$$\underline{-30 \quad -30} \qquad \text{Subtraction property of equality.}$$
$$x + 0 = 120 \qquad \text{Simplify.}$$
$$x = 120 \qquad \text{Simplify.}$$

The solution of the system is (120 points, 30 points).

Step 4 Look back.

Since 120 points + 30 points equals 150 total points and 120 points equals 4(30 points), the answer seems reasonable.

Step 5 Report the solution.

The instructor should assign 120 points to the short answer questions and 30 points to the multiple choice questions.

When there is *one* unknown amount in a problem, we often solve the problem by writing and solving *one* equation in *one* variable. In the next example, there are *two* unknown amounts. We solve the problem by writing and solving a system of *two* linear equations in *two* variables. Since the equations are in standard form, we use the elimination method to solve the system.

EXAMPLE 3 | The cost of four Blu-ray discs and 1 DVD is $113.45. The cost of three Blu-ray discs and 8 DVDs is $182.89. The cost of a case of 12 packages of microwave popcorn is $36. Find the cost of a Blu-ray disc. Find the cost of a DVD.

SOLUTION ▶ **Step 1 Understand the problem.**

The unknowns are the cost of a Blu-ray disc and the cost of a DVD. The cost of the microwave popcorn is extraneous information.

$$x = \text{cost per Blu-ray disc} \qquad y = \text{cost per DVD}$$

Step 2 Make a plan.

Write and solve a system of equations in which each equation represents the relationship of the number of Blu-ray discs, the number of DVDs, and the total cost. Both word equations are the same,

$$(\text{number of Blu-ray discs})\left(\frac{\text{cost}}{\text{1 Blu-ray disc}}\right) + (\text{number of DVDs})\left(\frac{\text{cost}}{\text{1 DVD}}\right)$$
$$= \text{total cost}$$

Step 3 Carry out the plan.

$$(4 \text{ Blu-ray discs})x + (1 \text{ DVD})y = \$113.45$$
$$(3 \text{ Blu-ray discs})x + (8 \text{ DVDs})y = \$182.89$$

Removing the units, the system is $\begin{aligned}4x + 1y &= 113.45 \\ 3x + 8y &= 182.89\end{aligned}$. Since both equations are in standard form, solve by elimination and eliminate the *y*-terms. (It does not matter which variable we choose to eliminate.)

Create *y*-terms that are opposites.

$$(-8)(4x + 1y) = (-8)(113.45) \qquad \text{Multiplication property of equality.}$$
$$-32x - 8y = -907.60 \qquad \text{Distributive property; simplify.}$$

The *y*-term in the other equation is already the opposite of $-8y$.

$$3x + 8y = 182.89 \qquad 8y \text{ is the opposite of } -8y.$$

Add the equations vertically to eliminate the y-terms. Solve for x.

$$-32x - 8y = -907.60 \quad \text{Add the equations vertically; the } y\text{-terms are opposites.}$$
$$+ \quad 3x + 8y = 182.89 \quad \text{Addition property of equality.}$$
$$\overline{\quad -29x + 0 = -724.71} \quad \text{Simplify.}$$
$$\frac{-29x}{-29} = \frac{-724.71}{-29} \quad \text{Simplify; division property of equality.}$$
$$x = \$24.99 \quad \text{Simplify; the solution of the system is (\$24.99, } y\text{).}$$

Replace x in either one of the original equations and solve for y.

$$4(\mathbf{24.99}) + 1y = 113.45 \quad \text{Replace } x \text{ in one of the equations with 24.99.}$$
$$\mathbf{99.96} + y = 113.45 \quad \text{Simplify.}$$
$$-99.96 \qquad -99.96 \quad \text{Subtraction property of equality.}$$
$$\overline{\quad \mathbf{0} + y = \mathbf{13.49}} \quad \text{Simplify.}$$
$$y = \$13.49 \quad \text{Simplify; include units.}$$

The solution of the system is (\$24.99, \$13.49).

Step 4 Look back.
Since $4(\$24.99) + 1(\$13.49)$ equals the total given in the problem, \$113.45, and $3(\$24.99) + 8(\$13.49)$ equals the total given in the problem, \$182.89, the answer seems reasonable. From this problem, we can learn that solving by elimination does not always require multiplying both equations by a constant. To end up with y-terms that were opposites in this system, we multiplied only one equation by a constant.

Step 5 Report the solution.
The cost per Blu-ray disc is \$24.99, and the cost per DVD is \$13.49.

Practice Problems

For problems 1–3, use a system of equations and the five steps.

1. The cost to make a product is \$40. The fixed overhead costs to make the product are \$1200 per month. The company sells each product for \$65. Find the number of products that must be made and sold each month to break even. Find the costs to make the products and the revenue from selling the products at the break-even point.

2. An investor has \$36,000. She wants to put twice as much in short-term investments as in long-term investments. Find the amount she should put in each kind of investment.

3. Two students ate lunch at a taco truck. Two tacos and a quesadilla cost \$6. Five tacos and two quesadillas cost \$13.25. Find the cost of a taco and the cost of a quesadilla.

Mixtures

A mixture includes two or more ingredients. If we know the total volume of the mixture and the percent of an ingredient in the mixture, we can find the volume of that ingredient.

Volume of an Ingredient in a Mixture

volume of an ingredient = (percent ingredient)(total volume of the mixture)

EXAMPLE 4   A drink box contains 200 mL of liquid that is 5% pineapple juice. Find the volume of pineapple juice in this box.

SOLUTION ▶

$$\text{volume of pineapple juice} = (\text{percent pineapple juice})(\text{total volume})$$
$$V = (0.05)(200\ \text{mL})$$
$$V = 10\ \text{mL}$$

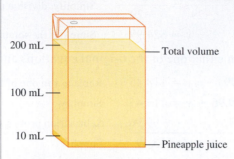

The drink box contains 10 mL of pineapple juice.

In the next example, to find the amount of orange juice in each drink, multiply the percent of juice in the drink by the volume of the drink.

EXAMPLE 5   Drink A is 5% orange juice. Drink B is 2% orange juice. Find the amount of each drink needed to make 15,000 gal of a new drink, Drink C, that is 3% orange juice. Round to the nearest whole number.

SOLUTION ▶ **Step 1 Understand the problem.**

The unknowns are the amount of Drink A and the amount of Drink B to mix together.

$x = $ amount of Drink A $y = $ amount of Drink B

Step 2 Make a plan.

Write and solve a system of equations in which one equation represents the volume of the drinks and the other equation represents the amount of orange juice in each drink. The word equations are

$$\text{amount A } + \text{ amount B } = \text{ amount C}$$
$$(\% \text{ orange})(\text{amount A}) + (\% \text{ orange})(\text{amount B}) = (\% \text{ orange})(\text{amount C})$$

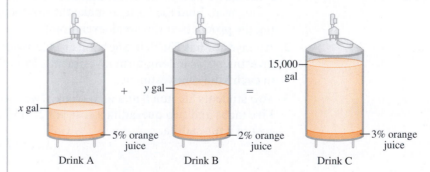

Step 3 Carry out the plan.

$$x + y = 15{,}000\ \text{gal}$$
$$0.05x + 0.02y = (0.03)(15{,}000\ \text{gal})$$

If we remove the units and simplify, the system is $\begin{aligned} x + y &= 15{,}000 \\ 0.05x + 0.02y &= 450 \end{aligned}$. Since both

equations are in standard form, solve by elimination and eliminate the *x*-terms. We need to multiply only one equation by a constant.

Create *x*-terms that are opposites.

$$(-0.05)(x + y) = (-0.05)(15{,}000)$$ Multiplication property of equality.

$$-0.05x - 0.05y = -750$$ Distributive property; simplify.

Add the equations vertically to eliminate the *x*-terms. Solve for *y*.

$$-0.05x - 0.05y = -750$$ Add the equations vertically; the *x*-terms are opposites.

$$+ \quad 0.05x + 0.02y = 450$$ Addition property of equality.

$$0 - 0.03y = -300$$ Simplify.

$$\frac{-0.03y}{-0.03} = \frac{-300}{-0.03}$$ Simplify; division property of equality.

$$y = 10{,}000 \text{ gal}$$ Simplify; the solution of the system is $(x, 10{,}000 \text{ gal})$.

Replace *y* in either one of the original equations and solve for *x*.

$$x + 10{,}000 = 15{,}000$$ Replace *y* in either of the equations.

$$\underline{-10{,}000 \quad -10{,}000}$$ Subtraction property of equality.

$$x + 0 = 5000$$ Simplify.

$$x = 5000 \text{ gal}$$ Simplify.

The solution of the system is (5000 gal Drink A, 10,000 gal Drink B).

Step 4 Look back.

Since 5000 gal + 10,000 gal equals 15,000 gal, the total amount is correct. In Drink A, $(0.05)(5000 \text{ gal})$ is 250 gal of orange juice. In Drink B, $(0.02)(10{,}000 \text{ gal})$ is 200 gal of orange juice. In Drink C, $(0.03)(15{,}000 \text{ gal})$ is 450 gal of orange juice, the same as the sum of the orange juice in Drink A and Drink B. The answer seems reasonable.

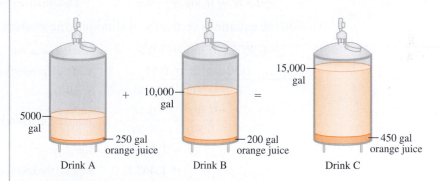

Step 5 Report the solution.

To make 15,000 gal of Drink C, mix together 5000 gal of Drink A and 10,000 gal of Drink B.

In stockrooms and factories, a weaker liquid solution is often made by diluting a stronger solution with a pure substance such as water. The percent of another substance in pure water is 0%.

EXAMPLE 6   A stock solution of hydrochloric acid sold by a chemical company is 30% acid. Find the amount of the stock solution and water to mix together to make 2 L of 8% acid. Round to the nearest thousandth.

SOLUTION ▶ **Step 1 Understand the problem.**

The unknowns are the amount of the stock solution and the amount of water to mix together.

$$x = \text{amount of stock solution} \qquad y = \text{amount of water}$$

Step 2 Make a plan.

Write a system of equations in which one equation represents the volume of the liquids and the other equation represents the amount of acid in each liquid. The word equations are:

amount stock solution + amount water = amount new solution

(% acid)(amount stock) + (% acid)(amount water) = (% acid)(amount new)

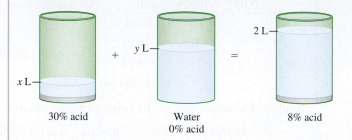

x L   30% acid   + y L   Water 0% acid   =   2 L   8% acid

Step 3 Carry out the plan.

$$x + y = 2 \text{ L}$$
$$0.30x + 0y = (0.08)(2 \text{ L})$$

If we remove the units and simplify, the system is $\begin{aligned} x + y &= 2 \\ 0.30x + 0y &= 0.16 \end{aligned}$. Since both

equations are in standard form, solve by elimination and eliminate the x-terms. We need to multiply only one equation by a constant.

Create x-terms that are opposites.

$(-0.30)(x + y) = (-0.30)(2)$	Multiplication property of equality.
$-0.30x - 0.30y = -0.6$	Distributive property; simplify.

Add the equations vertically to eliminate the x-terms. Solve for y.

$-0.30x - 0.30y = -0.6$	Add the equations vertically; the x-terms are opposites.
$+ \quad\;\; 0.30x + 0y = 0.16$	Addition property of equality.
$0 - 0.30y = -0.44$	Simplify.
$\dfrac{-0.30y}{-0.30} = \dfrac{-0.44}{-0.30}$	Simplify; division property of equality.
$y = 1.4666\ldots$	Simplify.
$y \approx \mathbf{1.467} \text{ L}$	Round; the solution of the system is $(x, 1.467 \text{ L})$.

Replace y in either one of the original equations and solve for x.

$x + \mathbf{1.467} = 2$	Replace y in one of the original equations.
$\underline{-1.467 \;\; -1.467}$	Subtraction property of equality.
$x + \mathbf{0} = \mathbf{0.533}$	Simplify.
$x = 0.533 \text{ L}$	Simplify.

The solution of the system is about $(0.533 \text{ L}, 1.467 \text{ L})$.

Step 4 Look back.

Since $0.533 \text{ L} + 1.467 \text{ L}$ equals 2 L, the total amount is correct. For the stock solution, $(0.30)(0.533 \text{ L})$ is about 0.16 L acid. The water contains no acid. For the new solution, $(0.08)(2 \text{ L})$ equals 0.16 L, the same as the sum of the acid in the stock solution and in the water. Although the water changes the volume of the final mixture, it does not change the amount of acid in the final mixture. The answer seems reasonable.

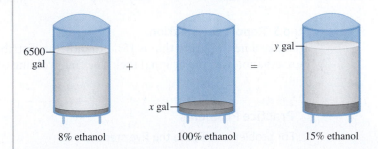

0.533 L ⌐   1.467 L ⌐   2 L ⌐

           + ⌐0.16 L   =
            ⌐   acid   ⌐0.16 L
            acid   acid
   30% acid   Water   8% acid
                        0% acid

Step 5 Report the solution.
To make 2 L of a solution that is 8% acid, mix together 0.533 L of concentrated acid and 1.467 L of water.

In Examples 5 and 6, the final amount of mixture is given. In the next example, this amount is not given. Instead, we know the amount of one of the solutions being mixed together. The pure substance in this problem is ethanol. Pure ethanol is 100% ethanol; in decimal form, 100% = 1.

EXAMPLE 7   A fuel distributor has 6500 gal of fuel that is 8% ethanol. Find the volume of pure ethanol to add to this original fuel to make a new fuel that is 15% ethanol, rounding to the nearest whole number. Find the total volume of the new fuel.

SOLUTION ▸ ### Step 1 Understand the problem.
The unknowns are the amount of pure ethanol and the amount of the new mixture.

$$x = \text{amount of pure ethanol} \qquad y = \text{amount of new mixture}$$

Step 2 Make a plan.
Write a system of equations in which one equation represents the volume of the liquids and the other equation represents the amount of ethanol in each liquid.
The word equations are

$$\text{amount original fuel} + \text{amount pure ethanol} = \text{amount new fuel}$$
$$(\% \text{ ethanol})(\text{amount original}) + (\% \text{ ethanol})(\text{amount pure}) = (\% \text{ ethanol})(\text{amount new})$$

6500 ⌐   y gal ⌐
 gal

       +   =

                    x gal ⌐

   8% ethanol   100% ethanol   15% ethanol

Step 3 Carry out the plan.
$$6500 \text{ gal} + x = y$$
$$(0.08)(6500 \text{ gal}) + 1x = 0.15y$$

If we remove the units, the system is $\begin{array}{c} 6500 + x = y \\ 520 + 1x = 0.15y \end{array}$. Since one of the equations is solved for y, solve by substitution. The substitution expression is $6500 + x$.

$$520 + 1x = 0.15(6500 + x) \qquad \text{Replace } y \text{ with the substitution expression, } 6500 + x.$$
$$520 + 1x = \mathbf{0.15} \cdot 6500 + \mathbf{0.15}x \qquad \text{Distributive property.}$$

$$520 + 1x = \mathbf{975} + 0.15x \qquad \text{Simplify.}$$

$$\underline{-520 \qquad\qquad -520} \qquad\qquad \text{Subtraction property of equality.}$$

$$\mathbf{0} + 1x = \mathbf{455} + 0.15x \qquad \text{Simplify.}$$

$$\underline{-0.15x \qquad\qquad -0.15x} \qquad \text{Subtraction property of equality.}$$

$$\mathbf{0.85}x = 455 + \mathbf{0} \qquad\qquad \text{Simplify.}$$

$$\frac{0.85x}{0.85} = \frac{455}{0.85} \qquad\qquad \text{Division property of equality.}$$

$$x = 535.2\ldots \qquad\qquad \text{Simplify. The solution of the system is } (535.2\ldots, y).$$

$$x \approx \mathbf{535} \text{ gal} \qquad\qquad \text{Round to the nearest whole number.}$$

$$6500 + x = y \qquad \text{To find } y, \text{ use either one of the original equations.}$$

$$6500 + \mathbf{535} = y \qquad \text{Replace } x \text{ with 535.}$$

$$7035 \text{ gal} = y \qquad \text{Simplify.}$$

The solution of the system is about (535 gal, 7035 gal).

Step 4 Look back.
Since 6500 gal + 535 gal equals 7035 gal, the total amount is correct. For the original fuel, $(0.08)(6500 \text{ gal})$ is 520 gal of ethanol, and this was mixed with 535 gal of pure ethanol for a total of 1055 gal of ethanol. For the new fuel, $(0.15)(7035 \text{ gal})$ is also 1055 gal of ethanol. The answer seems reasonable. When doing similar problems, we know that the percent of a substance in a pure substance is 100%; in decimal form, 100% = 1.

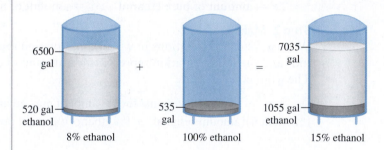

Step 5 Report the solution.
To make a new mixture that is 15% ethanol, the distributor should mix 535 gal of pure ethanol with the original fuel. The final volume of the new fuel will be 7035 gal.

Practice Problems

For problems 4–6, use the five steps.

4. Brass is an alloy of copper and zinc. Brass A is 65% copper. Brass B is 80% copper. Brass C is 68% copper. Find the amount of Brass A and Brass B needed to make 200 kg of Brass C.

5. Vinegar is 5% acetic acid. Find the amount of vinegar and water needed to make 8 gal of a solution that is 2% acetic acid. Round to the nearest tenth.

6. If 16 L of an antifreeze mixture contains 44% antifreeze, find the volume of pure antifreeze to add to this original mixture to make a new mixture that is 50% antifreeze. Find the total volume of the new mixture.

Distance, Rate, and Time

An average rate (speed) such as $\dfrac{60 \text{ mi}}{1 \text{ hr}}$ is distance divided by time, $\dfrac{\text{distance}}{\text{time}}$. The product of a rate and time is distance: $d = rt$. This equation can also be solved for t.

$$d = rt \qquad \text{distance} = (\text{rate})(\text{time})$$

$$\frac{d}{r} = \frac{rt}{r} \qquad \text{Division property of equality.}$$

$$\frac{d}{r} = t \qquad \text{Simplify.}$$

$$t = \frac{d}{r} \qquad \text{Rewrite with } t \text{ on the left.}$$

In the next example, both cars travel the same amount of time. To find the distance each car traveled, use $t = \dfrac{d}{r}$ to write two equations. Then use substitution to solve the system.

EXAMPLE 8 | Two cars enter a freeway traveling in opposite directions. The speed of the fast car is $\dfrac{65 \text{ mi}}{1 \text{ hr}}$. The speed of the slow car is $\dfrac{58 \text{ mi}}{1 \text{ hr}}$. Both cars have a full tank of gas. Find the distance each car is from the freeway entrance when the distance between the cars is 45 mi. Round to the nearest whole number.

SOLUTION ▶ **Step 1 Understand the problem.**

The unknowns are the distance each car is from the freeway entrance. The full tank of gas is extraneous information.

$$x = \text{distance of fast car} \qquad y = \text{distance of slow car}$$

Step 2 Make a plan.

Since the sum of the distances is the total distance, a word equation is

$$d_{\text{fast car}} + d_{\text{slow car}} = d_{\text{total}}$$

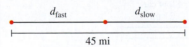

Both cars travel the same time, t. So $t_{\text{fast car}} = t_{\text{slow car}}$. Since $t = \dfrac{d}{r}$, the other word equation is $\dfrac{d_{\text{fast car}}}{r_{\text{fast car}}} = \dfrac{d_{\text{slow car}}}{r_{\text{slow car}}}$.

Step 3 Carry out the plan.

$$x + y = 45 \text{ mi}$$
$$\frac{x}{\dfrac{65 \text{ mi}}{1 \text{ hr}}} = \frac{y}{\dfrac{58 \text{ mi}}{1 \text{ hr}}}$$

Removing the units, the system is $\begin{aligned} x + y &= 45 \\ \dfrac{x}{65} &= \dfrac{y}{58} \end{aligned}$. If we solve $\dfrac{x}{65} = \dfrac{y}{58}$ for a variable, we can use substitution.

$$\frac{x}{65} = \frac{y}{58} \qquad \text{Solve for } x.$$

$$65 \cdot \frac{x}{65} = \frac{y}{58} \cdot 65 \qquad \text{Multiplication property of equality.}$$

$$x = \frac{65}{58}y \qquad \text{Simplify; the substitution expression is } \frac{65}{58}y.$$

$$x + y = 45 \qquad \text{The other equation in the system.}$$

$$\frac{65}{58}y + y = 45 \qquad \text{Replace } x \text{ with the substitution expression, } \frac{65}{58}y.$$

$$58\left(\frac{65}{58}y + y\right) = 58(45) \qquad \text{Clear fractions; multiplication property of equality.}$$

$$\mathbf{58} \cdot \frac{65}{58}y + \mathbf{58}y = \mathbf{2610} \qquad \text{Distributive property; simplify.}$$

$$65y + 58y = 2610 \qquad \text{Simplify.}$$

$$\mathbf{123}y = \mathbf{2610} \qquad \text{Combine like terms.}$$

$$\frac{123y}{123} = \frac{2610}{123} \qquad \text{Division property of equality.}$$

$$y = 21.21\ldots \qquad \text{Simplify; the solution of the system is } (x, 21.21\ldots).$$

$$y \approx \mathbf{21} \text{ mi} \qquad \text{Round to the nearest whole number.}$$

$$x + y = 45 \qquad \text{To find } x, \text{ use either one of the equations.}$$

$$x + \mathbf{21} = 45 \qquad \text{Replace } y \text{ with 21.}$$

$$\underline{-21 \quad -21} \qquad \text{Subtraction property of equality.}$$

$$x + \mathbf{0} = \mathbf{24} \qquad \text{Simplify.}$$

$$x = 24 \text{ mi} \qquad \text{Simplify.}$$

The solution of the system is about (24 mi, 21 mi).

Step 4 Look back.
Since 24 mi + 21 mi equals 45 mi, the total distance is correct. Since $\frac{24}{65} \approx 0.37$ and $\frac{21}{58} \approx 0.36$, both cars travel about the same amount of time. The difference is due to rounding. The answer seems reasonable.

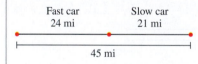

Step 5 Report the solution.
When the distance between the cars is 45 mi, the fast car will have traveled about 24 mi and the slow car will have traveled about 21 mi.

The **point of no return** is the location at which the flight time to the destination is the same as the flight time to return to the original airport. (This does not mean that the plane must always continue on to its destination in an emergency. It may be able to land at another airport.) In the next example, a plane is traveling from New York to London. To find the point of no return, write a system of two equations:

$$\text{time to NY} = \frac{\text{distance from NY}}{\text{rate}} \quad \text{and} \quad \text{time to London} = \frac{\text{distance from London}}{\text{rate}}$$

EXAMPLE 9   A jet flies 3458 mi from New York to London. The average speed of the jet in still air is $\frac{520 \text{ mi}}{1 \text{ hr}}$. The average tail wind is $\frac{65 \text{ mi}}{1 \text{ hr}}$. Find the distance from the point of no return to New York and the distance from the point of no return to London. Round to the nearest whole number.

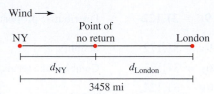

SOLUTION ▶ **Step 1 Understand the problem.**

The unknowns are the distance at the point of no return that the jet is from New York and the distance the jet is from London.

$$x = \text{distance from NY} \qquad y = \text{distance from London}$$

Step 2 Make a plan.

Since the sum of the distances is the total distance, a word equation is

$$d_{\text{NY}} + d_{\text{London}} = d_{\text{total}}$$

Wind →

| | Point of | |
| NY | no return | London |

$$d_{\text{NY}} \qquad d_{\text{London}}$$

3458 mi

The plane is the same time from both destinations, t. So $t_{\text{NY}} = t_{\text{London}}$. Since $t = \dfrac{d}{r}$, another word equation is

$$\frac{d_{\text{NY}}}{r_{\text{NY}}} = \frac{d_{\text{London}}}{r_{\text{London}}}$$

The speed of the plane if it returns to New York is its speed in still air minus the speed of the tail wind:

$$\frac{520 \text{ mi}}{1 \text{ hr}} - \frac{65 \text{ mi}}{1 \text{ hr}} = \frac{455 \text{ mi}}{1 \text{ hr}}$$

The speed of the plane if it continues on to London is its speed in still air plus the speed of the tail wind:

$$\frac{520 \text{ mi}}{1 \text{ hr}} + \frac{65 \text{ mi}}{1 \text{ hr}} = \frac{585 \text{ mi}}{1 \text{ hr}}$$

Step 3 Carry out the plan.

$$x + y = 3458 \text{ mi}$$

$$\frac{x}{\dfrac{455 \text{ mi}}{1 \text{ hr}}} = \frac{y}{\dfrac{585 \text{ mi}}{1 \text{ hr}}}$$

Removing the units, the system is $\begin{array}{c} x + y = 3458 \\ \dfrac{x}{455} = \dfrac{y}{585} \end{array}$. If we solve $\dfrac{x}{455} = \dfrac{y}{585}$ for a variable, we can use substitution.

$$\frac{x}{455} = \frac{y}{585}$$ Solve for x.

$$455 \cdot \frac{x}{455} = \frac{y}{585} \cdot 455$$ Multiplication property of equality.

$$x = \frac{455}{585} y$$ Simplify.

$$x = \frac{7 \cdot \cancel{65}}{9 \cdot \cancel{65}} y$$ Simplify into lowest terms.

$$x = \frac{7}{9} y$$ The substitution expression is $\frac{7}{9} y$.

$$x + y = 3458$$ The other equation in the system.

$$\frac{7}{9} y + y = 3458$$ Replace x with the substitution expression, $\frac{7}{9} y$.

$$9\left(\frac{7}{9} y + y\right) = 9(3458)$$ Clear fractions; multiplication property of equality.

$$9 \cdot \frac{7}{9} y + 9y = 31{,}122$$ Distributive property.

$$7y + 9y = 31{,}122$$ Simplify.

$$16y = 31{,}122$$ Combine like terms.

$$\frac{16y}{16} = \frac{31{,}122}{16}$$ Division property of equality.

$$y = 1945.1\ldots$$ Simplify.

$$y \approx 1945 \text{ mi}$$ Round.

$$x + y = 3458$$ To find x, use either one of the equations.

$$x + 1945 = 3458$$ Replace y with 1945.

$$\underline{-1945 \quad -1945}$$ Subtraction property of equality.

$$x + 0 = 1513$$ Simplify.

$$x = 1513 \text{ mi}$$ Simplify.

The solution of the system is about (1513 mi to NY, 1945 mi to London).

Step 4 Look back.
Since 1513 mi + 1945 mi equals 3458 mi, the total distance is correct. Since $\frac{1513}{455} \approx 3.325$ and $\frac{1945}{585} \approx 3.325$, both planes travel the same amount of time from the point of no return. Since a jet going with the wind travels faster, it makes sense that it can travel a longer distance (to London) in the same time that a jet going against the wind travels a shorter distance (to New York).

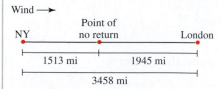

Step 5 Report the solution.
From the point of no return, the distance of the jet from New York is about 1513 mi, and the distance of the jet from London is about 1945 mi.

Practice Problems

7. Two cars left Des Moines, Iowa, on Interstate 80, traveling in opposite directions. The speed of the fast car is $\dfrac{75 \text{ mi}}{1 \text{ hr}}$. The speed of the slow car is $\dfrac{65 \text{ mi}}{1 \text{ hr}}$. One car has three passengers; the other car has two passengers. Find the distance each car is from Des Moines when the distance between the cars is 90 mi. Round to the nearest whole number.

8. A small twin engine plane flies 492 mi from Indianapolis to Harrisburg. The average speed of the plane in still air is $\dfrac{110 \text{ mi}}{1 \text{ hr}}$. The average tail wind is $\dfrac{60 \text{ mi}}{1 \text{ hr}}$. Find the distance from the point of no return to Indianapolis and the distance from the point of no return to Harrisburg. Round to the nearest whole number.

4.3 VOCABULARY PRACTICE

Match the term with its description.

1. $ax + by = c$
2. $y = mx + b$
3. The money made from selling a product
4. The product of rate and time
5. At this point, the cost of making the product is equal to the revenue from selling the product.
6. The location at which the time to reach the destination is the same as the time to return to the origin
7. An algebraic method for solving a system of equations in which we add the equations in the system
8. An algebraic method for solving a system of equations in which we replace a variable with an expression
9. A process in problem solving in which we state what the variable represents
10. When we use this method for solving a system of equations, the solution is an estimate.

A. assign the variable
B. break-even point
C. distance
D. elimination method
E. graphing method
F. point of no return
G. revenue
H. slope-intercept form of a linear equation
I. standard form of a linear equation
J. substitution method

4.3   Exercises

For exercises 1–22, use the five steps and a system of equations.

1. A professional fly tier uses about $0.45 in materials to tie a fly. Her fixed costs are $400 a week. Her average revenue per fly is $1.25. Find the number of flies that she must tie and sell per week to break even. Round up to the nearest fly. Find the revenue at the break-even point.

© Shane W. Thompson/Shutterstock

2. A home-based business sells pillar palm wax candles through a website. Each candle sells for $12. The cost to make each candle is $2. The fixed costs for the business are $500 per month. Find the number of candles that the business must make and sell per month to break even. Find the revenue at the break-even point.

3. In its business plan, a new bagel shop predicts that the average amount each customer will spend per visit is $5.35. The average cost of ingredients per customer will be $1.50. The overhead cost to run the bagel shop will be $2100 per week. Find the number of customers that must be served each week to break even. Round up to the nearest customer. Find the costs at the rounded break-even point.

4. In its business plan, a new restaurant predicts that the average amount each customer will spend per visit is $45. The average cost of ingredients per customer will be $22. The overhead cost to run the restaurant will be $12,000 per week. Find the number of customers that must be served each week to break even. Round up to the nearest customer. Find the costs at the rounded break-even point.

5. A taxpayer is putting $3500 into an IRA. Her investment strategy is to put three times as much in stocks as she does in certificates of deposit (CDs). Find the amount she should put in stocks and the amount she should put in certificates of deposit.

6. A consumer is shifting his debt on his credit cards. His total debt is $1950. He wants four times as much debt on his VISA® card than on his MasterCard®. Find the amount of debt that should be on his VISA card and the amount of debt that should be on his MasterCard.

7. In planning a 1500-calorie-per-day diet, a student wants to eat four times as much carbohydrate, including vegetables, as she eats of protein and fat. Find the number of calories in the diet that should be carbohydrates and the number that should be protein and fat.

8. In planning a 2400-calorie-per-day diet, a student wants to eat three times as much carbohydrate, including vegetables, as he eats of protein and fat. Find the number of calories in the diet that should be carbohydrate and the number that should be protein and fat.

9. The cost of five hamburgers and two orders of fries is $10.75. The cost of nine hamburgers and three orders of fries is $18.60. The cost of six hamburgers and four soft drinks is $13.86. Find the cost of a hamburger and the cost of an order of fries.

10. At the bookstore, the cost of four T-shirts and three sweatshirts is $178. The cost of two T-shirts and seven sweatshirts is $298. The cost of five T-shirts and three hats is $116. Find the cost of a T-shirt and the cost of a sweatshirt.

11. Cranberry juice drinks are a mixture of cranberry juice and water. Drink A is 27% cranberry juice. Drink B is 15% cranberry juice. Find the amount of each drink needed to make 2000 L of a new blend that is 20% cranberry juice. Round to the nearest whole number.

12. A juice blend is 30% guava juice. Another juice blend is 10% guava juice. Find the amount of each juice blend needed to make 500 L of a blend that is 18% guava juice.

13. Solution A is 20% acid. Find the amount of Solution A and the amount of pure water needed to make 5 L of a new solution that is 18% acid.

14. Solution B is 24% acid. Find the amount of Solution B and the amount of pure water needed to make 8 L of a new solution that is 5% acid. Round to the nearest tenth.

15. Copper and gold alloys are used to make distinctive jewelry. A rose-gold alloy is 25% copper. A red-gold alloy is 42% copper. Find the amount of the each alloy needed to make 80 g of a new alloy that is 30% copper. Round to the nearest tenth.

16. Solder is a metal alloy. Solder A is 63% tin. Solder B is 70% tin. Find the amount of each solder needed to make 15 g of a new solder that is 65% tin. Round to the nearest tenth.

17. Two light-rail trains leave an airport traveling in opposite directions. One train is traveling $\frac{40 \text{ mi}}{1 \text{ hr}}$. The other train is traveling $\frac{35 \text{ mi}}{1 \text{ hr}}$. Find the distance each train is from the airport when the total distance between the trains is 5 mi. Round to the nearest tenth.

18. A community college is located at a freeway exit. One student leaves campus traveling north at an average speed of $\frac{65 \text{ mi}}{1 \text{ hr}}$. Another student goes south in heavy traffic at an average speed of $\frac{40 \text{ mi}}{1 \text{ hr}}$. Find the distance each student is from campus when the total distance between their cars is 20 mi. Round to the nearest tenth.

19. Harry's Yard Service charges $10 per visit plus $7.25 per hour. Harriet's Landscape Care charges $15 per visit plus $5.25 per hour. Find the number of hours of yard service at which the costs of the service are the same for each company. Find the costs at the break-even point and round to the nearest hundredth.

20. Home Health Care charges $10 per visit plus $23 per hour. Eldercare charges $15.50 per visit plus $19 per hour. Find the number of hours at which the costs are the same for both companies. Find the costs at the break-even point and round to the nearest hundredth.

21. The average speed of a Boeing 777 in still air is $\frac{545 \text{ mi}}{1 \text{ hr}}$. The plane is flying 2256 mi from Los Angeles to Honolulu into an average headwind of $\frac{65 \text{ mi}}{1 \text{ hr}}$. Find the distance from the point of no return to Los Angeles. Find the distance from the point of no return to Honolulu. Round to the nearest whole number.

22. The average ground speed of a Boeing 767 in still air is $\frac{550 \text{ mi}}{1 \text{ hr}}$. The plane is flying 4800 mi from London to Seattle into an average headwind of $\frac{27 \text{ mi}}{1 \text{ hr}}$. Find the distance from the point of no return to London. Find the distance from the point of no return to Seattle. Round to the nearest whole number.

For exercises 23–54, use the five steps and a system of equations.

23. A feedlot manager adds soybean hulls to swine feed to prevent caking. The amount of hulls he uses depends on the humidity. The minimum amount of protein in swine feed is 23%. Feed A contains 20% soy hulls. Feed B contains 45% soy hulls. The manager needs 750 pounds of a new feed mixture that is 35% soy hulls. Find the number of pounds of Feed A and Feed B needed to make this mixture.

24. A department chair is estimating the cost of providing math classes. The cost of delivering a class is $1200 per class plus an additional $50 per student in the class. The revenue from student tuition is about $130 per student in the class. Of the students taking the class, about 78% receive financial aid. Find the number of students that need to be in a class so that the cost of delivering a class is the same as the revenues from student tuition. Round up to the nearest whole number.

25. A chef orders 50 lb of Chilean sea bass. For specials that evening in her restaurant, she creates an entrée that uses 6 oz of sea bass and will cost $29. She creates an appetizer that uses 4 oz of sea bass and will cost $8. She expects to sell twice as many entrées as appetizers. Only 30% of her customers order dessert. Find the number of appetizers and the number of entrées that she can make with 50 lb of sea bass.

26. The Old Babylonian period in Mesopotamia lasted from about 2000 BCE to 1600 BCE. The city of Babylon was located in what is now Iraq. A Babylonian clay tablet included a problem like this:

There are two fields whose total area is 1800 square yards. One produces grain at the rate of $\frac{2}{3}$ bushel per square yard. The other produces grain at the rate of $\frac{1}{2}$ bushel per square yard. If the total yield from both fields is 1100 bushels, what is the size of each field in square yards?

27. A landscape plan calls for a mixture of yellow daylilies ($4 each) and red daylilies ($5 each). The budget for daylilies is $340. The yellow daylilies are 36 in. tall. The plan calls for three times as many yellow daylilies as red daylilies. Find the number of yellow daylilies and the number of red daylilies to order.

28. A caterer is making 96 sandwiches for a luncheon for 50 men and 46 women. Ice tea, water, and coffee will be served. The organizers want three times as many ham sandwiches as vegetarian sandwiches. Find the number of ham sandwiches and the number of vegetarian sandwiches to order.

29. Sunflower meal is a by-product of extracting oil from sunflower seeds. No more than 25% of the grain ration fed to dairy cattle should be sunflower meal. The protein content of this meal depends on the amount of hull removed during processing. A feed company sells sunflower meal that is 34% protein, 28% protein, or 40% protein. A dairy farmer wants 2500 pounds of sunflower meal that is 30% protein. Find the amount of 34% protein meal and the amount of 28% protein meal needed to make this feed. Round to the nearest whole number.

30. Corn gluten meal is a by-product of wet milling corn to make cornstarch or corn syrup. A farmer can buy meal with a protein content of 40% or of 60%. A dairy farmer wants 1200 pounds of feed that is 45% protein. Find the amount of each kind of corn gluten meal needed to make this feed. Round to the nearest whole number.

© vesilvio/Shutterstock

31. By legal definition, the composition of *ice cream* must be greater than 10% milkfat. Some premium ice creams are 16% milkfat. *Ice milk* contains between 3% and 5% milkfat. *Light ice cream* contains between 8% and 10% milkfat. Blend A contains 5% milkfat. Blend B contains 13% milkfat. Find the number of gallons of each blend needed to make 3000 gallons of a new blend that is 9% milkfat.

32. A Household Hints website recommends making homemade shower cleaner by mixing vinegar and water. The cleaner is 1.6% acetic acid. The vinegar is 5% acetic acid. Find the amount of vinegar and the amount of water to mix to make 2 gallons of shower cleaner. Round to the nearest tenth.

33. Hydrochloric acid solution that is 32% hydrochloric acid by weight is sold in 525-lb drums. A solution used for cleaning steel is 22% hydrochloric acid by weight. Find the number of pounds of water to mix with 2100 lb of the acid solution in the drums to make this cleaning solution. Round to the nearest whole number.

© Steve Allen RF/Getty Images

34. In paper manufacturing, wood pulp is bleached with a solution that is 50% hydrogen peroxide. To save on shipping costs, manufacturers purchase 70% hydrogen peroxide solution. They then dilute it at the paper plant. Since shipping costs depend on weight, diluting the hydrogen peroxide at the plant reduces costs. Find the amount of water to add to 7000 lb of 70% hydrogen peroxide to make a new solution that is 50% hydrogen peroxide. Round to the nearest whole number.

35. A food co-op in Brooklyn requires that all members work at the store 13 times a year. A co-op member is making 30 lb of a mixture of cashews and peanuts. The final price of the mixture will be $5 per pound. Cashews cost $7.25 per pound. Peanuts cost $1.30 per pound. Find the number of pounds of cashews and peanuts needed to make this mixture. Round to the nearest tenth.

36. Chocolate-covered almonds cost $7.50 per pound. Chocolate-covered cashews cost $7.60 per pound. Chocolate-covered macadamia nuts cost $13.25 per pound. Find the number of pounds of macadamias and almonds needed to make 150 lb of a mixture that costs $10 per pound. Round to the nearest tenth.

37. Concentrated stop bath solution for film developing is 28% acetic acid. Find the amount of concentrated stop bath solution and the amount of water needed to make 1 qt of solution that is 2% acetic acid. Round to the nearest hundredth.

38. A lab technician is preparing organ tissue for mounting on slides. The tissue is cut so that the thinnest dimension is no greater than 5 mm. It is then placed in a 4% formaldehyde solution for at least 2 hr. The stock solution of formalin in the laboratory is 37% formaldehyde. Find the amount of stock formalin and the amount of water needed to make 75 mL of a 4% formaldehyde solution. Round to the nearest tenth.

For exercises 39 and 40, diesel fuel has a *cloud point temperature*. Below this temperature, waxy materials may appear in the fuel. This material can clog fuel filters. Number One (No. 1) Diesel fuel has a lower cloud point temperature than Number Two (No. 2) Diesel fuel.

39. For logging trucks working in the mountains in the winter, a petroleum jobber made a blended fuel that was 60% No. 2 Diesel and 40% No. 1 Diesel. When spring arrived, 9500 gal of this fuel had not been sold. It needs to be reblended so that it is 80% No. 2 Diesel. Find the amount of pure No. 2 Diesel fuel to add to the leftover blended fuel to make the new fuel.

© steve estvanik/Shutterstock

40. A petroleum jobber has 8500 gal of fuel that is 80% No. 2 Diesel fuel and 20% No. 1 Diesel fuel. He wants a new fuel that is 60% No. 2 Diesel fuel and 40% No. 1 Diesel fuel. Find the amount of pure No. 1 Diesel fuel to add to the 8500 gal to make this new fuel. Round to the nearest whole number.

41. A cell phone plan costs $39.99 per month. This plan allows for 450 minutes of out-of-network calls each month. Each additional minute outside the network costs $0.30. The same company has another plan with 900 minutes of out-of-network time for $59.99. Each additional minute outside the network costs $0.40. Find the number of total minutes outside the network a customer must use for the costs of the plans per month to be the same. Round to the nearest whole number.

42. A cell phone plan costs $39.99 per month. Each text message sent or received costs $0.15. A different plan costs $49.99 per month. Each text message sent or received costs $0.10. Find the number of text messages a customer must use for the costs of the plans per month to be the same. Round to the nearest whole number.

For exercises 43 and 44, a 14-karat gold alloy is 58.3% gold. A 20-karat gold alloy is 83.3% gold. An 18-karat gold alloy is 75% gold. Pure 100% gold is 24-karat.

43. A customer brings 10 oz of 14-karat gold alloy to a jeweler. Find the amount of 20-karat gold alloy to combine with the customer's alloy to make an alloy that is 18-karat gold. Round to the nearest tenth.

44. A jeweler is combining 10 oz of 20-karat gold with copper to make a new alloy that is 14-karat gold. Find the amount of copper needed. Round to the nearest tenth.

45. A solution is 35% antifreeze. Find the amount of this solution and the amount of pure antifreeze to mix to make 7 L of a solution that is 40% antifreeze. Round to the nearest tenth.

46. A saturated brine (salt) solution at 60°F is 26.5% salt. A curing tank contains 200 L of saturated brine solution. Find the amount of water to add to the tank to create a solution for curing salmon that is 15.8% salt. Round to the nearest whole number.

47. A bottle of concentrated insecticide contains 32% active ingredient. A homeowner needs 1 gal (128 ounces) of a mixture that is 5% active ingredient. Find the number of ounces of water and the number of ounces of concentrate needed to make this mixture.

48. The active ingredient in a broad-leaf weed killer is 2-4-dichlorophenoxyacetic acid (2,4-D). A bottle of concentrated 2,4-D contains 3.05% 2,4-D. The concentration of 2,4-D recommended for killing dandelions in a grass lawn is 0.2%. Find the number of ounces of water and the number of ounces of concentrate needed to make 2 gal of solution for killing dandelions. Round to the nearest ounce (1 gal = 128 oz).

49. A highway construction project requires 31,000 yd³ of concrete. A concrete company has two sizes of concrete mixer trucks, small and large. The concrete can be transported to the construction site with 1600 large loads and 1775 small loads, or it can be transported with 2000 large loads and 1250 small loads. Find the volume of the small truck and the volume of the large truck.

50. The Kentucky building code describes the charges for plan examinations and inspections required by the Department of Housing, Buildings and Construction. The charges for building a 600,000 ft^2 factory and a 7000 ft^2 daycare center are $72,910. The charges for building a 450,000 ft^2 factory and a 5000 ft^2 daycare center are $54,650. Find the cost per square foot for plan examinations and inspections for a factory and the cost per square foot for plan examinations and inspections for a daycare center. (*Source:* dhbc.ky.gov)

51. The price of a child ticket was $4. The price of an adult ticket was $9.50. The amount collected for the sale of 500 tickets was $3650. The amount of concession sales was $1950. Find the number of adult tickets and the number of child tickets sold.

52. The price of a child ticket was $4. The price of an adult ticket was $9.50. The amount collected for the sale of 500 tickets was $3787.50. The concessions stand sold 390 soft drinks. Find the number of adult tickets and the number of child tickets sold.

53. A recreational path along a river crosses two bridges to form a loop. The length of the path is 20 mi. A runner begins the trail, running at an average pace of 6 miles per hour. She runs this path four times a week. Thirty minutes later, a bicyclist starts the trail at the same place. The speed of the bicyclist is 18 miles an hour. Find the time it will take the bicyclist to catch up to the runner.

54. A tugboat leaves a port pushing barges filled with wheat down the Snake River toward Portland. Including the current, the barge is traveling at a speed of 9 miles per hour. Twenty minutes later, a powerboat leaves the port traveling at a speed of 24 miles per hour, including the current. Find the time it will take the powerboat to catch up to the barge.

55. Write a word problem that involves the selling of two kinds of tickets. You will need to include the price of each kind of ticket, the total number of tickets, and the total money collected from ticket sales. Use a system of equations to solve this problem.

56. Write a word problem that involves the mixing of two solutions to make a new solution. You will need to include the concentration of a substance in each solution, the volume of the new solution, and the concentration of the substance in the new solution. Use a system of equations to solve this problem.

Problem Solving: Practice and Review

Follow your instructors's guidelines for using the five steps as outlined in Section 1.5, p. 51.

57. The National Speakers Association in Tempe, Arizona, publishes *Speaker* magazine. A full-page four-color ad is $7\frac{1}{2}$ in. wide and 10 in. long. The cost to insert this ad in one issue is $1770. Find the price per square inch for the advertising. (*Source:* www.nsaspeaker.org)

58. Find the percent decrease in the number of Associate of Arts degrees awarded at Pima Community College from the 2007–2008 academic year to the 2009–2010 academic year. Round to the nearest tenth of a percent. (*Source:* www.pima.edu)

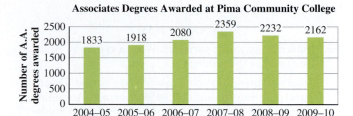

Associates Degrees Awarded at Pima Community College

59. Weeki Wachee Canoe and Kayak Rental® rents a two-seat canoe for $52 plus tax. One additional person with a weight less than 130 lb can sit in the middle for $12 plus tax. The sales tax rate is 6.5%. Find the cost for two adults and a child who weighs less than 130 lb to rent a canoe. (*Source:* www.floridacanoe.com)

60. Venture capital is needed to start a new company that is developing new "green" energy technology. Predict the percent of venture capital investments in 2013 that will be invested in energy technology. Round to the nearest percent.

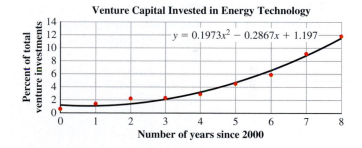

Venture Capital Invested in Energy Technology

$$y = 0.1973x^2 - 0.2867x + 1.197$$

Find the Mistake

For exercises 61–64, the completed problem has one mistake.
(a) Describe the mistake in words, or copy down the whole problem and highlight or circle the mistake.
(b) Do the problem correctly.

61. **Problem:** Assign the variables. Write a system of equations to solve the problem.

A silver alloy is 80% silver. Sterling silver alloy is 92.5% silver. Find the amount of each alloy needed to make 12 g of an alloy that is 88% silver.

Incorrect Answer:

x = number of grams of 80% silver alloy
y = number of grams of sterling silver alloy
$$x + y = 12$$
$$0.08x + 0.925y = 0.88(12)$$

62. **Problem:** Assign the variables. Write a system of equations to solve the problem.

A photographer offers two bargain packages for single-pose graduation pictures. Package A costs $52. It includes two 8-by-10 prints and four 5-by-7 prints. Package B costs $53. It includes one 8-by-10 print and six 5-by-7 prints. Find the cost of an 8-by-10 print. Find the cost of a 5-by-7 print.

Incorrect Answer: x = cost of 8-by-10

y = cost of 5-by-7

$$2x + 4y = 6(52)$$
$$1x + 6y = 7(53)$$

63. Problem: Assign the variables. Write a system of equations to solve the problem.

Dried porcini mushrooms cost $54.50 a pound. Dried chanterelle mushrooms cost $9.90 an ounce. An online gourmet supplier wants 40 lb of a blend of these mushrooms that will cost $6.60 an ounce. Find the needed amount of each mushroom.

Incorrect Answer: x = number of pounds of porcini

y = number of pounds of chanterelle

$$x + y = 40$$
$$54.50x + 9.90y = (40)(6.60)$$

64. Problem: Assign the variables. Write a system of equations to solve this problem.

A take-and-bake pizza with two toppings is $7. A take-and-bake pizza with one topping is $6. An organizer needs 64 pizzas. He wants three times as many pizzas with two toppings as pizzas with one topping. Find the number of pizzas to order.

Incorrect Answer: x = number of 2-topping pizzas

y = number of 1-topping pizzas

$$7x + 6y = 64$$
$$x = 3y$$

Review

65. Sketch the graph of a system of two linear equations with one solution.

66. Sketch the graph of a system of two linear equations with no solution.

67. Sketch the graph of a system of two linear equations with infinitely many solutions.

68. Write the standard form of a linear equation in two variables.

SUCCESS IN COLLEGE MATHEMATICS

69. When solving the system of equations $\begin{matrix} x - 4y = -1 \\ x + 7y = 10 \end{matrix}$ by elimination, a student multiplied each equation on both sides by a constant to create this equivalent system: $\begin{matrix} 2x - 8y = -2 \\ 2x + 14y = 20 \end{matrix}$. The student then added the equations and said that the sum was $0 + 6y = 18$. What mistake did the student make?

70. When using the elimination method to solve a system of equations, the equations are added together, and a variable is eliminated in the sum. For a variable to be eliminated, what has to be true about the signs of the terms that include this variable?

© kalewa/Shutterstock

A heart transplant patient needs to limit intake of fat, cholesterol, and sodium. For any three different foods, the amount of food that meets these requirements can be described with a system of three linear equations. In this section, we will learn how to solve these systems.

SECTION 4.4

Systems of Linear Equations in Three Variables

After reading the text, working the practice problems, and completing assigned exercises, you should be able to:

1. Identify a representation as a point, line, or plane.

2. Use an algebraic method to check whether an ordered triple is a solution of a system of three linear equations in three variables.

3. Use elimination to solve a system of three linear equations in three variables.

4. Use a system of three linear equations to solve an application problem.

Points, Lines, and Planes

We can represent an ordered pair (x, y) as a **point** on a graph. Each of the solutions of a linear equation in two variables, $ax + by = c$, is a point. The graph of a linear equation in two variables is the collection of these points which is a **line**.

The **general form of a linear equation in three variables** is $ax + by + cz = d$, where a, b, c, and d are real numbers and a, b, and c are not all 0. Each solution of a linear equation in three variables is an **ordered triple**, (x, y, z). The graph of an ordered triple is a **point**. The graph of a linear equation in three variables is the collection of these points, which is a **plane**. In geometry, a plane is an infinite two-dimensional space. For example, a piece of paper is part of a plane.

To graph an ordered triple, we need three axes in a three-dimensional space. All of the axes are perpendicular to each other.

We can use perspective to represent three dimensions on a two-dimensional piece of paper. Since we live in a three-dimensional visual world, we can also imagine these graphs. Picture the point that represents the ordered triple (3, 2, 4). From the origin, this point is 3 units in the positive x-direction, 2 units in the positive y-direction, and 4 units up in the positive z-direction (Figure 1).

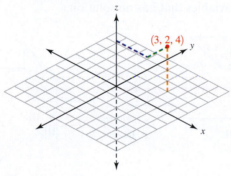

Figure 1

EXAMPLE 1 **(a)** Is the graph of $y = 3x + 2$ a point, a line, or a plane?

SOLUTION ▶   This is a linear equation in two variables; its graph is a line.

(b) Is the graph of $(2, 8)$ a point, a line, or a plane?

▶   This is an ordered pair; its graph is a point.

(c) Is the graph of $x + y + z = 10$ a point, a line, or a plane?

▶   This is a linear equation in three variables; its graph is a plane.

(d) Is the graph of $(2, 8, 3)$ a point, a line, or a plane?

▶   This is an ordered triple; its graph is a point.

The solution of a system of linear equations is an ordered pair(s) that is a solution of each equation. When we solve a system of two linear equations in two variables by graphing, the solution is the intersection point of the lines (Figure 2). If the lines represented by a system of linear equations do not intersect (Figure 3), then the system has no solution. If the lines represented by a system of linear equations are coinciding (Figure 4), then the system has infinitely many solutions.

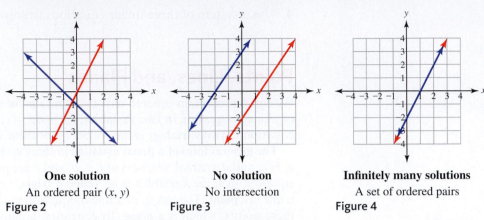

One solution

An ordered pair (x, y)

Figure 2

No solution

No intersection

Figure 3

Infinitely many solutions

A set of ordered pairs

Figure 4

We can represent a system of three linear equations in three variables with a graph of three planes. The solution of this system is the point(s) that all of the planes share.

EXAMPLE 2 (a) Describe the graph of a system of three linear equations in three variables that has one solution.

SOLUTION ►   If a system has one solution, the planes share a single point.

(b) Describe the graph of a system of three linear equations in three variables that has no solution.

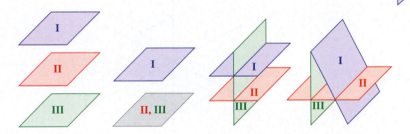

► If a system has no solution, the planes do not have any points in common. Two planes may share many points, but the system does not have a solution unless all three planes share at least one point.

(c) Describe the graph of a system of three linear equations in three variables that has infinitely many solutions.

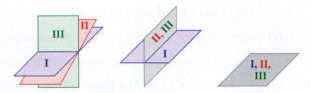

► If a system has infinitely many solutions, the planes may have a single line in common, or they may all be the same plane.

It is difficult to draw the plane that represents a linear equation in three variables. It is even more difficult to draw three planes. Instead of graphing, we can use an algebraic method to solve these systems of equations.

Ordered Triples

The solution of a system of three linear equations in three variables may be an ordered triple. To check the solution algebraically, replace the variables in each equation with the solution.

EXAMPLE 3 Use an algebraic method to check whether the ordered triple (1, 2, 3) is a solution of the system

$$x + y + z = 6$$
$$2x + y - z = 1$$
$$x - y - z = -4$$

SOLUTION ▶

$x + y + z = 6$	$2x + y - z = 1$	$x - y - z = -4$
$1 + 2 + 3 = 6$	$2(1) + 2 - 3 = 1$	$1 - 2 - 3 = -4$
$3 + 3 = 6$	$4 - 3 = 1$	$-1 - 3 = -4$
$6 = 6$ True.	$1 = 1$ True.	$-4 = -4$ True.

If the variables in each equation are replaced with the solution, each equation is true. This ordered triple is a solution of the system.

Elimination

In Section 4.2, we used the elimination method to solve a system of two linear equations in two variables. We can use this same strategy to solve a system of three linear equations in three variables.

Using Elimination to Solve a System of Linear Equations in Three Variables

1. Choose a variable to eliminate. Then choose two of the equations. Use the elimination method to eliminate this variable. The result will be an equation in only two variables.
2. Choose a different pair of equations. Use the elimination method to eliminate the same variable as in step 1. The result will be another equation in two variables.

3. Use elimination to solve the system of two equations created in steps 1 and 2. We now know the values of two of the variables in the ordered triple.

4. Replace the variables in one of the equations with the known values. Solve for the remaining variable.

5. To check, replace the variables in each equation with the solution.

EXAMPLE 4 **(a)** Solve
$$x + y + z = -1$$
$$2x + 3y + z = -11$$
$$x - 2y + 5z = 29$$
by elimination.

SOLUTION ▶ Choose a variable to eliminate, x. Eliminate x from the first two equations by multiplying one of the equations by a constant (-2) and adding the equations together.

$(-2)(x + y + z) = (-2)(-1)$	Multiplication property of equality.
$-2x - 2y - 2z = 2$	Simplify.

$-2x - 2y - 2z = 2$	Add the equations vertically; the x-terms are opposites.
$+\quad 2x + 3y + z = -11$	Addition property of equality.
$0 + y - z = -9$	Simplify.
$y - z = -9$	This equation has two variables, y and z.

Eliminate the same variable, x, from a different pair of equations.

$(-1)(x + y + z) = (-1)(-1)$	Multiplication property of equality.
$-x - y - z = 1$	Simplify.

$-x - y - z = 1$	Add the equations vertically; the x-terms are opposites.
$+\ x - 2y + 5z = 29$	Addition property of equality.
$0 - 3y + 4z = 30$	Simplify.
$-3y + 4z = 30$	This equation has two variables, y and z.

We now have a system of two equations in two variables, y and z. Solve this new system, $\begin{aligned} y - z &= -9 \\ -3y + 4z &= 30 \end{aligned}$, by elimination. We choose to eliminate y.

$3(y - z) = 3(-9)$	Multiplication property of equality.
$3y - 3z = -27$	Simplify.

$3y - 3z = -27$	Add the equations vertically; the y-terms are opposites.
$+\ -3y + 4z = 30$	Addition property of equality.
$0 + z = 3$	Simplify.
$z = 3$	Simplify; the solution of the system is $(x, y, 3)$.

Replace z with 3 in either of the two equations, $\begin{aligned} y - z &= -9 \\ -3y + 4z &= 30 \end{aligned}$.

$y - z = -9$	Choose one of the equations.
$y - 3 = -9$	Replace z with 3.
$\underline{+3 \quad +3}$	Addition property of equality.
$y + 0 = -6$	Simplify.
$y = -6$	Simplify; the solution of the system is $(x, -6, 3)$.

Replace y and z in any of the original equations,
$$x + y + z = -1$$
$$2x + 3y + z = -11, \text{ and}$$
$$x - 2y + 5z = 29$$
solve for x.

$$x + y + z = -1 \qquad \text{Choose an equation from the original system.}$$
$$x + (-6) + 3 = -1 \qquad \text{Replace } y \text{ with } -6 \text{ and } z \text{ with } 3.$$
$$x - 3 = -1 \qquad \text{Simplify.}$$
$$\underline{+3 \quad\; +3} \qquad \text{Addition property of equality.}$$
$$x + 0 = 2 \qquad \text{Simplify.}$$
$$x = 2 \qquad \text{Simplify.}$$

The solution of this system is $(2, -6, 3)$.

(b) Check.

$x + y + z = -1$	$2x + 3y + z = -11$	$x - 2y + 5z = 29$
$2 + (-6) + 3 = -1$	$2(2) + 3(-6) + 3 = -11$	$2 - 2(-6) + 5(3) = 29$
$-4 + 3 = -1$	$4 - 18 + 3 = -11$	$2 + 12 + 15 = 29$
$-1 = -1$	$-11 = -11$	$29 = 29$
True.	True.	True.

The ordered triple $(2, -6, 3)$ is a solution of each equation in the system and it is the solution of the system. So the planes that represent each equation intersect in a single point.

When we try to use elimination to solve a system of two linear equations with no solution, the result is a false equation. In the next example, we see similar results when we solve a system of three linear equations with no solution.

EXAMPLE 5 Solve
$$\begin{aligned} 2x + 3y + 5z &= 1 \\ 6x + 7y + 8z &= 12 \\ 10x + 13y + 18z &= 5 \end{aligned}$$
by elimination.

SOLUTION ▶ Choose two of the equations and then choose a variable to eliminate. In this example, we choose to eliminate y. Multiply each equation by a constant and then add the equations together.

$$(-7)(2x + 3y + 5z) = (-7)(1) \qquad \text{Multiplication property of equality.}$$
$$-14x - 21y - 35z = -7 \qquad \text{Simplify.}$$

$$3(6x + 7y + 8z) = 3(12) \qquad \text{Multiplication property of equality.}$$
$$18x + 21y + 24z = 36 \qquad \text{Simplify.}$$

$$\begin{aligned} -14x - 21y - 35z &= -7 \qquad \text{Add equations vertically; the } y\text{-terms are opposites.} \\ + \quad 18x + 21y + 24z &= 36 \qquad \text{Addition property of equality.} \\ \hline 4x + 0 - 11z &= 29 \qquad \text{Simplify.} \\ 4x - 11z &= 29 \qquad \text{This equation has two variables, } x \text{ and } z. \end{aligned}$$

Now eliminate the same variable, y, from a different pair of equations.

$$(-13)(6x + 7y + 8z) = (-13)(12) \qquad \text{Multiplication property of equality.}$$
$$-78x - 91y - 104z = -156 \qquad \text{Simplify.}$$

$$7(10x + 13y + 18z) = 7(5) \qquad \text{Multiplication property of equality.}$$
$$70x + 91y + 126z = 35 \qquad \text{Simplify.}$$

$$\begin{aligned} -78x - 91y - 104z &= -156 \qquad \text{Add equations vertically; the } y\text{-terms are opposites.} \\ + \quad 70x + 91y + 126z &= 35 \qquad \text{Addition property of equality.} \\ \hline -8x + 0 + 22z &= -121 \qquad \text{Simplify.} \\ -8x + 22z &= -121 \qquad \text{This equation has two variables, } x \text{ and } z. \end{aligned}$$

We now have a system of two equations in two variables, x and z. Solve this new system, $\begin{aligned} 4x - 11z &= 29 \\ -8x + 22z &= -121 \end{aligned}$, by elimination. We choose to eliminate z.

$2(4x - 11z) = 2(29)$	Multiplication property of equality.
$8x - 22z = 58$	Simplify.

$8x - 22z = 58$	Add the equations vertically; the z-terms are opposites.
$+\ -8x + 22z = -121$	Addition property of equality.
$0 + 0 = -63$	Simplify; this equation is false.

When the equations are added, the variables disappear. The new equation is false. This system has no solution; there is no ordered triple that is a solution of each of the equations. Since each equation is represented by a plane, this means that there is no point that is on all three planes. They have no point in common.

A graph of two coinciding lines represents a system of two linear equations with infinitely many solutions. When we use an algebraic method to solve the system, the variables disappear. The resulting equation is true. In the next example, we see similar results when we solve a system of linear equations in three variables that has infinitely many solutions.

EXAMPLE 6   Solve $\begin{aligned} x + y + z &= 10 \\ 2x + 3y + 4z &= 25 \\ 7x + 9y + 11z &= 80 \end{aligned}$ by elimination.

SOLUTION ▶ Choose two of the equations and then choose a variable to eliminate. In this example, we choose to eliminate x. Multiply one of the equations by a constant and then add the equations together.

$(-2)(x + y + z) = (-2)(10)$	Multiplication property of equality.
$-2x - 2y - 2z = -20$	Simplify.

$-2x - 2y - 2z = -20$	Add equations vertically; the x-terms are opposites.
$+\ \ \ 2x + 3y + 4z = 25$	Addition property of equality.
$0 + y + 2z = 5$	Simplify.
$y + 2z = 5$	This equation has only two variables, y and z.

Now eliminate the same variable, x, from a different pair of equations.

$(-7)(x + y + z) = (-7)(10)$	Multiplication property of equality.
$-7x - 7y - 7z = -70$	Simplify.

$-7x - 7y - 7z = -70$	Add equations vertically; the x-terms are opposites.
$+\ \ 7x + 9y + 11z = 80$	Addition property of equality.
$0 + 2y + 4z = 10$	Simplify.
$2y + 4z = 10$	Simplify; this equation has only two variables, y and z.

We now have a system of two equations in two variables, y and z. Solve this new system, $\begin{aligned} y + 2z &= 5 \\ 2y + 4z &= 10 \end{aligned}$, by elimination. We choose to eliminate y.

$(-2)(y + 2z) = (-2)(5)$	Multiplication property of equality.
$-2y - 4z = -10$	Simplify.

$-2y - 4z = -10$	Add equations vertically; y-terms and z-terms are opposites.
$+\ \ 2y + 4z = 10$	Addition property of equality.
$0 + 0 = 0$	Simplify; this equation is true.

This system has infinitely many solutions.

Since each equation is represented by a plane, this means that there is a line that is shared by all three planes. They have infinitely many points in common.

To describe the solution of Example 6 more specifically than just to say "infinitely many solutions," we use set-builder notation to describe how two of the variables depend on the third variable. The solution is in **parametric form**. For Example 6, we can use set-builder notation to describe how x and y depend on z.

When we eliminated x from $x + y + z = 10$ and $2x + 3y + 4z = 25$, the result was the equation $y + 2z = 5$. To find the relationship of y and z, solve this equation for y.

$y + 2z = 5$	This equation is a result of elimination.
$\underline{-2z \qquad -2z}$	Subtraction property of equality.
$y + 0 = 5 - 2z$	Simplify.
$y = -2z + 5$	Simplify; change order; this is the relationship of y and z.

To find the relationship of x and z, replace y in one of the original equations, $x + y + z = 10$, and solve for x.

$x + y + z = 10$	One of the original equations.
$x + (-2z + 5) + z = 10$	Since $y = -2z + 5$, substitute $-2z + 5$ for y.
$x - z + 5 = 10$	Simplify.
$\underline{\qquad -5 \quad -5}$	Subtraction property of equality.
$x - z + 0 = 5$	Simplify.
$\underline{+z \qquad +z}$	Addition property of equality.
$x + 0 = z + 5$	Simplify.
$x = z + 5$	Simplify; this is the relationship of x and z.

In parametric form, the solution of the system $\begin{matrix} x + y + z = 10 \\ 2x + 3y + 4z = 25 \\ 7x + 9y + 11z = 80 \end{matrix}$ is $\{(x, y, z) \mid x = z + 5, y = -2z + 5, z \text{ is a real number}\}$. This set-builder notation tells us that the solution of the system is the set of ordered triples, (x, y, z), such that x equals $z + 5$, y equals $-2z + 5$, and z is a real number.

> ### Using Elimination to Solve a System of Linear Equations in Three Variables with No Solution or Infinitely Many Solutions
>
> If the result of elimination is a false equation with no variables, the system has no solution.
>
> If the result of elimination is a true equation with no variables, the system has infinitely many solutions. Use set-builder notation to describe the equation more specifically.

Practice Problems

For problems 9–11,
(a) solve by elimination.
(b) if the solution is an ordered triple, check.

9. $\begin{aligned} x + 3y + 2z &= 11 \\ 7x - 2y + z &= -41 \\ 2x + 5y - 3z &= 4 \end{aligned}$

10. $\begin{aligned} 2x + 5y + z &= 9 \\ 3x + 4y - 16z &= -60 \\ -5x - 9y + 15z &= 51 \end{aligned}$

11. $\begin{aligned} x + y - 5z &= 24 \\ 2x + 3y + z &= 23 \\ 4x + 7y + 13z &= 20 \end{aligned}$

Applications

As we study a problem, we identify what is unknown. If there are three unknowns, our strategy for solving the problem may be to solve a system of three linear equations.

EXAMPLE 7 A port employs longshore workers (Lshore), crane operators, and ship clerks to unload container ships. The time needed to unload a ship depends on its size and the type of containers. Unloading Ship A takes 120 hr of longshore worker time, 30 hr of crane time, and 40 hr of clerk time. Ship B takes 150 hr of longshore worker time, 25 hr of crane time, and 50 hr of clerk time. Ship C takes 70 hr of longshore worker time, 15 hr of crane time, and 20 hr of clerk time. A port can provide 1990 hr of longshore time, 425 hr of crane time, and 650 hr of clerk time. How many of each kind of ship can be unloaded? (Assume that the goal is to use up all of the time for each kind of work.)

© Anneka/Shutterstock

SOLUTION ▶ **Step 1 Understand the problem.**
The unknowns are the number of each kind of ship that can be unloaded.

x = number of Ship A y = number of Ship B z = number of Ship C

Step 2 Make a plan.
Write a system of three equations in which one equation represents the longshore time, one equation represents the amount of crane time, and one equation represents the amount of clerk time. The word equations are

(Lshore A)(number A) + (Lshore B)(number B) + (Lshore C)(number C) = total Lshore time
(crane A)(number A) + (crane B)(number B) + (crane C)(number C) = total crane time
(clerk A)(number A) + (clerk B)(number B) + (clerk C)(number C) = total clerk time

Step 3 Carry out the plan.

$$\left(\frac{120 \text{ hr}}{1 \text{ Ship A}}\right)x + \left(\frac{150 \text{ hr}}{1 \text{ Ship B}}\right)y + \left(\frac{70 \text{ hr}}{1 \text{ Ship C}}\right)z = 1990 \text{ hr} \qquad \text{Longshore equation.}$$

$$\left(\frac{30 \text{ hr}}{1 \text{ Ship A}}\right)x + \left(\frac{25 \text{ hr}}{1 \text{ Ship B}}\right)y + \left(\frac{15 \text{ hr}}{1 \text{ Ship C}}\right)z = 425 \text{ hr} \qquad \text{Crane equation.}$$

$$\left(\frac{40 \text{ hr}}{1 \text{ Ship A}}\right)x + \left(\frac{50 \text{ hr}}{1 \text{ Ship B}}\right)y + \left(\frac{20 \text{ hr}}{1 \text{ Ship C}}\right)z = 650 \text{ hr} \qquad \text{Clerk equation.}$$

Remove the units.

$$120x + 150y + 70z = 1990$$
$$30x + 25y + 15z = 425$$
$$40x + 50y + 20z = 650$$

Choose a variable to eliminate, x, and choose two of the equations. Multiply one of the equations by a constant and add the equations together.

$$(-4)(30x + 25y + 15z) = (-4)(425) \qquad \text{Multiplication property of equality.}$$
$$-120x - 100y - 60z = -1700 \qquad \text{Simplify.}$$

$$120x + 150y + 70z = 1990 \qquad \text{Add equations vertically; the } x\text{-terms are opposites.}$$
$$+\ -120x - 100y - 60z = -1700 \qquad \text{Addition property of equality.}$$
$$\overline{0 + 50y + 10z = 290} \qquad \text{Simplify.}$$
$$\mathbf{50y + 10z = 290} \qquad \text{This equation has two variables, } y \text{ and } z.$$

When x is eliminated from two other equations, y is also eliminated.

$$(-3)(40x + 50y + 20z) = (-3)(650)$$ Multiplication property of equality.

$$-120x - 150y - 60z = -1950$$ Simplify.

$$120x + 150y + 70z = 1990$$ Add equations vertically.

$$+ \ -120x - 150y - 60z = -1950$$ Addition property of equality.

$$0 + 0 + 10z = 40$$ Simplify.

$$\frac{10z}{10} = \frac{40}{10}$$ Division property of equality.

$$z = 4 \qquad \text{Simplify; the solution of the system is } (x, y, 4 \text{ Ship C}).$$

$$50y + 10z = 290 \qquad \text{Use the equation from the first elimination.}$$

$$50y + 10(\mathbf{4}) = 290 \qquad \text{Replace } z \text{ with 4.}$$

$$50y + \mathbf{40} = 290 \qquad \text{Simplify.}$$

$$\underline{ -40 \quad -40} \qquad \text{Subtraction property of equality.}$$

$$50y + \mathbf{0} = 250 \qquad \text{Simplify.}$$

$$\frac{\mathbf{50y}}{50} = \frac{250}{50} \qquad \text{Division property of equality.}$$

$$y = 5 \qquad \text{Simplify; the solution of the system is } (x, 5 \text{ Ship B}, 4 \text{ Ship C}).$$

Choose one of the original equations, replace y and z, and solve for x.

$$40x + 50y + 20z = 650 \qquad \text{Choose an equation from the original system.}$$

$$40x + 50(\mathbf{5}) + 20(\mathbf{4}) = 650 \qquad \text{Replace } y \text{ with 5, and replace } z \text{ with 4.}$$

$$40x + \mathbf{330} = 650 \qquad \text{Simplify.}$$

$$\underline{ -330 \quad -330} \qquad \text{Subtraction property of equality.}$$

$$40x + \mathbf{0} = \mathbf{320} \qquad \text{Simplify.}$$

$$\frac{\mathbf{40x}}{40} = \frac{320}{40} \qquad \text{Division property of equality.}$$

$$x = 8 \qquad \text{Simplify.}$$

The solution of the system is $(8 \text{ Ship A}, 5 \text{ Ship B}, 4 \text{ Ship C})$.

Step 4 Look back.

Since $(120)(8) + (150)(5) + (70)(4) = 1990$ longshore hours,
$(30)(8) + (25)(5) + (15)(4) = 425$ crane hours, and
$(40)(8) + (50)(5) + (20)(4) = 650$ clerk hours, and since these times are equal to the number of hours given in the problem, the answer seems reasonable.

Step 5 Report the solution.

To use up all of the work hours, the port can unload 8 Ship A, 5 Ship B, and 4 Ship C.

In real life, problems involving scheduling and efficiency usually involve more than three variables. Business analysts use computers with specialized software to solve these systems. Computer scientists learn a great deal of mathematics to prepare them to design this kind of software.

Practice Problems

12. A dietician is selecting apples, oranges, or bananas for a 7-day lunch menu. He needs seven fruits. The total energy should equal 610 calories. The total calcium should equal 152 mg. An apple has 70 calories and 8 mg of calcium. An orange has 70 calories and 60 mg of calcium. A banana has 100 calories and 6 mg of calcium. Use a system of equations to find the number of apples, oranges, and bananas he should choose.

4.4 VOCABULARY PRACTICE

Match the term with its description.

1. A line has this many dimensions.
2. A cardboard box has this many dimensions.
3. A piece of paper has this many dimensions.
4. $ax + by + cz = d$
5. $ax + by = c$
6. (x, y, z)
7. (x, y)
8. An algebraic method for solving a system of equations
9. A process in problem solving in which we state what the variable represents
10. An infinite two-dimensional space

A. assign the variable
B. elimination method
C. linear equation in three variables
D. linear equation in two variables
E. one dimension
F. ordered pair
G. ordered triple
H. plane
I. three dimensions
J. two dimensions

4.4 Exercises

Follow your instructor's guidelines for showing your work.

For exercises 1–6, does the system of equations represented by the graph have one solution, no solution, or infinitely many solutions?

1. The three planes intersect in a single line.

2. Plane A intersects Plane B in a single line. Plane A intersects Plane C in a single line. Plane A and Plane C do not intersect.

3. The three planes intersect in a single point.

4. The three planes do not intersect and are parallel.

5. The intersection of the three planes is two different lines.

6. A single plane represents all of the equations.

For exercises 7–10, use an algebraic method to check whether the ordered triple is a solution of the system of equations.

7. $(5, 6, 4)$; $x + y - z = 7$
$$3x + 2y - 6z = 3$$
$$2x - 5y + z = -16$$

8. $(7, 2, 9)$; $8x + y - 3z = 31$
$$5x + 7y + z = 58$$
$$-9x + 10y + 2z = -25$$

9. $(2, 7, -3)$; $-4x + 3y + z = 10$
$$6x + 5y - 2z = 53$$
$$7x + y - z = 18$$

10. $(-4, 5, -8)$; $-10x + 3y + 2z = 39$
$$3x - 7y + 5z = -87$$
$$9x + 2y - z = 38$$

For exercises 11–26,
(a) solve by elimination. If the system has infinitely many solutions, it is sufficient to say so. Parametric form is not required.
(b) if the solution is an ordered triple, check.

11. $x + y + z = 8$
$$2x + y + 3z = 12$$
$$x + 2y + z = 13$$

12. $x + y + z = 9$
$$3x + y + 2z = 17$$
$$x + 2y + z = 13$$

13. $x + y + z = -2$
$$x + 3y + 2z = -9$$
$$5x + 5y + 4z = -5$$

14. $x + y + z = -2$
$$5x + 2y + z = 3$$
$$4x + 4y + 3z = -7$$

15. $x + y + z = 9$
$$2x + 3y + z = 12$$
$$3x + 4y + 2z = 20$$

16. $x + y + z = 13$
$5x + y + 2z = 28$
$6x + 2y + 3z = 40$

17. $x + y + z = 13$
$5x + 7y + 2z = 42$
$6x + 8y + 3z = 55$

18. $x + y + z = 16$
$3x + 8y + 2z = 64$
$4x + 9y + 3z = 80$

19. $4x + 3y + 2z = 1$
$8x + 5y - 3z = -7$
$-7x - 2y + z = -9$

20. $6x + 5y + 2z = 28$
$3x + 2y - 5z = -9$
$-5x - 4y + z = -13$

21. $-4x + 3y - 6z = -11$
$11x + 2y + 4z = 28$
$3x + 8y - 8z = 6$

22. $-6x + 5y - 2z = -12$
$7x + 3y + 9z = 45$
$-5x + 13y + 5z = 21$

23. $2x - 5y + 3z = -41$
$3x + 4y + 8z = 42$
$5x + 6y - 9z = 64$

24. $3x - 2y + 8z = -1$
$4x + 9y + 10z = 92$
$6x + 5y + 11z = 70$

25. $9x + 2y + 3z = -28$
$-2x + 7y - 8z = -16$
$5x + 16y - 13z = -50$

26. $8x + 5y + 11z = -20$
$7x - 3y + 2z = -3$
$22x - y + 15z = -20$

27. We can rewrite $y = 5$ as a linear equation in two variables: $0x + 1y = 5$. Rewrite $5x + 8y = 20$ as a linear equation in three variables.

28. We can rewrite $x = 3$ as a linear equation in two variables: $1x + 0y = 3$. Rewrite $11y + 3z = 15$ as a linear equation in three variables.

For exercises 29–34,
(a) solve by elimination. If the system has infinitely many solutions, it is sufficient to say so. Parametric form is not required.
(b) if the solution is an ordered triple, check.

29. $9x + 7y - 3z = 157$
$-4x + 2y + 5z = -31$
$5x + 8y = 112$

30. $4x + 5y - 6z = 47$
$-7x + 3y + 4z = -51$
$8x + 11y = 59$

31. $3x + 5y + 8z = 46$
$-4x + 2y + 3z = 7$
$11y + 3z = 15$

32. $4x + 13y + 5z = 14$
$9x + 3y + 8z = 38$
$7y + 2z = -4$

33. $3x + 4y - z = 20$
$7x + 2y + 5z = 21$
$x + 6y + 2z = 2$

34. $5x + 8y - z = 42$
$2x + 4y + 7z = -20$
$x + 12y + 3z = -5$

35. Write a system of linear equations in three variables in which the solution is the ordered triple $(4, 1, 6)$.

36. Write a system of linear equations in three variables in which the solution is the ordered triple $(3, 2, 9)$.

For exercises 37–38, the system of equations has infinitely many solutions. Use set-builder notation (parametric form) to describe the solutions.

37. $2x + 2y + z = 8$
$x + 2y + 2z = 6$
$4x + 6y + 5z = 20$

38. $2x + 2y + z = 6$
$x + 2y + 2z = 11$
$5x + 6y + 4z = 23$

For exercises 39–42, we can create a system of three linear equations in three variables with infinitely many solutions. We begin by deciding on an ordered triple. For example, we can choose $(3, 4, 5)$. We now write a system of two linear equations in three variables in which this ordered triple is a solution. One example is $x + y + z = 12$ and $2x + y + 3z = 25$. To get the third equation in the system, we add these two equations together: $3x + 2y + 4z = 37$.

39. Solve this system to show that it has infinitely many solutions.

40. Describe the solution in exercise 39 more specifically using set-builder notation (parametric form).

41. Write a different system of three linear equations in three variables that has infinitely many solutions.

42. Describe the solution of the system in exercise 41 more specifically using set-builder notation (parametric form).

For exercises 43–46, to create a system of three linear equations in three variables with no solution, begin by deciding on an ordered triple. For example, choose $(3, 4, 5)$. Now write a system of two linear equations in three variables in which this ordered triple is a solution. One example is $x + y + z = 12$ and $2x + y + 3z = 25$. To get the third equation in the system, add these two equations together: $3x + 2y + 4z = 37$. Now change the constant in the last equation: $3x + 2y + 4z = 36$. This change shifts the position of the planes enough that they have no point in common.

43. Solve this system to show that it has no solutions.

44. Rewrite the last equation with a constant other than 36 or 37. Solve to show that this system also has no solution.

45. Write a different system of three linear equations in three variables that has no solution.

46. Write another system of three linear equations in three variables that has no solution.

For exercises 47–54, use a system of three linear equations in three variables and the five steps.

47. A hiker is making a high-energy snack mix from roasted almonds, M&M's® milk chocolate candies, and raisins. Six cups of snack mix should have 4170 calories, 44.1 g of fiber, and 62 g of protein. One cup of roasted almonds has 825 calories, 16.3 g fiber, and 30.5 g protein. One cup of the candies has 1020 calories, 5.8 g fiber, and 9 g protein. One cup of raisins has 435 calories, 5.4 g fiber, and 4.5 g protein. Find the needed amount of each ingredient.

48. A dietitian is planning a lunch of low-salt soda crackers, low-salt chicken noodle soup, and skim milk for a heart transplant patient. The lunch should provide 7.4 g fat, 19 mg cholesterol, and 926 mg sodium. One soda cracker contains 0.4 g fat, no cholesterol, and 19 mg sodium. One can of chicken noodle soup contains 3 g fat, 9 mg cholesterol, and 530 mg sodium. One cup of skim milk contains 0.2 g fat, 5 mg cholesterol, and 103 mg sodium. Find the amount of each food that should be in the lunch.

49. The staff of a physical therapy clinic includes physical therapists, treatment aides, and exercise aides. Each day the clinic can provide 16 hr of therapist time, 23 hr of treatment aide time, and 27 hr of exercise aide time. At an average visit, patients with shoulder injuries need 0.25 hr of therapist time, 0.5 hr of treatment aide time, and 0.5 hr of exercise aide time. Patients with back injuries need 0.5 hr of therapist time, 1 hr of treatment aide time, and 0.5 hr of exercise aide time. Patients with knee injuries need 0.5 hr of therapist time, 0.25 hr of treatment aide time, and 1 hr of exercise aide time. How many of each kind of patient can the clinic treat with the available time?

50. A math tutoring center provides help using human tutors, video tutoring on televisions, and computer tutoring. With its equipment and staff, the center can provide 25 hr of human tutoring, 19 hr of video tutoring, and 33 hr of computer tutoring per day. On an average visit, a pre-algebra student needs 0.25 hr of human tutoring, 0.25 hr of video tutoring, and 0.5 hr of computer tutoring. An elementary algebra student needs 0.5 hr of human tutoring, 0.25 hr of video tutoring, and 0.25 hr of computer tutoring. An intermediate algebra student needs 0.5 hr of human tutoring, 0.25 hr of video tutoring, and 0.5 hr of computer tutoring. For how many of each kind of student can the center provide services with the available equipment and staff?

51. A small business makes small, medium, and large packs for dogs. With its equipment and employees, the business can provide 45 hr of cutting, 160 hr of construction, and 12 hr of shipping per week. A small dog pack requires 0.25 hr of cutting, 1 hr of construction, and 0.10 hr for shipping. A medium dog pack requires 0.25 hr of cutting, 1.25 hr of construction, and 0.10 hr for shipping. A large dog pack requires 0.5 hr of cutting, 1.5 hr of construc-

tion, and 0.10 hr for shipping. How many of each kind of pack can the business make per week with the available equipment and employees?

52. An organic farm makes apricot, blackberry, and cantaloupe jam. With its equipment and employees, the farm can provide 60 hr of peeling, 105 hr of seeding, and 35 hr of mixing per week. A batch of apricot jam requires 1 hr of peeling, 0.5 hr of seeding, and 0.25 hr of mixing. A batch of blackberry jam requires 0 hr of peeling, 1 hr of seeding, and 0.25 hr of mixing. A batch of cantaloupe jam requires 1 hr of peeling, 0.25 hr of seeding, and 0.25 hr of mixing. How many batches of each kind of jam can the farm make per week with the available equipment and employees?

53. A daughter received an inheritance of $60,000. Tax-free bonds pay 6% annual interest. A certificate of deposit pays 5% annual interest. The average annual return of a mutual stock fund is 11%. The daughter's goal for total return per year from these investments is $4300. She decides to put twice as much in the mutual fund as she puts into bonds. Find the amount she should put in each kind of investment.

54. A lottery winner decides to invest $90,000 of his prize money in mutual funds. The average annual return of a high-risk foreign stock fund is 14%. The average annual return of a low-risk utility fund is 6%. The average annual return of a health sector fund is 9%. The winner's goal for total return per year from these investments is $8000. He wants to put three times as much in health sector fund as he does in the utility fund. Find the amount he should put in each kind of investment.

Problem Solving: Practice and Review

Follow your instructor's guidelines for using the five steps as outlined in Section 1.5, p. 51.

55. The Bisbee mines are located near Tucson, Arizona. The value of metals taken from the mines is described in 1975 dollars. Between 1975 and 2008, the value of a dollar decreased by about 53%. Find the value of metals taken from the mine in 2008 dollars. Round to the nearest tenth of a billion. (*Source:* cost.jsc.nasa.gov/inflateGDP.html)

Before the Bisbee mines closed in 1975, the local mines produced metals valued at $6.1 billion (at 1975 price). This . . . came from the estimated production of 8,032,352,000 lbs of copper, 2,871,786 ounces of gold, 77,162,986 ounces of silver, 304,627,600 lbs of lead and 371,945,900 lbs of zinc! (*Source:* www.queenminetour.com)

56. A 2009 poll asked 2013 students about their cell phone use and 1002 parents about cell phone use by their children. The students were ages 13–18 years and were currently enrolled in grades 7–12. Eighty-four percent of the students had cell phones. Predict how many of the students in the poll admitted to cheating at least once with a cell phone. Round to the nearest whole number.

65% of students with cell phones say they use them during school. Only 23% of parents whose children have cell phones think that they use them during school. . . . More than a third of teens with cell phones (35%) admit to cheating at least once with them. (*Source:* www.commonsensemedia .org, June 18, 2009)

57. The bill from the surgeon for a lumbar fusion of three vertebrae in the spine that took 3 hr 30 min was $15,874. The bill from the physician assistant (PA) for the surgery was $2951. The insurance company paid $13,899 to the surgeon and $2768.64 to the PA. The patient paid the remaining amount. Find the percent of the total bills that the patient paid. Round to the nearest tenth.

58. The line of best fit for the relationship shown in the graph is $y = \left(\dfrac{0.46\%}{1 \text{ year}}\right)x + 7.06\%$.

a. Describe what the slope of the line of best fit represents.

b. Describe what the y-intercept of the line of best fit represents.

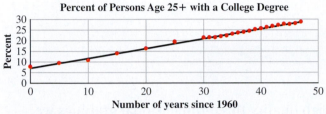

Percent of Persons Age 25+ with a College Degree

Source: www.census.gov, 2009

Find the Mistake

For exercises 59–62, the completed problem has one mistake.
(a) Describe the mistake in words, or copy down the whole problem and highlight or circle the mistake.
(b) Do the problem correctly.

59. Problem: Use an algebraic method to check whether $(2, -4, -7)$ is a solution of the system of equations.

$$x + y + z = -9$$
$$3x + 2y - 4z = 26$$
$$2x + 5y - z = -9$$

Incorrect Answer:

$$x + y + z = -9$$
$$2 + (-4) + (-7) = -9$$
$$-9 = -9 \quad \text{True.}$$

$$3x + 2y - 4z = 26$$
$$3(2) + 2(-4) - 4(-7) = 26$$
$$6 - 8 + 28 = 26$$
$$26 = 26 \quad \text{True.}$$

$$2x + 5y - z = -9$$
$$2(2) + 5(-4) - 7 = -9$$
$$4 - 20 - 7 = -9$$
$$-16 - 7 = -9 \quad \text{False.}$$

$(2, -4, -7)$ is *not* a solution of this system.

60. Problem: Assign the variables. Write a system of three equations in three variables to solve this problem.

An IRA account with $90,000 is being invested in stocks (average annual return 9%), bonds (average annual return 4%), and a real estate fund (average annual return 8%). The investor wants to put twice as much in real estate

as in stocks. The goal for annual return for the investments is $10,400. How much should be put in each kind of investment?

Incorrect Answer: $x =$ amount in stocks
$y =$ amount in bonds
$z =$ amount in real estate

$$x + y + z = \$90,000$$
$$0.09x + 0.04y + 0.08z = \$10,400$$
$$x = 2z$$

61. Problem: Assign the variables. Write a system of equations to solve this problem.

A small quilt company sells hand-made twin, full, and queen-sized quilts. With its equipment and employees, the business can provide 142 hr of cutting, 110 hr of sewing, and 271 hr of quilting per week. On average, a twin quilt requires 2 hr of cutting, 15 hr of sewing, and 20 hr of quilting. A full quilt requires 3 hr of cutting, 20 hr of sewing, and 30 hr of quilting. A queen quilt requires 3.5 hrs of cutting, 25 hr of sewing, and 40 hr of quilting. How many of each kind of quilt can the company make per week with the available equipment and employees?

Incorrect Answer: $x =$ twin quilts
$y =$ full quilts
$z =$ queen quilts

$$2x + 3y + 3.5z = 142$$
$$20x + 15y + 25z = 110$$
$$20x + 30y + 40z = 271$$

62. Problem: Solve $x + y + z = 9$
$x - y + z = 3$ by elimination.
$x + y - z = 1$

Incorrect Answer:

$$x + y + z = 9$$
$$\underline{+\ x - y + z = 3}$$
$$2x + 2z = 12$$

$$x + y + z = 9$$
$$\underline{+\ x + y - z = 1}$$
$$2x + 2y = 10$$

$$2x + 2z = 12$$
$$\underline{+\ -2x - 2y = -10}$$
$$0 + 0 = -2$$

This system has no solution.

Review

63. What is the reciprocal of 8?

64. What is the reciprocal of -8?

65. When we multiply a number by its reciprocal, what is the product?

66. The additive inverse of 7 is -7. What is the sum of a number and its additive inverse?

SUCCESS IN COLLEGE MATHEMATICS

67. In finding the slope of the line that passes through the points $(-3, -8)$ and $(-2, 6)$, a student used the slope formula, $m = \dfrac{y_2 - y_1}{x_2 - x_1}$, and incorrectly wrote

$$m = \frac{6 - (-8)}{-3 - (-2)}.$$

a. Describe the mistake that the student made.

b. What would you suggest that this student do to avoid making this mistake again?

68. In finding the slope of the line that passes through the points $(-3, -8)$ and $(-2, 6)$, a student used the slope formula, $m = \dfrac{y_2 - y_1}{x_2 - x_1}$, and wrote $m = \dfrac{6 - (-8)}{-2 - 3}$.

Simplifying, the student incorrectly wrote $m = \dfrac{-2}{-5}$ and then wrote $m = \dfrac{2}{5}$.

a. Describe the mistake that the student made.

b. What would you suggest that this student do to avoid making this mistake again?

© Angela Harburn/Shutterstock

Computers create the high quality three-dimensional graphics we expect in films and video games. These images are represented mathematically using matrices. In this section, we will use matrices to represent and solve a system of linear equations.

SECTION 4.5

Matrix Methods

After reading the text, working the practice problems, and completing assigned exercises, you should be able to:

1. Use an augmented matrix to represent a system of linear equations.

2. Use row operations to change an entry in a matrix to 0 or 1.

3. Solve a system of linear equations by rewriting its matrix in reduced row echelon form.

4. Use a system of three linear equations to solve an application problem.

Matrices

A **matrix** is an arrangement of numbers in rows and columns. The plural of matrix is **matrices**. This matrix has two rows and three columns. It is a 2×3 ("two by three") matrix. A matrix can represent the coefficients of a system of linear equations. Or an **augmented matrix** can represent the entire system. In an augmented matrix, a vertical line separates the coefficients of a system from its constants.

2 × 3 Matrix

$$\left[\begin{array}{rr|r} 4 & 5 & 1 \\ 3 & -2 & 8 \end{array}\right]$$

We often refer to the rows in a matrix using numbers, increasing from top to bottom. For efficiency, we also abbreviate the rows. For example, the top row, row 1, is r_1.

EXAMPLE 1   A system of linear equations in two variables is $\begin{array}{l} 3x + 2y = 26 \\ 5x - 6y = -22 \end{array}$. Write a

coefficients matrix and an augmented matrix for this system.

SOLUTION ▶   Coefficients matrix: $\begin{bmatrix} 3 & 2 \\ 5 & -6 \end{bmatrix}$ Augmented matrix: $\begin{bmatrix} 3 & 2 & 26 \\ 5 & -6 & -22 \end{bmatrix}$

If a variable is not included in an equation, its coefficient is 0. This 0 must be included in the matrix.

EXAMPLE 2   Represent $\begin{array}{l} x + y + 4z = 4 \\ x - y + z = 57 \\ 7x + 4y = 57 \end{array}$ with an augmented matrix.

SOLUTION ▶ $\begin{bmatrix} 1 & 1 & 4 & 4 \\ 1 & -1 & 1 & 57 \\ 7 & 4 & 0 & 57 \end{bmatrix}$

Practice Problems

For problems 1–2, represent each system of equations with an augmented matrix.

1. $\begin{array}{l} 2x + 7y = 58 \\ 3x - 5y = -53 \end{array}$ **2.** $\begin{array}{l} 2x + 7y + 3x = 24 \\ x + 8y - 2z = 6 \\ y - z = 10 \end{array}$

Row Operations

When using elimination to solve a system of linear equations, we often multiply one or both of the equations by a nonzero constant. Since a row in a matrix is a representation of an equation, we can also multiply each entry in a row of an augmented matrix by any real number except 0.

During elimination, we use the addition property of equality to add two equations together. Similarly, we can add two rows in a matrix together.

We can switch the positions of equations in a system of equations. This does not change the solution of the system. Similarly, we can switch the position of rows in a matrix. Doing these **row operations** does not change the solution of the system.

Row Operations That Do Not Change the Solution of a System of Equations

• Multiply the row by any real number except 0.

• Add two rows together.

• Switch the positions of rows in the matrix.

To change a number to 1, multiply it by its reciprocal.

EXAMPLE 3   A system of equations is represented by $\begin{bmatrix} 2 & 3 & 34 \\ 9 & -4 & 13 \end{bmatrix}$. Use a row operation to change the first entry in r_1 to 1.

SOLUTION ▶ To change 2 to 1, multiply row 1 by the reciprocal of 2, which is $\frac{1}{2}$.

$$\frac{1}{2}r_1 \qquad \text{The reciprocal of 2 is } \frac{1}{2}.$$

$$= \frac{1}{2}[2 \quad 3 \mid 34] \qquad r_1 = [2 \quad 3 \mid 34]$$

$$= \left[1 \quad \frac{3}{2} \mid 17\right] \qquad \text{Multiply each entry in the row by } \frac{1}{2}.$$

Use two row operations to change an entry to 0. In the next example, we first multiply row 1 by a constant and then add row 1 to row 2.

EXAMPLE 4 A system of equations is represented by $\begin{bmatrix} 1 & \frac{3}{2} & 17 \\ 9 & -4 & 13 \end{bmatrix}$.

(a) Use row operations to change the first entry in r_2 to 0.

SOLUTION ▶ The first entry in r_2 is 9. The opposite of 9 is -9. So multiply r_1 by -9 and then add the changed row to r_2.

$$-9r_1 \qquad \qquad r_1 = \left[1 \quad \frac{3}{2} \mid 17\right]$$

$$= -9\left[1 \quad \frac{3}{2} \mid 17\right] \qquad \text{Multiply: } -9r_1.$$

$$= \left[-9 \quad -\frac{27}{2} \mid -153\right] \qquad \text{Multiply each entry by } -9.$$

$$-9r_1 + r_2 \qquad \qquad \text{Add the rows together.}$$

$$= \left[-9 \quad -\frac{27}{2} \mid -153\right] + [9 \quad -4 \mid 13] \qquad r_2 = [9 \quad -4 \mid 13]$$

$$= \left[-9 \quad -\frac{27}{2} \mid -153\right] + \left[9 \quad -\frac{8}{2} \mid 13\right] \qquad -\frac{4}{1} \cdot \frac{2}{2} = -\frac{8}{2}$$

$$= \left[0 \quad -\frac{35}{2} \mid -140\right] \qquad \text{The first entry in } r_2 \text{ is now 0.}$$

(b) Rewrite the matrix.

▶ Although r_1 was used in the process of changing r_2, do not change it in the matrix. Write the original r_1.

$$\begin{bmatrix} 1 & \frac{3}{2} & 17 \\ 0 & -\frac{35}{2} & -140 \end{bmatrix}$$

Practice Problems

For problems 3–4, change the first entry in r_1 to 1. Rewrite the matrix.

3. $\begin{bmatrix} 4 & 8 & 12 \\ 3 & 6 & 15 \end{bmatrix}$ **4.** $\begin{bmatrix} -4 & 8 & 12 \\ 3 & 6 & 15 \end{bmatrix}$

For problems 5–6, change the first entry in r_2 to 0. Rewrite the matrix.

5. $\begin{bmatrix} 1 & 2 & 3 \\ 3 & 6 & 15 \end{bmatrix}$ **6.** $\begin{bmatrix} 1 & -2 & -3 \\ 3 & 6 & 15 \end{bmatrix}$

Reduced Row Echelon Form

An augmented matrix can represent a system of equations. To find the solution of the system, use row operations to rewrite the augmented matrix in **reduced row echelon form (rref)**.

> **Reduced Row Echelon Form (rref)**
> - The first entry in row 1 is 1.
> - Any row that includes only zeros is at the bottom of the matrix.
> - In any row that includes numbers other than 0, the first nonzero entry is a 1. This is a **leading 1**.
> - Each leading 1 is to the right of a leading 1 in any row above it. All of the other values in the column are 0.

EXAMPLE 5   Determine whether each matrix is in reduced row echelon form.

(a) $\begin{bmatrix} 1 & 0 & 0 & | & 4 \\ 0 & 0 & 1 & | & 3 \\ 0 & 1 & 0 & | & 9 \end{bmatrix}$

SOLUTION ▶ The leading 1 in row 3 is to the *left* of the leading 1 in row 2. The rules state that "each leading 1 is to the right of a leading 1 in any row above it." Since row 3 is below row 2, this matrix is *not* in reduced row echelon form.

(b) $\begin{bmatrix} 1 & 0 & 0 & | & 4 \\ 0 & 1 & 0 & | & 3 \\ \mathbf{0} & \mathbf{0} & \mathbf{0} & | & \mathbf{0} \end{bmatrix}$

▶ *Yes*, this matrix is in reduced row echelon form. Notice that the row that includes only zeros is at the bottom of the matrix.

To solve a system of equations, we can represent the system as an augmented matrix and then rewrite the matrix in reduced row echelon form.

EXAMPLE 6   The matrix is in reduced row echelon form. Identify the solution of the system of equations represented by the matrix.

(a) $\begin{bmatrix} 1 & 0 & | & 5 \\ 0 & 1 & | & -6 \end{bmatrix}$

SOLUTION ▶ This matrix represents the system of equations $\begin{matrix} 1x + 0y = 5 \\ 0x + 1y = -6 \end{matrix}$. Simplifying, $x = 5$ and $y = -6$. The solution is $(5, -6)$.

(b) $\begin{bmatrix} 1 & 0 & 0 & | & 9 \\ 0 & 1 & 0 & | & \dfrac{1}{2} \\ 0 & 0 & 1 & | & 3 \end{bmatrix}$

▶ This matrix represents the system of equations $\begin{matrix} 1x + 0y + 0z = 9 \\ 0x + 1y + 0z = \dfrac{1}{2} \\ 0x + 0y + 1z = 3 \end{matrix}$.

Simplifying, $x = 9$, $y = \dfrac{1}{2}$, and $z = 3$. The solution is $\left(9, \dfrac{1}{2}, 3\right)$.

EXAMPLE 7 The system $\begin{array}{l} x + 3y = 5 \\ 4x - 5y = 3 \end{array}$ is represented by $\begin{bmatrix} 1 & 3 & 5 \\ 4 & -5 & 3 \end{bmatrix}$. Solve this system by rewriting the matrix in reduced row echelon form.

(a) Change the first entry in r_1 to 1.

SOLUTION ▶   This entry is already 1.

(b) Change all the other entries in the first column to 0.

▶

$-4r_1 + r_2$

$= -4[1 \quad 3 \,|\, 5] + [4 \quad -5 \,|\, 3]$

$= [-4 \quad -12 \,|\, -20] + [4 \quad -5 \,|\, 3]$

$= [0 \quad -17 \,|\, -17]$

The other entry in the first column is 4. To change 4 to 0, multiply r_1 by -4 and add the result to r_2.

$\begin{bmatrix} 1 & 3 & 5 \\ 0 & -17 & -17 \end{bmatrix}$

Rewrite the matrix.

(c) Change the second entry in r_2 of the new matrix to 1. This will be the leading 1 in this row, and it will be to the right of the leading 1 above it.

▶

$-\dfrac{1}{17} r_2$

$= -\dfrac{1}{17}[0 \quad -17 \,|\, -17]$

$= [0 \quad 1 \,|\, 1]$

To change the second entry, -17, in r_2 to 1, multiply by the reciprocal of -17.

$\begin{bmatrix} 1 & 3 & 5 \\ 0 & 1 & 1 \end{bmatrix}$

Rewrite the matrix.

(d) Change the other entries in the second column to 0.

▶

$-3r_2 + r_1$

$= -3[0 \quad 1 \,|\, 1] + [1 \quad 3 \,|\, 5]$

$= [0 \quad -3 \,|\, -3] + [1 \quad 3 \,|\, 5]$

$= [1 \quad 0 \,|\, 2]$

The other entry in the second column of the new matrix is 3. To change 3 to 0, multiply r_2 by -3 and add the result to r_1.

$\begin{bmatrix} 1 & 0 & 2 \\ 0 & 1 & 1 \end{bmatrix}$

Rewrite the matrix.

(e) Identify the solution of the system.

▶   The matrix is in reduced row echelon form. It represents the system of equations $\begin{array}{l} 1x + 0y = 2 \\ 0x + 1y = 1 \end{array}$. Simplifying, $x = 2$ and $y = 1$. The solution of the system is $(2, 1)$.

Row operations often require arithmetic with fractions.

EXAMPLE 8 The system $\begin{array}{l} 3x + 4y = 11 \\ 5x - 6y = 12 \end{array}$ is represented by $\begin{bmatrix} 3 & 4 & 11 \\ 5 & -6 & 12 \end{bmatrix}$. Solve this system by rewriting the matrix in reduced row echelon form.

(a) Change the first entry in r_1 to 1.

SOLUTION ▶ $\dfrac{1}{3} r_1$

To change the first entry, 3, in r_1 to 1, multiply by the reciprocal of 3.

$= \dfrac{1}{3}[3 \quad 4 \,|\, 11]$

$= \left[1 \quad \dfrac{4}{3} \,\middle|\, \dfrac{11}{3}\right]$

$\begin{bmatrix} 1 & \dfrac{4}{3} & \dfrac{11}{3} \\ 5 & -6 & 12 \end{bmatrix}$

Rewrite the matrix.

(b) Change all the other entries in the first column to 0.

▶ $-5r_1 + r_2$

The other entry in the first column is 5. To change 5 to 0, multiply r_1 by -5 and add the result to r_2.

$$= -5\begin{bmatrix} 1 & \dfrac{4}{3} & \bigm| & \dfrac{11}{3} \end{bmatrix} + \begin{bmatrix} 5 & -6 & \bigm| & 12 \end{bmatrix}$$

$$= \begin{bmatrix} -5 & -\dfrac{20}{3} & \bigm| & -\dfrac{55}{3} \end{bmatrix} + \begin{bmatrix} 5 & -\dfrac{18}{3} & \bigm| & \dfrac{36}{3} \end{bmatrix}$$

$$= \begin{bmatrix} 0 & -\dfrac{38}{3} & \bigm| & -\dfrac{19}{3} \end{bmatrix}$$

$$\begin{bmatrix} 1 & \dfrac{4}{3} & \bigm| & \dfrac{11}{3} \\ 0 & -\dfrac{38}{3} & \bigm| & -\dfrac{19}{3} \end{bmatrix}$$

Rewrite the matrix.

(c) Change the second entry in r_2 to 1. This will be the leading 1 in this row, and it will be to the right of the leading 1 above it.

▶ $-\dfrac{3}{38}r_2$

To change the second entry, $-\dfrac{38}{3}$, in r_2 to 1, multiply by the reciprocal of $-\dfrac{38}{3}$.

$$= -\dfrac{3}{38}\begin{bmatrix} 0 & -\dfrac{38}{3} & \bigm| & -\dfrac{19}{3} \end{bmatrix}$$

$$= \begin{bmatrix} 0 & 1 & \bigm| & \dfrac{1}{2} \end{bmatrix}$$

$$\begin{bmatrix} 1 & \dfrac{4}{3} & \bigm| & \dfrac{11}{3} \\ 0 & 1 & \bigm| & \dfrac{1}{2} \end{bmatrix}$$

Rewrite the matrix.

(d) Change the other entries in the second column to 0.

▶ $-\dfrac{4}{3}r_2 + r_1$

The other entry in the second column of the new matrix is $\dfrac{4}{3}$. To change $\dfrac{4}{3}$ to 0, multiply r_2 by $-\dfrac{4}{3}$ and add the result to r_1.

$$= -\dfrac{4}{3}\begin{bmatrix} 0 & 1 & \bigm| & \dfrac{1}{2} \end{bmatrix} + \begin{bmatrix} 1 & \dfrac{4}{3} & \bigm| & \dfrac{11}{3} \end{bmatrix}$$

$$= \begin{bmatrix} 0 & -\dfrac{4}{3} & \bigm| & -\dfrac{2}{3} \end{bmatrix} + \begin{bmatrix} 1 & \dfrac{4}{3} & \bigm| & \dfrac{11}{3} \end{bmatrix}$$

$$= \begin{bmatrix} 1 & 0 & \bigm| & 3 \end{bmatrix}$$

$$\begin{bmatrix} 1 & 0 & \bigm| & 3 \\ 0 & 1 & \bigm| & \dfrac{1}{2} \end{bmatrix}$$

Rewrite the matrix.

The matrix is in reduced row echelon form.

(e) Identify the solution of the system.

▶ The solution of the system is $\left(3, \dfrac{1}{2}\right)$.

As the number of equations and variables increases, rewriting a matrix in reduced row echelon form is usually faster than solving a system of equations by elimination.

EXAMPLE 9 The system $\begin{aligned} 2x + y + z &= 21 \\ 3x + 2y - z &= 12 \\ 4x - y + 2z &= 24 \end{aligned}$ is represented by $\begin{bmatrix} 2 & 1 & 1 & \bigm| & 21 \\ 3 & 2 & -1 & \bigm| & 12 \\ 4 & -1 & 2 & \bigm| & 24 \end{bmatrix}$. Solve this system by rewriting the matrix in reduced row echelon form.

(a) Change the first entry in r_1 to 1.

SOLUTION ►

$$\frac{1}{2}r_1$$

$$= \frac{1}{2}[2 \ 1 \ 1 \,|\, 21]$$

$$= \left[1 \ \frac{1}{2} \ \frac{1}{2} \,\middle|\, \frac{21}{2}\right]$$

$$\begin{bmatrix} 1 & \frac{1}{2} & \frac{1}{2} & \frac{21}{2} \\ 3 & 2 & -1 & 12 \\ 4 & -1 & 2 & 24 \end{bmatrix}$$

Rewrite the matrix.

(b) Change all the other entries in the first column to 0.

$$-3r_1 + r_2$$

$$= -3\left[1 \ \frac{1}{2} \ \frac{1}{2} \,\middle|\, \frac{21}{2}\right] + [3 \ 2 \ -1 \,|\, 12]$$

$$= \left[-3 \ -\frac{3}{2} \ -\frac{3}{2} \,\middle|\, -\frac{63}{2}\right] + \left[3 \ \frac{4}{2} \ -\frac{2}{2} \,\middle|\, \frac{24}{2}\right]$$

$$= \left[0 \ \frac{1}{2} \ -\frac{5}{2} \,\middle|\, -\frac{39}{2}\right]$$

$$\begin{bmatrix} 1 & \frac{1}{2} & \frac{1}{2} & \frac{21}{2} \\ 0 & \frac{1}{2} & -\frac{5}{2} & -\frac{39}{2} \\ 4 & -1 & 2 & 24 \end{bmatrix}$$

Rewrite the matrix.

$$-4r_1 + r_3$$

$$= -4\left[1 \ \frac{1}{2} \ \frac{1}{2} \,\middle|\, \frac{21}{2}\right] + [4 \ -1 \ 2 \,|\, 24]$$

$$= [-4 \ -2 \ -2 \,|\, -42] + [4 \ -1 \ 2 \,|\, 24]$$

$$= [0 \ -3 \ 0 \,|\, -18]$$

$$\begin{bmatrix} 1 & \frac{1}{2} & \frac{1}{2} & \frac{21}{2} \\ 0 & \frac{1}{2} & -\frac{5}{2} & -\frac{39}{2} \\ 0 & -3 & 0 & -18 \end{bmatrix}$$

Rewrite the matrix.

(c) Change the second entry in r_2 of the new matrix to 1. This will be the leading 1 in this row, and it will be to the right of the leading 1 above it.

$$2r_2$$

$$= 2\left[0 \ \frac{1}{2} \ -\frac{5}{2} \,\middle|\, -\frac{39}{2}\right]$$

$$= [0 \ 1 \ -5 \,|\, -39]$$

$$\begin{bmatrix} 1 & \frac{1}{2} & \frac{1}{2} & \frac{21}{2} \\ 0 & 1 & -5 & -39 \\ 0 & -3 & 0 & -18 \end{bmatrix}$$

Rewrite the matrix.

(d) Change all the other entries in the second column to 0.

$$-\frac{1}{2}r_2 + r_1$$

$$= -\frac{1}{2}[0 \ 1 \ -5 \,|\, -39] + \left[1 \ \frac{1}{2} \ \frac{1}{2} \,\middle|\, \frac{21}{2}\right]$$

$$= \left[0 \ -\frac{1}{2} \ \frac{5}{2} \,\middle|\, \frac{39}{2}\right] + \left[1 \ \frac{1}{2} \ \frac{1}{2} \,\middle|\, \frac{21}{2}\right]$$

$$= [1 \ 0 \ 3 \,|\, 30]$$

$$\begin{bmatrix} 1 & 0 & 3 & 30 \\ 0 & 1 & -5 & -39 \\ 0 & -3 & 0 & -18 \end{bmatrix}$$

Rewrite the matrix.

$$3r_2 + r_3$$

$$= 3[0 \ 1 \ -5 \,|\, -39] + [0 \ -3 \ 0 \,|\, -18]$$

$$= [0 \ 3 \ -15 \,|\, -117] + [0 \ -3 \ 0 \,|\, -18]$$

$$= [0 \ 0 \ -15 \,|\, -135]$$

$$\begin{bmatrix} 1 & 0 & 3 & 30 \\ 0 & 1 & -5 & -39 \\ 0 & 0 & -15 & -135 \end{bmatrix}$$

Rewrite the matrix.

(e) Change the third entry in r_3 of the new matrix to 1. This will be the leading 1 in this row, and it will be to the right of the leading 1 above it.

$$-\frac{1}{15}r_3$$

$$= -\frac{1}{15}[0\ \ 0\ \ {-15}\ |\ {-135}]$$

$$= [0\ \ 0\ \ 1\,|\,9]$$

$$\begin{bmatrix} 1 & 0 & 3 & | & 30 \\ 0 & 1 & -5 & | & -39 \\ 0 & 0 & 1 & | & 9 \end{bmatrix}$$

Rewrite the matrix.

(f) Change all the other entries in the third column to 0.

$$-3r_3 + r_1$$

$$= -3[0\ \ 0\ \ 1\,|\,9] + [1\ \ 0\ \ 3\,|\,30]$$

$$= [\mathbf{0}\ \ \mathbf{0}\ \ {-\mathbf{3}}\,|\,{-\mathbf{27}}] + [1\ \ 0\ \ 3\,|\,30]$$

$$= [1\ \ 0\ \ 0\,|\,3]$$

$$\begin{bmatrix} \mathbf{1} & \mathbf{0} & \mathbf{0} & | & \mathbf{3} \\ 0 & 1 & -5 & | & -39 \\ 0 & 0 & 1 & | & 9 \end{bmatrix}$$

Rewrite the matrix.

$$5r_3 + r_2$$

$$= 5[0\ \ 0\ \ 1\,|\,9] + [0\ \ 1\ \ {-5}\,|\,{-39}]$$

$$= [\mathbf{0}\ \ \mathbf{0}\ \ \mathbf{5}\,|\,\mathbf{45}] + [0\ \ 1\ \ {-5}\,|\,{-39}]$$

$$= [0\ \ 1\ \ 0\,|\,6]$$

$$\begin{bmatrix} 1 & 0 & 0 & | & 3 \\ \mathbf{0} & \mathbf{1} & \mathbf{0} & | & \mathbf{6} \\ 0 & 0 & 1 & | & 9 \end{bmatrix}$$

Rewrite the matrix.

(g) Identify the solution of the system.

The solution of the system is the ordered triple $(3, 6, 9)$.

The system in Example 9 has one solution. A system of linear equations may instead have no solution or infinitely many solutions.

EXAMPLE 10 The system $\begin{aligned} 2x - y &= -7 \\ -4x + 2y &= 15 \end{aligned}$ is represented by $\begin{bmatrix} 2 & -1 & | & -7 \\ -4 & 2 & | & 15 \end{bmatrix}$. Solve this system by rewriting the matrix in reduced row echelon form.

(a) Change the first entry in r_1 to 1.

SOLUTION $$\frac{1}{2}r_1$$

$$= \frac{1}{2}[2\ \ {-1}\,|\,{-7}]$$

$$= \left[1\ \ {-\frac{1}{2}}\ \middle|\ {-\frac{7}{2}}\right]$$

$$\begin{bmatrix} 1 & -\dfrac{1}{2} & | & -\dfrac{7}{2} \\ -4 & 2 & | & 15 \end{bmatrix}$$

Rewrite the matrix.

(b) Change all the other entries in the first column to 0.

$$4r_1 + r_2$$

$$= 4\left[1\ \ {-\frac{1}{2}}\ \middle|\ {-\frac{7}{2}}\right] + [-4\ \ 2\,|\,15]$$

$$= [\mathbf{4}\ \ {-\mathbf{2}}\,|\,{-\mathbf{14}}] + [-4\ \ 2\,|\,15]$$

$$= [0\ \ 0\,|\,1]$$

$$\begin{bmatrix} 1 & -\dfrac{1}{2} & | & -\dfrac{7}{2} \\ \mathbf{0} & \mathbf{0} & | & \mathbf{1} \end{bmatrix}$$

Rewrite the matrix.

When we change the first entry in r_2 to 0, the second entry in r_2 is also 0. The bottom row represents an equation that cannot be true: $0x + 0y = 1$.

(c) Identify the solution of the system.

▶ This system has no solution.

In the next example, the system has infinitely many solutions. The bottom row in the matrix in reduced row echelon form is all zeros.

EXAMPLE 11 The system $\begin{aligned} 3x + 2y + z &= 13 \\ x + y + 2z &= 6 \\ 5x + 4y + 5z &= 25 \end{aligned}$ is represented by $\begin{bmatrix} 3 & 2 & 1 & | & 13 \\ 1 & 1 & 2 & | & 6 \\ 5 & 4 & 5 & | & 25 \end{bmatrix}$. Solve this system by rewriting the matrix in reduced row echelon form.

(a) Change the first entry in r_1 to 1.

SOLUTION ▶ We can multiply the first row by $\frac{1}{3}$, or we can switch the positions of r_1 and r_2. Switching rows does not change the solution of the system.

$\begin{bmatrix} 1 & 1 & 2 & | & 6 \\ 3 & 2 & 1 & | & 13 \\ 5 & 4 & 5 & | & 25 \end{bmatrix}$
Rewrite the matrix.

(b) Change all the other entries in the first column of the new matrix to 0.

▶ $-3r_1 + r_2$
$= -3[1\ 1\ 2\,|\,6] + [3\ 2\ 1\,|\,13]$
$= [-3\ -3\ -6\,|\,-18] + [3\ 2\ 1\,|\,13]$
$= [0\ -1\ -5\,|\,-5]$

$\begin{bmatrix} 1 & 1 & 2 & | & 6 \\ 0 & -1 & -5 & | & -5 \\ 5 & 4 & 5 & | & 25 \end{bmatrix}$
Rewrite the matrix.

$-5r_1 + r_3$
$= -5[1\ 1\ 2\,|\,6] + [5\ 4\ 5\,|\,25]$
$= [-5\ -5\ -10\,|\,-30] + [5\ 4\ 5\,|\,25]$
$= [0\ -1\ -5\,|\,-5]$

$\begin{bmatrix} 1 & 1 & 2 & | & 6 \\ 0 & -1 & -5 & | & -5 \\ 0 & -1 & -5 & | & -5 \end{bmatrix}$
Rewrite the matrix.

(c) Change the second entry in r_2 of the new matrix to 1. This will be the leading 1 in this row, and it will be to the right of the leading 1 above it.

▶ $-1r_2$
$= -1[0\ -1\ -5\,|\,-5]$
$= [0\ 1\ 5\,|\,5]$

$\begin{bmatrix} 1 & 1 & 2 & | & 6 \\ 0 & 1 & 5 & | & 5 \\ 0 & -1 & -5 & | & -5 \end{bmatrix}$
Rewrite the matrix.

(d) Change all the other entries in the second column to 0.

▶ $-1r_2 + r_1$
$= -1[0\ 1\ 5\,|\,5] + [1\ 1\ 2\,|\,6]$
$= [0\ -1\ -5\,|\,-5] + [1\ 1\ 2\,|\,6]$
$= [1\ 0\ -3\,|\,1]$

$\begin{bmatrix} 1 & 0 & -3 & | & 1 \\ 0 & 1 & 5 & | & 5 \\ 0 & -1 & -5 & | & -5 \end{bmatrix}$
Rewrite the matrix.

$r_2 + r_3$
$= [0\ 1\ 5\,|\,5] + [0\ -1\ -5\,|\,-5]$
$= [0\ 0\ 0\,|\,0]$

$\begin{bmatrix} 1 & 0 & -3 & | & 1 \\ 0 & 1 & 5 & | & 5 \\ 0 & 0 & 0 & | & 0 \end{bmatrix}$
Rewrite the matrix.

(e) Identify the solution of the system.

▶ This matrix is in reduced row echelon form. The bottom row represents an equation that is always true: $0x + 0y + 0z = 0$. The system of equations has infinitely many solutions.

We can describe the solution of Example 11 in a more specific way. The first row of the matrix in reduced row echelon form represents $1x + -3z = 1$. Solving for x, we have $x = 3z + 1$. The second row represents $1y + 5z = 5$. Solving for y, $y = -5z + 5$. The third equation tells us that when x and y depend on z in this way, z can be any real number. Using set-builder notation, the solution is $\{(x, y, z) \mid x = 3z + 1, y = -5z + 5, z \text{ is a real number}\}$. This solution is now in parametric form.

Practice Problems

For problems 7–9, solve each system by rewriting its matrix in reduced row echelon form.

7. $3x + 6y = 27$
$\quad\ 5x + 2y = 29$

8. $3x + 4y + z = 44$
$\quad\ x - 2y + z = -8$
$\quad\ 2x + 3y - 2z = 16$

9. $\quad\ 3x - y = 18$
$\quad -6x + 2y = -37$

Using Technology: Matrices and rref

EXAMPLE 12 Solve $\begin{array}{l} 2x + 3y = 7 \\ 9x + 11y = 34 \end{array}$ by rewriting $\left[\begin{array}{cc|c} 2 & 3 & 7 \\ 9 & 11 & 34 \end{array}\right]$ in reduced row echelon form.

Press 2nd x^{-1} to open the Matrix menu. If the calculator has done matrix work before, the rows × columns of the last matrix appears next to the name of the matrix. If the matrix memory on the calculator is empty, this will be blank.

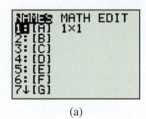

(a)

The last matrix in this calculator had 1 row and 1 column.

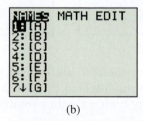

(b)

The matrix memory in this calculator is empty.

To work with choice 1, Matrix A, move the cursor to the right and highlight EDIT. Press ENTER . To enter a matrix with 2 rows and 3 columns, a 2 × 3 matrix, first change the row and column.

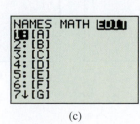

(c)   (d)

Type [2] as the new value for rows. Move the cursor to the columns; type [3]. Press [ENTER]. A matrix appears with a 0 in each position.

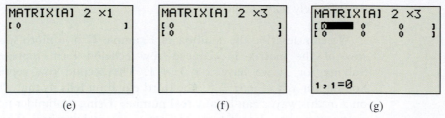

(e)   (f)   (g)

To enter the values from the matrix, type [2]. Press [ENTER]. A 2 appears in the first row, first column. The cursor moves to the first row, second column. Type [3].

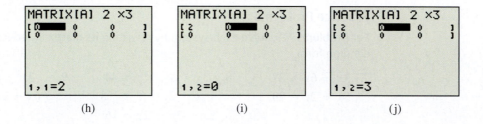

(h)   (i)   (j)

Press [ENTER]. A 3 appears in the first row, second column. Repeat this process until all of the values are in the matrix.

To find the reduced row echelon form of the matrix, return to the home screen by pressing [2nd] [MODE]. If necessary, press [CLEAR] to erase the screen. Now go back to the Matrix screen; press [2nd] [x^{-1}]. Move the cursor to the right to highlight MATH.

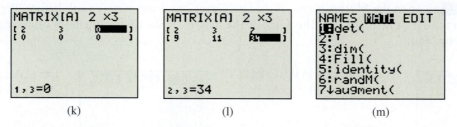

(k)   (l)   (m)

Move the cursor down, passing choice 7, until choice B appears, **rref (**. Press [ENTER]. **rref (** appears on the home screen. Press [2nd] [x^{-1}]. The highlighted choice is Matrix A. Press [ENTER]. Now **rref ([A]** appears on the screen. Close the expression by typing a right parenthesis, [ )].

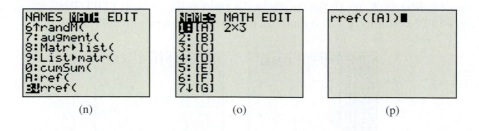

(n)   (o)   (p)

Press ENTER . The screen shows the reduced row echelon form: $\begin{bmatrix} 1 & 0 & 5 \\ 0 & 1 & -1 \end{bmatrix}$.

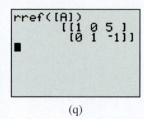

The solution of this system of equations is $(5, -1)$.

(q)

Practice Problems For problems 10–13,

(a) represent the system of equations with an argumented matrix.

(b) use a graphing calculator and **rref** to rewrite the matrix in reduced row echelon form.

(c) identify the solution.

10. $x + 3y = 5$
    $4x - 5y = 3$

11. $2x + y + z = 21$
    $3x + 2y - z = 12$
    $4x - y + 2z = 24$

12. $3x + 2y + z = 13$
    $x + y + 2z = 6$
    $5x + 4y + 5z = 25$

13. $2x - y = -7$
    $4x + 2y = 15$

4.5 VOCABULARY PRACTICE

Match the term with its description.

1. An arrangement of numbers in rows and columns
2. In this matrix, a vertical line separates the coefficients from the constants.
3. A system of equations is $\begin{matrix} 4x + y = 8 \\ 2x + 3y = 20 \end{matrix}$. An example of this kind of matrix is $\begin{bmatrix} 4 & 1 \\ 2 & 3 \end{bmatrix}$.
4. Switching the position of rows in a matrix is an example of this.
5. The product of this and any real number except 0 is 1.
6. The sum of a real number and this is 0.
7. The matrix $\begin{bmatrix} 1 & 0 & 0 & 4 \\ 0 & 1 & 0 & 9 \\ 0 & 0 & 1 & 3 \end{bmatrix}$ is in this form.
8. The first nonzero entry that is a 1 in a row
9. The type of solution of a system in which the bottom row of its augmented matrix in reduced row echelon form is all zeros to the left of the vertical line and not zero to the right of the vertical line
10. The type of solution of a system of equations represented by this matrix: $\begin{bmatrix} 1 & 3 & 6 \\ 0 & 0 & 0 \end{bmatrix}$

A. additive inverse
B. augmented matrix
C. coefficients matrix
D. infinitely many solutions
E. leading 1
F. matrix
G. no solution
H. reciprocal
I. reduced row echelon form
J. row operation

4.5 Exercises

Follow your instructor's guidelines for showing your work.

For exercises 1–6, represent each system of equations with an augmented matrix.

1. $3x + 8y = 19$
   $9x + 2y = 13$

2. $7x + 9y = 23$
   $6x + 5y = 17$

3. $2x + 5y + z = 18$
   $-6x + 5z = 26$
   $2x + 7y + 3z = 26$

4. $5x + 2y + z = 25$
   $-8x + 12z = -17$
   $7x + y - 4z = 26$

5. $x - y - z = -7$
   $x + y - z = 11$
   $x + y + z = 15$

6. $x - y - z = -11$
   $x + y - z = 3$
   $x + y + z = 21$

For exercises 7–12, use row operations to change the first entry in the first row to 1. Rewrite the matrix.

7. $\left[\begin{array}{cc|c} 6 & 12 & 24 \\ 9 & 3 & 17 \end{array}\right]$

8. $\left[\begin{array}{cc|c} 4 & 12 & 36 \\ 2 & 5 & 11 \end{array}\right]$

9. $\left[\begin{array}{ccc|c} -3 & 8 & 15 & 17 \\ -1 & 6 & 7 & 8 \\ 5 & 20 & 2 & 11 \end{array}\right]$

10. $\left[\begin{array}{ccc|c} -5 & 2 & 15 & 3 \\ 2 & -4 & 4 & 7 \\ 9 & 7 & 9 & 4 \end{array}\right]$

11. $\left[\begin{array}{cc|c} -1 & 5 & -13 \\ 4 & 1 & 11 \end{array}\right]$

12. $\left[\begin{array}{cc|c} -1 & 7 & -19 \\ 3 & 1 & 8 \end{array}\right]$

For exercises 13–22, use row operations to change the first entry in the second row to 0. Rewrite the matrix.

13. $\left[\begin{array}{cc|c} 1 & 2 & 10 \\ 3 & 5 & 23 \end{array}\right]$

14. $\left[\begin{array}{cc|c} 1 & 3 & 22 \\ 4 & 7 & 43 \end{array}\right]$

15. $\left[\begin{array}{ccc|c} 1 & -7 & 1 & -6 \\ -3 & 4 & 9 & -41 \\ 5 & 2 & 1 & -3 \end{array}\right]$

16. $\left[\begin{array}{ccc|c} 1 & -3 & 1 & -21 \\ -6 & 5 & 2 & 6 \\ 2 & 9 & 1 & 80 \end{array}\right]$

17. $\left[\begin{array}{cc|c} 1 & 4 & 3 \\ 3 & 2 & -\dfrac{11}{5} \\ 5 & 5 & \end{array}\right]$

18. $\left[\begin{array}{cc|c} 1 & 6 & 21 \\ 5 & 6 & \dfrac{9}{7} \\ 7 & 7 & \end{array}\right]$

19. $\left[\begin{array}{cc|c} 1 & 3 & 5 \\ -2 & -6 & 11 \end{array}\right]$

20. $\left[\begin{array}{cc|c} 1 & 4 & 9 \\ -3 & -12 & 17 \end{array}\right]$

21. $\left[\begin{array}{cc|c} 1 & 3 & 5 \\ 2 & 6 & 10 \end{array}\right]$

22. $\left[\begin{array}{cc|c} 1 & 4 & 9 \\ 3 & 12 & 27 \end{array}\right]$

23. Explain how to change the first entry in the first row of a matrix to 1. Assume that this entry is not 0.

24. The first entry in the first row of a matrix is 1. Explain how to change the first entry in the second row to 0.

For exercises 25–36, a matrix that represents a system of equations is in reduced row echelon form. Identify the solution of the system of equations. If the system has infinitely many solutions, it is sufficient to say so. Parametric form is not required.

25. $\left[\begin{array}{cc|c} 1 & 0 & -6 \\ 0 & 1 & 5 \end{array}\right]$

26. $\left[\begin{array}{cc|c} 1 & 0 & 11 \\ 0 & 1 & -23 \end{array}\right]$

27. $\left[\begin{array}{ccc|c} 1 & 0 & 0 & -8 \\ 0 & 1 & 0 & 4 \\ 0 & 0 & 1 & 9 \end{array}\right]$

28. $\left[\begin{array}{ccc|c} 1 & 0 & 0 & 3 \\ 0 & 1 & 0 & -2 \\ 0 & 0 & 1 & 10 \end{array}\right]$

29. $\left[\begin{array}{cc|c} 1 & 2 & -6 \\ 0 & 0 & 1 \end{array}\right]$

30. $\left[\begin{array}{cc|c} 1 & 8 & 4 \\ 0 & 0 & 1 \end{array}\right]$

31. $\left[\begin{array}{cc|c} 1 & 2 & -6 \\ 0 & 0 & 0 \end{array}\right]$

32. $\left[\begin{array}{cc|c} 1 & 8 & 4 \\ 0 & 0 & 0 \end{array}\right]$

33. $\left[\begin{array}{ccc|c} 1 & 0 & -2 & 6 \\ 0 & 1 & 1 & 8 \\ 0 & 0 & 0 & 1 \end{array}\right]$

34. $\begin{bmatrix} 1 & 0 & 9 & | & 8 \\ 0 & 1 & 1 & | & 4 \\ 0 & 0 & 0 & | & 1 \end{bmatrix}$

35. $\begin{bmatrix} 1 & 0 & -2 & | & 6 \\ 0 & 1 & 1 & | & 8 \\ 0 & 0 & 0 & | & 0 \end{bmatrix}$

36. $\begin{bmatrix} 1 & 0 & 9 & | & 8 \\ 0 & 1 & 1 & | & 4 \\ 0 & 0 & 0 & | & 0 \end{bmatrix}$

For exercises 37–52,
(a) represent the system of equations with an augmented matrix.
(b) solve the system by rewriting the matrix in reduced row echelon form. If there are infinitely many solutions, it is sufficient to say so. Parametric form is not required.
(c) if the solution is an ordered pair, check.

37. $3x + 5y = 1$
 $3x + y = 5$

38. $4x + 3y = 5$
 $4x + y = 7$

39. $2x - 3y = -13$
 $x + 3y = 7$

40. $2x - 5y = -16$
 $x + 4y = 5$

41. $-2x + 3y = 12$
 $2x + y = 4$

42. $-5x + 2y = 6$
 $5x + y = 3$

43. $-2x + 3y = 12$
 $2x + y = 20$

44. $-5x + 2y = -29$
 $5x + y = 38$

45. $4x + 3y = 25$
 $2x - y = -5$

46. $6x - 5y = -23$
 $3x - y = -1$

47. $4x - 5y = 4$
 $4x - 5y = 3$

48. $2x - 3y = 9$
 $2x - 3y = 1$

49. $3x + 5y = -14$
 $9x + 2y = 10$

50. $5x + 2y = 20$
 $10x + 3y = 45$

51. $2x - y = 6$
 $-4x + 2y = -12$

52. $3x - y = 8$
 $-6x + 2y = -16$

53. We can represent the system of equations in exercise 37 with two lines. Are these intersecting lines, parallel lines, or coinciding lines?

54. We can represent the system of equations in exercise 38 with two lines. Are these intersecting lines, parallel lines, or coinciding lines?

55. We can represent the system of equations in exercise 47 with two lines. Are these intersecting lines, parallel lines, or coinciding lines?

56. We can represent the system of equations in exercise 48 with two lines. Are these intersecting lines, parallel lines, or coinciding lines?

57. We can represent the system of equations in exercise 51 with two lines. Are these intersecting lines, parallel lines, or coinciding lines?

58. We can represent the system of equations in exercise 52 with two lines. Are these intersecting lines, parallel lines, or coinciding lines?

59. Use set-builder notation (parametric form) to write the solution for exercise 51.

60. Use set-builder notation (parametric form) to write the solution for exercise 52.

For exercises 61–72,
(a) represent the system of equations with an augmented matrix.
(b) solve the system by rewriting the matrix in reduced row echelon form. If there are infinitely many solutions, it is sufficient to say so. Parametric form is not required.
(c) if the solution is an ordered triple, check.

61. $x + y + z = 6$
 $2x + y + 2z = 9$
 $-2x - y + z = -3$

62. $x + y + z = 7$
 $3x + y + 3z = 19$
 $-3x - y + z = -3$

63. $3x + y + z = 9$
 $-3x + 2y + 3z = -8$
 $3x + 4y + 5z = 30$

64. $2x + y + z = 11$
 $-2x + 4y + z = 1$
 $2x + 6y + 3z = 25$

65. $x + y + z = 11$
 $2x - y + z = 5$
 $5x + 2y + 4z = 38$

66. $x + y + z = 10$
 $3x - y + z = 4$
 $3x + y + 2z = 17$

67. $2x + y + z = 6$
 $2x - y + 2z = 4$
 $6x + 2y + z = 22$

68. $2x + y + 2z = 8$
 $2x - y + z = 2$
 $8x + 2y + z = -10$

69. $x + y - 2z = 6$
$3x - y + z = 16$
$5x + y - 3z = 28$

70. $x + y + z = 22$
$2x - y + 4z = 6$
$4x + y + 6z = 50$

71. $2x + y + z = 14$
$-2x + 4y + z = -15$
$2y + z = -1$

72. $3x + y + 2z = 9$
$-3x + y + z = -13$
$y + 2z = -3$

73. Write the solution for exercise 69 using set-builder notation in parametric form.

74. Write the solution for exercise 70 using set-builder notation in parametric form.

For exercises 75–76, the variables are described using subscripts x_1, x_2, and x_3 rather than different letters.
(a) Represent the system of equations with an augmented matrix.
(b) Solve the system by rewriting the matrix in reduced row echelon form. If there are infinitely many solutions, it is sufficient to say so. Parametric form is not required.
(c) If the solution is an ordered triple, check.

75. $x_1 + x_2 + x_3 = 5$
$4x_1 - x_2 + 2x_3 = 3$
$2x_1 + x_2 + x_3 = 6$

76. $x_1 + x_2 - x_3 = 5$
$6x_1 + x_2 + 2x_3 = 13$
$2x_1 - x_2 + x_3 = -2$

For exercises 77–82,
(a) assign the variables.
(b) write a system of equations that represents the problem situation.
(c) represent the system with an augmented matrix.
(d) rewrite the matrix in reduced row echelon form.
(e) identify the solution of the system.
(f) use a sentence to report the solution to the problem.

77. A commercial cleaning solution of ammonia and water is 30% ammonia. Find the amount of this commercial solution and the amount of pure water needed to make 5 gallons of solution that is 12% ammonia.

78. A commercial bleach solution is 12% sodium hypochlorite. Household bleach is about 5% sodium hypochlorite. Find the amount of the commercial bleach solution and the amount of pure water needed to make 5 gallons of household bleach. Round to the nearest tenth.

79. The cost to make a product is $20. The fixed overhead costs to make the product are $1500 per month. The company sells each product for $40. Find the number of products that must be made and sold each month to break even. Find the revenue at the break-even point.

80. The cost to make a product is $25. The fixed overhead costs to make the product are $2000 per month. The company sells each product for $30. Find the number of products that must be made and sold each month to break even. Find the revenue at the break-even point.

81. At a food service, the cost of one burrito, one taco, and one soft drink is $5.10. The cost of two burritos, one taco, and two soft drinks is $8.90. The cost of one burrito, two tacos, and three soft drinks is $8.60. Find the price of each item.

82. At a ballpark, the cost of one hotdog, one candy bar, and one soft drink is $6. The cost of three hotdogs, no candy bars, and one soft drink is $9.50. The cost of one hotdog, two candy bars, and one soft drink is $7.50. Find the price of each item.

Problem Solving: Practice and Rewiew

Follow your instructor's guidelines for using the five steps as outlined in Section 1.5, p. 51.

83. In March 2008, Interior Secretary Dirk Kempthorne opened valves that allowed water to flow through Glen Canyon Dam and create an artificial flood through the Grand Canyon. Find the volume of the Empire State Building.

28,400 cubic feet per second was being released from Lake Powell, enough to fill the Empire State Building in 20 minutes. (*Source:* www.lmtribune.com, March 6, 2008)

84. Find the number of jobs in the clean energy economy. Round to the nearest thousand.

In 2007, 65 percent—501,551—of all jobs in the clean energy economy were in the category of Conservation and Pollution Mitigation, which includes the recycling industry. (*Source:* www.pewtrusts.org, June 2009)

85. A "square" trashcan is $19\frac{1}{2}$ in. wide, $19\frac{1}{2}$ in. long, and $27\frac{5}{8}$ in. high. Find its volume in gallons. Round to the nearest whole number (1 gal = 231 in.^{3}).

© AN NGUYEN/Shutterstock

86. Write a linear model that describes the relationship of the number of enrollees in Medicare and the number of years since 1990. Include the units in the model. Use the model to predict the number of enrollees in 2015. Round to the nearest tenth of a million.

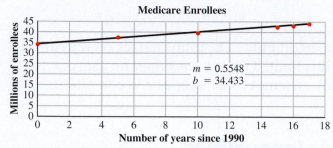

Medicare Enrollees

$m = 0.5548$
$b = 34.433$

Source: www. census. gov, 2009

Technology

For exercises 87–90,
(a) represent the system of equations with an augmented matrix.
(b) use a graphing calculator and rref to rewrite the matrix in reduced row echelon form.
(c) identify the solution.

87. $3x + 4y = -50$
$5x - 2y = 64$

88. $2x + 5y = -79$
$7x - 3y = 113$

89. $2x + 7y + 3z = -26$
$9x - 11y + 4z = 321$
$-4x + y - z = -68$

90. $18x + y - z = -14$
$11x + 3y - 2z = -19$
$9x + 4y = 36$

Find the Mistake

For exercises 91–94, the completed problem has one mistake.
(a) Describe the mistake in words, or copy down the whole problem and highlight or circle the mistake.
(b) Do the problem correctly.

91. Problem: Identify the solution of the system of equations

represented by this matrix: $\begin{bmatrix} 1 & 0 & 0 & | & 2 \\ 0 & 1 & 0 & | & 3 \\ 0 & 0 & 0 & | & 0 \end{bmatrix}$.

Incorrect Answer: The solution is $(2, 3, 0)$.

92. Problem: Identify the solution of the system of equations

represented by this matrix: $\begin{bmatrix} 1 & 0 & | & 8 \\ 0 & 0 & | & 15 \end{bmatrix}$.

Incorrect Answer: The solution is $(8, 15)$.

93. Problem: Change the first entry in row 1 to 1. Rewrite the

matrix: $\begin{bmatrix} -6 & 5 & | & 4 \\ 2 & 1 & | & 6 \end{bmatrix}$.

Incorrect Answer: Multiply row 1 by the reciprocal of -6, which is $\dfrac{1}{6}$, and rewrite the matrix:

$\begin{bmatrix} 1 & \dfrac{5}{6} & | & \dfrac{2}{3} \\ 2 & 1 & | & 6 \end{bmatrix}$.

94. Problem: Use an augmented matrix to represent the
$$3x + 4y - z = 22$$
system $2x - y - z = 7$.
$$3x + 8y = 31$$

Incorrect Answer: $\begin{bmatrix} 3 & 4 & -1 & | & 22 \\ 2 & -1 & -1 & | & 7 \\ 3 & 8 & 1 & | & 31 \end{bmatrix}$

Review

For exercises 95–98, graph the equation or inequality on an *xy*-coordinate system.

95. $2x - 9y = 27$

96. $2x - 9y \geq 27$

97. $y < -4$

98. $x = \dfrac{5}{2}$

SUCCESS IN COLLEGE MATHEMATICS

99. When adding $r_1 + r_2$ in the augmented matrix

$\begin{bmatrix} 1 & \dfrac{3}{5} & | & 18 \\ 4 & -7 & | & 23 \end{bmatrix}$, a student incorrectly said that

$\dfrac{3}{5} + -7 = -\dfrac{4}{5}$.

a. Describe the mistake that the student made.
b. Describe how to rewrite a whole number as a fraction with a denominator of 5.

Most credit cards have a credit limit. The total charges on the card cannot be greater than this limit. This is a *constraint* on charges. In this section, we will study systems of inequalities and constraints.

SECTION 4.6

Systems of Linear Inequalities

After reading the text, working the practice problems, and completing assigned exercises, you should be able to:

1. Graph a system of linear inequalities.
2. Identify the vertices of a bounded solution region.
3. Use a graphical or an algebraic method to determine whether an ordered pair is a solution of a system of inequalities.
4. Use linear inequalities to model the constraints in a problem situation.
5. Graph a system of linear constraints. Find the vertices of the feasible region.

A System of Inequalities

To graph a linear inequality (Section 1.8), first graph the boundary line. This line is either solid or dashed. Then, use a test point to determine where the graph should be shaded. The points in the **solution region** represent ordered pairs that are solutions of the inequality.

An inequality that includes a $>$ or $<$ sign is a **strict inequality**. The boundary line is dashed. The solution region is the shaded area and does not include the boundary line.

An inequality that includes a $\geq$ or $\leq$ sign is not strict. The boundary line is solid. The solution region includes the shaded area and the boundary line.

> **Graphing a Linear Inequality**
> 1. Determine whether the boundary line is solid or dashed. If the inequality is not strict, it includes a $\geq$ or $\leq$ sign, and the boundary line is a solid line. If the inequality is strict, it includes a $>$ or $<$ sign, and the boundary line is dashed.
> 2. Graph the boundary line.
> 3. Select a test point that is *not* on the boundary line. Replace the variables in the inequality with the coordinates of the test point. The resulting inequality will be true or false.
> 4. If the inequality is true, the test point is in the solution region. Shade the area on this side of the boundary line. If the inequality is not true, the test point is *not* in the solution region. Shade the area on the *other* side of the boundary line.

A **system of inequalities** is two or more inequalities. The solution region is the intersection of the solution regions of each inequality. So each point in the solution region of the system is in the solution region of every inequality in the system.

EXAMPLE 1 Graph the system $\begin{array}{c} y \le 2x - 7 \\ 6x + 5y \ge 30 \end{array}$. Label the solution region.

(a) Graph: $y \le 2x - 7$

SOLUTION ▶ The solution region of $y \le 2x - 7$ includes the solutions of $y = 2x - 7$ and $y < 2x - 7$. The graph of $y = 2x - 7$ is a line. This is the boundary of the solution region.

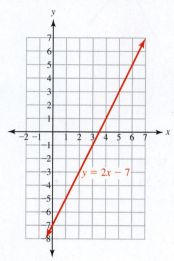

$y = 2x - 7$

To find the rest of the solution region, choose a test point. Replace the variables in the inequality with the coordinates from the test point. If the inequality is true, the test point is in the solution region. If the inequality is not true, the solution region is on the other side of the boundary line.

Test point: (0, 0)

$y \le 2x - 7$

$\mathbf{0} \le 2(\mathbf{0}) - 7$

$0 \le -7$ False.

The solution region does not include (0, 0).

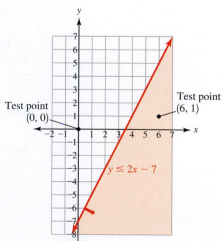

Test point (0, 0)

Test point (6, 1)

$y \le 2x - 7$

Check that the solution region is correct by testing a point in the region, (6, 1).

$y \le 2x - 7$

$\mathbf{1} \le 2(\mathbf{6}) - 7$

$1 \le 5$ True.   The shading is correct.

(b) Graph $6x + 5y \geq 30$ on the same coordinate system.

The solid boundary line is $6x + 5y = 30$.

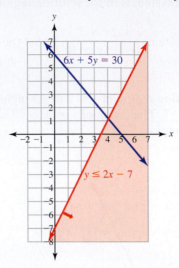

Test point: $(0, 0)$

$$6x + 5y \geq 30$$
$$6(0) + 5(0) \geq 30$$
$$0 \geq 30 \quad \text{False.}$$

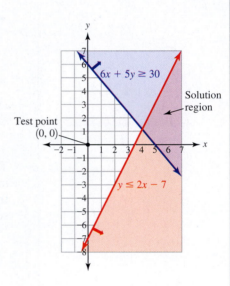

(c) Label the solution region.

The solution region does not include $(0, 0)$. It is on the other side of the boundary line. The **solution region** of the system is the area that is shaded by both inequalities.

In the next example, the boundary lines of the inequalities create a finite area that is the solution region. This is a **bounded solution region**.

EXAMPLE 2 Graph the system $\begin{array}{l} y \leq -x + 6 \\ x \geq 1 \\ y \geq -1 \end{array}$. Label the solution region.

(a) Graph: $y \leq -x + 6$

SOLUTION ▶ The solid boundary line is $y = -x + 6$. Use a test point to find the side of the graph that should be shaded. This shaded region together with the boundary line is the solution region of $y \leq -x + 6$.

Test point: $(0, 0)$

$y \leq -x + 6$

$0 \leq 0 + 6$

$0 \leq 6$ True.

The solution region includes $(0, 0)$; shade below the boundary line.

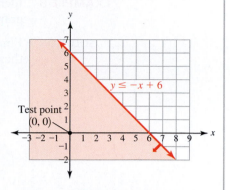

(b) Graph: $x \geq 1$

▶ The solid vertical boundary line is $x = 1$.

Test point: $(0, 0)$

$x \geq 1$

$0 \geq 1$ False.

The solution region for $x \geq 1$ does *not* include $(0, 0)$.

Shade to the right of the vertical line $x = 1$.

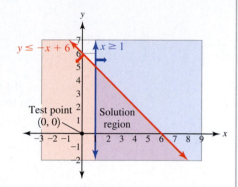

(c) Graph: $y \geq -1$

▶ The solid horizontal boundary line is $y = -1$.

Test point: $(0, 0)$

$y \geq -1$

$0 \geq -1$ True.

The solution region for $y \geq -1$ includes $(0, 0)$.

Shade above the horizontal line $y \geq -1$.

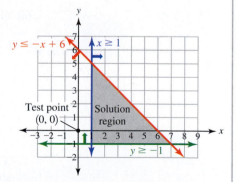

(d) Label the solution region

▶ The **bounded solution region** for the system $\begin{array}{l} y \leq -x + 6 \\ x \geq 1 \\ y \geq -1 \end{array}$ is the region that is

shaded by all three inequalities.

Each corner of a bounded solution region is a **vertex**. The plural of vertex is **vertices**.

EXAMPLE 3   Identify the vertices of the bounded solution region of the system of inequalities graphed in Example 2.

SOLUTION ▶ Find the corners of the bounded solution region and identify the coordinates of these points. The vertices are $(1, 5)$, $(1, -1)$, and $(7, -1)$.

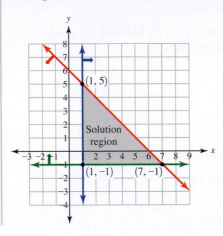

In the next example, the system of inequalities has no solution. There is no region on the graph that includes points from all three inequalities.

$$y > -\frac{1}{2}x + 3$$

EXAMPLE 4   Graph the system $y < -\frac{1}{2}x - 4$. Label the solution region.

$$y > 5$$

(a) Graph: $y > -\frac{1}{2}x + 3$

SOLUTION ▶   Test point: $(0, 0)$

$$y > -\frac{1}{2}x + 3$$

$$0 > -\frac{1}{2}(0) + 3 \quad \text{False.}$$

The shaded area does not include $(0, 0)$; shade above the boundary line.

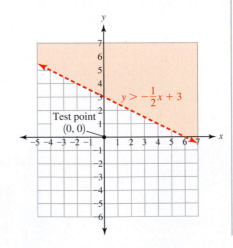

(b) Graph: $y < -\dfrac{1}{2}x - 4$

Test point: $(0, 0)$

$$y < -\frac{1}{2}x - 4$$

$$0 < -\frac{1}{2}(0) - 4 \quad \text{False.}$$

The shaded area does not include $(0, 0)$; shade below the boundary line.

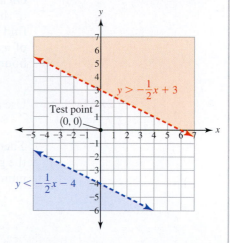

(c) Graph: $y > 5$

Test point: $(0, 0)$

$$y > 5$$

$$0 > 5 \quad \text{False.}$$

The shaded area does not include $(0, 0)$; shade above the horizontal boundary line.

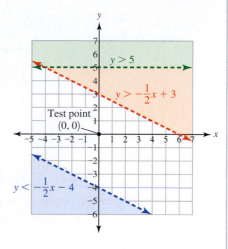

(d) Label the solution region.

Although the inequalities $y > 5$ and $y > -\dfrac{1}{2}x + 3$ share a solution region, the other inequality does not have any points in this region. This system of inequalities has no solution.

Although the test point for all three inequalities in Example 4 is $(0, 0)$, any point that is not on the boundary line can be used. We often use $(0, 0)$ because arithmetic with 0 is relatively easy.

Practice Problems

For problems 1–4,
(a) graph each system of inequalities. Show your work with test points. Label the solution region, if any.
(b) if the solution region is bounded, estimate the vertices.

1. $2x + 3y \leq 24$
$y \geq x + 1$
$x \geq 0$

2. $2x + 3y \leq 24$
$y \leq x + 1$
$y \geq 0$

3. $-x + y \leq 0$
$y \geq -2x + 6$
$x \leq 8$

4. $x \geq -2$
$x \leq 7$
$y \geq -3$
$y \leq 5$

Vertices

On a graph, the solution of a system of two linear equations is the intersection point of their graphs. We can use the algebraic methods of substitution and elimination to find the solution. These same methods can find the exact vertex of a solution region of a system of linear inequalities. A vertex is the intersection point of two of the boundary lines.

EXAMPLE 5   Use an algebraic method to find the vertices of the solution region of $\begin{aligned} 7x - 3y &\leq 21 \\ y &\geq -3x - 3 \\ y &\leq 0 \end{aligned}$.

There are three vertices; these are the points where the lines intersect. By looking at the graph, we can estimate the vertices. However, when the vertices are not whole numbers, we use an algebraic method to find their exact values.

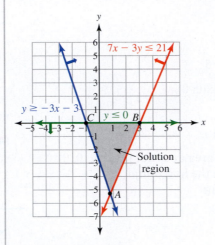

(a) Identify vertex A.

SOLUTION ▶   Vertex A is the intersection of $7x - 3y = 21$ and $y = -3x - 3$. To find the intersection point, solve this system of equations by substitution.

$$7x - 3y = 21$$
$$7x - 3(-3x - 3) = 21 \qquad \text{Replace } y \text{ with } -3x - 3.$$
$$7x - 3(-3x) - 3(-3) = 21 \qquad \text{Distributive property.}$$
$$7x + 9x + 9 = 21 \qquad \text{Simplify.}$$
$$16x + 9 = 21 \qquad \text{Simplify; combine like terms.}$$
$$\underline{ -9 \quad -9} \qquad \text{Subtraction property of equality.}$$
$$16x + 0 = 12 \qquad \text{Simplify.}$$
$$\frac{16x}{16} = \frac{12}{16} \qquad \text{Division property of equality.}$$
$$x = \frac{3}{4} \qquad \text{Vertex A is } \left(\frac{3}{4}, y\right).$$

$$y = -3x - 3 \qquad \text{To find } y, \text{ use either of the equations.}$$
$$y = -3\left(\frac{3}{4}\right) - 3 \qquad \text{Replace } x \text{ with } \frac{3}{4}.$$
$$y = -\frac{9}{4} - \frac{12}{4} \qquad \text{Simplify.}$$
$$y = -\frac{21}{4} \qquad \text{Simplify.}$$

Vertex A is $\left(\dfrac{3}{4}, -\dfrac{21}{4}\right)$.

(b) Identify vertex B.

▶ Vertex B is the intersection of $7x - 3y = 21$ and $y = 0$. To find the intersection point, solve this system of equations by substitution.

$7x - 3y = 21$	Solve by substitution.
$7x - 3(0) = 21$	Replace y with 0.
$7x - \mathbf{0} = 21$	Simplify.
$\dfrac{\mathbf{7x}}{7} = \dfrac{21}{7}$	Simplify; division property of equality.
$x = 3$	Simplify.

Since $y = 0$, vertex B is $(3, 0)$.

(c) Identify vertex C.

▶ Vertex C is the intersection of $y = -3x - 3$ and $y = 0$. To find the intersection point, solve this system of equations by substitution.

$y = -3x - 3$	Solve by substitution.
$0 = -3x - 3$	Replace y with 0.
$\underline{+3 \qquad\qquad +3}$	Addition property of equality.
$\mathbf{3} = -3x + \mathbf{0}$	Simplify.
$\dfrac{3}{-3} = \dfrac{\mathbf{-3x}}{-3}$	Simplify; division property of equality.
$-1 = x$	Simplify.

Since $y = 0$, vertex C is $(-1, 0)$.

The vertices of the solution region are $\left(\dfrac{3}{4}, -\dfrac{21}{4}\right)$, $(3, 0)$, and $(-1, 0)$.

Practice Problems

For problems 5–7,
(a) graph the system of inequalities. Label the solution region.
(b) use an algebraic method to find the vertices of the solution region.

5. $y \ge \dfrac{6}{5}x - 6$

$x + y \ge -5$

$y \le 6$

6. $-2x + y \ge 1$

$y \le -\dfrac{3}{4}x + 5$

$x \ge 0$

7. $x - y \le 4$

$y \ge 5x - 7$

$y \ge 0$

Identifying Solutions

If an ordered pair is a solution of a system of inequalities, the point it represents is in the solution region on the graph. If the variables in an inequality in the system are replaced by the solution, the inequality is true.

$$x \geq 0$$

EXAMPLE 6   The graph represents the system $2x + 5y \leq 10$.

$$y \geq \frac{8}{9}x - 4$$

(a) Use a graphical method to determine whether $(5, 0)$ is a solution of the system.

SOLUTION ▶   The point $(5, 0)$ is close to a boundary line. It appears that it is not in the solution region.

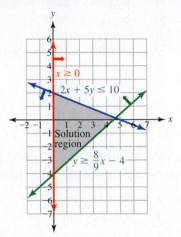

(b) Use an algebraic method to determine whether $(5, 0)$ is a solution of the system.

▶   Test $(5, 0)$ in each inequality. To be a solution of the system, it must be a solution of every inequality.

$x \geq 0$	$2x + 5y \leq 10$	$y \geq \dfrac{8}{9}x - 4$
$5 \geq 0$ True.	$2(5) + 5(0) \leq 10$	$0 \geq \dfrac{8}{9}(5) - 4$
	$10 \leq 10$ True.	$0 \geq \dfrac{4}{9}$ False.

Since $(5, 0)$ is not a solution of $y \geq \dfrac{8}{9}x - 4$, it is not a solution of this system of inequalities.

Practice Problems

For problems 8–9,
(a) graph the system.
(b) use a graphical method to determine whether the point is a solution of the system.
(c) use an algebraic method to determine whether the point is a solution of the system.

8. $(1, -5); x + y \leq 8$
$\qquad\quad y \geq 2x - 9$
$\qquad\quad x \geq 1$

9. $(3, -5); 4x + y \geq 10$
$\qquad\quad x \geq 1$
$\qquad\quad y \geq 8$

Constraints

In an application problem, a constraint limits the possible solutions. We can use a linear inequality to represent some constraints.

EXAMPLE 7 | Assign a variable and write a constraint that represents the number of street gangs linked to the arrests.

ICE agents nationwide have arrested more than 22,000 gang members and associates linked to more than 1,200 street gangs. (*Source:* www.dallasnews.com, June 14, 2011)

SOLUTION ▶ x = number of street gangs   Assign a variable.

$x > 1200$ street gangs   Write a constraint.

The solution region of the graph of a system of constraints is often called the **feasible region**. In many application problems, the variable(s) represents numbers that are greater than or equal to 0. Constraints ensure that the feasible region includes only these numbers.

EXAMPLE 8 | A student can spend no more than $225 this month on food and gas. She must spend at least $25 per month on gas. Write and graph three constraints that describe this situation. Label the feasible region, and estimate the vertices.

(a) Assign two variables.

SOLUTION ▶   Let x = amount spent on food and y = amount spent on gas.

(b) Write three constraints.

▶   The amount of money that can be spent is a constraint.

$x + y \le \$225$ The total spent is less than or equal to $225.

The minimum amount spent on food is a constraint. Since the student cannot spend a negative amount of money on food, the amount must be greater than or equal to 0.

$x \ge \$0$ The amount spent on food is greater than or equal to $0.

The minimum amount spent on gas is a constraint. The amount spent on gas must be greater than or equal to $25.

$y \ge \$25$ The amount spent on gas is greater than or equal to $25.

(c) Graph the constraints. Label the feasible region.

▶   Graph: $x + y \le \$225$

Graph the intercepts and one other point; solid boundary line.

x	y
0	$225
$225	0
$100	$125

Test point: ($0, $0)

$x + y \le \$225$

$\$0 + \$0 \le \$225$

$\$0 \le \$225$ True.

The point ($0, $0) is in the feasible region for this constraint.

Graph: $x \ge \$0$

To find which side of the solid vertical boundary line to shade, choose a test point other than ($0, $0). We choose ($100, $50).

Test point: ($100, $50)

$x \ge \$0$

$\$100 \ge \$0$ True.

The point ($100, $50) is in the feasible region for this constraint.

Graph: $y \geq \$25$

To find which side of the solid horizontal boundary line to shade, choose a test point.

Test point: ($100, $50)

$y \geq \$25$

$\mathbf{\$50} \geq \$25$ True.

The point ($100, $50) is in the feasible region for this constraint.

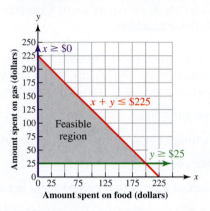

(d) Estimate the vertices of the feasible region.

▶   The feasible region is the area shaded by all of the constraints. It includes the boundary lines.

The estimated vertices are ($0, $25), ($0, $225), ($200, $25).

In the next example, the feasible region is limited to points with coordinates that are whole numbers. The potter can only sell whole cups or plates.

EXAMPLE 9   A pottery shop sells souvenir cups and plates. The glaze on the pottery is made of ash from the eruption of Mt. St. Helens. On the pottery wheel, a potter can make a cup in 5 min and a plate in 4 min. Each cup is made from 0.25 kg clay, and each plate is made from 0.8 kg clay. The potter has 8 hr (480 min) to make cups and plates and has 50 kg of clay. The potter must make at least 10 cups. Write and graph four constraints that describe this situation. Label the feasible region, and estimate the vertices.

(a) Assign two variables.

SOLUTION ▶   Let x = number of cups and y = number of plates.

(b) Write four constraints that describe this situation.

▶   The amount of time is a constraint. The total time must be less than or equal to 480 min.

$$\left(\frac{5 \text{ min}}{1 \text{ cup}}\right)x + \left(\frac{4 \text{ min}}{1 \text{ plate}}\right)y \leq 480 \text{ min} \qquad \text{time for cups + time for plates} \leq \text{total time}$$

The amount of clay is a constraint. The total amount must be less than or equal to 50 kg.

$$\left(\frac{0.25 \text{ kg}}{1 \text{ cup}}\right)x + \left(\frac{0.8 \text{ kg}}{1 \text{ plate}}\right)y \leq 50 \text{ kg} \qquad \text{clay for cups + clay for plates} \leq \text{total clay}$$

The minimum number of cups is a constraint. The number of cups must be greater than or equal to 10.

$x \geq 10$ cups   number of cups ≥ 10

The potter cannot make a negative number of plates. The number of plates must be greater than or equal to 0.

$y \geq 0$ plates   number of plates ≥ 0

(c) Graph the constraints. Label the feasible region.

▶ Graph: $\left(\dfrac{5 \text{ min}}{1 \text{ cup}}\right)x + \left(\dfrac{4 \text{ min}}{1 \text{ plate}}\right)y \leq 480$ min

Graph the intercepts and one other point; solid boundary line.

Test point: (0 cup, 0 plate)

x (cups)	y (plates)
0	120
96	0
50	57.5

$$\left(\dfrac{5 \text{ min}}{1 \text{ cup}}\right)x + \left(\dfrac{4 \text{ min}}{1 \text{ plate}}\right)y \leq 480 \text{ min}$$

$$\left(\dfrac{5 \text{ min}}{1 \text{ cup}}\right)(0 \text{ cup}) + \left(\dfrac{4 \text{ min}}{1 \text{ plate}}\right)(0 \text{ plate}) \leq 480 \text{ min}$$

$$0 \leq 480 \text{ min} \text{True.}$$

The point (0 cup, 0 plate) is in the feasible region for this inequality.

Graph: $\left(\dfrac{0.25 \text{ kg}}{1 \text{ cup}}\right)x + \left(\dfrac{0.8 \text{ kg}}{1 \text{ plate}}\right)y \leq 50$ kg

Graph the intercept and two other points; solid boundary line.

Test point: (0 cup, 0 plate)

x (cups)	y (plates)
0	62.5
200	0
100	31.25

$$\left(\dfrac{0.25 \text{ kg}}{1 \text{ cup}}\right)x + \left(\dfrac{0.8 \text{ kg}}{1 \text{ plate}}\right)y \leq 50 \text{ kg}$$

$$\left(\dfrac{0.25 \text{ kg}}{1 \text{ cup}}\right)(0 \text{ cup}) + \left(\dfrac{0.8 \text{ kg}}{1 \text{ plate}}\right)(0 \text{ plate}) \leq 50 \text{ kg}$$

$$0 \text{ kg} \leq 50 \text{ kg} \text{True.}$$

The point (0 cup, 0 plate) is in the feasible region for this constraint.

Graph: $x \geq 10$ cups

To find which side of the solid vertical boundary line to shade, choose a test point other than (0 cup, 0 plate). We choose (40 cups, 20 plates).

Test point: (40 cups, 20 plates)

$x \geq 10$ cups

40 cups ≥ 10 cups   True.

The point (40 cups, 20 plates) is in the feasible region for this constraint.

Graph: $y \geq 0$ plates

To find which side of the solid horizontal boundary line to shade, choose a test point.

Test point: (40 cups, 20 plates)

$y \geq 0$ plates

20 plates ≥ 0 plates   True.

The point (40 cups, 20 plates) is in the feasible region for this constraint.

The potter cannot sell part of a cup or plate. The only solutions are points with whole number coordinates. However, all of the points in the feasible region are shaded because it is not practical to mark each individual point with whole number coordinates. If the potter needs plates and cups to sell, it is unlikely that the potter will choose to make 10 cups and 0 plates. However, this is a feasible choice within the constraints.

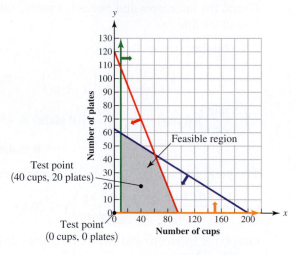

(d) Estimate the vertices of the feasible region.

The feasible region is the area that is shaded by all of the constraints. It includes the boundary lines.

The estimated vertices are (10 cups, 0 plates), (96 cups, 0 plates), (61 cups, 43 plates), and (10 cups, 60 plates).

In a finite mathematics or linear algebra class, the vertices of a feasible region are used to find the best or "optimal" solution for a problem situation. This process is called **optimization**.

Practice Problems

10. A furniture refinishing shop has a large number of wood chairs that need to be glued and sanded. It takes 10 min to glue a chair and 15 min to sand a chair. On a given day, an employee can work for 3 hr (180 min) gluing and sanding chairs. At least 8 chairs need to be glued. At least 4 chairs need to be sanded. Write and graph the constraints. Label the feasible region, and estimate the vertices, rounding the coordinates to the nearest whole number.
a. Assign two variables.
b. Write three constraints that describe this situation.
c. Graph the constraints. Label the feasible region.
d. Estimate the vertices.

Using Technology: Graphing a System of Linear Inequalities

Some calculators have four different shading patterns. These calculators can graph a maximum of four different inequalities with unique shading.

In the next example, we graph the inequalities one at a time. We graph the boundary line. We use a test point to find out whether the shading is "below" or "above" the line. We move the cursor to the left of the equals sign and choose the correct shading icon.

EXAMPLE 10 Graph $y \leq 3x$, $y \leq -2x + 10$, and $y \geq x - 8$. Estimate the vertices of the solution region.

To graph the boundary line of $y \leq 3x$, type its equation on the Y= screen. Press [ENTER]. Press [GRAPH]. The graph of the line appears.

With pencil and paper, use a test point to find which side of the line to shade.

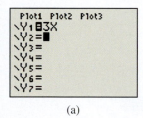
(a)

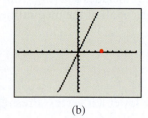
(b)

Choose a test point: $(4, 0)$.

$y \leq 3x$

$0 \leq 3(4)$

$0 \leq 12$ True.

To shade, move the cursor to the left of the equals sign. Press [ENTER] repeatedly until the "shaded below" icon appears. Press [GRAPH].

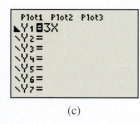
(c)

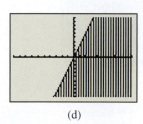
(d)

The shading includes the test point. This is the graph of $y \leq 3x$.

On the Y= screen, type the next boundary line equation, $y = -2x + 10$. Press [ENTER]. Press [GRAPH]. Another line appears on the graph.

With pencil and paper, use a test point to find which side of the line to shade.

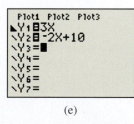
(e)

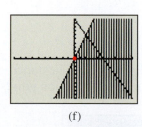
(f)

Choose a test point: $(0, 0)$.

$y \leq -2x + 10$

$0 \leq -2(0) + 10$

$0 \leq 10$ True.

To shade, move the cursor to the left of the equals sign. Press [ENTER] repeatedly until the "shaded below" icon appears. Press [GRAPH].

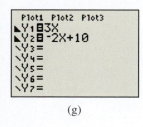
(g)

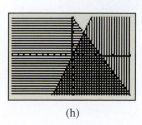
(h)

The shading includes the test point. Notice the solution region that is shaded by both inequalities.

On the Y= screen, type the next boundary equation, $y = x - 8$. Press ENTER . Press GRAPH .

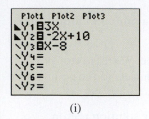

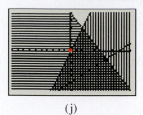

To add the shading,
choose a test point: $(0, 0)$.

$y \geq x - 8$

$0 \geq 0 - 8$ True.

(i)   (j)

To shade, move the cursor to the left of the equals sign. Press ENTER repeatedly until the "shaded above" icon appears. Press GRAPH .

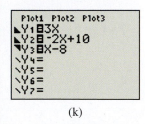

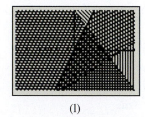

The solution region of the system appears. It includes shading from the graphs of all of the inequalities.

(k)   (l)

Before estimating the vertices, change the window to show the lower left vertex. Press WINDOW .
Change the Y_{min} value to -15. Press GRAPH .

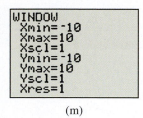

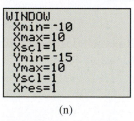

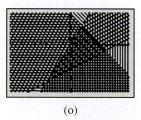

(m)   (n)   (o)

To find a vertex, go to the CALC menu. Choose 5: intersect. Press ENTER .

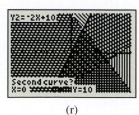

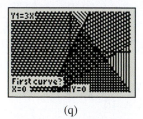

The shading makes it difficult to see the cursor. However, the location of the cursor is shown in the upper left-hand corner. Press ENTER again.

(p)   (q)

Press ENTER again. Move the cursor to the intersection point. Press ENTER . The upper vertex is $(2, 6)$. To find the other vertices, repeat this process for the other intersection points.

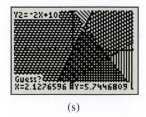

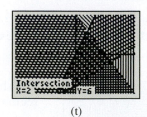

(r)   (s)   (t)

Practice Problems For problems 11–13,

  (a) graph the system of inequalities. Choose a window that shows all the vertices of the solution region. Sketch the graph; describe the window.

  (b) estimate the vertices of the solution region.

11. $y \leq 4x$
$y \geq x - 5$
$y \leq -2x + 12$

12. $x + y \leq 8$
$y \geq 2x - 9$
$y \leq 9$
$y \geq -3x - 9$

13. $2x + 3y \leq 24$
$y \leq x + 1$
$y \geq 0$

4.6 VOCABULARY PRACTICE

Match the term with its description.

1. $<$

2. $>$

3. $\leq$

4. $\geq$

5. We use this as we determine the shaded area on the graph of a linear inequality.

6. This line marks the edge of the solution region on the graph of a linear inequality.

7. On the graph of a linear inequality, this area includes all the points that represent solutions of the inequality.

8. On the graph of a constraint, this area includes all the points that represent the ordered pairs that are within the limits of the constraint.

9. A boundary line used to represent a strict inequality

10. A boundary line used to represent an inequality that is not strict

A. boundary line
B. dashed boundary line
C. feasible region
D. greater than
E. greater than or equal to
F. less than
G. less than or equal to
H. solid boundary line
I. solution region
J. test point

4.6 Exercises

Follow your instructor's guidelines for showing your work.

1. Explain how to determine whether the boundary line of the graph of an inequality should be solid or dashed.

2. Explain how to use a test point to determine which side of a boundary line to shade.

For exercises 3–22, graph the system of linear inequalities. Label the solution region.

3. $x - y \geq 6$
$x \geq 1$

4. $x + y \leq 12$
$x \geq 4$

5. $2x - 3y < 18$
$y < -x + 4$

6. $5x - 3y < 30$
$y < -x + 5$

7. $2x - 3y > 18$
$y < -x + 4$

8. $5x - 3y > 30$
$y < -x + 5$

9. $2x - 3y > 18$
$y > -x + 4$

10. $5x - 3y > 30$
$y > -x + 5$

11. $y > 3x - 7$
$y > -3x - 7$

12. $y > 2x - 8$
$y > -2x - 8$

13. $y \geq x$
$y \leq 6$
$x \geq 0$

14. $y \geq x$
$y \leq 5$
$x \geq 0$

15. $-9x + 2y < 18$
$-9x + 2y > -18$

16. $-5x + 2y < 10$
$-5x + 2y > -10$

17. $x \geq 3$
$x \leq 10$
$y \geq -6$
$y \leq 2$

18. $x \geq -4$
$x \leq 6$
$y \geq -5$
$y \leq 3$

19. $y > -\dfrac{1}{3}x$
$y < \dfrac{1}{3}x$
$x < 5$

20. $y > -\dfrac{1}{2}x$
$y < \dfrac{1}{2}x$
$x < 6$

21. $y \leq x$
$y \leq -x$
$y \geq -4$

22. $y \geq x$
$y \geq -x$
$y \leq 4$

For exercises 23–28, the graph represents a system of inequalities. Estimate the vertices of the solution region.

23.

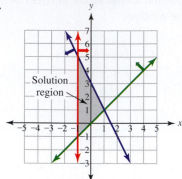

24.

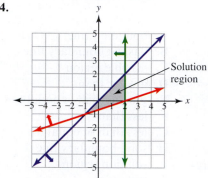

25.

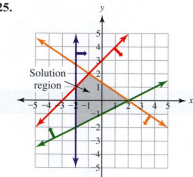

26.

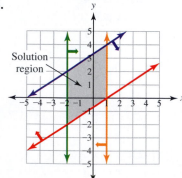

27.

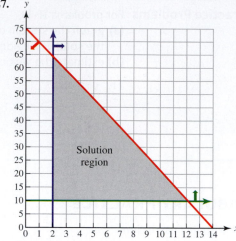

28.

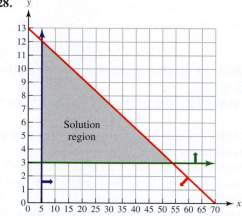

For exercises 29–50,
(a) graph the system of inequalities. Label the solution region.
(b) estimate the vertices of the solution region.

29. $x + 4y \le 8$
$x \ge 0$
$y \ge 0$

30. $x + 5y \le 15$
$x \ge 0$
$y \ge 0$

31. $y \le -2x + 9$
$x \ge 2$
$y \ge 1$

32. $y \le -2x + 12$
$x \ge 2$
$y \ge 2$

33. $y \ge 2x + 1$
$y \le -3x + 11$
$x \ge 0$

34. $y \ge 3x + 1$
$y \le -5x + 9$
$x \ge 0$

35. $-3x + y \le 3$
$3x + y \le 15$
$x \ge 0$
$y \ge 0$

36. $-4x + y \le 4$
$4x + y \le 12$
$x \ge 0$
$y \ge 0$

37. $y \le \dfrac{2}{3}x + 1$
$y \le -x + 11$
$y \ge 1$

38. $y \le \dfrac{3}{2}x - 3$
$y \le -x + 7$
$y \ge -3$

39. $y \ge \dfrac{2}{3}x + 1$
$y \le -x + 11$
$x \ge 0$

40. $y \ge \dfrac{3}{2}x - 3$
$y \le -x + 7$
$x \ge 0$

41.
$$x + y \le 60$$
$$800x + 500y \ge 16{,}000$$
$$y \le 60$$
$$y \ge 0$$
$$x \ge 0$$

42.
$$x + y \le 100$$
$$900x + 500y \ge 18{,}000$$
$$y \le 50$$
$$y \ge 0$$
$$x \ge 0$$

43.
$$x + y \le 80$$
$$400x + 300y \ge 12{,}000$$
$$y \le 40$$
$$y \ge 0$$

44.
$$x + y \le 70$$
$$500x + 200y \ge 10{,}000$$
$$x \ge 0$$
$$y \ge 0$$

45.
$$y \ge 2x$$
$$x + y \le 60$$
$$x \ge 0$$

46.
$$y \ge 4x$$
$$x + y \le 80$$
$$x \ge 0$$

47.
$$y \le 2x$$
$$x + y \le 60$$
$$y \ge 0$$

48.
$$y \le 4x$$
$$x + y \le 80$$
$$y \ge 0$$

49. $x \ge 20$
$$x \le 60$$
$$y \ge 10$$
$$y \le 45$$

50. $x \ge 15$
$$x \le 80$$
$$y \ge 5$$
$$y \le 20$$

For exercises 51–54,
(a) graph the system of inequalities.
(b) use an algebraic method to find the vertices of the solution region. (Use elimination or substitution to find the intersection point of each pair of boundary lines. Do not rewrite fractions as decimal numbers.)

51. $y \le -6x + 20$
$$y \ge 3x - 15$$
$$x \ge 0$$

52. $y \ge 4x - 9$
$$y \le -2x + 12$$
$$x \ge 0$$

53. $y \ge -3x + 6$
$$y \le -x + 6$$
$$y \ge 2x - 8$$

54. $y \ge -4x + 8$
$$y \le -x + 8$$
$$y \ge 3x - 9$$

For exercises 55–58, use an algebraic method to show that the ordered pair is *not* a solution of the system of inequalities.

55. $(2, -4); 2x - y \le 12$
$$y \ge 2x$$
$$x \ge 1$$

56. $(3, -5); 6x - y \le 23$
$$y \ge x$$
$$x \ge 2$$

57. $(3, 6); 3x + 4y \le 100$
$$20x + y \ge 50$$
$$y \le -5x + 20$$

58. $(-9, -4); 7x - 8y \ge -100$
$$x + 4y < 20$$
$$y \ge -3x + 2$$

For exercises 59–60, if a student receives a *subsidized loan*, the federal government pays the interest on the loan while the student is in school and for the first 6 months after the student leaves school.

59. In 2008, a first-year undergraduate student could borrow a maximum of $7500 in Stafford loans. No more than $3500 could be subsidized Stafford loans. Let x = amount of unsubsidized Stafford loans in the first year and y = amount of subsidized Stafford loans in the first year. (*Source:* www.ed.gov)
   a. Write four constraints that describe this situation.
   b. Graph the constraints. Label the feasible region.
   c. Estimate the vertices of the feasible region.

60. In 2008, a second-year undergraduate student could borrow a maximum of $8500 in Stafford loans. No more than $4500 could be subsidized Stafford loans. Let x = amount of unsubsidized Stafford loans in the second year and y = amount of subsidized Stafford loans in the second year. (*Source:* www.ed.gov)
   a. Write four constraints that describe this situation.
   b. Graph the constraints. Label the feasible region.
   c. Estimate the vertices of the feasible region.

For exercises 61–62, hog feed in the United States is made from grain and soybean meal. Grain provides 45–65% of the protein and 80–90% of the energy. Soybean meal provides the remaining 35–55% of the protein and 10–20% of the energy. (*Source:* www.ag.auburn.edu)

61. Let x = percent of protein provided by grain and y = percent of protein provided by soybean meal.
   a. Write five constraints that describe this situation.
   b. Graph the constraints. Label the feasible region.
   c. Estimate the vertices of the feasible region.

62. Let x = percent of energy provided by grain and y = percent of energy provided by soybean meal.
   a. Write five constraints that describe this situation.
   b. Graph the constraints. Label the feasible region.
   c. Estimate the vertices of the feasible region.

63. In some counties in Kansas, customers can buy alcohol by the drink at a business only if at least 30% of the revenue of the business is from the sale of food. Let x = percent of revenues from sale of alcohol and y = percent of revenues from sale of food.
   a. Write five constraints that describe this situation.
   b. Graph the constraints. Label the feasible region.
   c. Estimate the vertices of the feasible region.

64. The North Dakota Forest Service has a Cooperative Fire Assistance Grant Program. At least 10% of the total funding for a grant project must come from sources other than the federal government. No more than 90% of the total funding for the project may come from the federal government. Let x = percent of funds from sources other than the federal government and y = percent of funds

from the federal government. (*Source:* North Dakota Forest Service 2008)
a. Write five constraints that describe this situation.
b. Graph the constraints. Label the feasible region.
c. Estimate the vertices of the feasible region.

65. An investor can put up to $30,500 in her retirement plan. She wants at least four times as much in domestic investments as in foreign investments. She wants at least $2000 in foreign investments. Let x = amount in foreign investments and y = amount of domestic investments.
a. Write four constraints that describe this situation.
b. Graph the constraints. Label the feasible region.
c. Estimate the vertices of the feasible region.

66. An investor is reinvesting at least $18,000 in his retirement plan. He wants it divided between growth mutual funds and income mutual funds. He wants no more than three times as much in income funds as in growth funds. He wants no more than $5000 in growth funds. Let x = amount in growth funds and y = amount in income funds.
a. Write four constraints that describe this situation.
b. Graph the constraints. Label the feasible region.
c. Estimate the vertices of the feasible region.

67. An employee is buying packages of candy and jars of roasted peanuts for an open house. Each package of candy weighs 6 oz and costs $1.50. Each jar of roasted peanuts weighs 16 oz and costs $2.00. The maximum budget for candy and nuts for the open house is $120. The employee wants to buy no more than three times as many jars of roasted peanuts as packages of candy. He wants to buy at least 15 jars of peanuts. Let x = packages of candy and y = jars of roasted peanuts.
a. Write four constraints that describe this situation.
b. Graph the constraints. Label the feasible region.
c. Estimate the vertices of the feasible region.

68. An organic gardener is planting two kinds of beets, long season beets and baby beets. The maximum length of the row planted in beets is 150 ft. The gardener wants to plant at least twice the length in baby beets as in long season beets. Let x = feet of baby beets and y = feet of long season beets.
a. Write four constraints that describe this situation.
b. Graph the constraints. Label the feasible region.
c. Estimate the vertices of the feasible region.

Problem Solving: Practice and Review

Follow your instructor's guidelines for using the five steps as outlined in Section 1.5, p. 51.

69. Thioglycolic acid is the active ingredient in some solutions used for perming hair. A chemical supply company sells a solution that is 80% thioglycolic acid. The maxi-

mum concentration of thioglycolic acid recommended by the Cosmetic Ingredient Review is 15.2%. Use a system of equations to find the amount of 80% thioglycolic acid solution and the amount of pure water to mix together to make 3 L of 15.2% thioglycolic acid solution. Round to the nearest tenth.

70. Find the rate of patients contracting an infection during hospitalization per 1000 cases in 2006. Round to the nearest tenth.

In 2007, Pennsylvania hospitals reported that 27,949 patients contracted an infection during their hospitalization, a rate of 17.7 per 1,000 cases, which is a 7.8 percent decrease from . . . 2006. (*Source:* Pennsylvania Health Care Cost Containment Council, Jan. 2009)

71. Find the amount in today's dollars that Flagler spent on all of the sites he developed.

Henry Morrison Flagler gave a large portion of his life, and his money, to the State of Florida. Experts estimate that he spent about $50 million on all the sites he developed. Surprisingly, this would equate to about one-third of the whole value of Florida at the time. Two-fifths of this money went to the Key West Extension of the Overseas Railroad. It is estimated that, in today's dollars, this great project would have cost about $640 million. (*Source:* www.floridakeys.com)

72. Find the cost per acre for the parcel purchased in January 2008. Write the answer in place value notation. Round to the nearest whole number.

The undeveloped land is located . . . between the county's Punaluu and Whittington beach parks. The purchase of one, 234-acre parcel was completed in January 2008. The county used a $1.2 million state Legacy Land Conservation Program grant to help pay the $1.9 million cost. (*Source:* http://hawaiitribune-herald.com, June 19, 2009)

Technology

For exercises 73–76,
(a) use a graphing calculator to graph the system of inequalities. Sketch the graph. Use the same patterns for shading in the sketch as the calculator uses. Describe the window.
(b) estimate the vertices of the solution region.

73. $y \leq -x + 9$
$y \geq x - 9$
$y \leq 2x - 9$

74. $y \leq -\dfrac{3}{4}x + 9$
$y \geq 2x - 7$
$y \leq 3x + 4$
$y \geq 1$

75. $y \leq x + 9$
$y \geq 4x + 9$
$y \geq 3$

76. $y \leq -2x + 8$
$y \geq 3x - 6$
$y \geq -4$

Find the Mistake

For exercises 77–80, the completed problem has one mistake.
(a) Describe the mistake in words, or copy down the whole problem and highlight or circle the mistake.
(b) Do the problem correctly.

77. Problem: Graph $y \leq -x + 5$ and identify the solution region.
$$y \leq 3x - 9$$
$$x \geq 0$$
$$y \geq 0$$

Incorrect Answer:

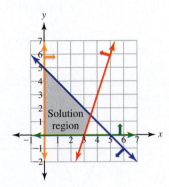

78. Problem: Identify the vertices of the solution region.
Incorrect Answer: The vertices are (0, 8) and (4, 0).

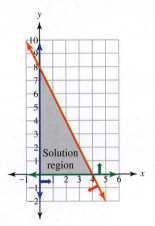

79. Problem: Graph: $3x - 5y < 15$
$$x \geq 0$$
$$y \leq 0$$

Incorrect Answer:

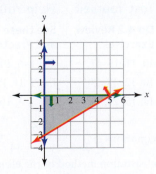

80. Problem: Graph: $x \geq 0$
$$y \geq 0$$
$$x \leq 2$$
$$y \leq 4$$

Incorrect Answer:

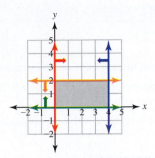

Review

For exercises 81–84, evaluate.

81. $\sqrt{64}$

82. $\sqrt{16}$

83. $\sqrt{-64}$

84. $\sqrt{25}$

SUCCESS IN COLLEGE MATHEMATICS

85. A student incorrectly graphed $3x - 5y < 15$ as shown.
   a. Describe the mistake that the student made in graphing the boundary line.
   b. Describe the mistake that the student made in shading.

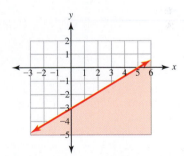

86. This student had incorrectly memorized that the shading for an inequality with a less than sign is always *under* the graphed line. Explain how the student should determine the shading of the graph of an inequality.

Study Plan for Review of Chapter 4

SECTION 4.1 **Systems of Linear Equations**

Ask Yourself	Test Yourself	Help Yourself
Can I...	Do 4.1 Review Exercises	See these Examples and Practice Problems
solve a system of linear equations by graphing?	3–8	Ex. 1–7, PP 1–8
use an algebraic method to determine whether an ordered pair is a solution of a system of linear equations?	1, 2	Ex. 8, PP 9, 10
use a graphical method to solve an application problem?	9	Ex. 9, PP 11, 12

4.1 Review Exercises

1. Use an algebraic method to determine whether $(-3, -5)$ is a solution of $\begin{array}{l} 2x - 7y = 29 \\ -x + 8y = -37 \end{array}$.

2. Use an algebraic method to determine whether $(1, 4)$ is a solution of $\begin{array}{l} 5x + y = 10 \\ y = -2x + 6 \end{array}$.

For exercises 3–6, solve by graphing.

3. $y = \dfrac{1}{2}x - 6$
   $2x - 4y = 24$

4. $y = 2x + 9$
   $6x - 3y = -21$

5. $x - 2y = -2$
   $3x + y = 29$

6. $y = 3x - 8$
   $x = 5$

7. Explain why $(2, 3)$ is not a solution of the system of equations represented by the graph.

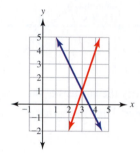

8. Explain why the system $\begin{array}{l} y = \dfrac{3}{4}x + 2 \\ y = \dfrac{3}{4}x - 9 \end{array}$ has no solution.

9. An investor will put a total of $20,000 in two mutual funds. Fund A is high risk. Fund B is medium risk. The investor decides to put four times as much in Fund A as in Fund B. Let x equal the amount of money put in Fund A, and let y equal the amount of money put in Fund B. To find the amount of money that should be put in each fund, solve the system $\begin{array}{l} x + y = \$20,000 \\ x = 4y \end{array}$ by graphing.

SECTION 4.2 **Algebraic Methods**

Ask Yourself	Test Yourself	Help Yourself
Can I...	Do 4.2 Review Exercises	See these Examples and Practice Problems
solve a system of two linear equations in two variables by substitution?	10, 11, 13–17	Ex. 1–6, PP 1–4
solve a system of linear equations in two variables by elimination?	10–12, 18–21	Ex. 7–11, PP 5–8
identify a system of equations with no solution or infinitely many solutions?	14, 15, 20, 21	Ex. 3, 4, 10, PP 2, 3, 7, 8

4.2 Review Exercises

10. A system of equations is $\begin{array}{l} 6x + 5y = 11 \\ 9x - 4y = 5 \end{array}$. Would you choose the substitution method or the elimination method to solve this system? Explain.

11. A system of equations is $\begin{array}{l} y = 9x - 2 \\ 3x - 5y = -34 \end{array}$. Would you choose the substitution method or the elimination method to solve this system? Explain.

12. A student is using the elimination method to solve the system of equations $\begin{array}{l} 4x - 3y = 2 \\ 5x + 8y = 26 \end{array}$. He chooses to multiply both sides of $4x - 3y = 2$ by a constant, 5. He chooses to multiply both sides of $5x + 8y = 26$ by a constant, -4. Identify the variable he is eliminating.

For exercises 13–17,
(a) solve by substitution.
(b) if the solution is an ordered pair, check.

13. $y = 2x - 15$
$\quad 7x + 4y = 45$

14. $y = \dfrac{3}{2}x + 4$
$\quad 3x - 2y = 2$

15. $y = -5x + 2$
$\quad 15x + 3y = 6$

16. $y = -4x + 9$
$\quad y = 3x + 6$

17. $\quad 2x + 3y = -21$
$\quad -7x - 10y = 74$

For exercises 18–21,
(a) use the elimination method to solve the system of equations.
(b) if the system has one solution, check.

18. $5x + 9y = -117$
$\quad -3x + 4y = 42$

19. $\dfrac{3}{4}x + \dfrac{5}{6}y = 11$
$\quad y = 8x - 58$

20. $\quad x - y = 15$
$\quad -2x + 2y = -30$

21. $2x - 3y = 5$
$\quad 4x - 6y = 24$

SECTION 4.3   Applications

Ask Yourself	Test Yourself	Help Yourself
Can I...	Do 4.3 Review Exercises	See these Examples and Practice Problems
write a system of linear equations that represents costs and revenue?	23	Ex. 1, PP 1
given the percent of an ingredient in a mixture and the total volume of the mixture, find the volume of the ingredient?	22, 25	Ex. 4–7, PP 4–6
write a formula that describes the relationship of distance, rate, and time?	28	Ex. 8, 9, PP 7, 8
use a system of linear equations to solve an application problem?	22–28	Ex. 1–3, 5–9, PP 1–8

4.3 Review Exercises

For exercises 22–28, use a system of equations and the five steps.

22. Find the amount of pure antifreeze and the amount of water needed to make 64 oz of a solution that is 80% antifreeze.

23. A photographer uses about $50 in materials to create a new framed print. Her fixed costs are $1200 a month. The average revenue per print is $110. Find the number of prints that need to be made and sold to break even. Find the revenue at the break-even point.

24. A librarian has $5000 to purchase new books. He wants to buy twice as much fiction as nonfiction. Find the amount to spend on fiction and the amount to spend on nonfiction. Round to the nearest whole number.

25. One salad dressing is 48% olive oil. Another salad dressing is 35% olive oil. Find the amount of each salad dressing needed to make 32 oz of dressing that is 40% olive oil. Round to the nearest whole number.

26. The cost of three wraps and two tacos is $17.25. The cost of two wraps and five tacos is $19.75. Find the cost of one wrap and the cost of one taco.

27. On a cross-country flight, 125 passengers checked baggage. Some passengers checked one piece of baggage for $15. Other passengers checked two pieces of baggage for $40. The total amount collected for checking baggage was $2475. Find the number of people who checked one piece of baggage. Find the number of passengers who checked two pieces of baggage.

28. A turtle traveling $0.5 \, \dfrac{\text{mi}}{\text{hr}}$ starts down a road. Twelve hours later, a rabbit traveling $25 \, \dfrac{\text{mi}}{\text{hr}}$ starts down the same road. Find the time in minutes when the rabbit will catch up to the turtle. Round to the nearest whole number. (Assume that both animals travel in a straight line at a constant speed and do not stop along the way.)

SECTION 4.4   Systems of Linear Equations in Three Variables

Ask Yourself	Test Yourself	Help Yourself
Can I...	Do 4.4 Review Exercises	See these Examples and Practice Problems
identify a graph as a point, a line, or a plane?	29	Ex. 1, 2, PP 1–6
use an algebraic method to determine whether an ordered triple is a solution of a system of three linear equations in three variables?	30	Ex. 3, PP 7, 8

Ask Yourself	Test Yourself	Help Yourself
Can I...	Do 4.4 Review Exercises	See these Examples and Practice Problems
solve a system of three linear equations in three variables by elimination?	31–34	Ex. 4–6, PP 9–11
identify a system of three linear equations in three variables that has no solution or infinitely many solutions?	32, 33	Ex. 5, 6, PP 10, 11
use a system of three linear equations to solve an application problem?	35	Ex.7, PP 12

4.4 Review Exercises

29. Identify the graph as a point, a line, or a plane.
   a. The graph in three dimensions of $ax + by + cz = d$ $(a \neq 0, b \neq 0, c \neq 0)$
   b. The graph of (x, y, z)
   c. The graph in two dimensions of $ax + by = c$ $(a \neq 0, b \neq 0)$

30. Use an algebraic method to show that $(1, -2, 4)$ is a solution of the system of equations.

$$2x + 5y + z = -4$$
$$x - 3y + 2z = 15$$
$$9x + 2y - z = 1$$

For exercises 31–34,
(a) solve by elimination. If there are infinitely many solutions, it is sufficient to say so. Parametric notation is not required.
(b) if the solution is an ordered triple, check.

31. $2x + 3y + z = 10$
$x - 5y + z = -17$
$8x + 2y - z = -40$

32. $x + 3y - z = 12$
$x + 6y + z = 3$
$3x + 12y - z = 27$

33. $3x + 4y + z = 10$
$x + y + z = 2$
$x + 2y - z = 9$

34. $2x + y + z = 10$
$3x + 2y + z = 17$
$x + 2z = 2$

35. A vineyard requires a yearly application of 900 lb nitrogen, 375 lb phosphate, and 675 lb potash. A 100-lb bag of 10-10-10 fertilizer contains 10 lb nitrogen, 10 lb phosphate, and 10 lb potash. A 100-lb bag of 18-4-12 fertilizer contains 18 lb nitrogen, 4 lb phosphate, and 12 lb potash. A 100-lb bag of 20-3-6 fertilizer contains 20 lb nitrogen, 3 lb phosphate, and 6 lb potash. Use the five steps and a system of three linear equations to find the number of bags of each kind of fertilizer to apply to the vineyard.

SECTION 4.5 Matrix Methods

Ask Yourself	Test Yourself	Help Yourself
Can I...	Do 4.5 Review Exercises	See these Examples and Practice Problems
represent a system of linear equations with an augmented matrix?	39, 40, 43–48	Ex. 1, 2, PP 1, 2
use row operations to change an entry in a matrix to 0 or 1?	41, 42	Ex. 3, 4, PP 3–6
determine whether a matrix is in reduced row echelon form?	43–48	Ex. 5
identify the solution of a system of linear equations that is represented by a matrix in reduced row echelon form?	36–38, 43–48	Ex. 6–10, PP 7–9
solve a system of linear equations by rewriting its matrix in reduced row echelon form?	43–48	Ex. 6–10, PP 7–9
identify a matrix in reduced row echelon form that represents a system of linear equations with no solution or infinitely many solutions?	37, 38, 45, 46	Ex. 9, 10, PP 9
use a system of three linear equations to solve an application problem?	47, 48	See Ex. 7 of Section 4.4, PP 12

4.5 Review Exercises

36. A matrix in reduced row echelon form is $\begin{bmatrix} 1 & 0 & -3 \\ 0 & 1 & 8 \end{bmatrix}$.

Identify the solution of this system of equations.

37. A system of linear equations in two variables has infinitely many solutions. If this system is represented by an augmented matrix and changed into reduced row echelon form, describe how to identify that the system has infinitely many solutions.

38. A system of linear equations in two variables has no solution. If this system is represented by an augmented matrix and changed into reduced row echelon form, describe how to identify that the system has no solution.

For exercises 39–40, represent the system of equations with an augmented matrix.

39. $2x + 5y = 1$
$-3x - y = 18$

40. $x + y + z = 10$
$3x + 2y - z = 1$
$3x + z = 13$

41. Change the first entry in row 1 to 1: $\begin{bmatrix} -8 & 5 & | & 17 \\ 9 & 1 & | & 39 \end{bmatrix}$.
Rewrite the matrix.

42. Change the first entry in row 2 to 0: $\begin{bmatrix} 1 & 2 & | & -9 \\ 4 & 3 & | & -11 \end{bmatrix}$.
Rewrite the matrix.

For exercises 43–46,
(a) represent the system with an augmented matrix.
(b) rewrite the matrix in reduced row echelon form.
(c) identify the solution of the system.

43. $2x + y = 33$
$4x - y = 57$

44. $x + y + z = 10$
$2x + y - z = 20$
$x + 2y + z = -14$

45. $3x - 4y = 8$
$-6x + 8y = -16$

46. $2x + y + z = 18$
$x + y - z = 5$
$5x + 3y + z = 40$

For exercises 47–48,
(a) assign the variables.
(b) write a system of equations that represents the problem situation.
(c) represent the system with an augmented matrix.
(d) rewrite the matrix in reduced row echelon form.
(e) identify the solution of the system.
(f) use a sentence to report the solution to the problem.

47. Soil scientists use the micro-Kjeldahl method to find the nitrogen in soil and plant materials. The method uses a 1% boric acid solution. A stockroom has 2.5 L of 4% boric acid solution. Find the amount of this boric acid solution and the amount of pure water needed to make 100 mL of a 1% boric acid solution.

48. The cost to make a product is $8. The fixed overhead costs to make the product are $3200 per month. The company sells each product for $24. Find the number of products that must be made and sold each month to break even. Find the revenue at the break-even point.

SECTION 4.6   **Systems of Linear Inequalities**

Ask Yourself	Test Yourself	Help Yourself
Can I . . .	**Do 4.6 Review Exercises**	**See these Examples and Practice Problems**
graph a system of linear inequalities and label the solution region?	50–52	Ex. 1, 2, 4, PP 1–7
estimate the vertices of the graph of a bounded solution region or find the vertices using an algebraic method?	49–52	Ex. 3, 5, PP 5–7
use an algebraic method to determine whether an ordered pair is a solution of a system of inequalities?	49	Ex. 6, PP 8, 9
write and/or graph a system of linear constraints and label the feasible region?	53	Ex. 7–9, PP 10
estimate the vertices of the feasible region of the graph of a system of linear constraints?	53	Ex. 8, 9, PP 10

4.6 Review Exercises

49. The graph represents the system $\begin{array}{l} y \geq x + 1 \\ y \leq 4 \\ y \geq -x + 4 \end{array}$.

a. Use the graph to estimate the vertices of the solution region.
b. Use an algebraic method to determine whether $(-3, 2)$ is a solution of the system.

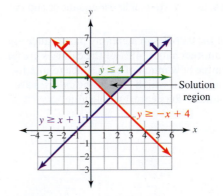

For exercises 50–52,
(a) graph the system of inequalities and label the solution region.
(b) estimate the vertices of the solution region.

50. $2x + 3y \leq 18$
$x \geq 3$
$y \geq 0$

51. $2x + y \leq 12$
$y \geq x$
$x \geq 0$

52. $y \leq -3x + 12$
$y \geq -3x + 6$
$x \geq 0$
$y \geq 0$

53. Jack is eating chicken stir-fry and a brown rice salad for dinner. Each serving of stir-fry contains 3 g of fat and 20 g of protein. Each serving of salad contains 6 g of fat and 5 g of protein. In this meal, Jack wants to eat at least 35 g of protein. He wants to eat no more than 21 g of fat. Let x = number of servings of stir-fry, and let y = number of servings of rice salad.

a. Write four constraints that describe this situation.
b. Graph the constraints. Label the feasible region.
c. Estimate the vertices of the feasible region.

Chapter 4 Test

1. Solve $\begin{array}{l} y = 2x + 6 \\ x + 4y = -12 \end{array}$ by graphing.

2. Solve $\begin{array}{l} 3x - 2y = 6 \\ y = \dfrac{3}{2}x + 5 \end{array}$ by graphing.

For problems 3 and 4,
(a) solve by substitution.
(b) if the solution is an ordered pair, check.

3. $y = 6x - 5$
$9x - 4y = -10$

4. $y = 9x - 3$
$18x - 2y = 6$

For problems 5 and 6,
(a) solve by elimination.
(b) if the solution is an ordered pair, check.

5. $2x + 7y = 18$
$3x - 9y = -51$

6. $6x + 8y = 9$
$5x - 6y = -2$

For problems 7 and 8, use a system of linear equations and the five steps.

7. Sixty percent of the calories in a creamy pasta sauce are fat calories. In another sauce, 48% of the calories are fat calories. Determine how much of each sauce to mix to create 32 oz of a sauce in which 52% of the calories are fat calories. Round to the nearest whole number.

8. A landscape architect is drawing plans for a 8.5-acre public park. She wants four times as much open space as planted space. Find the number of acres that should be open space and the number of acres that should be planted space.

9. **a.** Solve $\begin{array}{l} x + y + z = 10 \\ 2x + 2y + z = 22 \\ 2x + 4y - z = 44 \end{array}$ by elimination.

b. If the solution is an ordered triple, check.

10. A person on a diet is making a low-energy snack mix from raw broccoli, radishes, and cauliflower. The mix should have 106 calories, 9.2 g of fiber, and 8 g of protein. One cup of broccoli has 31 calories, 2.4 g fiber, and 2.6 g protein. One cup of radishes has 19 calories, 1.9 g fiber, and 0.8 g protein. One cup of cauliflower has 25 calories, 2.5 g fiber, and 2 g protein. Use a system of linear equations and the five steps to find the amount of broccoli, radishes, and cauliflower in this snack mix.

11. Solve the system $\begin{array}{l} 2x + y = 11 \\ 4x - y = 19 \end{array}$ by rewriting its matrix in reduced row echelon form.

12. **a.** Graph $\begin{array}{l} x + y \leq 8 \\ y \geq \dfrac{1}{3}x \\ x \geq 3 \end{array}$. Label the solution region.

b. Estimate the vertices of the solution region.

Rational Expressions, Equations, and Functions

SUCCESS IN COLLEGE MATHEMATICS

Making Connections

In Chapter 5, you will learn about rational expressions, equations, and functions. A rational number such as $\frac{5}{9}$ is a number that can be written as a fraction in which the numerator and denominator are integers. A rational expression such as $\frac{x^2 + 5x + 6}{x^2 - 9}$ can be written as a fraction in which the numerator and denominator are polynomials.

You already know how to do arithmetic with fractions, such as simplifying a fraction into lowest terms and adding or subtracting fractions with different denominators. You will use these same procedures to work with rational expressions, equations, and functions.

Look for examples in which a problem with rational numbers and a similar problem with rational expressions are done side by side. Work both problems, and compare the annotations for each line. Compare the procedures you know for doing fraction arithmetic with the procedures for doing algebra with rational expressions. Making connections between previous knowledge and new knowledge is an important part of learning and understanding algebra.

$$\frac{\text{integer}}{\text{integer}} = \frac{5}{9}$$

$$\frac{\text{polynomial}}{\text{polynomial}} = \frac{x^2 + 5x + 6}{x^2 - 9}$$

In common use, *rational* means something that is based on reason. In mathematics, a **rational number** is the *ratio* of two integers, such as $\frac{5}{9}$. We can also describe a rational number as the quotient of two integers. A **rational expression** is the quotient of two polynomials. In this section, we will study rational expressions.

Simplifying, Multiplying, and Dividing Rational Expressions

After reading the text, working the practice problems, and completing assigned exercises, you should be able to:

1. Simplify a rational expression.
2. Multiply rational expressions.
3. Divide rational expressions.

Simplifying Rational Expressions

We use the same principles to simplify rational expressions as we use to simplify rational numbers. We factor the numerator and denominator, look for common factors, and simplify fractions or rational expressions that are equal to 1.

> **Simplifying a Rational Number or Rational Expression into Lowest Terms**
> 1. Factor the numerator and denominator.
> 2. Find common factors in the numerator and denominator.
> 3. Simplify fractions or expressions that are equal to 1. (The product of a fraction or expression and 1 is equal to the fraction or expression.)
> 4. Multiply any remaining factors in the numerator; either multiply any remaining factors in the denominator or leave the denominator in factored form.
> 5. The number or expression is simplified if the numerator and denominator have no common factors.

When a rational number or expression has the same numerator and denominator, it is equal to 1. We often simplify by drawing a line through common factors in the numerator and denominator. We can do this because the common factors create a fraction that is equal to 1. When a number or expression is multiplied by 1, the value of the number or expression does not change. (The **multiplicative identity** of the real numbers is 1.)

EXAMPLE 1 | Simplify: $\dfrac{15}{24}$ and $\dfrac{x^2 + 5x + 6}{x^2 - 9}$

SOLUTION ▶

$$\frac{15}{24}$$

$$= \frac{5 \cdot 3}{8 \cdot 3} \qquad \text{Factor the numerator and denominator.}$$

$$= \frac{5 \cdot \cancel{3}}{8 \cdot \cancel{3}} \qquad \text{Find common factors.}$$

$$= \frac{5}{8} \cdot 1 \qquad \text{Simplify fractions that are equal to 1.}$$

$$= \frac{5}{8} \qquad \text{Simplify.}$$

$$\frac{x^2 + 5x + 6}{x^2 - 9}$$

$$= \frac{(x + 2)(x + 3)}{(x + 3)(x - 3)}$$

$$= \frac{(x + 2)(\cancel{x + 3})}{(\cancel{x + 3})(x - 3)}$$

$$= \frac{x + 2}{x - 3} \cdot 1$$

$$= \frac{x + 2}{x - 3}$$

Division by zero is undefined. In Example 1, the rational expression $\dfrac{(x + 2)(x + 3)}{(x + 3)(x - 3)}$ includes a variable in the denominator. If $x = 3$ or $x = -3$, the denominator is 0 and the expression is undefined. *In this section, we assume that all rational expressions are defined.* In other words, a variable cannot represent a number that results in a denominator of 0.

EXAMPLE 2   Simplify: $\dfrac{a^2 - 12a + 32}{a^2 - 16}$

SOLUTION ▶

$$\frac{a^2 - 12a + 32}{a^2 - 16}$$

$$= \frac{(a - 4)(a - 8)}{(a - 4)(a + 4)} \qquad \text{Factor the numerator and denominator.}$$

$$= \frac{(\cancel{a - 4})(a - 8)}{(\cancel{a - 4})(a + 4)} \qquad \text{Find common factors; } \frac{a - 4}{a - 4} = 1$$

$$= \frac{a - 8}{a + 4} \qquad \text{Simplify fractions that are equal to 1.}$$

In Example 2, after factoring and simplifying fractions that are equal to 1, the expression was simplified. In the next example, we finish simplifying by multiplying the remaining factors in the numerator and leaving the denominator factored. As we will see in Section 5.2, this makes it easier to add or subtract these expressions. Ask your instructor whether you should multiply factors in the denominator.

EXAMPLE 3   Simplify: $\dfrac{2w^2 + 14w - 36}{w^2 + 3w - 54}$

SOLUTION ▶

$$\frac{2w^2 + 14w - 36}{w^2 + 3w - 54}$$

$$= \frac{2(w^2 + 7w - 18)}{(w + 9)(w - 6)} \qquad \text{Factor the numerator and denominator.}$$

$$= \frac{2(\cancel{w + 9})(w - 2)}{(\cancel{w + 9})(w - 6)} \qquad \text{Factor the numerator completely; find common factors.}$$

$$= \frac{2(w - 2)}{w - 6} \qquad \text{Simplify fractions that are equal to 1; } \frac{w + 9}{w + 9} = 1$$

$$= \frac{2w - 4}{w - 6} \qquad \text{Multiply the factors in the numerator.}$$

When simplifying rational expressions, look for common *factors* in the numerator and denominator. When we simplify these factors, we are simplifying a fraction that is equal to 1. This is true only for *multiplication of factors*. We are using the same principles that we use in simplifying rational numbers.

Simplify Fractions That Are Equal to 1

Correct: $\dfrac{\cancel{2} \cdot 5}{\cancel{2} \cdot 3}$

$= \dfrac{5}{3}$

Correct: $\dfrac{\cancel{(x+3)}(x+1)}{\cancel{(x+3)}(x-2)}$

$= \dfrac{x+1}{x-2}$

Correct: $\dfrac{2+5}{2+3}$

$= \dfrac{7}{5}$

Wrong: $\dfrac{\cancel{x}+3}{\cancel{x}+1}$

$= \dfrac{3}{1}$

Wrong: $\dfrac{\cancel{6}}{x+\cancel{6}}$

$= \dfrac{1}{x}$

Wrong: $\dfrac{\cancel{2}+5}{\cancel{2}+3}$

$= \dfrac{5}{3}$

The factors $(4-p)$ and $(p-4)$ are not the same. However, $4-p = -1(-4+p)$. In the next example, to simplify $\dfrac{4-p}{p-4}$, we factor out -1 in the numerator, change the order of the factor to $p-4$, then simplify $\dfrac{p-4}{p-4}$.

EXAMPLE 4 Simplify: $\dfrac{4-p}{p-4}$

SOLUTION ▶ $\dfrac{4-p}{p-4}$

$= \dfrac{-1(-4+p)}{p-4}$ Factor out -1 in the numerator.

$= \dfrac{-1(p-4)}{p-4}$ Commutative property; rewrite $-4+p$ as $p-4$.

$= \dfrac{-1(\cancel{p-4})}{\cancel{p-4}}$ Find common factors.

$= -1$ Simplify fractions that are equal to 1; $\dfrac{p-4}{p-4} = 1$

Since the difference of squares factoring pattern is $a^2 - b^2 = (a-b)(a+b)$, $p^2 - 16 = (p-4)(p+4)$. This pattern also applies when the lead term is a number: $16 - p^2 = (4-p)(4+p)$.

EXAMPLE 5 Simplify: $\dfrac{16-p^2}{p^2-6p+8}$

SOLUTION ▶ $\dfrac{16-p^2}{p^2-6p+8}$

$= \dfrac{(4-p)(4+p)}{(p-4)(p-2)}$ Factor the numerator and denominator.

$= \dfrac{-1(-4+p)(4+p)}{(p-4)(p-2)}$ Factor; $4-p = -1(-4+p)$

$$= \frac{-1(p-4)(p+4)}{(p-4)(p-2)}$$ Commutative property.

$$= \frac{-1(p\!\!\!/-4)(p+4)}{(p\!\!\!/-4)(p-2)}$$ Find common factors.

$$= \frac{-1(p+4)}{p-2}$$ Simplify fractions that are equal to 1; $\dfrac{p-4}{p-4} = 1$

$$= \frac{-1p-1(4)}{p-2}$$ Multiply remaining factors; distributive property.

$$= \frac{-p-4}{p-2}$$ Simplify.

When the numerator and denominator of a rational expression are monomials, use the quotient rule of exponents, $\dfrac{x^m}{x^n} = x^{m-n}$, to simplify the variables. Use the relationship $x^{-n} = \dfrac{1}{x^n}$ to rewrite negative exponents as positive exponents.

EXAMPLE 6   Simplify: $\dfrac{36a^2b^6c}{48a^5b^2c^3}$

SOLUTION ▶ $\dfrac{36a^2b^6c}{48a^5b^2c^3}$

$$= \frac{3 \cdot \cancel{12} \cdot a^{2-5}b^{6-2}c^{1-3}}{4 \cdot \cancel{12}}$$ Factor; find common factors; quotient rule of exponents.

$$= \frac{3a^{-3}b^4c^{-2}}{4}$$ Simplify.

$$= \frac{3b^4}{4a^3c^2}$$ Rewrite with positive exponents; $a^{-3} = \dfrac{1}{a^3}; c^{-2} = \dfrac{1}{c^2}$

Practice Problems

For problems 1–4, simplify.

1. $\dfrac{x^2 - 9}{x^2 - 7x + 12}$ 2. $\dfrac{9 - w^2}{w^2 - 7w + 12}$

3. $\dfrac{2x^2 + 5x + 3}{2x^2 - 7x - 15}$ 4. $\dfrac{54p^5qr^9}{72p^8q^2r}$

Difference or Sum of Cubes in Rational Expressions

To factor a difference of squares or a sum or difference of cubes, use a pattern.

Patterns for Factoring Polynomials

Difference of Cubes Pattern $a^3 - b^3 = (a - b)(a^2 + ab + b^2)$

Sum of Cubes Pattern $a^3 + b^3 = (a + b)(a^2 - ab + b^2)$

EXAMPLE 7   Simplify: $\dfrac{8x^4 - xy^3}{4x^2y - y^3}$

SOLUTION ▶ $\dfrac{8x^4 - xy^3}{4x^2y - y^3}$

$= \dfrac{x(8x^3 - y^3)}{y(4x^2 - y^2)}$ Factor out the greatest common factor.

$= \dfrac{x((2x)^3 - y^3)}{y(4x^2 - y^2)}$ Rewrite numerator to match the pattern $a^3 - b^3$.

$= \dfrac{x(2x - y)((2x)^2 + 2xy + y^2)}{y(4x^2 - y^2)}$ Difference of cubes pattern; $(a - b)(a^2 + ab + b^2)$.

$= \dfrac{x(2x - y)(4x^2 + 2xy + y^2)}{y(4x^2 - y^2)}$ Simplify.

$= \dfrac{x(2x - y)(4x^2 + 2xy + y^2)}{y((2x)^2 - y^2)}$ Rewrite denominator to match the pattern $a^2 - b^2$.

$= \dfrac{x(2x - y)(4x^2 + 2xy + y^2)}{y(2x - y)(2x + y)}$ Difference of squares pattern; $(a - b)(a + b)$.

$= \dfrac{x(2x - y)(4x^2 + 2xy + y^2)}{y(2x - y)(2x + y)}$ Find common factors.

$= \dfrac{x(4x^2 + 2xy + y^2)}{y(2x + y)}$ Simplify.

$= \dfrac{4x^3 + 2x^2y + xy^2}{y(2x + y)}$ Multiply factors in the numerator.

Practice Problems

For problems 5–7, simplify.

5. $\dfrac{x^3 + 8}{x^2 + 5x + 6}$ **6.** $\dfrac{c^3 - 8}{c^2 - 4}$ **7.** $\dfrac{27n^3 + p^3}{15n^2 + 12n + 5np + 4p}$

Multiplying and Dividing Rational Expressions

We can use two different strategies to multiply rational numbers or expressions. Either we multiply the numerators and denominators and then simplify, or we first find common factors in the numerator and denominator, simplify fractions that are equal to 1, and then multiply the remaining factors. When multiplying rational expressions, we usually use the second strategy and begin by finding common factors.

Strategies for Multiplying Rational Numbers or Rational Expressions

1. Multiply numerators and multiply denominators.
2. Simplify the product.

Or

1. Identify common factors in the numerator and denominator.
2. Simplify rational numbers or expressions that are equal to 1.
3. Multiply any remaining factors in the numerator; either multiply the factors in the denominator or leave the denominator in factored form.

EXAMPLE 8 | Simplify: $\dfrac{15}{8} \cdot \dfrac{10}{9}$ and $\dfrac{x - 4}{x^2 - 6x} \cdot \dfrac{x^2 - 36}{x^2 - 4x}$

SOLUTION ▶

$\dfrac{15}{8} \cdot \dfrac{10}{9}$

$= \dfrac{3 \cdot 5}{2 \cdot 4} \cdot \dfrac{2 \cdot 5}{3 \cdot 3}$ Factor.

$= \dfrac{\cancel{3} \cdot 5}{2 \cdot 4} \cdot \dfrac{2 \cdot 5}{3 \cdot \cancel{3}}$ Find common factors.

$= \dfrac{25}{12}$ Simplify fractions equal to 1; multiply remaining factors.

$\dfrac{x - 4}{x^2 - 6x} \cdot \dfrac{x^2 - 36}{x^2 - 4x}$

$= \dfrac{x - 4}{x(x - 6)} \cdot \dfrac{(x - 6)(x + 6)}{x(x - 4)}$

$= \dfrac{\cancel{x - 4}}{x\cancel{(x - 6)}} \cdot \dfrac{\cancel{(x - 6)}(x + 6)}{x\cancel{(x - 4)}}$

$= \dfrac{x + 6}{x^2}$

If all of the factors in the numerator have matching factors in the denominator, multiply the factors in the numerator by 1. The numerator in the final simplified expression is then 1.

EXAMPLE 9 | Simplify: $\dfrac{x^2 + 4x}{2x^2 + 15x + 7} \cdot \dfrac{x^2 + 4x - 21}{x^3 + x^2 - 12x}$

SOLUTION ▶

$\dfrac{x^2 + 4x}{2x^2 + 15x + 7} \cdot \dfrac{x^2 + 4x - 21}{x^3 + x^2 - 12x}$

$= \dfrac{1 \cdot \cancel{x}\cancel{(x + 4)}}{(2x + 1)\cancel{(x + 7)}} \cdot \dfrac{\cancel{(x + 7)}\cancel{(x - 3)}}{\cancel{x}\cancel{(x - 3)}\cancel{(x + 4)}}$ Factor; find common factors.

$= \dfrac{1}{2x + 1}$ Simplify fractions equal to 1.

To divide rational numbers or rational expressions, rewrite division as multiplication by the reciprocal of the divisor: "invert and multiply."

> **Dividing Rational Numbers or Rational Expressions**
> 1. Rewrite division as multiplication by the reciprocal of the divisor.
> 2. Identify common factors in the numerators and denominators.
> 3. Simplify rational numbers or expressions that are equal to 1.
> 4. Multiply any remaining factors in the numerator; either multiply the factors in the denominator or leave the denominator in factored form.

By tradition, denominators in rational expressions are not negative. It is easier to add or subtract rational expressions in which the denominator is positive. In the next example, we rewrite $\dfrac{x + 9}{-x}$ as $\dfrac{-1(x + 9)}{x}$.

EXAMPLE 10 Simplify: $\dfrac{2x^2 - x - 6}{2x^2 + 3x} \div \dfrac{4 - x^2}{x^2 + 11x + 18}$

SOLUTION ▶

$$\dfrac{2x^2 - x - 6}{2x^2 + 3x} \div \dfrac{4 - x^2}{x^2 + 11x + 18}$$

$$= \dfrac{2x^2 - x - 6}{2x^2 + 3x} \cdot \dfrac{x^2 + 11x + 18}{4 - x^2} \qquad \text{Multiplication by reciprocal.}$$

$$= \dfrac{(x - 2)(2x + 3)}{x(2x + 3)} \cdot \dfrac{(x + 2)(x + 9)}{(2 - x)(2 + x)} \qquad \text{Factor.}$$

$$= \dfrac{(x - 2)(2x + 3)}{x(2x + 3)} \cdot \dfrac{(x + 2)(x + 9)}{-1(-2 + x)(2 + x)} \qquad \text{Factor out } -1.$$

$$= \dfrac{(x - 2)(2x + 3)}{x(2x + 3)} \cdot \dfrac{(x + 2)(x + 9)}{-1(x - 2)(x + 2)} \qquad \text{Commutative property.}$$

$$= \dfrac{\cancel{(x - 2)}\cancel{(2x + 3)}}{x\cancel{(2x + 3)}} \cdot \dfrac{\cancel{(x + 2)}(x + 9)}{-1\cancel{(x - 2)}\cancel{(x + 2)}} \qquad \text{Find common factors.}$$

$$= \dfrac{x + 9}{x(-1)} \qquad \text{Simplify fractions equal to 1.}$$

$$= \dfrac{(-1)(x + 9)}{x} \qquad \text{Positive denominator; } \dfrac{1}{-1} = \dfrac{-1}{1}$$

$$= \dfrac{-x - 9}{x} \qquad \text{Multiply remaining factors; distributive property.}$$

Practice Problems

For problems 8–11, simplify.

8. $\dfrac{k^2 + 8k + 15}{k^2 + 2k - 15} \cdot \dfrac{k^3 - k^2 - 6k}{k^2 + 5k + 6}$

9. $\dfrac{x^2 - x - 12}{x^2 + 5x + 6} \div \dfrac{x^2 + 4x}{x^2 + 2x}$

10. $\dfrac{5m^3 - 5m^2}{2m^2 - 9m - 18} \cdot \dfrac{2m + 3}{10m^2 - 10m}$

11. $\dfrac{3x^2y}{5p^3w^5} \div \dfrac{18x^5y^9}{25pw}$

Using Technology: Using a Graphing Calculator to Check

A graphing calculator cannot simplify a rational expression in one variable. However, we can use it to check our work. In Example 1, we simplified $\dfrac{x^2 + 5x + 6}{x^2 - 9}$ to an equivalent expression $\dfrac{x + 2}{x - 3}$. To check that this is correct, graph two functions: $y = \dfrac{x^2 + 5x + 6}{x^2 - 9}$ and $y = \dfrac{x + 2}{x - 3}$. If the answer is correct, the graphs will appear to be the same.

When typing the function on the Y= screen, put parentheses around the numerator and the denominator in the rational expression.

The graphs of both functions appear to be the same. Our simplifying is probably correct.

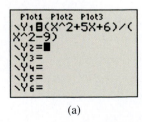

(a)

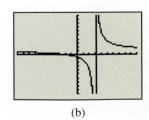

(b)

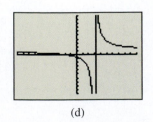

(c)

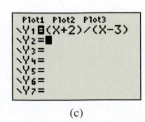

(d)

These two graphs are actually not exactly the same. Although the resolution on the calculator is not good enough to show it, there is a hole in the graph of $y = \dfrac{x^2 + 5x + 6}{x^2 - 9}$ at $x = -3$. As we will see in Section 5.5, the domain of a rational function must be restricted to prevent division by 0. However, the appearance of the actual graphs is close enough to confirm that we have simplified correctly.

Practice Problems For problems 12–14,
(a) use an algebraic method to simplify each expression.
(b) use a graphing calculator to check. Sketch the graph; describe the window.

12. $\dfrac{x^2 + 10x + 16}{x^2 - 4}$ 13. $\dfrac{x^2 - 9x + 18}{x^2 - 10x + 21}$ 14. $\dfrac{x^2 - 6x}{x^2 - 36}$

5.1 VOCABULARY PRACTICE

Match the term with its description.

1. The result of division
2. We can write each number in this set as a fraction in which the numerator and denominator are integers.
3. All numbers in this set are nonrepeating, nonterminating decimal numbers.
4. A polynomial with one term
5. When the numerator and denominator of an expression have no common factors
6. The quotient of two polynomials
7. This set of numbers includes the set of whole numbers and their opposites.
8. When we multiply these together, the result is a product.
9. This set of numbers is the union of the rational numbers and the irrational numbers.
10. Division by zero

A. factors
B. integers
C. irrational numbers
D. lowest terms
E. monomial
F. quotient
G. rational expression
H. rational numbers
I. real numbers
J. undefined

5.1 Exercises

Follow your instructor's guidelines for showing your work. Assume that all expressions in these exercises are defined. No variable represents a number that will cause division by zero.

For exercises 1–30, simplify. Do not rewrite improper fractions as mixed numbers.

1. $\dfrac{x^2 y^8 z}{x^7 y^2 z^3}$

2. $\dfrac{a^8 b^2 c}{a^3 b^6 c^4}$

3. $\dfrac{33 x^2 y^8 z}{45 x^7 y^2 z^3}$

4. $\dfrac{42 a^8 b^2 c}{54 a^3 b^6 c^4}$

5. $\dfrac{-14 hk}{28 h^4 k^5}$

6. $\dfrac{-24 xy}{48 x^3 y^7}$

7. $\dfrac{x^2 - 3x - 10}{x^2 + 5x + 6}$

8. $\dfrac{p^2 + 4p - 21}{p^2 + 9p + 14}$

9. $\dfrac{6y^2 - 5y - 4}{3y^2 + 14y - 24}$

10. $\dfrac{8n^2 + 10n - 3}{4n^2 + 19n - 5}$

11. $\dfrac{x^2 - 4x}{x^3 + 5x^2 + 6x}$

12. $\dfrac{c^2 - 5c}{c^3 + 6c^2 + 8c}$

13. $\dfrac{a^2 - 5a}{2a^2 - 3a - 35}$

14. $\dfrac{x^2 - 6x}{2x^2 - 3x - 54}$

15. $\dfrac{18 w^3 + 12 w^2}{36 w^6 - 18 w^5}$

16. $\dfrac{20 p^3 + 12 p^2}{80 p^8 - 20 p^7}$

17. $\dfrac{x^2 - 25}{x^2 - 8x + 15}$

18. $\dfrac{y^2 - 36}{y^2 - 14y + 48}$

19. $\dfrac{25 - x^2}{x^2 - 8x + 15}$

20. $\dfrac{36 - y^2}{y^2 - 14y + 48}$

21. $\dfrac{9 - b^2}{b^2 + 7b - 30}$

22. $\dfrac{49 - c^2}{c^2 - 5c - 14}$

23. $\dfrac{x^3 - 5x^2 + 6x}{x^2 - 2x}$

24. $\dfrac{d^3 - 7d^2 + 12d}{d^2 - 3d}$

25. $\dfrac{x^3 - 5x^2 + 6x}{2x - x^2}$

26. $\dfrac{d^3 - 7d^2 + 12d}{3d - d^2}$

27. $\dfrac{80 x^5 y^4}{100 x^6 y^3}$

28. $\dfrac{60 x^{11} y^{21}}{90 x^{12} y^{20}}$

29. $\dfrac{x^2 + 9x + 20}{x^2 + 6x + 8}$

30. $\dfrac{x^2 + 10x + 24}{x^2 + 12x + 32}$

31. Describe how the terms in a polynomial change when -1 is factored from the polynomial.

32. A student said that 5 divided by 0 equals 0. Use multiplication to show why this cannot be true.

For exercises 33–40, simplify.

33. $\dfrac{x^3 - y^3}{x^2 - y^2}$

34. $\dfrac{a^2 - b^2}{a^3 - b^3}$

35. $\dfrac{x^3 + y^3}{x^2 - y^2}$

36. $\dfrac{a^2 - b^2}{a^3 + b^3}$

37. $\dfrac{x^3 + 8}{x^2 + 7x + 10}$

38. $\dfrac{y^3 + 27}{2y^2 + y - 15}$

39. $\dfrac{27x^3 - 64y^6}{9x^2 - 16y^4}$

40. $\dfrac{8a^3 - 27b^6}{4a^2 - 9b^4}$

For exercises 41–58, simplify.

41. $\dfrac{c^2 + 10c + 21}{c^2 - 16} \cdot \dfrac{c^2 + 5c + 4}{c^2 + 8c + 7}$

42. $\dfrac{a^2 + 15a + 54}{a^2 - 9} \cdot \dfrac{a^2 + 10a + 21}{a^2 + 13a + 42}$

43. $\dfrac{x^2 + 2x - 15}{x^2 - 3x} \cdot \dfrac{x^2 + 9x}{x^2 + 15x + 54}$

44. $\dfrac{x^2 + 2x - 80}{x^2 - 8x} \cdot \dfrac{x^2 + 7x}{x^2 + 8x + 7}$

45. $\dfrac{6k^2 - k - 15}{3k^2 - 2k - 5} \cdot \dfrac{k^2 + 3k + 2}{2k^2 + 7k + 6}$

46. $\dfrac{6m^2 + 11m - 10}{3m^2 + m - 2} \cdot \dfrac{m^2 + 7m + 6}{2m^2 + 17m + 30}$

47. $\dfrac{x^3 + 2x^2 - 24x}{x^2 + 6x - 27} \cdot \dfrac{x^2 + x - 12}{x^3 - 16x}$

48. $\dfrac{y^3 + 3y^2 - 18y}{y^2 - 6y - 16} \cdot \dfrac{y^2 - 5y - 24}{y^3 - 9y}$

49. $\dfrac{z^2 - 16}{z^2 + 4z} \cdot \dfrac{z^2 - 10z}{z^2 - z - 12}$

50. $\dfrac{v^2 - 64}{v^2 + 8v} \cdot \dfrac{v^2 - 6v}{v^2 - v - 56}$

51. $\dfrac{16 - z^2}{z^2 + 4z} \cdot \dfrac{z^2 - 10z}{z^2 - z - 12}$

52. $\dfrac{64 - v^2}{v^2 + 8v} \cdot \dfrac{v^2 - 6v}{v^2 - v - 56}$

53. $\dfrac{6x^2 - 7x - 5}{3x^2 + 13x - 10} \cdot \dfrac{9x^2 - 4}{9x^2 - 9x - 10}$

54. $\dfrac{2x^2 - 3x - 5}{2x^2 - 15x + 28} \cdot \dfrac{4x^2 - 49}{4x^2 + 4x - 35}$

55. $\dfrac{6x^2 - 7x - 5}{3x^2 + 13x - 10} \cdot \dfrac{4 - 9x^2}{9x^2 - 9x - 10}$

56. $\dfrac{2x^2 - 3x - 5}{2x^2 - 15x + 28} \cdot \dfrac{49 - 4x^2}{4x^2 + 4x - 35}$

57. $\dfrac{5x^2 - 20x}{2y^2 - 4y} \cdot \dfrac{4y^8 - 8y^7}{25x^4 - 100x^3}$

58. $\dfrac{2x^2 - 4x}{8p^2 + 24p} \cdot \dfrac{4p^7 + 12p^6}{10x^5 - 20x^4}$

For exercises 59–74, simplify.

59. $\dfrac{x^2 + 9x + 20}{x^2 + 6x + 8} \div \dfrac{x^2 + 11x + 30}{x^2 + 3x + 2}$

60. $\dfrac{y^2 + 12y + 27}{y^2 + 6y + 5} \div \dfrac{y^2 + 7y + 12}{y^2 + 5y + 4}$

61. $\dfrac{w^2 - 4}{w^2 + 5w + 6} \div \dfrac{w^2 + 4w - 12}{w^2 + 10w + 24}$

62. $\dfrac{k^2 - 9}{k^2 + 5k + 6} \div \dfrac{k^2 + 2k - 15}{k^2 + 14k + 45}$

63. $\dfrac{4d^6 - 24d^5}{21f^9 + 21f^8} \div \dfrac{30d^{10} - 180d^9}{7f^2 + 7f}$

64. $\dfrac{8c^7 - 24c^6}{15a^6 + 30a^5} \div \dfrac{20c^{11} - 60c^{10}}{9a^2 + 18a}$

65. $\dfrac{x^2 - 25}{x^2 + 8x + 15} \div \dfrac{x^2 - 3x - 10}{x^2 + 9x + 18}$

66. $\dfrac{y^2 - 81}{y^2 + 11y + 18} \div \dfrac{y^2 - 8y - 9}{y^2 + 7y + 10}$

67. $\dfrac{25 - x^2}{x^2 + 8x + 15} \div \dfrac{x^2 - 3x - 10}{x^2 + 9x + 18}$

68. $\dfrac{81 - y^2}{y^2 + 11y + 18} \div \dfrac{y^2 - 8y - 9}{y^2 + 7y + 10}$

69. $\dfrac{20ab^2c^4}{21a^5bc} \div \dfrac{5a^8b^3c}{18abc^4}$

70. $\dfrac{30x^2yz^6}{55x^5y^4z} \div \dfrac{9x^2yz^9}{20xy^7}$

71. $\dfrac{3x^2 + 7x + 2}{3x^2 - 26x - 9} \div \dfrac{2x^2 - 5x - 12}{x^2 - 13x + 36}$

72. $\dfrac{4y^2 + 13y + 3}{4y^2 - 3y - 1} \div \dfrac{2y^2 - 7y - 30}{y^2 - 7y + 6}$

73. $\dfrac{a^2 + 4a - 21}{a^2 - a - 6} \div \dfrac{a^2 + 11a + 28}{a^2 + 6a + 8}$

74. $\dfrac{3c^2 - 14c - 24}{c^2 - 4c - 12} \div \dfrac{3c^2 - 23c - 36}{c^2 - 7c - 18}$

For exercises 75–82, simplify.

75. $\dfrac{7a^3b^7}{35a^2b^5} \cdot \dfrac{10b^2c^3}{6b^3c^5} \div \dfrac{ac^4}{3ac^7}$

76. $\dfrac{5a^2b^4}{15a^3b^3} \cdot \dfrac{6bc^3}{14bc^5} \div \dfrac{a^4c^5}{7a^6c^8}$

77. $\dfrac{z + 5}{z - 3} \div \dfrac{2z + 1}{2z + 3} \cdot \dfrac{2z^2 - z - 15}{2z^2 + 9z + 9}$

78. $\dfrac{y + 5}{y - 3} \div \dfrac{2y + 1}{3y + 2} \cdot \dfrac{3y^2 + 7y + 2}{3y^2 + 17y + 10}$

79. $\dfrac{x^2 - 5x - 6}{x^2 - 5x + 4} \cdot \dfrac{x^2 - 9x + 20}{x^2 - 6x - 7} \div \dfrac{x - 5}{x - 1}$

80. $\dfrac{x^2 - x - 6}{x^2 - 4x - 5} \cdot \dfrac{x^2 - 6x + 8}{x^2 - 5x + 6} \div \dfrac{x - 4}{x + 1}$

81. $\dfrac{2u-6}{2u-1} \cdot \dfrac{4u^2-1}{u^2-9} \cdot \dfrac{u^3+7u^2+12u}{u^3+u^2+3u}$

82. $\dfrac{3v-6}{3v-1} \cdot \dfrac{9v^2-1}{v^2-4} \cdot \dfrac{v^3+7v^2+10v}{v^3+v^2+5v}$

Problem Solving: Practice and Review

Follow your instructor's guidelines for following the five steps as outlined in Section 1.5, p. 51.

83. The U.S. Public Interest Research Group conducted a study of credit card marketing and practices. They surveyed 1584 students from 40 schools in 14 states. Find the number of students surveyed who reported that they had paid at least one late fee.

One in four respondents (25%) reported they had paid at least one late fee and 15% reported they had paid at least one over-the-limit fee. Over 6% of respondents reported that at least one card had been cancelled for non-payment. (*Source:* www.studentpirgs.org, March 2008)

84. Find the total amount received by Franklin Primary Health Care Center, Inc. in Alabama. Write the answer in millions of dollars; round to the nearest tenth of a million.

Franklin Primary received $710,000 in NAP funds for a new center in western Mobile County; $791,000 in IDS funds to increase service capacity; $1.4 million in Capital Improvement Program (CIP) funds for construction, renovation and equipment projects; and another $314,000 in CIP funds when it took over a health center in Baldwin County last fall. (*Source:* www.hhs.gov, June 16, 2011)

85. Find the percent of the additional patients who were uninsured. Round to the nearest percent.

When Franklin Primary sought the IDS funds, it proposed providing health services for 2,040 new patients, of whom 1,635 would be uninsured. However, the Recovery Act funds actually allowed Franklin Primary to serve 9,067 additional patients as of Dec. 31, 2010, of whom 3,376 were uninsured. (*Source:* www.hhs.gov, June 16, 2011)

86. A model of the Official National School Lunch Participation, y, is $y = \left(\dfrac{404{,}534 \text{ participants}}{1 \text{ year}}\right) x + 29{,}701{,}568$

participants, where x is the number of years since 2005. The Official National School Lunch Participation equals the average daily meals served divided by 0.927. Find the average daily meals served in 2011. Round to the nearest thousand. (*Source:* www.fns.usda.gov)

Technology

For exercises 87–90,
(a) simplify the expression.
(b) check your work with a graphing calculator using the method shown in this section. Sketch the graph; describe the window.

87. $\dfrac{x^2-36}{x^2-14x+48}$

88. $\dfrac{x^2-49}{x^2-2x-35}$

89. $\dfrac{2x+5}{6x^2+11x-10}$

90. $\dfrac{3x+1}{12x^2-5x-3}$

Find the Mistake

For exercises 91–94, the completed problem has one mistake.
(a) Describe the mistake in words, or copy down the whole problem and highlight or circle the mistake.
(b) Do the problem correctly.

91. Problem: Simplify: $\dfrac{x^2-4}{x^2-4x-12}$

Incorrect Answer: $\dfrac{x^2-4}{x^2-4x-12}$

$= \dfrac{\cancel{x^2}-\cancel{4}}{\cancel{x^2}-\cancel{4}x-12}$

$= \dfrac{1}{x-12}$

92. Problem: Simplify: $\dfrac{25-x^2}{x^2-3x-10}$

Incorrect Answer: $\dfrac{25-x^2}{x^2-3x-10}$

$= \dfrac{(\cancel{x-5})(x+5)}{(\cancel{x-5})(x+2)}$

$= \dfrac{x+5}{x+2}$

93. Problem: Simplify: $\dfrac{y^2-9y+20}{y^2+4y+3} \div \dfrac{y^2-3y-10}{y^2-2y-8}$

Incorrect Answer: $\dfrac{y^2-9y+20}{y^2+4y+3} \div \dfrac{y^2-3y-10}{y^2-2y-8}$

$= \dfrac{(\cancel{y-4})(y-5)}{(y+3)(y+1)} \cdot \dfrac{(y-5)\cancel{(y+2)}}{\cancel{(y-4)}\cancel{(y+2)}}$

$= \dfrac{(y-5)(y-5)}{(y+3)(y+1)}$

94. Problem: Simplify: $\dfrac{x^2-4x-45}{x^2-6x-27} \cdot \dfrac{x^2+12x+27}{x^2+14x+45}$

Incorrect Answer: $\dfrac{x^2-4x-45}{x^2-6x-27} \cdot \dfrac{x^2+12x+27}{x^2+14x+45}$

$= \dfrac{(x+9)(x-5)}{(x-9)\cancel{(x+3)}} \cdot \dfrac{\cancel{(x+3)}\cancel{(x+9)}}{(x+5)\cancel{(x+9)}}$

$= \dfrac{(x+9)(x-5)}{(x-9)(x+5)}$

Review

95. Write the prime factorization of 36.

96. Write the prime factorization of 48.

97. Use the prime factorization of 36 and 48 to find the least common multiple of 36 and 48.

98. Add $\dfrac{7}{36} + \dfrac{11}{48}$ using the least common denominator.

SUCCESS IN COLLEGE MATHEMATICS

99. A rational number or a rational expression that is simplified has no common factors in the numerator and denominator. Look at Example 1 in this section.
   a. What are the common factors in the numerator and denominator of $\dfrac{15}{24}$?
   b. What are the common factors in the numerator and denominator of $\dfrac{x^2 + 5x + 6}{x^2 - 9}$?

100. When simplifying the rational number $\dfrac{10}{6}$, we can rewrite the number as $\dfrac{2 \cdot 5}{2 \cdot 3}$. The simplified fraction is $\dfrac{5}{3}$. Sometimes we cross off the common factors like this: $\dfrac{\cancel{2} \cdot 5}{\cancel{2} \cdot 3}$. Explain why we can cross off these common factors.

$\frac{3}{8}$ pizza

To add rational numbers such as $\dfrac{3}{8}$ and $\dfrac{1}{10}$, we find a common denominator. To add rational expressions, we also find a common denominator. In this section, we will add and subtract rational expressions. Assume that all rational expressions are defined with denominators that do not equal 0.

$\frac{1}{10}$ pizza

SECTION 5.2

Adding and Subtracting Rational Expressions

After reading the text, working the practice problems, and completing assigned exercises, you should be able to:

1. Write the prime factorization of a rational number or polynomial.
2. Find the least common denominator of two rational expressions.
3. Rewrite a rational expression with a new denominator.
4. Add and subtract rational expressions.

Least Common Denominators

The set of whole numbers is $\{0, 1, 2, 3, 4, \ldots\}$. A **prime number** is a whole number greater than 1 with only two factors, itself and 1. The set of the first five prime numbers is $\{2, 3, 5, 7, 11\}$. By the fundamental theorem of arithmetic, we know that for any whole number greater than 1, there is only one combination of prime number factors with a product equal to this number. This expression is the **prime factorization** of the number.

> **The Fundamental Theorem of Arithmetic**
> Every whole number greater than 1 can be represented in exactly one way, other than rearrangement, as a product of one or more prime numbers.

> **Prime Factorization**
> 1. Choose any two factors of the number.
> 2. If a factor is not prime, rewrite it as the product of two numbers.
> 3. Continue this process until all of the factors are prime numbers.
> 4. Write the factors in increasing order.
> 5. Rewrite any repeated multiplications in exponential notation.

EXAMPLE 1   Write the prime factorization of 84.

SOLUTION ▶

$84 = 4 \cdot 21$ Choose any two factors of 84.

$84 = 2 \cdot 2 \cdot 3 \cdot 7$ If a factor is not prime, rewrite it as a product of two numbers.

$84 = 2^2 \cdot 3 \cdot 7$ Rewrite repeated multiplications in exponential notation; $2 \cdot 2 = 2^2$

The prime factorization of 84 is $2^2 \cdot 3 \cdot 7$.

We can use prime factorizations to find a least common denominator.

> **Finding Least Common Denominators of Rational Numbers and Rational Expressions**
> 1. Write the prime factorization of each denominator.
> 2. The least common denominator is the product of each unique factor with the greatest exponent on this factor in the prime factorizations.

Instead of using prime factorization, we could just multiply the denominators together to create a common denominator. However, this product might not be the smallest denominator. By choosing the greatest exponent for a factor that appears in either factorization, we find the *smallest* or *least* common denominator. Arithmetic with smaller factors is often easier.

EXAMPLE 2   For the rational numbers $\dfrac{3}{40}$ and $\dfrac{1}{50}$,

(a) write the prime factorization of each denominator.

SOLUTION ▶

40		50
$= 4 \cdot 10$	Choose any two factors.	$= 5 \cdot 10$
$= 2 \cdot 2 \cdot 2 \cdot 5$	Rewrite factors that are not prime.	$= 2 \cdot 5 \cdot 5$
$= 2^3 \cdot 5^1$	Use exponential notation.	$= 2^1 \cdot 5^2$

(b) find the least common denominator.

▶

The factors that appear in each factorization are 2 and 5. The greatest exponent with a base of 2 is 3; the greatest exponent with a base of 5 is 2.

$2^3 \cdot 5^2$ Write the product of each unique factor with the greatest exponent.

$= 8 \cdot 25$ Simplify.

$= 200$ Simplify.

In Example 2, the denominators are whole numbers. We can also use prime factorization to find a least common denominator that includes variables.

EXAMPLE 3 For the rational expressions $\dfrac{11}{52x^2y^3z}$ and $\dfrac{17}{40xyz^2}$,

(a) write the prime factorization of each denominator.

SOLUTION ▶

$$52x^2y^3z \qquad\qquad\qquad\qquad\qquad\qquad 40xyz^2$$

$\quad = 2 \cdot 2 \cdot 13 \cdot x^2y^3z$ Rewrite as product of primes. $= 2 \cdot 2 \cdot 2 \cdot 5 \cdot xyz^2$

$\quad = \mathbf{2^2 \cdot 13 \cdot x^2y^3z}$ Exponential notation. $= \mathbf{2^3 \cdot 5 \cdot xyz^2}$

(b) find the least common denominator.

▶ $2^3 \cdot 5 \cdot 13 \cdot x^2 \cdot y^3 \cdot z^2$ Choose each unique factor and the greatest exponent.

$\quad = \mathbf{8 \cdot 5 \cdot 13 \cdot x^2y^3z^2}$ Simplify.

$\quad = \mathbf{520x^2y^3z^2}$ Simplify.

In the next example, the denominators are trinomials.

EXAMPLE 4 For the rational expressions $\dfrac{17}{x^3 + 6x^2 + 9x}$ and $\dfrac{11}{x^2 + 5x + 6}$,

(a) write the prime factorization of each denominator.

SOLUTION ▶

$$x^3 + 6x^2 + 9x$$

$\quad = x(x^2 + 6x + 9)$ Factor out the greatest common factor.

$\quad = \mathbf{x(x + 3)(x + 3)}$ Factor the trinomial.

$\quad = \mathbf{x(x + 3)^2}$ Exponential notation.

$$x^2 + 5x + 6$$

$\quad = \mathbf{(x + 2)(x + 3)}$ Factor the trinomial.

(b) find the least common denominator.

▶ $\mathbf{x(x + 2)(x + 3)^2}$ Choose each unique factor and the greatest exponent.

Practice Problems

For problems 1–4,
(a) write the prime factorization of each denominator.
(b) find the least common denominator.

1. $\dfrac{3}{8}; \dfrac{1}{10}$ **2.** $\dfrac{5}{108xy^2}; \dfrac{7}{72x^2y}$ **3.** $\dfrac{3}{x^2 + 9x + 20}; \dfrac{2}{x^3 - 16x}$

4. $\dfrac{1}{2x^2 - 7x - 15}; \dfrac{3}{3x^2 - 14x - 5}$

Equivalent Expressions

When one large pizza is cut into 8 pieces and another large pizza is cut into 10 pieces, the pieces are not the same size. To add $\dfrac{3}{8}$ pizza $+ \dfrac{1}{10}$ pizza, we need to rewrite the fractions with a common denominator by multiplying each fraction by a fraction that is equal to 1. These new equivalent fractions describe pieces that are the same size.

EXAMPLE 5 | Rewrite $\dfrac{3}{8}$ and $\dfrac{1}{10}$ as equivalent fractions with a denominator of 40.

SOLUTION ▶

$\dfrac{3}{8}$ The least common denominator is 40. $\dfrac{1}{10}$

$= \dfrac{3}{8} \cdot \dfrac{5}{5}$ Multiply by a fraction equal to 1. $= \dfrac{1}{10} \cdot \dfrac{4}{4}$

$= \dfrac{15}{40}$ Multiply numerators; multiply denominators. $= \dfrac{4}{40}$

We use the same process to build an equivalent rational expression with a new denominator. To choose the fraction that is equal to 1, compare the denominator in the original expression with the new denominator. In the next example, the original denominator is $5x$. The new denominator is $10xy$. If we multiply $5x$ by $2y$, the result is $10xy$. So we multiply the expression by $\dfrac{2y}{2y}$, a fraction that is equal to 1.

EXAMPLE 6 | Rewrite $\dfrac{3}{5x}$ as an equivalent rational expression with the denominator $10xy$.

SOLUTION ▶

$\dfrac{3}{5x}$

$= \dfrac{3}{5x} \cdot \dfrac{2y}{2y}$ Multiply by a fraction equal to 1.

$= \dfrac{6y}{10xy}$ Multiply numerators; multiply denominators.

In the next example, one of the denominators of the original expressions is $52x^2y^3z$. To rewrite it as an equivalent expression with a denominator of $520x^2y^3z^2$, multiply the expression by $\dfrac{10z}{10z}$, a fraction that is equal to 1.

EXAMPLE 7 | Rewrite $\dfrac{3}{52x^2y^3z}$ and $\dfrac{5}{40xyz^2}$ with the denominator $520x^2y^3z^2$.

SOLUTION ▶

$\dfrac{3}{52x^2y^3z}$ $\dfrac{5}{40xyz^2}$

$= \dfrac{3}{52x^2y^3z} \cdot \dfrac{10z}{10z}$ Multiply by a fraction equal to 1. $= \dfrac{5}{40xyz^2} \cdot \dfrac{13xy^2}{13xy^2}$

$= \dfrac{30z}{520x^2y^3z^2}$ Multiply numerators; multiply denominators. $= \dfrac{65xy^2}{520x^2y^3z^2}$

Before multiplying by a fraction that is equal to 1, factor the denominators completely. After multiplying by 1, use the distributive property to multiply factors in the numerator. We usually leave the denominator factored.

EXAMPLE 8 Rewrite $\dfrac{3}{x^2 + 5x + 6}$ and $\dfrac{5}{x^3 + 6x^2 + 9x}$ with the denominator $x(x + 2)(x + 3)^2$.

SOLUTION ▶

$$\dfrac{3}{x^2 + 5x + 6}$$

$$= \dfrac{3}{(x + 3)(x + 2)}$$ Factor.

$$= \dfrac{3}{(x + 2)(x + 3)} \cdot \dfrac{x}{x} \cdot \dfrac{(x + 3)}{(x + 3)}$$ Multiply by 1.

$$= \dfrac{3x^2 + 9x}{x(x + 2)(x + 3)^2}$$ Distributive property.

$$\dfrac{5}{x^3 + 6x^2 + 9x}$$

$$= \dfrac{5}{x(x + 3)(x + 3)}$$

$$= \dfrac{5}{x(x + 3)(x + 3)} \cdot \dfrac{(x + 2)}{(x + 2)}$$

$$= \dfrac{5x + 10}{x(x + 2)(x + 3)^2}$$

Building an Equivalent Rational Expression with a New Denominator

1. Factor the denominator of the expression completely.
2. Identify the factor(s) needed to build the new denominator.
3. Multiply the expression by a fraction equal to 1 so that the denominator in the product is the new denominator.
4. Multiply factors in the numerator; leave the denominator factored.

Practice Problems

For problems 5–8,
(a) find the least common denominator of each pair of expressions.
(b) rewrite these expressions with this common denominator.

5. $\dfrac{3}{4}$ pizza and $\dfrac{1}{6}$ pizza

6. $\dfrac{5}{32xy^2}$ and $\dfrac{7}{72xy}$

7. $\dfrac{3}{x^2 - 6x + 8}$ and $\dfrac{2}{x^2 - 4x}$

8. $\dfrac{1}{6x^2 + 7x - 3}$ and $\dfrac{3}{3x^2 - x}$

Adding and Subtracting Rational Expressions

To add $\dfrac{15}{40}$ pizza and $\dfrac{3}{40}$ pizza, add the numerators. The denominators, which represent the size of the pieces of pizza, stay the same. The sum is $\dfrac{18}{40}$ pizza. Simplifying into lowest terms, the sum is $\dfrac{9}{20}$ pizza. We use the same process to add and subtract rational expressions.

EXAMPLE 9 Simplify: $\dfrac{7x}{(x + 2)(x + 3)} + \dfrac{14}{(x + 2)(x + 3)}$

SOLUTION ▶

$$\dfrac{7x}{(x + 2)(x + 3)} + \dfrac{14}{(x + 2)(x + 3)}$$ The denominators are the same.

$$= \dfrac{7x + 14}{(x + 2)(x + 3)}$$ Add the numerators.

$$= \dfrac{7\cancel{(x + 2)}}{\cancel{(x + 2)}(x + 3)}$$ Factor the numerator; find common factors.

$$= \dfrac{7}{x + 3}$$ Simplify.

> **Adding or Subtracting Rational Expressions with Like Denominators**
> 1. Combine numerators. The denominator stays the same.
> 2. If the numerator is not prime, factor the numerator. Simplify common factors.
> 3. Multiply any factors in the numerator; either multiply the factors in the denominator or leave it factored.

In the next example, the denominators in the expressions are not the same. Before adding, we need to find the least common denominator and rewrite both expressions as equivalent expressions with this new denominator.

EXAMPLE 10 Simplify: $\dfrac{7x}{x + 2} + \dfrac{x}{6}$

SOLUTION ▶

$\dfrac{7x}{x + 2} + \dfrac{x}{6}$ Least common denominator: $6(x + 2)$.

$= \dfrac{7x}{x + 2} \cdot \dfrac{6}{6} + \dfrac{x}{6} \cdot \dfrac{(x + 2)}{(x + 2)}$ Multiply by fractions equal to 1.

$= \dfrac{42x}{6(x + 2)} + \dfrac{x^2 + 2x}{6(x + 2)}$ The denominators are the same.

$= \dfrac{42x + x^2 + 2x}{6(x + 2)}$ Add the numerators.

$= \dfrac{x^2 + 44x}{6(x + 2)}$ Combine like terms.

$= \dfrac{x(x + 44)}{6(x + 2)}$ Factor the numerator; there are no common factors.

$= \dfrac{x^2 + 44x}{6(x + 2)}$ Multiply the remaining factors in the numerator.

> **Adding or Subtracting Rational Expressions with Unlike Denominators**
> 1. Factor the denominators completely.
> 2. Find the least common denominator.
> 3. Multiply by 1 to build equivalent expressions with the new denominator.
> 4. Combine numerators. The denominator stays the same.
> 5. Factor the numerator. Simplify common factors.
> 6. Multiply any factors in the numerator; either multiply the factors in the denominator or leave it factored.

EXAMPLE 11 Simplify: $\dfrac{a + 5}{2a^2 - 11a - 6} - \dfrac{3a - 7}{a^2 - 4a - 12}$

SOLUTION ▶

$\dfrac{a + 5}{2a^2 - 11a - 6} - \dfrac{3a - 7}{a^2 - 4a - 12}$ Find the least common denominator.

$= \dfrac{a + 5}{(2a + 1)(a - 6)} - \dfrac{3a - 7}{(a - 6)(a + 2)}$ Factor.

Least common denominator: $(2a + 1)(a - 6)(a + 2)$

$$= \frac{a + 5}{(2a + 1)(a - 6)} \cdot \frac{(a + 2)}{(a + 2)} - \frac{3a - 7}{(a - 6)(a + 2)} \cdot \frac{(2a + 1)}{(2a + 1)} \quad \text{Multiply by 1.}$$

$$= \frac{a^2 + 2a + 5a + 10}{(2a + 1)(a - 6)(a + 2)} - \frac{6a^2 + 3a - 14a - 7}{(2a + 1)(a - 6)(a + 2)} \quad \text{Distributive property.}$$

$$= \frac{a^2 + 7a + 10}{(2a + 1)(a - 6)(a + 2)} - \frac{6a^2 - 11a - 7}{(2a + 1)(a - 6)(a + 2)} \quad \text{Combine like terms.}$$

$$= \frac{a^2 + 7a + 10 - (6a^2 - 11a - 7)}{(2a + 1)(a - 6)(a + 2)} \quad \text{Combine numerators.}$$

$$= \frac{a^2 + 7a + 10 - 1(6a^2 - 11a - 7)}{(2a + 1)(a - 6)(a + 2)} \quad \text{Multiply by 1.}$$

$$= \frac{a^2 + 7a + 10 - 6a^2 + 11a + 7}{(2a + 1)(a - 6)(a + 2)} \quad \text{Distributive property.}$$

$$= \frac{-5a^2 + 18a + 17}{(2a + 1)(a - 6)(a + 2)} \quad \text{Combine like terms.}$$

Can the numerator, a quadratic trinomial, be factored? If the square root of the discriminant, $b^2 - 4ac$, is a whole number, the trinomial can be factored.

$b^2 - 4ac$

$= 18^2 - 4(-5)(17) \qquad a = -5, b = 18, c = 17$

$= 664 \qquad\qquad \sqrt{664} = 25.76\ldots,$ an irrational number.

Since the numerator is a prime polynomial and the numerator and denominator share no common factors, the expression is simplified.

When adding or subtracting more than two expressions, find a common denominator of all the expressions.

EXAMPLE 12 Simplify: $\dfrac{8}{x} + \dfrac{3}{x^2} - \dfrac{1}{6x}$

SOLUTION ▶

$$\frac{8}{x} + \frac{3}{x^2} - \frac{1}{6x} \qquad \text{Least common denominator: } 6x^2.$$

$$= \frac{8}{x} \cdot \frac{6x}{6x} + \frac{3}{x^2} \cdot \frac{6}{6} - \frac{1}{6x} \cdot \frac{x}{x} \qquad \text{Multiply by fractions equal to 1.}$$

$$= \frac{48x}{6x^2} + \frac{18}{6x^2} - \frac{x}{6x^2} \qquad \text{The denominators are the same.}$$

$$= \frac{48x + 18 - x}{6x^2} \qquad \text{Combine numerators.}$$

$$= \frac{47x + 18}{6x^2} \qquad \text{The numerator is prime.}$$

Since the numerator and denominator have no common factors, the expression is simplified.

To find the least common denominator, factor all of the expressions in the denominators completely.

EXAMPLE 13 | Simplify: $\dfrac{1}{p+1} - \dfrac{5p-2}{p^2-5p-6} + \dfrac{p}{p-6}$

SOLUTION ▶ $\dfrac{1}{p+1} - \dfrac{5p-2}{p^2-5p-6} + \dfrac{p}{p-6}$ Find the least common denominator.

$= \dfrac{1}{p+1} - \dfrac{5p-2}{(p-6)(p+1)} + \dfrac{p}{p-6}$ Factor denominators.

Least common denominator: $(p+1)(p-6)$

$= \dfrac{1}{p+1} \cdot \dfrac{(p-6)}{(p-6)} - \dfrac{5p-2}{(p-6)(p+1)} + \dfrac{p}{p-6} \cdot \dfrac{(p+1)}{(p+1)}$ Multiply by fractions equal to 1.

$= \dfrac{p-6}{(p+1)(p-6)} - \dfrac{5p-2}{(p+1)(p-6)} + \dfrac{p^2+p}{(p+1)(p-6)}$ Distributive property.

$= \dfrac{p-6-(5p-2)+p^2+p}{(p+1)(p-6)}$ Combine numerators.

$= \dfrac{p-6-1(5p-2)+p^2+p}{(p+1)(p-6)}$ Multiply by 1.

$= \dfrac{p-6-5p+2+p^2+p}{(p+1)(p-6)}$ Distributive property.

$= \dfrac{p^2-3p-4}{(p+1)(p-6)}$ Combine like terms.

$= \dfrac{(p-4)\cancel{(p+1)}}{\cancel{(p+1)}(p-6)}$ Factor the numerator.

$= \dfrac{p-4}{p-6}$ Simplify.

Practice Problems

For problems 9–12, simplify.

9. $\dfrac{2x^2+13x}{x^2-4x-45} + \dfrac{15}{x^2-4x-45}$

10. $\dfrac{7}{12hk^2} - \dfrac{1}{30h^2k}$

11. $\dfrac{2c}{c^2-36} - \dfrac{1}{c^2-6c}$

12. $\dfrac{y+4}{2y-10} - \dfrac{5}{y^2-25}$

5.2 VOCABULARY PRACTICE

Match the term with its description.

1. A quotient of polynomials
2. Division by zero
3. A whole number that is greater than 1. Its only factors are itself and 1.
4. A polynomial whose only factors are itself and 1
5. The result of multiplication
6. The result of division
7. When we multiply these together, the result is a product.
8. A rational expression in which the numerator and denominator have no factors in common
9. $ax^2 + bx + c$
10. $b^2 - 4ac$

A. discriminant
B. factors
C. prime number
D. prime polynomial
E. product
F. quotient
G. rational expression
H. rational expression in lowest terms
I. quadratic trinomial
J. undefined

5.2 Exercises

Follow your instructor's guidelines for showing your work. Assume that all expressions in these exercises are defined. No variable represents a number that will result in division by zero.

For exercises 1–10, write the prime factorization of each expression.

1. $48x^2y$

2. $56a^2b$

3. $231x^2 - 231x$

4. $130k - 130k^2$

5. $8c^3 - 18c$

6. $48x^3 - 27x$

7. $6f^2 - 78f + 240$

8. $5p^2 + 20p - 105$

9. $12y^2 + 62y + 80$

10. $30y^2 - 28y - 16$

11. In Chapter 2, we *factored polynomials completely.* In this section, we learned to write the *prime factorization of a polynomial.*
   a. Is this the same process?
   b. Explain.

12. Explain how to find the prime factorization of a whole number. Include an example in your explanation.

For exercises 13–26, find the least common denominator.

13. $\dfrac{2}{x^2 - 7x - 18}; \dfrac{7}{x^2 - 10x + 9}$

14. $\dfrac{5}{x^2 - 3x - 40}; \dfrac{6}{x^2 - 15x + 56}$

15. $\dfrac{3}{d^2 - 4d}; \dfrac{d}{3d - 12}$

16. $\dfrac{8}{5z + 10}; \dfrac{1}{z^2 + 2z}$

17. $\dfrac{1}{100a^2b}; \dfrac{1}{35ab^3c}$

18. $\dfrac{1}{72xy^2z}; \dfrac{1}{42xyz^3}$

19. $\dfrac{2}{x^3 - 9x}; \dfrac{1}{-x^3 + 9x}$

20. $\dfrac{8}{w^3 - 49w}; \dfrac{3}{-w^3 + 49w}$

21. $\dfrac{1}{k^2 + 7k - 30}; \dfrac{1}{9 - k^2}$

22. $\dfrac{1}{y^2 - 5y - 14}; \dfrac{1}{49 - y^2}$

23. $\dfrac{1}{y}; \dfrac{3}{y^2}; \dfrac{1}{8y}$

24. $\dfrac{1}{z^3}; \dfrac{3}{z}; \dfrac{1}{9z}$

25. $\dfrac{1}{x^2 - 4}; \dfrac{1}{x^2 - 6x + 8}; \dfrac{5}{x^2 - 2x - 8}$

26. $\dfrac{1}{a^2 - 9}; \dfrac{1}{a^2 - 10a + 21}; \dfrac{3}{a^2 - 4a - 21}$

27. Explain why it is necessary to use a common denominator when adding two rational expressions.

28. Write an example of a prime polynomial.

For exercises 29–80, simplify.

29. $\dfrac{8x}{(x + 2)(x - 6)} + \dfrac{16}{(x + 2)(x - 6)}$

30. $\dfrac{5x}{(x + 3)(x - 1)} + \dfrac{15}{(x + 3)(x - 1)}$

31. $\dfrac{4}{15ab^2} - \dfrac{7}{18a^2b}$

32. $\dfrac{2}{21xy^3} - \dfrac{3}{28x^2y}$

33. $\dfrac{9}{16y} - \dfrac{5}{24x} + 1$

34. $\dfrac{3}{16d} - \dfrac{5d}{12c} + 1$

35. $\dfrac{x^2}{(x + 5)(x + 9)} - \dfrac{4x + 45}{(x + 5)(x + 9)}$

36. $\dfrac{y^2}{(y + 6)(y + 8)} - \dfrac{2y + 48}{(y + 6)(y + 8)}$

37. $\dfrac{9d^2 - 20d - 40}{(7d + 8)(d - 3)} - \dfrac{2d^2 + 35d + 32}{(7d + 8)(d - 3)}$

38. $\dfrac{11m^2 - 30m - 9}{(8m + 3)(m - 14)} - \dfrac{3m^2 + 23m + 12}{(8m + 3)(m - 14)}$

39. $\dfrac{3x^2 + 10x}{(3x + 8)(x - 5)} + \dfrac{7x + 20}{(3x + 8)(x - 5)} - \dfrac{3x + 4}{(3x + 8)(x - 5)}$

40. $\dfrac{5x^2 + 20x}{(5x + 2)(x - 9)} + \dfrac{19x + 18}{(5x + 2)(x - 9)} - \dfrac{7x + 6}{(5x + 2)(x - 9)}$

41. $\dfrac{8n}{4n^2 - 81} + \dfrac{36}{4n^2 - 81}$

42. $\dfrac{6c}{9c^2 - 4} + \dfrac{4}{9c^2 - 4}$

43. $\dfrac{9k^2 + 6k}{6k^2 + 11k - 7} + \dfrac{44k + 53}{6k^2 + 11k - 7} - \dfrac{8k + 4}{6k^2 + 11k - 7}$

44. $\dfrac{9x^2 + 10x}{12x^2 + 17x - 5} + \dfrac{25x + 30}{12x^2 + 17x - 5} - \dfrac{5x + 5}{12x^2 + 17x - 5}$

45. $\dfrac{2}{x + 3} + \dfrac{1}{x - 4}$

46. $\dfrac{1}{x + 4} + \dfrac{2}{x - 2}$

47. $\dfrac{1}{x + 3} - \dfrac{1}{x - 4}$

48. $\dfrac{1}{x - 2} - \dfrac{1}{x + 4}$

49. $\dfrac{1}{x - 5} - \dfrac{1}{x + 3}$

50. $\dfrac{1}{x-2} - \dfrac{1}{x+3}$

51. $\dfrac{3x}{x+3} + \dfrac{5x}{x-5}$

52. $\dfrac{2x}{x-2} + \dfrac{3x}{x+3}$

53. $\dfrac{x}{x-6} - \dfrac{36}{x^2-6x}$

54. $\dfrac{z}{z-8} - \dfrac{64}{z^2-8z}$

55. $\dfrac{p-4}{4p^2-p} - \dfrac{4}{p}$

56. $\dfrac{w-2}{2w^2-w} - \dfrac{2}{w}$

57. $\dfrac{2x-3}{5x^2-3x} - \dfrac{1}{x}$

58. $\dfrac{5x-7}{8x^2-7x} - \dfrac{1}{x}$

59. $\dfrac{6y-1}{3y^2+y} + \dfrac{1}{y}$

60. $\dfrac{8z-5}{2z^2+5z} + \dfrac{1}{z}$

61. $\dfrac{2x+5}{2x+2} - \dfrac{9}{4x^2+14x+10}$

62. $\dfrac{2x+5}{2x+3} - \dfrac{4}{4x^2+16x+15}$

63. $\dfrac{4x+1}{x^2+x-30} + \dfrac{3x+2}{x^2-3x-10}$

64. $\dfrac{2x+5}{x^2+4x-21} + \dfrac{x+8}{x^2+5x-14}$

65. $\dfrac{c}{c^2+4c+3} - \dfrac{3}{c^2-4c-5}$

66. $\dfrac{2h-1}{h^2+h-6} - \dfrac{h+2}{h^2+5h+6}$

67. $\dfrac{1}{x^2-4} - \dfrac{3}{x^2-3x-10}$

68. $\dfrac{1}{x^2-9} - \dfrac{4}{x^2-2x-15}$

69. $\dfrac{3x}{x^2+7x+10} - \dfrac{2x}{x^2+6x+8}$

70. $\dfrac{4x}{x^2+8x+12} - \dfrac{3x}{x^2+7x+10}$

71. $\dfrac{x}{x+2} + \dfrac{x}{x+4} - \dfrac{x^2+7x+6}{x^2+6x+8}$

72. $\dfrac{x}{x+3} + \dfrac{x}{x+2} - \dfrac{x^2+3x+3}{x^2+5x+6}$

73. $\dfrac{3x^2-11x-1}{x^2-5x+4} - \dfrac{3x}{x-1} - \dfrac{3x}{x-4}$

74. $\dfrac{2x^2-3x-3}{x^2-5x+6} - \dfrac{2x}{x-3} - \dfrac{2x}{x-2}$

75. $\dfrac{2x^2-2x+2}{x^2-9} - \dfrac{x}{x-3} + \dfrac{2}{x+3}$

76. $\dfrac{2x^2+x+3}{x^2-4} - \dfrac{x}{x-2} + \dfrac{3}{x+2}$

77. $\dfrac{x}{x+3} + \dfrac{5}{x} - \dfrac{2x^2+3x+7}{x^2+3x}$

78. $\dfrac{x}{x+5} + \dfrac{4}{x} - \dfrac{2x^2+3x+18}{x^2+5x}$

79. $\dfrac{x}{2x^2-3x-2} + \dfrac{1}{2x^2-7x+6} - \dfrac{2x}{4x^2-4x-3}$

80. $\dfrac{x}{2x^2+3x-2} + \dfrac{1}{2x^2+9x+10} - \dfrac{2x}{4x^2+8x-5}$

Problem Solving: Practice and Review

Follow your instructor's guidelines for following the five steps as outlined in Section 1.5, p. 51.

81. See the table below. Find the difference between the percent of 16- to 19-year-olds and the percent of 20- to 24-year-olds who moved in 2006 because they were going to attend or leave college. Round to the nearest tenth of a percent.

Table for exercise 81

Age in years	16–19	20–24	25–29
Total population who moved in 2006 (in thousands)	2399	5904	5546
Reason for moving (in thousands)			
Wanted new or better home/apartment	397	867	839
Wanted better neighborhood/less crime	101	171	205
Wanted cheaper housing	141	389	327
Other housing reason	216	463	389
To attend or leave college	139	565	201
Change of climate	3	20	16
Health reasons	17	40	28
Natural disaster	33	62	58

Source: www.census.gov

For exercises 82–84, use the graph and polynomial model.

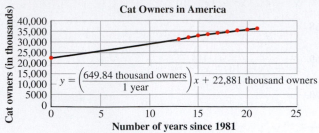

Cat Owners in America

$$y = \left(\frac{649.84 \text{ thousand owners}}{1 \text{ year}}\right) x + 22{,}881 \text{ thousand owners}$$

Source: http://petfoodinstitute.org

82. Use the model to find the number of Americans who will own cats in 2015. Report the answer in millions of Americans. Round to the nearest tenth.

83. What is the slope of this model? Describe what the slope represents.

84. What is the y-coordinate of the y-intercept of this model? Describe what the y-intercept represents.

Technology

For exercises 85–88,
(a) simplify the expression.
(b) check your work with a graphing calculator using the method shown in Section 5.1. Sketch the graph; describe the window.

85. $\dfrac{1}{x^2 - 1} + \dfrac{1}{x^2 + 4x + 3}$

86. $\dfrac{2x + 1}{x - 3} - \dfrac{11x + 2}{x^2 - x - 6}$

87. $\dfrac{3x - 7}{2x^2 + 7x} + \dfrac{1}{x}$

88. $\dfrac{3}{x - 3} - \dfrac{9}{x^2 - 3x}$

Find the Mistake

For exercises 89–92, the completed problem has one mistake.
(a) Describe the mistake in words, or copy down the whole problem and highlight or circle the mistake.
(b) Do the problem correctly.

89. **Problem:** Simplify: $\dfrac{x - 1}{x + 1} - \dfrac{4}{x^2 - 1}$

Incorrect Answer: $\dfrac{x - 1}{x + 1} - \dfrac{4}{x^2 - 1}$

$$= \frac{(x - 1)}{(x + 1)} \cdot \frac{(x - 1)}{(x - 1)} - \frac{4}{(x + 1)(x - 1)}$$

$$= \frac{x^2 - 1x - 1x - 1 - 4}{(x + 1)(x - 1)}$$

$$= \frac{x^2 - 2x - 5}{(x + 1)(x - 1)}$$

90. **Problem:** Simplify: $\dfrac{a}{a^2 - 5a - 24} - \dfrac{8}{a^2 - 5a - 24}$

Incorrect Answer: $\dfrac{a}{a^2 - 5a - 24} - \dfrac{8}{a^2 - 5a - 24}$

$$= \frac{a - 8}{a^2 - 5a - 24}$$

$$= \frac{a - 8}{(a - 8)(a + 3)}$$

$$= a + 3$$

91. **Problem:** Simplify: $\dfrac{3}{5xy^2} - \dfrac{1}{30x^2y}$

Incorrect Answer: $\dfrac{3}{5xy^2} - \dfrac{1}{30x^2y}$

$$= \left(\frac{3}{5xy^2}\right)\left(\frac{6x}{6x}\right) - \frac{1}{30x^2y}\left(\frac{y}{y}\right)$$

$$= \frac{3x}{30x^2y^2} - \frac{1}{30x^2y^2}$$

$$= \frac{3x - 1}{30x^2y^2}$$

92. **Problem:** Simplify: $\dfrac{4x}{x^2 - 16} - \dfrac{16}{x^2 - 16}$

Incorrect Answer: $\dfrac{4x}{x^2 - 16} - \dfrac{16}{x^2 - 16}$

$$= \frac{4x - 16}{x^2 - 16}$$

$$= \frac{4(x - 4)}{(x - 4)(x + 4)}$$

$$= \frac{\cancel{4}}{x + \cancel{4}}$$

$$= \frac{1}{x}$$

Review

For exercises 93–96, find the slope of the line that passes through the given points.

93. $\left(\dfrac{1}{8}, \dfrac{4}{9}\right); \left(\dfrac{7}{8}, \dfrac{5}{9}\right)$

94. $\left(\dfrac{1}{8}, \dfrac{1}{11}\right); \left(\dfrac{3}{8}, \dfrac{4}{11}\right)$

95. $\left(\dfrac{1}{4}, \dfrac{1}{3}\right); \left(\dfrac{1}{3}, \dfrac{1}{4}\right)$

96. $\left(\dfrac{1}{4}, \dfrac{1}{10}\right); \left(\dfrac{1}{2}, \dfrac{1}{5}\right)$

SUCCESS IN COLLEGE MATHEMATICS

97. When multiplying or dividing rational numbers, we do not need a common denominator. When adding or subtracting rational numbers with different denominators, we first find a common denominator and then rewrite each number as an equivalent fraction with this denominator. Explain why the denominators of fractions being added or subtracted have to be the same.

98. The fraction $\dfrac{3}{4}$ can be multiplied by $\dfrac{5}{5}$ and rewritten as an equivalent fraction with a different denominator: $\dfrac{3}{4} \cdot \dfrac{5}{5} = \dfrac{15}{20}$. Explain why we can multiply $\dfrac{3}{4}$ by $\dfrac{5}{5}$ without changing the value of the fraction.

$$m = \dfrac{\dfrac{1}{8} - \dfrac{1}{2}}{\dfrac{3}{8} - \dfrac{1}{4}}$$

A rational number or expression in which the numerator or denominator is itself a fraction is a **complex rational number** or **complex rational expression**. In this section, we will simplify these numbers and expressions.

SECTION 5.3

Complex Rational Expressions

After reading the text, working the practice exercises, and completing assigned exercises, you should be able to:

Simplify a complex rational number or complex rational expression.

Complex Rational Expressions

In Chapter 1, we used the slope formula, $m = \dfrac{y_2 - y_1}{x_2 - x_1}$, to find the slope of a line. In the next example, the coordinates of the points on the line, (x_1, y_1) and (x_2, y_2), are fractions.

EXAMPLE 1 Find the slope of the line that passes through $\left(\dfrac{1}{15}, \dfrac{2}{21}\right)$ and $\left(\dfrac{14}{15}, \dfrac{19}{21}\right)$.

SOLUTION ▶

$m = \dfrac{y_2 - y_1}{x_2 - x_1}$ The slope formula.

$m = \dfrac{\dfrac{19}{21} - \dfrac{2}{21}}{\dfrac{14}{15} - \dfrac{1}{15}}$ $(x_1, y_1) = \left(\dfrac{1}{15}, \dfrac{2}{21}\right); (x_2, y_2) = \left(\dfrac{14}{15}, \dfrac{19}{21}\right)$

$m = \dfrac{\dfrac{17}{21}}{\dfrac{13}{15}}$ Evaluate the numerator and denominator.

$m = \dfrac{17}{21} \div \dfrac{13}{15}$ Rewrite the fraction as a quotient.

$m = \dfrac{17}{21} \cdot \dfrac{15}{13}$ Rewrite as multiplication by the reciprocal of the divisor.

$m = \dfrac{17}{3 \cdot 7} \cdot \dfrac{3 \cdot 5}{13}$ Find common factors.

$m = \dfrac{85}{91}$ Simplify.

In Example 1, before simplifying, the numerator and denominator of the slope are fractions. Similarly, in a **complex rational expression**, there is at least one fraction in the numerator or denominator. To simplify, we rewrite the complex rational expression as a quotient of rational expressions.

Simplifying a Complex Rational Expression That Is a Quotient

1. Rewrite the complex rational expression as a quotient.
2. Rewrite division as multiplication by the reciprocal.
3. Factor.
4. Simplify common factors.
5. Multiply any remaining factors in the numerator; either multiply the remaining factors in the denominator or leave it in factored form.

EXAMPLE 2 Simplify: $\dfrac{\dfrac{a^2 + 3a}{5a}}{\dfrac{2a + 6}{15a}}$

SOLUTION ▶ $\dfrac{\dfrac{a^2 + 3a}{5a}}{\dfrac{2a + 6}{15a}}$

$= \dfrac{a^2 + 3a}{5a} \div \dfrac{2a + 6}{15a}$ Rewrite the complex fraction as a quotient.

$= \dfrac{a^2 + 3a}{5a} \cdot \dfrac{15a}{2a + 6}$ Rewrite division as multiplication by the reciprocal.

$= \dfrac{a(a + 3)}{5 \cdot a} \cdot \dfrac{3 \cdot 5 \cdot a}{2(a + 3)}$ Factor completely; find common factors.

$= \dfrac{3a}{2}$ Simplify.

If the numerator of a complex fraction is not itself a fraction, rewrite it as a fraction with a denominator of 1.

EXAMPLE 3 Simplify: $\dfrac{x + 5}{\dfrac{x^2 - 25}{2x}}$

SOLUTION ▶ $\dfrac{x + 5}{\dfrac{x^2 - 25}{2x}}$

$= \dfrac{\dfrac{x + 5}{1}}{\dfrac{x^2 - 25}{2x}}$ Rewrite the numerator as a fraction.

$= \dfrac{x + 5}{1} \div \dfrac{x^2 - 25}{2x}$ Rewrite the complex fraction as a quotient.

$= \dfrac{x + 5}{1} \cdot \dfrac{2x}{x^2 - 25}$ Rewrite division as multiplication by the reciprocal.

$= \dfrac{x + 5}{1} \cdot \dfrac{2x}{(x - 5)(x + 5)}$ Factor completely; find common factors.

$= \dfrac{2x}{x - 5}$ Simplify.

In the next example, each of the numerators and denominators must be factored.

EXAMPLE 4 Simplify: $\dfrac{\dfrac{h^2 + 5h - 36}{3h^2 + 2h - 8}}{\dfrac{h^2 + 7h - 18}{3h^2 - 4h}}$

SOLUTION ▶

$\dfrac{\dfrac{h^2 + 5h - 36}{3h^2 + 2h - 8}}{\dfrac{h^2 + 7h - 18}{3h^2 - 4h}}$

$= \dfrac{h^2 + 5h - 36}{3h^2 + 2h - 8} \div \dfrac{h^2 + 7h - 18}{3h^2 - 4h}$ Rewrite the complex fraction as a quotient.

$= \dfrac{h^2 + 5h - 36}{3h^2 + 2h - 8} \cdot \dfrac{3h^2 - 4h}{h^2 + 7h - 18}$ Rewrite division as multiplication by the reciprocal.

$= \dfrac{(h + 9)(h - 4)}{(3h - 4)(h + 2)} \cdot \dfrac{h(3h - 4)}{(h + 9)(h - 2)}$ Factor completely; find common factors.

$= \dfrac{h(h - 4)}{(h + 2)(h - 2)}$ Simplify.

$= \dfrac{h^2 - 4h}{(h + 2)(h - 2)}$ Multiply remaining factors in the numerator.

Practice Problems

For problems 1–6, simplify.

1. $\dfrac{\dfrac{4}{27}}{\dfrac{2}{15}}$

2. $\dfrac{\dfrac{x + 1}{x - 2}}{\dfrac{x - 1}{2x - 4}}$

3. $\dfrac{n + 8}{\dfrac{n^2 + 10n + 16}{n - 3}}$

4. $\dfrac{\dfrac{c^2 + c}{2c - 14}}{\dfrac{c^2 + 4c + 3}{c^2 - 4c - 21}}$

5. $\dfrac{\dfrac{y^2 - 4}{y^2 - y - 6}}{\dfrac{y^2 + 10y + 24}{y^2 + 3y - 18}}$

6. $\dfrac{\dfrac{3x - 12}{9x + 18}}{\dfrac{x^2 - 16}{x^2 + 8x + 16}}$

Complex Rational Expressions with Addition or Subtraction: Strategy 1

There are two different strategies for simplifying complex rational expressions that include addition or subtraction. In the first strategy, we begin by finding common denominators and complete the addition or subtraction.

EXAMPLE 5 | Evaluate: $\dfrac{\dfrac{2}{3}+\dfrac{1}{2}}{\dfrac{1}{8}+\dfrac{5}{6}}$

SOLUTION ▶
$\dfrac{\dfrac{2}{3}+\dfrac{1}{2}}{\dfrac{1}{8}+\dfrac{5}{6}}$
The least common denominator for the numerator is 6; the least common denominator for the denominator is 24.

$=\dfrac{\dfrac{2}{3}\left(\dfrac{2}{2}\right)+\dfrac{1}{2}\left(\dfrac{3}{3}\right)}{\dfrac{1}{8}\left(\dfrac{3}{3}\right)+\dfrac{5}{6}\left(\dfrac{4}{4}\right)}$
Multiply by fractions equal to 1.

$=\dfrac{\dfrac{4}{6}+\dfrac{3}{6}}{\dfrac{3}{24}+\dfrac{20}{24}}$
Multiply numerators; multiply denominators.

$=\dfrac{\dfrac{7}{6}}{\dfrac{23}{24}}$
Complete the addition.

$=\dfrac{7}{6}\div\dfrac{23}{24}$
Rewrite the complex fraction as a quotient.

$=\dfrac{7}{6}\cdot\dfrac{24}{23}$
Rewrite division as multiplication by the reciprocal.

$=\dfrac{7}{\cancel{6}}\cdot\dfrac{4\cdot\cancel{6}}{23}$
Find common factors.

$=\dfrac{28}{23}$
Multiply remaining factors.

Simplifying a Complex Rational Expression with Addition or Subtraction in the Numerator and/or Denominator (Strategy 1)

1. Find the least common denominator for the fractions in the numerator. Build equivalent fractions with this denominator. Add or subtract the fractions.

2. Find the least common denominator for the fractions in the denominator. Build equivalent fractions with this denominator. Add or subtract the fractions.

3. Rewrite the rational expression as a quotient.

4. Rewrite division as multiplication by the reciprocal of the divisor.

5. Factor; find common factors.

6. Multiply remaining factors in the numerator; either multiply the remaining factors in the denominator or leave it in factored form.

When the fractions include variables, we again begin by finding least common denominators and building equivalent fractions. Notice that we leave the denominators factored. This makes it easier to simplify in later steps.

EXAMPLE 6 Simplify: $\dfrac{\dfrac{1}{u} - \dfrac{1}{u+1}}{\dfrac{1}{u} + \dfrac{1}{u-1}}$

SOLUTION ▶

$\dfrac{\dfrac{1}{u} - \dfrac{1}{u+1}}{\dfrac{1}{u} + \dfrac{1}{u-1}}$

The least common denominator in the numerator is $u(u+1)$; the least common denominator in the denominator is $u(u-1)$.

$= \dfrac{\dfrac{1}{u}\left(\dfrac{u+1}{u+1}\right) - \dfrac{1}{u+1}\left(\dfrac{u}{u}\right)}{\dfrac{1}{u}\left(\dfrac{u-1}{u-1}\right) + \dfrac{1}{u-1}\left(\dfrac{u}{u}\right)}$

Multiply by fractions equal to 1.

$= \dfrac{\dfrac{u+1}{u(u+1)} - \dfrac{u}{u(u+1)}}{\dfrac{u-1}{u(u-1)} + \dfrac{u}{u(u-1)}}$

Multiply numerators; leave the denominators factored.

$= \dfrac{\dfrac{u+1-u}{u(u+1)}}{\dfrac{u-1+u}{u(u-1)}}$

Add numerators; the denominators do not change.

$= \dfrac{\dfrac{1}{u(u+1)}}{\dfrac{2u-1}{u(u-1)}}$

Simplify numerators.

$= \dfrac{1}{u(u+1)} \div \dfrac{2u-1}{u(u-1)}$

Rewrite the complex fraction as a quotient.

$= \dfrac{1}{u(u+1)} \cdot \dfrac{u(u-1)}{2u-1}$

Rewrite division as multiplication by the reciprocal.

$= \dfrac{1}{\cancel{u}(u+1)} \cdot \dfrac{\cancel{u}(u-1)}{2u-1}$

Find common factors.

$= \dfrac{u-1}{(u+1)(2u-1)}$

Multiply remaining factors in the numerator.

When using Strategy 1, rewrite whole numbers in the complex fraction as fractions with a denominator of 1.

EXAMPLE 7   Simplify: $\dfrac{2 + \dfrac{1}{x}}{\dfrac{1}{x+1} + 1}$

SOLUTION ▶

$\dfrac{2 + \dfrac{1}{x}}{\dfrac{1}{x+1} + 1}$ The common denominator for the numerator is x; the common denominator for the denominator is $x+1$.

$= \dfrac{\dfrac{2}{1} + \dfrac{1}{x}}{\dfrac{1}{x+1} + \dfrac{1}{1}}$ Rewrite whole numbers as fractions with denominators of 1.

$= \dfrac{\dfrac{2}{1}\left(\dfrac{x}{x}\right) + \dfrac{1}{x}}{\dfrac{1}{x+1} + \dfrac{1}{1}\left(\dfrac{x+1}{x+1}\right)}$ Multiply by fractions equal to 1.

$= \dfrac{\dfrac{2x}{x} + \dfrac{1}{x}}{\dfrac{1}{x+1} + \dfrac{x+1}{x+1}}$ Multiply numerators.

$= \dfrac{\dfrac{2x+1}{x}}{\dfrac{1+x+1}{x+1}}$ Add numerators; the denominators do not change.

$= \dfrac{\dfrac{2x+1}{x}}{\dfrac{x+2}{x+1}}$ Simplify.

$= \dfrac{2x+1}{x} \div \dfrac{x+2}{x+1}$ Rewrite the complex fraction as a quotient.

$= \dfrac{2x+1}{x} \cdot \dfrac{x+1}{x+2}$ Rewrite division as multiplication by the reciprocal.

$= \dfrac{(2x+1)(x+1)}{x(x+2)}$ Use parentheses for multiplication.

$= \dfrac{2x^2 + 3x + 1}{x(x+2)}$ Multiply remaining factors in the numerator.

Practice Problems

For problems 7–10, use Strategy 1 to evaluate or simplify.

7. $\dfrac{\dfrac{3}{4} + \dfrac{1}{2}}{\dfrac{1}{6} + \dfrac{2}{3}}$

8. $\dfrac{6 - \dfrac{4}{x+2}}{10 + \dfrac{7}{x+2}}$

9. $\dfrac{1 + \dfrac{3}{x}}{1 - \dfrac{9}{x^2}}$

10. $\dfrac{\dfrac{5}{3x-3} - \dfrac{1}{x+1}}{\dfrac{1}{x-1} + \dfrac{x}{x^2-1}}$

Complex Rational Expressions with Addition or Subtraction: Strategy 2

In Examples 6 and 7, we first combined the fractions in the numerator and/or the denominator of a complex fraction. In the next examples, we use a different strategy. We find the least common denominator of all of the individual fractions and multiply the numerator and denominator of the complex rational expression by this least common denominator. This clears the fractions in the complex rational expression. This is Strategy 2.

EXAMPLE 8 Evaluate: $\dfrac{\dfrac{2}{3} + \dfrac{1}{2}}{\dfrac{1}{8} + \dfrac{5}{6}}$

SOLUTION ▶ $\dfrac{\dfrac{2}{3} + \dfrac{1}{2}}{\dfrac{1}{8} + \dfrac{5}{6}}$ The least common denominator of 2, 3, 6, and 8 is 24.

$= \dfrac{\dfrac{2}{3} + \dfrac{1}{2} \cdot \dfrac{24}{24}}{\dfrac{1}{8} + \dfrac{5}{6}}$ Multiply by a fraction equal to 1.

$= \dfrac{\left(\dfrac{2}{3} + \dfrac{1}{2}\right) \cdot \dfrac{24}{1}}{\left(\dfrac{1}{8} + \dfrac{5}{6}\right) \cdot \dfrac{24}{1}}$ Use parentheses for multiplication; $\dfrac{24}{24} = \dfrac{\frac{24}{1}}{\frac{24}{1}}$

$= \dfrac{\dfrac{2}{3} \cdot \dfrac{24}{1} + \dfrac{1}{2} \cdot \dfrac{24}{1}}{\dfrac{1}{8} \cdot \dfrac{24}{1} + \dfrac{5}{6} \cdot \dfrac{24}{1}}$ Distributive property.

$= \dfrac{\dfrac{2}{3} \cdot \dfrac{3 \cdot 8}{1} + \dfrac{1}{2} \cdot \dfrac{2 \cdot 12}{1}}{\dfrac{1}{8} \cdot \dfrac{8 \cdot 3}{1} + \dfrac{5}{6} \cdot \dfrac{6 \cdot 4}{1}}$ Find common factors.

$= \dfrac{\dfrac{16}{1} + \dfrac{12}{1}}{\dfrac{3}{1} + \dfrac{20}{1}}$ Simplify.

$= \dfrac{16 + 12}{3 + 20}$ Simplify.

$= \dfrac{28}{23}$ Simplify; the simplified expression is the same as in Example 5, using Strategy 1.

> ## Simplifying a Complex Rational Expression with Addition or Subtraction in the Numerator and/or Denominator (Strategy 2)
>
> 1. Find the least common denominator of all of the individual fractions.
> 2. Multiply the complex rational expression by 1 (the numerator and denominator of the fraction that is equal to 1 are the least common denominator.)
> 3. Use the distributive property.
> 4. Factor; find common factors.
> 5. Simplify.

EXAMPLE 9 Simplify: $\dfrac{3 - \dfrac{9}{w - 4}}{5 + \dfrac{4}{w - 4}}$

SOLUTION ▶ $\dfrac{3 - \dfrac{9}{w - 4}}{5 + \dfrac{4}{w - 4}}$ The least common denominator is $w - 4$.

$= \left(\dfrac{3 - \dfrac{9}{w - 4}}{5 + \dfrac{4}{w - 4}}\right)\left(\dfrac{w - 4}{w - 4}\right)$ Multiply by a fraction equal to 1.

$= \left(\dfrac{\dfrac{3}{1} - \dfrac{9}{w - 4}}{\dfrac{5}{1} + \dfrac{4}{w - 4}}\right)\left(\dfrac{w - 4}{w - 4}\right)$ Rewrite whole numbers as fractions.

$= \left(\dfrac{\dfrac{3}{1} - \dfrac{9}{w - 4}}{\dfrac{5}{1} + \dfrac{4}{w - 4}}\right)\left(\dfrac{\dfrac{w - 4}{1}}{\dfrac{w - 4}{1}}\right)$ Write as a fraction with a denominator of 1.

$= \dfrac{\left(\dfrac{3}{1}\right)\left(\dfrac{w - 4}{1}\right) - \left(\dfrac{9}{w - 4}\right)\left(\dfrac{w - 4}{1}\right)}{\left(\dfrac{5}{1}\right)\left(\dfrac{w - 4}{1}\right) + \left(\dfrac{4}{w - 4}\right)\left(\dfrac{w - 4}{1}\right)}$ Distributive property; find common factors.

$= \dfrac{3(w - 4) - 9}{5(w - 4) + 4}$ Simplify.

$= \dfrac{3w - 12 - 9}{5w - 20 + 4}$ Distributive property.

$= \dfrac{3w - 21}{5w - 16}$ Combine like terms.

$= \dfrac{3(w - 7)}{5w - 16}$ Factor the numerator; no common factors.

$= \dfrac{3w - 21}{5w - 16}$ Multiply remaining factors in the numerator.

In the next example, the least common denominator has two factors.

EXAMPLE 10   Simplify: $\dfrac{2 + \dfrac{1}{x}}{\dfrac{1}{x+1} + 1}$

SOLUTION ▶

$\dfrac{2 + \dfrac{1}{x}}{\dfrac{1}{x+1} + 1}$ The least common denominator is $x(x+1)$.

$= \dfrac{2 + \dfrac{1}{x}}{\dfrac{1}{x+1} + 1} \cdot \dfrac{x(x+1)}{x(x+1)}$ Multiply by a fraction equal to 1.

$= \dfrac{\left(2 + \dfrac{1}{x}\right) \dfrac{x(x+1)}{1}}{\left(\dfrac{1}{x+1} + 1\right) \dfrac{x(x+1)}{1}}$ Write as a fraction with a denominator of 1.

$= \dfrac{\dfrac{2}{1} \cdot \dfrac{x(x+1)}{1} + \dfrac{1}{x} \cdot \dfrac{x(x+1)}{1}}{\dfrac{1}{x+1} \cdot \dfrac{x(x+1)}{1} + \dfrac{1}{1} \cdot \dfrac{x(x+1)}{1}}$ Distributive property.

$= \dfrac{\dfrac{2}{1} \cdot \dfrac{x(x+1)}{1} + \dfrac{1}{\cancel{x}} \cdot \dfrac{\cancel{x}(x+1)}{1}}{\dfrac{1}{\cancel{x+1}} \cdot \dfrac{\cancel{x(x+1)}}{1} + \dfrac{1}{1} \cdot \dfrac{x(x+1)}{1}}$ Find common factors.

$= \dfrac{2x(x+1) + 1(x+1)}{x + x(x+1)}$ Simplify.

$= \dfrac{2x^2 + 2x + x + 1}{x + x^2 + x}$ Distributive property.

$= \dfrac{2x^2 + 3x + 1}{x^2 + 2x}$ Combine like terms.

$= \dfrac{(2x+1)(x+1)}{x(x+2)}$ Factor; there are no common factors.

$= \dfrac{2x^2 + 3x + 1}{x(x+2)}$ Multiply factors in the numerator. The simplified expression is the same as in Example 7, using Strategy 1.

In the next example, the least common denominator has three factors.

EXAMPLE 11 | Simplify: $\dfrac{\dfrac{1}{a} + \dfrac{1}{a-1}}{\dfrac{1}{a} - \dfrac{1}{a+1}}$

SOLUTION ▶ $\dfrac{\dfrac{1}{a} + \dfrac{1}{a-1}}{\dfrac{1}{a} - \dfrac{1}{a+1}}$ 　　　　　The least common denominator is $a(a-1)(a+1)$.

$= \dfrac{\dfrac{1}{a} + \dfrac{1}{a-1}}{\dfrac{1}{a} - \dfrac{1}{a+1}} \cdot \dfrac{a(a-1)(a+1)}{a(a-1)(a+1)}$ 　　　Multiply by a fraction equal to 1.

$= \dfrac{\left(\dfrac{1}{a} + \dfrac{1}{a-1}\right) \cdot \dfrac{a(a-1)(a+1)}{1}}{\left(\dfrac{1}{a} - \dfrac{1}{a+1}\right) \cdot \dfrac{a(a-1)(a+1)}{1}}$ 　　　Write as a fraction with a denominator of 1.

$= \dfrac{\dfrac{1}{a} \cdot \dfrac{a(a-1)(a+1)}{1} + \dfrac{1}{a-1} \cdot \dfrac{a(a-1)(a+1)}{1}}{\dfrac{1}{a} \cdot \dfrac{a(a-1)(a+1)}{1} - \dfrac{1}{a+1} \cdot \dfrac{a(a-1)(a+1)}{1}}$ 　　　Distributive property.

$= \dfrac{\dfrac{\cancel{a}(a-1)(a+1)}{\cancel{a}} + \dfrac{a(\cancel{a-1})(a+1)}{\cancel{a-1}}}{\dfrac{\cancel{a}(a-1)(a+1)}{\cancel{a}} - \dfrac{a(a-1)(\cancel{a+1})}{\cancel{a+1}}}$ 　　　Find common factors.

$= \dfrac{(a-1)(a+1) + a(a+1)}{(a-1)(a+1) - a(a-1)}$ 　　　Simplify.

$= \dfrac{a(a+1) - 1(a+1) + a \cdot a + a(1)}{a(a+1) - 1(a+1) - a \cdot a - a(-1)}$ 　　　Distributive property.

$= \dfrac{a^2 + a - a - 1 + a^2 + a}{a^2 + a - a - 1 - a^2 + a}$ 　　　Simplify.

$= \dfrac{2a^2 + a - 1}{a - 1}$ 　　　Combine like terms.

$= \dfrac{(2a - 1)(a + 1)}{a - 1}$ 　　　Factor; there are no common factors.

$= \dfrac{2a^2 + a - 1}{a - 1}$ 　　　Multiply the factors in the numerator.

Practice Problems

For problems 11–14, use Strategy 2 to evaluate or simplify.

11. $\dfrac{\dfrac{3}{4} + \dfrac{1}{2}}{\dfrac{1}{6} + \dfrac{2}{3}}$ 　　
12. $\dfrac{6 - \dfrac{4}{x+2}}{10 + \dfrac{7}{x+2}}$ 　　
13. $\dfrac{1 + \dfrac{3}{x}}{1 - \dfrac{9}{x^2}}$ 　　
14. $\dfrac{\dfrac{5}{3x-3} - \dfrac{1}{x+1}}{\dfrac{1}{x-1} + \dfrac{x}{x^2-1}}$

5.3 VOCABULARY PRACTICE

Match the term with its description.

1. In $12 \div 3 = 4$, this is 3.

2. For the fractions $\dfrac{1}{4}, \dfrac{1}{6}$, and $\dfrac{1}{9}$, this is 36.

3. The value of a fraction in which the numerator and denominator are equal

4. A fraction in which the numerator and denominator have no common factors

5. For the fraction $\dfrac{x}{y}$, this is $\dfrac{y}{x}$.

6. Division by zero

7. $a(b + c) = ab + ac$

8. A rational expression in which the numerator and/or denominator are fractions

9. The result of division

10. The result of multiplication

A. complex rational expression
B. distributive property
C. divisor
D. least common denominator
E. lowest terms
F. one
G. product
H. quotient
I. reciprocal
J. undefined

5.3 Exercises

Follow your instructor's guidelines for showing your work.

For exercises 1–8, find the slope of the line that passes through the given points.

1. $(14, 8)(42, 15)$

2. $(15, 9)(42, 18)$

3. $\left(\dfrac{1}{4}, \dfrac{1}{3}\right)\left(\dfrac{3}{4}, \dfrac{7}{9}\right)$

4. $\left(\dfrac{2}{9}, \dfrac{1}{5}\right)\left(\dfrac{2}{3}, \dfrac{4}{5}\right)$

5. $\left(\dfrac{5}{3}, \dfrac{19}{10}\right)\left(\dfrac{13}{8}, \dfrac{31}{20}\right)$

6. $\left(\dfrac{11}{8}, \dfrac{7}{5}\right)\left(\dfrac{25}{16}, \dfrac{7}{4}\right)$

7. $\left(\dfrac{7}{9}, -\dfrac{4}{3}\right)\left(-\dfrac{5}{9}, -\dfrac{8}{3}\right)$

8. $\left(\dfrac{5}{8}, -\dfrac{3}{2}\right)\left(-\dfrac{3}{8}, -\dfrac{7}{2}\right)$

For exercises 9–76, evaluate or simplify.

9. $\dfrac{\dfrac{4}{5}}{\dfrac{3}{10}}$

10. $\dfrac{\dfrac{2}{3}}{\dfrac{5}{6}}$

11. $\dfrac{\dfrac{20}{3}}{\dfrac{5}{3}}$

12. $\dfrac{\dfrac{9}{5}}{\dfrac{3}{5}}$

13. $\dfrac{\dfrac{4q}{9}}{\dfrac{2q^2}{3}}$

14. $\dfrac{\dfrac{5r}{4}}{\dfrac{4}{2r^2}}$

15. $\dfrac{\dfrac{9m^2}{12}}{\dfrac{12}{15m}}$

16. $\dfrac{\dfrac{16n^2}{20}}{\dfrac{20}{12n}}$

17. $\dfrac{\dfrac{x}{x-2}}{\dfrac{x+4}{x+3}}$

18. $\dfrac{\dfrac{x-2}{x+1}}{\dfrac{x}{x-1}}$

19. $\dfrac{\dfrac{x+1}{x-3}}{\dfrac{x+3}{x-1}}$

20. $\dfrac{\dfrac{x-1}{x-2}}{\dfrac{x+2}{x+1}}$

21. $\dfrac{\dfrac{4x+32}{3x+9}}{\dfrac{8x+64}{15x+45}}$

22. $\dfrac{\dfrac{3x+12}{2x+6}}{\dfrac{9x+36}{8x+24}}$

23. $\dfrac{\dfrac{2y-6}{4y+16}}{\dfrac{12y+36}{5y+20}}$

24. $\dfrac{\dfrac{5z-5}{4z+12}}{\dfrac{10z+10}{7z+21}}$

25. $\dfrac{\dfrac{a^2+11a+24}{9a}}{\dfrac{a^2-64}{27a^3}}$

26. $\dfrac{\dfrac{b^2+6b+5}{6b}}{\dfrac{b^2-1}{24b^4}}$

27. $\dfrac{\dfrac{3a}{a^2+5a+6}}{\dfrac{9a^3}{a^2-9}}$

28. $\dfrac{\dfrac{2b}{b^2+6b+8}}{\dfrac{4b^4}{b^2-4}}$

29. $\dfrac{\dfrac{b^2-6b-27}{b^2-8b-9}}{\dfrac{b^2-b-12}{b^2-1}}$

30. $\dfrac{\dfrac{y^2-2y-24}{y^2-4y-12}}{\dfrac{y^2-4y-32}{y^2-4}}$

31. $\dfrac{\dfrac{2x^2-9x-5}{20x^3}}{\dfrac{2x^2-13x-7}{42x^8}}$

32. $\dfrac{\dfrac{3d^2-5d-12}{15d^4}}{\dfrac{3d^2-14d-24}{36d^7}}$

33. $\dfrac{\dfrac{h-4}{h^2+2h-24}}{\dfrac{h-3}{h^2+4h-21}}$

34. $\dfrac{\dfrac{d-6}{d^2+3d-54}}{\dfrac{d-5}{d^2-3d-10}}$

35. $\dfrac{\dfrac{x^2+3x+2}{x+2}}{x+1}$

36. $\dfrac{\dfrac{x^2-9}{x-3}}{x^2+4x+3}$

37. $\dfrac{\dfrac{1}{3}+\dfrac{1}{2}}{\dfrac{1}{2}+\dfrac{1}{7}}$

38. $\dfrac{\dfrac{1}{2}+\dfrac{1}{3}}{\dfrac{1}{3}+\dfrac{1}{5}}$

39. $\dfrac{\dfrac{1}{3}}{\dfrac{1}{2}+\dfrac{1}{7}}$

40. $\dfrac{\dfrac{1}{2}}{\dfrac{1}{3}+\dfrac{1}{5}}$

41. $\dfrac{\dfrac{2}{3}-\dfrac{1}{2}}{\dfrac{1}{3}+\dfrac{1}{6}}$

42. $\dfrac{\dfrac{2}{5}-\dfrac{1}{3}}{\dfrac{1}{5}+\dfrac{1}{3}}$

43. $\dfrac{\dfrac{5}{6}+\dfrac{1}{3}}{\dfrac{2}{3}-\dfrac{1}{12}}$

44. $\dfrac{\dfrac{4}{9}+\dfrac{3}{2}}{\dfrac{5}{9}-\dfrac{1}{6}}$

45. $\dfrac{\dfrac{3}{x-1}-4}{\dfrac{4}{x+2}-1}$

46. $\dfrac{\dfrac{1}{x-4}-2}{\dfrac{4}{x+5}-4}$

47. $\dfrac{6-\dfrac{3}{x+1}}{8+\dfrac{5}{x+1}}$

48. $\dfrac{9-\dfrac{2}{y+3}}{11+\dfrac{4}{y+3}}$

49. $\dfrac{1+\dfrac{4}{k}}{1-\dfrac{16}{k^2}}$

50. $\dfrac{1+\dfrac{6}{w}}{1-\dfrac{36}{w^2}}$

51. $\dfrac{x}{x+\dfrac{x}{x+2}}$

52. $\dfrac{x}{x+\dfrac{x}{x+3}}$

53. $\dfrac{\dfrac{1}{x}}{\dfrac{1}{x}+\dfrac{1}{x+2}}$

54. $\dfrac{\dfrac{1}{x}}{\dfrac{1}{x}+\dfrac{1}{x+3}}$

55. $\dfrac{\dfrac{1}{x}}{\dfrac{1}{x-2}+\dfrac{1}{x+2}}$

56. $\dfrac{\dfrac{1}{x}}{\dfrac{1}{x-3}+\dfrac{1}{x+3}}$

57. $\dfrac{\dfrac{1}{x+2}+\dfrac{1}{x}}{\dfrac{1}{x+2}}$

58. $\dfrac{\dfrac{1}{x+3}+\dfrac{1}{x}}{\dfrac{1}{x+3}}$

59. $\dfrac{\dfrac{1}{x+4}-\dfrac{1}{x}}{\dfrac{1}{x+4}}$

60. $\dfrac{\dfrac{1}{x+1}-\dfrac{1}{x}}{\dfrac{1}{x+1}}$

61. $\dfrac{\dfrac{4}{x-1}+\dfrac{2}{x+1}}{\dfrac{3}{x-1}+\dfrac{1}{x+1}}$

62. $\dfrac{\dfrac{5}{x+1}+\dfrac{3}{x-2}}{\dfrac{5}{x+1}+\dfrac{2}{x-2}}$

63. $\dfrac{\dfrac{2}{x-2}+\dfrac{3}{x+3}}{\dfrac{2}{x-2}+\dfrac{1}{x+3}}$

64. $\dfrac{\dfrac{2}{x+4}+\dfrac{5}{x-2}}{\dfrac{3}{x+4}+\dfrac{4}{x-2}}$

65. $\dfrac{\dfrac{3}{x-4}-\dfrac{1}{x+2}}{\dfrac{6}{x-4}+\dfrac{5}{x+2}}$

66. $\dfrac{\dfrac{3}{x-3}-\dfrac{4}{x+1}}{\dfrac{4}{x-3}+\dfrac{6}{x+1}}$

67. $\dfrac{\dfrac{7}{4x-8}-\dfrac{1}{x+2}}{\dfrac{1}{x-2}+\dfrac{x}{x^2-4}}$

68. $\dfrac{\dfrac{5}{4x-12}-\dfrac{1}{x+3}}{\dfrac{1}{x-3}+\dfrac{x}{x^2-9}}$

69. $\dfrac{x-y}{\dfrac{1}{y}-\dfrac{1}{x}}$

70. $\dfrac{x+y}{\dfrac{1}{x}+\dfrac{1}{y}}$

71. $\dfrac{\dfrac{1}{y}+\dfrac{1}{x}}{y+x}$

72. $\dfrac{\dfrac{1}{x}-\dfrac{1}{y}}{y-x}$

73. $\dfrac{\dfrac{c}{3}+\dfrac{d}{2}}{\dfrac{c}{2}+\dfrac{d}{7}}$

74. $\dfrac{\dfrac{a}{2}+\dfrac{b}{3}}{\dfrac{a}{3}+\dfrac{b}{5}}$

75. $\dfrac{\dfrac{3}{c}+\dfrac{2}{d}}{\dfrac{2}{c}+\dfrac{5}{d}}$

76. $\dfrac{\dfrac{2}{a}+\dfrac{3}{b}}{\dfrac{3}{a}+\dfrac{5}{b}}$

For exercises 77–80, these exercises are common in an introductory college course in electronics. Resistors restrict the flow of electrons in an electric circuit. The unit of resistance is an *ohm*. The formula $R = \dfrac{1}{\dfrac{1}{R_1}+\dfrac{1}{R_2}}$ describes the total resistance R in ohms of a parallel circuit with two resistors, R_1 and R_2.

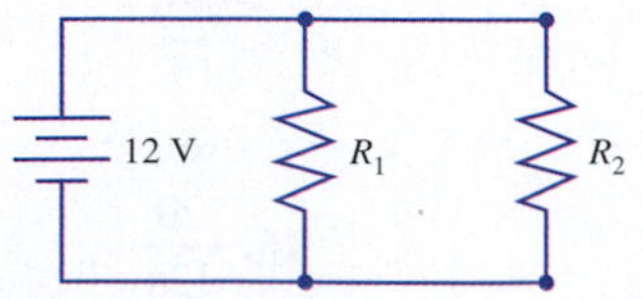

77. In a parallel circuit, Resistor R_1 has a resistance of 5 ohms and Resistor R_2 has a resistance of 20 ohms. Find the total resistance of the circuit.

78. In a parallel circuit, Resistor R_1 has a resistance of 3 ohms and Resistor R_2 has a resistance of 6 ohms. Find the total resistance of the circuit.

79. In a parallel circuit, Resistor R_1 has a resistance of 10 ohms and Resistor R_2 has a resistance of 10 ohms. Find the total resistance of the circuit.

80. In a parallel circuit, Resistor R_1 has a resistance of 8 ohms and Resistor R_2 has a resistance of 8 ohms. Find the total resistance of the circuit.

Problem Solving: Practice and Review

Follow your instructor's guidelines for following the five steps as outlined in Section 1.5, p. 51.

81. An American was born in 1956 and will reach full retirement age in 2021. His social security payments are based on his lifetime earnings. His first option is to retire at age 65 years and 10 months and receive $1000 a month for the rest of his life. His second option is to retire at age 62 (46 months before full retirement age) and receive $758 a month for the rest of his life. What age will he be in years and months when the payments he has received from either option are equal? Round up to the nearest month. (*Source:* www.ssa.gov)

82. The graph is from the Centers for Disease Control and Prevention website. A student said that a bar graph would be a better way to represent this information and that the lines between the data points should not be there. Do you agree or disagree? Explain.

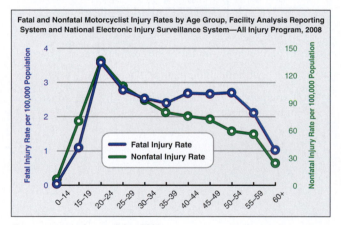

Fatal and Nonfatal Motorcyclist Injury Rates by Age Group, Facility Analysis Reporting System and National Electronic Injury Surveillance System—All Injury Program, 2008

Source: www.cdc.gov, May 2011

83. See the table below. Find the percent increase in associate degrees in health fields from the 1999–2000 academic year to the 2009–2010 academic year. Round to the nearest percent.

84. In April and May of 2011, Joliet, Illinois, residents, using 90-gal containers rather than bins, recycled 1683 tons. This was a 30% increase over the amount recycled in the same period in 2010, when bins were used. Find the amount recycled in the same period in 2010. Round to the nearest ten. (*Source:* www.chicagotribune.com, June 16, 2011)

Find the Mistake

For exercises 85–88, one part of simplifying a rational expression is completed.

(a) Describe the mistake in words, or copy down this part of the problem and highlight or circle the mistake.

(b) Do this part of the problem correctly.

85. Problem: In simplifying $\dfrac{\dfrac{2}{5} - \dfrac{1}{3}}{\dfrac{1}{5} + \dfrac{1}{3}}$, one of the steps in the

process is simplifying $\dfrac{1}{15} \div \dfrac{8}{15}$.

Incorrect Answer: $\dfrac{1}{15} \cdot \dfrac{15}{8}$

$$= \frac{1}{\cancel{15}} \cdot \frac{\cancel{15}}{8}$$

$$= 8$$

Table for exercise 83

Field of study	Number of associate degrees (1999–2000)	Number of associate degrees (2009–2010)
General Studies	187,454	284,758
Health	84,097	177,729
Business	97,831	116,895
Social/Behavorial Science	52,201	74,686
Vocations/Trades	44,747	51,227
Other	38,742	49,311
Protective Services	16,298	37,273
Humanities & Fine Arts	18,870	23,363
STEM	6,591	17,229
Education	8,226	17,048

Source: www.ccweek.com, June 13, 2011

86. Problem: In simplifying $\dfrac{2 + \dfrac{3}{x}}{3 + \dfrac{1}{x}}$, one of the steps is adding

the terms in the numerator and adding the terms in the denominator.

Incorrect Answer: $\dfrac{2 + \dfrac{3}{x}}{3 + \dfrac{1}{x}}$

$= \dfrac{\dfrac{5}{x}}{\dfrac{4}{x}}$

87. Problem: In simplifying $\dfrac{1 + \dfrac{7}{n-2}}{1 + \dfrac{3}{n+2}}$, one of the steps is

clearing the fractions.

Incorrect Answer: $\dfrac{1 + \dfrac{7}{n-2}}{1 + \dfrac{3}{n+2}} \cdot \dfrac{(n-2)(n+2)}{(n-2)(n+2)}$

$= \dfrac{1 + 7(n+2)}{1 + 3(n-2)}$

88. Problem: In simplifying $\dfrac{\dfrac{3}{x+2} + \dfrac{4}{x-2}}{\dfrac{2}{x+1} + \dfrac{3}{x-2}}$, one of the steps

is finding the least common denominator of all of the fractions.

Incorrect Answer: The least common denominator is $(x+2)(x+1)(x-2)^2$.

Review

For exercises 89–92,
(a) clear the fractions and solve.
(b) check.

89. $\dfrac{1}{3}x + 8 = 13$

90. $\dfrac{2}{3}x - 5 = -1$

91. $\dfrac{1}{3}x + \dfrac{8}{9} = \dfrac{26}{9}$

92. $\dfrac{2}{5}x - \dfrac{4}{7} = \dfrac{1}{3}$

SUCCESS IN COLLEGE MATHEMATICS

93. Rewrite the fraction $\dfrac{\dfrac{3}{4}}{\dfrac{5}{11}}$ as a quotient, and then rewrite

the quotient as multiplication by the reciprocal of the divisor.

94. In Example 9, we rewrite the complex fraction

$\dfrac{3 - \dfrac{9}{w-4}}{5 + \dfrac{4}{w-4}}$ as $\left(\dfrac{3 - \dfrac{9}{w-4}}{5 + \dfrac{4}{w-4}}\right)\left(\dfrac{w-4}{w-4}\right)$.

a. Explain why we multiply the complex fraction by $\dfrac{w-4}{w-4}$ without changing its value.
b. Explain why we choose to multiply by $\dfrac{w-4}{w-4}$ instead of by another fraction.

At some colleges and universities, teaching assistants grade exams for professors. In this section, we will use a rational equation to predict how long it will take two teaching assistants working together to grade a set of chemistry exams.

SECTION 5.4

Rational Equations in One Variable

After reading the text, working the practice problems, and completing assigned exercises, you should be able to:

1. Clear the fractions from a rational equation in one variable.

2. Solve a rational equation in one variable.

3. Use a rational equation to solve proportions, rate of work, and other application problems.

Rational Equations

In Section 1.3, we cleared fractions from equations to make them easier to solve. To **clear the fractions** from an equation, multiply both sides of the equation by the least common denominator and simplify common factors. The coefficients and constants in the equation are now integers.

EXAMPLE 1   Clear the fractions from $\dfrac{1}{7}x + \dfrac{3}{5} = \dfrac{8}{9}x$. Do not solve.

SOLUTION ▸ The least common denominator is $7 \cdot 5 \cdot 3^2$ which equals 315.

$$\frac{315}{1}\left(\frac{1}{7}x + \frac{3}{5}\right) = \frac{315}{1}\left(\frac{8}{9}x\right)$$ Multiplication property of equality.

$$\frac{315}{7}x + \frac{945}{5} = \frac{2520}{9}x$$ Distributive property.

$$\frac{45 \cdot \cancel{7}}{\cancel{7}}x + \frac{189 \cdot \cancel{5}}{\cancel{5}} = \frac{280 \cdot \cancel{9}}{\cancel{9}}x$$ Find common factors.

$$45x + 189 = 280x$$ Simplify; the fractions are cleared.

The equation in Example 1, $\dfrac{1}{7}x + \dfrac{3}{5} = \dfrac{8}{9}x$, is a polynomial equation in one variable. The variable is in the numerator. In the next example, we clear the fractions from a rational equation. A **rational equation** includes a variable in at least one denominator.

EXAMPLE 2 | Clear the fractions from $\dfrac{5}{x} - \dfrac{1}{4} = \dfrac{3}{x}$. Do not solve.

SOLUTION ▶

$$\dfrac{5}{x} - \dfrac{1}{4} = \dfrac{3}{x} \qquad \text{Least common denominator is } 4x.$$

$$\dfrac{4x}{1}\left(\dfrac{5}{x} - \dfrac{1}{4}\right) = \dfrac{4x}{1}\left(\dfrac{3}{x}\right) \qquad \text{Multiplication property of equality.}$$

$$\dfrac{20x}{x} - \dfrac{4x}{4} = \dfrac{12x}{x} \qquad \text{Distributive property; find common factors.}$$

$$20 - x = 12 \qquad \text{Simplify; the fractions are cleared.}$$

To find the least common denominator in the next example, the denominators need to be factored completely.

EXAMPLE 3 | Clear the fractions from $\dfrac{x}{x-3} + \dfrac{1}{x-5} = \dfrac{2}{x^2 - 8x + 15}$. Do not solve.

SOLUTION ▶

$$\dfrac{x}{x-3} + \dfrac{1}{x-5} = \dfrac{2}{x^2 - 8x + 15}$$

$$\dfrac{x}{x-3} + \dfrac{1}{x-5} = \dfrac{2}{(x-3)(x-5)} \qquad \text{Factor.}$$

The least common denominator is $(x-5)(x-3)$.

$$\dfrac{(x-5)(x-3)}{1}\left(\dfrac{x}{x-3} + \dfrac{1}{x-5}\right) = \dfrac{(x-5)(x-3)}{1} \cdot \dfrac{2}{(x-3)(x-5)}$$

$$\dfrac{(x-5)(x-3)x}{x-3} + \dfrac{(x-5)(x-3)1}{x-5} = \dfrac{(x-5)(x-3)\cdot 2}{(x-3)(x-5)} \qquad \text{Distributive property.}$$

$$x(x-5) + 1(x-3) = 2 \qquad \text{Simplify; the fractions are cleared.}$$

An equation includes an equals sign; an expression does not. We can only clear the fractions from an *equation*; we can never clear the fractions from an *expression*.

Clearing Fractions from a Rational Equation

1. Factor the denominators completely.

2. Identify the least common denominator.

3. Multiply both sides of the equation by the least common denominator (multiplication property of equality).

4. Simplify.

Practice Problems

For problems 1–3,
(a) identify the least common denominator.
(b) clear the fractions from each equation. Do not solve.

1. $\dfrac{5}{6}p - \dfrac{1}{8} = \dfrac{2}{15}$ **2.** $\dfrac{11}{2x} - \dfrac{2}{4x} = \dfrac{1}{8}$ **3.** $\dfrac{2x}{x-4} + \dfrac{2}{x-6} = \dfrac{4}{x^2 - 10x + 24}$

Solving Rational Equations and Extraneous Solutions

To solve a rational equation, clear the fractions and solve the new equation. In the next example, the new equation is a quadratic equation. To solve, use the zero product property: if $ab = 0$, then $a = 0$ or $b = 0$.

EXAMPLE 4   (a) Solve: $\dfrac{1}{2} - \dfrac{3}{x^2} = \dfrac{5}{2x}$

SOLUTION ▶

$$\dfrac{1}{2} - \dfrac{3}{x^2} = \dfrac{5}{2x}$$ Least common denoninator is $2x^2$.

$$\dfrac{2x^2}{1}\left(\dfrac{1}{2} - \dfrac{3}{x^2}\right) = \dfrac{2x^2}{1}\left(\dfrac{5}{2x}\right)$$ Multiplication property of equality.

$$\dfrac{2x^2}{2} - \dfrac{6x^2}{x^2} = \dfrac{2 \cdot x^2 \cdot 5}{2x}$$ Distributive property.

$$\dfrac{2x^2}{2} - \dfrac{6x^2}{x^2} = \dfrac{2 \cdot x \cdot x \cdot 5}{2x}$$ Find common factors.

$$x^2 - 6 = 5x$$ Simplify; the fractions are cleared.

$$\underline{ -5x \quad -5x}$$ Subtraction property of equality.

$$x^2 - 5x - 6 = 0$$ Standard form: $ax^2 + bx + c = 0$

$$(x - 6)(x + 1) = 0$$ Factor.

$$x - 6 = 0 \quad \text{or} \quad x + 1 = 0$$ Zero product property.

$$\underline{+6 \ +6 -1 \quad -1}$$ Property of equality.

$$x + 0 = 6 \quad \text{or} \quad x + 0 = -1$$ Simplify.

$$x = 6 \quad \text{or} \quad x = -1$$ Simplify.

(b) Check.

▶

Check: $x = 6$	Check: $x = -1$
$\dfrac{1}{2} - \dfrac{3}{(6)^2} = \dfrac{5}{2(6)}$	$\dfrac{1}{2} - \dfrac{3}{(-1)^2} = \dfrac{5}{2(-1)}$
$\dfrac{1}{2} - \dfrac{3}{36} = \dfrac{5}{12}$	$\dfrac{1}{2} - \dfrac{3}{1} = \dfrac{5}{-2}$
$\dfrac{18}{36} - \dfrac{3}{36} = \dfrac{15}{36}$	$\dfrac{1}{2} - \dfrac{6}{2} = -\dfrac{5}{2}$
$\dfrac{15}{36} = \dfrac{15}{36}$ True.	$-\dfrac{5}{2} = -\dfrac{5}{2}$ True.

In the next example, we clear the fractions and solve the resulting quadratic equation. However, one of the solutions of this new quadratic equation is not a solution of the original rational equation. It is an **extraneous solution**. It occurs because we cleared the fractions. A solution of the new equation (which has no variables in a denominator) may result in division by zero in the original equation. Since division by zero is undefined, it is not a solution of the original equation.

EXAMPLE 5   (a) Solve: $\dfrac{1}{x - 1} + \dfrac{2}{x} = \dfrac{x}{x - 1}$

SOLUTION ▶

$$\dfrac{1}{x - 1} + \dfrac{2}{x} = \dfrac{x}{x - 1}$$ Least common denominator is $x(x - 1)$.

$$\dfrac{x(x - 1)}{1}\left(\dfrac{1}{x - 1} + \dfrac{2}{x}\right) = \dfrac{x(x - 1)}{1}\left(\dfrac{x}{x - 1}\right)$$ Multiplication property of equality.

$$\frac{x(x-1)(1)}{x-1} + \frac{x(x-1)(2)}{x} = \frac{x(x-1)(x)}{x-1}$$ Distributive property; find common factors.

$$x + (x-1)(2) = x^2$$ Simplify.

$$x + 2x - 2 = x^2$$ Distributive property.

$$3x - 2 = x^2$$ Combine like terms.

$$\underline{-3x\ +2 \qquad\qquad -3x\ +2}$$ Properties of equality.

$$0 = x^2 - 3x + 2$$ Simplify; $ax^2 + bx + c = 0$

$$0 = (x-1)(x-2)$$ Factor.

$$x - 1 = 0 \quad \text{or} \quad x - 2 = 0$$ Zero product property.

$$\underline{+1\ +1} \qquad\qquad \underline{+2\ +2}$$ Addition property of equality.

$$x + 0 = 1 \quad \text{or} \quad x + 0 = 2$$ Simplify.

$$\cancel{x = 1} \quad \text{or} \quad x = 2$$ Simplify.

(b) Check.

Check: $x = 1$

$$\frac{1}{1-1} + \frac{2}{1} = \frac{1}{1-1}$$

$$\frac{1}{0} + \frac{2}{1} = \frac{1}{0} \qquad \text{Undefined; false.}$$

Check: $x = 2$

$$\frac{1}{2-1} + \frac{2}{2} = \frac{2}{2-1}$$

$$\frac{1}{1} + 1 = \frac{2}{1}$$

$$2 = 2 \quad \text{True.}$$

The solution of $\dfrac{1}{x-1} + \dfrac{2}{x} = \dfrac{x}{x-1}$ is $x = 2$. The other value, $x = 1$, is an **extraneous solution**. Although $x = 1$ is a solution of the quadratic equation $x^2 - 3x + 2 = 0$, it is not a solution of $\dfrac{1}{x-1} + \dfrac{2}{x} = \dfrac{x}{x-1}$. In the rational equation, $x = 1$ results in division by zero, which is undefined. Ask your instructor how you should mark extraneous solutions in your work.

We can identify extraneous solutions in a check because they result in division by zero. We can also identify possible extraneous solutions before we solve the equation. If every solution of a rational equation is extraneous, the equation has no solution.

EXAMPLE 6 | Solve: $\dfrac{3}{x-3} = \dfrac{x}{x-3} - \dfrac{3}{2}$

SOLUTION ▶ The denominator with a variable is $x - 3$. Since $3 - 3 = 0$, $x = 3$ results in division by 0. So if a solution is $x = 3$, it is extraneous.

$$\frac{3}{x-3} = \frac{x}{x-3} - \frac{3}{2}$$ Least common denominator is $2(x-3)$.

$$\frac{2(x-3)}{1}\left(\frac{3}{x-3}\right) = \frac{2(x-3)}{1}\left(\frac{x}{x-3} - \frac{3}{2}\right)$$ Multiplication property of equality.

$$\frac{2(x-3)(3)}{x-3} = \frac{2(x-3)(x)}{x-3} - \frac{2(x-3)(3)}{2}$$ Distributive property; find common factors.

$$6 = 2x - (x-3)(3)$$ Simplify.

$$6 = 2x - (3x - 9)$$ Distributive property.

$$6 = 2x - 1(3x - 9)$$ Multiply by 1.

$$6 = 2x - 3x + 9 \qquad \text{Distributive property.}$$

$$6 = -x + 9 \qquad \text{Combine like terms.}$$

$$\underline{-9 \qquad\qquad -9} \qquad \text{Addition property of equality.}$$

$$-3 = -x + 0 \qquad \text{Simplify.}$$

$$\frac{-3}{-1} = \frac{-x}{-1} \qquad \text{Division property of equality.}$$

$$3 = x \qquad \text{Simplify.}$$

The only solution of the equation $6 = 2x - (x - 3)(3)$ is $x = 3$. However, this solution results in division by zero in the original rational equation, $\dfrac{3}{x-3} = \dfrac{x}{x-3} - \dfrac{3}{2}$. Since $x = 3$ is extraneous, the equation has no solution.

Solving a Rational Equation

1. Clear the fractions.
2. Identify possible extraneous solutions that cause division by zero.
3. Solve the equation.
4. Check; identify extraneous solutions, and do not include them in the final answer. If an equation has no solution, write this.

EXAMPLE 7 Solve: $\dfrac{x}{x+5} + \dfrac{2x}{x+7} = \dfrac{x^2 + 2x - 7}{x^2 + 12x + 35}$

SOLUTION ▶ To find the least common denominator, factor the denominator of $\dfrac{x^2 + 2x - 7}{x^2 + 12x + 35}$. The least common denominator is $(x + 5)(x + 7)$. If a solution is $x = -5$ or $x = -7$, it is extraneous.

$$\frac{x}{x+5} + \frac{2x}{x+7} = \frac{x^2 + 2x - 7}{x^2 + 12x + 35}$$

$$\frac{x}{x+5} + \frac{2x}{x+7} = \frac{x^2 + 2x - 7}{(x+5)(x+7)} \qquad \text{Factor denominator.}$$

$$\frac{(x+5)(x+7)}{1}\left(\frac{x}{x+5} + \frac{2x}{x+7}\right) = \frac{(x+5)(x+7)}{1}\left(\frac{x^2 + 2x - 7}{(x+5)(x+7)}\right)$$

$$\frac{(x+5)(x+7)}{1}\left(\frac{x}{x+5} + \frac{2x}{x+7}\right) = x^2 + 2x - 7 \qquad \text{Simplify.}$$

$$\frac{(x+5)(x+7)}{1}\left(\frac{x}{x+5}\right) + \frac{(x+5)(x+7)}{1}\left(\frac{2x}{x+7}\right)$$

$$= x^2 + 2x - 7 \qquad \text{Distributive property.}$$

$$(x+7)x + (x+5)2x = x^2 + 2x - 7 \qquad \text{Simplify.}$$

$$x^2 + 7x + 2x^2 + 10x = x^2 + 2x - 7 \qquad \text{Distributive property.}$$

$$3x^2 + 17x = x^2 + 2x - 7 \qquad \text{Combine like terms.}$$

$$\underline{-x^2 - 2x + 7 \quad -x^2 - 2x + 7} \qquad \text{Properties of equality.}$$

$$2x^2 + 15x + 7 = 0 \qquad \text{Simplify; } ax^2 + bx + c = 0$$

$$(2x + 1)(x + 7) = 0 \qquad \text{Factor.}$$

$$2x + 1 = 0 \quad \text{or} \quad x + 7 = 0 \qquad \text{Zero product property.}$$

$$\underline{-1 \quad -1} \qquad \qquad \underline{-7 \quad -7} \qquad \text{Subtraction property of equality.}$$

$$2x + 0 = -1 \quad \text{or} \quad x + 0 = -7 \qquad \text{Simplify.}$$

$$\frac{2x}{2} = \frac{-1}{2} \quad \text{or} \quad \cancel{x = -7} \qquad \text{Simplify.}$$

$$x = -\frac{1}{2} \qquad \qquad \qquad \text{Simplify.}$$

The solution of $\dfrac{x}{x + 5} + \dfrac{2x}{x + 7} = \dfrac{x^2 + 2x - 7}{x^2 + 12x + 35}$ is $x = -\dfrac{1}{2}$.

Practice Problems

For problems 4–7,
(a) solve.
(b) check.

4. $\dfrac{3}{x} + \dfrac{2}{9} = \dfrac{1}{3}$ **5.** $\dfrac{2}{x} + \dfrac{x}{8} = \dfrac{x}{4}$ **6.** $\dfrac{1}{x + 1} - \dfrac{1}{x + 2} = \dfrac{1}{2}$

7. $\dfrac{2a}{a - 1} + \dfrac{a - 5}{a^2 - 1} = 1$

Proportions

A **ratio** such as $\dfrac{6 \text{ people}}{100 \text{ people}}$ or a **rate** such as $\dfrac{30 \text{ mi}}{1 \text{ hr}}$ compares two measurements. An equation with two equivalent ratios or rates such as $\dfrac{30 \text{ mi}}{1 \text{ hr}} = \dfrac{60 \text{ mi}}{2 \text{ hr}}$ is a **proportion**. In the next example, the problem includes a *known ratio* from a study, $\dfrac{1 \text{ baby with sickle cell disease}}{500 \text{ babies}}$. To find the *unknown ratio*, $\dfrac{N}{609{,}550 \text{ babies}}$, we write and solve a proportion, *known ratio = unknown ratio*.

EXAMPLE 8   In 2009, about 609,550 African-American babies were born. Predict the number of these babies who had sickle cell disease. Round to the nearest ten.

In sickle cell disease, the red blood cells become hard and sticky and look like a C-shaped farm tool called a "sickle". The sickle cells die early, which causes a constant shortage of red blood cells. Also, when they travel through small blood vessels, they get stuck and clog the blood flow. This can cause pain and other serious problems. . . . [Sickle cell disease] occurs among about 1 of every 500 Black or African-American births and among about 1 out of every 36,000 Hispanic-American births. (*Source:* www.cdc.gov, June 13, 2011)

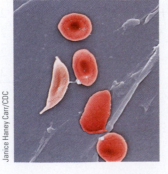

Janice Haney Carr/CDC

SOLUTION ▶ **Step 1 Understand the problem.**

The unknown is the number of African-American babies born in 2009 with sickle cell disease. The number of Hispanic-American babies with sickle cell disease is extraneous information.

N = number of babies born in 2009 with sickle cell disease

Step 2 Make a plan.

In a proportion, unknown rate = known rate. The rate is

$$\frac{\text{number of babies with sickle cell disease}}{\text{total number of babies}}.$$

Step 3 Carry out the plan.

known rate = unknown rate	Word equation.
$\dfrac{\text{1 baby with sickle cell disease}}{500 \text{ babies}} = \dfrac{N}{609{,}550 \text{ babies}}$	Proportion.
$\dfrac{1}{500} = \dfrac{N}{609{,}550}$	To solve, remove the units.
$609{,}550 \cdot \dfrac{1}{500} = \cancel{609{,}550} \cdot \dfrac{N}{\cancel{609{,}550}}$	Multiplication property of equality.
$1219.1 = N$	Simplify.
$\mathbf{1220 \approx N}$	Round.
$1220 \textbf{ babies with disease} \approx N$	Include units.

Step 4 Look back.

The known rate should equal the unknown rate. Since $\dfrac{1}{500} = 0.002$ and $\dfrac{1220}{609{,}550} = 0.002\ldots$, the answer seems reasonable.

Step 5 Report the solution.

Of 609,550 African-American babies born in 2009, about 1220 babies had sickle cell disease.

Practice Problems

For problems 8–9, use a proportion and the five steps.

8. A city has a population of 60,325 people. Predict how many of these people will have a diagnosis of Tourette syndrome in their lifetime. Round to the nearest whole number.

   Tourette syndrome (TS) is an inheritable, childhood-onset neurologic disorder marked by persistent multiple motor tics and at least one vocal tic. . . . Based on data from the 2007 National Survey of Children's Health, the estimated prevalence of a lifetime diagnosis of TS by parent report was 3.0 per 1,000. A diagnosis of TS was almost three times as likely for boys as girls. (*Source:* www.cdc.gov, June 5, 2009)

9. In 2005, there were at least 13,288,000 Medicare patients discharged from the hospital. Predict how many of these patients returned to the hospital within 30 days. (*Source:* www.cdc.gov, Dec. 2007)

   Millions of patients each year leave the hospital only to return within weeks or months for lack of proper follow-up care. One in five Medicare patients, for example, returns to the hospital within 30 days. Over all, readmissions cost the federal government an estimated $17 billion a year. (*Source:* www.nytimes.com, May 5, 2009)

Rate of Work

A typist works at a rate of $\dfrac{60 \text{ words}}{1 \text{ min}}$. A data entry operator works at a rate of $\dfrac{250 \text{ keys}}{1 \text{ min}}$. These rates describe the amount of work done in a certain time. Other rates of work describe how long it takes to *complete a job*. For example, we do not describe the number of teeth a dental hygienist cleans per minute. Instead, we use a rate that describes how long it takes to clean an entire mouth, such as $\dfrac{1 \text{ mouth}}{45 \text{ min}}$.

The fraction of a job completed is equal to the product of the rate of work and the time worked.

> ### Fraction of a Job Completed
>
> fraction of a job completed $=$ (rate of work)(time worked)

EXAMPLE 9 The work rate of a dental hygienist is $\dfrac{1 \text{ mouth}}{45 \text{ min}}$. Find the fraction of the job completed after the hygienist works 20 min.

SOLUTION ▶ fraction of job completed

$= (\text{rate of work})(\text{time})$

$= \left(\dfrac{1 \text{ mouth}}{45 \, \cancel{\text{min}}}\right)(20 \, \cancel{\text{min}})$

$= \dfrac{4}{9} \text{ mouth}$

In some jobs, more than one person can work together, and their individual rates of work do not change.

> ### Fraction of a Job Completed by Two People, A and B (Working Independently)
>
> (rate for A)(time A works) $+$ (rate for B)(time B works)
> $\quad =$ fraction of job completed

EXAMPLE 10 Worker A can wash the windows in an apartment in 45 min. Worker B can wash the same windows in 30 min. If they work together, how long will it take them to wash the windows? (Assume that their rates of work do not change when they work together.)

SOLUTION ▶ **Step 1 Understand the problem.**
The unknown is the amount of time it will take the workers to wash the windows working together.

$T = $ amount of time to wash the windows

Step 2 Make a plan.
Worker A's rate is $\dfrac{1 \text{ job}}{45 \text{ min}}$, and Worker B's rate is $\dfrac{1 \text{ job}}{30 \text{ min}}$. Both work the same amount of time, T. Since the time of each worker equals (rate of work)(time), a word equation is (rate A)(time A) $+$ (rate B)(time B) $=$ 1 job or

$$\left(\dfrac{1 \text{ job}}{45 \text{ min}}\right)T + \left(\dfrac{1 \text{ job}}{30 \text{ min}}\right)T = 1 \text{ job}.$$

Step 3 Carry out the plan.

$$\frac{1}{45}T + \frac{1}{30}T = 1 \qquad \text{Remove the units; the least common denominator is 90.}$$

$$\frac{90}{1}\left(\frac{1}{45}T + \frac{1}{30}T\right) = 90(1) \qquad \text{Clear fractions; multiplication property of equality.}$$

$$\frac{90}{45}T + \frac{90}{30}T = 90 \qquad \text{Distributive property; simplify.}$$

$$2T + 3T = 90 \qquad \text{Simplify.}$$

$$5T = 90 \qquad \text{Combine like terms.}$$

$$\frac{5T}{5} = \frac{90}{5} \qquad \text{Division property of equality.}$$

$$T = 18 \text{ min} \qquad \text{Simplify; the units are minutes.}$$

Step 4 Look back.
When Worker A washes windows for 18 min, $\frac{18}{45}$ job, or $\frac{2}{5}$ job, is done. When Worker B washes windows for 18 min, $\frac{18}{30}$ job, or $\frac{3}{5}$ job, is done. Since $\frac{2}{5}$ job $+ \frac{3}{5}$ job equals 1 finished job, the answer seems reasonable. This problem shows that clearing the fractions from an equation allows us to avoid adding fractions with relatively large denominators.

Step 5 Report the solution.
It will take the two workers 18 min to wash the windows working together.

In writing equations, the simplified units on the left side must be the same as the simplified units on the right side. In the next example, we change the units of time into minutes.

EXAMPLE 11 Two students are working as graders for a chemistry professor. Student A can grade the tests for a general chemistry class in 3 hr. Student B can grade the tests for the class in 2 hr 30 min. If they work together, how long will it take them to grade the tests? Write the answer as a decimal number in minutes; round to the nearest whole number. (Assume that their rate of work does not change when they work together.)

SOLUTION ▶ **Step 1 Understand the problem.**
The unknown is the amount of time it will take them to grade the tests working together.

T = amount of time to grade the tests

Step 2 Make a plan.
Change the time into minutes and write a rate for each grader.

Student A		**Student B**
3 hr	Time working alone.	2 hr + 30 min
$= \frac{3\text{ hr}}{1} \cdot \frac{60 \text{ min}}{1 \text{ hr}}$	Change units.	$= \frac{2 \text{ hr}}{1} \cdot \frac{60 \text{ min}}{1 \text{ hr}} + 30 \text{ min}$
$= 180 \text{ min}$	Simplify.	$= 150 \text{ min}$
$\frac{1 \text{ job}}{180 \text{ min}}$	Rate.	$\frac{1 \text{ job}}{150 \text{ min}}$

Both students work the same amount of time, T. A word equation is

$$(\text{rate A})(\text{time A}) + (\text{rate B})(\text{time B}) = 1 \text{ job}$$

Step 3 Carry out the plan.

$$\left(\frac{1 \text{ job}}{180 \text{ min}}\right)T + \left(\frac{1 \text{ job}}{150 \text{ min}}\right)T = 1 \text{ job} \qquad \text{Algebraic equation.}$$

$$\frac{1}{180}T + \frac{1}{150}T = 1 \qquad \text{Remove units; least common denominator is 900.}$$

$$\frac{900}{1}\left(\frac{1}{180}T + \frac{1}{150}T\right) = 900(1) \qquad \text{Multiplication property of equality.}$$

$$\frac{900}{180}T + \frac{900}{150}T = 900 \qquad \text{Distributive property; simplify.}$$

$$5T + 6T = 900 \qquad \text{Simplify.}$$

$$11T = 900 \qquad \text{Combine like terms.}$$

$$\frac{11T}{11} = \frac{900}{11} \qquad \text{Division property of equality.}$$

$$T = 81.81\ldots \qquad \text{Rewrite as a decimal number.}$$

$$T \approx 82 \text{ min} \qquad \text{Round to the nearest whole number.}$$

Step 4 Look back.
When Student A grades for 82 min, $\frac{82}{180}$ job, or about 0.46 job, is done. When Student B grades for 82 min, $\frac{82}{150}$ job, or about 0.55 job, is done. Since 0.46 job + 0.55 job equals about 1 finished job, the answer seems reasonable. From this problem, we can learn that the denominator of a rate should not be a mixed measurement with more than 1 unit.

Step 5 Report the solution.
It will take the graders about 82 min to grade the tests working together.

Sometimes work is done and undone at the same time. For example, a person can be putting hotdog buns in a warmer at a concession stand while another person is taking hotdog buns out to serve to customers or a faucet can be filling a sink at the same time that an open drain is emptying the sink.

EXAMPLE 12 When the drain is closed, it takes 175 seconds (s) for a faucet to fill an empty sink. After 120 s of filling an empty sink with the drain closed, the drain is opened, and water begins to drain out of the sink at the same time that the water is pouring into it. After 300 s of simultaneously filling and draining, the sink is full. Find the time for the drain to empty a full sink. Round to the nearest whole number. (Drains normally will empty a sink faster than it can be filled by a faucet; this drain has a partial clog.)

SOLUTION ▶ **Step 1 Understand the problem.**
The unknown is the amount of time for the drain to empty a full sink.

x = amount of time for the drain to empty the sink

Step 2 Make a plan.
Write a rate for filling the sink and a rate for draining the sink.

$$\text{fill rate: } \frac{1 \text{ sink}}{175 \text{ s}} \qquad \text{drain rate: } \frac{1 \text{ sink}}{x}$$

For the first 120 s, water is going into the sink. The drain is closed. For the next 300 s, water is going into the sink and draining out of the sink at the same time. Since the water is draining out of the sink more slowly than it is going into the sink, eventually the sink is full. A word equation is

amount filled 120 s + amount filled 300 s − amount drained 300 s = 1 full sink

We can rewrite the word equation as

$$(\text{fill rate})(\text{time}_1) + (\text{fill rate})(\text{time}_2) - (\text{drain rate})(\text{time}_2) = 1 \text{ full sink}$$

Step 3 Carry out the plan.

$$\left(\frac{1 \text{ sink}}{175 \text{ s}}\right)(120 \text{ s}) + \left(\frac{1 \text{ sink}}{175 \text{ s}}\right)(300 \text{ s}) - \left(\frac{1 \text{ sink}}{x}\right)(300 \text{ s}) = 1 \text{ full sink}$$

$$\left(\frac{1}{175}\right)120 + \left(\frac{1}{175}\right)(300) - \left(\frac{1}{x}\right)(300) = 1 \qquad \text{Remove units.}$$

$$\left(\frac{24 \cdot \cancel{5}}{35 \cdot \cancel{5}}\right) + \left(\frac{12 \cdot \cancel{25}}{7 \cdot \cancel{25}}\right) - \frac{300}{x} = 1 \qquad \text{Find common factors.}$$

$$\frac{24}{35} + \frac{12}{7} - \frac{300}{x} = 1 \qquad \text{Simplify; least common denominator is } 35x.$$

$$\frac{35x}{1}\left(\frac{24}{35} + \frac{12}{7} - \frac{300}{x}\right) = 35x(1) \qquad \text{Multiplication property of equality.}$$

$$\frac{(35x)(24)}{35} + \frac{(35x)(12)}{7} + \frac{(35x)(-300)}{x} = 35x(1) \qquad \text{Distributive property.}$$

$$24x + 60x - 10{,}500 = 35x \qquad \text{Simplify.}$$

$$84x - 10{,}500 = 35x \qquad \text{Combine like terms.}$$

$$\underline{-84x \qquad\qquad -84x} \qquad \text{Subtraction property of equality.}$$

$$0 - 10{,}500 = -49x \qquad \text{Simplify.}$$

$$\frac{-10{,}500}{-49} = \frac{-49x}{-49} \qquad \text{Division property of equality.}$$

$$214.2\ldots = x \qquad \text{Simplify.}$$

$$214 \text{ s} \approx x \qquad \text{Round; include the units.}$$

Step 4 Look back.

The faucet is filling the sink for 120 s plus 300 s, a total of 420 s; $\left(\dfrac{1 \text{ sink}}{175 \text{ s}}\right)(420 \text{ s})$ equals about 2.4 sinks of water. At the end when the sink is full, 1.4 sinks of water must have gone down the drain. Since $\left(\dfrac{1 \text{ sink}}{214 \text{ s}}\right)(300 \text{ s})$ drains about 1.4 sink, the drain rate of $\dfrac{1 \text{ sink}}{214 \text{ s}}$ seems reasonable.

Step 5 Report the solution.

It takes about 214 s for the drain to empty the sink.

Practice Problems

For problems 10–11, use the five steps. Assume that the rates of work do not change when people are working together.

10. A restaurant provides sandwiches each day for a U.S. Forest Service wildfire crew. Stan can make all the sandwiches by himself in 1 hr 10 min. Araceli can make the sandwiches by herself in 55 min. Find the time it will take Stan and Araceli working together to make the sandwiches. Round to the nearest minute.

11. A diaper service provides clean cotton diapers for its customers. Alaina washes, dries, sorts by size, folds, and stacks diapers. It takes 3 hr for her to fill the storage shelves with clean diapers. Chris takes diapers from the storage shelves and fills orders for delivery to customers. At the beginning of the day, the shelves are empty. Alaina works by herself for 1 hr, putting diapers on the shelves. Chris then comes to work and begins to take diapers off the shelves. After Chris and Alaina work together for 4 hr, the shelves are full. Find the rate at which Chris empties the shelves.

Other Applications

EXAMPLE 13 | The initial markup is the average markup (percent) required on all of a retailer's products so that selling the products covers the cost of all of the items and incidental expenses and returns a reasonable profit. The formula for initial markup, I, is $I = \dfrac{E + R + P}{F + R}$, where E is operating expenses, R is total price reductions, P is expected profit, and F is forecasted net sales. If a business is planning an initial markup of 50% on forecasted net sales of $60,000, operating expenses of $15,000, and an expected profit of $6000, find the amount of price reductions.

SOLUTION ▶ **Step 1 Understand the problem.**
The unknown is the amount of price reductions.

R = amount of price reductions

Step 2 Make a plan.
Use the formula $I = \dfrac{E + R + P}{F + R}$, where $I = 0.50$, $E = \$15{,}000$, $P = \$6000$, and $F = \$60{,}000$.

Step 3 Carry out the plan.

$$0.50 = \frac{\$15{,}000 + R + \$6000}{\$60{,}000 + R} \qquad I = \frac{E + R + P}{F + R}$$

$$(\$60{,}000 + R)(0.50) = \frac{(\cancel{\$60{,}000 + R})}{1}\left(\frac{\$15{,}000 + R + \$6000}{\cancel{\$60{,}000 + R}}\right) \quad \begin{array}{l}\text{Clear}\\\text{fractions.}\end{array}$$

$$(\$60{,}000 + R)(0.50) = \mathbf{\$15{,}000 + R + \$6000} \qquad \text{Simplify.}$$

$$(\$60{,}000)(\mathbf{0.50}) + R(\mathbf{0.50}) = \$15{,}000 + R + \$6000 \qquad \text{Distributive property.}$$

$$\mathbf{\$30{,}000} + 0.50R = R + \mathbf{\$21{,}000} \qquad \text{Simplify.}$$

$$\underline{-0.50R \quad -0.50R} \qquad \text{Subtraction property of equality.}$$

$$\mathbf{\$30{,}000 + 0} = \mathbf{0.50R} + \$21{,}000 \qquad \text{Simplify.}$$

$$\underline{-\$21{,}000 \qquad\qquad -\$21{,}000} \qquad \text{Subtraction property of equality.}$$

$$\mathbf{\$9000} = 0.50R + \mathbf{0} \qquad \text{Simplify.}$$

$$\frac{\$9000}{0.50} = \frac{\mathbf{0.50R}}{0.50} \qquad \text{Division property of equality.}$$

$$\$18{,}000 = R \qquad \text{Simplify.}$$

Step 4 Look back.
The ratio $\dfrac{\$15{,}000 + R + \$6000}{\$60{,}000 + R}$ should equal 50%. Since

$$\frac{\$15{,}000 + \mathbf{\$18{,}000} + \$6000}{\$60{,}000 + \mathbf{\$18{,}000}} = \frac{\$39{,}000}{\$78{,}000} \qquad \text{and} \qquad \frac{\$39{,}000}{\$78{,}000} = 0.5$$

the answer seems reasonable.

Step 5 Report the solution.
The price reductions equal $18,000.

Practice Problems

For problem 12, use the five steps.

12. The *capitalization ratio*, R, of a company is a percent in decimal form that is a measure of how much a company is leveraged. The formula is $R = \dfrac{D}{D + E}$, where D is the amount of long-term debt and E is the amount of stockholders' equity. The balance sheet of a company shows stockholders' equity of $13,302 million. Find the amount of long-term debt that the company can hold and have a capitalization ratio of 32%. Round to the nearest million.

5.4 VOCABULARY PRACTICE

Match the term with its description.

1. If $ab = 0$, then $a = 0$ or $b = 0$.

2. Division by zero

3. Two equations that may look different but that have the same solution

4. The product of the rate of work and the time worked

5. This compares two measurements with the same units.

6. This compares two measurements with different units.

7. An equation in which two rates are equal

8. $a(b + c) = ab + ac$

9. We can multiply both sides of an equation by any number except 0 without changing the solution of the equation.

10. In a rational equation, replacing the variable with this results in division by zero.

A. distributive property

B. equivalent equations

C. extraneous solution

D. fraction of a job completed

E. multiplication property of equality

F. proportion

G. rate

H. ratio

I. undefined

J. zero product property

5.4 Exercises

Follow your instructor's guidelines for showing your work.

For exercises 1–12, clear the fractions. Use the distributive property and combine like terms to simplify. Do not solve.

1. $\dfrac{1}{3}x + \dfrac{4}{21} = \dfrac{5}{8}$

2. $\dfrac{1}{5}y + \dfrac{3}{35} = \dfrac{2}{9}$

3. $\dfrac{1}{3x} + \dfrac{4}{21} = \dfrac{5}{8}$

4. $\dfrac{1}{5y} + \dfrac{3}{35} = \dfrac{2}{9}$

5. $\dfrac{1}{3x} + \dfrac{4}{21} = \dfrac{5}{8x^2}$

6. $\dfrac{1}{5y} + \dfrac{3}{35} = \dfrac{2}{9y^2}$

7. $\dfrac{2}{x - 5} + \dfrac{5}{x + 2} = \dfrac{3}{8}$

8. $\dfrac{4}{y - 9} + \dfrac{3}{y + 1} = \dfrac{5}{6}$

9. $\dfrac{2}{x - 5} - \dfrac{5}{x + 2} = \dfrac{3}{8}$

10. $\dfrac{4}{y-9} - \dfrac{3}{y+1} = \dfrac{5}{6}$

11. $\dfrac{2}{x-5} - \dfrac{5}{x+2} = \dfrac{3}{x^2-3x-10}$

12. $\dfrac{4}{y-9} - \dfrac{3}{y+1} = \dfrac{5}{y^2-8y-9}$

13. Explain the purpose of clearing the denominators from a rational equation.

14. Explain how to find the least common multiple of the denominators of a rational equation.

For exercises 15–62,
(a) solve.
(b) check.

15. $\dfrac{2}{5}x + 18 = -31$

16. $\dfrac{3}{8}y + 24 = -40$

17. $\dfrac{3}{10}p + \dfrac{3}{9} = \dfrac{2}{5}p$

18. $\dfrac{4}{9}w + \dfrac{3}{4} = \dfrac{1}{6}w$

19. $\dfrac{4}{7} + \dfrac{3}{x} = \dfrac{9}{14}$

20. $\dfrac{8}{9} + \dfrac{2}{f} = \dfrac{25}{18}$

21. $\dfrac{2}{k} + \dfrac{3}{5} = \dfrac{9}{10}$

22. $\dfrac{2}{x} + \dfrac{4}{5} = \dfrac{11}{12}$

23. $\dfrac{2}{c} - \dfrac{5}{6} = \dfrac{7}{c}$

24. $\dfrac{2}{a} - \dfrac{3}{7} = \dfrac{11}{a}$

25. $\dfrac{1}{z} - \dfrac{5}{z} = \dfrac{1}{9}$

26. $\dfrac{1}{y} - \dfrac{3}{y} = \dfrac{1}{8}$

27. $\dfrac{6}{5x} - \dfrac{1}{x} = 12$

28. $\dfrac{4}{9x} - \dfrac{1}{x} = 15$

29. $\dfrac{1}{w} - \dfrac{8}{w^2} = -\dfrac{1}{49}$

30. $\dfrac{1}{p} - \dfrac{5}{p^2} = \dfrac{1}{36}$

31. $\dfrac{1}{2} - \dfrac{4}{p^2} = \dfrac{7}{2p}$

32. $\dfrac{1}{3} - \dfrac{5}{y^2} = \dfrac{2}{3y}$

33. $\dfrac{3}{a+4} = \dfrac{6}{a-8}$

34. $\dfrac{5}{d+9} = \dfrac{7}{d+3}$

35. $\dfrac{2}{x+4} + \dfrac{4}{x} = \dfrac{x}{x+4}$

36. $\dfrac{2}{h+10} + \dfrac{4}{h} = \dfrac{h}{h+10}$

37. $\dfrac{1}{a-1} + \dfrac{3}{a} = \dfrac{a}{a-1}$

38. $\dfrac{1}{c-1} + \dfrac{5}{c} = \dfrac{c}{c-1}$

39. $\dfrac{7}{x-3} = \dfrac{x}{x-3} - \dfrac{3}{x}$

40. $\dfrac{3}{x+2} = \dfrac{x}{x+2} - \dfrac{4}{x}$

41. $\dfrac{6}{m-6} = \dfrac{m}{m-6} - \dfrac{4}{m}$

42. $\dfrac{2}{b-2} + \dfrac{8}{b} = \dfrac{b}{b-2}$

43. $\dfrac{7}{w+2} - \dfrac{20}{w} = \dfrac{w}{w+2}$

44. $\dfrac{y}{y+12} + \dfrac{4}{y} = \dfrac{-10}{y+12}$

45. $\dfrac{3c}{c+2} = -\dfrac{3}{c+2} + \dfrac{10}{c}$

46. $\dfrac{2z}{z-8} = -\dfrac{6}{z-8} - \dfrac{5}{z}$

47. $\dfrac{8n}{n+1} = 6 - \dfrac{8}{n+1}$

48. $\dfrac{6v}{v-2} = 5 + \dfrac{12}{v-2}$

49. $\dfrac{3p}{p^2-9} + \dfrac{1}{p-3} = \dfrac{3}{p-3}$

50. $\dfrac{7q}{q^2-36} + \dfrac{2}{q-6} = \dfrac{8}{q-6}$

51. $\dfrac{9}{x+4} = \dfrac{6}{x-4} + \dfrac{2x}{x^2-16}$

52. $\dfrac{10}{y+7} = \dfrac{8}{y-7} + \dfrac{3y}{y^2-49}$

53. $\dfrac{8}{x+4} = \dfrac{6}{x-4} + \dfrac{2x}{x^2-16}$

54. $\dfrac{11}{y+7} = \dfrac{8}{y-7} + \dfrac{3y}{y^2-49}$

55. $\dfrac{x}{x+1} + \dfrac{x}{x+4} = \dfrac{x^2+7x+3}{x^2+5x+4}$

56. $\dfrac{x}{x+3} + \dfrac{x}{x+2} = \dfrac{x^2+3x+3}{x^2+5x+6}$

57. $\dfrac{3x}{x-1} + \dfrac{3x}{x-4} = \dfrac{3x^2 - 11x - 1}{x^2 - 5x + 4}$

58. $\dfrac{2x}{x-3} + \dfrac{2x}{x-2} = \dfrac{2x^2 - 3x - 3}{x^2 - 5x + 6}$

59. $\dfrac{x}{x-3} - \dfrac{2}{x+3} = \dfrac{2x^2 - 2x + 2}{x^2 - 9}$

60. $\dfrac{x}{x-2} - \dfrac{3}{x+2} = \dfrac{2x^2 + x + 3}{x^2 - 4}$

61. $\dfrac{x}{x+3} = \dfrac{2x^2 + 3x + 7}{x^2 + 3x} - \dfrac{5}{x}$

62. $\dfrac{x}{x+5} = \dfrac{2x^2 + 3x + 18}{x^2 + 5x} - \dfrac{4}{x}$

63. A rational equation has no variables in the denominator.
   a. Can this equation have extraneous solutions?
   b. Explain.

64. A rational equation has one variable in the denominator.
   a. Can this equation have extraneous solutions?
   b. Explain.

For exercises 65–74, use the five steps from Section 1.5 and a proportion.

65. The Seattle Baby Diaper Service adds bleach to the wash water for dirty diapers. It uses two cups of bleach for 250 lb of diapers. Find the amount of bleach needed to wash 1350 lb of diapers.

© William McKellar/Getty Images

66. In 2009, seven out of ten daycare centers in Alabama were unlicensed. Predict how many of 135 daycare centers were licensed. Round to the nearest whole number. (*Source:* www.arc.org, April 2009)

For exercises 67–68, a digester at the Top Deck Dairy in Iowa decomposes about 17,000 gal of cow manure *each day*. This decomposition process produces methane gas. The methane gas is sent through a pipe to an electricity-generating plant. *Each year*, this process produces 864,000 kilowatt·hour of electricity. This is enough to meet the energy needs of 100 homes. (*Source:* www.alliantenergy.com)

67. Find the number of gallons of cow manure needed to produce 100,000 kilowatt·hour of electricity. Round to the nearest thousand.

68. Find the number of gallons of cow manure needed to meet the daily energy needs of 275 homes. Round to the nearest thousand.

69. Predict the number of the 86.7 million uninsured individuals who were uninsured for nine months or more. Round to the nearest million.

86.7 million people under the age of 65 went without health insurance for some or all of the two-year period from 2007 to 2008. One out of three people . . . under the age of 65 were uninsured for some or all of 2007–2008. Of the 86.7 million uninsured individuals, three in five . . . were uninsured for nine months or more. Nearly three-quarters . . . were uninsured for six months or more. (*Source:* www.familiesusa.org, March 2009)

70. Predict the number of babies born in 2007 whose mothers drank alcohol while pregnant. Round to the nearest thousand.

The preliminary estimate of births in 2007 rose 1 percent to 4,317,119, the highest number of births ever registered for the United States. (*Source:* www.cdc.gov, March 18, 2009)

The number of women who drink alcohol while pregnant has not changed substantially over time, according to a 15-year study by the Centers for Disease Control and Prevention. Approximately 1 in 8 women drank alcohol while pregnant, the study says. Alcohol consumption during pregnancy is a risk factor for poor birth outcomes including fetal alcohol syndrome, birth defects, and low birth weight. (*Source:* www.cdc.gov, June 3, 2009)

71. At a community college, 8032 students who registered for spring classes did not have health insurance. This represented two out of five students who registered for spring classes. Find the number of students who registered for spring classes.

72. According to the Centers for Disease Control and Prevention, a total of 3,397,615 deaths occurred in the United States in 2004. Predict how many of these deaths were caused by injuries. Round to the nearest thousand.

Injuries caused 1 out of 14 deaths in the United States in 2004 (1), including 3 out of 4 deaths among adolescents and young adults. In 2005, one in nine people in the United States sought medical attention for an injury. (*Source:* www .cdc.gov, March 2008)

73. According to the American Red Cross, four out of five Americans do not know that home fires are the most common disaster in the United States. Use this information to write a word problem that can be solved using a proportion. (*Source:* www.crossnet.org, Feb. 13, 2006)

74. According to a 2009 survey, three out of five employers maintained 401(k) retirement plans for their employees, despite the economic crisis. Use this information to write a word problem that can be solved using a proportion. (*Source:* www.worldatwork.org, March 17, 2009)

For exercises 75–84, use the five steps. Assume that the rates of work do not change when people are working together.

75. Charles can weed an entire garden in 50 min. Annie can weed the same garden in 35 min. If they work together, find the time in minutes to do this job. Round to the nearest whole number.

76. Miguel can restock all the vending machines in the student union in 35 min. Carl can do the same job in 48 min. If they work together, find the time in minutes to do this job. Round to the nearest minute.

77. Nina and Abbie are graders for an anatomy class. Nina can grade a set of exams in 3 hr. Nina and Abbie start grading a set of exams. Abbie works for 20 min. She then leaves to proctor an exam for a professor. It takes Nina 1 hr 20 min more to finish grading the set of exams. Find the time in hours and minutes for Abbie to grade a set of exams. Round to the nearest minute.

78. Jonas and Ehsan are graders for a psychology class. Jonas can grade a set of psychology exams in 3 hr. He grades a set of exams for 25 min by himself. Then Ehsan arrives and works with Jonas until the set of exams is graded. It takes them 70 more minutes to finish. Find the time in hours and minutes for Ehsan to grade a set of exams. Round to the nearest minute.

79. An investment adviser produces a monthly newsletter. She can print all of the newsletters on her personal laser printer in 80 min. She can print all of the newsletters on the network office printer in 45 min. If she uses both printers, find the time in minutes to print the newsletters. Round to the nearest whole number.

80. A mother and a grandmother are quilting the top of a queen bedspread for a wedding gift. The mother can quilt a queen bedspread in about 50 hr. The grandmother can quilt the same top in about 65 hr. If they work together, find the time in hours and minutes to do this job. Round to the nearest minute.

81. Bob opens the drain on his hot tub. The tub empties. Bob cleans the tub. He puts the garden hose in the tub and turns it on. It normally takes 75 min for the hose to fill the tub. After 15 min, Bob realizes that he forgot to close the drain. After he goes outside and closes the drain, it takes an additional 68 min to fill the tub. Find the drain rate of the tub. Round to the nearest minute.

82. A garden hose fills a hot tub in 75 min. It takes 2 hr 10 min for the tub to drain. Bob opens the drain of the tub. The tub empties. Bob cleans the tub. He puts the garden hose in the tub and turns it on. Since Bob did not close the drain, the hose is pouring water into the tub, and the drain is letting water drain out. Find the time in minutes that it takes the tub to fill with the drain open. Round to the nearest whole number.

83. Pat and Chris are washing cars at a car dealership. Pat can wash all of the cars on the lot in 4 hr. Pat works by himself for 40 min. Chris then joins him. They work together for 90 min to finish washing all of the cars. Find the time in minutes for Chris to wash all of the cars. Round to the nearest minute.

84. A restaurant makes fresh salsa daily using fresh tomatoes, peppers, onions, cilantro, and spices. Azar can prepare all the vegetables needed for the salsa in 90 min. Azar works by herself for 35 min. Ed then joins her. They work together for 30 more min to finish the job. Find the time in minutes for Ed to prepare all of the vegetables. Round to the nearest whole number.

For exercises 85–86,
(a) use the information to write a word problem.
(b) solve the problem.

85. Ben can dust the entire house in 1 hr 15 min. Sarah can dust the entire house in 38 min.

86. Anders can shovel the entire driveway after a snowfall of 4 in. in 65 min. Hans can shovel the entire driveway after a snowfall of 4 in. in 80 min.

For exercises 87–88, different materials have different *resistances* to the movement of electrons. In a parallel circuit with two resistors, the total resistance is calculated using the formula $R_T = \dfrac{R_1 \cdot R_2}{R_1 + R_2}$. The total resistance in the circuit is R_T, the resistance of one resistor is R_1, and the resistance of the other resistor is R_2. The unit of measurement of resistance is ohms. These problems are common in introductory college electronics classes and physics classes.

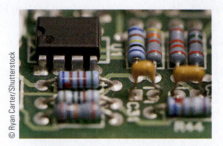

© Ryan Carter/Shutterstock

87. A parallel circuit has a total resistance of 5 ohms. One of the resistors in the circuit is 6 ohms. Find the resistance of the other resistor.

88. A parallel circuit has a total resistance of 6 ohms. One of the resistors in the circuit is 15 ohms. Find the resistance of the other resistor.

For exercises 89–90, a thin converging lens bends light rays. Each light ray passes through the focal point. The thin lens equation shows the relationship of O (the distance of the object from the lens), I (the distance of the image from the lens), and f (the focal length): $\dfrac{1}{O} + \dfrac{1}{I} = \dfrac{1}{f}$. These problems are common in introductory college physics classes.

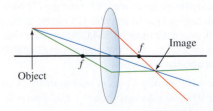

89. The focal length of a converging lens is 24.6 cm. The object is 61.1 cm from the lens. Find the distance of the image from the lens. Round to the nearest tenth.

90. The focal length of a converging lens is 24.6 cm. The object is 34.2 cm from the lens. Find the distance of the image from the lens. Round to the nearest tenth.

For exercises 91–92, baseball team managers use the On-Base Percentage O to compare player performance. In the formula $O = \dfrac{H + W + P}{A + W + H + S}$, the number of hits is H, the number of walks is W, the number of times the player was hit by a pitch is P, the number of times the player was at bat is A, and the number of sacrifice fly balls hit by the player is S. O is rounded to the nearest thousandth.

91. In six seasons, Albert Pujols of the St. Louis Cardinals had 3489 at-bats. His on-base percentage was 0.419. He walked 493 times. He was hit by a pitch 48 times. He hit 31 sacrifice flies. Find his number of hits. Round to the nearest whole number.

92. In eight seasons, Lance Berkman of the Houston Astros had 3687 at-bats. His on-base percentage was 0.416. He walked 690 times. He was hit by a pitch 45 times. He hit 36 sacrifice flies. Find his number of hits. Round to the nearest whole number.

For exercises 93–94, an NBA offensive statistic is Total Field Shooting Percentage, P. In the formula $P = \dfrac{F_m + T_m}{F_A + T_A}$, the total field shooting percentage is P, the number of field goals (2 points) made is F_m, the number of three-pointers made is T_m, the number of field goals attempted is F_A, and the number of three-pointers attempted is T_A.

93. In 11 seasons, Kobe Bryant had a total field shooting percentage of 0.434. He attempted 14,019 field goals and made 6343 field goals. He made 885 three-pointers. How many three-pointers did he attempt? Round to the nearest whole number.

94. In 15 seasons, Shaquille O'Neal had a total field shooting percentage of 0.579. He attempted 17,057 field goals. He made 1 three-pointer and attempted 20 three-pointers. How many field goals did he make? Round to the nearest whole number.

Problem Solving: Practice and Review

Follow your instructor's guidelines for using the five steps as outlined in Section 1.5, p. 51.

95. Find the percent decrease in the number of tuberculosis cases per 100,000 people. Round to the nearest tenth of a percent.

There were an estimated 9.27 million new cases of tuberculosis worldwide in 2007. . . . Although this figure represents an increase from 9.24 million in 2006, the world population has also grown, making the number of cases per capita a more useful measure of the problem; this figure peaked in 2004 at 142 per 100,000 and fell to 139 per 100,000 in 2007. (*Source:* www.nejm.org, June 4, 2009)

96. Dr. Heather Henson-Ramsey studied the effect of the insecticide malathion on earthworms. In this research, she measured how rapidly malathion breaks down in soil. The model $y = \left(\dfrac{-0.2953 \text{ ppm}}{1 \text{ hr}}\right)x + 34.78 \text{ ppm}$ describes the concentration in parts per million of malathion in soil, y, after x hr. (*Source:* Henson-Ramsey et al., Development of a Dynamic Pharmacokinetic Model to Estimate Bioconcentration of 2 Xenobiotics in Earthworms, *Environmental Modeling and Assessment*, 2007)

Courtesy Heather Henson-Ramsey

a. State the slope of this model. Include units of measurement. Describe what the slope represents.

b. State the y-intercept of this model. Include units of measurement. Describe what the y-intercept represents.

c. Find the concentration of malathion in the soil after 3 hr. Round to the nearest hundredth.

97. The five causes of death in the graph "accounted for 81% of all deaths from injuries in 2005." Find the number of deaths from injuries in 2005 that were not a result of these causes. Round to the nearest hundred.

U.S. Top Causes of Death from Injuries in 2005

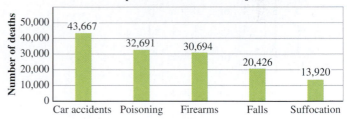

Source: www.usatoday.com, April 2, 2008

98. In fiscal year 2008, 720 million pounds of trash and 12,050 tons of recyclable material were collected in Ada County, Idaho. This was about a 16.3% decrease in trash and a 3.3% increase in recyclable material compared to fiscal year 2007. Find the total amount of trash and recyclable material collected in fiscal year 2007. Round to the nearest tenth of a million pounds (1 ton = 2000 pounds). (*Source:* Ada County Annual Report, Fiscal Year 2007–2008)

Find the Mistake

For exercises 99–102, the completed problem has one mistake.

(a) Describe the mistake in words, or copy down the whole problem and highlight or circle the mistake.

(b) Do the problem correctly.

99. **Problem:** Solve: $\dfrac{3}{x-4} = \dfrac{x+1}{x+4}$

Incorrect Answer:

$$\frac{3}{x-4} = \frac{x+1}{x+4}$$

$$\frac{3}{x-4} = \frac{\cancel{x}+1}{\cancel{x}+4}$$

$$\frac{3}{x-4} = \frac{1}{4}$$

$$4(x-4)\frac{3}{x-4} = 4(x-4)\frac{1}{4}$$

$$12 = x-4$$

$$16 = x$$

100. **Problem:** Solve: $\dfrac{-2x}{x-1} + \dfrac{x+3}{x^2-1} = 1$

Incorrect Answer:

$$\frac{(x-1)(x+1)}{1}\left(\frac{-2x}{x-1}\right) +$$

$$(x-1)(x+1)\left(\frac{x+3}{(x-1)(x+1)}\right) = \frac{(x-1)(x+1)}{1}(1)$$

$$-2x(x+1) + (x+3) = (x-1)(x+1)$$

$$-2x^2 - 2x + x + 3 = x^2 - 1$$

$$0 = 3x^2 + x - 4$$

$$0 = (3x+4)(x-1)$$

$$\begin{array}{ccc} 0 = 3x+4 & \text{or} & 0 = x-1 \\ -4 \quad\quad -4 & & +1 \quad\quad +1 \\ \hline \dfrac{-4}{3} = \dfrac{3x}{3} & \text{or} & 1 = x \\ x = -\dfrac{4}{3} & \text{or} & x = 1 \end{array}$$

101. **Problem:** Jose can muck out the barn in 50 min. Rob can muck out the barn in 44 min. Find the time in minutes for them to do the job working together. Assume that they work at the same speed alone or together.

Incorrect Answer: $t =$ time to muck out the barn

$$t = \frac{1}{2}(50 \text{ min}) + \frac{1}{2}(44 \text{ min})$$

$$t = 25 \text{ min} + 22 \text{ min}$$

$$t = 47 \text{ minutes}$$

102. **Problem:** Clear the fractions from $\dfrac{2}{x-2} - \dfrac{5}{x+2} = 1$.

Incorrect Answer:

$$\frac{2}{x-2} - \frac{5}{x+2} = 1$$

$$\frac{(x-2)(x+2)}{1} \cdot \frac{2}{x-2} - \frac{(x-2)(x+2)}{1} \cdot \frac{5}{x+2} = 1$$

$$2(x+2) - 5(x-2) = 1$$

$$2x + 4 - 5x + 10 = 1$$

$$-3x + 14 = 1$$

Review

103. Use interval notation to represent the domain of $f(x) = |x+5| - 9$.

104. Use interval notation to represent the domain of $g(x) = x^2 + 5x + 6$.

105. If $f(x) = 4x^2 - 7x - 3$, find $f(-2)$.

106. Use an algebraic method to find any real zeros of $f(x) = x^2 - 6x + 8$.

SUCCESS IN COLLEGE MATHEMATICS

107. To clear fractions from a linear equation in one variable, we multiply both sides of the equation by a common denominator of all of the fractions in the equation. Explain why multiplying both sides of the equation by a common denominator clears the fractions.

108. When we clear the fractions from a rational equation that includes at least one variable in a denominator and then solve the remaining equation, a solution may be *extraneous*. Explain why extraneous solutions can occur when solving a rational equation.

© Denis Tabler/Shutterstock

Thomas Young (1773–1829) is famous for his contributions to mathematics, physics, and medicine. He developed a rule of thumb for medication dosages for children that is an example of a rational function, $C(x) = \dfrac{Ax}{x + 12}$. In this section, we will study this and other rational functions.

SECTION 5.5 Rational Functions

After reading the text, working the practice problems, and completing assigned exercises, you should be able to:

1. Use a number line graph to represent the domain of a rational function.
2. Use interval notation to represent the domain of a rational function.
3. Write the equation of the vertical asymptote(s) of a rational function.
4. Predict where a hole will appear in the graph of a rational function.
5. Evaluate a rational function.
6. Use an algebraic method to find any real zeros of a rational function.

The Domain of a Rational Function

The domain of a function is the set of its input values. For some rational functions, the domain is a **union** of two sets. The notation for union is ∪.

EXAMPLE 1 What set does the interval represent? Graph the set on a number line graph.

(a) $(-\infty, \infty)$

SOLUTION ▶ The set of real numbers.

<—+—+—◄—+—+—+—►—+—+—► x
−4 −3 −2 −1　0　1　2　3　4

(b) $[2, \infty)$

▶ The set of real numbers greater than or equal to 2.

<—+—+—[—+—+—►—+—+—► x
　0　1　2　3　4　5　6　7

(c) $(-\infty, 0)$

▶ The set of real numbers less than 0.

<—+—+—◄—+—+—)—+—+—► x
−5 −4 −3 −2 −1　0　1　2

(d) $(-\infty, 0) \cup (0, \infty)$

▶ The set of real numbers less than 0 and the set of real numbers greater than 0. Zero is not an element in the set.

<—+—+—◄—+—+—✕—+—►—+—+—► x
−5 −4 −3 −2 −1　0　1　2　3　4　5

The function rule of a rational function is the quotient of two polynomials: $f(x) = \dfrac{\text{polynomial}}{\text{polynomial}}$. If the polynomial in the denominator includes a variable, the domain of the function may need to be **restricted** to prevent division by zero.

EXAMPLE 2 **(a)** Use a number line graph to represent the domain of $f(x) = \dfrac{1}{x}$.

SOLUTION ▶ Since an input of 0 causes division by 0 and division by 0 is undefined, 0 must be restricted from the domain.

$$\xleftarrow{\hspace{0.3cm}}\overset{\displaystyle\times}{\underset{-5\ -4\ -3\ -2\ -1\quad 0\quad 1\ 2\ 3\ 4\ 5}{\rule{5cm}{0pt}}}\xrightarrow{\hspace{0.3cm}} x$$

(b) Use interval notation to represent the domain of $f(x) = \dfrac{1}{x}$.

▶ $(-\infty, 0) \cup (0, \infty)$

In the next example, the polynomial in the denominator has two factors. Two input values result in a factor that equals 0. Since division by zero is undefined, these input values must be restricted from the domain.

EXAMPLE 3 **(a)** Use a number line graph to represent the domain of $f(x) = \dfrac{1}{x^2 - 5x - 6}$.

SOLUTION ▶ $f(x) = \dfrac{1}{x^2 - 5x - 6}$

$f(x) = \dfrac{1}{(x - 6)(x + 1)}$ Factor the denominator.

$(x - 6)(x + 1) = 0$ Identify input values that result in division by 0.

$x - 6 = 0$ or $x + 1 = 0$ Zero product property.

$\underline{+6\ \ +6}$ $\underline{-1\ \ -1}$ Properties of equality.

$x + 0 = 6$ or $x + 0 = -1$ Simplify.

$x = 6$ or $x = -1$ These input values cause division by 0.

The domain is the set of real numbers *except* -1 and 6.

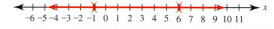

(b) Use interval notation to represent the domain of $f(x) = \dfrac{1}{x^2 - 5x - 6}$.

▶ $(-\infty, -1) \cup (-1, 6) \cup (6, \infty)$

Practice Problems

For problems 1–3,
(a) use a number line graph to represent the domain of the function.
(b) use interval notation to represent the domain of each function.

1. $f(x) = \dfrac{5}{x - 2}$ **2.** $h(x) = \dfrac{x}{x^2 - 5x - 24}$

3. $y = \dfrac{6x + 7}{x^2 - x - 6} + \dfrac{5}{x + 2}$

The Graph of a Rational Function

To prevent division by zero, one or more real numbers may be restricted from the domain of a rational function. The graph of a rational function is affected by these restrictions.

In the next example, we study the rational function $y = \dfrac{-5}{x+6}$. The domain of a function is the set of the inputs of the function. Since an input value of -6 results in division by 0, this value is restricted from the domain. To better understand this rational function, we look at the input values that are close to $x = -6$ and their outputs.

EXAMPLE 4 | A rational function is $y = \dfrac{-5}{x+6}$.

(a) Use interval notation to represent the domain.

SOLUTION ▶ If $x = -6$, the denominator is equal to 0. Because division by 0 is undefined, restrict -6 from the domain. The domain is $(-\infty, -6) \cup (-6, \infty)$.

(b) Describe what happens to the output values before and after the restricted input value, $x = -6$.

x (input value)	y (output value)	Numerator of $\dfrac{-5}{x+6}$	Denominator of $\dfrac{-5}{x+6}$
-6.3	16.7	-5	-0.3
-6.1	50	-5	-0.1
-6.001	5000	-5	-0.001
-6.0001	50,000	-5	-0.0001
-6	undefined	-5	0
-5.9999	$-50,000$	-5	0.0001
-5.999	-5000	-5	0.001
-5.9	-50	-5	0.1
-5.7	-16.7	-5	0.3

Some values are rounded.

▶ The numerator of the output value is always -5. In the table of ordered pairs, we see that as the input value, x, slowly increases and approaches -6, the denominator is also a negative number, approaching 0. Since the quotient of two negative numbers is positive, the output value is positive. Since the numerator is always -5 and the denominator is approaching 0, the output value is increasing rapidly and is very very positive.

When the input value is just greater than -6, the numerator is still a negative number but the denominator is now positive. Since the quotient of a negative and a positive number is a negative number, the output value is now very very negative.

(c) The graph represents $y = \dfrac{-5}{x+6}$. Write the equation of the vertical asymptote of the graph.

As the input value, x, increases and approaches the restricted input value of -6, the graph of the function approaches but never intersects the vertical line represented by the equation $x = -6$. This line is a **vertical asymptote**. As the input value, x, decreases and approaches the restricted input value $x = -6$, the graph again approaches but never intersects the same vertical line. The equation of the vertical asymptote of this graph is $x = -6$.

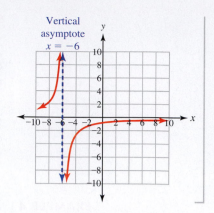

In the next example, the graph again has a vertical asymptote. Unlike the function in Example 4, there is no "jump" from a very very large positive output value to a very very negative output value. In the table of ordered pairs, we see that the output values are very very negative both before and after the restricted input value, $x = 4$.

EXAMPLE 5 A rational function is $y = \dfrac{x^2 - 2x - 15}{x^2 - 8x + 16}$.

(a) Use interval notation to represent the domain.

SOLUTION ▸ To find input values that result in division by 0, factor the denominator:

$$y = \frac{x^2 - 2x - 15}{(x - 4)(x - 4)}$$

If $x = 4$, the denominator is equal to 0. Because division by 0 is undefined, restrict 4 from the domain. The domain is $(-\infty, 4) \cup (4, \infty)$.

(b) Describe what happens to the output values before and after the restricted input value, $x = 4$.

x (input value)	$y = \dfrac{x^2 - 2x - 15}{(x - 4)(x - 4)}$ (output value)	Numerator of $\dfrac{x^2 - 2x - 15}{(x - 4)(x - 4)}$	Denominator of $\dfrac{x^2 - 2x - 15}{(x - 4)(x - 4)}$
3.7	-96.8	-8.71	0.09
3.8	-204	-8.16	0.04
3.9	-759	-7.59	0.01
3.99	$-71,000\dots$	-7.06	0.0001
4	undefined	-7	0
4.01	$-69,000$	-6.94	0.0001
4.1	-639	-6.39	0.01
4.2	-144	-5.76	0.04
4.3	-56.8	-5.11	0.09

Some values are rounded.

▸ As the input, x, increases and approaches the restricted input value of 4, the numerator is a negative number. The denominator is a positive number that is approaching 0. Since the quotient of a negative number and a positive number is negative, the output value, y, is a negative number. As the denominator approaches 0, the value of the output value, y, becomes very very negative.

When the input value is just greater than the restricted input value, 4, the numerator is still a negative number and the denominator is still a positive number. The value of the fraction continues to be very very negative.

(c) The graph represents $y = \dfrac{x^2 - 2x - 15}{x^2 - 8x + 16}$. Write the equation of the vertical asymptote of the graph.

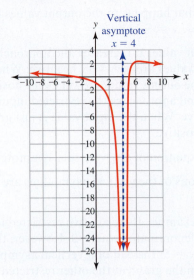

▶ As the input value, x, increases and approaches the restricted input value of $x = 4$, the graph of the function approaches but never intersects the vertical line represented by the equation $x = 4$. As the input value, x, decreases and approaches the restricted input value $x = 4$, the graph again approaches but never intersects the same vertical line. The equation of the **vertical asymptote** is $x = 4$.

If the domain of a function has two restrictions, its graph may have two vertical asymptotes.

EXAMPLE 6 The graph represents $y = \dfrac{6x + 7}{x^2 - x - 6}$.

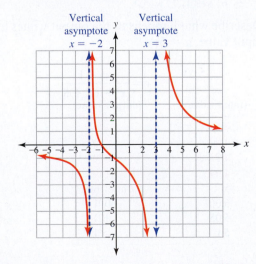

(a) Use interval notation to represent the domain.

SOLUTION ▶ To find input values that result in division by 0, factor the denominator:

$$y = \frac{6x + 7}{(x - 3)(x + 2)}$$

If $x = 3$ or $x = -2$, the denominator is equal to 0. Because division by 0 is undefined, restrict -2 and restrict 3 from the domain. The domain is $(-\infty, -2) \cup (-2, 3) \cup (3, \infty)$.

(b) Describe what happens to the output values before and after the restricted input values, $x = -2$ and $x = 3$.

▶ As the input value, x, increases and approaches the restricted input value -2, the output values are very very negative. When the input value is just greater than -2, the output values are very very positive. As the input value, x, increases and approaches the restricted input value 3, the output values are very very negative. When the input value is just greater than 3, the output values are very very positive.

(c) Write the equation of any vertical asymptotes.

▶ The equations of the vertical asymptotes are $x = -2$ and $x = 3$.

In the next example, there are two restricted input values in the domain. If $x = 1$ or $x = 5$, the denominator equals 0. However, there is only one vertical asymptote on the graph, $x = 5$. Instead of a vertical asymptote, there is a "hole" (also called a *discontinuity*) in the graph at the other restricted input value, $x = 1$.

EXAMPLE 7 A rational function is $y = \dfrac{x^2 - 1}{x^2 - 6x + 5}$.

(a) Use interval notation to represent the domain.

SOLUTION ▶ To find input values that result in division by 0, factor the denominator:

$$y = \frac{x^2 - 1}{(x - 5)(x - 1)}$$

If $x = 5$ or $x = 1$, the denominator is equal to 0. Because division by 0 is undefined, restrict 5 and restrict 1 from the domain. The domain is $(-\infty, 1) \cup (1, 5) \cup (5, \infty)$.

(b) Describe what happens to the output values before and after the restricted input value, $x = 5$.

x (input value)	$y = \dfrac{x^2 - 1}{x^2 - 6x + 5}$ (output value)
4.5	−11
4.9	−59
4.95	−119
4.995	−1199
5	undefined
5.001	6001
5.05	121
5.1	61

▶ As the input value, x, increases and gets closer to 5, the outputs get very very negative. The input $x = 5$ is restricted from the domain. When the input values are just greater than 5, the output values are very very large positive numbers. On the graph, we will see a vertical asymptote at this input value.

(c) Describe what happens to the output values before and after the restricted input value, $x = 1$.

x (input value)	$y = \dfrac{x^2 - 1}{x^2 - 6x + 5}$ (output value)
0.7	$-0.39...$
0.8	$-0.42...$
0.9	$-0.46...$
0.95	$-0.48...$
1	undefined
1.1	$-0.53...$
1.2	$-0.57...$
1.3	$-0.62...$

▶ As the input value, x, increases and gets closer to 1, the outputs get more negative. However, when the input values are just greater than 1, the output values are only slightly more negative.

The only effect on the graph of restricting 1 from the domain is the hole in the graph. This happens because there cannot be a point with an x-coordinate of 1, since an input of 1 in this function results in division by zero.

(d) The graph represents $y = \dfrac{x^2 - 1}{x^2 - 6x + 5}$. Write the equation of the vertical asymptote of the graph, and identify the location of the hole on the graph.

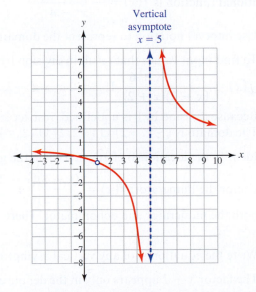

▶ As the input value, x, increases and approaches the restricted input value of $x = 5$, the graph of the function approaches but never intersects the vertical line represented by the equation $x = 5$. As the input value, x, decreases and approaches $x = 5$, the function approaches but never intersects the same vertical line. The equation of the vertical asymptote is $x = 5$.

As the input value, x, increases and approaches the restricted input value of $x = 1$, the output values are not very very negative or very very positive. However, since there is no point with an input value, x, of 1, there is a **hole** in the graph at $x = 1$.

To predict whether the graph will have a hole or an asymptote associated with an input value restricted from the domain, compare the numerators and denominators. If we factor the numerator as well as the denominator, the function is $y = \dfrac{(x - 1)(x + 1)}{(x - 5)(x - 1)}$. Since the factor $x - 1$ is in both the numerator and the denominator, there is a hole in the graph at the restricted input value $x = 1$. Since the factor $x - 5$ does not appear in the numerator, there is a vertical asymptote in the graph at the restricted input value $x = 5$.

Identifying Vertical Asymptotes and Holes in Rational Functions

1. Factor the numerator and denominator.

2. Identify input values that are restricted from the domain because they result in division by 0.

3. If the same factor is in the numerator at least as many times as it is in the denominator, there will be a *hole* in the graph at the input value that results in this factor being equal to 0.

4. If the factor is found in the denominator more times than in the numerator, there will be a *vertical asymptote* in the graph at the input value that results in this factor being equal to 0.

EXAMPLE 8 A rational function is $f(x) = \dfrac{x^2 + 5x + 6}{x^2 - 4}$.

(a) Use interval notation to represent the domain.

SOLUTION ▶ To find input values that result in division by 0, factor the denominator:

$f(x) = \dfrac{x^2 + 5x + 6}{(x - 2)(x + 2)}$. If $x = 2$ or $x = -2$, the denominator is equal to 0.

Because division by 0 is undefined, restrict 2 and restrict -2 from the domain. The domain is $(-\infty, -2) \cup (-2, 2) \cup (2, \infty)$.

(b) Identify the location(s) of any hole(s) on the graph.

▶ Factor the numerator: $f(x) = \dfrac{(x + 3)(x + 2)}{(x - 2)(x + 2)}$. The factor $x + 2$ is present in both the numerator and denominator. There is a hole at the restricted value $x = -2$.

(c) Write the equation(s) of any vertical asymptote(s) on the graph.

▶ The factor $x - 2$ appears only in the denominator. The equation of the vertical asymptote is $x = 2$.

Practice Problems

For problems 4–8,
(a) use interval notation to represent the domain of the function.
(b) write the equation of any vertical asymptotes on the graph.
(c) identify the location of any holes on the graph.

4.

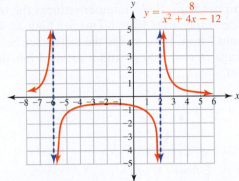

$$y = \frac{8}{x^2 + 4x - 12}$$

5.

$$y = \frac{5x - 15}{x^2 + 4x - 21}$$

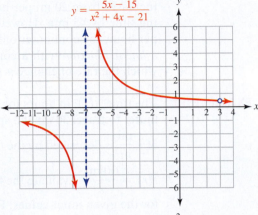

6.

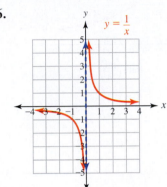

$$y = \frac{1}{x}$$

7. $f(x) = \dfrac{x^2 - 1}{x^2 + 9x + 14}$

8. $f(x) = \dfrac{x^2 - 1}{x^2 + 9x + 8}$

Evaluating a Rational Function

To evaluate a rational function, replace the variable with the input value. If the input value causes division by zero, the function is undefined.

EXAMPLE 9   A function is $f(x) = \dfrac{1}{x}$.

(a) Evaluate: $f\left(\dfrac{1}{200}\right)$

SOLUTION ▶ $f\left(\dfrac{1}{200}\right) = \dfrac{1}{\dfrac{1}{200}}$ Replace x with $\dfrac{1}{200}$.

$f\left(\dfrac{1}{200}\right) = 1 \div \dfrac{1}{200}$ Rewrite fraction as division.

$f\left(\dfrac{1}{200}\right) = (1)\left(\dfrac{200}{1}\right)$ Multiply by reciprocal of divisor.

$f\left(\dfrac{1}{200}\right) = 200$ Simplify.

(b) Evaluate: $f(0)$

▶ $f(0) = \dfrac{1}{0}$ Undefined.

In the next example, the rational function is a model of the relationship between the rate (speed) of a commuter and the time needed to travel 30 mi. If the speed increases from 20 mi per hour to 40 mi per hour, the commuter reduces the traveling time by 45 min. However, if the speed increases from 60 mi per hour to 80 mi per hour, the commuter only saves about 7 min. Here we see the difference between a *linear function* with a constant average rate of change and a *rational function* in which the rate of change is not constant.

EXAMPLE 10

The time, $t(r)$, in hours for a commuter to travel 30 mi at a speed of $r \dfrac{\text{mi}}{\text{hr}}$ is described by the function $t(r) = \dfrac{30 \text{ mi}}{r}$. Evaluate the function for the given input values. Round to the nearest hundredth.

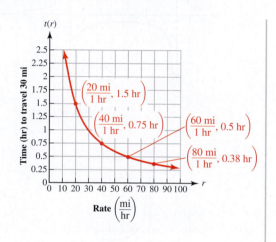

(a) Evaluate: $t\left(\dfrac{20 \text{ mi}}{1 \text{ hr}}\right)$

SOLUTION ▶ $t\left(\dfrac{\textbf{20 mi}}{\textbf{1 hr}}\right) = \dfrac{30 \text{ mi}}{\dfrac{20 \text{ mi}}{1 \text{ hr}}}$

$t\left(\dfrac{20 \text{ mi}}{1 \text{ hr}}\right) = \textbf{1.5 hr}$ (90 min)

(b) Evaluate: $t\left(\dfrac{40 \text{ mi}}{1 \text{ hr}}\right)$

▶ $t\left(\dfrac{\textbf{40 mi}}{\textbf{1 hr}}\right) = \dfrac{30 \text{ mi}}{\dfrac{40 \text{ mi}}{1 \text{ hr}}}$

$t\left(\dfrac{40 \text{ mi}}{1 \text{ hr}}\right) = \textbf{0.75 hr}$ (45 min)

(c) Evaluate: $t\left(\dfrac{60 \text{ mi}}{1 \text{ hr}}\right)$

▶ $t\left(\dfrac{\textbf{60 mi}}{\textbf{1 hr}}\right) = \dfrac{30 \text{ mi}}{\dfrac{60 \text{ mi}}{1 \text{ hr}}}$

$t\left(\dfrac{60 \text{ mi}}{1 \text{ hr}}\right) = \textbf{0.5 hr}$ (30 min)

(d) Evaluate: $t\left(\dfrac{80 \text{ mi}}{1 \text{ hr}}\right)$

▶ $t\left(\dfrac{\textbf{80 mi}}{\textbf{1 hr}}\right) = \dfrac{30 \text{ mi}}{\dfrac{80 \text{ mi}}{1 \text{ hr}}}$

$t\left(\dfrac{80 \text{ mi}}{1 \text{ hr}}\right) = \textbf{0.375 hr}$ (22.5 min)

Notice that when the speed increases from $\dfrac{20 \text{ mi}}{1 \text{ hr}}$ to $\dfrac{40 \text{ mi}}{1 \text{ hr}}$, the commuter saves 45 min in commuting time, and when the speed increases from $\dfrac{60 \text{ mi}}{1 \text{ hr}}$ to $\dfrac{80 \text{ mi}}{1 \text{ hr}}$, the commuter saves 7.5 min in commuting time.

Practice Problems

9. Evaluate $f(10)$ if $f(x) = \dfrac{x + 5}{x - 5}$.

10. Evaluate $h(-4)$ if $h(x) = \dfrac{x^2 - 10x + 16}{x^2 - 5x - 24}$.

11. Evaluate $g(3)$ if $g(x) = \dfrac{x}{x^2 - 9}$. 12. Evaluate $f\left(\dfrac{1}{300}\right)$ if $f(x) = \dfrac{1}{2x}$.

13. A company wants to make \$500,000 from the sale of an item. The relationship of number of products, n, and product price, p, is $p = \dfrac{\$500{,}000}{n}$. If 2500 products are sold, find the product price needed to produce this revenue.

14. Young's formula is $C(x) = \dfrac{(\text{adult dose})(x)}{x + 12}$ where x is the age of a child between 1 year and 12 years of age and $C(x)$ is the estimated amount of medication to give to a child. The adult dose of a medication is 50 mg. Use Young's formula to find the dose of the same medication for a child who is 8 years old. (*Note:* This formula is from the nineteenth century; medication for children is now determined more precisely.)

A Real Zero of a Rational Function

A real **zero of a function** is an input value that corresponds to an output value of 0. On a graph, a real zero is the x-coordinate of the x-intercept. To find the exact value of a real zero using an algebraic method, replace the output value in the function rule with 0.

EXAMPLE 11 Use an algebraic method to find any real zeros of $f(x) = \dfrac{7x - 3}{x - 1}$.

SOLUTION ▶

$$f(x) = \frac{7x - 3}{x - 1}$$

$$0 = \frac{7x - 3}{x - 1} \qquad \text{Replace } f(x), \text{ the output value, with 0.}$$

$$(x - 1)(0) = \frac{(x - 1)}{1}\left(\frac{7x - 3}{x - 1}\right) \qquad \text{Clear fractions; multiplication property of equality.}$$

$$0 = 7x - 3 \qquad \text{Simplify.}$$

$$\underline{+3 \qquad\qquad +3} \qquad \text{Addition property of equality.}$$

$$3 = 7x + 0 \qquad \text{Simplify.}$$

$$\frac{3}{7} = \frac{7x}{7} \qquad \text{Division property of equality.}$$

$$\frac{3}{7} = x \qquad \text{A real zero of this function is } x = \dfrac{3}{7}.$$

The real zero of this function is the *x*-coordinate of the *x*-intercept of its graph, $\left(\frac{3}{7}, 0\right)$.

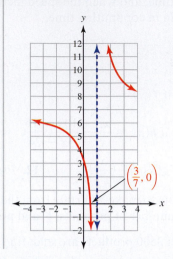

In the next example, the graph of the function does not intersect the *x*-axis. Since there is no input value that corresponds to an output value of 0, there is no *x*-intercept, and the function does not have a real zero.

EXAMPLE 12   A rational function is $f(x) = \dfrac{1}{x + 2}$.

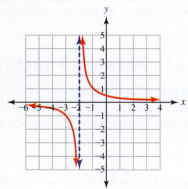

(a) Use an algebraic method to show that this function does not have a real zero.

SOLUTION ▶

$$f(x) = \frac{1}{x + 2}$$

$$0 = \frac{1}{x + 2}$$ Replace $f(x)$, the output value, with 0.

$$(x + 2)(0) = \frac{(x + 2)}{1}\left(\frac{1}{x + 2}\right)$$ Clear fractions; multiplication property of equality.

$$0 = \frac{(x + 2)}{1}\left(\frac{1}{x + 2}\right)$$ Simplify; find common factors.

$$0 = 1 \quad \text{False.}$$ Simplify; this equation has no solution.

Since this equation has no solution, there is no input value that corresponds to an output value of 0. This function has no real zero. Notice that the graph of the function has no x-intercept. A real zero of a function is the x-coordinate of the x-intercept.

(b) Use interval notation to represent the domain of this function.

▸ If $x = -2$, the denominator is equal to 0. Because division by 0 is undefined, restrict -2 from the domain. The domain is $(-\infty, -2) \cup (-2, \infty)$.

(c) Explain why this function has no real zeros.

▸ A rational expression can equal 0 only if the numerator equals 0. If the numerator of the function rule cannot equal 0, the function has no real zero.

In Example 12, the denominator of the function rule can equal 0, and the domain is restricted. In the next example, the denominator of the function rule cannot equal 0. The domain is the set of real numbers, $(-\infty, \infty)$.

EXAMPLE 13 | A rational function is $y = \dfrac{5}{x^2 + 1}$.

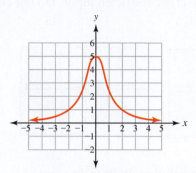

(a) Use interval notation to represent the domain.

SOLUTION ▸ The denominator equals zero when $x^2 = -1$. The square of a real number cannot be negative. So the denominator will never be zero. The domain is $(-\infty, \infty)$. No restrictions are needed.

(b) Explain why this function cannot have a real zero.

▸ A zero of a function is the input value that corresponds to an output value of 0. For this function to have an output of 0, the numerator must be 0. The numerator of this function is always 5. So this function cannot have a zero that is a real number. The graph of the function does not have an x-intercept.

Real Zeros and Rational Functions

- A real zero is an input value that corresponds to an output value of 0.
- On the graph of a rational function, a real zero is the x-coordinate of the x-intercept.
- The *numerator* of the rule of a rational function determines whether the function has any real zeros. If the numerator cannot equal 0, the function has no real zeros. If a function has no real zeros, its graph does not intersect the x-axis.

Practice Problems

For problems 15–18,
(a) use interval notation to represent the domain of the function.
(b) use an algebraic method to find any real zeros.

15. $f(x) = \dfrac{x + 5}{x - 5}$ **16.** $h(x) = \dfrac{x^2 - 10x + 16}{x^2 - 5x - 24}$ **17.** $g(x) = \dfrac{x}{x^2 - 9}$

18. $y = \dfrac{5}{x + 2}$

Using Technology: Graphing a Rational Function

With some graphing calculators, the graph of a rational function appears to include a vertical line at a restricted input value. This is not part of the graph.

The correct sketch of the graph has a dashed line at the restricted input. Do not connect the dashed line to the graph of the function.

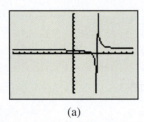   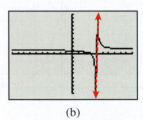

(a)   (b)

The screen of a graphing calculator has limited resolution. It often does not show the hole in a graph at a restricted input. Use an algebraic method to predict the existence of a hole, and show the location of the hole with an open circle on the sketch of the graph.

EXAMPLE 14 Graph: $y = \dfrac{x + 2}{x^2 - 4}$

This function is equivalent to $y = \dfrac{x + 2}{(x + 2)(x - 2)}$. The restricted input values are $x = -2$ and $x = 2$. There is a vertical asymptote at $x = 2$. Because the numerator and denominator both include the factor $x + 2$, there is a hole at $x = -2$.

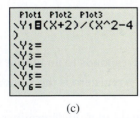   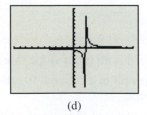   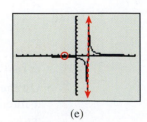

(c)   (d)   (e)

Practice Problems For problems 19–21,
(a) graph each function. Sketch the graph, including any vertical asymptotes or holes. Describe the window.
(b) use interval notation to represent the domain.
(c) write the equation of any vertical asymptotes.
(d) identify the location of any holes.
(e) identify any real zeros of the function.

19. $f(x) = \dfrac{2x + 4}{x - 3}$ **20.** $h(x) = \dfrac{9}{x^2 - 3x - 10}$ **21.** $g(x) = \dfrac{3x - 12}{x^2 - 2x - 8}$

5.5 Vocabulary Practice

Match the term with its description.

1. An input value that results in an output value of zero
2. The product of rate and time
3. The ratio of two integers
4. $f(x) = \dfrac{\text{polynomial}}{\text{polynomial}}$
5. Undefined
6. The set of the inputs of a function
7. The set of the outputs of a function
8. A relation in which each input corresponds to exactly one output
9. Finding the output value of a function for a given input value
10. $\cup$

A. distance
B. division by zero
C. domain
D. evaluating a function
E. function
F. range
G. rational function
H. rational number
I. real zero of a function
J. union

5.5 Exercises

Follow your instructor's guidelines for showing your work.

For exercises 1–10,
(a) use a number line to represent the domain of the function.
(b) use interval notation to represent the domain of the function.

1. $f(x) = \dfrac{5}{x}$

2. $g(x) = \dfrac{4}{x}$

3. $y = \dfrac{5}{x + 1}$

4. $y = \dfrac{7}{x + 4}$

5. $y = \dfrac{x + 2}{x + 1}$

6. $y = \dfrac{x + 3}{x + 4}$

7. $f(t) = \dfrac{t - 4}{t^2 + 10t + 24}$

8. $f(t) = \dfrac{t - 3}{t^2 + 13t + 40}$

9. $f(t) = \dfrac{t - 4}{t^2 - 10t + 24}$

10. $f(t) = \dfrac{t - 3}{t^2 - 13t + 40}$

For exercises 11–40,
(a) use interval notation to represent the domain of the function.
(b) write the equation of any vertical asymptotes.
(c) identify the location of any holes on the graph.

11. $f(x) = \dfrac{3}{2x^2 - 6x}$

12. $g(x) = \dfrac{2}{3x^2 - 9x}$

13. $g(x) = \dfrac{5x - 4}{10x^2 + 7x - 12}$

14. $k(x) = \dfrac{2x + 5}{6x^2 + 5x - 25}$

15. $R(x) = \dfrac{3x}{2x^2 - 6x}$

16. $g(x) = \dfrac{2x}{3x^2 - 9x}$

17. $f(x) = \dfrac{x^2 + 12x + 36}{x^2 - x - 30}$

18. $h(x) = \dfrac{x^2 + 18x + 81}{x^2 - 11x + 18}$

19. $f(x) = \dfrac{2x}{x^3 - 3x^2 - 28x}$

20. $f(x) = \dfrac{3x}{x^3 - 5x^2 - 24x}$

21. $f(x) = \dfrac{3}{x - 2}$

22. $f(x) = \dfrac{-5}{x + 3}$

23. $f(x) = \dfrac{12}{x^2 - 9}$

24. $f(x) = \dfrac{12}{x^2 - 16}$

25. $h(x) = \dfrac{x - 3}{x^2 - x - 6}$

26. $h(x) = \dfrac{x - 4}{x^2 - 2x - 8}$

27. $g(x) = \dfrac{-10}{x^2 + 3}$

28. $g(x) = \dfrac{-4}{x^2 + 1}$

29. $g(x) = \dfrac{10}{x^2 + 3}$

30. $g(x) = \dfrac{4}{x^2 + 1}$

31. $f(x) = \dfrac{x}{4}$

32. $f(x) = \dfrac{x}{2}$

33. $y = \dfrac{10x}{x^2 + 1}$

34. $y = -\dfrac{6x}{x^2 + 1}$

35. $y = \dfrac{7}{x^2 + 6x + 9}$

36. $y = \dfrac{2}{x^2 - 8x + 16}$

37. $f(x) = \dfrac{x^2 + 5x + 6}{x + 3}$

38. $f(x) = \dfrac{x^2 + 7x + 10}{x + 2}$

39. $f(x) = \dfrac{20}{x^3 + 8x^2 + 12x}$

40. $f(x) = \dfrac{50}{x^3 + 9x^2 + 18x}$

For exercises 41–46, each table of ordered pairs includes the inputs and outputs near a restricted input.

(a) Use this information to decide whether the graph of this function has an asymptote or a hole at this input.

(b) Explain your decision.

41.

x	y
1.2	−3.75
1.4	−5
1.6	−7.5
1.8	−15
2	undefined
2.2	15
2.4	7.5
2.6	5

42.

x	y
1.2	0.722
1.4	0.730
1.6	0.736
1.8	0.744
2	undefined
2.2	0.756
2.4	0.762
2.6	0.767

43.

x	y
1.2	−0.67
1.4	−0.59
1.6	−0.52
1.8	−0.46
2	undefined
2.2	−0.34
2.4	−0.30
2.6	−0.25

44.

x	y
1.2	−6.25
1.4	−8.3
1.6	−12.5
1.8	−25
2	undefined
2.2	25
2.4	12.5
2.6	8.3

45.

x	y
−3.5	−6
−3	−7.5
−2.5	−10
−2	−15
−1.5	−30
−1	undefined
−0.5	30
0	15
0.5	10

46.

x	y
−5.5	−80
−5	−100
−4.5	−133.33
−4	−200
−3.5	−400
−3	undefined
−2.5	400
−2	200
−1.5	133.33

For exercises 47–50, evaluate $f(x) = \dfrac{5}{x}$ for the given input value.

47. $f(-10)$

48. $f(-20)$

49. $f\left(-\dfrac{1}{10}\right)$

50. $f\left(\dfrac{1}{10}\right)$

For exercises 51–58, evaluate the function for the given input value.

51. $f(x) = \dfrac{x}{x-6}; f(6)$

52. $f(x) = \dfrac{x}{x-5}; f(5)$

53. $f(x) = \dfrac{x}{x-6}; f(-6)$

54. $f(x) = \dfrac{x}{x-5}; f(-5)$

55. $f(x) = \dfrac{x^2 - 4x - 45}{x^2 - 9}; f(0)$

56. $f(x) = \dfrac{x^2 - 5x - 36}{x^2 - 64}; f(0)$

57. $f(x) = \dfrac{x^2 - 4x - 45}{x^2 - 9}; f(3)$

58. $f(x) = \dfrac{x^2 - 5x - 36}{x^2 - 64}; f(8)$

For exercises 59–62, the lateral surface area of a can (not including the ends) is 39 in.². The radius of the can, r, depends on the height of the can, h: $r = \dfrac{39\ \text{in.}^2}{2\pi h}$.

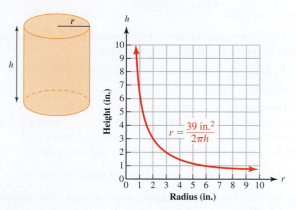

59. If the height is 4.8 in., find the radius. Round to the nearest tenth.

60. If the height is 3 in., find the radius. Round to the nearest tenth.

61. Use the graph to estimate the radius when the height is 2 in.

62. Use the graph to estimate the radius when the height is 6 in.

For exercises 63–66, the time in hours, h, to travel 180 miles depends on the speed, r, in miles per hour: $h = \dfrac{180\ \text{mi}}{r}$.

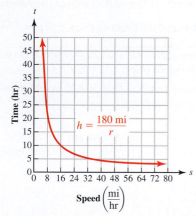

63. At a speed of $\dfrac{65\ \text{mi}}{1\ \text{hr}}$, find the time to travel 180 mi. Round to the nearest tenth.

64. At a speed of $\dfrac{30\ \text{mi}}{1\ \text{hr}}$, find the time to travel 180 mi.

65. Use the graph to estimate the time to travel 180 mi at a speed of $24\ \dfrac{\text{mi}}{\text{hr}}$. Round to the nearest whole number.

66. Use the graph to estimate the time to travel 180 mi at a speed of $8\ \dfrac{\text{mi}}{\text{hr}}$. Round to the nearest whole number.

For exercises 67–68, a new refrigerator costs $520. It uses 479 kilowatt·hour of electrical power a year. In 2006, the cost of a kilowatt·hour in Miami, Florida, was $0.11. The average annual cost in dollars, $C(n)$, to own and use the refrigerator for n years is $C(n) = \dfrac{520 + (479)(0.11)n}{n}$. Assuming that the costs stay the same, find the average annual cost for someone in Miami to own and use this refrigerator for

67. 4 years

68. 10 years

For exercises 69–70, a dog from an animal shelter costs $150. This includes the cost of licensing and vet care. It costs $300 per year to feed and care for the dog. The average annual cost in dollars, $C(n)$, to own and care for the dog for n years is $C(n) = \dfrac{150 + 300n}{n}$. Assuming that the cost of caring for the dog stays the same, find the average annual cost of owning and caring for this dog for

69. 4 years

70. 10 years

71. Think of an item that involves an annual expense. Describe it. Find its initial cost. Estimate the annual cost of using or owning it. Use this information to write a function that predicts the average annual cost, $C(n)$, of owning and using this item for n years.

72. Use the function in exercise 71 to predict the average annual cost of owning and using this item for 5 years.

For exercises 73–74, a company wants to collect $600,000 from selling a product. The relationship of the number of products sold, n, and the product price, p, is $p = \dfrac{\$600,000}{n}$.

73. The company predicts that 20,000 products will be sold. Find the product price.

74. The company predicts that 30,000 products will be sold. Find the product price.

For exercises 75–80,
(a) use an algebraic method to find any real zero(s).
(b) if the function does not have any real zeros, explain why.

75. $f(x) = \dfrac{x - 3}{x + 9}$

76. $g(x) = \dfrac{x - 7}{x + 14}$

77. $f(x) = \dfrac{5}{x + 9}$

78. $g(x) = \dfrac{8}{x + 14}$

79. $y = \dfrac{x^2 - x - 6}{x + 1}$

80. $y = \dfrac{x^2 - 2x - 24}{x + 3}$

Problem Solving: Practice and Review

Follow your instructor's guidelines for using the five steps as outlined in Section 1.5, p. 51.

81. In 2008 and 2009, many employees working in industries that depended on new car sales lost their jobs. Many

of these people lived in Michigan. Use the model of the relationship between the number of people receiving food assistance, y, and the months since October 2008, x, to predict the number of people receiving assistance in August, 2009. Round to the nearest thousand.

People Receiving Food Assistance in Michigan

$y = 1555x^2 + 12,994x + 1,316,443$

Number of months since October 2008

Source: www.mich.gov

82. A standard showerhead uses 2.5 gal of water per minute. A low-flow showerhead uses 1.4 gal per minute. A student takes a shower every day. The average length of the shower is 5 min. Find the difference in the gallons of water used per year with these showerheads (1 year = 365 days). Round to the nearest whole number.

For exercises 83–84, a formula for the weight of copper pipe is $W = (0.2537)(A^2 - B^2)L$. The outside diameter of the pipe in inches is A, the inside diameter of the pipe in inches is B, the length of the pipe in inches is L, and the weight of the pipe in pounds is W.

© kotomiti/Shutterstock

83. The formula includes a constant, 0.2537, without any units. Find the units of this constant in this formula.

84. A copper pipe has a nominal size of $\dfrac{1}{2}$ in. The nominal size is used for identification and often does not correspond to the actual measurement. Its actual outside diameter is 0.625 in., and its actual inside diameter is 0.527 in. Find the weight of 20 ft of this pipe. Round to the nearest tenth.

Technology

For exercises 85–88, use a graphing calculator to graph each function.
(a) Sketch the graph. Include any holes or vertical asymptotes. Describe the window.
(b) Use interval notation to represent the domain.
(c) Write the equation of any vertical asymptotes.
(d) Identify the location of any holes.
(e) Find any real zeros.

85. $y = \dfrac{10}{x^2 - 3x - 18}$

86. $g(x) = \dfrac{12}{x^2 + 3x - 28}$

87. $f(x) = \dfrac{10x - 60}{x^2 - 3x - 18}$

88. $g(x) = \dfrac{12x - 48}{x^2 + 3x - 28}$

Find the Mistake

For exercises 89–92, the completed problem has one mistake.
(a) Describe the mistake in words, or copy down the whole problem and highlight or circle the mistake.
(b) Do the problem correctly.

89. **Problem:** Use interval notation to state the domain of

$$f(x) = \dfrac{x + 4}{x^2 + 7x + 12}.$$

Incorrect Answer: $f(x) = \dfrac{x + 4}{x^2 + 7x + 12}$

$$f(x) = \dfrac{x + 4}{(x + 4)(x + 3)}$$

$$f(x) = \dfrac{x + 4}{(x + 4)(x + 3)}$$

$$f(x) = \dfrac{1}{x + 3}$$

The domain is $(-\infty, -3) \cup (-3, \infty)$.

90. **Problem:** Find any real zeros of the function $f(x) = \dfrac{6}{x + 2}$.

Incorrect Answer: $x + 2 = 0$

$$\dfrac{-2 \quad -2}{x = -2}$$

A real zero of this function is $x = -2$.

91. **Problem:** Find any real zeros of the function

$$f(x) = \dfrac{x + 1}{x^2 - 1}.$$

Incorrect Answer: If $x = -1$, the numerator of the function is 0. So the real zero of this function is $x = -1$.

92. **Problem:** Write the equations of any vertical asymptotes of the function $f(x) = \dfrac{4x + 12}{x^2 + 5x + 6}$.

Incorrect Answer: $f(x) = \dfrac{4x + 12}{x^2 + 5x + 6}$

$$f(x) = \dfrac{4x + 12}{(x + 3)(x + 2)}$$

$$x + 3 = 0 \quad \text{or} \quad x + 2 = 0$$

$$\dfrac{-3 \quad -3}{x = -3} \qquad \dfrac{-2 \quad -2}{x = -2}$$

The equations of the vertical asymptotes of this function are $x = -3$ and $x = -2$.

Review

93. A function is $f(x) = 6x$. As the input value increases, does the output value increase or decrease?

94. A function is $f(x) = -6x$. As the input value increases, does the output value increase or decrease?

95. a. Graph: $f(x) = 6x$
    b. Identify the slope of the graph.
    c. Identify the y-intercept of the graph.

96. a. Graph: $f(x) = -6x$
    b. Identify the slope of the graph.
    c. Identify the y-intercept of the graph.

SUCCESS IN COLLEGE MATHEMATICS

97. Division by zero is undefined. So the rational number $\dfrac{4}{0}$ is undefined, and $4 \div 0$ is undefined. Explain why $\dfrac{4}{0}$ cannot equal 0.

98. The domain of the rational function $y = \dfrac{2}{x - 4}$ cannot include 4. Explain why.

In many jobs, our pay depends on how many hours we work. We make more money if we work more hours. This is an example of a *direct variation*. In this section, we will study functions that are *direct variations* and *inverse variations*.

SECTION 5.6

Variation

After reading the text, working the practice problems, and completing assigned exercises, you should be able to:

1. Write and evaluate a function that is a direct variation.
2. Write and evaluate a function that is an inverse variation.
3. Graph a function that is a direct or inverse variation.
4. Determine whether a function is a joint or combined variation and find the constant of proportionality.
5. Identify whether the relationship between two variables in a formula is direct or inverse.
6. Solve a formula for a variable.

Direct Variation

A home health care worker earns \$7.80 an hour. A function that describes the relationship of her gross earnings, y, and the number of hours she works, x, is $y = \left(\dfrac{\$7.80}{1 \text{ hr}} \right) x$. This function is a **direct variation**. As the input changes, the output changes at a constant rate. This rate is the slope of the graph of the function (Figure 1). It is also the **constant of proportionality**, k.

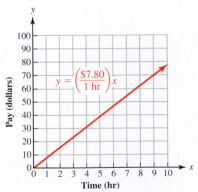

Figure 1

> **Direct Variation**
>
> $$f(x) = kx \qquad \text{or} \qquad y = kx$$
>
> The constant of proportionality is k; $k > 0$. As the input value increases, the output value increases. As the input value decreases, the output value decreases. This is a linear function.

Given an input value and its corresponding output value, we can find the constant of proportionality, k, of a direct variation.

EXAMPLE 1   The relationship of the input value, x, and the output value, y, is a direct variation. When $x = 2$, $y = 12$.

(a) Find the constant of proportionality, k.

SOLUTION ▶

$y = kx$ A direct variation.

$12 = k(2)$ $x = 2$; $y = 12$

$\dfrac{12}{2} = \dfrac{k(2)}{2}$ Division property of equality.

$6 = k$ Simplify.

(b) Write a function that describes this direct variation.

▶

$y = kx$ A direct variation.

$y = \mathbf{6}x$ $k = 6$

(c) Evaluate this function to find y when $x = 3$.

▶

$y = 6x$

$y = 6(\mathbf{3})$ $x = 3$

$y = \mathbf{18}$ Simplify.

(d) Use slope-intercept graphing to graph this function.

▶

The y-intercept is $(0, 0)$. The slope is $\dfrac{6}{1}$.

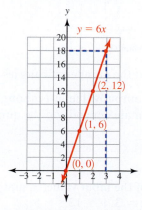

(e) Use the graph to find y when x is 3.

▶

Draw a vertical dashed line up from 3 on the x-axis. When the dashed line intersects the graph, pivot and draw a horizontal dashed line to the y-axis. Since this horizontal dashed line intersects the y-axis at 18, when $x = 3$, $y = 18$.

In the next example, the constant of proportionality, k, includes units.

EXAMPLE 2 The relationship between the volume in tablespoons (T), x, and the calories (cal) contained in chocolate chips, y, is a direct variation. Twelve tablespoons of chocolate chips have 840 cal.

(a) Find the constant of proportionality for this direct variation.

SOLUTION ▶
$$y = kx \qquad \text{A direct variation.}$$
$$840 \text{ cal} = k(12 \text{ T}) \qquad y = 840 \text{ cal}; x = 12 \text{ T}$$
$$\frac{840 \text{ cal}}{12 \text{ T}} = \frac{k(\cancel{12 \text{ T}})}{\cancel{12 \text{ T}}} \qquad \text{Division property of equality.}$$
$$\frac{70 \text{ cal}}{1 \text{ T}} = k \qquad \text{Simplify.}$$

(b) Write a function that describes this direct variation.

$$y = \left(\frac{70 \text{ cal}}{1 \text{ T}}\right)x \qquad y = kx$$

(c) Use this function to find the calories in 32 T of chocolate chips. (This is the standard amount used in a chocolate chip cookie recipe.)

$$y = \left(\frac{70 \text{ cal}}{1 \cancel{T}}\right)(32 \cancel{T}) \qquad \text{Input, } x, \text{ is 32 T.}$$
$$y = 2240 \text{ cal} \qquad \text{Simplify.}$$

(d) Graph this function.

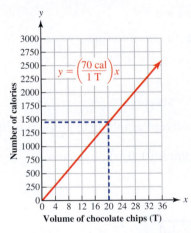

(e) Use the graph to find the number of calories in 20 T of chocolate chips.

When $x = 20$ T, the output value, y, is about 1400 cal. (This is an estimate; the accuracy of this estimate depends on the quality of the graph.)

Practice Problems

1. The relationship of the input value, x, and the output value, y, is a direct variation. When $x = 3$, $y = 12$.
   a. Find the constant of proportionality, k.
   b. Write a function that describes this direct variation.
   c. Evaluate this function to find y when $x = 5$.
   d. Graph this function.
   e. Use the graph to find y when $x = 2$.

2. The relationship of the miles driven, x, and the amount in dollars that the IRS considers deductible business costs, y, is a direct variation. For driving 1000 mi for business, the IRS allows a $500 deductible business cost. (*Source:* www.irs.gov)

a. Find the constant of proportionality, k. Include the units.

b. Write a function that describes this direct variation.

c. Evaluate this function to find the deductible business cost for driving 4500 mi.

d. Graph this function.

e. Use the graph to find the deductible business cost for driving 4000 mi.

Inverse Variation

In a cheese factory, bricks of cheese move on conveyor belts. A function showing the relationship of the time in seconds between each brick of cheese, x, and the number of bricks that the conveyer belt moves per hour, y, is

$$y = \frac{\dfrac{3600 \text{ brick} \cdot \text{s}}{1 \text{ hr}}}{x}$$

This function is an **inverse variation**. As the input value (the time between bricks) increases, the output value (the number of bricks moved by the conveyer belt) decreases (Figure 2).

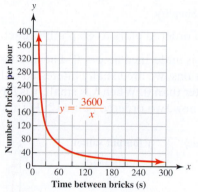

Figure 2

The constant of proportionality, k, is the product of the input value and the output value. The units of k in inverse variations are often complicated. When writing the function, the units for k are often removed: $y = \dfrac{3600}{x}$.

Inverse Variation

$$f(x) = \frac{k}{x} \qquad \text{or} \qquad y = \frac{k}{x}$$

The constant of proportionality is k; $k > 0$. As the input value increases, the output value decreases. As the input value decreases, the output value increases. This is a rational function.

EXAMPLE 3   The relationship of the input value, x, and the output value, y, is an inverse variation. When $x = 2$, $y = 12$.

(a) Find the constant of proportionality, k.

SOLUTION ▶ $y = \dfrac{k}{x}$ An inverse variation.

$12 = \dfrac{k}{2}$ $x = 2$; $y = 12$

$2 \cdot 12 = 2 \cdot \dfrac{k}{2}$ Multiplication property of equality.

$24 = k$ Simplify.

(b) Write a function that describes this inverse variation.

▶ $y = \dfrac{k}{x}$ An inverse variation.

$y = \dfrac{24}{x}$ $k = 24$

(c) Evaluate this function to find y when $x = 3$.

▶ $y = \dfrac{24}{x}$

$y = \dfrac{24}{3}$ $x = 3$

$y = 8$ Simplify.

(d) Use a table of ordered pairs to graph this function.

▶   The function is undefined at $x = 0$. The line $x = 0$ is a vertical asymptote. With inverse variations, we are usually interested only in the graph for input values that are greater than 0. When x is small, y is large. As x increases, y decreases. We already know two ordered pairs: $(2, 12)$ and $(8, 3)$.

x (input value)	y (output value)
0	undefined
1	24
2	12
3	8
4	6
8	3
12	2

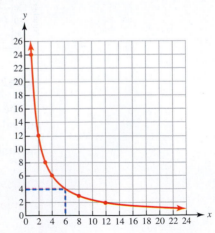

(e) Use the graph to find y when $x = 6$.

▶   Draw a vertical dashed line up from 6 on the x-axis. When the dashed line intersects the graph, pivot and draw a horizontal dashed line to the y-axis. This horizontal dashed line intersects the y-axis at 4, so when $x = 6$, $y = 4$.

In the next example, the constant of proportionality for an inverse variation includes units.

EXAMPLE 4   The relationship between the speed of a car, x, and the time it travels, y, over a fixed distance is an inverse variation. The car travels at a speed of $\dfrac{45\,\text{mi}}{1\,\text{hr}}$ for 0.75 hr.

(a) Find the constant of proportionality for this inverse variation.

SOLUTION ▶

$$y = \frac{k}{x} \qquad\qquad \text{An inverse variation.}$$

$$0.75\ \text{hr} = \frac{k}{\dfrac{45\ \text{mi}}{1\ \text{hr}}} \qquad\qquad y = 0.75\ \text{hr};\ x = \frac{45\,\text{mi}}{1\,\text{hr}}$$

$$\left(\frac{45\,\text{mi}}{1\,\text{hr}}\right)(0.75\,\text{hr}) = \left(\frac{45\,\text{mi}}{1\,\text{hr}}\right)\!\left(\frac{k}{\dfrac{45\,\text{mi}}{1\,\text{hr}}}\right) \qquad \text{Multiplication property of equality.}$$

$$33.75\ \text{mi} = k \qquad\qquad \text{Simplify.}$$

(b) Write a function that describes this inverse variation.

$$y = \frac{33.75\ \text{mi}}{x} \qquad y = \frac{k}{x}$$

(c) Use this function to find the time *in minutes* required to travel this distance at a speed of $\dfrac{60\,\text{mi}}{1\,\text{hr}}$. Round to the nearest whole number.

$$y = \frac{33.75\ \text{mi}}{\dfrac{60\ \text{mi}}{1\ \text{hr}}} \qquad\qquad x = \frac{60\,\text{mi}}{1\,\text{hr}}$$

$$y = 33.75\ \text{mi} \div \frac{60\ \text{mi}}{1\ \text{hr}} \qquad\qquad \text{Rewrite the fraction as division.}$$

$$y = (33.75\,\text{mi})\left(\frac{1\ \text{hr}}{60\ \text{mi}}\right) \qquad\qquad \text{Rewrite as multiplication by the reciprocal.}$$

$$y = 0.5625\ \text{hr} \qquad\qquad \text{Simplify.}$$

$$y = \frac{(0.5625\,\text{hr})}{1}\left(\frac{60\ \text{min}}{1\ \text{hr}}\right) \qquad\qquad \text{Change hours into minutes; multiply by 1.}$$

$$y = 33.75\ \text{min} \qquad\qquad \text{Simplify.}$$

$$y \approx 34\ \text{min} \qquad\qquad \text{Round.}$$

(d) Use a table of ordered pairs to graph $y = \dfrac{33.75 \text{ mi}}{x}$.

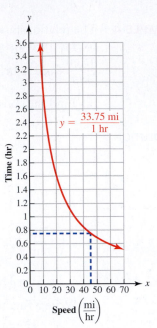

$x \left(\dfrac{\text{mi}}{\text{hr}} \right)$	y (hr)
0	undefined
10	3.38
20	1.69
30	1.13
40	0.84
50	0.68
60	0.56
70	0.48

(e) Use the graph to find the time required to travel this distance at a speed of $\dfrac{45 \text{ mi}}{1 \text{ hr}}$. Round to the nearest tenth.

At a speed of $\dfrac{45 \text{ mi}}{1 \text{ hr}}$, it will take about 0.75 hr to travel this distance. (This is an estimate; the accuracy of the estimate depends on the quality of the graph.)

Practice Problems

3. The relationship of the input value, x, and the output value, y, is an inverse variation. When $x = 3$, $y = 12$.
  a. Find the constant of proportionality, k.
  b. Write a function that describes this inverse variation.
  c. Evaluate this function to find y when $x = 4$.
  d. Graph this function.
  e. Use the graph to find y when $x = 2$.

4. The relationship of the length of a rectangle, x, and its width, y, for a fixed area is an inverse variation. When the length of the rectangle is 10 in., the width is 4 in.
  a. Find the constant of proportionality, k. Include the units.
  b. Write a function that describes this inverse variation.
  c. Evaluate this function to find the width of this rectangle if the length is 8 in.
  d. Graph this function.
  e. Use the graph to find the width of this rectangle when the length is 5 in.

Joint and Combined Variation

A formula is $y = kxz$, where x, y, and z are variables and k is a constant of proportionality. The value of y is directly proportional to both x and z. This is an example of a **joint variation**. A change in the value of x or z produces a directly proportional change in y.

EXAMPLE 5 The kinetic energy of an object varies jointly with its mass and with the square of its velocity (speed). If y is the kinetic energy, m is the mass, and v is the speed, $y = kmv^2$. A bowling ball has a mass of 5 kg, a velocity of $7 \frac{m}{s}$, and a kinetic energy of $122.5 \frac{kg \cdot m^2}{s^2}$.

(a) Find the constant of proportionality for this joint variation.

SOLUTION ▶

$$y = kmv^2 \qquad \text{A joint variation.}$$

$$122.5 \frac{kg \cdot m^2}{s^2} = k(5 \text{ kg})\left(7 \frac{m}{s}\right)^2 \qquad \text{Replace } y, m, \text{ and } v.$$

$$122.5 \frac{kg \cdot m^2}{s^2} = k\left(\frac{245 \text{ kg} \cdot m^2}{s^2}\right) \qquad \text{Simplify.}$$

$$\frac{122.5 \dfrac{kg \cdot m^2}{s^2}}{245 \dfrac{kg \cdot m^2}{s^2}} = \frac{k\left(245 \dfrac{kg \cdot m^2}{s^2}\right)}{245 \dfrac{kg \cdot m^2}{s^2}} \qquad \text{Division property of equality.}$$

$$0.5 = k \qquad \text{Simplify.}$$

(b) Write a function that describes this joint variation.

▶ $y = \mathbf{0.5}mv^2 \qquad y = kxz^2$

(c) Evaluate this function to find the kinetic energy of a ball bearing with a mass of 0.01 kg and a velocity of $6.0 \frac{m}{s}$.

▶ $$y = 0.5mv^2$$

$$y = (0.5)(\mathbf{0.01 \text{ kg}})\left(\mathbf{6.0} \frac{\mathbf{m}}{\mathbf{s}}\right)^2 \qquad \text{Replace } x \text{ and } z.$$

$$y = \frac{\mathbf{0.18 \text{ kg} \cdot m^2}}{\mathbf{s^2}} \qquad \text{Simplify.}$$

In the formula $y = \dfrac{kx}{z}$, the variable y is *directly* proportional to x and *inversely* proportional to z. This is an example of a **combined variation**.

EXAMPLE 6 The volume of a gas is directly proportional to its temperature and is inversely proportional to its pressure. If V is the volume in liters, T is the temperature in kelvins (K), and P is the pressure in atmospheres (atm), $V = \dfrac{kT}{P}$. When the temperature is 300 K and the pressure is 1 atm, the volume is about 24.63 L.

(a) Find the constant of proportionality for this combined variation.

SOLUTION ▶

$$V = \frac{kT}{P} \qquad \text{A combined variation.}$$

$$24.63 \text{ L} = \frac{k(300 \text{ K})}{1 \text{ atm}} \qquad \text{Replace } V, T, \text{ and } P.$$

$$(1 \text{ atm})(24.63 \text{ L}) = (1 \text{ atm}) \cdot \frac{k(300 \text{ K})}{1 \text{ atm}} \qquad \text{Multiplication property of equality.}$$

$$24.63 \text{ L} \cdot \text{atm} = k(300 \text{ K}) \qquad \text{Simplify.}$$

$$\frac{24.63 \text{ L} \cdot \text{atm}}{300 \text{ K}} = \frac{k(300 \text{ K})}{300 \text{ K}} \qquad \text{Division property of equality.}$$

$$\frac{0.0821 \text{ L} \cdot \text{atm}}{1 \text{ K}} = k \qquad \text{Simplify.}$$

(b) Write a function that describes this combined variation.

$$V = \frac{\left(\dfrac{0.0821 \text{ L} \cdot \text{atm}}{1 \text{ K}}\right)T}{P} \qquad y = \frac{kx}{z}$$

(c) Evaluate this function to find the volume of the same amount of gas at a pressure of 2.00 atm and temperature of 310 K. Round to the nearest tenth.

$$V = \frac{\left(\dfrac{0.0821 \text{ L} \cdot \text{atm}}{1 \text{ K}}\right)T}{P}$$

$$V = \frac{\left(\dfrac{0.0821 \text{ L} \cdot \text{atm}}{1 \text{ K}}\right)(310 \text{ K})}{(2 \text{ atm})} \qquad \text{Replace } k, T, \text{ and } P.$$

$$V = 12.7255 \text{ L} \qquad \text{Simplify.}$$

$$V \approx 12.7 \text{ L} \qquad \text{Round.}$$

This relationship of time, pressure, and volume of a gas is a topic in introductory chemistry classes. Although it may seem awkward to include the units, they are an important check for reasonability. Instructors in these classes will expect you to include them as you solve problems.

Practice Problems

5. A joint variation is $y = kxz$.
    a. If $x = 20$, $z = 10$, and $y = 400$, find k.
    b. Write a function that represents this joint variation.
    c. Evaluate this function when $x = 30$ and $z = 15$.

6. A combined variation is $y = \dfrac{kx}{z}$.
    a. If $x = 20$, $z = 10$, and $y = 18$, find k.
    b. Write a function that represents this joint variation.
    c. Evaluate this function when $x = 10$ and $z = 5$.

Variation and Formulas

Many formulas include more than two variables. We can describe the relationship of two of the variables when the other variables are constant.

EXAMPLE 7 The formula $V = LWH$ describes the relationship between the volume of a box and its length L, width W, and height H.

(a) Is the relationship between V and L a direct variation or an inverse variation?

SOLUTION ▶ Imagine this equation as a balanced scale. We assume that the values of W and H cannot change. If L increases, what must happen to V so that the scale remains balanced?

$$\frac{V}{\blacktriangle} = LWH$$

If L increases, the scale will tip to the right. To balance the scale, V must also increase. This is a **direct variation**.

(b) Assume that V and W are constant. Is the relationship between L and H a direct variation or an inverse variation?

▶ Imagine this equation as a balanced scale. We assume that the values of V and W cannot change. If L increases, what must happen to H so that the scale remains balanced?

$$\frac{V}{\blacktriangle} = LWH$$

If L increases, the scale will tip to the right. The only way to balance this scale is for H to decrease. This is an **inverse variation**.

Practice Problems

For problems 7–9, a truck has a recommended standard size set of tires. Some owners choose to put a nonstandard set of tires on a truck. The effect of this change is described by the formula $T = \dfrac{336\,gm}{R}$. In this formula, T is the tire diameter, g is the gear ratio, m is the speed, and R is the revolutions of the tires per minute.

7. Is the relationship between T and R a direct variation or an inverse variation?

8. Is the relationship between g and m a direct variation or an inverse variation?

9. Is the relationship between T and m a direct variation or an inverse variation?

Solving for a Variable

Spreadsheet software programs allow us to evaluate formulas quickly. To enter the formula, we must solve it for a variable.

EXAMPLE 8 | The maximum package measurement formula at UPS is $L + 2W + 2H = 165$, where L is the length, W is the width, and H is the height. All measurements are in inches. Solve this formula for W.

SOLUTION ▶
$$L + 2W + 2H = 165$$
$$\underline{-L \qquad -2H \qquad -L\ -2H} \qquad \text{Subtraction property of equality.}$$
$$0 + 2W + 0 = 165 - L - 2H$$
$$\frac{2W}{2} = \frac{165 - L - 2H}{2} \qquad \text{Simplify; division property of equality.}$$
$$W = \frac{165 - L - 2H}{2} \qquad \text{Simplify.}$$

In the next example, we are solving for a variable that is in two unlike terms. To isolate the variable, we factor.

EXAMPLE 9 The formula for the future value of an investment earning simple interest is $A = P + PRT$, where the future value is A, the principal (original amount invested) is P, the annual interest rate in decimal form is R, and the time of investment in years is T. Solve this formula for P.

SOLUTION ▶

$$A = P + PRT \qquad \text{P is in two unlike terms.}$$

$$A = P(1 + RT) \qquad \text{Distributive property; factor out P.}$$

$$\frac{A}{(1 + RT)} = \frac{P\cancel{(1 + RT)}}{\cancel{(1 + RT)}} \qquad \text{Division property of equality.}$$

$$\frac{A}{1 + RT} = P \qquad \text{Simplify.}$$

In some formulas, the variable we want to solve for is in the denominator. We first clear the fractions.

EXAMPLE 10 One version of the *thin lens* formula is $\dfrac{1}{f} = \dfrac{1}{d_o} + \dfrac{1}{d_i}$. The focal length of the lens is f, the distance of the object from the lens is d_o, and the distance of the image from the lens is d_i. To calculate the focal length, solve this formula for f.

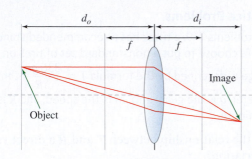

SOLUTION ▶ To isolate f, first clear the fractions. Since the least common denominator is $fd_o d_i$, multiply both sides of the formula by $fd_o d_i$.

$$\frac{1}{f} = \frac{1}{d_o} + \frac{1}{d_i}$$

$$(fd_o d_i)\left(\frac{1}{f}\right) = (fd_o d_i)\left(\frac{1}{d_o} + \frac{1}{d_i}\right) \qquad \text{Multiplication property of equality.}$$

$$(\cancel{f}d_o d_i)\left(\frac{1}{\cancel{f}}\right) = (f\cancel{d_o}d_i)\left(\frac{1}{\cancel{d_o}}\right) + (fd_o\cancel{d_i})\left(\frac{1}{\cancel{d_i}}\right) \qquad \text{Distributive property; find common factors.}$$

$$d_o d_i = fd_i + fd_o \qquad \text{Simplify.}$$

$$d_o d_i = f(d_i + d_o) \qquad \text{Factor.}$$

$$\frac{d_o d_i}{(d_i + d_o)} = \frac{f\cancel{(d_i + d_o)}}{\cancel{(d_i + d_o)}} \qquad \text{Division property of equality.}$$

$$\frac{d_o d_i}{d_i + d_o} = f \qquad \text{Simplify.}$$

Practice Problems

10. The ideal gas law is $PV = nRT$. Solve this formula for R.

11. A formula for finding compression ratio is $C = \dfrac{V_1 + V_2}{V_2}$. Solve for V_2.

12. The Friedewald equation is used to estimate LDL cholesterol in milligrams per deciliter: $L = C - H - \dfrac{T}{5}$. Solve for T. (*Source:* www.healthharvard.edu)

5.6 VOCABULARY PRACTICE

Match the function with the term. Terms may be used more than once.

1. $y = \dfrac{k}{x}$

2. $y = kx$

3. $d = \left(60 \dfrac{\text{miles}}{\text{hour}}\right)t$

4. $t = \dfrac{125 \text{ miles}}{r}$

5. $F = m\left(32 \dfrac{\text{meters}}{\text{second}^2}\right)$

6. $C = 2\pi r$

7. $f(x) = 120x$

8. $P = \dfrac{nRT}{V}$, where n, R, and T are constant

9. $P = \dfrac{nRT}{V}$, where n, R, and V are constant

10. $W = \dfrac{A}{L}$, where A is constant

A. direct variation
B. inverse variation

5.6 Exercises

Follow your instructor's guidelines for showing your work.

1. The relationship of the input value, x, and the output value, y, is a *direct variation*. When $x = 2$, $y = 6$.
   a. Find the constant of proportionality, k.
   b. Write a function that describes this direct variation.
   c. Evaluate this function to find y when $x = 4$.
   d. Graph this function.
   e. Use the graph to find y when $x = 5$.

2. The relationship of the input value, x, and the output value, y, is a *direct variation*. When $x = 2$, $y = 8$.
   a. Find the constant of proportionality, k.
   b. Write a function that describes this direct variation.
   c. Evaluate this function to find y when $x = 3$.
   d. Graph this function.
   e. Use the graph to find y when $x = 5$.

3. The relationship of the input value, x, and the output value, y, is a *direct variation*. When $x = 1$, $y = 6$.
   a. Find the constant of proportionality, k.
   b. Write a function that describes this direct variation.
   c. Evaluate this function to find y when $x = 4$.
   d. Graph this function.
   e. Use the graph to find y when $x = 2$.

4. The relationship of the input value, x, and the output value, y, is a *direct variation*. When $x = 1$, $y = 4$.
   a. Find the constant of proportionality, k.
   b. Write a function that describes this direct variation.
   c. Evaluate this function to find y when $x = 2$.
   d. Graph this function.
   e. Use the graph to find y when $x = 3$.

5. The relationship between the amount in tablespoons (T) of ground coffee, x, and the amount in ounces (oz) of coffee brewed, y, is a *direct variation*. According to a gourmet coffee company, a customer should use 14 T of ground coffee to brew 42 oz of coffee.
   a. Find the constant of proportionality. Include the units of measurement.
   b. Write a function that describes this relationship.
   c. Evaluate this function to find the amount of coffee that can be brewed from 48 T of ground coffee.
   d. Graph the function.
   e. Use the graph to predict the amount of coffee that can be brewed from 32 T of coffee. Use dashed lines on the graph to show your work.

6. The relationship between the number of servings of Wish-Bone Thousand Island® salad dressing, x, and the amount of sodium in milligrams (mg) in the dressing, y, is a *direct variation*. Seven servings of dressing contain 2100 mg of sodium. (*Source:* www.wish-bone.com)
   a. Find the constant of proportionality. Include the units of measurement.
   b. Write a function that describes this relationship.
   c. Evaluate this function to find the amount of sodium in a bottle of salad dressing that contains 16 servings.
   d. Graph the function.
   e. Use the graph to find the amount of sodium in three servings of salad dressing. Use dashed lines on the graph to show your work.

7. The relationship of the input value, x, and the output value, y, is an *inverse variation*. When $x = 2$, $y = 6$.
   a. Find the constant of proportionality, k.
   b. Write a function that describes this inverse variation.
   c. Evaluate this function to find y when $x = 4$.
   d. Graph this function.
   e. Use the graph to find y when $x = 3$.

8. The relationship of the input value, x, and the output value, y, is an *inverse variation*. When $x = 3$, $y = 6$.
   a. Find the constant of proportionality, k.
   b. Write a function that describes this inverse variation.
   c. Evaluate this function to find y when $x = 9$.
   d. Graph this function.
   e. Use the graph to find y when $x = 2$.

9. The relationship of the input value, x, and the output value, y, is an *inverse variation*. When $x = 2$, $y = 10$.
   a. Find the constant of proportionality, k.
   b. Write a function that describes this inverse variation.
   c. Evaluate this function to find y when $x = 5$.
   d. Graph this function.
   e. Use the graph to find y when $x = 4$.

10. The relationship of the input value, x, and the output value, y, is an *inverse variation*. When $x = 4$, $y = 5$.
    a. Find the constant of proportionality, k.
    b. Write a function that describes this inverse variation.
    c. Evaluate this function to find y when $x = 10$.
    d. Graph this function.
    e. Use the graph to find y when $x = 5$.

11. The relationship of the pressure in atmospheres (atm) of a gas, x, and the volume in liters (L) of a gas, y, is an *inverse variation*. When the pressure is 1.20 atm, the volume of a gas is 18.7 L.
    a. Find the constant of variation. Include the units of measurement.
    b. Write a function that describes this relationship.
    c. Evaluate this function to find the volume of a gas that is at 1.40 atm of pressure. Round to the nearest tenth.
    d. Use at least seven ordered pairs to graph the function.
    e. Use the graph to find the volume of a gas that is at 2 atm of pressure. Use dashed lines on the graph to show your work.

12. The relationship of the cross-sectional area in circular mils (CM) of a fixed length of copper wire, x, and the resistance in ohms of the wire, y, is an *inverse variation*.

When the cross-sectional area of copper wire is 1022 CM, its resistance is 0.51 ohm.
    a. Find the constant of variation. Include the units of measurement.
    b. Write a function that describes this relationship.
    c. Evaluate this function to find the resistance of copper wire that has a cross-sectional area of 2533 CM. Round to the nearest hundredth.
    d. Use at least seven ordered pairs to graph the function.
    e. Use the graph to find the resistance of a copper wire that has a cross-sectional area of 1624 CM.

13. The relationship between the radius in centimeters (cm) of a circle, x, and the circumference of the circle in centimeters, y, is a *direct variation*. When the radius of a circle is 10 cm, the circumference of this circle is 62.8 cm.

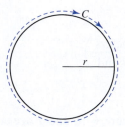

    a. Find the constant of proportionality. Include the units of measurement.
    b. Write a function that describes this relationship.
    c. Evaluate this function to find the circumference of a circle with a radius of 20 cm.
    d. Graph the function.
    e. Use the graph to find the circumference of a circle with a radius of 15 cm. Show your work using dashed lines on the graph.

14. The relationship between the acceleration in $\dfrac{\text{meter}}{\text{second}^2}$ experienced by an object, x, and the force in $\dfrac{\text{meter} \cdot \text{kilogram}}{\text{second}^2}$ applied to the object, y, is a *direct variation*. When the acceleration is $\dfrac{0.5 \text{ m}}{1 \text{ s}^2}$, the force applied is $\dfrac{2.5 \text{ m} \cdot \text{kg}}{1 \text{ s}^2}$.
    a. Find the constant of proportionality. Include the units of measurement.
    b. Write a function that describes this relationship.
    c. Evaluate this function to find the force applied when the acceleration is $\dfrac{2 \text{ m}}{1 \text{ s}^2}$.
    d. Graph the function.
    e. Use the graph to estimate the force when the acceleration is $\dfrac{1 \text{ m}}{1 \text{ s}^2}$. Show your work using dashed lines on the graph.

15. If the price per share of a company's stock is constant, the relationship between the earnings per share, x, and the financial ratio "price to earnings," y, is an *inverse variation*. When the earnings per share of a company is $2.83, its price to earnings ratio is 15.7.
    a. Find the constant of proportionality. Include the units of measurement.
    b. Write a function that describes this relationship.

c. If the price to earnings ratio is 14, evaluate this function to find the earnings per share. Round to the nearest hundredth.

16. If the annual credit sales are constant, the relationship of the accounts receivable, x, and the financial ratio "receivables turnover," y, is an *inverse variation*. When the accounts receivable of a company is $184 million, the receivables turnover ratio is 17.4.
   a. Find the constant of proportionality. Include the units of measurement.
   b. Write a function that describes this relationship.
   c. If the accounts receivable are $200 million, evaluate this function to find the receivables turnover ratio. Round to the nearest tenth.

17. The relationship of the cross-sectional area in square meters (m^2) of copper wire used for household circuits, x, and the current flow in ohms through the wire, y, is an *inverse variation*. For a wire with a cross-sectional area of 3.14×10^{-6} m^2, the resistance is 5.4×10^{-3} ohm.
   a. Find the constant of proportionality. Use scientific notation. Include the units of measurement. Round the mantissa to the nearest ten-thousandth.
   b. Write a function that describes this relationship.
   c. Evaluate this function to find the resistance for a wire with a cross-sectional area of 2.05×10^{-6} m^2. Round the mantissa to the nearest tenth.

18. If the force acting on an object is constant, the relationship of the mass in kilograms of the object, x, and the acceleration in $\dfrac{\text{meter}}{\text{second}^2}$ of the object, y, is an *inverse variation*. For a car with a mass of 1000 kg, the acceleration is $\dfrac{4 \text{ m}}{1 \text{ s}^2}$.
   a. Find the constant of proportionality. Include the units of measurement.
   b. Write a function that describes this relationship.
   c. Evaluate this function to find the acceleration of a car with a mass of 1500 kg. Round to the nearest tenth.

19. Two students often order pizza delivered to their homes. One student believes that the relationship between the distance of the pizza place to a home in miles, y, and the time in minutes to deliver a pizza to a home, x, is an *inverse variation*. It takes 45 min to deliver a pizza to his house. His house is 8 mi away from the pizza place.
   a. Find the constant of proportionality. Include the units of measurement.
   b. Write a function that describes this relationship.
   c. Do you think that this relationship is usually true? Explain.

20. A student believes that the relationship between the percent score she earns on a test, y, and the time in minutes she spends studying the night before the test, x, is an *inverse variation*. She spent 45 min studying for a test. Her grade on the test was 65%.
   a. Find the constant of proportionality. Include the units of measurement.

b. Write a function that describes this relationship.
   c. Do you think that this relationship is usually true? Explain.

21. A spring hangs from a horizontal support. An object hangs from the bottom of the spring. The relationship of the distance in centimeters (cm) that the spring stretches, x, is *directly proportional* to the mass in grams (g) of the object, y. When the mass of an object is 50 g, the stretch in the spring is 4.0 cm.

© Andy Crawford/Getty Images

   a. Find the constant of proportionality. Include the units of measurement.
   b. Write a function that describes this relationship.
   c. Evaluate this function to find the stretch of the spring when the mass of the object is 85 g.

22. The relationship of the length in meters (m) of a gold wire, x, and the resistance in ohms, y, is a *direct variation*. If the length of the wire is 8.0 m, the resistance is 0.06 ohm.
   a. Find the constant of proportionality. Include the units of measurement.
   b. Write a function that describes this relationship.
   c. Evaluate this function to find the resistance when the length of the wire is 3 m. Round to the nearest hundredth.

23. The relationship of the number of general admission tickets sold, x, and the total ticket receipts, y, is a *direct variation*. When 11,000 tickets are sold, the total ticket receipts are $495,000.
   a. Find the constant of proportionality. Include the units of measurement.
   b. Write a function that describes this relationship.
   c. Evaluate this function to find the total receipts from the sale of 7575 tickets.
   d. What does k represent in this function?

24. The relationship between the assessed value of property, x, and the tax owed, y, is a *direct variation*. In a county in New Jersey, a property had an assessed value of $150,000. The property tax bill on this property was $5637.
   a. Find the constant of proportionality. Include the units of measurement. Round to the nearest whole number.
   b. Write a function that describes this relationship.
   c. Evaluate this function to find the tax owed for a property with an assessed value of $185,000. Round to the nearest whole number.
   d. What does k represent in this function?

25. The relationship between the time in hours that a chartered jet travels, x, and the total cost to charter a jet, y, is a *direct variation*. The travel time for a round-trip flight from New York to Nassau in the Bahamas is 5.8 hr. The round-trip cost is $18,238.
    a. Find the constant of proportionality. Include the units of measurement. Round to the nearest hundredth.
    b. Write a function that describes this relationship.
    c. Find the cost to charter a jet for a round-trip with a travel time of 12 hr. Round to the nearest whole number
    d. What does k represent in this function?

26. The relationship between the number of miles driven on a freeway, x, and the cost of gasoline, y, is a *direct variation*. For a trip of 400 freeway miles, the total cost is $60.
    a. Find the constant of proportionality. Include the units of measurement.
    b. Write a function that describes this relationship.
    c. Evaluate this function to find the cost to drive 225 mi on the freeway in this car.
    d. What does k represent in this function?

27. The relationship between the speed in miles per hour $\left(\dfrac{\text{mi}}{\text{hr}}\right)$ of a car, x, and the time it travels in hours (hr), y, over a fixed distance is an *inverse variation*. A car travels at a speed of $\dfrac{35 \text{ mi}}{1 \text{ hr}}$ for 0.5 hr.
    a. Find the constant of proportionality. Include the units of measurement.
    b. Write a function that describes this relationship.
    c. Evaluate this function to find the time in minutes required to travel this distance at a speed of $\dfrac{50 \text{ mi}}{1 \text{ hr}}$.

28. The relationship between number of people who charter a jet to fly from New York to Chicago, x, and the cost per person, y, is an *inverse variation*. When two people charter the jet, the cost per person is $4300.
    a. Find the constant of proportionality. Include the units of measurement.
    b. Write a function that describes this relationship.
    c. The price of a first class round-trip ticket between New York and Chicago is $1010. A business is flying important customers from New York to Chicago. When the group of customers gets large enough, it is cheaper to charter a jet than to fly the customers on commercial airlines in first class. Find how many customers are needed to make it cheaper to choose the charter for the flight rather than commercial airlines in first class.

29. In radiography, a grid is used to reduce the effect of X-ray scattering on the final X-ray. The relationship of the distance in micrometers of the interspace, x, on the grid and the grid ratio, y, is an *inverse variation*. When the interspace distance on a grid is 300 micrometers, the grid ratio is 8. Write a function that describes this variation.

30. When the radiation is constant, the relationship between the current in milliamps in an X-ray tube, x, and the sec-

onds of exposure time, y, is an *inverse variation*. When a radiographic procedure uses a current of 600 milliamps, the exposure time is 0.2 s. Write a function that describes this variation.

31. A joint variation is $y = kxz$. If $x = 15$, $z = 2$, and $y = 45$, find k.

32. A joint variation is $y = kxz$. If $x = 8$, $z = 3$, and $y = 54$, find k.

33. A combined variation is $y = \dfrac{kx}{z}$. If $x = 30$, $z = 60$, and $y = 8$, find k.

34. A combined variation is $y = \dfrac{kx}{z}$. If $x = 90$, $z = 150$, and $y = 12$, find k.

35. The formula $R = k\dfrac{IE}{h}$ shows the relationship of the amount of radiation R received by a patient during a CT scan, the beam intensity I, the average beam energy E, and the thickness of the slice h.
    a. Is this an example of a joint variation or a combined variation? Explain.
    b. Is the radiation directly or inversely related to the beam intensity?
    c. Is the radiation directly or inversely related to the thickness of the slice?

36. Centripetal force causes objects to move in a curved path. The formula for centripetal force is $F_c = \dfrac{kv^2}{r}$, where F_c is the centripetal force, v is the velocity of the object, and r is the radius of the curve.
    a. Is this an example of a joint variation or a combined variation? Explain.
    b. Is the centripetal force directly or inversely related to the velocity?
    c. Is the centripetal force directly or inversely related to the radius?

37. The formula $D = kB^2S$ shows the relationship of the piston displacement in a car engine D, the bore of the cylinder B, and the length of the stroke S.

© GTS Production/Shutterstock

    a. Is this an example of a joint variation or a combined variation?
    b. Explain.

38. The formula $L = kR^2T^4$ shows the relationship of the luminosity of a star L, the radius of the star R, and its temperature T.
    a. Is this an example of a joint variation or a combined variation?
    b. Explain.

39. The formula $f = \dfrac{c}{\lambda}$ shows the relationship between the frequency of light f, the wavelength of light λ, and the speed of light c. Assume that c is constant. Is the relationship between f and λ a direct variation or an inverse variation?

40. The formula $R = \dfrac{V}{I}$ shows the relationship between the resistance R, voltage V, and current I in an electric circuit. Assume that V is constant. Is the relationship between R and I a direct variation or an inverse variation?

41. The formula $R = \dfrac{dVP}{F}$ shows the relationship between fluid velocity V, pipe diameter P, fluid density d, fluid viscosity F, and the Reynolds number R. Assume that d, V, and F are constant. Is the relationship between R and P a direct variation or an inverse variation?

42. The formula $M = \dfrac{ER}{168G}$ shows the relationship between tire revolutions per minute E, radius of the tire R, the final gear ratio G, and the speed of the motorcycle in miles per hour M. Assume that E and G are constant. Is the relationship between M and R a direct variation or an inverse variation?

43. For the formula $R = \dfrac{dVP}{F}$, assume that d, V, and P are constant. Is the relationship between R and F a direct variation or an inverse variation?

44. For the formula $M = \dfrac{ER}{168G}$, assume that E and R are constant. Is the relationship between M and G a direct variation or an inverse variation?

45. The formula $C = \dfrac{P_m P_i}{TF}$ is used to find the cost of insurance for machinery. Solve this formula for F.

46. The formula $N = \dfrac{12F}{CR}$ is used to find the number of teeth in a circular saw. Solve this formula for R.

© Alena Brozova/Shutterstock

47. A formula is $C = \dfrac{P_m P_i}{TF}$. Assume that C, P_m, and P_i are constant. Is the relationship between F and T a direct variation or an inverse variation?

48. A formula is $N = \dfrac{12F}{CR}$. Assume that F and N are constant. Is the relationship between C and R a direct variation or an inverse variation?

49. In chemistry, the ideal gas law is $PV = nRT$. Solve this equation for T.

50. Solve $PV = nRT$ for n.

51. A formula is $PV = nRT$. Assume that P, V, and R are constant. Is the relationship between T and n a direct variation or an inverse variation?

52. A formula is $PV = nRT$. Assume that P, R, and T are constant. Is the relationship between n and V a direct variation or an inverse variation?

53. The maximum package measurement formula at UPS is $L + 2W + 2H = 165$. Solve this formula for H. (*Source:* www.ups.com)

54. The formula for finding the surface area of a single ridge greenhouse is $A = 2LH + 2WH + 2LR + GW$. Solve for R.

© Pavel Svoboda/Shutterstock

55. Solve $A = 2LH + 2WH + 2LR + GW$ for L.

56. Solve $A = 2LH + 2WH + 2LR + GW$ for W.

57. The formula for fielding average in baseball is $F = \dfrac{P + A}{P + A + E}$. Solve this formula for E. (*Source:* www.baseball-almanac.com)

58. The formula for on-base percentage in baseball is $O = \dfrac{H + W + P}{A + W + P + S}$. Solve this formula for S. (*Source:* www.baseball-almanac.com)

59. Solve $F = \dfrac{P + A}{P + A + E}$ for A.

60. Solve $O = \dfrac{H + W + P}{A + W + P + S}$ for W.

61. In baseball, the formula for total base percentage is $T = \dfrac{H + W + P}{A + W + P}$. Solve this formula for W. (*Source:* www.baseball-almanac.com)

62. Solve $T = \dfrac{H + W + P}{A + W + P}$ for P.

63. The thin lens formula is $\dfrac{1}{f} = \dfrac{1}{d_o} + \dfrac{1}{d_i}$. Solve this formula for d_o.

64. Solve $\dfrac{1}{f} = \dfrac{1}{d_o} + \dfrac{1}{d_i}$ for d_i.

65. The formula $G = \dfrac{N^2 L}{239}\left(1 + \dfrac{1}{2A}\right)$ shows the

relationship between maximum piston acceleration, crankshaft speed, stroke length, and ratio of connecting rod length to stroke. Solve this formula for A.

66. The formula $R = \dfrac{P - O}{O} - \dfrac{M_2 - M_1}{M_1}$ shows the

relationship between the market adjusted returns, the closing price on the first trading day, the offer price in the prospectus, the closing value of a selected market index on that date, and the closing value of a selected market index on the date before listing. Solve this formula for M_1.

67. In weaving, Ashenhurst's formula predicts the maximum

sett of a yarn: $S = \dfrac{DE}{E + I}$. The number of ends of warp

in one repeat of the pattern is E, the number of times the weft passes from one side of the fabric to the other is I, and the thread diameter number is E. Solve this formula for E. (*Source:* www.faena.medievaltextiles.org)

68. Solve $S = \dfrac{DE}{E + I}$ for I.

69. The formula $R = k\dfrac{L}{A}$ shows the relationship between

the resistance in a wire, the constant of resistivity of the material, the length of the wire, and the cross-sectional area of the wire. Solve this formula for A.

70. Solve $R = k\dfrac{L}{A}$ for L.

71. The summer simmer index, S, describes the combined effect of air temperature, T, and relative humidity, H. $S = 1.98[T - (0.55 - 0.0055H)(T - 58)] - 56.83$. Solve this formula for T. (*Source:* www.gorhamschaffler.com)

72. Solve $S = 1.98\,[T - (0.55 - 0.0055H)(T - 58)] - 56.83$ for H.

73. The formula $H = \dfrac{P}{h - 3P}$ describes the amount of heat,

H, added to a room by a motor consuming P watts of electrical power with an efficiency of h. Solve this formula for P.

74. Solve $H = \dfrac{P}{h - 3P}$ for h.

Problem Solving: Practice and Review

Follow your instructor's guidelines for using the five steps as outlined in Section 1.5, p. 51.

For exercises 75–76, the graph represents the relationship of the number of elementary and secondary teachers from 1980 to 2009 in the United States, y, and the number of years since 1980, x.

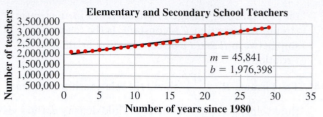

Source: nces.ed.gov

75. Write a function that represents the lines of best fit. Include the units of measurement.

76. Use the function to find the number of these teachers in the United States in 2015. Round to the nearest thousand.

77. Use a proportion to find the number out of 200,000 Americans aged 20 years and over who have elevated blood pressure. Round to the nearest hundred.

About one in six Americans aged 20 years and over has elevated blood pressure, and one in four has hypertension. (*Source:* www.cdc.gov/nchs, Feb. 25, 2009)

78. An employer may be required by court order to *garnish* an employee's wages to pay a debt, back taxes, or child support. For ordinary garnishments (not for child support, bankruptcy, or taxes), the weekly amount is limited "to the lesser of 25 percent of disposable earnings or the amount by which disposable earnings are greater than 30 times the federal minimum hourly wage. . . . The federal minimum wage is $7.25 per hour effective July 24, 2009." The disposable earnings of an employee are $300 a week. Find the maximum amount that can be garnished from this employee's wages per week for an ordinary debt. (*Source:* www.dol.gov/elaws/elg)

Technology

For exercises 79–82,

(a) graph the function in an appropriate window. Sketch the graph; describe the window.

(b) identify the function as a direct or an inverse variation.

79. The cost in dollars, y, of buying x lb of hamburger is described

by the function $y = \left(\dfrac{\$2.99}{1\text{ lb}}\right)x$.

80. The cost in dollars, y, of buying x lb of bananas is described by the function

$$y = \left(\frac{\$0.79}{1 \text{ lb}}\right)x.$$

81. A buyer has \$15,000 saved for a down payment. The relationship of the price of the house in dollars, y, and the percent (in decimal form) that the down payment is of the price of the house, x, is described by the function

$$y = \frac{\$15,000}{x}.$$

82. The relationship of the length in centimeters (cm) of the base of a triangle, x, and the height of the triangle in centimeters, y, for a triangle with a fixed area of 25 cm² is described by the function $y = \dfrac{50 \text{ cm}^2}{x}$.

Find the Mistake

For exercises 83–86, the completed problem has one mistake.
(a) Describe the mistake in words, or copy down the whole problem and highlight or circle the mistake.
(b) Do the problem correctly.

83. Problem: Solve $y = \dfrac{x_1 + x_2}{x_2}$ for x_2.

Incorrect Answer: $y = \dfrac{x_1 + x_2}{x_2}$

$$x_2 y = \frac{(x_1 + x_2)}{x_2} \cdot x_2$$

$$x_2 y = x_1 + x_2$$

$$\underline{-x_1 \quad -x_1}$$

$$x_2 y - x_1 = x_2$$

$$x_2 = x_2 y - x_1$$

84. Problem: Creatinine clearance is a medical test used to assess kidney function. The formula is $C = \dfrac{UV}{1440P}$. Solve this formula for P.

Incorrect Answer: $C = \dfrac{UV}{1440P}$

$$\left(\frac{1440}{UV}\right)C = \left(\frac{1440}{UV}\right)\left(\frac{UV}{1440P}\right)$$

$$\frac{1440C}{UV} = P$$

85. Problem: Solve $\dfrac{1}{x} + \dfrac{1}{y} = \dfrac{1}{z}$ for y.

Incorrect Answer: $\dfrac{1}{x} + \dfrac{1}{y} = \dfrac{1}{z}$

$$\frac{x}{1} + \frac{y}{1} = \frac{z}{1}$$

$$x + y = z$$

$$\underline{-x \qquad\quad -x}$$

$$y = z - x$$

86. Problem: The calories burned per minute depend on the kind of exercise. A woman wants to burn 250 calories (cal) per day with exercise. The relationship of the $\dfrac{\text{calories}}{\text{minute}}$ required by the exercise, x, and the time (min) of exercise, y, is described by the function $y = \dfrac{250 \text{ cal}}{x}$. The woman can schedule 30 min per day for exercise. Find how many calories per minute the exercise must use. Round to the nearest tenth.

Incorrect Answer: $y = \dfrac{250 \text{ cal}}{x}$

$$y = \frac{250 \text{ cal}}{30 \text{ min}}$$

$$y \approx 8.3 \frac{\text{cal}}{\text{min}}$$

Review

87. Use long division to find $428 \div 4$.

88. Use long division to find $428 \div 6$. Do not write in additional zeros after the decimal point. Write the remainder as a fraction, $\dfrac{\text{remainder}}{\text{divisor}}$, and simplify.

89. Identify the degree of the polynomial expression $8x + 2$.

90. Identify the degree of the polynomial expression 50.

SUCCESS IN COLLEGE MATHEMATICS

91. A direct variation is a function that can be written in the form $y = kx$, where k is the constant of proportionality. The graph of a direct variation is a line with a slope of k. What is the y-intercept of this line?

92. An inverse variation is a function that can be written in the form $y = \dfrac{k}{x}$, where k is the constant of proportionality and $k > 0$. Explain why the output of this function can never equal 0.

$$
\begin{array}{r}
x - 3 \\
x - 2 \overline{)\smash{x^2 - 5x + 6}} \\
\underline{-(x^2 - 2x)} \\
-3x + 6 \\
\underline{-(-3x + 6)} \\
0
\end{array}
$$

In Chapter 2, we learned how to add, subtract, and multiply polynomials. We are now ready to divide polynomials. We again assume that all expressions are defined. No denominators equal 0.

SECTION 5.7

Division of Polynomials; Synthetic Division

After reading the text, working the practice problems, and completing assigned exercises, you should be able to:

1. Divide a monomial by a monomial.

2. Divide a polynomial by a monomial.

3. Use long division to divide a polynomial by a binomial.

4. Use synthetic division to divide a polynomial by a binomial.

5. Use long division, synthetic division, or function evaluation to determine if a binomial is a factor of a polynomial.

6. Use the remainder theorem and synthetic division to evaluate a polynomial function.

Division by a Monomial

We use the same language to describe the division of polynomials as we use to describe the division of numbers. In $10 \div 2 = 5$, the **dividend** is 10, the **divisor** is 2, and the **quotient** is 5.

We use a variety of notations and language to indicate division. Appropriate ways to say the notation $10x \div 2$, $\dfrac{10x}{2}$, or $2\overline{)10x}$ in words include "divide $10x$ by 2," "simplify $10x$ divided by 2," "simplify $10x$ over 2," and "find the quotient of $10x$ and 2." When we are dividing whole numbers, we say "evaluate" rather than "simplify."

EXAMPLE 1   **(a)** Simplify: $\dfrac{38x^{10}y^5}{2x^3y}$

SOLUTION ▶

$\dfrac{38x^{10}y^5}{2x^3y}$ The quotient of two monomials: $(38x^{10}y^5) \div (2x^3y)$

$= \dfrac{19 \cdot \cancel{2} \cdot x^{10-3}y^{5-1}}{\cancel{2}}$ Simplify; quotient rule of exponents.

$= 19x^7y^4$ Simplify.

(b) Identify the dividend, divisor, and quotient.

▶   The dividend is $38x^{10}y^5$. The divisor is $2x^3y$. The quotient is $19x^7y^4$.

To add fractions with a common denominator, we combine the numerators and leave the denominator the same. We can write $\dfrac{5}{11}$ as $\dfrac{2}{11} + \dfrac{3}{11}$. We use this same idea to divide a polynomial by a monomial.

EXAMPLE 2 (a) Simplify: $\dfrac{20x^4 + 5x^3 - 35x - 5}{5x}$

SOLUTION ▶

$\dfrac{20x^4 + 5x^3 - 35x - 5}{5x}$ dividend $\overline{\hspace{1.5cm}}$ divisor

$= \dfrac{20x^4}{5x} + \dfrac{5x^3}{5x} - \dfrac{35x}{5x} - \dfrac{5}{5x}$ Rewrite as addition and subtraction of separate terms.

$= 4x^3 + x^2 - 7 - \dfrac{1}{x}$ Simplify each term.

(b) Is the quotient a polynomial?

▶   No, one of the terms has a variable in the denominator.

Dividing a Polynomial by a Monomial

1. Write the division as a fraction: $\dfrac{\text{dividend}}{\text{divisor}}$.

2. Rewrite as addition and/or subtraction of separate terms.

3. Simplify.

Practice Problems

For problems 1–3,
(a) identify the dividend and divisor.
(b) simplify.

1. $\dfrac{21a^6 b^{11} c}{7a^2 bc}$ 2. $\dfrac{15x^3 - 27x^2 + 6x - 9}{3x}$ 3. $\dfrac{21a^6 b^{11} c - 28a^5 bc + 35}{7a^2 bc}$

Division by a Polynomial

We can use the process of **long division** to find the quotient of whole numbers.

EXAMPLE 3 (a) Use long division to find the quotient: $358 \div 2$.

SOLUTION ▶

$2\overline{)358}$

$\begin{array}{r} 1 \\ 2\overline{)358} \end{array}$ Look at the first digit in the dividend, 3. What should we multiply the divisor by so that the product is the largest number that is less than or equal to 3? The answer is 1.

$\begin{array}{r} 1 \\ 2\overline{)358} \\ -2 \\ \hline 1 \end{array}$ Multiply the divisor by 1: $1(2) = 2$. Write this under the first digit in the dividend. Subtract. The difference is 1.

$\begin{array}{r} 1 \\ 2\overline{)358} \\ -2 \downarrow \\ \hline 15 \end{array}$ Bring down the 5 from the dividend. What should we multiply 2 by so that the product is the largest number that is less than or equal to 15? The answer is 7.

$$\begin{array}{r} 17 \\ 2\overline{)358} \\ \underline{-2} \\ 15 \\ \underline{-14} \\ 1 \end{array}$$

Multiply the divisor by 7. Write this product below 15. Subtract. The difference is 1.

$$\begin{array}{r} 17 \\ 2\overline{)358} \\ \underline{-2} \\ 15 \\ \underline{-14} \\ 18 \end{array}$$

Bring down the 8 from the dividend. What should we multiply 2 by so that the product is the largest number that is less than or equal to 18? The answer is 9.

$$\begin{array}{r} 179 \\ 2\overline{)358} \\ \underline{-2} \\ 15 \\ \underline{-14} \\ 18 \\ \underline{-18} \\ 0 \end{array}$$

Multiply the divisor by 9. Write this product below 18. Subtract. The difference is 0. There is no remainder. The quotient is 179.

(b) Check.

▶

$$(\text{quotient})(\text{divisor}) = \text{dividend}$$
$$(179)(2) = 358$$
$$\mathbf{358} = 358 \quad \text{True.} \quad \text{Evaluate; the quotient is correct.}$$

We can also use long division to find the quotient of two polynomials. The terms in the dividend and the divisor should be in **descending order**. If there is a constant, it is the last term.

EXAMPLE 4 **(a)** Use long division to find the quotient: $(x^2 - 5x + 6) \div (x - 2)$.

SOLUTION ▶

$$\begin{array}{r} x \\ x - 2\overline{)x^2 - 5x + 6} \end{array}$$

Look at the first term in the dividend, x^2. What should we multiply the divisor by so that the first term in the product is x^2? The answer is x.

$$\begin{array}{r} x \\ x - 2\overline{)x^2 - 5x + 6} \\ \underline{-(x^2 - 2x)} \\ 0 - 3x \end{array}$$

Multiply the divisor by x: $x(x - 2) = x^2 - 2x$. Subtract. The difference is $-3x$.

$$\begin{array}{r} x - 3 \\ x - 2\overline{)x^2 - 5x + 6} \\ \underline{-(x^2 - 2x)} \\ -3x + 6 \end{array}$$

Now bring down the 6 from the dividend. What should we multiply x by to equal $-3x$? The answer is -3.

$$\begin{array}{r} x - 3 \\ x - 2\overline{)x^2 - 5x + 6} \\ \underline{-(x^2 - 2x)} \\ -3x + 6 \\ \underline{-(-3x + 6)} \\ 0 \end{array}$$

Multiply the divisor by -3: $-3(x - 2) = -3x + 6$. Subtract. The difference is 0. There is no remainder. The quotient of $x^2 - 5x + 6$ and $x - 2$ is $x - 3$.

(b) Check.

$$(\text{quotient})(\text{divisor}) = \text{dividend}$$

$(x - 3)(x - 2) = x^2 - 5x + 6$	
$x(x - 2) - 3(x - 2) = x^2 - 5x + 6$	Distributive property.
$x^2 - 2x - 3x + 6 = x^2 - 5x + 6$	Simplify.
$x^2 - 5x + 6 = x^2 - 5x + 6$ True.	Simplify; the quotient is correct.

Practice Problems

For problems 4–6,
(a) use long division.
(b) check.

4. $(x^2 + 12x + 20) \div (x + 2)$ **5.** $(x^2 + 8x + 15) \div (x + 3)$

6. $(15x^2 + 14x - 8) \div (5x - 2)$

Remainders

Quotients are written with remainders. In the next example, we see how a quotient of long division can be written with a whole number remainder, using a mixed number or using a decimal number, and how to check the quotient.

EXAMPLE 5   **(a)** Use long division to find the quotient: $91 \div 2$.

SOLUTION ▶

$$\begin{array}{r} 45 \\ 2\overline{)91} \\ -8 \\ \hline 11 \\ -10 \\ \hline 1 \end{array}$$

We can write this quotient with a remainder as 45R(1), as the mixed number $45\frac{1}{2}$, where the fraction in the mixed number is $\dfrac{\text{remainder}}{\text{divisor}}$, or we can continue the long division by writing a decimal point in the dividend and as many zeros as necessary after the decimal point. If we do this, the quotient is 45.5.

(b) Check the mixed number quotient.

A mixed number can be written with a plus sign: $45\frac{1}{2}$ represents the sum $45 + \frac{1}{2}$.

$$(\text{divisor})(\text{quotient}) = \text{dividend}$$

$$(2)\left(45 + \frac{1}{2}\right) = 91$$

$$2(45) + 2\left(\frac{1}{2}\right) = 91$$

$$90 + 1 = 91$$

$$91 = 91 \quad \text{True.}$$

Since the product of the divisor and the quotient equals the dividend, the quotient is correct.

The quotient of two polynomials may include a remainder. In long division of two polynomials, divide until the degree of the remainder is less than the degree of the divisor. (The degree of a polynomial in one variable is the value of the largest exponent. The degree of a constant such as 5 is 0 because 5 equals $5x^0$.) Write the quotient of a polynomial that includes a remainder with a plus sign. The last term in the quotient is written as $\dfrac{\text{remainder}}{\text{dividend}}$.

EXAMPLE 6 (a) Use long division to find the quotient: $(x^2 + 8x + 10) \div (x + 2)$.

SOLUTION ▶

$$x + 2\overline{\smash{)}x^2 + 8x + 10}$$
$$x$$

Look at the first term in the dividend, x^2. What should we multiply the divisor by so that the first term in the product is x^2? The answer is x.

$$\begin{array}{r} x \\ x + 2\overline{\smash{)}x^2 + 8x + 10} \\ -(x^2 + 2x) \\ \hline 6x \end{array}$$

Multiply the divisor by x: $x(x + 2) = x^2 + 2x$. Subtract this product from the dividend. The difference is $6x$.

$$\begin{array}{r} x + 6 \\ x + 2\overline{\smash{)}x^2 + 8x + 10} \\ -(x^2 + 2x) \downarrow \\ \hline 6x + 10 \end{array}$$

Now bring down the 10 from the dividend. What should we multiply the divisor by to equal $6x$? The answer is 6.

$$\begin{array}{r} x + 6 \\ x + 2\overline{\smash{)}x^2 + 8x + 10} \\ -(x^2 + 2x) \\ \hline 6x + 10 \\ -(6x + 12) \\ \hline -2 \end{array}$$

Multiply the divisor by 6: $6(x + 2) = 6x + 12$. Subtract this product from the dividend. The difference is -2. This is the remainder. Since $-2 = -2x^0$, the degree of the remainder is 0. The degree of the divisor, $x^1 + 2$, is 1. Since the degree of the remainder is less than the degree of the divisor, the division is complete.

The quotient is $x + 6 + \left(\dfrac{-2}{x + 2}\right)$.

(b) Check.

▶

$$(\text{divisor})(\text{quotient}) = \text{dividend}$$

$$(x + 2)\left(x + 6 + \left(\frac{-2}{x + 2}\right)\right) = x^2 + 8x + 10$$

$$(x + 2)(x) + (x + 2)(6) + (x + 2)\left(\frac{-2}{x + 2}\right) = x^2 + 8x + 10$$

$$x^2 + 2x + 6x + 12 - 2 = x^2 + 8x + 10$$

$$x^2 + 8x + 10 = x^2 + 8x + 10 \quad \text{True.}$$

Since the product of the quotient and divisor equals the dividend, the quotient is correct.

The number 407 is 4 hundreds + 0 tens + 7 ones. The 0 is a **placeholder**. In polynomial division, we also need placeholders. In the next example, we rewrite the polynomial $8x^3 + 10x + 1$ as $8x^3 + 0x^2 + 10x + 1$.

EXAMPLE 7 (a) Use long division to find the quotient: $(8x^3 + 10x + 1) \div (2x + 3)$.

SOLUTION ▶

$$2x + 3\overline{\smash{)}8x^3 + 0x^2 + 10x + 1}$$

The term $0x$ is used as a placeholder.

$$\begin{array}{r} 4x^2 \\ 2x + 3\overline{\smash{)}8x^3 + 0x^2 + 10x + 1} \end{array}$$

Look at the first term in the dividend, $8x^3$. What should we multiply the divisor by so that the first term in the product is $8x^3$? The answer is $4x^2$.

$$\begin{array}{r} 4x^2 \\ 2x + 3\overline{\smash{)}8x^3 + 0x^2 + 10x + 1} \\ -(8x^3 + 12x^2) \\ \hline -12x^2 \end{array}$$

Multiply the divisor by $4x^2$: $4x^2(2x + 3) = 8x^3 + 12x^2$. Subtract this product from the dividend. The difference is $-12x^2$.

$$
\begin{array}{r}
4x^2 - 6x \\
2x + 3{\overline{\smash{\big)}\,8x^3 + 0x^2 + 10x + 1}} \\
-(8x^3 + 12x^2) \\
\hline
-12x^2 + 10x
\end{array}
$$

Now bring down the $10x$ from the dividend. What should we multiply the divisor by to equal $-12x^2$? The answer is $-6x$.

$$
\begin{array}{r}
4x^2 - 6x \\
2x + 3{\overline{\smash{\big)}\,8x^3 + 0x^2 + 10x + 1}} \\
-(8x^3 + 12x^2) \\
\hline
-12x^2 + 10x \\
-(-12x^2 - 18x) \\
\hline
28x
\end{array}
$$

Multiply the divisor by $-6x$: $-6x(2x + 3) = -12x^2 - 18x$. Subtract this product from the dividend. The difference is $28x$.

$$
\begin{array}{r}
4x^2 - 6x + 14 \\
2x + 3{\overline{\smash{\big)}\,8x^3 + 0x^2 + 10x + 1}} \\
-(8x^3 + 12x^2) \\
\hline
-12x^2 + 10x \\
-(-12x^2 - 18x) \\
\hline
28x + 1
\end{array}
$$

Now bring down the 1 from the dividend. What should we multiply the divisor by to equal $28x$? The answer is 14.

$$
\begin{array}{r}
4x^2 - 6x + 14 \\
2x + 3{\overline{\smash{\big)}\,8x^3 + 0x^2 + 10x + 1}} \\
-(8x^3 + 12x^2) \\
\hline
-12x^2 + 10x \\
-(-12x^2 - 18x) \\
\hline
28x + 1 \\
-(28x + 42) \\
\hline
-41
\end{array}
$$

Multiply the divisor by 14: $14(2x + 3) = 28x + 42$. Subtract this product from the dividend. The difference is -41. This is the remainder. The degree of the remainder is 0. The degree of the divisor, $2x^1 + 3$, is 1. Since the degree of the remainder is less than the degree of the divisor, the division is complete.

The quotient is $4x^2 - 6x + 14 + \left(\dfrac{-41}{2x + 3} \right)$.

(b) Check.

$$(\text{divisor})(\text{quotient}) = \text{dividend}$$

$$(2x + 3)\left(4x^2 - 6x + 14 + \left(\dfrac{-41}{2x + 3}\right)\right) = 8x^3 + 10x + 1$$

$$8x^3 + 12x^2 - 12x^2 - 18x + 28x + 42 + (-41) = 8x^3 + 10x + 1$$

$$8x^3 + 0 + 10x + 1 = 8x^3 + 10x + 1$$

$$8x^3 + 10x + 1 = 8x^3 + 10x + 1 \quad \text{True.}$$

Since the product of the quotient and the divisor equals the dividend, the quotient is correct.

Practice Problems

For problems 7–8,
(a) use long division to find the quotient.
(b) check.

7. $(x^2 + 8x + 17) \div (x + 2)$ **8.** $(5x^3 + 7x^2 + 3) \div (x + 4)$

Synthetic Division

The process of synthetic division is an **algorithm**. An algorithm is a procedure, often written as a series of steps. One of the advantages of the synthetic division algorithm is that it involves only multiplication and addition. This allows us to complete the algorithm quickly. We will use synthetic division only when the polynomial divisor is in the form $x - k$ and k is an integer.

Synthetic Division of Polynomials

1. Rewrite the division of polynomials in synthetic division format.

   All polynomials must be in descending order.

   If necessary, use 0 as a coefficient to create a placeholder term.

   When the divisor is $x - k$, the synthetic division test value is k.

   When the divisor is $x + k$, the synthetic division test value is $-k$.

   The synthetic division dividend includes no variables.

2. Leave an empty row below the dividend. Draw a horizontal line below this row. Bring down the first number below this line.

3. Multiply the test value by the first number. Write this product below the second number, above the line.

4. Add the second number and the product from step 3. Write the sum below the line.

5. Multiply this sum by the test value. Write this product below the third number, above the line.

6. Add the third number and the last product. Write the sum below the line.

7. Repeat steps 5 and 6 for any remaining numbers in the dividend.

8. To rewrite the quotient with variables, the final number below the line is the remainder. The other numbers are the coefficients of the polynomial, in descending order.

EXAMPLE 8   Use synthetic division to find the quotient: $(2x^2 + 3x + 4) \div (x - 1)$.

SOLUTION ▶ **Test value** ⟶ $\underline{1|}$ 2 3 4

   $\overline{}$

   2

The divisor, $x - 1$, is in the form $x - k$, so the test value is 1. Leave an empty row. Draw a horizontal line. Bring down the 2 below the line.

$\underline{1|}$ 2 3 4

       2

   2 **5**

Multiply the test value by the first number: $(1)(2) = 2$. Write this product, 2, below the 3. Add the numbers in this column: $3 + 2 = 5$. Write this sum below the line.

$\underline{1|}$ 2 3 4

      2 **5**

   2 5 **9**

Multiply the test value by the sum: $(1)(5) = 5$. Write 5 below the 4. Add the numbers in this column, $4 + 5 = 9$. Write this sum below the line.

$\underline{1|}$ 2 3 4

      2 5

   2 5 | 9

The division is finished. The last number in the bottom row is the remainder. Draw a vertical line between the quotient and the remainder.

The quotient is $2x + 5 + \dfrac{9}{x - 1}$.   Rewrite the quotient with variables.

A goal of synthetic division of polynomials is to discover if the divisor is a factor of a polynomial. If the remainder is 0, the divisor is a factor of the dividend. In Example 8, the remainder is 9. So $x - 1$ is *not* a factor of $2x^2 + 3x + 4$. In the next example, since synthetic division results in a remainder of 0, the divisor is a factor of the dividend.

EXAMPLE 9   (a) Use synthetic division to find the quotient: $(2x^2 + 11x + 15) \div (x + 3)$.

SOLUTION ▶

$$\text{Test value} \longrightarrow \underline{-3|} \quad 2 \quad 11 \quad 15$$

$$\text{_____}$$

$$2$$

The test value is -3. Leave an empty row. Draw a horizontal line. Bring down the 2 below the line.

$$\underline{-3|} \quad 2 \quad 11 \quad 15$$
$$\qquad \quad -6$$
$$\text{_____}$$
$$2 \quad 5$$

Multiply the test value by the first number: $(-3)(2) = -6$. Write this product, -6 below the 11. Add the numbers in this column: $11 + (-6) = 5$. Write the sum below the line.

$$\underline{-3|} \quad 2 \quad 11 \quad 15$$
$$\qquad \quad -6 \; -15$$
$$\text{_____}$$
$$2 \quad 5 \quad 0$$

Multiply the test value by the second number, $(-3)(5) = -15$. Write -15 below the 15. Add the numbers in this column: $15 + (-15) = 0$.

$$\underline{-3|} \quad 2 \quad 11 \quad 15$$
$$\qquad \quad -6 \; -15$$
$$\text{_____}$$
$$2 \quad 5 \;|\; 0$$

The division is finished. The remainder is 0. Draw a vertical line between the quotient and the remainder.

The quotient is $2x + 5$.   Rewrite the quotient with variables.

(b) Is $x + 3$ a factor of $2x^2 + 11x + 15$?

▶   Since the remainder is 0, $x + 3$ is a factor.

We can use long division or synthetic division of polynomials to find out if a divisor is a factor of a dividend. We can also find this information by evaluating a function. If $f(\text{test value}) = 0$, the divisor is a factor of the function rule.

EXAMPLE 10   Evaluate $f(-3)$ to determine whether $x + 3$ is a factor of $f(x) = 2x^2 + 11x + 15$.

SOLUTION ▶

$$f(x) = 2x^2 + 11x + 15$$

$$f(-3) = 2(-3)^2 + 11(-3) + 15$$ Replace x with -3; the test value of the divisor $x + 3$.

$$f(-3) = 18 - 33 + 15$$ Simplify.

$$f(-3) = 0$$ Simplify.

Since $f(\text{test value}) = 0$, $x + 3$ is a factor of $2x^2 + 11x + 15$.

In the next example, we use synthetic division and evaluate a function to find out whether $x + 5$ is a factor of $3x^2 - 11x - 20$.

EXAMPLE 11   (a) Use synthetic division to find out whether $x + 5$ is a factor of $3x^2 - 11x - 20$.

SOLUTION ▶

$$\text{Test value} \longrightarrow \underline{-5|} \quad 3 \quad -11 \quad -20$$

$$\text{_____}$$

$$3$$

The test value is -5. Bring down the 3.

$$\underline{-5|} \quad 3 \quad -11 \quad -20$$
$$\qquad \quad -15$$
$$\text{_____}$$
$$3 \quad -26$$

Multiply the test value by the first number: $(-5)(3) = -15$. Write -15 below the -11. Add the numbers in this column: $-11 + (-15) = -26$.

$$\begin{array}{r|rrr} -5 & 3 & -11 & -20 \\ & & -15 & \mathbf{130} \\ \hline & 3 & -26 & \mathbf{110} \end{array}$$

Multiply the test value by the second number: $(-5)(-26) = 130$. Write 130 below the -20. Add the numbers in this column: $-20 + 130 = 110$. Write the sum below the line.

$$\begin{array}{r|rrr} -5 & 3 & -11 & -20 \\ & & -15 & 130 \\ \hline & 3 & -26 & | \ 110 \end{array}$$

The division is finished. The last number in the bottom row is the remainder. Draw a vertical line between the quotient and the remainder.

The quotient is $3x - 26 + \dfrac{110}{x + 5}$. Since there is a remainder, $x + 5$ is not a factor of $3x^2 - 11x - 20$.

(b) Evaluate a function to find out whether $x + 5$ is a factor of $3x^2 - 11x - 20$.

$$f(x) = 3x^2 - 11x - 20$$

$$f(-5) = 3(-5)^2 - 11(-5) - 20 \qquad \text{Replace } x \text{ with } -5, \text{ the test value of the divisor } x + 5.$$

$$f(-5) = 75 + 55 - 20 \qquad \text{Simplify.}$$

$$f(-5) = 110 \qquad \text{Simplify.}$$

Since $f(-5)$ does not equal 0, $x + 5$ is not a factor of the polynomial.

In Example 11, we see that the remainder of synthetic division and the result of evaluation of the function are the same number. We can use one process to check the other.

EXAMPLE 12   A student uses synthetic division to find that the quotient of $(x^2 - x - 72) \div (x + 9)$ is $x - 10 + \dfrac{18}{x + 9}$. Evaluate a polynomial function to check that the remainder of this division is correct.

SOLUTION ▶

$$f(x) = x^2 - x - 72 \qquad \text{The function rule is the dividend.}$$

$$f(-9) = (-9)^2 - (-9) - 72 \qquad \text{Replace } x \text{ with } -9, \text{ the test value of } x + 9.$$

$$f(-9) = 81 + 9 - 72 \qquad \text{Simplify.}$$

$$f(-9) = 18 \qquad \text{Simplify.}$$

The result of evaluating the function, 18, is the same as the numerator of the remainder found using synthetic division. The quotient found using synthetic division is most likely correct.

Practice Problems

For problems 9–10,
(a) use synthetic division to find the quotient.
(b) check the result by evaluating a function.

9. $(x^2 + 12x + 20) \div (x + 2)$
10. $(x^2 + 9x + 15) \div (x + 3)$

For problems 11–12, evaluate a polynomial function to determine whether the divisor is a factor of the dividend.

11. $(x^2 + 12x + 20) \div (x + 2)$
12. $(x^2 + 9x + 15) \div (x + 3)$

The Remainder Theorem

Another use of synthetic division is for evaluating polynomial functions. The remainder theorem describes the relationship between the remainder of a division and evaluating a polynomial function.

> ### The Remainder Theorem
>
> If $f(x)$ is a polynomial, then the remainder of $f(x) \div (x - c)$ is $f(c)$.

If we use synthetic division to divide $f(x)$ by $x - c$, the remainder of this division equals $f(c)$. In Example 11, we used synthetic division to find the quotient of $(3x^2 - 11x - 20) \div (x + 5)$. The remainder is 110.

$$
\begin{array}{r|rrr}
-5 & 3 & -11 & -20 \\
   &   & -15 & 130 \\
\hline
   & 3 & -26 & 110
\end{array}
$$

According to the remainder theorem, for $f(x) = 3x^2 - 11x - 20$, $f(-5)$ should equal 110. We can verify this by evaluating $f(-5)$.

EXAMPLE 13   Evaluate $f(-5)$ for $f(x) = 3x^2 - 11x - 20$.

SOLUTION ▶

$$f(x) = 3x^2 - 11x - 20$$
$$f(\mathbf{-5}) = 3(\mathbf{-5})^2 - 11(\mathbf{-5}) - 20 \qquad \text{Replace the variable with } -5.$$
$$f(-5) = \mathbf{110} \qquad \text{Simplify.}$$

$f(-5)$ is equal to the remainder of $(3x^2 - 11x - 20) \div (x + 5)$.

With functions such as $f(x) = 3x^2 - 11x - 20$ and inputs such as -5, it is quicker just to replace the variables in the function and simplify than to do synthetic division. However, for more complicated functions and nonreal inputs, this might not be true. Then the remainder theorem can be a quicker way to evaluate a function.

> ### Practice Problems
>
> For problems 13–14,
> **(a)** use synthetic division to find the quotient.
> **(b)** use the remainder theorem to evaluate the function.
>
> **13.** $x^2 + 4x + 13 \div (x + 2)$; $f(x) = x^2 + 4x + 13$; $f(-2)$
> **14.** $x^2 - 6x + 14 \div (x - 3)$; $f(x) = x^2 - 6x + 14$; $f(3)$

Using Technology: Using a Graphing Calculator to Check

We can use a graphing calculator to check some quotients. In Example 6, we saw that $(x^2 + 8x + 10) \div (x + 2) = x + 6 - \dfrac{2}{x + 2}$. To check, we graph two functions: $y = \dfrac{x^2 + 8x + 10}{x + 2}$ and $y = x + 6 - \dfrac{2}{x + 2}$. If the graphs appear to be the same, the quotient is probably correct. (There is a hole in both graphs when $x = -2$ but the resolution of the calculator does not show it.)

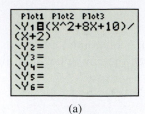

(a)

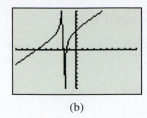

(b)

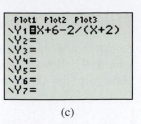

(c)

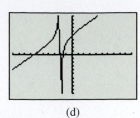

(d)

Put parentheses around the numerator and the denominator.

Put parentheses around the denominator, $x + 2$.

The graphs appear to be the same. The quotient is probably correct.

Practice Problems For problems 15–17,
(a) use long division to find the quotient.
(b) use a graphing calculator to check.

15. $\dfrac{x^2 - 11x + 18}{x - 2}$ **16.** $\dfrac{x^2 - 11x + 18}{x + 2}$ **17.** $\dfrac{x^2 - 11x + 18}{x - 9}$

5.7 VOCABULARY PRACTICE

Match the term with its description.

1. An algorithm used for dividing a polynomial by a linear polynomial
2. The product of the quotient and the divisor
3. The result of division
4. The product of this and a variable is an example of a term.
5. In the expression $3x + 5$, 5 is an example of this.
6. The result of multiplication
7. A polynomial in one variable in which the terms are arranged with the exponents on the variable in order from largest to smallest
8. A polynomial with only one term
9. What the dividend is divided by
10. What are multiplied together to get a product

A. coefficient
B. constant
C. descending order
D. dividend
E. divisor
F. factors
G. monomial
H. product
I. quotient
J. synthetic division

5.7 Exercises

Follow your instructor's guidelines for showing your work. For all exercises, assume that the expression is defined; the denominator does not equal 0.

For exercises 1–4,
(a) identify the dividend.
(b) identify the divisor.

1. $\dfrac{x^2 + 6x}{x}$

2. $x^2 + 9x + 20$ divided by $x + 5$

3. $\dfrac{108x^6y^3z}{48xy^9}$

4. $x - 6\overline{)x^2 - 12x + 36}$

For exercises 5–14, simplify.

5. $\dfrac{102a^8bc^2}{54ab^6c}$

6. $\dfrac{105h^4km^6}{60hk^7m}$

7. $\dfrac{54ab^6c}{102a^8bc^2}$

8. $-\dfrac{60hk^7m}{105h^4km^6}$

9. $\dfrac{12p^6 + 18p^4 - 21p + 15}{3p}$

10. $\dfrac{75y^4 + 20y^3 - 100y + 30}{5y}$

11. $\dfrac{20x^2y^4 - 18xy^2 + 45xy}{9xy}$

12. $\dfrac{41g^3h^3 - 24gh^2 + 42gh}{6gh}$

13. $\dfrac{20x^2y^4z^3 - 18xy^2z + 45xy}{xy}$

14. $\dfrac{41g^3h^3k - 24gh^2k + 42gh}{gh}$

For exercises 15–54,
(a) use long division to find the quotient.
(b) check.

15. $(y^2 + 9y + 18) \div (y + 6)$

16. $(a^2 + 9a + 20) \div (a + 5)$

17. $(y^2 + 9y + 18) \div (y + 3)$

18. $(a^2 + 9a + 20) \div (a + 4)$

19. $(y^2 - 9y + 18) \div (y - 6)$

20. $(a^2 - 9a + 20) \div (a - 5)$

21. $(c^2 + 4c - 21) \div (c + 7)$

22. $(d^2 + 3d - 40) \div (d + 8)$

23. $(c^2 + 4c - 21) \div (c - 3)$

24. $(d^2 + 3d - 40) \div (d - 5)$

25. $(c^2 + 4c - 21) \div (c + 3)$

26. $(d^2 + 3d - 40) \div (d + 5)$

27. $(2x^2 - 11x - 21) \div (x - 7)$

28. $(2p^2 - 7p - 30) \div (p - 6)$

29. $(2x^2 - 11x - 21) \div (x + 7)$

30. $(2p^2 - 7p - 30) \div (p + 6)$

31. $(2x^2 - 11x - 21) \div (2x + 3)$

32. $(2p^2 - 7p - 30) \div (2p + 5)$

33. $(2x^2 - 11x - 21) \div (2x - 3)$

34. $(2p^2 - 7p - 30) \div (2p - 5)$

35. $(5b^2 + 47b + 18) \div (b + 9)$

36. $(7x^2 + 45x + 18) \div (x + 6)$

37. $(5b^2 + 47b + 18) \div (5b + 2)$

38. $(7x^2 + 45x + 18) \div (7x + 3)$

39. $\dfrac{8z^2 - 10z - 3}{4z + 1}$

40. $\dfrac{10m^2 - 33m - 7}{5m + 1}$

41. $\dfrac{20w^2 + 12w - 11}{10w + 11}$

42. $\dfrac{22c^2 + 15c - 13}{11c + 13}$

43. $\dfrac{25x^2 + 20x + 4}{5x + 2}$

44. $\dfrac{36y^2 + 60y + 25}{6y + 5}$

45. $\dfrac{x^3 + 8x^2 + 22x + 21}{x + 3}$

46. $\dfrac{k^3 + 10k^2 + 21k + 40}{k + 8}$

47. $\dfrac{x^3 + 8x^2 + 22x + 21}{x^2 + 5x + 7}$

48. $\dfrac{k^3 + 10k^2 + 21k + 40}{k^2 + 2k + 5}$

49. $\dfrac{x^4 - x^2 - 56}{x^2 + 7}$

50. $\dfrac{z^4 - 4z^2 - 45}{z^2 + 5}$

51. $\dfrac{a^4 - 6a^3 + 3a^2 - 12a + 2}{a^2 + 2}$

52. $\dfrac{w^4 - 7w^3 + 4w^2 - 21w + 3}{w^2 + 3}$

53. $\dfrac{b^8 - 6b^5 + 2b^3 - 12}{b^3 - 6}$

54. $\dfrac{c^8 - 5c^5 + 4c^3 - 20}{c^3 - 5}$

For exercises 55–74,
(a) use synthetic division to find the quotient.
(b) check the result by evaluating a function.

55. $\dfrac{x^2 + 9x + 18}{x + 6}$

56. $\dfrac{x^2 + 9x + 20}{x + 5}$

57. $\dfrac{x^2 + 9x + 18}{x + 3}$

58. $\dfrac{x^2 + 9x + 20}{x + 4}$

59. $\dfrac{y^2 + 4y - 21}{y + 7}$

60. $\dfrac{y^2 + 3y - 40}{y + 8}$

61. $\dfrac{c^2 + 4c - 21}{c + 3}$

62. $\dfrac{d^2 + 3d - 40}{d + 5}$

63. $\dfrac{2p^2 - 15p - 27}{p + 9}$

64. $\dfrac{2m^2 - 15m - 50}{m + 10}$

65. $\dfrac{2p^2 - 15p - 27}{p - 9}$

66. $\dfrac{2m^2 - 15m - 50}{m - 10}$

67. $\dfrac{4x^2 + 19x + 21}{x - 3}$

68. $\dfrac{6x^2 + 17x + 10}{x - 2}$

69. $\dfrac{4x^2 + 19x + 21}{x + 3}$

70. $\dfrac{6x^2 + 17x + 10}{x + 2}$

71. $\dfrac{6x^3 - 19x^2 - 18x - 8}{x - 4}$

72. $\dfrac{8x^3 - 37x^2 - 11x - 20}{x - 5}$

73. $\dfrac{6x^3 - 19x^2 - 18x - 8}{x + 4}$

74. $\dfrac{8x^3 - 37x^2 - 11x - 20}{x + 5}$

For exercises 75–82, evaluate a polynomial function to determine whether the divisor is a factor of the dividend.

75. $(x^2 + 15x + 54) \div (x + 9)$

76. $(x^2 + 15x + 56) \div (x + 8)$

77. $(x^2 + 15x + 54) \div (x - 9)$

78. $(x^2 + 15x + 56) \div (x - 8)$

79. $(2x^2 - 15x - 8) \div (x - 8)$

80. $(3x^2 - 17x - 6) \div (x - 6)$

81. $(x^3 + 2x^2 - 8x - 21) \div (x - 3)$

82. $(x^3 + 4x^2 - 41x - 20) \div (x - 5)$

For exercises 83–88, use synthetic division to find the quotient and the remainder theorem to evaluate the function.

83. $f(x) = x^2 + 9x + 19;\ f(-2)$

84. $f(x) = x^2 + 9x + 19;\ f(-3)$

85. $f(x) = x^2 + 9x + 19;\ f(2)$

86. $f(x) = x^2 + 9x + 19;\ f(3)$

87. $f(x) = x^3 + 5x^2 - x + 1;\ f(4)$

88. $f(x) = x^3 + 7x^2 - x + 3;\ f(2)$

Problem Solving: Practice and Review

Follow your instructor's guidelines for using the five steps as outlined in Section 1.5, p. 51.

89. Find the number of LCD TV shipments in the first quarter of 2008. Round to the nearest thousand.

Despite continued economic weakness, North American consumers continued their love affair with flat panel TVs. . . . Quarter 1 2009 LCD TV shipments surged 23% [compared to first quarter 2008] to 7,200,000 units. (*Source:* www .displaysearch.com, May 11, 2009)

90. Find the proposed reduction in property taxes for a home with an appraised value of $235,750. Round to the nearest hundredth.

Capitol Heights [Maryland] will reduce the current real property tax rate of 41.2 cents per $100 of assessed value to . . . 40.1 cents. (*Source:* www.gazette.net, June 4, 2009)

91. The Phoebe A. Hearst Museum of Anthropology in California has about 4 million items. However, a lack of facilities and modern records keeps most of the collection closed to the public. In 2010, efforts began to digitize the catalog and raise money to construct a new building. Find the percent of the annual budget of the museum in 2008–2009 that was paid by sources other than the State of California.

The current operating budget is $3 million, of which $1.275 million was paid by the state in the 2008–9 fiscal year, down about $50,000 from the previous year. (*Source:* www .nytimes.com, Aug. 14, 2010)

92. The function $y = \left(\dfrac{0.485\%}{1 \text{ year}}\right)x + 21.094\%$ describes the relationship of the percent of Americans age 25 and over with a bachelor's degree, y, with the number of years since 1991, x. Predict the percent of Americans who will have earned a bachelor's degree or higher in 2012. Round to the nearest tenth of a percent. (*Source:* nces.ed.gov)

Technology

For exercises 93–96,
(a) find the quotient using either long division or synthetic division.
(b) check your work with a graphing calculator using the method shown in this section. Sketch the graph; describe the window.

93. $(x^2 + 5x - 24) \div (x + 8)$

94. $(x^2 + 7x - 18) \div (x + 9)$

95. $(x^2 + 5x - 24) \div (x - 8)$

96. $(x^2 + 7x - 18) \div (x - 9)$

Find the Mistake

For exercises 97–100, the completed problem has one mistake.
(a) Describe the mistake in words or copy down the whole problem and highlight or circle the mistake.
(b) Do the problem correctly.

97. Problem: Find the quotient: $(20x^3 + 4x^2 - 2x) \div (4x)$.

Incorrect Answer: $\dfrac{20x^3 + 4x^2 - 2x}{4x}$

$$= 5x^2 + 4x^2 - 2x$$

98. Problem: Use long division to find the quotient: $(x^2 - 8x + 12) \div (x - 6)$.

Incorrect Answer:

$$
\begin{array}{r}
x - 14 \\
x - 6{\overline{\smash{\big)}\,x^2 - 8x + 12}} \\
\underline{x^2 - 6x} \\
-14x + 12 \\
\underline{-14x + 84} \\
-72
\end{array}
$$

The quotient is $x - 14 + \left(\dfrac{-72}{x - 6}\right)$.

99. Problem: Use long division to find the quotient:
$(x^2 - 13x - 24) \div (x - 8)$.

Incorrect Answer:

$$
\begin{array}{r}
x - 21 \\
x - 8 \overline{) x^2 - 13x - 24} \\
-(x^2 - 8x) \\
\hline
-21x - 24 \\
-(-21x + 168) \\
\hline
-144
\end{array}
$$

The quotient is $x - 21 + \dfrac{-144}{x - 8}$.

100. Problem: Use synthetic division to find the quotient:
$(x^2 - 2x - 8) \div (x + 2)$.

Incorrect Answer:

$$
\begin{array}{r|rrr}
2 & 1 & -2 & -8 \\
  &   & 2 & 0 \\
\hline
  & 1 & 0 & -8
\end{array}
$$

The quotient is $x + \dfrac{-8}{x + 2}$.

Review

101. Graph: $y = -\dfrac{3}{4}x + 10$

102. a. Solve: $2(4x + 1) - 17 = 5x + 9$
    b. Check.

103. a. Solve: $5x^2 + 3x - 8 = 0$
    b. Check.

104. Find any real zero(s) of $f(x) = 6x^2 - 17x - 88$.

SUCCESS IN COLLEGE MATHEMATICS

105. Explain how to check the quotient of dividing a whole number by another whole number when the quotient does not include a remainder.

106. Explain how to check the quotient of dividing a whole number by another whole number when the quotient includes a remainder.

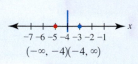

$(-\infty, -4)(-4, \infty)$

In Section 2.6, we found boundary values for absolute value inequalities and then used them to describe intervals on the real number line. We then tested values to find the intervals that included solutions of the inequality. In this section, we will use the same process to solve rational inequalities.

SECTION 5.8

Rational Inequalities

After reading the text, working the practice problems, and completing assigned exercises, you should be able to:

1. Find the boundary values of a rational inequality.

2. Solve a rational inequality.

3. Use interval notation or a number line graph to represent the solution of a rational inequality.

Boundary Values

The solution of the linear equation $2x = 6$ is $x = 3$. If we replace x in the equation with any other number, the equation is not true. In contrast, the linear inequality $2x > 6$ has many solutions. All real numbers greater than 3 are solutions of this inequality. The set that includes all of the solutions of the inequality is its **solution set**. The number 3 is a **boundary value** of this inequality. The boundary value divides the set of real numbers into intervals. If a number in an interval is a solution of the inequality, all of the numbers in the interval are solutions.

In Chapter 1, we used the properties of inequality to solve linear inequalities. We can instead find the boundary values of the inequalities, use the boundary values to divide the set of real numbers into intervals, and choose a number in each interval to check if the interval includes solutions of the inequality.

EXAMPLE 1 Solve the linear inequality $-2x + 6 < 14$.

(a) Find any real boundary values.

SOLUTION ▶ To find the boundary values, solve the linear equation $-2x + 6 = 14$.

$$-2x + 6 = 14$$

$$\underline{\quad -6 \quad -6 \quad}$$ Subtraction property of equality.

$$-2x + 0 = 8$$ Simplify.

$$\frac{-2x}{-2} = \frac{8}{-2}$$ Division property of equality.

$$x = -4$$ Simplify.

(b) Use the boundary values to solve the inequality.

▶ The boundary values divide the set of real numbers into intervals. Choose a number in each interval, and check whether it is a solution of the inequality.

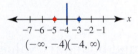

$(-\infty, -4)(-4, \infty)$

Interval: $(-\infty, -4)$	**Interval:** $(-4, \infty)$
Test: $x = -5$	**Test:** $x = -3$
$-2x + 6 < 14$	$-2x + 6 < 14$
$-2(-5) + 6 < 14$	$-2(-3) + 6 < 14$
$16 < 14$ False.	$12 < 14$ True.

Since $x = -5$ is not a solution of the inequality, none of the numbers in the interval $(-\infty, -4)$ are solutions. Since $x = -3$ is a solution of the inequality, all of the numbers in the interval $(-4, \infty)$ are solutions.

(c) Use a number line graph and interval notation to represent the solution.

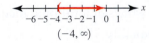

$(-4, \infty)$

To solve a rational inequality, we follow a similar process. To find the boundary values for a strict inequality ($>$ or $<$), rewrite the rational inequality as an equation and solve it. Identify any values that result in division by 0.

EXAMPLE 2 Solve the rational inequality $\dfrac{2x}{x - 2} > 1$.

(a) Find any real boundary values.

SOLUTION ▶ If $x = 2$, the denominator equals 0. Since division by 0 is undefined, $x = 2$ is a boundary value.

To find other boundary values, solve the rational equation $\dfrac{2x}{x-2} = 1$.

$$\frac{2x}{x-2} = 1$$

$$\left(\frac{x-2}{1}\right)\frac{2x}{x-2} = (x-2)1 \qquad \text{Clear fractions; multiplication property of equality.}$$

$$2x = x - 2 \qquad \text{Simplify.}$$

$$\underline{-x \quad -x} \qquad \text{Subtraction property of equality.}$$

$$x = 0 - 2 \qquad \text{Simplify.}$$

$$x = -2 \qquad \text{Simplify.}$$

(b) Use the boundary values to solve the inequality.

The boundary values divide the set of real numbers into intervals. Choose a number in each interval, and check whether it is a solution of the inequality.

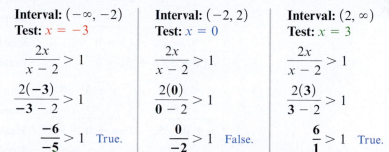

$(-\infty, -2) \quad (-2, 2) \quad (2, \infty)$

Interval: $(-\infty, -2)$	**Interval:** $(-2, 2)$	**Interval:** $(2, \infty)$
Test: $x = -3$	**Test:** $x = 0$	**Test:** $x = 3$
$\dfrac{2x}{x-2} > 1$	$\dfrac{2x}{x-2} > 1$	$\dfrac{2x}{x-2} > 1$
$\dfrac{2(-3)}{-3-2} > 1$	$\dfrac{2(0)}{0-2} > 1$	$\dfrac{2(3)}{3-2} > 1$
$\dfrac{-6}{-5} > 1$ True.	$\dfrac{0}{-2} > 1$ False.	$\dfrac{6}{1} > 1$ True.

Since $x = -3$ is a solution of the inequality, all of the numbers in the interval $(-\infty, -2)$ are solutions. Since $x = 0$ is not a solution of the inequality, none of the numbers in the interval $(-2, 2)$ are solutions. Since $x = 3$ is a solution of the inequality, all of the numbers in the interval $(2, \infty)$ are solutions.

(c) Use a number line graph and interval notation to represent the solution.

$(-\infty, -2) \quad \cup \quad (2, \infty)$

Practice Problems

For problems 1–2,
(a) find any real boundary values.
(b) use the boundary values to solve.
(c) use a number line graph and interval notation to represent the solution.

1. $\dfrac{3x}{x-6} < 1$ **2.** $\dfrac{3x}{x-6} > 1$

More Rational Inequalities

In the next example, there are three boundary values.

EXAMPLE 3 | Solve: $\dfrac{1}{3} - \dfrac{5}{x^2} < \dfrac{2}{3x}$

(a) Find any real boundary values.

SOLUTION ▶ If $x = 0$, the denominator equals 0. Since division by 0 is undefined, $x = 0$ is a boundary value.

To find other boundary values, solve the rational equation $\dfrac{1}{3} - \dfrac{5}{x^2} = \dfrac{2}{3x}$.

$$\dfrac{1}{3} - \dfrac{5}{x^2} = \dfrac{2}{3x}$$

$$\left(\dfrac{3x^2}{1}\right)\left(\dfrac{1}{3} - \dfrac{5}{x^2}\right) = \left(\dfrac{3x^2}{1}\right)\dfrac{2}{3x} \qquad \text{Clear fractions; multiplication property of equality.}$$

$$3x^2\left(\dfrac{1}{3}\right) + 3x^2\left(-\dfrac{5}{x^2}\right) = 2x \qquad \text{Distributive property; simplify.}$$

$$x^2 - 15 = 2x \qquad \text{Simplify.}$$

$$\underline{\quad\quad -2x \; -2x \quad\quad} \qquad \text{Subtraction property of equality.}$$

$$x^2 - 2x - 15 = 0 \qquad \text{Simplify.}$$

$$(x - 5)(x + 3) = 0 \qquad \text{Factor.}$$

$$x - 5 = 0 \quad \text{or} \quad x + 3 = 0 \qquad \text{Zero product property.}$$

$$\underline{+5 \;\; +5} \qquad\quad \underline{-3 \quad -3} \qquad \text{Property of equality.}$$

$$x + 0 = 5 \qquad\quad x + 0 = -3 \qquad \text{Simplify.}$$

$$x = 5 \quad \text{or} \qquad x = -3 \qquad \text{Simplify.}$$

(b) Use the boundary values to solve the inequality.

▶ The boundary values divide the set of real numbers into intervals. Choose a number in each interval, and check whether it is a solution of the inequality.

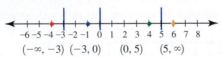

$$(-\infty, -3) \; (-3, 0) \quad (0, 5) \quad (5, \infty)$$

Interval: $(-\infty, -3)$	**Interval:** $(-3, 0)$
Test: $x = -4$	**Test:** $x = -1$
$\dfrac{1}{3} - \dfrac{5}{x^2} < \dfrac{2}{3x}$	$\dfrac{1}{3} - \dfrac{5}{x^2} < \dfrac{2}{3x}$
$\dfrac{1}{3} - \dfrac{5}{(-4)^2} < \dfrac{2}{3(-4)}$	$\dfrac{1}{3} - \dfrac{5}{(-1)^2} < \dfrac{2}{3(-1)}$
$\dfrac{1}{3} - \dfrac{5}{16} < \dfrac{2}{-12}$	$\dfrac{1}{3} - \dfrac{5}{1} < -\dfrac{2}{3}$
$\dfrac{16}{48} - \dfrac{15}{48} < -\dfrac{8}{48}$	$\dfrac{1}{3} - \dfrac{15}{3} < -\dfrac{2}{3}$
$\dfrac{1}{48} < -\dfrac{8}{48}$ False.	$-\dfrac{14}{3} < -\dfrac{2}{3}$ True.

Interval: $(0, 5)$
Test: $x = 4$

$$\frac{1}{3} - \frac{5}{x^2} < \frac{2}{3x}$$

$$\frac{1}{3} - \frac{5}{(4)^2} < \frac{2}{3(4)}$$

$$\frac{1}{3} - \frac{5}{16} < \frac{2}{12}$$

$$\frac{16}{48} - \frac{15}{48} < \frac{1}{6}$$

$$\frac{1}{48} < \frac{1}{6} \qquad \text{True.}$$

Interval: $(5, \infty)$
Test: $x = 6$

$$\frac{1}{3} - \frac{5}{x^2} < \frac{2}{3x}$$

$$\frac{1}{3} - \frac{5}{(6)^2} < \frac{2}{3(6)}$$

$$\frac{1}{3} - \frac{5}{36} < \frac{2}{18}$$

$$\frac{12}{36} - \frac{5}{36} < \frac{4}{36}$$

$$\frac{7}{36} < \frac{4}{36} \qquad \text{False.}$$

Since $x = -4$ is not a solution of the inequality, none of the numbers in the interval $(-\infty, -3)$ are solutions. Since $x = -1$ and $x = 4$ are solutions of the inequality, all of the numbers in the intervals $(-3, 0)$ and $(0, 5)$ are solutions. Since $x = 6$ is not a solution of the inequality, none of the numbers in the interval $(5, \infty)$ are solutions. Since the only value that results in division by 0 is 0 and 0 is not included in any of the intervals that are solutions, we do not need to make any further restrictions.

(c) Use a number line graph and interval notation to represent the solution.

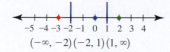

$$(-3, 0) \cup \quad (0, 5)$$

When an inequality is not strict, we also test the boundary values to see whether they are solutions of the inequality.

EXAMPLE 4 Solve: $\dfrac{x - 1}{x + 2} \geq 0$

(a) Find any real boundary values.

SOLUTION ▶ If $x = -2$, the denominator equals 0. Since division by 0 is undefined, $x = -2$ is a boundary value.
  To find other boundary values, solve the rational equation $\dfrac{x - 1}{x + 2} = 0$.

$$\frac{x - 1}{x + 2} = 0$$

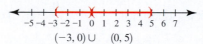   Clear fractions; multiplication property of equality.

$$x - 1 = 0 \qquad \text{Simplify.}$$

$$\underline{+ 1 + 1} \qquad \text{Addition property of equality.}$$

$$x + 0 = 1 \qquad \text{Simplify.}$$

$$x = 1 \qquad \text{Simplify.}$$

(b) Use the boundary values to solve the inequality.

The boundary values divide the set of real numbers into intervals. Choose a number in each interval, and check whether it is a solution of the inequality.

$$(-\infty, -2)(-2, 1)(1, \infty)$$

Interval: $(-\infty, -2)$	**Interval:** $(-2, 1)$	**Interval:** $(1, \infty)$
Test: $x = -3$	**Test:** $x = 0$	**Test:** $x = 2$
$\dfrac{x-1}{x+2} \geq 0$	$\dfrac{x-1}{x+2} \geq 0$	$\dfrac{x-1}{x+2} \geq 0$
$\dfrac{-3-1}{-3+2} \geq 0$	$\dfrac{0-1}{0+2} \geq 0$	$\dfrac{2-1}{2+2} \geq 0$
$\dfrac{-4}{-1} \geq 0$ True.	$-\dfrac{1}{2} \geq 0$ False.	$\dfrac{1}{4} \geq 0$ True.

Since $x = -3$ is a solution of the inequality, all of the numbers in the interval $(-\infty, -2)$ are solutions. Since $x = 0$ is not a solution of the inequality, none of the numbers in the interval $(-2, 1)$ are solutions. Since $x = 2$ is a solution of the inequality, all of the numbers in the interval $(1, \infty)$ are solutions.

Since the inequality $\dfrac{x-1}{x+2} \geq 0$ is not strict, we need to also test the boundary values to see whether they are solutions.

Test: $x = -2$	**Test:** $x = 1$
$\dfrac{x-1}{x+2} \geq 0$	$\dfrac{x-1}{x+2} \geq 0$
$\dfrac{-2-1}{-2+2} \geq 0$	$\dfrac{1-1}{1+2} \geq 0$
$\dfrac{-3}{0} \geq 0$ Undefined.	$\dfrac{0}{3} \geq 0$ True.

Since $x = -2$ results in division by 0, it is not a solution. However, $x = 1$ is a solution of the inequality.

(c) Use a number line graph and interval notation to represent the solution.

$$(-\infty, -2) \ \cup \ [1, \infty)$$

Solving a Rational Inequality

1. Find the boundary values of the inequality.

2. Use the boundary values to divide the set of real numbers into intervals.

3. Choose a number from each interval. If this number is a solution of the inequality, every number in the interval is a solution.

4. If the inequality is not strict, test the boundary value(s) to see whether it is a solution.

The only boundary values of some rational inequalities are those that cause division by 0.

EXAMPLE 5   **(a)** Solve: $\dfrac{3}{x^2 + 9x + 8} \leq 0$

SOLUTION   Factor the denominator. The inequality is $\dfrac{3}{(x+8)(x+1)} \leq 0$. Since $x = -8$ and $x = -1$ cause division by 0, these are boundary values.

To find other boundary values, solve $\dfrac{3}{x^2 + 9x + 8} = 0.$

$$\frac{3}{x^2 + 9x + 8} = 0$$

$$\left(\frac{\cancel{x^2 + 9x + 8}}{1}\right)\left(\frac{3}{\cancel{x^2 + 9x + 8}}\right) = (x^2 + 9x + 8)0 \qquad \text{Multiplication property of equality.}$$

$$3 = 0 \quad \text{False.} \qquad \text{Simplify.}$$

Since this equation has no solution, the only boundary values are $x = -8$ and $x = -1$.

The boundary values divide the set of real numbers into intervals. Choose a number in each interval, and check whether it is a solution of the inequality.

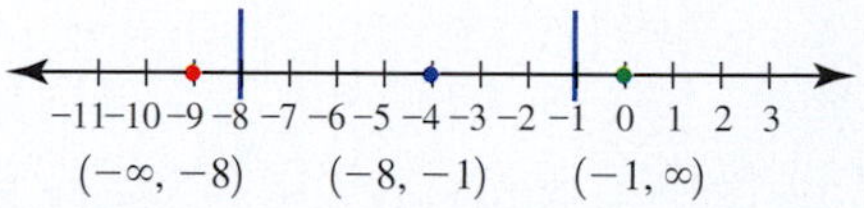
−11 −10 −9 −8 −7 −6 −5 −4 −3 −2 −1 0 1 2 3
(−∞, −8)
(−8, −1)
(−1, ∞)

Interval: $(-\infty, -8)$
Test: $x = -9$

$$\frac{3}{x^2 + 9x + 8} \le 0$$

$$\frac{3}{(\mathbf{-9})^2 + 9(\mathbf{-9}) + 8} \le 0$$

$$\frac{\mathbf{3}}{\mathbf{8}} \le 0 \quad \text{False.}$$

Interval: $(-8, -1)$
Test: $x = -4$

$$\frac{3}{x^2 + 9x + 8} \le 0$$

$$\frac{3}{(\mathbf{-4})^2 + 9(\mathbf{-4}) + 8} \le 0$$

$$\frac{\mathbf{3}}{\mathbf{-12}} \le 0 \quad \text{True.}$$

Interval: $(-1, \infty)$
Test: $x = 0$

$$\frac{3}{x^2 + 9x + 8} \le 0$$

$$\frac{3}{(\mathbf{0})^2 + 9(\mathbf{0}) + 8} \le 0$$

$$\frac{\mathbf{3}}{\mathbf{8}} \le 0 \quad \text{False.}$$

Since $x = -9$ is not a solution of the inequality, none of the numbers in the interval $(-\infty, -8)$ are solutions. Since $x = -4$ is a solution of the inequality, all of the numbers in the interval $(-8, -1)$ are solutions. Since $x = 0$ is not a solution of the inequality, none of the numbers in the interval $(-1, \infty)$ are solutions.

Since the inequality $\dfrac{3}{x^2 + 9x + 8} \le 0$ is not strict, the boundary values could be solutions. However, since we already know that $x = -8$ and $x = -1$ result in division by 0, they cannot be solutions.

(b) Use a number line graph and interval notation to represent the solution.

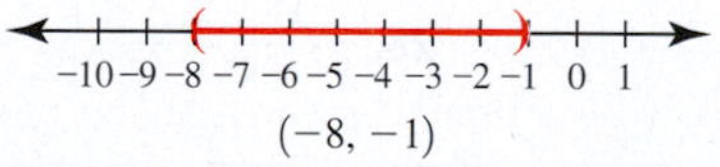
−10 −9 −8 −7 −6 −5 −4 −3 −2 −1 0 1
(−8, −1)

Practice Problems

For problems 3–6,
(a) solve the inequality.
(b) use interval notation to represent the solution.

3. $\dfrac{x^2 - 9}{x + 3} \ge 0$ **4.** $\dfrac{2}{x + 4} > \dfrac{1}{x - 3}$

5. $\dfrac{2}{x + 4} \le \dfrac{1}{x - 3}$ **6.** $\dfrac{x^2 - 9}{x + 4} \ge 0$

5.8 VOCABULARY PRACTICE

Match the term with its description.

1. Division by zero
2. $\geq$
3. $<$
4. $(-3, 5)$
5. $[-3, 5]$
6. $(-\infty, \infty)$
7. The quotient of two polynomials
8. The quotient of two integers
9. If a fraction is this, the numerator and denominator have no common factors.
10. If $ab = 0$, then $a = 0$ or $b = 0$.

A. greater than or equal to
B. lowest terms
C. rational expression
D. rational number
E. set of real numbers
F. strict inequality
G. this interval includes the endpoints
H. this interval does not include the endpoints
I. undefined
J. zero product property

5.8   Exercises

Follow your instructor's guidelines for showing your work.

For exercises 1–8, find any real boundary values.

1. $\dfrac{1}{x} > 4$

2. $\dfrac{1}{x} > 5$

3. $\dfrac{x - 2}{x + 4} \leq 3$

4. $\dfrac{2x - 4}{x + 2} \leq 4$

5. $\dfrac{2}{x^2 + 7x + 10} < 0$

6. $\dfrac{3}{x^2 - 2x + 1} \leq 0$

7. $\dfrac{x + 2}{x} > 0$

8. $\dfrac{x + 3}{x} > 0$

For exercises 9–18, use interval notation to represent the number line graph.

9.
   -8 -7 -6 -5 -4 -3 -2 -1 0 1 2

10.
   -9 -8 -7 -6 -5 -4 -3 -2 -1 0

11.
   -10 -9 -8 -7 -6 -5 -4 -3 -2 -1

12.
   -6 -5 -4 -3 -2 -1 0

13.
   10   12   14   16

14.
   10   12   14   16   18   20

15.
   -7 -6 -5 -4 -3 -2 -1 0 1 2 3 4

16.
   -9 -8 -7 -6 -5 -4 -3 -2 -1 0 1 2 3

17.
   -2 -1 0 1 2 3 4 5 6

18.
   -6 -5 -4 -3 -2 -1 0 1 2 3 4

For exercises 19–30,
(a) find any boundary values.
(b) use the boundary values to solve the inequality.
(c) use interval notation to represent the solution.

19. $\dfrac{5x}{4x - 4} > 1$

20. $\dfrac{6x}{5x - 10} > 1$

21. $\dfrac{5x}{4x - 4} \leq 1$

22. $\dfrac{6x}{5x - 10} \leq 1$

23. $\dfrac{2}{x^2 - 9} \geq 0$

24. $\dfrac{3}{x^2 - 16} \geq 0$

25. $\dfrac{3}{x - 4} \geq \dfrac{2}{x + 7}$

26. $\dfrac{4}{x - 5} \leq \dfrac{3}{x + 9}$

27. $\dfrac{x}{x - 9} < 10$

28. $\dfrac{x}{x - 8} < 9$

29. $\dfrac{2}{x^2 + 7x + 10} > 0$

30. $\dfrac{5}{x^2 + 6x + 8} > 0$

Problem Solving: Practice and Review

Follow your instructor's guidelines for using the five steps as outlined in Section 1.5, p. 51.

31. A headline in the *Seattle Times* read, "Some UW nursing students face 43% tuition increase." Find the correct percent increase in tuition. Round to the nearest percent.

    Gillian Ehrlich, a student in the UW doctoral family-nurse-practitioner program, will see her tuition rise from $2,600 a year to $26,532. (*Source:* www.seattletimes.nwsource.com, June 6, 2009)

32. A tower in a small wind energy system in Carroll County will be the maximum height allowed. The distance of the guy cables from the base of the tower is the guy radius. This will be three-quarters of the tower height. Find the length of each guy cable.

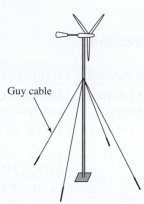

Guy cable

May 6, 2008—The Board of County Commissioners today unanimously voted to allow small wind energy systems in Carroll County. It is believed to be the first such vote in any Maryland county. An amendment to the county's zoning ordinance permits the power-generating wind turbines as an accessory use in all zoning districts, as long as certain requirements are met. To be considered a small wind energy system, its capacity must be no more than 50 kilowatts and its tower cannot exceed 150 feet. The allowable height varies, depending on the size of the property. (*Source:* ccgovernment.carr.org)

There are two basic types of towers: self-supporting (free standing) and guyed. Most home wind power systems use a guyed tower. . . . However, because the guy radius must be one-half to three-quarters of the tower height, guyed towers require enough space to accommodate them. (*Source:* www .energysavers.gov)

33. Find the total U.S. expenditures for physician and clinical services. Round to the nearest billion.

Physician practices report that overall the costs of interacting with insurance plans is $31 billion annually and 6.9 percent of all U.S. expenditures for physician and clinical services. (*Source:* weill.cornell.edu, May 14, 2009)

34. A homeowner is building a rectangular patio in his backyard. He wants the patio to be twice as long as it is wide. He has enough patio bricks to cover an area of 750 ft². Find the width and length of the rectangular patio. Round to the nearest tenth.

Find the Mistake

For exercises 35–38, the completed problem has one mistake.
(a) Describe the mistake in words, or copy down the whole problem and highlight or circle the mistake.
(b) Do the problem correctly.

35. Problem: Find the boundary values for the inequality
$$\frac{5}{x + 9} > 0.$$
Incorrect Answer: Since the equation $\frac{5}{x + 9} = 0$ has no solution, the inequality has no boundary values.

36. Problem: Solve $\frac{x}{x + 3} < 4$. Use interval notation to represent the solution.

Incorrect Answer: To find boundary values, solve
$$\frac{x}{x + 3} = 4.$$

$$\left(\frac{x + 3}{1}\right)\left(\frac{x}{x + 3}\right) = (x + 3)(4)$$

$$x = 4x + 12$$
$$-3x = 12$$
$$x = -4$$

Interval: $(-\infty, -4)$	**Interval:** $(-1, \infty)$
Test: $x = -5$	**Test:** $x = 0$
$\dfrac{x}{x + 3} < 4$	$\dfrac{x}{x + 3} < 4$
$\dfrac{-5}{-5 + 3} < 4$	$\dfrac{0}{0 + 3} < 4$
$\dfrac{5}{2} < 4$ True.	$0 < 4$ True.

The solution is $(-\infty, -1) \cup (-1, \infty)$.

37. Problem: Use interval notation to represent the number line graph.

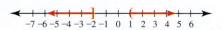

-7 -6 -5 -4 -3 -2 -1 0 1 2 3 4 5 6

Incorrect Answer: $(-\infty, -2) \cup (1, \infty)$

38. Problem: Find the boundary values for the inequality
$$\frac{4}{x^2 - 9x} < 0$$
Incorrect Answer: The equation $\frac{4}{x^2 - 9x} = 0$ has no solution. We can factor the denominator:
$$\frac{4}{(x - 3)(x + 3)} < 0.$$ Because they cause division by zero, $x = 3$ and $x = -3$ are boundary values.

Review

For exercises 39–42, evaluate.

39. 4^3 **40.** 4^{-3}

41. 4^0 **42.** 4^{-1}

SUCCESS IN COLLEGE MATHEMATICS

43. The solution of the linear inequality $7x < 14$ is $x < 2$. What is the boundary value of this linear inequality?

44. Explain why, when solving a rational inequality, we first identify its boundary values.

Study Plan for Review of Chapter 5

In these exercises, assume that all variables in expressions represent nonnegative real numbers. Use interval notation to represent domains and ranges.

SECTION 5.1 Simplifying, Multiplying, and Dividing Rational Expressions

Ask Yourself	Test Yourself	Help Yourself
Can I . . .	**Do 5.1 Review Exercises**	**See these Examples and Practice Problems**
explain how to determine if a rational expression is simplified?	1, 2, 7–9	Ex. 1–7, PP 1–7
simplify a rational expression?		
multiply rational expressions?	3, 4	Ex. 8, 9, PP 8, 10
divide rational expressions?	5, 6	Ex. 10, PP 9, 11

5.1 Review Exercises

For exercises 1–6, simplify.

1. $\dfrac{-42a^2bc^9}{102a^7b^4c^2}$

2. $\dfrac{6x^2 - 13x - 28}{3x^2 + 22x + 24}$

3. $\dfrac{x^2 - 4x - 21}{2x^2 + 3x + 1} \cdot \dfrac{2x^2 - 7x - 4}{x^2 - 11x + 28}$

4. $\dfrac{x^2 + x}{x^3 + 4x^2 + 3x} \cdot \dfrac{x^2 + x - 6}{x^2 - 1}$

5. $\dfrac{21w^2k^5}{8h^4} \div \dfrac{9w^{10}k^8}{24h}$

6. $\dfrac{3x^2 + 28x + 9}{3x^2 + 6x - 189} \div \dfrac{3x^2 - 5x - 2}{x^2 - 4}$

7. Explain why $\dfrac{x(x - 1)(x + 2)}{(x + 3)(x + 2)}$ is not simplified.

8. An expression is $\dfrac{(7 - a)(a + 3)}{(a - 7)(a + 2)}$. Describe the next step in simplifying this expression.

9. Explain why $\dfrac{4x^{-2}y^7}{9}$ is not simplified.

SECTION 5.2 Adding and Subtracting Rational Expressions

Ask Yourself	Test Yourself	Help Yourself
Can I . . .	**Do 5.2 Review Exercises**	**See these Examples and Practice Problems**
write the prime factorization of a polynomial?	10, 11	Ex. 1–4, PP 1–4
find the least common denominator of rational numbers or expressions?	12, 13	Ex. 2–4, PP 1–4
rewrite a rational expression as an equivalent expression with a different denominator?	14, 15	Ex. 6–8, PP 5–8
add or subtract rational expressions with the same denominator?	16	Ex. 9, PP 9
add or subtract rational expressions with different denominators?	17–20	Ex. 10–12, PP 10–12

5.2 Review Exercises

For exercises 10–11, write the prime factorization of the expression.

10. $48a^2bc^3$

11. $x^3 + 4x^2 - 45x$

For exercises 12–13, find the least common denominator.

12. $\dfrac{5}{x^3 - 9x}; \dfrac{8}{2x^2 + x - 15}$

13. $\dfrac{9}{24h^2k}; \dfrac{8}{36hkm}$

14. Rewrite $\dfrac{2}{15x^2y}$ with a denominator of $105x^3y^4$.

15. Rewrite $\dfrac{5}{3x^2 + 22x - 16}$ with a denominator of $(3x - 2)(x + 8)(x - 1)$.

For exercises 16–20, simplify.

16. $\dfrac{z + 3}{2z + 8} + \dfrac{z + 7}{2z + 8}$

17. $\dfrac{3}{2h} - \dfrac{h}{14}$

18. $\dfrac{3p + 1}{p^2 - 1} + \dfrac{p}{p + 1}$

19. $\dfrac{x + 27}{x^2 - 2x - 15} + \dfrac{2x - 9}{x^2 + x - 6}$

20. $\dfrac{c}{c + 2} - \dfrac{c}{c - 2}$

SECTION 5.3   Complex Rational Expressions

Ask Yourself	Test Yourself	Help Yourself
Can I...	**Do 5.3 Review Exercises**	**See these Examples and Practice Problems**
simplify a complex rational expression?	21	Ex. 2–4, PP 2-6
simplify a complex rational expression that includes addition or subtraction in the numerator and/or denominator using Strategy 1 or Strategy 2?	22, 23	Ex. 5–11, PP 7–14

5.3 Review Exercises

For exercises 21–23, simplify.

21. $\dfrac{\dfrac{2x^2 + 13x + 15}{x^2 - 25}}{\dfrac{2x^2 + 3x}{x^2 + 7x}}$

22. $\dfrac{1 - \dfrac{7}{x + 1}}{\dfrac{4}{x + 1} + 1}$

23. $\dfrac{3 - \dfrac{1}{d}}{9 - \dfrac{1}{d^2}}$

SECTION 5.4   Rational Equations in One Variable

Ask Yourself	Test Yourself	Help Yourself
Can I...	**Do 5.4 Review Exercises**	**See these Examples and Practice Problems**
clear the fractions from an equation in one variable?	24–26, 29	Ex. 1–3, PP 1–3
use the zero product property to solve a polynomial equation?	24–26	Ex. 4, 5, PP 5–7
solve a rational equation in one variable?	24–26, 29	Ex. 4–7, PP 4–7
identify an extraneous solution?	24, 25, 28	Ex. 5–7, PP 7
explain why some rational equations have an extraneous solution(s)?		
use a proportion to solve an application problem?	30	Ex. 8, PP 8–9
use a rational equation to solve an application problem involving rates of work?	31	Ex. 10–12, PP 10, 11
use a rational equation to solve an application problem other than involving rates of work?	32	Ex. 13, PP 12

5.4 Review Exercises

For exercises 24–26,
(a) solve.
(b) check.

24. $\dfrac{k}{k + 4} + \dfrac{3}{k - 2} = \dfrac{18}{k^2 + 2k - 8}$

25. $\dfrac{2}{x - 2} + \dfrac{9}{x} = \dfrac{x}{x - 2}$

26. $\dfrac{1}{3} - \dfrac{2}{x^2} = \dfrac{5}{3x}$

27. Explain why some rational equations have an extraneous solution.

28. To solve $\dfrac{x}{x - 6} = \dfrac{x}{x - 6} + 36$, a student cleared the fractions and correctly solved the equation $x = x + 36(x - 6)$. His solution was $x = 6$. Explain why this is not a solution of the equation.

29. A student is solving $\dfrac{5}{a + 4} = \dfrac{12a^2 + 19}{a^2 + 7a + 12} - \dfrac{3}{a + 3}$. After factoring, the equation is $\dfrac{5}{a + 4} = \dfrac{12a^2 + 19}{(a + 3)(a + 4)} - \dfrac{3}{a + 3}$. Describe what the student should do next.

For exercises 30–31, use the five steps.

30. Predict the number of 10,225 people who are willing to pay more to buy a gas-electric hybrid vehicle. Round to the nearest hundred.

    Only one out of four people are willing to pay more to buy a gas-electric hybrid vehicle compared to a conventional car, according to a new online poll. (*Source:* www.usatoday.com, May 17, 2009)

31. Two caterers are preparing 15 dozen stuffed mushrooms as appetizers for a big party. From past experience, they know that Janisa can prepare the mushrooms in 1 hour and 30 minutes. Omar needs 1 hour and 50 minutes to

prepare the mushrooms. If they work together, find the amount of time in minutes to do this job. Round to the nearest whole number.

32. Super elevation is the percent tilt (banking) on a roadway that helps a car stay on the road as it goes around a curve. The relationship of the radius R of a curve in meters, the side friction factor F, the designed speed V for the curve in kilometers per hour, and the percent super elevation E in decimal form is $R = \dfrac{V^2}{127.5E + 127.5F}$. A curve has a radius of 390 m, a design speed of 90 $\dfrac{\text{km}}{\text{hr}}$, and a side friction factor of 0.13. Find the needed super elevation. Round to the nearest tenth of a percent. (*Source:* www .aseansec.org)

© DNY59 RF/Getty Images RF

SECTION 5.5   Rational Functions

Ask Yourself	Test Yourself	Help Yourself
Can I . . .	**Do 5.5 Review Exercises**	**See these Examples and Practice Problems**
write a definition of *function, domain,* and *range*?	33–41, 48–50	Ex. 1–8, 13, PP 1–8
use a number line graph or interval notation to represent the domain of a rational function?		
write the equations of any vertical asymptotes on the graph of a rational function?	33–41, 46, 47	Ex. 4–8, PP 4–8
predict the location of any holes in the graph of a rational function?	33–41, 47	Ex. 7, 8, PP 4–8
evaluate a rational function?	42, 43	Ex. 9, 10, PP 9–14
use an algebraic method to identify any real zeros of a rational function?	44, 45	Ex. 11–13, PP 15–18

5.5 Review Exercises

For exercises 33–41,
(a) use interval notation to represent the domain of the function.
(b) write the equation of any vertical asymptotes.
(c) identify the location of any holes on the graph.

33. $y = \dfrac{4}{x}$

34. $y = \dfrac{7}{x + 24}$

35. $f(x) = \dfrac{7}{2x^2 - x - 45}$

36. $g(x) = \dfrac{x + 3}{x^2 - 9}$

37. $f(x) = \dfrac{x + 6}{x^2 - 7x - 18}$

38. $g(x) = \dfrac{x - 4}{x^2 - 16}$

39. $y = \dfrac{12}{x^2 - 10x + 25}$

40. $h(x) = \dfrac{x^2 + x - 12}{x^2 + 7x - 30}$

41. $y = \dfrac{3}{x^2 + 5x}$

42. In radiography, scattered X-rays can blur the image. A grid has sections of material that absorbs X-rays alter-

nated with sections of material that allow X-rays to pass. The width of the absorbing material in micrometers is w. The function $P(w) = \dfrac{w}{w + 350 \text{ micrometers}}$ describes the relationship of the width of the grid strip, w, to the percent of X-rays absorbed by the grid strip, $P(w)$. Find $P(50 \text{ micrometers})$.

43. A rational function is $b(x) = \dfrac{1}{x + 2}$. Evaluate $b(-2)$.

44. Use an algebraic method to find the real zero(s) of $h(x) = \dfrac{7x + 2}{x - 9}$.

45. Explain why $f(x) = \dfrac{6}{x + 2}$ has no real zeros.

46. Explain why the graph of $f(x) = \dfrac{6}{x + 2}$ has one vertical asymptote.

47. Explain why the graph of $f(x) = \dfrac{x + 2}{x^2 - 4}$ has one vertical asymptote and one hole.

48. Explain why the domain of $f(x) = \dfrac{2}{x^2 + 1}$ is $(-\infty, \infty)$, without any restrictions.

49. Explain why the domain of $f(x) = \dfrac{2}{x - 6}$ is $(-\infty, 6) \cup (6, \infty)$, without any additional restrictions.

50. A student said that the domain of $f(x) = \dfrac{x(x + 4)}{x(x - 5)}$ is $(-\infty, 5) \cup (5, \infty)$ and that there is no restriction at $x = 0$ because the function simplifies to $f(x) = \dfrac{x + 4}{x - 5}$. Explain what is wrong with this answer.

SECTION 5.6 **Variation**

Ask Yourself

Can I...	Test Yourself	Help Yourself
	Do 5.6 Review Exercises	See these Examples and Practice Problems
identify a function that is a direct variation?	52, 54, 58	Ex. 1, 2, PP 1, 2
find the constant of proportionality for a direct variation, write and graph a function that describes this variation, and evaluate the function?		
identify a function that is an inverse variation?	51, 55, 59	Ex. 3, 4, PP 3, 4
find the constant of proportionality for an inverse variation, write and graph a function that describes this variation, and evaluate the function?		
find the constant of proportionality for a joint variation, write a function that describes this variation, and evaluate the function?	56	Ex. 5, PP 5
find the constant of proportionality for a combined variation, write a function that describes this variation, evaluate the function?	57	Ex. 6, PP 6
identify whether the relationship between two variables in a formula is direct or inverse?	53, 56–59	Ex. 7, PP 7–9
solve a formula for a variable?	53	Ex. 8–10, PP 10–12

5.6 Review Exercises

51. The relationship of water pressure and lung volume of a scuba diver is described by the function

$$y = \dfrac{35 \text{ lb} \cdot \text{qt}}{\dfrac{\text{in.}^2}{x}}.$$ In this function, y is the air capacity of the lungs in quarts. The pressure of the water in pounds per square inch (psi) is x.

  a. Is this function a direct variation or an inverse variation? Explain.

  b. Find the lung volume in quarts of this diver when the water pressure is $\dfrac{29.4 \text{ lb}}{\text{in.}^2}$. Round to the nearest tenth.

  c. Use a table of ordered pairs to graph this function.

  d. Use the graph to find the lung volume of this diver when the pressure is $\dfrac{44 \text{ lb}}{\text{in.}^2}$.

52. Hot water flows from the Yampah Spring into the pools of the Hot Springs Lodge and Pool in Glenwood Springs, Colorado. The amount of water discharged by the spring in gallons, y, for x days is described by the function

$$y = \left(\dfrac{3{,}500{,}000 \text{ gal}}{1 \text{ day}} \right) x.$$

  a. Is this function a direct variation or an inverse variation? Explain.

b. Find the amount of water output by the spring in 5 days.

c. Use a table of ordered pairs to graph this function.

d. Use the graph to find the amount of water discharged by the spring after 4 days.

53. The breathing time a scuba diver gets from a particular cylinder of air can be estimated with the formula

$$T = \frac{V(C - P)}{CR}.$$

a. Is the relationship between T and R a direct variation or an inverse variation?

b. Is the relationship between T and V a direct variation or an inverse variation?

c. Solve for P.

d. Solve for C.

54. The relationship between the number of scrapbooking cardstock pages in a stack, x, and the thickness in inches of the stack, y, is a direct variation. The thickness of 20 pages is 0.24 in.

© Amy Walters/Shutterstock

a. Find the constant of proportionality for this direct variation.

b. Write a function that describes this direct variation.

55. If the area of a rectangle is fixed, the relationship between the length of the rectangle, x, and the width of the rectangle, y, is an inverse variation. When the length is 10 cm, the width is 4 cm.

a. Find the constant of proportionality for this inverse variation.

b. Write a function that describes this inverse variation.

56. A joint variation is $y = kxz$, where $x = 50$, $z = 2$, and $y = 25$.

a. Find k.

b. Is the relationship between x and y direct or inverse?

c. Is the relationship between z and y direct or inverse?

d. If $x = 80$ and $z = 4$, find y.

57. A combined variation is $y = \dfrac{kx}{z}$, where $x = 50$, $z = 2$, and $y = 75$.

a. Find k.

b. Is the relationship between x and y direct or inverse?

c. Is the relationship between z and y direct or inverse?

d. If $x = 80$ and $z = 6$, find y.

58. For a bouquet of roses, the charge depends on the number of roses. If the number of roses increases, the price also increases. Is this an example of a direct or inverse variation?

59. As the length of a board increases, the force needed to break the board decreases. Is this an example of a direct or inverse variation?

SECTION 5.7 Division of Polynomials; Synthetic Division

Ask Yourself	**Test Yourself**	**Help Yourself**
Can I . . .	Do 5.7 Review Exercises	See these Examples and Practice Problems
identify the dividend, divisor, and quotient in a division problem?	62, 63	Ex. 1–5, PP 1–10
check the quotient of a division problem?		
divide a monomial by a monomial?	60	Ex. 1, PP 1
divide a polynomial by a monomial?	61	Ex. 2, PP 2, 3
use long division to divide a polynomial by a binomial?	62, 63	Ex. 4, 6, 7, PP 4–8
use synthetic division to divide a polynomial by a binomial?	64, 65	Ex. 8, 9, 11, PP 9, 10
use synthetic division to determine whether a divisor is a factor of a polynomial?	66	Ex. 11
evaluate a polynomial function to check whether the remainder of a quotient is correct?	65	Ex. 12, PP 9, 10
use synthetic division and the remainder theorem to evaluate a polynomial function?	68	Ex. 13, PP 13, 14

5.7 Review Exercises

For exercises 60–61, simplify.

60. $\dfrac{54h^4k^2}{75hk^9}$

61. $\dfrac{12x^3 + 30x^2 - 16x + 8}{2x^2}$

For exercises 62–63,
(a) use long division to find the quotient.
(b) check the solution.

62. $(3x^2 + 23x + 30) \div (x + 6)$

63. $(3x^2 + 23x + 30) \div (x - 6)$

64. Use synthetic division to find the quotient:
$(3x^2 + 23x + 30) \div (x + 6)$.

65. a. Use synthetic division to find the quotient:
$(3x^2 + 23x + 30) \div (x - 6)$.
b. Check the remainder in this quotient by evaluating a polynomial function.

66. Use synthetic division to determine whether $x - 6$ is factor of $9x^2 - 53x - 6$.

67. Determine whether $x - 6$ is a factor of $9x^2 - 53x - 6$ by evaluating a function.

68. A function is $f(x) = 2x^2 + 15x + 9$. Use synthetic division and the remainder theorem to evaluate $f(-6)$.

SECTION 5.8   **Rational Inequalities**

Ask Yourself	Test Yourself	Help Yourself
Can I . . .	Do 5.8 Review Exercises	See these Examples and Practice Problems
find the real boundary values of a rational inequality?	69, 70	Ex. 1–5, PP 1–6
solve a rational inequality?	69, 70	Ex. 2–5, PP 1–6
use a number line graph or interval notation to represent the solution of a rational inequality?	69, 70	Ex. 2–5, PP 1–6

5.8 Review Exercises

69. a. Find the boundary values for $\dfrac{x}{2x + 10} > 3$.
b. Use the boundary values to find the solution.
c. Use a number line graph to represent the solution.
d. Use interval notation to represent the solution.

70. a. Solve: $\dfrac{40}{x + 9} \le 2$
b. Use a number line graph to represent the solution.
c. Use interval notation to represent the solution.

Chapter 5 Test

For problems 1–7, simplify.

1. $\dfrac{x^2 - 6x}{x^3 - 4x^2 - 12x}$

2. $\dfrac{x^3 - 4x^2 - 12x}{4x}$

3. $\dfrac{2x^2 + 19x + 9}{x^3 - 81x} \div \dfrac{2x^2 - 11x - 6}{x^2 - 6x}$

4. $\dfrac{y^2 - 4}{2y^2 - 11y - 6} \cdot \dfrac{2y^2 + 7y + 3}{y^2 - y - 12}$

5. $\dfrac{x}{x^2 + 5x + 6} - \dfrac{2}{x^2 + 3x + 2}$

6. $\dfrac{x}{x^2 + 9x + 18} + \dfrac{5}{x^2 + 10x + 21}$

7. $\dfrac{4x - 1}{6x^2 - x} - \dfrac{1}{x}$

For problems 8–9,
(a) solve.
(b) check.

8. $\dfrac{3}{x - 5} + \dfrac{1}{x + 5} = \dfrac{2}{x^2 - 25}$

9. $\dfrac{x}{x - 5} + \dfrac{5}{x} = \dfrac{11}{6}$

10. Solve $7P - 4Q = 8AQ + 3PR$ for P.

11. Find any real zeros of $f(x) = \dfrac{x + 3}{x - 7}$.

12. Explain why the domain of the function of $f(x) = \dfrac{x + 3}{x - 7}$ is not the set of real numbers.

13. A rational function is $f(x) = \dfrac{x^2 - 25}{x^2 + 7x + 10}$.
a. Write the equation of any vertical asymptotes of the graph of this function.
b. Identify the location of any holes.

14. The relationship between the number of sheep in a pasture, S, and the length of the grass in the pasture, L, is described by the equation $L = \dfrac{k}{S}$.

   a. Is the length of the grass directly or inversely related to the number of sheep in the pasture?

   b. If the number of sheep increases, does the length of the grass increase or decrease?

   c. If the number of sheep decreases, does the length of the grass increase or decrease?

15. A formula is $Z = \dfrac{4HW}{KL}$.

   a. If W increases and H, L, and K are kept constant, what will happen to Z?

   b. If K increases and H, L, and W are kept constant, what will happen to Z?

   c. Are L and Z directly related or inversely related?

   d. Are H and W directly related or inversely related?

   e. Solve this formula for H.

For problems 16–17, use the five steps from Section 5.1.

16. Chris can set all of the tables in a banquet hall in 32 min. Lucia can set all of the tables in a banquet hall in 47 min.

Find the time in minutes for them to do the job together. Round to the nearest whole number. (Assume that they work at the same rate together and apart.)

17. According to a 2009 survey, U.S. employers planned to decrease salary budgets but still give raises to three out of four employees. Predict how many of 4580 employees received a raise in 2009. Round to the nearest hundred. (*Source:* www.worldatwork.org, Feb. 3, 2009)

18. Simplify: $\dfrac{12x^4 + 9x^3 - x^2 - 4x + 8}{12x}$

19. Use long division to divide $2x^2 + 13x + 21$ by $x + 3$.

20. Use synthetic division to divide $5x^2 - 27x - 56$ by $x - 7$.

21. A polynomial function is $f(x) = x^2 + 13x + 20$. Use synthetic division and the remainder theorem to evaluate $f(-3)$.

22. Solve $\dfrac{2x}{x + 8} \le 4$. Use interval notation to represent the solution.

Radical Expressions, Equations, and Functions

SUCCESS IN COLLEGE MATHEMATICS

Keys for Success

To succeed in a mathematics class such as Intermediate Algebra, you should plan to read your textbook, attend classes, take notes and use them, and review for tests. However, doing these activities alone may not always be enough. *What you believe about yourself and your actions also makes a difference.*

You need to believe that you are in control of yourself and your actions. You can make choices, even in difficult circumstances. You are not a victim without options. The only person responsible for your success is you. Avoid blaming others and their actions for your difficulties. Instead, make choices to limit the effect that the actions of others can have on you.

You need to believe that persistence is an important part of success. Success does not come in one giant step. Instead, success happens as a series of small steps. You need to believe that continued hard work will pay off, one day at a time. Expect that some topics will be more difficult than others and that these topics will require more time to learn. Although our culture often celebrates instant success and gratification, the truth is that most achievements result from persistent, planned effort.

You need to believe that your success may require the assistance and support of others. Asking for help does not mean that you do not have what it takes to succeed. Instead, it shows that you are in control and know that you need assistance to be successful.

The English word **radical** comes from the Latin word *radicalis*, "having roots." In biology, a *radical* is something that grows from a *root* or the base of a stem. A dandelion plant has radical leaves. In medicine, a *radical* surgery removes the *root* of a disease. In mathematics, a *radical* is another word for the *root* of a number. In this section, we begin our study of radicals.

SECTION 6.1 Introduction to Radicals

After reading the text, working the practice problems, and completing assigned exercises, you should be able to:

1. Identify the index and radicand of a radical.

2. Evaluate a radical expression.

3. Simplify a radical expression.

Radicals

To *square* a number, we multiply it by itself: $3^2 = 9$. To find the **square root** of a number, we undo the process of squaring. Since $(3)(3) = 9$ and $(-3)(-3) = 9$, the square roots of 9 are 3 and –3. The positive square root of a number is its **principal square root**. Since the notation for principal square root is $\sqrt{}$, $\sqrt{9} = 3$. The principal square root of 9 is 3.

EXAMPLE 1   **(a)** What are the square roots of 16?

SOLUTION ▶   The principal square root of 16 is 4. $4 \cdot 4 = 16$

The other square root of 16 is -4. $(-4)(-4) = 16$

(b) Evaluate: $\sqrt{16}$

▶ $\sqrt{16}$ What is the principal square root of 16?

$= 4$ $(4)(4) = 16$

Since the product of a real number and itself is always positive or 0, the *square* of a real number cannot be a negative number. For example, there is no real number that can be multiplied by itself and result in a product of –16. So *the square root of a negative number is not a real number.* We will learn more about the square roots of negative numbers in Chapter 7.

The expression 2^3 equals $2 \cdot 2 \cdot 2$ or 8. The expression $(-2)^3$ equals $(-2)(-2)(-2)$ or –8. To find the **principal cube root** of a number, undo the process of cubing. The principal cube root may be a positive or a negative number. Since the product of three negative numbers is a negative number, *the principal cube root of a negative number is a negative real number.* Since the notation for a principal cube root is $\sqrt[3]{}$, $\sqrt[3]{8} = 2$ and $\sqrt[3]{-8} = -2$.

EXAMPLE 2   **(a)** Evaluate: $\sqrt{64}$

SOLUTION ▶ $\sqrt{64}$ What is the principal square root of 64?

$= 8$ $(8)(8) = 64$

(b) Evaluate: $\sqrt{-64}$ What is the principal square root of -64?

▶ $\sqrt{-64}$ is not a real number.   No real number multiplied by itself is negative.

(c) Evaluate: $-\sqrt{64}$

▶ $-\sqrt{64}$ What is the opposite of the principal square root of 64?

        $= -8$ $(8)(8) = 64$

(d) Evaluate: $\sqrt[3]{64}$

▶ $\sqrt[3]{64}$ What is the principal cube root of 64?

        $= 4$ $(4)(4)(4) = 64$

(e) Evaluate: $\sqrt[3]{-64}$

▶ $\sqrt[3]{-64}$ What is the principal cube root of –64?

        $= -4$ $(-4)(-4)(-4) = -64$

Square Roots

Each real number greater than 0 has two real square roots. The positive square root is the **principal square root**. The notation for a principal square root is $\sqrt{\ }$. The square root of a negative number is not a real number. The square root of 0 is 0.

Cube Roots

Each real number has one real cube root, the **principal cube root**. The notation for a principal cube root is $\sqrt[3]{\ }$. The principal cube root of a negative number is a negative number. The principal cube root of a positive number is a positive number. The cube root of 0 is 0.

The expression $\sqrt[3]{-64}$ is a cube root. It is also a **radical**. The number -64 is the **radicand** and the number 3 is the **index**. The index of a square root is 2, although the 2 is usually not included in notation for a principal square root, $\sqrt{\ }$.

Radical, Index, and Radicand

Where $\sqrt[n]{x}$ is a radical, n is the index and x is the radicand.

EXAMPLE 3   Identify the index and the radicand. Evaluate.

(a) $\sqrt[5]{-32}$ Index: 5; radicand: -32.

SOLUTION ▶ $= -2$ $(-2)(-2)(-2)(-2)(-2) = -32$

(b) $\sqrt{144}$ Index: 2; radicand: 144.

▶ $= 12$ $(12)(12) = 144$

(c) $\sqrt[3]{\dfrac{27}{343}}$ Index: 3; radicand: $\dfrac{27}{343}$.

▶ $= \dfrac{3}{7}$ $\left(\dfrac{3}{7}\right)\left(\dfrac{3}{7}\right)\left(\dfrac{3}{7}\right) = \dfrac{27}{343}$

In Example 3, each of the simplified radicals is a rational number (a number that can be written as a fraction in which the numerator and denominator are integers). In the next example, the radical is an irrational number. Although there are methods for evaluating irrational square roots without a calculator, we most often use a calculator.

EXAMPLE 4 | Evaluate $\sqrt{37}$. Round to the nearest hundredth.

SOLUTION ▶ $\sqrt{37}$ Use a calculator.

$= 6.082762\ldots$ This irrational number does not repeat or terminate.

≈ 6.08 Round.

When using a formula with radicals, round irrational numbers to the given place value.

EXAMPLE 5 | The SMOG (Simple Measure of Gobbledygook) readability formula for finding the grade level, G, of a book or article is $G = \sqrt{p} + 3$. The total number of polysyllabic (three or more syllables) words in a selection of 30 sentences is p. Find the reading grade level of an article in which the 30 sentences include 38 polysyllabic words. Round to the nearest tenth.

SOLUTION ▶ $G = \sqrt{p} + 3$

$G = \sqrt{38} + 3$ Replace p with 38.

$G = 6.164\ldots + 3$ Use a calculator; $\sqrt{38}$ is irrational.

$G \approx 9.2$ Round to the nearest tenth.

The reading level of this article is grade 9.2 (freshman in high school).

Practice Problems

For problems 1–7,
(a) identify the index and radicand in each radical.
(b) evaluate the radical. If the number is irrational, round to the nearest hundredth.

1. $\sqrt[3]{27}$ **2.** $\sqrt[4]{81}$ **3.** $\sqrt{\dfrac{49}{144}}$ **4.** $\sqrt{-100}$ **5.** $\sqrt[3]{-125}$

6. $\sqrt{5}$ **7.** $-\sqrt{169}$

Perfect Squares, Perfect Cubes, and Perfect Powers

When a square root is a rational number, the radicand is a **perfect square**. When a cube root is a rational number, the radicand is a **perfect cube**.

Perfect Square Numbers

The principal square root of a number that is a perfect square is a rational number. If a number is not a perfect square, its principal square root is an irrational number.

Perfect Squares   **Not Perfect Squares**

16 because $\sqrt{16} = 4$ 11 because $\sqrt{11} = 3.31662\ldots$

$\dfrac{9}{16}$ because $\sqrt{\dfrac{9}{16}} = \dfrac{3}{4}$ $\dfrac{10}{13}$ because $\sqrt{\dfrac{10}{13}} = 0.877058\ldots$

> ### Perfect Cube Numbers
>
> The principal cube root of a number that is a perfect cube is a rational number. If a number is not a perfect cube, its principal cube root is an irrational number.
>
> | **Perfect Cubes** | **Not Perfect Cubes** |
> |---|---|
> | 27 because $\sqrt[3]{27} = 3$ | 5 because $\sqrt[3]{5} = 1.70997\ldots$ |

We can extend these ideas to other exponents. But instead of using visual images such as a square and a cube, we just call them perfect powers. For example, 16 is a perfect fourth power because $2^4 = 16$ and $\sqrt[4]{16} = 2$. The first step in simplifying a radical is to find whole number factors of the radicand that are perfect powers. A list of frequently used factors is included here for reference.

Base, n	2	3	4	5	6	7	8	9	10	11	12	13	14	15
Perfect Square n^2	4	9	16	25	36	49	64	81	100	121	144	169	196	225
Perfect Cube n^3	8	27	64	125	216	343	512	729	1000					
Perfect 4th Power n^4	16	81	256	625										
Perfect 5th Power n^5	32	243	1024	3125										
Perfect 6th Power n^6	64	729	4096											

EXAMPLE 6   Find the greatest whole number factor that is a perfect square.

(a) 28

SOLUTION ▶   Perfect squares less than 28: 4, 9, 16, 25

Factors of 28: 1, 2, 4, 7, 14, 28

The greatest whole number factor of 28 that is a perfect square is 4.

(b) 50

▶   Perfect squares less than 50: 4, 9, 16, 25, 36, 49

Factors of 50: 1, 2, 5, 10, 25, 50

The greatest whole number factor of 50 that is a perfect square is 25.

(c) 180

▶   Perfect squares less than 180: 4, 9, 16, 25, 36, 49, 64, 81, 100, 121, 144, 169

Factors of 180: 1, 2, 3, 4, 5, 6, 9, 10, 12, 15, 18, 20, 30, 36, 45, 60, 90, 180

The greatest whole number factor of 180 that is a perfect square is 36.

Practice Problems

For problems 8–11, find the greatest whole number factor that is a perfect square.

8. 72   **9.** 54   **10.** 3000   **11.** 98

For problems 12–15, find the greatest whole number factor that is a perfect cube.

12. 72   **13.** 54   **14.** 3000   **15.** 128

Radicals with Variables

The notation $\sqrt{}$ refers only to the principal square root, a positive number or 0.

EXAMPLE 7   **(a)** What are the square roots of 16?

SOLUTION ▶   The square roots of 16 are 4 and -4.

(b) Evaluate: $\sqrt{16}$

▶ $\sqrt{16}$
$= 4$

(c) Simplify $\sqrt{x^2}$ when x is a real number.

▶   If $x \geq 0$, then $\sqrt{x^2}$ is x.

If $x < 0$, then $\sqrt{x^2} \neq x$ because the principal square root must be positive and x is a negative number. However, $\sqrt{x^2}$ does equal $|x|$, since $|x|$ is a positive number.

$\sqrt{x^2}$ x is a real number; it can be positive or negative.

$= |x|$ The absolute value of x is greater than or equal to 0.

(d) Simplify $\sqrt{(x-8)^2}$ when x is a real number.

▶ $\sqrt{(x-8)^2}$

$= |x-8|$ The $\sqrt{}$ notation represents only the positive square root.

Assumptions About Variables in Radicands

If x is a real number, $\sqrt{x^2} = |x|$.

If x is a real number greater than or equal to 0, $\sqrt{x^2} = x$.

If x and k are real numbers, $\sqrt{(x+k)^2} = |x+k|$.

For the rest of this section, we will assume that all factors in radicands represent real numbers that are greater than or equal to 0.

For now, we will assume that factors in radicands represent real numbers that are greater than or equal to 0. In Chapter 7, we will change this assumption so that we can extend our knowledge about radicals to the solutions of quadratic equations.

To simplify radicands that include exponential expressions, we use the power rule of exponents, $(x^m)^n = x^{mn}$. Since $(a^4)^3 = a^{12}$, we can simplify a cube root such as $\sqrt[3]{a^{12}}$ by rewriting it as $\sqrt[3]{(a^4)^3}$. Since $\sqrt[3]{(a^4)^3} = a^4$, $\sqrt[3]{a^{12}} = a^4$. This same relationship applies to any radical with a radicand that is a perfect power exponential expression. *The exponent in the simplified expression equals the original exponent in the radicand divided by the index.*

EXAMPLE 8   **(a)** Simplify: $\sqrt{x^{10}}$

SOLUTION ▶ $\sqrt[2]{x^{10}}$

$= \sqrt[2]{(x^5)^2}$ Power rule of exponents; $x^{10} = (x^5)^2$

$= x^5$ Simplify.

(b) Simplify: $\sqrt[3]{x^6}$

▶ $\sqrt[3]{x^6}$

$= \sqrt[3]{(x^2)^3}$ Power rule of exponents; $x^6 = (x^2)^3$

$= x^2$ Simplify.

(c) Simplify: $\sqrt[11]{x^{22}}$

$\sqrt[11]{x^{22}}$

$= \sqrt[11]{(x^2)^{11}}$ Power rule of exponents; $x^{22} = (x^2)^{11}$

$= x^2$ Simplify.

(d) Simplify: $\sqrt[79]{x^{79}}$

$\sqrt[79]{x^{79}}$

$= \sqrt[79]{(x^1)^{79}}$ Power rule of exponents; $x^{79} = (x^1)^{79}$

$= x$ Simplify.

Exponential Expressions That Are Perfect Powers

Exponential Expressions That Are Perfect Squares   A variable raised to a power is a perfect square if its exponent is a multiple of 2.

Exponential Expressions That Are Perfect Cubes   A variable raised to a power is a perfect cube if its exponent is a multiple of 3.

Exponential Expressions That Are Perfect nth Powers   A variable raised to a power is a perfect nth power if its exponent is a multiple of n and n is a whole number greater than or equal to 2.

Practice Problems

For problems 16–19, simplify. Assume that all variables represent nonnegative real numbers.

16. $\sqrt{x^8}$ **17.** $\sqrt[4]{y^{20}}$ **18.** $\sqrt[3]{z^{18}}$ **19.** $\sqrt[10]{a^{10}}$

Simplifying Radical Expressions

The notation for multiplication of radicals is either a multiplication dot or writing the radicals next to each other: $\sqrt{x} \cdot \sqrt{y} = \sqrt{x}\sqrt{y}$. The **product rule of radicals** describes how to multiply and simplify radicals.

Product Rule of Radicals

If a and b are real numbers greater than or equal to 0 and n is an even whole number greater than 0, $\sqrt[n]{ab} = \sqrt[n]{a} \cdot \sqrt[n]{b}$ and $\sqrt[n]{a} \cdot \sqrt[n]{b} = \sqrt[n]{ab}$.

If a and b are real numbers and n is an odd whole number greater than 1, $\sqrt[n]{ab} = \sqrt[n]{a} \cdot \sqrt[n]{b}$ and $\sqrt[n]{a} \cdot \sqrt[n]{b} = \sqrt[n]{ab}$.

To simplify a square root, use the product rule of radicals to rewrite it as the product of two or more radicals with radicands that are perfect squares. Then simplify each radical.

EXAMPLE 9   Simplify: $\sqrt{36x^{10}}$

SOLUTION $\sqrt{36x^{10}}$ Find perfect square factors of the radicand; 36, x^{10}.

$= \sqrt{36}\sqrt{x^{10}}$ Product rule of radicals.

$= 6x^5$ Simplify.

In Example 6, we identified the greatest perfect square factor of whole numbers. In the next example, the radicand is not a perfect square. However, some of the *factors* of the radicand are perfect squares. Since 25 is a perfect square, we use the product rule of radicals to rewrite $\sqrt{50}$ as $\sqrt{25} \cdot \sqrt{2}$. We do not rewrite $\sqrt{50}$ as $\sqrt{10} \cdot \sqrt{5}$ because 10 and 5 are not perfect squares.

EXAMPLE 10 Simplify: $\sqrt{50}$

SOLUTION ▶

$\sqrt{50}$ Find perfect square factors of the radicand, 25.

$= \sqrt{25}\sqrt{2}$ Product rule of radicals.

$= 5\sqrt{2}$ Simplify.

When simplifying, find the *greatest* perfect square factor of the radicand. Otherwise, the radical will have to be simplified again.

EXAMPLE 11 Simplify: $\sqrt{500}$

SOLUTION ▶

$\sqrt{500}$	$\sqrt{500}$	Find perfect square factors of the radicand: 100 or 25.
$= \sqrt{100}\sqrt{5}$	$= \sqrt{25}\sqrt{20}$	Product rule of radicals.
$= 10\sqrt{5}$ Simplify.	$= 5\sqrt{4}\sqrt{5}$	Not simplified; 4 is a perfect square factor of 20.
	$= 5 \cdot 2\sqrt{5}$	Simplify.
	$= 10\sqrt{5}$	Simplify.

When we choose the greatest perfect square factor, 100, we have to simplify only once. When we choose a smaller perfect square factor, 25, we have to simplify twice.

In the next example, to simplify $\sqrt{c^7}$, we need to rewrite it as the product of two radicals. The radicand of one of the radicals should be the largest perfect square factor of c^7. The exponent of the largest perfect square factor is the *largest multiple of 2* that is less than or equal to 7. Since 6 is the largest multiple of 2 that is less than or equal to 7, we rewrite $\sqrt{c^7}$ as $\sqrt{c^6}\sqrt{c}$. After simplifying, we use the product rule of radicals to multiply the remaining radicals: $\sqrt{c}\sqrt{d} = \sqrt{cd}$.

EXAMPLE 12 Simplify: $\sqrt{c^7dw^8}$

SOLUTION ▶

$\sqrt{c^7dw^8}$

$= \sqrt{c^7}\sqrt{d}\sqrt{w^8}$ Find perfect square factors of the radicand: c^6, w^8.

$= \sqrt{c^6}\sqrt{c}\sqrt{d}\sqrt{w^8}$ Product rule of radicals.

$= c^3 \cdot \sqrt{c} \cdot \sqrt{d} \cdot w^4$ Simplify.

$= c^3w^4\sqrt{cd}$ Product rule of radicals.

An expression is not simplified until every exponent in the radicand is less than the index.

EXAMPLE 13 Simplify: $\sqrt{x^5y^2z^9}$

SOLUTION ▶

$\sqrt{x^5y^2z^9}$

$= \sqrt{x^5}\sqrt{y^2}\sqrt{z^9}$ Find perfect square factors of the radicand: x^4, y^2, z^8.

$= \sqrt{x^4}\sqrt{x}\sqrt{y^2}\sqrt{z^8}\sqrt{z}$ Product rule of radicals.

$= x^2 \cdot \sqrt{x} \cdot y \cdot z^4 \cdot \sqrt{z}$ Simplify.

$= x^2yz^4\sqrt{xz}$ Product rule of radicals; exponents in the radicand are less than the index, 2.

In the next example, we find the greatest perfect square factor of the number and of each exponential expression.

EXAMPLE 14   Simplify: $\sqrt{500h^{15}k^9}$

SOLUTION ▶

$\sqrt{500h^{15}k^9}$ — Find perfect square factors of the radicand: 100, h^{14}, k^8.

$= \sqrt{500}\sqrt{h^{15}}\sqrt{k^9}$ — Product rule of radicals.

$= \sqrt{100}\sqrt{5}\sqrt{h^{14}}\sqrt{h}\sqrt{k^8}\sqrt{k}$ — Product rule of radicals.

$= 10h^7k^4\sqrt{5hk}$ — Product rule of radicals; exponents in the radicand are less than the index, 2.

To simplify a cube root, look for factors that are perfect cubes. The expression is not simplified until every exponent in the radicand is less than the index, 3.

EXAMPLE 15   Simplify: $\sqrt[3]{-1024x^{13}z^2}$

SOLUTION ▶

$\sqrt[3]{-1024x^{13}z^2}$ — Find perfect cube factors of the radicand: 512, x^{12}.

$= \sqrt[3]{-1024}\sqrt[3]{x^{13}}\sqrt[3]{z^2}$ — Product rule of radicals.

$= \sqrt[3]{-512}\sqrt[3]{2}\sqrt[3]{x^{12}}\sqrt[3]{x}\sqrt[3]{z^2}$ — Product rule of radicals.

$= -8 \cdot \sqrt[3]{2} \cdot x^4 \cdot \sqrt[3]{x} \cdot \sqrt[3]{z^2}$ — Simplify.

$= -8x^4\sqrt[3]{2xz^2}$ — Product rule of radicals; exponents in the radicand are less than the index, 3.

The exponent on an exponential expression that is a perfect fifth power is a multiple of 5. When a fifth root is simplified, all of the exponents in the radicand are less than 5.

EXAMPLE 16   Simplify: $\sqrt[5]{729h^4k^{12}}$

SOLUTION ▶

$\sqrt[5]{729h^4k^{12}}$ — Find perfect fifth factors of the radicand: 243, k^{10}.

$= \sqrt[5]{729}\sqrt[5]{h^4}\sqrt[5]{k^{12}}$ — Product rule of radicals.

$= \sqrt[5]{243}\sqrt[5]{3}\sqrt[5]{h^4}\sqrt[5]{k^{10}}\sqrt[5]{k^2}$ — Product rule of radicals.

$= 3 \cdot \sqrt[5]{3} \cdot \sqrt[5]{h^4} \cdot k^2 \cdot \sqrt[5]{k^2}$ — Simplify.

$= 3k^2\sqrt[5]{3h^4k^2}$ — Product rule of radicals; exponents in the radicand are less than the index, 5.

Before starting to simplify a radical, refer to or build a list of perfect powers for the index of the root. Then look for the largest number in the list that is a factor of the radicand.

EXAMPLE 17   Simplify: $\sqrt[4]{405x^4y^2}$

SOLUTION ▶

$\sqrt[4]{405x^4y^2}$ — Refer to or build a list of perfect fourth powers.

$= \sqrt[4]{405}\sqrt[4]{x^4}\sqrt[4]{y^2}$ — Find perfect fourth factors of the radicand: 81, x^4.

$= \sqrt[4]{81}\sqrt[4]{5}\sqrt[4]{x^4}\sqrt[4]{y^2}$ — Product rule of radicals.

$= 3 \cdot \sqrt[4]{5} \cdot x \cdot \sqrt[4]{y^2}$ — Simplify.

$= 3x\sqrt[4]{5y^2}$ — Product rule of radicals; exponents in the radicand are less than the index, 4.

In the next example, the radicand is a polynomial. To simplify, first factor the polynomial. We cannot rewrite a radical in which the radicand is a polynomial as a sum of radicals. *There is no "sum rule" for radicals*: $\sqrt{x^2 + 12x + 36} \neq \sqrt{x^2} + \sqrt{12x} + \sqrt{36}$.

EXAMPLE 18 | Simplify: $\sqrt{x^2 + 12x + 36}$

SOLUTION ▶ $\sqrt{x^2 + 12x + 36}$ A quadratic trinomial: $ax^2 + bx + c$.

$= \sqrt{(x + 6)(x + 6)}$ Factor.

$= x + 6$ Simplify.

In this section, we are assuming that all factors in radicands are greater than or equal to 0. If we did not make this assumption, we would need to use an absolute value to ensure that the result of simplifying a principal square root is greater than or equal to 0. Without the assumption, the expression simplifies to $|x + 6|$.

Simplifying Radicals

Assume that all factors that are variables or variable expressions in radicands represent nonnegative real numbers.

1. Rewrite the radical as a product of two or more radicals in which each radicand is a constant or a single variable: $\sqrt[n]{ab} = \sqrt[n]{a}\sqrt[n]{b}$.

2. If the radicand is not a perfect power, rewrite each factor as a product of two radicals. Choose radicands that are the largest perfect powers.

3. Simplify. When a radical is simplified, the radicand has no perfect power factors; every exponent in the radicand is less than the index.

4. Use the product rule of radicals to combine any remaining radicals into a single radical: $\sqrt[n]{a}\sqrt[n]{b} = \sqrt[n]{ab}$.

Practice Problems

For problems 20–25, simplify.

20. $\sqrt{92x^4y^9}$ **21.** $\sqrt{625b^3c}$ **22.** $\sqrt[3]{54x^8}$ **23.** $\sqrt[4]{405a^{10}b^3}$

24. $\sqrt[3]{-24z^6}$ **25.** $\sqrt{4x^2 + 4x + 1}$

6.1 VOCABULARY PRACTICE

Match the term with its description.

1. The result of multiplication
2. The 3 in $\sqrt[3]{x}$ is an example of this.
3. $x^m \cdot x^n = x^{m+n}$
4. $(x^m)^n = x^{mn}$
5. The square root of a nonnegative real number that is a perfect square is this.
6. The square root of a nonnegative real number that is not a perfect square is this.
7. The positive square root of a positive number
8. $\sqrt[n]{a} \cdot \sqrt[n]{b} = \sqrt[n]{ab}$
9. The entire expression $\sqrt[3]{x}$ is an example of this.
10. The x in $\sqrt[3]{x}$ is an example of this.

A. index
B. irrational number
C. power rule of exponents
D. principal square root
E. product
F. product rule of exponents
G. product rule of radicals
H. radical
I. radicand
J. rational number

6.1 Exercises

Follow your instructor's guidelines for showing your work. Assume that all factors in radicands in these exercises represent nonnegative real numbers.

For exercises 1–4,
(a) identify the index.
(b) identify the radicand.

1. $\sqrt{5}$

2. $\sqrt{7}$

3. $\sqrt[3]{x}$

4. $\sqrt[3]{y}$

For exercises 5–20, evaluate. Round irrational numbers to the nearest tenth.

5. $\sqrt{81}$

6. $\sqrt{9}$

7. $\sqrt[3]{-512}$

8. $\sqrt[3]{-343}$

9. $\sqrt[4]{1296}$

10. $\sqrt[4]{4096}$

11. $\sqrt{11}$

12. $\sqrt{7}$

13. $-\sqrt{60}$

14. $-\sqrt{50}$

15. $\sqrt{-21}$

16. $\sqrt{-17}$

17. $\sqrt{\dfrac{100}{49}}$

18. $\sqrt{\dfrac{25}{64}}$

19. $\sqrt{\dfrac{1}{64}}$

20. $\sqrt{\dfrac{1}{25}}$

21. Since $\sqrt{\dfrac{1}{100}} = \dfrac{1}{10}$ and $\dfrac{1}{100} = 0.01$, what is $\sqrt{0.01}$? Write the answer as a decimal number.

22. Since $\sqrt{\dfrac{1}{10,000}} = \dfrac{1}{100}$ and $\dfrac{1}{10,000} = 0.0001$, what is $\sqrt{0.0001}$? Write the answer as a decimal number.

23. Think about the square root of a fraction between 0 and 1. Is this square root greater than or less than the original fraction?

24. Think about the square root of a whole number greater than 1. Is this square root greater than or less than the original number?

25. A second story window is 25 ft above the ground. A bush is growing next to the house. Because of this bush, the base of a ladder must be 10 ft away from the house. The minimum length of ladder needed to reach the window equals $\sqrt{(25 \text{ ft})^2 + (10 \text{ ft})^2}$. Find the minimum length of the ladder. Round *up* to the nearest whole number.

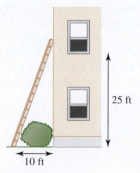

25 ft

10 ft

26. A first baseman throws from first base to third base. The distance of the throw equals $\sqrt{(90 \text{ ft})^2 + (90 \text{ ft})^2}$. Find the distance of the throw. Round to the nearest whole number.

90 ft 90 ft 90 ft 90 ft

For exercises 27–28, the formula $G = \sqrt{p} + 3$ is used to find the reading grade level of a document. The reading grade level is *G*. The number of polysyllabic words in 30 selected sentences from the document is *p*. Find the reading grade level of a document with the given number of polysyllabic words. Round to the nearest tenth.

27. 65 polysyllabic words

28. 51 polysyllabic words

For exercises 29–58, simplify.

29. $\sqrt{100x^6}$

30. $\sqrt{49a^6}$

31. $\sqrt{200x^6}$

32. $\sqrt{98a^6}$

33. $\sqrt{200x^7}$

34. $\sqrt{98a^7}$

35. $\sqrt{48h^5k^2}$

36. $\sqrt{27x^5y^2}$

37. $\sqrt{128abc^2}$

38. $\sqrt{50jkm^4}$

39. $\sqrt{288x^{13}}$

40. $\sqrt{512p^{15}}$

41. $\sqrt{80xy}$

42. $\sqrt{45bc}$

43. $\sqrt[12]{x^{12}}$

44. $\sqrt[14]{z^{14}}$

45. $\sqrt[24]{x^{24}}$

46. $\sqrt[32]{x^{32}}$

47. $\sqrt[24]{x^{48}}$

48. $\sqrt[32]{x^{64}}$

49. $\sqrt[7]{a^{21}}$

50. $\sqrt[8]{h^{24}}$

51. $\sqrt[3]{40a^6b^7}$

52. $\sqrt[3]{250p^9q^4}$

53. $\sqrt[3]{81g^2h}$

54. $\sqrt[3]{48ab^2}$

55. $\sqrt[3]{-128x^{12}y^{16}}$

56. $\sqrt[3]{-108a^{15}b^7}$

57. $\sqrt[3]{-512x^{21}y^3}$

58. $\sqrt[3]{-343x^{24}y^3}$

For exercises 59–66, simplify.

59. $\sqrt[4]{32a^5b^8}$

60. $\sqrt[4]{162c^7d^4}$

61. $\sqrt[4]{256r^{15}w^{13}}$

62. $\sqrt[4]{1296b^7c^9}$

63. $\sqrt[4]{243h^3k^2}$

64. $\sqrt[4]{80c^2d^3}$

65. $\sqrt[4]{128x^{12}y^9}$

66. $\sqrt[4]{64w^{20}z^{13}}$

For exercises 67–72, simplify.

67. $\sqrt[8]{256x^{15}}$

68. $\sqrt[7]{128y^{12}}$

69. $\sqrt[9]{x^{31}}$

70. $\sqrt[11]{x^{27}}$

71. $\sqrt[100]{a^{201}}$

72. $\sqrt[50]{v^{51}}$

For exercises 73–86, simplify.

73. $\sqrt{(x + 5)^2}$

74. $\sqrt{(x + 6)^2}$

75. $\sqrt{a^2 - 8a + 16}$

76. $\sqrt{b^2 - 14b + 49}$

77. $\sqrt{z^2 - 2z + 1}$

78. $\sqrt{c^2 + 2c + 1}$

79. $\sqrt{4x^2 + 12x + 9}$

80. $\sqrt{4x^2 + 20x + 25}$

81. $\sqrt{4h^2 - 36h + 81}$

82. $\sqrt{4k^2 - 12k + 9}$

83. $\sqrt{9x^2 - 42x + 49}$

84. $\sqrt{9y^2 - 48y + 64}$

85. $\sqrt[3]{(x - 5)^3}$

86. $\sqrt[3]{(x + 4)^3}$

Problem Solving: Practice and Review

Follow your instructor's guidelines for using the five steps as outlined in Section 1.5, p. 51.

87. A security system includes four cameras and a digital video recorder (DVR) with color monitor. A small business owner can watch the monitor at the office or through the Internet and can retrieve stored information from the DVR. The cost of the system is an initial $99 setup fee and a maintenance fee of $99 per month.

  a. Write a linear model that describes the cost to use the system, y, for x months.

  b. Find the cost to use the system for 5 years.

88. Choose whole numbers from 1 to 10 to complete the headline for the article: ___ out of ___ Employees Can Expect a Raise.

  Depending on where they work, 77 percent of the workforce can still expect to receive base pay raises. (*Source:* www.worldatwork.org, Feb. 3, 2009)

89. Find the cost of attending CSU as a commuter student in 2009.

  College costs have grown much faster than wages. In 2009, the average wage for retail salespeople was $10.90 an hour. A student working 12 weeks over the summer would earn $5200, enough to pay only 20% of the current cost of a year at UC, and 40% of the cost of attending CSU as a commuter student. (*Source:* www.cpec.ca.gov, March 2011)

90. A recommended ratio of water to fish for an aquarium is 4 gal of water per inch of fish. For example, if a person wants to keep 6 fish that are each 1 in. long, the minimum size of their aquarium is 24 gal. Find the number of inches of fish recommended for an aquarium that is 48 in. long, 18 in. wide, and 25 in. high (1 ft³ ≈ 7.481 gal). Round to the nearest tenth.

Find the Mistake

For exercises 91–94, the completed problem has one mistake.

(a) Describe the mistake in words, or copy down the whole problem and highlight or circle the mistake.

(b) Do the problem correctly.

91. **Problem:** Simplify: $\sqrt{32x^{25}}$

  Incorrect Answer: $\sqrt{32x^{25}}$

$$= \sqrt{32}\sqrt{x^{25}}$$
$$= \sqrt{16}\sqrt{2}\sqrt{x^{25}}$$
$$= 4x^5\sqrt{2}$$

92. **Problem:** Simplify: $\sqrt[3]{32x}$

  Incorrect Answer: $\sqrt[3]{32x}$

$$= \sqrt{32}\sqrt{x}$$
$$= \sqrt{16}\sqrt{2}\sqrt{x}$$
$$= 4\sqrt{2x}$$

93. **Problem:** Simplify: $\sqrt{x^2 + 10x + 25}$

  Incorrect Answer: $\sqrt{x^2 + 10x + 25}$

$$= \sqrt{x^2} + \sqrt{10x} + \sqrt{25}$$
$$= x + 5 + \sqrt{10x}$$

94. **Problem:** Simplify: $\sqrt{72x^5}$

  Incorrect Answer: $\sqrt{72x^5}$

$$= \sqrt{72}\sqrt{x^5}$$
$$= \sqrt{4}\sqrt{18}\sqrt{x^4}\sqrt{x}$$
$$= 2x^2\sqrt{18x}$$

Review

For exercises 95–97, simplify.

95. $3x - 9y - 5x + 17y$

96. $2(3x - 9)$

97. $3(2x - 9) - (7x + 4)$

98. Write the distributive property.

SUCCESS IN COLLEGE MATHEMATICS

99. When a student begins a math class, the instructor assumes that the student already knows the material taught in previous classes. The student is at the "starting line" of the class. When you began this class, did you think that you were at the starting line?

100. Students may transfer from other colleges with math courses that have different content. Placement tests are not perfect. When a student does not know material from previous courses, it can be difficult to understand new material. Instead of feeling confident and in control, the student may feel that it is impossible to succeed. What do you think a student in this position should do?

© mashe/Shutterstock

Radishes and beets are both roots. However, they are not the same kind of root. They are not "like roots." In this section, we will add and subtract radicals. Only radicals that are "like radicals" can be combined.

SECTION 6.2

Adding, Subtracting, Multiplying, and Simplifying Radical Expressions

After reading the text, working the practice problems, and completing assigned exercises, you should be able to:

1. Add and subtract radical expressions.

2. Multiply radical expressions.

Adding and Subtracting Radicals

To simplify the expression $3x + 5x + 2y - 9y$, we combine **like terms**. The simplified expression is $8x - 7y$. To add and subtract radical expressions, we combine **like radicals**. Like radicals have the same radicands and the same index.

EXAMPLE 1   Simplify: $3\sqrt{x} + 5\sqrt{x} + 2\sqrt{y} - 9\sqrt{y}$

SOLUTION ▶
$$3\sqrt{x} + 5\sqrt{x} + 2\sqrt{y} - 9\sqrt{y}$$
$$= 8\sqrt{x} - 7\sqrt{y} \qquad \text{Combine like radicals.}$$

In the next example, the radicals are not the same. However, after each radical is simplified, there are like radicals that can be combined.

EXAMPLE 2   Simplify: $-\sqrt{18} + \sqrt{50} - \sqrt{162}$

SOLUTION ▶
$$-\sqrt{18} + \sqrt{50} - \sqrt{162} \qquad \text{Find perfect square factors of the radicands: 9, 25, 81.}$$
$$= -\sqrt{9}\sqrt{2} + \sqrt{25}\sqrt{2} - \sqrt{81}\sqrt{2} \qquad \text{Product rule of radicals.}$$
$$= -3\sqrt{2} + 5\sqrt{2} - 9\sqrt{2} \qquad \text{Simplify.}$$
$$= -7\sqrt{2} \qquad \text{Combine like radicals.}$$

The radicands of like radicals may include both numbers and variables.

EXAMPLE 3 Simplify: $\sqrt{6a^3} + \sqrt{24a^3} + \sqrt{54a^3}$

SOLUTION ▶

$\sqrt{6a^3} + \sqrt{24a^3} + \sqrt{54a^3}$

$= \sqrt{6}\sqrt{a^3} + \sqrt{24}\sqrt{a^3} + \sqrt{54}\sqrt{a^3}$ Find perfect square factors of the radicands: a^2, 4, 9.

$= \sqrt{6}\sqrt{a^2}\sqrt{a} + \sqrt{4}\sqrt{6}\sqrt{a^2}\sqrt{a} + \sqrt{9}\sqrt{6}\sqrt{a^2}\sqrt{a}$ Product rule of radicals.

$= \sqrt{6}\cdot a \cdot \sqrt{a} + 2\cdot\sqrt{6}\cdot a \cdot \sqrt{a} + 3\cdot\sqrt{6}\cdot a \cdot \sqrt{a}$ Simplify.

$= a\sqrt{6a} + 2a\sqrt{6a} + 3a\sqrt{6a}$ Product rule of radicals.

$= 6a\sqrt{6a}$ Combine like radicals.

Adding and Subtracting Radicals

Assume that all variables represent nonnegative real numbers.

1. Simplify each radical. When a radical is simplified, the radicand has no perfect power factors; every exponent in the radicand is less than the index.

2. Combine like radicals. Like radicals have the same radicand and the same index.

In Examples 2 and 3, we combined square roots. We use the same process to combine cube roots. Find perfect cube factors in each radicand, use the product rule of radicals to rewrite each cube root as a product of cube roots, simplify, and combine like radicals.

EXAMPLE 4 Simplify: $\sqrt[3]{128} - \sqrt[3]{54} + \sqrt[3]{2}$

SOLUTION ▶

$\sqrt[3]{128} - \sqrt[3]{54} + \sqrt[3]{2}$ Find perfect cube factors of the radicands: 64, 27.

$= \sqrt[3]{64}\sqrt[3]{2} - \sqrt[3]{27}\sqrt[3]{2} + \sqrt[3]{2}$ Product rule of radicals.

$= 4\sqrt[3]{2} - 3\sqrt[3]{2} + 1\sqrt[3]{2}$ Simplify; rewrite $\sqrt[3]{2} = 1\sqrt[3]{2}$.

$= 2\sqrt[3]{2}$ Combine like radicals.

In working with radicals, it is important to write neatly. The difference between an exponent in an expression and the index of a radical must be clear.

EXAMPLE 5 Simplify: $\sqrt[3]{16x^{13}} + \sqrt[3]{54x^{13}}$

SOLUTION ▶

$\sqrt[3]{16x^{13}} + \sqrt[3]{54x^{13}}$

$= \sqrt[3]{16}\sqrt[3]{x^{13}} + \sqrt[3]{54}\sqrt[3]{x^{13}}$ Find perfect cube factors of the radicands: 8, 27, x^{12}.

$= \sqrt[3]{8}\cdot\sqrt[3]{2}\cdot\sqrt[3]{x^{12}}\cdot\sqrt[3]{x} + \sqrt[3]{27}\cdot\sqrt[3]{2}\cdot\sqrt[3]{x^{12}}\cdot\sqrt[3]{x}$ Product rule of radicals.

$= 2\cdot\sqrt[3]{2}\cdot x^4\cdot\sqrt[3]{x} + 3\cdot\sqrt[3]{2}\cdot x^4\cdot\sqrt[3]{x}$ Simplify.

$= 2x^4\sqrt[3]{2x} + 3x^4\sqrt[3]{2x}$ Product rule of radicals.

$= 5x^4\sqrt[3]{2x}$ Combine like radicals.

In the next example, the radicals have an index of 4. It may be helpful to refer to or make a list of perfect fourth powers: $2^4 = 16$, $3^4 = 81$, $4^4 = 256$, $5^4 = 625$, $6^4 = 1296$.

EXAMPLE 6 | Simplify: $\sqrt[4]{162x^5} + \sqrt[4]{32x^5} - \sqrt[4]{1250x^5}$

SOLUTION ▶

$\sqrt[4]{162x^5} + \sqrt[4]{32x^5} - \sqrt[4]{1250x^5}$

$= \sqrt[4]{162}\,\sqrt[4]{x^5} + \sqrt[4]{32}\,\sqrt[4]{x^5} - \sqrt[4]{1250}\,\sqrt[4]{x^5}$ Find perfect fourth powers: $81x^4$, 16, 625.

$= \sqrt[4]{81}\cdot\sqrt[4]{2}\cdot\sqrt[4]{x^4}\cdot\sqrt[4]{x} + \sqrt[4]{16}\cdot\sqrt[4]{2}\cdot\sqrt[4]{x^4}\cdot\sqrt[4]{x} - \sqrt[4]{625}\cdot\sqrt[4]{2}\cdot\sqrt[4]{x^4}\cdot\sqrt[4]{x}$

$= \mathbf{3}\cdot\sqrt[4]{2}\cdot\mathbf{x}\cdot\sqrt[4]{x} + \mathbf{2}\cdot\sqrt[4]{2}\cdot\mathbf{x}\cdot\sqrt[4]{x} - \mathbf{5}\cdot\sqrt[4]{2}\cdot\mathbf{x}\cdot\sqrt[4]{x}$ Simplify.

$= 3x\sqrt[4]{2x} + 2x\sqrt[4]{2x} - 5x\sqrt[4]{2x}$ Product rule.

$= \mathbf{5x\sqrt[4]{2x}} - 5x\sqrt[4]{2x}$ Simplify.

$= 0$ Combine like radicals.

Practice Problems

For problems 1–5, simplify.

1. $\sqrt{45} + \sqrt{80} - \sqrt{5}$ **2.** $2\sqrt{50x} - 5\sqrt{8x}$ **3.** $3p\sqrt{20} + 4\sqrt{5p^2}$

4. $\sqrt[3]{128} + \sqrt[3]{54}$ **5.** $\sqrt[4]{16x} + \sqrt[4]{81x}$

Multiplying Radicals

Using the product rule of radicals, $\sqrt[n]{ab} = \sqrt[n]{a}\cdot\sqrt[n]{b}$, we can rewrite a radical as a product of radicals. To multiply radicals, we reverse the rule, $\sqrt[n]{a}\cdot\sqrt[n]{b} = \sqrt[n]{ab}$, and simplify the product. In a simplified expression, the radical is usually written last.

EXAMPLE 7 | Simplify: $\sqrt{3h}\,\sqrt{7h}$

SOLUTION ▶

$\sqrt{3h}\,\sqrt{7h}$

$= \sqrt{3h\cdot 7h}$ Product rule of radicals.

$= \sqrt{21h^2}$ Simplify.

$= \sqrt{21}\,\sqrt{h^2}$ Product rule of radicals.

$= \sqrt{21}\mathbf{h}$ Simplify.

$= h\sqrt{21}$ Commutative property; write the radical last.

In the next example, the radicand of the product is 315. The perfect squares less than 315 are 4, 9, 16, 25, 36, 49, 64, 81, 100, 121, 144, 169, 196, 225, 256, and 289. The only perfect square factor that is a factor of 315 is **9**.

EXAMPLE 8 | Simplify: $\sqrt{15}\,\sqrt{21}$

SOLUTION ▶

$\sqrt{15}\,\sqrt{21}$

$= \sqrt{15\cdot 21}$ Product rule of radicals.

$= \sqrt{315}$ Simplify; find perfect square factors of the radicand: 9.

$= \sqrt{9}\,\sqrt{35}$ Product rule of radicals.

$= 3\sqrt{35}$ Simplify.

Finding factors in a list of perfect squares can take some time. A different way to simplify is to look for perfect square factors of each radical and simplify *before* finishing the multiplication.

EXAMPLE 9 Simplify: $\sqrt{15}\sqrt{21}$

SOLUTION ▶ $\sqrt{15}\sqrt{21}$

$= \sqrt{15 \cdot 21}$ Product rule of radicals.

$= \sqrt{3 \cdot 5 \cdot 3 \cdot 7}$ Factor each number in the radicand.

$= \sqrt{3^2 \cdot 35}$ 3^2 is a perfect square, 9.

$= \sqrt{3^2}\sqrt{35}$ Product rule of radicals.

$= 3\sqrt{35}$ Simplify.

Although the individual radicals $\sqrt{15}$ and $\sqrt{6}$ cannot be simplified, $\sqrt{15 \cdot 6}$ can be simplified because both 15 and 6 have a factor of 3.

EXAMPLE 10 Simplify: $7\sqrt{15x} \cdot 2\sqrt{6}$

SOLUTION ▶ $7\sqrt{15x} \cdot 2\sqrt{6}$

$= 7 \cdot 2 \cdot \sqrt{15x \cdot 6}$ Product rule of radicals.

$= 14\sqrt{3 \cdot 5 \cdot x \cdot 2 \cdot 3}$ Factor each number in the radicand; 3^2 is a perfect square, 9.

$= 14\sqrt{3^2}\sqrt{10x}$ Product rule of radicals.

$= 14 \cdot 3 \cdot \sqrt{10x}$ Simplify.

$= 42\sqrt{10x}$ Simplify; product rule of radicals.

Multiplying Radicals

Assume that all variables represent nonnegative real numbers.

Strategy 1

1. Use the product rule of radicals to rewrite the expression as a single radical. Multiply the radicands.

2. Simplify.

Strategy 2

1. Use the product rule of radicals to rewrite the expression as a single radical. Do not multiply the radicands.

2. Factor each radicand; look for perfect powers.

3. Simplify.

The distributive property, $a(b + c) = ab + ac$, also applies to multiplication with radicals. We can multiply two radical expressions with different radicands. However, we cannot add or subtract radical expressions with different radicands. In the next example, the final expression is $3\sqrt{2xy} + 27\sqrt{x}$. Since the radicands are not the same, this expression is simplified.

EXAMPLE 11 Simplify: $3\sqrt{x}(\sqrt{2y} + 9)$

SOLUTION ▶ $3\sqrt{x}(\sqrt{2y} + 9)$

$= 3\sqrt{x} \cdot \sqrt{2y} + 3\sqrt{x} \cdot 9$ Distributive property.

$= 3\sqrt{2xy} + 27\sqrt{x}$ Product rule of radicals; these are not like radicals and cannot be combined.

In multiplying an expression such as $2\sqrt{5p} \cdot 3\sqrt{7p}$, multiply the whole numbers, $2 \cdot 3$, and use the product rule of radicals to multiply $\sqrt{5p}\sqrt{7p}$.

EXAMPLE 12 | Simplify: $2\sqrt{5p}(3\sqrt{7p} - 4\sqrt{10})$

SOLUTION ▶

$2\sqrt{5p}(3\sqrt{7p} - 4\sqrt{10})$

$= 2\sqrt{5p}\cdot 3\sqrt{7p} + 2\sqrt{5p}\cdot(-4\sqrt{10})$ Distributive property.

$= 2\cdot 3\cdot\sqrt{5p\cdot 7p} + 2(-4)\sqrt{5p\cdot 10}$ Product rule of radicals.

$= 6\sqrt{5\cdot 7\cdot p^2} - 8\sqrt{5\cdot p\cdot 2\cdot 5}$ Factor radicands. Find perfect square factors.

$= 6\sqrt{35}\sqrt{p^2} - 8\sqrt{5^2}\sqrt{2p}$ Product rule of radicals.

$= 6\cdot\sqrt{35}\cdot p - 8\cdot 5\cdot\sqrt{2p}$ Simplify.

$= 6p\sqrt{35} - 40\sqrt{2p}$ Simplify; these are not like radicals and cannot be combined.

In Chapter 2, we used the distributive property to multiply binomials. In the next example, we use it to multiply expressions that include two radicals. There are four radicals in the final expression that are not like radicals and cannot be combined.

EXAMPLE 13 | Simplify: $(\sqrt{x} - 9\sqrt{y})(\sqrt{3p} - \sqrt{w})$

SOLUTION ▶

$(\sqrt{x} - 9\sqrt{y})(\sqrt{3p} - \sqrt{w})$

$= (\sqrt{x} - 9\sqrt{y})(\sqrt{3p} - 1\sqrt{w})$ Rewrite $-\sqrt{w}$ as $-1\sqrt{w}$.

$= \sqrt{x}(\sqrt{3p} - 1\sqrt{w}) - 9\sqrt{y}(\sqrt{3p} - 1\sqrt{w})$ Distributive property.

$= \sqrt{x}\sqrt{3p} + \sqrt{x}(-1\sqrt{w}) - 9\sqrt{y}\sqrt{3p} - 9\sqrt{y}(-1\sqrt{w})$

$= \sqrt{3px} - \sqrt{wx} - 9\sqrt{3py} + 9\sqrt{wy}$ Product rule of radicals; these are not like radicals and cannot be combined.

In the next example, the factors are the same except that the signs are opposite. These factors are a **conjugate pair**. Their product does not include any radicals. In the next section, we will use conjugate pairs to rationalize denominators.

EXAMPLE 14 | Simplify: $(3\sqrt{x} - \sqrt{7})(3\sqrt{x} + \sqrt{7})$

SOLUTION ▶

$(3\sqrt{x} - \sqrt{7})(3\sqrt{x} + \sqrt{7})$

$= 3\sqrt{x}(3\sqrt{x} + \sqrt{7}) - \sqrt{7}(3\sqrt{x} + \sqrt{7})$ Distributive property.

$= 3\sqrt{x}\cdot 3\sqrt{x} + 3\sqrt{x}\cdot\sqrt{7} - \sqrt{7}\cdot 3\sqrt{x} - \sqrt{7}\cdot\sqrt{7}$ Distributive property.

$= 9x + 3\sqrt{7x} - 3\sqrt{7x} - 7$ Product rule of radicals; simplify.

$= 9x + 0 - 7$ Simplify.

$= 9x - 7$ Simplify; the expression has no radicals.

Practice Problems

For problems 6–10, simplify.

6. $2\sqrt{21}\cdot 10\sqrt{14}$ 7. $\sqrt{3x}(\sqrt{6x} + \sqrt{5})$ 8. $3\sqrt{x}(\sqrt{6x} + \sqrt{5})$

9. $\sqrt[3]{9c}(\sqrt[3]{3c} + \sqrt[3]{6c^2})$ 10. $(\sqrt{10h} + \sqrt{15})(\sqrt{2h} - \sqrt{21})$

6.2 VOCABULARY PRACTICE

Match the term with its description.

1. $\sqrt[n]{a} \cdot \sqrt[n]{b} = \sqrt[n]{ab}$
2. $a(b + c) = ab + ac$
3. $x^m \cdot x^n = x^{m+n}$
4. The positive square root of a positive number
5. Terms that have exactly the same variable(s) with exactly the same exponents
6. The expression under a radical sign
7. The 3 in $\sqrt[3]{x}$ is an example of this.
8. $(x^m)^n = x^{mn}$
9. A real number that is greater than or equal to 0
10. A number that is an element of the set $\{\ldots, -2, -1, 0, 1, 2, 3, \ldots\}$

A. distributive property
B. index
C. integer
D. like terms
E. nonnegative real number
F. power rule of exponents
G. principal square root
H. product rule of exponents
I. product rule of radicals
J. radicand

6.2 Exercises

Follow your instructor's guidelines for showing your work. Assume that all radicands and bases of exponential expressions represent nonnegative real numbers.

For exercises 1–38, simplify. If the expression is already simplified, write "already simplified."

1. $6\sqrt{3} + 5\sqrt{3}$
2. $7\sqrt{11} + 2\sqrt{11}$
3. $-7\sqrt{5} + 3\sqrt{5}$
4. $-4\sqrt{5} + 9\sqrt{5}$
5. $\sqrt{21} - 15\sqrt{21}$
6. $\sqrt{35} - 6\sqrt{35}$
7. $6\sqrt[3]{x^2} + 7\sqrt[3]{x^2}$
8. $5\sqrt[4]{y^3} + 9\sqrt[4]{y^3}$
9. $3w\sqrt{5} + 9w\sqrt{5} - w\sqrt{5}$
10. $2p\sqrt{6} + 13p\sqrt{6} - p\sqrt{6}$
11. $10\sqrt{11} + 2\sqrt{13}$
12. $21\sqrt{3} + 15\sqrt{5}$
13. $\sqrt{20} + \sqrt{5}$
14. $\sqrt{72} + \sqrt{2}$
15. $\sqrt{20x} + \sqrt{5x}$
16. $\sqrt{72v} + \sqrt{2v}$
17. $\sqrt{20x^2} + \sqrt{5x^2}$
18. $\sqrt{72v^2} + \sqrt{2v^2}$
19. $\sqrt{300} + \sqrt{27}$
20. $\sqrt{72} + \sqrt{8}$
21. $\sqrt[3]{54} + \sqrt[3]{16}$
22. $\sqrt[3]{375} + \sqrt[3]{24}$
23. $5\sqrt[3]{81x} - \sqrt[3]{375x}$
24. $6\sqrt[3]{40x^2} - \sqrt[3]{320x^2}$
25. $\sqrt[4]{16a^3} - \sqrt[4]{81a^3}$
26. $\sqrt[4]{256y} + \sqrt[4]{16y}$
27. $\sqrt[3]{-16k^2} - \sqrt[3]{54k^2}$
28. $\sqrt[3]{-72m^2} - \sqrt[3]{243m^2}$
29. $4\sqrt{12} - 3\sqrt{27}$
30. $6\sqrt{75} - 8\sqrt{48}$
31. $\sqrt[5]{64x^7} + \sqrt[5]{486x^7}$
32. $\sqrt[5]{64a^9} + \sqrt[5]{2048a^9}$
33. $\sqrt[5]{-64x^7} + \sqrt[5]{486x^7}$
34. $\sqrt[5]{-64a^9} + \sqrt[5]{2048a^9}$

35. $\sqrt[4]{80p^4} + \sqrt[4]{1280p^4}$
36. $\sqrt[4]{64w^4} + \sqrt[4]{324w^4}$
37. $\sqrt[4]{80p^4} + \sqrt[4]{1280p^{13}}$
38. $\sqrt[4]{32w^4} + \sqrt[4]{162w^{15}}$
39. Explain why $\sqrt{-25}$ is not a real number.
40. Explain why $\sqrt[4]{-16}$ is not a real number.
41. a. Simplify: $\sqrt{xy^2}$
    b. Simplify: $\sqrt{x^2y}$
    c. Does $\sqrt{xy^2} = \sqrt{x^2y}$?
42. Explain why we cannot combine the terms in the expression $5\sqrt{xy} + 8\sqrt[3]{xy}$.
43. We know that $\sqrt{5}\sqrt{5} = 5$.
    a. Does $(\sqrt[3]{5})(\sqrt[3]{5}) = 5$?
    b. Does $(\sqrt[3]{5})(\sqrt[3]{5})(\sqrt[3]{5}) = 5$?
44. We know that $\sqrt{7}\sqrt{7} = 7$.
    a. Does $(\sqrt[3]{7})(\sqrt[3]{7}) = 7$?
    b. Does $(\sqrt[3]{7})(\sqrt[3]{7})(\sqrt[3]{7}) = 7$?

For exercises 45–64, simplify.

45. $\sqrt{15}\sqrt{6}$
46. $\sqrt{21}\sqrt{14}$
47. $\sqrt{20}\sqrt{30}$
48. $\sqrt{30}\sqrt{15}$
49. $\sqrt{14x}\sqrt{42x}$
50. $\sqrt{10a}\sqrt{30a}$
51. $\sqrt{21p}\sqrt{10p}$
52. $\sqrt{35w}\sqrt{22w}$
53. $\sqrt[3]{2}\sqrt[3]{12}$
54. $\sqrt[3]{4}\sqrt[3]{6}$
55. $\sqrt[3]{9}\sqrt[3]{9}$
56. $\sqrt[3]{25}\sqrt[3]{25}$
57. $\sqrt[4]{x^3}\sqrt[4]{x}$
58. $\sqrt[4]{n}\sqrt[4]{n^3}$
59. $\sqrt[5]{n^3}\sqrt[5]{n^3}$
60. $\sqrt[5]{d^4}\sqrt[5]{d^3}$
61. $\sqrt[101]{w^{49}}\sqrt[101]{w^{71}}$
62. $\sqrt[101]{w^{53}}\sqrt[101]{w^{61}}$
63. $\sqrt[17]{a^5}\sqrt[17]{a^{11}}$
64. $\sqrt[19]{a^{11}}\sqrt[19]{a^7}$

For exercises 65–88, simplify.

65. $\sqrt{3}(\sqrt{2} + 4\sqrt{3})$
66. $\sqrt{2}(\sqrt{5} + 6\sqrt{2})$

67. $\sqrt{2x}(\sqrt{x} - \sqrt{5x})$

68. $\sqrt{3h}(\sqrt{h} - \sqrt{7h})$

69. $\sqrt{2x}(\sqrt{6x} - \sqrt{10x})$

70. $\sqrt{3h}(\sqrt{15h} - \sqrt{21h})$

71. $(x + \sqrt{2})(x + \sqrt{2})$

72. $(b + \sqrt{5})(b + \sqrt{5})$

73. $(x + \sqrt{2})(x - \sqrt{2})$

74. $(b + \sqrt{5})(b - \sqrt{5})$

75. $(x - \sqrt{2})(x - \sqrt{2})$

76. $(b - \sqrt{5})(b - \sqrt{5})$

77. $(\sqrt{x} - 2)(\sqrt{x} - 2)$

78. $(\sqrt{b} - 2)(\sqrt{b} - 2)$

79. $(\sqrt{3h} - \sqrt{2k})(\sqrt{6h} + \sqrt{10k})$

80. $(\sqrt{6m} + \sqrt{15p})(\sqrt{2m} - \sqrt{3p})$

81. $(\sqrt{5a} - \sqrt{10a})(\sqrt{15a} - \sqrt{14a})$

82. $(\sqrt{3w} - \sqrt{21w})(\sqrt{6w} - \sqrt{35w})$

83. $(3\sqrt{z} - 5)(3\sqrt{z} + 5)$

84. $(9\sqrt{x} - 8)(9\sqrt{x} + 8)$

85. $(6\sqrt{2x} + 1)(6\sqrt{2x} + 1)$

86. $(8\sqrt{3x} + 2)(8\sqrt{3x} + 2)$

87. $(\sqrt[3]{4c})(\sqrt[3]{4c})(\sqrt[3]{4c})$

88. $(\sqrt[3]{7y})(\sqrt[3]{7y})(\sqrt[3]{7y})$

Problem Solving: Practice and Review

Follow your instructor's guidelines for using the five steps as outlined in Section 1.5, p. 51.

89. An incorrect headline for an article is "Three out of four dream of a new career whilst on vacation." Rewrite the headline using whole numbers less than ten.

A survey . . . has revealed that 81% of people dream about new career options whilst on vacation but only 26% of those have ever acted on this dream. (*Source:* www.ehotelier.com, May 27, 2009)

90. At an Arizona nursery, a large saguaro cactus costs $145 for the first 3 feet and $85 for each additional foot.
a. Write a linear model if y is the total cost of a cactus that is x feet high.
b. Find the cost to buy a 12-foot saguaro cactus.

91. Find the percent increase in the number of heroin-related deaths.

The number of heroin-related deaths has skyrocketed from five in 2008 to 18 last year. (*Source:* www.stltoday.com, June 22, 2011)

92. The Las Vegas Motor Speedway was completed in 1996. It seats 142,000 people. The 1.5-mi track is a D-shaped oval. The length of the frontstretch is 2275 ft. The length of the backstretch is 1572 ft. A NASCAR race at the Speedway is 267 laps. Find the percent of the total distance in this race that is driven on a curve, not on the backstretch or frontstretch (1 mi = 5280 ft). Round to the nearest tenth of a percent.

Find the Mistake

For exercises 93–96, the completed problem has one mistake.
(a) Describe the mistake in words, or copy down the whole problem and highlight or circle the mistake.
(b) Do the problem correctly.

93. Problem: Simplify: $\sqrt{72x^5} + \sqrt{200x^5}$

Incorrect Answer:
$$\sqrt{72x^5} + \sqrt{200x^5}$$
$$= \sqrt{72}\sqrt{x^5} + \sqrt{200}\sqrt{x^5}$$
$$= \sqrt{9}\sqrt{8}\sqrt{x^4}\sqrt{x} + \sqrt{25}\sqrt{8}\sqrt{x^4}\sqrt{x}$$
$$= 3x^2\sqrt{8x} + 5x^2\sqrt{8x}$$
$$= 8x^2\sqrt{8x}$$

94. Problem: Simplify: $\sqrt[3]{16x^4} + \sqrt[3]{64x^4}$

Incorrect Answer: $\sqrt[3]{16x^4} + \sqrt[3]{64x^4}$
$$= \sqrt{16}\sqrt{x^4} + \sqrt{64}\sqrt{x^4}$$
$$= 4x^2 + 8x^2$$
$$= 12x^2$$

95. Problem: Simplify: $\sqrt{3}(\sqrt{2x} - 8)$

Incorrect Answer: $\sqrt{3}(\sqrt{2x} - 8)$
$$= \sqrt{6x} - 8\sqrt{3}$$

96. Problem: Simplify: $\sqrt{32x^{16}} + \sqrt{18x^{16}}$

Incorrect Answer: $\sqrt{32x^{16}} + \sqrt{18x^{16}}$
$$= \sqrt{32}\sqrt{x^{16}} + \sqrt{18}\sqrt{x^{16}}$$
$$= \sqrt{2}\sqrt{16}\sqrt{x^{16}} + \sqrt{9}\sqrt{2}\sqrt{x^{16}}$$
$$= 4x^4\sqrt{2} + 3x^4\sqrt{2}$$
$$= 7x^4\sqrt{2}$$

Review

97. Write the quotient rule of exponents.

For exercises 98–100, simplify.

98. $\dfrac{x^2y^7}{x^6y^3}$

99. $\dfrac{12x^2y^9}{40x^8y^4}$

100. $\dfrac{x^3y^8}{4x^9y^{20}}$

SUCCESS IN COLLEGE MATHEMATICS

101. A student is self-confident and is sure that he will get an A on the first test. Although he did all of the homework assignments, he did not do anything else to prepare. His grade on the first test was a D. He tells his friends that the test wasn't fair and that he can't learn from this professor.
a. Who does the student believe is responsible for his failure?
b. Self-confidence on its own does not ensure success. What else does this student need to do to be successful?

$$1.41421\overline{)1.00000}^{\;\;?}$$

$$2\overline{)1.41421}^{\;\;?}$$

The top division problem on the left is more difficult than the bottom division problem. This is one reason that in this section, we will **rationalize denominators**. We again assume that factors in radicands and bases of exponential expressions represent nonnegative real numbers.

| **SECTION 6.3** | # Dividing Radical Expressions and Conjugates |

After reading the text, working the practice problems, and completing assigned exercises, you should be able to:

1. Simplify a radical expression that includes division.
2. Rationalize the numerator or denominator of a fraction.
3. Identify the conjugate of an expression.

Simplifying Radicals

To simplify radicals in which the radicand is a fraction, use the quotient rule of radicals.

> **Quotient Rule of Radicals**
>
> If a, b are real numbers, $b \neq 0$, and n is a whole number greater than 1,
>
> $$\sqrt[n]{\frac{a}{b}} = \frac{\sqrt[n]{a}}{\sqrt[n]{b}}.$$

Using the quotient rule of radicals, we can rewrite $\sqrt{\dfrac{17x^3}{16y^2}}$ as $\dfrac{\sqrt{17x^3}}{\sqrt{16y^2}}$ and then simplify each of the radicals.

EXAMPLE 1 Simplify: $\sqrt{\dfrac{17x^3}{16y^2}}$

SOLUTION ▶

$$\sqrt{\frac{17x^3}{16y^2}}$$

$$= \frac{\sqrt{17x^3}}{\sqrt{16y^2}} \qquad \text{Quotient rule of radicals.}$$

$$= \frac{\sqrt{17}\sqrt{x^3}}{\sqrt{16}\sqrt{y^2}} \qquad \text{Find perfect square factors of the radicands: } x^2, 16, y^2.$$

$$= \frac{\sqrt{17} \cdot \sqrt{x^2} \cdot \sqrt{x}}{\sqrt{16} \cdot \sqrt{y^2}} \qquad \text{Product rule of radicals.}$$

$$= \frac{x\sqrt{17x}}{4y} \qquad \text{Simplify; product rule of radicals.}$$

In the next example, we simplify the radicand *before* using the quotient rule of radicals.

EXAMPLE 2   Simplify: $\sqrt{\dfrac{30a^4b^2}{24a}}$

SOLUTION ▶

$\sqrt{\dfrac{30a^4b^2}{24a}}$

$= \sqrt{\dfrac{30a^{4-1}b^2}{24}}$ Quotient rule of exponents.

$= \sqrt{\dfrac{30a^3b^2}{24}}$ Simplify.

$= \sqrt{\dfrac{\mathbf{6} \cdot 5 \cdot a^3b^2}{\mathbf{6} \cdot 4}}$ Factor: find common factors.

$= \sqrt{\dfrac{5a^3b^2}{4}}$ Simplify.

$= \dfrac{\sqrt{5a^3b^2}}{\sqrt{4}}$ Quotient rule of radicals.

$= \dfrac{\sqrt{5}\sqrt{a^3}\sqrt{b^2}}{\sqrt{4}}$ Find perfect square factors of the radicands: a^2, b^2.

$= \dfrac{\sqrt{5}\sqrt{a^2}\sqrt{a}\sqrt{b^2}}{\sqrt{4}}$ Product rule of radicals.

$= \dfrac{\sqrt{5} \cdot a \cdot \sqrt{a} \cdot b}{2}$ Simplify.

$= \dfrac{ab\sqrt{5a}}{2}$ Product rule of radicals.

Practice Problems

For problems 1–4, simplify.

1. $\sqrt{\dfrac{9h^5}{100k^2}}$ 2. $\sqrt{\dfrac{10a^6c^5}{32ac^2}}$

3. $\sqrt[3]{\dfrac{27x^{11}}{64y^6}}$ 4. $\sqrt[4]{\dfrac{625c^5}{16c^{12}}}$

Rationalizing a Denominator

Before calculators, dividing a whole number by an irrational number such as $\sqrt{2}$ was almost always done by long division. Long division by a rounded irrational number is more difficult than long division by a whole number. This is one reason that *a simplified expression traditionally has no radicals in the denominator.*

EXAMPLE 3 | The expression $\dfrac{1}{\sqrt{2}} \approx 1 \div 1.41421$, and the expression $\dfrac{\sqrt{2}}{2} \approx 1.41421 \div 2.$

Use long division to find the quotients: $1 \div 1.41421$ and $1.41421 \div 2.$

SOLUTION ▶

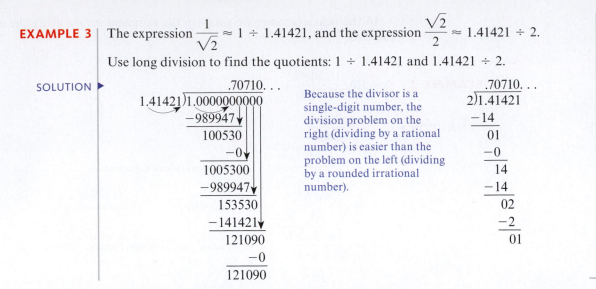

Because the divisor is a single-digit number, the division problem on the right (dividing by a rational number) is easier than the problem on the left (dividing by a rounded irrational number).

To rewrite $\dfrac{1}{\sqrt{2}}$ as an equivalent fraction without a radical in the denominator, multiply by a fraction equal to 1. This process is **rationalizing the denominator**.

EXAMPLE 4 | Rationalize the denominator of $\dfrac{1}{\sqrt{2}}$.

SOLUTION ▶

$\dfrac{1}{\sqrt{2}}$ A radical is in the denominator; the denominator is irrational.

$= \dfrac{1}{\sqrt{2}} \cdot \dfrac{\sqrt{2}}{\sqrt{2}}$ Multiply by a fraction equal to 1, $\dfrac{\sqrt{2}}{\sqrt{2}} = 1.$

$= \dfrac{\sqrt{2}}{\sqrt{2} \cdot \sqrt{2}}$ Product rule of radicals.

$= \dfrac{\sqrt{2}}{2}$ Simplify; $\sqrt{2}\sqrt{2} = 2$; the denominator has no radicals.

In the next example, the denominator is $\sqrt{x}$. Assuming that any radical in the denominator is irrational, multiply by a fraction equal to 1: $\dfrac{\sqrt{x}}{\sqrt{x}}$. Multiplying by 1 does not change the value of the fraction.

EXAMPLE 5 | Rationalize the denominator of $\dfrac{5y\sqrt{2}}{\sqrt{x}}$.

SOLUTION ▶

$\dfrac{5y\sqrt{2}}{\sqrt{x}}$ A radical is in the denominator.

$= \dfrac{5y\sqrt{2}}{\sqrt{x}} \cdot \dfrac{\sqrt{x}}{\sqrt{x}}$ Multiply by a fraction equal to 1, $\dfrac{\sqrt{x}}{\sqrt{x}} = 1.$

$= \dfrac{5y\sqrt{2 \cdot x}}{\sqrt{x}\sqrt{x}}$ Product rule of radicals.

$= \dfrac{5y\sqrt{2x}}{x}$ Simplify; $\sqrt{x}\sqrt{x} = x$

If the radicand is a rational number, the product of a square root and itself is a rational number: $\sqrt{2}\sqrt{2} = 2$. However, to have a product that is a rational number, we need to multiply a cube root by itself *three* times: $\sqrt[3]{2} \cdot \sqrt[3]{2} \cdot \sqrt[3]{2} = 2$. In the next example, the denominator is a cube root. To rationalize the denominator, multiply by a fraction that is equal to 1 *two* times so that the denominator is the product of three cube roots.

EXAMPLE 6 Rationalize the denominator of $\dfrac{6}{\sqrt[3]{x}}$.

SOLUTION ▶

$\dfrac{6}{\sqrt[3]{x}}$ A radical is in the denominator.

$= \dfrac{6}{\sqrt[3]{x}} \cdot \dfrac{\sqrt[3]{x}}{\sqrt[3]{x}} \cdot \dfrac{\sqrt[3]{x}}{\sqrt[3]{x}}$ Multiply twice by a fraction equal to 1, $\dfrac{\sqrt[3]{x}}{\sqrt[3]{x}} = 1$.

$= \dfrac{6\sqrt[3]{x^2}}{x}$ Simplify; $\sqrt[3]{x} \cdot \sqrt[3]{x} \cdot \sqrt[3]{x} = x$; the denominator has no radicals.

We can also rationalize the denominator by multiplying *once* by a fraction equal to 1, $\dfrac{\sqrt[3]{x^2}}{\sqrt[3]{x^2}}$. Both strategies result in the same answer.

$\dfrac{6}{\sqrt[3]{x}}$

$= \dfrac{6}{\sqrt[3]{x}} \cdot \dfrac{\sqrt[3]{x^2}}{\sqrt[3]{x^2}}$ Multiply by a fraction equal to 1, $\dfrac{\sqrt[3]{x^2}}{\sqrt[3]{x^2}} = 1$.

$= \dfrac{6\sqrt[3]{x^2}}{\sqrt[3]{x^3}}$ Simplify.

$= \dfrac{6\sqrt[3]{x^2}}{x}$ Simplify; the denominator has no radicals.

Simplifying a Radical with a Radicand That Is a Quotient

Assume that all variables represent nonnegative real numbers.

1. Simplify the radicand.

2. Use the quotient rule of radicals to rewrite the expression as the quotient of two radicals.

3. Simplify each radical.

4. If the denominator includes a radical, rationalize the denominator by multiplying by a fraction equal to 1. A rational denominator does not include a radical.

When rationalizing denominators, remember that the product of a square root and itself is the radicand of the square root. For example, $\sqrt{3} \cdot \sqrt{3} = 3$ and $\sqrt{2y}\sqrt{2y} = 2y$.

EXAMPLE 7 | Simplify: $\sqrt{\dfrac{15x}{32y}}$

SOLUTION ▶ $\sqrt{\dfrac{15x}{32y}}$

$= \dfrac{\sqrt{15x}}{\sqrt{32y}}$ Quotient rule of radicals.

$= \dfrac{\sqrt{15x}}{\sqrt{16}\sqrt{2y}}$ Product rule of radicals.

$= \dfrac{\sqrt{15x}}{4\sqrt{2y}}$ Simplify; a radical is in the denominator.

$= \dfrac{\sqrt{15x}}{4\sqrt{2y}} \cdot \dfrac{\sqrt{2y}}{\sqrt{2y}}$ Multiply by a fraction equal to 1, $\dfrac{\sqrt{2y}}{\sqrt{2y}} = 1$.

$= \dfrac{\sqrt{15x \cdot 2y}}{4 \cdot 2y}$ Simplify; $\sqrt{2y} \cdot \sqrt{2y} = 2y$

$= \dfrac{\sqrt{30xy}}{8y}$ Simplify; the denominator has no radicals.

Practice Problems

For problems 5–9, rationalize the denominator.

5. $\dfrac{x + \sqrt{5}}{\sqrt{3}}$ **6.** $\dfrac{5}{\sqrt[3]{2x}}$ **7.** $\dfrac{6}{\sqrt[4]{y}}$ **8.** $\sqrt{\dfrac{5a}{c}}$ **9.** $\sqrt{\dfrac{21p}{50w}}$

Using a Conjugate to Rationalize a Denominator

The conjugate of $\sqrt{p} + 5$ is $\sqrt{p} - 5$. The conjugate of $\sqrt{p} - 5$ is $\sqrt{p} + 5$. The expressions $\sqrt{p} + 5$ and $\sqrt{p} - 5$ are a **conjugate pair**.

> **Conjugate**
>
> The conjugate of $a + b$ is $a - b$. The conjugate of $a - b$ is $a + b$. The expressions $a + b$ and $a - b$ are a **conjugate pair**.

When we multiply an expression that includes a square root by its conjugate, there are no radicals in the product.

EXAMPLE 8 | **(a)** Multiply $\sqrt{p} + 5$ by its conjugate.

SOLUTION ▶ $(\sqrt{p} + 5)(\sqrt{p} - 5)$ The conjugate of $\sqrt{p} + 5$ is $\sqrt{p} - 5$.

$= (\sqrt{p})(\sqrt{p} - 5) + 5(\sqrt{p} - 5)$ Distributive property.

$= \sqrt{p}\sqrt{p} + \sqrt{p}(-5) + 5\sqrt{p} + 5(-5)$ Distributive property.

$= p - 5\sqrt{p} + 5\sqrt{p} - 25$ Simplify; product rule of radicals.

$= p - 25$ Simplify; the expression has no radicals.

(b) Multiply $\sqrt{5x} - \sqrt{y}$ by its conjugate.

$$(\sqrt{5x} - \sqrt{y})(\sqrt{5x} + \sqrt{y}) \qquad \text{The conjugate of } \sqrt{5x} - \sqrt{y} \text{ is } \sqrt{5x} + \sqrt{y}.$$

$$= \sqrt{5x}(\sqrt{5x} + \sqrt{y}) - \sqrt{y}(\sqrt{5x} + \sqrt{y}) \qquad \text{Distributive property.}$$

$$= \sqrt{5x}\sqrt{5x} + \sqrt{5x}\sqrt{y} - \sqrt{y}\sqrt{5x} - \sqrt{y}\sqrt{y} \qquad \text{Distributive property.}$$

$$= 5x + \sqrt{5xy} - \sqrt{5xy} - y \qquad \text{Simplify; product rule of radicals.}$$

$$= 5x - y \qquad \text{Simplify; the expression has no radicals.}$$

In the next example, to rationalize the denominator, we again multiply by a fraction equal to 1, $\dfrac{\sqrt{x} + 3}{\sqrt{x} + 3}$. *The numerator and the denominator of the fraction equal to 1 are the conjugate of the denominator,* $\sqrt{x} - 3$.

EXAMPLE 9 Rationalize the denominator of $\dfrac{6}{\sqrt{x} - 3}$.

SOLUTION $\blacktriangleright$

$$\dfrac{6}{\sqrt{x} - 3} \qquad \text{A radical is in the denominator.}$$

$$= \dfrac{6}{\sqrt{x} - 3} \cdot \dfrac{(\sqrt{x} + 3)}{(\sqrt{x} + 3)} \qquad \text{Multiply by a fraction equal to 1, } \dfrac{\sqrt{x} + 3}{\sqrt{x} + 3} = 1.$$

$$= \dfrac{6\sqrt{x} + 6 \cdot 3}{\sqrt{x}(\sqrt{x} + 3) - 3(\sqrt{x} + 3)} \qquad \text{Distributive property.}$$

$$= \dfrac{6\sqrt{x} + 18}{\sqrt{x}\sqrt{x} + \sqrt{x} \cdot 3 - 3\sqrt{x} - 3 \cdot 3} \qquad \text{Simplify; distributive property.}$$

$$= \dfrac{6\sqrt{x} + 18}{x - 9} \qquad \text{Simplify; product rule of radicals; the denominator has no radicals.}$$

Rationalizing a Denominator That Includes a Square Root

Multiply by a fraction equal to 1.

1. If the denominator is a radical or the product of a constant and a radical, multiply by a fraction in which the numerator and the denominator are the radical.

2. If the denominator is a sum or difference, multiply by a fraction in which the numerator and denominator are the conjugate of the sum or difference.

EXAMPLE 10 **(a)** Rationalize the denominator of $\dfrac{3}{5\sqrt{z}}$.

SOLUTION $\blacktriangleright$

$$\dfrac{3}{5\sqrt{z}} \qquad \text{A radical is in the denominator.}$$

$$= \dfrac{3}{5\sqrt{z}} \cdot \dfrac{\sqrt{z}}{\sqrt{z}} \qquad \text{Multiply by a fraction equal to 1, } \dfrac{\sqrt{z}}{\sqrt{z}} = 1.$$

$$= \dfrac{3\sqrt{z}}{5z} \qquad \text{Simplify; product rule of radicals; the denominator has no radicals.}$$

(b) Rationalize the denominator of $\dfrac{3}{5 + \sqrt{z}}$.

$$\frac{3}{5 + \sqrt{z}}$$ — A radical is in the denominator.

$$= \frac{3}{(5 + \sqrt{z})} \cdot \frac{(5 - \sqrt{z})}{(5 - \sqrt{z})}$$ — Multiply by a fraction equal to 1, $\dfrac{5 - \sqrt{z}}{5 - \sqrt{z}} = 1$.

$$= \frac{3 \cdot 5 - 3\sqrt{z}}{5(5 - \sqrt{z}) + \sqrt{z}(5 - \sqrt{z})}$$ — Distributive property.

$$= \frac{15 - 3\sqrt{z}}{25 - 5\sqrt{z} + 5\sqrt{z} - \sqrt{z}\sqrt{z}}$$ — Distributive property.

$$= \frac{15 - 3\sqrt{z}}{25 - z}$$ — Simplify; product rule of radicals; the denominator has no radicals.

In the next example, both the numerator and denominator have two terms. In multiplying the original radical expression by the conjugate, put parentheses around the expressions and use the distributive property. The conjugate of $\sqrt{6x} - 9$ is $\sqrt{6x} + 9$.

EXAMPLE 11 Simplify: $\dfrac{\sqrt{3x} + \sqrt{7}}{\sqrt{6x} - 9}$

SOLUTION ▶

$$\frac{\sqrt{3x} + \sqrt{7}}{\sqrt{6x} - 9}$$ — A radical is in the denominator.

$$= \frac{(\sqrt{3x} + \sqrt{7})}{(\sqrt{6x} - 9)} \cdot \frac{(\sqrt{6x} + 9)}{(\sqrt{6x} + 9)}$$ — Multiply by a fraction equal to 1, $\dfrac{\sqrt{6x} + 9}{\sqrt{6x} + 9} = 1$.

$$= \frac{\sqrt{3x}(\sqrt{6x} + 9) + \sqrt{7}(\sqrt{6x} + 9)}{\sqrt{6x}(\sqrt{6x} + 9) - 9(\sqrt{6x} + 9)}$$ — Distributive property.

$$= \frac{\sqrt{3x}\sqrt{6x} + \sqrt{3x} \cdot 9 + \sqrt{7}\sqrt{6x} + \sqrt{7} \cdot 9}{\sqrt{6x}\sqrt{6x} + \sqrt{6x} \cdot 9 - 9\sqrt{6x} - 9 \cdot 9}$$ — Distributive property.

$$= \frac{\sqrt{3 \cdot x \cdot 2 \cdot 3 \cdot x} + 9\sqrt{3x} + \sqrt{42x} + 9\sqrt{7}}{6x + 9\sqrt{6x} - 9\sqrt{6x} - 81}$$ — Simplify; product rule of radicals.

$$= \frac{\sqrt{3 \cdot 3}\sqrt{x \cdot x}\sqrt{2} + 9\sqrt{3x} + \sqrt{42x} + 9\sqrt{7}}{6x + 0 - 81}$$ — Simplify; product rule of radicals.

$$= \frac{3x\sqrt{2} + 9\sqrt{3x} + \sqrt{42x} + 9\sqrt{7}}{6x - 81}$$ — Simplify; the denominator has no radicals.

A simplified expression has no radical in the denominator. However, we sometimes ignore this rule if we need a numerator to be rational. In future math classes, you may sometimes be asked to rationalize a numerator instead of a denominator.

Practice Problems

For problems 10–12,
(a) identify the conjugate of the denominator.
(b) rationalize the denominator.

10. $\dfrac{\sqrt{x}}{\sqrt{y} - 5}$ **11.** $\dfrac{5}{\sqrt{x} - \sqrt{y}}$ **12.** $\dfrac{2\sqrt{3a}}{\sqrt{3a} + \sqrt{5b}}$

6.3 VOCABULARY PRACTICE

Match the term with its description. A term may be used more than once.

1. $\sqrt[n]{\dfrac{a}{b}} = \dfrac{\sqrt[n]{a}}{\sqrt[n]{b}}$

2. $\sqrt[n]{a} \cdot \sqrt[n]{b} = \sqrt[n]{ab}$

3. A number that can be written as a fraction in which the numerator and denominator are integers

4. A number that cannot be written as a fraction in which the numerator and denominator are integers

5. $a + b$ and $a - b$

6. $\dfrac{x^m}{x^n} = x^{m-n}$

7. $x - \sqrt{y}$ and $x + \sqrt{y}$

8. A fraction in which the numerator and denominator have no common factors

9. In this fraction, the numerator is less than the denominator.

10. In this fraction, the numerator is greater than or equal to the denominator.

A. conjugate pair
B. improper fraction
C. irrational number
D. lowest terms
E. product rule of radicals
F. proper fraction
G. quotient rule of exponents
H. quotient rule of radicals
I. rational number

6.3 Exercises

Follow your instructor's guidelines for showing your work. Assume that radicands and bases of exponential expressions represent nonnegative real numbers.

For exercises 1–24, simplify.

1. $\sqrt{\dfrac{36w^8}{49p^6}}$

2. $\sqrt{\dfrac{81p^{10}}{4q^2}}$

3. $\sqrt{\dfrac{100m}{81p^2}}$

4. $\sqrt{\dfrac{25y}{144a^{10}}}$

5. $\sqrt[3]{\dfrac{27k^6}{8h^{12}}}$

6. $\sqrt[3]{\dfrac{64z^9}{125m^6}}$

7. $\sqrt[3]{\dfrac{17b^4}{27m^{18}}}$

8. $\sqrt[3]{\dfrac{15c^5}{64d^{21}}}$

9. $\sqrt{\dfrac{72a^7b}{25c^2d^4}}$

10. $\sqrt{\dfrac{20h^9k}{81x^4y^2}}$

11. $\sqrt{\dfrac{108x^5y^3}{49z^2}}$

12. $\sqrt{\dfrac{147a^7b^5}{36c^2}}$

13. $\sqrt{\dfrac{24x^5y}{32x}}$

14. $\sqrt{\dfrac{6x^5y}{48x}}$

15. $\sqrt{\dfrac{6a^4b^3}{50}}$

16. $\sqrt{\dfrac{10a^3b^4}{72}}$

17. $\sqrt{\dfrac{xy}{100}}$

18. $\sqrt{\dfrac{ab}{121}}$

19. $\sqrt[3]{\dfrac{15h^5}{24h}}$

20. $\sqrt[3]{\dfrac{20k^5}{32k}}$

21. $\sqrt[4]{\dfrac{5x^5}{16x}}$

22. $\sqrt[4]{\dfrac{7a^6}{16a^2}}$

23. $\sqrt[5]{\dfrac{24b^4}{96}}$

24. $\sqrt[5]{\dfrac{21u^3}{84}}$

For exercises 25–78, simplify. The denominator must be rational.

25. $\dfrac{5}{\sqrt{x}}$

26. $\dfrac{8}{\sqrt{y}}$

27. $\dfrac{5}{3\sqrt{x}}$

28. $\dfrac{8}{9\sqrt{y}}$

29. $\dfrac{5}{3 + \sqrt{x}}$

30. $\dfrac{8}{9 + \sqrt{y}}$

31. $\dfrac{5}{3 - \sqrt{x}}$

32. $\dfrac{8}{9 - \sqrt{y}}$

33. $\dfrac{5\sqrt{x}}{3 - \sqrt{x}}$

34. $\dfrac{8\sqrt{y}}{9 - \sqrt{y}}$

35. $\dfrac{7}{\sqrt[3]{2}}$

36. $\dfrac{6}{\sqrt[3]{5}}$

37. $\dfrac{9}{\sqrt[4]{a}}$

38. $\dfrac{3}{\sqrt[4]{m}}$

39. $\dfrac{\sqrt{k}}{3 + \sqrt{k}}$

40. $\dfrac{\sqrt{h}}{4 + \sqrt{h}}$

41. $\dfrac{\sqrt{x} + 9}{\sqrt{5}}$

42. $\dfrac{\sqrt{y} + 8}{\sqrt{6}}$

43. $\dfrac{5}{x - 3\sqrt{y}}$

44. $\dfrac{2}{p - 4\sqrt{y}}$

45. $\dfrac{3\sqrt{x}}{\sqrt{2x} + 9}$

46. $\dfrac{5\sqrt{w}}{\sqrt{7w} + 1}$

47. $\dfrac{\sqrt{x} + \sqrt{y}}{\sqrt{x} - \sqrt{y}}$

48. $\dfrac{\sqrt{x} - \sqrt{y}}{\sqrt{x} + \sqrt{y}}$

49. $\dfrac{\sqrt{3}}{x - \sqrt{3}}$

50. $\dfrac{\sqrt{5}}{w - \sqrt{5}}$

51. $\dfrac{\sqrt{3}}{\sqrt{x} - \sqrt{3}}$

52. $\dfrac{\sqrt{5}}{\sqrt{w} - \sqrt{5}}$

53. $\dfrac{\sqrt[4]{h}}{\sqrt[4]{3h}}$

54. $\dfrac{\sqrt[4]{p}}{\sqrt[4]{5p}}$

55. $\dfrac{3}{2\sqrt{a} - 5\sqrt{b}}$

56. $\dfrac{7}{3\sqrt{c} - 4\sqrt{d}}$

57. $\dfrac{3\sqrt{a}}{2\sqrt{a} - 5\sqrt{b}}$

58. $\dfrac{7\sqrt{c}}{3\sqrt{c} - 4\sqrt{d}}$

59. $\dfrac{\sqrt{3a}}{2\sqrt{a} - 5\sqrt{b}}$

60. $\dfrac{\sqrt{7c}}{3\sqrt{c} - 4\sqrt{d}}$

61. $\dfrac{\sqrt{15a} - \sqrt{7}}{\sqrt{6a} - 5}$

62. $\dfrac{\sqrt{14c} - \sqrt{5}}{\sqrt{6c} - 3}$

63. $\sqrt{\dfrac{2}{5x}}$

64. $\sqrt{\dfrac{3}{7x}}$

65. $\sqrt{\dfrac{27p}{32q^3}}$

66. $\sqrt{\dfrac{20x}{63y^5}}$

67. $\sqrt[3]{\dfrac{27x}{2}}$

68. $\sqrt[3]{\dfrac{64y}{5}}$

69. $\sqrt[3]{\dfrac{8c^3}{5}}$

70. $\sqrt[3]{\dfrac{125d^3}{2}}$

71. $\sqrt[4]{\dfrac{3a^3}{4a}}$

72. $\sqrt[4]{\dfrac{3c^2}{25c}}$

73. $\sqrt{\dfrac{x}{3yz}}$

74. $\sqrt{\dfrac{a}{5bc}}$

75. $\dfrac{7}{\sqrt[9]{x}}$

76. $\dfrac{11}{\sqrt[8]{y}}$

77. $\sqrt[5]{\dfrac{9x}{2y}}$

78. $\sqrt[5]{\dfrac{7a}{3b}}$

79. The difference of two squares pattern is $(a - b)(a + b) = a^2 - b^2$. Use this pattern to multiply $(\sqrt{x} - \sqrt{y})(\sqrt{x} + \sqrt{y})$.

80. A perfect square trinomial pattern is $(a + b)(a + b) = a^2 + 2ab + b^2$. Use this pattern to multiply $(\sqrt{x} + \sqrt{y})(\sqrt{x} + \sqrt{y})$.

81. The quotient rule of radicals states, "If a, b are real numbers, $b \neq 0$, and n is a whole number greater than 0, $\sqrt[n]{\dfrac{a}{b}} = \dfrac{\sqrt[n]{a}}{\sqrt[n]{b}}$." Explain why b cannot equal 0.

82. Explain why the quotient rule of exponents cannot be used to simplify $\dfrac{x^5}{y^3}$.

83. The fraction $\dfrac{3}{2}$ is a rational number, but $\dfrac{\sqrt{3}}{2}$ is not a rational number. Explain why.

84. Describe the difference between a rational number and an irrational number.

Problem Solving: Practice and Review

Follow your instructor's guidelines for using the five steps as outlined in Section 1.5, p. 51.

85. Heidi's Events and Catering requires a minimum of one server for every 25 guests. The cost per server is $18.50 per hour. (*Source:* www.heidisevents.com)
    a. Write a linear model that describes the cost of servers, y, for x guests for 4 hours. Include the units of measurement.
    b. Use this model to find the cost for servers for a 4-hr reception for 300 guests.

86. A pharmacy technician works fifty 40-hour weeks per year. Use the hourly earnings in the chart to predict her annual salary.

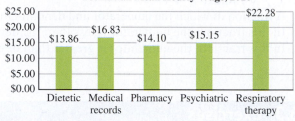

Technician Mean Hourly Wage, 2010

Source: www.bls.gov, May 2010

87. A model of the number of degrees awarded by technical colleges in South Carolina is $D(x) = 8.44x^2 + 123.15x + 4817.4$. The number of years since 1997 is x. The number of degrees awarded is $D(x)$. Use this model to predict the number of degrees awarded in 2008. Round to the nearest hundred.

88. A Boeing 757 is flying at an altitude of 37,000 ft. The average ground speed of the plane is 550 miles per hour. (The ground speed is the speed of the aircraft relative to the ground. It is the sum of the aircraft air speed and the speed of the wind.) Find the time for this plane to fly 1500 mi. Report the time in hours and minutes. Round to the nearest minute.

Find the Mistake

For exercises 89–92, the completed problem has one mistake.
(a) Describe the mistake in words, or copy down the whole problem and highlight or circle the mistake.
(b) Do the problem correctly.

89. **Problem:** Simplify: $\dfrac{\sqrt{2}}{\sqrt{8x}}$

    Incorrect Answer: $\dfrac{\sqrt{2}}{\sqrt{8x}}$

    $= \dfrac{\sqrt{2}}{\sqrt{8x}} \cdot \dfrac{\sqrt{8x}}{\sqrt{8x}}$

    $= \dfrac{\sqrt{16x}}{8x}$

    $= \dfrac{\sqrt{16}\sqrt{x}}{8x}$

    $= \dfrac{4\sqrt{x}}{8x}$

90. Problem: Simplify: $\dfrac{5}{\sqrt{z}+9}$

Incorrect Answer: $\dfrac{5}{\sqrt{z}+9}$

$$= \frac{5}{\sqrt{z}+9} \cdot \frac{\sqrt{z}+9}{\sqrt{z}+9}$$

$$= \frac{5\sqrt{z}+5\cdot 9}{\sqrt{z}\sqrt{z}+\sqrt{z}\cdot 9+9\sqrt{z}+9\cdot 9}$$

$$= \frac{5\sqrt{z}+45}{z^2+18\sqrt{z}+81}$$

91. Problem: Simplify: $\sqrt[4]{\dfrac{5}{2x}}$

Incorrect Answer: $\sqrt[4]{\dfrac{5}{2x}}$

$$= \frac{\sqrt[4]{5}}{\sqrt[4]{2x}}$$

$$= \frac{\sqrt[4]{5}}{\sqrt[4]{2x}} \cdot \frac{\sqrt[4]{2x}}{\sqrt[4]{2x}}$$

$$= \frac{\sqrt[4]{10x}}{2x}$$

92. Problem: Simplify: $\sqrt{\dfrac{45p}{32q}}$

Incorrect Answer: $\sqrt{\dfrac{45p}{32q}}$

$$= \frac{\sqrt{45p}}{\sqrt{32q}}$$

$$= \frac{\sqrt{45p}}{\sqrt{32q}} \cdot \frac{\sqrt{32q}}{\sqrt{32q}}$$

$$= \frac{\sqrt{1440pq}}{32q}$$

Review

For exercises 93–96, simplify. All exponents must be positive.

93. $(x^3)^2$ **94.** x^3x^2 **95.** $\dfrac{x^5}{x^3}$ **96.** $\dfrac{x^2}{x^7}$

SUCCESS IN COLLEGE MATHEMATICS

97. A student has been missing class. At midterm, she is barely passing the course and is struggling to keep up. When her instructor asks her why she has been missing class, she says, "I can't help it. My roommate stays up late with her friends. I sleep right through my alarm because I'm so tired."
   a. Who does the student believe is responsible for her difficulties?
   b. The student needs to make choices that will allow her to be successful. What do you think she should do?

$$\sqrt[3]{x^2} = x^{\frac{2}{3}}$$

The expression $\sqrt[3]{x^2}$ is in radical notation. In exponential notation, this expression is written $x^{\frac{2}{3}}$. In this section, we will use both notations. Throughout this section, we will assume that factors in radicands and bases of exponential expressions represent nonnegative real numbers.

SECTION 6.4

Rational Exponents

After reading the text, working the practice problems, and completing assigned exercises, you should be able to:

1. Rewrite a radical expression as an expression with rational exponents.
2. Rewrite an expression with rational exponents as a radical.
3. Simplify an expression that includes rational exponents.
4. Simplify a radical expression by rewriting it as an expression with rational exponents.

Rational Exponents

We can rewrite a radical as an exponential expression.

> **Radicals and Exponential Expressions**
>
> If m and n are whole numbers and $n > 1$, $\sqrt[n]{x^m} = x^{\frac{m}{n}}$.

EXAMPLE 1 | Show that $\sqrt{9} = 9^{\frac{1}{2}}$.

SOLUTION ▶ Multiply each expression by itself.

$\sqrt{9}\sqrt{9}$	$9^{\frac{1}{2}} \cdot 9^{\frac{1}{2}}$ The bases are the same.
$= 3 \cdot 3$ Simplify.	$= 9^{\left(\frac{1}{2}+\frac{1}{2}\right)}$ Product rule of exponents.
$= 9$ Simplify.	$= 9^{1}$ Simplify.
	$= 9$ Simplify.

Since $\left(\sqrt{9}\right)^2 = 9$ and $\left(9^{\frac{1}{2}}\right)^2 = 9$, we know that $\sqrt{9} = 9^{\frac{1}{2}}$.

In some situations, it is easier to work with radicals. In other situations, it is easier to work with rational exponents. So it is important to be able to change from one notation into the other.

EXAMPLE 2 | **(a)** Rewrite $\sqrt[3]{x^2}$ as an exponential expression.

SOLUTION ▶ $\sqrt[3]{x^2}$

$= x^{\frac{2}{3}}$ $\sqrt[n]{x^m} = x^{\frac{m}{n}}$

(b) Rewrite $\sqrt[5]{p}$ as an exponential expression.

▶ $\sqrt[5]{p}$

$= \sqrt[5]{p^1}$ $m = 1, n = 5$

$= p^{\frac{1}{5}}$ $\sqrt[n]{x^m} = x^{\frac{m}{n}}$

(c) Rewrite $\sqrt{c}$ as an exponential expression.

▶ $\sqrt{c}$

$= \sqrt[2]{c^1}$ $m = 1, n = 2$

$= c^{\frac{1}{2}}$ $\sqrt[n]{x^m} = x^{\frac{m}{n}}$

In rewriting a radical as an exponential expression, exponents may be improper fractions but must be in lowest terms. Since rational numbers can be written as fractions, exponents that are fractions are often called **rational exponents**.

EXAMPLE 3   (a) Rewrite $\sqrt{x^3 y^{10}}$ as an exponential expression.

SOLUTION ▶ $\sqrt{x^3 y^{10}}$

$= \sqrt{x^3}\sqrt{y^{10}}$ Product rule of radicals.

$= x^{\frac{3}{2}} y^{\frac{10}{2}}$ Exponential expressions.

$= x^{\frac{3}{2}} y^5$ Simplify; $y^{\frac{10}{2}} = y^5$

(b) Rewrite $\sqrt[3]{x^3 y^{10}}$ as an exponential expression.

▶ $\sqrt[3]{x^3 y^{10}}$

$= \sqrt[3]{x^3}\sqrt[3]{y^{10}}$ Product rule of radicals.

$= x^{\frac{3}{3}} y^{\frac{10}{3}}$ Exponential expressions.

$= xy^{\frac{10}{3}}$ Simplify; $x^{\frac{3}{3}} = x^1$

We can also rewrite exponential expressions with rational exponents in radical notation. If the radicand includes perfect square factors, simplify.

EXAMPLE 4   (a) Rewrite $144^{\frac{1}{2}}$ as a radical. Simplify.

SOLUTION ▶ $144^{\frac{1}{2}}$

$= \sqrt{144}$ $x^{\frac{m}{n}} = \sqrt[n]{x^m}$

$= 12$ Simplify.

(b) Rewrite $5^{\frac{2}{3}}$ as a radical. Simplify.

▶ $5^{\frac{2}{3}}$

$= \sqrt[3]{5^2}$ $x^{\frac{m}{n}} = \sqrt[n]{x^m}$

$= \sqrt[3]{25}$ Simplify; $5^2 = 25$

Practice Problems

For problems 1–3,
(a) rewrite the radical as an exponential expression.
(b) simplify.

1. $\sqrt{x^4 y^9}$ **2.** $\sqrt[3]{x^4 y^9}$ **3.** $\sqrt[4]{x^4 y^{12}}$

For problems 4–5,
(a) rewrite the exponential expression as a radical.
(b) simplify.

4. $16^{\frac{1}{2}}$ **5.** $6^{\frac{2}{3}}$

Products with Rational Exponents

We can use the product rule of exponents, $x^m \cdot x^n = x^{m+n}$, to simplify expressions that include rational exponents. In a simplified expression, all fractions are in lowest terms.

EXAMPLE 5   Simplify: $x^{\frac{2}{9}} y^{\frac{1}{4}} x^{\frac{1}{9}}$

SOLUTION ▶

$$x^{\frac{2}{9}} y^{\frac{1}{4}} x^{\frac{1}{9}}$$

$$= x^{\left(\frac{2}{9}+\frac{1}{9}\right)} y^{\frac{1}{4}} \qquad \text{Product rule of exponents.}$$

$$= x^{\frac{3}{9}} y^{\frac{1}{4}} \qquad \text{Simplify; } \frac{3}{9} \text{ is not in lowest terms.}$$

$$= x^{\frac{1 \cdot 3}{3 \cdot 3}} y^{\frac{1}{4}} \qquad \text{Find common factors.}$$

$$= x^{\frac{1}{3}} y^{\frac{1}{4}} \qquad \text{Simplify.}$$

We can add only fractions with a common denominator. In the next example, the rational exponents have different denominators. We change the denominators into the common denominator by multiplying by a fraction equal to 1, just as we would in adding any fractions.

EXAMPLE 6   Simplify: $x^{\frac{5}{24}} x^{\frac{1}{60}}$

SOLUTION ▶

$$x^{\frac{5}{24}} x^{\frac{1}{60}} \qquad \text{The least common denominator of 24 and 60 is 120.}$$

$$= x^{\left(\frac{5}{24}\right)\left(\frac{5}{5}\right)} x^{\left(\frac{1}{60}\right)\left(\frac{2}{2}\right)} \qquad \text{Multiply each exponent by a fraction equal to 1.}$$

$$= x^{\frac{25}{120}} x^{\frac{2}{120}} \qquad \text{Simplify.}$$

$$= x^{\left(\frac{25}{120}+\frac{2}{120}\right)} \qquad \text{Product rule of exponents.}$$

$$= x^{\frac{27}{120}} \qquad \text{Simplify; } \frac{27}{120} \text{ is not in lowest terms.}$$

$$= x^{\frac{9 \cdot 3}{40 \cdot 3}} \qquad \text{Find common factors.}$$

$$= x^{\frac{9}{40}} \qquad \text{Simplify.}$$

When an exponent is raised to another power, use the power rule of exponents, $(x^m)^n = x^{m \cdot n}$, and multiply the exponents.

EXAMPLE 7 | Simplify $\left(a^{\frac{3}{4}}\right)^{\frac{2}{9}}$. Rewrite the simplified expression in radical notation.

SOLUTION ▶

$$\left(a^{\frac{3}{4}}\right)^{\frac{2}{9}}$$

$$= a^{\left(\frac{3}{4}\right)\left(\frac{2}{9}\right)} \qquad \text{Power rule of exponents.}$$

$$= a^{\left(\frac{3}{2 \cdot 2}\right)\left(\frac{2}{3 \cdot 3}\right)} \qquad \text{Find common factors.}$$

$$= a^{\frac{1}{6}} \qquad \text{Simplify.}$$

$$= \sqrt[6]{a} \qquad \text{Rewrite in radical notation; } a^{\frac{1}{6}} = \sqrt[6]{a}$$

The next example is an expression that is a fourth root of a cube root. As a radical, it may seem very difficult to simplify. However, we can rewrite the expression with rational exponents and then use the exponent rules to simplify.

EXAMPLE 8 | Simplify $\sqrt[4]{\sqrt[3]{x^2}}$. Write the final expression as a radical.

SOLUTION ▶

$$\sqrt[4]{\sqrt[3]{x^2}} \qquad \text{The fourth root of the cube root of } x^2.$$

$$= \sqrt[4]{x^{\frac{2}{3}}} \qquad \text{Rewrite the inner radical; } \sqrt[3]{x^2} = x^{\frac{2}{3}}$$

$$= \left(x^{\frac{2}{3}}\right)^{\frac{1}{4}} \qquad \text{Rewrite the remaining radical; } \sqrt[4]{n} = n^{\frac{1}{4}}$$

$$= x^{\left(\frac{2}{3}\right)\left(\frac{1}{4}\right)} \qquad \text{Power rule of exponents.}$$

$$= x^{\left(\frac{2}{3}\right)\left(\frac{1}{2 \cdot 2}\right)} \qquad \text{Find common factors.}$$

$$= x^{\frac{1}{6}} \qquad \text{Simplify.}$$

$$= \sqrt[6]{x} \qquad \text{Rewrite as a radical; } \sqrt[6]{x^1} = x^{\frac{1}{6}}$$

Practice Problems

For problems 6–8, simplify.

6. $z^{\frac{2}{5}}z^{\frac{1}{5}}$ **7.** $x^{\frac{5}{18}}x^{\frac{7}{24}}$ **8.** $x^{\frac{1}{8}}y^{\frac{1}{9}}x^{\frac{3}{20}}y^{\frac{2}{15}}$

For problems 9–12,
(a) simplify.
(b) write the final expression as a radical.

9. $\left(h^{\frac{2}{9}}\right)^{\frac{3}{10}}$ **10.** $\left(x^{\frac{3}{8}}\right)^{\frac{8}{3}}$ **11.** $\left(p^{\frac{1}{2}}\right)^2$ **12.** $\sqrt{\sqrt[5]{z^3}}$

Rational Exponents and Division

To simplify a quotient with rational exponents, use the quotient rule of exponents $\dfrac{x^m}{x^n} = x^{m-n}$.

EXAMPLE 9 Simplify: $\dfrac{x^{\frac{11}{12}}}{x^{\frac{5}{12}}}$

SOLUTION ▶

$\dfrac{x^{\frac{11}{12}}}{x^{\frac{5}{12}}}$

$= x^{\left(\frac{11}{12} - \frac{5}{12}\right)}$ Quotient rule of exponents.

$= x^{\frac{6}{12}}$ Simplify; $\dfrac{6}{12}$ is not in lowest terms.

$= x^{\frac{6 \cdot 1}{6 \cdot 2}}$ Find common factors.

$= x^{\frac{1}{2}}$ Simplify.

To subtract rational exponents, the denominators of the exponents must be the same.

EXAMPLE 10 Simplify: $\dfrac{8x^{\frac{7}{9}}}{60x^{\frac{2}{5}}}$

SOLUTION ▶

$\dfrac{8x^{\frac{7}{9}}}{60x^{\frac{2}{5}}}$ The least common denominator of 9 and 5 is 45.

$= \dfrac{8x^{\left(\frac{7}{9} - \frac{2}{5}\right)}}{60}$ Quotient rule of exponents.

$= \dfrac{8x^{\left(\frac{7}{9} \cdot \frac{5}{5} - \frac{2}{5} \cdot \frac{9}{9}\right)}}{60}$ Multiply by fractions equal to 1.

$= \dfrac{8x^{\left(\frac{35}{45} - \frac{18}{45}\right)}}{60}$ Simplify.

$= \dfrac{8x^{\frac{17}{45}}}{60}$ Simplify.

$= \dfrac{2 \cdot 4 \cdot x^{\frac{17}{45}}}{15 \cdot 4}$ Find common factors of 8 and 60.

$= \dfrac{2x^{\frac{17}{45}}}{15}$ Simplify.

As expressions get more complicated, it is often easier to first simplify the expression inside the parentheses as much as possible.

EXAMPLE 11 Simplify: $\left(\dfrac{x^{\frac{5}{6}}}{x^{\frac{1}{3}}}\right)^{\frac{1}{4}}$

SOLUTION ▶

$\left(\dfrac{x^{\frac{5}{6}}}{x^{\frac{1}{3}}}\right)^{\frac{1}{4}}$ The least common denominator of 6 and 3 is 6.

$$= \left(x^{\frac{5}{6} - \frac{1}{3}} \right)^{\frac{1}{4}} \qquad \text{Quotient rule of exponents.}$$

$$= \left(x^{\frac{5}{6} - \frac{1}{3} \cdot \frac{2}{2}} \right)^{\frac{1}{4}} \qquad \text{Multiply by a fraction equal to 1; } \frac{2}{2} = 1$$

$$= \left(x^{\frac{5}{6} - \frac{2}{6}} \right)^{\frac{1}{4}} \qquad \text{Simplify.}$$

$$= \left(x^{\frac{3}{6}} \right)^{\frac{1}{4}} \qquad \text{Simplify; } \frac{3}{6} \text{ is not in lowest terms.}$$

$$= \left(x^{\frac{1 \cdot 3}{2 \cdot 3}} \right)^{\frac{1}{4}} \qquad \text{Find common factors.}$$

$$= \left(x^{\frac{1}{2}} \right)^{\frac{1}{4}} \qquad \text{Simplify.}$$

$$= x^{\left(\frac{1}{2} \right)\left(\frac{1}{4} \right)} \qquad \text{Power rule of exponents.}$$

$$= x^{\frac{1}{8}} \qquad \text{Simplify.}$$

Since a simplified expression has only positive exponents, in the next example we rewrite $x^{\frac{1}{6}} y^{-\frac{1}{6}}$ as $\dfrac{x^{\frac{1}{6}}}{y^{\frac{1}{6}}}$.

Negative Exponents

If x and m are real numbers and x is not equal to 0, $x^{-m} = \dfrac{1}{x^m}$ and $\dfrac{1}{x^{-m}} = x^m$.

EXAMPLE 12 Rewrite $\dfrac{\sqrt[3]{x^2 y}}{\sqrt{xy}}$ as an exponential expression. Simplify.

SOLUTION ▶

$$\frac{\sqrt[3]{x^2 y}}{\sqrt{xy}}$$

$$= \frac{x^{\frac{2}{3}} y^{\frac{1}{3}}}{x^{\frac{1}{2}} y^{\frac{1}{2}}} \qquad \text{Rewrite as an exponential expression; } \sqrt[n]{x^m} = x^{\frac{m}{n}}$$

$$= x^{\left(\frac{2}{3} - \frac{1}{2} \right)} y^{\left(\frac{1}{3} - \frac{1}{2} \right)} \qquad \text{Quotient rule of exponents; least common denominator is 6.}$$

$$= x^{\left(\frac{2}{3} \cdot \frac{2}{2} - \frac{1}{2} \cdot \frac{3}{3} \right)} y^{\left(\frac{1}{3} \cdot \frac{2}{2} - \frac{1}{2} \cdot \frac{3}{3} \right)} \qquad \text{Multiply by fractions equal to 1.}$$

$$= x^{\left(\frac{4}{6} - \frac{3}{6} \right)} y^{\left(\frac{2}{6} - \frac{3}{6} \right)} \qquad \text{Simplify.}$$

$$= x^{\frac{1}{6}} y^{-\frac{1}{6}} \qquad \text{Simplify; } -\frac{1}{6} \text{ is a negative exponent.}$$

$$= \frac{x^{\frac{1}{6}}}{y^{\frac{1}{6}}} \qquad \text{Rewrite with only positive exponents; } x^{-m} = \frac{1}{x^m}$$

Simplified Radical Expressions and Exponential Expressions

If a radical expression is simplified,

- the factors in the radicand have no perfect power factors.
- the exponents in the radicand are less than the index of the radical.
- there are no radicals in a denominator.

If an exponential expression with rational exponents is simplified,

- there is only one term with a given variable.
- the exponents are in lowest terms (but may be improper).
- the exponents are positive.
- the coefficient is in lowest terms.

Practice Problems

For problems 13–16, simplify. Rational exponents may be improper but must be in lowest terms.

13. $\dfrac{x^{\frac{5}{7}}}{x^{\frac{3}{8}}}$

14. $\left(\dfrac{n^{\frac{4}{5}}}{n^{\frac{1}{3}}}\right)^{\frac{1}{8}}$

15. $\dfrac{21x^{\frac{1}{8}}y^{\frac{5}{11}}}{35x^{\frac{2}{9}}y^{\frac{1}{4}}}$

16. $\dfrac{w^2}{w^{\frac{1}{8}}}$

For problems 17–18, rewrite as an exponential expression. Simplify.

17. $\dfrac{\sqrt[5]{x^4}}{\sqrt[7]{x^3}}$

18. $\dfrac{\sqrt[8]{x}}{\sqrt[5]{x^4}}$

6.4 VOCABULARY PRACTICE

Match the term with its description.

1. $x^m \cdot x^n = x^{m+n}$

2. $\dfrac{x^m}{x^n} = x^{m-n}$

3. $(x^m)^n = x^{m \cdot n}$

4. The number 3 in $\sqrt[3]{5x}$

5. The expression $5x$ in $\sqrt[3]{5x}$

6. A number that cannot be written as a fraction in which the numerator and denominator are integers

7. $\{\ldots, -3, -2, -1, 0, 1, 2, 3, \ldots\}$

8. In this fraction, the numerator is less than the denominator.

9. In this fraction, the numerator is greater than or equal to the denominator.

10. In $x^{\frac{a}{b}}$, $\dfrac{a}{b}$ is this.

A. improper fraction
B. index
C. integers
D. irrational number
E. power rule of exponents
F. product rule of exponents
G. proper fraction
H. quotient rule of exponents
I. radicand
J. rational exponent

6.4 Exercises

Follow your instructor's guidelines for showing your work. Assume that radicands and bases of exponential expressions represent nonnegative real numbers.

For exercises 1–24, rewrite the radical expression as an exponential expression. Exponents must be in lowest terms but may be improper.

1. $\sqrt{x}$

2. $\sqrt{y}$

3. $\sqrt{x^3}$

4. $\sqrt{y^5}$

5. $\sqrt[3]{x^2 y}$

6. $\sqrt[3]{xy^2}$

7. $\sqrt[4]{x^3}$

8. $\sqrt[4]{y^3}$

9. $\sqrt[4]{x^2}$

10. $\sqrt[4]{y^2}$

11. $\sqrt[5]{x}$

12. $\sqrt[5]{y}$

13. $\sqrt[99]{x}$

14. $\sqrt[97]{y}$

15. $\sqrt[20]{a^4}$

16. $\sqrt[20]{a^5}$

17. $\sqrt[24]{d^3}$

18. $\sqrt[24]{d^4}$

19. $\sqrt[3]{h^4}$

20. $\sqrt[3]{k^4}$

21. $\sqrt[4]{a^6 b^8}$

22. $\sqrt[4]{a^8 b^{10}}$

23. $\sqrt[60]{x^{48} y}$

24. $\sqrt[60]{xy^{24}}$

For exercises 25–30, rewrite the expression as a single radical.

25. $m^{\frac{3}{4}}$

26. $p^{\frac{5}{6}}$

27. $b^{\frac{1}{3}} c^{\frac{2}{3}}$

28. $a^{\frac{1}{4}} b^{\frac{3}{4}}$

29. $a^{\frac{1}{4}} b^{\frac{5}{4}}$

30. $a^{\frac{1}{5}} b^{\frac{6}{5}}$

For exercises 31–32,
(a) rewrite the rational exponents with the least common denominator.
(b) rewrite the expression as a single radical.

31. $x^{\frac{1}{3}} y^{\frac{1}{4}}$

32. $a^{\frac{1}{6}} b^{\frac{1}{4}}$

33. A web page of a manufacturer of steel pipe includes Manning's Formula for pipe discharge, $Q = \dfrac{1.486A}{n} R^{\frac{2}{3}} S^{\frac{1}{2}}$. Rewrite this formula in radical notation.

34. A scientific paper about improving drug patches includes the formula for drug effusion, $R = 200\left(\dfrac{DT}{\pi h^2}\right)^{\frac{1}{2}}$. Rewrite this formula in radical notation.

For exercises 35–56, simplify. Write the final expression in exponential notation.

35. $x^6 x^2$

36. $y^5 y^2$

37. $a^{\frac{1}{3}} \cdot a^{\frac{1}{3}}$

38. $u^{\frac{1}{5}} \cdot u^{\frac{2}{5}}$

39. $x^{\frac{1}{6}} \cdot x^{\frac{1}{6}}$

40. $y^{\frac{1}{8}} \cdot y^{\frac{1}{8}}$

41. $x^{\frac{2}{9}} \cdot x^{\frac{1}{5}}$

42. $y^{\frac{1}{8}} \cdot y^{\frac{3}{5}}$

43. $\left(x^6\right)^2$

44. $\left(y^5\right)^2$

45. $\left(x^{\frac{1}{2}}\right)^{\frac{1}{4}}$

46. $\left(y^{\frac{1}{6}}\right)^{\frac{1}{3}}$

47. $\left(x^{\frac{2}{3}}\right)^{\frac{3}{8}}$

48. $\left(y^{\frac{3}{4}}\right)^{\frac{2}{3}}$

49. $\left(x^{\frac{2}{3}}\right)^0$

50. $\left(y^{\frac{2}{5}}\right)^0$

51. $\left(x^{\frac{3}{5}}\right)^{\frac{10}{21}}$

52. $\left(y^{\frac{5}{6}}\right)^{\frac{12}{25}}$

53. $\left(x^{\frac{3}{8}}\right)^2$

54. $\left(y^{\frac{5}{9}}\right)^3$

55. $x^{\frac{2}{9}} x^{\frac{3}{5}} x^{\frac{1}{6}}$

56. $y^{\frac{3}{10}} y^{\frac{1}{4}} y^{\frac{2}{15}}$

For exercises 57–60,
(a) rewrite the expression in exponential notation. Simplify.
(b) write the final expression as a radical.

57. $\sqrt[6]{\sqrt[3]{x^2}}$

58. $\sqrt[8]{\sqrt[3]{x^2}}$

59. $\sqrt[5]{\sqrt[3]{x}}$

60. $\sqrt[8]{\sqrt[3]{x}}$

For exercises 61–76, simplify. Exponents must be positive and in lowest terms. They may be improper fractions.

61. $\dfrac{9p^8}{12p^3}$

62. $\dfrac{15n^9}{40n^3}$

63. $\dfrac{20a^3}{15a^5}$

64. $\dfrac{15z^2}{10z^9}$

65. $\dfrac{a^{\frac{3}{8}}}{a^{\frac{1}{8}}}$

66. $\dfrac{b^{\frac{3}{10}}}{b^{\frac{1}{10}}}$

67. $\dfrac{a^{\frac{1}{8}}}{a^{\frac{3}{8}}}$

68. $\dfrac{b^{\frac{1}{10}}}{b^{\frac{3}{10}}}$

69. $\dfrac{a^{\frac{1}{5}}}{a^{\frac{3}{7}}}$

70. $\dfrac{b^{\frac{3}{10}}}{b^{\frac{4}{7}}}$

71. $\dfrac{x^{\frac{3}{5}} y^{\frac{4}{5}}}{x^{\frac{2}{7}} y^{\frac{1}{7}}}$

72. $\dfrac{x^{\frac{5}{9}} y^{\frac{7}{8}}}{x^{\frac{1}{2}} y^{\frac{3}{11}}}$

73. $\dfrac{15m^{\frac{1}{3}}}{63m^{\frac{9}{10}}}$

74. $\dfrac{24k^{\frac{1}{4}}}{64k^{\frac{5}{9}}}$

75. $\dfrac{4x^{\frac{13}{20}}}{20x^{\frac{7}{8}}}$

76. $\dfrac{6y^{\frac{7}{15}}}{54y^{\frac{17}{20}}}$

For exercises 77–82,
(a) rewrite as an exponential expression.
(b) simplify.

77. $\dfrac{\sqrt[5]{x^3 y}}{\sqrt{xy}}$

78. $\dfrac{\sqrt[4]{x^3 y}}{\sqrt{xy}}$

79. $\dfrac{\sqrt[3]{ab}}{\sqrt[5]{a^2 b^3}}$

80. $\dfrac{\sqrt[5]{hk}}{\sqrt[6]{h^2 k^3}}$

81. $\dfrac{\sqrt{144 pw}}{\sqrt[3]{125 pw}}$

82. $\dfrac{\sqrt{100 ab}}{\sqrt[3]{216 ab}}$

For exercises 83–88, simplify.

83. $\left(\dfrac{x^{\frac{2}{3}}}{x^{\frac{1}{4}}}\right)^{\frac{1}{2}}$

84. $\left(\dfrac{y^{\frac{3}{5}}}{y^{\frac{3}{8}}}\right)^{\frac{1}{2}}$

85. $\dfrac{x^{\frac{7}{8}} x^{\frac{1}{2}}}{x^{\frac{2}{9}}}$

86. $\dfrac{y^{\frac{7}{9}} y^{\frac{1}{3}}}{y^{\frac{1}{10}}}$

87. $\dfrac{x^4}{\left(x^{\frac{3}{4}}\right)^{\frac{1}{2}}}$

88. $\dfrac{y^6}{\left(y^{\frac{5}{7}}\right)^{\frac{1}{3}}}$

Problem Solving: Practice and Review

Follow your instructor's guidelines for using the five steps as outlined in Section 1.5, p. 51.

89. Predict how many of 3350 children living in the neighborhoods tested have elevated lead levels. Round to the nearest whole number.

   When the soil sampling results came in last month, the lead contamination . . . was traced to peeling paint, not the plant. Of the children tested from those neighborhoods [on Staten Island], about 7 out of 1,000 had elevated lead levels, health department data show, compared with a citywide average of 4.5 out of 1,000. (*Source:* www.nytimes.com, April 10, 2010)

90. The average distance traveled by Americans during the Memorial Day weekend in 2011 was expected to be 792 mi, a 27% increase over the average distance traveled by Americans during the same holiday weekend in 2010. Find the average distance traveled in 2010. Round to the nearest whole number. (*Source:* www.aaanewsroom.net, May 19, 2011)

91. As of June 15, 2010, California had recorded 910 cases of whooping cough as compared to 219 cases in the same period in 2009. Find the percent increase in the number of cases of whooping cough. Round to the nearest percent. (*Source:* www.usatoday.com, June 25, 2010)

92. A small goat cheese company sells a 4-ounce package of chevre for $3. The cost of making the cheese is $1.85 per package. If the overhead is $5000 per month, find the number of packages of chevre that need to be sold to break even (costs = revenue). Round to the nearest whole number.

Find the Mistake

For exercises 93–96, the completed problem has one mistake.
(a) Describe the mistake in words, or copy down the whole problem and highlight or circle the mistake.
(b) Do the problem correctly.

93. **Problem:** Simplify: $\left(x^{\frac{2}{7}}\right)^{\frac{3}{4}}$

   Incorrect Answer: $\left(x^{\frac{2}{7}}\right)^{\frac{3}{4}}$

   $= x^{\left(\frac{2}{7}+\frac{3}{4}\right)}$

   $= x^{\left(\frac{2}{7}\cdot\frac{4}{4}+\frac{3}{4}\cdot\frac{7}{7}\right)}$

   $= x^{\left(\frac{8}{28}+\frac{21}{28}\right)}$

   $= x^{\frac{29}{28}}$

94. **Problem:** Simplify: $\dfrac{a^{\frac{1}{6}}}{a^{\frac{5}{6}}}$

   Incorrect Answer: $\dfrac{a^{\frac{1}{6}}}{a^{\frac{5}{6}}}$

   $= a^{-\frac{4}{6}}$

   $= a^{\left(-\frac{\cancel{2}\cdot 2}{\cancel{2}\cdot 3}\right)}$

   $= a^{-\frac{2}{3}}$

95. **Problem:** Simplify: $x^{\frac{2}{7}}\cdot x^{\frac{3}{7}}$

   Incorrect Answer: $x^{\frac{2}{7}}\cdot x^{\frac{3}{7}}$

   $= x^{\left(\frac{2}{7}\right)\left(\frac{3}{7}\right)}$

   $= x^{\frac{6}{49}}$

96. **Problem:** Simplify: $\dfrac{x^{\frac{9}{10}}}{x^{\frac{3}{10}}}$

   Incorrect Answer: $\dfrac{x^{\frac{9}{10}}}{x^{\frac{3}{10}}}$

   $= x^{\left(\frac{9}{10}-\frac{3}{10}\right)}$

   $= x^{\frac{6}{10}}$

Review

For exercises 97–99,
(a) solve the equation.
(b) check the solution(s).

97. $3(2x - 9) + 1 = 4x + 20$

98. $|x + 3| - 28 = -5$

99. $2x^2 - 7x - 15 = 0$

100. State the zero product property.

SUCCESS IN COLLEGE MATHEMATICS

101. A student has a family, including small children. For about two weeks, members of his family have a flu virus. He misses a few days of class, he has much less time to study and do homework, and his grade on the first test is a D. What do you think the student should do in this situation?

A business needs inventory of products ready to sell to customers. However, it is expensive to store unsold products in a warehouse. In this section, we will use a radical equation to predict the amount of inventory a business should order.

SECTION 6.5 Radical Equations

After reading the text, working the practice problems, and completing assigned exercises, you should be able to:

1. Solve a radical equation that includes one radical.
2. Identify an extraneous solution of a radical equation.
3. Solve a radical equation that includes two radicals.
4. Solve application problems using radical equations.

Solving Radical Equations with One Radical

In Chapter 1, we used the properties of equalities to solve linear equations in one variable, such as $2x + 5 = 25$. In Chapter 2, we used the zero product property to solve quadratic equations that can be factored, such as $x^2 + 5x + 6 = 0$. In Chapter 5, we solved rational equations by clearing the fractions and isolating the variable, knowing that a solution might be extraneous. Now we need a new strategy for solving equations that include a variable in a radicand. In the next example, we isolate the radical and raise each side of the equation to a power that is equal to the index, 2.

EXAMPLE 1   (a) Solve: $\sqrt{x + 2} = 5$

SOLUTION ▶

$\sqrt{x + 2} = 5$ — The radical is isolated.

$(\sqrt{x + 2})^2 = 5^2$ — The index of the radicand is 2; raise each side to the second power.

$x + 2 = 25$ — Simplify; $(\sqrt{x + 2})^2 = x + 2$

$\underline{-2 \quad -2}$ — Subtraction property of equality.

$x + 0 = 23$ — Simplify.

$x = 23$ — Simplify.

(b) Check.

$$\sqrt{x + 2} = 5 \qquad \text{Check in the original equation.}$$
$$\sqrt{23 + 2} = 5 \qquad \text{Replace the variable with the solution, 23.}$$
$$\sqrt{25} = 5 \qquad \text{Simplify.}$$
$$5 = 5 \quad \text{True.} \qquad \text{Simplify; since } 5 = 5 \text{ is true, the solution is correct.}$$

If the equation includes a cube root, isolate the cube root and raise each side of the equation to a power that is equal to the index, 3.

EXAMPLE 2 **(a)** Solve: $\sqrt[3]{2x + 8} + 15 = 21$

SOLUTION ▶

$$\sqrt[3]{2x + 8} + 15 = 21 \qquad \text{The radical is not isolated.}$$
$$\underline{\qquad\quad -15 \quad -15} \qquad \text{Subtraction property of equality.}$$
$$\sqrt[3]{2x + 8} + 0 = 6 \qquad \text{Simplify.}$$
$$\left(\sqrt[3]{2x + 8}\right)^3 = 6^3 \qquad \text{The index of the radicand is 3; raise each side to the third power.}$$
$$2x + 8 = 216 \qquad \text{Simplify; } \left(\sqrt[3]{2x + 8}\right)^3 = 2x + 8$$
$$\underline{\quad -8 \qquad -8} \qquad \text{Subtraction property of equality.}$$
$$2x + 0 = 208 \qquad \text{Simplify.}$$
$$\frac{2x}{2} = \frac{208}{2} \qquad \text{Division property of equality.}$$
$$x = 104 \qquad \text{Simplify.}$$

(b) Check.

$$\sqrt[3]{2x + 8} + 15 = 21 \qquad \text{Check in the original equation.}$$
$$\sqrt[3]{2(104) + 8} + 15 = 21 \qquad \text{Replace the variable with the solution, 104.}$$
$$\sqrt[3]{208 + 8} + 15 = 21 \qquad \text{Simplify.}$$
$$\sqrt[3]{216} + 15 = 21 \qquad \text{Simplify.}$$
$$6 + 15 = 21 \qquad \text{Evaluate the cube root.}$$
$$21 = 21 \quad \text{True.} \qquad \text{Simplify; since } 21 = 21 \text{ is true, the solution is correct.}$$

Solving a Radical Equation with One Radical

1. Isolate the radical.

2. Raise both sides to a power equal to the index of the radical.

3. Simplify. Solve the equation.

4. Check the solution(s).

When we clear the fractions from a *rational equation* and then solve, the solutions of the new equation may result in division by zero in the original equation. Such solutions are **extraneous**. For example, the solution $x = 1$ is an extraneous solution of $\frac{1}{x - 1} + \frac{2}{x} = \frac{x}{x - 1}$. Replacing x with 1 results in division by zero, which is undefined. When we are solving *radical equations*, the process of raising both sides of the equation to a power can also result in an extraneous solution(s).

EXAMPLE 3 (a) Solve: $x = \sqrt{2x + 3}$

SOLUTION ▶

$x = \sqrt{2x + 3}$ The radical is isolated.

$x^2 = \left(\sqrt{2x + 3}\right)^2$ The index of the radicand is 2; raise both sides to the second power.

$x^2 = 2x + 3$ Simplify; this is a quadratic equation.

$\underline{-2x - 3 \quad -2x - 3}$ Subtraction property of equality.

$x^2 - 2x - 3 = 0$ Standard form: $ax^2 + bx + c = 0$

$(x - 3)(x + 1) = 0$ Factor.

$x - 3 = 0$ or $x + 1 = 0$ Zero product property.

$\underline{+3 \quad +3} \qquad \underline{-1 \quad -1}$ Properties of equality.

$x = 3$ or $\cancel{x = -1}$ Simplify.

(b) Check.

▶

Check: $x = 3$

$x = \sqrt{2x + 3}$ Check in the original equation.

$3 = \sqrt{2(3) + 3}$ Replace the variable with each solution.

$3 = \sqrt{9}$ Simplify.

$3 = 3$ True.

Check: $x = -1$

$x = \sqrt{2x + 3}$

$-1 = \sqrt{2(-1) + 3}$

$-1 = \sqrt{1}$

$-1 = 1$ False.

Since $3 = 3$ is true, $x = 3$ is a solution of $x = \sqrt{2x + 3}$. Since $-1 = 1$ is false, $x = -1$ is an extraneous solution. The only solution of $x = \sqrt{2x + 3}$ is $x = 3$.

Why are some solutions extraneous? A principal square root is greater than or equal to 0. So the solution of $x = \sqrt{2x + 3}$ is also greater than or equal to 0. When we square both sides, we create a quadratic equation, $x^2 = 2x + 3$, with two solutions. The process of squaring creates an equation with a solution that is not a solution of the original equation. Such solutions are extraneous. Checking the solution of a radical equation can reveal arithmetic or algebra mistakes *and* any extraneous solutions. Ask your instructor how to mark extraneous solutions.

EXAMPLE 4 (a) Solve: $x + 2 = \sqrt{-2x - 5}$

SOLUTION ▶

$x + 2 = \sqrt{-2x - 5}$ The radical is isolated.

$(x + 2)^2 = \left(\sqrt{-2x - 5}\right)^2$ The index of the radicand is 2; raise both sides to the second power.

$(x + 2)(x + 2) = -2x - 5$ Simplify.

$x(x + 2) + 2(x + 2) = -2x - 5$ Distributive property.

$x^2 + 2x + 2x + 4 = -2x - 5$ Distributive property.

$x^2 + 4x + 4 = -2x - 5$ Simplify; this is a quadratic equation.

$\underline{+2x + 5 \qquad +2x + 5}$ Addition property of equality.

$x^2 + 6x + 9 = 0$ Standard form: $ax^2 + bx + c = 0$

$(x + 3)(x + 3) = 0$ Factor.

$x + 3 = 0$ Zero product property.

$\underline{-3 \quad -3}$ Subtraction property of equality.

$\cancel{x = -3}$ Simplify.

(b) Check.

$$x + 2 = \sqrt{-2x - 5}$$ Check in the original equation.

$$-3 + 2 = \sqrt{(-2)(-3) - 5}$$ Replace variable.

$$-1 = \sqrt{6 - 5}$$ Simplify.

$$-1 = \sqrt{1}$$ Simplify.

$$-1 = 1$$ False.

Since $-1 = 1$ is false, $x = -3$ is an extraneous solution of $x + 2 = \sqrt{-2x - 5}$. This equation has no solution.

Practice Problems

For problems 1–5,
(a) solve.
(b) check.

1. $\sqrt{3x} + 12 = 24$ **2.** $\sqrt[3]{x + 8} = 7$ **3.** $x = \sqrt{4x + 5}$

4. $x + 3 = \sqrt{16x - 12}$ **5.** $x - 3 = \sqrt{-5x + 15}$

Solving Radical Equations with Two Radicals

To solve equations with two radicals, eliminate both radicals from the equation.

Solving a Radical Equation with Two Radicals

1. Use the properties of equality to rewrite the equation with one of the radicals isolated.

2. Raise both sides to the power that is equal to the index of the radical.

3. Simplify.

4. If a radical remains, isolate this radical. Again, raise both sides to a power equal to the index. Simplify.

5. Solve the equation.

6. Check the solution(s).

EXAMPLE 5   **(a)** Solve: $\sqrt{2x + 6} = 1 + \sqrt{x + 4}$

SOLUTION $\sqrt{2x + 6} = 1 + \sqrt{x + 4}$ The radicals are on opposite sides.

$\left(\sqrt{2x + 6}\right)^2 = \left(1 + \sqrt{x + 4}\right)^2$ The index is 2; raise both sides to the second power.

$2x + 6 = \left(1 + \sqrt{x + 4}\right)\left(1 + \sqrt{x + 4}\right)$ Simplify; $\left(\sqrt{2x + 6}\right)^2 = 2x + 6$

$2x + 6 = 1 + 1\sqrt{x + 4} + 1\sqrt{x + 4} + x + 4$ Distributive property.

$2x + 6 = x + 2\sqrt{x + 4} + 5$ Simplify.

$\underline{-x\ -5\quad -x\qquad\qquad\quad -5}$ Subtraction property of equality.

$x + 1 = 0 + 2\sqrt{x + 4} + 0$ Simplify; the radical is isolated.

$(x + 1)^2 = \left(2\sqrt{x + 4}\right)^2$ Raise both sides to the second power.

$(x + 1)(x + 1) = 4(x + 4)$ Simplify; $\left(2\sqrt{x + 4}\right)^2 = 2 \cdot 2\left(\sqrt{x + 4}\right)^2$

$x^2 + x + x + 1 = 4x + 16$ Distributive property.

$$x^2 + 2x + 1 = 4x + 16 \qquad \text{Simplify.}$$

$$\underline{-4x - 16 \quad -4x \quad -16} \qquad \text{Subtraction property of equality.}$$

$$x^2 - 2x - 15 = 0 \qquad \text{Standard form: } ax^2 + bx + c = 0$$

$$(x - 5)(x + 3) = 0 \qquad \text{Factor.}$$

$$x - 5 = 0 \quad \text{or} \quad x + 3 = 0 \qquad \text{Zero product property.}$$

$$\underline{+5 \ +5} \qquad \underline{-3 \ -3} \qquad \text{Properties of equality.}$$

$$x = 5 \quad \text{or} \quad \cancel{x = -3} \qquad \text{Simplify.}$$

(b) Check.

▶   Check: $x = 5$ Check: $x = -3$

$$\sqrt{2x + 6} = 1 + \sqrt{x + 4} \qquad\qquad \sqrt{2x + 6} = 1 + \sqrt{x + 4}$$

$$\sqrt{2(5) + 6} = 1 + \sqrt{5 + 4} \qquad\qquad \sqrt{2(-3) + 6} = 1 + \sqrt{-3 + 4}$$

$$\sqrt{16} = 1 + \sqrt{9} \qquad\qquad\qquad \sqrt{0} = 1 + \sqrt{1}$$

$$4 = 4 \quad \text{True.} \qquad\qquad\qquad\quad 0 = 2 \quad \text{False.}$$

Since $x = -3$ is an extraneous solution, the only solution of $\sqrt{2x + 6} = 1 + \sqrt{x + 4}$ is $x = 5$.

Practice Problems

For problems 6–7,
(a) solve.
(b) check.

6. $\sqrt{2x + 3} = \sqrt{x + 2} + 2$ **7.** $\sqrt{4 - p} + \sqrt{p + 6} = 4$

An equation may include rational exponents instead of radicals, $\sqrt[n]{x^m} = x^{\frac{m}{n}}$. To solve, isolate the expression with the rational exponent and raise both sides of the equation to a power that is equal to the denominator of the rational exponent. To simplify, use the power rule of exponents, $(x^m)^n = x^{m \cdot n}$.

EXAMPLE 6   **(a)** Solve: $(3n + 12)^{\frac{1}{2}} + 4 = 10$

SOLUTION ▶

$$(3n + 12)^{\frac{1}{2}} + 4 = 10$$

$$\underline{\qquad\qquad -4 \quad -4} \qquad \text{Subtraction property of equality.}$$

$$(3n + 12)^{\frac{1}{2}} + 0 = 6 \qquad \text{Simplify.}$$

$$\left((3n + 12)^{\frac{1}{2}}\right)^2 = 6^2 \qquad \text{The denominator of the rational exponent is 2; raise both sides to the second power.}$$

$$(3n + 12)^{\left(\frac{1}{2}\right)(2)} = 36 \qquad \text{Power rule of exponents; simplify.}$$

$$3n + 12 = 36 \qquad \text{Simplify; } \left(\frac{1}{2}\right)(2) = 1; (3n + 12)^1 = 3n + 12$$

$$\underline{\qquad -12 \ -12} \qquad \text{Subtraction property of equality.}$$

$$3n + 0 = 24 \qquad \text{Simplify.}$$

$$\frac{3n}{3} = \frac{24}{3} \qquad \text{Division property of equality.}$$

$$n = 8 \qquad \text{Simplify.}$$

(b) Check.

$$(3n + 12)^{\frac{1}{2}} + 4 = 10 \qquad \text{Check in the original equation.}$$

$$(3(\mathbf{8}) + 12)^{\frac{1}{2}} + 4 = 10 \qquad \text{Replace the variable with the solution, 8.}$$

$$(\mathbf{36})^{\frac{1}{2}} + 4 = 10 \qquad \text{Simplify.}$$

$$6 + 4 = 10 \qquad \text{Simplify; } 36^{\frac{1}{2}} = \sqrt{36}; \sqrt{36} = 6$$

$$\mathbf{10} = 10 \quad \text{True.} \qquad \text{Simplify; since } 10 = 10 \text{ is true, the solution is correct.}$$

Practice Problems

For problems 8–9,
(a) solve.
(b) check.

8. $(2n + 6)^{\frac{1}{2}} + 3 = 11$ **9.** $(5n - 20)^{\frac{1}{2}} - 4 = 6$

Applications

When a retailer sells products, new products need to be available in *inventory* to replace the sold products. Some retailers use the formula $Q = \sqrt{\dfrac{2DC_o}{C_hC_p}}$ to predict the **ideal order quantity**. The ideal number of products to order for inventory is Q, the average annual demand for the product is D, the cost of placing the order is C_o, the annual cost to store the product per dollar of value in a back room or warehouse is C_h, and the cost of each product is C_p.

EXAMPLE 7   A retailer expects to sell 2000 exercise bikes this year. The cost per bike is \$450. The charge by the supplier for placing an order is \$500. The holding cost is \$0.25 per dollar value of bikes stored in the warehouse. Use the formula $Q = \sqrt{\dfrac{2DC_o}{C_hC_p}}$ to find the ideal number of bikes to order. Round to the nearest whole number.

SOLUTION ▶ **Step 1 Understand the problem.**
The unknown is the ideal number of exercise bikes to order.

     Q = ideal number of exercise bikes

Step 2 Make a plan.
Use the formula $Q = \sqrt{\dfrac{2DC_o}{C_hC_p}}$. Since the constant in the formula, 2, does not include units, the simplified units for the other measurements will not be the units for the answer, bikes. Do not include units when replacing the variables in the formula.

Step 3 Carry out the plan.

$$Q = \sqrt{\dfrac{2DC_o}{C_hC_p}} \qquad \text{The formula for ideal order quantity, } Q.$$

$$Q = \sqrt{\dfrac{2(\mathbf{2000})(\mathbf{500})}{(\mathbf{0.25})(\mathbf{450})}} \qquad D = 2000 \text{ bikes; } C_p = \$450; C_h = \$0.25; C_o = \$500$$

$$Q = 133.33\ldots \qquad \text{Simplify.}$$

$$Q \approx \mathbf{133 \text{ exercise bikes}} \qquad \text{Round; this is the ideal number of bikes to order.}$$

Step 4 Look back.

If the retailer expected to sell 1500 bikes,

$$Q = \sqrt{\frac{2(\mathbf{1500})(500)}{(0.25)(450)}} \approx 115 \text{ bikes}$$

If the retailer expected to sell 2500 bikes,

$$Q = \sqrt{\frac{2(\mathbf{2500})(500)}{(0.25)(450)}} \approx 149 \text{ bikes}$$

Since 2000 bikes is between 1500 bikes and 2500 bikes and the ideal order quantity of 133 bikes is between 115 bikes and 149 bikes, the answer seems reasonable. From this problem, we learn that if the constant in a formula does not include units, the simplified units may not make sense.

Step 5 Report the solution.

The ideal order quantity is 133 exercise bikes per order.

Practice Problems

10. A retailer expects to sell 6250 blenders this year. The charge by the supplier for placing an order is $400. The holding cost is $0.25 per dollar value of the blenders stored in the warehouse. The cost for each blender is $48. Find the ideal number of blenders in an order. Round to the nearest whole number.

11. A formula for finding the length of a spiral (helix) is $L = \sqrt{(\pi D)^2 + H^2}$. In this formula, L is the length of material needed to make one loop of the helix, D is the diameter of the helix, and H is the distance between loops of the helix. An engineer is building a spiral rebar cage with a diameter of 5 ft 3 in. and a distance between loops of 6 in. Find the length of rebar in inches needed to make 10 loops of this helix. Round to the nearest hundred.

Using Technology: Solving an Equation with a Graphing Calculator

The solution of $x + 5 = 8$ is $x = 3$. We can find this solution by isolating x. We subtract 5 from each side of the equation. We can also estimate this solution by graphing two linear equations, $y = x + 5$ and $y = 8$. The x-coordinate of the intersection of the graphs is the solution, $x = 3$.

EXAMPLE 8 Solve $x + 5 = 8$ by graphing.

Put the calculator in a standard window. Press [Y=]. Type both equations. Press [GRAPH]. Go to the CALC menu. Choose 5: intersect. The cursor is on the graph of $y = x + 5$. It asks "First curve?" Press [ENTER]. (Notice that the calculator calls the graph of any equation, even a straight line, a "curve.")

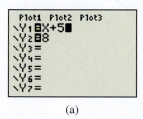

(a)

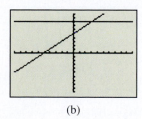

(b)

(c)

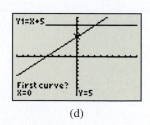

(d)

The cursor has moved to the graph of $y = 8$. Since this is the second "curve," press ENTER. Now the calculator asks "Guess?" Move the cursor to the intersection point. Press ENTER. The coordinates of the intersection point appear.

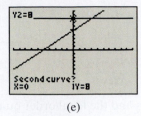
(e)

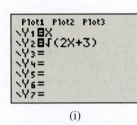
(f)

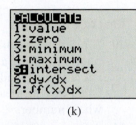
(g)

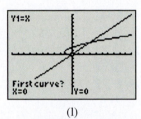
(h)

The solution of the equation is the x-coordinate of the intersection point. In this example, the solution of $x + 5 = 8$ is $x = 3$.

EXAMPLE 9   Solve: $x = \sqrt{2x + 3}$

Find the intersection point of the graphs of $y = x$ and $y = \sqrt{2x + 3}$.

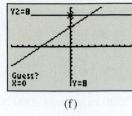
(i)

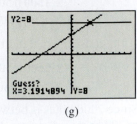
(j)

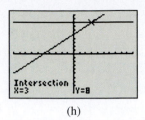
(k)

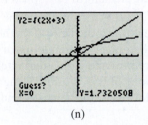
(l)

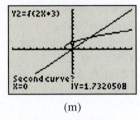
(m)

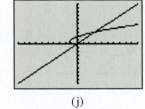
(n)

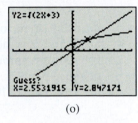
(o)

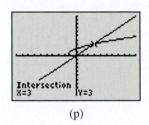
(p)

The solution of this equation is $x = 3$. This is the same solution we found algebraically in Example 3 in this section. Notice that the extraneous solution $x = -1$ does not appear on the graph; the lines do not intersect at $x = -1$.

Practice Problems   For problems 12–15,
    (a) use the graphical method to solve each equation. Sketch the graph; describe the window.
    (b) identify the solution.

    12. $2x - 8 = -3x + 2$ **13.** $x^2 - 2x = 8$
    14. $\sqrt{x + 9} = 2$ **15.** $x = \sqrt{5x}$

6.5 VOCABULARY PRACTICE

Match the term with its description.

1. $\sqrt{x}$ is an example of this.
2. $\sqrt{ab} = \sqrt{a}\sqrt{b}$
3. $\sqrt{\dfrac{a}{b}} = \dfrac{\sqrt{a}}{\sqrt{b}}$
4. A number that can be written as a fraction in which the numerator and denominator are integers
5. In the expression $\sqrt[3]{4x}$, $4x$ is an example of this.
6. In the expression $\sqrt[3]{4x}$, 3 is an example of this.
7. $\dfrac{x^m}{x^n} = x^{m-n}$
8. $x^m \cdot x^n = x^{m+n}$
9. A number that cannot be written as a fraction in which the numerator and denominator are integers
10. $\sqrt{x} - 3$ and $\sqrt{x} + 3$ are examples of this.

A. conjugate pair
B. index
C. irrational number
D. product rule of exponents
E. product rule of radicals
F. quotient rule of exponents
G. quotient rule of radicals
H. radical
I. radicand
J. rational number

6.5 Exercises

Follow your instructor's guidelines for showing your work.

For exercises 1–48,
(a) solve.
(b) check.

1. $\sqrt{x} + 3 = 11$
2. $\sqrt{z} + 2 = 9$
3. $\sqrt{p} = -3$
4. $\sqrt{y} = -8$
5. $-\sqrt{x} = -4$
6. $-\sqrt{x} = -12$
7. $\sqrt[3]{m} = 5$
8. $\sqrt[3]{c} = 2$
9. $\sqrt[3]{d} = -9$
10. $\sqrt[3]{w} = -6$
11. $\sqrt[4]{x} = 3$
12. $\sqrt[4]{h} = 5$
13. $\sqrt[5]{3h} = 3$
14. $\sqrt[5]{2z} = 2$
15. $-3\sqrt{x} = 15$
16. $-8\sqrt{f} = 24$
17. $\sqrt{y + 4} = 3$
18. $\sqrt{x + 8} = 5$
19. $\sqrt{y - 4} = 3$
20. $\sqrt{x - 8} = 5$
21. $\sqrt[3]{x + 4} = 6$
22. $\sqrt[3]{h + 3} = 7$
23. $\sqrt[4]{x + 1} = 2$
24. $\sqrt[4]{x + 2} = 8$
25. $\sqrt[4]{x + 1} = 2$
26. $\sqrt[4]{x + 2} = 8$
27. $-5\sqrt{x - 1} = -30$
28. $-3\sqrt{y + 2} = -21$
29. $\sqrt{2p - 3} + 4 = 7$
30. $\sqrt{7d + 1} + 2 = 10$
31. $\sqrt{4 - 2k} = 6$
32. $\sqrt{3 - 2x} = 7$
33. $2\sqrt{5 - x} = 18$
34. $2\sqrt{4 - x} = 12$
35. $\sqrt{x + 72} = x$
36. $\sqrt{x + 56} = x$
37. $\sqrt{x + 2} = x + 2$
38. $\sqrt{h - 5} = h - 5$
39. $y - 1 = \sqrt{8y - 23}$
40. $k + 3 = \sqrt{15k - 5}$
41. $w = \sqrt{5w}$
42. $x = \sqrt{7x}$
43. $-w = \sqrt{5w}$
44. $-x = \sqrt{7x}$

45. $\sqrt{2p - 3} = 5 - 2p$
46. $\sqrt{7y + 29} = y + 3$
47. $x - 4 = \sqrt{-5x + 20}$
48. $a - 6 = \sqrt{-7a + 50}$

For exercises 49–54,
(a) solve.
(b) check.

49. $\sqrt{p + 5} - \sqrt{p} = 1$
50. $\sqrt{x + 12} - \sqrt{x} = 2$
51. $\sqrt{7 - x} - 6 = \sqrt{x + 11}$
52. $\sqrt{m + 3} + 1 = \sqrt{m - 8}$
53. $\sqrt{a - 5} + \sqrt{a + 6} = 11$
54. $\sqrt{y - 4} + \sqrt{y + 7} = 11$

For exercises 55–68,
(a) solve.
(b) check.

55. $2p^{\frac{1}{2}} = 10$
56. $5k^{\frac{1}{2}} = 10$
57. $(y - 4)^{\frac{1}{2}} + 6 = 9$
58. $(x - 8)^{\frac{1}{2}} + 10 = 15$
59. $y^{\frac{1}{2}} + 4 = 3$
60. $x^{\frac{1}{2}} + 8 = 5$
61. $(5x - 3)^{\frac{1}{2}} + 8 = 9$
62. $(7r - 3)^{\frac{1}{2}} + 2 = 5$
63. $\sqrt{n + 12} = 2 + \sqrt{n}$
64. $\sqrt{p + 40} = 4 + \sqrt{p}$
65. $(x + 3)^{\frac{1}{2}} + (x - 1)^{\frac{1}{2}} = 2$
66. $(x + 4)^{\frac{1}{2}} + (x - 1)^{\frac{1}{2}} = 5$
67. $(x + 6)^{\frac{1}{2}} = 4 - (x - 2)^{\frac{1}{2}}$
68. $(x + 1)^{\frac{1}{2}} = 2 - (x - 3)^{\frac{1}{2}}$

For exercises 69–70, a minimum skid speed formula is $S = \sqrt{30Dfn}$. The average distance of the skid marks in feet is D. The drag factor of the road surface is f. The percent braking efficiency in decimal form is n. The minimum speed of the car when the wheels lock up and the car begins to skid in miles per hour is S.

69. An asphalt road surface has a drag factor of 0.75. All four wheels of the car brake during a skid. The braking efficiency is 0.50 (perhaps these brakes should be replaced). Predict the distance of the skid marks when the minimum speed of the car is 35 miles per hour. Round to the nearest whole number.

70. A concrete road surface has a drag factor of 0.80. All four wheels of the car brake during a skid. The braking efficiency is 0.50. Predict the distance of the skid marks when the minimum speed of the car is 50 miles per hour. Round to the nearest whole number.

For exercises 71–74, a maximum hull speed formula is $S = 1.34\sqrt{L}$. The maximum speed in knots that a single-hull sailboat can gain from wind power is S. The length in feet of the hull of the boat at the waterline is L (1 knot = 1.152 miles per hour).

71. A Hunter 216 single-hull sailboat is 21.5 ft long. The length of the hull at the waterline is 18.75 ft. Find the maximum speed of the boat under sail in miles per hour. Round to the nearest tenth.

72. A Hunter 27X single-hull sailboat is 27 ft long. The length of the hull at the waterline is 23.58 ft. Find the maximum speed of the boat under sail in miles per hour. Round to the nearest tenth.

73. Determine the length at the waterline of a single-hull sailboat that will have a maximum speed of 8 knots. Round to the nearest tenth.

74. Determine the length at the waterline of a single-hull sailboat that will have a maximum speed of 5 knots. Round to the nearest tenth.

For exercises 75–76, a doctor may use the body surface area of a patient to determine a dosage of chemotherapy. The Mosteller body surface area formula is $A = \sqrt{\dfrac{HW}{3600}}$. The patient's height in centimeters is H, the patient's weight in kilograms is W, and the patient's body surface area in square meters is A (1 in. = 2.54 cm; 1 kg ≈ 2.2 lb).

75. Find the body surface area in square meters of a patient who is 6 ft tall and weighs 170 lb. Round to the nearest hundredth.

76. Find the body surface area in square meters of a patient who is 5 ft tall and weighs 105 lb. Round to the nearest hundredth.

For exercises 77–78, the low height of a broadcast antenna for a college FM radio station in a flat area limits its reception area. The formula $D = 1.415\sqrt{H}$ finds the maximum range of signal reception in miles for listeners without elevated antennas, D, when H is the height of an FM antenna in feet.

77. Find the height of antenna required for a maximum range of signal reception of 12 mi. Round to the nearest whole number.

78. Find the height of antenna required for a maximum range of signal reception of 25 mi. Round to the nearest whole number.

For exercises 79–82, students in introductory physics classes often study the motion of a pendulum. The period of a pendulum is the time it takes for the pendulum to complete one swing, returning to its original position, $T = 2\pi\sqrt{\dfrac{L}{32\,\frac{\text{ft}}{\text{s}^2}}}$. The period in seconds is T. The length of the pendulum in feet is L.

79. Find the length in *feet and inches* of a pendulum with a period of 8 s. Round to the nearest inch.

80. Find the length in *feet and inches* of a pendulum with a period of 2 s. Round to the nearest inch.

81. A grandfather clock kit includes a 38-in. pendulum. Predict the time of the period of this pendulum in seconds. Round to the nearest hundredth (1ft = 12 in.).

82. A grandfather clock kit includes a 42-in. pendulum. Predict the time of the period of this pendulum in seconds. Round to the nearest hundredth (1ft = 12 in.).

For exercises 83–84, the mean (sometimes called the average) is a measure of *central tendency*. It is one way to estimate the "center" of a collection of test scores. The standard deviation is a measure of *dispersion*. It is one way to describe the amount that the scores are different from the mean. The formula for finding the standard deviation of a sample is $s = \sqrt{\dfrac{\text{the sum of } (X_n - M)^2}{N - 1}}$, where s is the standard deviation, X_n is each test score, M is the average test score, and N is the number of test scores.

83. Ten students took a math test. Their scores were 78, 76, 88, 92, 70, 64, 86, 86, 90, and 80. So $X_1 = 78$, $X_2 = 76$, and so on.
  a. Find the mean score, M. The mean is the sum of the scores divided by the number of scores.
  b. For each score X, find the difference of it and the mean, $X - M$. Square each difference. Add these ten "squared differences" together to find the sum of $(X_n - M)^2$.
  c. Use the formula to find s. Round to the nearest tenth.

84. A student measured the mass in grams of eight mice at the beginning of an experiment. Their masses were 40 g, 43 g, 38 g, 39 g, 39 g, 40 g, 32 g, and 41 g.
  a. Find the mean weight, M.
  b. For each weight X, find the difference of it and the mean, $X - M$. Square each difference. Add these eight "squared differences" together. This result is the sum of $(X_n - M)^2$.
  c. Use the formula to find s. Round to the nearest tenth.

For exercises 85–86, an Australian company charges license fees for the use of its images in books or magazines. The price in Australian dollars depends on the number of copies made, according to the function $C = 100 + \sqrt{n}$, where n is the number of copies in the print run and C is in Australian dollars.

85. Find the license fees charged for an image that will appear in 2000 copies of a magazine. Round to the nearest hundredth.

86. Find the license fees charged for an image that will appear in 20,000 copies of a magazine. Round to the nearest hundredth.

For exercises 87–88, use the ideal order quantity formula from Example 7.

87. A retailer expects to sell 10,000 gadgets this year. The charge by the supplier for placing an order is $400. The holding cost is $0.25 per dollar value of gadgets stored in the warehouse. The cost for each gadget is $75. Find the ideal number of gadgets in an order. Round to the nearest whole number.

88. Use the ideal number of gadgets per order found in exercise 87 to estimate how many orders for gadgets the retailer will place this year. Round to the nearest whole number.

Problem Solving: Practice and Review

Follow your instructor's guidelines for using the five steps as outlined in Section 1.5, p. 51.

89. Oakite-33 is a liquid acidic compound used to prepare metals for painting. It will gradually dissolve ordinary steel tanks, washing machines, or piping. The rate of metal loss for steel in Oakite-33 at 5% by volume, 29°C, is 0.123 in. per year. (*Source:* www.chemetalloakite.com)
   a. Write a linear model that describes the loss in metal, y, in x years.
   b. Find the time in years for Oakite-33 to dissolve through a steel tank with walls that are 1-in. thick. Round to the nearest tenth.

90. An estimate of production costs for growing sweet onions in the Columbia Valley in Washington State is $2758 per acre. The average yield of onions is $\dfrac{32 \text{ tons}}{1 \text{ acre}}$. In 2007, the wholesale price for onions was $\dfrac{\$22.40}{100 \text{ lb}}$.
   (*Source:* Washington State University Extension office)
   a. Write a revenue function, $R(x)$, if x is acres planted in onions.
   b. Write a cost function, $C(x)$, if x is acres planted in onions.
   c. Write a profit function, $P(x)$, by finding the difference of $R(x)$ and $C(x)$.
   d. Use the profit function to estimate the profit if 140 acres are planted in onions.

91. The Fan Cost Index™ for Major League Baseball games equals the price of two average-price adult tickets, two

average-price tickets for children, two small draft beers, four small soft drinks, four regular hotdogs, parking for one car, two game programs and two of the least expensive adult-size adjustable caps. Find the Fan Cost Index in 2010. Round to the nearest hundredth.

The total price to take a family of four to a game [in 2011] increased by 2.0 percent to $197.35, according to Team Marketing Report's exclusive 2011 Fan Cost Index. (*Source:* www.teammarketing.com, April 2011)

92. The graph shows the line of best fit for the relationship between the number of years since 1996, x, and the number of students enrolled in the public schools of Colorado.
   a. Write a linear model for this relationship. Include the units of measurement.
   b. Use this model to predict the enrollment in 2013. Round to the nearest thousand.

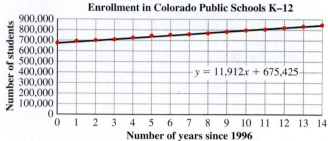

Enrollment in Colorado Public Schools K–12

$y = 11{,}912x + 675{,}425$

Source: Colorado Department of Education, Feb 28, 2011

Technology

For exercises 93–96,
(a) use a graphing calculator to solve each equation. Choose a window that shows the point of intersection. Sketch the graph; describe the window.
(b) identify the solution(s).

93. $\sqrt{2x + 18} = 4$

94. $\sqrt{3x + 24} = 6$

95. $\sqrt{2x + 18} = \sqrt{5x}$

96. $\sqrt{3x + 24} = \sqrt{9x}$

Find the Mistake

For exercises 97–100, the completed problem has one mistake.
(a) Describe the mistake in words, or copy down the whole problem and highlight or circle the mistake.
(b) Do the problem correctly.

97. Problem: Solve: $\sqrt[3]{x} + 2 = 8$

   Incorrect Answer: $\sqrt[3]{x} + 2 = 8$
   $$\left(\sqrt[3]{x} + 2\right)^3 = 8$$
   $$x + 2 = 8$$
   $$x = 6$$

98. Problem: Solve: $\sqrt{3x - 12} = 15$

Incorrect Answer: $\sqrt{3x - 12} = 15$

$$(\sqrt{3x - 12})^2 = 15^2$$
$$3x - 12 = 30$$
$$\underline{+12 \quad +12}$$
$$3x + 0 = 42$$
$$\frac{3x}{3} = \frac{42}{3}$$
$$x = 14$$

99. Problem: Solve: $x - 1 = \sqrt{5x - 9}$

Incorrect Answer: $x - 1 = \sqrt{5x - 9}$

$$x^2 - 1^2 = (\sqrt{5x - 9})^2$$
$$x^2 - 1 = 5x - 9$$
$$x^2 - 5x + 8 = 0$$

The discriminant of $x^2 - 5x + 8$ is $b^2 - 4ac = (-5)^2 - 4(1)(8) = -7$. Since $-7 < 0$, $x^2 - 5x + 8$ is a prime polynomial. We don't know a method for solving this equation.

100. Problem: Solve: $x + 3 = \sqrt{-2x - 3}$

Incorrect Answer: $x + 3 = \sqrt{-2x - 3}$

$$(x + 3)^2 = (\sqrt{-2x - 3})^2$$
$$x^2 + 6x + 9 = -2x - 3$$
$$x^2 + 8x + 12 = 0$$
$$(x + 6)(x + 2) = 0$$
$$x + 6 = 0 \quad \text{or} \quad x + 2 = 0$$
$$x = -6 \text{ or} \qquad x = -2$$

Review

101. Use interval notation to represent the domain of $f(x) = x^2 - 5x - 6$.

102. Use a table of ordered pairs to graph $f(x) = x^2 - 5x - 6$.

103. Use an algebraic method to find the zero(s) of $f(x) = x^2 - 5x - 6$.

104. The vertex of the graph of a quadratic function is $(0, 5)$. The graph opens up. Use interval notation to write the domain of the function.

SUCCESS IN COLLEGE MATHEMATICS

105. A student thinks that she cannot solve application problems. She skips them on her homework, assuming that she cannot do them correctly. She tells her friends that she can't do these problems and that she will never be able to do them. What do you think this student could do to change this situation?

© Dennis Hallinan/Getty Images

A function is a set of ordered pairs in which each input value corresponds to exactly one output value. If the function rule includes a radical, it is a radical function. The radical function $f(d) = \sqrt{32.8d}$ describes the relationship of the input value, d, the distance traveled by a car moving at a constant acceleration in a drag race, and the output value, $f(d)$, the speed of the car. In this section, we will study this and other radical functions.

SECTION 6.6

Radical Functions

After reading the text, working the practice problems, and completing assigned exercises, you should be able to:

1. Use a table of ordered pairs to graph a radical function.

2. Use translation to graph a radical function.

3. Use an algebraic method to find the domain of a radical function.

4. Evaluate a radical function.

5. Use the graph of a radical function to estimate its range.

6. Use an algebraic or graphical method to find a real zero of a radical function.

Graphing a Radical Function

The square root of a negative number is not a real number. In graphing $y = \sqrt{x}$ with a table of ordered pairs, the values for x must be greater than or equal to 0.

EXAMPLE 1   **(a)** Use a table of ordered pairs to graph $y = \sqrt{x}$. Round irrational output values to the nearest hundredth.

SOLUTION ▶   Since the square root of a negative number is not a real number, the smallest value of x is 0. When x is not a perfect square, y is an irrational number.

x	y
0	0
4	2
10	≈ 3.16
16	4

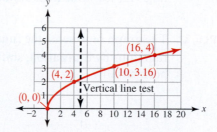

(b) What does the vertical line test tell us about the set of ordered pairs represented by this graph?

▶   Since each vertical line crosses the graph in at most one point, this graph represents a function.

A function such as $y = \sqrt{x}$ is a **radical function**. The input values of a function with a square root must be restricted so that the radicand is greater than or equal to 0. Since the cube root of a negative number is a real number, there are no similar restrictions on the input values for a cube root.

EXAMPLE 2   Use a table of ordered pairs to graph $f(x) = \sqrt[3]{x}$. Round irrational output values to the nearest hundredth.

SOLUTION ▶   If an input value is not a perfect cube, the output value is irrational.

x	y
-8	-2
-1	-1
0	0
1	1
4	≈ 1.59
8	2

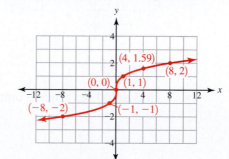

Practice Problems

For problems 1–3, use a table of ordered pairs to graph each function.

1. $f(x) = \sqrt{x} - 2$ **2.** $f(x) = \sqrt[3]{x} + 1$ **3.** $y = -\sqrt{x}$

Graphing by Translation

In Example 1, we used a table of ordered pairs to graph $f(x) = \sqrt{x}$. The graph of $f(x) = \sqrt{x} + 3$ is the same curve. However, it is shifted vertically up 3 units. If a constant is added *outside the radical*, the graph shifts vertically up or down. If a constant is added *inside the radical*, the graph shifts horizontally right or left. These shifts are vertical and horizontal **translations**.

The advantage to using translation is that only the basic graph needs to be built with a table of ordered pairs. Graphs with vertical or horizontal translations are just copies of the basic graph shifted to a new location. Graphing by translation takes far less time than building another table of ordered pairs.

EXAMPLE 3 | The graph represents the function $f(x) = \sqrt{x}$. Graph each of the following functions by translation, shifting the graph of $f(x) = \sqrt{x}$ vertically and/or horizontally.

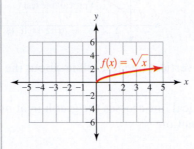

(a) $f(x) = \sqrt{x} + 3$

SOLUTION ▶ Vertical shift up 3 units.

(b) $f(x) = \sqrt{x} - 3$

Vertical shift down 3 units.

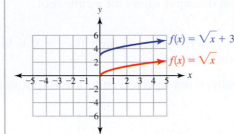

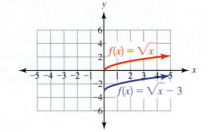

(c) $f(x) = \sqrt{x - 3}$

▶ Horizontal shift right 3 units.

(d) $f(x) = \sqrt{x + 3}$

Horizontal shift left 3 units.

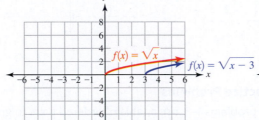

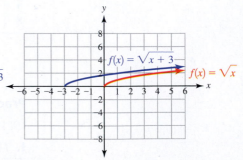

(e) $f(x) = \sqrt{x + 2} - 4$

Horizontal shift left 2 units.
Vertical shift down 4 units.

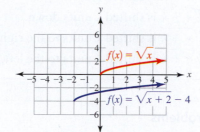

(f) $f(x) = \sqrt{x - 1} + 2$

Horizontal shift right 1 unit.
Vertical shift up 2 units.

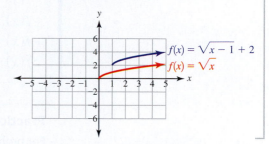

Why does the graph shift to the left when we add a constant inside the radical? Since we are adding, it seems as though the graph should move in the positive direction. However, think about the input value needed for the output value to be 0. For $0 = \sqrt{x}$, $x = 0$. For $0 = \sqrt{x + 3}$, $x = -3$. So when we add 3 inside the radical, the "first point" of the graph shifts to the left (Figure 1).

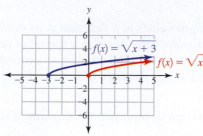

Figure 1

We can do similar translations with functions that include a cube root.

EXAMPLE 4 The graph represents the function $f(x) = \sqrt[3]{x}$. Graph each of the following functions by translation, shifting the graph of $f(x) = \sqrt[3]{x}$ vertically or horizontally.

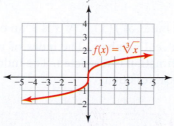

(a) $f(x) = \sqrt[3]{x} + 2$

SOLUTION ▶ Vertical shift up 2 units.

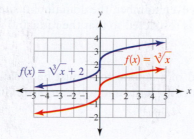

(b) $f(x) = \sqrt[3]{x + 2}$

Horizontal shift left 2 units.

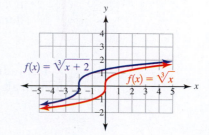

Graphing Radical Functions by Translation

If $f(x) = \sqrt[n]{x}$, where n is a whole number greater than or equal to 2 and $k > 0$,

$f(x) = \sqrt[n]{x} + k$ is shifted k units up.

$f(x) = \sqrt[n]{x} - k$ is shifted k units down.

$f(x) = \sqrt[n]{x - k}$ is shifted k units to the right.

$f(x) = \sqrt[n]{x + k}$ is shifted k units to the left.

Practice Problems

For problems 4–6, refer to $f(x) = \sqrt{x}$ (Example 3) or $f(x) = \sqrt[3]{x}$ (Example 4) and translation. Graph each function by translation, shifting the graph vertically or horizontally.

4. $f(x) = \sqrt{x} + 1$ **5.** $f(x) = \sqrt[3]{x} - 2$ **6.** $y = \sqrt{x + 5} - 2$

Domain

The set of the input values of a function is its domain. The domain of a *polynomial function* or an *absolute value function* is the set of real numbers, $(-\infty, \infty)$. To prevent division by zero, we restrict the domain of a *rational function* in which a variable is in the denominator. The domains of some *radical functions* must also be restricted.

EXAMPLE 5   Use interval notation to represent the domain of the function.

(a) $f(x) = \sqrt[3]{x}$

SOLUTION ▶   The cube root of any real number is a real number. The domain is $(-\infty, \infty)$.

(b) $g(x) = \sqrt{x}$

▶   The square root of a real number less than 0 is not a real number. The inputs must be real numbers that are greater than or equal to 0. The domain of this function is $[0, \infty)$.

In Example 5, the index of $f(x) = \sqrt[3]{x}$ is 3, an odd number. The domain of this function is the set of real numbers. The index of $g(x) = \sqrt{x}$ is 2, an even number. The domain of this function is restricted. The index of the radical helps us identify the domain of a radical function.

Domains of Radical Functions

For $f(x) = \sqrt[n]{x}$, where n is a whole number greater than or equal to 2,

- if n is an odd number, the domain is the set of real numbers, $(-\infty, \infty)$.
- if n is an even number, the domain is a subset of the real numbers. The domain is restricted so that the radicand is greater than or equal to 0.

The radicand of a square root must be greater than or equal to 0. To find the domain algebraically, write and solve an inequality, radicand ≥ 0.

EXAMPLE 6   Use an algebraic method to find the domain of $h(x) = \sqrt{-2x + 6} + 7$, and use interval notation to represent it.

SOLUTION ▶ The radicand, $-2x + 6$, must be greater than or equal to 0. Ignore the 7 in the function rule as it does not affect the value of the radicand.

$$-2x + 6 \geq 0 \qquad \text{The radicand } \geq 0.$$
$$\underline{\quad -6 \;\; -6 \quad} \qquad \text{Subtraction property of inequality.}$$
$$-2x + 0 \geq -6 \qquad \text{Simplify.}$$
$$\frac{-2x}{-2} \leq \frac{-6}{-2} \qquad \text{Division property of inequality; reverse the inequality sign.}$$
$$x \leq 3 \qquad \text{The domain is the set of real numbers less than or equal to 3.}$$

The domain is $(-\infty, 3]$.

Some radical functions are models of real-life situations. In the next example, the maximum ocean depth restricts the domain.

EXAMPLE 7   The function $V = \sqrt{\left(\dfrac{9.8 \text{ m}}{1 \text{ s}^2}\right)d}$ represents the relationship between the wave speed of a tsunami, V, and the depth of the ocean, d, where V is in meters per second and the depth is in meters.

(a) Use interval notation to represent the domain of this function in an ocean with a maximum depth of 3500 m.

SOLUTION ▶   [0 m, 3500 m]

(b) Explain why this is a reasonable domain.

▶ Each input value is an ocean depth. Since the depth must be greater than or equal to 0 m and less than or equal to 3500 m, the domain is reasonable.

Evaluating a Radical Function

To evaluate a function for a given input, find the corresponding output value. If the output is a radical, simplify but do not evaluate radicals that are irrational numbers.

EXAMPLE 8 | Evaluate $f(9)$ when $f(x) = \sqrt{12x}$.

SOLUTION ▶

$f(x) = \sqrt{12x}$

$f(9) = \sqrt{12 \cdot 9}$ Replace x with the input value, 9.

$f(9) = \sqrt{108}$ Multiply inside the radicand.

$f(9) = \sqrt{36}\sqrt{3}$ 36 is a perfect square factor of 108; product rule of radicals.

$f(9) = 6\sqrt{3}$ Simplify; do not evaluate the square root.

Road signs called *chevrons* mark highway curves. In the next example, we evaluate a radical function that finds the distance between chevron markers. Many function rules like this one found in manuals and reports do not include units of measurement.

Courtesy of the Author

EXAMPLE 9 | The function $f(R) = 3\sqrt{R - 50}$ represents the relationship between the radius of a curve, R, and the recommended distance between chevron signs, $f(R)$. Find the recommended distance (in feet) between chevrons on a curve with a radius of 700 ft. Round to the nearest whole number.

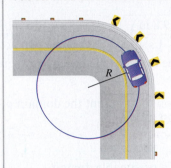

SOLUTION ▶ **Step 1 Understand the problem.**
The unknown is the recommended distance between chevrons on a curve with a radius of 700 ft. The unknown is the output, $f(R)$, of a function f.

$f(R)$ = recommended distance between chevrons

Step 2 Make a plan.
Evaluate the function $f(R) = 3\sqrt{R - 50}$.

Step 3 Carry out the plan.

$f(R) = 3\sqrt{R - 50}$ The function does not include units.

$f(700) = 3\sqrt{700 - 50}$ Replace R with 700.

$f(700) = 3\sqrt{650}$ Simplify.

$f(700) = 3(25.495\dots)$ Evaluate the square root.

$f(700) \approx 76$ ft   Round; the units of distance are feet.

Step 4 Look back.
If the radius of the curve is 600 ft, the distance between signs is 70 ft. If the radius of the curve is 800 ft, the distance between signs is 82 ft.

$f(600) = 3\sqrt{600 - 50}$ $f(800) = 3\sqrt{800 - 50}$

$f(600) \approx 70$ ft $f(800) \approx 82$ ft

Since the input value of 700 ft is between 600 ft and 800 ft and the output value of 76 ft is between 70 ft and 82 ft, the answer seems reasonable.

Step 5 Report the solution.
The recommended distance between the chevrons is 76 ft.

Practice Problems

11. Evaluate $f(12)$ when $f(x) = \sqrt{20x}$.

12. The function $f(d) = \sqrt{32.8d}$ represents the relationship between speed, $f(d)$, and distance, d, at a constant acceleration. A car is stopped at the starting line of a straight and level drag strip. When the traffic light turns green, the car accelerates forward at a constant rate. Find the speed of the car in feet per second after it has traveled 100 ft. Round to the nearest whole number.

13. The function $f(A) = 2\sqrt{\dfrac{A}{\pi}}$ represents the relationship of the area of a circle, A, and its diameter, $f(A)$. Use this function to find the diameter of a circular horse corral with an area of 750 ft^2. Round to the nearest whole number.

Range

Since the radicand of a square root must be greater than or equal to 0, the smallest input value of $f(x) = \sqrt{x}$ is 0. The value of $f(0)$ is 0. This output value is included in the range; the closed interval begins with a bracket. The function has no maximum output value.

EXAMPLE 10   Use interval notation to represent the range.

(a) $f(x) = \sqrt{x}$

SOLUTION ▶   The domain of this function is $[0, \infty)$. The minimum *input* value is 0. Since $f(0) = 0$, the minimum *output* value is 0. There is no maximum output value. As the graph confirms, the range is $[0, \infty)$.

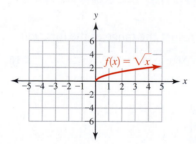

(b) $g(x) = \sqrt{x} + 3$

▶   The domain of this function is $[0, \infty)$. The minimum *input* value is 0. Since $g(0) = 3$, the minimum *output* value is 3. As the graph confirms, the range is $[3, \infty)$.

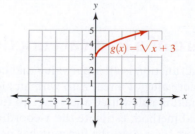

(c) $h(x) = \sqrt{x + 4} + 3$

▶ The domain of this function is $[-4, \infty)$. The minimum *input* value is -4. Since $h(-4) = 3$, the minimum *output* value is 3. As the graph confirms, the range is $[3, \infty)$.

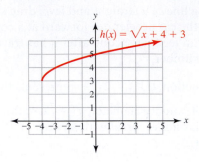

(d) $c(x) = -\sqrt{x}$

▶ The domain of this function is $[0, \infty)$. The graph shows that the minimum input value of 0 corresponds to the *maximum* output value. Since $c(0) = 0$, the maximum output value is 0. As the graph confirms, the range is $(-\infty, 0]$.

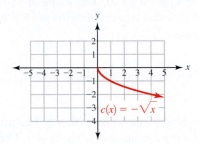

Practice Problems

For problems 14–16, use interval notation to represent the range of the function.

14. $y = \sqrt{x} + 5$

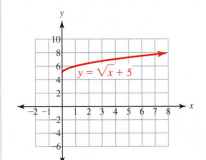

15. $y = \sqrt{x} - 4$

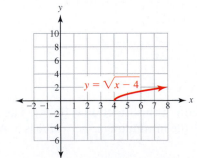

16. $y = -\sqrt{x} + 1$

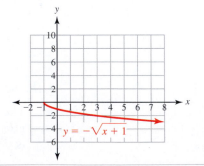

A Real Zero of a Radical Function

A real zero of a function is an input value that corresponds to an output value of 0. To find the exact value of a real zero with an algebraic method, replace the output value in the function rule with 0 and solve. To find an estimated value of a real zero with a graphical method, estimate the x-coordinate of an x-intercept. If the graph does not intersect the x-axis, the function has no real zeros.

EXAMPLE 11 A function is $f(x) = \sqrt{x} - 3$.

(a) Use an algebraic method to find any real zeros.

SOLUTION ▶

$f(x) = \sqrt{x} - 3$	The original function.
$0 = \sqrt{x} - 3$	Replace $f(x)$, the output value, with 0.
$\underline{+3 \qquad\qquad +3}$	Addition property of equality.
$3 = \sqrt{x} + 0$	Simplify.
$3^2 = (\sqrt{x})^2$	Raise both sides to the second power.
$9 = x$	A real zero of this function is $x = 9$.

Since $x = 9$ could be an extraneous solution of the original equation, check by evaluating $f(9)$. If 9 is a real zero of the function, $f(9)$ will equal 0.

$f(9) = \sqrt{9} - 3$	Replace x with the input value, 9.
$f(9) = 3 - 3$	Evaluate the square root.
$f(9) = 0$	The output value is 0 when the input value is 9.

Since $f(9) = 0$, 9 is a real zero of $f(x) = \sqrt{x} - 3$.

(b) Use a graphical method to estimate any real zeros.

▶ Estimate the x-intercept: $(9, 0)$. The estimated real zero is the x-coordinate of this x-intercept, $x = 9$.

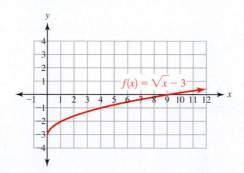

If the graph of a radical function does not intersect the x-axis, it does not have any real zeros.

EXAMPLE 12 Use an algebraic method to show that $y = -\sqrt{x} - 1$ has no real zeros.

SOLUTION ▶

$y = -\sqrt{x} - 1$	
$0 = -\sqrt{x} - 1$	Replace the output variable, y, with 0.
$\underline{+1 \qquad\qquad +1}$	Addition property of equality.
$1 = -1\sqrt{x} + 0$	Simplify; $-\sqrt{x} = -1\sqrt{x}$
$\dfrac{1}{-1} = \dfrac{-1\sqrt{x}}{-1}$	Division property of equality.
$-1 = \sqrt{x}$	Simplify; this is a radical equation.
$(-1)^2 = (\sqrt{x})^2$	Solve the radical equation; raise both sides to the second power.
$\cancel{1 = x}$	Simplify.

Check.

$$y = -\sqrt{x} - 1 \qquad \text{Check in the original function.}$$

$$0 = -\sqrt{1} - 1 \qquad \text{Replace the variables.}$$

$$0 = -1 - 1 \qquad \text{Evaluate the square root.}$$

$$0 = -2 \quad \text{False.} \qquad \text{Simplify; since } 0 = -2 \text{ is false, } -1 \text{ is an extraneous solution of}$$
$$0 = -\sqrt{x} - 1.$$

Since $0 = -\sqrt{x} - 1$ has no solution, the function $y = -\sqrt{x} - 1$ has no real zeros. The graph of the function confirms this; the graph does not have an x-intercept.

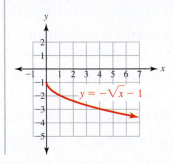

Practice Problems

For problems 17–18,
(a) use an algebraic method to find any real zeros.
(b) check.

17. $f(x) = \sqrt{x} - 8$ **18.** $f(x) = 3\sqrt{x} - 6$

For problems 19–21, use the graph to estimate any real zeros of the function.

19.

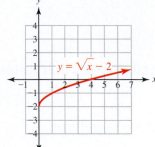

20.

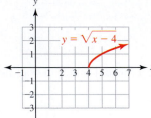

21.

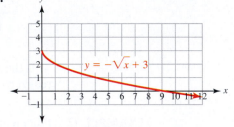

Using Technology: The Range and Real Zero of a Radical Function

EXAMPLE 13   Estimate any real zeros of $f(x) = \sqrt{x} - 5$. Round to the nearest hundredth.

Press WINDOW. Change the window settings to $[-1, 15, 1, -5, 5, 1]$. Go to the Y= screen. Type in the function, enclosing the radicand in parentheses. Go to the CALC menu. Choose 2: zero. Press ENTER. The calculator asks

"Left Bound?" However, the graph of the function begins at the x-intercept. We cannot move the cursor to the left of the x-intercept. So we cannot use this method to estimate the real zero.

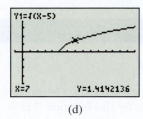 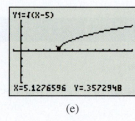 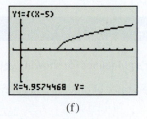

(a)   (b)   (c)

Instead, press GRAPH. Press TRACE. The function rule and the current coordinates of the cursor appear on the screen. Use the left arrow key to move the cursor as close as possible to the x-intercept. The estimated value of the real zero is $x = 5.12. \ldots$ If we move the cursor farther to the left, there will be no y-value. This value of x, $4.95. \ldots$, is not in the domain of this function.

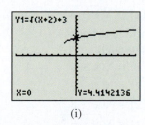 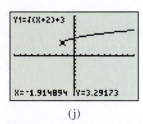

(d)   (e)   (f)

To represent the range of a function, we need the minimum or maximum output value. If we use the minimum or maximum command in the CALC window, we must find left and right bounds. If this is not possible, again use TRACE.

EXAMPLE 14 Estimate the range of $f(x) = \sqrt{x + 2} + 3$. Round to the nearest hundredth.

Reset the window to Standard. Press Y=. Type in the function. Press GRAPH. Press TRACE. Move the cursor as close as possible to the minimum output value so that a value for Y is still visible. This means that the cursor is still on the graph of the function. The estimated minimum output value is $3.29. \ldots$ Using this method, we estimate that the range is $[3.29, \infty)$.

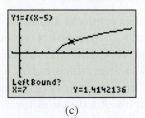

(g)   (h)   (i)   (j)

Practice Problems   For problems 22–25,
         (a) graph each function. Choose a window that shows the real zero or minimum output value. Sketch the graph; describe the window.
         (b) estimate the real zero.
         (c) estimate the minimum output value.
         (d) use interval notation to represent the range.
         If necessary, round the real zero or minimum to the nearest hundredth.

         22. $f(x) = \sqrt{x - 3}$ **23.** $h(x) = \sqrt{8x + 5} - 4$
         24. $g(x) = \sqrt{-x + 7}$ **25.** $y = \sqrt{-x + 7} - 6$

6.6 VOCABULARY PRACTICE

Match the term with its description.

1. An input value that results in an output value of zero
2. A relation in which each input corresponds to exactly one output
3. In the expression $\sqrt{7x}$, $7x$ is an example of this.
4. This test is used to determine whether a graph represents a function.
5. In this fraction, the numerator is greater than or equal to the denominator.
6. The set of the inputs of a function
7. The set of the outputs of a function
8. In $\sqrt[n]{x}$, n is an example of this.
9. A whole number that is divisible by 2
10. A whole number that is not divisible by 2

A. domain
B. even number
C. function
D. improper fraction
E. index
F. odd number
G. radicand
H. range
I. real zero of a function
J. vertical line test

6.6 Exercises

Follow your instructor's guidelines for showing your work.

For exercises 1–20, use a table of ordered pairs or translation to graph the function.

1. $f(x) = \sqrt{x} + 1$
2. $g(x) = \sqrt{x} + 4$
3. $f(x) = \sqrt{x} - 1$
4. $g(x) = \sqrt{x} - 4$
5. $f(x) = \sqrt{x+1}$
6. $g(x) = \sqrt{x+4}$
7. $f(x) = \sqrt{x-1}$
8. $g(x) = \sqrt{x-4}$
9. $f(x) = \sqrt[3]{x} + 4$
10. $g(x) = \sqrt[3]{x} + 5$
11. $f(x) = \sqrt[3]{x} - 4$
12. $g(x) = \sqrt[3]{x} - 5$
13. $f(x) = -\sqrt{x}$
14. $h(x) = -\sqrt[3]{x}$
15. $f(x) = -\sqrt{x} + 5$
16. $h(x) = -\sqrt[3]{x} + 5$
17. $f(x) = \sqrt{x+3} - 2$
18. $f(x) = \sqrt{x-3} + 2$
19. $f(x) = \sqrt[3]{x-4} + 2$
20. $f(x) = \sqrt[3]{x+4} - 2$

For exercises 21–30, use an algebraic method to find the domain, and use interval notation to represent it.

21. $f(x) = \sqrt{x+1}$
22. $g(x) = \sqrt{x+9}$
23. $f(x) = \sqrt{x-1}$
24. $g(x) = \sqrt{x-9}$
25. $f(x) = \sqrt{x-6}$
26. $g(x) = \sqrt{x-8}$
27. $f(x) = \sqrt{-x+4}$
28. $g(x) = \sqrt{-x+2}$
29. $f(x) = \sqrt{x+14}$
30. $g(x) = \sqrt{x+12}$

31. Explain why the domain of $f(x) = \sqrt{x}$ is not the set of real numbers.

32. Explain why the domain of $g(x) = -\sqrt{x}$ is not the set of real numbers.

33. Explain why the domain of $f(x) = \sqrt[3]{x}$ is the set of real numbers.

34. Explain why the domain of $f(x) = \sqrt[3]{-x}$ is the set of real numbers.

35. The function $f(H) = \frac{1}{6}\sqrt{73H}$ represents the relationship between the height, H, of an adult male patient in meters and the patient's body surface area, $f(H)$, in square meters. The patient weighs 73 kg. (*Source: Clinical Anatomy*, Vol. 18, No. 2, 2005)

a. Use interval notation to represent a reasonable domain for this function (1 m ≈ 3.28 ft; 1 kg ≈ 2.2 lb).

b. Explain why this domain is reasonable.

36. The function $f(H) = \frac{1}{6}\sqrt{68H}$ represents the relationship between the height, H, of an adult female patient in meters and the patient's body surface area, $f(H)$, in square meters. The patient weighs 68 kg. (*Source: Clinical Anatomy*, Vol. 18, No. 2, 2005)

a. Use interval notation to represent a reasonable domain for this function (1 m ≈ 3.28 ft; 1 kg ≈ 2.2 lb).

b. Explain why this domain is reasonable.

37. Evaluate $f(24)$ when $f(x) = \sqrt{3x}$.
38. Evaluate $f(30)$ when $f(x) = \sqrt{5x}$.
39. Evaluate $f(-20)$ when $f(x) = \sqrt{-5x}$.
40. Evaluate $f(-2)$ when $f(x) = \sqrt{-50x}$.

41. If the length of one leg of a right triangle is 2 in., the function $c = \sqrt{(2 \text{ in.})^2 + b^2}$ represents the relationship between the length of the other leg, b, and the length of the hypotenuse, c. Find the length of the hypotenuse if the length of the other leg is 5 in. Round to the nearest tenth.

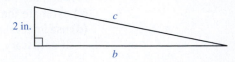

42. The function $N = 6000\sqrt{n}$ represents the relationship between the number of items that should be stocked in warehouses (the inventory), N, and the number of warehouse locations, n. Find the number of items that should be stocked if a supplier is using four warehouses.

43. The function $P = \dfrac{1}{2(0.31)}\sqrt{\dfrac{T}{0.006}}$ represents the relationship between the pitch, P, in hertz of a piano string and the tension, T, of the string in newtons (N). Find the pitch of this piano string with a tension of 650 N. Round to the nearest whole number.

© Mark McClare/Shutterstock

44. The function $T = \sqrt{\dfrac{2h}{9.8}}$ represents the relationship between the time, T, in seconds that it takes for a rescue package to hit the ground and the height, h, of the rescue plane above the ground in meters. Find the time it takes a package to hit the ground when the height of the plane is 100 m. Round to the nearest tenth.

45. If an object travels away from a planet at a fast enough speed, the *escape velocity*, the force of gravity cannot bring it back. The function $V = (4.56 \times 10^{-4})\sqrt{M}$ represents the relationship between the mass of a planet, M, in kilograms and the escape velocity, V, in $\dfrac{\text{meters}}{\text{second}}$. The mass of the earth is 5.98×10^{24} kg. Find the escape velocity for the earth. Write the answer in scientific notation. Round the mantissa to the nearest hundredth.

46. The function $V = \sqrt{\dfrac{(6.67 \times 10^{-11})(5.98 \times 10^{24})}{(d + 6.38 \times 10^6)}}$ represents the relationship between the speed, V, in $\dfrac{\text{meters}}{\text{second}}$ of a satellite and the distance of the satellite from the surface of the earth, d, in meters. Find the speed of a satellite that is in orbit 1.5×10^6 m above the surface of the earth. Write the answer in scientific notation; round the mantissa to the nearest hundredth.

For exercises 47–48, the function $f(n) = \sqrt{n} + 3$ represents the relationship of the SMOG reading grade level $f(n)$ of a book or article and the number, n, of polysyllabic words in a selection. The selection includes 10 sentences near the beginning of the article, 10 sentences in the middle, and 10 sentences near the end. (A polysyllabic word has three or more syllables.) Include repetitions of the same word.

47. A teacher selected 30 sentences from an article. He counted 70 polysyllabic words in these sentences. Find the SMOG reading grade level of this article. Round to the nearest tenth.

48. Find the SMOG reading grade level of a book or article assigned in one of your classes. Include the title of the book or article and the number of polysyllabic words. Round to the nearest tenth.

For exercises 49–58, use interval notation to represent
(a) the domain of the function represented by the graph.
(b) the range of the function represented by the graph.

49.

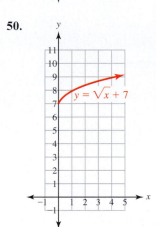

$y = \sqrt{x} + 6$

50.

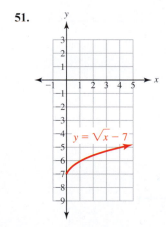

$y = \sqrt{x} + 7$

51.

$y = \sqrt{x} - 7$

52.

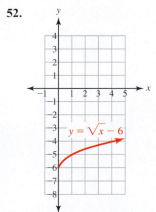

$y = \sqrt{x} - 6$

53.

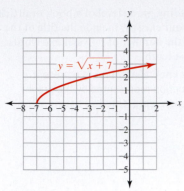

54.

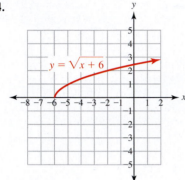

55.

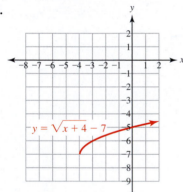

56.

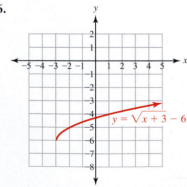

57.

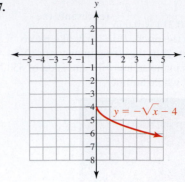

58.

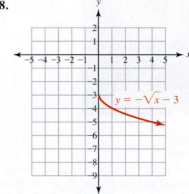

For exercises 59–68, use an algebraic method to find any real zeros of the function.

59. $f(x) = \sqrt{x + 1}$ **60.** $g(x) = \sqrt{x + 3}$

61. $f(x) = \sqrt{x - 1}$ **62.** $g(x) = \sqrt{x - 3}$

63. $f(x) = \sqrt{x} - 6$ **64.** $g(x) = \sqrt{x} - 4$

65. $f(x) = \sqrt{-x + 4}$ **66.** $g(x) = \sqrt{-x + 2}$

67. $f(x) = \sqrt[3]{x} + 2$ **68.** $g(x) = \sqrt[3]{x} + 4$

For exercises 69–72, use the graph to estimate any real zeros of the function.

69.

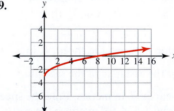

70.

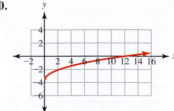

71.

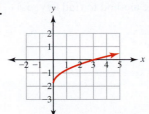

72.

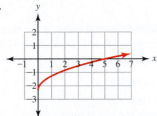

For exercises 73–90, use interval notation to represent the domain of the function. These are a mixture of constant, linear, quadratic, cubic, absolute value, rational, and radical functions.

73. $f(x) = x^2 + 5x + 4$

74. $f(x) = x^2 + 4x + 3$

75. $f(x) = \dfrac{x^2 + 5x + 4}{x^2 - 16}$

76. $f(x) = \dfrac{x^2 + 4x + 3}{x^2 - 9}$

77. $y = 9$

78. $y = -2$

79. $g(x) = \dfrac{7}{x^2 + 3}$

80. $g(x) = \dfrac{9}{x^2 + 2}$

81. $y = \sqrt{3x - 21}$

82. $y = \sqrt{4x - 20}$

83. $y = \dfrac{5}{6}x - 1$

84. $y = \dfrac{3}{4}x - 2$

85. $y = \dfrac{2}{x - 2}$

86. $y = \dfrac{3}{x - 3}$

87. $f(x) = |x - 6| + 8$

88. $f(x) = |x - 5| + 4$

89. $y = \sqrt{9 - x}$

90. $y = \sqrt{6 - x}$

Problem Solving: Practice and Review

Follow your instructor's guidelines for using the five steps as outlined in Section 1.5, p.51.

91. A circular sinkhole near Plant City, Florida, has a diameter of 10 ft and a depth of 14 ft. The homeowner is going to fill the sinkhole with gravel. Gravel is sold in cubic yards. Each cubic yard of gravel weighs about 2700 lb. Estimate the weight of the gravel needed to fill the hole (1 yd = 3 ft; volume of a circular cylinder is $V = \pi r^2 h$; $\pi \approx 3.14$). (*Source:* www.dep.state.fl.us/geology, www.eugenesand.com)

92. Find the amount generated by the current parking fees at Austin Community College. Round to the nearest hundred.

Parking fees are among the student charges that could rise . . . the change would generate about $755,000 annually, 68 percent more than the current fee. (*Source:* www.statesman.com, June 8, 2009)

93. A New York City resident smokes two packs of cigarettes a day. Find the total amount of tax this smoker will pay in a year of buying cigarettes (365 days = 1 year).

Starting tomorrow, New York City smokers will have to pay $9 or more for a pack of cigarettes. . . . New York City cigarettes became the most expensive in the nation last June, and will remain the priciest tomorrow, when the federal tax jumps from 39 cents to $1.01 per pack, pushing the total tax to $5.26 per pack. (*Source:* www.nyc.gov, March 31, 2009)

94. The population in the United States living in counties that are located on a coast is increasing.
  a. Use the information on the graph to write a linear model, including units.
  b. Predict the average number of people per square mile of land area in coastal counties in 2020. Round to the nearest whole number.

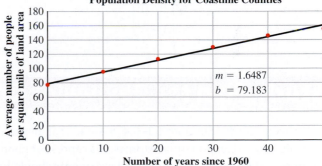

Source: www.census.gov, May 2010

Technology

For exercises 95–98,
(a) use a graphing calculator to graph each function. Choose a window that shows any real zeros. Sketch the graph; describe the window.
(b) use interval notation to represent the domain.
(c) use interval notation to represent the range.
(d) estimate any real zeros of the function.

95. $f(x) = 3\sqrt{x} - 6$

96. $g(x) = 2\sqrt{x} - 8$

97. $f(x) = \sqrt{-x} + 4$

98. $g(x) = \sqrt{-x} + 2$

Find The Mistake

For exercises 99–102, the completed problem has one mistake.
(a) Describe the mistake in words, or copy down the whole problem and highlight or circle the mistake.
(b) Do the problem correctly.

99. **Problem:** Use an algebraic method to find the domain of $f(x) = \sqrt{x - 8} + 2$, and use interval notation to represent it.

 Incorrect Answer: $x - 8 + 2 \geq 0$

$$x - 6 \geq 0$$
$$x \geq 6$$
$$[6, \infty)$$

100. **Problem:** Find the domain of $f(x) = \sqrt[3]{x + 1}$, and use interval notation to represent it.

 Incorrect Answer: $x + 1 \geq 0$

$$x \geq -1$$
$$[-1, \infty)$$

101. **Problem:** Use interval notation to represent the range of the function shown in the graph.

 Incorrect Answer: The range extends from negative infinity to 3. The range is $(-\infty, 3]$.

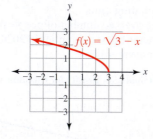

102. **Problem:** Use an algebraic method to find any real zeros of $f(x) = \sqrt{x} + 3$.

 Incorrect Answer: $0 = \sqrt{x} + 3$

$$-3 = \sqrt{x}$$
$$(-3)^2 = (\sqrt{x})^2$$
$$9 = x$$

The real zero of this function is $x = 9$.

Review

103. Write the equation in slope-intercept form of the line that passes through the points $(8, -7)$ and $(20, -4)$.

104. Solve $|x + 3| - 8 = 15$. Check the solutions.

105. Factor $6x^3 - x^2 - 15x$ completely.

106. Use the zero product property to solve $x^2 - 10x + 16 = 0$. Check the solution(s).

SUCCESS IN COLLEGE MATHEMATICS

107. How do you get extra help when you need it?

Study Plan for Review of Chapter 6

In the exercises for Sections 6.1–6.4, assume that all variables or variable expressions in radicands and variable bases of exponential expressions represent nonnegative real numbers. Use interval notation to represent domains and ranges. All exponents should be positive and in lowest terms.

SECTION 6.1   **Introduction to Radicals**

Ask Yourself	Test Yourself	Help Yourself
Can I . . .	Do 6.1 Review Exercises	See these Examples and Practice Problems
identify both square roots of a real number?	1–10, 20	Ex. 1–5, PP 1–7
identify the notation for the principal square root?		
explain why the square root of a negative number is not a real number?		
explain why the principal cube root of a negative number is a real number?		
identify the index and the radicand of a radical?		
evaluate a radical?		
create a list of perfect squares, perfect cubes, perfect fourths, and perfect fifths?	11–19	Ex. 6, PP 8–15
find the greatest whole number factor of a number that is a perfect power?		

Ask Yourself	Test Yourself	Help Yourself
Can I . . .	**Do 6.1 Review Exercises**	**See these Examples and Practice Problems**
write the product rule of radicals?	11–19	Ex. 7–18, PP 16–24
simplify a radical expression?		
simplify a square root in which the radicand is a perfect square trinomial?	15	Ex. 18, PP 25

6.1 Review Exercises

1. What are the square roots of 100?
2. Explain why $\sqrt{100} = 10$ but does not equal -10.
3. Explain why the square root of a negative number is not a real number.
4. Explain why the cube root of a positive number and the cube root of a negative number are real numbers.

For exercises 5–6, identify the index and radicand of each radical.

5. $\sqrt{x-4}$
6. $\sqrt[3]{x+2} - 6$

For exercises 7–10, evaluate. If the number is irrational, round to the nearest hundredth.

7. $\sqrt{64}$
8. $\sqrt[5]{-243}$
9. $\sqrt{35}$
10. $\sqrt{-49}$

For exercises 11–18, simplify.

11. $\sqrt[3]{x^6}$
12. $\sqrt[4]{y^{35}}$
13. $\sqrt{50a^6b^7c}$
14. $\sqrt[3]{-54x^8y^9z^2}$
15. $\sqrt{9p^2 + 6p + 1}$
16. $\sqrt[75]{x^{75}}$
17. $\sqrt[4]{48n^9p^{12}}$
18. $\sqrt{21u^5z}$
19. A student simplified $\sqrt{32x^3y^2}$ to $2xy\sqrt{8x}$. This expression is not completely simplified. Explain why.
20. A student said that $\sqrt[4]{4} = 1$ because $1 + 1 + 1 + 1 = 4$. Explain what is wrong with this thinking.

SECTION 6.2 **Adding, Subtracting, Multiplying, and Simplifying Radical Expressions**

Ask Yourself	Test Yourself	Help Yourself
Can I . . .	**Do 6.2 Review Exercises**	**See these Examples and Practice Problems**
identify like radicals?	21, 22	Ex. 1, PP 1–5
add or subtract like radicals?	23–28	Ex. 1, PP 1–5
simplify radicals?	24–28	Ex. 2–6, PP 1–5
use the product rule of radicals to multiply radicals?	29–31	Ex. 7–10, PP 6–10
use the distributive property to multiply a radical by a sum or difference?	32–34	Ex. 11–14, PP 7–10

6.2 Review Exercises

21. Explain why $\sqrt{6x}$ and $\sqrt{5y}$ are not like radicals.
22. Explain why $\sqrt{7a}$ and $\sqrt[3]{7a}$ are not like radicals.
23. Simplify $-\sqrt{u} + 9\sqrt{u} - 15\sqrt{u}$.

For exercises 24–28, simplify.

24. $\sqrt{48} + \sqrt{27}$
25. $\sqrt{54k} + \sqrt{600k}$
26. $3\sqrt{128x^3} + 5x\sqrt{18x}$
27. $\sqrt[3]{40xy} - \sqrt[3]{135xy} + 6\sqrt[3]{5xy}$
28. $\sqrt[4]{324a^4b^5} - \sqrt[4]{64a^4b^5}$

For exercises 29–31, simplify.

29. $\sqrt{15xy} \cdot \sqrt{10x}$
30. $\sqrt{21ab} \cdot \sqrt{42ab}$
31. $\sqrt[3]{2p^2} \cdot \sqrt[3]{2p} \cdot \sqrt[3]{2p}$

For exercises 32–34, simplify.

32. $3\sqrt{c}(\sqrt{4c} + 15c)$
33. $(\sqrt{h} - \sqrt{5})(\sqrt{h} + \sqrt{5})$
34. $(\sqrt{2a} - 3)(\sqrt{6a} - 5)$

Ask Yourself	Test Yourself	Help Yourself
Can I . . .	Do 6.3 Review Exercises	See these Examples and Practice Problems
write the quotient rule of radicals?	35–37, 42	Ex. 1, 2, 7, PP 1–4, 8, 9
simplify a radical in which the radicand is a quotient?		
identify an expression in which the denominator is not rational?	38	Ex. 4–6, 9–11, PP 5–12
rationalize the denominator of an expression in which the denominator is a radical or the product of a number and a radical?	40, 42–44	Ex. 4–7, 10, PP 5–9
identify the conjugate of an expression?	39	Ex. 8–11, PP 10–12
rationalize the denominator of an expression in which the denominator is a sum or difference?	41	Ex. 9–11 PP 10–12

6.3 Review Exercises

For exercises 35–37, simplify.

35. $\sqrt{\dfrac{x^5}{4z^6}}$

36. $\sqrt{\dfrac{a^4}{b^{12}}}$

37. $\sqrt{\dfrac{98w^5}{81p^2}}$

38. Choose the expression in which the denominator is rationalized.

 a. $\dfrac{3}{\sqrt{x}}$ b. $\dfrac{\sqrt{x}}{\sqrt{x}+3}$

 c. $\dfrac{\sqrt{x}+3}{x-\sqrt{3}}$ d. $\dfrac{\sqrt{x}+3}{x}$

39. What is the conjugate of $\sqrt{x}-\sqrt{5}$?

For exercises 40–44, simplify. The denominator must be rational.

40. $\dfrac{\sqrt{2p}+3}{\sqrt{5p}}$

41. $\dfrac{\sqrt{5p}}{\sqrt{2p}+3}$

42. $\sqrt{\dfrac{50x^4y^2}{20xy}}$

43. $\dfrac{5x}{\sqrt[3]{6x}}$

44. $\dfrac{5x}{\sqrt[4]{6x}}$

Ask Yourself	Test Yourself	Help Yourself
Can I . . .	Do 6.4 Review Exercises	See these Examples and Practice Problems
rewrite a radical expression as an expression with rational exponents?	45, 47, 59–61	Ex. 2, 3, PP 1–3
rewrite an expression with rational exponents as a radical?	46	Ex. 4, 7, PP 4, 5, 9–12
write the product rule of exponents, the quotient rule of exponents, and the power rule of exponents?	47–50, 52–58	Ex. 5–12 , PP 6–16
simplify an expression that includes rational exponents?		
find a common denominator for two rational exponents and rewrite the exponents with this denominator?	51	Ex. 6, 10–12 PP 6–8, 13–18
simplify a radical expression by rewriting it as an expression with rational exponents?	59–61	Ex. 8, 12 PP 12, 17, 18

6.4 Review Exercises

45. Rewrite $\sqrt[11]{d^5}$ as an exponential expression.

46. Rewrite $a^{\frac{23}{24}}$ as a radical.

47. Rewrite $\sqrt[3]{x^5 y^{12}}$ as an exponential expression and simplify.

48. The expression $a^{\frac{12}{10}}$ is not simplified. Explain why.

49. The expression $a^{\frac{3}{4}} b^{-\frac{1}{8}}$ is not simplified. Explain why.

50. The expression $a^{\frac{5}{4}}$ is simplified. The equivalent expression $\sqrt[4]{a^5}$ is not simplified. Explain why.

51. Fractions can be added, subtracted, multiplied, and divided. For which of these operations do you need to find a common denominator?

For exercises 52–58, simplify.

52. $a^{\frac{1}{8}} a^{\frac{1}{8}} a^{\frac{5}{8}}$

53. $x^{\frac{1}{4}} x^{\frac{2}{5}}$

54. $\left(a^{\frac{2}{3}}\right)^{\frac{3}{5}}$

55. $\dfrac{c^{\frac{3}{4}}}{c^{\frac{2}{9}}}$

56. $\dfrac{4x^{\frac{3}{8}}}{36x^{\frac{1}{2}}}$

57. $\dfrac{12x^{\frac{1}{2}} z^{\frac{2}{5}}}{80xz^{\frac{3}{7}}}$

58. $\left(\dfrac{a^{\frac{2}{3}} b}{ab^{\frac{1}{4}}}\right)^{\frac{1}{2}}$

For exercises 59–61,
(a) rewrite the radical as an exponential expression.
(b) simplify. Leave the final answer in exponential notation.

59. $\sqrt{\dfrac{x^{14}}{z^{11}}}$

60. $\dfrac{\sqrt[4]{p^2 w}}{\sqrt{pw}}$

61. $\sqrt[3]{\sqrt{x^3 y^4}}$

SECTION 6.5   **Radical Equations**

Ask Yourself	Test Yourself	Help Yourself
Can I . . .	Do 6.5 Review Exercises	See these Examples and Practice Problems
solve a radical equation that includes one radical?	66–70	Ex. 1–4, PP 1–5
identify an extraneous solution of a radical equation?	62–65, 67, 70	Ex. 3–5, PP 3, 5
solve a radical equation that includes more than one radical?	71, 72	Ex. 5, PP 6, 7
solve an equation that includes rational exponents?	73, 74	Ex. 6, PP 8, 9
use a radical equation to solve an application problem?	75	Ex. 7, PP 10, 11

6.5 Review Exercises

62. Describe how to identify an extraneous solution of a radical equation.

63. A student solved $x + 4 = \sqrt{38 - x}$ and said that the solutions were $x = 2$ and $x = -11$. Explain why $x = -11$ is an extraneous solution.

64. a. Can a radical equation with one cube root have an extraneous solution?
b. Explain why or why not.

65. Explain why the equation $\sqrt{3x - 4} = -5$ has no solution.

For exercises 66–74,
(a) solve.
(b) check.

66. $\sqrt{2x + 9} = 8$

67. $x + 3 = \sqrt{-2x - 7}$

68. $\sqrt[3]{x + 7} = 4$

69. $\sqrt[3]{x + 6} = -2$

70. $x - 3 = \sqrt{10x + 89}$

71. $\sqrt{2k + 3} - \sqrt{k + 1} = 1$

72. $\sqrt{y^2 + 8} = 2\sqrt{2y - 1}$

73. $(5n - 4)^{\frac{1}{2}} - 2 = 9$

74. $(x + 7)^{\frac{1}{2}} - 1 = (x + 9)^{\frac{1}{2}}$

75. The volume swept by the piston in an engine cylinder is called its *displacement*. Displacement depends on the *bore* (diameter) of the cylinder and the *stroke* (vertical distance of the piston). If B is the bore (in.), D is the displacement (in.^{3}), S is the stroke (in.), and C is the number of cylinders in the engine, then $B = \sqrt{\dfrac{4D}{SC\pi}}$. Find D for an eight-cylinder engine with a bore of 4.312 in. and a stroke of 3.50 in. Round to the nearest whole number.

SECTION 6.6 Radical Functions

Ask Yourself	Test Yourself	Help Yourself
Can I . . .	**Do 6.6 Review Exercises**	**See these Examples and Practice Problems**
use a table of ordered pairs to graph a radical function?	76, 77	Ex. 1, 2, PP 1–3
use translation to graph a radical function?	78, 79, 81	Ex. 3, 4, PP 4–6
find the domain of a radical function?	86, 87	Ex. 5–7, PP 7–11
evaluate a radical function?	82	Ex. 8, 9, PP 11–13
identify the domain or range of a radical function, given its graph?	80, 84, 85	Ex. 10, PP 14–16
use an algebraic method to find a real zero of a radical function?	83	Ex. 11, 12, PP 17, 18
estimate any real zeros of a radical function, given its graph?	79, 80	Ex. 11, PP 19–21

6.6 Review Exercises

For exercises 76–77, use a table of ordered pairs to graph the function.

76. $y = \sqrt{x} + 4$

77. $f(x) = \sqrt[3]{x} - 2$

78. The graph represents the function $y = \sqrt{x} + 2$. Use this graph and translation to graph $y = \sqrt{x} + 1$.

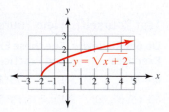

79. The graph represents the function $y = -\sqrt{x} + 2$.
   a. Use a graphical method to find any real zeros of this function.
   b. Use this graph and translation to graph $y = -\sqrt{x} + 3$.

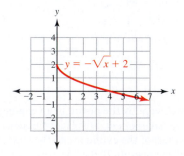

80. a. Explain why the graph at the top of the next column shows that the function $f(x) = \sqrt{x} + 1$ does not have a real zero.
   b. Use interval notation to represent the domain of this function.
   c. Use interval notation to represent the range of this function.

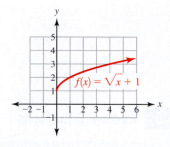

81. A radical function is $y = \sqrt{x} + 5$. Describe the difference between the graph of this function and the graph of the function $y = \sqrt{x} + 8$.

82. A boat anchored in 32 ft of water has a swinging radius that depends on the length of the anchor line. The function $R = 35 + \sqrt{L^2 - 37^2}$ represents the relationship between the swinging radius, R, in feet and the length of the anchor line, L, in feet. Find the swinging radius if the length of the anchor line is 100 ft. Round to the nearest whole number.

For exercises 83–85, the function represented by the graph is $g(x) = \sqrt{12x + 24} - 6$.

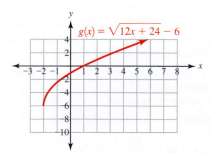

83. Use an algebraic method to find any real zeros of this function.

84. Use interval notation to represent the domain of this function.

85. Use interval notation to represent the range of this function.

For exercises 86–87, use an algebraic method to find the domain of the function, and use interval notation to represent it.

86. $f(x) = \sqrt{2x - 14}$

87. $g(x) = \sqrt{-2x + 8} - 15$

88. a. The notation for the principal square root, $\sqrt{\ }$, represents only the nonnegative square root of a real number. For example, $\sqrt{16} = 4$. If the notation $\sqrt{\ }$ represented both the positive and the negative square root of a real root, would $f(x) = \sqrt{x}$ be a function?

b. Explain why or why not.

Chapter 6 Test

Assume that all variables or variable expressions in radicands and all variable bases in exponential expressions represent nonnegative real numbers. Use interval notation to represent domains and ranges. All exponents should be positive and in lowest terms. All denominators should be rational.

For exercises 1–16, simplify.

1. $\sqrt{162a^6b^5c}$

2. $\sqrt[3]{72x^{10}y^9z^2}$

3. $\sqrt{\dfrac{75x^3y^5}{49x^8}}$

4. $8\sqrt{11} + 3\sqrt{99} - \sqrt{11}$

5. $\sqrt{180} + \sqrt{320}$

6. $\sqrt{42d}\,\sqrt{21cd}$

7. $\sqrt{3w}(\sqrt{6w} - \sqrt{5})$

8. $(5 - \sqrt{6})(\sqrt{15} + \sqrt{21})$

9. $\sqrt[5]{256x^9y^5}$

10. $\sqrt[51]{x^{51}}$

11. $x^{\frac{1}{9}}x^{\frac{5}{9}}$

12. $x^{\frac{4}{9}}x^{\frac{2}{5}}$

13. $\left(a^{\frac{2}{9}}\right)^{\frac{3}{8}}$

14. $\dfrac{x^{\frac{3}{4}}y^{\frac{8}{9}}}{x^{\frac{1}{3}}y^{\frac{2}{5}}}$

15. $\dfrac{2}{\sqrt{x} + 5}$

16. $\sqrt{\dfrac{32a^5}{3}}$

17. Rewrite $\sqrt[9]{x^2}$ as an exponential expression.

18. Rewrite $k^{\frac{3}{7}}$ as a radical.

19. Rationalize the denominator:

a. $\dfrac{5\sqrt{a} - 2}{4\sqrt{z}}$

b. $\dfrac{3\sqrt{p}}{\sqrt{p} - \sqrt{w}}$

20. Use an algebraic method to find the domain, and use interval notation to represent it.

a. $f(x) = \sqrt{x + 7}$

b. $f(x) = \sqrt{-6x - 30}$

21. a. Solve: $x + 3 = \sqrt{16x - 12}$

b. Check.

22. The graph represents $f(x) = \sqrt{x + 4} - 2$.

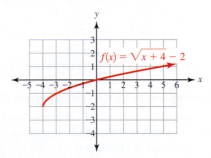

a. Estimate any real zeros of this function.

b. Use interval notation to represent the domain.

c. Use interval notation to represent the range.

d. Use this graph and translation to graph $f(x) = \sqrt{x + 4} + 1$.

23. Use an algebraic method to find any real zeros of $y = \sqrt{x + 2} - 6$.

24. The function $P = \dfrac{1}{0.65}\sqrt{\dfrac{T}{3.83 \times 10^{-4}}}$ represents the relationship of the tension of a violin string, T, in newtons and the pitch produced by the string, P, in hertz. Find the pitch when the tension is 70 N. Round to the nearest whole number.

Cumulative Review Chapters 4–6

Follow your instructor's guidelines for showing your work.

1. Use graphing to solve $\begin{array}{l} y = -2x + 2 \\ 2x + 3y = -6 \end{array}$.

2. **a.** Use substitution to solve $\begin{array}{l} y = 4x - 7 \\ 6x + 5y = -22 \end{array}$.

   b. Check.

3. Use elimination to solve $\begin{array}{l} 9x + 7y = 5 \\ 4x + 3y = 1 \end{array}$.

4. A system of two linear equations in two variables has no solution. Describe its graph.

5. A system of two linear equations in two variables has infinitely many solutions. Describe its graph.

6. **a.** Use elimination or substitution to solve

   $2x + 3y + z = 14$

   $x - 4y + z = -6$.

   $3x + 4y + z = 10$

   b. Check.

7. Solve $\begin{array}{l} 3x + 5y = 42 \\ 2x - y = 2 \end{array}$ by rewriting a matrix in reduced row echelon form.

8. **a.** Graph: $\begin{array}{l} y \geq 2x - 6 \\ y \leq -x + 6 \\ x \geq 0 \end{array}$

   b. Use the graph to estimate the vertices of the solution region.

For exercises 9–10, use the five steps from Section 1.5.

9. Find the amount of steel that is 12% nickel and the amount of steel that is 17% nickel to combine to make 2500 lb of steel that is 14% nickel.

10. A gardener is adding a total of 50 lb of sand and compost to a vegetable garden. She wants three times as much compost as sand. Find out how much sand and how much compost to add to the garden.

For exercises 11–15, simplify. Assume that all expressions in these exercises are defined. No variable represents a number that will cause division by zero.

11. $\dfrac{x^3 - 16x}{2x^2 - 5x - 12}$

12. $\dfrac{8 - 2a^2}{2a^2 - 10a + 12}$

13. $\dfrac{36a^2 + 12a + 1}{18a^2 + 15a + 2} \div \dfrac{6a^2 - 17a - 3}{3a^2 - 16a - 12}$

14. $\dfrac{1}{x - 1} + \dfrac{x}{x^2 - 1}$

15. $\dfrac{n}{n^2 + 4n + 3} - \dfrac{3}{n^2 - 4n - 5}$

For exercises 16–25, simplify. Assume that all factors in radicands are nonnegative.

16. $\sqrt{240a^3bc^2}$

17. $\sqrt{242a} + \sqrt{98a} - \sqrt{8a}$

18. $\sqrt{\dfrac{128x}{25y}}$

19. $(4\sqrt{x} - \sqrt{5})(4\sqrt{x} + \sqrt{5})$

20. $\sqrt{\dfrac{27x^3}{32y^3}}$

21. $\dfrac{\sqrt{8}}{\sqrt{n} - \sqrt{6}}$

22. $\sqrt[3]{54x^5y^6}$

23. $\sqrt[4]{48h^5k^4}$

24. $\dfrac{8w^{\frac{7}{9}}}{36w^{\frac{2}{5}}}$

25. $\left(9x^{\frac{1}{2}}y^{\frac{3}{4}}\right)^2$

26. Rewrite $\sqrt[11]{x^{10}}$ as an exponential expression.

27. Rewrite $x^{\frac{4}{5}}$ as a radical.

28. What is the conjugate of $\sqrt{x} + \sqrt{3}$?

29. Describe the difference between a rational number and an irrational number.

30. Use long division to find the quotient: $(x^2 - 6x + 8) \div (x - 4)$.

31. Use synthetic division to find the quotient: $(x^2 + 7x + 12) \div (x + 3)$.

For exercises 32–36,
(a) solve.
(b) check.

32. $\dfrac{1}{x} - \dfrac{4}{x^2} = \dfrac{1}{25}$

33. $\dfrac{4}{x - 4} = \dfrac{x}{x - 4} - \dfrac{4}{3}$

34. $\sqrt{4x - 20} = 10$

35. $x - 2 = \sqrt{2x - 1}$

36. $\sqrt{2x - 4} = 1 + \sqrt{x - 1}$

37. **a.** Solve: $\dfrac{2}{x^2 + 6x + 8} < 0$

   b. Use interval notation and a graph to represent the solution.

38. Solve $\dfrac{3}{x} + \dfrac{2}{y} = \dfrac{5}{z}$ for y.

39. A formula is $4AZ = \dfrac{6K}{11B}$. Assume that A and K are constant.

    a. Is the relationship between Z and B a direct variation or an inverse variation?

    b. Explain.

40. The relationship between the cost of a product, x, and the sales tax owed, y, is a direct variation. A product costs $470. The sales tax in Lewiston, Idaho, on this product is $30.55.

    a. Find k, the constant of proportionality.

    b. Write a function that represents this direct variation.

    c. Evaluate this function to find the sales tax on a product that costs $850.

    d. What does k represent in this function?

41. In the formula $d = \sqrt{(x_1 - x_2)^2 + (y_1 - y_2)^2}$, d is the distance between two points (x_1, y_1) and (x_2, y_2). Find the distance between $(5, -6)$ and $(2, -10)$.

For exercises 42–43, use the five steps.

42. According to a Rasmussen Report national telephone survey, three out of four Americans trust their own judgment more on economic issues facing the nation than the judgment of the average member of Congress. Find the number out of 1,250,000 Americans who trust their own judgment more. (*Source:* www.rasmussenreports.com, June 25, 2009)

43. A security officer can open up all the classrooms in a large building in 35 min. A different security officer can do the same job in 28 min. Find the time it will take them to do this job working together. Assume that the rate of work does not change when the officers work together. Round to the nearest whole number.

For exercises 44–46,
(a) use interval notation to represent the domain.
(b) write the equation of any vertical asymptotes.
(c) identify the location of any holes on the graph.

44. $g(x) = \dfrac{4x}{x - 9}$

45. $f(x) = \dfrac{x^2 + 2x - 15}{x^2 - 9}$

46. $y = \dfrac{3}{x^2 + 1}$

47. Evaluate $f(-2)$ when $f(x) = \dfrac{x - 4}{x^2 - 6x + 8}$.

48. Use an algebraic method to find any real zeros of

$$g(x) = \dfrac{x^2 + x - 20}{x + 3}.$$

For exercises 49–50,
(a) use a table of ordered pairs to graph the function.
(b) use interval notation to represent the domain.
(c) use interval notation to represent the range.
(d) use an algebraic method to find any real zeros.

49. $f(x) = \sqrt{x} + 2$

50. $g(x) = \sqrt{x} - 3$

51. For a regular cylinder with a volume of 16 in.^{3}, the function $R = \sqrt{\dfrac{16 \text{ in.}^3}{\pi h}}$ represents the relationship of the height, h, and the radius, R. Find the radius of a regular cylinder with a height of 12 in. Round to the nearest hundredth.

Quadratic Expressions, Equations, and Functions

7

Planning for the Future

One or more mathematics courses are required for most degrees or certificates. Mathematics courses often have prerequisites which may be other math courses, a high enough score on a placement test, or placement by an instructor. Different programs require different courses in mathematics. Intermediate Algebra may be the last math course you have to take, or it may be a prerequisite for your next course.

If you are earning an associate's or a bachelor's degree, you may need to complete a group of courses in general education, sometimes called the General Education Core. One or more courses in mathematics are often required to complete the General Education Core.

To reach your college and career goals or if you are undecided about your future, you need to register for the right courses in the correct sequence. It is important to talk to an advisor before registration begins. If you are undecided about your major or career goal, some math classes allow you more career options than others. If you know your career goal, your advisor can help you choose the math course or courses that are required by your program and, if necessary, that fulfill the math course requirement in the General Education Core. To be successful in the future, it is important to look and plan ahead.

© Brittany Courville/Shutterstock

A power plant produces alternating current (AC) electricity. To describe alternating current, we use the number i. We also need this number to find the solution of many polynomial equations. In this section, we will learn how to do arithmetic with i.

Complex Numbers

After reading the text, working the practice problems, and completing assigned exercises, you should be able to:

1. Simplify a square root or cube root in which the radicand is less than zero.

2. Multiply square roots when at least one of the square roots has a radicand that is less than zero.

3. Identify whether a number is an element of the set of complex numbers, imaginary numbers, real numbers, rational numbers, irrational numbers, integers, or whole numbers.

4. Add, subtract, or multiply complex numbers.

5. Write a complex number in $a + bi$ form.

6. Simplify i^p where p is a nonzero whole number.

The Number i

The square root of -1 is not a *real* number. However, it is a number. Mathematicians define the value of i^2 as $i^2 = -1$. So $i = \sqrt{-1}$. The set of **imaginary numbers** includes i. Other imaginary numbers are multiples of i such as $3i$. In daily life, *imaginary* often means "make-believe" or "pretend." But imaginary numbers are not make-believe. They exist.

To simplify a square root in which the radicand is negative, rewrite it as a product of a square root and $\sqrt{-1}$. Then $\sqrt{-1}$ simplifies to i.

EXAMPLE 1 Simplify: $\sqrt{-2}$

SOLUTION ▶ $\qquad \sqrt{-2}$

$\qquad = \sqrt{2}\,\sqrt{-1}$ Rewrite as the product of a square root and $\sqrt{-1}$.

$\qquad = \sqrt{2}\,i$ Simplify. This is an imaginary number; $\sqrt{-1} = i$

Using the commutative property, we can rewrite this expression as $i\sqrt{2}$. Ask your instructor whether you should write this expression as $\sqrt{2}i$ or as $i\sqrt{2}$. If you write $\sqrt{2}i$, be sure that the i is *not* under the radical sign.

In simplifying square roots, the product rule of radicals applies only to radicands that are greater than or equal to 0. So if the radicand is negative, the first step in simplifying is to rewrite it as a product of a square root and $\sqrt{-1}$.

> ### Product Rule of Radicals
> If a and b are *real numbers greater than or equal to 0* and n is an even whole number greater than 0, $\sqrt[n]{ab} = \sqrt[n]{a} \cdot \sqrt[n]{b}$ and $\sqrt[n]{a} \cdot \sqrt[n]{b} = \sqrt[n]{ab}$.

EXAMPLE 2 | Simplify: $\sqrt{-32x^2}$

SOLUTION ▶

$\sqrt{-32x^2}$	Rewrite as the product of a square root and $\sqrt{-1}$.
$= \sqrt{32x^2}\sqrt{-1}$	Find perfect square factors: 16, x^2.
$= \sqrt{16}\sqrt{2}\sqrt{x^2}\,i$	Product rule of radicals.
$= 4 \cdot \sqrt{2} \cdot x \cdot i$	Simplify.
$= 4x\sqrt{2}\,i$	Change the order; commutative property.

> ### Simplifying Square Roots
> Assume that all variables represent nonnegative real numbers.
>
> 1. If the radicand is less than 0, rewrite the radical as a product of a square root and $\sqrt{-1}$. Simplify; $\sqrt{-1} = i$.
> 2. Simplify the remaining square root. When a square root is simplified, the radicand has no perfect square factors; every exponent in the radicand is less than 2.
> 3. Combine any remaining radicals into a single radical.

The square root of a negative number is not a real number. However, the cube root of a negative number is a real number. When cube roots are simplified, the result is not an imaginary number.

EXAMPLE 3 | Simplify: $\sqrt[3]{-16x^5y^3}$

SOLUTION ▶

$\sqrt[3]{-16x^5y^3}$	Find perfect cube factors: -8, x^3, y^3.
$= \sqrt[3]{-8}\sqrt[3]{2}\sqrt[3]{x^3}\sqrt[3]{x^2}\sqrt[3]{y^3}$	Product rule of radicals.
$= -2xy\sqrt[3]{2x^2}$	Simplify.

Practice Problems
For problems 1–4, simplify.

1. $\sqrt{-9}$ 2. $\sqrt{-18x^2}$ 3. $\sqrt{-128x^9y^{11}}$ 4. $\sqrt[3]{-128x^9y^{11}}$

Multiplication

When multiplying square roots with negative radicands, first rewrite each square root as the product of a radical and $\sqrt{-1}$.

EXAMPLE 4 Simplify: $\sqrt{15}\sqrt{-10}$

SOLUTION ▶

$\sqrt{15}\sqrt{-10}$

$= \sqrt{15}\sqrt{10}\sqrt{-1}$ Rewrite as the product of a square root and $\sqrt{-1}$.

$= \sqrt{15}\sqrt{10}\cdot i$ Simplify; $\sqrt{-1} = i$

$= \sqrt{15\cdot 10}\,i$ Product rule of radicals.

$= \sqrt{3\cdot 5\cdot 2\cdot 5}\,i$ Find common factors.

$= \sqrt{5^2}\sqrt{3\cdot 2}\,i$ Product rule of radicals.

$= 5\sqrt{6}\,i$ Simplify.

In simplifying square roots, the product rule of radicals applies only to radicands that are greater than or equal to 0. So if the radicand is negative, *the first step in simplifying is always to rewrite it as a product of a square root and $\sqrt{-1}$*. Notice what happens to the sign of the simplified expression below if this step is not done first.

Correct:	**Wrong:**
$\sqrt{-9}\cdot\sqrt{-4}$	$\sqrt{-9}\cdot\sqrt{-4}$
$= \sqrt{9}\sqrt{-1}\sqrt{4}\sqrt{-1}$	$= \sqrt{36}$
$= \sqrt{9}\cdot i\cdot\sqrt{4}\cdot i$	$= 6$ The sign is positive. This answer is not correct.
$= \sqrt{36}\cdot i^2$	
$= \sqrt{36}(-1)$	
$= 6(-1)$	
$= -6$ The sign is negative.	

EXAMPLE 5 Simplify: $\sqrt{-2x}\sqrt{-6x}$

SOLUTION ▶

$\sqrt{-2x}\sqrt{-6x}$

$= \sqrt{2x}\sqrt{-1}\sqrt{6x}\sqrt{-1}$ Rewrite as the product of a square root and $\sqrt{-1}$.

$= \sqrt{2x}\cdot i\cdot\sqrt{6x}\cdot i$ Simplify; $\sqrt{-1} = i$

$= \sqrt{2x}\sqrt{6x}\,i^2$ $i\cdot i = i^2$

$= -1\sqrt{2x}\sqrt{6x}$ Simplify; $i^2 = -1$

$= -1\sqrt{2x\cdot 6x}$ Product rule of radicals.

$= -1\sqrt{2\cdot x\cdot 2\cdot 3\cdot x}$ Find common factors in the radicand.

$= -1\sqrt{2^2}\sqrt{x^2}\sqrt{3}$ Product rule of radicals.

$= -1\cdot 2\cdot x\cdot\sqrt{3}$ Simplify.

$= -2x\sqrt{3}$ Simplify.

In the next example, notice that the product of two square roots with negative radicands is not imaginary.

EXAMPLE 6 Simplify: $\sqrt{-6abc}\sqrt{-15ab}$

SOLUTION ▶

$\sqrt{-6abc}\sqrt{-15ab}$

$= \sqrt{6abc}\sqrt{-1}\sqrt{15ab}\sqrt{-1}$ Rewrite as the product of a square root and $\sqrt{-1}$.

$= \sqrt{6abc}\cdot i\cdot\sqrt{15ab}\cdot i$ Simplify; $\sqrt{-1} = i$

$= \sqrt{6abc}\sqrt{15ab}\cdot i^2$ $i\cdot i = i^2$

$$- -1\sqrt{6abc}\,\sqrt{15ab} \qquad \text{Simplify; } i^2 = -1$$
$$= -1\sqrt{6abc\cdot15ab} \qquad \text{Product rule of radicals.}$$
$$= -1\sqrt{2\cdot3\cdot abc\cdot3\cdot5\cdot ab} \qquad \text{Find common factors.}$$
$$= -1\sqrt{3^2}\,\sqrt{a^2}\,\sqrt{b^2}\,\sqrt{2\cdot5\cdot c} \qquad \text{Product rule of radicals.}$$
$$= -3ab\sqrt{10c} \qquad \text{Simplify.}$$

Practice Problems

For problems 5–8, simplify.

5. $\sqrt{-6}\,\sqrt{-10}$ **6.** $\sqrt{-7a}\,\sqrt{-5b}$ **7.** $\sqrt{-6hk}\cdot\sqrt{-18h^2k}$
8. $\sqrt{-14p^5}\cdot\sqrt{22p^7}$

The Set of Complex Numbers

In Section 1.1, we identified subsets of the real numbers.

The Set of Real Numbers and Subsets

Whole Numbers	$\{0, 1, 2, 3, \ldots\}$
Integers	$\{\ldots, -3, -2, -1, 0, 1, 2, \ldots\}$
Rational Numbers	Numbers that can be written as a fraction in which the numerator and denominator are integers. If a and b are integers and $b \neq 0$, $\dfrac{a}{b}$ is a rational number.
Irrational Numbers	Numbers that cannot be written as a fraction in which the numerator and denominator are integers; nonrepeating, nonterminating decimal numbers
Real Numbers	The rational and irrational numbers

Numbers such as $3i$ are not real numbers; they are elements of the set of imaginary numbers. This set includes the number i and its multiples. Since $\sqrt{-36}$ simplifies to $6i$, it is also an imaginary number.

The sum of a real number and an imaginary number is a **complex number**. The elements in the set of **complex numbers** are numbers that can be written in the form $a + bi$, where a and b are real numbers. Since an imaginary number can be written in the form $0 + bi$, every imaginary number is an element of the set of complex numbers. Since a real number can be written in the form $a + 0i$, every real number is an element of the set of complex numbers.

Although $3 + 8i$ is a complex number, it is not an element of the set of imaginary numbers or of the set of real numbers.

EXAMPLE 7 (a) Rewrite the real number 17 in $a + bi$ form.

SOLUTION ▶ 17

$= 17 + 0i$

(b) Rewrite the imaginary number $4i$ in $a + bi$ form.

▶ $4i$

$= 0 + 4i$

We can visualize the set of complex numbers as a set with many subsets (Figure 1). Each set that is inside the rectangle of another set is a subset of that set. If the rectangles of two sets do not overlap, the sets are **disjoint** and have no elements in common. In the diagram, the size of the rectangle does not imply the number of elements in the set.

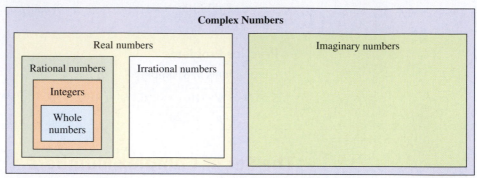

Figure 1

EXAMPLE 8 Put a checkmark or X in the box if the number is an element of the set at the top of the column.

SOLUTION ▶

	Complex numbers	Imaginary numbers	Real numbers	Rational numbers	Irrational numbers	Integers	Whole numbers
3	X		X	X		X	X
$\sqrt{16}$	X		X	X		X	X
-7	X		X	X		X	
$\sqrt{2}$	X		X		X		
π	X		X		X		
$\dfrac{5}{8}$	X		X	X			
0.1	X		X	X			
$4i$	X	X					
$\sqrt{-64}$	X	X					
$7 + 2i$	X						

- The set of whole numbers is $\{0, 1, 2, 3, \ldots\}$, so 3 and $\sqrt{16}$ (which simplifies to 4) are whole numbers.
- The set of integers is $\{\ldots, -3, -2, 1, 0, 1, \ldots\}$, so 3, $\sqrt{16}$, and -7 are integers.
- An irrational number cannot be written as a fraction in which the numerator and denominator are integers; $\sqrt{2}$ and π are irrational numbers.
- A rational number can be written as a fraction in which the numerator and denominator are integers so 3, $\sqrt{16}$, -7, $\dfrac{5}{8}$, and 0.1 are rational numbers.
- The real numbers include the rational and irrational numbers and can be written in the form $a + 0i$ so 3, $\sqrt{16}$, -7, $\dfrac{5}{8}$, 0.1, $\sqrt{2}$, and π are real numbers.
- The imaginary numbers can be written in the form $0 + bi$, so $4i$ and $\sqrt{-64}$ (which simplifies to $8i$) are imaginary numbers.
- Since all of the numbers can be written in the form $a + bi$, they are all complex numbers.

Practice Problems

9. Put a checkmark or X in the box if the number is an element of the set at the top of the column.

	Complex numbers	Imaginary numbers	Real numbers	Rational numbers	Irrational numbers	Integers	Whole numbers
$6 + 2i$							
6							
$\sqrt{6}$							
$6i$							
-6							
2							
$\sqrt{25}$							
$\sqrt{-25}$							

Arithmetic with Complex Numbers

To add two complex numbers, add the real numbers and add the imaginary numbers. The sum is in $a + bi$ form.

EXAMPLE 9 Simplify: $(6 + 2i) + (10 - 8i)$

SOLUTION ▶ $(6 + 2i) + (10 - 8i)$

$= 6 + 10 + 2i - 8i$ Commutative property.

$= 16 - 6i$ Simplify; $6 + 10 = 16$; $2i - 8i = -6i$

In physics classes, students may analyze alternating current flowing through a circuit. They use complex numbers to describe the voltage of the current. The magnitude of the voltage is the real part of the complex number. The phase shift of the voltage is the imaginary part of the complex number.

EXAMPLE 10 Simplify: $(9.8298 + 6.8829i)$ volts $+ (9.6442 - 8.7735i)$ volts

SOLUTION ▶ $(9.8298 + 6.8829i)$ volts $+ (9.6442 - 8.7735i)$ volts

$= [(9.8298 + 9.6442) + (6.8829 - 8.7735)i]$ **volts**   Commutative property.

$= (\mathbf{19.4740 - 1.8906}i)$ volts   Simplify.

When subtracting complex numbers, use the distributive property.

EXAMPLE 11 Simplify: $(6 + 2i) - (10 - 8i)$

SOLUTION ▶ $(6 + 2i) - (10 - 8i)$

$= (6 + 2i) - \mathbf{1}(10 - 8i)$ Multiply by 1; $-(10 - 8i) = -1(10 - 8i)$

$= 6 + 2i - \mathbf{1}(10) - \mathbf{1}(-8i)$ Distributive property.

$= 6 + 2i - \mathbf{10 + 8}i$ Simplify.

$= -4 + 10i$ Simplify; $6 - 10 = -4$; $2i + 8i = 10i$

The product of complex numbers may include i^2. Simplify by rewriting i^2 as -1.

EXAMPLE 12 | Simplify: $3i(7 - 5i)$

SOLUTION ▶

$$3i(7 - 5i)$$

$= \mathbf{3i}(7) + \mathbf{3i}(-5i)$	Distributive property.
$= 21i - 15i^2$	Simplify.
$= 21i - 15(-1)$	Simplify; $i^2 = -1$
$= 21i + \mathbf{15}$	Simplify.
$= 15 + 21i$	Write in $a + bi$ form.

In Example 1, the simplified expression is $\sqrt{2}\,i$. In the next example, the simplified expression is $47 + 16i$. These are complex numbers but they are not real numbers. We often call them nonreal numbers.

EXAMPLE 13 | Simplify: $(6 - 7i)(2 + 5i)$

SOLUTION ▶

$$(6 - 7i)(2 + 5i)$$

$= \mathbf{6}(2 + 5i) - \mathbf{7i}(2 + 5i)$	Distributive property.
$= \mathbf{6}(2) + \mathbf{6}(5i) - \mathbf{7i}(2) - \mathbf{7i}(5i)$	Distributive property.
$= 12 + 30i - 14i - 35i^2$	Simplify.
$= 12 + \mathbf{16i} - 35(-1)$	Simplify; $i^2 = -1$
$= 12 + 16i + \mathbf{35}$	Simplify.
$= \mathbf{47} + 16i$	Simplify; write in $a + bi$ form.

In Chapter 5, we learned that the conjugate of $a + b$ is $a - b$. When we multiply an expression that includes a radical by its conjugate, the product does not include any radicals. When we multiply a complex number by its conjugate, the product does not include any imaginary numbers. It is a real number.

> ### Complex Conjugates
>
> The complex conjugate of $a + bi$ is $a - bi$. The complex conjugate of $a - bi$ is $a + bi$. The product of a conjugate pair of complex numbers is a real number.

In the next example, we multiply $6 + 2i$ by its complex conjugate, $6 - 2i$. The product is a real number, 40.

EXAMPLE 14 | Simplify: $(6 + 2i)(6 - 2i)$

SOLUTION ▶

$$(6 + 2i)(6 - 2i)$$

$= \mathbf{6}(6 - 2i) + \mathbf{2i}(6 - 2i)$	Distributive property.
$= \mathbf{6}(6) + \mathbf{6}(-2i) + \mathbf{2i}(6) + \mathbf{2i}(-2i)$	Distributive property.
$= 36 - 12i + 12i - 4i^2$	Simplify.
$= 36 - \mathbf{0} - 4(-1)$	Simplify; $i^2 = -1$
$= 36 + \mathbf{4}$	Simplify.
$= 40$	Simplify. The product is a real number.

Practice Problems

For problems 10–14, simplify.

10. $(7 - 5i) + (3 - 8i)$ **11.** $(7 - 5i) - (3 - 8i)$ **12.** $5i(3 - 8i)$
13. $(7 - 5i)(3 - 8i)$ **14.** $(7 - 5i)(7 + 5i)$
15. Identify the complex conjugate of $9 + 2i$.
16. Identify the complex conjugate of $4 - 6i$.

Writing a Complex Number in $a + bi$ Form

We can write a fraction as a sum of two fractions. We can also write a complex number such as $\dfrac{3 + 5i}{11}$ as a sum of two fractions. We do this so that the complex number is in $a + bi$ form.

$$\frac{8}{11}$$

$$= \frac{3 + 5}{11}$$

$$= \frac{3}{11} + \frac{5}{11}$$

$$\frac{3 + 5i}{11}$$

$$= \frac{3}{11} + \frac{5i}{11} \qquad \text{Rewrite as the sum of two fractions.}$$

$$= \frac{3}{11} + \frac{5}{11}i \qquad \text{Rewrite in } a + bi \text{ form.}$$

EXAMPLE 15 | Rewrite $\dfrac{6 + 2i}{15}$ in $a + bi$ form.

SOLUTION ▶

$$\frac{6 + 2i}{15} \qquad \text{This number is not in } a + bi \text{ form.}$$

$$= \frac{6}{15} + \frac{2}{15}i \qquad \text{Rewrite the number as two fractions.}$$

$$= \frac{2 \cdot 3}{5 \cdot 3} + \frac{2}{15}i \qquad \text{Find common factors.}$$

$$= \frac{2}{5} + \frac{2}{15}i \qquad \text{Simplify; this number is in } a + bi \text{ form.}$$

In the next example, the number i is in the denominator. To rewrite this number in $a + bi$ form, we multiply by a fraction that is equal to 1, $\dfrac{i}{i}$.

EXAMPLE 16 | Rewrite $\dfrac{8}{i}$ in $a + bi$ form.

SOLUTION ▶

$$\frac{8}{i} \qquad \text{This number is not in } a + bi \text{ form.}$$

$$= \frac{8}{i} \cdot \frac{i}{i} \qquad \text{Multiply by a fraction equal to 1; } \frac{i}{i} = 1$$

$$= \frac{8i}{i^2} \qquad \text{Simplify.}$$

$$= \frac{8i}{-1} \qquad \text{Simplify; } i^2 = -1$$

$$= -8i \qquad \text{This is equivalent to } 0 + -8i. \text{ The expression is now in } a + bi \text{ form.}$$

Although we say that the form of a complex number should be $a + bi$, a number in the form $a - bi$ is also acceptable. The expressions are equivalent; $a - bi = a + (-bi)$.

EXAMPLE 17 | Rewrite $\dfrac{2+i}{3i}$ in $a+bi$ form.

SOLUTION ▶

$\dfrac{2+i}{3i}$	This number is not in $a+bi$ form.
$=\dfrac{2+i}{3i}\cdot\dfrac{i}{i}$	Multiply by a fraction equal to 1; $\dfrac{i}{i}=1$
$=\dfrac{2i+i^2}{3i^2}$	Use the distributive property in the numerator.
$=\dfrac{2i+(-1)}{3(-1)}$	Simplify; $i^2=-1$
$=\dfrac{2i-1}{-3}$	The denominator is real, but the number is not in $a+bi$ form.
$=\dfrac{2i}{-3}+\left(\dfrac{-1}{-3}\right)$	Rewrite as two fractions with the same denominator.
$=-\dfrac{2}{3}i+\dfrac{1}{3}$	Simplify; rewrite $\dfrac{2i}{-3}$ as $-\dfrac{2}{3}i$.
$=\dfrac{1}{3}-\dfrac{2}{3}i$	Change the order to write in $a+bi$ form.

In the next example, the denominator of the expression is $6+2i$. To rewrite the expression in $a+bi$ form, we multiply by a fraction that is equal to 1, $\dfrac{6-2i}{6-2i}$. The conjugate of $6+2i$ is $6-2i$.

EXAMPLE 18 | Rewrite $\dfrac{5}{6+2i}$ in $a+bi$ form.

SOLUTION ▶

$\dfrac{5}{6+2i}$	This number is not in $a+bi$ form.
$=\dfrac{5}{(6+2i)}\cdot\dfrac{(6-2i)}{(6-2i)}$	$6-2i$ is the conjugate of $6+2i$; $\dfrac{6-2i}{6-2i}=1$
$=\dfrac{5(6)+5(-2i)}{6(6-2i)+2i(6-2i)}$	Distributive property.
$=\dfrac{30-10i}{6(6)+6(-2i)+2i(6)+2i(-2i)}$	Simplify; distributive property.
$=\dfrac{30-10i}{36-12i+12i-4i^2}$	Simplify.
$=\dfrac{30-10i}{36+0-4(-1)}$	Simplify; $-12i+12i=0$; $i^2=-1$
$=\dfrac{30-10i}{36+4}$	Simplify.
$=\dfrac{30-10i}{40}$	Simplify.
$=\dfrac{30}{40}-\dfrac{10}{40}i$	Rewrite in $a+bi$ form.
$=\dfrac{3\cdot10}{4\cdot10}-\dfrac{1\cdot10}{4\cdot10}i$	Find common factors.
$=\dfrac{3}{4}-\dfrac{1}{4}i$	Simplify; this number is in $a+bi$ form.

Practice Problems

For problems 17–20, rewrite each number in $a + bi$ form.

17. $\dfrac{8 + 9i}{18}$ **18.** $\dfrac{7}{2i}$ **19.** $\dfrac{7}{2 + i}$ **20.** $\dfrac{7}{2 - 3i}$

Powers of i

To simplify an expression such as i^{15}, we need to think about multiplying -1 by itself p times.

What if we multiply (-1) by itself an *even* number of times? The product $(-1)(-1)$ is 1. The product is 1 whenever p is an *even* nonzero whole number: $(-1)^p = 1$.

What if we multiply (-1) by itself an *odd* number of times? The product $(-1)(-1)(-1)$ is -1. The product is -1 whenever p is an *odd* nonzero whole number: $(-1)^p = -1$.

EXAMPLE 19 **(a)** Simplify: $(-1)^4$

SOLUTION ▶
$$(-1)^4 \qquad\qquad \text{The exponent, } p, \text{ is even.}$$
$$= (-1)(-1)(-1)(-1) \quad \text{Rewrite as multiplication.}$$
$$= 1 \qquad\qquad \text{The product is 1.}$$

(b) Simplify: $(-1)^5$

▶
$$(-1)^5 \qquad\qquad \text{The exponent, } p, \text{ is odd.}$$
$$= (-1)(-1)(-1)(-1)(-1) \quad \text{Rewrite as multiplication.}$$
$$= -1 \qquad\qquad \text{The product is } -1.$$

The power rule of exponents is $(x^m)^n = x^{mn}$. If we can write an expression so that $x^m = i^2$, then $x^m = -1$.

EXAMPLE 20 Simplify.

SOLUTION ▶
(a) i^3 **(b)** i^4 **(c)** i^5 **(d)** i^6 **(e)** i^7

$= (i^2)i$ $= (i^2)^2$ $= (i^2)^2 i$ $= (i^2)^3$ $= (i^2)^3 i$

$= -1i$ $= (-1)^2$ $= (-1)^2 i$ $= (-1)^3$ $= (-1)^3 i$

 $= 1$ $= 1i$ $= -1$ $= -1i$

If we continue to simplify i^p, where p is a nonzero whole number, a pattern appears. Simplified, $i^p = i, -i, 1,$ or -1. Each position on the circle in Figure 2 is equal to one of these values. As we move clockwise, p increases, and the values change from i to -1 to $-i$ to 1, and then the pattern repeats again.

i	$i^2 = -1$	$i^3 = -1i$	$i^4 = 1$	
$i^5 = i$	$i^6 = -1$	$i^7 = -1i$	$i^8 = 1$	
$i^9 = i$	$i^{10} = -1$	$i^{11} = -1i$	$i^{12} = 1$	
$i^{13} = i$	$i^{14} = -1$	$i^{15} = -1i$	$i^{16} = 1$	

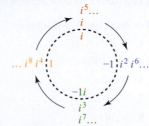

Figure 2

When p is large, use the power rule of exponents, $(x^m)^n = x^{mn}$, to rewrite the expression i^p with a base of i^2.

EXAMPLE 21 (a) Simplify: i^{340}

SOLUTION ▶

i^{340}	340 is an even number; $340 \div 2 = 170$
$= (i^2)^{170}$	Rewrite with a base of i^2; $(2)(170) = 340$
$= (-1)^{170}$	Simplify; $i^2 = -1$
$= 1$	170 is even; -1 raised to an even power $= 1$; $(-1)^{170} = 1$

(b) Simplify: i^{723}

i^{723}	723 is an odd number.
$= i^{722} \cdot i^1$	Rewrite as a product; 722 is even; $722 \div 2 = 361$
$= (i^2)^{361} i$	Rewrite with a base of i^2; $(2)(361) = 722$
$= (-1)^{361} i$	Simplify; $i^2 = -1$
$= -1i$	361 is odd; -1 raised to an odd power $= -1$; $(-1)^{361} = -1$

Practice Problems

For problems 21–26, simplify.

21. i^9 **22.** i^{19} **23.** i^{14} **24.** i^{203} **25.** i^{204} **26.** i^{205}

7.1 VOCABULARY PRACTICE

Match the term with its description.

1. This set of numbers is the union of the rational numbers and the irrational numbers.
2. All numbers in this set can be written as a fraction in which the numerator and denominator are integers.
3. All numbers in this set are nonrepeating, nonterminating decimal numbers.
4. The numbers in the set $\{0, 1, 2, 3, \dots\}$
5. The numbers in the set $\{\dots, -3, -2, -1, 0, 1, 2, \dots\}$
6. The set of all numbers that can be written in the form $a + bi$, where a and b are real numbers
7. $4 + 3i$ and $4 - 3i$ are examples of these
8. The expression under a radical
9. $a(b + c) = ab + ac$
10. This set of numbers only includes i and real number multiples of i.

A. complex conjugates
B. complex numbers
C. distributive property
D. imaginary numbers
E. integers
F. irrational numbers
G. radicand
H. rational numbers
I. real numbers
J. whole numbers

7.1 Exercises

Follow your instructor's guidelines for showing your work. Assume that all variables in radicands in these exercises represent nonnegative real numbers.

For exercises 1–34, simplify.

1. $\sqrt{-25}$

2. $\sqrt{-36}$

3. $\sqrt{-100x}$

4. $\sqrt{-49z}$

5. $\sqrt{-6a^2}$

6. $\sqrt{-10b^2}$

7. $\sqrt{-18a^2b}$

8. $\sqrt{-72u^2z}$

9. $\sqrt{50xy^3}$

10. $\sqrt{98xy^3}$

11. $\sqrt{-50xy^3}$

12. $\sqrt{-98xy^3}$

13. $\sqrt{-14}\sqrt{21}$

14. $\sqrt{-30}\sqrt{50}$

15. $\sqrt{-14}\sqrt{-21}$

16. $\sqrt{-30}\sqrt{-50}$

17. $\sqrt{-3w}\sqrt{6w}$

18. $\sqrt{-5z}\sqrt{10z}$

19. $\sqrt{-3w}\sqrt{-6w}$

20. $\sqrt{-5z}\sqrt{-10z}$

21. $\sqrt{ab}\sqrt{-ab}$

22. $\sqrt{cd}\sqrt{-cd}$

23. $\sqrt{-30ab}\sqrt{-42ab}$

24. $\sqrt{-39cd}\sqrt{-26cd}$

25. $\sqrt{108ab}\sqrt{-15ac}$

26. $\sqrt{125hk}\sqrt{-30hp}$

27. $\sqrt[3]{-32c^4d}$

28. $\sqrt[3]{-40w^5z}$

29. $\sqrt[3]{-32c^4d}\sqrt[3]{-24c^2d}$

30. $\sqrt[3]{-40w^5z}\sqrt[3]{50wz^4}$

31. $\sqrt{3^2 - 4(2)(8)}$

32. $\sqrt{3^2 - 4(8)(1)}$

33. $\sqrt{3^2 - 4(2)(-8)}$

34. $\sqrt{3^2 - 4(8)(-1)}$

35. Explain why the square root of a negative number cannot be a real number.

36. The square root of a negative number is not a real number. However, the cube root of a negative number is a real number. Explain why.

37. Copy the table below. Put a checkmark or X in the box if the number is an element of the set at the top of the column.

38. Copy the table below. Put a checkmark or X in the box if the number is an element of the set at the top of the column.

For exercises 39–82, simplify. The answer must be in $a + bi$ form.

39. $(5 + 3i) + (6 + 9i)$ **40.** $(11 + 8i) + (2 + 4i)$

41. $(5 - 3i) + (6 - 9i)$ **42.** $(11 - 8i) + (2 - 4i)$

43. $(5 - 3i) - (6 - 9i)$ **44.** $(11 - 8i) - (2 - 4i)$

45. $(5 - 3i) - (6 + 9i)$ **46.** $(11 - 8i) - (2 + 4i)$

47. $6(3 - 9i)$ **48.** $8(2 - 5i)$

49. $(8 + 5i)(3 + 6i)$ **50.** $(3 + 2i)(4 + 7i)$

51. $(8 + 5i)(3 - 6i)$ **52.** $(3 + 2i)(4 - 7i)$

53. $(8 - 5i)(3 - 6i)$ **54.** $(3 - 2i)(4 - 7i)$

55. $\dfrac{12 + 9i}{54}$ **56.** $\dfrac{15 + 6i}{63}$

57. $\dfrac{10 - 18i}{90}$ **58.** $\dfrac{12 - 20i}{60}$

59. $\dfrac{5}{i}$ **60.** $\dfrac{3}{i}$

61. $\dfrac{7}{3i}$ **62.** $\dfrac{5}{3i}$

63. $\dfrac{2}{5 + 3i}$ **64.** $\dfrac{3}{6 + 3i}$

65. $\dfrac{6}{3 - 4i}$ **66.** $\dfrac{8}{4 - 5i}$

67. $\dfrac{6i}{3 - 4i}$ **68.** $\dfrac{8i}{4 - 6i}$

69. $\dfrac{2i}{5 + 2i}$ **70.** $\dfrac{3i}{4 + 3i}$

71. $\dfrac{3 + 2i}{5 - 4i}$ **72.** $\dfrac{4 + 3i}{8 - 6i}$

73. $\dfrac{3 + 2i}{6 + 4i}$ **74.** $\dfrac{4 + 3i}{8 + 6i}$

75. i^{15} **76.** i^{23}

77. i^{24} **78.** i^{16}

79. i^{134} **80.** i^{126}

81. i^{333} **82.** i^{341}

83. The voltage drops in an AC circuit are $3.1470 + 17.177i$ volts, $-18.797 + 3.0850i$ volts, and $135.65 - 20.262i$ volts. Find the sum of these voltages.

84. The voltage drops in an AC circuit are $15 + 26.6564i$ volts, $-9.7298 - 6.8829i$ volts, and $9.7442 - 19.7735i$ volts. Find the sum of these voltages.

85. Impedance in an AC circuit is the sum of the resistance (measured in real numbers) and inductive reactance (measured with imaginary numbers). A circuit has two resistors. Each causes 5 ohms of resistance in the circuit. It also has two inductors. Each inductor causes $3.7699i$ ohms of reactance. Find the total impedance in this circuit.

86. An AC circuit has three resistors. Each causes 5 ohms of resistance in the circuit. It also has three inductors. Each inductor causes $3.7699i$ ohms of reactance. Find the total impedance in this circuit.

Table for exercise 37

	Complex numbers	Imaginary numbers	Real numbers	Rational numbers	Irrational numbers	Integers	Whole numbers
$\sqrt{5}$							
$\sqrt{-5}$							
$5i$							
$3 - 7i$							
9							
0.4							

Table for exercise 38

	Complex numbers	Imaginary numbers	Real numbers	Rational numbers	Irrational numbers	Integers	Whole numbers
$15 + i$							
$\sqrt{26}$							
$\sqrt{-16}$							
$\sqrt{16}$							
$\dfrac{3}{11}$							
$-8i$							

Problem Solving: Practice and Review

Follow your instructor's guidelines for using the five steps as outlined in Section 1.5, p. 51.

87. Find the poverty level income for a family of four. Round to the nearest whole number.

Any child at a participating school may purchase a meal through the National School Lunch Program. Children from families with incomes at or below 130 percent of the poverty level are eligible for free meals. Those with incomes between 130 percent and 185 percent of the poverty level are eligible for reduced-price meals, for which students can be charged no more than 40 cents. (For the period July 1, 2010, through June 30, 2011, 130 percent of the poverty level is $28,665 for a family of four; 185 percent is $40,793.) (*Source:* www.fns .usda.gov, fiscal year 2010–2011)

88. Find the amount of donations to the City University of New York in the 2002 fiscal year. Round to the nearest tenth of a million.

At the City University of New York ... donations reached $279 million last year, a 39 percent increase over the previous year and three and a half times the total in the 2002 fiscal year. (*Source:* www.nytimes.com, Feb. 20, 2008)

89. Find the difference in the average cost per person for the 2000 Census and the projected cost per person for the 2010 Census. Round to the nearest hundredth.

Source: www.census.gov

90. The diameter of a circular above-ground pool is 27 ft. The height of the steel wall is 54 in. A family will fill the pool until the water is 4 in. below the top of the steel walls. Find the gallons of water needed to fill the pool (1 gal = 231 in.^{3}; $\pi \approx 3.14$). Round to the nearest hundred.

Technology

For exercises 91–94, use a calculator to simplify. Many graphing calculators do arithmetic with numbers that include i. Look for the symbol i above a key on the calculator.

91. $8i \cdot 4i$ **92.** $6i \cdot 5i$

93. $53i + 21i$ **94.** $71i + 18i$

Find the Mistake

For exercises 95–98, the completed problem has one mistake.

(a) Describe the mistake in words, or copy down the whole problem and highlight or circle the mistake.

(b) Do the problem correctly.

95. Problem: Rewrite $\dfrac{16}{2 + 8i}$ in $a + bi$ form.

Incorrect Answer: $\dfrac{16}{2 + 8i}$

$= \dfrac{16}{2} + \dfrac{16}{8i}$

$= 8 + \dfrac{2}{i}$

$= 8 - 2i$

96. Problem: Simplify: $\sqrt[3]{-81x^4y^6}$

Incorrect Answer: $\sqrt[3]{-81x^4y^6}$

$= \sqrt[3]{81x^4y^6} \sqrt[3]{-1}$

$= \sqrt[3]{81x^4y^6}\,i$

$= \sqrt[3]{27}\,\sqrt[3]{3x}\,\sqrt[3]{x^3}\,\sqrt[3]{y^6}\,i$

$= 3xy^2\sqrt[3]{3x}\,i$

97. Problem: Simplify: $\sqrt{-10xy}\,\sqrt{-15xy}$

Incorrect Answer: $\sqrt{-10xy}\,\sqrt{-15xy}$

$= \sqrt{150x^2y^2}$

$= \sqrt{25}\,\sqrt{6}\,\sqrt{x^2}\,\sqrt{y^2}$

$= 5xy\sqrt{6}$

98. Problem: Simplify: $\dfrac{8}{3 - 7i}$

Incorrect Answer: $\dfrac{8}{3 - 7i}$

$= \dfrac{8}{(3 - 7i)} \cdot \dfrac{(3 + 7i)}{(3 + 7i)}$

$= \dfrac{24 + 56i}{9 + 21i - 21i - 49i^2}$

$= \dfrac{24 + 56i}{9 - 49(-1)}$

$= \dfrac{24 + 56i}{9 - 49}$

$= \dfrac{24 + 56i}{-40}$

$= \dfrac{24}{-40} + \dfrac{56}{-40}\,i$

$= -\dfrac{3}{5} - \dfrac{7}{5}\,i$

Review

99. Write the zero product property.

For exercises 100–102, use the zero product property to solve.

100. $x^2 - 15x + 54 = 0$

101. $4x^2 - 25 = 0$

102. $6x^2 + 17x + 5 = 0$

SUCCESS IN COLLEGE MATHEMATICS

103. Who is your advisor?

104. What is the best way to make an appointment to meet with your advisor?

© Mike Flippo/Shutterstock

In baseball, the distance from third base to first base is a solution of a quadratic equation. In this section, we will learn several different strategies for solving quadratic equations.

SECTION 7.2

Solving Quadratic Equations

After reading the text, working the practice problems, and completing assigned exercises, you should be able to:

1. Solve a quadratic equation in standard form, $ax^2 + bx + c = 0$, in which $b = 0$.
2. Use the Pythagorean theorem to solve application problems.
3. Use the zero product property to solve a quadratic equation.

Solving a Quadratic Equation in One Variable

The standard form of a quadratic equation in one variable is $ax^2 + bx + c = 0$. The coefficients a, b, and c are real numbers. The **lead coefficient**, a, cannot equal 0. The **degree** of a quadratic equation in one variable equals the value of its largest exponent, 2. To solve quadratic equations in which $b = 0$, we use principal square roots. The notation for the principal square root is the square root sign, $\sqrt{}$.

EXAMPLE 1 **(a)** What are the square roots of 25?

SOLUTION ▶ Since $5 \cdot 5 = 25$ and $(-5)(-5) = 25$, the square roots of 25 are 5 and -5.

(b) Evaluate: $\sqrt{25}$

▶ Since the square root notation, $\sqrt{}$, represents the principal square root, $\sqrt{25} = 5$.

In Chapter 6, we assumed that all factors in radicands including variables were greater than or equal to 0. With this assumption, the principal square root of x^2 equals x: $\sqrt{x^2} = x$. However, if the variable represents a negative number, the principal square root is a positive number, and this assumption is not true. For example, when $x = -5$, $\sqrt{(-5)^2}$ equals 5, not -5.

When solving quadratic equations, we cannot assume that the variables represent real numbers that are greater than or equal to 0. So we use the definition of absolute value to simplify a principal square root: $\sqrt{x^2} = |x|$.

> **Absolute Value**
>
> For any real number x,
>
> if $x \geq 0$, then $|x| = x$.
> if $x < 0$, then $|x| = -x$.

We can use $\sqrt{x^2} = |x|$ to solve quadratic equations such as $x^2 = 9$, in which $b = 0$. Begin by taking the square root of each side. Since $\sqrt{x^2} = |x|$, the result is an absolute value equation with two solutions. In the first case, when $x \geq 0$, the solution is x. In the second case, when $x < 0$, the solution is $-x$.

EXAMPLE 2 (a) Solve: $x^2 = 9$

SOLUTION ▶

$$x^2 = 9 \qquad \text{The } ax^2 \text{ term, } 1x^2, \text{ is isolated.}$$
$$\sqrt{x^2} = \sqrt{9} \qquad \text{Take the square root of each side.}$$
$$|x| = 3 \qquad \text{Simplify; } \sqrt{x^2} = |x|$$

To solve absolute value equations, use the definition of absolute value to write two equations: $|x| = x$ or $|x| = -x$. Solve each equation. In this example, $|x| = 3$ is rewritten as two equations: $x = 3$ or $-x = 3$.

First case: $x = 3$ **Second case:** $-x = 3$

$$\frac{-x}{-1} = \frac{3}{-1} \qquad \text{Division property of equality.}$$
$$x = -3 \qquad \text{Simplify.}$$

Two solutions: $x = 3$ or $x = -3$.

(b) Check.

▶

Check: $x = 3$ Check: $x = -3$

$x^2 = 9$ Check in original equation. $x^2 = 9$

$3^2 = 9$ Replace variable with solution. $(-3)^2 = 9$

$9 = 9$ True.   Simplify; the solutions are correct. $9 = 9$ True.

Plus-minus notation $(\pm)$ is another way to represent the solutions of $x^2 = 9$. In this notation, the solutions of $x^2 = 9$ are $x = \pm 3$. To say this notation aloud, say, "x equals 3 or x equals -3" or say, "x equals plus or minus 3."

In solving quadratic equations, we often do not include the work with absolute value and immediately use $\pm$ notation.

EXAMPLE 3 Solve: $w^2 = 27$

SOLUTION ▶

$$w^2 = 27$$
$$\sqrt{w^2} = \sqrt{27} \qquad \text{Take the square root of both sides.}$$
$$w = \pm\sqrt{27} \qquad \text{Simplify; } \pm \text{ notation.}$$
$$w = \pm\sqrt{9}\sqrt{3} \qquad \text{Find perfect square factors; product rule of radicals.}$$
$$w = \pm 3\sqrt{3} \qquad \text{Simplify.}$$

The $\pm$ notation represents two solutions: $w = 3\sqrt{3}$ or $w = -3\sqrt{3}$.

Before taking the square root of both sides, the ax^2 term should be isolated, and the lead coefficient, a, should equal 1.

Solving a Quadratic Equation, $ax^2 + bx + c = 0$, When $b = 0$

1. Use the addition or subtraction property of equality to isolate ax^2.

2. Use the multiplication or division property of equality so that $a = 1$.

3. Take the square root of both sides.

4. Show the work in solving the absolute value equation or use $\pm$ notation.

5. Check each solution in the original equation.

In the next example, to solve the quadratic equation $2x^2 - 12 = 0$, we use the addition property of equality to isolate the ax^2 term and then use the division property of equality so that the lead coefficient, a, equals 1.

EXAMPLE 4   (a) Solve: $2x^2 - 12 = 0$.

SOLUTION ▶

$2x^2 - 12 = 0$	The ax^2 term is not isolated.
$\underline{+12 \ +12}$	Addition property of equality.
$2x^2 + 0 = 12$	Simplify.
$2x^2 = 12$	Simplify; the ax^2 term is isolated.
$\dfrac{2x^2}{2} = \dfrac{12}{2}$	Division property of equality.
$x^2 = 6$	Simplify; $a = 1$
$\sqrt{x^2} = \sqrt{6}$	Take the square root of each side.
$x = \pm\sqrt{6}$	Simplify; use $\pm$ notation.

(b) Check.

▶

Check: $x = \sqrt{6}$

		Check: $x = -\sqrt{6}$
$2x^2 - 12 = 0$	Check in original equation.	$2x^2 - 12 = 0$
$2(\sqrt{6})^2 - 12 = 0$	Replace variable with the solution.	$2(-\sqrt{6})^2 - 12 = 0$
$2(6) - 12 = 0$	Simplify.	$2(6) - 12 = 0$
$12 - 12 = 0$	Simplify.	$12 - 12 = 0$
$0 = 0$ True.	Simplify; the solutions are correct.	$0 = 0$ True.

In the next example, the solutions of the quadratic equation are imaginary numbers.

EXAMPLE 5   (a) Solve: $p^2 + 32 = 0$

SOLUTION ▶

$p^2 + 32 = 0$	We need to isolate p^2.
$\underline{-32 \quad -32}$	Subtraction property of equality.
$p^2 + 0 = -32$	Simplify.
$\sqrt{p^2} = \sqrt{-32}$	Take the square root of both sides.
$p = \pm\sqrt{-32}$	Simplify; use $\pm$ notation.
$p = \pm\sqrt{32}\,\sqrt{-1}$	Rewrite as the product of a radical and $\sqrt{-1}$.
$p = \pm\sqrt{16}\,\sqrt{2}\,i$	Product rule of radicals.
$p = \pm4\sqrt{2}\,i$	Simplify.

(b) Check.

▶

Check: $p = 4\sqrt{2}\,i$

	Check: $p = -4\sqrt{2}\,i$
$p^2 + 32 = 0$	$p^2 + 32 = 0$
$(4\sqrt{2}\,i)^2 + 32 = 0$	$(-4\sqrt{2}\,i)^2 + 32 = 0$
$(4\sqrt{2}\,i)(4\sqrt{2}\,i) + 32 = 0$	$(-4\sqrt{2}\,i)(-4\sqrt{2}\,i) + 32 = 0$
$16 \cdot 2 \cdot i^2 + 32 = 0$	$16 \cdot 2 \cdot i^2 + 32 = 0$
$(32)(-1) + 32 = 0$	$(32)(-1) + 32 = 0$
$-32 + 32 = 0$	$-32 + 32 = 0$
$0 = 0$ True.	$0 = 0$ True.

Both solutions are correct.

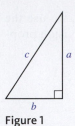

Figure 1

In the next example, we use the Pythagorean theorem for right triangles, $a^2 + b^2 = c^2$. As shown in Figure 1, a **right triangle** has a 90-degree angle. The side opposite the right angle is the **hypotenuse** with length c. The lengths of the other two sides (the **legs**) are a and b. Notice that a, b, and c represent different things in the standard form of a quadratic equation, $ax^2 + bx + c = 0$, and in the Pythagorean theorem, $a^2 + b^2 = c^2$.

EXAMPLE 6 | A baseball diamond has four right angles. The distance from each base to the next is 90 ft. Find the distance from first base to third base. Round to the nearest whole number.

SOLUTION ▶ **Step 1 Understand the problem.**
The unknown is the distance from first base to third base.

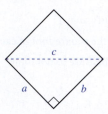

$c =$ distance from first base to third base

Step 2 Make a plan.
Since the bottom half of a baseball diamond forms a right triangle and the distance from first base to third base is the hypotenuse, use the Pythagorean theorem, $a^2 + b^2 = c^2$, to find the length of the hypotenuse.

Step 3 Carry out the plan.

$$a^2 + b^2 = c^2 \qquad \text{The Pythagorean theorem for right triangles.}$$
$$(\mathbf{90\ ft})^2 + (\mathbf{90\ ft})^2 = c^2 \qquad a = 90\ \text{ft};\ b = 90\ \text{ft}$$
$$\mathbf{8100\ ft^2} + \mathbf{8100\ ft^2} = c^2 \qquad \text{Simplify.}$$
$$\mathbf{16{,}200\ ft^2} = c^2 \qquad \text{Simplify; this is a quadratic equation.}$$
$$\sqrt{16{,}200\ \text{ft}^2} = \sqrt{c^2} \qquad \text{Take the square root of both sides.}$$
$$\pm 127.27\ldots\ \text{ft} = c \qquad \text{Simplify; use } \pm \text{ notation.}$$
$$c \approx 127 \quad \text{or} \quad c \approx -127 \qquad \text{Round.}$$

Although the equation $(90\ \text{ft})^2 + (90\ \text{ft})^2 = c^2$ has two solutions, the distance between the bases must be positive. The only solution that is an answer to the problem is $c \approx 127$ ft.

Step 4 Look back.
The distance from third base to first base should be longer than the distance from first base to home plate (90 ft). However, the distance should not be twice as long (180 ft). Since the solution of 127 ft is between 90 ft and 180 ft, it seems reasonable.

Step 5 Report the solution.
The distance from first base to third base is about 127 ft.

Practice Problems

For problems 1–4,
(a) solve each equation.
(b) check.

1. $x^2 = 25$ **2.** $b^2 + 49 = 0$ **3.** $k^2 - 56 = 0$ **4.** $m^2 + 63 = 0$
5. Rewrite $x = \pm 8i$ as two separate solutions.

6. Rewrite $x = \sqrt{5}$ or $x = -\sqrt{5}$ using $\pm$ notation.

7. Football players run wind sprints on a diagonal from one corner to the opposite corner on the other end of the field. Including the end zones, the field is 120 yd long and 160 ft wide. Find the distance that the players run (1 yd = 3 ft). Round to the nearest foot.

The Zero Product Property

In Section 2.4, we used the zero product property to solve quadratic equations with rational solutions.

> **Zero Product Property**
>
> If a and b are real numbers and $a \cdot b = 0$, then $a = 0$ or $b = 0$.

To solve the quadratic equation in the next example, we factor and use the zero product property. The work in factoring by the ac method is not shown. However, you should show any work you do to factor.

EXAMPLE 7   **(a)** Use the zero product property to solve $10x^2 + 7x - 12 = 0$.

SOLUTION ▶

$$10x^2 + 7x - 12 = 0 \qquad a = 10, b = 7, c = -12, ac = -120$$

$$(5x - 4)(2x + 3) = 0 \qquad \text{Factor.}$$

$$5x - 4 = 0 \quad \text{or} \quad 2x + 3 = 0 \qquad \text{Zero product property.}$$

$$\underline{+4 \quad +4} \qquad \qquad \underline{-3 \quad -3} \qquad \text{Properties of equality.}$$

$$5x + 0 = 4 \qquad \qquad 2x + 0 = -3$$

$$\frac{5x}{5} = \frac{4}{5} \qquad \qquad \frac{2x}{2} = \frac{-3}{2} \qquad \text{Division property of equality.}$$

$$x = \frac{4}{5} \quad \text{or} \quad x = -\frac{3}{2} \qquad \text{Simplify.}$$

(b) Check.

Check: $x = \dfrac{4}{5}$ 　　　　　　　 Check: $x = -\dfrac{3}{2}$

$$10x^2 + 7x - 12 = 0 \qquad\qquad 10x^2 + 7x - 12 = 0$$

$$10\left(\frac{4}{5}\right)^2 + \frac{7}{1}\left(\frac{4}{5}\right) - 12 = 0 \qquad\qquad 10\left(-\frac{3}{2}\right)^2 + \frac{7}{1}\left(-\frac{3}{2}\right) - 12 = 0$$

$$\frac{10}{1}\left(\frac{16}{25}\right) + \frac{28}{5} - 12 = 0 \qquad\qquad \frac{10}{1}\left(\frac{9}{4}\right) - \frac{21}{2} - 12 = 0$$

$$\frac{\cancel{5} \cdot 2 \cdot 16}{\cancel{5} \cdot 5} + \frac{28}{5} - 12 = 0 \qquad\qquad \frac{\cancel{2} \cdot 5 \cdot 9}{\cancel{2} \cdot 2} - \frac{21}{2} - 12 = 0$$

$$\frac{32}{5} + \frac{28}{5} - \frac{60}{5} = 0 \qquad\qquad \frac{45}{2} - \frac{21}{2} - \frac{24}{2} = 0$$

$$\frac{60}{5} - \frac{60}{5} = 0 \qquad\qquad \frac{24}{2} - \frac{24}{2} = 0$$

$$0 = 0 \quad \text{True.} \qquad\qquad 0 = 0 \quad \text{True.}$$

Both solutions are correct.

> **Solving a Quadratic Equation When $ax^2 + bx + c$ Can Be Factored**
>
> 1. Write the quadratic equation in standard form, $ax^2 + bx + c = 0$.
> 2. Factor $ax^2 + bx + c$.
> 3. Use the zero product property to write equations in which each factor equals 0.
> 4. Solve these equations.
> 5. Check.

For a triangle with area A, base b, and height h, $A = \frac{1}{2}bh$. If we know the area and the relationship of b and h, we can write a quadratic equation in either b or h.

EXAMPLE 8 The height of a triangle is 2 in. longer than its base. The area of the triangle is 60 in.^{2}. Use a quadratic equation in one variable to find the height and base of the triangle.

SOLUTION ▶ **Step 1 Understand the problem.**
There are two unknowns, the height and the base of the triangle. Since we know the relationship of the base and height, we can assign one unknown for the base and write an expression for the height in this variable.

$$b = \text{base}$$

Step 2 Make a plan.
The height equals the base plus 2 in.: $b + 2$ in. The formula for the area of a triangle is $A = \frac{1}{2}bh$. If we replace h with $b + 2$ in. and A with 60 in.^{2}, the result is a quadratic equation. If this equation can be factored, we can use the zero product property to solve it.

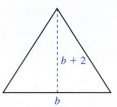

Step 3 Carry out the plan.

$$60 \text{ in.}^2 = \frac{1}{2}b(b + 2 \text{ in.}) \qquad A = \frac{1}{2}bh; \ h = b + 2 \text{ in.}$$

$$60 = \frac{1}{2}b(b + 2) \qquad \text{Remove the units.}$$

$$60 = \frac{1}{2}b \cdot b + \frac{1}{2}b \cdot 2 \qquad \text{Distributive property.}$$

$$60 = \frac{1}{2}b^2 + b \qquad \text{Simplify; this is a quadratic equation.}$$

$$2(60) = 2\left(\frac{1}{2}b^2 + b\right) \qquad \text{Multiplication property of equality; clear the fractions.}$$

$$120 = 2 \cdot \frac{1}{2}b^2 + 2b \qquad \text{Distributive property.}$$

$$120 = b^2 + 2b \qquad \text{Simplify.}$$

$$\frac{-120 \qquad\qquad -120}{0 = b^2 + 2b - 120} \qquad \text{Subtraction property of equality.}$$

$$0 = b^2 + 2b - 120 \qquad \text{Simplify. Standard form; } ax^2 + bx + c = 0$$

$$0 = (b + 12)(b - 10) \qquad \text{Factor.}$$

$$b + 12 = 0 \quad \text{or} \quad b - 10 = 0 \qquad \text{Zero product property.}$$

$$\frac{-12 \quad -12 \qquad\qquad +10 \ +10}{b + 0 = -12 \qquad\qquad b + 0 = 10} \qquad \text{Properties of equality.}$$

$$b + 0 = -12 \qquad\qquad b + 0 = 10 \qquad \text{Simplify.}$$

$$b = -12 \quad \text{or} \qquad b = 10 \qquad \text{Simplify.}$$

Since distance is positive, the only useful solution is $b = 10$ in. Since the height is 2 in. more than the base, $h = 12$ in.

Step 4 Look back.

The product $\frac{1}{2}(10 \text{ in.})(12 \text{ in.})$ is 60 in². Since this is the area of the triangle in the problem, the solution seems reasonable.

Step 5 Report the solution.

The base of the triangle is 10 in. The height is 12 in.

Practice Problems

For problems 8–10,
(a) solve.
(b) check.

8. $x^2 - 19x + 84 = 0$ **9.** $6y^2 - y - 12 = 0$ **10.** $p^2 - 56p = 0$

11. The length of a rectangle is 3 cm more than twice its width. The area of the rectangle is 152 cm². Use a quadratic equation to find the length and width of the rectangle.

12. The height of a triangle is 4 in. longer than its base. The area of the triangle is 48 in.². Use a quadratic equation to find the height and base of the triangle.

7.2 VOCABULARY PRACTICE

Match the term with its description.

1. $ax^2 + bx + c = 0$ when $a \neq 0$

2. $A = \frac{1}{2}bh$

3. The expression under a radical

4. If $ab = 0$, then $a = 0$ or $b = 0$.

5. $\sqrt{x^2}$ is equivalent to this.

6. For a right triangle in which c is the length of the hypotenuse and a and b are the lengths of the other two sides, $a^2 + b^2 = c^2$.

7. The set of numbers that can be written as fractions in which the numerator and denominator are integers

8. The largest exponent in a polynomial in one variable determines this.

9. $A = LW$

10. In a quadratic equation in standard form, a is this.

A. absolute value of x
B. area of a rectangle
C. area of a triangle
D. degree of a polynomial in one variable
E. lead coefficient
F. Pythagorean theorem
G. radicand
H. rational numbers
I. standard form of a quadratic equation in one variable
J. zero product property

7.2 Exercises

Follow your instructor's guidelines for showing your work.

1. What is the standard form of a quadratic equation in one variable?

2. What is the degree of a quadratic equation in one variable?

For exercises 3–8, an equation is given with two solutions. Verify that each solution is correct by doing a check.

3. $x^2 + 8x + 12 = 0$
   a. Check: $x = -6$
   b. Check: $x = -2$

4. $p^2 + 10p - 24 = 0$
   a. Check: $p = 2$
   b. Check: $p = -12$

5. $4z^2 - 4z - 7 = 0$
   a. Check: $z = \frac{1}{2} + \sqrt{2}$
   b. Check: $z = \frac{1}{2} - \sqrt{2}$

6. $2x^2 + 4x - 2 = 0$
   a. $x = -1 + \sqrt{2}$
   b. $x = -1 - \sqrt{2}$

7. $16a^2 - 8a + 13 = 0$
   a. $a = \frac{1}{4} - \frac{\sqrt{3}}{2}i$
   b. $a = \frac{1}{4} + \frac{\sqrt{3}}{2}i$

8. $25w^2 - 20w + 9 = 0$
   a. $w = \frac{2}{5} - \frac{\sqrt{5}}{5}i$
   b. $w = \frac{2}{5} + \frac{\sqrt{5}}{5}i$

For exercises 9–24,
(a) solve.
(b) check.

9. $x^2 = 25$
10. $z^2 = 49$
11. $x^2 = -25$
12. $z^2 = -49$
13. $2x^2 = 108$
14. $2x^2 = 126$
15. $3d^2 = -468$
16. $4z^2 = -592$
17. $h^2 - 36 = 0$
18. $k^2 - 100 = 0$
19. $h^2 + 36 = 0$
20. $k^2 + 100 = 0$
21. $p^2 - 243 = 0$
22. $w^2 - 320 = 0$
23. $x^2 + 735 = 0$
24. $x^2 + 200 = 0$

25. A second story window is 30 ft above the ground. A bush is growing next to the house. Because of this bush, the base of a ladder on the ground must be 12 ft away from the house. Use the Pythagorean theorem to find the

minimum length of ladder needed to reach the window. Round *up* to the nearest foot.

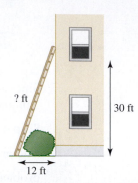

26. In exercise 25, we do not follow the usual rules for rounding. Instead, we round up to the nearest foot. Explain why.

For exercises 27–28, use the diagram.

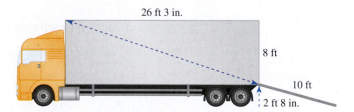

26 ft 3 in.
8 ft
10 ft
2 ft 8 in.

27. The floor of the truck is 2 ft 8 in. above the ground. The ramp connecting the rear of the truck to the ground is 10 ft long. Use the Pythagorean theorem to find the distance in feet and inches from the end of the ramp to the rear of the truck. Round to the nearest inch.

28. Find the length in feet and inches of the diagonal in the diagram. Round to the nearest inch.

For exercises 29–30, the door opening of this moving van is 5 ft 11 in. wide by 5 ft 6 in. high.

5 ft 11 in.
5 ft 6 in.

29. Find the length of the diagonal on this van. Round to the nearest inch.

30. A California king-size box spring is 72 in. wide, 84 in. long, and 8 in. thick.
   a. Will this box spring fit through the rear door of the moving van?
   b. Explain.

31. A triangle is drawn on an xy-coordinate system. The three vertices of the triangle are at point A (1, 1), point B (1, 10), and point C (6, 1). The distances between these vertices have no units.
   a. Draw an xy-coordinate system. Plot the points A, B, and C. Connect the points with line segments to create a triangle.
   b. Find the distance between point A and point B.
   c. Find the distance between point A and point C.
   d. The distance between point B and point C is equal to $\sqrt{(10-1)^2 + (6-1)^2}$. Find this distance. Round to the nearest tenth.

32. Use the Pythagorean theorem to explain why the distance in exercise 31d is equal to $\sqrt{(10-1)^2 + (6-1)^2}$.

33. We can describe a point on a coordinate system as (x_1, y_1) and another point as (x_2, y_2). A third point can be (x_1, y_2). If these points are the vertices of a right triangle, use the Pythagorean theorem to create a formula for finding the distance between (x_1, y_1) and (x_2, y_2).

34. A triangle is drawn on an xy-coordinate system. The three vertices of the triangle are at point A (0, 0), point B (0, 8), and point C (9, 0).
   a. Draw an xy-coordinate system. Plot the points A, B, and C. Connect the points with line segments to create a triangle.
   b. Find the distance between point A and point B. This distance will have no units.
   c. Find the distance between point A and point C.
   d. The distance between point B and point C is equal to $\sqrt{(8-0)^2 + (9-0)^2}$. Find this distance. Round to the nearest tenth.

For exercises 35–42, the only factors of a prime polynomial are itself and 1. In Section 2.3, we used the discriminant, $b^2 - 4ac$, to predict whether a quadratic trinomial is prime. If a, b, and c have no common factors, and the square root of the discriminant is irrational or is a negative number, the trinomial is prime.
(a) Identify a, b, and c for each quadratic trinomial.
(b) Use the discriminant to determine whether the trinomial is prime.

35. $x^2 - 6x - 7$

36. $x^2 - 8x - 9$

37. $10x^2 + 33x + 27$

38. $10x^2 - 33x + 20$

39. $9h^2 - 35h + 20$

40. $16k^2 + 3k - 6$

41. $7u^2 + 3u + 1$

42. $9v^2 + 2v + 1$

For exercises 43–68,
(a) use the zero product property to solve.
(b) check.

43. $x^2 + 13x + 22 = 0$

44. $x^2 + 15x + 26 = 0$

45. $w^2 - 8w + 12 = 0$

46. $p^2 - 10p + 16 = 0$

47. $z^2 - 15z - 54 = 0$

48. $h^2 - 14h - 51 = 0$

49. $z^2 + 15z - 54 = 0$

50. $h^2 + 14h - 51 = 0$

51. $2x^2 + x - 15 = 0$

52. $8x^2 - 18x + 7 = 0$

53. $6p^2 + 13p - 5 = 0$

54. $6q^2 + q - 12 = 0$

55. $21x^2 - 10x + 1 = 0$

56. $18x^2 - 11x + 1 = 0$

57. $6m^2 + 7m - 5 = 0$

58. $6n^2 + 2n - 4 = 0$

59. $7z^2 - 19z - 6 = 0$

60. $5w^2 + 12w - 9 = 0$

61. $4x^2 + 12x + 9 = 0$

62. $9x^2 + 12x + 4 = 0$

63. $9v^2 - 6v + 1 = 0$

64. $25u^2 - 10u + 1 = 0$

65. $10x^2 + 9x = -2$

66. $6x^2 + 17x = -12$

67. $-x^2 + 13x = 36$

68. $-x^2 + 15x = 50$

For exercises 69–72, use a quadratic equation to find the length and width of the rectangle.

69. The width of a rectangle is 2 ft less than the length. Its area is 360 ft^2.

70. The width of a rectangle is 3 in. less than the length. Its area is 180 in.^{2}.

71. The length of a rectangle is 4 cm more than twice the width. Its area is 448 cm^2.

72. The length of a rectangle is 8 ft less than twice the width. Its area is 330 ft^2.

For exercises 73–74, use a quadratic equation to find the base and height of the triangle.

73. The base of a triangle is 4 in. longer than its height. Its area is 160 in.^{2}.

74. The base of a triangle is 3 in. shorter than its height. Its area is 275 in.^{2}.

For exercises 75–76, a trapezoid is a geometric figure with four sides. Two of the sides are parallel, b_1 and b_2. The formula for the area of a trapezoid is $A = \dfrac{1}{2}(b_1 + b_2)h$.

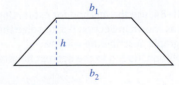

75. One base of a trapezoid, b_2, is twice as long as the other base, b_1. The height is 4 cm less than b_1. Its area is 90 cm^2. Use a quadratic equation to find b_1, b_2, and h.

76. One base of a trapezoid, b_2, is three times as long as the other base, b_1. The height is 2 cm less than b_1. The area is 96 cm^2. Use a quadratic equation to find b_1, b_2, and h.

Problem Solving: Practice and Review

Follow your instructor's guidelines for using the five steps as outlined in Section 1.5, p. 51.

77. The annual production cost, including harvesting, of organic blueberries in Santa Barbara County, California, is \$63,581 per acre. An average yield per acre is 14,000 pounds of blueberries. The grower can sell the blueberries for \$7.50 per pound. (*Source:* coststudies.ucdavis.edu, 2007)
    a. Write a revenue function, $R(x)$, that finds the revenue in dollars from the sale of x lb of blueberries.
    b. Write a cost function, $C(x)$, that finds the cost in dollars to grow x lb of blueberries. Round to the nearest hundredth.
    c. Write a profit function, $P(x)$, that finds the profit in dollars of growing and selling x lb of blueberries.
    d. Find the profit (or loss) from growing and selling 10 acres of blueberries.

78. Find the percent increase in foreclosures from 2005 to 2007. Round to the nearest percent.

    In 2007, there were 15,000 foreclosure filings in the city [New York], an increase from 7,000 in 2005. (*Source:* www.nytimes.com, Jan. 14, 2009)

79. Report the information about deaths from firearms in the United States in 2005 as a rate, $\dfrac{1 \text{ death}}{x \text{ min}}$. Round to the nearest minute (1 year = 365 days).

    Firearms were used to kill 30,143 people in the United States in 2005, the most recent year with complete data from the Centers for Disease Control and Prevention. (*Source:* www.nejm.org, March 19, 2008)

80. A *light-year* is the distance that light travels in a vacuum in 1 year. One light-year equals about 5.9 trillion miles.

    The Swift [satellite] detected an explosion from deep space that was so powerful that its afterglow was briefly visible to the naked eye. . . . The explosion was so far away that it took its light . . . 7.5 billion years to reach Earth! In fact, the explosion took place so long ago that Earth had not yet come into existence. (*Source:* www.nasa.gov, March 21, 2008)
    a. Rewrite 7.5 billion light-years in scientific notation.
    b. Write the relationship of 1 light-year and miles in scientific notation.
    c. Convert the distance of the exploding star from the earth into miles. Write in scientific notation. Round the mantissa to the nearest tenth.

Technology

For exercises 81–84, use the method for estimating the solution of an equation, as shown in Section 6.5, to solve the equation.
(a) Sketch the graph; describe the window.
(b) State the solution. Round to the nearest hundredth.

81. $x^2 + 4x = 5$

82. $x^2 + 5x = 6$

83. $2x^2 - 3x = 20$

84. $2x^2 - 7x = 30$

Find the Mistake

For exercises 85–88, the completed problem has one mistake.
(a) Describe the mistake in words, or copy down the whole problem and highlight or circle the mistake.
(b) Do the problem correctly.

85. **Problem:** Solve: $x^2 + 20 = 0$

    Incorrect Answer: $x^2 + 20 = 0$
    $$\dfrac{-20 \quad -20}{x^2 + 0 = -20}$$
    $$\sqrt{x^2} = \sqrt{-20}$$
    $$x = \sqrt{20}\,\sqrt{-1}$$
    $$x = \sqrt{4}\,\sqrt{5}\,i$$
    $$x = 2\sqrt{5}\,i$$

86. **Problem:** Solve: $x^2 + 7x + 12 = 2$

    Incorrect Answer: $x^2 + 7x + 12 = 2$
    $$(x + 3)(x + 4) = 2$$
    $$x + 3 = 0 \quad \text{or} \quad x + 4 = 0$$
    $$\dfrac{-3 \quad -3}{x = -3} \quad \text{or} \quad \dfrac{-4 \quad -4}{x = -4}$$

87. **Problem:** Solve: $x^2 + 18 = 0$

    Incorrect Answer: $x^2 + 18 = 0$
    $$x^2 = -18$$
    $$\sqrt{x^2} = \pm\sqrt{-18}$$
    $$x = \pm\sqrt{9}\,\sqrt{2}\,\sqrt{-1}$$
    $$x = \pm 3\sqrt{2}\,i$$

88. **Problem:** Solve: $-x^2 = 9$

    Incorrect Answer: $-x^2 = 9$
    $$\sqrt{-x^2} = \sqrt{9}$$
    $$-x = \pm\sqrt{9}$$
    $$-x = \pm 3$$
    $$-x = 3 \quad \text{or} \quad -x = -3$$
    $$\dfrac{-x}{-1} = \dfrac{3}{-1} \quad \text{or} \quad \dfrac{-x}{-1} = \dfrac{-3}{-1}$$
    $$x = -3 \quad \text{or} \quad x = 3$$

Review

For exercises 89–92, evaluate $\left(\dfrac{1}{2}b\right)^2$ for the given value of b.

89. $b = 8$

90. $b = 3$

91. $b = -3$

92. $b = -9$

SUCCESS IN COLLEGE MATHEMATICS

93. Are you required to meet with your advisor before you register for the next term?

94. Explain how to register for classes at your college.

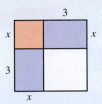

The area of a square is equal to the product of its length and width. The area of the large square in the diagram is $(x + 3)(x + 3)$ or $x^2 + 6x + 9$. This is a perfect square trinomial. In this section, we will learn how to solve a quadratic equation by "completing" a perfect square trinomial.

SECTION 7.3 | Completing the Square

After reading the text, working the practice problems, and completing assigned exercises, you should be able to:

1. Use completing the square to solve a quadratic equation in which $a = 1$.

2. Use completing the square to solve a quadratic equation in which $a \neq 1$.

Completing the Square

If we cannot factor the quadratic expression in a quadratic equation, we cannot use the zero product property to solve the equation. However, we can solve these equations by "completing the square." To understand this method, we begin with a visual model of a square.

EXAMPLE 1 | Add a constant to change $x^2 + 6x$ into a perfect square trinomial.

SOLUTION ▶ A visual model of x^2 is a square. The length of each side is x.

A visual model of $6x$ is a rectangle. The width is x. The length is 6. We can divide this rectangle into two equal parts. The width of each new rectangle is x. The length is 3.

We arrange these three pieces, x^2, $3x$, and $3x$, to make a partial square. To complete the square, we need one more piece. This piece is itself a square. Its length is 3, and its width is 3. Its area is 9.

We began with $x^2 + 6x$. To complete the square, we add 9. The polynomial is now a perfect square trinomial, $x^2 + 6x + 9$.

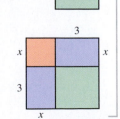

In Example 1, we completed the square by adding a constant and used geometric reasoning to find the value of the constant. We can also use algebraic reasoning to complete the square.

EXAMPLE 2   Find the term needed to complete the square for the polynomial expression $ax^2 + bx$ when $a = 1$.

SOLUTION ▶ If $a = 1$, then $ax^2 + bx = x^2 + bx$. Using guess and check factoring, the factors will be $(x + \underline{\quad})(x + \underline{\quad})$. For the expression to be a perfect square, both factors have to be the same. If the missing term is $\frac{1}{2}b$, then the factors are $\left(x + \frac{1}{2}b \right)\left(x + \frac{1}{2}b \right)$.

$$\left(x + \frac{1}{2}b \right)\left(x + \frac{1}{2}b \right)$$

$$= \mathbf{x}\left(x + \frac{1}{2}b \right) + \frac{1}{2}\mathbf{b}\left(x + \frac{1}{2}b \right) \qquad \text{Distributive property.}$$

$$= \mathbf{x} \cdot \mathbf{x} + \mathbf{x} \cdot \frac{1}{2}b + \frac{1}{2}\mathbf{b} \cdot \mathbf{x} + \frac{1}{2}\mathbf{b} \cdot \frac{1}{2}b \qquad \text{Distributive property.}$$

$$= x^2 + \frac{1}{2}bx + \frac{1}{2}bx + \left(\frac{1}{2}b \right)^2 \qquad \text{Simplify.}$$

$$= x^2 + bx + \left(\frac{1}{2}b \right)^2 \qquad \text{Combine like terms.}$$

This is a perfect square polynomial; it equals $\left(x + \frac{1}{2}b \right)^2$. So to complete the square for $x^2 + bx$, add the term $\left(\frac{1}{2}b \right)^2$.

EXAMPLE 3   **(a)** Add a constant to change $x^2 + 8x$ into a perfect square trinomial.

SOLUTION ▶

$$\left(\frac{1}{2}b \right)^2 \qquad \text{For } x^2 + 8x, a = 1 \text{ and } b = 8.$$

$$= \left(\frac{1}{2} \cdot \mathbf{8} \right)^2 \qquad \text{Replace } b \text{ with 8.}$$

$$= \mathbf{4}^2 \qquad \text{Simplify.}$$

$$= 16 \qquad \text{Simplify.}$$

$x^2 + 8x + 16$ is a perfect square trinomial.

(b) Check by factoring.

▶ $x^2 + 8x + 16$

$$= (x + 4)(x + 4) \qquad \text{Factor.}$$

$$= (x + 4)^2 \qquad \text{This trinomial is a perfect square.}$$

In a geometric model, the length of a rectangle, b, is always a positive number. In an algebraic model, the value of b can be positive or negative.

EXAMPLE 4   Add a constant to change $p^2 - 12p$ into a perfect square trinomial.

SOLUTION ▸

$$\left(\frac{1}{2}b\right)^2 \qquad \text{For } p^2 - 12p,\ a = 1 \text{ and } b = -12.$$

$$= \left(\frac{1}{2} \cdot (\mathbf{-12})\right)^2 \qquad \text{Replace } b \text{ with } -12.$$

$$= (\mathbf{-6})^2 \qquad \text{Simplify.}$$

$$= 36$$

$p^2 - 12p + 36$ is a perfect square trinomial.

In the next example, the constant that completes the square is a fraction.

EXAMPLE 5   Add a constant to change $w^2 + 7w$ into a perfect square trinomial.

SOLUTION ▸

$$\left(\frac{1}{2}b\right)^2 \qquad \text{For } w^2 + 7w,\ a = 1 \text{ and } b = 7.$$

$$= \left(\frac{1}{2} \cdot (\mathbf{7})\right)^2 \qquad \text{Replace } b \text{ with } 7.$$

$$= \left(\frac{\mathbf{7}}{\mathbf{2}}\right)^2 \qquad \text{Simplify.}$$

$$= \frac{49}{4} \qquad \text{Simplify.}$$

$w^2 + 7w + \dfrac{49}{4}$ is a perfect square trinomial.

Practice Problems

For problems 1–5, add a constant to change the expression into a perfect square trinomial.

1. $x^2 + 10x$ **2.** $z^2 - 20z$ **3.** $k^2 + 2k$ **4.** $u^2 - 3u$ **5.** $w^2 - 9w$

Solving a Quadratic Equation by Completing the Square

If a quadratic equation cannot be factored, we can solve it by *completing the square*.

Solving a Quadratic Equation, $ax^2 + bx + c = 0$, by Completing the Square ($a = 1, b \neq 0$)

1. Isolate the terms that have variables on one side of the equation.

2. Add $\left(\dfrac{1}{2}b\right)^2$ to both sides of the equation.

3. Factor the perfect square trinomial.

4. Take the square root of both sides.

5. Simplify; use $\pm$ notation.

6. Rewrite as two equations; solve each equation.

7. Check the solution(s) in the original equation.

In the next example, we use completing the square to solve the equation $x^2 + 8x = -12$. To complete the square, we add 16 to the left side of the equation. We must also add 16 to the right side of the equation.

EXAMPLE 6 Solve $x^2 + 8x = -12$ by completing the square.

SOLUTION ▶

$$x^2 + 8x = -12 \qquad b = 8;\ \left(\tfrac{1}{2}b\right)^2 = \left(\tfrac{1}{2}(8)\right)^2 = 16$$

$$\underline{+16 \quad +16} \qquad \text{Complete the square; addition property of equality.}$$

$$x^2 + 8x + 16 = 4 \qquad \text{Simplify.}$$

$$(x + 4)(x + 4) = 4 \qquad \text{Factor the perfect square trinomial.}$$

$$\sqrt{(x + 4)(x + 4)} = \sqrt{4} \qquad \text{Take the square root of both sides.}$$

$$x + 4 = \pm 2 \qquad \text{Simplify; use } \pm \text{ notation.}$$

$$x + 4 = 2 \quad \text{or} \quad x + 4 = -2 \qquad \text{Rewrite as two equations.}$$

$$\underline{-4 \ -4} \qquad \qquad \underline{-4 \quad -4} \qquad \text{Subtraction property of equality.}$$

$$x + 0 = -2 \quad \text{or} \quad x + 0 = -6 \qquad \text{Simplify.}$$

$$x = -2 \quad \text{or} \qquad x = -6 \qquad \text{Simplify; two rational solutions.}$$

Before completing the square, isolate the terms that include a variable.

EXAMPLE 7 Solve $x^2 - 14x + 40 = 0$ by completing the square.

SOLUTION ▶

$$x^2 - 14x + 40 = 0 \qquad \text{Isolate } x^2 - 14x.$$

$$\underline{-40 \quad -40} \qquad \text{Subtraction property of equality.}$$

$$x^2 - 14x + 0 = -40 \qquad \text{Simplify.}$$

$$x^2 - 14x = -40 \qquad \text{Simplify; } b = -14;\ \left(\tfrac{1}{2}b\right)^2 = \left(\tfrac{1}{2}(-14)\right)^2 = 49$$

$$\underline{+49 \quad +49} \qquad \text{Complete the square; addition property of equality.}$$

$$x^2 - 14x + 49 = 9 \qquad \text{Simplify.}$$

$$(x - 7)(x - 7) = 9 \qquad \text{Factor the perfect square trinomial.}$$

$$\sqrt{(x - 7)(x - 7)} = \sqrt{9} \qquad \text{Take the square root of both sides.}$$

$$x - 7 = \pm 3 \qquad \text{Simplify; use } \pm \text{ notation.}$$

$$x - 7 = 3 \quad \text{or} \quad x - 7 = -3 \qquad \text{Rewrite as two equations.}$$

$$\underline{+7 \ +7} \qquad \qquad \underline{+7 \quad +7} \qquad \text{Addition property of equality.}$$

$$x + 0 = 10 \quad \text{or} \quad x + 0 = 4 \qquad \text{Simplify.}$$

$$x = 10 \quad \text{or} \qquad x = 4 \qquad \text{Simplify; two rational solutions.}$$

In the next example, we need to solve the equation $x + 3 = \pm\sqrt{19}$. To isolate the variable, we subtract 3 from both sides. Write the -3 to the *left* of the $\pm$ notation.

EXAMPLE 8 **(a)** Solve $x^2 + 6x - 10 = 0$ by completing the square.

SOLUTION ▶

$$x^2 + 6x - 10 = 0 \qquad \text{Isolate } x^2 + 6x.$$

$$\underline{+10 \ +10} \qquad \text{Addition property of equality.}$$

$$x^2 + 6x + 0 = 10 \qquad \text{Simplify; } b = 6;\ \left(\tfrac{1}{2}b\right)^2 = \left(\tfrac{1}{2}\cdot 6\right)^2 = 9$$

$$\underline{+9 \ +9} \qquad \text{Complete the square; addition property of equality.}$$

$$x^2 + 6x + 9 = 19 \qquad \text{Simplify.}$$

$$(x + 3)(x + 3) = 19 \qquad \text{Factor the perfect square trinomial.}$$

$$\sqrt{(x + 3)(x + 3)} = \sqrt{19} \qquad \text{Take the square root of both sides.}$$

$$x + 3 = \pm\sqrt{19} \qquad \text{Simplify; use } \pm \text{ notation.}$$

$$\underline{ -3 \quad -3} \qquad \text{Subtraction property of equality.}$$

$$x + 0 = -3 \pm \sqrt{19} \qquad \text{Simplify; write } -3 \text{ to the left of } \pm.$$

$$x = -3 \pm \sqrt{19} \qquad \text{Simplify; two irrational solutions.}$$

$$x = -3 + \sqrt{19} \quad \text{or} \quad x = -3 - \sqrt{19} \qquad \text{The solutions written as two equations with "or."}$$

(b) Check one of the solutions, $x = -3 + \sqrt{19}$.

$$x^2 + 6x - 10 = 0 \qquad \text{Original equation.}$$

$$\left(-3 + \sqrt{19}\right)^2 + 6\left(-3 + \sqrt{19}\right) - 10 = 0 \qquad \text{Replace } x \text{ with } -3 + \sqrt{19}.$$

$$\left(-3 + \sqrt{19}\right)\left(-3 + \sqrt{19}\right) + 6(-3) + 6\sqrt{19} - 10 = 0 \qquad \text{Distributive property.}$$

$$-3\left(-3 + \sqrt{19}\right) + \sqrt{19}\left(-3 + \sqrt{19}\right) - 18 + 6\sqrt{19} - 10 = 0 \qquad \text{Distributive property; simplify.}$$

$$-3(-3) - 3\sqrt{19} + \sqrt{19}(-3) + \sqrt{19}\sqrt{19} - 28 + 6\sqrt{19} = 0 \qquad \text{Distributive property; simplify.}$$

$$9 - 3\sqrt{19} - 3\sqrt{19} + 19 - 28 + 6\sqrt{19} = 0 \qquad \text{Simplify.}$$

$$9 + 19 - 28 - 3\sqrt{19} - 3\sqrt{19} + 6\sqrt{19} = 0 \qquad \text{Change order.}$$

$$28 - 28 - 6\sqrt{19} + 6\sqrt{19} = 0 \qquad \text{Combine like terms.}$$

$$0 = 0 \qquad \text{Simplify; since } 0 = 0 \text{ is true, the solution is correct.}$$

To simplify a square root with a negative radicand, use $\sqrt{-1} = i$.

EXAMPLE 9   **(a)** Solve $u^2 + 4u + 9 = 0$ by completing the square.

SOLUTION ▶

$$u^2 + 4u + 9 = 0 \qquad \text{We need to isolate } u^2 + 4u.$$

$$\underline{ -9 \quad -9} \qquad \text{Subtraction property of equality.}$$

$$u^2 + 4u + 0 = -9 \qquad \text{Simplify; } b = 4; \left(\frac{1}{2}b\right)^2 = \left(\frac{1}{2}\cdot 4\right)^2 = 4$$

$$\underline{ +4 \quad +4} \qquad \text{Complete the square; addition property of equality.}$$

$$u^2 + 4u + 4 = -5 \qquad \text{Simplify.}$$

$$(u + 2)(u + 2) = -5 \qquad \text{Factor the perfect square trinomial.}$$

$$\sqrt{(u + 2)(u + 2)} = \sqrt{-5} \qquad \text{Take the square root of both sides.}$$

$$u + 2 = \pm\sqrt{-5} \qquad \text{Simplify; use } \pm \text{ notation.}$$

$$u + 2 = \pm\sqrt{5}\sqrt{-1} \qquad \text{Rewrite as the product of a square root and } \sqrt{-1}.$$

$$u + 2 = \pm\sqrt{5}i \qquad \text{Simplify; } \sqrt{-1} = i$$

$$\underline{ -2 \quad -2} \qquad \text{Subtraction property of equality.}$$

$$u + 0 = -2 \pm \sqrt{5}i \qquad \text{Simplify.}$$

$$u = -2 \pm \sqrt{5}i \qquad \text{Simplify; two nonreal solutions.}$$

(b) Check one of the solutions, $u = -2 - \sqrt{5}i$.

$u^2 + 4u + 9 = 0$	Original equation.
$(-2 - \sqrt{5}i)^2 + 4(-2 - \sqrt{5}i) + 9 = 0$	Replace x.
$(-2 - \sqrt{5}i)(-2 - \sqrt{5}i) + 4(-2) + 4(-\sqrt{5}i) + 9 = 0$	Distributive property.
$-2(-2 - \sqrt{5}i) - \sqrt{5}i(-2 - \sqrt{5}i) - 8 - 4\sqrt{5}i + 9 = 0$	Distributive property.
$-2(-2) - 2(-\sqrt{5}i) - \sqrt{5}i(-2) - \sqrt{5}i(-\sqrt{5}i) + 1 - 4\sqrt{5}i = 0$	Distributive property.
$4 + 2\sqrt{5}i + 2\sqrt{5}i + 5i^2 + 1 - 4\sqrt{5}i = 0$	Distributive property.
$4 + 2\sqrt{5}i + 2\sqrt{5}i + 5(-1) + 1 - 4\sqrt{5}i = 0$	$i^2 = -1$
$4 - 5 + 1 + 2\sqrt{5}i + 2\sqrt{5}i - 4\sqrt{5}i = 0$	Change order.
$-1 + 1 + 4\sqrt{5}i - 4\sqrt{5}i = 0$	Simplify.
$0 = 0$	Simplify; since $0 = 0$ is true, the solution is correct.

If b is not an even number, $\left(\frac{1}{2}b\right)^2$ is a fraction. This means that the last term in the perfect square trinomial is a fraction. Using a pattern can make it easier to factor these trinomials.

Patterns for Factoring Polynomials

Perfect Square Trinomial Pattern $a^2 + 2ab + b^2 = (a + b)(a + b)$

Perfect Square Trinomial Pattern $a^2 - 2ab + b^2 = (a - b)(a - b)$

In the next example, we need to rewrite $w^2 + 5w + \frac{25}{4}$ so that it matches the factoring pattern $a^2 + 2ab + b^2$. Matching the first term, $a = w$. Matching the last term, $b = \sqrt{\frac{25}{4}} = \frac{5}{2}$. To rewrite the middle term, $5w$, to match the pattern $2ab$, multiply by a fraction equal to 1, $\frac{2}{2}$.

EXAMPLE 10   Solve $w^2 + 5w - 9 = 0$ by completing the square.

SOLUTION

$w^2 + 5w - 9 = 0$	We need to isolate $w^2 + 5w$.
$\underline{ +9 \quad +9}$	Addition property of equality.
$w^2 + 5w + 0 = 9$	Simplify.
$w^2 + 5w = 9$	$b = 5; \left(\frac{1}{2}b\right)^2 = \left(\frac{1}{2} \cdot 5\right)^2 = \frac{25}{4}$
$\underline{ +\frac{25}{4} \quad +\frac{25}{4}}$	Complete the square; addition property of equality.
$w^2 + 5w + \frac{25}{4} = 9 + \frac{25}{4}$	Simplify.

$$w^2 + 5w + \frac{25}{4} = \frac{36}{4} + \frac{25}{4}$$ A common denominator is 4; $\left(\frac{9}{1}\right)\left(\frac{4}{4}\right) = \frac{36}{4}$

$$w^2 + 5w + \frac{25}{4} = \frac{61}{4}$$ Simplify.

$$w^2 + 2w\left(\frac{5}{2}\right) + \left(\frac{5}{2}\right)^2 = \frac{61}{4}$$ Rewrite to match the pattern $a^2 + 2ab + b^2$; $5w = \frac{2}{2} \cdot \frac{5}{1} \cdot \frac{w}{1} = 2 \cdot w \cdot \frac{5}{2}$

$$\left(w + \frac{5}{2}\right)\left(w + \frac{5}{2}\right) = \frac{61}{4}$$ Factor; $a^2 + 2ab + b^2 = (a + b)(a + b)$

$$\sqrt{\left(w + \frac{5}{2}\right)\left(w + \frac{5}{2}\right)} = \sqrt{\frac{61}{4}}$$ Take the square root of both sides.

$$w + \frac{5}{2} = \pm\sqrt{\frac{61}{4}}$$ Simplify; use $\pm$ notation.

$$w + \frac{5}{2} = \pm\frac{\sqrt{61}}{\sqrt{4}}$$ Quotient rule of radicals.

$$w + \frac{5}{2} = \pm\frac{\sqrt{61}}{2}$$ Evaluate the radical in the denominator.

$$-\frac{5}{2} \qquad -\frac{5}{2}$$ Subtraction property of equality.

$$w + 0 = -\frac{5}{2} \pm \frac{\sqrt{61}}{2}$$ Simplify.

$$w = -\frac{5}{2} \pm \frac{\sqrt{61}}{2}$$ Simplify; two irrational solutions.

In Example 10, since both fractions in the solution have the same denominator, we can rewrite the solution as a single fraction. In $\pm$ notation, the solutions are $w = \dfrac{-5 \pm \sqrt{61}}{2}$. However, if the solutions are nonreal numbers, leave the solution in $a + bi$ form and do not write it as a single fraction.

Practice Problems

6. a. Solve $x^2 + 10x = -21$ by completing the square.
   b. Check.

For problems 7–9, solve by completing the square.

7. $z^2 - 12z + 27 = 0$ **8.** $k^2 + 2k + 8 = 0$ **9.** $z^2 + 3z - 9 = 0$

Completing the Square When $a \neq 1$

In the next example, the **lead coefficient**, a, does not equal 1. Each term in the equation must be divided by a before completing the square. This changes the original values of b and c in the equation. Complete the square using this new value for b in $\left(\frac{1}{2}b\right)^2$.

EXAMPLE 11 Solve $4x^2 + 3x - 20 = 0$ by completing the square.

SOLUTION ▶

$$4x^2 + 3x - 20 = 0$$ — The lead coefficient, a, is 4.

$$\frac{4x^2}{4} + \frac{3x}{4} - \frac{20}{4} = \frac{0}{4}$$ — Change a to 1; division property of equality.

$$1x^2 + \frac{3}{4}x - 5 = 0$$ — Simplify. The lead coefficient is now 1.

$$\underline{\phantom{x^2 + \frac{3}{4}x + 0} +5 \ +5}$$ — Isolate $x^2 + \frac{3}{4}x$; addition property of equality.

$$x^2 + \frac{3}{4}x + 0 = 5$$ — $b = \frac{3}{4}; \left(\frac{1}{2}b\right)^2 = \left(\frac{1}{2} \cdot \frac{3}{4}\right)^2 = \frac{9}{64}$

$$\underline{ +\frac{9}{64} \ +\frac{9}{64}}$$ — Complete the square; addition property of equality.

$$x^2 + \frac{3}{4}x + \frac{9}{64} = 5 + \frac{9}{64}$$ — Simplify; the common denominator is 64.

$$x^2 + \frac{3}{4}x + \frac{9}{64} = \frac{5}{1} \cdot \frac{64}{64} + \frac{9}{64}$$ — Multiply by a fraction equal to 1; $\frac{64}{64} = 1$

$$x^2 + \frac{3}{4}x + \frac{9}{64} = \frac{320}{64} + \frac{9}{64}$$ — Simplify.

$$x^2 + \frac{3}{4}x + \frac{9}{64} = \frac{329}{64}$$ — Simplify.

$$x^2 + 2x\left(\frac{3}{8}\right) + \left(\frac{3}{8}\right)^2 = \frac{329}{64}$$ — Rewrite to match the pattern $a^2 + 2ab + b^2$; $\frac{3}{4}x = \frac{2}{2} \cdot \frac{3}{4} \cdot x = 2 \cdot x \cdot \frac{3}{8}$

$$\left(x + \frac{3}{8}\right)\left(x + \frac{3}{8}\right) = \frac{329}{64}$$ — Factor; $a^2 + 2ab + b^2 = (a + b)(a + b)$

$$\sqrt{\left(x + \frac{3}{8}\right)\left(x + \frac{3}{8}\right)} = \sqrt{\frac{329}{64}}$$ — Take the square root of both sides.

$$x + \frac{3}{8} = \pm\sqrt{\frac{329}{64}}$$ — Simplify; use $\pm$ notation.

$$x + \frac{3}{8} = \pm\frac{\sqrt{329}}{\sqrt{64}}$$ — Quotient rule of radicals.

$$x + \frac{3}{8} = \pm\frac{\sqrt{329}}{8}$$ — Evaluate the radical in the denominator.

$$\underline{-\frac{3}{8} \qquad\qquad -\frac{3}{8}}$$ — Subtraction property of equality.

$$x + 0 = -\frac{3}{8} \pm \frac{\sqrt{329}}{8}$$ — Simplify; write the rational number first.

$$x = -\frac{3}{8} \pm \frac{\sqrt{329}}{8}$$ — Simplify.

$$x = \frac{-3 \pm \sqrt{329}}{8}$$ — Another way to write this solution.

In the next example, the radicand of one of the square roots is a negative number.

EXAMPLE 12 | Solve $2w^2 + 10w + 19 = 0$ by completing the square.

SOLUTION ▶

$$2w^2 + 10w + 19 = 0$$ The lead coefficient, a, is 2.

$$\frac{2w^2}{2} + \frac{10w}{2} + \frac{19}{2} = \frac{0}{2}$$ Change a to 1; division property of equality.

$$w^2 + 5w + \frac{19}{2} = 0$$ Simplify: the lead coefficient is now 1.

$$\begin{array}{c} -\frac{19}{2} \quad -\frac{19}{2} \\ \hline w^2 + 5w + 0 = -\frac{19}{2} \end{array}$$ Isolate $w^2 + 5w$; subtraction property of equality.

$\left(\frac{1}{2}b\right)^2 = \left(\frac{1}{2} \cdot 5\right)^2 = \frac{25}{4}$

$$\begin{array}{c} +\frac{25}{4} \quad +\frac{25}{4} \\ \hline w^2 + 5w + \frac{25}{4} = -\frac{19}{2} + \frac{25}{4} \end{array}$$ Complete the square; addition property of equality.

Simplify; common denominator is 4.

$$w^2 + 5w + \frac{25}{4} = -\frac{19}{2} \cdot \frac{2}{2} + \frac{25}{4}$$ Multiply by a fraction equal to 1; $\frac{2}{2} = 1$

$$w^2 + 5w + \frac{25}{4} = -\frac{38}{4} + \frac{25}{4}$$ Simplify.

$$w^2 + 5w + \frac{25}{4} = -\frac{13}{4}$$ Simplify.

$$w^2 + 2w\left(\frac{5}{2}\right) + \left(\frac{5}{2}\right)^2 = -\frac{13}{4}$$ Rewrite to match pattern $a^2 + 2ab + b^2$; $5w = \frac{2}{2} \cdot \frac{5}{1} \cdot \frac{w}{1} = 2 \cdot w \cdot \frac{5}{2}$

$$\left(w + \frac{5}{2}\right)\left(w + \frac{5}{2}\right) = -\frac{13}{4}$$ Factor; $a^2 + 2ab + b^2 = (a+b)(a+b)$

$$\sqrt{\left(w + \frac{5}{2}\right)\left(w + \frac{5}{2}\right)} = \sqrt{-\frac{13}{4}}$$ Take the square root of both sides.

$$\sqrt{\left(w + \frac{5}{2}\right)\left(w + \frac{5}{2}\right)} = \sqrt{\frac{13}{4}}\sqrt{-1}$$ Rewrite as the product of a square root and $\sqrt{-1}$.

$$\sqrt{\left(w + \frac{5}{2}\right)\left(w + \frac{5}{2}\right)} = \sqrt{\frac{13}{4}}\,i$$ Simplify; $\sqrt{-1} = i$

$$w + \frac{5}{2} = \pm\sqrt{\frac{13}{4}}\,i$$ Simplify; use $\pm$ notation.

$$w + \frac{5}{2} = \pm\frac{\sqrt{13}}{\sqrt{4}}\,i$$ Quotient rule of radicals.

$$w + \frac{5}{2} = \pm\frac{\sqrt{13}}{2}\,i$$ Simplify.

$$\begin{array}{c} -\frac{5}{2} \quad -\frac{5}{2} \\ \hline w + 0 = -\frac{5}{2} \pm \frac{\sqrt{13}}{2}\,i \end{array}$$ Subtraction property of equality.

Simplify.

$$w = -\frac{5}{2} \pm \frac{\sqrt{13}}{2}\,i$$ Simplify.

Since the solutions need to stay in $a + bi$ form, do not write them as a single fraction.

Practice Problems

For problems 10–12, solve by completing the square.

10. $4p^2 + 8p - 7 = 0$ **11.** $2x^2 - 7x - 8 = 0$ **12.** $3x^2 + x + 2 = 0$

7.3 VOCABULARY PRACTICE

Match the term with its description.

1. $ax^2 + bx + c = 0$ when $a \neq 0$
2. $a^2 + 2ab + b^2$
3. The expression under a radical
4. $\sqrt{-1}$
5. $\sqrt[n]{ab} = \sqrt[n]{a} \cdot \sqrt[n]{b}$
6. $\sqrt[n]{\dfrac{a}{b}} = \dfrac{\sqrt[n]{a}}{\sqrt[n]{b}}$
7. If $ab = 0$, then either $a = 0$ or $b = 0$.
8. Real numbers that are not rational numbers
9. Complex numbers, $a + bi$, in which $a = 0$ and $b \neq 0$
10. The lead coefficient of a quadratic equation in one variable in standard form

A. a
B. i
C. imaginary numbers
D. irrational numbers
E. perfect square trinomial
F. product rule of radicals
G. quotient rule of radicals
H. radicand
I. standard form of a quadratic equation in one variable
J. zero product property

7.3 Exercises

Follow your instructor's guidelines for showing your work.

1. What is the degree of a quadratic equation in one variable?

2. What is the standard form of a quadratic equation in one variable?

For exercises 3–10, add a constant to change the expression into a perfect square trinomial.

3. $x^2 + 12x$

4. $x^2 + 20x$

5. $u^2 - 6u$

6. $w^2 - 8w$

7. $z^2 + 3z$

8. $k^2 + 5k$

9. $x^2 - 11x$

10. $x^2 - 13x$

For exercises 11–34,
(a) solve by completing the square.
(b) check one of the solutions.

11. $x^2 + 6x + 5 = 0$

12. $x^2 + 4x + 3 = 0$

13. $n^2 - 8n + 12 = 0$

14. $m^2 - 6m + 8 = 0$

15. $j^2 + 8j - 17 = 0$

16. $k^2 + 10k - 13 = 0$

17. $j^2 + 8j + 17 = 0$

18. $k^2 + 10k + 26 = 0$

19. $x^2 - 2x + 7 = 0$

20. $h^2 - 2h + 6 = 0$

21. $x^2 - 2x - 3 = 0$

22. $h^2 - 2h - 8 = 0$

23. $a^2 + 3a + 2 = 0$

24. $b^2 + 5b + 4 = 0$

25. $x^2 + 9x - 10 = 0$

26. $x^2 + 7x - 8 = 0$

27. $x^2 - 11x = -3$

28. $x^2 - 13x = -5$

29. $z^2 = z - 4$

30. $p^2 = p - 5$

31. $d^2 = 6d - 9$

32. $v^2 = 8v - 16$

33. $u^2 + 10u = -25$

34. $c^2 + 12c = -36$

For exercises 35–60, solve by completing the square.

35. $4x^2 - 4x - 15 = 0$

36. $4x^2 - 8x - 5 = 0$

37. $4x^2 + 4x + 1 = 0$

38. $4x^2 + 12x + 9 = 0$

39. $2x^2 - 5x - 3 = 0$

40. $2x^2 - 7x - 4 = 0$

41. $2x^2 - 5x = -3$

42. $2x^2 + 9x = -4$

43. $2x^2 = -10x - 5$

44. $2x^2 = -12x - 7$

45. $3w^2 + 9w - 8 = 0$

46. $3p^2 + 9p - 4 = 0$

47. $2x^2 + 5x - 11 = 0$

48. $2x^2 + 7x - 13 = 0$

49. $2x^2 + 5x + 11 = 0$

50. $2x^2 + 7x + 13 = 0$

51. $2d^2 + 25d + 72 = 0$

52. $2f^2 + 19f + 42 = 0$

53. $3x^2 + 2x + 1 = 0$

54. $5x^2 + 3x + 1 = 0$

55. $3x^2 + 2x = 1$

56. $5x^2 + 4x = 1$

57. $2x^2 = 6x + 39$

58. $2x^2 = 6x + 29$

59. $2x^2 - 8x + 15 = 0$

60. $2x^2 - 4x + 15 = 0$

For exercises 61–80,
(a) solve using any of the methods from this chapter.
(b) check.

61. $x^2 + 24 = 0$

62. $x^2 + 28 = 0$

63. $p^2 + 6p - 27 = 0$

64. $w^2 + 4w - 12 = 0$

65. $0 = x^2 - 6x + 3$

66. $0 = x^2 - 8x + 3$

67. $c^2 = 4c - 11$

68. $d^2 = 6d - 10$

69. $q^2 = 8^2 + 5^2$

70. $z^2 = 11^2 + 5^2$

71. $2x^2 + 6x = 13$

72. $2x^2 + 10x = 13$

73. $z^2 - 64z = 0$

74. $p^2 - 25p = 0$

75. $z^2 - 64 = 0$

76. $p^2 - 25 = 0$

77. $3k^2 + 6k + 4 = 0$

78. $4h^2 + 8h + 11 = 0$

79. $p^2 = -32$

80. $n^2 = -48$

Problem Solving: Practice and Review

Follow your instructor's guidelines for using the five steps as outlined in Section 1.5, p. 51.

81. Find the percent of the guild's voting members who voted to accept the package. Round to the nearest percent.

Members of the Newspaper Guild at *The Boston Globe* narrowly rejected a proposed package of wage and compensation cuts . . . the guild said its members had rejected the package by a vote of 277 to 265—meaning 542 of the guild's 670 members had voted. (*Source:* www.nytimes.com, June 6, 2009)

82. The speed limit on Interstate 84 is 65 miles per hour for cars and 55 miles per hour for trucks. Assume that a car and a truck are both traveling at the speed limit from Mt. Hood Community College in Portland to Pendleton, Oregon. Find the difference in time in hours and minutes that it takes for the car and the truck to travel the 201 miles. Round to the nearest minute.

83. The troy system of measurement is used to measure weight. The basic unit in this system is the **grain**. The **troy ounce** is now used only for measuring precious metals. One troy ounce equals 480 grains. One grain equals $\dfrac{1}{7000}$ lb or 0.065 g. If the price of gold is \$1001.40 per troy ounce, find the price of gold per gram. Round to the nearest hundredth. (*Source:* www.troy-ounce .com)

84. In March 2008, the Federal Reserve Board approved a decrease in the discount rate at some Federal Reserve Banks from 3.50 percent per year to 3.25 percent per year. An investment bank needs a loan of \$25,000,000. Find the difference in one *day* of interest at these rates (1 year = 365 days). Round to the nearest whole number. (*Source:* www.federalreserve.gov)

Technology

For exercises 85–88, use a graphing calculator and the "intersection method" (Section 6.5) to solve the equation. One of the graphed lines, $y = 0$, is the *x*-axis; it will not be visible on the screen of the calculator. However, the calculator will recognize it and find its intersection with the other curve.
(a) Sketch the graph; describe the window.
(b) Identify the solution. Round to the nearest thousandth.

85. $x^2 - x - 6 = 0$

86. $2x^2 - 11x - 6 = 0$

87. $x^2 - 7 = 0$

88. $x^2 - 3x - 11 = 0$

Find the Mistake

For exercises 89–92, the completed problem has one mistake.
(a) Describe the mistake in words, or copy down the whole problem and highlight or circle the mistake.
(b) Do the problem correctly.

89. Problem: Solve $x^2 + 6x - 10 = 0$ by completing the square.

Incorrect Answer: $x^2 + 6x - 10 = 0$

$$\dfrac{ +10 \ +10}{x^2 + 6x = 10}$$

$$x^2 + 6x + 9 = 10 + 9$$

$$(x + 3)(x + 3) = 19$$

$$\sqrt{(x + 3)(x + 3)} = \sqrt{19}$$

$$x + 3 = \sqrt{19}$$

$$\dfrac{-3 \quad -3}{x = -3 + \sqrt{19}}$$

90. Problem: Solve $2x^2 + 16x + 9 = 0$ by completing the square.

Incorrect Answer: $2x^2 + 16x + 9 = 0$

$$\frac{-9 \quad -9}{2x^2 + 16x = -9}$$

$$2x^2 + 16x + 64 = -9 + 64$$

$$2x^2 + 16x + 64 = 55$$

$$(x + 8)(x + 8) = 55$$

$$\sqrt{(x + 8)(x + 8)} = \pm\sqrt{55}$$

$$x + 8 = \pm\sqrt{55}$$

$$\frac{-8 \quad -8}{x + 0 = -8 \pm \sqrt{55}}$$

$$x = -8 + \sqrt{55} \quad \text{or} \quad x = -8 - \sqrt{55}$$

91. Problem: Solve $x^2 + 2x + 8 = 0$ by completing the square.

Incorrect Answer: $x^2 + 2x + 8 = 0$

$$\frac{-8 \quad -8}{x^2 + 2x = -8}$$

$$x^2 + 2x + 1 = -8 + 1$$

$$(x + 1)(x + 1) = -7$$

$$\sqrt{(x + 1)(x + 1)} = \sqrt{-7}$$

$$x + 1 = \pm\sqrt{-7}$$

$$\frac{-1 \quad -1}{x + 0 = -1 \pm \sqrt{-7}}$$

92. Problem: Solve $2x^2 + 8x + 15 = 0$ by completing the square.

Incorrect Answer: $2x^2 + 8x + 15 = 0$

$$2x^2 + 8x = -15$$

$$\frac{2x^2}{2} + \frac{8x}{2} = -15$$

$$x^2 + 4x + 4 = -15 + 4$$

$$(x + 2)(x + 2) = -11$$

$$\sqrt{(x + 2)(x + 2)} = \sqrt{-11}$$

$$x + 2 = \pm\sqrt{11}\sqrt{-1}$$

$$x + 2 = \pm\sqrt{11}i$$

$$\frac{-2 \quad -2}{x + 0 = -2 \pm \sqrt{11}i}$$

$$x = -2 + \sqrt{11}i \quad \text{or} \quad x = -2 - \sqrt{11}i$$

Review

For exercises 93–96, evaluate $\sqrt{b^2 - 4ac}$ for the given values of a, b, and c.

93. $a = 1, b = -3, c = 2$

94. $a = 2, b = -3, c = 2$

95. $a = 2, b = 6, c = 10$

96. $a = 2, b = 6, c = -10$

SUCCESS IN COLLEGE MATHEMATICS

97. Part of planning ahead is identifying long-term goals so that you are motivated to work on short-term goals. Does your college have a center for career counseling? If so, where is it located?

98. Do you think that you need more career counseling? Explain.

The time it takes for a hammer to fall to the ground from a roof is the solution of a quadratic equation. In this section, we develop the quadratic formula. We then use this formula and quadratic equations to solve application problems.

SECTION 7.4

Quadratic Formula

After reading the text, working the practice problems, and completing assigned exercises, you should be able to:

1. Use the quadratic formula to solve a quadratic equation.
2. Use the discriminant to describe the solutions of a quadratic equation.
3. Use a quadratic equation to solve an application problem.

The Quadratic Formula

In Section 7.3, we solved quadratic equations by completing the square. Now we will use completing the square to solve $ax^2 + bx + c = 0$. The result is the **quadratic formula**. Be careful not to confuse the a, b, and c that represent the coefficients and constant in the quadratic equation with the a and b in the factoring pattern for a perfect square trinomial.

EXAMPLE 1 Solve $ax^2 + bx + c = 0$ by completing the square.

SOLUTION ▶

$$ax^2 + bx + c = 0$$

$$\frac{a}{a}x^2 + \frac{b}{a}x + \frac{c}{a} = 0 \qquad \text{So that } a = 1, \text{ divide each term by } a.$$

$$1x^2 + \frac{b}{a}x + \frac{c}{a} = 0 \qquad \text{Simplify; the lead coefficient is 1.}$$

$$\qquad\qquad\quad -\frac{c}{a} \quad -\frac{c}{a} \qquad \text{Subtraction property of equality.}$$

$$x^2 + \frac{b}{a}x + 0 = \frac{-c}{a} \qquad \text{Simplify; } x^2 + \frac{b}{a}x \text{ is isolated.}$$

$$x^2 + \frac{b}{a}x + \frac{b^2}{4a^2} = \frac{-c}{a} + \frac{b^2}{4a^2} \qquad \text{Complete the square; } \left(\frac{1}{2}\cdot\frac{b}{a}\right)^2 = \frac{b^2}{4a^2}$$

$$x^2 + 2x\left(\frac{b}{2a}\right) + \left(\frac{b}{2a}\right)^2 = \frac{-c}{a} + \frac{b^2}{4a^2} \qquad \text{Rewrite to match the pattern } a^2 + 2ab + b^2.$$

$$\left(x + \frac{b}{2a}\right)\left(x + \frac{b}{2a}\right) = \frac{-c}{a} + \frac{b^2}{4a^2} \qquad \text{Factor; } a^2 + 2ab + b^2 = (a+b)(a+b)$$

$$\left(x + \frac{b}{2a}\right)\left(x + \frac{b}{2a}\right) = \frac{-c}{a} \cdot \frac{4a}{4a} + \frac{b^2}{4a^2}$$

Multiply by a fraction equal to 1; $\frac{4a}{4a} = 1$

$$\left(x + \frac{b}{2a}\right)\left(x + \frac{b}{2a}\right) = \frac{-4ac}{4a^2} + \frac{b^2}{4a^2}$$

Simplify; the fractions have the same denominator.

$$\left(x + \frac{b}{2a}\right)\left(x + \frac{b}{2a}\right) = \frac{b^2 - 4ac}{4a^2}$$

Combine into a single fraction.

$$\sqrt{\left(x + \frac{b}{2a}\right)\left(x + \frac{b}{2a}\right)} = \sqrt{\frac{b^2 - 4ac}{4a^2}}$$

Take the square root of both sides.

$$x + \frac{b}{2a} = \pm\sqrt{\frac{b^2 - 4ac}{4a^2}}$$

Simplify; use $\pm$ notation.

$$x + \frac{b}{2a} = \pm\frac{\sqrt{b^2 - 4ac}}{\sqrt{4a^2}}$$

Quotient rule of radicals.

$$x + \frac{b}{2a} = \pm\frac{\sqrt{b^2 - 4ac}}{2a}$$

Simplify the denominator.

$$\underset{\displaystyle -\frac{b}{2a}}{} \quad \underset{\displaystyle -\frac{b}{2a}}{}$$

Subtraction property of equality.

$$x + 0 = -\frac{b}{2a} \pm \frac{\sqrt{b^2 - 4ac}}{2a}$$

Simplify.

$$x = \frac{-b \pm \sqrt{b^2 - 4ac}}{2a}$$

Simplify. Combine the fractions.

The solutions of $ax^2 + bx + c = 0$ are $x = \dfrac{-b \pm \sqrt{b^2 - 4ac}}{2a}$. This is the **quadratic formula**.

The Quadratic Formula

For the equation $ax^2 + bx + c = 0$ where a, b, and c are real numbers and $a \neq 0$,

$$x = \frac{-b \pm \sqrt{b^2 - 4ac}}{2a}$$

In the next example, we use the quadratic formula to solve a quadratic equation. We do not have to complete the square. We already did that work in creating the quadratic formula. Ask your instructor how to write solutions when they are irrational or include i. Some instructors prefer that the solutions be written using $\pm$ notation. Others prefer that the solutions be written using "or."

EXAMPLE 2   **(a)** Use the quadratic formula to solve $6x^2 + 5x - 9 = 0$.

SOLUTION ▶ $6x^2 + 5x - 9 = 0$ $a = 6, b = 5, c = -9$

$$x = \frac{-b \pm \sqrt{b^2 - 4ac}}{2a}$$

The quadratic formula.

$$x = \frac{-5 \pm \sqrt{5^2 - 4(6)(-9)}}{2(6)}$$

Replace a, b, and c.

$$x = \frac{-5 \pm \sqrt{25 + 216}}{12} \qquad \text{Simplify.}$$

$$x = \frac{-5 \pm \sqrt{241}}{12} \qquad \text{Simplify.}$$

(b) Check one of the solutions: $x = \dfrac{-5 + \sqrt{241}}{12}$

$$6x^2 + 5x - 9 = 0 \qquad \text{Original equation.}$$

$$6\left(\frac{-5 + \sqrt{241}}{12}\right)^2 + 5\left(\frac{-5 + \sqrt{241}}{12}\right) - 9 = 0 \qquad \text{Replace variables.}$$

$$6\left(\frac{-5 + \sqrt{241}}{12}\right)\left(\frac{-5 + \sqrt{241}}{12}\right) + \frac{-25 + 5\sqrt{241}}{12} - 9 = 0 \qquad \text{Distributive property.}$$

$$6\left(\frac{25 - 5\sqrt{241} - 5\sqrt{241} + 241}{144}\right) + \frac{-25 + 5\sqrt{241}}{12} - 9 = 0 \qquad \text{Distributive property.}$$

$$6\left(\frac{266 - 10\sqrt{241}}{6 \cdot 24}\right) + \frac{-25 + 5\sqrt{241}}{12} - 9 = 0 \qquad \text{Combine like terms.}$$

$$\frac{266 - 10\sqrt{241}}{24} + \frac{-25 + 5\sqrt{241}}{12} - 9 = 0 \qquad \text{Simplify.}$$

$$\frac{266 - 10\sqrt{241}}{24} + \left(\frac{2}{2}\right)\left(\frac{-25 + 5\sqrt{241}}{12}\right) - 9 = 0 \qquad \text{Multiply by 1.}$$

$$\frac{266 - 10\sqrt{241}}{24} + \frac{-50 + 10\sqrt{241}}{24} - 9 = 0 \qquad \text{Distributive property.}$$

$$\frac{266 - 10\sqrt{241} - 50 + 10\sqrt{241}}{24} - 9 = 0 \qquad \text{Combine fractions.}$$

$$\frac{216}{24} - 9 = 0 \qquad \text{Simplify.}$$

$$9 - 9 = 0 \qquad \text{Simplify.}$$

$$0 = 0 \qquad \text{True.}$$

Since $0 = 0$ is true, the solution is correct.

When b is negative, include the negative sign from the quadratic formula and the negative sign for b.

EXAMPLE 3 | Use the quadratic formula to solve $2x^2 - 11x - 6 = 0$.

SOLUTION ▶

$$2x^2 - 11x - 6 = 0 \qquad a = 2, b = -11, c = -6$$

$$x = \frac{-(-11) \pm \sqrt{(-11)^2 - 4(2)(-6)}}{2(2)} \qquad \text{Replace } a, b, \text{ and } c \text{ in } x = \frac{-b \pm \sqrt{b^2 - 4ac}}{2a}.$$

$$x = \frac{11 \pm \sqrt{121 + 48}}{4} \qquad \text{Simplify.}$$

$$x = \frac{11 \pm \sqrt{169}}{4} \qquad \text{Simplify.}$$

$$x = \frac{11 \pm 13}{4} \qquad \text{Evaluate the square root.}$$

$$x = \frac{11 + 13}{4} \quad \text{or} \quad x = \frac{11 - 13}{4} \qquad \text{Write the solutions using ``or.''}$$

$$x = \frac{24}{4} \quad \text{or} \quad x = \frac{-2}{4} \qquad \text{Simplify the numerators.}$$

$$x = \frac{4 \cdot 6}{4 \cdot 1} \quad \text{or} \quad x = \frac{-1 \cdot 2}{2 \cdot 2} \qquad \text{Find common factors.}$$

$$x = 6 \quad \text{or} \quad x = -\frac{1}{2} \qquad \text{Simplify.}$$

Since the solutions of this equation are rational numbers, we could have instead factored and used the zero product property to find the solutions.

The solutions of some quadratic equations are not real numbers. Write these solutions in $a + bi$ form. Notice that a and b are used in different ways in the quadratic formula and in describing a complex number.

EXAMPLE 4 Use the quadratic formula to solve $3x^2 - 3x + 8 = 0$.

SOLUTION ▶

$$3x^2 - 3x + 8 = 0 \qquad a = 3, b = -3, c = 8.$$

$$x = \frac{-(-3) \pm \sqrt{(-3)^2 - 4(3)(8)}}{2(3)} \qquad \text{Replace } a, b, \text{ and } c \text{ in } x = \frac{-b \pm \sqrt{b^2 - 4ac}}{2a}.$$

$$x = \frac{3 \pm \sqrt{9 - 96}}{6} \qquad \text{Simplify.}$$

$$x = \frac{3 \pm \sqrt{-87}}{6} \qquad \text{Simplify.}$$

$$x = \frac{3 \pm \sqrt{87}\sqrt{-1}}{6} \qquad \text{Rewrite as the product of a square root and } \sqrt{-1}.$$

$$x = \frac{3 \pm \sqrt{87}i}{6} \qquad \text{Simplify; } \sqrt{-1} = i$$

$$x = \frac{3}{6} \pm \frac{\sqrt{87}}{6}i \qquad \text{Rewrite in } a + bi \text{ form.}$$

$$x = \frac{1 \cdot 3}{2 \cdot 3} \pm \frac{\sqrt{87}}{6}i \qquad \text{Find common factors.}$$

$$x = \frac{1}{2} \pm \frac{\sqrt{87}}{6}i \qquad \text{Simplify.}$$

In the next example, we again write the solutions using "or" so that the fractions can be simplified into lowest terms more easily.

EXAMPLE 5 Use the quadratic formula to solve $5x^2 + 6x - 3 = 0$.

SOLUTION ▶

$$5x^2 + 6x - 3 = 0 \qquad a = 5, b = 6, c = -3$$

$$x = \frac{-6 \pm \sqrt{6^2 - 4(5)(-3)}}{2(5)} \qquad \text{Replace } a, b, \text{ and } c \text{ in } x = \frac{-b \pm \sqrt{b^2 - 4ac}}{2a}.$$

$$x = \frac{-6 \pm \sqrt{36 + 60}}{10} \qquad \text{Simplify.}$$

$$x = \frac{-6 \pm \sqrt{96}}{10} \qquad \text{Simplify.}$$

$$x = \frac{-6 \pm \sqrt{16}\sqrt{6}}{10} \qquad \text{Product rule of radicals.}$$

$$x = \frac{-6 \pm 4\sqrt{6}}{10}$$ Simplify.

$$x = -\frac{6}{10} + \frac{4\sqrt{6}}{10} \quad \text{or} \quad x = -\frac{6}{10} - \frac{4\sqrt{6}}{10}$$ Write as two equations.

$$x = -\frac{3 \cdot 2}{5 \cdot 2} + \frac{2 \cdot 2\sqrt{6}}{2 \cdot 5} \quad \text{or} \quad x = -\frac{3 \cdot 2}{5 \cdot 2} - \frac{2 \cdot 2\sqrt{6}}{2 \cdot 5}$$ Find common factors.

$$x = -\frac{3}{5} + \frac{2\sqrt{6}}{5} \quad \text{or} \quad x = -\frac{3}{5} - \frac{2\sqrt{6}}{5}$$ Simplify.

We can also use the quadratic formula to solve equations in which $b = 0$ or $c = 0$.

EXAMPLE 6 | Use the quadratic formula to solve $7x^2 - 9 = 0$.

SOLUTION ▶ $\quad 7x^2 - 9 = 0 \qquad\qquad\qquad a = 7, b = 0, c = -9$

$$x = \frac{-0 \pm \sqrt{0^2 - 4(7)(-9)}}{2(7)}$$ Replace a, b, and c in $x = \dfrac{-b \pm \sqrt{b^2 - 4ac}}{2a}$.

$$x = \frac{\pm\sqrt{252}}{14}$$ Simplify.

$$x = \frac{\pm\sqrt{36}\sqrt{7}}{14}$$ Product rule of radicals.

$$x = \frac{\pm 6\sqrt{7}}{14}$$ Simplify.

$$x = \pm\frac{2 \cdot 3\sqrt{7}}{2 \cdot 7}$$ Find common factors.

$$x = \pm\frac{3\sqrt{7}}{7}$$ Simplify.

Practice Problems

For problems 1–6, use the quadratic formula to solve.

1. $2x^2 + 5x - 13 = 0$ **2.** $4x^2 + 3x + 7 = 0$ **3.** $4x^2 + 3x - 7 = 0$
4. $x^2 - 3x + 2 = 0$ **5.** $x^2 - 14x + 49 = 0$ **6.** $3x^2 + 10 = 0$

The Discriminant

In Section 2.3, we learned that the **discriminant** of the quadratic expression $ax^2 + bx + c$ is $b^2 - 4ac$. We used the discriminant to predict whether a quadratic trinomial expression was prime. Now we see that the discriminant is the *radicand* of the square root in the quadratic formula: $x = \dfrac{-b \pm \sqrt{b^2 - 4ac}}{2a}$. We can use the discriminant to determine the number of unique solutions of a quadratic equation and whether the solutions are rational, irrational, or nonreal numbers.

When the discriminant is equal to 0, the square root in the quadratic formula, $\sqrt{b^2 - 4ac}$, is 0. The quadratic equation has one rational solution.

EXAMPLE 7 A quadratic equation is $x^2 + 6x + 9 = 0$.

(a) Find the discriminant, $b^2 - 4ac$.

SOLUTION ▶

$b^2 - 4ac$	$a = 1, b = 6, c = 9$
$= (6)^2 - 4(1)(9)$	Replace a, b, and c.
$= 36 - 36$	Simplify.
$= 0$	Simplify. The discriminant is equal to 0.

(b) Use the quadratic formula to solve the equation.

▶

$x^2 + 6x + 9 = 0$	$a = 1, b = 6, c = 9$
$x = \dfrac{-6 \pm \sqrt{(6)^2 - 4(1)(9)}}{2(1)}$	$x = \dfrac{-b \pm \sqrt{b^2 - 4ac}}{2a}$
$x = \dfrac{-6 \pm \sqrt{0}}{2}$	Simplify; the discriminant is 0.
$x = -\dfrac{6}{2}$	Simplify.
$x = -3$	Simplify.

The discriminant is equal to 0; the quadratic equation has one rational solution.

When the discriminant is a perfect square that is greater than 0, the square root in the quadratic formula, $\sqrt{b^2 - 4ac}$, is an integer. The quadratic equation has two rational solutions.

EXAMPLE 8 A quadratic equation is $2x^2 + 5x - 7 = 0$.

(a) Find the discriminant, $b^2 - 4ac$.

SOLUTION ▶

$b^2 - 4ac$	$a = 2, b = 5, c = -7$
$= (5)^2 - 4(2)(-7)$	Replace a, b, and c.
$= 25 + 56$	Simplify.
$= 81$	Simplify. The discriminant is a perfect square.

(b) Use the quadratic formula to solve the equation.

▶

$2x^2 + 5x - 7 = 0$	$a = 2, b = 5, c = -7$
$x = \dfrac{-5 \pm \sqrt{5^2 - 4(2)(-7)}}{2(2)}$	$x = \dfrac{-b \pm \sqrt{b^2 - 4ac}}{2a}$
$x = \dfrac{-5 \pm \sqrt{81}}{4}$	Simplify; the discriminant is 81.
$x = \dfrac{-5 \pm 9}{4}$	Simplify.
$x = \dfrac{-5 + 9}{4}$ or $x = \dfrac{-5 - 9}{4}$	Write as two equations.
$x = 1$ or $x = -\dfrac{14}{4}$	Simplify.
$x = -\dfrac{7 \cdot 2}{2 \cdot 2}$	Find common factors.
$x = -\dfrac{7}{2}$	Simplify.

The discriminant is greater than 0 and is a perfect square; the quadratic equation has two rational solutions.

When the discriminant is greater than 0 but is not a perfect square, the square root in the quadratic formula, $\sqrt{b^2 - 4ac}$, is an irrational number. The quadratic equation has two irrational solutions.

EXAMPLE 9 A quadratic equation is $x^2 - 7x - 2 = 0$.

(a) Find the discriminant, $b^2 - 4ac$.

SOLUTION ▶
$$b^2 - 4ac \qquad\qquad a = 1, b = -7, c = -2$$
$$= (-7)^2 - 4(1)(-2) \qquad \text{Replace } a, b, \text{ and } c.$$
$$= 49 + 8 \qquad\qquad \text{Simplify.}$$
$$= 57 \qquad\qquad \text{Simplify; the discriminant is not a perfect square.}$$

(b) Use the quadratic formula to solve the equation.

▶
$$x^2 - 7x - 2 = 0 \qquad\qquad a = 1, b = -7, c = -2$$
$$x = \frac{-(-7) \pm \sqrt{(-7)^2 - 4(1)(-2)}}{2(1)} \qquad x = \frac{-b \pm \sqrt{b^2 - 4ac}}{2a}$$
$$x = \frac{7 \pm \sqrt{57}}{2} \qquad\qquad \text{Simplify; the discriminant is 57.}$$

The discriminant is greater than 0 but is not a perfect square; the quadratic equation has two irrational solutions.

When the discriminant is less than 0, the square root in the quadratic formula, $\sqrt{b^2 - 4ac}$, is an imaginary number. The quadratic equation has two nonreal solutions.

EXAMPLE 10 A quadratic equation is $3x^2 + x + 9 = 0$.

(a) Find the discriminant, $b^2 - 4ac$.

SOLUTION ▶
$$b^2 - 4ac \qquad\qquad a = 3, b = 1, c = 9$$
$$= (1)^2 - 4(3)(9) \qquad \text{Replace } a, b, \text{ and } c.$$
$$= 1 - 108 \qquad\qquad \text{Simplify.}$$
$$= -107 \qquad\qquad \text{Simplify. The discriminant is less than 0.}$$

(b) Use the quadratic formula to solve the equation.

▶
$$3x^2 + x + 9 = 0 \qquad\qquad a = 3, b = 1, c = 9$$
$$x = \frac{-1 \pm \sqrt{(1)^2 - 4(3)(9)}}{2(3)} \qquad x = \frac{-b \pm \sqrt{b^2 - 4ac}}{2a}$$
$$x = \frac{-1 \pm \sqrt{-107}}{6} \qquad\qquad \text{Simplify; the discriminant is } -107.$$
$$x = \frac{-1 \pm \sqrt{107}\sqrt{-1}}{6} \qquad\qquad \text{Rewrite as the product of a square root and } \sqrt{-1}.$$
$$x = \frac{-1 \pm \sqrt{107}\,i}{6} \qquad\qquad \text{Simplify.}$$
$$x = -\frac{1}{6} \pm \frac{\sqrt{107}}{6}\,i \qquad\qquad \text{Write in the form } a + bi.$$

The discriminant is less than 0; the quadratic equation has two nonreal solutions.

> **Using the Discriminant, $b^2 - 4ac$, to Describe the Solutions of a Quadratic Equation, $ax^2 + bx + c = 0$**
>
> If $b^2 - 4ac = 0$, the quadratic equation has one rational solution.
>
> If $b^2 - 4ac > 0$ and is a perfect square, the quadratic equation has two rational solutions.
>
> If $b^2 - 4ac > 0$ and is not a perfect square, the quadratic equation has two irrational solutions.
>
> If $b^2 - 4ac < 0$, the quadratic equation has two nonreal solutions.

Practice Problems

For problems 7–12,
(a) find the discriminant.
(b) use the discriminant to identify the number of solutions.
(c) identify the solutions as rational, irrational, or nonreal numbers.

7. $2x^2 + 5x - 13 = 0$ **8.** $4x^2 + 3x + 7 = 0$ **9.** $4x^2 + 3x - 7 = 0$
10. $x^2 - 3x + 2 = 0$ **11.** $x^2 - 14x + 49 = 0$ **12.** $3x^2 + 10 = 0$

Applications

We can use quadratic equations to describe many situations. However, we can seldom factor these equations. In application problems with quadratic equations, the quadratic formula is often the best method for finding the solution.

Ignoring any air resistance, the equation $d = v_i t + \dfrac{1}{2} at^2$ describes the motion of a falling object. The distance the object falls is d, the initial velocity (speed) of the object is v_i, the acceleration of the object caused by the gravity of the earth is a, and t is time. If we know d, v_i, and a, then $d = v_i t + \dfrac{1}{2} at^2$ is a quadratic equation in one variable, t.

EXAMPLE 11   According to OSHA safety rules, workers must not rest tools such as hammers on a roof. The initial vertical speed, v_i, of a hammer as it falls off a roof is $\dfrac{8.0 \text{ ft}}{1 \text{ s}}$. The acceleration caused by gravity, a, is $\dfrac{32 \text{ ft}}{1 \text{ s}^2}$. Find the time it takes for the hammer to fall 28 ft and hit the ground. Round to the nearest tenth.

SOLUTION ▶ **Step 1 Understand the problem.**
The unknown is the time in seconds that it takes the hammer to hit the ground.

$t = $ time

Step 2 Make a plan.

Replace d, v_i, and a in the formula $d = v_i t + \dfrac{1}{2} at^2$. Use the quadratic formula to solve for t. Since t is the variable in the equation, the quadratic formula is

$$t = \frac{-b \pm \sqrt{b^2 - 4ac}}{2a}.$$

Step 3 Carry out the plan.

Replace the variables in the formula, $d = v_i t + \frac{1}{2}at^2$.

$$d = v_i t + \frac{1}{2}at^2 \qquad \text{The formula for motion of a falling object.}$$

$$\mathbf{28\ ft} = \left(\frac{\mathbf{8.0\ ft}}{\mathbf{1\ s}}\right)t + \frac{1}{2}\left(\frac{\mathbf{32\ ft}}{\mathbf{1\ s^2}}\right)t^2 \qquad \text{Replace } d, v_i, \text{ and } a.$$

$$28 = 8.0t + 16t^2 \qquad \text{Remove units; simplify.}$$

$$\frac{-28 \qquad\qquad\qquad -28}{0 = 16t^2 + 8.0t - 28} \qquad \begin{array}{l}\text{Subtraction property of equality.}\\ \text{Simplify; the equation is in standard form}\\ ax^2 + bx + c = 0.\end{array}$$

Use the quadratic formula to solve $0 = 16t^2 + 8.0t - 28$.

$$t = \frac{-\mathbf{8.0} \pm \sqrt{(\mathbf{8.0})^2 - 4(\mathbf{16})(-\mathbf{28})}}{2(\mathbf{16})} \qquad \begin{array}{l} t = \frac{-b \pm \sqrt{b^2 - 4ac}}{2a};\\ a = 16, b = 8.0, c = -28 \end{array}$$

$$t = \frac{-8.0 \pm \sqrt{\mathbf{1856}}}{\mathbf{32}} \qquad \text{Simplify.}$$

$$t = \frac{-8.0 \pm \mathbf{43.08}\ldots}{32} \qquad \begin{array}{l}\text{Evaluate the square root.}\\ \text{Do not round.}\end{array}$$

$$t = \frac{-8.0 + 43.08\ldots}{32} \quad \text{or} \quad t = \frac{-8.0 - 43.08}{32} \qquad \begin{array}{l}\text{Rewrite as two separate}\\ \text{equations.}\end{array}$$

$$t = \mathbf{1.096}\ldots \quad \text{or} \quad t = -\mathbf{1.59}\ldots \qquad \text{Simplify.}$$

$$t \approx \mathbf{1.1} \qquad \text{or} \quad t \approx -\mathbf{1.6} \qquad \text{Round to the nearest tenth.}$$

Since elapsed time is a positive number, the only reasonable solution is $t \approx 1.1$ s.

Step 4 Look back.

Since $\left(\frac{8.0\ ft}{1\ s}\right)(1.1\ s) + \frac{1}{2}\left(\frac{32\ ft}{1\ s^2}\right)(1.1\ s)^2$ equals 28.16 ft and this is close to the actual distance of 28 ft, the answer seems reasonable. Also, to count the 60 seconds in one minute, we say "one-one thousand, two-one thousand. . . ," each "one thousand" representing one second of time passed. From experience, it seems reasonable that the hammer will hit the ground in the time that a person can say "one-one thousand," or about 1 s.

Step 5 Report the solution.

The hammer hits the ground in about 1.1 s.

The area of a rectangle is the product of its length and width. If we know the area and the perimeter of a rectangle, we can write a quadratic equation in either L or W.

EXAMPLE 12

In Lewiston, Idaho, many people own livestock that they keep in pastures next to their homes. The cost of a 100-ft roll of wire horse fence is $187. A horse owner buys 10 rolls of fencing. Use a quadratic equation in one variable to find the length and width in feet of a one-acre (43,560 ft^2) rectangular pasture surrounded by this fencing. All of the fencing must be used. Round to the nearest whole number.

SOLUTION ▶ **Step 1 Understand the problem.**

There are two unknowns, the length and the width of the pasture. To use a quadratic equation in one variable to find the unknowns, we need to solve the formula for perimeter for a variable. If we choose to solve for L, the unknown is W:

$W =$ width.

Since each roll is 100 ft and there are 10 rolls, the perimeter of the fence is 1000 ft.

Step 2 Make a plan.

The formula for the perimeter of a rectangle is $P = 2W + 2L$. Solve the formula for perimeter for L. The formula for the area of a rectangle is $A = LW$. Replace L in the area formula with the result of solving the perimeter formula, and replace A with the area of the pasture, 43,560 ft². The result is a quadratic equation in W that can be solved by using the quadratic formula.

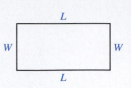

Step 3 Carry out the plan.

Solve $P = 2W + 2L$ for L.

$P = 2W + 2L$	Formula for perimeter.
1000 ft $= 2W + 2L$	Replace P with 1000 ft.
$\dfrac{-2W \qquad -2W}{}$	Subtraction property of equality.
$1000 \text{ ft} - 2W = 0 + 2L$	Simplify.
$\dfrac{1000 \text{ ft}}{2} - \dfrac{2W}{2} = \dfrac{2L}{2}$	Division property of equality.
$500 \text{ ft} - W = L$	Simplify.

```
       500 ft − W
    ┌───────────────┐
  W │               │ W
    └───────────────┘
       500 ft − W
```

Replace L in the area formula with $500 \text{ ft} - W$, and replace A with 43,560 ft².

$A = LW$	
43,560 ft² $= (500 \text{ ft} - W)W$	Replace A with 43,560 ft² and L with $500 \text{ ft} - W$.
$43,560 = (500 - W)W$	Remove units.
$43,560 = 500W - W^2$	Distributive property.
$\dfrac{-500W + W^2 \quad -500W \quad + W^2}{}$	Simplify.
$W^2 - 500W + 43,560 = 0$	Standard form, $a = 1, b = -500, c = 43,560$

Use the quadratic formula to solve for W. Since the variable in the equation is W, the quadratic formula is $W = \dfrac{-b \pm \sqrt{b^2 - 4ac}}{2a}$.

$W = \dfrac{-(-500) \pm \sqrt{(-500)^2 - 4(1)(43,560)}}{2(1)}$	$W = \dfrac{-b \pm \sqrt{b^2 - 4ac}}{2a}$; $a = 1; b = -500; c = 43,560$
$W = \dfrac{500 \pm \sqrt{75,760}}{2}$	Simplify.
$W = \dfrac{500 \pm 275.2\ldots}{2}$	Evaluate the square root; do not round.
$W = \dfrac{500 + 275.2\ldots}{2}$ or $W = \dfrac{500 - 275.2\ldots}{2}$	Rewrite as two separate equations.
$W = 387.6\ldots$ or $W = 112.3\ldots$	Simplify.
$W \approx 388 \text{ ft}$ or $W \approx 112 \text{ ft}$	Round; replace the units.

The first solution matches a diagram in which the length is shorter than the width. Since $L = 500 \text{ ft} - W$, then $L = 500 \text{ ft} - 388 \text{ ft}$, which simplifies to

$L = 112$ ft. The second solution matches the diagram we used to visualize this problem in which the length is longer than the width: $L = 500$ ft $- 112$ ft, which simplifies to $L = 388$ ft.

Step 4 Look back.

If we round 388 ft to 400 ft and 112 ft to 100 ft, the area of the rectangle is (400 ft)(100 ft), which equals 40,000 ft². Since this is close to the actual area of 43,560 ft², the answer seems reasonable.

Step 5 Report the solution.

The pasture should be about 388 ft long and 112 ft wide.

Practice Problems

13. A light rail train traveling at $\dfrac{14.66 \text{ ft}}{1 \text{ s}}$ is accelerating at a constant rate of $\dfrac{2.5 \text{ ft}}{1 \text{ s}^2}$. Find the time in seconds that it takes the train to travel 1 mi by solving the quadratic equation $5280 \text{ ft} = \left(\dfrac{14.66 \text{ ft}}{1 \text{ s}}\right)t + \dfrac{1}{2}\left(\dfrac{2.5 \text{ ft}}{1 \text{ s}^2}\right)t^2$.

    Round to the nearest whole number.

14. A rectangle has an area of 77 ft². The rectangle is 4 ft longer than it is wide. Use a quadratic equation to find the length and width.

15. Acetic acid reacts with ethanol to make water and ethyl acetate. In an experiment, 0.10 mole of acetic acid reacts with 0.15 mole of ethanol. The moles of ethyl acetate made by this reaction, x, is one of the solutions of the quadratic equation $3x^2 - x + 0.060 = 0$. Find the number of moles of ethyl acetate made in this reaction. Round to the nearest hundredth. (One of the solutions is unreasonable because the amount of ethyl acetate made is more than the combined amount of acetic acid and ethanol.)

7.4 VOCABULARY PRACTICE

Match the term with its description.

1. The degree of a quadratic equation in one variable
2. $b^2 - 4ac$
3. The product of the length and the width of a rectangle
4. $x = \dfrac{-b \pm \sqrt{b^2 - 4ac}}{2a}$
5. $ax^2 + bx + c = 0, a \neq 0$
6. A fraction in which the numerator and denominator have no common factors
7. In the term $3x$, 3 is an example of this.
8. The method for solving a quadratic equation that we used to find the quadratic formula
9. The solution of a quadratic equation with a discriminant that is greater than zero belongs to this set of numbers
10. The distance around a rectangle

A. area of a rectangle
B. coefficient
C. completing the square
D. discriminant
E. fraction in lowest terms
F. perimeter of a rectangle
G. quadratic formula
H. real numbers
I. standard form of a quadratic equation in one variable
J. two

7.4 Exercises

Follow your instructor's guidelines for showing your work.

For exercises 1–12,
(a) find the discriminant.
(b) use the discriminant to identify the number of solutions.
(c) identify the solutions as rational, irrational, or nonreal numbers.

1. $6p^2 + 13p - 5 = 0$
2. $6q^2 + q - 12 = 0$
3. $2x^2 - 10x + 5 = 0$
4. $2x^2 - 12x + 7 = 0$
5. $2w^2 + 5w + 11 = 0$
6. $2d^2 + 7d + 13 = 0$
7. $9x^2 + 12x + 4 = 0$
8. $4x^2 + 12x + 9 = 0$
9. $5x^2 - x = 0$
10. $3x^2 - x = 0$
11. $x^2 + 49 = 0$
12. $x^2 + 36 = 0$

13. If the discriminant is equal to 0, a quadratic equation has only one solution. Explain why.

14. If the discriminant is less than 0, the solutions of a quadratic equation are always nonreal numbers. Explain why these solutions cannot be real numbers.

15. If the discriminant is a perfect square greater than 0, a quadratic equation has two solutions that are rational numbers. Explain why these solutions cannot be irrational numbers.

16. If the discriminant is not a perfect square and it is greater than 0, a quadratic equation has two solutions that are irrational numbers. Explain why these solutions cannot be rational numbers.

For exercises 17–42, use the quadratic formula to solve.

17. $7x^2 + 9x - 2 = 0$
18. $11x^2 + 7x - 3 = 0$
19. $7x^2 + 9x + 2 = 0$
20. $11x^2 + 8x - 3 = 0$
21. $7x^2 + 9x + 20 = 0$
22. $11x^2 + 7x + 3 = 0$
23. $4w^2 + 6w - 9 = 0$
24. $3d^2 + 8d - 12 = 0$
25. $x^2 + 3x + 1 = 0$
26. $x^2 + 5x + 1 = 0$
27. $x^2 + 3x + 2 = 0$
28. $x^2 + 5x + 6 = 0$
29. $3m^2 - 16 = 0$
30. $5v^2 - 49 = 0$
31. $6r^2 + r + 8 = 0$
32. $8m^2 + m + 6 = 0$

33. $9x^2 + 12x + 4 = 0$
34. $4x^2 + 12x + 9 = 0$
35. $x^2 + 8x = 0$
36. $x^2 + 12x = 0$
37. $x^2 - 8x = 0$
38. $x^2 - 12x = 0$
39. $x^2 + 8 = 0$
40. $x^2 + 12 = 0$
41. $x^2 - 8 = 0$
42. $x^2 - 12 = 0$

For exercises 43–46,
(a) use the quadratic formula to solve.
(b) solve by completing the square.
(c) solve by factoring and using the zero product property.

43. $x^2 - 22x + 117 = 0$
44. $x^2 - 20x + 96 = 0$
45. $6x^2 + 7x - 20 = 0$
46. $3x^2 + 17x - 56 = 0$

For exercises 47–48,
(a) use the quadratic formula to solve.
(b) solve by factoring and using the zero product property.
(c) solve by isolating the x^2 term and taking the square root of both sides.

47. $x^2 - 169 = 0$
48. $4x^2 - 25 = 0$

For exercises 49–52, use a quadratic equation and the quadratic formula to solve the application problem.

49. A 2009 Toyota Corolla is traveling about $\dfrac{29.3 \text{ ft}}{1 \text{ s}}$ on a freeway entrance ramp. To merge, the driver applies a constant acceleration of $\dfrac{8.5 \text{ ft}}{1 \text{ s}^2}$. Use the equation $d = v_i t + \dfrac{1}{2} at^2$ to find the time for the car to travel the 1000 ft to the freeway. Round to the nearest tenth.

50. A 2009 Pontiac G8 GXP is traveling about $\dfrac{29.3 \text{ ft}}{1 \text{ s}}$ on a freeway entrance ramp. To merge, the driver applies a constant acceleration of $\dfrac{17.6 \text{ ft}}{1 \text{ s}^2}$. Use the equation $d = v_i t + \dfrac{1}{2} at^2$ to find the time for the car to travel the 1000 ft to the freeway. Round to the nearest tenth.

51. A quilter receives a gift of 4.5 yd of antique lace. She wants to make a wall hanging and trim it using all of this lace. The area of the wall hanging will be 1620 in.^{2}. Find the length and width of the wall hanging (1 yd = 36 in.).

52. A homeowner will use 55 ft of decorative edging for a rectangular cement patio. The area of the patio will be 186 ft^2. Find the length and width of the patio.

Exercises 53 and 54 are typical exercises in beginning college chemistry classes. The unit for the amount of a substance is a mole.

53. The amount in moles of phosphoruspentachloride gas (PCl_5) at the equilibrium point of a reaction is the solution of the quadratic equation $x^2 + 0.198x - 0.0076 = 0$. Find this amount. Round to the nearest thousandth. (The amount cannot be less than 0.)

54. The amount in moles of nitrous oxide gas at the equilibrium point of a reaction is two times the solution of the quadratic equation $3.99x^2 + 0.0048x - 0.0021 = 0$. Find this amount. Round to the nearest thousandth. (The amount cannot be less than 0.)

Exercises 55–56 are typical exercises in a beginning college class in electronics. Resistors restrict the flow of electrons in an electric circuit. The unit of resistance is an ohm.

55. Two resistors in series, R_1 and R_2, have a total resistance of 20 ohms. The resistance of R_2 is one of the solutions of the equation $x^2 - 20x + 75 = 0$. The other solution of the equation is the resistance of R_1. Find the value of each resistor in ohms.

56. Two resistors in series, R_1 and R_2, have a total resistance of 3000 ohms. The resistance of R_2 is one of the solutions of $x^2 - 3000x + 2,187,500 = 0$. The other solution is the resistance of R_1. Find the value of each resistor in ohms.

For exercises 57–58, use the Pythagorean theorem $a^2 + b^2 = c^2$, the relationship distance = (speed)(time), and the quadratic formula to solve.

57. Interstate 70 meets Highway 71 in Colorado. From this junction, one car travels east at $\dfrac{75 \text{ mi}}{1 \text{ hr}}$ on Interstate 70. One hour later, another car leaves this junction traveling south at $\dfrac{60 \text{ mi}}{1 \text{ hr}}$. When the cars are 200 mi apart, how long has the first car been traveling? Write the time in hours and minutes. Round to the nearest minute.

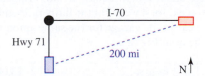

58. Interstate 35 meets Highway 64 in Oklahoma. From this junction, one car travels north at $\dfrac{75 \text{ mi}}{1 \text{ hr}}$ on Interstate 35. One hour later, another car leaves this junction traveling west at $\dfrac{55 \text{ mi}}{1 \text{ hr}}$. When the cars are 150 mi apart, how long has the first car been traveling? Write the time in hours and minutes. Round to the nearest minute.

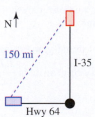

For exercises 59–64, use any method from this chapter to solve the quadratic equation.

59. A mine tour includes a vertical shaft that is 275 ft deep. A tourist drops a rock into the shaft. $\left(\text{The initial velocity of the rock is } \dfrac{0 \text{ ft}}{1 \text{ s}}. \right)$ The acceleration on the rock caused by gravity is $\dfrac{32 \text{ ft}}{1 \text{ s}^2}$. Use the formula $d = v_i t + \dfrac{1}{2} at^2$ to find the time it takes a rock to drop to the bottom of the shaft. Round to the nearest tenth.

60. A wind turbine project in Washington state includes towers that are 50 m high. A wrench is dropped from the top of a tower. $\left(\text{The initial velocity of the wrench is } \dfrac{0 \text{ m}}{\text{s}}. \right)$ The acceleration on the wrench caused by gravity is $\dfrac{9.8 \text{ m}}{1 \text{ s}^2}$. Use the formula $d = v_i t + \dfrac{1}{2} at^2$ to find the time it takes the wrench to hit the ground. Round to the nearest tenth.

61. The speed of a tour boat in still water is $\dfrac{11 \text{ mi}}{1 \text{ hr}}$. The boat travels 18 miles upstream on the Columbia River against the current. After a stop at a historic site, the boat returns downstream. The total time spent traveling on the water was 3.5 hr. The speed of the current is one of the solutions of $18(11 + x) + 18(11 - x) = 3.5(11 + x)(11 - x)$. Find the speed of the current. Round to the nearest tenth.

62. The speed of a bicyclist on flat ground in still air is $\dfrac{13 \text{ mi}}{1 \text{ hr}}$. The bicyclist rides 18 mi on a flat road against the wind. After a rest stop, the bicyclist returns with the wind. The total time spent biking was 4 hr. The change in the speed of the bicyclist caused by the wind is one of the solutions of $18(13 + x) + 18(13 - x) = 4(13 + x)(13 - x)$. Find the change in speed caused by the wind. Round to the nearest tenth.

63. A drinking fountain spouts water approximately in the shape of a parabola. The times at which the water reaches a height of 0.30 ft are the solutions of the equation $0.3 = 5.65t - 16t^2$, where t is in seconds. Find these values for t. Round to the nearest hundredth.

64. A punter kicks a football at a 30-degree angle. The path of the football is approximately in the shape of a parabola. The times at which the football reaches a height of 25.0 ft are the solutions of the equation $25 = 45t - 16t^2$, where t is in seconds. Find these values for t. Round to the nearest hundredth.

For exercises 65–66, a polygon is a closed figure in a plane such as a square or triangle. The sides of a polygon never intersect. Exactly two sides connect at any vertex. A diagonal is a line from vertex to vertex that is not a side. The number of diagonals is equal to d where $d = \frac{1}{2}n^2 - \frac{3}{2}n$ and n is the number of sides.

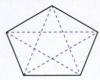

65. A polygon has 20 diagonals. How many sides does it have?

66. A polygon has 54 diagonals. How many sides does it have?

For exercises 67–68, a right triangle has a 90-degree angle. The side opposite the 90-degree angle is the *hypotenuse*. The length of the hypotenuse is c. The other sides are legs. The lengths of the legs are a and b. The Pythagorean theorem describes the relationship of the legs and hypotenuse: $a^2 + b^2 = c^2$.

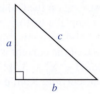

67. The hypotenuse of a right triangle is 15 in. One leg is 3 in. shorter than the other leg. Find the length of each leg.

68. The hypotenuse of a right triangle is 155 cm. One leg is 31 cm longer than the other leg. Find the length of each leg.

For exercises 69–70, a developer is dividing a large, trapezoidal-shaped lot into two smaller lots with equal areas. A line drawn parallel to the north and south edges divides the lots.

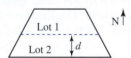

69. The length of the north edge of the lot is 160 ft. The south edge is 200 ft. The lot is 400 ft deep. The distance of the dividing line from the south edge is one of the solutions of $d^2 - 4000d + 720,000 = 0$. Find this distance. Round to the nearest tenth.

70. The length of the north edge of the lot is 110 ft. The south edge is 150 ft. The lot is 300 ft deep. The distance of the dividing line from the south edge is one of the solutions of $d^2 - 2250d + 292,500 = 0$. Find this distance. Round to the nearest tenth.

For exercises 71–72, in the formula $d = \dfrac{V^2}{2fg}$, the minimum stopping distance in feet for a car is d. The speed of the car in feet per second is V. The acceleration caused by gravity, g, is $32\dfrac{ft}{s^2}$. The friction coefficient of the road and tire is f. The friction coefficient between a dry road and a car with good tires is 0.8.

71. The stopping distance of a car with good tires on a dry road is 151.25 ft. Find the speed of the car.

72. A car with good tires is traveling $\dfrac{30\ mi}{1\ hr}$ on a dry road. Find the stopping distance of the car. Round to the nearest foot. (Change the speed into units of feet per second; 1 mi = 5280 ft)

For exercises 73–74, when the cost of making a product is equal to the revenue received from selling the product, the company is "breaking even."

73. The cost in dollars of making x products is $0.2x^2 + 25x + 150$. The revenue in dollars received from selling x products is $70x$. Find the two values of x at which the company breaks even. Round to the nearest whole number.

74. The cost in dollars of making x products is $0.1x^2 + 25x + 150$. The revenue in dollars received from selling x products is $60x$. Find the two values of x at which the company breaks even. Round to the nearest whole number.

Problem Solving: Practice and Review

Follow your instructor's guidelines for using the five steps as outlined in Section 1.5, p. 51.

75. The All-Alaska Sweepstakes is a winner-take-all 408-mi dogsled race from Nome to Candle in Alaska. There is one prize of $100,000 for the winning musher and team. There were 16 mushers entered for the March 2008 race. On March 25, 2008, the price of one troy ounce of gold was $933. Find the percent of the prize that was paid for by entry fees. Round to the nearest percent.

Mushers can enter the 100th anniversary Sweepstakes race until midnight on Sunday, March 16. The entry fee is $2,000 plus an ounce of gold. (*Source:* www.allalaskasweepstakes .org, March 15, 2008)

76. Find the difference in the cost to park a car in the parking garage under Union Square in San Francisco for 8 hr, 5 days a week, for 4 weeks at the daily rates and the cost to park a car in the garage for the same amount of time paying the monthly rate.

0–1 hours	$3.00
1–2 hours	$7.00
2–3 hours	$9.00
3–4 hours	$13.00
4–5 hours	$17.00
5–6 hours	$21.00
6–7 hours	$22.00
7–24 hour maximum	$31.00
Regular monthly parking	$360.00

Source: www.unionsquareshop.com/ parking.html

77. The relationship of density, mass, and volume is $D = \dfrac{M}{V}$. The average density of granite is $\dfrac{2.75\ g}{1\ cm^3}$. Find the volume in cubic centimeters of a granite curling stone that weighs 42 lb (1 kg ≈ 2.2 1b). Round to the nearest hundred.

A curling stone is of circular shape, having a circumference no greater than 91.44 cm (36 in.), a height no less than 11.43 cm (4.5 in.), and a weight, including handle and bolt, no greater than 19.96 kg (44 lb) and no less than 17.24 kg (38 lb). (*Source:* www.worldcurling.org)

© Bork/Shutterstock

78. A model of the number of children under age 18 in the United States living with a single parent, $N(x)$, in x years since 1970 is $N(x) = \left(\dfrac{353{,}866 \text{ children}}{1 \text{ year}}\right)x + 9{,}084{,}929$ children. Predict how many children will live with a single parent in 2015. Round to the nearest thousand. (*Source:* www.cdc.gov)

Find the Mistake

For exercises 79–82, the completed problem has one mistake.
(a) Describe the mistake in words, or copy down the whole problem and highlight or circle the mistake.
(b) Do the problem correctly.

79. Problem: Use the discriminant to describe the solutions of $x^2 - 6x + 3 = 0$.

Incorrect Answer: $b^2 - 4ac$

$$= -6^2 - 4(1)(3)$$
$$= -36 - 12$$
$$= -48$$

Since the discriminant is less than 0, this equation has two nonreal solutions.

80. Problem: Use the quadratic formula to solve $x^2 - 5x = 0$.

Incorrect Answer: $x = \dfrac{-b \pm \sqrt{b^2 - 4ac}}{2a}$

$$x = \dfrac{-5 \pm \sqrt{(-5)^2 - 4(1)(0)}}{2(1)}$$
$$x = \dfrac{-5 \pm \sqrt{25}}{2}$$
$$x = \dfrac{-5 + 5}{2} \quad \text{or} \quad x = \dfrac{-5 - 5}{2}$$
$$x = 0 \quad \text{or} \quad x = -5$$

81. Problem: Use the quadratic formula to solve $x^2 + 3x + 11 = 0$.

Incorrect Answer: $x = \dfrac{-b \pm \sqrt{b^2 - 4ac}}{2a}$

$$x = \dfrac{-3 \pm \sqrt{3^2 - 4(1)(11)}}{2(1)}$$
$$x = \dfrac{-3 \pm \sqrt{-35}}{2}$$

82. Problem: Use the quadratic formula to solve $2x^2 + 5x + 4 = 0$.

Incorrect Answer: $x = \dfrac{-b \pm \sqrt{b^2 - 4ac}}{2a}$

$$x = \dfrac{-5 \pm \sqrt{5^2 - 4(1)(4)}}{2(1)}$$
$$x = \dfrac{-5 \pm \sqrt{9}}{2}$$
$$x = \dfrac{-5 + \sqrt{9}}{2} \quad \text{or} \quad x = \dfrac{-5 - \sqrt{9}}{2}$$
$$x = -1 \quad \text{or} \quad x = -4$$

Review

83. Use an algebraic method to find the zero of $f(x) = 3x + 12$.

84. Use interval notation to describe the domain of $f(x) = x^2$.

85. Explain the purpose of the vertical line test.

86. Use an algebraic method to find any real zeros of $f(x) = |x + 6| + 9$.

SUCCESS IN COLLEGE MATHEMATICS

87. If you have a program, major, or career goal selected, describe it. If you are undecided, describe what you are most interested in.

© pakul54/Shutterstock

A function is a set of ordered pairs in which each input value corresponds to exactly one output value. The function $f(x) = -466x^2 + 60,412x + 4,908,164$ describes the relationship of the number of years since 2000, x, and the projected population of Minnesota, $f(x)$. In this section, we will study this and other quadratic functions.

SECTION 7.5

Quadratic Functions

After reading the text, working the practice problems, and completing assigned exercises, you should be able to:

1. Given the graph of a quadratic function, identify its vertex, axis of symmetry, minimum or maximum output value, domain, range, and real zeros.

2. Use an algebraic method to find the real zeros of a quadratic function.

3. Use the discriminant to describe the real zeros of a quadratic function.

4. Identify the vertex, axis of symmetry, minimum or maximum output value, domain, and range of a quadratic function arranged in vertex form.

5. Evaluate a quadratic function.

The Graph of a Quadratic Function

A quadratic function is a second-degree polynomial function. The standard form of a quadratic function is $y = ax^2 + bx + c$, where $a, b,$ and c are real numbers and $a \neq 0$. The domain of a quadratic function, like that of all polynomial functions, is the set of real numbers, $(-\infty, \infty)$.

The graph of a quadratic function is a parabola (Figure 1). In Section 3.3, we used a table of ordered pairs to graph quadratic functions. The vertex is the "top" or "bottom" point of the parabola. When the parabola opens up, the y-coordinate of the vertex is the **minimum** output value. The range is then the set of real numbers greater than or equal to the minimum output value. In interval notation, the range is $[\text{minimum output value}, \infty)$. When the parabola opens down, the y-coordinate of the vertex is the **maximum** output value. The range is then the set of real numbers less than or equal to the maximum output value. In interval notation, the range is $(-\infty, \text{maximum output value}]$.

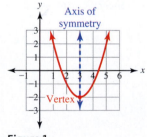
Figure 1

The **axis of symmetry** is the vertical line that passes through the vertex. Each "arm" of the parabola is symmetric about this vertical line. Since the axis of symmetry is not part of the function, it is drawn on the graph of the function as a dashed line.

When we use a graph to identify the vertex of a parabola, the vertex is an estimate. The accuracy of the estimate depends on the quality of the graph.

EXAMPLE 1 The graph represents the quadratic function
$f(x) = x^2 + 2$.

SOLUTION **(a)** Estimate the vertex.

▶ $(0, 2)$

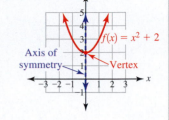

(b) Identify the direction in which the graph opens.

▶ Opens up

(c) Estimate the minimum output value.

▶ 2

(d) Write the equation of the estimated axis of symmetry.

▶ $x = 0$

(e) Use interval notation to represent the domain.

▶ $(-\infty, \infty)$

(f) Use interval notation to represent the estimated range.

▶ $[2, \infty)$

In the next example, the parabola opens down. The y-coordinate of the vertex is the maximum output value.

EXAMPLE 2 The graph represents the quadratic function
$y = -2x^2 + 12x - 17$.

SOLUTION **(a)** Estimate the vertex.

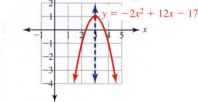

▶ $(3, 1)$

(b) Identify the direction in which the graph opens.

▶ Opens down

(c) Estimate the maximum output value.

▶ 1

(d) Write the equation of the estimated axis of symmetry.

▶ $x = 3$

(e) Use interval notation to represent the domain.

▶ $(-\infty, \infty)$

(f) Use interval notation to represent the estimated range.

▶ $(-\infty, 1]$

Practice Problems

For problems 1–3, the graph represents a quadratic function.
(a) Estimate the vertex.
(b) Identify the direction that the graph opens.
(c) Estimate the minimum or maximum output value.
(d) Write the equation of the estimated axis of symmetry.
(e) Use interval notation to represent the domain.
(f) Use interval notation to represent the estimated range.

1.

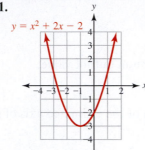

$y = x^2 + 2x - 2$

2.

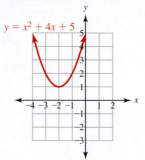

$y = x^2 + 4x + 5$

3.

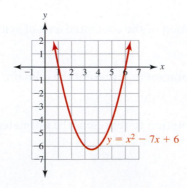

$y = -x^2 + 10x - 28$

Real Zeros of a Quadratic Function

A **zero** of a function is an input value that corresponds to an output value of 0. The x-coordinate of any x-intercept of the graph of a quadratic function is a real zero. If the graph does not have an x-intercept, the function has no zeros that are real numbers.

EXAMPLE 3 Use the graph of $y = x^2 - 7x + 6$ to estimate its real zeros.

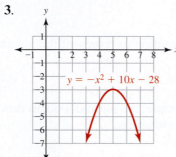

$y = x^2 - 7x + 6$

SOLUTION ▶ The x-intercepts appear to be $(1, 0)$ and $(6, 0)$. The estimated real zeros of the function are $x = 1$ and $x = 6$.

To use an algebraic method to find the exact value of a real zero, replace the output value with 0 and solve the quadratic equation for the input value. If the solutions are nonreal numbers, the function has no real zeros. To solve, factor and use the zero product property, use completing the square, or use the quadratic formula. In the next example, the polynomial looks as though it may be difficult to factor, so we use the quadratic formula.

EXAMPLE 4   The graph represents $f(x) = 6x^2 - 13x - 63$. Use an algebraic method to find any real zeros of this function.

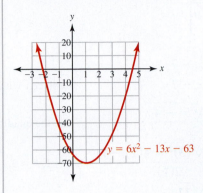

SOLUTION ▶

$$f(x) = 6x^2 - 13x - 63$$

$$0 = 6x^2 - 13x - 63 \qquad \text{Replace } f(x) \text{ with 0.}$$

$$x = \frac{-b \pm \sqrt{b^2 - 4ac}}{2a} \qquad \text{Use the quadratic formula to solve.}$$

$$x = \frac{-(-13) \pm \sqrt{(-13)^2 - 4(6)(-63)}}{2(6)} \qquad a = 6, b = -13, c = -63$$

$$x = \frac{13 \pm \sqrt{1681}}{12} \qquad \text{Simplify.}$$

$$x = \frac{13 \pm 41}{12} \qquad \text{Evaluate the radical.}$$

$$x = \frac{13 + 41}{12} \quad \text{or} \quad x = \frac{13 - 41}{12} \qquad \text{Rewrite as two equations.}$$

$$x = \frac{54}{12} \quad \text{or} \quad x = \frac{-28}{12} \qquad \text{Simplify.}$$

$$x = \frac{9 \cdot 6}{2 \cdot 6} \quad \text{or} \quad x = -\frac{7 \cdot 4}{3 \cdot 4} \qquad \text{Find common factors.}$$

$$x = \frac{9}{2} \quad \text{or} \quad x = -\frac{7}{3} \qquad \text{Simplify.}$$

This function has two rational zeros, $x = \dfrac{9}{2}$ and $x = -\dfrac{7}{3}$. Since the zeros are rational, we could have solved $0 = 6x^2 - 13x - 63$ by factoring and using the zero product property. The factored equation is $0 = (2x - 9)(3x + 7)$.

In the next example, the zeros of the function are irrational numbers.

EXAMPLE 5 The graph represents $y = x^2 + 3x - 7$. Use an algebraic method to find any real zeros of this function.

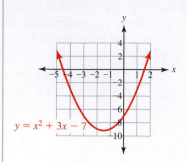

$y = x^2 + 3x - 7$

SOLUTION ▶

$$y = x^2 + 3x - 7$$

$$0 = x^2 + 3x - 7 \qquad \text{Replace } y \text{ with } 0.$$

$$x = \frac{-b \pm \sqrt{b^2 - 4ac}}{2a} \qquad \text{Use the quadratic formula to solve.}$$

$$x = \frac{-3 \pm \sqrt{(3)^2 - 4(1)(-7)}}{2(1)} \qquad a = 1, b = 3, c = -7$$

$$x = \frac{-3 \pm \sqrt{37}}{2} \qquad \text{Simplify.}$$

This function has two irrational zeros, $x = \dfrac{-3 + \sqrt{37}}{2}$ and $x = \dfrac{-3 - \sqrt{37}}{2}$.

To find the real zeros of a quadratic function, we solve a quadratic equation in standard form, $ax^2 + bx + c = 0$. The discriminant of this equation is $b^2 - 4ac$. We can use the discriminant to identify the number of real zeros and whether they are rational or irrational numbers.

Using the Discriminant, $b^2 - 4ac$, to Describe the Zeros of a Quadratic Function, $y = ax^2 + bx = 0$

If $b^2 - 4ac = 0$, the quadratic function has one rational zero. The graph of the function has one x-intercept.

If $b^2 - 4ac > 0$ and is a perfect square, the quadratic function has two rational zeros. The graph of the function has two x-intercepts.

If $b^2 - 4ac > 0$ and is not a perfect square, the quadratic function has two irrational zeros. The graph of the function has two x-intercepts.

If $b^2 - 4ac < 0$, the quadratic function has no real zeros. The graph of the function has no x-intercepts.

EXAMPLE 6 The graph represents $f(x) = x^2 + 6x + 10$.

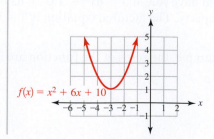

$f(x) = x^2 + 6x + 10$

(a) Find the discriminant.

SOLUTION ▶   For $f(x) = x^2 + 6x + 10$, $a = 1$, $b = 6$, and $c = 10$.

$b^2 - 4ac$ The discriminant.

$= (6)^2 - 4(1)(10)$ Replace a, b, and c.

$= -4$ Simplify.

(b) Identify the number of real zeros.

▶   Since the discriminant is less than 0, this function has no real zeros. Notice that the graph of the function has no x-intercepts.

A real zero of a quadratic function is the x-coordinate of the x-intercept. If we know the direction in which the graph of a quadratic function opens and the vertex, we can identify the number of x-intercepts and the number of real zeros.

EXAMPLE 7   **(a)** The vertex of the graph of a quadratic function is $(3, 2)$, and it opens up. Identify the number of real zeros of this function.

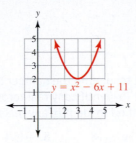

SOLUTION ▶   The vertex is above the x-axis. Since the graph opens up, it does not intersect the x-axis and has no x-intercepts. The function has no real zeros.

(b) The vertex of the graph of a quadratic function is $(3, 2)$, and it opens down. Identify the number of real zeros of this function.

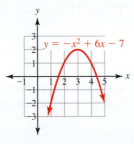

▶   The vertex is above the x-axis. Since the graph opens down, it intersects the x-axis twice and has two x-intercepts. The function has two real zeros.

(c) The vertex of the graph of a quadratic function is $(3, 0)$, and it opens up. Identify the number of real zeros of this function.

▶   The vertex is on the x-axis. The graph has one x-intercept. The function has one real zero.

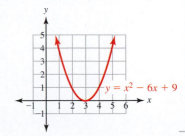

Practice Problems

For problems 4–6, use the graph of the function to estimate any real zeros.

4.
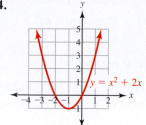
$y = x^2 + 2x$

5.
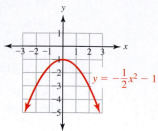
$y = -\dfrac{1}{2}x^2 - 1$

6.
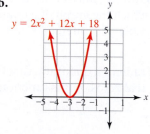
$y = 2x^2 + 12x + 18$

For problems 7–9,
(a) find the discriminant.
(b) use the discriminant to identify the number of real zeros.
(c) use the discriminant to identify any real zeros as rational or irrational numbers.
(d) use an algebraic method to identify any real zeros.

7. $f(x) = -2x^2 + 15x - 18$ **8.** $p(x) = 3x^2 + 5x - 9$

9. $w(x) = x^2 - 3x + 10$

10. Use the vertex and the direction that the graph of a function opens to identify the number of real zeros.

    a. vertex: $(-4, 5)$; opens up   **b.** vertex: $(-4, 5)$; opens down
    c. vertex: $(-4, 0)$; opens up   **d.** vertex: $(6, -2)$; opens up
    e. vertex: $(6, -2)$; opens down

Vertex Form

If we write a quadratic function in vertex form, we can identify the vertex, the axis of symmetry, and the maximum or minimum output value.

> **Vertex Form of a Quadratic Equation in One Variable**
>
> $$y = a(x - h)^2 + k \qquad a, h, \text{ and } k \text{ are real numbers, } a \neq 0$$
>
> The vertex is (h, k). The axis of symmetry is $x = h$.
>
> If $a > 0$, the parabola opens up. The minimum output value is k.
>
> If $a < 0$, the parabola opens down. The maximum output value is k.

The value of a determines the relative "height" of the graph. When a is positive, the parabola opens up. As $|a|$ increases, the graph of the parabola changes. The graph gets "taller." *For a given input, the output of the tallest graph is greater than the output of the shortest graph.*

EXAMPLE 8 Identify the tallest parabola and the shortest parabola.

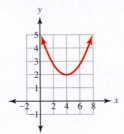

A: $y = 0.2(x - 4)^2 + 2$

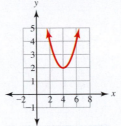

B: $y = 0.5(x - 4)^2 + 2$

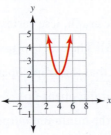

C: $y = 1(x - 4)^2 + 2$

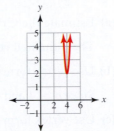

D: $y = 10(x - 4)^2 + 2$

SOLUTION ▶ The vertex of all of the graphs is (4, 2). To find the tallest graph, evaluate a given input value, such as $x = 6$. The value of $f(6)$ for the tallest graph will be greater than $f(6)$ for the other graphs. We can also identify a and compare the value of $|a|$. The value of $|a|$ is greatest for the tallest graph and is least for the shortest graph.

Graph	Function	a	$\|a\|$	$f(6)$	
A	$y = 0.2(x - 4)^2 + 2$	0.2	0.2	**2.8**	Shortest
B	$y = 0.5(x - 4)^2 + 2$	0.5	0.5	**4**	
C	$y = 1(x - 4)^2 + 2$	1	1	**6**	
D	$y = 10(x - 4)^2 + 2$	10	10	**42**	Tallest

When a is negative, the parabola opens down. As $|a|$ increases, the graph again changes. Even though the parabola opens down, we say that the parabola in Figure 5 is *taller* than the parabolas in Figures 2, 3, and 4.

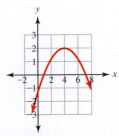

Figure 2
$y = -0.2(x - 4)^2 + 2$

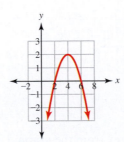

Figure 3
$y = -0.5(x - 4)^2 + 2$

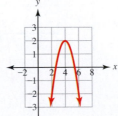

Figure 4
$y = -1(x - 4)^2 + 2$

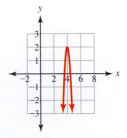

Figure 5
$y = -10(x - 4)^2 + 2$

In vertex form, the ordered pair (h, k) is the vertex of the function. When the parabola opens up, k is the minimum output value. When the parabola opens down, k is the maximum output value.

EXAMPLE 9 The quadratic function represented by the graph, $g(x) = 4(x - 3)^2 + 5$, is in vertex form, $y = a(x - h)^2 + k$.

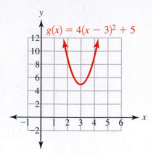

(a) Estimate the vertex from the graph. Use vertex form to identify the exact vertex.

SOLUTION ▶   The estimated vertex is $(3, 5)$. The exact vertex (h, k) is $(3, 5)$.

(b) Use vertex form to confirm that the graph opens up.

▶   Since $a = 4$ and $4 > 0$, the graph opens up.

(c) Use vertex form to identify the minimum output value.

▶   The exact minimum output value is $k, 5$.

(d) Use vertex form to write the equation of the axis of symmetry.

▶   The equation of the axis of symmetry is $x = h$: $x = 3$.

(e) Use interval notation to represent the domain.

▶ $(-\infty, \infty)$

(f) Use interval notation to represent the range.

▶ $[5, \infty)$

In vertex form, $y = a(x - h)^2 + k$, the value of h does not include the $-$ sign to its left. In the next example, the function is not in vertex form. We rewrite $(x + h)$ as $(x - (-h))$.

EXAMPLE 10 The graph represents the quadratic function $g(x) = -7(x + 2)^2 + 1$.

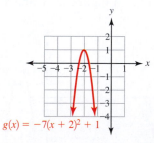

SOLUTION ▶   **(a)** Estimate the vertex from the graph. Use vertex form to identify the exact vertex.

▶   The estimated vertex is $(-2, 1)$. Since vertex form is $y = a(x - h)^2 + k$, $g(x) = -7(x + 2)^2 + 1$ is *not* in vertex form. Rewrite in vertex form by rewriting $+2$ as $-(-2)$: $g(x) = -7(x - (-2))^2 + 1$. The exact vertex (h, k) is $(-2, 1)$.

(b) Use vertex form to confirm that the graph opens down.

▶   Since $a = -7$ and $-7 < 0$, the graph of this function opens down.

(c) Use vertex form to identify the maximum output value.

▶ The maximum output value, k, is 1.

(d) Use vertex form to write the equation of the axis of symmetry.

▶ The axis of symmetry, $x = h$, is $x = -2$.

(e) Use interval notation to represent the domain.

▶ $(-\infty, \infty)$

(f) Use interval notation to represent the range.

▶ $(-\infty, 1]$

If we know the coordinates of the vertex and the value of a, we can write the equation of a quadratic function in vertex form.

EXAMPLE 11 | Use the value of a and the coordinates of the vertex to complete the function.

SOLUTION | **(a)** $a = 3$; vertex: $(7, 1)$

▶ $\quad y = a(x - h)^2 + k \qquad$ Vertex form.

$\quad y = 3(x - 7)^2 + 1 \qquad a = 3; h = 7, k = 1$

(b) $a = -4$; vertex: $(-6, 9)$

▶ $\quad y = a(x - h)^2 + k \qquad$ Vertex form.

$\quad y = -4(x - (-6))^2 + 9 \qquad a = -4; h = -6, k = 9$

$\quad y = -4(x + \mathbf{6})^2 + 9 \qquad$ Simplify.

(c) $a = 1$; vertex: $(2, -8)$

▶ $\quad y = a(x - h)^2 + k \qquad$ Vertex form.

$\quad y = 1(x - 2)^2 + (-8) \qquad a = 1; h = 2, k = -8$

$\quad y = (x - 2)^2 - 8 \qquad$ Simplify.

Practice Problems

For problems 11–12,
(a) use vertex form to identify the vertex.
(b) use vertex form to identify whether the graph opens up or down.
(c) use vertex form to identify the maximum or minimum output value.
(d) use vertex form to write the equation of the axis of symmetry.
(e) use interval notation to represent the domain.
(f) use interval notation to represent the range.
(g) predict the number of real zeros.

11. $y = 5(x - 8)^2 + 1$ $\qquad$ **12.** $y = -5(x + 7)^2 - 6$

13. Use the value of a and the coordinates of the vertex to complete the function.
$\quad$ **a.** $a = 3$; vertex: $(6, 2)$ $\qquad$ **b.** $a = -2$; vertex: $(-8, 4)$
$\quad$ **c.** $a = -1$; vertex: $(5, -10)$

Evaluating a Quadratic Function

To evaluate a quadratic function, replace the variable in the function rule with the input value and simplify. When the input value is a negative number, use parentheses.

EXAMPLE 12 A quadratic function is $f(x) = x^2 - 6x - 15$. Evaluate $f(-4)$.

SOLUTION ▶

$$f(x) = x^2 - 6x - 15$$
$$f(\mathbf{-4}) = (\mathbf{-4})^2 - 6(\mathbf{-4}) - 15 \qquad \text{Replace } x \text{ with } -4.$$
$$f(-4) = \mathbf{16 + 24} - 15 \qquad \text{Simplify.}$$
$$f(-4) = \mathbf{25} \qquad \text{Simplify.}$$

In the next example, the quadratic function is a model and it does not include any units. We determine the units from the description of the relationship represented by the model.

EXAMPLE 13 The function $f(x) = -466x^2 + 60{,}412x + 4{,}908{,}164$ describes the relationship of the number of years since 2000, x, and the projected population of Minnesota, $f(x)$. Find the projected population in 2018. Round to the nearest thousand.

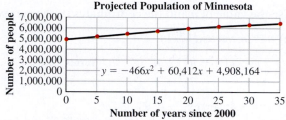

Projected Population of Minnesota

$y = -466x^2 + 60{,}412x + 4{,}908{,}164$

Source: www.lmic.state.mn.us/

SOLUTION ▶

$$f(x) = -466x^2 + 60{,}412x + 4{,}908{,}164$$
$$f(\mathbf{18}) = -466(\mathbf{18})^2 + 60{,}412(\mathbf{18}) + 4{,}908{,}164 \qquad \begin{array}{l}2018 - 2000 = 18 \text{ years;}\\ \text{replace } x \text{ with } 18.\end{array}$$
$$f(18) = \mathbf{-150{,}984 + 1{,}087{,}416} + 4{,}908{,}164 \qquad \text{Simplify.}$$
$$f(18) = \mathbf{5{,}844{,}596} \qquad \text{Simplify.}$$
$$f(18) \approx \mathbf{5{,}845{,}000 \text{ people}} \qquad \text{Round; include units.}$$

Practice Problems

14. A quadratic function is $x^2 - 6x + 7$.
    a. Evaluate: $f(3)$ **b.** Evaluate: $f(-3)$

15. A quadratic function is $-x^2 + 8x - 15$.
    a. Evaluate: $f(6)$ **b.** Evaluate: $f(-6)$

16. A car with good tires is traveling on a dry road. The function $D = \dfrac{2.15x^2}{51.2}$ estimates its stopping distance in feet, D, when x is the speed of the car in miles per hour. Estimate the stopping distance when the speed of a car is $\dfrac{55 \text{ mi}}{1 \text{ hr}}$. Round to the nearest whole number.

17. The quadratic function $C(x) = 0.062x^2 + 2.16x + 20.95$ is a model of local government spending on criminal justice from 1982 to 2006. In this function, x is the years since 1982 and $C(x)$ is the amount in billions of dollars. Assume that the amount of spending continues to grow at this rate. Use this function to estimate the justice expenditures in 2013. Round to the nearest tenth of a billion. (*Source:* bjs.ujp-usdoj.gov, Dec. 2008)

7.5 Vocabulary Practice

Match the term with its description.

1. $y = ax^2 + bx + c, a \neq 0$
2. $y = a(x - h)^2 + k, a \neq 0$
3. The vertical line that passes through the vertex of a parabola
4. A set of ordered pairs in which each input corresponds to only one output
5. The set of the inputs of a function
6. The set of the outputs of a function
7. $b^2 - 4ac$
8. The shape of the graph of a quadratic function
9. The ordered pair whose y-coordinate is the maximum or minimum output value of a quadratic function
10. An input value that results in an output value of 0

A. axis of symmetry
B. discriminant
C. domain
D. function
E. parabola
F. range
G. standard form of a quadratic function
H. vertex
I. vertex form of a quadratic function
J. real zero of a function

7.5 Exercises

Follow your instructor's guidelines for showing your work.

1. Describe the shape of a graph that represents a quadratic function.

2. The axis of symmetry passes through the vertex of a parabola. Although not perfectly symmetric, the body of a goat also has a horizontal and a vertical axis of symmetry. Describe or show an axis of a symmetry on a goat.

© Laura Bracken

For exercises 3–10, use a table of ordered pairs to graph the quadratic function. (To review, refer to Section 3.3.)

3. $g(x) = x^2 - 8$
4. $g(x) = x^2 - 6$
5. $y = x^2 - 10x + 23$
6. $y = x^2 - 12x + 33$
7. $f(x) = -x^2 + 3$
8. $f(x) = -x^2 + 4$
9. $y = -x^2 - 8x - 11$
10. $y = -x^2 - 4x + 2$

For exercises 11–22, the graph represents a quadratic function.
(a) Estimate the vertex.
(b) Identify the direction in which the graph opens.
(c) Estimate the minimum or maximum output value.
(d) Write the equation of the estimated axis of symmetry.
(e) Use interval notation to represent the domain.
(f) Use interval notation to represent the estimated range.

11.

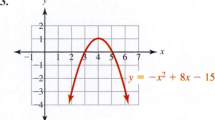

$y = x^2 - 8x + 17$

12.

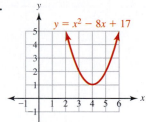

$y = x^2 - 10x + 27$

13.

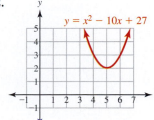

$y = -x^2 + 8x - 15$

14.

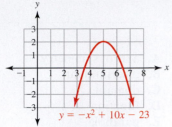

$y = -x^2 + 10x - 23$

15.

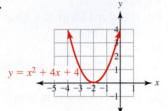

$y = x^2 + 4x + 4$

16.

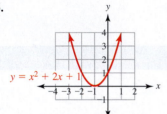

$y = x^2 + 2x + 1$

17.

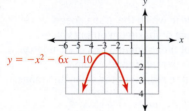

$y = -x^2 - 6x - 10$

18.

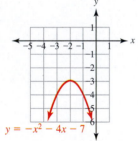

$y = -x^2 - 4x - 7$

19.

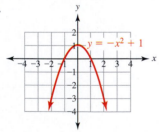

$y = -x^2 + 1$

20.

$y = -x^2 - 1$

21.

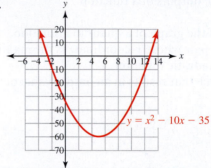

$y = x^2 - 10x - 35$

22.

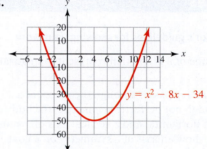

$y = x^2 - 8x - 34$

For exercises 23–28, the graph represents a quadratic function. Estimate any real zeros of the function.

23.

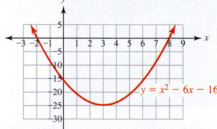

$y = x^2 - 6x - 16$

24.

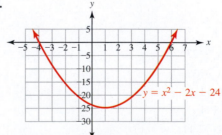

$y = x^2 - 2x - 24$

25.

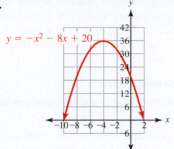

$y = -x^2 - 8x + 20$

26.

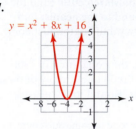

$y = -x^2 + 2x + 8$

27.

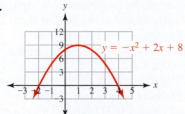

$y = x^2 + 8x + 16$

28.

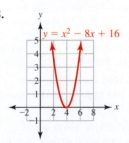

$y = x^2 - 8x + 16$

For exercises 29–42,

(a) find the discriminant.

(b) use the discriminant to identify the number of real zeros.

(c) use the discriminant to identify the real zeros as rational or irrational numbers.

(d) if the function has any real zeros, use an algebraic method to find them.

29. $f(x) = x^2 - 4x - 21$

30. $g(x) = x^2 - 4x - 45$

31. $w(x) = 6x^2 - 3x + 5$

32. $p(x) = 10x^2 - 7x + 3$

33. $q(x) = 3x^2 - 9x + 2$

34. $f(x) = 3x^2 - 7x + 1$

35. $p(x) = -2x^2 + 5x - 4$

36. $w(x) = -3x^2 + 7x - 8$

37. $f(x) = x^2 + 8x + 16$

38. $f(x) = x^2 - 8x + 16$

39. $y = x^2 + 12$

40. $y = x^2 + 20$

41. $y = x^2 - 12$

42. $y = x^2 - 20$

For exercises 43–50,

(a) the graph of a quadratic function has the given vertex and opens in the given direction. Identify the number of real zeros of the function.

(b) explain.

43. $(5, 6)$; down

44. $(4, 9)$; down

45. $(-3, -6)$; down

46. $(-9, -4)$; down

47. $(0, 0)$; up

48. $(0, 0)$; down

49. $(-7, 1)$; up

50. $(-8, 2)$; up

51. We can use the discriminant to predict the number of solutions of a quadratic equation in one variable. We can also use the discriminant to predict the number of real zeros of a quadratic function. Describe the difference between a quadratic *equation* in one variable and a quadratic *function*.

52. We can use the discriminant to predict whether the real solutions of a quadratic equation in one variable are rational or irrational. We can also use the discriminant to predict whether the zeros of a quadratic function are rational or irrational. Describe the difference between a quadratic *equation* in one variable and a quadratic *function*.

For exercises 53–68,

(a) use vertex form to identify the vertex.

(b) use vertex form to identify whether the graph opens up or down.

(c) use vertex form to identify the maximum or minimum output value.

(d) use vertex form to write the equation of the axis of symmetry.

(e) use interval notation to represent the domain.

(f) use interval notation to represent the range.

53. $y = -4(x - 2)^2 + 5$

54. $y = -5(x - 3)^2 + 7$

55. $g(x) = 4(x - 2)^2 - 5$

56. $w(x) = 5(x - 3)^2 - 7$

57. $y = 4(x + 2)^2 - 5$

58. $y = 5(x + 3)^2 - 7$

59. $y = 3(x - 5)^2$

60. $y = 3(x - 4)^2$

61. $g(x) = -5x^2$

62. $p(x) = -6x^2$

63. $f(x) = (x + 2)^2$

64. $f(x) = (x - 3)^2$

65. $y = -10(x - 8)^2 - 1$

66. $y = -8(x - 7)^2 - 2$

67. $g(x) = -(x + 3)^2$

68. $g(x) = -(x + 1)^2$

For exercises 69–74, use the value of *a* and the coordinates of the vertex to complete the function.

69. $a = 2$; vertex: $(5, -9)$

70. $a = 4$; vertex: $(3, -15)$

71. $a = 2$; vertex: $(-5, 9)$

72. $a = 4$; vertex: $(-3, 15)$

73. $a = 2$; vertex: $(0, 0)$

74. $a = 4$; vertex: $(0, 0)$

75. The function $N(x) = 93.232x^2 + 1499.6x + 7090.8$ describes the relationship of the number of years since 2003, *x,* and the number of Starbucks Coffee Company stores open at fiscal year end, $N(x)$. If this relationship continues, find the number of Starbucks Coffee Company stores in 2011. Round to the nearest hundred. (*Source:* investor.starbucks.com, 2008)

76. The function $A(x) = 816.29x^2 + 187.26x + 2895.4$ describes the relationship of the number of years since 2004, *x,* and the number of organic acres of hay produced in Washington state, $A(x)$. If this relationship continues, find the number of acres in 2012. Round to the nearest hundred. (*Source:* csanr.wsu.edu, Feb. 2009)

77. The function $D(x) = 0.0225x^2 - 0.16x + 6.0201$ describes the relationship of the number of years since 1980, *x,* and the number of people in millions with diagnosed diabetes in the United States, $D(x)$. If this relationship continues, find the number of people with diagnosed diabetes in 2012. Round to the nearest tenth of a million. (*Source:* www.cdc.gov, Oct. 2008)

78. The function $N(x) = 1347.7x^2 + 2702.3x + 102{,}171$ describes the relationship of the number of years since 1991, *x,* and the number of nursing students enrolled in U.S. baccalaureate nursing programs, $N(x)$. If this relationship continues, find the number of students in 2014. Round to the nearest thousand. (*Source:* www.cdc.gov, 2008)

79. The Alico Building in Waco, Texas, was once the tallest skyscraper west of the Mississippi River. The height to the observation tower is 282 ft. The function $d(t) = 16t^2$ describes the relationship of the time that an object falls in seconds, *t,* and the distance in feet an object will fall from the observation tower in this time (neglecting air resistance), $d(t)$. If a coin falls from the observation tower, will it have hit the ground after 3 s? Explain.

80. The Can Manufacturers Institute publishes a manual that includes the dimensions of standard can sizes.

The function $V(d) = 4.6875\pi\left(\dfrac{d}{2}\right)^2$ describes the

relationship of the outside diameter of a can, *d,* and the volume of the can in cubic inches, $V(d)$. The diameter of a No. 1 Tall can is 3.0625 in. The diameter of a No. 2.5 can is 4.0625 in. Find the difference in volume of these cans. Round to the nearest tenth.

Problem Solving: Practice and Review

Follow your instructor's guidelines for using the five steps as outlined in Section 1.5, p. 51.

81. In the election for the president of Mexico in 2006, candidate López Obrador lost to Felipe Calderón by a margin of six tenths of 1 percent of the total number of certified votes. Find the total vote. Round to the nearest thousand. (*Source:* global.nytimes.com, Feb. 4, 2009)

On Sept. 5, the Federal Electoral Judicial Tribunal certified Calderón's victory by a margin of 233,831 votes. (*Source:* www.washingtonpost.com, July 2006)

82. In March 2011, retail sales of video game systems, games and accessories totaled $1.47 billion. This is a 4% decrease from the retail sales in March 2010. Find the retail sales in March 2010. Round to the nearest hundredth of a billion. (*Source:* www.huffingtonpost.com, April 15, 2011)

83. According to the U.S. Census, the population in 1900 was about 76,094,000 people. Predict how many of these people died from typhoid fever. Round to the nearest thousand.

A century ago, it was common for food to come into contact with human sewage, picking up germs. For instance, in 1900, typhoid fever killed 31 people out of 100,000. As sanitation improved and such diseases largely disappeared, new ailments, many associated with animal waste, took their place. (*Source:* www.nytimes.com, April 11, 2009)

84. Find the number of U.S. adults who claim that they know someone personally who has experienced domestic violence.

Approximately 33 million or 15% of all U.S. adults admit that they were a victim of domestic violence. Furthermore, 6 in 10 (U.S) adults claim that they know someone personally who has experienced domestic violence. (*Source:* www.ndvh .org, 2006)

Technology

For exercises 85–88, graph the function on a graphing calculator. Choose a window that shows the intercepts and any maximum or minimum value. Round irrational numbers to the nearest hundredth.

(a) Sketch the graph; describe the window.

(b) Use the calculator to identify any real zero(s).

(c) Use the calculator to identify any minimum or maximum.

(d) Use interval notation to represent the domain.

(e) Use interval notation to represent the range.

(f) Use the value command to evaluate the function when $x = 2$.

85. $f(x) = -4x^2 - 12x - 5$

86. $g(x) = -8x^2 - 22x - 9$

87. $h(x) = 3x^2 - 54x + 228$

88. $k(x) = 3x^2 - 48x + 178$

Find the Mistake

For exercises 89–92, the completed problem has one mistake.
(a) Describe the mistake in words, or copy down the whole problem and highlight or circle the mistake.
(b) Do the problem correctly.

89. Problem: Identify the vertex of the graph of
$f(x) = 3(x + 2)^2 - 9$.

Incorrect Answer: The vertex is $(2, -9)$.

90. Problem: Use an algebraic method to find any real zeros
of $g(x) = x^2 + 3x + 5$.

Incorrect Answer: $0 = x^2 + 3x + 5$

$$x = \frac{-3 \pm \sqrt{3^2 - 4(1)(5)}}{2(1)}$$

$$x = \frac{-3 \pm \sqrt{-11}}{2}$$

$$x = \frac{-3 \pm \sqrt{11}\,i}{2}$$

Real zeros: $x = -\dfrac{3}{2} + \dfrac{\sqrt{11}}{2}\,i,\ x = -\dfrac{3}{2} - \dfrac{\sqrt{11}}{2}\,i$

91. Problem: The function $N(x) = -0.0343x^2 - 0.8289x + 57.083$ describes the relationship between the number of years since 1997, x, and the rate of colon cancer per 100,000 people in the United States, $N(x)$. If this relationship continues, find the rate of colon cancer per 100,000 people in the United States in 2015. Round to the nearest tenth. (*Source:* Surveillance, Epidemiology, and End Results (SEER) Program of the National Cancer Institute, July 2011)

Incorrect Answer: $N(x) = -0.0343x^2 - 0.8289x + 57.083$

$$N(2015) = -0.0343(2015)^2 - 0.8289(2015) + 57.083$$

$$N(2015) \approx -140{,}878.9$$

Since the rate is negative, the model predicts that there will be no colon cancer in the United States in 2015.

92. Problem: Use the graph of the function to estimate any real zeros.

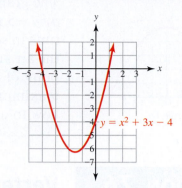

Incorrect Answer: The real zeros of the function appear to be $(-4, 0)$ and $(1, 0)$.

Review

93. Change $y = 6x - 9$ to shift its graph up 5 units.

94. Change $y = \dfrac{2}{3}x$ to shift its graph down 7 units.

95. Change $y - 6 = 3(x + 2)$ to shift the graph right 10 units and down 3 units.

96. Change $y + 9 = 3(x + 7)$ to shift its graph to the left 5 units and up 4 units.

SUCCESS IN COLLEGE MATHEMATICS

97. An important part of planning for success in college is choosing the right courses in the right sequence. Do you have to take another math class after you complete Intermediate Algebra? If so, what are its name and course number?

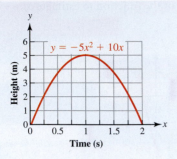

The graph represents the quadratic function $y = -5x^2 + 10x$, where $a = -5$ and $b = 10$. This function represents the relationship of the time in seconds, x, that a football thrown at a 30-degree angle is in the air and the height in meters, y, of the football above its original point of release. The vertex of the graph of this function includes the maximum output value of the function. In this section, we will learn how to use a and b to find the maximum output value.

SECTION 7.6

Vertex Form of a Quadratic Function

After reading the text, working the practice problems, and completing assigned exercises, you should be able to:

1. Change a quadratic function in vertex form to cause a horizontal and/or vertical shift in its graph.

2. Given a quadratic equation in standard form ($a = 1$), rewrite it in vertex form.

3. Given a quadratic equation in standard form ($a \neq 1$), rewrite it in vertex form.

4. Given a quadratic equation in standard form, use a, b, and c or a and b to identify the vertex of its graph.

Translation

In Section 3.6, we studied horizontal and vertical shifts in the graphs of linear functions. When a linear equation is in slope-intercept form, $y = mx + b$, a change in the value of b causes a vertical shift in the graph of the function.

EXAMPLE 1 The linear function represented by the graph is $y = -3x + 2$.

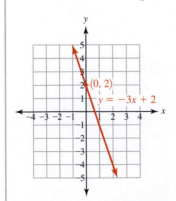

(a) Change this function to shift its graph down 5 units.

SOLUTION ▶ The function is in slope-intercept form, $y = mx + b$, and $b = 2$. To shift the graph down 5 units, subtract 5 from b: $2 - 5 = -3$. The new function is $y = -3x - 3$.

(b) Use translation to graph the new function.

> The graph of this function is the same as the graph of $y = -3x + 2$, shifted down 5 units. The y-intercept is $(0, -3)$, and the slope is $-\dfrac{3}{1}$.

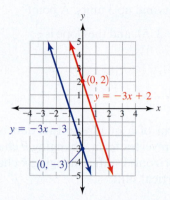

When a linear equation is in point-slope form, $y - y_1 = m(x - x_1)$, a change in the value of x_1 causes a horizontal shift in the graph. A change in the value of y_1 causes a vertical shift in the graph.

Translation of Linear Functions

If a linear function is in slope-intercept form, $y = mx + b$, changing b causes a vertical shift in the position of its graph.

If a linear function is in point-slope form, $y - y_1 = m(x - x_1)$, changing x_1 causes a horizontal shift in its graph. Changing y_1 causes a vertical shift in its graph.

EXAMPLE 2   The linear function represented by the graph is $y - 2 = 3(x - 4)$.

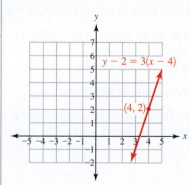

(a) Identify: (x_1, y_1)

SOLUTION ▶ $(4, 2)$

(b) Change the function to shift its graph to the left 7 units and up 3 units.

> To shift the graph to the left 7 units, subtract 7 from x_1: $4 - 7 = -3$. To shift the graph up 3 units, add 3 to y_1: $2 + 3 = 5$. Since the graph of the shifted line includes the point $(-3, 5)$ and the slope remains the same, the new function is $y - 5 = 3(x - (-3))$. Simplifying, the function is $y - 5 = 3(x + 3)$.

(c) Use translation to graph the new function.

▶ The graph of this function is the same as the graph of $y - 2 = 3(x - 4)$, shifted to the left 7 units and up 3 units. A point on the line is $(-3, 5)$, and the slope is $\frac{3}{1}$.

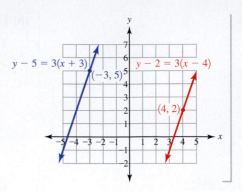

The vertex form of a quadratic function is $y = a(x - h)^2 + k$. When a quadratic function is in vertex form, changing h changes the x-coordinate of the vertex and results in a horizontal shift. Changing k changes the y-coordinate of the vertex and results in a vertical shift.

EXAMPLE 3   The quadratic function represented by the graph is $y = (x - 2)^2 + 1$.

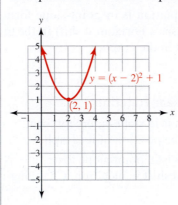

(a) Identify: (h, k)

SOLUTION ▶ $(2, 1)$

(b) Change the function to shift its graph to the right 3 units and down 5 units.

▶ To shift the graph to the right 3 units, add 3 to h: $2 + 3 = 5$. To shift the graph down 5 units, subtract 5 from k: $1 - 5 = -4$. The vertex of the shifted graph is $(5, -4)$. The new function is $y = (x - 5)^2 - 4$.

(c) Use translation to graph the new function.

▶ The graph of this function is the same as the graph of $y = (x - 2)^2 + 1$, shifted to the right 3 units and down 5 units. The vertex is $(5, -4)$.

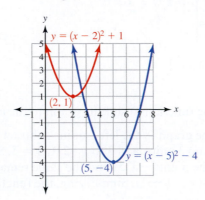

> **Translating Quadratic Functions in Vertex Form, $y = a(x - h)^2 + k$**
>
> 1. Identify (h, k).
> 2. Change h to cause a horizontal shift in the graph. Change k to cause a vertical shift in the graph.

EXAMPLE 4   The quadratic function represented by the graph is $y = (x + 4)^2 + 3$.

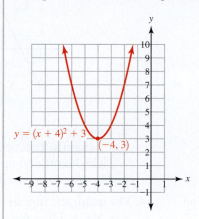

(a) Identify: (h, k)

SOLUTION ▶   In vertex form, this function is $y = (x - (-4))^2 + 3$. The vertex is $(-4, 3)$.

(b) Change the function to shift its graph to the left 1 unit and up 3 units.

▶   To shift the graph to the left 1 unit, subtract 1 from h: $-4 - 1 = -5$. To shift the graph up 3 units, add 3 to k: $3 + 3 = 6$. The vertex of the shifted graph is $(-5, 6)$. The new function is $y = (x - (-5))^2 + 6$. Simplifying, the function is $y = (x + 5)^2 + 6$.

(c) Use translation to graph the new function.

▶   The graph of this function is the same as the graph of $y = (x + 4)^2 + 3$, shifted to the left 1 unit and up 3 units. The vertex is $(-5, 6)$.

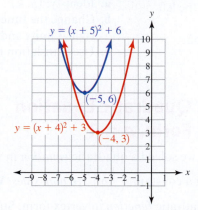

Practice Problems

1. The linear function represented by the graph is $y = 2x - 7$.

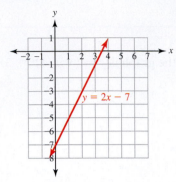

a. Change this function to shift its graph up 6 units.

b. Use translation to graph the new function.

2. The linear function represented by the graph is $y - 5 = 2(x - 1)$.

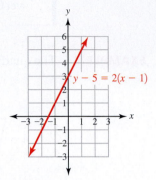

a. Identify: (x_1, y_1)

b. Change the function to shift its graph to the right 3 units and up 11 units.

c. Use translation to graph the new function.

3. The quadratic function represented by the graph is $y = 5(x - 8)^2 + 1$.

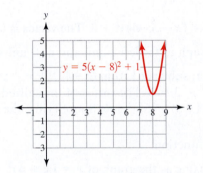

a. Identify: (h, k)

b. Change the function to shift its graph to the left 5 units and down 6 units.

c. Use translation to graph the new function.

4. The quadratic function represented by the graph is $y = 5(x + 7)^2 - 6$.

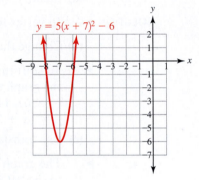

a. Identify: (h, k)

b. Change the function to shift its graph to the right 3 units and up 2 units.

c. Use translation to graph the new function.

Writing a Quadratic Equation in Vertex Form, $a = 1$

In Section 7.3, we solved a quadratic *equation* in which $a = 1$ by completing the square, adding a constant, $\left(\dfrac{1}{2}b\right)^2$, to *both sides*. We can use completing the square to rewrite a quadratic *function* in vertex form. Since we cannot add $\left(\dfrac{1}{2}b\right)^2$ to *both sides* of the function, we *add and subtract* $\left(\dfrac{1}{2}b\right)^2$ on the *same side* of the function. Since $\left(\dfrac{1}{2}b\right)^2 - \left(\dfrac{1}{2}b\right)^2 = 0$, this does not change the function.

EXAMPLE 5   A quadratic function in standard form is $f(x) = x^2 + 6x - 7$.

(a) Use completing the square to rewrite this function in vertex form.

SOLUTION ▶

$f(x) = x^2 + 6x - 7$	$a = 1, b = 6, c = -7$
$f(x) = (x^2 + 6x) - 7$	Group the first two terms.
$f(x) = (x^2 + 6x + \underline{}) - \underline{} - 7$	Complete the square; add and subtract a number.
$f(x) = (x^2 + 6x + \mathbf{9}) - \mathbf{9} - 7$	Complete the square; $\left(\frac{1}{2}b\right)^2 = 9$
$f(x) = (x^2 + 6x + 9) - \mathbf{16}$	Simplify.
$f(x) = \mathbf{(x + 3)(x + 3)} - 16$	Factor $x^2 + 6x + 9$.
$f(x) = \mathbf{(x + 3)^2} - 16$	Use exponential notation.
$f(x) = (x - (-3))^2 - 16$	Rewrite in vertex form; $3 = -(-3)$

(b) Identify the vertex, (h, k).

▶   The vertex is $(-3, -16)$.

(c) Write the equation of the axis of symmetry.

▶   The equation of the axis of symmetry is $x = -3$.

(d) Identify whether the parabola opens up or down.

▶   Since $a = 1$ and $1 > 0$, the parabola opens up.

(e) Identify the maximum or minimum output value.

▶   Since the parabola opens up, it is a minimum, -16.

(f) Use interval notation to represent the domain.

▶   The domain of a quadratic function is the set of real numbers, $(-\infty, \infty)$.

(g) Use interval notation to represent the range.

▶   The minimum output value is -16. The range is $[-16, \infty)$.

(h) Identify the number of real zeros.

▶   Since the vertex is below the x-axis and the graph opens up, the graph has two x-intercepts and the function has two real zeros.

The pattern for factoring perfect square trinomials is $a^2 + 2ab + b^2 = (a + b)(a + b)$. In the next example, we need to factor a perfect square trinomial in which the constant is a fraction, $x^2 + 5x + \frac{25}{4}$. Rewriting to match the pattern $a^2 + 2ab + b^2$, the trinomial is $(x)^2 + 2(x)\left(\frac{5}{2}\right) + \left(\frac{5}{2}\right)^2$. The factors are $(a + b)(a + b)$ or $\left(x + \frac{5}{2}\right)\left(x + \frac{5}{2}\right)$.

EXAMPLE 6   A quadratic function in standard form is $y = x^2 + 5x + 3$.

(a) Use completing the square to rewrite this function in vertex form.

SOLUTION ▶

$y = x^2 + 5x + 3$	$a = 1, b = 5, c = 3$
$y = (x^2 + 5x) + 3$	Group the first two terms.
$y = (x^2 + 5x + \underline{}) - \underline{} + 3$	Complete the square; add and subtract a number.
$y = \left(x^2 + 5x + \frac{25}{4}\right) - \frac{25}{4} + 3$	Complete the square; $\left(\frac{1}{2}b\right)^2 = \frac{25}{4}$

$$y = \left(x^2 + 5x + \frac{25}{4}\right) - \frac{25}{4} + \frac{3}{1} \cdot \frac{4}{4} \qquad \text{Multiply by a fraction equal to 1; } \frac{4}{4} = 1$$

$$y = \left(x^2 + 5x + \frac{25}{4}\right) - \frac{25}{4} + \frac{12}{4} \qquad \text{Simplify.}$$

$$y = \left(x^2 + 5x + \frac{25}{4}\right) - \frac{13}{4} \qquad \text{Simplify.}$$

$$y = \left(x + \frac{5}{2}\right)\left(x + \frac{5}{2}\right) - \frac{13}{4} \qquad \begin{array}{l}\text{Use a pattern to factor;}\\ a^2 + 2ab + b^2 = (a + b)(a + b)\end{array}$$

$$y = \left(x + \frac{5}{2}\right)^2 - \frac{13}{4} \qquad \text{Use exponential notation.}$$

$$y = \left(x - \left(-\frac{5}{2}\right)\right)^2 - \frac{13}{4} \qquad \text{Rewrite in vertex form; } \frac{5}{2} = -\left(-\frac{5}{2}\right)$$

(b) Identify the vertex, (h, k).

▶ The vertex is $\left(-\frac{5}{2}, -\frac{13}{4}\right)$.

Writing a Quadratic Function, $y = ax^2 + bx + c$, in Vertex Form, $y = a(x - h)^2 + k$, When $a = 1$

1. Group the first two terms, $x^2 + bx$, in parentheses: $y = (x^2 + bx) + c$.

2. Add $\left(\frac{1}{2}b\right)^2$ inside the parentheses; subtract $\left(\frac{1}{2}b\right)^2$ outside the parentheses. The function is now

$$y = \left(x^2 + bx + \left(\frac{1}{2}b\right)^2\right) - \left(\frac{1}{2}b\right)^2 + c$$

3. Factor the trinomial in the parentheses, and rewrite in exponential notation; simplify outside of the parentheses.

When the value of b is a negative number, the value of $\left(\frac{1}{2}b\right)^2$ is a positive number. In the next example, since $\left(\frac{1}{2}b\right)^2 = 25$, we add 25 inside the parentheses and subtract 25 outside the parentheses.

EXAMPLE 7 A quadratic function in standard form is $y = x^2 - 10x - 8$.

(a) Use completing the square to rewrite this function in vertex form.

SOLUTION ▶

$$y = x^2 - 10x - 8 \qquad a = 1, b = -10, c = -8$$

$$y = (x^2 - 10x) - 8 \qquad \text{Group the first two terms.}$$

$$y = (x^2 - 10x + \underline{\quad}) - \underline{\quad} - 8 \qquad \text{Complete the square; add and subtract a number.}$$

$$y = (x^2 - 10x + \mathbf{25}) - \mathbf{25} - 8 \qquad \text{Complete the square; } \left(\frac{1}{2}b\right)^2 = 25$$

$$y = (x^2 - 10x + 25) - \mathbf{33} \qquad \text{Simplify.}$$

$$y = (\mathbf{x} - \mathbf{5})(\mathbf{x} - \mathbf{5}) - 33 \qquad \text{Factor.}$$

$$y = (x - 5)^2 - 33 \qquad \text{Use exponential notation.}$$

(b) Identify the vertex, (h, k).

▶ The vertex is $(5, -33)$.

Practice Problems

For problems 5–8,
(a) use completing the square to rewrite each function in vertex form.
(b) identify the vertex.
(c) write the equation of the axis of symmetry.
(d) use interval notation to represent the domain.
(e) use interval notation to represent the range.

5. $y = x^2 + 8x - 21$ **6.** $f(x) = x^2 - 12x + 7$
7. $f(x) = x^2 + 7x + 3$ **8.** $y = x^2 + 3x - 11$

In the next example, the value of a in $y = ax^2 + bx + c$ is not 1. To rewrite this function in vertex form, we begin by grouping the first two terms and factoring out a.

EXAMPLE 8 | A quadratic function in standard form is $y = 2x^2 + 12x + 17$.

(a) Use completing the square to rewrite this function in vertex form.

SOLUTION ▶

$y = 2x^2 + 12x + 17$	$a = 2, b = 12, c = 17$
$y = (2x^2 + 12x) + 17$	Group the first two terms.
$y = 2(x^2 + 6x) + 17$	Factor out 2.
$y = 2(x^2 + 6x + \underline{}) - \underline{} + 17$	Complete the square; add and subtract a number.

To complete the square, we add $\left(\dfrac{1}{2}b\right)^2$, which equals 9, inside the parentheses.

However, since the terms in the parentheses are multiplied by 2, we need to subtract 18 outside the parentheses.

$y = 2(x^2 + 6x + \mathbf{9}) - \mathbf{2 \cdot 9} + 17$	
$y = 2(x^2 + 6x + 9) - \mathbf{18} + 17$	$\left(\dfrac{1}{2}b\right)^2 = 9; 2(9) = 18$
$y = 2(x^2 + 6x + 9) - \mathbf{1}$	Simplify.
$y = 2(x + 3)(x + 3) - 1$	Factor.
$y = 2(x + 3)^2 - 1$	Use exponential notation.
$y = 2(x - (-3))^2 - 1$	Rewrite $+3$ as $-(-3)$.

(b) Identify the vertex, (h, k).

▶   The vertex is $(-3, -1)$.

In the next example, the value of a in $y = ax^2 + bx + c$ is a negative number, -3. To rewrite this function in vertex form, we begin by grouping the first two terms and factoring out -3.

EXAMPLE 9 | A quadratic function in standard form is $y = -3x^2 + 5x + 9$.

(a) Use completing the square to rewrite this function in vertex form.

SOLUTION ▶

$y = -3x^2 + 5x + 9$	$a = -3, b = 5, c = 9$
$y = (-3x^2 + 5x) + 9$	Group the first two terms.

To factor out -3, we need to find a number whose product with -3 is 5:
$-3(x^2 + \underline{}x)$.

Since $-3\left(-\dfrac{5}{3}\right) = 5$, the factoring is $-3\left(x^2 - \dfrac{5}{3}x\right)$.

$$y = -3\left(x^2 - \frac{5}{3}x\right) + 9 \qquad \text{Factor out } -3.$$

$$y = -3\left(x^2 - \frac{5}{3}x + \underline{}\right) - \underline{} + 9 \qquad \begin{array}{l}\text{Complete the square; add and} \\ \text{subtract a number.}\end{array}$$

To complete the square, we add $\left(\dfrac{1}{2}b\right)^2$, which equals $\dfrac{25}{36}$, inside the parentheses. However, since the terms in the parentheses are multiplied by -3, we need to subtract $-3\left(\dfrac{25}{36}\right)$ outside the parentheses.

$$y = -3\left(x^2 - \frac{5}{3}x + \frac{\mathbf{25}}{\mathbf{36}}\right) - (-3)\left(\frac{\mathbf{25}}{\mathbf{36}}\right) + 9 \qquad \left(\frac{1}{2}b\right)^2 = \frac{25}{36}; -3\left(\frac{25}{36}\right) = -\frac{25}{12}$$

$$y = -3\left(x^2 - \frac{5}{3}x + \frac{25}{36}\right) + \frac{\mathbf{25}}{\mathbf{12}} + 9 \qquad \text{Simplify.}$$

$$y = -3\left(x^2 - \frac{5}{3}x + \frac{25}{36}\right) + \frac{25}{12} + \frac{9}{1}\cdot\frac{\mathbf{12}}{\mathbf{12}} \qquad \text{Multiply by a fraction equal to 1; } \frac{12}{12} = 1$$

$$y = -3\left(x^2 - \frac{5}{3}x + \frac{25}{36}\right) + \frac{25}{12} + \frac{\mathbf{108}}{\mathbf{12}} \qquad \text{Simplify.}$$

$$y = -3\left(x^2 - \frac{5}{3}x + \frac{25}{36}\right) + \frac{\mathbf{133}}{\mathbf{12}} \qquad \text{Simplify.}$$

The pattern for factoring perfect square trinomials is $a^2 - 2ab + b^2 = (a - b)(a - b)$. Rewriting $\left(x^2 - \dfrac{5}{3}x + \dfrac{25}{36}\right)$ to match the pattern $a^2 - 2ab + b^2$, the trinomial is $(x)^2 - 2(x)\left(\dfrac{5}{6}\right) + \left(\dfrac{5}{6}\right)^2$. The factors are $(a - b)(a - b)$ or $\left(x - \dfrac{5}{6}\right)\left(x - \dfrac{5}{6}\right)$.

$$y = -3\left(x - \frac{\mathbf{5}}{\mathbf{6}}\right)\left(x - \frac{\mathbf{5}}{\mathbf{6}}\right) + \frac{133}{12} \qquad \begin{array}{l}\text{Use a pattern to factor;} \\ a^2 - 2ab + b^2 = (a - b)(a - b)\end{array}$$

$$y = -3\left(x - \frac{5}{6}\right)^{\mathbf{2}} + \frac{133}{12} \qquad \text{Simplify.}$$

(b) Identify the vertex, (h, k).

▶ The vertex is $\left(\dfrac{5}{6}, \dfrac{133}{12}\right)$.

(c) Write the equation of the axis of symmetry.

▶ The equation of the axis of symmetry is $x = \dfrac{5}{6}$.

(d) Identify the direction in which the parabola opens.

▶ Since $a = -3$ and $-3 < 1$, the parabola opens down.

(e) Identify the maximum or minimum output value.

▶ Since the parabola opens down, the output value is a maximum, $\dfrac{133}{12}$.

(f) Use interval notation to represent the domain.

▶ The domain of a quadratic function is the set of real numbers, $(-\infty, \infty)$.

(g) Use interval notation to represent the range.

▶ The maximum output value is $\dfrac{133}{12}$. The range is $\left(-\infty, \dfrac{133}{12}\right]$.

(h) Identify the number of real zeros.

▶ Since the vertex is above the x-axis and the graph opens down, the graph has two x-intercepts and the function has two real zeros.

Writing a Quadratic Function, $y = ax^2 + bx + c$, in Vertex Form, $y = a(x - h)^2 + k$, When $a \neq 1$

1. Group the first two terms, $ax^2 + bx$, in parentheses: $y = (ax^2 + bx) + c$.
2. Factor out a from the first two terms.
3. Add the term to complete the square inside the parentheses; subtract the product of a and the term to complete the square outside the parentheses.
4. Factor the trinomial in the parentheses and rewrite in exponential notation; simplify outside of the parentheses.

Practice Problems

For problems 9–10, use completing the square to rewrite the function in vertex form.

9. $y = 2x^2 + 8x + 9$ **10.** $y = -4x^2 + 12x + 3$

For problems 11–12,
(a) use completing the square to rewrite the function in vertex form.
(b) identify the vertex, (h, k).
(c) write the equation of the axis of symmetry.
(d) identify the direction in which the parabola opens.
(e) identify the maximum or minimum output value.
(f) use interval notation to represent the domain.
(g) use interval notation to represent the range.
(h) identify the number of real zeros.

11. $y = 3x^2 + 2x + 8$ **12.** $y = -5x^2 + 9x + 2$

Using a, b, and c to Identify the Vertex of $y = ax^2 + bx + c$

In Section 7.4, we used completing the square to solve $ax^2 + bx + c = 0$. The result was the quadratic formula. We can also use completing the square to rewrite the quadratic function $y = ax^2 + bx + c$ in vertex form.

EXAMPLE 10 | The standard form of a quadratic function is $y = ax^2 + bx + c$.

(a) Use completing the square to rewrite this function in vertex form.

SOLUTION ▶

$y = ax^2 + bx + c$

$y = (ax^2 + bx) + c$ Group the first two terms.

$y = a\left(x^2 + \dfrac{b}{a}x\right) + c$ Factor out a from the first two terms.

$y = a\left(x^2 + \dfrac{b}{a}x + \underline{\ \ }\right) - \underline{\ \ } + c$ Complete the square; add and subtract a number.

$y = a\left(x^2 + \dfrac{b}{a}x + \dfrac{b^2}{4a^2}\right) - a\left(\dfrac{b^2}{4a^2}\right) + c$ $\left(\dfrac{1}{2}\cdot\dfrac{b}{a}\right)^2 = \dfrac{b^2}{4a^2}$

$y = a\left(x^2 + \dfrac{b}{a}x + \dfrac{b^2}{4a^2}\right) - \dfrac{b^2}{4a} + c$ Simplify.

$y = a\left(x^2 + \dfrac{b}{a}x + \dfrac{b^2}{4a^2}\right) + c - \dfrac{b^2}{4a}$ Change order; commutative property.

$y = a\left(x^2 + \dfrac{b}{a}x + \dfrac{b^2}{4a^2}\right) + \dfrac{c}{1}\cdot\dfrac{4a}{4a} - \dfrac{b^2}{4a}$ Multiply by a fraction equal to 1; $\dfrac{4a}{4a} = 1$

$y = a\left(x^2 + \dfrac{b}{a}x + \dfrac{b^2}{4a^2}\right) + \dfrac{4ac}{4a} - \dfrac{b^2}{4a}$ Simplify.

$y = a\left(x^2 + \dfrac{b}{a}x + \dfrac{b^2}{4a^2}\right) + \dfrac{4ac - b^2}{4a}$ Combine fractions.

$y = a\left(x + \dfrac{b}{2a}\right)\left(x + \dfrac{b}{2a}\right) + \dfrac{4ac - b^2}{4a}$ Factor.

$y = a\left(x + \dfrac{b}{2a}\right)^2 + \dfrac{4ac - b^2}{4a}$ Use exponential notation.

$y = a\left(x - \left(-\dfrac{b}{2a}\right)\right)^2 + \dfrac{4ac - b^2}{4a}$ Rewrite; $\dfrac{b}{2a} = -\left(-\dfrac{b}{2a}\right)$

(b) Identify the vertex, (h, k).

▶ The vertex is $\left(-\dfrac{b}{2a}, \dfrac{4ac - b^2}{4a}\right)$.

Vertex (h, k) of a Quadratic Function in Standard Form $y = ax^2 + bx + c$

$$(h, k) = \left(-\dfrac{b}{2a}, \dfrac{4ac - b^2}{4a}\right)$$

We can use the values of a, b, and c from a quadratic function and $(h, k) = \left(-\dfrac{b}{2a}, \dfrac{4ac - b^2}{4a}\right)$ to find the vertex of its graph.

EXAMPLE 11 | Use the values of a, b, and c to identify the vertex of $y = 2x^2 - 12x + 11$.

SOLUTION ▶

$y = 2x^2 - 12x + 11$ $a = 2; b = -12; c = 11$

(h, k) The vertex.

$= \left(\dfrac{-(-12)}{2(2)}, \dfrac{4(2)(11) - (-12)^2}{4(2)} \right)$ $\left(-\dfrac{b}{2a}, \dfrac{4ac - b^2}{4a} \right)$; replace a, b and c.

$= \left(\dfrac{12}{4}, \dfrac{88 - 144}{8} \right)$ Simplify.

$= (3, -7)$ Simplify.

We can instead find the x-coordinate of the vertex, $-\dfrac{b}{2a}$, replace x in the original function with this value, and solve to find the y-coordinate of the vertex.

EXAMPLE 12 | Use the values of a and b to identify the vertex of $y = 2x^2 - 12x + 11$.

SOLUTION ▶

$y = 2x^2 - 12x + 11$ $a = 2; b = -12; c = 11$

$h = -\dfrac{b}{2a}$ h is the x-coordinate of the vertex.

$h = -\dfrac{(-12)}{2(2)}$ Replace a and b.

$h = 3$ Simplify; the x-coordinate of the vertex.

$y = 2x^2 - 12x + 11$ The original function.

$y = 2(3)^2 - 12(3) + 11$ Replace x with the x-coordinate of the vertex, 3.

$y = 2(9) - 36 + 11$ Simplify.

$y = -7$ Simplify; the y-coordinate of the vertex.

The vertex of this function is $(3, -7)$.

In the next example, we use $h = -\dfrac{b}{2a}$ to find the maximum height of a football pass.

EXAMPLE 13 | The quadratic function $y = -5x^2 + 10x$ represents the relationship of the time in seconds, x, that a football thrown at a 30-degree angle is in the air and the height in meters, y, of the football above its original release point. Use the values of a and b to find the maximum height of the pass above where the quarterback released the ball.

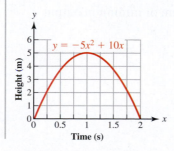

SOLUTION ▶ $y = -5x^2 + 10x$ $a = -5, b = 10$

$h = -\dfrac{b}{2a}$ h is the x-coordinate of the vertex.

$h = -\dfrac{10}{2(-5)}$ Replace a and b.

$h = 1$ Simplify; the x-coordinate of the vertex.

$y = -5x^2 + 10x$ The original function.

$y = -5(1)^2 + 10(1)$ Replace x with the x-coordinate of the vertex, 1.

$y = -5 + 10$ Simplify.

$y = 5\,\text{m}$ Simplify; the unit of height is meters.

The maximum height of the football is 5 m above where the quarterback released the ball.

Practice Problems

13. Use the values of a, b, and c to identify the vertex of $y = 2x^2 + 6x + 15$.
14. Use the values of a and b to identify the vertex of $y = x^2 + 6x - 7$.
15. The quadratic function $y = -5x^2 + 3x$ describes the relationship of the time in seconds, x, that a football thrown at a 10-degree angle is in the air and the height in meters, y, of the football above its original release point. Use the values of a and b to find the maximum height of the pass above where the quarterback released the ball.

7.6 Vocabulary Practice

Match the term with its description.

1. $y = ax^2 + bx + c, a \neq 0$
2. $y = a(x - h)^2 + k, a \neq 0$
3. The vertical line that passes through the vertex of a parabola
4. $h = -\dfrac{b}{2a}$
5. In $y = ax^2 + bx + c$, a is this.
6. Shifting the position of a graph of a function by changing h and/or k
7. $b^2 - 4ac$
8. The shape of the graph of a quadratic function
9. The ordered pair whose y-coordinate is the maximum or minimum output value of a quadratic function
10. An input value that results in an output value of 0

A. axis of symmetry
B. discriminant
C. lead coefficient
D. parabola
E. standard form of a quadratic function
F. translation
G. vertex
H. vertex form of a quadratic function
I. x-coordinate of the vertex of a quadratic function
J. real zero of a function

7.6 Exercises

Follow your instructor's guidelines for showing your work.

1. The linear function represented by the graph is
$y = x - 4$.

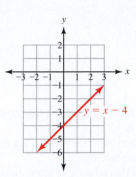

   a. Change this function to shift its graph up 1 unit.
   b. Use translation to graph the new function.

2. The linear function represented by the graph is
$y = x + 2$.

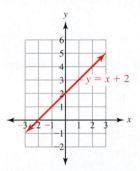

   a. Change this function to shift its graph up 1 unit.
   b. Use translation to graph the new function.

3. The linear function represented by the graph is
$y - 1 = 2(x - 4)$.

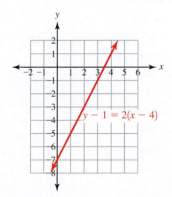

   a. Identify (x_1, y_1).
   b. Change the function to shift its graph to the left
3 units and down 5 units.
   c. Use translation to graph the new function.

4. The linear function represented by the graph is
$y - 4 = 3(x - 2)$.

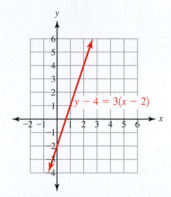

   a. Identify (x_1, y_1).
   b. Change the function to shift its graph to the left
3 units and down 5 units.
   c. Use translation to graph the new function.

5. The quadratic function represented by the graph is
$y = 2(x - 1)^2 + 3$.

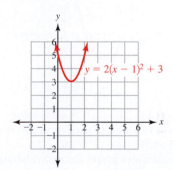

   a. Identify (h, k).
   b. Change the function to shift its graph to the right
4 units and down 5 units.
   c. Use translation to graph the new function.

6. The quadratic function represented by the graph is
$y = 2(x - 4)^2 + 1$.

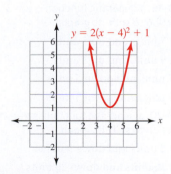

   a. Identify (h, k).
   b. Change the function to shift its graph to the right 4
units and down 5 units.
   c. Use translation to graph the new function.

7. The quadratic function represented by the graph is $y = (x + 2)^2 - 4$.

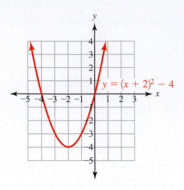

a. Identify (h, k).
b. Change the function to shift its graph to the left 3 units and up 1 unit.
c. Use translation to graph the new function.

8. The quadratic function represented by the graph is $y = (x + 3)^2 - 2$.

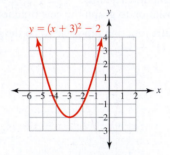

a. Identify (h, k).
b. Change the function to shift its graph to the left 1 unit and up 3 units.
c. Use translation to graph the new function.

For exercises 9–12, identify the vertex, (h, k) of the quadratic function.

9. $y = (x - 5)^2 + 7$

10. $y = (x - 3)^2 + 9$

11. $y = (x + 6)^2 - 7$

12. $y = (x + 4)^2 - 3$

For exercises 13–16, a quadratic function is $y = 3(x - 4)^2 + 9$. Change the function to shift its graph:

13. to the right 7 units and up 5 units.

14. to the right 9 units and up 3 units.

15. to the left 7 units and down 5 units.

16. to the left 9 units and down 3 units.

For exercises 17–20, a quadratic function is $y = 2(x + 6)^2 - 7$. Change the function to shift its graph:

17. to the right 8 units and down 9 units.

18. to the right 10 units and down 20 units.

19. to the left 8 units and up 9 units.

20. to the left 10 units and up 20 units.

For exercises 21–26, the value of the lead coefficient, a, in a quadratic function is 1.
(a) write the equation in vertex form of the quadratic function with the given vertex.
(b) change the function to shift its graph to the left 2 units and up 15 units.

21. $(9, 1)$

22. $(8, 1)$

23. $(-6, 3)$

24. $(-5, 2)$

25. $(-12, -8)$

26. $(-15, -6)$

For exercises 27–32, the value of the lead coefficient, a, in a quadratic function is -2.
(a) write the equation in vertex form of the quadratic function with the given vertex.
(b) change the function to shift its graph to the right 5 units and down 7 units.

27. $(-2, -1)$

28. $(-3, -1)$

29. $(0, 0)$

30. $(0, 4)$

31. $(-5, 0)$

32. $(-9, 0)$

For exercises 33–56, use completing the square to rewrite the function in vertex form.

33. $y = x^2 + 6x + 15$

34. $y = x^2 + 10x + 18$

35. $y = x^2 - 10x + 15$

36. $y = x^2 - 6x + 18$

37. $w(x) = x^2 + 20x - 65$

38. $p(x) = x^2 + 30x - 71$

39. $y = x^2 + 5x - 11$

40. $y = x^2 + 7x - 13$

41. $d(t) = t^2 + 9t + 8$

42. $v(t) = t^2 + 3t + 4$

43. $f(t) = t^2 - 15t - 30$

44. $f(x) = x^2 - 13x - 28$

45. $y = x^2 - x + 4$

46. $y = x^2 - x + 6$

47. $f(x) = x^2 + 12x$

48. $g(x) = x^2 + 16x$

49. $h(x) = x^2 - 10x$

50. $f(x) = x^2 - 20x$

51. $y = x^2 + x$

52. $y = x^2 - x$

53. $y = x^2 + 2x - 11$

54. $y = x^2 + 2x - 13$

55. $y = x^2 + 7x$

56. $y = x^2 + 5x$

For exercises 57–62,

(a) use completing the square to rewrite the function in vertex form.

(b) identify the vertex.

(c) write the equation of the axis of symmetry.

(d) identify the direction in which parabola opens.

(e) identify the minimum or maximum output value.

(f) use interval notation to represent the domain.

(g) use interval notation to represent the range.

(h) describe the number and type of any real zeros.

57. $y = x^2 - 20x - 12$

58. $y = x^2 - 18x - 15$

59. $y = x^2 + 5x + 8$

60. $y = x^2 + 7x + 16$

61. $f(x) = x^2 - 6x$

62. $g(x) = x^2 - 8x$

For exercises 63–72, use completing the square to rewrite the function in vertex form.

63. $y = 2x^2 + 12x + 51$

64. $y = 2x^2 - 24x + 48$

65. $y = 2x^2 - 16x + 30$

66. $y = 2x^2 - 20x + 21$

67. $f(x) = 2x^2 - 9x + 3$

68. $g(x) = 2x^2 - 5x + 8$

69. $y = 3x^2 - 21x + 2$

70. $y = 3x^2 - 15x + 5$

71. $y = -x^2 + 20x + 42$

72. $y = -x^2 + 18x + 30$

For exercises 73–76,

(a) use completing the square to rewrite the function in vertex form.

(b) identify the vertex, (h, k).

(c) write the equation of the axis of symmetry.

(d) identify the direction in which the parabola opens.

(e) identify the maximum or minimum output value.

(f) use interval notation to represent the domain.

(g) use interval notation to represent the range.

(h) identify the number of real zeros.

73. $y = -x^2 - 20x + 42$

74. $y = -x^2 - 18x + 30$

75. $y = 4x^2 + 12x - 25$

76. $y = 4x^2 + 20x - 32$

For exercises 77–78, use the values of a, b, and c to identify the vertex.

77. $y = 4x^2 + 10x - 5$

78. $y = 4x^2 + 14x - 11$

For exercises 79–82, use the values of a and b to identify the vertex.

79. $y = 5x^2 + 30x + 10$

80. $y = 5x^2 + 20x + 15$

81. $y = -5x^2 + 30x + 10$

82. $y = -5x^2 + 20x + 15$

83. The quadratic function $y = -5x^2 + 14x$ describes the relationship of the time in seconds, x, that a football thrown at a 45-degree angle is in the air and the height in meters, y, of the football above its original release point. Use the values of a and b to find the maximum height of the pass above where the quarterback released the ball.

84. The quadratic function $y = -5x^2 + 8x$ describes the relationship of the time in seconds, x, that a football thrown at a 25-degree angle is in the air and the height in meters, y, of the football above its original release point. Use the values of a and b to find the maximum height of the pass above where the quarterback released the ball.

Problem Solving: Practice and Review

Follow your instructor's guidelines for using the five steps as outlined in Section 1.5, p. 51.

85. Many evergreen trees naturally lose their deep green summer color because of shorter day lengths and a change in the angle of the sun. To sell these trees for Christmas, they are sprayed green. The colorant costs $13 per gallon. One gallon of colorant is diluted with 29 gallons of water to make 30 gallons of spray. Thirty gallons of spray can color 120 Christmas trees. The cost of hiring workers to spray the trees is $7.25 per hour. A worker can spray 40 trees per hour. Find the cost of spraying 1000 trees. Round to the nearest whole number. (*Source:* www.ext.vt.edu)

86. The cost of flying coach round-trip from Dallas to Los Angeles is $256. The same ticket can be purchased for 32,500 frequent flyer miles plus $49. The cost of flying first class round-trip on the same plane is $1538. The same ticket can be purchased for 62,600 frequent flyer miles plus $49. Find the difference in the value of 1000 frequent flyer miles used by the passenger who flies coach and by the passenger who flies first-class. Round to the nearest hundredth.

87. In October 2008, a 20-lb piece of metal fell from the sky and crushed the passenger side of a rental car in Jersey City. Police believe the metal came from an airplane. Ignoring air resistance, the velocity V of an object that free-falls to the ground from an altitude of d *feet* can be found by using the formula $V = \sqrt{2\left(32\dfrac{\text{ft}}{\text{s}^2}\right)d}$. If the piece of metal fell from a plane flying at 40,000 ft, find its velocity as it hit the ground *in miles per hour*. Round to the nearest whole number. (*Source:* www.nj.com, Oct. 21, 2008)

88. Find the percent increase in the number of viewers of Sunday night football games from 2009 to 2010. Round to the nearest percent.

Sunday night football games have increased to an average of 20.8 million viewers in 2010, up from 18.9 million in 2009 and 16.5 million in 2008. (*Source:* money.cnn.com, Feb. 3, 2011)

Technology

For exercises 89–92, graph the function on a graphing calculator. Choose a window that shows the intercepts and any maximum or minimum value. Round irrational numbers to the hundredths place.
(a) Sketch the graph; describe the window.
(b) Use the calculator to identify any real zeros.
(c) Use the calculator to identify any minimum or maximum output value.
(d) Use interval notation to represent the domain.
(e) Use interval notation to represent the range.
(f) Use the value command to evaluate the function when $x = 2$.

89. $f(x) = -2(x - 3)^2 + 8$

90. $g(x) = 3(x + 5)^2 - 2$

91. $y = (x - 4)^2$

92. $y = -(x - 6)^2 - 2$

Find the Mistake

For exercises 93–96, the completed problem has one mistake.
(a) Describe the mistake in words, or copy down the whole problem and highlight or circle the mistake.
(b) Do the problem correctly.

93. Problem: Use completing the square to rewrite $y = x^2 + 6x + 11$ in vertex form.

Incorrect Answer: $y = x^2 + 6x + 11$
$$y = (x^2 + 6x + 9) + 11$$
$$y = (x + 3)(x + 3) + 11$$
$$y = (x + 3)^2 + 11$$

94. Problem: Use completing the square to rewrite $y = 2x^2 + 8x + 5$ in vertex form.

Incorrect Answer: $y = 2x^2 + 8x + 5$
$$y = (x^2 + 8x + 16) - 16 + 11$$
$$y = (x + 4)(x + 4) - 5$$
$$y = (x + 4)^2 - 5$$

95. Problem: A function is $y = (x - 5)^2 - 7$. Change the function to shift its graph to the right 4 units and down 2 units.

Incorrect Answer: $y = (x - 1)^2 - 9$

96. Problem: Use the values of a and b to identify the vertex of $y = x^2 - 8x - 6$.

Incorrect Answer: $a = 1, b = -8$

$$h = -\frac{b}{2a}$$
$$h = \frac{-8}{2(1)}$$
$$h = -4$$
$$y = x^2 - 8x - 6$$
$$y = (-4)^2 - 8(-4) - 6$$
$$y = 16 + 32 - 6$$
$$y = 42$$

The vertex is $(-4, 42)$.

Review

For exercises 97–100,
(a) solve the inequality.
(b) use a number line graph to represent the solution.

97. $3x - 12 > 21$

98. $-3x - 12 \geq 21$

99. $-2x + 10 \geq 18$

100. $2x + 10 < 18$

SUCCESS IN COLLEGE MATHEMATICS

101. A college catalog, usually available on-line, is a valuable resource for planning ahead. It often includes degree requirements, program plans, and course descriptions. To show that you know how to access the catalog, choose a course that you plan to take in the future at your college. Find its course description in the college catalog. Copy this course description here.

To find an *exact* amount of fencing, we solve an equation. To find the *minimum* amount of fencing, we solve an inequality. In this section, we will solve quadratic inequalities.

SECTION 7.7

Quadratic Inequalities

After reading the text, working the practice problems, and completing assigned exercises, you should be able to:

1. Find the boundary values of a quadratic inequality.

2. Solve a quadratic inequality.

3. Use interval notation or a graph to represent the solution of a quadratic inequality.

4. Use a quadratic inequality to solve an application problem.

Boundary Values

The solution of the linear equation $3x = 12$ is $x = 4$. If we replace x in the equation with any other number, the equation is not true. In contrast, the linear inequality $3x > 12$ has infinitely many solutions. All real numbers greater than 4 are solutions of this inequality. The set that includes all of the solutions of the inequality is its **solution set**. The number 4 is a **boundary value** of this inequality. The boundary value divides the set of real numbers into intervals. If a number in an interval is a solution of the inequality, all of the numbers in the interval are solutions.

In Chapter 1, we used the properties of inequality to solve linear inequalities. We can instead find the boundary values of the inequalities, use the boundary values to divide the set of real numbers into intervals, and choose a number in each interval to check whether the interval includes solutions of the inequality.

EXAMPLE 1   Solve the linear inequality $3x - 9 > 5x + 17$.

(a) Find any real boundary values.

SOLUTION ▶   To find the boundary values, solve the linear equation $3x - 9 = 5x + 17$.

$$3x - 9 = 5x + 17$$

$\underline{-5x \qquad\quad -5x}$	Subtraction property of equality.
$-2x - 9 = 0 + 17$	Simplify.
$\underline{\quad +9 \qquad\quad +9}$	Addition property of equality.
$-2x + 0 = 26$	Simplify.
$\dfrac{-2x}{-2} = \dfrac{26}{-2}$	Division property of equality.
$x = -13$	Simplify.

(b) Use the boundary values to solve the inequality.

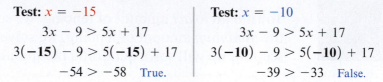

$$(-\infty, -13)(-13, \infty)$$

The boundary values divide the set of real numbers into intervals. Choose a number in each interval and check whether it is a solution of the inequality.

Test: $x = -15$

$$3x - 9 > 5x + 17$$
$$3(-15) - 9 > 5(-15) + 17$$
$$-54 > -58 \quad \text{True.}$$

Test: $x = -10$

$$3x - 9 > 5x + 17$$
$$3(-10) - 9 > 5(-10) + 17$$
$$-39 > -33 \quad \text{False.}$$

Since $x = -15$ is a solution of the inequality, all of the numbers in the interval $(-\infty, -13)$ are solutions. Since $x = -10$ is not a solution of the inequality, none of the numbers in the interval $(-13, \infty)$ are solutions.

(c) Use a number line graph and interval notation to represent the solution.

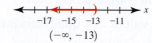

$$(-\infty, -13)$$

To solve a quadratic inequality, we follow a similar process.

EXAMPLE 2 Solve the quadratic inequality $x^2 + 7x + 10 > 0$.

(a) Find any real boundary values.

SOLUTION To find the boundary values, solve the quadratic equation $x^2 + 7x + 10 = 0$.

$$x^2 + 7x + 10 = 0$$
$$(x + 5)(x + 2) = 0 \qquad \text{Factor.}$$
$$x + 5 = 0 \quad \text{or} \quad x + 2 = 0 \qquad \text{Zero product property.}$$
$$\underline{ -5 \quad -5} \qquad \underline{ -2 \quad -2} \qquad \text{Subtraction property of equality.}$$
$$x + 0 = -5 \quad \text{or} \quad x + 0 = -2 \qquad \text{Simplify.}$$
$$x = -5 \quad \text{or} \qquad x = -2 \qquad \text{Simplify.}$$

(b) Use the boundary values to solve the inequality.

$$(-\infty, -5)(-5, -2)(-2, \infty)$$

The boundary values divide the set of real numbers into intervals. Choose a number in each interval, and check whether it is a solution of the inequality.

Test: $x = -6$

$$x^2 + 7x + 10 > 0$$
$$(-6)^2 + 7(-6) + 10 > 0$$
$$36 - 42 + 10 > 0$$
$$4 > 0$$
$$\text{True.}$$

Test: $x = -3$

$$x^2 + 7x + 10 > 0$$
$$(-3)^2 + 7(-3) + 10 > 0$$
$$9 - 21 + 10 > 0$$
$$-2 > 0$$
$$\text{False.}$$

Test: $x = -1$

$$x^2 + 7x + 10 > 0$$
$$(-1)^2 + 7(-1) + 10 > 0$$
$$1 - 7 + 10 > 0$$
$$4 > 0$$
$$\text{True.}$$

Since $x = -6$ is a solution of the inequality, all of the numbers in the interval $(-\infty, -5)$ are solutions. Since $x = -3$ is not a solution of the inequality, none of the numbers in the interval $(-5, -2)$ are solutions. Since $x = -1$ is a solution of the inequality, all of the numbers in the interval $(-2, \infty)$ are solutions.

(c) Use a number line graph and interval notation to represent the solution.

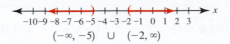

$$(-\infty, -5) \quad \cup \quad (-2, \infty)$$

Practice Problems

For problems 1–5,
(a) find the boundary values.
(b) use the boundary values to solve the inequality.
(c) use a number line graph to represent the solution.
(d) use interval notation to represent the solution.

1. $2x - 12 > 18$ **2.** $-2x - 12 > 18$ **3.** $x^2 + 4x - 45 > 0$
4. $x^2 + 4x - 21 < 0$ **5.** $x^2 - 3x > 0$

More Quadratic Inequalities

In the next example, the boundary values of $x^2 - 6 \le 0$ are irrational numbers. When checking a number in each interval, choose rational numbers to make the arithmetic easier. Since the inequality sign is $\le$, the inequality is not strict, and we must check the boundary values as well as a value in each interval.

EXAMPLE 3 **(a)** Solve: $x^2 - 6 \le 0$

SOLUTION ▶ To find the boundary values, solve $x^2 - 6 = 0$.

$$x^2 - 6 = 0$$

$\underline{+6 \quad +6}$	Addition property of equality.
$x^2 + 0 = 6$	Simplify.
$\sqrt{x^2} = \sqrt{6}$	Take the square root of both sides.
$x = \pm\sqrt{6}$	Simplify; use $\pm$ notation.
$x = \sqrt{6} \quad \text{or} \quad x = -\sqrt{6}$	The boundary values.

The boundary values divide the real numbers into three intervals: $(-\infty, -\sqrt{6})$, $(-\sqrt{6}, \sqrt{6})$, and $(\sqrt{6}, \infty)$. Since $\sqrt{6} \approx 2.4$, the intervals are approximately $(-\infty, -2.4)$, $(-2.4, 2.4)$, and $(2.4, \infty)$.

$$(-\infty, -\sqrt{6}) \quad (\sqrt{6}, \sqrt{6}) \quad (\sqrt{6}, \infty)$$

Test: $x = -3$	**Test:** $x = 0$	**Test:** $x = 3$
$x^2 - 6 \le 0$	$x^2 - 6 \le 0$	$x^2 - 6 \le 0$
$(-3)^2 - 6 \le 0$	$(0)^2 - 6 \le 0$	$(3)^2 - 6 \le 0$
$9 - 6 \le 0$	$0 - 6 \le 0$	$9 - 6 \le 0$
$3 \le 0$ False.	$-6 \le 0$ True.	$3 \le 0$ False.

Since $x = 0$ is a solution of the inequality, the numbers in the interval $(-\sqrt{6}, \sqrt{6})$ are solutions of the inequality. Since $x^2 - 6 \le 0$ is not strict, we need to also test the boundary values to see whether they are solutions.

Test: $x = \sqrt{6}$	**Test:** $x = -\sqrt{6}$
$x^2 - 6 \leq 0$	$x^2 - 6 \leq 0$
$(\sqrt{6})^2 - 6 \leq 0$	$(-\sqrt{6})^2 - 6 \leq 0$
$6 - 6 \leq 0$	$6 - 6 \leq 0$
$0 \leq 0$ True.	$0 \leq 0$ True.

The values $x = \sqrt{6}$ and $x = -\sqrt{6}$ are part of the solution set of the inequality.

(b) Use interval notation to represent the solution.

▶ Including the boundary values and the interval $(-\sqrt{6}, \sqrt{6})$, the solution of this inequality is $[-\sqrt{6}, \sqrt{6}]$.

In the next example, we solve a quadratic inequality that has only one solution. Notice that the trinomial in the inequality is a perfect square and the inequality is not strict.

EXAMPLE 4   **(a)** Solve: $x^2 + 6x + 9 \leq 0$.

SOLUTION ▶   To find the boundary values, solve $x^2 + 6x + 9 = 0$.

$$x^2 + 6x + 9 = 0$$
$$(x + 3)(x + 3) = 0 \qquad \text{Factor; } x^2 + 6x + 9 \text{ is a perfect square.}$$
$$x + 3 = 0 \quad \text{or} \quad x + 3 = 0 \qquad \text{Zero product property.}$$
$$\underline{-3 \quad -3} \qquad\qquad \underline{-3 \quad -3} \qquad \text{Subtraction property of equality.}$$
$$x = -3 \qquad\qquad x = -3 \qquad \text{The only boundary value is } x = -3.$$

The boundary value divides the real numbers into two intervals: $(-\infty, -3)$ and $(-3, \infty)$.

```
◄──┼──┼──●──┼──┼──┼──┼──●──┼──┼──► x
  -6 -5 -4 -3 -2 -1  0  1  2
  (-∞, -3)(-3, ∞)
```

Test: $x = -4$	**Test:** $x = 0$
$x^2 + 6x + 9 \leq 0$	$x^2 + 6x + 9 \leq 0$
$(-4)^2 + 6(-4) + 9 \leq 0$	$(0)^2 + 6(0) + 9 \leq 0$
$16 - 24 + 9 \leq 0$	$0 + 0 + 9 \leq 0$
$1 \leq 0$ False.	$9 \leq 0$ False.

Since neither $x = -4$ nor $x = 0$ is a solution of the inequality, the numbers in the intervals $(-\infty, -3)$ and $(-3, \infty)$ are not solutions. Since $x^2 + 6x + 9 \leq 0$ is not strict, we need to also test the boundary value, -3, to see whether it is a solution.

Test: $x = -3$

$$x^2 + 6x + 9 \leq 0$$
$$(-3)^2 + 6(-3) + 9 \leq 0$$
$$9 - 18 + 9 \leq 0$$
$$0 \leq 0 \qquad \text{True.}$$

The only solution of this inequality is the boundary value $x = -3$.

(b) Use a number line graph and interval notation to represent the solution.

```
◄──┼──┼──┼──┼──●──┼──┼──┼──┼──► x
  -7 -6 -5 -4 -3 -2 -1  0  1
        [-3, -3]
```

A boundary value must be a real number. In the next example, since the solutions of the quadratic equation are nonreal numbers, they are not boundary values. In this situation, the solution of the inequality is either the set of real numbers or the inequality has no solution.

EXAMPLE 5   Solve: $x^2 + 3x + 7 \geq 0$

SOLUTION ▶ To find the boundary values, solve $x^2 + 3x + 7 = 0$. Since $x^2 + 3x + 7$ is a prime quadratic trinomial, use the quadratic formula.

$$x^2 + 3x + 7 = 0 \qquad\qquad\qquad a = 1, b = 3, c = 7$$

$$x = \frac{-3 \pm \sqrt{(3)^2 - 4(1)(7)}}{2(1)} \qquad\qquad x = \frac{-b \pm \sqrt{b^2 - 4ac}}{2a}$$

$$x = \frac{-3 \pm \sqrt{9 - 28}}{2} \qquad\qquad\qquad \text{Simplify.}$$

$$x = \frac{-3 \pm \sqrt{-19}}{2} \qquad\qquad\qquad \text{Simplify.}$$

$$x = -\frac{3}{2} + \frac{\sqrt{19}}{2}i \quad \text{or} \quad x = -\frac{3}{2} - \frac{\sqrt{19}}{2}i \qquad \text{The solutions are nonreal.}$$

This inequality has no real boundary values. So either all real numbers are solutions or no real number is a solution. Test any real number in the original inequality. Since arithmetic with 0 is relatively easy, choose $x = 0$.

Test: $x = 0$

$$x^2 + 3x + 7 \geq 0$$
$$(0)^2 + 3(0) + 7 \geq 0$$
$$0 + 0 + 7 \geq 0$$
$$7 \geq 0 \quad \text{True.}$$

Since a real number, 0, is a solution of the inequality, then all real numbers are solutions. The solution of this inequality is the set of real numbers: $(-\infty, \infty)$.

The solution of the inequality in Example 5, $x^2 + 3x + 7 \geq 0$, is the set of real numbers. In the next example, we change the inequality sign. The inequality $x^2 + 3x + 7 \leq 0$ has no real boundary values and no solution.

EXAMPLE 6   Solve: $x^2 + 3x + 7 \leq 0$

SOLUTION ▶
$$x^2 + 3x + 7 = 0 \qquad\qquad\qquad \text{To find the boundary values, solve.}$$

$$x = -\frac{3}{2} + \frac{\sqrt{19}}{2}i \quad \text{or} \quad x = -\frac{3}{2} - \frac{\sqrt{19}}{2}i \qquad \text{This equation is solved in Example 5.}$$

This inequality has no real boundary values. Either all real numbers are solutions or no real number is a solution. Test any real number.

Test: $x = 0$

$$x^2 + 3x + 7 \leq 0$$
$$(0)^2 + 3(0) + 7 \leq 0$$
$$0 + 0 + 7 \leq 0$$
$$7 \leq 0 \quad \text{False.}$$

This inequality has no real solution.

Solving a Quadratic Inequality

1. Find the boundary values of the inequality.
2. Use the boundary values to divide the set of real numbers into intervals.
3. Choose a number from each interval. If this number is a solution of the inequality, every number in the interval is a solution.
4. If the inequality is not strict, test the boundary values to see whether they are solutions.
5. If the inequality has no real boundary values, either it has no solution or its solution is the set of real numbers. Choose a real number. If this number is a solution of the inequality, then every real number is a solution. If this number is not a solution, then the inequality has no real solution.

Practice Problems

For problems 6–12,
(a) solve the inequality.
(b) use interval notation to represent the solution.

6. $2x^2 + x - 3 \leq 0$ 7. $x^2 - 5 \geq 0$ 8. $x^2 + 3x - 10 \geq 0$
9. $x^2 + 3x - 10 \leq 0$ 10. $x^2 + 8x + 16 \geq 0$
11. $x^2 + 8x + 16 < 0$ 12. $x^2 + 8x + 16 \leq 0$

Applications

A quadratic inequality in one variable can be a model of a problem situation.

EXAMPLE 7   A construction site covers a city block with an area of 237,600 ft². If 2000 ft of temporary fence must surround a rectangular area that is at least 175,000 square feet, find the possible widths of this rectangle. Round to the nearest whole number.

SOLUTION ▶ **Step 1 Understand the problem.**
Although this situation is best described by an inequality rather than an equality, this problem is similar to Example 12 in Section 7.4. We know the perimeter and the area of the rectangle. If we use the formula for perimeter, $P = 2W + 2L$, to write an expression for the length in L, the unknown is the length, L.

$$2W + 2L = 2000 \text{ ft} \qquad \text{The perimeter is 2000 ft; solve for } L.$$

$$\underline{-2W \qquad\qquad\qquad -2W} \qquad \text{Subtraction property of equality.}$$

$$2L = 2000 \text{ ft} - 2W \qquad \text{Simplify.}$$

$$\frac{2L}{2} = \frac{2000 \text{ ft}}{2} - \frac{2W}{2} \qquad \text{Division property of equality.}$$

$$L = 1000 \text{ ft} - W \qquad \text{Simplify.}$$

$$L = 1000 - W \qquad \text{Remove the units.}$$

Step 2 Make a plan.
The area is greater than (length)(width). If we replace L with $1000 - W$ and A with 175,000 ft², the result is a quadratic inequality. To find the boundary values of this inequality, we can use the quadratic formula to solve for W.

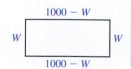

Step 3 Carry out the plan.

$$\text{area} \geq 175{,}000 \text{ ft}^2 \qquad \text{Minimum area is } 175{,}000 \text{ ft}^2.$$

$$(\text{length})(\text{width}) \geq 175{,}000 \qquad \text{Area} = (\text{length})(\text{width}); \text{ remove units.}$$

$$(1000 - W)W \geq 175{,}000 \qquad \text{Replace } L \text{ with } 1000 - W.$$

$$1000W - W^2 \geq 175{,}000 \qquad \text{Distributive property.}$$

To find the boundary values, solve $1000W - W^2 = 175{,}000$. To ensure that rounding does not cause a solution that is too small, round the minimum solution *up* to the nearest whole number and the maximum solution *down* to the nearest whole number.

$$1000W - W^2 = 175{,}000 \qquad \text{Not in standard form.}$$

$$\underline{ -175{,}000 \quad -175{,}000} \qquad \text{Subtraction property of equality.}$$

$$1000W - W^2 - 175{,}000 = 0 \qquad \text{Simplify.}$$

$$-W^2 + 1000W - 175{,}000 = 0 \qquad \text{Change order; write in standard form.}$$

$$W = \frac{-1000 \pm \sqrt{(1000)^2 - 4(-1)(-175{,}000)}}{2(-1)} \qquad x = \frac{-b \pm \sqrt{b^2 - 4ac}}{2a}$$

$$W = \frac{-1000 \pm \sqrt{300{,}000}}{-2} \qquad \text{Simplify.}$$

$$W = \frac{-1000 + \sqrt{300{,}000}}{-2} \quad \text{or} \quad W = \frac{-1000 - \sqrt{300{,}000}}{-2} \qquad \text{Rewrite using "or" instead of } \pm.$$

$$W \approx 227 \text{ ft} \quad \text{or} \quad W \approx 773 \text{ ft} \qquad \text{Simplify; round.}$$

The width cannot be less than 0 ft or greater than the length of the fence, 2000 ft. The boundary values create three intervals: $(0, 227)$, $(227, 773)$, and $(773, 2000)$.

Test: $W = 200$

$$1000W - W^2 \geq 175{,}000$$
$$1000(200) - (200)^2 \geq 175{,}000$$
$$160{,}000 \geq 175{,}000 \quad \text{False.}$$

Test: $W = 500$

$$1000W - W^2 \geq 175{,}000$$
$$1000(500) - (500)^2 \geq 175{,}000$$
$$250{,}000 \geq 175{,}000 \quad \text{True.}$$

Test: $W = 800$

$$1000W - W^2 \geq 175{,}000$$
$$1000(800) - (800)^2 \geq 175{,}000$$
$$160{,}000 \geq 175{,}000 \quad \text{False.}$$

The solution is $227 \leq x \leq 773$. Replacing the units, $227 \text{ ft} \leq x \leq 773 \text{ ft}$.

Step 4 Look back.

The minimum width is 227 ft. Since the perimeter is 2000 ft, the length is 773 ft. The minimum area is $(227 \text{ ft})(773 \text{ ft}) = 175{,}471 \text{ ft}^2$. This is greater than the minimum area of $175{,}000 \text{ ft}^2$. The maximum width is 773 ft. The length is then 227 ft. The area is the same, $175{,}471 \text{ ft}^2$. Since the area produced by a width between 227 ft and 773 ft is greater than $175{,}000 \text{ ft}^2$, the answer seems reasonable.

Step 5 Report the solution.

The width of the rectangle should be between 227 ft and 773 ft.

Practice Problems

13. The organizers of an artisan's fair use stakes and a tape with colored flags to mark the edges of a rectangular area. Each roll of tape is 113 ft long. The organizers have 15 rolls. They need at least 150,000 ft² for the fair. Use a quadratic inequality in one variable to find the possible widths for this rectangle. Round to the nearest foot.

7.7 VOCABULARY PRACTICE

Match the term with its description.

1. $ax^2 + bx + c = 0$
2. If $ab = 0$, then $a = 0$ or $b = 0$.
3. $x = \dfrac{-b \pm \sqrt{b^2 - 4ac}}{2a}$
4. A polynomial expression with three terms
5. $\sqrt{-1}$
6. $\sqrt{6}$
7. $\dfrac{3}{4}$
8. $a^2 + 2ab + b^2 = (a + b)(a + b)$
9. $(-\infty, \infty)$
10. $(0, \infty)$

A. i
B. irrational number
C. perfect square trinomial factoring pattern
D. quadratic formula
E. rational number
F. set of real numbers
G. set of real numbers greater than 0
H. standard form of a quadratic equation in one variable
I. trinomial
J. zero product property

7.7 Exercises

Follow your instructor's guidelines for showing your work.

For exercises 1–4,
(a) find the boundary value.
(b) use the boundary value to solve the inequality.
(c) use a number line graph to represent the solution.

1. $8x - (x + 5) > 2x + 30$
2. $9x - (x - 15) > 3x + 20$
3. $-4x - 16 \geq -24$
4. $-3x - 15 \geq -24$

For exercises 5–20, find any real boundary values.

5. $x^2 - 4 > 0$
6. $x^2 - 9 > 0$
7. $x^2 + 10x + 16 > 0$
8. $x^2 + 12x + 32 > 0$
9. $x^2 - 3x - 108 < 0$
10. $x^2 - 5x - 84 < 0$
11. $6x^2 + 7x - 20 \geq 0$
12. $6x^2 + 17x - 45 \geq 0$

13. $3x^2 + x - 5 \leq 0$
14. $2x^2 + x - 7 \leq 0$
15. $x^2 - 13 < 0$
16. $x^2 - 17 < 0$
17. $x^2 + 15 > 0$
18. $x^2 + 3 > 0$
19. $x^2 + 10x + 25 \geq 0$
20. $x^2 + 12x + 36 \geq 0$

21. A quadratic inequality has no real boundary values. Explain what this tells us about the solution of this inequality.

22. Explain how to find the boundary value(s) for a quadratic inequality.

For exercises 23–28, use interval notation to represent the set graphed on the number line.

23.

24.

25.

26.

27.

28.

For exercises 29–52,
(a) find the boundary value(s).
(b) use the boundary value(s) to solve the inequality.
(c) use a number line graph to represent the solution.
(d) use interval notation to represent the solution.

29. $x^2 - 3x - 40 > 0$

30. $x^2 - 5x - 14 > 0$

31. $x^2 - 3x - 40 < 0$

32. $x^2 - 5x - 14 < 0$

33. $9x^2 - 12x - 5 \leq 0$

34. $4x^2 - 4x - 63 \leq 0$

35. $9x^2 - 12x - 5 \geq 0$

36. $4x^2 - 4x - 63 \geq 0$

37. $x^2 + 8x < 0$

38. $x^2 + 3x < 0$

39. $x^2 + 8x > 0$

40. $x^2 + 3x > 0$

41. $x^2 + 18x + 81 \geq 0$

42. $x^2 + 20x + 100 \geq 0$

43. $x^2 + 18x + 81 > 0$

44. $x^2 + 20x + 100 > 0$

45. $x^2 + 18x + 81 < 0$

46. $x^2 + 20x + 100 < 0$

47. $x^2 + 18x + 81 \leq 0$

48. $x^2 + 20x + 100 \leq 0$

49. $3x^2 + 4x + 9 \geq 0$

50. $5x^2 + 4x + 11 \geq 0$

51. $3x^2 + 4x + 9 \leq 0$

52. $5x^2 + 4x + 11 \leq 0$

53. Write a quadratic inequality with no real boundary values that has no solution.

54. Write a quadratic inequality with no real boundary values that has infinitely many solutions.

55. Write a quadratic inequality with one real boundary value whose solution is a single real number.

56. Write a quadratic inequality with one real boundary value that has no solution.

For exercises 57–70,
(a) solve the quadratic inequality.
(b) use interval notation to represent the solution.

57. $x^2 + 6x - 16 \leq 0$

58. $x^2 + 2x - 63 \leq 0$

59. $80x^2 + 6x - 2 \geq 0$

60. $66x^2 - 7x - 3 \geq 0$

61. $2x^2 - 9x < 0$

62. $3x^2 - 8x < 0$

63. $4x^2 + 20x + 25 > 0$

64. $9x^2 + 12x + 4 > 0$

65. $-x^2 + 5x - 10 < 0$

66. $-x^2 + 8x - 19 < 0$

67. $-x^2 - 12x + 36 < 0$

68. $-x^2 - 10x + 25 < 0$

69. $x^2 - 12x + 36 \leq 0$

70. $x^2 - 8x + 16 \leq 0$

For exercises 71–76, use a quadratic inequality in one variable to solve the problem. For rounding, see Example 7.

71. An animal rescue organization receives a donation of 20 rolls of fencing. Each roll is 100 ft long. The organization wants to create a rectangular pasture for donkeys with an area of *at least* 2 acres using all of this fencing (1 acre = 43,560 ft^2). Find the possible widths for this rectangle.

72. Every year, community volunteers organize an egg hunt for children under the age of 5. They use boundary tape to mark off the rectangle in which they hide the eggs. They have 5 rolls of boundary tape. Each roll is 36 yd long. The area of the rectangle must be *at least* 15,000 ft^2. If all the tape is used, find the possible widths for this rectangle.

73. A moving company has enclosed trailers for rent. The rectangular door opening on a trailer is 3 ft 9 in. wide and 4 ft high. Find the possible widths of a sheet of plywood that can fit into this trailer. Round down to the nearest inch.

74. The table below lists the standard sizes of box springs. A student can rent a 6-ft by 12-ft trailer with a rectangular door opening that is 5 ft high and 4 ft 10 in. wide, or she can save some money and rent a smaller trailer with a rectangular door opening that is 4 ft high and 3 ft 9 in. wide. She has a queen-size box spring.

Single (twin)	39 in. × 75 in.
Double (full)	54 in. × 75 in.
Queen	60 in. × 80 in.
King	76 in. × 80 in.

a. Find the trailer that she should rent.
b. Explain.

75. The function $C(x) = 600 + 90x$ models the cost in dollars of making x products. The function $R(x) = 200x - x^2$ models the revenue received in dollars for selling x products. At the break-even point, $C(x) = R(x)$. If the company makes a profit, $R(x) > C(x)$. Find the number of products to make and sell so that the company makes a profit. Make sure that any rounding does not create an unprofitable situation.

76. The function $C(x) = 7500 + 20x$ models the cost in dollars of making x products. The function $R(x) = 200x - x^2$ models the revenue received in dollars for selling x products. At the break-even point, $C(x) = R(x)$. If the company makes a profit, $R(x) > C(x)$. Find the number of products to make and sell so that the company makes a profit. Make sure that any rounding does not create an unprofitable situation.

Problem Solving: Practice and Review

Follow your instructor's guidelines for using the five steps as outlined in Section 1.5, p. 51.

77. The function $y = 0.00821x^2 + 0.05x + 4.77$ is a model of the percent of American adolescents ages 13–19 who are overweight, y, for x years since 1966. Predict the percent of overweight adolescents in 2013. Round to the nearest percent. (*Source*: www.cdc.gov, 2007)

78. The formula for the volume, V, of a sphere is $V = \frac{4}{3}\pi r^3$, where r is the radius of the sphere. The formula for the volume, V, of a cone is $V = \frac{1}{3}\pi r^2 h$, where r is the radius of the cone and h is the height of the cone. At an ice cream store, a customer orders a single-scoop ice cream cone. The diameter of the cone is 1.9 in. Its height is 4.5 in. The cone is filled with ice cream and is topped with a half-sphere of ice cream with a diameter of 1.9 in. Find the total volume of ice cream. Round to the nearest whole number.

© Ledo/Shutterstock

For exercises 79–80, the Durango and Silverton Narrow Gauge Railroad uses coal-fired steam locomotives to power its excursion train. The round trip from Durango to Silverton is 90 mi. Each trip requires six short tons of coal.

© David Gaylor/Shutterstock

A million-dollar effort to reduce the carbon footprint of the D&SNG is under way, and the railroad is working with CarbonZero . . . to plant 2,587 trees this year on the train route between Durango and Silverton. (*Source*: www .durangoherald.com, May 3, 2008)

Carbon dioxide (CO_2) forms during coal combustion when one atom of carbon (C) unites with two atoms of oxygen (O) from the air. . . . For example, coal with a carbon content

of 78 percent and a heating value of 14,000 Btu per pound emits about 204.3 pounds of carbon dioxide per million Btu when completely burned. Complete combustion of 1 short ton (2,000 pounds) of this coal will generate about 5,720 pounds (2.86 short tons) of carbon dioxide. (*Source*: www .eia.doe.gov)

79. Assume that the coal burned by the railroad is similar to the coal described in the government report. Find the weight of carbon dioxide produced by each round trip of the railroad from Durango to Silverton. Round to the nearest hundred.

80. The train operates daily from the beginning of May to the beginning of November, completing at least 180 round-trips. Assume that the coal burned by the railroad is similar to the coal described in the government report. Find the weight of carbon dioxide produced by 180 round-trips. Round to the nearest thousand.

Technology

For exercises 81–82,
(a) graph both functions in the given window. Sketch the graph; describe the window.
(b) use the **intersect** command to find the number of products to make and sell so that the company makes a profit. Do not let rounding create an unprofitable situation.

81. The function $C(x) = 90x + 600$ models the cost in dollars of making x products. The function $R(x) = 200x - x^2$ models the revenue received in dollars for selling x products. At the break-even point, $C(x) = R(x)$. If the company makes a profit, $R(x) > C(x)$. Window: $[-100, 170, 20; -10{,}000, 20{,}000, 5000]$

82. The function $C(x) = 20x + 7500$ models the cost in dollars of making x products. The function $R(x) = 200x - x^2$ models the revenue received in dollars for selling x products. At the break-even point, $C(x) = R(x)$. If the company makes a profit, $R(x) > C(x)$. Window: $[-50, 170, 20; -2500, 12{,}000, 2500]$

Find the Mistake

For exercises 83–86, the completed problem has one mistake.
(a) Describe the mistake in words, or copy down the whole problem and highlight or circle the mistake.
(b) Do the problem correctly.

83. **Problem:** Solve $x^2 + 6x + 11 > 0$. Use interval notation to describe the solution.

    Incorrect Answer: To find boundary values, solve $x^2 + 6x + 11 = 0$.

    $$x = \frac{-6 \pm \sqrt{6^2 - 4(1)(11)}}{2(1)}$$

    $$x = -3 \pm \sqrt{2}i$$

    Since the boundary values are not real, the inequality has no solution.

84. Problem: Solve: $x^2 + 18x + 81 > 0$

Incorrect Answer: To find the boundary values, solve $x^2 + 18x + 81 = 0$.

$$x = \frac{-18 \pm \sqrt{18^2 - 4(1)(81)}}{2(1)}$$

$$x = \pm\frac{18}{2}$$

$$x = \pm 9$$

Test: $x = -10$

$$x^2 + 18x + 81 > 0$$
$$(-10)^2 + 18(-10) + 81 > 0$$
$$1 > 0 \quad \text{True.}$$

Test: $x = 0$

$$x^2 + 18x + 81 > 0$$
$$(0)^2 + 18(0) + 81 > 0$$
$$81 > 0 \quad \text{True.}$$

Test: $x = 10$

$$x^2 + 18x + 81 > 0$$
$$(10)^2 + 18(10) + 81 > 0$$
$$361 > 0 \quad \text{True.}$$

The solution of this inequality is the set of real numbers, $(-\infty, \infty)$.

85. Problem: Use interval notation to represent the set graphed on the number line.

$$-2\ -1\ \ 0\ \ 1\ \ 2\ \ 3\ \ 4\ \ 5\ \ 6\ \ 7\ \ 8\ \ 9\ \ 10$$

Incorrect Answer: $[-\infty, 3] \cup (5, \infty)$

86. Problem: Find the boundary values for the quadratic inequality $-x^2 + 8 \leq 2$.

Incorrect Answer: $-x^2 + 8 \leq 2$

$$\frac{-2\ -2}{-x^2 + 6 \leq 0}$$

To find the boundary values, solve $-x^2 + 6 = 0$.

$$-x^2 + 6 = 0$$
$$-x(x - 6) = 0$$
$$-x = 0 \qquad x - 6 = 0$$
$$\frac{-x}{-1} = \frac{0}{-1} \qquad \frac{+6\ +6}{x + 0 = 6}$$
$$x = 0 \qquad x = 6$$

The boundary values are $x = 0$ and $x = 6$.

Review

87. Explain how to simplify a fraction into lowest terms.

88. Explain how to find the common denominator for two fractions.

89. In fraction arithmetic, we add, subtract, multiply, and divide. For which of these operations do the fractions need to have a common denominator?

90. Explain how to divide two fractions.

SUCCESS IN COLLEGE MATHEMATICS

91. Part of planning for success is planning your course schedule. Some students may want a break from math and do not want to take math next term. However, students who take their next math course as soon as possible usually are more successful than students who do not. Explain why you think that this is true.

Study Plan for Review of Chapter 7

In these exercises, assume that all variables in *expressions* represent nonnegative real numbers. Use interval notation to represent domains and ranges.

SECTION 7.1 Complex Numbers

Ask Yourself	Test Yourself	Help Yourself
Can I...	**Do 7.1 Review Exercises**	**See these Examples and Practice Problems**
simplify a square root or a cube root in which the radicand is less than 0?	2–4	Ex. 1–3, PP 1–4
identify whether a number is an element of the set of complex numbers, imaginary numbers, real numbers, rational numbers, irrational numbers, integers, or whole numbers?	1, 6, 7	Ex. 8, PP 9
multiply square roots in which at least one of the radicands is less than 0?	8	Ex. 4–6, PP 5–8
add or subtract complex numbers?	9, 10	Ex. 9–11, PP 10, 11
multiply complex numbers?	10–12, 21	Ex. 12–14, PP 12–14
identify the complex conjugate of a complex number?	20	Ex. 14, 18, PP 15, 16
rewrite a complex number in $a + bi$ form?	5, 13–16	Ex. 7, 15–18, PP 17–20
simplify the expression i^p, where p is a nonzero whole number?	17–19	Ex. 20, 21, PP 21–26

7.1 Review Exercises

1. Copy the table below. Put a checkmark or X in the box if the number is an element of the set.

	Complex numbers	Imaginary numbers	Real numbers	Rational numbers	Irrational numbers	Integers	Whole numbers
$3i$							
$7-15i$							
$\dfrac{3}{5}$							
6							
$\sqrt{11}$							
-41							

For exercises 2–5, simplify.

2. $\sqrt{-75xy^3z^4}$

3. $\sqrt[3]{-128p^4q^7}$

4. $\sqrt{6^2 - 4(9)(3)}$

5. $\dfrac{3}{2i}$

6. Identify the numbers that are elements of the set of imaginary numbers.
   a. $\sqrt{-8}$
   b. $\sqrt[3]{-8}$
   c. $7i$
   d. $6 + 7i$
   e. 12
   f. $\dfrac{2}{i}$

7. a. Is every real number a complex number?
   b. Is every imaginary number a complex number?
   c. Is every complex number a real number?

8. a. Simplify: $\sqrt{-15a^3b}\,\sqrt{-20ab^2}$
   b. Simplify: $\sqrt{-7xy}\,\sqrt{14x}$

For exercises 9–19, simplify. The final answer must be in $a + bi$ form.

9. $(12 + 3i) - (7 - 2i)$

10. $4(8 - 7i) - 3(9 - 2i)$

11. $(7 - 3i)(4 + 9i)$

12. $-6i(5 - i)$

13. $-\dfrac{2}{5i}$

14. $\dfrac{6 - 7i}{2 + 3i}$

15. $\dfrac{6 + i}{2}$

16. $\dfrac{9 - i}{i}$

17. i^3

18. i^{121}

19. i^{1050}

20. a. What is the complex conjugate of $1 - i$?
    b. What is the complex conjugate of $9 + 6i$?

21. A complex number is $3 + i$. Find a complex number to multiply it by so that the product is the real number 10.

SECTION 7.2 Solving Quadratic Equations

Ask Yourself	Test Yourself	Help Yourself
Can I...	Do 7.2 Review Exercises	See these Examples and Practice Problems
solve a quadratic equation in which $b = 0$ and check the solution(s)? rewrite a solution in $\pm$ notation as two separate solutions?	22–24	Ex. 2–5, PP 1–6
write the Pythagorean theorem? identify a right triangle? use the Pythagorean theorem to solve an application problem?	25, 26	Ex. 6, PP 7
write the zero product property? use the zero product property to solve a quadratic equation?	27–31	Ex. 7, 8, PP 8–10
use a quadratic equation to solve an application problem that involves a geometric figure such as a triangle or rectangle?	32	Ex. 6, 8, PP 7, 11, 12

7.2 Review Exercises

For exercises 22–24,
(a) solve the equation.
(b) check.

22. $h^2 + 45 = 0$

23. $p^2 = 98$

24. $3x^2 = -135$

25. Explain how you know that a diagonal drawn between opposite corners of a window is the hypotenuse of a right triangle.

26. A carpenter is cutting wood to make a rafter for a gable roof. The altitude, a, of the roof is 5 ft 4 in. The total span of the roof is 23 ft 10 in. The rafter must have an overhang of 10 in. Use the Pythagorean theorem to find the length of the rafter in feet and inches. Round to the nearest inch.

For problems 27–29, use the zero product property to solve.

27. $6x^2 + 13x - 5 = 0$

28. $x^2 - 5x = 0$

29. $3x^2 + 6x = 0$

30. Explain why the zero product property cannot be used to solve $x^2 + 9x + 13 = 0$.

31. Write a quadratic equation in standard form with solutions $x = 2$ and $x = 5$ that can be solved using the zero product property.

32. The width of a rectangle is 3 ft less than the length. Its area is 108 ft². Use a quadratic equation to find the length and width.

SECTION 7.3 Completing the Square

Ask Yourself	Test Yourself	Help Yourself
Can I...	Do 7.3 Review Exercises	See these Examples and Practice Problems
add a constant to change an expression into a perfect square trinomial?	33, 34	Ex. 1–12, PP 1–12
factor a perfect square trinomial?	35–37	Ex. 6–12, PP 6–12
solve a quadratic equation by completing the square when $a = 1$?	43–45	Ex. 6–10, PP 6–9
solve a quadratic equation by completing the square when $a \neq 1$?	46–48	Ex. 11, 12, PP 10–12

7.3 Review Exercises

33. What constant should be added to change $x^2 + 26x$ into a perfect square trinomial?

34. What constant should be added to change $x^2 + 9x$ into a perfect square trinomial?

For exercises 35–37, factor.

35. $x^2 - 18x + 81$ **36.** $x^2 + 20x + 100$

37. $x^2 + 7x + \dfrac{49}{4}$

38. Rewrite $x = -7 \pm \sqrt{5}\,i$ as two separate equations.

39. Rewrite $x = 3 + \sqrt{2}$ or $x = 3 - \sqrt{2}$ using $\pm$ notation.

40. Rewrite $x = \dfrac{3 \pm \sqrt{5}}{2}$ as two separate equations.

41. A solution of $2x^2 - 9x - 35 = 0$ is $x = -\dfrac{5}{2}$. Check this solution

42. A solution of $x^2 - x + 10 = 0$ is $x = \dfrac{1}{2} - \dfrac{\sqrt{39}}{2}i$. Check this solution.

For exercises 43–48, solve by completing the square.

43. $h^2 + 9h + 14 = 0$

44. $x^2 - 6x + 11 = 0$

45. $z^2 - 5z + 2 = 0$

46. $4x^2 + 16x + 15 = 0$

47. $2x^2 - 6x + 11 = 0$

48. $2x^2 + 13x - 7 = 0$

SECTION 7.4 **Quadratic Formula**

Ask Yourself	Test Yourself	Help Yourself
Can I . . .	**Do 7.4 Review Exercises**	**See these Examples and Practice Problems**
write the quadratic formula?	49–52, 55	Ex. 2–6, PP 1–6
use the quadratic formula to solve a quadratic equation and check the solution(s)?		
find the discriminant of a quadratic equation and use the discriminant to describe the solutions?	49–54	Ex. 7–10, PP 7–12
use a quadratic equation and the quadratic formula to solve an application problem?	56–57	Ex. 11, 12, PP 13–15

7.4 Review Exercises

For exercises 49–52,
(a) find the discriminant.
(b) use the discriminant to describe the solutions of the equation.
(c) use the quadratic formula to find the exact solutions.

49. $3x^2 - 7x + 4 = 0$

50. $5x^2 - 9x + 10 = 0$

51. $8x^2 + 17x - 6 = 0$

52. $5x^2 + 7x = 0$

53. Explain why a quadratic equation with a discriminant that equals 0 has only one solution.

54. Explain why a quadratic equation with a discriminant that is less than 0 has nonreal solutions.

55. Identify the solution(s) of the quadratic equation $ax^2 + bx + c = 0$.

56. On a wet road, a Ford Expedition brakes at an average acceleration of $\dfrac{-22 \text{ ft}}{1 \text{ s}^2}$. When its initial speed is $\dfrac{88 \text{ ft}}{1 \text{ s}}$, its stopping distance is 176 ft. Use the formula $d = v_i t + \dfrac{1}{2}at^2$, where d is distance, v_i is initial speed, a is acceleration, and t is time, to find the time in seconds that it will take the Expedition to stop. Round to the nearest whole number.

57. The area of a rectangle is 104 ft². The width of the rectangle is 5 ft less than the length. Use a quadratic equation in one variable to find the length and width of this rectangle.

SECTION 7.5   **Quadratic Functions**

Ask Yourself	Test Yourself	Help Yourself
Can I . . .	**Do 7.5 Review Exercises**	**See these Examples and Practice Problems**
identify the domain of a polynomial function?	58–63	Ex. 1, 2, PP 1–3
identify the vertex, the equation of the axis of symmetry, the minimum or maximum output value, the domain, and the range of a quadratic function, given its graph?		
use the graph to estimate the real zeros of a quadratic function?	60, 64–66	Ex. 3–6, PP 4–9
use an algebraic method to find the real zeros of a quadratic function?		
write the standard form of a quadratic function?		
use the discriminant to identify the number of zeros of a quadratic function and identify whether they are rational, irrational, or nonreal numbers?		
given the vertex and the direction that a parabola opens, identify the number of real zeros?	67–72	Ex. 7, PP 10–11
write the vertex form of a quadratic function?	67–70	Ex. 9, 10, PP 11, 12
identify the vertex, the equation of the axis of symmetry, the minimum or maximum output value, the domain, and the range of a quadratic function written in vertex form?		
write a quadratic function in vertex form, given the vertex and the value of a?	71, 72	Ex. 11, PP 13
evaluate a quadratic function?	73, 74	Ex. 12, 13, PP 14–17

7.5 Review Exercises

For exercises 58–63, use the graph of the function.

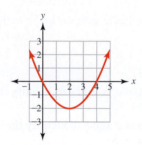

58. Identify the vertex.

59. Write the equation of the axis of symmetry.

60. Estimate any real zeros.

61. Identify the minimum or maximum output value.

62. Use interval notation to represent the domain.

63. Use interval notation to represent the range.

For exercises 64–66,
(a) find the discriminant.
(b) use the discriminant to identify the number of real zeros.
(c) use an algebraic method to find the exact value of any real zeros.

64. $y = x^2 + 5x - 36$

65. $f(x) = x^2 + 4x + 15$

66. $y = x^2 - 8x + 16$

For exercises 67–70, copy and complete the table below for the given function.

67. $f(x) = (x - 4)^2 + 9$

68. $y = (x + 6)^2 - 21$

69. $y = -2(x - 4)^2$

70. $y = -(x + 1)^2 + 50$

71. A quadratic function has a vertex of $(-2, 7)$ and $a = 1$.
   a. Write the equation of this function in vertex form.
   b. Identify the direction in which the parabola opens.
   c. Identify the number of real zeros.

Table for exercises 67–70

	Vertex	Opens up or down?	Domain	Range	Axis of symmetry	Number of real zeros
67.						
68.						
69.						
70.						

72. A quadratic function has a vertex of $(-2, 7)$ and $a = -3$.
  a. Write the equation of this function in vertex form.
  b. Identify the direction in which the parabola opens.
  c. Identify the number of real zeros.

73. The function $P(x) = 0.0838x^2 + 12.403x + 84.998$ represents the relationship of the number of years since

1994, x, and the number of prescription drugs in millions delivered by mail order per year, $P(x)$. Find the number of prescriptions delivered by mail order in 2012. Round to the nearest million. (*Source:* www.census.gov, 2009)

74. If $f(x) = x^2 - x - 11$, evaluate $f(-3)$.

SECTION 7.6 Vertex Form of a Quadratic Function

Ask Yourself	Test Yourself	Help Yourself
Can I . . .	**Do 7.6 Review Exercises**	**See these Examples and Practice Problems**
change a linear function in slope-intercept form or point-slope form to cause a horizontal and/or vertical shift in its graph?	81, 82	Ex. 1–4, PP 1–4
change a quadratic function in vertex form to cause a horizontal and/or a vertical shift in its graph?		
use completing the square to rewrite a quadratic function in vertex form $(a = 1)$?	75, 76	Ex. 5–7, PP 5–8
use completing the square to rewrite a quadratic function in vertex form $(a \neq 1)$?	77, 78	Ex. 8–10, PP 9–12
identify the vertex of a quadratic function written in vertex form?	75–79	Ex. 5–10, PP 11, 12
use a, b, and c or use a and b to identify the vertex of a quadratic function?	83–85	Ex. 11–13, PP 13–15

7.6 Review Exercises

For exercises 75–78,
(a) use completing the square to rewrite the equation in vertex form.
(b) identify the vertex.

75. $y = x^2 - 8x + 11$

76. $f(x) = x^2 + 5x + 13$

77. $y = 2x^2 - 20x + 44$

78. $f(x) = 3x^2 + 6x + 11$

79. A student said that the vertex of the graph of the function $y = 2(x + 3)^2 - 9$ is $(3, -9)$. Explain what is wrong with this answer.

80. Factor 5 out of the expression $10x^2 - 3x$.

For exercises 81–82,
(a) identify the vertex of the function.
(b) change the function to shift the graph to the right 7 units and up 2 units.
(c) change the function to shift the graph to the left 7 units and down 2 units.

81. $y = 4(x - 5)^2 + 8$

82. $g(x) = (x + 3)^2 - 6$

83. Use the values of a, b, and c to identify the vertex of $g(x) = x^2 + 6x + 1$.

84. Use the values of a and b to identify the vertex of $y = 2x^2 - 16x + 37$.

85. A quadratic function is $y = x^2 - 14x + 50$.
  a. Use the values of a and b to identify the vertex.
  b. Write the equation of the axis of symmetry.
  c. Identify the minimum or maximum output value.
  d. Use interval notation to represent the domain.
  e. Use interval notation to represent the range.

SECTION 7.7 Quadratic Inequalities

Ask Yourself	Test Yourself	Help Yourself
Can I . . .	**Do 7.7 Review Exercises**	**See these Examples and Practice Problems**
identify the boundary values of a quadratic inequality in one variable?	86–91	Ex. 2–7, PP 1–12
solve a quadratic inequality in one variable?	86–91	Ex. 2–7, PP 3–12
use a number line graph and/or interval notation to represent the solution of a quadratic inequality?	86–90	Ex. 2–5, PP 3–12

Ask Yourself	Test Yourself	Help Yourself
Can I...	**Do 7.7 Review Exercises**	**See these Examples and Practice Problems**
identify a quadratic inequality with a solution that is the set of real numbers?	90	Ex. 5, PP 10
identify a quadratic inequality with no solution?	91	Ex. 6, PP 11
use a quadratic inequality to solve an application problem?	92	Ex. 7, PP 13

7.7 Review Exercises

For exercises 86–91,
(a) identify the boundary values.
(b) solve the inequality.
(c) use interval notation to represent the solution.

86. $x^2 + 8x + 15 \geq 0$

87. $x^2 - 4x - 7 < 0$

88. $x^2 - 14x + 49 \leq 0$

89. $x^2 - 18x + 81 > 0$

90. $x^2 + 2x + 8 > 0$

91. $x^2 + 3x + 5 < 0$

92. At a salvage store, a homeowner finds 300 tiles that are 1 in. square. He wants to use these tiles to create a border around a rectangular area in his kitchen. He wants the area of the rectangle to be at least 1600 in.^{2}. Find the possible widths for this rectangle. Round to the nearest inch.

Chapter 7 Test

1. Copy the chart below. If a number is an element of the set, mark the box with a checkmark or an X.

		Complex numbers	Imaginary numbers	Real numbers	Rational numbers	Irrational numbers	Integers	Whole numbers
a.	-17							
b.	$2 - 8i$							
c.	$\sqrt{14}$							
d.	$-6i$							
e.	$\dfrac{2}{5}$							
f.	8							
g.	$\sqrt{-5}$							

For exercises 2–8, simplify. The final answer must be in $a + bi$ form.

2. $\dfrac{2}{6 - 7i}$

3. $\dfrac{9 + 5i}{17}$

4. $\dfrac{4 + 5i}{3i}$

5. $\sqrt{-108}$

6. $\sqrt{-30x^3}\sqrt{-42y}$

7. $(4 - 6i) - (2 + 9i)$

8. $(4 - 6i)(2 + 9i)$

9. Solve $x^2 - 18x - 40 = 0$ by
   a. factoring and using the zero product property.
   b. completing the square.
   c. using the quadratic formula.

10. Solve each quadratic equation using the method of your choice.
   a. $x^2 + 63 = 0$
   b. $3x^2 - 5x + 8 = 0$
   c. $x^2 - 16x = 0$

11. A rectangle has an area of 204.6 m^2 and a length that is 4.1 m more than its width. Find its length and width.

12. A quadratic equation is $3x^2 - 5x - 7 = 0$.
   a. Find the discriminant.
   b. Use the discriminant to identify the number of solutions.
   c. Identify the solutions as rational, irrational, or nonreal numbers.

13. Copy and complete the table below. Use interval notation to represent the domain and range.

14. Change the function $y = 4(x - 9)^2 - 3$ to shift its graph to the left 8 units and up 2 units.

15. Use completing the square to rewrite $y = x^2 + 16x - 31$ in vertex form.

16. Use completing the square to rewrite $y = 2x^2 + 10x + 11$ in vertex form.

17. The function $C(x) = 63.49x^2 + 130.81x + 14{,}814$ describes the relationship between the number of years since 1990, x, and the amount of compensation and pension, $C(x)$, in millions of dollars. Find the amount of compensation and pension in 2013. Round to the nearest tenth of a *billion*. (*Source:* www.census.gov, 2009)

18. Solve: $x^2 - 9x - 22 \geq 0$

Table for exercise 13

Function	Vertex	Domain	Range	Axis of symmetry	Opens up or down?	Number of real zeros
$y = 4(x - 1)^2 - 6$						
$y = -2(x + 5)^2 - 3$						

Exponential and Logarithmic Expressions, Equations, and Functions

8

SUCCESS IN COLLEGE MATHEMATICS

Final Exams

The questions on a comprehensive final exam can ask about anything you have learned from the beginning of the term. The final exam assesses what you remember and contributes to your final grade. It also gives you the chance to review everything you've learned. When you review, the connections between concepts in different chapters may become much clearer. This review process may also help transfer your knowledge more permanently into your long-term memory.

The Success in College Mathematics exercises in each section of this chapter suggest some ways to review for the final exam.

As the end of the term nears, you may be tired. You may have big projects or papers due in other classes. Perhaps you feel overwhelmed and just want to be done. However, you are not done. You need to prepare for your final exam. This means that you need to build time into your schedule now to review for your final exam.

© A. L. Spangler/Shutterstock

At a stand at a farmer's market, **revenue** is the money received for selling products. **Costs** include the money needed to make the product. **Profit** is the difference between revenue and costs. If the output of one function is revenue and the output of another function is costs, the difference of the output of the two functions is the profit. In this section, we will learn how to add, subtract, multiply, and compose functions.

SECTION 8.1

Operations with Functions and One-to-One Functions

After reading the text, working the practice problems, and completing assigned exercises, you should be able to:

1. Add or subtract functions.

2. Multiply two functions.

3. Compose two functions.

4. Determine whether a function is a one-to-one function.

Addition and Subtraction of Functions

To add or subtract functions with rules, add or subtract the function rules. If the functions being added are polynomial functions, their sum is also a polynomial function.

EXAMPLE 1 If $f(x) = 3x - 9$ and $g(x) = 12x + 1$,

(a) find $f(x) + g(x)$.

SOLUTION

$$f(x) + g(x) = (3x - 9) + (12x + 1) \qquad \text{Add function rules.}$$
$$f(x) + g(x) = \mathbf{15x - 8} \qquad \text{Simplify.}$$

(b) use interval notation to represent the domain of $f(x) + g(x)$.

Since $f(x) + g(x) = 15x - 8$ is a polynomial function, its domain is the set of real numbers, $(-\infty, \infty)$.

(c) find $g(x) + f(x)$.

$$g(x) + f(x) = (12x + 1) + (3x - 9) \qquad \text{Add function rules.}$$
$$g(x) + f(x) = \mathbf{15x - 8} \qquad \text{Simplify.}$$

(d) use interval notation to represent the range of $g(x) + f(x)$.

Since $g(x) + f(x) = 15x - 8$ is a linear function, its range is the set of real numbers, $(-\infty, \infty)$.

(e) Is addition of functions commutative?

Since changing the order does not change the sum, addition of functions is commutative: $f(x) + g(x) = g(x) + f(x)$.

698

When subtracting functions, enclose the function rule being subtracted in parentheses.

EXAMPLE 2   If $f(x) = 2x^2 + 5x - 16$ and $g(x) = x^2 - 3x - 2$,

(a) find $f(x) - g(x)$.

SOLUTION ▶

$f(x) - g(x) = 2x^2 + 5x - 16 - (x^2 - 3x - 2)$	Subtract function rules.
$f(x) - g(x) = 2x^2 + 5x - 16 - 1(x^2 - 3x - 2)$	Multiply by 1.
$f(x) - g(x) = 2x^2 + 5x - 16 - x^2 + 3x + 2$	Distributive property.
$f(x) - g(x) = x^2 + 8x - 14$	Simplify.

(b) find $g(x) - f(x)$.

▶

$g(x) - f(x) = x^2 - 3x - 2 - (2x^2 + 5x - 16)$	Subtract function rules.
$g(x) - f(x) = x^2 - 3x - 2 - 1(2x^2 + 5x - 16)$	Multiply by 1.
$g(x) - f(x) = x^2 - 3x - 2 - 2x^2 - 5x + 16$	Distributive property.
$g(x) - f(x) = -x^2 - 8x + 14$	Simplify.

(c) use interval notation to represent the range of $f(x) - g(x)$.

▶ Since the functions are quadratic, to find the range, use completing the square to rewrite the function in vertex form, $y = a(x - h)^2 + k$. Use a to identify whether the graph of the function opens up or down. The minimum or maximum output value is k.

$f(x) - g(x) = x^2 + 8x - 14$	
$f(x) - g(x) = (x^2 + 8x) - 14$	Group $x^2 + 8x$.
$f(x) - g(x) = (x^2 + 8x + 16) - 16 - 14$	Add and subtract a number.
$f(x) - g(x) = (x + 4)(x + 4) - 30$	Factor; simplify.
$f(x) - g(x) = (x + 4)^2 - 30$	Exponential notation.

Since $a = 1$ and $1 > 0$, the graph opens up. The minimum output value is -30. The range of the function is $[-30, \infty)$.

(d) Is subtraction of functions commutative?

▶ Since changing the order does change the difference, subtraction of functions is not commutative: $f(x) - g(x) \neq g(x) - f(x)$.

When adding or subtracting radical functions, simplify by combining like radicals.

EXAMPLE 3   If $f(x) = 3\sqrt{x + 5}$ and $g(x) = 7\sqrt{x + 5}$,

(a) find $f(x) - g(x)$.

SOLUTION ▶

$f(x) - g(x) = 3\sqrt{x + 5} - 7\sqrt{x + 5}$	Subtract function rules.
$f(x) - g(x) = -4\sqrt{x + 5}$	Simplify.

(b) use interval notation to represent the domain of $f(x) - g(x)$.

▶ The square root of a number less than 0 is not a real number. The domain must be restricted so that the radicand is greater than or equal to 0.

$x + 5 \geq 0$	The radicand must be greater than or equal to 0.
$\underline{-5 \quad -5}$	Subtraction property of equality.
$x + 0 \geq -5$	Simplify.
$x \geq -5$	Simplify.

The domain of $f(x) - g(x)$ is $[-5, \infty)$.

When adding or subtracting rational functions, the function rules must have a common denominator.

EXAMPLE 4 If $f(x) = \dfrac{8}{x^2 - 4x}$ and $g(x) = \dfrac{2}{x}$,

(a) find $f(x) + g(x)$.

SOLUTION ▶

$f(x) + g(x) = \dfrac{8}{x^2 - 4x} + \dfrac{2}{x}$ Add function rules.

$f(x) + g(x) = \dfrac{8}{x(x - 4)} + \dfrac{2}{x}$ Factor denominator.

The least common denominator is $x(x - 4)$.

$f(x) + g(x) = \dfrac{8}{x(x - 4)} + \dfrac{2}{x} \cdot \dfrac{x - 4}{x - 4}$ Multiply by a fraction equal to 1.

$f(x) + g(x) = \dfrac{8}{x(x - 4)} + \dfrac{2x - 8}{x(x - 4)}$ Distributive property.

$f(x) + g(x) = \dfrac{2x}{x(x - 4)}$ Add numerators; find common factors.

$f(x) + g(x) = \dfrac{2}{x - 4}$ Simplify.

(b) use interval notation to represent the domain of $f(x) + g(x)$.

▶ The domain of this function must be restricted to prevent division by 0. Looking back at the original functions, $f(x)$ and $g(x)$, we find that an input value of 0 or 4 results in division by 0. So the domain of $f(x) + g(x)$ is $(-\infty, 0) \cup (0, 4) \cup (4, \infty)$.

In the next example, we write a **revenue function**, $R(x)$, and a **cost function**, $C(x)$. The **profit function** is the difference of these functions: $P(x) = R(x) - C(x)$.

EXAMPLE 5 A family grows 5 acres of sweet corn to sell in bags at a farmer's market. Each bag holds 1 dozen ears (1 dozen = 12 ears). The price of each bag is $3.50. The cost of growing the corn is $500 per acre. Each acre produces about 1000 dozen ears of sweet corn. (*Source:* extension.missouri.edu)

(a) Write a revenue function, $R(x)$, that finds the revenue from the sale of x bags of corn.

SOLUTION ▶

$R(x) = \left(\dfrac{\$3.50}{1 \text{ bag}} \right) x$ The revenue per bag sold is $3.50.

(b) Change the units of cost from dollars per acre to dollars per bag.

▶ $\left(\dfrac{\$500}{1 \text{ acre}} \right) \left(\dfrac{1 \text{ acre}}{1000 \text{ dozen ears}} \right) \left(\dfrac{1 \text{ dozen ears}}{1 \text{ bag}} \right)$ Multiply by fractions equal to 1.

$= \dfrac{\$0.50}{1 \text{ bag}}$ Simplify.

(c) Write a cost function, $C(x)$, that finds the cost to grow x bags of corn.

▶ $C(x) = \left(\dfrac{\$0.50}{1 \text{ bag}} \right) x$ The cost to grow a bag is 0.50.

(d) Write a profit function, $P(x)$, that finds the profit from growing and selling x bags of corn.

$$P(x) = R(x) - C(x) \qquad \text{profit} = \text{revenue} - \text{cost}$$

$$P(x) = \left(\frac{\$3.50}{1 \text{ bag}}\right)x - \left(\frac{\$0.50}{1 \text{ bag}}\right)x \qquad \text{Replace } R(x) \text{ and } C(x).$$

$$P(x) = \left(\frac{\$3.00}{1 \text{ bag}}\right)x \qquad \text{Simplify.}$$

(e) Describe the meaning of the profit function.

This profit function shows that the family makes a profit of $3 on every bag of corn they sell.

Practice Problems

For problems 1–3, $f(x) = 8x - 2$ and $g(x) = 5x + 9$. Find:

1. $f(x) + g(x)$ **2.** $f(x) - g(x)$ **3.** $g(x) - f(x)$

4. If $f(x) = 6\sqrt{x - 2}$ and $g(x) = 9\sqrt{x - 2}$, find $f(x) - g(x)$.

5. If $f(x) = \dfrac{2x + 1}{x - 1}$ and $g(x) = \dfrac{5}{x^2 - 1}$, find $f(x) + g(x)$.

6. If the family in Example 5 decreases the price of corn to $2 a bag and reduces the costs of growing the corn to $450 an acre, write a profit function, $P(x)$ that finds the profit in dollars of growing and selling x bags of corn.

7. The cost for a company to make a product is $25 each. The company charges wholesale buyers $37 for each product.
    a. Write a revenue function, $R(x)$, that finds the revenue from the sale of x products.
    b. Write a cost function, $C(x)$, that finds the cost to make x products.
    c. Write a profit function, $P(x)$, that finds the profit from making and selling x products.

8. The company in practice problem 7 has a profit goal of $55,000 for sale of these products. Use the profit function to find the number of products, x, that must be made and sold to make this profit. Round up to the nearest product.

Multiplication of Functions

To multiply polynomial functions, use the distributive property and the product rule of exponents, $x^m \cdot x^n = x^{m+n}$.

EXAMPLE 6   If $f(x) = 3x + 10$ and $g(x) = 6x^3$,

(a) find $f(x) \cdot g(x)$.

SOLUTION ▶ $f(x) \cdot g(x) = (3x + 10)(6x^3)$

       $f(x) \cdot g(x) = 3x(6x^3) + 10(6x^3)$ Distributive property.

       $f(x) \cdot g(x) = 18x^4 + 60x^3$ Simplify; product rule of exponents.

(b) find $g(x) \cdot f(x)$.

       $g(x) \cdot f(x) = (6x^3)(3x + 10)$

       $g(x) \cdot f(x) = 6x^3(3x) + 6x^3(10)$ Distributive property.

       $g(x) \cdot f(x) = 18x^4 + 60x^3$ Simplify.

(c) Is multiplication of functions commutative?

▶ Since changing the order does not change the product, multiplication of functions is commutative: $f(x) \cdot g(x) = g(x) \cdot f(x)$.

In the next example, the functions f and g are linear functions. The product of f and g is a quadratic function.

EXAMPLE 7 **(a)** If $f(x) = 3x + 10$ and $g(x) = 6x - 5$, find $f(x) \cdot g(x)$.

SOLUTION ▶

$f(x) \cdot g(x) = (3x + 10)(6x - 5)$

$f(x) \cdot g(x) = 3x(6x - 5) + 10(6x - 5)$ Distributive property.

$f(x) \cdot g(x) = 3x(6x) + 3x(-5) + 10(6x) + 10(-5)$ Distributive property.

$f(x) \cdot g(x) = 18x^2 - 15x + 60x - 50$ Simplify.

$f(x) \cdot g(x) = 18x^2 + 45x - 50$ Simplify.

(b) Use interval notation to represent the domain of $f(x)$, $g(x)$, and $f(x) \cdot g(x)$.

▶ Since all of these functions are polynomial functions and the domain of a polynomial function is the set of real numbers, the domain of each function is $(-\infty, \infty)$.

Practice Problems

For problems 9–11, $f(x) = 4x$, $g(x) = 7x - 2$, $h(x) = 8x + 3$, and $k(x) = x^2 + 10x + 25$. Find:

9. $f(x) \cdot g(x)$ **10.** $g(x) \cdot h(x)$ **11.** $f(x) \cdot k(x)$

Composition of Functions

We can add, subtract, or multiply functions. Addition, subtraction, and multiplication are **operations**. Another operation is **composition**. In composition, the output of one function is the input of another function. There are two notations for composition of two functions f and g: $f \circ g$ and $f(g(x))$, said, "f of g of x." For both of these notations, the output of g is the input value for f.

We can visualize composition by using two "function machines."

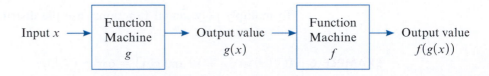

EXAMPLE 8 **(a)** Find $f(g(x))$ when $g(x) = 5x + 2$ and $f(x) = 6x - 1$.

SOLUTION ▶

$f(g(x)) = f(5x + 2)$ The input is $g(x)$, $5x + 2$.

$f(g(x)) = 6(5x + 2) - 1$ Replace x in $f(x)$ with $5x + 2$.

$f(g(x)) = 30x + 12 - 1$ Distributive property.

$f(g(x)) = 30x + 11$ Simplify.

(b) Use interval notation to represent the domain and range of $f(g(x))$.

▶ When we compose two polynomial functions, the result is a new polynomial function. In this example, $f(g(x)) = 30x + 11$ is a linear function. Its domain and its range are the set of real numbers, $(-\infty, \infty)$.

If we change the order of composition in Example 8, the result is not the same. Although there are functions for which we can change the order of composition and the result is the same, this is not true for all functions. Just one example in which the result is not the same means that composition for functions is *not* commutative.

EXAMPLE 9 Find $g(f(x))$ when $g(x) = 5x + 2$ and $f(x) = 6x - 1$.

SOLUTION ▶

$$g(f(x)) = g(6x - 1) \qquad \text{The input is } f(x), 6x - 1.$$
$$g(f(x)) = 5(6x - 1) + 2 \qquad \text{Replace } x \text{ in } g(x) \text{ with } 6x - 1.$$
$$g(f(x)) = \mathbf{30x - 5} + 2 \qquad \text{Distributive property.}$$
$$g(f(x)) = 30x - 3 \qquad \text{Simplify; } g(f(x)) \neq f(g(x))$$

To compose two radical functions, first rewrite the radicals as exponential expressions. Assume that the variables represent real numbers greater than or equal to 0.

EXAMPLE 10 Find $h(g(x))$ when $g(x) = \sqrt{x}$ and $h(x) = \sqrt[3]{x^2}$. Write the final answer in radical notation.

SOLUTION ▶ Since $\sqrt[n]{x^m} = x^{\frac{m}{n}}$, $g(x) = x^{\frac{1}{2}}$ and $h(x) = x^{\frac{2}{3}}$.

$$h(g(x)) = h\left(x^{\frac{1}{2}}\right) \qquad \text{The input is } g(x), x^{\frac{1}{2}}.$$

$$h(g(x)) = \left(x^{\frac{1}{2}}\right)^{\frac{2}{3}} \qquad \text{Replace } x \text{ in } h(x) \text{ with } x^{\frac{1}{2}}.$$

$$h(g(x)) = x^{\left(\frac{1}{2}\right)\left(\frac{2}{3}\right)} \qquad \text{Power rule of exponents.}$$

$$h(g(x)) = x^{\frac{1}{3}} \qquad \text{Simplify.}$$

$$h(g(x)) = \sqrt[3]{x} \qquad \text{Rewrite in radical notation.}$$

In the next example, the functions created by composition are quadratic.

EXAMPLE 11 Two functions are $g(x) = x^2 + 2x + 1$ and $f(x) = 7x - 4$.

(a) Find: $f(g(x))$

SOLUTION ▶

$$f(g(x)) = f(x^2 + 2x + 1) \qquad \text{The input is } g(x), x^2 + 2x + 1.$$
$$f(g(x)) = 7(x^2 + 2x + 1) - 4 \qquad \text{Replace } x \text{ in } f(x) \text{ with } x^2 + 2x + 1.$$
$$f(g(x)) = \mathbf{7x^2 + 14x + 7} - 4 \qquad \text{Distributive property.}$$
$$f(g(x)) = 7x^2 + 14x + 3 \qquad \text{Simplify.}$$

(b) Find: $g(f(x))$

▶

$$g(f(x)) = g(7x - 4) \qquad \text{The input is } f(x), 7x - 4.$$
$$g(f(x)) = (7x - 4)^2 + 2(7x - 4) + 1 \qquad \text{Replace } x \text{ in } g(x) \text{ with } 7x - 4.$$
$$g(f(x)) = \mathbf{(7x - 4)(7x - 4)} + 14x - 8 + 1 \qquad (7x - 4)^2 = (7x - 4)(7x - 4)$$
$$g(f(x)) = \mathbf{49x^2 - 28x - 28x + 16} + 14x - 7 \qquad \text{Distributive property; simplify.}$$
$$g(f(x)) = 49x^2 - 42x + 9 \qquad \text{Simplify.}$$

A spreadsheet uses data (the input values) and a formula (the function rule). For each input value, the spreadsheet uses the formula to calculate an output value. Since this output value can then be an input value for another formula, a spreadsheet can do composition of functions. A spreadsheet that finds the payroll for a business often uses composition.

EXAMPLE 12 | A fast-food restaurant pays its workers $8 per hour. A worker's *gross pay* is the product of this pay rate and the number of hours worked. The employer must withhold 7.65% of the gross pay for taxes.

(a) Write a function in which the gross pay in dollars, $g(x)$, depends on the hours worked, x.

SOLUTION ▶ $\quad g(x) = \left(\dfrac{\$8}{1 \text{ hr}}\right)x \qquad x = \text{ hours worked}; g(x) = \text{ gross pay}$

(b) Write a function in which the amount on the paycheck after taxes, $p(x)$, depends on the gross pay, x.

▶ $\quad p(x) = x - 0.0765x \qquad x = \text{ gross pay}; p(x) = \text{ amount on paycheck}$

(c) Find $p(g(x))$. Round to the nearest hundredth.

▶ $\quad p(g(x)) = p\left(\left(\dfrac{\$8}{1 \text{ hr}}\right)x\right) \qquad$ The input is $g(x)$, $\dfrac{\$8}{1 \text{ hr}}x$.

$\quad p(g(x)) = \left(\dfrac{\$8}{1 \text{ hr}}\right)x - 0.0765\left(\left(\dfrac{\$8}{1 \text{ hr}}\right)x\right) \qquad$ Replace x in $p(x)$ with $\dfrac{\$8}{1 \text{ hr}}x$.

$\quad p(g(x)) = \left(\dfrac{\$8}{1 \text{ hr}}\right)x - \left(\dfrac{\$0.612}{1 \text{ hr}}\right)x \qquad$ Simplify.

$\quad p(g(x)) = \left(\dfrac{\$7.388}{1 \text{ hr}}\right)x \qquad$ Simplify.

$\quad p(g(x)) \approx \left(\dfrac{\$7.39}{1 \text{ hr}}\right)x \qquad$ Round.

(d) Describe the meaning of $p(g(x))$.

▶ The function $p(g(x))$ describes the relationship of the earnings of the worker after withholding for taxes and the number of hours worked. After taxes, this worker makes about $7.39 per hour. Notice that the input value, x, represents two different things. For $g(x)$, the input value is the number of hours that an employee works. The output of $g(x)$ is the gross pay, in dollars. This output of $g(x)$ is now an input for $p(x)$. The final output of $p(x)$ is the final amount on the paycheck (after withholding for taxes), in dollars.

Practice Problems

For problems 12–14,
(a) find $f(g(x))$.
(b) find $g(f(x))$.

12. $f(x) = 8x - 2; g(x) = 5x + 9$
13. $f(x) = \sqrt[3]{x^2}; g(x) = \sqrt{x}$
14. $f(x) = x^2 - 8x + 2; g(x) = 9x + 1$

15. If $R(x) = \left(\dfrac{\$8.99}{1 \text{ lb}}\right)x$ and $S(x) = x - 0.30x$, find $S(R(x))$. Round to the nearest hundredth.

One-to-One Functions

A **function** is a set of ordered pairs in which each input corresponds to exactly one output. No two ordered pairs have the same input value. However, two or more ordered pairs may have the same output value. If a vertical line crosses the graph of a relation in more than one point, at least two ordered pairs on the graph have the same input. So the graph does not represent a function. This is the **vertical line test**.

A **one-to-one function** is a special kind of function in which each output corresponds to exactly one input. No two ordered pairs have the same output value. Since it is a function, no two ordered pairs have the same input value.

EXAMPLE 13 Determine whether each function is a one-to-one function.

(a) $f = \{(0, 1), (2, 3), (5, 11)\}$

SOLUTION ▶ Yes, the outputs of this function are 1, 3, and 11. No output is the same as another output.

(b) $h = \{(\text{pear, green}), (\text{banana, yellow}), (\text{apple, green})\}$

▶ No, two ordered pairs have the same output value, green.

If every horizontal line crosses the graph of a function in *at most one* point, no two ordered pairs on the graph have the same output, and the graph represents a one-to-one function. This is the **horizontal line test**.

EXAMPLE 14 Use the horizontal line test to determine whether the graph represents a one-to-one function.

(a)
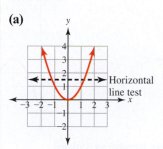

(b)
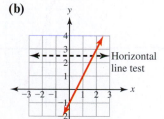

(c)
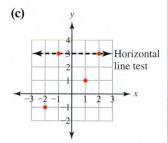

SOLUTION ▶ At least one horizontal line crosses the graph in more than one point. This is not a one-to-one function.

Every horizontal line crosses the graph in at most one point. This is a one-to-one function.

At least one horizontal line crosses the graph in more than one point. This is not a one-to-one function.

Vertical and Horizontal Line Tests

Vertical Line Test   If every vertical line crosses the graph of a relation in at most one point, the graph represents a function.

Horizontal Line Test   If every horizontal line crosses the graph of a function in at most one point, the graph represents a one-to-one function.

A bar graph can represent some functions. The height of each bar represents a single output value.

EXAMPLE 15 | Determine whether the bar graph represents a one-to-one function.

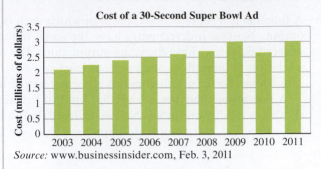

Cost of a 30-Second Super Bowl Ad

Source: www.businessinsider.com, Feb. 3, 2011

SOLUTION ▶ If a function is one-to-one, each output value corresponds to exactly one input value. No input or output values repeat.

The years on the horizontal axis are the input values. None of the input values repeat. The height of each bar shows an output value. At least two of the bars (2009 and 2011) have the same heights. The output value of $3 million repeats. So this bar graph does not represent a one-to-one function.

Practice Problems

For problems 16–22, determine whether the function represented by roster notation or a graph is one-to-one.

16. {(Ecuador, Quito), (Peru, Lima), (Argentina, Buenos Aires)}

17. {(Wambach, forward), (Morgan, forward), (Solo, goalkeeper), (Sauerbrunn, defender)}

18.

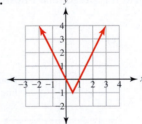

19.

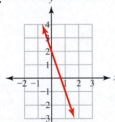

20.

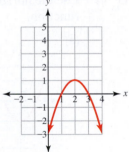

21.

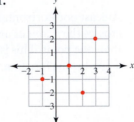

22.

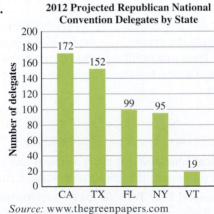

2012 Projected Republican National Convention Delegates by State

Source: www.thegreenpapers.com

8.1 VOCABULARY PRACTICE

Match the term with its description.

1. A test used to determine whether a graph of a function represents a one-to-one function
2. A test used to determine whether a graph represents a function
3. The expense of making a product
4. Changing the order of multiplication or addition does not change the result.
5. $x^m \cdot x^n = x^{m+n}$
6. A function in which each output corresponds to exactly one input
7. The amount of money made by selling a product
8. The difference of revenue and costs
9. Using the output of one function as the input of the next function
10. A relation

A. commutative property
B. composition
C. costs
D. horizontal line test
E. one-to-one function
F. product rule of exponents
G. profit
H. revenue
I. set of ordered pairs
J. vertical line test

8.1 Exercises

Follow your instructor's guidelines for showing your work.

For exercises 1–12, $f(x) = x + 2$, $g(x) = 4x$, $h(x) = x + 5$, and $k(x) = 3x$. Find:

1. $f(x) + g(x)$
2. $h(x) + k(x)$
3. $g(x) - f(x)$
4. $k(x) - h(x)$
5. $f(x) - g(x)$
6. $h(x) - k(x)$
7. $f(x) \cdot g(x)$
8. $h(x) \cdot k(x)$
9. $f(g(x))$
10. $h(k(x))$
11. $g(f(x))$
12. $k(h(x))$

For exercises 13–24, $f(x) = -3x - 4$, $g(x) = -6x - 12$, $h(x) = -2x - 5$, and $k(x) = -4x - 7$. Find:

13. $f(x) + g(x)$
14. $h(x) + k(x)$
15. $g(x) - f(x)$
16. $k(x) - h(x)$
17. $f(x) - g(x)$
18. $h(x) - k(x)$
19. $f(x) \cdot g(x)$
20. $h(x) \cdot k(x)$
21. $f(g(x))$
22. $h(k(x))$
23. $g(f(x))$
24. $k(h(x))$

For exercises 25–28, $f(x) = 8x + 2$ and $g(x) = x^2 + 5x - 4$. Find:

25. $f(x) + g(x)$
26. $g(x) + f(x)$
27. $f(g(x))$
28. $g(f(x))$

For exercises 29–30, $f(x) = 3x + 6$ and $g(x) = x^2 + 2x - 9$. Find:

29. $g(f(x))$
30. $f(g(x))$

For exercises 31–32, $f(x) = 7x^2$ and $g(x) = 3x - 2$. Find:

31. $f(g(x))$
32. $g(f(x))$

For exercises 33–34, $f(x) = 5x^2$ and $g(x) = 2x - 3$. Find:

33. $g(f(x))$
34. $f(g(x))$

For exercises 35–36, $f(x) = \frac{3}{4}x - 2$ and $g(x) = -4x^2$. Find:

35. $f(g(x))$
36. $g(f(x))$

For exercises 37–38, $f(x) = \frac{4}{5}x - 7$ and $g(x) = -5x^2$. Find:

37. $g(f(x))$
38. $f(g(x))$

For exercises 39–44, $f(x) = \sqrt[3]{x^2}$, $g(x) = \sqrt[4]{x^3}$, $h(x) = \sqrt[6]{x^5}$, and $k(x) = \sqrt[5]{x^3}$. Assume that x is a real number greater than or equal to 0. Find:

39. $f(g(x))$
40. $h(k(x))$
41. $g(f(x))$
42. $k(h(x))$
43. $f(h(x))$
44. $g(k(x))$

For exercises 45–48, $f(x) = \frac{x}{x - 8}$ and $g(x) = \frac{64}{x^2 - 8x}$. Find:

45. $f(x) + g(x)$
46. $g(x) + f(x)$
47. $f(x) - g(x)$
48. $g(x) - f(x)$

49. Addition of functions is commutative. Explain what this means.

50. Composition of functions is not commutative. Explain what this means.

For exercises 51–52, $f(x) = 6x + 3$ and $g(x) = 8x - 2$.

51. a. Find: $f(g(x))$
    b. Use interval notation to represent the domain of $f(g(x))$.
    c. Use interval notation to represent the range of $f(g(x))$.

52. a. Find: $g(f(x))$
    b. Use interval notation to represent the domain of $g(f(x))$.
    c. Use interval notation to represent the range of $f(g(x))$.

For exercises 53–54, $f(x) = x^2 + 5$ and $g(x) = x + 4$.

53. a. Find: $f(g(x))$
   b. Use interval notation to represent the domain of $f(g(x))$.
   c. Use interval notation to represent the range of $f(g(x))$.

54. a. Find: $g(f(x))$
   b. Use interval notation to represent the domain of $g(f(x))$.
   c. Use interval notation to represent the range of $f(g(x))$.

For exercises 55–56, $f(x) = x - 5$ and $g(x) = x^2 + 3$.

55. a. Find: $f(g(x))$
   b. Use interval notation to represent the domain of $f(g(x))$.
   c. Use interval notation to represent the range of $f(g(x))$.

56. a. Find: $g(f(x))$
   b. Use interval notation to represent the domain of $g(f(x))$.
   c. Use interval notation to represent the range of $f(g(x))$.

For exercises 57–58, $f(x) = 8\sqrt{3x + 12}$ and $g(x) = 5\sqrt{3x + 12}$.

57. a. Find: $f(x) + g(x)$
   b. Use interval notation to represent the domain of $f(x) + g(x)$.

58. a. Find: $f(x) - g(x)$
   b. Use interval notation to represent the domain of $f(x) - g(x)$.

For exercises 59–60, $f(x) = \sqrt{-3x + 15}$ and $g(x) = 9\sqrt{-3x + 15}$.

59. a. Find: $f(x) - g(x)$
   b. Use interval notation to represent the domain of $f(x) - g(x)$.

60. a. Find: $f(x) + g(x)$
   b. Use interval notation to represent the domain of $f(x) + g(x)$.

For exercises 61–62, $f(x) = 7\sqrt{-2x + 6}$ and $g(x) = 5\sqrt{-2x + 6}$.

61. a. Find: $f(x) - g(x)$
   b. Use interval notation to represent the domain of $f(x) - g(x)$.

62. a. Find: $g(x) - f(x)$
   b. Use interval notation to represent the domain of $g(x) - f(x)$.

63. If $f(x) = \dfrac{x}{x - 6}$ and $g(x) = \dfrac{36}{x^2 - 6x}$,
   a. find $f(x) - g(x)$.
   b. use interval notation to represent the domain of $f(x) - g(x)$.

64. If $f(x) = \dfrac{x}{x - 7}$ and $g(x) = \dfrac{49}{x^2 - 7x}$,
   a. find $f(x) - g(x)$.
   b. use interval notation to represent the domain of $f(x) - g(x)$.

For exercises 65 and 66, the income statement of a company includes the total revenue and the net income. *Total revenue* is the amount of money a business brought in during the time period covered by the income statement. *Net income* is the income that a company has after subtracting costs, other expenses, and taxes from the total revenue.

65. The table shows part of an annual income statement for Harley-Davidson, Inc. What were the costs, other expenses, and taxes during this time? (*Source:* www .hoovers.com)

Total revenue	$6,143.0 million
Net income	$933.8 million

66. The table shows part of an annual income statement for Nike, Inc. What were the costs, other expenses, and taxes during this time? (*Source:* www.hoovers.com)

Total revenue	$16,325.9 million
Net income	$1,491.5 million

67. It costs $2.30 to make a product. The selling price of the product is $15.99.
   a. Write a revenue function, $R(x)$, that finds the revenue from the sale of x products.
   b. Write a cost function, $C(x)$, that finds the costs to make x products.
   c. Write a profit function, $P(x)$, that finds the profit from making and selling x products.
   d. Evaluate the profit function if a company makes and sells 40,000 products.

68. It costs $4.50 to make a product. The selling price of the product is $10.99.
   a. Write a revenue function, $R(x)$, that finds the revenue from the sale of x products.
   b. Write a cost function, $C(x)$, that finds the costs to make x products.
   c. Write a profit function, $P(x)$, that finds the profit from making and selling x products.
   d. Evaluate the profit function if a company makes and sells 400,000 products.

69. An estimate of the cost of growing pumpkins is $814 per acre. The yield of pumpkins is $\dfrac{22{,}000 \text{ lb}}{1 \text{ acre}}$. The wholesale price of pumpkins is $\dfrac{\$0.08}{1 \text{ lb}}$. (*Source:* www.agmrc.org)
   a. Write a revenue function, $R(x)$, that finds the revenue from the sale of x acres of pumpkins.
   b. Write a cost function, $C(x)$, that finds the costs to grow x acres of pumpkins.
   c. Write a profit function, $P(x)$, that finds the profit from growing and selling x acres of pumpkins.
   d. Evaluate the profit function if 5 acres of pumpkins are grown and sold.

70. A quilting service charges customers $\dfrac{\$0.06}{1 \text{ in.}^2}$. The service pays its quilters $\dfrac{\$12}{1 \text{ hr}}$. To quilt 9000 in.² takes about 30 hr.

   a. Write a revenue function, $R(x)$, that finds the revenue of quilting x square inches.

   b. Write a cost function, $C(x)$, that finds the cost of quilting x square inches.

   c. Write a profit function, $P(x)$, that finds the profit from quilting x square inches.

   d. Evaluate the profit function for quilting a bed quilt that is 80 in. long and 90 in. wide.

71. If $R(x) = \left(\dfrac{\$4.99}{1 \text{ lb}}\right)x$ and $S(x) = x - 0.20x$, find $S(R(x))$.
Round to the nearest hundredth.

72. If $R(x) = \left(\dfrac{\$3.99}{1 \text{ lb}}\right)x$ and $S(x) = x - 0.25x$, find $S(R(x))$.
Round to the nearest hundredth.

73. If $R(x) = \left(\dfrac{\$500}{1 \text{ sale}}\right)x$ and $S(x) = x + 0.20x$, find $S(R(x))$.

74. If $R(x) = \left(\dfrac{\$50}{1 \text{ sale}}\right)x$ and $S(x) = x + 0.04x$, find $S(R(x))$.

75. A discount store in Seattle is having a Back to College Sale with many items discounted 15%. The function $P(x) = x - 0.15x$ represents the relationship of the original price, x, and the price of the item on sale, $P(x)$. The sales tax rate in Seattle is 9.5%. The function $C(x) = x + 0.095x$ describes the cost of an item including the sales tax. A student is buying a desk with an original price of $149.99. Find $C(P(\$149.99))$. Round to the nearest hundredth. (*Source:* dor.wa.gov)

76. A store in New York City is having an end-of-season clearance sale with many items discounted 40%. The function $P(x) = x - 0.40x$ represents the relationship of the original price, x, and the price of the item on sale, $P(x)$. The sales tax rate in New York City is 8.875%. The function $C(x) = x + 0.08875x$ describes the cost of an item including the sales tax. A student is buying a bookcase with an original price of $135. Find $C(P(\$135))$. Round to the nearest hundredth. (*Source:* www.nyc.gov)

77. Describe the difference between a relation and a function.

78. Describe the difference between a function and a one-to-one function.

79. Create a set with six ordered pairs that is a one-to-one function.

80. Create a set with six ordered pairs that is a function but is not a one-to-one function.

81. Explain why the horizontal line test can be used to determine whether a graph represents a one-to-one function.

82. Explain why the vertical line test can be used to determine whether a graph represents a function.

83. The graph represents the function $y = |x - 2| - 1$.

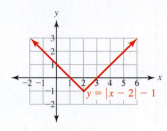

   a. Copy the graph, and use the horizontal line test.

   b. Is this function one-to-one?

84. The graph represents the function $y = |x - 3| - 2$.

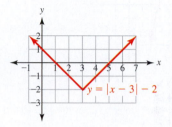

   a. Copy the graph, and use the horizontal line test.

   b. Is this function one-to-one?

85. Does the graph represent a one-to-one function?

   a.

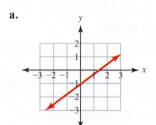

   b.

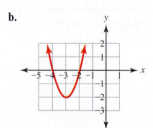

   c.

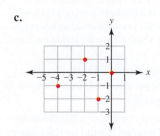

86. Does the graph represent a one-to-one function?

a.

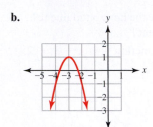

b.

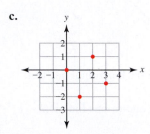

c.

y

Problem Solving: Practice and Review

Follow your instructor's guidelines for using the five steps outlined in Section 1.5, p. 51.

87. Find the total number of American drivers on the road. Round to the nearest million.

The 2011 GMAC Insurance National Drivers Test results revealed that . . . 36.9 million American drivers—roughly 18 percent—would not pass the written drivers test if taken today. (*Source:* www.gmacinsurance.com, May 26, 2011)

For exercises 88–89, use the information below.

Their stories, archived in a state database and detailed in hundreds of confidential files obtained by The Oregonian, show that one of every five [developmentally disabled] clients in state-licensed foster or group homes have been victims of at least one serious instance of abuse or neglect during the past seven years. . . . Now, the state spends $134 million a year to house 4,200 residents in 611 foster homes, which are operated by individuals, and 544 group homes, which are most often run by nonprofits. (*Source:* www.oregonian.com, Nov. 4, 2007)

88. Find the approximate number of residents who have been victims of at least one serious instance of abuse or neglect during the past 7 years.

89. Find the average cost per year to house a developmentally disabled resident in Oregon. Round to the nearest hundred.

90. The graph represents the relationship of the number of years since 1998, x, and the amount of gross domestic product (GDP) of the United States in millions of dollars, y.

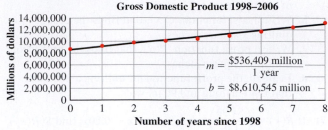

Source: money.cnn.com, Dec. 1, 2008

a. Write a linear function that is a model of this relationship.

b. Use this function to find the GDP in 2011. Report the answer in *trillions* of dollars (1 trillion = 1,000,000 million). Round to the nearest tenth.

c. For an economic downturn to be called a recession, the gross domestic product must decline for two consecutive quarters. In December 2008, the National Bureau of Economic Research said that the United States had been in a recession since December 2007. Is this model a good predictor of the GDP in 2011? Explain.

Find the Mistake

For exercises 91–94, the completed problem has one mistake.
(a) Describe the mistake in words, or copy down the whole problem and highlight or circle the mistake.
(b) Do the problem correctly.

91. Problem: Find $f(g(x))$ when $f(x) = -7x + 2$ and $g(x) = 5x - 9$.

Incorrect Answer: $f(g(x)) = 5(-7x + 2) - 9$
$$f(g(x)) = -35x + 10 - 9$$
$$f(g(x)) = -35x + 1$$

92. Problem: Find $f(g(x))$ when $f(x) = 4x + 9$ and $g(x) = -3x + 11$.

Incorrect Answer: $f(g(x)) = 4(-3x + 11)$
$$f(g(x)) = -12x + 44$$

93. Problem: Find $f(x) - g(x)$ when $f(x) = -7x + 2$ and $g(x) = 5x - 9$.

Incorrect Answer: $f(x) - g(x) = -7x + 2 - 5x - 9$
$$f(x) - g(x) = -12x - 7$$

94. Problem: The cost to make a product is $3.85. The selling price of the product is $29.99. Write a revenue function, $R(x)$, that predicts the revenue from the sale of x products.

Incorrect Answer: $R(x) = \left(\dfrac{\$3.85}{1 \text{ product}}\right) x$

Review

For exercises 95–98,
(a) solve the equation.
(b) check the solution(s).

95. $|x| = 8$

96. $|x - 9| = 15$

97. $|3x - 9| - 6 = 15$

98. $|3x + 5| + 6 = 6$

SUCCESS IN COLLEGE MATHEMATICS

99. Complete one of the cumulative reviews in the textbook. As you complete and check the answers for the exercises in these reviews, classify each question into three types: A, B, and C. Mark as an "A" any exercise that you can answer correctly without looking anything up. You know this material and do not need to spend any more time reviewing it. Mark as a "B" any exercise that you can answer correctly after you refresh your memory by looking at an example. You need to do a few more exercises like this one before the final exam. Mark as a "C" any exercise that you cannot do at all or that you do incorrectly. You will need to go back, reread the section of the text in which this question was discussed, and do some practice problems from this section.

A function is a relation in which each input corresponds to exactly one output. When the input of a function is an exponent, it is an **exponential function**. A model of the growth of the number of open source computer software projects is an exponential function. In this section, we will study this and other exponential functions.

SECTION 8.2

Exponential Functions

After reading the text, working the practice problems, and completing assigned exercises, you should be able to:

1. Identify the domain, range, and y-intercept of an exponential function.

2. Determine whether an exponential function is one-to-one.

3. Evaluate an exponential function.

4. Identify increasing and decreasing exponential functions.

Exponential Functions

In an exponential function, the input value is an exponent. For example, $y = 2^x$ is an exponential function. The base of this function is 2. The exponent is the input value, x. The output value is y.

EXAMPLE 1 | The graph and the table of ordered pairs are representations of the exponential function $y = 2^x$.

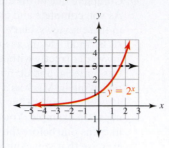

x	y
-30	9.3×10^{-10}
-20	9.5×10^{-7}
-10	9.8×10^{-4}
0	1
10	1024
20	1.0×10^6
30	1.1×10^9

(a) Use interval notation to represent the domain.

SOLUTION ▶   Since no real number input values cause division by zero or result in output values that are not real numbers, the domain is $(-\infty, \infty)$.

(b) Use interval notation to represent the range.

▶   As we see in the table, as x increases, y also increases. As x decreases, y gets closer to 0 but never equals 0. The range is $(0, \infty)$.

(c) Identify the y-intercept.

▶ $y = 2^x$

$y = 2^{\mathbf{0}}$ The x-coordinate of the y-intercept is 0.

$y = \mathbf{1}$ The y-intercept is $(0, 1)$.

For any exponential function $y = b^x$ with a base b, $b^0 = 1$. So the y-intercept of any exponential function $y = b^x$ is $(0, 1)$.

(d) Identify the x-intercept.

▶   The y-coordinate of an x-intercept is 0. Since the range of an exponential function $y = b^x$ does not include 0, there is no x-intercept. The graph does not intersect the x-axis.

(e) Is this function one-to-one?

▶   Since the graph passes the horizontal line test, it is a one-to-one function. Each output value corresponds to exactly one input value.

(f) In function notation, $y = 2^x$ is $f(x) = 2^x$. Evaluate $f(4)$.

▶ $f(x) = 2^x$

$f(\mathbf{4}) = 2^{\mathbf{4}}$ Replace x with 4.

$f(4) = \mathbf{16}$ Simplify.

In Example 1, the base of the function is greater than 1. As the input increases, the output also increases. This is an **increasing** function. In the next example, the base of the function is greater than 0 and less than 1. As the input increases, the output decreases. It is a **decreasing** function.

EXAMPLE 2   The graph and the table of ordered pairs are representations of the exponential function $y = 0.5^x$.

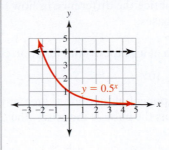

x	y
-10	1024
0	1
10	9.8×10^{-4}
20	9.5×10^{-7}
30	9.3×10^{-10}
40	9.1×10^{-13}
50	8.9×10^{-16}

(a) Use interval notation to represent the domain.

SOLUTION ▶   Since no real number input values cause division by zero or result in output values that are not real numbers, the domain is $(-\infty, \infty)$.

(b) Use interval notation to represent the range.

▶   As we see in the table, as x increases, y decreases. As x increases, y gets closer to 0 but never equals 0. The range is $(0, \infty)$.

(c) Identify the y-intercept.

▶ $y = 0.5^x$

$y = 0.5^0$ The x-coordinate of the y-intercept is 0.

$y = \mathbf{1}$ The y-intercept is (0, 1).

(d) Identify the x-intercept.

▶   The y-coordinate of an x-intercept is 0. Since the range of an exponential function $y = b^x$ does not include 0, there is no x-intercept. The graph does not intersect the x-axis.

(e) Is this function one-to-one?

▶   Since the graph passes the horizontal line test, it is a one-to-one function. Each output value corresponds to exactly one input value.

(f) In function notation $y = 0.5^x$ is $f(x) = 0.5^x$. Evaluate $f(4)$.

▶ $f(x) = 0.5^x$

$f(\mathbf{4}) = 0.5^{\mathbf{4}}$ Replace x with 4.

$f(4) = \mathbf{0.0625}$ Simplify.

(g) Is this function increasing or decreasing?

▶   As the input values increase, the output values decrease. This is a decreasing function.

Exponential Functions

Exponential functions in the form $y = b^x$ or $f(x) = b^x$, where b is a real number, $b > 0$, and $b \neq 1$, are one-to-one functions. The domain is $(-\infty, \infty)$ and the range is $(0, \infty)$. The y-intercept is (0, 1). There is no x-intercept. If $0 < b < 1$, the function is decreasing. If $b > 1$, the function is increasing.

Changing the value of the base, b, changes the graph of an exponential function. In the next examples, the table shows the input and output values for functions with different bases. Notice the difference in how fast the output values increase or decrease for different bases.

EXAMPLE 3 | Describe how the graph of an exponential function changes when the base is changed.

(a) $y = 2^x$, $y = 5^x$, $y = 10^x$

SOLUTION ▶ The base in each exponential function is greater than 1. In both the tables and the graphs, we see that as the base, b, increases, the outputs *increase* at a faster rate.

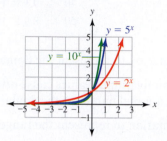

$y = 2^x$		$y = 5^x$		$y = 10^x$	
x	y	x	y	x	y
0	1	0	1	0	1
1	2	1	5	1	10
2	4	2	25	2	100
3	8	3	125	3	1000

(b) $y = 0.2^x$, $y = 0.5^x$, $y = 0.7^x$

▶ The base in each exponential function is between 0 and 1. In both the tables and the graphs, we see that as b increases, the outputs *decrease* at a slower rate.

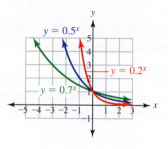

$y = 0.2^x$		$y = 0.5^x$		$y = 0.7^x$	
x	y	x	y	x	y
−4	625	−4	16	−4	4.16
−3	125	−3	8	−3	2.92
−2	25	−2	4	−2	2.04
−1	5	−1	2	−1	1.43
0	1	0	1	0	1
1	0.2	1	0.5	1	0.7

Some output values are rounded.

Practice Problems

For problems 1–2,
(a) complete the table of ordered pairs for each function.
(b) draw a graph of the function.
(c) use interval notation to represent the domain.
(d) use interval notation to represent the range.
(e) identify the *y*-intercept.
(f) is the function one-to-one?
(g) is the function increasing or decreasing?

1. $y = 3^x$

x	y
−20	2.9×10^{-10}
−10	
0	
10	
20	
30	
40	1.2×10^{15}

2. $y = 0.2^x$

x	y
−20	9.5×10^{13}
−10	
0	
10	
20	
30	
40	1.1×10^{-28}

3. Imagine graphing $y = 15^x$ on the same coordinate system as $y = 3^x$. For which function do the outputs increase faster?

4. Imagine graphing $y = 0.7^x$ on the same coordinate system as $y = 0.2^x$. For which function do the outputs decrease faster?

The Irrational Number *e*

We cannot write an irrational number as a fraction with integers in the numerator and denominator. For example, $\sqrt{3}$ and π are irrational numbers. An approximate value of π is 3.14. Another irrational number, 2.718. . . , is named with the letter e. Most scientific calculators include an e^x key.

EXAMPLE 4 | A function is $f(x) = e^x$. Evaluate $f(3)$. Round to the nearest hundredth.

SOLUTION ▶

$f(x) = e^x$ The base is the irrational number e, 2.718. . . .

$f(\mathbf{3}) = e^{\mathbf{3}}$ The input value is 3.

$f(3) = \mathbf{20.08553. . .}$ Use a calculator.

$f(3) \approx \mathbf{20.09}$ Round.

The graph in Figure 1 represents the functions $y = 2^x$, $y = e^x$, and $y = 3^x$. Since $e > 1$, $y = e^x$ is an increasing function. Like other exponential functions, its domain is $(-\infty, \infty)$, its range is $(0, \infty)$, and the y-intercept of its graph is $(0, 1)$.

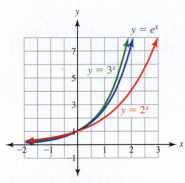

Figure 1

> **Practice Problems**
>
> For problems 5–8, evaluate $f(x) = e^x$ for the given input values. Round irrational numbers to the nearest hundredth.
>
> **5.** $f(2)$ **6.** $f(-2)$ **7.** $f(1)$ **8.** $f(0)$

Exponential Functions and the Order of Operations

Some exponential functions are in the form $y = A(b^x)$, where A is a real number that is not equal to 1.

EXAMPLE 5 | An exponential function is $f(x) = 4(3^x)$. Evaluate $f(2)$.

SOLUTION ▶

$f(x) = 4(3^x)$

$f(\mathbf{2}) = 4(3^{\mathbf{2}})$ Replace x with 2.

$f(2) = 4(\mathbf{9})$ Evaluate the exponential expression.

$f(2) = \mathbf{36}$ Simplify.

When evaluating an exponential function, do not round until the end.

EXAMPLE 6 | An exponential function is $y = 2000(1.08)^x$. Evaluate this function when $x = 5$. Round to the nearest hundredth.

SOLUTION ▶

$y = 2000(1.08)^x$

$y = 2000(1.08)^{\mathbf{5}}$ Replace x with 5.

$y = 2000(\mathbf{1.46. . .})$ Evaluate the exponential expression. Do not round.

$y = \mathbf{2938.656. . .}$ Simplify.

$y \approx \mathbf{2938.66}$ Round.

In the next example, the exponential expression is multiplied by a constant, 7.1511. The exponent also includes a constant, 0.0499.

EXAMPLE 7 The exponential function $f(x) = (7.1511)(e^{0.0499x})$ represents the relationship of the number of months since January 1995, x, and the number of open source projects, $f(x)$. Evaluate this function to predict the number of open source projects in January 2013. Round to the nearest thousand. (*Source:* Deshpande, A., and Riehle, D., *The Total Growth of Open Source*, 2008)

SOLUTION ▶ Since there are 12 months in a year and the difference between 2013 and 1995 is 18 years, $x = 216$.

$$f(x) = (7.1511)(e^{0.0499x})$$
$$f(\mathbf{216}) = (7.1511)(e^{(0.0499)(\mathbf{216})}) \qquad \text{Replace } x \text{ with 216.}$$
$$f(216) = (7.1511)(e^{\mathbf{10.7784}}) \qquad \text{Follow the order of operations.}$$
$$f(216) = (7.1511)(\mathbf{47{,}973.3 \ldots}) \qquad \text{Follow the order of operations.}$$
$$f(216) = \mathbf{343{,}061.9 \ldots} \qquad \text{Follow the order of operations.}$$
$$f(216) \approx \mathbf{343{,}000 \text{ projects}} \qquad \text{Round.}$$

The functions $f(x) = b^x$ and $f(x) = Ab^x$ have the same domain and range. However, the y-intercepts are not the same. The y-intercept of $f(x) = b^x$ is $(0, 1)$. The y-intercept of $f(x) = Ab^x$ is $(0, A)$.

$$f(x) = Ab^x \qquad \text{An exponential function.}$$
$$f(\mathbf{0}) = Ab^{\mathbf{0}} \qquad \text{The } x\text{-coordinate of the } y\text{-intercept is 0. Replace } x \text{ with 0.}$$
$$f(0) = A(\mathbf{1}) \qquad \text{Simplify.}$$
$$f(0) = A \qquad \text{Simplify. The } y\text{-intercept is } (0, A).$$

EXAMPLE 8 Compare the domain, range, and y-intercept of $f(x) = 3^x$ and $f(x) = 2(3^x)$.

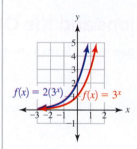

SOLUTION ▶

$f(x) = 3^x$	$f(x) = 2(3^x)$
Domain: $(-\infty, \infty)$	**Domain:** $(-\infty, \infty)$
Range: $(0, \infty)$	**Range:** $(0, \infty)$
y-intercept: $(0, 1)$	**y-intercept:** $(0, 2)$

The domain and range of the functions are the same. The y-intercepts are different. The function $f(x) = 2(3^x)$ is in the form $f(x) = Ab^x$. The y-intercept, $(0, A)$, is $(0, 2)$.

Practice Problems

 9. Evaluate $f(2)$ when $f(x) = 3(4^x)$.

 10. When $x = 2$, evaluate $y = 3000(4^x)$.

 11. Identify the y-intercept of $y = 4^x$.

 12. Identify the y-intercept of $y = 5(4^x)$.

 13. Evaluate the function in Example 7 to predict the number of open source projects in January 2014. Round to the nearest thousand.
 (*Source:* Deshpande, A., and Riehle, D., *The Total Growth of Open Source*, 2008)

Using Technology: Evaluating e^x and Graphing Exponential Functions

Different scientific calculators use different keystrokes to evaluate an exponential expression with a base of e. Find e^x on your calculator. It may have its own key, or it may be printed above the **LN** key.

After pressing the appropriate key(s), if the entry line of the calculator shows $e^{\wedge}($, then type in the exponent, type in the closing parenthesis,), and press **ENTER** or **=**.

If the entry line of the calculator shows 2.718281828, press **CLEAR**. Type the exponent first; now press the appropriate keys for e^x.

EXAMPLE 9 Graph $y = e^x$. Find the output value when the input value is 2.

Go to the Y= screen. Look for e^x above the **LN** key. Press **2nd**, **LN**. The screen shows $e^{\wedge}($. To type in the variable, x, press the **X,T,θ,n** key. Close the parentheses by pressing **)**. Press **GRAPH**.

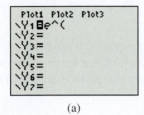
(a)

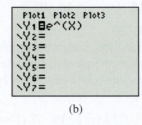
(b)

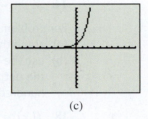
(c)

It looks as though the graph touches the x-axis and stops when x is approximately equal to -3. However, we know that an exponential function does not have an x-intercept. The range of this function is $(0, \infty)$. We also know that the graph does not stop. The domain of this function is $(-\infty, \infty)$. There are output values for every real number. We can see this trend if we change the window of the graph.

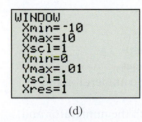
(d)

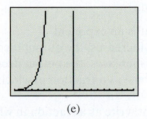
(e)

The input values are the same as in a standard window. The maximum output value is $y = 0.01$. Now we see that the graph does not stop when $x = -3$. As x gets very small, the y values get closer and closer to 0 but y *never equals* 0.

EXAMPLE 10 Graph $f(x) = 3^x$ in a standard window. Find $f(-8)$. Round the mantissa to the nearest hundredth.

Type the function in the Y= screen. Press **GRAPH**. To evaluate the function, go to the CALC screen. The **Value** command is highlighted.

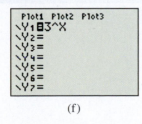
(f)

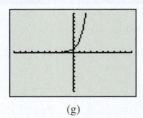
(g)

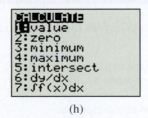
(h)

Press ENTER . The cursor is next to X=. Type the input value, −8. Press ENTER .

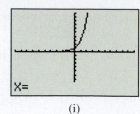 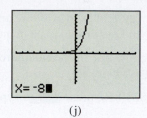 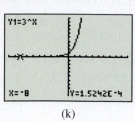

(i)   (j)   (k)

When $x = -8$, $f(x) = 1.5242\ldots \times 10^{-4}$. Rounding, $f(x) \approx 1.52 \times 10^{-4}$.

Practice Problems For problems 14–17, use a calculator to evaluate. Round any irrational numbers to the nearest ten-thousandth.

14. 2^3 **15.** e^3 **16.** 10^{-2} **17.** e^{-2}

For problems 18–20,
(a) graph the function. Sketch the graph; describe the window.
(b) use the **Value** command to evaluate the function for the given input value. If the output value is in scientific notation, round the mantissa to the nearest hundredth.

18. $f(x) = 3^x$; $f(5)$ **19.** $f(x) = 5^x$; $f(-8)$ **20.** $f(x) = 0.2^x$; $f(4)$

8.2 VOCABULARY PRACTICE

Match the term with its description.

1. A function in which the input is an exponent
2. As the input to this function increases, the output increases.
3. As the input to this function increases, the output decreases.
4. A real number that we can write as a fraction in which the numerator and denominator are integers
5. A real number that we cannot write as a fraction in which the numerator and denominator are integers
6. The set of the inputs of a function
7. The set of the outputs of a function
8. A function in which each output corresponds to exactly one input
9. The approximate value of this irrational number is 3.14.
10. The approximate value of this irrational number is 2.72.

A. decreasing function
B. domain
C. e
D. exponential function
E. increasing function
F. irrational number
G. one-to-one function
H. π
I. range
J. rational number

8.2 Exercises

Follow your instructor's guidelines for showing your work.

For exercises 1–6,
(a) identify the base of the function.
(b) complete the table of ordered pairs.
(c) graph the function.
(d) use interval notation to represent the domain.
(e) use interval notation to represent the range.
(f) identify the y-intercept.
(g) identify the function as increasing or decreasing.

x	y
−20	
−10	
0	
10	
20	
30	
40	

1. $y = 4^x$
2. $y = 5^x$
3. $y = 0.4^x$
4. $y = 0.5^x$
5. $y = e^x$
6. $y = 2e^x$

7. A student is solving a word problem in which he is trying to find energy. When he assigns the variable, he wants to use e = energy. Explain why we usually do not use e as a variable.

8. Another important irrational number, pronounced "phi" or "phee," is $\phi = 1.618.\dots$ This is the golden ratio of art and architecture. A student wrote $\phi = 1.618$. What mistake did the student make?

9. Imagine graphing $y = 2^x$ on the same coordinate system as $y = 6^x$. For which function do the outputs increase faster?

10. Imagine graphing $y = 4^x$ on the same coordinate system as $y = 10^x$. For which function do the outputs increase faster?

11. Imagine graphing $y = 0.3^x$ on the same coordinate system as $y = 0.8^x$. For which function do the outputs decrease faster?

12. Imagine graphing $y = 0.2^x$ on the same coordinate system as $y = 0.7^x$. For which function do the outputs decrease faster?

For exercises 13–20, identify the function as increasing or decreasing.

13. $f(x) = 0.6^x$

14. $f(x) = 0.8^x$

15. $g(x) = 15^x$

16. $g(x) = 23^x$

17. $y = e^x$

18. $y = 10^x$

19. $y = 6(2^x)$

20. $y = 9(3^x)$

21. For the function $y = 1^x$,
   a. complete the table of ordered pairs.
   b. is this function one-to-one?
   c. is this an exponential function?

x	y
−2	
−1	
0	
1	
2	
3	
4	

22. For the function $y = (-1)^x$,
   a. complete the table of ordered pairs.
   b. is this function one-to-one?
   c. is this an exponential function?

x	y
−2	
−1	
0	
1	
2	
3	
4	

23. For the function $y = 2^x$,
   a. complete the table of ordered pairs. Write the outputs as fractions.
   b. as x decreases, what happens to y?
   c. does y ever equal 0?

x	y
0	1
−1	
−2	
−3	
−4	
−5	
−6	

24. For the function $y = 3^x$,
   a. complete the table of ordered pairs. Write the outputs as fractions.
   b. as x decreases, what happens to y?
   c. does y ever equal 0?

x	y
0	1
−1	
−2	
−3	
−4	
−5	
−6	

For exercises 25–26, the graphs represent the functions $y = 3^x$ and $y = 6^x$.

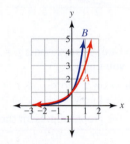

25. a. Which graph, A or B, represents $y = 3^x$?
    b. Explain how you know.

26. a. Which graph, A or B, represents $y = 6^x$?
    b. Explain how you know.

For exercises 27–28, the graphs represent the functions $y = 0.3^x$ and $y = 0.6^x$.

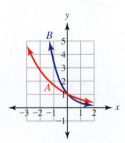

27. a. Which graph, A or B, represents $y = 0.3^x$?
    b. Explain how you know.

28. a. Which graph, A or B, represents $y = 0.6^x$?
    b. Explain how you know.

For exercises 29–30, the graphs represent the functions $y = 0.4^x$ and $y = 7^x$.

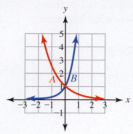

29. a. Which graph, *A* or *B*, represents $y = 0.4^x$?
   b. Explain how you know.

30. a. Which graph, *A* or *B*, represents $y = 7^x$?
   b. Explain how you know.

For exercises 31–32, the graphs represent the functions $y = e^x$ and $y = 2^x$.

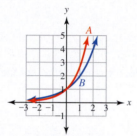

31. a. Which graph, *A* or *B*, represents $y = e^x$?
   b. Explain how you know.

32. a. Which graph, *A* or *B*, represents $y = 2^x$?
   b. Explain how you know.

For exercises 33–34, the graphs represent the functions $y = e^x$ and $y = 4^x$.

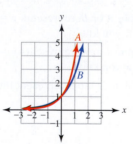

33. a. Which graph, *A* or *B*, represents $y = e^x$?
   b. Explain how you know.

34. a. Which graph, *A* or *B*, represents $y = 4^x$?
   b. Explain how you know.

For exercises 35–36, the graphs represent the functions $y = 2^x$ and $y = 5(2^x)$.

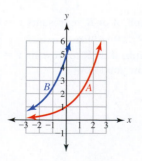

35. a. Which graph, *A* or *B*, represents $y = 2^x$?
   b. Explain how you know.

36. a. Which graph, *A* or *B*, represents $y = 5(2^x)$?
   b. Explain how you know.

For exercises 37–38, the graphs represent the functions $y = 0.9^x$ and $y = 5(0.9^x)$.

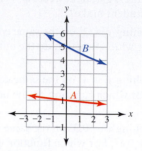

37. a. Which graph, *A* or *B*, represents $y = 0.9^x$?
   b. Explain how you know.

38. a. Which graph, *A* or *B*, represents $y = 5(0.9^x)$?
   b. Explain how you know.

39. The *y*-intercept of an exponential function in the form $y = b^x$, where $b > 0$ and $b \neq 1$, is $(0, 1)$. Explain why.

40. What is the value of any base except 0 that is raised to the 0 power?

41. A function is $y = 12(7^x)$. What is the *y*-intercept of the graph of this function?

42. A function is $y = 13(5^x)$. What is the *y*-intercept of the graph of this function?

For exercises 43–56, evaluate the function for the given input value. If the answer is irrational, round to the nearest thousandth.

43. $f(x) = 11^x$; $f(3)$

44. $f(x) = 7^x$; $f(2)$

45. $f(x) = 2(11^x)$; $f(3)$

46. $f(x) = 2(7^x)$; $f(2)$

47. $g(x) = 0.9^x$; $g(3)$

48. $g(x) = 0.8^x$; $g(2)$

49. $g(x) = 4(0.9^x)$; $g(3)$

50. $g(x) = 3(0.8^x)$; $g(2)$

51. $h(x) = 1000(1.05^x)$; $h(2)$

52. $h(x) = 1000(1.04^x)$; $h(3)$

53. $A(x) = 2549(1.15^x)$; $A(1)$

54. $A(x) = 3575(1.15^x)$; $A(1)$

55. $f(x) = e^x$; $f(3)$

56. $f(x) = e^x$; $f(4)$

For exercises 57–58, the exponential function $f(x) = (2 \text{ million})(e^{0.0464x})$ describes the relationship of the number of months since January 1995, x, and the number of open source lines of code, $f(x)$.

57. Evaluate this function to predict the number of open source lines of code in January 2012. Round to the nearest million. (*Source:* Deshpande, A., and Riehle, D., *The Total Growth of Open Source,* 2008)

58. Evaluate this function to predict the number of open source lines of code in January 2011. Round to the nearest million. (*Source:* Deshpande, A., and Riehle, D., *The Total Growth of Open Source,* 2008)

59. A linear function is $f(x) = 2x$. An exponential function is $g(x) = 2^x$. For both functions, as the input value increases, the output value also increases. If $x > 1$, for which function does the output value increase at a faster rate?

60. A linear function is $f(x) = 3x$. An exponential function is $g(x) = 3^x$. For both functions, as the input value increases, the output value also increases. If $x > 1$, for which function does the output value increase at a faster rate?

61. A function is $y = 1^x$.
   a. Is this function one-to-one?
   b. Explain.

62. A function is $y = 1^x$.
   a. Is this an exponential function?
   b. Explain.

For exercises 63–66, $f(x) = 3(5^x)$.

63. Use interval notation to represent the domain.

64. Use interval notation to represent the range.

65. Identify the y-intercept.

66. Identify the x-intercept.

For exercises 67–70, $f(x) = 2(6^x)$.

67. Use interval notation to represent the range.

68. Use interval notation to represent the domain.

69. Identify the x-intercept.

70. Identify the y-intercept.

For exercises 71–74, $f(x) = 6(0.5^x)$.

71. Use interval notation to represent the domain.

72. Use interval notation to represent the range.

73. Identify the y-intercept.

74. Identify the x-intercept.

For exercises 75–78, $f(x) = 3(0.2^x)$.

75. Use interval notation to represent the range.

76. Use interval notation to represent the domain.

77. Identify the x-intercept.

78. Identify the y-intercept.

For exercises 79–80, the function $D(x) = (4.99 \times 10^{12})e^{0.07x}$ represents the relationship of the number of years since 2000, x, and the total public debt in the United States in dollars, $D(x)$. (*Source:* www.treasurydirect.gov)

79. Find the public debt in 2012 in trillions of dollars. Round to the nearest hundredth of a trillion.

80. Find the public debt in 2013. Write in scientific notation; round the mantissa to the nearest hundredth.

Problem Solving: Practice and Review

Follow your instructor's guidelines for using the five steps as outlined in Section 1.5, p. 51.

81. The U.S. Census Bureau estimates that the population of Washington, D.C., in 2007 was 588,292 people. Predict the number of people living with AIDS in Washington, D.C., in 2007. Round to the nearest hundred.

The overall rate of persons living with AIDS increased from 1,395.5 in 2003 to 1,724.2 cases per 100,000 population in 2007, a 23.6% increase. (*Source:* www.dchealth.dc.gov, 2008)

82. In October 2010, 82,083 students were eligible for free and reduced-price lunches and breakfasts in the ten Southeast Valley districts. Find the number of students who received free and reduced price lunches in October 2009. Round to the nearest ten.

[From Oct. 2009 to Oct. 2010] the 10 Southeast Valley districts have had a 15 percent increase in the number of students eligible for free and reduced-price lunches and breakfasts. (*Source:* www.azcentral.com, June 7, 2011)

83. The shipping costs for an order from an on-line nursery include $10.35 for the first bundle of daylily plants plus $3.00 per additional bundle.
   a. Write a linear function that describes the relationship between the total number of bundles ordered, x, and the total shipping cost, y.
   b. Use this function to find the cost of shipping 15 bundles.

84. The two graphs represent the same model of the growth of employees for x years since 2000. The increase in the number of employees from 2000 to 2006 looks greater in the second graph. Explain why.

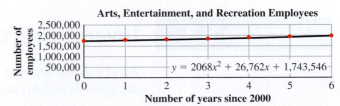

Source: www.census.gov, 2009

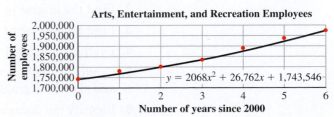

Source: www.census.gov, 2009

Technology

For exercises 85–88, use a graphing calculator to graph each function.

(a) Sketch the graph; describe the window.
(b) Use interval notation to represent the domain.
(c) Use interval notation to represent the range.
(d) Use the **Value** command to evaluate the function when $x = 2$.

85. $y = 4^x$

86. $y = 5^x$

87. $y = 0.4^x$

88. $y = 0.5^x$

Find the Mistake

For exercises 89–92, the completed problem has one mistake.
(a) Describe the mistake in words, or copy down the whole problem and highlight or circle the mistake.
(b) Do the problem correctly.

89. Problem: Identify the domain and range of the function $y = 3^x$.

Incorrect Answer: The domain is $(-\infty, \infty)$. The range is $[0, \infty)$.

90. Problem: What is the x-intercept of the function $y = 3^x$?

Incorrect Answer: The x-intercept is $(0, 1)$.

91. Problem: A function is $f(x) = 6^x$. Find $f(-2)$.

Incorrect Answer: $f(x) = 6^x$
$$f(-2) = 6^{-2}$$
$$f(-2) = -36$$

92. Problem: A function is $f(x) = 2(6^x)$. Identify the y-intercept of this function.

Incorrect Answer: The y-intercept is $(0, 1)$.

Review

93. Find $f(g(x))$ when $f(x) = 3x - 6$ and $g(x) = \frac{1}{3}x + 2$.

94. Graph: $y = x$

95. Graph $y = 3x - 6$ and $y = \frac{1}{3}x + 2$ on the same coordinate system.

96. Find $g(f(x))$ when $f(x) = 3x - 6$ and $g(x) = \frac{1}{3}x + 2$.

SUCCESS IN COLLEGE MATHEMATICS

97. In exercise 99 of Section 8.1, you completed a cumulative review. For the exercises in that review that you marked with a "C," find the section in the book in which this material was first taught. Re-read the section. Do a practice problem from this section that is similar to the "C" exercise. Since all of the answers to the practice problems are in the back of the book, you can check your answers.

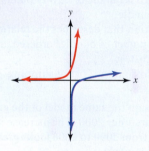

If $f(x)$ and $g(x)$ are inverses, then each undoes the other. If we input x into f, the output is $f(x)$. If $f(x)$ is then input into g, the output is once again x. The inverse of an exponential function is a **logarithmic function**. In this section, we will study logarithmic functions.

SECTION 8.3

Inverse Functions and Logarithmic Functions

After reading the text, working the practice problems, and completing assigned exercises, you should be able to:

1. Verify that two functions are inverses.

2. Find the inverse of a linear function.

3. Find the inverse of an exponential function, $y = b^x$.

4. Evaluate a logarithm.

5. Complete a table of ordered pairs for an exponential function and its inverse, and sketch its graph.

6. Identify the domain, range, and x-intercept of a logarithmic function.

Inverse Functions

If $f(x)$ and $g(x)$ are inverses, then each function undoes the other.

$$x \xrightarrow{\;f\;} f(x) \xrightarrow{\;g\;} x$$
$$\text{Input} \qquad\quad \text{Output} \qquad\quad \text{Output}$$

$$x \xrightarrow{\;g\;} g(x) \xrightarrow{\;f\;} x$$
$$\text{Input} \qquad\quad \text{Output} \qquad\quad \text{Output}$$

When x is the input for the function, f, the output is $f(x)$. When $f(x)$ is then input into the function, g, the output is once again x.

Inverse Functions

The functions f and g are inverses if and only if:

- f is a one-to-one function and g is a one-to-one function.
- the domain of f is the range of g and the domain of g is the range of f.
- $f(g(x)) = x$ and $g(f(x)) = x$.

EXAMPLE 1   Verify that $f(x) = 2x + 5$ and $g(x) = \dfrac{1}{2}x - \dfrac{5}{2}$ are inverse functions.

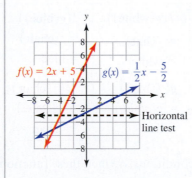

(a) Show that both functions are one-to-one.

SOLUTION ▶   Since each graph passes the horizontal line test, both functions are one-to-one.

(b) Show that the domain of one function is the range of the other function.

▶   The domain of a linear function is $(-\infty, \infty)$, and the range of a linear function is $(-\infty, \infty)$. Since both $f(x)$ and $g(x)$ are linear functions, the domain of one function is the range of the other.

(c) Show that $f(g(x)) = x$ and $g(f(x)) = x$.

▶
$$f(g(x))$$
$$= f\left(\frac{1}{2}x - \frac{5}{2}\right)$$
$$= 2\left(\frac{1}{2}x - \frac{5}{2}\right) + 5$$
$$= 2\left(\frac{1}{2}x\right) + 2\left(\frac{-5}{2}\right) + 5$$
$$= x - 5 + 5$$
$$= x$$

$$g(f(x))$$
$$= g(2x + 5)$$
$$= \frac{1}{2}(2x + 5) - \frac{5}{2}$$
$$= \frac{1}{2}(2x) + \frac{1}{2}(5) - \frac{5}{2}$$
$$= x + \frac{5}{2} - \frac{5}{2}$$
$$= x$$

Since both functions are one-to-one, the domain of one function is the range of the other function, and the compositions $f(g(x))$ and $g(f(x))$ both equal x, these two functions are inverses.

In the next example, each function is a set of ordered pairs written in roster notation.

EXAMPLE 2 Verify that $f = \{(\text{flag, black}), (\text{hat, white}), (\text{chair, blue})\}$ and $g = \{(\text{black, flag}), (\text{white, hat}), (\text{blue, chair})\}$ are inverse functions.

(a) Show that both functions are one-to-one.

SOLUTION ▶ Every input in a one-to-one function has an unique output. In f and g, no two ordered pairs have the same input value. No two ordered pairs have the same output value. These functions are one-to-one.

(b) Show that the domain of one function is the range of the other function.

▶ The domain of f, $\{\text{flag, hat, chair}\}$, is equal to the range of g, $\{\text{flag, hat, chair}\}$. The range of f, $\{\text{black, white, blue}\}$, is equal to the domain of g, $\{\text{black, white, blue}\}$.

(c) Show that $f(g(x)) = x$ and $g(f(x)) = x$.

▶ Since these functions do not have a rule, we need to show that $f(g(\text{input})) = \text{input}$ and $g(f(\text{input})) = \text{input}$.

$f(g(\text{black}))$	$f(g(\text{white}))$	$f(g(\text{blue}))$
$= f(\text{flag})$	$= f(\text{hat})$	$= f(\text{chair})$
$= \text{black}$	$= \text{white}$	$= \text{blue}$

$g(f(\text{flag}))$	$g(f(\text{hat}))$	$g(f(\text{chair}))$
$= g(\text{black})$	$= g(\text{white})$	$= g(\text{blue})$
$= \text{flag}$	$= \text{hat}$	$= \text{chair}$

Since the three conditions are met, these functions are inverses.

If a function is not one-to-one, it does not have an inverse. For example, the graph of a quadratic function (Figure 1) is a parabola. The graph fails the horizontal line test, and the function is not one-to-one. It does not have an inverse. The graph of an exponential function is one-to-one (Figure 2). It passes the horizontal line test. Later in this section, we will define a logarithmic function as the inverse of an exponential function.

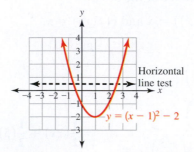

Figure 1

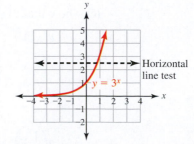

Figure 2

Practice Problems

1. Verify that $f(x) = -2x + 5$ and $g(x) = -\dfrac{1}{2}x + \dfrac{5}{2}$ are inverse functions.

2. Verify that $f = \{(6, 0), (3, 9), (2, 4)\}$ and $g = \{(0, 6), (9, 3), (4, 2)\}$ are inverse functions.

3. Explain why the function $y = x^2 + 3$ does not have an inverse.

Finding the Inverse of a Linear Function

To find the inverse of a linear function, we switch the variables and solve for y. We are changing the input of the original function to the output of the inverse function and the output of the original function to the input of the inverse function.

EXAMPLE 3 Find the inverse of $y = 7x - 4$.

SOLUTION ▶

$$y = 7x - 4 \qquad \text{The original function.}$$

$$x = 7y - 4 \qquad \text{Switch the variables.}$$

$$\underline{+4 \qquad\qquad +4} \qquad \text{Addition property of equality.}$$

$$x + 4 = 7y + 0 \qquad \text{Simplify.}$$

$$\frac{x}{7} + \frac{4}{7} = \frac{7y}{7} \qquad \text{Division property of equality.}$$

$$y = \frac{1}{7}x + \frac{4}{7} \qquad \text{Rewrite in slope-intercept form.}$$

The inverse of $y = 7x - 4$ is $y = \dfrac{1}{7}x + \dfrac{4}{7}$.

The notation for the inverse of a function $f(x)$ is $f^{-1}(x)$, which is said, "f inverse of x." The -1 represents the word inverse. To find the inverse of a function written in function notation, first replace $f(x)$ with y. After finding the inverse, replace y with $f^{-1}(x)$.

Finding the Inverse of a Linear Function

1. If the function is written in $f(x)$ notation, replace $f(x)$ with y.

2. For a function $y = mx + b$, switch the variables and rewrite the function as $x = my + b$.

3. Solve for y. Rewrite in slope-intercept form, $y = mx + b$.

4. If the original function is written in $f(x)$ notation, replace y with $f^{-1}(x)$.

EXAMPLE 4 Find the inverse of $f(x) = -\dfrac{2}{3}x + 6$.

SOLUTION ▶

$$f(x) = -\frac{2}{3}x + 6 \qquad \text{The original function.}$$

$$y = -\frac{2}{3}x + 6 \qquad \text{Replace } f(x) \text{ with } y.$$

$$x = -\frac{2}{3}y + 6 \qquad \text{Switch the variables.}$$

$$\underline{\qquad -6 \qquad\qquad -6\qquad} \qquad \text{Subtraction property of equality.}$$

$$x - 6 = -\frac{2}{3}y + \mathbf{0} \qquad \text{Simplify.}$$

$$\left(-\frac{3}{2}\right)(x - 6) = \left(-\frac{3}{2}\right)\left(-\frac{2}{3}y\right) \qquad \text{Multiplication property of equality.}$$

$$-\frac{3}{2}x - \frac{3}{2}(-6) = y \qquad \text{Distributive property; simplify.}$$

$$-\frac{3}{2}x + \mathbf{9} = y \qquad \text{Simplify.}$$

$$y = -\frac{3}{2}x + 9 \qquad \text{Write in slope-intercept form.}$$

$$f^{-1}(x) = -\frac{3}{2}x + 9 \qquad \text{Replace } y \text{ with } f^{-1}(x).$$

The inverse of $f(x) = -\frac{2}{3}x + 6$ is $f^{-1}(x) = -\frac{3}{2}x + 9$.

Practice Problems

4. Find the inverse of $y = \frac{3}{5}x + 8$.

5. Find the inverse of $f(x) = -9x - 4$.

Logarithmic Functions

For an exponential function $y = b^x$, the input x is an exponent. A function that is the inverse of this function *must have an output that is an exponent*. The inverse of an exponential function is a **logarithmic function**, $y = \log_b(x)$, where b is the base of the logarithm. For example, the inverse of $y = 10^x$ is $y = \log_{10}(x)$. The inverse of $f(x) = \log_4(x)$ is $f^{-1}(x) = 4^x$.

Logarithmic Functions

Logarithmic functions in the form $y = \log_b(x)$ or $f(x) = \log_b(x)$, where b is a real number, $b > 0$, and $b \neq 1$, are one-to-one functions. The domain is $(0, \infty)$ and the range is $(-\infty, \infty)$. There is no y-intercept. The x-intercept is $(1, 0)$.

The function $y = \log_b(x)$ is the inverse of the function $y = b^x$ and the function $f^{-1}(x) = \log_b(x)$ is the inverse of the function $f(x) = b^x$.

EXAMPLE 5 A function is $f(x) = 5^x$.

(a) What is $f^{-1}(x)$?

SOLUTION ▶ $\qquad f^{-1}(x) = \log_5(x) \qquad$ $f^{-1}(x)$ is a logarithmic function with a base of 5.

(b) Use interval notation to represent the domain and range of $f(x) = 5^x$.

▶ Since $f(x) = 5^x$ is an exponential function, its domain is $(-\infty, \infty)$ and its range is $(0, \infty)$.

(c) Use interval notation to represent the domain and range of $f^{-1}(x) = \log_5(x)$.

▶ Since $f^{-1}(x)$ and $f(x)$ are inverses, the domain of $f^{-1}(x)$ is the same as the range of $f(x)$, $(0, \infty)$. The range of $f^{-1}(x)$ is the same as the domain of $f(x)$, $(-\infty, \infty)$.

(d) Identify the y-intercept of $f(x)$ and $f^{-1}(x)$.

▶ The y-intercept of $f(x) = 5^x$ is $(0, 1)$. Since $f^{-1}(x) = \log_5(x)$ is a logarithmic function, it has no y-intercept.

(e) Identify the x-intercept of $f(x)$ and $f^{-1}(x)$.

▶ Since $f(x) = 5^x$ is an exponential function, it has no x-intercept. The x-intercept of $f^{-1}(x) = \log_5(x)$ is $(1, 0)$.

In the next example, we evaluate a logarithmic function by rewriting it as an exponential function. *The output of a logarithmic function is an exponent.*

Evaluating a Logarithm

If $y = \log_b(x)$, then $b^y = x$.

The output of a logarithmic function, y, is an exponent.

The input of a logarithmic function, x, is the argument.

If logarithm = $\log_{\text{base}}(\text{argument})$, then base^{logarithm} = argument.

A logarithm is an exponent.

EXAMPLE 6   A function is $y = \log_4(x)$. Find y when $x = 16$.

SOLUTION ▶

$y = \log_4(x)$

$y = \log_4(16)$ The argument (input) is 16. The base is 4. The exponent is y.

$4^y = 16$ The logarithm, y, is an exponent.

$y = 2$ $4^2 = 16$

A base 10 logarithm is a **common logarithm**. We often omit the "10" when writing a common logarithm. Instead of writing $\log_{10}(x)$, we write $\log(x)$. This is described as "log, base 10, of x" or "common log of x."

EXAMPLE 7   A function is $f(x) = \log(x)$. Evaluate $f(1000)$.

SOLUTION ▶

$f(x) = \log(x)$

$f(\mathbf{1000}) = \log(\mathbf{1000})$ The base is 10; base^{logarithm} = argument; $10^{\text{logarithm}} = 1000$

$f(1000) = \mathbf{3}$ $10^3 = 1000$

Sometimes we write a logarithm without using function notation. Instead of "Find $f(1)$ when $f(x) = \log_3(x)$," we write, "evaluate $\log_3(1)$."

EXAMPLE 8   **(a)** Evaluate: $\log_3(1)$

SOLUTION ▶

$y = \log_3(1)$ The input value is 1. The base is 3.

$3^y = 1$ A logarithm is an exponent.

$y = 0$ $3^0 = 1$

(b) Evaluate: $\log\left(\dfrac{1}{10}\right)$

$y = \log\left(\dfrac{1}{10}\right)$ The input value is $\dfrac{1}{10}$. The base is 10.

$10^y = \dfrac{1}{10}$ A logarithm is an exponent; base^{logarithm} = argument.

$y = -1$ $10^{-1} = \dfrac{1}{10^1} = \dfrac{1}{10}$

(c) Evaluate: $\log(10)$

$y = \log(10)$ The input value is 10. The base is 10.

$10^y = 10$ A logarithm is an exponent; base^{logarithm} = argument.

$y = 1$ $10^1 = 10$

(d) Evaluate: $\log(1)$

$y = \log(1)$ The input value is 1. The base is 10.

$10^y = 1$ A logarithm is an exponent; base^{logarithm} = argument.

$y = 0$ Solve by inspection; $10^0 = 1$

(e) Evaluate: $\log_2\left(\dfrac{1}{8}\right)$

$y = \log_2\left(\dfrac{1}{8}\right)$ The input value is 1. The base is 2.

$2^y = \dfrac{1}{8}$ A logarithm is an exponent; base^{logarithm} = argument.

$y = -3$ $2^{-3} = \dfrac{1}{2^3}$ and $\dfrac{1}{2^3} = \dfrac{1}{8}$

(f) Evaluate: $\log_6\left(\dfrac{1}{6}\right)$

$y = \log_6\left(\dfrac{1}{6}\right)$ The input value is 1. The base is 6.

$6^y = \dfrac{1}{6}$ A logarithm is an exponent; base^{logarithm} = argument.

$y = -1$ $6^{-1} = \dfrac{1}{6^1}$ and $\dfrac{1}{6^1} = \dfrac{1}{6}$

For inverse functions, the range of one function is the domain of the other function. Since the range of an exponential function is $(0, \infty)$, the domain of its inverse, a logarithmic function is also $(0, \infty)$. The input values for a logarithmic function must be greater than 0. *The logarithm of a negative number is not a real number; the logarithm of 0 is undefined.*

EXAMPLE 9   **(a)** Evaluate: $\log(-3)$

SOLUTION ▶   The domain of $y = \log(x)$ does not include -3; $\log(-3)$ is not a real number.

(b) Evaluate: $\log_4(0)$

The domain of $y = \log_4(x)$ does not include 0; $\log_4(0)$ is undefined.

Practice Problems

For problems 6–10, evaluate.

6. $\log_5(25)$ **7.** $\log_2\left(\dfrac{1}{64}\right)$ **8.** $\log(1000)$ **9.** $\log_8(1)$ **10.** $\log(-7)$

Graphs

At first glance, the typical human face is symmetric about a line drawn down through the nose. Each side of the face is the same—there is an eye and an eyebrow and half of a nose and a cheek and half of a mouth on each side of the line. If we use imagination to fold the face right on this line, the eyes would fold on top of each other and edges of the nose and mouth would match perfectly. The line drawn through the middle of the face is an **axis of symmetry**.

The graphs of inverse functions are also symmetric. However, the axis of symmetry is the line $y = x$. Because the axis of symmetry is not part of the graph of the functions, we draw it as a dashed line.

The functions $f(x) = 2x + 5$ and $g(x) = \dfrac{1}{2}x - \dfrac{5}{2}$ are inverses. If we fold the graph of these functions (Figure 3) on the axis of symmetry, the line $y = x$, the points of the two lines coincide.

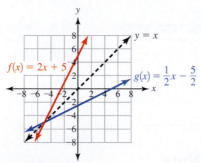

Figure 3

The functions $y = \log(x)$ and $y = 10^x$ are also inverses. In a table of ordered pairs, we see that the *inputs of one function* are the *outputs of its inverse*. In other words, the domain of a function is the range of its inverse.

$y = 10^x$		$y = \log(x)$	
x	y	x	y
-2	$\dfrac{1}{100}$	$\dfrac{1}{100}$	-2
-1	$\dfrac{1}{10}$	$\dfrac{1}{10}$	-1
0	1	1	0
1	10	10	1
2	100	100	2
3	1000	1000	3

The input values for $y = 10^x$ correspond to the output values for $y = \log(x)$. The output values for $y = \log(x)$ correspond to the input values of $y = 10^x$.

EXAMPLE 10 Sketch the graph of $y = 10^x$ and $y = \log(x)$. Graph the axis of symmetry, $y = x$, as a dashed line.

SOLUTION ▸ It is difficult to graph these functions precisely on standard graph paper because the output values vary from $\dfrac{1}{100}$ to 100. What is most important is to show that the graphs are mirror images around the axis of symmetry, the line $y = x$, that the y-intercept of $y = 10^x$ is $(0, 1)$ and that the x-intercept of $y = \log(x)$ is $(1, 0)$.

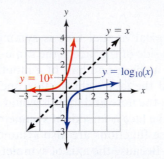

The graph of $y = \log(x)$ (Figure 4) does not intersect the y-axis. Since the domain of $y = \log(x)$ is $(0, \infty)$ and 0 is not an input value, the graph does not intersect the y-axis. Moving left from 1 down to 0, the output values decrease rapidly.

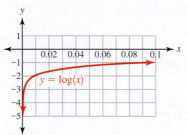

Figure 4

Practice Problems

For problems 11–12,

(a) complete the table of ordered pairs for the inverse.
(b) use interval notation to represent the domain of the logarithmic function.
(c) use interval notation to represent the range of the logarithmic function.
(d) identify the x-intercept of the logarithmic function.
(e) identify the y-intercept of the logarithmic function.

11.

$y = 3^x$		$y = \log_3(x)$	
x	y	x	y
-2	$\dfrac{1}{9}$	$\dfrac{1}{9}$	
-1	$\dfrac{1}{3}$	$\dfrac{1}{3}$	
0	1	1	
1	3	3	
2	9	9	
3	27	27	

12.

$y = 4^x$		$y = \log_4(x)$	
x	y	x	y
-2	$\dfrac{1}{16}$	$\dfrac{1}{16}$	
-1	$\dfrac{1}{4}$	$\dfrac{1}{4}$	
0	1	1	
1	4	4	
2	16	16	
3	64	64	

8.3 VOCABULARY PRACTICE

Match the term with its description.

1. A function in which each output corresponds to exactly one input
2. A function in which the input is an exponent
3. A function in which the output is an exponent
4. The domain of an exponential function, $y = b^x$
5. A test used to verify that a graph represents a function
6. A test used to verify that a graph represents a one-to-one function
7. The domain of a logarithmic function, $y = \log_b(x)$
8. A number that can be written as a fraction in which the numerator and denominator are integers
9. A number that cannot be written as a fraction in which the numerator and denominator are integers
10. A logarithm with a base that is equal to 10

A. common logarithm
B. exponential function
C. horizontal line test
D. irrational number
E. logarithmic function
F. one-to-one function
G. rational number
H. set of real numbers
I. set of real numbers greater than 0
J. vertical line test

8.3 Exercises

Follow your instructor's guidelines for showing your work.

For exercises 1–12, $f(x) = 4x - 9$, $g(x) = \dfrac{1}{4}x + \dfrac{9}{4}$, $h(x) = 3x + 7$, and $k(x) = \dfrac{1}{3}x - \dfrac{7}{3}$.

1. Graph $f(x)$, $g(x)$, and $y = x$ on the same coordinate system. Draw $y = x$ as a dashed line.

2. Graph $h(x)$, $k(x)$, and $y = x$ on the same coordinate system. Draw $y = x$ as a dashed line.

3. Are the graphs of $f(x)$ and $g(x)$ symmetric about the line $y = x$?

4. Are the graphs of $h(x)$ and $k(x)$ symmetric about the line $y = x$?

5. Show that $f(g(x)) = x$.

6. Show that $h(k(x)) = x$.

7. Identify the domain and range of $f(x)$ and the domain and range of $g(x)$.

8. Identify the domain and range of $h(x)$ and the domain and range of $k(x)$.

9. Is the domain of $f(x)$ the same as the range of $g(x)$?

10. Is the domain of $h(x)$ the same as the range of $k(x)$?

11. Is the domain of $g(x)$ the same as the range of $f(x)$?

12. Is the domain of $k(x)$ the same as the range of $h(x)$?

For exercises 13–16, find the inverse of the function. Write the inverse in slope-intercept form.

13. $y = 6x - 5$

14. $y = 3x - 8$

15. $f(x) = -\dfrac{4}{5}x + 10$

16. $f(x) = -\dfrac{3}{4}x + 5$

For exercises 17–18, to review graphing quadratic functions, go to Section 3.3.

17. a. Use a table of ordered pairs to graph $f(x) = x^2 + 5x + 6$.
    b. Does this function have an inverse?
    c. Explain.

18. a. Use a table of ordered pairs to graph $f(x) = x^2 + 4x + 8$.
    b. Does this function have an inverse?
    c. Explain.

For exercises 19–32, $f(x) = 2^x$ and $g(x) = 3^x$.

19. Use interval notation to represent the domain of $f(x)$.

20. Use interval notation to represent the domain of $g(x)$.

21. Use interval notation to represent the range of $f(x)$.

22. Use interval notation to represent the range of $g(x)$.

23. Identify the y-intercept of $f(x)$.

24. Identify the y-intercept of $g(x)$.

25. Identify the x-intercept of $f(x)$.

26. Identify the x-intercept of $g(x)$.

27. Find $f^{-1}(x)$.

28. Find $g^{-1}(x)$.

29. Use interval notation to represent the domain of $f^{-1}(x)$.

30. Use interval notation to represent the domain of $g^{-1}(x)$.

31. Use interval notation to represent the range of $f^{-1}(x)$.

32. Use interval notation to represent the range of $g^{-1}(x)$.

For exercises 33–46, $f(x) = \log_7(x)$ and $g(x) = \log_4(x)$.

33. Use interval notation to represent the domain of $f(x)$.

34. Use interval notation to represent the domain of $g(x)$.

35. Use interval notation to represent the range of $f(x)$.

36. Use interval notation to represent the range of $g(x)$.

37. Identify the y-intercept of $f(x)$.

38. Identify the y-intercept of $g(x)$.

39. Identify the x-intercept of $f(x)$.

40. Identify the x-intercept of $g(x)$.

41. Find: $f^{-1}(x)$

42. Find: $g^{-1}(x)$

43. Use interval notation to represent the domain of $f^{-1}(x)$.

44. Use interval notation to represent the domain of $g^{-1}(x)$.

45. Use interval notation to represent the range of $f^{-1}(x)$.

46. Use interval notation to represent the range of $g^{-1}(x)$.

For exercises 47–74, evaluate. Do not use the logarithm keys on a calculator.

47. $\log_7(49)$

48. $\log_3(81)$

49. $\log_2(128)$

50. $\log_9(81)$

51. $\log_6(1)$

52. $\log_5(1)$

53. $\log_3(-3)$

54. $\log_2(-2)$

55. $\log(10,000)$

56. $\log(100,000)$

57. $\log_4(4)$

58. $\log_8(8)$

59. $\log(100)$

60. $\log(10)$

61. $\log_6(6)$

62. $\log_7(7)$

63. $\log\left(\dfrac{1}{10}\right)$

64. $\log\left(\dfrac{1}{100}\right)$

65. $\log_5\left(\dfrac{1}{5}\right)$

66. $\log_4\left(\dfrac{1}{4}\right)$

67. $\log_2(64)$

68. $\log_2(32)$

69. $\log_3(0)$

70. $\log_5(0)$

71. $\log_6\left(\dfrac{1}{36}\right)$

72. $\log_4\left(\dfrac{1}{16}\right)$

73. $\log(-1)$

74. $\log(-4)$

For exercises 75–80, $f(x) = 3^x$ and $g(x) = 2^x$.

75. Find: $f^{-1}(x)$

76. Find: $g^{-1}(x)$

77. Complete the table of ordered pairs for $f(x)$ and $f^{-1}(x)$.

$f(x) = 3^x$		$f^{-1}(x)$	
x	y	x	$y = f^{-1}(x)$
-2		-2	
-1		-1	
0		0	
1		1	
2		2	
3		3	

78. Complete the table of ordered pairs for $g(x)$ and $g^{-1}(x)$.

$g(x) = 2^x$		$g^{-1}(x)$	
x	y	x	$y = g^{-1}(x)$
-2		-2	
-1		-1	
0		0	
1		1	
2		2	
3		3	

79. Sketch the graphs of $f(x)$, $f^{-1}(x)$, and $y = x$ (dashed line) on the same coordinate system.

80. Sketch the graphs of $g(x)$, $g^{-1}(x)$, and $y = x$ (dashed line) on the same coordinate system.

For exercises 81–86, $f(x) = \log_7(x)$ and $g(x) = \log_5(x)$.

81. Find: $f^{-1}(x)$

82. Find: $g^{-1}(x)$

83. Complete the table of ordered pairs for $f(x)$ and $f^{-1}(x)$.

$f(x) = \log_7(x)$		$f^{-1}(x)$	
x	$y = f(x)$	x	$y = f^{-1}(x)$
$\dfrac{1}{49}$		$\dfrac{1}{49}$	
$\dfrac{1}{7}$		$\dfrac{1}{7}$	
1		1	
7		7	
49		49	
343		343	

84. Complete the table of ordered pairs for $g(x)$ and $g^{-1}(x)$.

$g(x) = \log_5 x$		$g^{-1}(x)$	
x	$y = g(x)$	x	$y = g^{-1}(x)$
$\dfrac{1}{25}$		$\dfrac{1}{25}$	
$\dfrac{1}{5}$		$\dfrac{1}{5}$	
1		1	
5		5	
25		25	
125		125	

85. Sketch the graphs of $f(x), f^{-1}(x)$, and $y = x$ (dashed line) on the same coordinate system.

86. Sketch the graphs of $g(x), g^{-1}(x)$ and $y = x$ (dashed line) on the same coordinate system.

87. Explain why the graph of $y = \log_2(x)$ does not have a y-intercept.

88. Explain why the graph of $y = 2^x$ does not have an x-intercept.

89. Explain why the graph of $y = 3^x$ does not have an x-intercept.

90. Explain why the graph of $y = \log_5(x)$ does not have a y-intercept.

Problem Solving: Practice and Review

Follow your instructor's guidelines for using the five steps as outlined in Section 1.5, p. 51.

91. The function $P(t) = \left(\dfrac{75{,}267 \text{ people}}{1 \text{ year}}\right)t + 2{,}018{,}357$ people is a model of the population $P(t)$ of Nevada from 2000 to 2008. The number of years since 2000 is t. Use the function to predict the population of Nevada in 2014. Round to the nearest thousand. (*Source:* www.census .gov)

92. Find the total volume of one case of Fruit D'Licious juice concentrate.

Our most affordable juice concentrate solution! Easy to use and easy to store, one case of the **Fruit D'Licious** 7:1 ratio [7 parts water: 1 part concentrate] juice concentrate makes **12 gallons** of delicious, high quality juice drink! (*Source:* www.daycarejuice.com)

93. The New York City skyline includes thousands of wood water tanks, many built by the Rosenwach Tank Company. They are circular cylinders with a conical roof. Water is pumped up from city water lines and stored in the tank. Gravity provides water with adequate pressure to residents in upper floors. The diameter of a tank is 13 ft. Find the weight of the water when the depth of water in the tank is 8 ft. (1 ft$^3 \approx 7.48$ gal; 1 gal $\approx$ 8.33 lb; $V = \pi r^2 h$; $\pi \approx 3.14$.) Round to the nearest hundred.

A saw trims the ends for a 12-foot-high tank. . . . This tank's diameter is 13 feet. . . . When complete, this tank will hold 8500 gallons of water. (*Source:* www.ny1.com, May 22, 2006)

© SVLuma/Shutterstock

94. A pressure of about 0.43 lb per square inch (psi) is needed to raise a column of water 1 ft. The pressure in a municipal water system is 60 psi. Each story in a building is 10 ft high. How many stories can the water be lifted? Round *down* to the nearest whole number.

Find the Mistake

For exercises 95–98, the completed problem has one mistake.
(a) Describe the mistake in words, or copy down the whole problem and highlight or circle the mistake.
(b) Do the problem correctly.

95. Problem: A function is $f(x) = 5^x$. What is $f^{-1}(x)$?
Incorrect Answer: If $f(x) = 5^x$, then $f^{-1}(x) = 5^{-x}$.

96. Problem: A function is $y = \log_2(x)$. Find y when $x = 16$.
Incorrect Answer: Since $16 = (2)(8), y = 8$.

97. Problem: Evaluate: $\log(-100)$
Incorrect Answer: Since $\log(100) = 2, \log(-100) = -2$.

98. Problem: Evaluate: $\log_2(8)$
Incorrect Answer: Since $8 \div 2 = 4, \log_2(8) = 4$.

Review

For exercises 99–102, a concert ticket vendor charges $76 for each ticket and a service fee for each order of $11.50.

99. Write a function that represents the relationship of the number of tickets purchased, x, and the total cost of the order, y.

100. What does the slope in this function represent?

101. What does the y-intercept in this function represent?

102. Use the function to find the total cost of an order of six tickets.

SUCCESS IN COLLEGE MATHEMATICS

103. Build yourself a practice test using exercises from chapter reviews that match your "B" and "C" questions from the cumulative review. Although you can take the time to write these out, you can also just build a list of problems with page numbers and exercise numbers. Complete this practice test. Since many of the answers to the chapter reviews are in the back of the book, you can check most of the answers.

In the past, scientists, mathematicians, and students used slide rules or tables to identify the logarithms that are irrational numbers. Today, we most often use a calculator. In this section, we will evaluate logarithms that are irrational numbers, including logarithms with a base of *e*.

© huibvisser/Shutterstock

SECTION 8.4   Natural Logarithms and Logarithm Rules

After reading the text, working the practice problems, and completing assigned exercises, you should be able to:

1. Use a calculator to evaluate a logarithmic function and explain why the logarithm is reasonable.

2. Identify the base of a natural logarithm.

3. Use the change of base formula to evaluate a logarithm.

4. Use logarithms to multiply whole numbers.

5. Use the rules of logarithms to rewrite an expression with a single logarithm.

6. Use the rules of logarithms to rewrite an expression with more than one logarithm.

Using a Calculator to Evaluate Logarithms

A logarithm is an exponent. In Section 8.3, we studied logarithms that are rational numbers. However, many logarithms are irrational. We will evaluate irrational logarithms with a calculator.

> **Using Technology: Evaluating a Base 10 Logarithm**
>
> Most scientific calculators can evaluate logs in base 10.
>
> Look for the LOG command printed on a key or above it. The order of keys pressed varies with different calculators. Press LOG.
>
> If the entry line of the calculator shows **log(**, then type in the argument, type in the closing parenthesis,), and press ENTER or =.
>
> If the entry line of the calculator shows *error*, press CLEAR. Type the number to be logged; press LOG.

Practice Problems

For problems 1–4, use a calculator to evaluate. Round any irrational numbers to the nearest ten-thousandth.

1. $\log(10{,}000)$ **2.** $\log(50)$ **3.** $\log(1)$ **4.** $\log(200)$

EXAMPLE 1   **(a)** Use a scientific calculator to evaluate $\log(75)$. Round to the nearest ten-thousandth.

SOLUTION ▶ $\log(75)$

$= 1.87506\ldots$ Evaluate with a scientific calculator.

≈ 1.8751 Round.

(b) Explain why this is a reasonable answer.

▶   Since $10^1 = 10$ and $10^2 = 100$ and 75 is between 10 and 100, we expect $\log(75)$ to be between 1 and 2. So 1.8751 is a reasonable value for $\log(75)$.

In Section 8.2, we used the irrational number e as a base in exponential functions ($e \approx 2.72$). The inverse of an exponential function with a base of e is a logarithm with a base of e. A base e logarithm is a **natural logarithm**. Instead of writing $\log_e(x)$, we often write $\ln(x)$, saying, "the natural logarithm of x" or "the log, base e, of x." Most scientific calculators can evaluate natural logarithms.

EXAMPLE 2   **(a)** Use a scientific calculator to evaluate $\ln(9)$. Round to the nearest ten-thousandth.

SOLUTION ▶ $\ln(9)$

$= 2.19722\ldots$ Evaluate with a scientific calculator.

≈ 2.1972 Round.

(b) Explain why this is a reasonable answer.

▶   The irrational number $e = 2.718\ldots \approx 2.7$. Since $2.7^2 = 7.29$ and $2.7^3 = 19.683$, we expect $\ln(9)$ to be between 2 and 3. Since 2.1972 is between 2 and 3, it is a reasonable value for $\ln(9)$.

Most scientific calculators are programmed to find only logs in base 10 and base e. To find logarithms in bases other than e or 10, use the change of base formula.

Change of Base Formula for Logarithms

$$\log_b(x) = \frac{\log_a(x)}{\log_a(b)}$$

where a and b are real numbers greater than 0 and not equal to 1 and x is a real number greater than 0

Using the change of base formula, we can rewrite a logarithm in a base other than 10 or e as a ratio of logs in base 10 or base e. So $\log_2(11) = \dfrac{\log(11)}{\log(2)}$ or $\log_2(11) = \dfrac{\ln(11)}{\ln(2)}$. It does not matter whether we choose to use common (base 10) logarithms or natural (base e) logarithms, the answer is the same.

EXAMPLE 3 (a) Evaluate $\log_2(30)$ using the change of base formula and base 10 logarithms. Round to the nearest ten-thousandth.

SOLUTION ▶

$\log_2(30)$

$= \dfrac{\log(30)}{\log(2)}$ $\log_b(x) = \dfrac{\log_a(x)}{\log_a(b)}; a = 10, b = 2, x = 30$

$= \dfrac{1.4771\ldots}{0.3010\ldots}$ Evaluate each base 10 logarithm. Do not round.

$= 4.90689\ldots$ Divide.

≈ 4.9069 Round.

(b) Evaluate $\log_2(30)$ using the change of base formula and base e logarithms. Round to the nearest ten-thousandth.

▶

$\log_2(30)$

$= \dfrac{\ln(30)}{\ln(2)}$ $\log_b(x) = \dfrac{\log_a(x)}{\log_a(b)}; a = e, b = 2, x = 30$

$= \dfrac{3.4011\ldots}{0.6931\ldots}$ Evaluate each base e logarithm. Do not round.

$= 4.90689\ldots$ Divide.

≈ 4.9069 Round. This is the same result as in part (a).

(c) Explain why this is a reasonable answer.

▶

Since $2^4 = 16$ and $2^5 = 32$, we expect $\log_2(30)$ to be between 4 and 5. So 4.9069 is a reasonable value for $\log_2(30)$.

Practice Problems

For problems 5–10, evaluate. Round to the nearest ten-thousandth.

5. $\log(60)$ **6.** $\ln(60)$ **7.** $\log_2(60)$ **8.** $\log_3(60)$

9. $\log_4(1)$ **10.** $\log_4(-8)$

Logarithms and Arithmetic

In the United States, most people use the method shown here (Figure 1) to find the product of 113,598 and 24,601. However, we can also estimate the product of two numbers using base 10 logarithms and the product rule of exponents, $x^m \cdot x^n = x^{m+n}$.

$$
\begin{array}{r}
113598 \\
\times\ 24601 \\
\hline
113598 \\
000000 \\
681588 \\
454392 \\
227196 \\
\hline
2794624398
\end{array}
$$

Figure 1

EXAMPLE 4 Use base 10 logarithms to multiply 113,598 and 24,601.

SOLUTION ▶ $\log(113{,}598) = 5.0553\ldots$ so $10^{5.0553\ldots} = 113{,}598$

                 $\log(24{,}601) = 4.3909\ldots$ so $10^{4.3909\ldots} = 24{,}601$

$$(113,598)(24,601)$$

$= (10^{5.0553\ldots})(10^{4.3909\ldots})$	Rewrite the whole numbers as exponential expressions.
$= 10^{(5.0553\ldots + 4.3909\ldots)}$	Product rule of exponents.
$= 10^{9.4463\ldots}$	Simplify.
$\approx 2,794,624,398$	Use a calculator to evaluate the power of 10.

It is much quicker to add two logarithms than to multiply two large numbers. Before calculators, every math or science book included several pages of tables of logarithms. Mathematicians constructed these tables using pencil and paper arithmetic. In the 1970s, the price of calculators dropped to a level that allowed some college students to purchase them. (In 1974, a calculator that did arithmetic, had a square root key, but did not evaluate logarithms or evaluate a power of ten cost $110.) Now calculators are so inexpensive that tables of logarithms are no longer included in math textbooks.

Practice Problems

11. Multiply: $(3942)(275)$

12. Find $\log(3942)$ and $\log(275)$. Round to the nearest ten-thousandth.

13. Use the logarithms in problem 12 to estimate the product of $(3942)(275)$.

Power Rule of Logarithms

We use the power rule of exponents to simplify exponential expressions.

Power Rule of Exponents

$$(x^m)^n = x^{mn}$$

When raising an exponential expression to a power, multiply the exponents and keep the same base; x is a real number; m and n are integers; x and mn cannot both be 0.

Since a logarithm is an exponent, the power rule also applies to logarithms. However, we use logarithm notation and restrict the variables to match the restrictions for logarithms.

Power Rule of Logarithms

$$\log_b(m^p) = p\log_b(m)$$

When raising a logarithm to a power, multiply the power and the logarithm. The base, b, is a real number greater than 0 that is not equal to 1; m is a real number greater than 0; and p is a real number.

EXAMPLE 5 | Use the power rule of logarithms to rewrite $\log(10^x)$ as a product. Simplify.

SOLUTION ▶

$\log(10^x)$	The base of this logarithm is 10.
$= x\log(10)$	Power rule of logarithms.
$= x \cdot 1$	$\log_{10}(10) = 1$ because $10^1 = 10$.
$= x$	Simplify.

In the next example, the exponent is a product. Notice that since $e^1 = e$ and the base of the natural logarithm is e, $\ln(e) = 1$.

EXAMPLE 6 | Use the power rule of logarithms to rewrite $\ln(e^{3x})$ as a product. Simplify.

SOLUTION ▶

$\ln(e^{3x})$	The base of this logarithm is e.
$= \mathbf{3x} \cdot \ln(e)$	Power rule of logarithms.
$= 3x \cdot \mathbf{1}$	$\log_e(e) = 1$ because $e^1 = e$.
$= 3x$	Simplify.

---

log(10) and ln(e)

Since $10^1 = 10$, $\log_{10}(10) = 1$ and $\log(10) = 1$.
Since $e^1 = e$, $\log_e(e) = 1$ and $\ln(e) = 1$.

We can also use the power rule of logarithms to rewrite a product as a single logarithm.

EXAMPLE 7 | Use the power rule of logarithms to rewrite $9 \log(x)$ as a single logarithm.

SOLUTION ▶

$9 \log(x)$	
$= \log(x^9)$	Power rule of logarithms.

Practice Problems

For problems 14–17, use the power rule of logarithms to rewrite the expression as a product and simplify.

14. $\ln(e^x)$ **15.** $\log(10^{2x})$ **16.** $\log_5(5^x)$ **17.** $\log(10^{4x})$

For problems 18–20, use the power rule of logarithms to rewrite the expression as a single logarithm.

18. $2 \log(x)$ **19.** $6 \log(y)$ **20.** $7x \ln(y)$

The Product Rule and the Quotient Rule of Logarithms

We use the product rule of exponents to simplify exponential expressions.

Product Rule of Exponents

$$x^m \cdot x^n = x^{m+n}$$

When multiplying exponential expressions with the same base, add the exponents and keep the same base; x is a real number; m and n are integers; both x and $m + n$ are not 0.

Since a logarithm is an exponent, the product rule also applies to logarithms. However, we use logarithm notation and restrict the variables to match the restrictions for logarithms.

Product Rule of Logarithms

$$\log_b(m) + \log_b(n) = \log_b(m \cdot n)$$

If the bases are the same, the sum of the logarithms equals the logarithm of the product. The base, b, is a real number greater than 0 that is not equal to 1; m and n are real numbers greater than 0.

EXAMPLE 8   Use the product rule of logarithms to rewrite $\log_4(2) + \log_4(8)$ as a single logarithm. Evaluate.

SOLUTION ▶ $\log_4(2) + \log_4(8)$

$= \log_4(2 \cdot 8)$	Product rule of logarithms.
$= \log_4(\mathbf{16})$	Simplify.
$= 2$	Evaluate; $4^2 = 16$ or use the change of base formula.

We can also use the product rule of logarithms to rewrite a single logarithm as a sum of logarithms.

EXAMPLE 9   Use the product rule of logarithms to rewrite $\log(xy)$ as a sum of logarithms.

SOLUTION ▶ $\log(xy)$

$= \log(x) + \log(y)$ Product rule of logarithms.

We use the quotient rule of exponents to simplify exponential expressions.

Quotient Rule of Exponents

$$\frac{x^m}{x^n} = x^{m-n} \qquad x \neq 0$$

When dividing exponential expressions with the same base, subtract the exponents and keep the same base; x is a real number (except 0); m and n are integers.

Since a logarithm is an exponent, the quotient rule also applies to logarithms. However, we use logarithm notation and restrict the variables to match the restrictions for logarithms.

Quotient Rule of Logarithms

$$\log_b(m) - \log_b(n) = \log_b\left(\frac{m}{n}\right)$$

If the bases are the same, the difference of the logarithms is the logarithm of the quotient. The base, b, is a real number greater than 0 that is not equal to 1; m and n are real numbers greater than 0.

EXAMPLE 10   (a) Use the quotient rule of logarithms to rewrite $\ln(0.50) - \ln(0.42)$ as a single logarithm.

SOLUTION ▶

$\ln(0.50) - \ln(0.42)$ $\log_b(m) - \log_b(n)$

$= \ln\left(\dfrac{0.50}{0.42}\right)$ Quotient rule of logarithms.

(b) Evaluate. Round to the nearest ten-thousandth.

▶

$\ln\left(\dfrac{0.50}{0.42}\right)$

$= \ln(\mathbf{1.19047...})$ Divide.

$= 0.17435...$ Evaluate the logarithm.

≈ 0.1744 Round.

We can also use the quotient rule of logarithms to rewrite a single logarithm as a difference of logarithms.

EXAMPLE 11   Use the quotient rule of logarithms to rewrite $\log_4\left(\dfrac{x}{y}\right)$ as a difference of logarithms.

SOLUTION ▶

$\log_4\left(\dfrac{x}{y}\right)$

$= \log_4(x) - \log_4(y)$ Quotient rule of logarithms.

Practice Problems

For problems 21–24,
(a) use the product or quotient rule of logarithms to rewrite as a single logarithm.
(b) evaluate. Round irrational numbers to the nearest ten-thousandth.

21. $\log(100) + \log(10)$ **22.** $\ln(45) + \ln(2)$
23. $\log(100) - \log(10)$ **24.** $\ln(45) - \ln(9)$

For problems 25–26, use the product or quotient rule of logarithms to rewrite as a sum or difference of logarithms.

25. $\log_5\left(\dfrac{c}{d}\right)$ **26.** $\log_2(cd)$

Using More Than One Logarithm Rule

We can use the logarithm rules to rewrite some expressions that include more than one logarithm as a single logarithm or to rewrite a single logarithm as an expression with more than one logarithm.

EXAMPLE 12   Use the logarithm rules to rewrite $5\log(x) - 3\log(y)$ as a single logarithm.

SOLUTION ▶

$5\log(x) - 3\log(y)$

$= \log(x^5) - \log(y^3)$ Power rule of logarithms.

$= \log\left(\dfrac{x^5}{y^3}\right)$ Quotient rule of logarithms.

When rewriting an expression with a single logarithm as an expression with more than one logarithm, do not rewrite exponential expressions using the power rule of logarithms until the last step.

EXAMPLE 13   Use the logarithm rules to rewrite $\ln\left(\dfrac{x^3 y}{z}\right)$ as an expression with more than one logarithm.

SOLUTION ▶

$$\ln\left(\dfrac{x^3 y}{z}\right)$$

$$= \ln(x^3 y) - \ln(z) \qquad \text{Quotient rule of logarithms.}$$

$$= \mathbf{\ln(x^3) + \ln(y)} - \ln(z) \qquad \text{Product rule of logarithms.}$$

$$= \mathbf{3}\ln(x) + \ln(y) - \ln(z) \qquad \text{Power rule of logarithms.}$$

In Section 8.7, we will rewrite an expression with more than one logarithm as a single logarithm so that we can solve a logarithmic equation.

EXAMPLE 14   Use the logarithm rules to rewrite $6\log(x) - 3\log(y) + \log(z)$ as a single logarithm.

SOLUTION ▶

$$6\log(x) - 3\log(y) + \log(z)$$

$$= \log(x^6) - \log(y^3) + \log(z) \qquad \text{Power rule of logarithms.}$$

$$= \mathbf{\log\left(\dfrac{x^6}{y^3}\right)} + \log(z) \qquad \text{Quotient rule of logarithms.}$$

$$= \log\left(\dfrac{x^6 z}{y^3}\right) \qquad\qquad \text{Product rule of logarithms.}$$

Practice Problems

For problems 27–28, use the logarithm rules to rewrite the expression with a single logarithm.

27. $8\log(x) + \log(y) - 4\log(z)$ **28.** $\ln(x) - 2\ln(y) + 3\ln(z)$

For problems 29–30, use the logarithm rules to rewrite the expression with more than one logarithm.

29. $\log\left(\dfrac{x^5 y^3}{z^2}\right)$ **30.** $\ln\left(\dfrac{x}{y^2 z^4}\right)$

Using Technology: Graphing Logarithmic Functions

EXAMPLE 15 Graph $y = \log(x)$. Find the output value when the input value is 2.

Go to the Y= screen. Press ▢ LOG ▢. The screen shows **log(**.

To type in the variable, x, press the ▢ X,T,θ,n ▢ key. Close the parentheses by pressing ▢) ▢. Press ▢ GRAPH ▢.

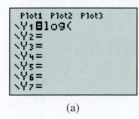

(a)

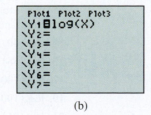

(b)

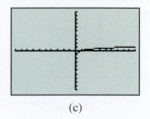

(c)

It looks as though the graph touches the y-axis and stops when x is approximately equal to 0. However, we know that a logarithmic function does not have a y-intercept. The domain of this function is $(0, \infty)$. We can see this trend if we look at a table of the graph, beginning at $x = -0.5$, which is not in the domain, with an interval between input values of 0.1.

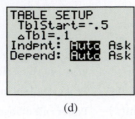

(d)

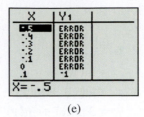

(e)

The calculator prints "ERROR" for x-values that are not in the domain. The first entry in the table is $\log(0.1) = -1$.

Practice Problems   For problems 31–32,
(a) graph the function. Sketch the graph; describe the window.
(b) use the **Value** command to evaluate the function for the given input value. Round to the nearest ten-thousandth.

31. $f(x) = \ln(x); f(5)$ **32.** $f(x) = \log(2x); f(3)$

8.4 VOCABULARY PRACTICE

Match the term with its description.

1. A function in which the output is an exponent
2. A function in which the input is an exponent
3. A number that can be written as a fraction in which the numerator and denominator are integers
4. $\log_b(x) = \dfrac{\log_a(x)}{\log_a(b)}$
5. $\{\ldots, -2, -1, 0, 1, 2, 3, \ldots\}$
6. $\log_b(m) - \log_b(n) = \log_b\left(\dfrac{m}{n}\right)$
7. $\log_b(m) + \log_b(n) = \log_b(m \cdot n)$
8. $\log_b(m^p) = p \log_b(m)$
9. A logarithm with a base equal to e
10. A logarithm with a base equal to 10

A. change of base formula
B. common logarithm
C. exponential function
D. logarithmic function
E. natural logarithm
F. power rule of logarithms
G. product rule of logarithms
H. quotient rule of logarithms
I. rational number
J. set of integers

8.4 Exercises

Follow your instructor's guidelines for showing your work.

For exercises 1–12, use a calculator to evaluate the logarithm. Round to the nearest ten-thousandth.

1. $\log(45)$
2. $\log(65)$
3. $\ln(45)$
4. $\ln(65)$
5. $\log(3 \times 10^{-8})$
6. $\log(5 \times 10^{-11})$
7. $\log(6 \times 10^2)$
8. $\log(4 \times 10^3)$
9. $\log(0.3)$
10. $\log(0.6)$
11. $\ln(7)$
12. $\ln(6)$

For exercises 13–24, use the change of base formula and a calculator to evaluate the logarithm. Round to the nearest ten-thousandth.

13. $\log_2(40)$
14. $\log_2(30)$
15. $\log_9(38)$
16. $\log_9(40)$
17. $\log_5(100)$
18. $\log_5(200)$
19. $\log_6(100)$
20. $\log_6(200)$
21. $\log_3(21)$
22. $\log_3(18)$
23. $\log_4(12)$
24. $\log_4(20)$

25. Explain why $\log(140)$ is between 2 and 3.

26. Explain why $\log(7500)$ is between 3 and 4.

27. Explain why $\log_5(400)$ is between 3 and 4.

28. Explain why $\log_7(250)$ is between 2 and 3.

29. Explain why $\ln(3)$ is between 1 and 2.

30. Explain why $\ln(4)$ is between 1 and 2.

31. Use logarithms to estimate $(8)(7)$. Round to the nearest whole number.

32. Use logarithms to estimate $(6)(8)$. Round to the nearest whole number.

33. Use logarithms to estimate $(42)(53)$. Round to the nearest whole number.

34. Use logarithms to estimate $(29)(71)$. Round to the nearest whole number.

For exercises 35–38, use the power rule of logarithms to rewrite the expression as a product and simplify.

35. $\ln(e^{2x})$

36. $\ln(e^{3x})$

37. $\log_4(4^x)$

38. $\log_3(3^x)$

For exercises 39–42, use the power rule of logarithms to rewrite each expression as a product.

39. $\log(x^4)$

40. $\log(x^5)$

41. $\log(y^2)$

42. $\log(y^3)$

For exercises 43–46, use the power rule of logarithms to rewrite the expression as a single logarithm.

43. $8 \log(x)$

44. $9 \log(w)$

45. $2x \ln(y)$

46. $8x \ln(y)$

47. In the power rule of exponents, the base must be a real number. In the power rule of logarithms, the base must be greater than 0 and not equal to 1. Explain why we must include these additional restrictions.

48. In the power rule of exponents, each exponent must be a real number. In the power rule of logarithms, $\log_b(m^p) = p \log_b(m)$, m must be greater than 0. Explain why.

For exercises 49–56, use the product or quotient rule of logarithms to rewrite each expression as a single logarithm and evaluate. Round irrational numbers to the nearest ten-thousandth.

49. $\log(7) + \log(5)$

50. $\log(6) + \log(5)$

51. $\ln(18) + \ln(2)$

52. $\ln(15) + \ln(3)$

53. $\log(8) - \log(4)$

54. $\log(12) - \log(3)$

55. $\ln(15) - \ln(3)$

56. $\ln(16) - \ln(4)$

For exercises 57–64, use the product or quotient rule of logarithms to rewrite each expression as a single logarithm.

57. $\log_5(x) + \log_5(y)$

58. $\log_2(c) + \log_2(d)$

59. $\log_5(c) - \log_5(d)$

60. $\log_2(u) - \log_2(w)$

61. $\log(x) + \log(y) + \log(z)$

62. $\ln(x) + \ln(y) + \ln(z)$

63. $\log(x) - \log(y) + \log(z)$

64. $\log(x) + \log(y) - \log(z)$

65. Describe the restrictions on the base in the product rule of logarithms.

66. Describe the restrictions on the base in the quotient rule of logarithms.

For exercises 67–78, use the logarithm rules to rewrite the expression as a single logarithm.

67. $4 \log_5(x) + \log_5(y)$

68. $3 \log_2(x) + \log_2(y)$

69. $\log(x) - \log(y) - \log(z)$

70. $\ln(x) - \ln(y) - \ln(z)$

71. $4 \log_5(x) - \log_5(y)$

72. $8 \log_4(x) - \log_4(y)$

73. $3 \log_5(x) + \log_5(y) - \log_5(z)$

74. $7 \log_2(x) + \log_2(y) - \log_2(z)$

75. $\log(6x) - 2 \log(y) + \log(z)$

76. $\log(9x) - 5 \log(y) + \log(z)$

77. $4 \ln(x) + 8 \ln(x) - 9 \ln(x)$

78. $2 \ln(x) + 7 \ln(x) - 4 \ln(x)$

For exercises 79–84, use the logarithm rules to rewrite the expression with more than one logarithm.

79. $\log\left(\dfrac{xy}{z}\right)$

80. $\log\left(\dfrac{cd}{k}\right)$

81. $\log\left(\dfrac{x^2 y^5}{z^8}\right)$

82. $\log\left(\dfrac{x^3 y^7}{z^2}\right)$

83. $\ln\left(\dfrac{x}{y^8 z^2}\right)$

84. $\ln\left(\dfrac{x}{y^4 z^6}\right)$

Problem Solving: Practice and Review

Follow your instructor's guidelines for using the five steps as outlined in Section 1.5, p. 51.

For exercises 85–87, the graph represents the relationship of the number of years since 2001, x, and the percent of low birth-weight live births in the United States, y.

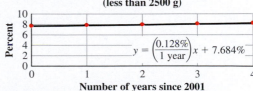

Percent of Low Birthweight Live Births (less than 2500 g)

$y = \left(\dfrac{0.128\%}{1 \text{ year}}\right)x + 7.684\%$

Number of years since 2001

Source: www.cdc.gov

85. Is the percent of live births that were low birthweight during this time period growing exponentially? Explain.

86. Identify the average rate of change in the percent of live births that were low birthweight.

87. Use the function to find the percent of live births that will be low birthweight in 2012. Round to the nearest tenth of a percent.

88. In early 2011, Netflix offered subscribers a plan that included unlimited streaming video and one DVD at a time for $9.99 per month. In July 2011, Netflix separated the plans and began charging $7.99 per month for unlimited streaming video and $7.99 per month for one DVD at a time. For a customer who continued to want unlimited streaming video and one DVD by mail, find the percent increase in cost. Round to the nearest percent. (*Source:* blog.netflix.com, July 12, 2011)

Technology

For exercises 89–92,
(a) graph the function. Sketch the graph; describe the window.
(b) use the **Value** command to evaluate the function for the given input value. Round to the nearest ten-thousandth.

89. $f(x) = \log(4x)$; $f(3)$

90. $f(x) = \log(5x)$; $f(3)$

91. $f(x) = \ln(3x)$; $f(5)$

92. $f(x) = \ln(2x)$; $f(10)$

Find the Mistake

For exercises 93–96, the completed problem has one mistake.
(a) Describe the mistake in words, or copy down the whole problem and highlight or circle the mistake.
(b) Do the problem correctly.

93. Problem: Use the product or quotient rule of logarithms to rewrite $\log_4(8) + \log_4(12)$ as a single logarithm and evaluate. Round irrational numbers to the nearest ten-thousandth.

Incorrect Answer: $\log_4(8) + \log_4(12)$

$$= \log_4(20)$$

$$= \frac{\log(20)}{\log(4)}$$

$$= \frac{1.3010\ldots}{0.6020\ldots}$$

$$\approx 2.1610$$

94. Problem: Use the product or quotient rule of logarithms to rewrite $\log(12) - \log(6)$ as a single logarithm and evaluate. Round irrational numbers to the nearest ten-thousandth.

Incorrect Answer: $\log(12) - \log(6)$

$$= \frac{\log(12)}{\log(6)}$$

$$= \frac{1.0791\ldots}{0.7781\ldots}$$

$$\approx 1.3869$$

95. Problem: Use the logarithm rules to rewrite $3\log(x) - 5\log(y) - 7\log(z)$ as a single logarithm.

Incorrect Answer: $3\log(x) - 5\log(y) - 7\log(z)$

$$= \log(x^3) - \log(y^5) - \log(z^7)$$

$$= \log\left(\frac{x^3 z^7}{y^5}\right)$$

96. Problem: Use the logarithm rules to rewrite $2\log(x) - 9\log(y)$ as a single logarithm.

Incorrect Answer: $2\log(x) - 9\log(y)$

$$= \log(2x) - \log(9y)$$

$$= \log\left(\frac{2x}{9y}\right)$$

Review

For exercises 97–100, graph each inequality on an *xy*-coordinate system.

97. $x > 3$

98. $y \le 2$

99. $3x - 7y > 14$

100. $y > -\dfrac{3}{4}x + 6$

SUCCESS IN COLLEGE MATHEMATICS

101. Look at the vocabulary practices for Chapters 1–6. List the terms that you do not know. Then review these terms using flashcards or another memorizing method.

Cockroaches lay many eggs. As new generations are born, they also reproduce. The total number of cockroaches increases rapidly. For a while, the rate of reproduction is exponential. To predict the number of cockroaches in the future, we can evaluate an exponential function. In this section, we will solve applications using exponential and logarithmic functions.

SECTION 8.5

Applications

After reading the text, working the practice problems, and completing assigned exercises, you should be able to:

1. Solve application problems by evaluating an exponential or logarithmic function.

2. Solve application problems by using a formula that includes logarithms or exponential expressions.

Applications of Exponential Functions and Formulas

An exponential function $y = A(b^{kt})$ can represent the relationship of the input value of time, t, and the output value, y. The original amount when $t = 0$ is A. Changing b or k changes the rate of exponential increase or decrease. In the next example, we use an exponential function to model the growth of tuition.

EXAMPLE 1 A college increases tuition once a year in the fall. The function $y = \$4500(1.08)^t$ represents the relationship of the number of years since fall 2010, t, and the semester tuition. The chief financial officer of the college expects that tuition will follow this model for 6 years.

(a) Identify A, b, and k.

SOLUTION ▶ For $y = \$4500(1.08)^t$, $A = \$4500$, $b = 1.08$, and $k = 1$.

(b) Find the semester tuition in fall 2010.

▶ $y = \$4500(1.08)^t$

$y = \$4500(1.08)^0$ In 2010, $t = 0$ years; replace t with 0.

$y = \$4500(\mathbf{1})$ Simplify.

$y = \mathbf{\$4500}$ Simplify.

(c) Predict the semester tuition in fall 2015. Round to the nearest whole number.

▶ $y = \$4500(1.08)^t$

$y = \$4500(1.08)^5$ In 2015, $t = 5$ years; replace t with 5.

$y = \$4500(\mathbf{1.46932\ldots})$ Follow the order of operations.

$y \approx \mathbf{\$6612}$ Round.

(d) Describe a reasonable domain for this function.

▶ Time is not negative. Tuition changes once a year. We expect the model to apply for 6 years. A reasonable domain is {0 year, 1 year, 2 years, 3 years, 4 years, 5 years, 6 years}.

(e) Describe a reasonable range for this function. Round to the nearest whole number.

▶ The output values correspond to the input values in the domain. The range is {$4500, $4860, $5249, $5669, $6122, $6612, $7141}.

In the next example, the process of adding interest earned to an investment so that the combined amount also earns interest is an example of exponential growth. The formula for the future value of the investment, A, is $A = P\left(1 + \dfrac{r}{n}\right)^{nt}$ where the original amount of the investment is the principal, P. The annual interest rate in decimal form is r. The number of times per year that the interest is calculated and added to the original investment (the number of compoundings) is n. The time in years that the money is invested is t.

EXAMPLE 2 | The annual interest rate paid on an investment of $2000 is 5.5%. Find the future value of the investment in 3 years if the interest is compounded 365 days a year. Round to the nearest hundredth.

SOLUTION ▶ **Step 1 Understand the problem.**
The unknown is the future value.

$$A = \text{future value}$$

Step 2 Make a plan.

Use the formula for future value, $A = P\left(1 + \dfrac{r}{n}\right)^{nt}$. The principal, P, is $2000; the annual interest rate, r, is 0.055; the number of compoundings per year, n, is 365; and the time, t, is 3 years.

Step 3 Carry out the plan.

$A = P\left(1 + \dfrac{r}{n}\right)^{nt}$	Formula for future value.
$A = 2000\left(1 + \dfrac{0.055}{365}\right)^{(365)(3)}$	Replace variables.
$A = 2000(1.00015\ldots)^{1095}$	Follow the order of operations.
$A = 2000(1.179\ldots)$	Follow the order of operations.
$A = \$2358.756\ldots$	Follow the order of operations.
$A \approx \$2358.76$	Round.

Step 4 Look back.
The formula for simple interest (no compounding) is $I = PRT$. Without compounding, the interest at 5.5% for 3 years is ($2000)(0.055)(3), which equals $330, and the future value is $2330. Compounding should result in a future value that is a little larger than $2330. So $2358.76 seems reasonable.

Step 5 Report the solution.
The future value of the investment is $2358.76.

When the interest is continuously added to the principal, this is **continuous compounding**. The future value formula for continuous compounding is $A = Pe^{rt}$, where the base is the irrational number e.

EXAMPLE 3 | The annual interest rate paid on an investment of $2000 is 5.5%. Find the future value of the investment in 3 years if the interest is compounded continuously. Round to the nearest hundredth.

SOLUTION ▶ **Step 1 Understand the problem.**
The unknown is the future value.

A = future value

Step 2 Make a plan.
Use the formula for future value, $A = Pe^{rt}$. The principal, P, is $2000. The annual interest rate is r, 0.055. The time, t, is 3 years.

Step 3 Carry out the plan.

$A = Pe^{rt}$	Formula for future value.
$A = 2000e^{(0.055)(3)}$	Replace variables.
$A = 2000e^{0.165}$	Follow the order of operations.
$A = 2000(1.179\ldots)$	Follow the order of operations.
$A = \$2358.786\ldots$	Follow the order of operations.
$A \approx \$2358.79$	Round.

Step 4 Look back.
Continuously compounding results in only a slightly higher future value ($0.03) than compounding 365 days per year. Most banks and credit cards compound interest daily (365 days a year).

Step 5 Report the solution.
The future value of the investment is $2358.79.

When using formulas, avoid rounding until the end of the problem. Rounding at each step will change the final answer. On the left, we find the future value of $2000 invested for 3 years at 5.5%, compounded monthly, rounding only once at the end. On the right is the same problem, with rounding in each step to the nearest thousandth. Rounding increases the future value.

Rounded once at the end:

$$A = \$2000\left(1 + \frac{0.055}{12}\right)^{(12)(3)}$$

$$A = \$2000(1 + 0.00458\ldots)^{36}$$

$$A = \$2000(1.1789\ldots)$$

$$A = \$2357.897\ldots$$

$$A \approx \$2358$$

Rounded at each step to the nearest thousandth:

$$A = \$2000\left(1 + \frac{0.055}{12}\right)^{(12)(3)}$$

$$A \approx \$2000(1 + 0.005)^{36}$$

$$A \approx \$2000(1.197)$$

$$A \approx \$2394$$

When undergoing radioactive decay, an unstable radioactive atom such as uranium-235 emits energy and changes into a stable atom such as lead. For a given sample of uranium-235, the number of uranium-235 atoms in the sample decreases exponentially and the number of atoms of lead increases exponentially. In the next example, we evaluate a function that is a model of the exponential decrease of the number of krypton-85 atoms emitted at the Three Mile Island Unit 2. The function is in the form $y = A(b^{kt})$, where $k < 0$.

EXAMPLE 4 | In 1979, the rupture of fuel elements at Three Mile Island Unit 2 released about 50,000 curies of radioactive krypton-85. The function $N(t) = 50,000e^{-0.063t}$ represents the relationship of the time in years since 1979, t, and the number of curies of krypton-85 remaining from the accident, $N(t)$. Evaluate the function to find the amount of krypton-85 remaining in 2012. Round to the nearest hundred. (*Source:* www.ead.anl.gov)

SOLUTION ▶ **Step 1 Understand the problem.**
The input value is the difference between 2012 and 1979, which equals 33 years. The base in the function is e. The value of k is less than 0, -0.063.

Step 2 Make a plan.
Evaluate $N(33 \text{ years})$ for the function $N(t) = 50,000e^{-0.063t}$.

Step 3 Carry out the plan.

$$N(t) = 50,000e^{-0.063t}$$
$$N(33) = 50,000e^{(-0.063)(33)} \qquad \text{Replace } t \text{ with 33.}$$
$$N(33) = 50,000e^{-2.079} \qquad \text{Follow the order of operations.}$$
$$N(33) = 50,000(0.125\ldots) \qquad \text{Follow the order of operations.}$$
$$N(33) = 6252.7\ldots \text{ curies} \qquad \text{Follow the order of operations.}$$
$$N(33) \approx 6300 \text{ curies} \qquad \text{Round.}$$

Step 4 Look back.
If $k = -0.06$, the rate of decrease is slower; $N(33) = 50,000e^{(-0.06)(33)}$, which is about 6903 curies. If $k = -0.07$, the rate of decrease is higher; $N(33) = 50,000e^{(-0.07)(33)}$, which is about 4963 curies. Since the solution of 6300 curies is between 4963 curies and 6903 curies, it seems reasonable.

Step 5 Report the solution.
In 2012, about 6300 curies of radioactive krypton-85 gas remained in the atmosphere from the accident.

Practice Problems

1. The function $y = 18^{4t}$ is a model of exponential population growth of German cockroaches. In this function, y is the population of cockroaches in t years. Find the number of cockroaches after 6 months (6 months = 0.5 year). This model assumes that none of the cockroaches die and that all females reproduce.

2. Find the future value of an investment of $2500 in 2 years if the annual interest is 6.5%, compounded monthly. Round to the nearest hundredth.

3. Find the future value of an investment of $2500 in 2 years if the annual interest is 6.5%, compounded continuously. Round to the nearest hundredth.

Applications of Logarithms

If the pH of a solution is greater than 7, the solution is *basic*. The pH of most shampoos is around 8. If the pH is less than 7, it is *acidic*. The pH of most hair conditioners is around 4. If the pH of a solution equals 7, it is *neutral*. Pure water is neutral. The pH of a solution depends on the concentration of hydronium ions, H_3O^+, in a solution. The notation for concentration of hydronium ions in moles per liter is $\left[H_3O^+\right]$. We use a base 10 logarithm to define the pH of a solution.

> **pH of a Solution**
>
> $$\text{pH} = -\log[\text{H}_3\text{O}^+] \quad \text{or} \quad y = -\log(x)$$
>
> where x is the concentration of hydronium ions in moles per liter and y is the pH

EXAMPLE 5   The concentration of hydronium ions in a solution is 4.5×10^{-8} mole per liter. Find the pH of the solution. Round to the nearest tenth.

SOLUTION ▸

$$\text{pH} = -\log[\text{H}_3\text{O}^+]$$

$$\text{pH} = -\log(\mathbf{4.5 \times 10^{-8}}) \qquad \text{Replace } [\text{H}_3\text{O}^+] \text{ with } 4.5 \times 10^{-8}.$$

$$\text{pH} = -(\mathbf{-7.34\ldots}) \qquad \text{Evaluate the logarithm.}$$

$$\text{pH} = \mathbf{7.34\ldots} \qquad \text{Simplify.}$$

$$\text{pH} \approx \mathbf{7.3} \qquad \text{Round.}$$

An electrochemical cell such as a car battery changes chemical energy into electrical energy. Chemists use the **Nernst equation**, $E = E^0 - \dfrac{0.05915}{z}\log\!\left(\dfrac{[\text{reduced ions}]}{[\text{oxidized ions}]}\right)$, to find the electrical potential of a cell. In college chemistry classes, students learn how to complete a Nernst equation. In the next example, we solve an already completed Nernst equation.

EXAMPLE 6   Find the potential, E, in volts of an electrochemical cell when

$$E = 0.15 - \frac{0.05915}{2}\log\!\left(\frac{0.40}{0.10}\right). \text{ Round to the nearest hundredth.}$$

SOLUTION ▸

$$E = 0.15 - \frac{0.05915}{2}\log\!\left(\frac{0.40}{0.10}\right) \qquad \text{The Nernst equation for a reaction with tin.}$$

$$E = 0.15 - \frac{0.05915}{2}\log(\mathbf{4}) \qquad \text{Simplify inside the parentheses.}$$

$$E = 0.15 - \frac{0.05915}{2}(\mathbf{0.60205\ldots}) \qquad \text{Evaluate the logarithm. Do not round.}$$

$$E = 0.15 - \mathbf{0.0178\ldots} \qquad \text{Simplify.}$$

$$E \approx \mathbf{0.13 \text{ volt}} \qquad \text{Simplify. Round. The units are volts.}$$

The intensity of a sound, I, is a measure of its energy in watts per square meter, $\left(\dfrac{\text{watt}}{\text{m}^2}\right)$. The formula for loudness in decibels is $L = 10\log\!\left(\dfrac{I}{1 \times 10^{-12}}\right)$.

EXAMPLE 7   The intensity of sound in the front row of a rock concert is about 1.1 watts per square meter. Find the loudness of this sound in decibels. Round to the nearest whole number.

SOLUTION ▸

$$L = 10\log\!\left(\frac{I}{1 \times 10^{-12}}\right)$$

$$L = 10\log\!\left(\frac{\mathbf{1.1}}{1 \times 10^{-12}}\right) \qquad \text{Replace } I \text{ with } 1.1.$$

$$L = 10\log(\mathbf{1.1 \times 10^{12}}) \qquad \text{Simplify.}$$

$$L = 10(\mathbf{12.041\ldots}) \qquad \text{Evaluate the logarithm.}$$

$$L \approx \mathbf{120 \text{ decibels}} \qquad \text{Simplify. Round; the units are decibels.}$$

Practice Problems

4. The concentration of hydronium ions in a solution is 2.4×10^{-5} mole per liter. Find the pH of this solution. Round to the nearest tenth.
5. The intensity of sound from the earbud of an iPod is 3.2×10^{-1} watt per square meter. Find the loudness of this sound in decibels. Round to the nearest whole number.
6. Use the completed Nernst equation, $E = 0.769 - \dfrac{0.05915}{1} \log\left(\dfrac{0.30}{0.20}\right)$, to find the potential in volts of a reaction involving iron. Round to the nearest hundredth.

Using Technology: Solving an Exponential Equation with a Graphing Calculator

To estimate the solution of the exponential equation $3^x = 8$, find the intersection point of the graphs of $y = 3^x$ and $y = 8$. The solution is the x-coordinate of this point. This is the same process used in Section 6.5 to solve a radical equation.

EXAMPLE 8 Use a graphing calculator to solve $3^x = 8$. Round to the nearest hundredth.

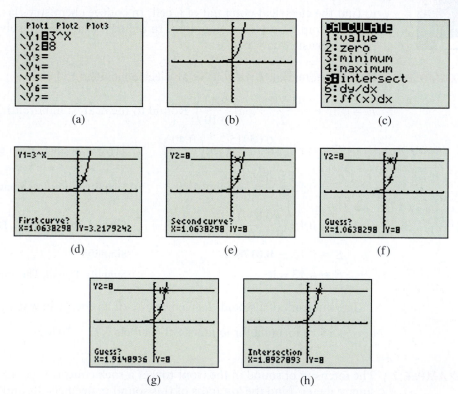

The intersection point is $(1.892\ldots, 8)$. Rounding to the nearest hundredth, the solution is $x \approx 1.89$.

Practice Problems For problems 7–10, use the graphical method to solve each equation. Choose a window that shows the intersection point.
(a) Sketch the graph; describe the window.
(b) Identify the solution, rounding to the nearest hundredth.

7. $2^x = 5$ 8. $3^x = 15$ 9. $6^{2x} = 20$ 10. $4^x = 16$

8.5 VOCABULARY PRACTICE

Match the term with its description.

1. Unit of measurement of loudness
2. Unit of measurement of sound intensity
3. The original amount of an investment
4. $-\log[\text{H}_3\text{O}^+]$
5. The pH of this solution equals 7
6. The pH of this solution is less than 7
7. The pH of this solution is greater than 7
8. The base of a common logarithm
9. The base of a natural logarithm
10. $A = Pe^{rt}$

A. acidic solution
B. basic solution
C. decibel
D. e
E. future value of an investment with continuous compounding
F. neutral solution
G. pH
H. principal
I. 10
J. watt per square meter

8.5 Exercises

Follow your instructor's guidelines for showing your work.

For exercises 1–4, an investment of $5000 earns an interest rate of 6% for 3 years. Find the future value, rounded to the nearest hundredth, if the interest is compounded:

1. annually
2. monthly
3. continuously
4. daily (1 year = 365 days)

For exercises 5–8, $5000 is invested at an interest rate of 8% for 3 years. Find the future value, rounded to the nearest hundredth, if the interest is compounded:

5. monthly
6. annually
7. daily (1 year = 365 days)
8. continuously

For exercises 9–12, credit card companies make money by charging interest on the balance owed after the due date. A student has a balance of $1200 on his credit card. Assume that the student makes no more purchases with this card. Round to the nearest hundredth.

9. The annual interest rate on the credit card account is 14.5%, compounded daily. Find the balance on the account after 30 days.
10. The annual interest rate on the credit card is 11%, compounded daily. Find the balance on the account after 30 days.
11. The annual interest rate on the credit card is 14.5%, compounded daily. After 30 days, the student makes the minimum payment of $20. Find the balance on the account after 60 days.
12. The annual interest rate on the credit card is 11%, compounded daily. After 30 days, the student makes the minimum payment of $20. Find the balance on the account after 60 days.

For exercises 13–16, a company sold 15,000 units of a product in 2007 and expects a 19% increase in the number of units sold each year for the next 3 years. The function

$D(t) = 15,000(1.19^t)$ represents the relationship of the years since 2007, t, and the number of units sold, $D(t)$.

13. Predict the number of units of this product sold in 2009. Round to the nearest whole number.
14. Predict the number of units of this product sold in 2010. Round to the nearest whole number.
15. Describe a reasonable domain (set of inputs) for this function.
16. Describe a reasonable range (set of outputs) for this function. Round to the nearest whole number.

For exercises 17–20, some bacteria reproduce asexually by binary fission (dividing in half). At the beginning, the bacteria may not reproduce. This is the *lag phase*. Next, the population grows exponentially. Then the population levels off in a *stationary phase*. Depending on the environment, the population may then decline.

17. A population of 100,000 bacteria begins exponential growth. The function $y = (100,000)(2^t)$ represents the number of bacteria, y, after t hr. Predict the population of the bacteria after 3 hr.
18. A population of 50,000 bacteria begins exponential growth. The function $y = (50,000)(2^t)$ represents the number of bacteria, y, after t hr. Predict the population of the bacteria after 3 hr.
19. A population of 100,000 bacteria begins exponential growth. The function $y = (100,000)(2^t)$ represents the number of bacteria, y, after t 30-min intervals. Predict the population of the bacteria after 3 hr.
20. A population of 50,000 bacteria begins exponential growth. The function $y = (50,000)(2^t)$ represents the number of bacteria, y, after t 30-min intervals. Predict the population of the bacteria after 3 hr.

For exercises 21–22, the function $y = 3^{6t}$ is a model of the exponential population growth of house mice, where t is time in *years*, and y is the population of mice.

21. Predict the population of mice after 6 *months*.

22. Predict the population of mice after 9 *months*. Round to the nearest whole number.

For exercises 23–24, the function $N(t) = 1e^{0.0722t}$ is a model of the exponential population growth of an earwig (*Euborellia annulata*) in laboratory conditions, where t is the time in *days* and $N(t)$ is the population of earwigs. (*Source:* N. Nonci, *Indonesian Journal of Agricultural Science,* 2005)

23. Predict the population of earwigs after 60 days. Round to the nearest whole number.

24. Predict the population of earwigs after 90 days. Round to the nearest whole number.

For exercise 25–28, the ratio of radioactive carbon-14 atoms and stable carbon-12 atoms in the atmosphere and in the tissues of living plants and animals is $\dfrac{1 \text{ atom carbon-14}}{1 \times 10^{12} \text{ atoms carbon-12}}$. Once death occurs, no new carbon atoms are absorbed. The carbon-14 atoms in the tissues release radiation and become stable nitrogen-14 atoms. For an initial sample of 2 g of carbon-14, the function $N(t) = 2e^{(-1.2097 \times 10^{-4})(t)}$ represents the relationship of the number of years after death, t, and the remaining amount of carbon-14 atoms, $N(t)$, in grams.

25. Find the amount of carbon-14 atoms remaining after 5730 years. Round to the nearest tenth.

26. Find the amount of carbon-14 atoms remaining after 11,460 years. Round to the nearest tenth.

27. The half-life of a radioactive substance is the time it takes for one-half of a sample to change into a stable atom. Use the results of exercise 25 to estimate the half-life of carbon-14.

28. The half-life of a radioactive substance is the time it takes for one-half of a sample to change into a stable atom. The half-life of radium is 1602 years. A sample contains 12 g of radium. Find the time needed for the sample to contain 3 g of radium.

For exercises 29–30, Canadian meteorologists use the *humidex* to measure the combined effect of temperature and humidity. The humidex is calculated using the formula

$$H = T + (0.5555)\left[\left(6.11e^{(5417)\left(\frac{1}{273.16} - \frac{1}{273+D}\right)}\right) - 10\right]$$

The humidex is H. The air temperature in degrees Celsius is T. The dew point temperature in degrees Celsius is D. The humidex has no units. If the humidex is greater than 45, the risk of heat stroke is very high.

29. On June 9, 2008, the high temperature in New York City was 96 °F (35.6 °C). The dew point temperature was 67 °F (19.4 °C). Find the humidex for these conditions. Round to the nearest tenth. (*Source:* www.wunderground.com)

30. On June 9, 2008, the high temperature in Philadelphia was 98 °F (36.7 °C). The dew point temperature was 70 °F (21.1 °C). Find the humidex for these conditions. Round to the nearest tenth. (*Source:* www.wunderground.com)

For exercises 31–34, a spectrophotometer can measure the amount of light transmitted through a solution. The Beer-Lambert law represents the relationship of the concentration of a dissolved substance and the amount of light that passes through it: $T = e^{-kbc}$. The concentration of the substance in moles per liter is c. The percent of light transmitted (in decimal form) is T. The molar absorbtivity constant for the substance is k. The width of the sample in centimeters is b.

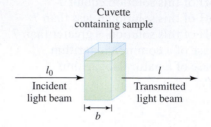

Cuvette containing sample

l_0 Incident light beam

l Transmitted light beam

b

31. The concentration of a solution of copper sulfate is $\dfrac{0.16 \text{ mole}}{1 \text{ L}}$. The absorbtivity constant for copper sulfate is $\dfrac{20 \text{ L}}{1 \text{ mole} \cdot \text{cm}}$. The width of the sample is 1 cm. Predict the percent of light transmitted through this sample. Round to the nearest tenth of a percent.

32. The concentration of a protein in a solution is $\dfrac{0.8 \text{ mole}}{1 \text{ L}}$. The absorbtivity constant for this protein is $\dfrac{0.8382 \text{ L}}{1 \text{ mole} \cdot \text{cm}}$. The width of the sample is 1 cm. Predict the percent of light transmitted through this sample. Round to the nearest tenth of a percent.

33. The concentration of a solution of copper sulfate is $\dfrac{0.32 \text{ mole}}{1 \text{ L}}$. The absorbtivity constant for copper sulfate is $\dfrac{20 \text{ L}}{1 \text{ mole} \cdot \text{cm}}$. The width of the sample is 1 cm. Predict the percent of light transmitted through this sample. Round to the nearest hundredth of a percent.

34. The concentration of a protein in a solution is $\dfrac{0.2 \text{ mole}}{1 \text{ L}}$. The absorbtivity constant for this protein is $\dfrac{0.8382 \text{ L}}{1 \text{ mole} \cdot \text{cm}}$. The width of the sample is 1 cm. Predict the percent of light transmitted through this sample. Round to the nearest hundredth of a percent.

For exercises 35–36, since a child's kidney is large compared to total body surface area, children often need more medication than adults. The relationship of the child dose and the adult dose is $F = (W^{-0.338})(5.69 - 3.87e^{-0.0822A})$. The weight of the child in *kilograms* is W. The age of the child in *months* is A. The multiplier of the adult dose is F. (*Source:* W. Hayton, "Maturation and Growth of Renal Function: Dosing Renally Cleared Drugs in Children," 2000)

35. Find the multiplier of an adult dose for a 3-year-old child who weighs 18 kg. Round to the nearest hundredth.

36. Find the multiplier of an adult dose for a 3-year-old child who weighs 14 kg. Round to the nearest hundredth.

For exercise 37–40, the Arrhenius equation describes the relationship between the rate constant for a reaction, k, and the temperature, T. The units of temperature are Kelvins. The Arrhenius equation for a reaction is $k = (1 \times 10^{12})e^{-\frac{76,100}{8.314T}}$.

37. Find k when the temperature is 429 Kelvin. Round the mantissa to the nearest hundredth.

38. Find k when the temperature is 389 Kelvin. Round the mantissa to the nearest hundredth.

39. As the temperature increases, does the rate constant, k, increase or decrease? Explain.

40. As the temperature decreases, does the rate constant, k, increase or decrease? Explain.

For exercises 41–42, a dose of 2 g of an antibiotic is administered by intravenous bolus. The function $C = 140e^{-0.095t}$ represents the relationship of the number of hours after the drug is given, t, and the concentration of the drug in the bloodstream in milligrams per milliliter, C. (*Source:* Bjork, et al., *The Prostate*, April 2001)

41. Find the concentration of the antibiotic after 30 min. Round to the nearest tenth.

42. Find the concentration of the antibiotic after 3 hr. Round to the nearest tenth.

For exercises 43–44, the effectiveness of a treatment for prostate cancer was determined by testing the blood serum level of human glandular kallikrein 2 (hK2). The function $y = (1.03776)^{-t}$ represents the relationship of the time in days after beginning treatment, t, and the concentration of hK2 in nanograms per milliliter, y.

43. Find the concentration of hK2 after 3 days of treatment. Round to the nearest hundredth.

44. Find the concentration of hK2 after 14 days of treatment. Round to the nearest hundredth.

For exercises 45–48, find the pH of a solution with the given concentration of hydronium ions. Round to the nearest tenth.

45. $\dfrac{3.7 \times 10^{-9} \text{ mole}}{1 \text{ L}}$

46. $\dfrac{5.3 \times 10^{-4} \text{ mole}}{1 \text{ L}}$

47. $\dfrac{9.2 \times 10^{-3} \text{ mole}}{1 \text{ L}}$

48. $\dfrac{4.9 \times 10^{-10} \text{ mole}}{1 \text{ L}}$

For exercises 49–52, solve the given Nernst equation to find the cell potential in volts. Round to the nearest tenth.

49. $E = 0.563 - \dfrac{0.05915}{2} \log\left(\dfrac{1.0}{1 \times 10^{-4}}\right)$

50. $E = 1.78 - \dfrac{0.05915}{6} \log\left(\dfrac{1 \times 10^{-6}}{1 \times 10^{-2}}\right)$

51. $E = 1.44 - \dfrac{0.05915}{1} \log\left(\dfrac{3.3 \times 10^{-2}}{1.3 \times 10^{-4}}\right)$

52. $E = 0.34 - \dfrac{0.05915}{2} \log\left(\dfrac{1.0}{2 \times 10^{-1}}\right)$

For exercises 53–56, find the loudness in decibels of the sound with the given intensity. Round to the nearest whole number.

53. shotgun blast: $\dfrac{1.05 \times 10^{5} \text{ watt}}{1 \text{ m}^2}$

54. inside subway: $\dfrac{2.5 \times 10^{-3} \text{ watt}}{1 \text{ m}^2}$

55. whisper: $\dfrac{3 \times 10^{-10} \text{ watt}}{1 \text{ m}^2}$

56. space shuttle launch: $\dfrac{1 \times 10^{7} \text{ watt}}{1 \text{ m}^2}$

57. For a reaction with glucose, $\ln\left(\dfrac{0.05}{0.1}\right) = (-8.8 \times 10^{-5})t$. Find the time in seconds, t. Round to the nearest whole number.

58. For a reaction with cyclobutane, $\ln\left(\dfrac{0.01}{0.1}\right) = (-9.2 \times 10^{-3})t$. Find the time in seconds, t. Round to the nearest whole number.

For exercises 59–60, the formula

$k = -\ln(R - 0.008t) + (4 - 3.5R)\left(\dfrac{V}{W}\right)$ is used to judge the

effectiveness of kidney dialysis. The blood urea is R, the time of dialysis in hours is t, the volume of blood filtered in liters is V, and the postdialysis weight of the patient in kilograms is W. For kidney dialysis to be effective, k must be at least 1.2.

59. Find k when $R = 0.3$, $t = 3$ hr, $V = 3.97$ L, and $W = 70$ kg. Round to the nearest tenth.

60. Find k when $R = 0.3$, $t = 3.5$ hr, $V = 3.3$ L, and $W = 71$ kg. Round to the nearest tenth.

For exercises 61–62, a proposed formula for determining annual dues for the International Federation of University Women is $x = 2\ln(G) - \ln(M)$. The gross per capita national income is G. The number of members in a national federation or organization is M. The dues per member, D, depend on x. (*Source:* www.ifuw.org)

If $x < 7$, $D = 8$ Swiss francs.

If $7 \le x < 8.5$, $D = 10.5$ Swiss francs.

If $8.5 \le x < 10$, $D = 13$ Swiss francs.

If $10 \le x < 11.5$, $D = 15.5$ Swiss francs.

If $x \ge 11.5$, $D = 18$ Swiss francs.

61. El Salvador has 150 members. Its gross per capita national income is $2110. Find the proposed dues per member from this country.

62. Finland has 1175 members. Its gross per capita national income is $23,890. Find the proposed dues per member from this country.

For exercises 63–64, when planning a new development, the need for new roads and traffic lights must be determined. The predicted number of vehicle trips, T, caused by a new regional shopping center depends on the size of its gross leasable area, A, in thousands of square feet: $T = e^{(0.756)\ln(A) + 5.25}$.

63. Predict the number of new vehicle trips for a regional shopping center with an area of 300 thousand ft^2. Round to the nearest thousand.

64. Predict the number of new vehicle trips for a regional shopping center with an area of 400 thousand ft². Round to the nearest thousand.

For exercises 65–66, the interest on a car loan for P dollars is compounded monthly. The annual interest rate in decimal form is r. The borrower pays a monthly payment, A. A formula for the number of payments, n, needed to pay off the loan and interest is

$$n = \frac{-\ln\left(1 - \dfrac{Pr}{12A}\right)}{\ln\left(1 + \dfrac{r}{12}\right)}$$

65. A lender charges 6% annual interest for a car loan for $18,000. A buyer can afford a maximum payment of $300. Find the number of payments needed to pay off this loan. Round up to the nearest whole number.

66. A lender charges 6 % annual interest for a car loan for $18,000. A buyer can afford a maximum payment of $350. Find the number of payments needed to pay off this loan. Round up to the nearest whole number.

For exercises 67–68, teams that play an on-line strategy war game are assigned a *team rating* each week. When the team rating is greater than 1500, the relationship of the individual player points P and the team rating T is represented by the formula

$$T = \frac{\ln\left(\dfrac{2894 - P}{259P}\right)}{-0.0025}$$

67. If an individual player earns 1200 points, find his or her team rating. Round to the nearest whole number.

68. If an individual player earns 1500 points, find his or her team rating. Round to the nearest whole number.

Problem Solving: Practice and Review

Follow your instructor's guidelines for using the five steps as outlined in Section 1.5, p. 51.

69. The population of a city in a high-income country includes 25,500 women who do *not* take oral contraceptives. Predict how many of these women will get ovarian cancer before age 75.

In high-income countries, 10 years use of oral contraceptives was estimated to reduce ovarian cancer incidence before age 75 from 1.2 to 0.8 per 100 users and mortality from 0.7 to 0.5 per 100; for every 5000 woman-years of use, about two ovarian cancers and one death from the disease before age 75 are prevented. (*Source:* www.thelancet.com, Jan. 26, 2008)

For exercises 70–71, the graph represents a linear model of the population in Texas, y, in x years since 1960.

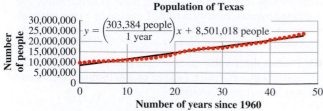

Population of Texas

Source: www.ojp.usdoj.gov/bjs

70. What is the average rate of change in the number of people living in Texas?

71. Predict the population of Texas in 2015. Round to the nearest thousand.

72. The cost to make a product is $4. The overhead per month to make the product is $5500. The product sells for $12. Assume that the company sells all of the product it makes. How many products must be made and sold per month for the company to break even? Round to the nearest whole number.

Technology

73. A function for finding the future value $f(x)$ of an investment x is $f(x) = 100\left(1 + \dfrac{0.04}{12}\right)^{(12x)}$.

    a. Graph this function in the window [0, 10, 1, 0, 150, 15]. Sketch the graph; describe the window. (When typing the function, put 12x in parentheses.)

    b. Use the **Value** command to find $f(5$ years). Round to the nearest hundredth.

74. A function for finding future value $f(x)$ of an investment x is $f(x) = 100\left(1 + \dfrac{0.04}{365}\right)^{(365x)}$.

    a. Graph this function in the window [0, 10, 1, 0, 150, 15]. Sketch the graph; describe the window. (When typing the function, put 365x in parentheses.)

    b. Use the **Value** command to find $f(5$ years). Round to the nearest hundredth.

75. Write an application problem that can be solved with the information in exercise 73.

76. Write an application problem that can be solved with the information in exercise 74.

Find the Mistake

For exercises 77–80, the completed problem has one mistake.
(a) Describe the mistake in words, or copy down the whole problem and highlight or circle the mistake.
(b) Do the problem correctly.

77. **Problem:** An investment of $10,000 earns an interest rate of 5.75%, compounded monthly. Find the future value in 3 years of this investment. Round to the nearest hundredth.

Incorrect Answer: $A = P\left(1 + \dfrac{r}{n}\right)^{nt}$

$$A = 10,000\left(1 + \frac{5.75}{12}\right)^{36}$$

$$A = \$13,201,278,960$$

78. **Problem:** An investment of $750 earns an interest rate of 6%, compounded daily (365 days per year). Find the future value in 2 years of this investment. Round to the nearest hundredth.

Incorrect Answer: $A = P\left(1 + \dfrac{r}{n}\right)^{nt}$

$$A = 750\left(1 + \frac{0.06}{365}\right)^{2}$$

$$A \approx \$750.25$$

79. Problem: The concentration of hydronium ions in a solution is 3.2×10^{-4} mole per liter. Find the pH of the solution. Round to the nearest tenth.

Incorrect Answer:
$$pH = -\log[H_3O^+]$$
$$pH = -\log(3.2 \times 10^{-4})$$
$$pH = -3.494\ldots$$
$$pH \approx -3.5$$

80. Problem: A population of 10,000 bacteria begins exponential growth. The function $y = (10,000)(2^n)$ describes the number of bacteria, y, after n 30-min intervals. Predict the population of the bacteria after 4 hr.

Incorrect Answer:
$$y = (10,000)(2^n)$$
$$y = (10,000)(2^4)$$
$$y = (10,000)(16)$$
$$y = 160,000 \text{ bacteria}$$

Review

81. Write the power rule of exponents.

82. Simplify: $(x^2 y^3)^4$

83. If $10^x = 100$, what is x?

84. If $5^x = 1$, what is x?

SUCCESS IN COLLEGE MATHEMATICS

85. Using the second test from this term, identify the questions on which you missed points. It might be that you have learned this material since you took this test. Or you still might not be able to do some questions. Copy the questions that you cannot do on a fresh piece of paper to create a practice test. Complete this test. You might need to go to a tutoring center or visit your instructor during office hours to make sure that you have done them correctly.

The *doubling time* for a population of mice is the time needed for the population to double. We can find the doubling time by solving an exponential equation. In this section, we will learn how to solve exponential equations and find doubling times and half-lives.

SECTION 8.6

Exponential Equations

After reading the text, working the practice problems, and completing assigned exercises, you should be able to:

1. Solve an exponential equation in which the variable is an exponent.

2. Solve an application using an exponential equation.

3. Given a function that is a model of exponential growth, find the doubling time.

4. Given a function that is a model of exponential decrease, find the half-life.

Exponential Equations

In an exponential equation, the variable is an exponent. To solve an exponential equation, take a base 10 logarithm or a base e logarithm of each side and use the power rule of logarithms to isolate the variable.

> **Power Rule of Logarithms**
>
> $$\log_b(m^p) = p \log_b(m)$$
>
> When raising a logarithm to a power, multiply the power and the logarithm. The base, b, is a real number greater than 0 that is not equal to 1; m is a real number greater than 0; and p is a real number.

EXAMPLE 1   Solve: $10^x = 100$

SOLUTION ▶

$$10^x = 100$$ The variable is an exponent.

$$\log(10^x) = \log(100)$$ Take the logarithm (base 10) of each side.

$$x \log(10) = \log(100)$$ Power rule of logarithms.

$$\frac{x \log(10)}{\log(10)} = \frac{\log(100)}{\log(10)}$$ Division property of equality.

$$x = \frac{\log(100)}{\log(10)}$$ Simplify.

$$x = \frac{2}{1}$$ Evaluate the logarithms.

$$x = 2$$ Simplify.

Solving an Exponential Equation in One Variable (the variable is an exponent)

1. Isolate the exponential expression.
2. Take the logarithm of each side.
3. Use the power rule to rewrite the exponential expression as a product.
4. Use the properties of equality to isolate the variable.
5. Check.

In the next example, the solution is irrational. Do not round until the end.

EXAMPLE 2   (a) Solve $5^x = 18$. Round to the nearest ten-thousandth.

SOLUTION ▶

$$5^x = 18$$ The variable is an exponent.

$$\log(5^x) = \log(18)$$ Take the logarithm (base 10) of each side.

$$x \log(5) = \log(18)$$ Power rule of logarithms.

$$\frac{x \log(5)}{\log(5)} = \frac{\log(18)}{\log(5)}$$ Division property of equality.

$$x = \frac{\log(18)}{\log(5)}$$ Simplify.

$$x = \frac{1.25527\ldots}{0.69897\ldots}$$ Evaluate the logarithms.

$$x = 1.79588\ldots$$ Simplify.

$$x \approx 1.7959$$ Round.

(b) Check.

▶

$$5^x = 18$$ Use the original equation.

$$5^{1.7959} = 18$$ Replace the variable with the rounded solution.

$$18.0003 \approx 18 \quad \text{True.}$$ The small difference is caused by rounding.

When solving an exponential equation, we take the logarithm of both sides. A logarithm of any base can be used. However, since $\ln(e) = 1$, we usually solve exponential equations in which the base is e by taking the natural logarithm of both sides.

EXAMPLE 3 | **(a)** Solve $e^{3x} = 21$. Round to the nearest ten-thousandth.

SOLUTION ▶

$e^{3x} = 21$	The variable is an exponent.
$\ln(e^{3x}) = \ln(21)$	Take the logarithm (base e) of each side.
$3x \ln(e) = \ln(21)$	Power rule of logarithms.
$\dfrac{3x \ln(e)}{3 \ln(e)} = \dfrac{\ln(21)}{3 \ln(e)}$	Division property of equality.
$x = \dfrac{\ln(21)}{3 \ln(e)}$	Simplify; $\dfrac{3x \ln(e)}{3 \ln(e)} = x$
$x = \dfrac{3.04452\ldots}{3(1)}$	Evaluate the logarithms.
$x = 1.01484\ldots$	Evaluate.
$x \approx 1.0148$	Round.

(b) Check.

$e^{3x} = 21$		Use the original equation.
$e^{3(1.0148)} = 21$		Replace the variable with the rounded solution.
$e^{3.0444} = 21$		Multiply the exponents.
$20.9974 \approx 21$	True.	The small difference is caused by rounding.

In the next example, we isolate the exponential expression before taking the logarithm of both sides.

EXAMPLE 4 | Solve $4000e^x = 12{,}000$. Round to the nearest ten-thousandth.

SOLUTION ▶

$4000e^x = 12{,}000$	The variable is an exponent.
$\dfrac{4000e^x}{4000} = \dfrac{12{,}000}{4000}$	Isolate e^x; division property of equality.
$e^x = 3$	Simplify.
$\ln(e^x) = \ln(3)$	Take the logarithm (base e) of each side.
$x \ln(e) = \ln(3)$	Power rule of logarithms.
$\dfrac{x \ln(e)}{\ln(e)} = \dfrac{\ln(3)}{\ln(e)}$	Division property of equality.
$x = \dfrac{\ln(3)}{\ln(e)}$	Simplify.
$x = \dfrac{1.09861\ldots}{1}$	Evaluate the logarithms; $\ln(e) = 1$
$x = 1.09861\ldots$	Simplify.
$x \approx 1.0986$	Round.

Practice Problems

For problems 1–4,
(a) solve. Round any irrational solutions to the nearest ten-thousandth.
(b) check.

1. $10^x = 1000$ **2.** $4^x = 20$ **3.** $6^{2x} = 100$ **4.** $200(e^x) = 800$

Applications of Exponential Equations

In Section 8.5, we used the exponential function $y = \$4500(1.08)^t$ as a model of the annual tuition at a college beginning in the fall of 2010. We found the value of y for different values of t. Given y, we can solve an exponential equation to find t.

EXAMPLE 5 A college increases tuition once a year in the fall. The function $y = \$4500(1.08)^t$ describes the relationship of the number of years since fall 2010, t, and the semester tuition. The chief financial officer of the college expects that tuition will follow this model for 6 years. Find the time when the semester tuition is greater than $5500.

SOLUTION ▶ **Step 1 Understand the problem.**
The unknown is the time when the semester tuition is greater than $5500.

$$t = \text{time}$$

Step 2 Make a plan.
Replace y in the function $y = \$4500(1.08)^t$ with $5500, and solve for t.

Step 3 Carry out the plan.

$$y = \$4500(1.08)^t$$

$$\mathbf{\$5500} = \$4500(1.08)^t \qquad \text{Replace } y \text{ with } \$5500; \text{ the variable is an exponent.}$$

$$\frac{\$5500}{\$4500} = \frac{\$4500(1.08)^t}{\$4500} \qquad \text{Isolate } 1.08^t; \text{ division property of equality.}$$

$$\mathbf{1.22\ldots} = 1.08^t \qquad \text{Simplify; do not round.}$$

$$\mathbf{\log(1.2222\ldots)} = \mathbf{\log(1.08^t)} \qquad \text{Take the logarithm (base 10) of both sides.}$$

$$\log(1.2222\ldots) = t\log(1.08) \qquad \text{Power rule of logarithms.}$$

$$\frac{\log(1.2222\ldots)}{\log(1.08)} = \frac{t\log(1.08)}{\log(1.08)} \qquad \text{Division property of equality.}$$

$$\frac{\log(1.2222\ldots)}{\log(1.08)} = \mathbf{t} \qquad \text{Simplify.}$$

$$\mathbf{2.6074\ldots} = t \qquad \text{Evaluate the logarithms; simplify.}$$

Since t is the number of years since fall 2010, the semester tuition is more than $5500 in fall 2013 ($t = 3$ years).

Step 4 Look back.
When $t = 2$ years, $y = \$4500(1.08^2) \approx \$5249$. When $t = 3$ years, $y = \$4500(1.08^3) \approx \$5669$. So it is reasonable that the tuition will be more than $5500 in 3 years.

Step 5 Report the solution.
The semester tuition will be more than $5500 in fall 2013.

In the next example, we again solve an exponential equation to find time, t.

EXAMPLE 6 An investor has $5000 to invest at an annual interest rate of 6%, compounded daily (365 days per year). Find the time in years and months needed for the investment to have a future value of $10,000. Round up to the nearest month.

SOLUTION ▶ **Step 1 Understand the problem.**
The unknown is the time when the future value of the investment is $10,000.

$$t = \text{time}$$

Step 2 Make a plan.

Using the formula $A = P\left(1 + \dfrac{r}{n}\right)^{nt}$, replace A with \$10,000, P with \$5000, r with 0.06, and n with 365. Solve for t.

Step 3 Carry out the plan.

$$10{,}000 = 5000\left(1 + \frac{0.06}{365}\right)^{365t}$$ Replace variables; the variable is an exponent.

$$\frac{10{,}000}{5000} = \frac{5000\left(1 + \dfrac{0.06}{365}\right)^{365t}}{5000}$$ Isolate $\left(1 + \dfrac{0.06}{365}\right)^{365t}$; division property of equality.

$$2 = \left(1 + \frac{0.06}{365}\right)^{365t}$$ Simplify.

$$\log(2) = \log\left(1 + \frac{0.06}{365}\right)^{365t}$$ Take the logarithm (base 10) of each side.

$$\log(2) = 365t \cdot \log\left(1 + \frac{0.06}{365}\right)$$ Power rule of logarithms.

$$\frac{\log(2)}{365 \cdot \log\left(1 + \dfrac{0.06}{365}\right)} = \frac{365t \cdot \log\left(1 + \dfrac{0.06}{365}\right)}{365 \cdot \log\left(1 + \dfrac{0.06}{365}\right)}$$ Division property of equality.

$$\frac{\log(2)}{365 \cdot \log\left(1 + \dfrac{0.06}{365}\right)} = t$$ Simplify.

$$\frac{0.30102\ldots}{0.02605\ldots} = t$$ Evaluate logarithms; do not round.

$$11.553\ldots \text{ year} = t$$ Simplify; the units of time are years.

$$11 \text{ years} + (0.553\ldots \text{ year})\left(\frac{12 \text{ months}}{1 \text{ year}}\right) = t$$ Convert 0.533... year into months.

$$11 \text{ years } \textbf{7 months} \approx t$$ Simplify; round up to the nearest month.

Step 4 Look back.

If $t = 10$ years, $y = 5000\left(1 + \dfrac{0.06}{365}\right)^{(365)(10)}$, which is about \$9110. If $t = 12$ years, $y = 5000\left(1 + \dfrac{0.06}{365}\right)^{(365)(12)}$, which is about \$10,272. Since the future value goal of \$10,000 is between \$9110 for 10 years and \$10,272 for 12 years, a time of 11 years 7 months seems reasonable.

Step 5 Report the solution.

The future value will be \$10,000 in about 11 years 7 months.

Practice Problems

5. The function $P = 85(1.10^t)$ describes the relationship of the time in years, t, and the population of squirrels in a study area. Find the time in years and months it will take for the population to increase to 170 squirrels. Round *up* to the nearest month.

6. An investor has \$1500 to invest at an annual interest rate of 5.5%, compounded monthly. Find the time in years and months required for the investment to have a future value of \$5000. Round *up* to the nearest month.

Doubling Time and Half-Life

The exponential function $y = Ab^{kt}$ describes the relationship of the time, t, and the output, y. The original amount of the output is A. When the original amount has doubled, $y = 2A$. To find the **doubling time**, we solve $2A = Ab^{kt}$. Notice that we do not need to know the original amount, A, to find the doubling time.

$$y = A(b^{kt})$$

$$2A = A(b^{kt}) \qquad \text{To find the doubling time, replace } y \text{ with } 2A.$$

$$\frac{2A}{A} = \frac{A(b^{kt})}{A} \qquad \text{Division property of equality.}$$

$$2 = b^{kt} \qquad \text{Simplify; this is the doubling time equation.}$$

Doubling Time Equation

$$2 = b^{kt} \qquad \text{where } b > 0, b \neq 1, \text{ and } k \text{ is a real number greater than 0}$$

EXAMPLE 7   The function $y = D(1.19^t)$ describes the relationship of the time in years, t, and the number of products sold per year. Find the time it will take for sales of this product to double. Round up to the nearest whole number.

SOLUTION ▶ **Step 1 Understand the problem.**
The unknown is the doubling time, t. We do not need to know the original amount of products sold, D, to find the doubling time.

$$t = \text{time}$$

Step 2 Make a plan.
Use the doubling time equation, $2 = b^{kt}$, with $b = 1.19$ and $k = 1$. Solve for t.

Step 3 Carry out the plan.

$$2 = b^{kt} \qquad \text{Doubling time equation.}$$

$$2 = \mathbf{1.19}^{(1t)} \qquad b = 1.19, k = 1$$

$$\mathbf{\log(2)} = \mathbf{\log(1.19^t)} \qquad \text{Take the logarithm (base 10) of both sides.}$$

$$\log(2) = t\log(1.19) \qquad \text{Power rule of logarithms.}$$

$$\frac{\log(2)}{\log(1.19)} = \frac{t\log(1.19)}{\log(1.19)} \qquad \text{Division property of equality.}$$

$$\frac{\log(2)}{\log(1.19)} = t \qquad \text{Simplify.}$$

$$\mathbf{3.984\ldots \text{ years}} = t \qquad \text{Evaluate the logarithms and divide; replace the units.}$$

$$\mathbf{4 \text{ years}} \approx t \qquad \text{Round.}$$

Step 4 Look back.
If we replace D with any value such as 1000 units, replace t with the proposed doubling time, 4 years, and simplify, then the value of y should be about $2(1000 \text{ units})$, or 2000 units. Since $y = 1000(1.19^4)$ equals about 2005 units, the answer seems reasonable.

Step 5 Report the solution.
The time for the sales of the product to double is about 4 years.

For an exponential decrease, the function $y = Ab^{-kt}$ describes the relationship of the time, t, and the output, y. The original amount of the output is A. When the output is one-half of the original amount, $y = 0.5A$. To find the **half-life** (the time for the original amount to decrease by one-half), we solve $0.5 = Ab^{-kt}$.

$$y = Ab^{-kt}$$

$$\mathbf{0.5A} = Ab^{-kt} \qquad \text{Replace } y \text{ with } 0.5A.$$

$$\frac{0.5A}{A} = \frac{Ab^{-kt}}{A} \qquad \text{Division property of equality.}$$

$$0.5 = b^{-kt} \qquad \text{Simplify; this is the half-life equation.}$$

Half-Life Equation

$$0.5 = b^{-kt} \qquad \text{where } b > 0, b \neq 1, \text{ and } k \text{ is a real number greater than } 0$$

In the next example, the base of the function is the irrational number e. To solve the half-life equation, we take the natural logarithm of both sides.

EXAMPLE 8 The function $y = Ae^{-0.063t}$ describes the relationship of the time in years, t, and the amount of krypton-85 atoms in a sample. Find the half-life of krypton-85. Round to the nearest whole number.

SOLUTION ▶ **Step 1 Understand the problem.**
The unknown is the half-life, t. We do not need to know the original amount of krypton-85 atoms, A, to find the half-life.

$$t = \text{time}$$

Step 2 Make a plan.
Use the half-life equation, $0.5 = b^{-kt}$, with $b = e$ and $k = -0.063$. Solve for t.

Step 3 Carry out the plan.

$$0.5 = b^{-kt} \qquad \text{Doubling time equation.}$$

$$0.5 = e^{-0.063t} \qquad b = e, k = -0.063$$

$$\mathbf{ln}(0.5) = \mathbf{ln}(e^{-0.063t}) \qquad \text{Take the logarithm (base } e\text{) of both sides.}$$

$$\ln(0.5) = \mathbf{-0.063}t \cdot \ln(e) \qquad \text{Power rule of logarithms.}$$

$$\ln(0.5) = -0.063t \cdot 1 \qquad \text{Simplify; } \ln(e) = 1$$

$$\frac{\ln(0.5)}{-0.063} = \frac{-0.063t}{-0.063} \qquad \text{Division property of equality.}$$

$$\frac{\ln(0.5)}{-0.063} = \boldsymbol{t} \qquad \text{Simplify.}$$

$$\mathbf{11.0 \ldots \text{ years}} = t \qquad \text{Evaluate the logarithm; simplify. The units are years.}$$

$$\mathbf{11 \text{ years}} \approx t \qquad \text{Round.}$$

Step 4 Look back.
If we replace A with any value such as 100 g, replace t with the proposed half-life, 11 years, and simplify, then the value of y should be about $0.5(100 \text{ g})$, or 50 g. Since $y = \mathbf{100}e^{(-0.063)(\mathbf{11})}$ equals about 50 g, the answer seems reasonable.

Step 5 Report the solution.
The half-life of krypton-85 is about 11 years.

Practice Problems

7. The function $y = A(e^{0.0114t})$ describes the relationship of the time in years, t, and world population, y. Assuming that this rate of population growth does not change, find the doubling time of world population. Round up to the nearest whole number.

8. Radioactive xenon-133 gas is used to evaluate lung function. The function $y = Ae^{-0.0132t}$ describes the relationship of the time in days, t, and the amount of xenon-133 in a sample. Find the half-life of xenon-133. Round up to the nearest whole number.

8.6 VOCABULARY PRACTICE

Match the term with its description. You may use a term more than once.

1. The original amount of an investment
2. A function in which the output value is an exponent
3. $\log_b(m^p) = p \log_b(m)$
4. The base of a common logarithm
5. The base of a natural logarithm
6. The process of adding interest earned to the principal
7. A rounded value of this irrational number is 2.72
8. $(x^m)^n = x^{mn}$
9. $A = P\left(1 + \dfrac{r}{n}\right)^{nt}$
10. A real number that cannot be written as a fraction in which the numerator and denominator are integers

A. compounding
B. e
C. future value formula
D. irrational number
E. logarithmic function
F. power rule of exponents
G. power rule of logarithms
H. principal
I. 10

8.6 Exercises

Follow your instructor's guidelines for showing your work.

For exercises 1–4, use the power rule of logarithms to rewrite each expression as a product. Do not evaluate the logarithm.

1. $\log(4^x)$
2. $\log(5^x)$
3. $\log(4^{8x})$
4. $\log(5^{7x})$

5. In the power rule of exponents, the base must be a real number. In the power rule of logarithms, the base must be greater than 0 and not equal to 1. Explain why we must include these additional restrictions.

6. In the power rule of exponents, each exponent must be a real number. In the power rule of logarithms, $\log_b(m^p) = p \log_b(m)$, m must be greater than 0. Explain why.

For exercises 7–14,
(a) solve. Round any irrational solutions to the nearest ten-thousandth.
(b) check.

7. $36 = 6^x$
8. $49 = 7^x$
9. $100{,}000 = 10^x$
10. $1{,}000{,}000 = 10^x$
11. $4^x = 40$
12. $3^x = 30$
13. $e^x = 15$
14. $e^x = 24$

15. A student incorrectly solved the equation $5^x = 30$ as $x = 6$. Explain why the solution is unreasonable.

16. A student incorrectly solved the equation $6^x = 30$ as $x = 5$. Explain why the solution is unreasonable.

For exercises 17–24, solve. Round any irrational solutions to the nearest ten-thousandth.

17. $625 = 5^{2t}$
18. $64 = 2^{2t}$
19. $54 = 6^{2t}$
20. $108 = 4^{2t}$
21. $5^{3x} = 750$
22. $9^{3x} = 810$
23. $e^{4x} = 80$
24. $e^{5x} = 95$

25. To solve $e^x = 15$, a student took the natural logarithm of both sides. She then used the power rule of logarithms. Could she have used common logarithms instead of natural logarithms? Explain.

26. To solve $e^x = 15$, a student took the natural logarithm of both sides. He then used the power rule of logarithms. What is the advantage of using the natural logarithm rather than a common logarithm? (If you are not sure, try it both ways to discover the advantage.)

For exercises 27–42, solve. Round any irrational solutions to the nearest ten-thousandth.

27. $5000 = 2500(5^x)$

28. $8000 = 4000(3^x)$

29. $500 = 200(4^x)$

30. $700 = 300(6^x)$

31. $12{,}000 = 6500(1.06^t)$

32. $14{,}000 = 7500(1.04^t)$

33. $12{,}000 = 2000\left(1 + \dfrac{0.06}{4}\right)^{4t}$

34. $15{,}000 = 5000\left(1 + \dfrac{0.05}{4}\right)^{4t}$

35. $12{,}000 = 2000\left(1 + \dfrac{0.06}{12}\right)^{12t}$

36. $15{,}000 = 5000\left(1 + \dfrac{0.05}{12}\right)^{12t}$

37. $12{,}000 = 2000\left(1 + \dfrac{0.06}{365}\right)^{365t}$

38. $15{,}000 = 5000\left(1 + \dfrac{0.05}{365}\right)^{365t}$

39. $2 = e^{0.05t}$

40. $2 = e^{0.06t}$

41. $0.5 = e^{-0.05t}$

42. $0.5 = e^{-0.06t}$

43. If $3000 is invested at an annual interest rate of 3%, compounded 365 times a year (daily), find the time in years and months when the future value is $4500. Round up to the nearest month.

44. If $2000 is invested at an annual interest rate of 3%, compounded 12 times a year (monthly), find the time in years and months when the future value is $3750. Round up to the nearest month.

45. If $2000 is invested at an annual interest rate of 5%, compounded 12 times a year (monthly), find the time in years and months when the future value is $3750. Round up to the nearest month.

46. If $3000 is invested at an annual interest rate of 4%, compounded 365 times a year (daily), find the time in years and months when the future value is $4900. Round up to the nearest month.

47. The population of a county in 2008 was 535,000 people. The function $y = 535{,}000(1.09)^t$ represents the relationship of the number of years since 2008, t, and the population, y. Find when the population of the county will be 755,000 people. Round to the nearest whole number.

48. The population of a town in 2008 was 45,000 people. The function $y = 45{,}000(1.06)^t$ represents the relationship of the number of years since 2008, t, and the population, y. Find when the population of the town will be 60,000 people. Round to the nearest whole number.

For exercises 49–52, Gordon Moore, a cofounder of Intel Corporation, developed a model for the number of transistors that can be placed on an integrated circuit at a minimum component cost. The Moore Model is $y = 2300(e^{0.35t})$, where t is the years since 1971, and the number of transistors is y. Find the number of transistors in the given year. Round to the nearest million. (*Source:* www.intel.com)

49. 2000

50. 2010

51. In 2009, Intel released a new integrated circuit (computer chip) for desktop computers with 820 million transistors. Does this output follow the Moore model?

52. In 2006, Intel released the Intel Core 2 Duo chip for desktop computers with 291 million transistors. Does this output follow the Moore model?

For exercises 53–54, the function $P = 43{,}764(e^{-0.03d})$ represents the relationship of the depth in meters of sediments in Lake Baikal, d, and the number of oligochaete worms in each square meter of sediments, P. Find the depth at which a square meter of sediments contains the given number of worms. Round to the nearest whole number. (*Source:* Martin, et al., *Hydrobiologia*, 1999)

53. 5000 worms

54. 1000 worms

For exercises 55–56, scientists studied the possible cross-pollination of corn plants by genetically engineered corn plants planted *upwind*. The function $p = 27.67(e^{-0.4098d})$ represents the relationship of the distance in meters between the normal corn and the genetically engineered corn, d, and the percent of the normal corn that was cross-pollinated by the genetically engineered corn, p. (*Source:* B. L. Ma, "Frequency of Pollen Drift in Genetically Engineered Corn," 2005)

55. If the percent of cross-fertilization is 2%, find the distance of the normal corn from the genetically engineered corn. Round to the nearest tenth.

56. If the percent of cross-fertilization is 5%, find the distance of the normal corn from the genetically engineered corn. Round to the nearest tenth.

For exercises 57–58, scientists studied the possible cross-pollination of corn plants by genetically engineered corn plants planted *downwind*. The function $p = 15.38(e^{-0.6488d})$ represents the relationship of the distance in meters between the normal corn and the genetically engineered corn, d, and the percent of the normal corn that was cross-pollinated by the genetically engineered corn, p. (*Source:* B. L. Ma, "Frequency of Pollen Drift in Genetically Engineered Corn," 2005)

57. If the percent of cross-fertilization is 2%, find the distance of the normal corn from the genetically engineered corn. Round to the nearest tenth.

58. If the percent of cross-fertilization is 5%, find the distance of the normal corn from the genetically engineered corn. Round to the nearest tenth.

For exercises 59–62, the function $N = 2.09(2.7^t)$ represents the relationship of the time in years since 2002, t, and the net income of a real estate investment company in millions of dollars, N.

59. Find the number of years when the annual net income of this company will be $5902 million. Round to the nearest whole number.

60. Find the number of years when the annual net income of this company will be $43,031 million. Round to the nearest whole number.

61. Find the annual net income in 2015 in *billions* of dollars. Round to the nearest thousandth.

62. Find the annual net income in 2014 in *billions* of dollars. Round to the nearest thousandth.

For exercises 63–66, the function $C = 140(e^{-0.095t})$ represents the relationship of the concentration of an antibiotic in the blood in milliliters per milliliter, C, and the time since the antibiotic was given in hours, t.

63. Find the time when the concentration is $\dfrac{105 \text{ mg}}{1 \text{ mL}}$. Round to the nearest whole number.

64. Find the time when the concentration is $\dfrac{116 \text{ mg}}{1 \text{ mL}}$. Round to the nearest whole number.

65. Find the concentration of the antibiotic after 4 hr. Round to the nearest whole number.

66. Find the concentration of the antibiotic after 5 hr. Round to the nearest whole number.

For exercises 67–70, the function $y = 3^{6t}$ represents the relationship of the population of mice, y, and the time in years, t.

67. Find the time for the population to reach 1000 mice. Round to the nearest whole number.

68. Find the time for the population to reach 1,000,000 mice. Round to the nearest whole number.

69. Find the doubling time for the population of mice in *weeks* (1 year = 52 weeks). Round to the nearest tenth.

70. Find the population of mice after 11 *months*. Round to the nearest whole number.

71. The function $y = 535,000(1.09^t)$ represents the relationship of the time in years, t, and the population of a country, y. Find the doubling time of the population. Round to the nearest whole number.

72. The function $y = 45,000(1.06^t)$ represents the relationship of the time in years, t, and the population of a country, y. Find the doubling time of the population. Round to the nearest whole number.

73. The function $C = 140(e^{-0.095t})$ represents the relationship of the time after an antibiotic is given in hours, t, and the concentration of the antibiotic in the blood in milligrams per milliliter, C. Find the half-life of the antibiotic in hours and minutes. Round to the nearest minute.

74. The function $C = 140(e^{-0.1925t})$ represents the relationship of the time after an antibiotic is given in hours, t, and the concentration of the antibiotic in the blood in milligrams per milliliter, C. Find the half-life of the antibiotic in hours and minutes. Round to the nearest minute.

For exercises 75–76, the half-life of prostate-specific antigen (PSA) may indicate the aggressiveness of a given prostate cancer. The formula $D = \dfrac{t \ln(2)}{\ln(P_2) - \ln(P_1)}$ represents the relationship between the time between PSA measurements in years, t, the first PSA measurement in nanograms per milliliter, P_1, the second PSA measurement in nanograms per milliliter, P_2, and the doubling time in years, D.

75. The first PSA reading for a patient was 0.40 nanograms per milliliter. Thirteen months later (1.08 years), his PSA reading was 0.52 nanograms per milliliter. Find the PSA doubling time for this patient in years and months. Round to the nearest month.

76. The first PSA reading for a patient was 0.40 nanograms per milliliter. Thirteen months later (1.08 years), his PSA reading was 0.6 nanograms per milliliter. Find the PSA doubling time for this patient in years and months. Round to the nearest month.

For exercises 77–78, the rule of 72 is used to estimate doubling time for investments that are compounded annually at annual interest rates from 6% to 10%. To find the doubling time in years, divide 72 by the annual interest rate in percent form.

77. Use the rule of 72 to estimate the doubling time for an investment of $1000 that is invested at 8% compounded annually.

78. Use the rule of 72 to estimate the doubling time for an investment of $1000 that is invested at 6% compounded annually.

79. Use $A = P\left(1 + \dfrac{r}{n}\right)^{nt}$ to find the doubling time in years and months for the investment in exercise 77. Round to the nearest month.

80. Use $A = P\left(1 + \dfrac{r}{n}\right)^{nt}$ to find the doubling time in years and months for the investment in exercise 78. Round to the nearest month.

81. Iridium-192 is a radioactive element often used in "seeds" planted within the body to treat cancer. The function $y = A(e^{-0.0094t})$ represents the relationship of the time in days, t, and the amount of iridium-192, y. Find the half-life of iridium-192 in days and hours. Round to the nearest hour.

82. Copper-64 is a radioactive element used in positron emission tomography (PET) scanning. The function $y = A(e^{-0.05458t})$ represents the relationship of the time in hours, t, and the amount of copper-64, y. Find the half-life of copper-64 in hours and minutes. Round to the nearest minute.

Problem Solving: Practice and Review

Follow your instructor's guidelines for using the five steps as outlined in Section 1.5, p. 51.

83. A small Midwest regional trucking company has 20 semi-trailer trucks. At a speed of 70 miles per hour, each truck travels an average of 5.5 miles per gallon of diesel. At a speed of 65 miles per hour, each truck travels an average of 6.0 mi per gallon of diesel. On average, a truck travels 2800 mi per week, 52 weeks a year. If the cost of diesel is $\dfrac{\$4.03}{1 \text{ gal}}$, how much will the company save in fuel costs per year if the trucks travel at 65 mi per hour instead of 70 mi per hour? Round to the nearest hundred. (*Source:* Gary Kooiker, professional semi-trailer truck driver)

84. In Hernando County, Florida, 4 out of 25 employees have a job in education or health services. In 2006, there were 40,008 employees in the county. Find the number of these employees with jobs in education or health services. Round to the nearest hundred. (*Source:* www.eflorida.com)

85. Fertilizer A contains 7% phosphoric pentoxide. Fertilizer B contains 52% phosphoric pentoxide. Find the amount of each fertilizer to mix to make 25,000 lb of a new fertilizer that contains 32% phosphoric pentoxide. Round to the nearest hundred.

86. On June 15, 2009, the New York City Council asked the New York State Legislature to raise the city sales tax from 4 percent to 4.5 percent. This increase is predicted to raise $518 million in the next fiscal year. Find the amount of total taxable sales in billions of dollars needed to generate an increase of $518 million. Round to the nearest billion. (*Source:* www.nytimes.com, June 15, 2009)

Technology

For exercises 87–90, use a graphing calculator and intersection to solve the equation (see Section 6.5).
(a) Sketch the graph; describe the window.
(b) Identify the solution. If the solution is irrational, round to the nearest ten-thousandth.

87. $2^x = 32$

88. $3^x = 27$

89. $4^x = 30$

90. $2^x = 30$

Find The Mistake

For exercises 91–94, the completed problem has one mistake.
(a) Describe the mistake in words, or copy down the whole problem and highlight or circle the mistake.
(b) Do the problem correctly.

91. Problem: Solve $6^x = 7776$. Round to the nearest ten-thousandth.

Incorrect Answer: $6^x = 7776$

$$\ln(6^x) = \ln(7776)$$
$$x \ln(6) = \ln(7776)$$
$$x = \ln\left(\frac{7776}{6}\right)$$
$$x \approx 7.1670$$

92. Problem: Solve $0.5 = e^{-0.04t}$. Round to the nearest ten-thousandth.

Incorrect Answer: $0.5 = e^{-0.04t}$

$$\log(0.5) = \log(e^{-0.04t})$$
$$\log(0.5) = -0.04t \cdot \log(e)$$
$$\frac{\log(0.5)}{-0.04 \cdot \log(e)} = t$$
$$\frac{-0.3010\ldots}{(-0.04)(1)} = t$$
$$7.5257 \approx t$$

93. Problem: The radioactive element californium-252 is used in cancer treatment. The function $y = A\left(e^{-0.2626t}\right)$ represents the relationship of the time in years, t, and the amount of californium-252, y. Find the half-life of californium-252. Round to the nearest tenth.

Incorrect Answer: $2 = e^{-0.2626t}$

$$\ln(2) = \ln(e^{-0.2626t})$$
$$\ln(2) = (-0.2626t)\ln(e)$$
$$\frac{0.6931\ldots}{(-0.2626)(1)} = t$$
$$-2.6 \text{ years} \approx t$$

94. Problem: An investor has $3000 to invest at an annual interest rate of 5%, compounded 12 times a year (monthly). The investor needs a future value of $12,000. How long should the money be invested? Round to the nearest tenth of a year.

Incorrect Answer: $12{,}000 = 3000(1 + 0.05)^{12t}$

$$\frac{12{,}000}{3000} = (1 + 0.05)^{12t}$$
$$4 = 1.05^{12t}$$
$$\ln(4) = \ln(1.05^{12t})$$
$$\ln(4) = 12t \cdot \ln(1.05)$$
$$\frac{\ln(4)}{12 \ln(1.05)} = t$$
$$2.4 \text{ years} \approx t$$

Review

95. Write the quotient rule of exponents.

96. Simplify: $\dfrac{x^4 y^3}{x^6 y}$

97. Write the product rule of exponents.

98. Simplify: $(x^5 y^6)(x^3 y^2)$

SUCCESS IN COLLEGE MATHEMATICS

99. Describe your plan for studying for the final exam. Include when you will study and what you will do.

Leaf blowers are loud. In Palo Alto, California, gas-powered leaf blowers are banned, and electric leaf blowers can be used only during certain hours. The intensity of the sound of a leaf blower is measured in watts per square meter. In this section, we will find this intensity by solving a logarithmic equation.

© jocicalek/Shutterstock

SECTION 8.7

Logarithmic Equations

After reading the text, working the practice problems, and completing assigned exercises, you should be able to:

1. Solve a logarithmic equation with one logarithm.

2. Solve a logarithmic equation with more than one logarithm.

3. Solve an application using a logarithmic equation.

Logarithmic Equations

When we find the logarithm of a number, this number is the **argument**. In the equation $\log_3(9) = 2$, the base is 3, the logarithm is 2, and the argument is 9. We need a method to solve equations such as $\log_2(x) = 3$, in which the argument is a variable.

> **Solving a Logarithmic Equation in One Variable (the variable is the argument)**
>
> **1.** Identify the base, logarithm, and argument: $\log_{\text{base}}(\text{argument}) = \text{logarithm}$.
> **2.** Rewrite the equation in exponential form: $\text{base}^{\text{logarithm}} = \text{argument}$.
> **3.** Solve by isolating the variable.
> **4.** Check.

EXAMPLE 1   Solve: $\log_2(x) = 3$

SOLUTION ▶

$\log_2(x) = 3$ Base = 2, logarithm = 3, argument = x

$2^3 = x$ $\text{base}^{\text{logarithm}} = \text{argument}$

$\mathbf{8} = x$ The solution of the equation.

In the next example, the solution is irrational.

EXAMPLE 2   Solve $\ln(x) = 4$. Round the solution to the nearest ten-thousandth.

SOLUTION ▶

$\ln(x) = 4$ Base = e, logarithm = 4, argument = x

$e^4 = x$ $\text{base}^{\text{logarithm}} = \text{argument}$

$54.59815\ldots = x$ The solution is irrational.

$\mathbf{54.5982} \approx x$ Round.

Since the domain of $y = \log_b(x)$ is $(0, \infty)$, the logarithm of a negative number is not a real number. However, the *solution* of a logarithmic equation can be a negative number.

EXAMPLE 3 (a) Solve: $\log_4(-2x) = 3$

SOLUTION ►

$$\log_4(-2x) = 3 \qquad \text{Base} = 4, \text{logarithm} = 3, \text{argument} = -2x$$

$$4^3 = -2x \qquad \text{base}^{\text{logarithm}} = \text{argument}$$

$$64 = -2x \qquad \text{Evaluate; } 4^3 = 64$$

$$\frac{64}{-2} = \frac{-2x}{-2} \qquad \text{Division property of equality.}$$

$$-32 = x \qquad \text{Simplify.}$$

(b) Check.

$$\log_4(-2x) = 3 \qquad\qquad \text{Check in the original equation.}$$

$$\log_4(-2(-32)) = 3 \qquad\qquad \text{Replace the variable, } x, \text{ with the solution.}$$

$$\log_4(64) = 3 \qquad\qquad \text{Evaluate; } (-2)(-32) = 64$$

$$\frac{\log(64)}{\log(4)} = 3 \qquad\qquad \text{Change of base formula.}$$

$$\frac{1.80617\ldots}{0.60205\ldots} = 3 \qquad\qquad \text{Evaluate the logarithms.}$$

$$3 = 3 \quad \text{True.} \qquad \text{The solution is correct.}$$

In the next example, the equation includes the logarithm of a quotient.

EXAMPLE 4 Solve $\log\left(\dfrac{x}{100}\right) = -3.5$. Round to the nearest ten-thousandth.

SOLUTION ►

$$\log\left(\frac{x}{100}\right) = -3.5$$

$$10^{-3.5} = \frac{x}{100} \qquad\qquad \text{base}^{\text{logarithm}} = \text{argument}$$

$$0.00031622\ldots = \frac{x}{100} \qquad\qquad \text{Simplify.}$$

$$(100)(0.00031622\ldots) = (100)\left(\frac{x}{100}\right) \qquad \text{Multiplication property of equality.}$$

$$0.031622\ldots = x \qquad\qquad \text{Simplify}$$

$$0.0316 \approx x \qquad\qquad \text{Round.}$$

Practice Problems

For problems 1–4,
(a) solve each equation. If the solution is irrational, round to the nearest ten-thousandth.
(b) check.

1. $\log_4(x) = 5$ 2. $\ln(x) = 3$ 3. $\log(-3x) = 2.1$ 4. $\log\left(\dfrac{x}{10}\right) = 2.5$

Equations with More Than One Logarithm

To solve an equation that includes a sum of logarithms, use the product rule of logarithms to rewrite the sum as a single logarithm.

> ## Product Rule of Logarithms
>
> $$\log_b(m) + \log_b(n) = \log_b(m \cdot n)$$
>
> If the bases are the same, the sum of the logarithms equals the logarithm of the product. The base, b, is a real number greater than 0 that is not equal to 1; m and n are real numbers greater than 0.

EXAMPLE 5 (a) Solve: $\log_2(x) + \log_2(4) = 5$

SOLUTION ▶

$\log_2(x) + \log_2(4) = 5$	The equation has more than one logarithm.
$\log_2(4x) = 5$	Product rule of logarithms.
$2^5 = 4x$	$\text{base}^{\text{logarithm}} = \text{argument}$
$32 = 4x$	Evaluate 2^5.
$\dfrac{32}{4} = \dfrac{4x}{4}$	Division property of equality.
$8 = x$	Simplify.

(b) Check.

$\log_2(x) + \log_2(4) = 5$	The original equation.
$\log_2(8) + \log_2(4) = 5$	Replace the variable x, with the solution.
$\dfrac{\log(8)}{\log(2)} + \dfrac{\log(4)}{\log(2)} = 5$	Change of base formula.
$\dfrac{0.90308\ldots}{0.30102\ldots} + \dfrac{0.602059\ldots}{0.30102\ldots} = 5$	Evaluate logarithms; do not round.
$3 + 2 = 5$	Divide.
$5 = 5$ True.	The solution is correct.

To solve an equation that includes a difference of logarithms, use the quotient rule of logarithms to rewrite the difference as a single logarithm.

> ## Quotient Rule of Logarithms
>
> $$\log_b(m) - \log_b(n) = \log_b\left(\frac{m}{n}\right)$$
>
> If the bases are the same, the difference of the logarithms is the logarithm of the quotient. The base, b, is a real number greater than 0 that is not equal to 1; m and n are real numbers greater than 0.

EXAMPLE 6 Solve $\ln(x) - \ln(0.5) = 6$. Round to the nearest ten-thousandth.

SOLUTION ▶

$\ln(x) - \ln(0.5) = 6$	The equation has more than one logarithm.
$\ln\left(\dfrac{x}{0.5}\right) = 6$	Quotient rule of logarithms.
$e^6 = \dfrac{x}{0.5}$	$\text{base}^{\text{logarithm}} = \text{argument}$
$(0.5)(e^6) = (0.5)\left(\dfrac{x}{0.5}\right)$	Multiplication property of equality.
$(0.5)(e^6) = x$	Simplify.

$$(0.5)(403.42879\ldots) = x \quad \text{Follow the order of operations; do not round.}$$
$$201.71439\ldots = x \quad \text{Simplify.}$$
$$201.7144 \approx x \quad \text{Round.}$$

In the next example, we use the power rule and the quotient rule of logarithms to rewrite the equation with a single logarithm.

> **Power Rule of Logarithms**
>
> $$\log_b(m^p) = p\log_b(m)$$
>
> When raising a logarithm to a power, multiply the power and the logarithm. The base, b, is a real number greater than 0 that is not equal to 1; m is a real number greater than 0; and p is a real number.

EXAMPLE 7 | Solve $4\log(x) - \log(x^3) = 5.2$. Round to the nearest whole number.

SOLUTION ▶

$$4\log(x) - \log(x^3) = 5.2$$
$$\log(x^4) - \log(x^3) = 5.2 \quad \text{Power rule of logarithms.}$$
$$\log\left(\frac{x^4}{x^3}\right) = 5.2 \quad \text{Quotient rule of logarithms.}$$
$$\log(x) = 5.2 \quad \text{Simplify.}$$
$$10^{5.2} = x \quad \text{base}^{\text{logarithm}} = \text{argument}$$
$$158{,}489.3\ldots = x \quad \text{Simplify.}$$
$$158{,}489 \approx x \quad \text{Round.}$$

Practice Problems

For problems 5–8,
(a) solve. Round irrational solutions to the nearest ten-thousandth.
(b) check.

5. $\log_6(3) + \log_6(x) = 4$ 6. $\log(x) - \log(5) = 2$
7. $\log(x^6) - 5\log(x) = 4$ 8. $\ln(6x) - \ln(3) = 1.2$

Applications of Logarithms

Given a pH, we can solve the equation $\text{pH} = -\log[\text{H}_3\text{O}^+]$ and find the hydronium ion concentration, $[\text{H}_3\text{O}^+]$.

EXAMPLE 8 | The pH of a solution is 8.5. Find the hydronium ion concentration in moles per liter. Use scientific notation. Round the mantissa to the nearest tenth.

SOLUTION ▶

$$\text{pH} = -\log[\text{H}_3\text{O}^+]$$
$$8.5 = -\log[\text{H}_3\text{O}^+] \quad \text{Replace pH with 8.5.}$$
$$\frac{8.5}{-1} = \frac{-\log[\text{H}_3\text{O}^+]}{-1} \quad \text{Division property of equality.}$$
$$-8.5 = \log[\text{H}_3\text{O}^+] \quad \text{Simplify; the logarithm is isolated.}$$
$$10^{-8.5} = [\text{H}_3\text{O}^+] \quad \text{base}^{\text{logarithm}} = \text{argument}$$
$$3.16 \times 10^{-9} = [\text{H}_3\text{O}^+] \quad \text{Evaluate the logarithm; use scientific notation.}$$
$$\frac{3.2 \times 10^{-9}\,\text{mole}}{1\,\text{L}} \approx [\text{H}_3\text{O}^+] \quad \text{Round; the units are moles per liter.}$$

In Section 8.5, we used the formula $L = 10 \log\left(\dfrac{I}{1 \times 10^{-12}}\right)$ to find the loudness of a sound, L. In the next example, we use the formula to find the intensity of a sound, I.

EXAMPLE 9 The loudness of a leaf blower is 115 decibels. Find the intensity of this sound in watts per square meter. Write the answer in scientific notation.

SOLUTION ▶

$$L = 10 \log\left(\frac{I}{1 \times 10^{-12}}\right)$$

$$115 = (10)\log\left(\frac{I}{1 \times 10^{-12}}\right) \qquad \text{Do not use units;}$$
replace L with 115.

$$\frac{115}{10} = \frac{(10)\log\left(\dfrac{I}{1 \times 10^{-12}}\right)}{10} \qquad \text{Isolate } \log\left(\dfrac{I}{1 \times 10^{-12}}\right);$$
division property
of equality.

$$11.5 = \log\left(\frac{I}{1 \times 10^{-12}}\right) \qquad \text{Simplify.}$$

$$10^{11.5} = \frac{I}{1 \times 10^{-12}} \qquad \text{base}^{\text{logarithm}} = \text{argument}$$

$$3.16227\ldots \times 10^{11} = \frac{I}{1 \times 10^{-12}} \qquad \text{Simplify; scientific notation.}$$

$$(1 \times 10^{-12})(3.16227\ldots \times 10^{11}) = (1 \times 10^{-12})\left(\frac{I}{1 \times 10^{-12}}\right) \qquad \begin{array}{l}\text{Multiplication}\\\text{property of equality.}\end{array}$$

$$3.162\ldots \times 10^{-1} = I \qquad \text{Simplify.}$$

$$\frac{3.2 \times 10^{-1} \text{ watt}}{1 \text{ m}^2} \approx I \qquad \text{Round; the units are watts per square meter.}$$

An important formula for describing chemical reactions is $\ln\left(\dfrac{A_t}{A_0}\right) = -kt$, where the concentration of a substance in moles per liter at the beginning of a reaction is A_0, the concentration of the substance in moles per liter when the time is t is A_t, and k is the rate constant for the reaction. The unit of measurement for k is the reciprocal of the unit of time. Since the units of concentration for A_t and A_0 are the same, the units on the left side of the formula equal 1. Since the product of a unit and its reciprocal is 1, the units on the right side also equal 1. Since the units simplify to 1, they are not included in the next example.

EXAMPLE 10 In a reaction of hydrogen peroxide in which the amount of hydrogen peroxide decreases over time, the original concentration of hydrogen peroxide is $\dfrac{0.1 \text{ mole}}{1 \text{ L}}$ and $k = \dfrac{0.012}{\text{s}}$. Find the concentration of hydrogen peroxide after 60 s. Round to the nearest hundredth.

SOLUTION ▶

$$\ln\left(\frac{A_t}{A_0}\right) = -kt$$

$$\ln\left(\frac{A_t}{0.1}\right) = -(0.012)(60) \qquad \text{Do not use units; replace variables.}$$

$$\ln\left(\frac{A_t}{0.1}\right) = -0.72 \qquad \text{Simplify.}$$

$$e^{-0.72} = \frac{A_t}{0.1} \qquad \text{base}^{\text{logarithm}} = \text{argument}$$

$$(0.1)e^{-0.72} = (0.1)\frac{A_t}{0.1} \qquad \text{Multiplication property of equality.}$$

$$(0.1)(\mathbf{0.4867\ldots}) = A_t \qquad \text{Simplify.}$$

$$\mathbf{0.04867\ldots} = A_t \qquad \text{Simplify.}$$

$$\frac{\mathbf{0.05 \ mole}}{\mathbf{1 \ L}} \approx A_t \qquad \text{Round; the units are moles per liter.}$$

Practice Problems

9. The pH of a solution is 4.2. Find the concentration of hydronium ions in this solution in moles per liter. Use scientific notation; round the mantissa to the nearest tenth.

10. The loudness of a motorcycle is 85 decibels. Find the intensity of this sound in watts per square meter. Use scientific notation; round the mantissa to the nearest tenth.

11. The original concentration of dinitrogen pentoxide is $\dfrac{4.5 \times 10^{-3} \ \text{mole}}{1 \ \text{L}}$ and $k = \dfrac{0.146}{1 \ \text{s}}$. Find the concentration of dinitrogen pentoxide after 60 s. Use scientific notation; round the mantissa to the nearest hundredth.

Using Technology: Solving a Logarithmic Equation

To solve an equation in one variable by graphing, we rewrite the equation as two functions. The solution is the x-coordinate of the intersection point of the graphs of the functions (Section 6.5). We can use this same strategy to solve a logarithmic equation such as $\log(x) = 2.5$.

EXAMPLE 11 Solve $\log(x) = 2.5$ by graphing. Round to the nearest tenth.

In a standard window, the horizontal line $y = 2.5$ and the curve $y = \log(x)$ do not intersect. Zooming out once still does not show the point of intersection.

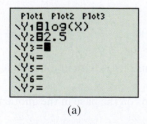

(a)   (b)   (c)   (d)

To see the intersection point, adjust the window by changing values in the window screen. Keep adjusting until the intersection point is clearly visible. The solution of the equation is the x-coordinate of the intersection point, rounded to the nearest tenth, $x \approx 316.2$.

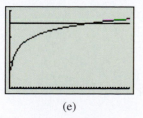

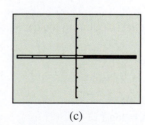
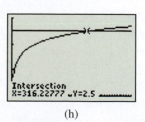

(e)   (f)   (g)   (h)

Practice Problems For problems 12–14, solve using the graphical method.
(a) Sketch the graph; describe the window.
(b) Identify the solution; round to the nearest tenth.

12. $\log(x) = 1.6$ **13.** $\log(x) = 3.9$ **14.** $\log(x) = -1$

8.7 VOCABULARY PRACTICE

Match the term with its description.

1. $\log_b(x) = \dfrac{\log_a(x)}{\log_a(b)}$

2. $-\log[H_3O^+]$

3. $\log_b(m) - \log_b(n) = \log_b\left(\dfrac{m}{n}\right)$

4. $\log_b(m^p) = p\log_b(m)$

5. $\log_b(m) + \log_b(n) = \log_b(mn)$

6. Unit of measurement of loudness

7. Unit of measurement of sound intensity

8. $(x^m)^n = x^{mn}$

9. $x^m \cdot x^n = x^{m+n}$

10. $\dfrac{x^m}{x^n} = x^{m-n}$

A. change of base formula
B. decibel
C. pH
D. power rule of exponents
E. power rule of logarithms
F. product rule of exponents
G. product rule of logarithms
H. quotient rule of exponents
I. quotient rule of logarithms
J. watts per square meter

8.7 Exercises

Follow your instructor's guidelines for showing your work.

1. A logarithmic equation is $\log(x) = 3$. Identify the base of the logarithm.

2. A logarithmic equation is $\ln(x) = 2$. Identify the base of the logarithm.

For exercises 3–22, solve. Round irrational solutions to the nearest ten-thousandth.

3. $\log_5(x) = 2$

4. $\log_4(x) = 2$

5. $\log(x) = 3.5$

6. $\log(x) = 4.2$

7. $\ln(x) = 3.5$

8. $\ln(x) = 4.2$

9. $7 = \log_2(x)$

10. $3 = \log_6(x)$

11. $\log_6(2x) = 3$

12. $\log_8(2x) = 3$

13. $\log_6(-2x) = 3$

14. $\log_8(-2x) = 3$

15. $\ln(3x) = 1$

16. $\ln(5x) = 1$

17. $\log(5x) = -3$

18. $\log(2x) = -1$

19. $\log(x) = -3$

20. $\log(x) = -1$

21. $\log_9(x) = -2$

22. $\log_8(x) = -2$

For exercises 23–34, rewrite the expression as a single logarithm. Simplify but do not evaluate the logarithms.

23. $\log_5(x) + \log_5(2)$

24. $\log_2(7) + \log_2(x)$

25. $\log_5(x) - \log_5(2)$

26. $\log_2(7) - \log_2(x)$

27. $\log_5(x) + \log_5(2) - \log_5(9)$

28. $\log_2(7) + \log_2(x) - \log_2(3)$

29. $4\log_5(x) + \log_5(2)$

30. $3\log_2(x) + \log_2(7)$

31. $4\log_5(x) - \log_5(2)$

32. $3\log_2(x) - \log_2(7)$

33. $\log(6x) - \log(x^2) + \log(3)$

34. $\log(10x^2) - \log(x^3) + \log(5)$

35. Describe the restrictions on the base in the product rule of logarithms.

36. Describe the restrictions on the base in the quotient rule of logarithms.

For exercises 37–58,
(a) solve. If the solution is irrational, round to the nearest ten-thousandth.
(b) check.

37. $\log_6(x) + \log_6(2) = 4$

38. $\log_3(x) + \log_3(9) = 4$

39. $\log_6(x) - \log_6(2) = 4$

40. $\log_3(x) - \log_3(9) = 4$

41. $\log_6(216) - \log_6(x) = 4$

42. $\log_3(27) - \log_3(x) = 4$

43. $\log(4) + \log(x) = 2$

44. $\log(5) + \log(x) = 3$

45. $\log(x) - \log(4) - \log(3) = 2$

46. $\log(x) - \log(4) - \log(2) = 3$

47. $\ln(x) + \ln(5) = 3$

48. $\ln(x) + \ln(7) = 4$

49. $\log(x^7) - 6\log(x) = 2$

50. $\log(x^8) - 7\log(x) = 3$

51. $\log_4(8x) - \log_4(4) = 5$

52. $\log_2(12x) - \log_2(3) = 5$

53. $\log(5x) + \log(5) + \log(2) = 3$

54. $\log(2x) + \log(5) + \log(2) = 3$

55. $\log(100) - \log(1000) = 2x$

56. $\log(1000) - \log(10,000) = 2x$

57. $\log_4(16) + \log_3(9) = 4x$

58. $\log_2(16) + \log_5(25) = 3x$

For exercises 59–68, rewrite the expression as a sum or difference of logarithms. Do not evaluate the logarithms.

59. $\log(4x)$

60. $\log(6x)$

61. $\log(4x)^2$

62. $\log(6x)^2$

63. $\log\left(\dfrac{x}{4}\right)$

64. $\log\left(\dfrac{x}{6}\right)$

65. $\log\left(\dfrac{x^2}{4}\right)$

66. $\log\left(\dfrac{x^3}{6}\right)$

67. $\log\left(\dfrac{I}{1 \times 10^{-12}}\right)$

68. $\ln\left(\dfrac{A_t}{A_0}\right)$

For exercises 69–72, evaluate. Write the number in scientific notation, and round the mantissa to the nearest tenth.

69. $10^{-8.3}$

70. $10^{-7.9}$

71. $10^{-4.2}$

72. $10^{-5.1}$

73. The pH of a solution is 8.9. Find the hydronium ion concentration in moles per liter. Use scientific notation. Round the mantissa to the nearest tenth.

74. The pH of a solution is 9.2. Find the hydronium ion concentration in moles per liter. Use scientific notation. Round the mantissa to the nearest tenth.

75. The loudness of a dial tone is 81 decibels. Find the intensity of this sound in watts per square meter. Write in scientific notation. Round the mantissa to the nearest tenth.

76. The loudness of a train whistle is 91 decibels. Find the intensity of this sound in watts per square meter. Round the mantissa to the nearest tenth.

77. The loudness of a power mower is 107 decibels. Find the intensity of this sound in watts per square meter. Round the mantissa to the nearest tenth.

78. The loudness of a power saw is 109 decibels. Find the intensity of this sound in watts per square meter. Round the mantissa to the nearest tenth.

79. The original concentration of benzenediazonomium chloride is $\dfrac{0.2 \text{ mole}}{1 \text{ L}}$ and $k = \dfrac{0.0651}{1 \text{ min}}$. Find the concentration of benzenediazonomium chloride after 30 min. Round to the nearest hundredth.

80. The original concentration of benzenediazonomium chloride is $\dfrac{0.2 \text{ mole}}{1 \text{ L}}$ and $k = \dfrac{0.0651}{1 \text{ min}}$. Find the concentration of benzenediazonomium chloride after 15 min. Round to the nearest hundredth.

81. The original concentration of methyl acetate is $\dfrac{0.3 \text{ mole}}{1 \text{ L}}$ and $k = \dfrac{3.36 \times 10^{-3}}{1 \text{ min}}$. Find the concentration of methyl acetate after 60 min. Round to the nearest hundredth.

82. The original concentration of methyl acetate is $\dfrac{0.3 \text{ mole}}{1 \text{ L}}$ and $k = \dfrac{3.36 \times 10^{-3}}{1 \text{ min}}$. Find the concentration of methyl acetate left after 90 min. Round to the nearest hundredth.

Problem Solving: Practice and Review

Follow your instructor's guidelines for using the five steps as outlined in Section 1.5, p. 51.

83. Find the percent increase in the amount that executive pay is compared to the income of an average worker.

In 1970, average executive pay at the nation's top companies was 28 times average worker income. . . . By 2005, executive pay had jumped to 158 times that of the average worker. (*Source:* seattletimes.nwsource.com, June 24, 2011)

84. Find the number of jobs in the project area in 2007. Round to the nearest thousand.

Since 2007, employment in the project area has declined by nearly 66,000 jobs, about 14 percent. Loss of employment has been seen across the country, but the 14 percent decrease within the project area is above the national average and higher than the Chicago metro area average of 7 percent for the same period. (*Source:* www.elginoharewestbypass.org, June 30, 2011)

85. The function $P(t) = 9,925,640(e^{0.01603t})$ represents the relationship of the number of years since 2011, t, and the population of Somalia, $P(t)$. Find the population of Somalia in 2015. Round to the nearest thousand. (*Source:* www.cia.gov)

86. According to a report by the Pew Foundation, the Millennial generation includes people in the United States born after 1980. In 2010, there were about 50 million people in this generation. Use a proportion to find the number of Millennials who were married with children at home.

Just one-in-five Millennials is currently married . . . and just one-in-eight . . . is married with children at home. (*Source:* www.pewsocialtrends.org, Feb. 2010.)

Technology

For exercises 87–90, use a graphing calculator to solve. Adjust the window as needed.
(a) Sketch the graph; describe the window.
(b) Identify the solution. If it is not an integer, round to the nearest ten-thousandth.

87. $\log(2x) = 0.5$

88. $\log(4x) = 0.8$

89. $\ln(2x) = 4$

90. $\ln(4x) = 3$

Find the Mistake

For exercises 91–94, the completed problem has one mistake.
(a) Describe the mistake in words, or copy down the whole problem and highlight or circle the mistake.
(b) Do the problem correctly.

91. Problem: Solve: $\log_4(x) = 3$

   Incorrect Answer: $\log_4(x) = 3$

$$x = 3^4$$
$$x = 81$$

92. Problem: Solve: $\log(x) + \log(4) = 2$

   Incorrect Answer: $\log(x) + \log(4) = 2$

$$\log(x + 4) = 2$$
$$10^2 = x + 4$$
$$100 = x + 4$$
$$96 = x$$

93. Problem: Rewrite $\log(x) - \log(2)$ as a single logarithm.

   Incorrect Answer: $\log(x) - \log(2)$

$$= \frac{\log(x)}{\log(2)}$$

94. Problem: Solve: $\log_3(x) = -2$

   Incorrect Answer: No real number solution. Logarithms are not negative numbers.

Review

95. Find the slope of the line that passes through $(-8, 1)$ and $(4, -3)$.

96. Write the equation in slope-intercept form of the line that passes through $(5, 2)$ and $(2, 17)$.

97. Graph: $y = -\frac{3}{4}x + 7$

98. Graph: $3x - 8y = 24$

SUCCESS IN COLLEGE MATHEMATICS

99. Name the two concepts, topics, or procedures that you think that you need to review the most before the final exam.

100. On a scale of 1 to 10, where 1 is almost no anxiety and 10 is overwhelming anxiety, rate your anxiety about taking the final exam.

Study Plan for Review of Chapter 8

SECTION 8.1   Operations with Functions and One-to-One Functions

Ask Yourself	Test Yourself	Help Yourself
Can I ...	**Do 8.1 Review Exercises**	**See these Examples and Practice Problems**
add or subtract functions?	1, 2, 5	Ex. 1–4, PP 1–5
write a revenue function, cost function, and profit function?	11	Ex. 5, PP 6–8
multiply two functions?	3	Ex. 6, 7, PP 9–11
compose two functions?	4, 6, 7	Ex. 8–12, PP 12–15
determine whether a function in roster notation is one-to-one	8	Ex. 13, PP 16, 17
determine whether a graph represents a one-to-one function?	9, 10	Ex. 14, 15, PP 18–22

8.1 Review Exercises

For exercises 1–4, $f(x) = 7x + 6$ and $g(x) = 3x - 5$. Find:

1. $f(x) + g(x)$

2. $f(x) - g(x)$

3. $f(x) \cdot g(x)$

4. $f(g(x))$

5. If $f(x) = \dfrac{x}{x + 6}$ and $g(x) = \dfrac{8}{2x + 12}$, find $f(x) + g(x)$.

6. a. Find $h(k(x))$ if $h(x) = x - 2$ and $k(x) = x^2 + 8x + 4$.
   b. Rewrite $h(k(x))$ in vertex form.
   c. Use interval notation to represent the domain of $h(k(x))$.
   d. Use interval notation to represent the range of $h(k(x))$.

7. If $f(x) = \sqrt{x}$ and $g(x) = \sqrt[5]{x^3}$, find $g(f(x))$.

8. Write a function with four ordered pairs in roster notation that is one-to-one.

9. A horizontal line intersects the graph of a function in two points. Explain why this shows that the function is not one-to-one.

10. The function $y = 3$ is not one-to-one. Explain why.

11. The cost function for a business making a product is $C(x) = \$4000x + \$150{,}000$, where x is the number of products and $C(x)$ is the cost of making x products. The revenue function for this product is $R(x) = \$2100x$, where $R(x)$ is the revenue collected from selling x products.
   a. Write a profit function for this product.
   b. Will this company make a profit if they make and sell this product? Explain.

SECTION 8.2   Exponential Functions

Ask Yourself	Test Yourself	Help Yourself
Can I ...	**Do 8.2 Review Exercises**	**See these Examples and Practice Problems**
identify the domain and range of an exponential function?	12, 13	Ex. 1, 2, 8, PP 1, 2
identify the y-intercept of an exponential function?	12, 13, 17	Ex. 1, 2, 8, PP 1, 2, 11, 12
evaluate an exponential function?	12, 13, 15	Ex. 1, 2, 4–7, PP 5–8, 9, 10, 13
describe how the graph of an exponential function changes when the base is changed?	14	Ex. 3, PP 3, 4
identify an exponential function that is increasing?	12, 13, 19	Ex. 3, PP 1
identify an exponential function that is decreasing?	12, 13	Ex. 2, 3, PP 2
write an approximate value for the number *e*?	16	Ex. 4

8.2 Review Exercises

For exercises 12–13,
(a) use interval notation to represent the domain of the function.
(b) use interval notation to represent the range of the function.
(c) identify the y-intercept.
(d) is this function one-to-one?
(e) evaluate $f(3)$.
(f) is this function increasing or decreasing?
(g) evaluate $f(0)$.

12. $f(x) = 7^x$

13. $f(x) = 0.7^x$

14. Describe how the graph of $y = 6^x$ is different from the graph of:
 a. $y = 3^x$
 b. $y = 10^x$
 c. $y = 0.6^x$

15. When $f(x) = e^x$, find $f(4)$. Round to the nearest ten-thousandth.

16. Is *e* greater than or less than π?

17. Explain why the y-intercept of the exponential function $y = b^x$ is $(0, 1)$.

18. Does the graph of an exponential function pass the horizontal line test?

19. Identify the exponential functions that are increasing.
 a. $y = 3^x$
 b. $y = 0.3^x$
 c. $f(x) = e^x$
 d. $f(x) = \pi^x$
 e. $f(x) = \left(\dfrac{1}{5}\right)^x$

SECTION 8.3   Inverse Functions and Logarithmic Functions

Ask Yourself	Test Yourself	Help Yourself
Can I ...	**Do 8.3 Review Exercises**	**See these Examples and Practice Problems**
verify that two functions are inverses?	20–22	Ex. 1, 2, PP 1–3
identify the inverse of a linear, exponential, or logarithmic function?	23, 24	Ex. 3–5, PP 4, 5
use $f^{-1}(x)$ notation?	23–25	Ex 5
evaluate a logarithm?	26, 27	Ex 6–9, PP 6–10
graph an exponential function and its inverse and the axis of symmetry?	28	Ex. 10
identify the domain, range, x-intercept, and y-intercept of an exponential or logarithmic function?	28, 29	Ex. 5, 10, PP 11, 12

8.3 Review Exercises

20. Verify that $f(x) = -\frac{1}{5}x + \frac{3}{5}$ and $g(x) = -5x + 3$ are inverse functions.

21. Explain why a quadratic function cannot have an inverse.

22. A function is $y = 3$. Does this function have an inverse? Explain.

23. Find the inverse of $f(x) = -3x + 8$. Use $f^{-1}(x)$ notation.

24. Identify the inverse of the function. Use $f^{-1}(x)$ notation.
   a. $f(x) = 20^x$
   b. $f(x) = \log_{17}(x)$

25. If $f(x)$ is one-to-one and $f(3) = 8$, evaluate $f^{-1}(8)$.

26. Identify the base of each logarithm.
   a. $\log_{11}(x)$
   b. $\log(x)$
   c. $\ln(x)$
   d. natural logarithm
   e. common logarithm

27. Evaluate. Do not use a calculator.
   a. $\log_4(256)$
   b. $\log(1)$
   c. $\log_3(9)$
   d. $\log(0)$
   e. $\log(100)$

28. a. Complete the table of ordered pairs for $y = 2^x$.

$y = 2^x$	
x	y
-2	
-1	
0	
1	
2	
3	

b. Use the output values from the table for $y = 2^x$ to complete the table of ordered pairs for $y = \log_2(x)$.

$y = \log_2(x)$	
x	y
	-2
	1
	0
	1
	2
	3

c. Sketch a graph of $y = 2^x$, $y = \log_2(x)$, and the axis of symmetry.

d. Use interval notation to represent the domain of $y = 2^x$.

e. Use interval notation to represent the range of $y = \log_2(x)$.

f. Use interval notation to represent the range of $y = 2^x$.

g. Use interval notation to represent the domain of $y = \log_2(x)$.

29. A function is $f(x) = \log_{20}(x)$.
   a. Use interval notation to represent the domain.
   b. Use interval notation to represent the range.
   c. What is the x-intercept of the graph that represents this function?
   d. Is this function one-to-one?

SECTION 8.4 **Natural Logarithms and Logarithm Rules**

Ask Yourself	Test Yourself	Help Yourself
Can I . . .	**Do 8.4 Review Exercises**	**See these Examples and Practice Problems**
use a calculator to evaluate a logarithm and explain why the logarithm is reasonable?	30	Ex. 1, 2, PP 1–10
use the change of base formula to evaluate a logarithm?	30	Ex. 3, PP 7–10
use logarithms to multiply whole numbers?	31	Ex. 4, PP 11–13
use the rules of logarithms to rewrite an expression with a single logarithm?	33–36	Ex. 7, 8, 10, 12, 14, PP 18–24, 27, 28
use the rules of logarithms to rewrite an expression with more than one logarithm?	32	Ex. 5, 6, 9, 11, 13, PP 14–17, 25, 26, 29, 30

8.4 Review Exercises

30. Evaluate. Round irrational numbers to the nearest ten-thousandth.
   a. $\ln(5)$
   b. $\log(41)$
   c. $\log_3(-4)$
   d. $\log_3(18)$

31. a. Multiply 732 and 45. Do not use a calculator.
   b. Use logarithms to multiply 732 and 45. Use a calculator to evaluate logarithms and exponential expressions.

32. Identity the expression(s) that are equivalent to $\log(2x)^3$.
  a. $3 \log(2x)$
  b. $\log 3(2x)$
  c. $(\log(2x)) \cdot 3$
  d. $\log(2^3 \cdot x^3)$
  e. $\log(8x^3)$

For exercises 33–36, rewrite each expression as a single logarithm. Do not evaluate the logarithms.

33. $\log_2(9) + \log_2(x)$

34. $\log_2(x) - \log_2(9)$

35. $5 \log(x) - \log(y) + 6 \log(z)$

36. $3 \log_8(x) - \log_8(2)$

SECTION 8.5 Applications

Ask Yourself	Test Yourself	Help Yourself
Can I...	**Do 8.5 Review Exercises**	**See these Examples and Practice Problems**
use an exponential function or formula to solve an application problem?	37, 40, 41	Ex. 1–4, PP 1–3
use a function or formula that includes a logarithm to solve an application problem?	38, 39, 42, 43	Ex. 5–7, PP 4–6

8.5 Review Exercises

37. The function $y = 20(0.85^t)$ represents the relationship of the current age of a child minus 5 years, t, and the percent of children at this age who wet the bed, y. Find the percent of 10-year-old children who wet the bed. Round to the nearest percent.

38. The International Organization for Standardization (ISO) uses a logarithmic scale and a linear scale for measuring the speed of film. The function
$$y = 3 \log_2\left(\frac{128}{100}x\right)$$
represents the relationship between the linear-scale speed, x, and the log-scale speed, y. Find the log-scale speed of a film with a linear-scale speed of 200.

39. The hydronium ion concentration of a solution is 3.8×10^{-10} mole per liter. Find the pH of this solution. Round to the nearest tenth.

40. Find the future value of an investment of $7600 that is invested at 6%, compounded monthly, for 3 years. Round to the nearest hundredth.

41. Find the future value of an investment of $7600 that is invested at 6%, compounded annually, for 3 years. Round to the nearest hundredth.

42. At a distance of 10 m from a running diesel truck, the intensity of sound is about $\dfrac{2 \times 10^{-3} \text{ watt}}{1 \text{ m}^2}$. Find the loudness of this sound in decibels. Round to the nearest whole number.

43. The function $M_w = \dfrac{2}{3} \log(M_o) - 6.05$ compares the seismic moment in newton · meters of an earthquake, M_o, and the moment magnitude (no units), M_w. The Sumatra earthquake of December 26, 2004, which caused a catastrophic tsunami, had a seismic moment of about 4.4×10^{22} newton · meters. Find the moment magnitude of this earthquake. Round to the nearest tenth. (*Source:* Richardson, E., www.e-education.psu.edu)

SECTION 8.6 Exponential Equations

Ask Yourself	Test Yourself	Help Yourself
Can I...	**Do 8.6 Review Exercises**	**See these Examples and Practice Problems**
solve an exponential equation?	44–47	Ex. 1–4, PP 1–4
use an exponential equation to solve an application problem?	48	Ex. 5, 6, PP 5, 6
write the doubling time equation?	49, 51	Ex. 7, PP 7
find the doubling time of a function?		
write the half-life equation?	50, 52	Ex. 8, PP 8
find the half-life of a function?		

8.6 Review Exercises

For exercises 44–47,
(a) solve. Round irrational solutions to the nearest ten-thousandth.
(b) check.

44. $20 = 10^x$

45. $20 = e^x$

46. $12 = 5^x$

47. $100 = 3^{2x}$

48. An investor has $2000 to invest at an annual interest rate of 6.3%, compounded monthly. Find the time that the money must be invested in years and months so that it has a future value of $3500. Round to the nearest month.

49. The function $y = A(e^{0.0103t})$ represents the relationship of the time in days, t, and the size of a large cell lung cancer tumor, y. Find the doubling time for this tumor. Round to the nearest whole number.

50. The function $y = A(e^{-0.002t})$ represents the relationship of the time of a chemical reaction in minutes, t, and the amount of dinitrogen pentoxide remaining, y. Find the half-life of this reaction. Round to the nearest whole number.

51. The doubling time for $2000 invested at 5% interest compounded daily is 13 years 10 months, rounded to the nearest month. Find the doubling time for $3500 invested at 5% interest compounded daily, rounded to the nearest month.

52. Iodine-123 is used in nuclear medicine imaging, especially for studies of the thyroid. It takes 13.2335 hr for a 5 g sample of the radioactive isotope iodine-123 to decay into a sample that only contains 2.5 g. Find the half-life of a 4 g sample of iodine-123. (*Source:* www.nist.gov; Sampson, Charles. *Textbook of Radiopharmacy, Theory and Practice*. vol. 3. Luxembourg: Gordon and Breach, 1994.)

SECTION 8.7	**Logarithmic Equations**

Ask Yourself	Test Yourself	Help Yourself
Can I . . .	**Do 8.7 Review Exercises**	**See these Examples and Practice Problems**
solve a logarithmic equation with one logarithm?	53–59	Ex. 1–4, PP 1–4
solve a logarithmic equation with more than one logarithm?	60, 61	Ex. 5–7, PP 5–8
use a logarithmic equation to solve an application problem?	62–64	Ex. 8–10, PP 9–11

8.7 Review Exercises

For exercises 53–56, rewrite the logarithmic equation as an equivalent exponential equation. Do not solve.

53. $1.2 = \log(x)$

54. $1.2 = \ln(x)$

55. $1.2 = \log_5(x)$

56. $\log_{base}(\text{argument}) = \text{logarithm}$

For exercises 57–59,
(a) solve. Round irrational solutions to the nearest ten-thousandth.
(b) check.

57. $\log_7(x) = 2.1$

58. $\ln(x) = 4$

59. $\log(x) = 3.5$

For exercises 60–61,
(a) solve. Round irrational solutions to the nearest ten-thousandth.
(b) check.

60. $\log_3(x) + \log_3(6) = 2$

61. $5\ln(x) - \ln(x^4) = 2$

62. The pH of a solution is 9.1. Find the concentration of hydronium ions in this solution in moles per liter. Use scientific notation. Round the mantissa to the nearest tenth.

63. The loudness of a dog sneeze is 65 decibels. Find the intensity of this sound in watts per square meter. Write the answer in scientific notation. Round the mantissa to the nearest tenth.

64. The function $M_w = \frac{2}{3}\log(M_o) - 6.05$ compares the seismic moment of in newton·meters of an earthquake, M_o, and the moment magnitude (no units), M_w. An earthquake with a moment magnitude of 7.0 struck Haiti on January 12, 2010. Find the seismic moment of this earthquake. Use scientific notation; round the mantissa to the nearest tenth. (*Source:* www.usgs.gov)

Chapter 8 Test

Round all irrational numbers to the nearest ten-thousandth.

1. Explain why $y = 3^x$ is a function.

2. Verify that $f(x) = 2x + 2$ and $g(x) = \frac{1}{2}x - 1$ are inverses.

For problems 3–12, $f(x) = 7^x$.

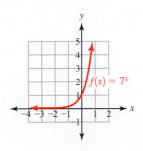

3. Use interval notation to represent the domain of this function.

4. Use interval notation to represent the range of this function.

5. Is this function one-to-one? Explain.

6. Is this function an increasing or decreasing function? Explain.

7. Describe how the graph of this function is different from the graph of $g(x) = 2^x$.

8. Describe how the graph of this function is different from the graph of $h(x) = 0.7^x$.

9. Evaluate: $f(-2)$

10. What is $f^{-1}(x)$?

11. Identify the y-intercept of the graph of this function.

12. Identify the x-intercept of the graph of this function.

For problems 13–17, evaluate. Round to the nearest ten-thousandth.

13. $\ln(168)$

14. $\log_2(14)$

15. $\log(-100)$

16. e^3

17. $\log(17)$

For problems 18–22, $f(x) = \ln(x)$.

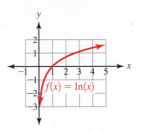

18. Use interval notation to represent the domain of this function.

19. Use interval notation to represent the range of this function.

20. Is this function one-to-one? Explain.

21. Identify the x-intercept of the graph of this function.

22. Identify the y-intercept of the graph of this function.

23. Solve: $5^x = 70$

24. Solve: $\log(x) + \log(4) = 5$

25. The formula for future value of an investment is $A = P\left(1 + \dfrac{r}{n}\right)^{nt}$. Find the future value of an investment of $7000 that is invested at 4%, compounded daily, for 3 years. Round to the nearest hundredth.

26. The function $P = 760e^{-0.145H}$ represents the relationship of the height above the earth's surface in kilometers, H, and the atmospheric pressure in millimeters of mercury, P. Find the atmospheric pressure at a height of 20 km. Round to the nearest hundredth.

27. Is the function in problem 26 an increasing function or a decreasing function? Explain.

28. The definition of pH is $\text{pH} = -\log[H_3O^+]$. The pH of a solution is 4.2. Find the concentration of H_3O^+ in moles per liter. Use scientific notation. Round the mantissa to the nearest tenth.

29. The function $P(t) = 45{,}000(1.15^t)$ represents the relationship of the time in years, t, and the population of a country, $P(t)$. Find the doubling time of this population. Round to the nearest whole number.

Conic Sections and Systems of Nonlinear Equations

9

Why Study Mathematics?

As you study for your math class, you may sometimes think, "When am I ever going to use this?" You might not be able to see the application of the mathematics to your career, your daily life, or your responsibilities as a citizen. Mathematics that is directly useful is often called applied mathematics. Mathematics that is not directly useful—at least not yet—is often called pure mathematics.

In some majors, students need to learn pure mathematics in order to know enough to do applied mathematics. Engineers cannot solve very useful differential equations without knowing the basics of algebra and calculus. Students majoring in the social sciences cannot understand the statistics used in their field unless they know the basics of algebra and probability.

However, even mathematics that is not directly useful is the result of human thought and creativity, just like art, literature, or music. We study the work of our ancestors as we seek to understand what it means to be human and to celebrate human accomplishments. *Mathematics is an amazing human accomplishment.*

In this chapter, you will study conic sections. Today we use our knowledge of conic sections to predict orbits, build telescopes, make "eyes" for robots, and collect solar energy. We also celebrate conic sections as evidence of human ingenuity and critical thinking.

781

© Philip Scalia /Alamy

The Thomas Viaduct crosses the Patapsco River near Baltimore. Completed in 1835, it still carries trains across the river. The railbed is supported by arches. The top of each arch is a semicircle. In this section, we begin our work with conic sections by learning about circles and semicircles.

SECTION 9.1

Distance Formula, Midpoint Formula, and Circles

After reading the text, working the practice problems, and completing assigned exercises, you should be able to:

1. Use the distance formula to find the distance between two points.

2. Use the midpoint formula to find the midpoint of a line segment.

3. Write the equation of a circle in standard form.

4. Given the equation of a circle, identify the center and radius and sketch its graph.

5. Use the equation of a circle or semicircle as a model.

The Distance and Midpoint Formulas

In Chapter 7, we used the Pythagorean theorem to find the distance from first base to third base on a baseball diamond. For a right triangle, $a^2 + b^2 = c^2$, where c is the hypotenuse opposite the right angle (Figure 1). Each of the corners of the triangle is a **vertex**. To use the Pythagorean theorem to develop a formula for finding the distance d between any two points, we solve the Pythagorean theorem for c.

Figure 1

$$a^2 + b^2 = c^2 \qquad \text{Pythagorean theorem.}$$
$$\sqrt{a^2 + b^2} = \sqrt{c^2} \qquad \text{Take the square root of both sides.}$$
$$\sqrt{a^2 + b^2} = \pm c \qquad \text{Simplify; use } \pm \text{ notation.}$$
$$c = \sqrt{a^2 + b^2} \qquad \text{Since distance is positive, use only the positive solution.}$$

A line segment connecting any two points on a coordinate system, (x_1, y_1) and (x_2, y_2), is the hypotenuse of a right triangle (Figure 2). The vertical distance, a, between (x_1, y_1) and (x_2, y_2) is $|y_2 - y_1|$. The horizontal distance, b, is $|x_2 - x_1|$. The distance between the two points is d.

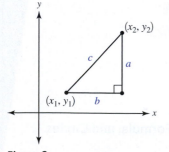

Figure 2

$$c = \sqrt{a^2 + b^2} \qquad \text{Distance is always positive.}$$
$$d = \sqrt{(|x_2 - x_1|)^2 + (|y_2 - y_1|)^2} \qquad \text{Replace } a, b, \text{ and } c.$$
$$d = \sqrt{(x_2 - x_1)^2 + (y_2 - y_1)^2} \qquad \text{Since } (x_2 - x_1)^2 > 0 \text{ and } (y_2 - y_1)^2 > 0, \text{ remove absolute value.}$$

This is the **distance formula**.

The Distance Formula

The distance, d, between two points P and Q represented by the ordered pairs (x_1, y_1) and (x_2, y_2) is

$$d = \sqrt{(x_2 - x_1)^2 + (y_2 - y_1)^2}$$

EXAMPLE 1   Use the distance formula to find the exact distance between $(-5, 9)$ and $(-2, 3)$. Do not round.

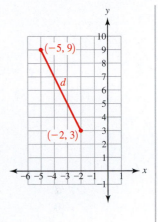

SOLUTION ▶

$d = \sqrt{(x_2 - x_1)^2 + (y_2 - y_1)^2}$ $(x_1, y_1) = (-5, 9); (x_2, y_2) = (-2, 3)$

$d = \sqrt{(\mathbf{-2} - (\mathbf{-5}))^2 + (\mathbf{3} - \mathbf{9})^2}$ Replace variables.

$d = \sqrt{(\mathbf{3})^2 + (\mathbf{-6})^2}$ Simplify.

$d = \sqrt{\mathbf{45}}$ Simplify.

$d = \sqrt{\mathbf{9}}\sqrt{\mathbf{5}}$ Product rule of radicals.

$d = \mathbf{3}\sqrt{\mathbf{5}}$ Simplify.

On a line segment connecting two points, (x_1, y_1), and (x_2, y_2), the **midpoint** is the point that is the same distance from (x_1, y_1) and (x_2, y_2).

Midpoint Formula

The midpoint (x_m, y_m) of two points (x_1, y_1) and (x_2, y_2) is

$$(x_m, y_m) = \left(\frac{x_1 + x_2}{2}, \frac{y_1 + y_2}{2} \right)$$

EXAMPLE 2   Find the midpoint between $(-5, 9)$ and $(-2, 3)$.

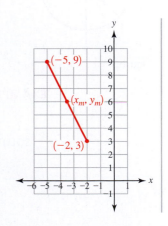

SOLUTION ▶

$$(x_m, y_m) = \left(\frac{x_1 + x_2}{2}, \frac{y_1 + y_2}{2}\right) \qquad (x_1, y_1) = (-5, 9); (x_2, y_2) = (-2, 3)$$

$$(x_m, y_m) = \left(\frac{-5 + (-2)}{2}, \frac{9 + 3}{2}\right) \qquad \text{Replace the variables.}$$

$$(x_m, y_m) = \left(-\frac{7}{2}, 6\right) \qquad \text{Simplify.}$$

Practice Problems

For problems 1–3,
(a) graph the given points, and draw a line segment between the points.
(b) use the distance formula to find the exact distance between the points.

1. $(-9, 4)$ and $(-2, 7)$ **2.** $(6, 5)$ and $(2, 5)$ **3.** $(2, 8)$ and $(6, 10)$

For problems 4–6,
(a) graph the given points and draw a line segment between the points.
(b) use the midpoint formula to find the midpoint.

4. $(-9, 4)$ and $(-2, 7)$ **5.** $(6, 5)$ and $(2, 5)$ **6.** $(2, 8)$ and $(6, 10)$

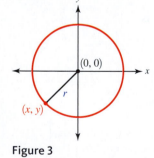

Conic Sections and Circles

The intersection of a horizontal plane and a vertical cone is a **circle**. A circle is an example of a **conic section**.

> ### Circle
> A circle is the points (x, y) in a plane that are the same distance r from a fixed point, (h, k). The fixed point is the **center**. The distance r is the **radius**.

To develop the standard form of the equation of a circle with a center at the origin $(0, 0)$, $x^2 + y^2 = r^2$ (Figure 3), we use the distance formula.

$$d = \sqrt{(x_2 - x_1)^2 + (y_2 - y_1)^2} \qquad \text{Distance formula.}$$

$$r = \sqrt{(x - 0)^2 + (y - 0)^2} \qquad \text{Replace variables.}$$

$$r = \sqrt{x^2 + y^2} \qquad \text{Simplify.}$$

$$r^2 = x^2 + y^2 \qquad \text{Square both sides.}$$

$$x^2 + y^2 = r^2 \qquad \text{Standard form of the equation of a circle, center } (0, 0).$$

Figure 3

EXAMPLE 3 | The center of a circle is $(0, 0)$, and its radius is 6.

(a) Write the equation of this circle in standard form.

SOLUTION ▶

$$x^2 + y^2 = r^2 \qquad \text{Standard form, center at origin.}$$

$$x^2 + y^2 = 6^2 \qquad \text{Replace } r \text{ with 6.}$$

$$x^2 + y^2 = 36 \qquad \text{Simplify.}$$

(b) Sketch the graph.

Rather than building a table of ordered pairs and graphing many points, we often graph the center, graph one or more points on the circle such as $(0, 6)$, $(6, 0)$, $(0, -6)$, and $(-6, 0)$, and draw a "sketch" of a circle through the points. A compass can create a more accurate sketch.

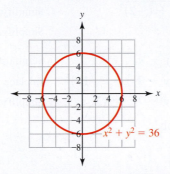

We can also use the distance formula to write an equation for a circle with a center at (h, k) (Figure 4).

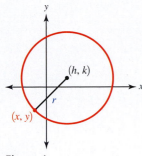

Figure 4

$$d = \sqrt{(x_2 - x_1)^2 + (y_2 - y_1)^2} \qquad \text{Distance formula.}$$
$$r = \sqrt{(x - h)^2 + (y - k)^2} \qquad \text{Replace variables.}$$
$$r^2 = (x - h)^2 + (y - k)^2 \qquad \text{Square both sides.}$$
$$(x - h)^2 + (y - k)^2 = r^2 \qquad \text{Standard form of the equation of a circle; center } (h, k).$$

Standard Forms of the Equation of a Circle

$$(x - h)^2 + (y - k)^2 = r^2 \qquad \text{center at } (h, k); \text{ radius, } r; r > 0$$
$$x^2 + y^2 = r^2 \qquad \text{center at origin; radius, } r; r > 0$$

The radius of a circle is a distance; r^2 cannot be a negative number. Since the radius of a circle is *always a number greater than 0*, $(x - 3)^2 + (y - 4)^2 = -25$ is *not* the equation of a circle.

EXAMPLE 4 | Write the equation of a circle with a radius of 3 and a center at $(-9, 1)$.

SOLUTION ▶
$$(x - h)^2 + (y - k)^2 = r^2 \qquad \text{Standard form, center at } (h, k).$$
$$(x - (-9))^2 + (y - 1)^2 = 3^2 \qquad \text{Replace variables; } h = -9, k = 1, r = 3$$
$$(x + 9)^2 + (y - 1)^2 = 9 \qquad \text{Simplify.}$$

The radius of a circle is the distance between the center and any point on the circle. To find the equation of a circle given the center and one point on the circle, first use the distance formula to find the radius. Then replace the variables in the standard form of the equation of the circle.

EXAMPLE 5   The center of a circle is $(2, 7)$. A point on the circle is $(-1, 5)$.

(a) Write the equation of this circle in standard form.

SOLUTION ▶

$$d = \sqrt{(x_2 - x_1)^2 + (y_2 - y_1)^2} \qquad \text{Distance formula.}$$

$$r = \sqrt{(-1 - 2)^2 + (5 - 7)^2} \qquad \text{Replace variables; } (x_1, y_1) = (2, 7), (x_2, y_2) = (-1, 5).$$

$$r = \sqrt{(-3)^2 + (-2)^2} \qquad \text{Simplify.}$$

$$r = \sqrt{13} \qquad \text{Simplify.}$$

$$(x - h)^2 + (y - k)^2 = r^2 \qquad \text{Standard form of the equation of a circle.}$$

$$(x - 2)^2 + (y - 7)^2 = (\sqrt{13})^2 \qquad \text{Replace variables; } (h, k) = (2, 7).$$

$$(x - 2)^2 + (y - 7)^2 = 13 \qquad \text{Simplify.}$$

(b) Sketch the graph.

▶

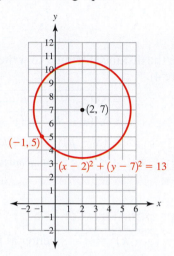

Writing the Equation of a Circle in Standard Form, Given (h, k) and a Point on the Circle

1. Use the distance formula, the center, and the other point to find the radius, r.
2. Use the standard form of the equation of a circle, $(x - h)^2 + (y - k)^2 = r^2$. Replace (h, k) with the center, and replace r with the radius.

In standard form, the values of h and k do not include the $-$ sign.

EXAMPLE 6   **(a)** The equation of a circle is $(x - 3)^2 + (y - 4)^2 = 25$. Identify the center, (h, k).

SOLUTION ▶   This equation is in standard form, $(x - h)^2 + (y - k)^2 = r^2$. The center is (h, k): $(3, 4)$.

(b) The equation of a circle is $(x + 6)^2 + (y)^2 = 25$. Identify the center, (h, k).

▶   Rewritten in standard form, the equation is $(x - (-6))^2 + (y - 0)^2 = 25$. The center is (h, k): $(-6, 0)$.

Another form of the equation of a circle is the **general form** of the equation of a circle, $x^2 + y^2 + Cx + Dy + E = 0$, where C, D, and E are integers. To rewrite the equation in standard form, we group like variables, complete perfect square trinomials, and factor.

EXAMPLE 7   (a) Rewrite $x^2 + y^2 - 6x + 8y - 5 = 0$ in standard form.

SOLUTION ▶

$$x^2 + y^2 - 6x + 8y - 5 = 0$$

$$\underline{+5 \ +5}$$ Addition property of equality.

$$x^2 + y^2 - 6x + 8y + \mathbf{0} = \mathbf{5}$$ Simplify.

$$(\mathbf{x^2 - 6x}) + (\mathbf{y^2 + 8y}) = 5$$ Simplify; group like variables.

To rewrite $(x^2 - 6x)$ as a perfect square trinomial, we add $\left(\dfrac{1}{2}b^2\right)$, or 9, to both sides of the equation. To rewrite $(y^2 + 8y)$ as a perfect square trinomial, we add $\left(\dfrac{1}{2}b^2\right)$, or 16, to both sides of the equation.

$$(x^2 - 6x + 9) + (y^2 + 8y + 16) = 5 + 9 + 16$$ Addition property of equality.

$$(x - 3)(x - 3) + (y + 4)(y + 4) = 30$$ Factor trinomials; simplify.

$$(x - 3)^2 + (y + 4)^2 = (\sqrt{30})^2$$ Use exponential notation.

$$(x - 3)^2 + (y - (-4))^2 = (\sqrt{30})^2$$ Rewrite in standard form.

(b) Identify the center, (h, k), and the radius, r.

▶   The center is $(3, -4)$, and the radius is $\sqrt{30}$.

Writing the Equation of a Circle in Standard Form, Given an Equation of a Circle in General Form, $x^2 + y^2 + Cx + Dy + E = 0$

1. Use the properties of equality to isolate the terms with variables.

2. Change the order, and use parentheses to group like variables.

3. Complete the square in the first group by adding $\left(\dfrac{1}{2}C\right)^2$ to the group; add $\left(\dfrac{1}{2}C\right)^2$ to the other side of the equation.

4. Complete the square in the second group by adding $\left(\dfrac{1}{2}D\right)^2$ to the group; add $\left(\dfrac{1}{2}D\right)^2$ to the other side of the equation.

5. Factor each group; use exponential notation.

6. Rewrite the constant on the right side of the equation as an exponential expression with an exponent of 2.

Practice Problems

7. The center of a circle is $(0, 0)$. Its radius is 7. Write its equation in standard form.

8. The center of a circle is $(-2, 3)$. A point on the circle is $(5, -4)$. Write its equation in standard form.

9. The equation of a circle is $x^2 + y^2 + 12x - 8y - 12 = 0$.
   a. Rewrite this equation in standard form.
   b. Identify the center, (h, k).
   c. Identify the radius, r.

Applications

In the next example, the equation of a circle is a model for the semicircular arch of a bridge. A semicircle is one-half of a full circle.

EXAMPLE 8 The Tunkhannock Viaduct in northeastern Pennsylvania is supported by semicircular arches. The diameter of each arch is about 180 ft.

(a) Write an equation that is a model of the semicircle. Assume that the origin of the xy-coordinate system is at the lower left edge of the semicircular arch.

SOLUTION ▶ Since the diameter is 180 ft, the radius is 90 ft. On a coordinate system, place the circle with its center (h, k) at $(90\text{ ft}, 0\text{ ft})$.

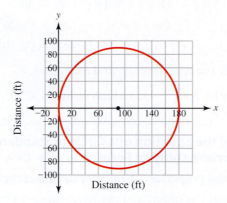

$$(x - h)^2 + (y - k)^2 = r^2 \qquad \text{Standard form of a circle.}$$
$$(x - 90\text{ ft})^2 + (y - 0\text{ ft})^2 = (90\text{ ft})^2 \qquad \text{Replace variables.}$$
$$(x - 90)^2 + y^2 = 8100 \qquad \text{Simplify; remove units.}$$
$$\underline{-(x - 90)^2 \qquad\qquad -(x - 90)^2} \qquad \text{Subtraction property of equality.}$$
$$y^2 = 8100 - (x - 90)^2 \qquad \text{Simplify.}$$
$$\sqrt{y^2} = \sqrt{8100 - (x - 90)^2} \qquad \text{Square root both sides.}$$
$$y = \pm\sqrt{8100 - (x - 90)^2} \qquad \text{Simplify.}$$

For the upper semicircle, the values of y are positive. So the equation of the upper semicircle is $y = \sqrt{8100\text{ ft}^2 - (x - 90\text{ ft})^2}$.

(b) Use this equation to find the height of the arch 10 ft to the right of the center. Round to the nearest tenth.

▶ The center of the arch is $(90\text{ ft}, 0\text{ ft})$. The point we need to find is $(100\text{ ft}, y_1)$, where y_1 represents the height of the arch 10 ft to the right of the center.

$$y = \sqrt{8100\text{ ft}^2 - (x - 90\text{ ft})^2} \qquad \text{Equation of semicircle.}$$
$$y_1 = \sqrt{8100\text{ ft}^2 - (100\text{ ft} - 90\text{ ft})^2} \qquad \text{Replace variables.}$$
$$y_1 = \sqrt{8100\text{ ft}^2 - 100\text{ ft}^2} \qquad \text{Simplify.}$$
$$y_1 = \sqrt{8000\text{ ft}^2} \qquad \text{Simplify.}$$
$$y_1 \approx 89.4\text{ ft} \qquad \text{Evaluate and round.}$$

The height of the arch 10 ft to the right of center is about 89.4 ft.

Practice Problems

10. Completed in 1835, the Thomas Viaduct near Baltimore has eight semi-circular arches. The radius of each arch is about 58 ft. Write an equation that describes the upper semicircle that forms each arch. Assume that the origin of the xy-coordinate system is at the lower left corner of the semicircle.

11. A satellite is in a circular orbit above the equator at an altitude of 800 mi. The radius of the earth is about 3963 mi. Assume that the center of the earth is at the origin of an xy-coordinate system. Write an equation that is a model of this circular orbit.

Using Technology: Graphing a Circle

We can graph a function on a graphing calculator. However, a circle fails the vertical line test and is not a function. To graph a circle on a graphing calculator, rewrite its equation as two separate functions.

EXAMPLE 9 Graph: $(x - 3)^2 + (y - 5)^2 = 4$

$$(x - 3)^2 + (y - 5)^2 = 4 \qquad \text{Equation of a circle.}$$

$$(y - 5)^2 = 4 - (x - 3)^2 \qquad \text{Isolate the expression with } y.$$

$$\sqrt{(y - 5)^2} = \sqrt{4 - (x - 3)^2} \qquad \text{Square root of both sides.}$$

$$y - 5 = \pm\sqrt{4 - (x - 3)^2} \qquad \text{Simplify.}$$

$$y = 5 \pm\sqrt{4 - (x - 3)^2} \qquad \text{Solve for } y.$$

$$y = 5 + \sqrt{4 - (x - 3)^2} \quad \text{or} \quad y = 5 - \sqrt{4 - (x - 3)^2} \qquad \text{Rewrite using "or."}$$

Begin with a standard window. Press [Y=]. Type both equations. Press [GRAPH].

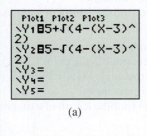

(a)

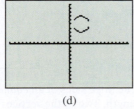

(b)

$[-10, 10, 1, -10, 10, 1]$

In a standard window, each axis extends from -10 to 10. However, the distances between the tick marks on each axis are not the same. This stretches the graph out in the horizontal direction.

To change the window so that the distance between adjacent tick marks on each axis is the same, press [ZOOM]; choose **zsquare**.

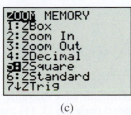

(c)

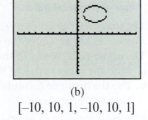

(d)

$[-15.16, -15.16, 1, -10, 10, 1]$

The graph of each function is a semicircle. This calculator cannot show the complete graph of each semicircle.

Practice Problems For problems 12–13, graph. Begin in a standard window, and then use **zsquare** to change the window. Sketch the graph; describe the window.

12. $(x - 5)^2 + (y - 2)^2 = 9$ **13.** $(x - 5)^2 + (y + 2)^2 = 9$

9.1 VOCABULARY PRACTICE

Match the term with its description.

1. In a right triangle where c is the length of the hypotenuse and a and b are the lengths of its legs, $a^2 + b^2 = c^2$.
2. The point where two sides of a triangle intersect
3. $d = \sqrt{(x_2 - x_1)^2 + (y_2 - y_1)^2}$
4. $\left(\dfrac{x_1 + x_2}{2}, \dfrac{y_1 + y_2}{2}\right)$
5. A 90-degree angle
6. A triangle that includes a 90-degree angle
7. The distance from the center of a circle to a point on the circle
8. A collection of points in a plane that are all the same positive distance r from a fixed point
9. The intersection of a plane and a cone; a circle is an example of this.
10. The side opposite the 90-degree angle in a triangle

A. circle
B. conic section
C. distance formula
D. hypotenuse
E. midpoint formula
F. Pythagorean theorem
G. radius of a circle
H. right angle
I. right triangle
J. vertex

9.1 Exercises

Follow your instructor's guidelines for showing your work.

1. The two legs of a right triangle measure 6 in. and 9 in. Find the exact length of the hypotenuse. Do not round.

2. The two legs of a right triangle measure 7 in. and 11 in. Find the exact length of the hypotenuse. Do not round.

For exercises 3–10, use the distance formula to find the exact distance between the given points.

3. $(8, 5)$; $(5, 1)$ 4. $(15, 8)$; $(12, 4)$

5. $(2, 5)$; $(-10, -11)$ 6. $(-3, 1)$; $(9, 17)$

7. $(12, 21)$; $(3, 6)$ 8. $(15, 10)$; $(23, 22)$

9. $(0, 10)$; $(6, 0)$ 10. $(0, 12)$; $(10, 0)$

11. The y-coordinates of each point on a horizontal line are the same. Use the distance formula to find the distance on a horizontal line between the points (x_1, y_1) and (x_2, y_1).

12. The x-coordinates of each point on a vertical line are the same. Use the distance formula to find the distance on a vertical line between the points (x_1, y_1) and (x_1, y_2).

13. The lower left vertex of a square is the point $(1, 3)$. The upper left vertex of the square is the point $(1, 8)$.
    a. Find the other two vertices of this square.
    b. Use the distance formula to find the length of a diagonal of this square.

14. The lower left vertex of a square is the point $(2, 5)$. The upper left vertex of the square is the point $(2, 13)$.
    a. Find the other two vertices of this square.
    b. Use the distance formula to find the length of a diagonal of this square.

15. The lower right vertex of a *rectangle* is the point $(10, -3)$. The upper left vertex of the rectangle is the point $(1, 1)$. The upper right vertex of the rectangle is $(10, 1)$.
    a. Find the other vertex of this rectangle.
    b. Use the distance formula to find the length of a diagonal of this rectangle.

16. The lower left vertex of a *rectangle* is the point $(2, -7)$. The upper left vertex of the rectangle is the point $(2, -1)$. The upper right vertex of the rectangle is $(5, -1)$.
    a. Find the other vertex of this rectangle.
    b. Use the distance formula to find the length of a diagonal of this rectangle.

17. The three sides of an *equilateral triangle* have the same length. Two of the vertices of the triangle are $(1, 4)$ and $(5, 9)$. Find the perimeter of this triangle.

18. The three sides of an *equilateral triangle* have the same length. Two of the vertices of the triangle are $(3, 11)$ and $(7, 15)$. Find the perimeter of this triangle.

For exercises 19–22, find the midpoint of the line segment with the given endpoints.

19. $(1, 4)$; $(5, 9)$ 20. $(3, 11)$; $(7, 15)$

21. $(-8, -2)$; $(-20, 20)$ 22. $(-6, -12)$; $(-30, 30)$

For exercises 23–24, an *isosceles* triangle has two sides of equal length and two angles of equal measure. The third side is the base. The given points are the vertices of the base of the triangle. Find the midpoint of the base.

23. $(2, 5); (2, 13)$ **24.** $(3, 8); (3, 22)$

For exercises 25–30, write the equation in standard form of the circle with the given center and radius.

25. Center: $(0, 0)$; radius: 8

26. Center: $(0, 0)$; radius: 10

27. Center: $(3, 1)$; radius: 9

28. Center: $(11, 2)$; radius: 15

29. Center: $(-8, 2)$; radius: 20

30. Center: $(-3, 6)$; radius: 8

For exercises 31–34, sketch the graph.

31. $(x - 5)^2 + (y - 6)^2 = 4$ **32.** $(x - 4)^2 + (y - 3)^2 = 5$

33. $(x + 1)^2 + (y - 3)^2 = 6$ **34.** $(x + 2)^2 + (y - 4)^2 = 3$

For exercises 35–40, write the equation of the circle in standard form.

35. Center: $(-4, 5)$; point: $(3, 8)$

36. Center: $(-6, 9)$; point: $(2, 5)$

37. Center: $(-2, -3)$; point: $(4, -9)$

38. Center: $(-8, -1)$; point: $(3, -7)$

39. Center: $(0, 0)$; point: $(2, 6)$

40. Center: $(0, 0)$; point: $(8, 1)$

For exercises 41–54,
(a) rewrite the equation of the circle in standard form.
(b) identify the center, (h, k).
(c) identify the radius, r.

41. $x^2 + y^2 - 4x + 12y - 9 = 0$

42. $x^2 + y^2 - 12x + 4y - 9 = 0$

43. $x^2 + y^2 + 10x - 6y - 2 = 0$

44. $x^2 + y^2 + 6x - 10y - 2 = 0$

45. $x^2 + y^2 + 8x - 2y - 10 = 0$

46. $x^2 + y^2 + 4x - 20y - 8 = 0$

47. $x^2 + y^2 - 8y - 9 = 0$

48. $x^2 + y^2 - 14y - 15 = 0$

49. $x^2 + y^2 - 18x - 4 = 0$

50. $x^2 + y^2 - 16x - 12 = 0$

51. $x^2 - 6x + 9 + y^2 - 8y + 16 = 12$

52. $x^2 - 10x + 25 + y^2 - 4y + 4 = 16$

53. $x^2 + 16x + y^2 + 4y + 68 = 12$

54. $x^2 + 6x + y^2 + 2y + 10 = 14$

For exercises 55–58, solve the equation for y.

55. $(x - 3)^2 + y^2 = 15$

56. $(x - 2)^2 + y^2 = 18$

57. $(x + 6)^2 + y^2 = 49$

58. $(x + 4)^2 + y^2 = 16$

For exercises 59–60, the radius of the earth is about 3963 mi or 6378 km. To write the equation of this orbit, place the origin of the coordinate system at the center of the earth.

59. Write the equation of the circular orbit of a satellite that is 22,241 mi above the earth's surface.

60. Write the equation of the circular orbit of a satellite that is 35,786 km above the earth's surface.

For exercises 61–62, the time for a satellite to complete one orbit is an orbital period, P. For a satellite in a circular orbit,

$P = 2\pi \sqrt{\dfrac{r^3}{\mu}}$, where P is in seconds, r is the radius of the orbit (including the radius of the earth, 3963 mi or 6378 km), and the gravitational parameter is μ ("mu"). For a satellite to remain over the same location on the earth's surface, it must be in a geostationary orbit, and its period must be 86,400 s (24 hr). ($\pi = 3.14$.)

61. The gravitational parameter is $\dfrac{95{,}629 \text{ mi}^3}{s^2}$.
   a. Solve the formula for r.
   b. Find the radius in miles for a satellite in a geostationary orbit. Round to the nearest whole number.
   c. Find the height of the satellite above the earth in miles.

62. The gravitational parameter is $\dfrac{398{,}601 \text{ km}^3}{s^2}$.
   a. Solve the formula for r.
   b. Find the radius in kilometers for a satellite in a geostationary orbit. Round to the nearest whole number.
   c. Find the height of the satellite above the earth in kilometers.

63. The radius of the outer edge of a semicircular window is 14.5 in. Write an equation that is a model of this window. Assume that the origin is at the lower left corner of the window.

© Niels Quist/Shutterstock

64. The radius of the outer edge of a semicircular window is 25.5 in. Write an equation that is a model of this window. Assume that the origin is at the lower left corner of the window.

65. For the window in exercise 63, find the vertical distance between the top and bottom edge at a point that is 4 in. to the right of the midpoint of the bottom edge. Round to the nearest tenth.

66. For the window in exercise 64, find the vertical distance between the top and bottom edge at a point that is 4 in. to the right of the midpoint of the bottom edge. Round to the nearest tenth.

67. Jardin House is a skyscraper in Hong Kong that includes 1748 circular windows. The diameter of each window is 1.8 m. Write an equation that is a model of a window. Assume that the origin of the xy-coordinate system is a point on the left edge of the window.

68. St. John's in the Wilderness Church in White Bear Lake, Minnesota, has a circular stained glass window. The diameter of this window is 9 ft. Write an equation that is a model of this window. Assume that the origin of the xy-coordinate system is a point on the edge of the window.

Problem Solving: Practice and Review

Follow your instructor's guidelines for using the five steps as outlined in Section 1.5, p. 51.

69. According to the Food and Agriculture Association of the United Nations, a person is affected by hunger if less than 1800 calories of food is available per day. On June 20, 2009, the U.S. Census Bureau estimated that the population of the world was 6,787,822,164 people. Predict the number of these people who were affected by hunger. Round to the nearest million.

Because of war, drought, political instability, high food prices and poverty, hunger now affects one in six people, by the United Nations' estimate. (*Source:* www.ap.org, June 19, 2009)

70. The National Assessment of Educational Progress in the Arts found that 56% of eighth-graders in 2008 could identify a saxophone as the instrument playing a melody. In 1997, 66% of these students could do this. Find the percent decrease from 1997 to 2008 in the number of students who could do this. Round to the nearest percent. (*Source:* nces.ed.gov, June 15, 2009)

71. The function $y = 10.565e^{-0.243x}$ represents the relationship of the number of years since 1997, x, and the number of acute hepatitis A cases per 100,000 people, y. Find the number of acute hepatitis A cases per 100,000 people in 2013. Round to the nearest tenth. (*Source:* www.cdc.gov, 2008)

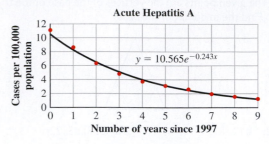

72. Find the original asking price of Ennis House.

Frank Lloyd Wright's Ennis House . . . one of the master's most celebrated residential designs and one of Los Angeles' most revered architectural landmarks, has sold . . . for about $4.5 million, 70% less than its original asking price. (*Source:* www.latimes.com, July 17, 2011)

Technology

For exercises 73–76,
(a) write the equations of the semicircles.
(b) sketch the graph; describe the window.

73. $(x - 2)^2 + (y - 6)^2 = 9$

74. $(x + 4)^2 + (y - 2)^2 = 16$

75. $(x - 2)^2 + (y - 4)^2 = 25$

76. $(x + 4)^2 + (y - 2)^2 = 36$

Find the Mistake

For exercises 77–80, the completed problem has one mistake.
(a) Describe the mistake in words, or copy down the whole problem and highlight or circle the mistake.
(b) Do the problem correctly.

77. Problem: The equation of a circle is $(x + 4)^2 + (y - 7)^2 = 9$. Identify the center.

Incorrect Answer: The center is $(4, -7)$.

78. Problem: The equation of a circle is $(x + 4)^2 + (y - 7)^2 = 9$. Find the radius.

Incorrect Answer: The radius of the circle is 9.

79. Problem: The equation of a circle is $x^2 + y^2 - 4x - 14y - 11 = 0$. Find the radius.

Incorrect Answer: $x^2 + y^2 - 4x - 14y - 11 = 0$

$$\underline{ +11 \ \ +11}$$
$$x^2 + y^2 - 4x - 14y + 0 = 11$$
$$(x^2 - 4x + 4) + (y^2 - 14y + 49) = 11$$
$$(x - 2)^2 + (y - 7)^2 = 11$$

The radius is $\sqrt{11}$.

80. Problem: Find the distance between $(2, 8)$ and $(-3, 1)$.

Incorrect Answer: $d = (-3 - 2)^2 + (1 - 8)^2$

$$d = 25 + 49$$
$$d = 74$$

Review

81. **a.** Rewrite $y = x^2 + 12x - 9$ in vertex form.
b. Identify the vertex of the parabola that represents this function.

82. Solve $x^2 + 24x + 8 = 0$ by completing the square.

83. Use the quadratic formula to solve $x^2 + 24x + 8 = 0$.

84. Solve $x^2 + 10x + 16 = 0$ by factoring and using the zero product property.

85. In college classes, you may sometimes think that the topic you are studying will not be useful to you in your future career. How do you motivate yourself to study such topics?

© jiawangkun/Shutterstock

President's Park includes the White House, Lafayette Park, and the Ellipse. The Ellipse is located just south of the White House fence. It is open to the public and hosts demonstrations as well as Ultimate Frisbee competitions. In this section, we will study this and other ellipses.

SECTION 9.2

Ellipses

After reading the text, working the practice problems, and completing assigned exercises, you should be able to:

1. Write the equation of an ellipse in standard form.

2. Sketch the graph of an ellipse.

3. Identify the vertices, center, major axis, and minor axis of an ellipse.

4. Use the equation of an ellipse as a model.

Conic Sections and Ellipses

The intersection of a vertical cone and a horizontal plane is a circle. If we tilt the plane, the intersection may be an **ellipse**. An ellipse is another example of a conic section.

> **Ellipse**
>
> An ellipse includes all the points (x, y) in a plane such that the sum of the distances from two fixed points is a constant. Each of the fixed points is a **focus** of the ellipse. (The plural of focus is *foci*.) The line containing the foci is the **major axis**. The midpoint of the line segment between the foci is the **center**. The line that is perpendicular to the major axis and passes through the center is the **minor axis**. The **vertices** are the points of intersection of the ellipse with either axis.
>
>

The plural of ellipse is *ellipses*. Although it is pronounced similarly, *ellipses* is not the same word as *ellipsis*. An *ellipsis* is the series of three dots that we use to show that a pattern or number continues. For example, the number 2.5757. . . includes an ellipsis.

To create an ellipse on paper, draw a major axis and a minor axis, perpendicular to each other (Figure 1). Mark the center. Select a length for the distance between the center and each focus. Mark the foci on the major axis. Put the paper on top of a piece of cardboard. Cut a piece of string that is longer than twice the distance between the foci. Use thumbtacks to anchor the ends of the string at the foci. Use a pencil to pull the string tight. Keeping the string tight, allow the pencil to trace an ellipse. Because the string is tight, the sum of the distances from the foci to any point on the ellipse is always the same.

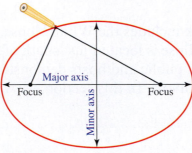

Figure 1

In creating an ellipse for a formal garden, the "gardener's method" is used. The thumbtacks are replaced with stakes and the string with rope. Instead of a pencil, a shovel or chalk is used to mark the edge of the ellipse.

The equation of an ellipse represents the relationship of each point on the ellipse (x, y) with the distance of the vertices from the center. The distance from the center to a vertex on the major axis is a. The distance from the center to a vertex on the minor axis is b. For any ellipse, $a > b > 0$. The equation for an ellipse in which the major axis is horizontal is different from the equation for an ellipse in which the major axis is vertical.

Standard Form of the Equation of an Ellipse, Center at the Origin (0, 0)

Distance between vertices: major axis = $2a$, minor axis = $2b$, $a > b > 0$

Major axis is horizontal:

$$\frac{x^2}{a^2} + \frac{y^2}{b^2} = 1$$

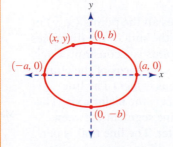

Major axis is vertical:

$$\frac{x^2}{b^2} + \frac{y^2}{a^2} = 1$$

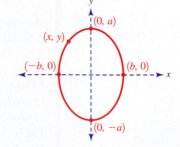

EXAMPLE 1 | The graph represents an ellipse. The center is $(0, 0)$. The distance between the vertices on the horizontal major axis is 8. The distance between the vertices on the vertical minor axis is 6.

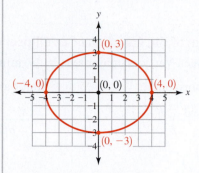

(a) Identify a.

SOLUTION ▶

$2a = 8$ Distance between vertices on major axis.

$\dfrac{2a}{2} = \dfrac{8}{2}$ Division property of equality.

$a = 4$ Simplify.

(b) Identify b.

▶

$2b = 6$ Distance between vertices on minor axis.

$\dfrac{2b}{2} = \dfrac{6}{2}$ Division property of equality.

$b = 3$ Simplify.

(c) Identify the vertices on each axis.

▶   Major axis: $(4, 0)$ and $(-4, 0)$; minor axis: $(0, 3)$ and $(0, -3)$

(d) Write the equation of this ellipse in standard form.

▶

$\dfrac{x^2}{a^2} + \dfrac{y^2}{b^2} = 1$ Standard form when major axis is horizontal.

$\dfrac{x^2}{(4)^2} + \dfrac{y^2}{(3)^2} = 1$ $a = 4; b = 3$

$\dfrac{x^2}{16} + \dfrac{y^2}{9} = 1$ Simplify.

We can also write the equation of an ellipse with a center at (h, k). Given the vertices on both axes, we can find a and b. If the major axis is horizontal, the distance between the vertices on this axis is the difference of the x-coordinates of the vertices. Since the distance between vertices also equals $2a$, we can solve for a. Since the minor axis is vertical, the distance between the vertices is the difference of the y-coordinates of the vertices. Since the distance between vertices also equals $2b$, we can solve for b.

If the major axis is vertical, the distance between its vertices is the difference in the y-coordinates of the vertices. The distance between the vertices on the horizontal minor axis is the difference in the x-coordinates of the vertices.

Standard Form of the Equation of an Ellipse, Center at (h, k)

Distance Between Vertices: major axis $= 2a$, minor axis $= 2b$, $a > b > 0$

Major axis is horizontal:

$$\frac{(x-h)^2}{a^2} + \frac{(y-k)^2}{b^2} = 1$$

Major axis is vertical:

$$\frac{(x-h)^2}{b^2} + \frac{(y-k)^2}{a^2} = 1$$

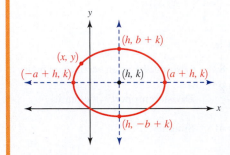

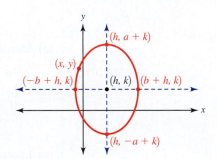

EXAMPLE 2   The vertices on the vertical major axis of an ellipse are $(2, 7)$ and $(2, -5)$. The vertices on the horizontal minor axis are $(6, 1)$ and $(-2, 1)$.

(a) Identify a.

SOLUTION ▶

$2a = 7 - (-5)$	Distance between vertices on major axis $= 2a$.
$2a = \mathbf{12}$	Simplify.
$\dfrac{2a}{2} = \dfrac{12}{2}$	Division property of equality.
$a = 6$	Simplify.

(b) Identify b.

▶

$2b = 6 - (-2)$	Distance between vertices on minor axis $= 2b$.
$2b = \mathbf{8}$	Simplify.
$\dfrac{2b}{2} = \dfrac{8}{2}$	Division property of equality.
$b = 4$	Simplify.

(c) Identify the center.

▶

The major axis is a vertical line passing through $(2, 7)$ and $(2, -5)$. The x-coordinate of every point on the line is 2, and its equation is $x = 2$. The minor axis is a horizontal line passing through $(6, 1)$ and $(-2, 1)$. The y-coordinate of every point on the line is 1, and its equation is $y = 1$. The center of the ellipse is at the intersection of the major and minor axis: $(2, 1)$.

(d) Write the equation of the ellipse.

$$\frac{(x-h)^2}{b^2} + \frac{(y-k)^2}{a^2} = 1 \qquad \text{Standard form; vertical major axis; center } (h, k).$$

$$\frac{(x-2)^2}{16} + \frac{(y-1)^2}{36} = 1 \qquad a = 6; b = 4; (h, k) = (2, 1)$$

(e) Sketch the graph.

Rather than building a table of ordered pairs to graph an ellipse, we often graph the vertices and the center. Then we just sketch the ellipse through the vertices.

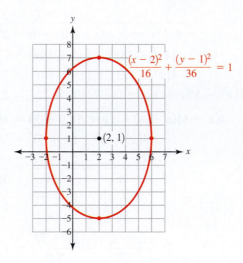

$$\frac{(x-2)^2}{16} + \frac{(y-1)^2}{36} = 1$$

Practice Problems

For problems 1–2,
(a) identify the vertices on the major axis.
(b) identify the vertices on the minor axis.
(c) write the equation of the ellipse in standard form.
(d) sketch the graph.

1. The center of an ellipse is $(0, 0)$. The distance between the vertices on its horizontal major axis is 10. The distance between the vertices on its vertical minor axis is 6.

2. The center of an ellipse is $(2, 3)$. The distance between the vertices on its horizontal major axis is 10. The distance between the vertices on its vertical minor axis is 6.

3. The vertices on the horizontal major axis of an ellipse are $(0, 3)$ and $(12, 3)$. The vertices on the vertical minor axis are $(6, 5)$ and $(6, 1)$.
   a. Write the equation of the major axis.
   b. Write the equation of the minor axis.
   c. Identify its center, (h, k).
   d. Write the equation of the ellipse in standard form.
   e. Sketch the graph.

Completing the Square

In Section 9.1, we used completing the square to rewrite the general form of the equation of a *circle* in standard form. Given the equation of an ellipse in general form, $Ax^2 + By^2 - Cx + Dy + E = 0$, we use the same process to rewrite it in standard form.

EXAMPLE 3   The equation of an ellipse is $4x^2 + y^2 - 48x + 10y - 27 = 0$.

(a) Rewrite this equation in standard form.

SOLUTION ▶

$$4x^2 + y^2 - 48x + 10y - 27 = 0$$

$$\underline{\hspace{4cm} +27 \quad +27} \qquad \text{Addition property of equality.}$$

$$4x^2 + y^2 - 48x + 10y + \mathbf{0} = \mathbf{27} \qquad \text{Simplify.}$$

$$(\mathbf{4x^2 - 48x}) + (y^2 + 10y) = 27 \qquad \text{Simplify. Group like variables.}$$

$$4(x^2 - 12x) + (y^2 + 10y) = 27 \qquad \text{The lead coefficient must equal 1.}$$

$$4(x^2 - 12x + 36) + (y^2 + 10y + 25) = 27 + 4(36) + 25 \qquad \text{Complete squares.}$$

$$4(x - 6)(x - 6) + (y + 5)(y + 5) = \mathbf{196} \qquad \text{Factor trinomials; simplify.}$$

$$4(x - 6)^2 + (y + 5)^2 = 196 \qquad \text{Exponential notation.}$$

$$\frac{4(x - 6)^2}{196} + \frac{(y + 5)^2}{196} = \frac{196}{196} \qquad \text{Division property of equality.}$$

$$\frac{(x - 6)^2}{49} + \frac{(y + 5)^2}{196} = 1 \qquad \text{Simplify.}$$

$$\frac{(x - 6)^2}{7^2} + \frac{(y + 5)^2}{14^2} = 1 \qquad \begin{array}{l}\text{Exponential notation}\\\text{in denominators.}\end{array}$$

$$\frac{(x - 6)^2}{7^2} + \frac{(y - (-5))^2}{14^2} = 1 \qquad \text{Rewrite } (y + 5) \text{ as } (y - (-5)).$$

(b) Identify the center.

▶   The center, (h, k), is $(6, -5)$.

(c) Write the equation of the major axis and the minor axis.

▶   The length of the major axis in an ellipse, a, is greater than the length of the minor axis, b. Since the form of the ellipse matches $\dfrac{(x - h)^2}{b^2} + \dfrac{(y - k)^2}{a^2} = 1$, the major axis is a vertical line passing through the center, $(6, -5)$, and its equation is $x = 6$. The minor axis is a horizontal line passing through the center, and its equation is $y = -5$.

(d) Identify the vertices on the major axis.

▶   Since $a = 14$, the y-coordinates of the vertices on the major vertical axis are 14 greater than or 14 less than the center: $(6, -5 + \mathbf{14})$ and $(6, -5 - \mathbf{14})$. Simplified, the vertices are $(6, \mathbf{9})$ and $(6, \mathbf{-19})$.

(e) Identify the vertices on the minor axis.

▶   Since $b = 7$, the x-coordinates of the vertices on the minor horizontal axis are 7 greater than or 7 less than the center: $(6 + \mathbf{7}, -5)$ and $(6 - \mathbf{7}, -5)$. Simplified, the vertices are $(\mathbf{13}, -5)$ and $(\mathbf{-1}, -5)$.

(f) Sketch the graph.

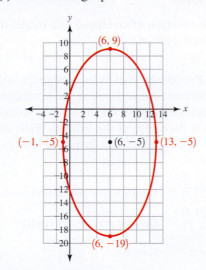

Writing the Equation of an Ellipse in Standard Form, Given an Equation in General Form, $Ax^2 + By^2 + Cx + Dy + E = 0$

1. Use a property of equality to isolate the terms with variables.

2. Change the order, and use parentheses to group like variables.

3. If $A \neq 1$ or $B \neq 1$, factor this constant out of the group.

4. Complete the square in the first group by adding $\left(\frac{1}{2} \cdot \frac{C}{A}\right)^2$ to the group; add $A\left(\frac{1}{2} \cdot \frac{C}{A}\right)^2$ to the other side of the equation, and simplify.

5. Complete the square in the second group by adding $\left(\frac{1}{2} \cdot \frac{D}{B}\right)^2$ to the group; add $B\left(\frac{1}{2} \cdot \frac{D}{B}\right)^2$ to the other side of the equation, and simplify.

6. Factor each group; use exponential notation.

7. Use the division property of equality to rewrite the constant on one side of the equation as 1.

8. Rewrite the denominators as exponential expressions with an exponent of 2.

Practice Problems

For problems 4–5,
(a) rewrite the equation in standard form.
(b) identify the center.
(c) write the equation of the major axis
(d) write the equation of the minor axis.
(e) identify the vertices on the major axis.
(f) identify the vertices on the minor axis.
(g) sketch the graph.

4. $4x^2 + y^2 + 24x - 4y + 36 = 0$ **5.** $x^2 + 4y^2 + 6x - 16y + 21 = 0$

Applications

We can use the equation of an ellipse as a model for ellipses such as planetary orbits or the Ellipse in President's Park.

EXAMPLE 4 The Ellipse in President's Park is located just south of the White House grounds. The distance between the vertices on the horizontal major axis is about 1058 ft. The distance between the vertices on the vertical minor axis is about 903 ft. Assume that the origin is the center of the Ellipse. Write an equation that is a model of the Ellipse. (*Source:* Kimberling, "The Shape and History of the Ellipse in Washington, D.C.")

(a) Identify a.

SOLUTION ▶
$$2a = 1058 \text{ ft} \qquad \text{Distance between vertices on horizontal major axis.}$$
$$\frac{2a}{2} = \frac{1058 \text{ ft}}{2} \qquad \text{Division property of equality}$$
$$a = 529 \text{ ft} \qquad \text{Simplify.}$$

(b) Identify b. Write as a decimal number.

▶
$$2b = 903 \text{ ft} \qquad \text{Distance between vertices on vertical minor axis.}$$
$$\frac{2b}{2} = \frac{903 \text{ ft}}{2} \qquad \text{Division property of equality}$$
$$b = 451.5 \text{ ft} \qquad \text{Simplify.}$$

(c) Write the equation in standard form that is a model of this ellipse. Place the origin $(0, 0)$ at the center of the ellipse.

▶
$$\frac{x^2}{a^2} + \frac{y^2}{b^2} = 1 \qquad \text{Horizontal major axis; vertical minor axis.}$$
$$\frac{x^2}{(\mathbf{529 \text{ ft}})^2} + \frac{y^2}{(\mathbf{451.5 \text{ ft}})^2} = 1 \qquad \text{Replace } a \text{ and } b.$$

Practice Problems

6. Kepler's first law of planetary motion states that the orbit of a planet is an ellipse. For Mars, the distance between the two vertices on the major axis is 3.04 AU. (An AU is an astronomical unit of distance that is approximately equal to the average distance between the earth and the sun, 93 million miles.) The distance between the two vertices on the minor axis is 3.02 AU. Assume that the origin is the center of Mars's elliptical orbit. Write an equation in standard form that is a model of this orbit.

7. A builder designed an elliptical deck for a homeowner in the Pacific Northwest. The distance between two vertices on the major axis is 24 ft. The distance between two vertices on the minor axis is 20 ft. Assume that the origin is at the center of the deck. Write an equation in standard form that is a model of the outer edge of the deck.

9.2 VOCABULARY PRACTICE

Match the term with its description.

1. $\dfrac{(x - h)^2}{b^2} + \dfrac{(y - k)^2}{a^2} = 1$

2. The intersection of a plane and a cone; circles and ellipses are examples of this.
3. A collection of points such that the sum of the distances from two fixed points is a constant
4. The fixed points on the major axis of an ellipse such that the sum of the distances from these fixed points to any point on the ellipse is a constant
5. The line that includes the foci of an ellipse
6. The midpoint of the line segment between the foci of an ellipse
7. The line that is perpendicular to the major axis and passes through the center
8. A point of intersection of an ellipse with the major or minor axis
9. The distance from the center of an ellipse to a vertex on the major axis
10. The distance from the center of an ellipse to a vertex on the minor axis

A. a
B. b
C. center of an ellipse
D. conic section
E. ellipse
F. foci
G. major axis
H. minor axis
I. standard form of the equation of an ellipse
J. vertex

9.2 Exercises

Follow your instructor's guidelines for showing your work.

For exercises 1–8, the center, the distance between vertices on the major axis, and the distance between vertices on the minor axis are given.
(a) Identify the vertices of the ellipse.
(b) Write the equation of the ellipse in standard form.
(c) Sketch a graph of the ellipse.

1. $(0, 0)$; horizontal major axis $= 12$; vertical minor axis $= 10$
2. $(0, 0)$; horizontal major axis $= 10$; vertical minor axis $= 8$
3. $(0, 0)$; vertical major axis $= 12$; horizontal minor axis $= 10$
4. $(0, 0)$; vertical major axis $= 10$; horizontal minor axis $= 8$
5. $(2, -1)$; vertical major axis $= 10$; horizontal minor axis $= 8$
6. $(3, -2)$; vertical major axis $= 6$; horizontal minor axis $= 4$
7. $(-4, 3)$; horizontal major axis $= 14$; vertical minor axis $= 8$
8. $(-3, 4)$; horizontal major axis $= 16$; vertical minor axis $= 10$

For exercises 9–16, sketch a graph of the ellipse that represents the equation.

9. $\dfrac{x^2}{9} + \dfrac{y^2}{16} = 1$

10. $\dfrac{x^2}{9} + \dfrac{y^2}{25} = 1$

11. $\dfrac{x^2}{16} + \dfrac{y^2}{4} = 1$

12. $\dfrac{x^2}{25} + \dfrac{y^2}{9} = 1$

13. $\dfrac{(x - 3)^2}{4} + \dfrac{(y + 2)^2}{16} = 1$

14. $\dfrac{(x - 1)^2}{9} + \dfrac{(y + 3)^2}{25} = 1$

15. $\dfrac{(x + 1)^2}{36} + \dfrac{(y - 3)^2}{9} = 1$

16. $\dfrac{(x + 2)^2}{49} + \dfrac{(y - 4)^2}{16} = 1$

For exercises 17–22, the vertices of an ellipse are given.
(a) Write the equation of the major axis.
(b) Write the equation of the minor axis.
(c) Identify the center.
(d) Write the equation in standard form.

17. Horizontal major axis: $(6, 4)$, $(16, 4)$; vertical minor axis: $(11, 5)$, $(11, 3)$
18. Horizontal major axis: $(2, 7)$, $(14, 7)$; vertical minor axis: $(8, 12)$, $(8, 2)$
19. Horizontal major axis: $(-14, -1)$, $(0, -1)$; vertical minor axis: $(-7, 5)$, $(-7, -7)$
20. Horizontal major axis: $(-20, -3)$, $(-2, -3)$; vertical minor axis: $(-11, 5)$, $(-11, -11)$
21. Vertical major axis: $(3, 15)$, $(3, -5)$; horizontal minor axis: $(1, 5)$, $(5, 5)$
22. Vertical major axis: $(4, 20)$, $(4, 2)$; horizontal minor axis: $(1, 11)$, $(7, 11)$

23. In the definition of an ellipse, $a > b > 0$. Why do you think that a and b cannot be equal?

24. In astronomy, the line segment from the center of the ellipse to a vertex on the major axis is called the **semimajor axis**. Explain why you think the prefix *semi* is used to describe this line segment.

For exercises 25–32, rewrite the equation of the ellipse in standard form.

25. $x^2 + 9y^2 - 4x - 54y + 76 = 0$
26. $x^2 + 16y^2 - 6x - 64y + 57 = 0$
27. $x^2 + 4y^2 + 2x - 16y + 1 = 0$
28. $4x^2 + 9y^2 + 16x - 18y - 11 = 0$

29. $4x^2 + 9y^2 + 24x + 18y + 9 = 0$

30. $9x^2 + 16y^2 + 72x + 32y + 16 = 0$

31. $4x^2 + y^2 + 6y - 91 = 0$

32. $4x^2 + y^2 + 4y - 140 = 0$

For exercises 33–34,
(a) identify the center of the ellipse with the given equation.
(b) is the major axis horizontal or vertical?

33. $\dfrac{(x - 21)^2}{100} + \dfrac{(y + 13)^2}{144} = 1$

34. $\dfrac{(x - 15)^2}{121} + \dfrac{(y + 11)^2}{169} = 1$

For exercises 35–36,
(a) identify the vertices on the major axis.
(b) identify the vertices on the minor axis.

35. $\dfrac{(x - 6)^2}{400} + \dfrac{(y + 8)^2}{196} = 1$ **36.** $\dfrac{(x - 5)^2}{225} + \dfrac{(y + 9)^2}{100} = 1$

For exercises 37–38,
(a) identify the vertices on the major axis.
(b) identify the vertices on the minor axis.

37. $\dfrac{(x + 7)^2}{64} + \dfrac{(y - 3)^2}{16} = 1$ **38.** $\dfrac{(x + 9)^2}{81} + \dfrac{(y - 8)^2}{25} = 1$

For exercises 39–40,
(a) write the equation of the major axis.
(b) write the equation of the minor axis.

39. $\dfrac{(x + 4)^2}{25} + \dfrac{(y - 1)^2}{36} = 1$

40. $\dfrac{(x + 1)^2}{36} + \dfrac{(y - 3)^2}{49} = 1$

41. The floor of the Blue Room in the White House is an ellipse. The length of the major axis is about 40 ft. The length of the minor axis is about 29 ft. If the center of the ellipse is $(0, 0)$, write an equation in standard form that is a model of this ellipse.

42. A doctor is taking a biopsy of a lesion on a patient's arm that may be malignant (cancerous). This removal is an *excision*. To minimize scarring, the excision should be elliptical. The major axis should be parallel to skin tension lines. The major axis should be 3 times the length of the minor axis. If the minor axis is 2 mm and the center is $(0, 0)$, write an equation in standard form that is a model of the ellipse.

43. The design of an elliptical coffee table is based on the famous elliptical window over the main entrance of Mount Vernon. The length of the major axis is 48 in. The length of the minor axis is 32 in. If the center of the ellipse is $(0, 0)$, write an equation in standard form that is a model of this ellipse.

44. The pupil of a goat is elliptical, and the major axis of the ellipse is horizontal. During the day, the length of the major axis is 16 mm and the length of the minor axis is

4 mm. If the center of the ellipse is $(0, 0)$, write an equation in standard form that is a model of this ellipse.

Courtesy of the Author

For exercises 45–48, the astronomical unit (AU) is a unit of distance that is approximately equal to the average distance between the earth and the sun, 93 million miles.

45. The orbit of Mercury is an ellipse. The distance between two vertices on the major axis of the orbit is 0.774 AU. The distance between two vertices on the minor axis of the orbit is 0.757 AU. Assume that the origin $(0, 0)$ is the center of this orbit. Write an equation in standard form that is a model of this orbit.

46. The orbit of Pluto is an ellipse. The distance between two vertices on the major axis of the orbit is 496 AU. The distance between two vertices on the minor axis of the orbit is 480 AU. Assume that the origin $(0, 0)$ is the center of this orbit. Write an equation in standard form that is a model of this orbit.

For exercises 47–48, Kepler's third law of planetary motion is $P = \dfrac{2\pi\sqrt{a^3}}{k}$. The period of an orbiting planet is P, the time it takes to complete one orbit. The length of the semimajor axis, a, is the distance in astronomical units between a vertex on the major axis and the center. Use the approximate value of 3.14 for π.

47. The value of k is $\dfrac{0.017202\ \text{AU}^{\frac{3}{2}}}{1\ \text{day}}$. The length of the semimajor axis of the earth is 1 AU. Find the period of earth's orbit in days. Round to the nearest whole number.

48. The value of k is $\dfrac{0.017202\ \text{AU}^{\frac{3}{2}}}{1\ \text{day}}$. The length of the semimajor axis of Saturn is 29.42 AU. Find the period of Saturn's orbit in days. Round to the nearest whole number.

For exercises 49–50, a *semiellipse* is one-half of an ellipse, divided along the major axis. The transom at the top of the door is a semiellipse.

© Robert Naratham/Shutterstock

(a) Write an equation in standard form that is a model of the *entire ellipse*. Assume that the center of the ellipse is at the midpoint of the bottom of the pictured transom.
(b) Solve this equation for *y*.
(c) Write an equation that is a model of the semielliptical transom.

49. The length of the major axis of a semielliptical window is 66 in. If the window were a full ellipse, the length of the minor axis would be 32 in.

50. The length of the major axis of a semielliptical window is 78 in. If the window were a full ellipse, the length of the minor axis would be 32 in.

Problem Solving: Practice and Review

Follow your instructor's guidelines for using the five steps as outlined in Section 1.5, p. 51.

51. In 2007, about 213 GW (gigawatts) of electric power were produced by coal-fired power plants. Find the amount of carbon dioxide per year produced by these plants. (1 MW = 10^6 watts; 1 GW = 10^9 watts). (*Source:* www.eia.doe.gov, Jan. 21, 2009)

One 500-MW [megawatt] coal-fired power plant produces approximately 3 million tons/year of carbon dioxide. (*Source:* www.pewclimate.org)

52. The final cost of a car including sales tax of 8.25%, is $17,500. The price of the car was marked down 12% from its original price. Find the original price. Round to the nearest whole number.

53. Find the consumption of electricity by state of Hawaii agencies in 2005. Round to the nearest million.

As a result of the Lead by Example program, State agencies consumed 661 million kWh of electricity in Fiscal Year 2009. This represents a 5.8 percent drop from Fiscal Year 2008 (saving $10 million), and a 2.5 percent drop from the baseline year Fiscal Year 2005. This is the first time since the Lead by Example initiative started that consumption has gone below the Fiscal Year 2005 baseline numbers. (*Source:* hawaii.gov, 2010 Annual Report)

54. A poll was conducted of a random sample of 1362 Dallas residents. About one in five believe that it is common for the police to use excessive force, use offensive language, or break laws or police rules. Find the number of people polled who think it is *not* common for the police to behave in these ways. Round to the nearest whole number. (*Source:* www.rand.org, 2008)

Technology

For exercises 55–58, in Section 9.1, we graphed a circle on a graphing calculator by graphing the equations of its semicircles. Use this same procedure to graph an ellipse.
(a) Write the equations of the semiellipses.
(b) Sketch the graph; describe the window.

55. $\dfrac{x^2}{16} + \dfrac{y^2}{25} = 1$

56. $\dfrac{x^2}{9} + \dfrac{y^2}{36} = 1$

57. $\dfrac{x^2}{25} + \dfrac{y^2}{16} = 1$

58. $\dfrac{x^2}{36} + \dfrac{y^2}{9} = 1$

Find the Mistake

For exercises 59–62, the completed problem has one mistake.
(a) Describe the mistake in words, or copy down the whole problem and highlight or circle the mistake.
(b) Do the problem correctly.

59. Problem: Identify the center of the graph of $\dfrac{(x-3)^2}{4^2} + \dfrac{(y+4)^2}{5^2} = 1.$

Incorrect Answer: The center is $(-3, 4)$.

60. Problem: Identify the length of the major axis of $\dfrac{(x+3)^2}{4} + \dfrac{(y-9)^2}{9} = 1.$

Incorrect Answer: Since $a = 2$, the length of the major axis is 4.

61. Problem: Identify the length of the major axis of $\dfrac{(x-2)^2}{4} + \dfrac{(y-1)^2}{9} = 1.$

Incorrect Answer: Since $a = 9$, the length of the major axis is 18.

62. Problem: Identify the major axis of $\dfrac{(x-3)^2}{4^2} + \dfrac{(y+4)^2}{5^2} = 1.$

Incorrect Answer: The major axis is $y = -4$.

Review

For exercises 63–66, solve by completing the square.

63. $x^2 + 4x + 9 = 0$

64. $2x^2 + 6x + 15 = 0$

65. $2x^2 - 5x + 15 = 0$

66. $x^2 - 7x + 3 = 0$

SUCCESS IN COLLEGE MATHEMATICS

67. Some people think that studying mathematics, even mathematics that is not directly useful, helps students learn to be persistent even when problems are difficult. They think that students learn to try another strategy when their first idea does not work. Do you think that studying mathematics has helped you become more persistent? Explain.

When signals strike a parabolic satellite dish, each signal parallel to the axis of symmetry reflects back to the same point. This point, the **focus**, is the receiver of the satellite dish. In this section, we will study parabolas as conic sections. Some parabolas do not represent functions.

SECTION 9.3

Parabolas

After reading the text, working the practice problems, and completing assigned exercises, you should be able to:

1. Write the equation of a parabola.
2. Graph a parabola.
3. Given a quadratic function in vertex form, identify the vertex, the equation of the axis of symmetry, the direction in which the graph opens, the domain, and the range.
4. Use the equation of a parabola as a model.

Conic Sections and Parabolas

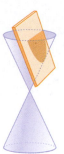

In Section 9.2, we saw that the intersection of a cone and a tilted plane can be an ellipse. If we tilt the plane differently, the intersection is instead a **parabola**. A parabola is another example of a conic section.

> **Parabola**
> A parabola includes all the points (x, y) in a plane such that the distance from (x, y) to a fixed point is equal to the distance from (x, y) to a fixed line. The fixed point is the **focus**. The fixed line is the **directrix**. The line that includes the focus and is perpendicular to the directrix is the **axis of symmetry**. The intersection of the parabola and the axis of symmetry is the **vertex**.

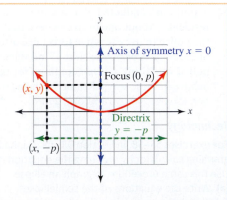

The distance between (x, y) on the parabola and the focus is equal to the distance between (x, y) and the directrix. In a parabola with a vertex at $(0, 0)$ and a focus on either the x-axis or y-axis, the vertical or horizontal distance of the focus from the origin is p.

Forms of the Equation of a Parabola, Vertex at (0, 0), Focus on the *x*- or *y*-Axis, $p > 0$					
Vertex	Focus	Directrix	Equation	Axis of symmetry	Opens
$(0, 0)$	$(0, p)$	$y = -p$	$x^2 = 4py$	$x = 0$	Up
$(0, 0)$	$(0, -p)$	$y = p$	$x^2 = -4py$	$x = 0$	Down
$(0, 0)$	$(p, 0)$	$x = -p$	$y^2 = 4px$	$y = 0$	Right
$(0, 0)$	$(-p, 0)$	$x = p$	$y^2 = -4px$	$y = 0$	Left

Given the equation of a parabola, a sketch of its graph can be drawn using the directrix, focus, vertex, the axis of symmetry, and a point on each of the arms of the parabola.

EXAMPLE 1 The equation of a parabola is $y^2 = -8x$.

(a) Identify the form of the equation.

SOLUTION ▶ The form of this equation is $y^2 = -4px$.

(b) Identify the vertex of the graph.

▶ The vertex is $(0, 0)$.

(c) Identify the direction in which the graph opens and the axis of symmetry.

▶ It opens left. The axis of symmetry is $y = 0$.

(d) Find two other points on the graph.

▶ Choose two values for the squared variable in the original equation, y. If possible, choose values so that y^2 is divisible by -8. In this example, good choices are $y = 4$ and $y = -4$. Replace y with these values, and solve for x.

$$
\begin{array}{c|c}
y^2 = -8x & y^2 = -8x \\
(4)^2 = -8x & (-4)^2 = -8x \\
16 = -8x & 16 = -8x \\
\dfrac{16}{-8} = \dfrac{-8x}{-8} & \dfrac{16}{-8} = \dfrac{-8x}{-8} \\
x = -2 & x = -2
\end{array}
$$

Two points on the parabola are $(-2, 4)$ and $(-2, -4)$.

(e) Identify p, the focus, and the equation of the directrix.

▶ For a parabola in this form, the focus is $(-p, 0)$, and the equation of the directrix is $x = p$. The equation of this parabola is $y^2 = -8x$. The form of this parabola is $y^2 = -4px$. Since $y^2 = y^2$, we can solve for p.

$$
\begin{aligned}
y^2 &= y^2 \\
-8x &= -4px && \text{Replace } y^2. \\
\dfrac{-8x}{-4x} &= \dfrac{-4px}{-4x} && \text{Division property of equality.} \\
2 &= p && \text{Simplify.}
\end{aligned}
$$

Since $p = 2$, the focus is $(-2, 0)$, and the equation of the directrix is $x = 2$.

(f) Sketch the graph. Label the focus, the vertex, the axis of symmetry, and the directrix.

▶ Graph the vertex $(0, 0)$, the points $(-2, 4)$ and $(-2, -4)$, the focus, the axis of symmetry, and the directrix. Then sketch the rest of the parabola, opening to the left.

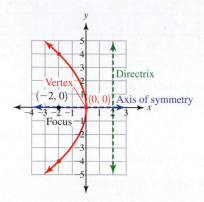

(g) Does this graph represent a function?

▶ Since the graph fails the vertical line test, it does not represent a function.

Sketching the Graph of a Parabola with Vertex (0, 0), Given Its Equation

1. Identify the form of the equation, the vertex, and the direction in which the parabola opens.
2. Find two other points on the parabola by choosing numbers to replace the squared variable in the equation of the parabola and solving for the other variable.
3. Identify the focus and directrix.
4. Graph the vertex, the two points, the focus, and the directrix. Sketch the graph of the parabola.

Given the vertex, the axis of symmetry, and a point on the graph, we can write the equation of a parabola.

EXAMPLE 2 The vertex of a parabola is $(0, 0)$, the axis of symmetry is $x = 0$, and another point on the parabola is $(8, -2)$.

(a) Identify the direction in which the graph opens and the form of its equation.

SOLUTION ▶ Since the axis of symmetry is the vertical line $x = 0$ (the y-axis), this parabola opens up or down (not right or left). Since the vertex is $(0, 0)$ and the point $(8, -2)$ is below the vertex, the parabola must open down. The form of a parabola that opens down is $x^2 = -4py$.

(b) Identify p, the focus, and the equation of the directrix.

▶
$$x^2 = -4py \qquad \text{Form of a parabola opening down.}$$
$$8^2 = -4p(-2) \qquad \text{A point is } (8, -2); \text{ replace } x \text{ with 8 and } y \text{ with } -2.$$
$$64 = 8p \qquad \text{Simplify.}$$
$$\frac{64}{8} = \frac{8p}{8} \qquad \text{Division property of equality.}$$
$$8 = p \qquad \text{Simplify.}$$

Since $p = 8$, the parabola opens down, and the focus is directly below the vertex, the focus is $(0, -8)$. The equation of the directrix is $y = 8$.

(c) Write the equation of this parabola.

$$x^2 = -4py \qquad \text{Form of a parabola opening down.}$$
$$x^2 = -4(\mathbf{8})(y) \qquad \text{Replace } p \text{ with 8.}$$
$$x^2 = \mathbf{-32}y \qquad \text{Simplify.}$$

(d) Sketch the graph. Label the point $(8, -2)$, the vertex, the axis of symmetry, the focus, and the directrix.

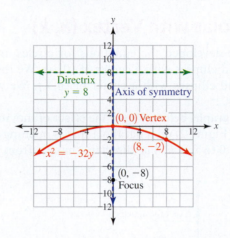

(e) Does this graph represent a function?

Since the graph passes the vertical line test, it represents a function.

Writing the Equation of a Parabola with Vertex (0, 0), Given the Equation of the Axis of Symmetry and One Other Point, (x_1, y_1)

1. Identify the direction in which the parabola opens and the form of its equation.

2. Replace x and y in the equation with the coordinates of (x_1, y_1); solve for p.

3. Replace p, x_1, and y_1 in the equation; simplify.

Practice Problems

For problems 1–2,
(a) identify the form of this equation.
(b) identify the vertex of the graph.
(c) identify the direction in which the graph opens and the axis of symmetry.
(d) find two other points on the graph.
(e) identify p, the focus, and write the equation of the directrix.
(f) sketch the graph, labeling the focus, the vertex, the axis of symmetry, and the directrix.
(g) does this equation represent a function?

1. $x^2 = 16y$ 　　　**2.** $y^2 = 24x$

3. The vertex of a parabola is $(0, 0)$, the axis of symmetry is the x-axis, and the graph includes the point $(5, 10)$.

    a. Identify the direction in which the graph opens.

    b. Identify the form of this equation.

    c. Identify the focus, and write the equation of the directrix.

    d. Write the equation of this parabola.

    e. Sketch the graph, labeling the focus, the vertex, the axis of symmetry, and the directrix.

    f. Does this equation represent a function?

Parabolas with Vertex (h, k)

When we developed the equations for circles and ellipses, we began by studying circles and ellipses with a center of $(0, 0)$. We then extended our understanding to circles and ellipses with a center of (h, k). We will follow this same process with parabolas.

In the graphs below, we see the effect of moving the vertex from $(0, 0)$ (Figure 1) to (h, k) (Figure 2) on a parabola that is opening up. The axis of symmetry changes from $x = 0$ to $x = h$. The directrix changes from $y = -p$ to $y = k - p$.

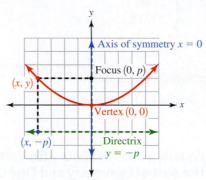

Figure 1

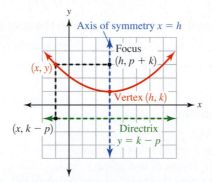

Figure 2

Forms of the Equation of a Parabola, Vertex at (h, k), Focus on a Line Parallel to the x- or y-Axis, $p > 0$					
Vertex	Focus	Directrix	Equation	Axis of symmetry	Opens
(h, k)	$(h, k + p)$	$y = k - p$	$(x - h)^2 = 4p(y - k)$	$x = h$	Up
(h, k)	$(h, k - p)$	$y = k + p$	$(x - h)^2 = -4p(y - k)$	$x = h$	Down
(h, k)	$(h + p, k)$	$x = h - p$	$(y - k)^2 = 4p(x - h)$	$y = k$	Right
(h, k)	$(h - p, k)$	$x = h + p$	$(y - k)^2 = -4p(x - h)$	$y = k$	Left

EXAMPLE 3 The vertex of a parabola is $(3, -2)$, the axis of symmetry is $y = -2$, and the graph includes the point $(4, 6)$.

(a) Identify the direction in which the graph opens and the form of its equation.

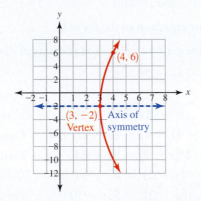

SOLUTION ▶ Since the axis of symmetry is the horizontal line $y = -2$, this parabola opens right or left. Since the vertex is $(3, -2)$ and the point $(4, 6)$ is to the right of the vertex, the parabola must open to the right. The form of a parabola opening to the right is $(y - k)^2 = 4p(x - h)$.

(b) Identify p, the focus, and the equation of the directrix.

▶ Replace h and k in $(y - k)^2 = 4p(x - h)$ with the coordinates of the vertex, and replace x and y with $(4, 6)$. Solve for p.

$$(y - k)^2 = 4p(x - h)$$ Form of a parabola opening to the right.

$$(\mathbf{6} - (\mathbf{-2}))^2 = 4p(\mathbf{4} - 3)$$ Replace variables: $h = 3; k = -2, y = 6, x = 4$

$$64 = 4p(\mathbf{1})$$ Simplify.

$$\frac{64}{4} = \frac{4p}{4}$$ Division property of equality.

$$16 = p$$ Simplify.

Since $p = 16$ and the parabola opens right, the focus is 16 units directly to the right of the vertex, $(3, -2)$. The focus is $(19, -2)$. The directrix is a vertical line passing through the point that is 16 units to the left of the vertex. The equation of the directrix is $x = -13$.

(c) Write the equation of this parabola.

▶ $$(y - k)^2 = 4p(x - h)$$ Form of a parabola opening to the right.

$$(y - (-2))^2 = 4(\mathbf{16})(x - 3)$$ Replace p with 16.

$$(y + \mathbf{2})^2 = \mathbf{64}(x - 3)$$ Simplify.

Writing the Equation of a Parabola with Vertex (h, k), Given the Equation of the Axis of Symmetry and One Other Point

1. Identify the direction in which the parabola opens and the form of its equation.

2. Replace h and k with the coordinates of the vertex; solve for p.

3. Replace h, k, and p in the equation; simplify.

In the next example, we use completing the square to rewrite an equation in one of the forms of a parabola.

EXAMPLE 4 The equation of a parabola is $x^2 - 4x - 8y + 28 = 0$.

(a) Rewrite this equation in the form $(x - h)^2 = 4p(y - k)$.

SOLUTION ▶ Since x is the only variable that is squared, isolate the terms that include x.

$$x^2 - 4x - 8y + 28 = 0$$

$$\underline{+8y \;\; -28 \;\; +8y \;\; -28} \qquad \text{Properties of equality.}$$

$x^2 - 4x = 8y - 28$	x^2 and $4x$ are isolated.
$(x^2 - 4x + 4) = 8y - 28 + 4$	Complete the square.
$(x - 2)(x - 2) = 8y - 24$	Factor; simplify.
$(x - 2)^2 = 8y - 24$	Use exponential notation.
$(x - 2)^2 = 8(y - 3)$	Factor.
$(x - 2)^2 = 4(2)(y - 3)$	Rewrite 8 as 4(2); this is $4p$.

(b) Identify the direction in which the graph opens.

▶ The graph of a parabola in the form $(x - h)^2 = 4p(y - k)$ opens up.

(c) Identify the vertex.

▶ The vertex, (h, k), of $(x - 2)^2 = 4(2)(y - 3)$ is $(2, 3)$.

(d) Write the equation of the axis of symmetry.

▶ The equation of the axis of symmetry, $x = h$, is $x = 2$.

(e) Identify p and the focus.

▶ The focus is $(h, k + p)$, $(2, 3 + 2)$, which simplifies to $(2, 5)$.

(f) Write the equation of the directrix.

▶ The equation of the directrix is $y = k - p$, $y = 3 - 2$, which simplifies to $y = 1$.

(g) Find two other points on the parabola.

▶ Choose two values for the squared variable, x, and solve for y. If possible, the expression $(x - 2)^2$ should be divisible by 8.

Choose: $x = 6$		Choose: $x = -2$	
$(x - 2)^2 = 8(y - 3)$	Equation of the parabola.	$(x - 2)^2 = 8(y - 3)$	
$(6 - 2)^2 = 8(y - 3)$	Replace x.	$(-2 - 2)^2 = 8(y - 3)$	
$4^2 = 8(y - 3)$	Simplify.	$(-4)^2 = 8(y - 3)$	
$16 = 8(y - 3)$	Simplify.	$16 = 8(y - 3)$	
$\dfrac{16}{8} = \dfrac{8(y - 3)}{8}$	Division property of equality.	$\dfrac{16}{8} = \dfrac{8(y - 3)}{8}$	
$2 = y - 3$	Simplify.	$2 = y - 3$	
$\underline{+3 \qquad +3}$	Addition property of equality.	$\underline{+3 \qquad +3}$	
$5 = y + 0$	Simplify.	$5 = y + 0$	
$5 = y$	Simplify.	$5 = y$	

Two points on the parabola are $(6, 5)$ and $(-2, 5)$.

(h) Sketch the graph. Label the axis of symmetry, the focus, the vertex, and the directrix.

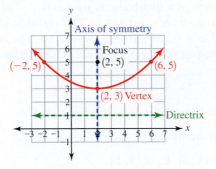

(i) Does this graph represent a function?

Yes, the graph passes the vertical line test.

Writing the Equation of a Parabola by Completing the Square

1. Identify the variable that is raised to the second power. Isolate the terms with this variable, and use parentheses to group them.

2. Complete the square; add the same number to both sides of the equation.

3. Factor the square; write in exponential notation.

4. Factor out the greatest common factor of the expression on the other side of the equation.

5. Rewrite the greatest common factor as a product of 4 and a constant.

Practice Problems

4. The vertex of a parabola is $(1, 4)$. The axis of symmetry is $y = 4$. Its graph includes the point $(3, 8)$.
   a. Identify the direction the graph opens.
   b. Identify the form of the equation.
   c. Identify p, the focus, and the equation of the directrix.
   d. Write the equation of this parabola.

5. The equation of a parabola is $y^2 - 4y + 12x - 56 = 0$.
   a. Rewrite this equation in the form $(y - k)^2 = -4p(x - h)$.
   b. Identify the direction in which the graph opens.
   c. Identify the vertex.
   d. Write the equation of the axis of symmetry.
   e. Identify p and the focus.
   f. Write the equation of the directrix.
   g. Find two other points on the parabola.
   h. Sketch the graph. Label the axis of symmetry, focus, vertex, and directrix.
   i. Does this graph represent a function?

Quadratic Functions

The graphs of some parabolas represent functions. These graphs open up or down. Parabolas that open left or right do not pass the vertical line test and are not functions.

> **Vertex Form of a Quadratic Equation in One Variable**
>
> Vertex form of a quadratic equation is $y = a(x - h)^2 + k$, where a, h, and k are real numbers and $a \neq 0$. The vertex is (h, k). The axis of symmetry is $x = h$. If $a > 0$, the parabola opens up. The *minimum* output value is k. If $a < 0$, the parabola opens down. The *maximum* output value is k.

EXAMPLE 5 | A quadratic function is $y = 3(x - 4)^2 + 5$.

(a) Identify the vertex.

SOLUTION ▶   The vertex, (h, k), is $(4, 5)$.

(b) Write the equation of the axis of symmetry.

▶   The equation of the axis of symmetry, $x = h$, is $x = 4$.

(c) Identify the direction in which the graph opens.

▶   Since $a = 3$ and $3 > 0$, the graph opens up.

(d) Use interval notation to represent the domain.

▶   Like all polynomial functions, the domain of a quadratic function is the set of real numbers, $(-\infty, \infty)$.

(e) Use interval notation to represent the range.

▶   The vertex is $(4, 5)$, and the graph opens up. The minimum output value is 5. The range is $[5, \infty)$.

Practice Problems

For problems 6–7,
(a) identify the vertex.
(b) write the equation of the axis of symmetry.
(c) identify the direction in which the graph opens.
(d) use interval notation to represent the domain.
(e) use interval notation to represent the range.

6. $y = 2(x - 1)^2 - 3$ **7.** $y = -2(x + 4)^2 + 1$

Applications

If we spin a parabola around its axis of symmetry, we create a "parabolic dish" such as a parabolic mirror or a satellite dish. Any ray of light or other energy that is parallel to the axis of symmetry hits the dish and reflects to the focus. This concentrates a satellite signal or solar energy at one location.

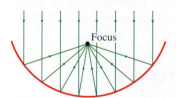

Focus

EXAMPLE 6 When too many people in an area use wood for heat or cooking, deforestation can occur. Parabolic solar cookers are an alternative to heating with wood. The heat generated is enough to boil water and fry meat. A cooking pot is placed at the focus of a parabolic mirror. Rays of sunlight that are parallel to the axis of symmetry hit the mirror and reflect to the focus. The parabolic mirror of a cooker is 20 ft across at its edge. It is 6 ft deep. Find the distance of the focus from the vertex in feet and inches.

© Jonathan Noden-Wilkinson/Shutterstock

SOLUTION ▶ Put the vertex of the dish at $(0, 0)$ with the mirror opening up. The equation is $x^2 = 4py$. The focus is $(0, p)$. Since the mirror is 20 ft across (a radius of 10 ft) and 6 ft deep, a point on the edge of the parabola is $(10 \text{ ft}, 6 \text{ ft})$.

$x^2 = 4py$	Parabola opening up.
$(\mathbf{10 \text{ ft}})^2 = 4p(\mathbf{6 \text{ ft}})$	Replace x with 10 ft and y with 6 ft.
$\mathbf{100 \text{ ft}^2} = (\mathbf{24 \text{ ft}})p$	Simplify.
$\dfrac{100 \text{ ft}^2}{24 \text{ ft}} = \dfrac{(24 \text{ ft})}{24 \text{ ft}}p$	Division property of equality.
$\dfrac{25}{6} \text{ ft} = p$	Simplify.
$4\dfrac{1}{6} \text{ ft} = p$	Rewrite as a mixed number.
$\mathbf{4 \text{ ft } 2 \text{ in.}} = p$	Change units; $\dfrac{1}{6} \text{ ft} \cdot \dfrac{12 \text{ in.}}{\text{ft}} = 2 \text{ in.}$

The focus is 4 ft 2 in. from the vertex on the axis of symmetry.

In a flashlight, the light source is at the focus. Light rays that strike the parabolic mirror reflect parallel to the axis of symmetry.

EXAMPLE 7 Designers of a small flashlight want the depth of its parabolic mirror at the center to be 1 in. The light source will be 0.5 in. away from the vertex of the mirror. What should the diameter of the mirror be? Round to the nearest tenth.

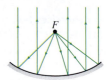

SOLUTION ▶ Put the vertex of the reflector at $(0, 0)$, opening up. The equation is $x^2 = 4py$. The focus is $(0, 0.5 \text{ in.})$. Since the reflector is 1 in. deep, a point on the edge of the parabola is $(x, 1 \text{ in.})$. The value of x is the radius of the mirror. To find the diameter, double this radius.

$x^2 = 4py$	Parabola opening up.
$x^2 = 4(\mathbf{0.5 \text{ in.}})(\mathbf{1 \text{ in.}})$	Replace p with 0.5 in. and y with 1 in.
$x^2 = \mathbf{2 \text{ in.}^2}$	Simplify.
$\sqrt{x^2} = \pm\sqrt{2} \text{ in.}$	Take the square root of both sides.
$x \approx 1.4 \text{ in.}$	Round; distances are positive.

Since the radius is about 1.4 in., the diameter should be about 2.8 in.

Practice Problems

8. The parabolic reflector of a satellite dish is 4 ft across at its edge. It is 1.5 ft deep. Find the distance of the focus from the vertex in feet and inches.

9. The parabolic mirror of a searchlight is 3 ft deep at the center. The focus is 1.5 ft from the vertex. What should the diameter of the mirror be? Write the answer in feet and inches. Round to the nearest tenth of an inch.

9.3 VOCABULARY PRACTICE

Match the term with its description.

1. A collection of points such that the distance from any point to the directrix is equal to the distance from the point to the focus	**A.** axis of symmetry
2. A set of ordered pairs	**B.** conic section
3. A set of ordered pairs in which each input corresponds to exactly one output	**C.** directrix
4. The intersection of a plane and a cone; circles, ellipses, and parabolas are examples of this.	**D.** function
5. The distance from a point on a parabola to this line is the same as the distance from the point to the focus.	**E.** parabola
6. $y = a(x - h)^2 + k$	**F.** parallel lines
7. The line that includes the focus of a parabola and is perpendicular to the directrix	**G.** perpendicular lines
8. The intersection of a parabola and its axis of symmetry	**H.** relation
9. Lines that never intersect	**I.** vertex
10. Lines that intersect and form a 90-degree angle	**J.** vertex form of a quadratic function

9.3 Exercises

Follow your instructor's guidelines for showing your work.

For exercises 1–4,
(a) identify the vertex.
(b) write the equation of the axis of symmetry.
(c) identify the direction in which the parabola opens.

1. $y = 3(x - 4)^2 - 5$

2. $y = 5(x - 6)^2 - 9$

3. $y = -3(x + 4)^2 - 5$

4. $y = -5(x + 6)^2 - 9$

For exercises 5–20,
(a) identify the form of the equation.
(b) identify the vertex.
(c) identify the direction in which the graph opens and the axis of symmetry.
(d) find two other points on the graph.
(e) identify p, the focus, and the equation of the directrix.
(f) sketch the graph, labeling the focus, the vertex, the directrix, and the axis of symmetry.
(g) does the graph represent a function?

5. $x^2 = 32y$

6. $x^2 = 4y$

7. $y^2 = 4x$

8. $y^2 = 32x$

9. $x^2 = -40y$

10. $x^2 = -20y$

11. $y^2 = -20x$

12. $y^2 = -40x$

13. $y^2 = 24x$

14. $y^2 = 12x$

15. $x^2 = -28y$

16. $x^2 = -36y$

17. $y^2 = -32x$

18. $y^2 = -28x$

19. $x^2 = 8y$

20. $x^2 = 16y$

For exercises 21–36, a parabola has the given axis of symmetry and point; its vertex is $(0, 0)$.
(a) Identify the direction that the graph opens and the form of its equation.
(b) Identify the focus and the equation of the directrix.
(c) Write the equation of this parabola.
(d) Sketch the graph, labeling the focus, the vertex, the directrix, and the axis of symmetry.
(e) Does this graph represent a function?

21. $y = 0$; $(6, 12)$

22. $y = 0$; $(3, 6)$

23. $x = 0$; $(14, -7)$

24. $x = 0$; $(18, -9)$

25. $y = 0$; $(-8, 16)$

26. $y = 0$; $(-7, 14)$

27. $x = 0$; $(4, 2)$

28. $x = 0$; $(4, 1)$

29. $x = 0$; $(16, 8)$

30. $x = 0$; $(4, 4)$

31. $y = 0$; $(4, 4)$

32. $y = 0$; $(2, 8)$

33. $x = 0$; $(20, -10)$

34. $x = 0$; $(10, -5)$

35. $y = 0$; $(-5, 10)$

36. $y = 0$; $(-10, 20)$

For exercises 37–48, write the equation of the parabola with the given vertex, point on the parabola, and axis of symmetry.

37. Vertex: $(1, 1)$; point: $(3, 0)$; $x = 1$

38. Vertex: $(2, 2)$; point: $(6, 0)$; $x = 2$

39. Vertex: $(5, 8)$; point: $(11, 9)$; $x = 5$

40. Vertex: $(3, 7)$; point: $(11, 9)$; $x = 3$

41. Vertex: $(5, 8)$; point: $(8, 20)$; $y = 8$

42. Vertex: $(3, 7)$; point: $(8, 17)$; $y = 7$

43. Vertex: $(-2, 5)$; point: $(-7, 25)$; $y = 5$

44. Vertex: $(-6, 1)$; point: $(-13, 15)$; $y = 1$

45. Vertex: $(-4, -2)$; point: $(14, -11)$; $x = -4$

46. Vertex: $(-7, -8)$; point: $(1, -9)$; $x = -7$

47. Vertex: $(5, 3)$; point: $(6, 6)$; $y = 3$

48. Vertex: $(7, 11)$; point: $(10, 2)$; $y = 11$

For exercises 49–56, sketch the graph. Label the directrix, focus, vertex, and axis of symmetry on the graph.

49. $(y - 3)^2 = 12(x - 1)$

50. $(y - 2)^2 = 16(x - 3)$

51. $(x - 5)^2 = -20(y + 4)$

52. $(x + 2)^2 = -32(y - 4)$

53. $(y + 2)^2 = -12(x - 5)$

54. $(y + 3)^2 = -8(x - 1)$

55. $(x + 1)^2 = 8(y + 6)$

56. $(x + 2)^2 = 12(y + 5)$

For exercises 57–66,

(a) rewrite the equation in one of the forms of a parabola.

(b) identify the direction in which the graph opens.

(c) identify the vertex.

(d) write the equation of the axis of symmetry.

(e) identify the focus.

(f) write the equation of the directrix.

(g) find two other points on the parabola.

(h) sketch the graph. Label the axis of symmetry, focus, vertex, and directrix.

(i) does the graph represent a function?

57. $x^2 - 12x + 8y + 28 = 0$

58. $x^2 - 10x - 12y + 49 = 0$

59. $x^2 + 6x - 8y + 1 = 0$

60. $x^2 + 2x - 8y - 23 = 0$

61. $y^2 + 4x - 8y + 8 = 0$

62. $y^2 - 16x + 4y + 20 = 0$

63. $y^2 + 20x - 4y + 4 = 0$

64. $y^2 + 8x - 6y + 9 = 0$

65. $x^2 - 12y - 60 = 0$

66. $x^2 - 4y - 24 = 0$

For exercises 67–74,

(a) identify the vertex, (h, k).

(b) write the equation of the axis of symmetry.

(c) identify the direction in which the graph opens.

(d) use interval notation to represent the domain.

(e) use interval notation to represent the range.

67. $y = 2(x - 3)^2 - 6$

68. $y = 3(x - 2)^2 - 5$

69. $y = -4(x + 1)^2 - 9$

70. $y = -6(x + 2)^2 - 7$

71. $y = 5(x + 8)^2 + 2$

72. $y = 7(x + 2)^2 + 1$

73. $y = x^2 - 4$

74. $y = x^2 - 6$

75. The diameter of the parabolic reflector of a small satellite dish is 18 in. The depth is 2 in. Find the distance on the axis of symmetry from the focus to the vertex.

76. The diameter of the parabolic reflector of a large satellite dish is 36 ft. The depth is 4.5 ft. Find the distance on the axis of symmetry from the focus to the vertex.

77. The parabolic mirror of a flashlight is 3 in. deep at the center. The focus is 1 in. from the vertex. What should the diameter of the mirror be? Round to the nearest tenth.

78. The parabolic mirror of a flashlight is 4 cm deep at the center. The focus is 2 cm from the vertex. What should the diameter of the mirror be? Round to the nearest tenth.

Problem Solving: Practice and Review

Follow your instructor's guidelines for using the five steps as outlined in Section 1.5, p. 51.

79. Find the number of vehicle repossessions in 2006. Write in place value notation. Round to the nearest thousand.

The number of vehicle repossessions is expected to rise 5% this year. That's after it jumped 12% to 1.67 million nationally in 2008. . . . That followed a 9% increase in 2007. (*Source:* www .usatoday.com, Feb. 27, 2009)

80. In a study of 38,480 children, two out of twenty-five children had a food allergy. Of these children, 38.7% had a history of severe reactions. Find the number of children with food allergies who had a history of severe reactions. Round to the nearest whole number. (*Source:* Gupta, *Pediatrics,* July 1, 2011)

81. The enrollment at a college in 2010 was 3578 students. The dean of admissions predicts that enrollment will, on average, grow at a linear rate of 125 students per year.
    a. Write a linear model that represents the relationship of the number of years since 2010, x, and the enrollment at the college, y.
    b. Find the enrollment at the college in 2015.

82. The perimeter of a rectangle is 82 in. Its area is 378 in.^{2}. Use a quadratic equation in one variable to find the length and width of the rectangle.

Technology

For exercises 83–86,
(a) solve the equation for y.
(b) graph the equation. Choose a window that shows the vertex and both arms of the parabola. Sketch the graph; describe the window.

83. $x^2 - 6x - 12y + 21 = 0$

84. $x^2 - 8x + 8y + 32 = 0$

85. $x^2 + 8x + 2y - 8 = 0$

86. $x^2 + 10x - 4y + 33 = 0$

Find the Mistake

For exercises 87–90, the completed problem has one mistake.
(a) Describe the mistake in words, or copy down the whole problem and highlight or circle the mistake.
(b) Do the problem correctly.

87. **Problem:** The equation of a parabola is $(y + 3)^2 = 12(x - 1)$. Identify the vertex.

    Incorrect Answer: The vertex is $(-3, 1)$.

88. **Problem:** The equation of a parabola is $(y + 3)^2 = 12(x - 1)$. Find the focus.

    Incorrect Answer: The focus is $(0, 3)$.

89. **Problem:** The equation of a parabola is $(y + 1)^2 = -20(x - 3)$. Write the equation of the axis of symmetry.

    Incorrect Answer: The axis of symmetry is $x = 3$.

90. **Problem:** The equation of a parabola is $(y - 4)^2 = -16(x - 1)$. Identify the direction in which the parabola opens.

    Incorrect Answer: Because $p = -4$, this parabola opens down.

Review

91. Write the slope-intercept form of a linear equation in two variables.

92. Identify the slope and the y-intercept of the graph of $y = -\dfrac{3}{4}x$.

93. Identify the slope and y-intercept of the graph of $y = \dfrac{3}{4}x$.

94. **a.** Are the lines $y = -\dfrac{3}{4}x$ and $y = \dfrac{3}{4}x$ perpendicular?
    b. Explain.

SUCCESS IN COLLEGE MATHEMATICS

95. Some people think that studying mathematics helps students develop their ability to solve problems. Do you think that your work in solving problems in mathematics can help you solve problems in other areas of your life or career? Explain.

96. Is logical thought an important skill in the career that you hope to have? Explain.

© Neo Edmund/Shutterstock

The Hubble telescope is a Cassegrain reflecting telescope. It contains a parabolic mirror and a hyperbolic mirror. In this section, we will learn about hyperbolas and hyperbolic reflectors.

SECTION 9.4

Hyperbolas

After reading the text, working the practice problems, and completing assigned exercises, you should be able to:

1. Write the equation of a hyperbola.

2. Given its equation, identify the foci, vertices, and transverse axis of a hyperbola.

3. Use asymptotes to sketch the graph of a hyperbola.

4. Use the equation of a hyperbola as a model.

Conic Sections and Hyperbolas

In Section 9.1, we saw that the intersection of a cone and a horizontal plane is a circle. If the plane is vertical and it intersects two cones, the intersection is instead a **hyperbola**. A hyperbola is another example of a **conic section**.

> ### Hyperbola
>
> A hyperbola includes all the points (x, y) in a plane such that the absolute value of the difference of the distances from two fixed points is a constant. These points form two separate curves, called **branches**. Each of the fixed points is a focus of the hyperbola. The line containing the foci is the **transverse axis**. The midpoint of the line segment between the foci is the **center**. The line that is perpendicular to the transverse axis and passes through the center is the **conjugate axis**. The **vertices** are the points of intersection of the hyperbola with the transverse axis.

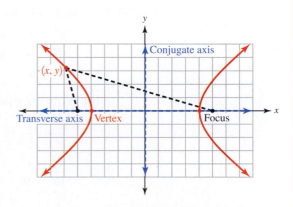

A *hyperbola* is a conic section. Although it sounds similar, a *hyperbole* is not a conic section; it is a term for an outrageous claim used for effect. Describing someone's feet by saying that they are as "big as a barge" is a hyperbole. Descriptions such as "slower than molasses" and "faster than lightning" are also hyperboles.

Forms of the Equation of a Hyperbola, Center at $(0, 0)$, $p > a$, $b^2 = p^2 - a^2$				
Center	**Foci**	**Vertices**	**Transverse axis**	**Equation**
$(0, 0)$	$(-p, 0)$ and $(p, 0)$	$(-a, 0)$ and $(a, 0)$	$y = 0$	$\dfrac{x^2}{a^2} - \dfrac{y^2}{b^2} = 1$
$(0, 0)$	$(0, -p)$ and $(0, p)$	$(0, -a)$ and $(0, a)$	$x = 0$	$\dfrac{y^2}{a^2} - \dfrac{x^2}{b^2} = 1$

Given the center, a focus, a vertex, and the equation of the transverse axis, we can use the relationship $b^2 = p^2 - a^2$ to write the equation of a hyperbola.

EXAMPLE 1 A hyperbola is centered at $(0, 0)$, the transverse axis is $y = 0$, a focus is $(5, 0)$, and a vertex is $(4, 0)$.

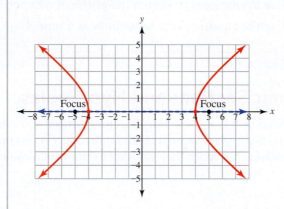

(a) Identify b^2.

SOLUTION ▶ Since a focus is $(5, 0)$, $p = 5$. Since a vertex is $(4, 0)$, $a = 4$.

$$b^2 = p^2 - a^2$$
$$b^2 = \mathbf{5}^2 - \mathbf{4}^2 \qquad p = 5; a = 4$$
$$b^2 = \mathbf{9} \qquad \text{Simplify.}$$

(b) Write the equation of the hyperbola.

▶ $$\frac{x^2}{a^2} - \frac{y^2}{b^2} = 1 \qquad \text{Transverse axis: } y = 0$$

$$\frac{x^2}{\mathbf{16}} - \frac{y^2}{\mathbf{9}} = 1 \qquad a^2 = 16; b^2 = 9$$

The foci and vertices are on the transverse axis of a hyperbola. Given a focus and vertex, we can identify the transverse axis.

EXAMPLE 2 The foci of a hyperbola are $(0, 6)$ and $(0, -6)$. One vertex of the hyperbola is $(0, 3)$.

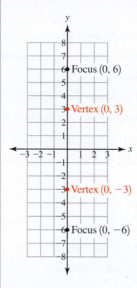

(a) Identify the transverse axis.

SOLUTION ▶ The foci and vertices are on the transverse axis. Since $(0, 6)$, $(0, -6)$, and $(0, 3)$ are points on the line $x = 0$ (the y-axis), this is the transverse axis.

(b) Identify p.

▶ Since a focus is $(0, 6)$, $p = 6$.

(c) Identify a.

▶ Since a vertex is $(0, 3)$, $a = 3$.

(d) Identify b^2.

▶ $b^2 = p^2 - a^2$
$b^2 = \mathbf{6}^2 - \mathbf{3}^2$ $p = 6; a = 3$
$b^2 = \mathbf{27}$ Simplify.

(e) Write the equation of this hyperbola.

▶ $\dfrac{y^2}{a^2} - \dfrac{x^2}{b^2} = 1$ Transverse axis: $x = 0$

$\dfrac{y^2}{\mathbf{9}} - \dfrac{x^2}{\mathbf{27}} = 1$ $a^2 = 9; b^2 = 27$

Given the equation of a hyperbola, we can identify the foci and vertices.

EXAMPLE 3 The equation of a hyperbola is $\dfrac{x^2}{25} - \dfrac{y^2}{64} = 1$.

(a) Identify the form of this equation and the transverse axis.

SOLUTION ▶ $\dfrac{x^2}{a^2} - \dfrac{y^2}{b^2} = 1$ Transverse axis: $y = 0$

(b) Identify p and the foci.

$$b^2 = p^2 - a^2$$
$$64 = p^2 - 25 \qquad a^2 = 25;\ b^2 = 64$$
$$\underline{+25 \qquad\quad +25} \qquad \text{Addition property of equality.}$$
$$\mathbf{89} = p^2 + \mathbf{0} \qquad \text{Simplify.}$$
$$\sqrt{89} = \sqrt{p^2} \qquad \text{Square root both sides.}$$
$$\pm\sqrt{89} = p \qquad \text{Simplify; use} \pm \text{notation.}$$

Since distance is positive, the only solution we use is $p = \sqrt{89}$. Since the transverse axis is $y = 0$, the foci are $(-p, 0)$ and $(p, 0)$ or $(-\sqrt{89}, 0)$ and $(\sqrt{89}, 0)$.

(c) Identify a and the vertices.

$$a^2 = 25$$
$$\sqrt{a^2} = \sqrt{25} \qquad \text{Square root both sides.}$$
$$a = \pm 5 \qquad \text{Simplify; use} \pm \text{notation.}$$

Since the transverse axis is $y = 0$, the vertices are $(-a, 0)$ and $(a, 0)$ or $(-5, 0)$ and $(5, 0)$.

Practice Problems

1. The center of a hyperbola is $(0, 0)$. The transverse axis is $x = 0$. A focus is $(0, -7)$. A vertex is $(0, -3)$.
   a. Identify the form of the equation of this hyperbola.
   b. Identify a, p, and b.
   c. Write the equation of this hyperbola.
2. The vertices of a hyperbola are $(-12, 0)$ and $(12, 0)$. A focus is $(-14, 0)$.
   a. Identify the form of the equation of this hyperbola.
   b. Identify a, p, and b.
   c. Write the equation of this hyperbola.
3. The equation of a hyperbola is $\dfrac{y^2}{16} - \dfrac{x^2}{100} = 1$.
   a. Identify the form of this equation and the transverse axis.
   b. Identify a, b, and p.
   c. Identify the foci.
   d. Identify the vertices.

Using Asymptotes to Graph a Hyperbola

In Section 5.5, we saw that a restriction in the domain of a rational function can result in a vertical asymptote on its graph. The graph of the rational function gets close to, but never touches, the vertical line that passes through the x-axis at the restricted value (Figure 1).

The graph of a hyperbola has two asymptotes. When the center of the hyperbola is $(0, 0)$, the asymptotes are *diagonal lines* with opposite slopes that pass through $(0, 0)$.

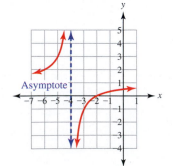

Figure 1

Asymptotes of a Hyperbola, Center at (0, 0), $p > a$, $b^2 = p^2 - a^2$

Center	Foci vertices	Transverse axis	Equation	Equation of asymptotes
$(0, 0)$	$(-p, 0)$ and $(p, 0)$ $(-a, 0)$ and $(a, 0)$	$y = 0$	$\dfrac{x^2}{a^2} - \dfrac{y^2}{b^2} = 1$	$y = -\dfrac{b}{a}x$ and $y = \dfrac{b}{a}x$
$(0, 0)$	$(0, -p)$ and $(0, p)$ $(0, -a)$ and $(0, a)$	$x = 0$	$\dfrac{y^2}{a^2} - \dfrac{x^2}{b^2} = 1$	$y = -\dfrac{a}{b}x$ and $y = \dfrac{a}{b}x$

EXAMPLE 4 The equation of a hyperbola is $\dfrac{x^2}{4} - \dfrac{y^2}{1} = 1$.

(a) Identify a and b.

SOLUTION ▶ Since the form of the hyperbola is $\dfrac{x^2}{a^2} - \dfrac{y^2}{b^2} = 1$, $a^2 = 4$ and $b^2 = 1$. Since a and b are greater than 0, $a = 2$ and $b = 1$.

(b) Write the equations of the asymptotes.

▶ For this form of a hyperbola, the equations of the asymptotes are $y = -\dfrac{b}{a}x$ and $y = \dfrac{b}{a}x$. Since $a = 2$ and $b = 1$, the equations are $y = -\dfrac{1}{2}x$ and $y = \dfrac{1}{2}x$.

(c) Graph the vertices and asymptotes. Sketch the graph of the hyperbola.

▶ Because it is not part of the graph of the hyperbola, an asymptote is a dashed line. Sketch the graph of the hyperbola. Each branch passes through its vertex. It approaches the asymptote but never touches it.

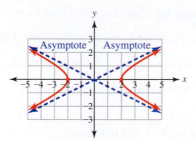

Practice Problems

For problems 4–6,
(a) identify a and b.
(b) write the equations of the asymptotes.
(c) graph the asymptotes and sketch the graph of the hyperbola.

4. $\dfrac{x^2}{9} - \dfrac{y^2}{16} = 1$ **5.** $\dfrac{y^2}{9} - \dfrac{x^2}{16} = 1$ **6.** $\dfrac{y^2}{25} - \dfrac{x^2}{9} = 1$

Applications

A hyperbolic mirror is one branch of a hyperbola, rotated around the transverse axis (Figure 2). Imagine a ray of light that is on the same line as the focus of this branch of the hyperbola. When the light strikes the mirror, it reflects back towards the focus of the other branch of the hyperbola. An eyepiece or some other receiver is at this other focus.

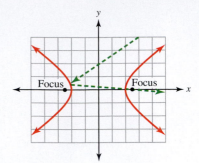

Figure 2

In a Cassegrain telescope, a parabolic mirror collects the light rays entering the telescope. These light rays reflect back to the focus of the parabola. One focus of a hyperbolic mirror is at the focus of the parabola. The light rays strike the hyperbolic mirror and reflect back through a hole in the parabolic mirror to the other focus of the hyperbola. The eyepiece or receiver is at that focus (Figure 3).

Focus of parabolic mirror and of hyperbolic mirror

Focus of hyperbolic mirror and receiver or eyepiece

Figure 3

EXAMPLE 5 The focus of a parabolic mirror in a Cassegrain telescope is 6000 mm from its vertex. For the hyperbola that is rotated to create the hyperbolic mirror, a is 2958 mm, and p is 3042 mm.

(a) Find b^2.

SOLUTION ▶

$$b^2 = p^2 - a^2 \qquad \text{Use the values of } p \text{ and } a \text{ to find } b.$$

$$b^2 = (3042 \text{ mm})^2 - (2958 \text{ mm})^2 \qquad p = 3042 \text{ mm}; a = 2958 \text{ mm}$$

$$b^2 = 504{,}000 \text{ mm}^2$$

(b) Write an equation that describes this hyperbola.

▶ Choose a coordinate system for the hyperbola and its equation. When the foci are placed on the x-axis, the equation of the hyperbola is $\dfrac{x^2}{a^2} - \dfrac{y^2}{b^2} = 1$. After a and b have been replaced, the equation is

$$\frac{x^2}{8{,}749{,}764 \text{ mm}^2} - \frac{y^2}{504{,}000 \text{ mm}^2} = 1$$

Another application of hyperbolas is LORAN navigation. A ship navigator receives signals from broadcasting stations run by the Coast Guard. The location of the ship is point P on a hyperbola. Any two broadcasting stations are point A and point B and are the foci of the hyperbola. From the definition of a hyperbola, we know that the difference of the distance between A and P and the distance between A and B is a constant. Similarly, the ship receives a signal from a third station at point C. This allows the navigator to construct another hyperbola. The location of the ship is the intersection of the two hyperbolas. However, with the advances in Global Positioning System (GPS) technology, LORAN is not as commonly used. Russia is the only large country still using LORAN.

Practice Problems

7. A Czech company sells hyperbolic mirrors used in security systems. The surface of the mirror is created by rotating one branch of a hyperbola around its axis. The equation of the hyperbola that is rotated is $\dfrac{x^2}{789 \text{ mm}^2} - \dfrac{y^2}{548 \text{ mm}^2} = 1$. Assume that the hyperbola is centered at the origin and the transverse axis is $y = 0$. Find the distance between the vertex of the mirror and the focus. Round to the nearest tenth.

9.4 VOCABULARY PRACTICE

Match the term with its description.

1. The line that includes the foci of a hyperbola
2. The midpoint of the line segment between the foci of a hyperbola
3. The intersection of a plane and a cone; circles and ellipses are examples of this.
4. The two separate curves that are the graph of a hyperbola
5. $\dfrac{x^2}{a^2} - \dfrac{y^2}{b^2} = 1$
6. $\dfrac{x^2}{a^2} + \dfrac{y^2}{b^2} = 1$
7. $x^2 + y^2 = r^2$
8. $(x - h)^2 = 4p(y - k)$
9. The points of intersection of a hyperbola with the transverse axis
10. The absolute value of the difference of the distances from any point on a hyperbola to these fixed points is a constant.

A. branches
B. center of a hyperbola
C. conic section
D. equation of a circle
E. equation of an ellipse
F. equation of a hyperbola
G. equation of a parabola
H. foci
I. transverse axis
J. vertices

9.4 Exercises

Follow your instructor's guidelines for showing your work. The center of every hyperbola in these exercises is (0, 0).

For exercises 1–8, the vertex, focus, and the equation of the transverse axis are given.
(a) Identify a, p, and b.
(b) Write the equation of the hyperbola.

1. focus $(8, 0)$; vertex $(3, 0)$; $y = 0$

2. focus $(7, 0)$; vertex $(2, 0)$; $y = 0$

3. focus $(0, 8)$; vertex $(0, 3)$; $x = 0$

4. focus $(0, 7)$; vertex $(0, 2)$; $x = 0$

5. focus $(-11, 0)$; vertex $(6, 0)$; $y = 0$

6. focus $(-14, 0)$; vertex $(9, 0)$; $y = 0$

7. focus $(0, -7)$; vertex $(0, 4)$; $x = 0$

8. focus $(0, 2)$; vertex $(0, 1)$; $x = 0$

For exercises 9–16,
(a) identify a, p, and b.
(b) write the equation of the hyperbola.

9. foci $(-5, 0)$, $(5, 0)$; vertex $(3, 0)$

10. foci $(-13, 0)$, $(13, 0)$; vertex $(12, 0)$

11. foci $(0, -5)$, $(0, 5)$; vertex $(0, 3)$

12. foci $(0, -5)$, $(0, 5)$; vertex $(0, 4)$

13. vertices $(0, -8)$, $(0, 8)$; focus $(0, 10)$

14. vertices $(0, -12)$, $(0, 12)$; focus $(0, 15)$

15. vertices $(-24, 0)$, $(24, 0)$; focus $(26, 0)$

16. vertices $(-24, 0)$, $(24, 0)$; focus $(25, 0)$

For exercises 17–20, describe the intersection of a plane and cone(s) that results in the given conic section.

17. hyperbola

18. ellipse

19. parabola

20. circle

For exercises 21–28,
(a) identify the transverse axis.
(b) identify the foci. If a coordinate is irrational, do not round.
(c) identify the vertices. If a coordinate is irrational, do not round.

21. $\dfrac{y^2}{25} - \dfrac{x^2}{64} = 1$

22. $\dfrac{y^2}{81} - \dfrac{x^2}{36} = 1$

23. $\dfrac{x^2}{25} - \dfrac{y^2}{64} = 1$

24. $\dfrac{x^2}{81} - \dfrac{y^2}{36} = 1$

25. $\dfrac{x^2}{169} - \dfrac{y^2}{21} = 1$

26. $\dfrac{x^2}{196} - \dfrac{y^2}{37} = 1$

27. $\dfrac{y^2}{13} - \dfrac{x^2}{11} = 1$

28. $\dfrac{y^2}{15} - \dfrac{x^2}{17} = 1$

29. A student said that he could write the equation of a hyperbola given one focus and one vertex. Explain what is wrong with his thinking.

30. A student said that she could write the equation of a hyperbola given the transverse axis and the vertex. Explain what is wrong with her thinking.

For exercises 31–36,
(a) identify a.
(b) identify p.
(c) identify b.

31. foci: $(-45, 0)$, $(45, 0)$; vertices: $(-36, 0)$, $(36, 0)$

32. foci: $(-20, 0)$, $(20, 0)$; vertices: $(-16, 0)$, $(16, 0)$

33. foci: $(0, -125)$, $(0, 125)$; vertices: $(0, -100)$, $(0, 100)$

34. foci: $(0, -80)$, $(0, 80)$; vertices: $(0, -64)$, $(0, 64)$

35. foci: $(0, -8)$, $(0, 8)$; vertices: $(0, -5)$, $(0, 5)$

36. foci: $(-12, 0)$, $(12, 0)$; vertices: $(-7, 0)$, $(7, 0)$

For exercises 37–44, write the equations of the asymptotes of each hyperbola.

37. $\dfrac{y^2}{121} - \dfrac{x^2}{169} = 1$

38. $\dfrac{y^2}{100} - \dfrac{x^2}{81} = 1$

39. $\dfrac{x^2}{16} - \dfrac{y^2}{4} = 1$

40. $\dfrac{x^2}{100} - \dfrac{y^2}{4} = 1$

41. $\dfrac{x^2}{169} - \dfrac{y^2}{21} = 1$

42. $\dfrac{x^2}{196} - \dfrac{y^2}{37} = 1$

43. $\dfrac{y^2}{13} - \dfrac{x^2}{9} = 1$

44. $\dfrac{y^2}{19} - \dfrac{x^2}{16} = 1$

For exercises 45–52,
(a) write the equations of the asymptotes.
(b) graph the asymptotes and sketch the graph of the hyperbola.

45. $\dfrac{x^2}{36} - \dfrac{y^2}{25} = 1$

46. $\dfrac{x^2}{49} - \dfrac{y^2}{36} = 1$

47. $\dfrac{y^2}{36} - \dfrac{x^2}{25} = 1$

48. $\dfrac{y^2}{49} - \dfrac{x^2}{36} = 1$

49. $\dfrac{y^2}{25} - \dfrac{x^2}{64} = 1$

50. $\dfrac{y^2}{4} - \dfrac{x^2}{16} = 1$

51. $\dfrac{x^2}{16} - \dfrac{y^2}{49} = 1$

52. $\dfrac{x^2}{1} - \dfrac{y^2}{4} = 1$

53. A Cassegrain telescope has a parabolic mirror and a hyperbolic mirror. For the hyperbola that is rotated to create the hyperbolic mirror, a is 575 mm and p is 625 mm. Write an equation that describes this hyperbola.

54. A Cassegrain telescope has a parabolic mirror and a hyperbolic mirror. For the hyperbola that is rotated to create the hyperbolic mirror, a is 720 mm and p is 800 mm. Write an equation that describes this hyperbola.

Problem Solving: Practice and Review

Follow your instructor's guidelines for using the five steps as outlined in Section 1.5, p. 51.

55. In July 2011, *Harry Potter and the Deathly Hallows, Part 2* set a record for weekend ticket sales in the United States and Canada. Find the revenue for weekend ticket sales a year earlier. Round to the nearest million.

    Revenue for the weekend for the top 12 films rose 46 percent to $250.7 million from a year earlier. (*Source:* www.sfgate .com, July 18, 2011)

56. The bar graph represents the relationship between the type of job (the input value) and the median annual wage (the output value.) Is this relationship a function? Explain.

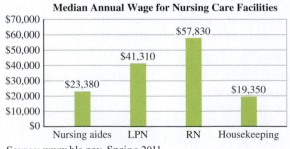

Median Annual Wage for Nursing Care Facilities

Source: www.bls.gov, Spring 2011

57. A study of the Dallas Municipal Courts system found a backlog of 114,000 cases in June 2009. This backlog was growing at a rate of 11%, compounded annually. Assuming no changes in the court, predict the backlog of cases in 5 years. Round to the nearest thousand. (*Source:* www .dallascityhall.com/council_briefings, June 17, 2009)

58. Find the percent increase in total charitable contributions from 2009 to 2010. Round to the nearest tenth of a percent.

Giving USA Foundation™ and its research partner, the Center on Philanthropy at Indiana University, today announced that total charitable contributions from American individuals, corporations and foundations were an estimated $290.89 billion in 2010, up from a revised estimate of $280.30 billion for 2009. (*Source:* www.aafrc.org, June 20, 2011)

Technology

For exercises 59–62, to graph the equation of a circle, solve for *y* and graph the equations of the two semicircles (Section 9.1). To graph a hyperbola, follow the same process.
(a) Write the equation of each branch of the hyperbola.
(b) Graph both equations. Sketch the graph; describe the window.

59. $\dfrac{x^2}{16} - \dfrac{y^2}{25} = 1$

60. $\dfrac{x^2}{9} - \dfrac{y^2}{36} = 1$

61. $\dfrac{y^2}{16} - \dfrac{x^2}{25} = 1$

62. $\dfrac{y^2}{9} - \dfrac{x^2}{36} = 1$

Find the Mistake

For exercises 63–66, the completed problem has one mistake.
(a) Describe the mistake in words, or copy down the whole problem and highlight or circle the mistake.
(b) Do the problem correctly.

63. Problem: Identify the vertices of the graph of $\dfrac{x^2}{16} - \dfrac{y^2}{25} = 1$.

Incorrect Answer: Since $a = 16$, the vertices are $(-16, 0)$ and $(16, 0)$.

64. Problem: Identify the foci of the graph of $\dfrac{x^2}{16} - \dfrac{y^2}{25} = 1$.

Incorrect Answer: Since $p^2 = 25$, the foci are $(-5, 0)$ and $(5, 0)$.

65. Problem: The foci of a hyperbola are $(0, -11)$ and $(0, 11)$. A vertex is $(0, -6)$. Identify the transverse axis.

Incorrect Answer: Since the foci are on the *y*-axis, the transverse axis is $y = 0$.

66. Problem: Write the equation of a hyperbola with foci $(0, -3)$ and $(0, 3)$ and a vertex $(0, -2)$.

Incorrect Answer: Since $a = 2$ and $b = 3$, the equation is $\dfrac{y^2}{4} - \dfrac{x^2}{9} = 1$.

Review

67. Describe and sketch the graph of a system of two linear equations with one solution.

68. Describe and sketch the graph of a system of two linear equations with no solution.

69. Describe and sketch the graph of a system of two linear equations with infinitely many solutions.

70. Use the elimination method to solve the system of equations: $\begin{array}{l} 3x + 8y = 38 \\ -5x + 7y = 18 \end{array}$.

SUCCESS IN COLLEGE MATHEMATICS

71. Some people think that studying mathematics helps students develop their ability to think logically and to use diagrams. Appendix 3, on conic sections, includes the steps for developing the equation of an ellipse, a parabola, and a hyperbola. All of these equations begin with the distance between two or more points and then, step by step, are changed into the final equation.
a. What is the formula for finding the distance between two points?
b. In developing the formula for the equation of an ellipse, the left side of the equation uses the coordinates of three points that we can call point *A*, point *B*, and point *C*. Identify the coordinates of each of these points.

Net Sales for Amazon.com

$y = 0.8786x^2 + 2.1557x + 11.077$

Number of years since 2006

Source: retailsales.com

Financial statements for retail companies usually include an entry for net sales. We can develop and graph a system of equations that are models of the net sales for two different companies. The intersection point of the graphs represents when the net sales are equal. If one of the equations is not a linear model, the system is a nonlinear system. In this section, we will study systems of nonlinear equations.

SECTION 9.5

Systems of Nonlinear Equations in Two Variables

After reading the text, working the practice problems, and completing assigned exercises, you should be able to:

1. Use graphing to solve a system of nonlinear equations.
2. Use substitution to solve a system of nonlinear equations.
3. Use elimination to solve a system of nonlinear equations.
4. Use a system of nonlinear equations as a model.

Nonlinear Systems of Equations

We can write a **linear equation** in two variables in the form $ax + by = c$ where a and b are not both 0. The graph of a linear equation is a straight line. In Chapter 4, we studied systems of linear equations that included two or more equations. The intersection of the graphs that represent these equations is the solution of the system. If the lines intersect in one point, the system has one solution. If the lines are parallel, the system has no solution. If the lines coincide, the system has infinitely many solutions. If we graph the system, we can estimate the solution. To find the exact solution, we use the algebraic methods of substitution or elimination.

In Sections 9.1–9.4, we learned about the equations of conic sections. Since the graphs of circles, ellipses, parabolas, and hyperbolas are not straight lines, equations of conic sections are **nonlinear equations**. A system of nonlinear equations in two variables includes at least one nonlinear equation. As with linear systems, we can solve these systems by graphing or with the algebraic methods of substitution and elimination.

EXAMPLE 1 | Solve $\begin{aligned} x^2 &= 8y \\ y &= 2x - 6 \end{aligned}$ by graphing.

(a) Graph: $x^2 = 8y$

SOLUTION ▶ The form of this equation is $x^2 = 4py$. Its graph is a parabola that opens up with vertex $(0, 0)$. To sketch the graph, we need to find two more points on the parabola.

$$x^2 = 8y \qquad \text{Replace } y \text{ with any value.}$$
$$x^2 = 8(2) \qquad \text{Choose 2 because } \sqrt{16} = 4.$$
$$x^2 = 16 \qquad \text{Simplify.}$$
$$\sqrt{x^2} = \sqrt{16} \qquad \text{Take the square root of both sides.}$$
$$x = \pm 4 \qquad \text{Simplify; use } \pm \text{ notation.}$$

Two other points on the parabola are $(4, 2)$ and $(-4, 2)$.

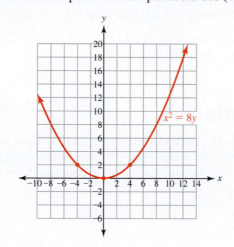

(b) Graph: $y = 2x - 6$

  ▶   Slope-intercept graphing: $m = \dfrac{2}{1}$ and $b = -6$

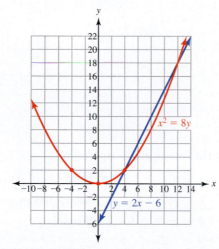

(c) Estimate the solutions of this system.

  ▶   The graphs appear to intersect at $(4, 2)$ and $(12, 18)$. These ordered pairs are the two solutions of this system.

(d) Check the solutions.

  ▶   Check: $(4, 2)$

$$x^2 = 8y \qquad\qquad y = 2x - 6$$
$$4^2 = 8(2) \qquad\qquad 2 = 2(4) - 6$$
$$16 = 16 \quad \text{True.} \qquad 2 = 2 \quad \text{True.}$$

Check: $(12, 18)$

$x^2 = 8y$	$y = 2x - 6$
$12^2 = 8(18)$	$18 = 2(12) - 6$
$144 = 144$ True.	$18 = 18$ True.

Practice Problems

For problems 1–2, solve by graphing.

1. $x^2 = 4y$
$\quad y = 2x - 3$

2. $(x - 3)^2 + (y - 2)^2 = 2$
$\quad y = x - 3$

Substitution

We can also use the substitution method (Section 4.2) to solve a system of nonlinear equations. In substitution, we solve one of the equations for a variable and then replace this variable in the other equation with the "substitution expression."

EXAMPLE 2   Solve $\begin{aligned} x^2 &= 8y \\ 2x - y &= 6 \end{aligned}$ by substitution.

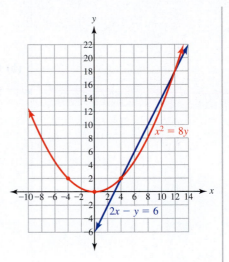

(a) Identify the substitution expression.

SOLUTION ▶

$2x - y = 6$	Isolate a variable; choose y.
$\underline{-2x \qquad\quad -2x}$	Subtraction property of equality.
$0 - y = -2x + 6$	Simplify.
$\dfrac{-y}{-1} = \dfrac{-2x}{-1} + \dfrac{6}{-1}$	Division property of equality.
$y = 2x - 6$	Simplify.

The substitution expression is $2x - 6$.

(b) Replace y in the other equation with the substitution expression, $2x - 6$. Solve for x.

▶

$x^2 = 8y$	
$x^2 = 8(2x - 6)$	Replace y with the substitution expression, $2x - 6$.
$x^2 = 16x - 48$	Distributive property.
$\underline{-16x + 48 \quad -16x + 48}$	Properties of equality.
$x^2 - 16x + 48 = 0$	Simplify.
$(x - 12)(x - 4) = 0$	Factor.

$$x - 12 = 0 \quad \text{or} \quad x - 4 = 0 \qquad \text{Zero product property.}$$

$$\underline{+12 \ +12 \qquad\qquad +4 \ +4} \qquad \text{Properties of equality.}$$

$$x + 0 = 12 \quad \text{or} \quad x + 0 = 4 \qquad \text{Simplify.}$$

$$x = 12 \quad \text{or} \qquad x = 4 \qquad \text{Simplify.}$$

The solutions are $(12, ?)$ and $(4, ?)$.

(c) Find the other coordinate of the solutions.

$$x^2 = 8y \qquad \text{Use either of the equations.}$$

$$(\mathbf{12})^2 = 8y \qquad \text{Replace } x \text{ with } 12.$$

$$\mathbf{144} = 8y \qquad \text{Simplify.}$$

$$\frac{144}{8} = \frac{8y}{8} \qquad \text{Division property of equality.}$$

$$18 = y \qquad \text{Simplify.}$$

One solution of the system is $(12, 18)$.

$$x^2 = 8y \qquad \text{Use either of the equations.}$$

$$(\mathbf{4})^2 = 8y \qquad \text{Replace } x \text{ with } 4.$$

$$\mathbf{16} = 8y \qquad \text{Simplify.}$$

$$\frac{16}{8} = \frac{8y}{8} \qquad \text{Division property of equality.}$$

$$2 = y \qquad \text{Simplify.}$$

The other solution of the system is $(4, 2)$. The solutions $(12, 18)$ and $(4, 2)$ correspond to the intersection points on the graph of the system.

In the next example, we use the quadratic formula to solve a quadratic equation in standard form, $ax^2 + bx + c = 0$.

EXAMPLE 3 | Solve $\begin{aligned} x^2 + y^2 &= 25 \\ 3x - 2y &= 6 \end{aligned}$ by substitution.

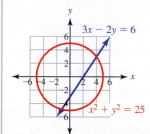

(a) Identify the substitution expression.

SOLUTION ▶

$$3x - 2y = 6 \qquad\qquad\qquad \text{Solve for } y.$$

$$\underline{-3x \qquad\qquad -3x} \qquad\qquad \text{Subtraction property of equality.}$$

$$\mathbf{0} - 2y = -\mathbf{3x} + 6 \qquad\qquad \text{Simplify.}$$

$$\frac{-2y}{-2} = \frac{-3x}{-2} + \frac{6}{-2} \qquad\qquad \text{Division property of equality.}$$

$$y = \frac{3}{2}x - 3 \qquad\qquad\qquad \text{Simplify.}$$

The substitution expression is $\dfrac{3}{2}x - 3$.

(b) Replace y in the other equation with the substitution expression, $\dfrac{3}{2}x - 3$. Solve for x.

$$x^2 + y^2 = 25$$

$$x^2 + \left(\frac{3}{2}x - 3\right)^2 = 25 \qquad \text{Replace } y \text{ with the substitution}$$
$$\text{expression, } \tfrac{3}{2}x - 3.$$

$$x^2 + \left(\frac{3}{2}x - 3\right)\left(\frac{3}{2}x - 3\right) = 25 \qquad \text{Rewrite as multiplication.}$$

$$x^2 + \frac{9}{4}x^2 - \frac{9}{2}x - \frac{9}{2}x + 9 = 25 \qquad \text{Distributive property.}$$

$$4\left(x^2 + \frac{9}{4}x^2 - \frac{9}{2}x - \frac{9}{2}x + 9\right) = 4(25) \qquad \text{Clear fractions; multiplication property of equality.}$$

$$4x^2 + 9x^2 - 18x - 18x + 36 = 100 \qquad \text{Simplify.}$$

$$13x^2 - 36x + 36 = 100 \qquad \text{Simplify.}$$

$$\underline{\qquad\qquad -100 \ -100} \qquad \text{Subtraction property of equality.}$$

$$13x^2 - 36x - 64 = 0 \qquad \text{Simplify; } a = 13, b = -36, c = -64$$

$$x = \frac{-(-36) \pm \sqrt{(-36)^2 - 4(13)(-64)}}{2(13)} \qquad \text{Quadratic formula: } x = \frac{-b \pm \sqrt{b^2 - 4ac}}{2a}$$

$$x = \frac{36 \pm \sqrt{4624}}{26} \qquad \text{Simplify radicand.}$$

$$x = \frac{36 \pm 68}{26} \qquad \text{Evaluate radical.}$$

$$x = \frac{36 + 68}{26} \quad \text{or} \quad x = \frac{36 - 68}{26} \qquad \text{Rewrite as two solutions.}$$

$$x = 4 \quad \text{or} \quad x = -\frac{16}{13} \qquad \text{Simplify.}$$

The solutions are $(4, ?)$ and $\left(-\dfrac{16}{13}, ?\right)$.

(c) Find the other coordinate of the solutions.

$$3x - 2y = 6 \qquad \text{Use either of the equations.}$$
$$3(4) - 2y = 6 \qquad \text{Replace } x \text{ with 4.}$$
$$12 - 2y = 6 \qquad \text{Simplify.}$$
$$\underline{-12 \qquad\qquad -12} \qquad \text{Subtraction property of equality.}$$
$$0 - 2y = -6 \qquad \text{Simplify.}$$
$$\frac{-2y}{-2} = \frac{-6}{-2} \qquad \text{Division property of equality.}$$
$$y = 3 \qquad \text{Simplify.}$$

One solution of the system is $(4, 3)$.

$$3x - 2y = 6 \qquad \text{Use either of the equations.}$$
$$3\left(-\frac{16}{13}\right) - 2y = 6 \qquad \text{Replace } x \text{ with } -\frac{16}{13}.$$
$$-\frac{48}{13} - 2y = 6 \qquad \text{Simplify.}$$
$$13\left(-\frac{48}{13} - 2y\right) = 13(6) \qquad \text{Clear fractions; multiplication property of equality.}$$

$$13\left(\frac{-48}{13}\right) + 13(-2y) = 78 \qquad \text{Distributive property; simplify.}$$

$$-48 - 26y = 78 \qquad \text{Simplify.}$$

$$\underline{+48 \qquad\qquad +48} \qquad \text{Addition property of equality.}$$

$$0 - 26y = 126 \qquad \text{Simplify.}$$

$$\frac{-26y}{-26} = \frac{126}{-26} \qquad \text{Division property of equality.}$$

$$y = -\frac{63}{13} \qquad \text{Simplify.}$$

The other solution of the system is $\left(-\dfrac{16}{13}, -\dfrac{63}{13}\right)$. These solutions correspond to the intersection points on the graph of the system.

In the next example, the nonlinear system has no real solution. The graphs of the equations do not intersect. When we solve by substitution, the solutions are not real numbers.

EXAMPLE 4 | Solve $\begin{array}{l} x^2 - y = 0 \\ y = x - 3 \end{array}$ by substitution.

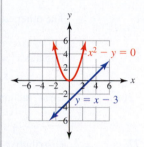

(a) Identify the substitution expression.

SOLUTION ▶ Since $y = x - 3$ is solved for y, the substitution expression is $x - 3$.

(b) Replace y in the other equation with the substitution expression, $x - 3$. Solve for x.

$$x^2 - y = 0 \qquad \text{The other equation.}$$

$$x^2 - (x - 3) = 0 \qquad \text{Replace } y \text{ with the substitution expression, } x - 3.$$

$$x^2 - x + 3 = 0 \qquad \text{Distributive property; } a = 1, b = -1, c = 3$$

$$x = \frac{-(-1) \pm \sqrt{(-1)^2 - 4(1)(3)}}{2(1)} \qquad \text{Quadratic formula; } x = \frac{-b \pm \sqrt{b^2 - 4ac}}{2a}$$

$$x = \frac{1 \pm \sqrt{-11}}{2} \qquad \text{Simplify the radicand.}$$

$$x = \frac{1 \pm \sqrt{11}\sqrt{-1}}{2} \qquad \text{Rewrite the radical as a product.}$$

$$x = \frac{1}{2} \pm \frac{\sqrt{11}}{2}i \qquad \text{Simplify.}$$

Since the values of x are nonreal numbers, this system has no real solution. The graphs of these equations do not intersect.

In the next example, both equations in the system are conic sections.

EXAMPLE 5 Solve $\begin{aligned} \dfrac{x^2}{24} + \dfrac{y^2}{72} &= 1 \\ x^2 &= 2y \end{aligned}$ by substitution.

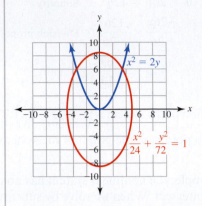

(a) Identify the substitution expression.

SOLUTION ▶ Since $x^2 = 2y$ is solved for x^2, the substitution expression is $2y$.

(b) Replace x^2 in the other equation with the substitution expression, $2y$. Solve for y.

$$\dfrac{x^2}{24} + \dfrac{y^2}{72} = 1 \qquad \text{The other equation.}$$

$$\dfrac{\mathbf{2y}}{24} + \dfrac{y^2}{72} = 1 \qquad \text{Replace } x^2 \text{ with the substitution expression.}$$

$$72\left(\dfrac{2y}{24} + \dfrac{y^2}{72}\right) = 72(1) \qquad \text{Clear fractions; multiplication property of equality.}$$

$$72\left(\dfrac{2y}{24}\right) + 72\left(\dfrac{y^2}{72}\right) = 72 \qquad \text{Distributive property.}$$

$$6y + y^2 = 72 \qquad \text{Simplify.}$$

$$\underline{ -72 \quad -72} \qquad \text{Subtraction property of equality.}$$

$$y^2 + 6y - 72 = 0 \qquad \text{Simplify.}$$

$$(y - 6)(y + 12) = 0 \qquad \text{Factor.}$$

$$y - 6 = 0 \quad \text{or} \quad y + 12 = 0 \qquad \text{Zero product property.}$$

$$\underline{+6 \ +6} \qquad \underline{-12 \quad -12} \qquad \text{Properties of equality.}$$

$$y + 0 = 6 \qquad y + 0 = -12 \qquad \text{Simplify.}$$

$$y = 6 \quad \text{or} \qquad y = -12 \qquad \text{Simplify.}$$

The solutions are $(?, 6)$ and $(?, -12)$.

(c) Find the other coordinate of the solutions.

$$x^2 = 2y \qquad \text{Use either of the equations.}$$

$$x^2 = 2(-12) \qquad \text{Replace } y \text{ with } -12.$$

$$x^2 = -24 \qquad \text{Simplify.}$$

$$\sqrt{x^2} = \sqrt{-24} \qquad \text{Square root both sides.}$$

$$x = \pm\sqrt{-24} \qquad \text{Simplify; use } \pm \text{ notation.}$$

$$x = \pm\sqrt{24}\sqrt{-1} \qquad \text{Rewrite the radical as a product.}$$

$$x = \pm\sqrt{24}\,i \qquad \text{Simplify.}$$

Since the values of x are nonreal numbers, this is not a real solution of the system.

$$x^2 = 2y \qquad \text{Use either of the equations.}$$
$$x^2 = 2(\mathbf{6}) \qquad \text{Replace } y \text{ with 6.}$$
$$x^2 = \mathbf{12} \qquad \text{Simplify.}$$
$$\sqrt{x^2} = \sqrt{12} \qquad \text{Square root both sides.}$$
$$x = \pm\sqrt{12} \qquad \text{Simplify.}$$
$$x = \pm\sqrt{\mathbf{4}}\sqrt{\mathbf{3}} \qquad \text{Product rule of radicals.}$$
$$x = \pm\mathbf{2}\sqrt{3} \qquad \text{Simplify.}$$

One solution of the system is $(2\sqrt{3},\ 6)$. The other solution is $(-2\sqrt{3},\ 6)$. These correspond to the two intersection points on the graph of the system.

Practice Problems

For problems 3–5, solve by substitution.

3. $x^2 = 4y$ **4.** $x^2 + y^2 = 16$

    $y = 2x - 3$ $3x + y = 4$

5. $\dfrac{x^2}{24} + \dfrac{y^2}{48} = 1$

    $x^2 = 4y$

Elimination

We can also use the elimination method to find the solution of a nonlinear system of equations. In the next example, we choose to eliminate the y-terms. The system has four solutions.

EXAMPLE 6   Solve $\begin{array}{l} 4x^2 + y^2 = 57 \\ x^2 + y^2 = 30 \end{array}$ by elimination.

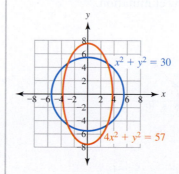

(a) Eliminate one of the variables, y, and solve for x.

SOLUTION ▶   To create y-terms that are opposites, multiply one of the equations by -1.

$$-1(x^2 + y^2) = (-1)30 \qquad \text{Multiplication property of equality.}$$
$$-x^2 - y^2 = -30 \qquad \text{Simplify.}$$

Add the equations together. Solve for x.

$$
\begin{array}{ll}
4x^2 + y^2 = 57 & \text{To eliminate the y-terms, add the equations.} \\
\underline{+ -x^2 - y^2 = -30} & \text{Addition property of equality.} \\
3x^2 + 0 = 27 & \text{Simplify.} \\
\dfrac{3x^2}{3} = \dfrac{27}{3} & \text{Division property of equality.} \\
x^2 = 9 & \text{Simplify.} \\
\sqrt{x^2} = \sqrt{9} & \text{Square root both sides.} \\
x = \pm 3 & \text{Simplify; use } \pm \text{ notation.}
\end{array}
$$

The solutions are $(-3, ?)$ and $(3, ?)$.

(b) Find the other coordinate of the solutions.

$$
\begin{array}{ll}
x^2 + y^2 = 30 & \text{Choose either equation.} \\
(3)^2 + y^2 = 30 & \text{Replace } x \text{ with 3.} \\
9 + y^2 = 30 & \text{Simplify.} \\
\underline{-9 \qquad\quad -9} & \text{Subtraction property of equality.} \\
0 + y^2 = 21 & \text{Simplify.} \\
\sqrt{y^2} = \sqrt{21} & \text{Square root both sides.} \\
y = \pm\sqrt{21} & \text{Simplify.} \\
y = \sqrt{21} \quad \text{or} \quad y = -\sqrt{21} &
\end{array}
$$

If we replace the variable x in $x^2 + y^2 = 30$ with -3, the solutions are the same, since $(3)^2$ and $(-3)^2$ both equal 9. The solutions are $\left(-3, -\sqrt{21}\right)$, $\left(3, -\sqrt{21}\right)$, $\left(-3, \sqrt{21}\right)$, and $\left(3, \sqrt{21}\right)$. These correspond to the four intersection points on the graph of the system.

In the next example, the system of nonlinear equations includes a hyperbola and a parabola. There are two solutions.

EXAMPLE 7   Solve $\begin{aligned} x^2 - 2y^2 &= 32 \\ x^2 &= 16y \end{aligned}$ by elimination.

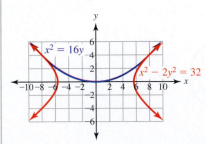

(a) Eliminate one of the variables, x, and solve for y.

SOLUTION ▶   To create x-terms that are opposites, multiply one of the equations by -1.

$$
\begin{array}{ll}
(-1)x^2 = (-1)16y & \text{Multiplication property of equality.} \\
-x^2 = -16y & \text{Simplify.}
\end{array}
$$

Add the equations together. Solve for y.

$$x^2 - 2y^2 = 32 \qquad \text{Add the equations together.}$$
$$+ \; -x^2 + 0y^2 = -16y \qquad \text{Addition property of equality.}$$
$$\overline{\;\; 0 - 2y^2 = -16y + 32 \;\;} \qquad \text{Combine like terms.}$$
$$+2y^2 \qquad +2y^2 \qquad \text{Properties of equality.}$$
$$\overline{\;\; 0 = 2y^2 - 16y + 32 \;\;} \qquad \text{Simplify.}$$
$$0 = 2(y^2 - 8y + 16) \qquad \text{Factor out the greatest common factor, 2.}$$
$$0 = 2(y - 4)(y - 4) \qquad \text{Factor the trinomial.}$$

$$y - 4 = 0 \qquad \text{Zero product property.}$$
$$\underline{+4 \;\; +4} \qquad \text{Addition property of equality.}$$
$$y = 4 \qquad \text{Simplify.}$$

The solutions are $(?, 4)$.

(b) Find the other coordinate of the solution.

$$x^2 = 16y \qquad \text{Use one of the original equations.}$$
$$x^2 = 16(4) \qquad \text{Replace } y \text{ with 4.}$$
$$x^2 = 64 \qquad \text{Multiply.}$$
$$\sqrt{x^2} = \sqrt{64} \qquad \text{Square root both sides.}$$
$$x = \pm 8 \qquad \text{Simplify.}$$

The solutions are $(8, 4)$ and $(-8, 4)$. These correspond to the two intersection points on the graph of the system.

When solving by elimination, clear any fractions so that the coefficients being added are integers.

EXAMPLE 8 Solve
$$\frac{x^2}{32} + \frac{y^2}{8} = 1$$
$$x^2 + y^2 = 20$$
by elimination.

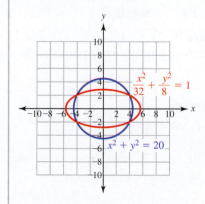

(a) Eliminate one of the variables, x, and solve for y.

SOLUTION ▶ Clear the fractions from $\dfrac{x^2}{32} + \dfrac{y^2}{8} = 1$.

$$\frac{x^2}{32} + \frac{y^2}{8} = 1$$

$$32\left(\frac{x^2}{32} + \frac{y^2}{8}\right) = 32(1) \qquad \text{Multiplication property of equality.}$$

$$32\left(\frac{x^2}{32}\right) + 32\left(\frac{y^2}{8}\right) = 32 \qquad \text{Distributive property; simplify.}$$

$$x^2 + 4y^2 = 32 \qquad \text{Simplify; the coefficients are integers.}$$

To create x-terms that are opposites, multiply this equation by -1.

$$-1(x^2 + 4y^2) = -1(32) \qquad \text{Create } x\text{-terms that are opposites.}$$

$$-x^2 - 4y^2 = -32 \qquad \text{Simplify.}$$

Add the equations together. Solve for y.

$$\begin{array}{ll} -x^2 - 4y^2 = -32 & \text{Add the equations together.} \\ +\quad x^2 + y^2 = 20 & \text{Addition property of equality.} \\ \hline 0 - 3y^2 = -12 & \text{Simplify.} \end{array}$$

$$\frac{-3y^2}{-3} = \frac{-12}{-3} \qquad \text{Division property of equality.}$$

$$y^2 = 4 \qquad \text{Simplify.}$$

$$\sqrt{y^2} = \sqrt{4} \qquad \text{Square root both sides.}$$

$$y = \pm 2 \qquad \text{Simplify.}$$

(b) Find the other coordinates of the solutions.

Since $(2)^2 = 4$ and $(-2)^2 = 4$, we need to solve only $x^2 + (2)^2 = 20$.

$$x^2 + y^2 = 20 \qquad \text{Use one of the original equations.}$$

$$x^2 + (\mathbf{2})^2 = 20 \qquad \text{Replace } y \text{ with 2.}$$

$$x^2 + \mathbf{4} = 20 \qquad \text{Simplify.}$$

$$\begin{array}{ll} \underline{ -4 \quad -4} & \text{Subtraction property of equality.} \\ x^2 + \mathbf{0} = \mathbf{16} & \text{Simplify.} \end{array}$$

$$\sqrt{x^2} = \sqrt{16} \qquad \text{Square root both sides.}$$

$$x = \pm 4 \qquad \text{Simplify.}$$

The solutions are $(4, 2)$, $(-4, 2)$, $(4, -2)$, and $(-4, -2)$. These correspond to the four intersection points on the graph of the system.

Practice Problems

For problems 6–8, solve by elimination.

6. $5x^2 + y^2 = 70$
$\quad\; x^2 + y^2 = 54$

7. $x^2 = 4y$
$\quad x^2 - y^2 = 4$

8. $\dfrac{x^2}{8} + \dfrac{y^2}{72} = 1$
$\quad x^2 + y^2 = 40$

Applications

In the next example, the solution of a system of equations is the time when two companies have the same net sales and the amount of net sales. Although the coordinates of the solutions are integers, many solutions of applications include coordinates that are not integers.

EXAMPLE 9   The function $y = 115x + 600$ represents the relationship of the number of years since 2001, x, and the net sales of a company, y, in millions of dollars. The function $y = 57.5x^2 - 460x + 1520$ represents the relationship of the number of years since 2001, x, and the net sales of a different company, y, in millions of dollars. Find the time(s) when the two companies have the same net sales and the amount of net sales at that time.

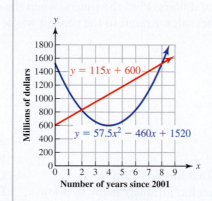

SOLUTION   ▶   Use the substitution method to solve. Since $y = 115x + 600$ is solved for y, the substitution expression is $115x + 600$.

$$y = 57.5x^2 - 460x + 1520 \qquad \text{The other equation.}$$

$$\mathbf{115x + 600} = 57.5x^2 - 460x + 1520 \qquad \text{Replace } y \text{ with the substitution expression.}$$

$$\underline{-115x\ -600 \qquad\qquad\quad -115x\quad -600} \qquad \text{Subtraction property of equality.}$$

$$0 = 57.5x^2 - \mathbf{575x + 920} \qquad \text{Simplify.}$$

$$x = \frac{-(-575) \pm \sqrt{(-575)^2 - 4(57.5)(920)}}{2(57.5)} \qquad \text{Use the quadratic formula.}$$

$$x = \frac{575 \pm \sqrt{119{,}025}}{115} \qquad \text{Simplify.}$$

$$x = \frac{575 \pm \mathbf{345}}{115} \qquad \text{Evaluate the radical.}$$

$$x = \frac{575 + 345}{115} \quad \text{or} \quad x = \frac{575 - 345}{115} \qquad \text{Rewrite as two separate solutions.}$$

$$x = \mathbf{8} \qquad\quad \text{or} \quad x = \mathbf{2} \qquad \text{Simplify.}$$

To find the net sales, find the other coordinates of the solutions.

$$y = 115x + 600 \qquad \text{Use either equation.}$$
$$y = 115(\mathbf{2}) + 600 \qquad \text{Replace } x \text{ with 2.}$$
$$y = \mathbf{\$830\ million} \qquad \text{Simplify; the units are millions of dollars.}$$

$$y = 115x + 600 \qquad \text{Use either equation.}$$
$$y = 115(\mathbf{8}) + 600 \qquad \text{Replace } x \text{ with 8.}$$
$$y = \mathbf{\$1520\ million} \qquad \text{Simplify; the units are millions of dollars.}$$

In 2003 ($x = 2$), the net sales for both companies were $830 million. In 2009 ($x = 8$), the net sales for both companies were $1520 million ($1.52 billion).

Practice Problems

9. The function $y = 30x + 120$ represents the relationship of the number of years since 1998, x, and the net sales of a company, y, in millions of dollars. The function $y = 4.4x^2 - 35.5x + 181$ represents the relationship of the number of years since 1998, x, and the sales of a different company, y, in millions of dollars. Find the time(s) when the companies had the same amount of net sales. Round to the nearest whole number.

9.5 VOCABULARY PRACTICE

Match the term with its description.

1. $ax + by = c$
2. (x, y)
3. The graph of a system of two linear equations that has no solution
4. The graph of a system of two linear equations that has infinitely many solutions
5. $\dfrac{x^2}{a^2} + \dfrac{y^2}{b^2} = 1$
6. $\dfrac{x^2}{a^2} - \dfrac{y^2}{b^2} = 1$
7. $(x - h)^2 = 4p(y - k)$
8. $x^2 + y^2 = r^2$
9. An algebraic method for solving a nonlinear system of equations
10. The result of finding a solution of a nonlinear system of equations by graphing

A. coinciding lines
B. equation of a circle
C. equation of an ellipse
D. equation of a hyperbola
E. equation of a line
F. equation of a parabola
G. estimate
H. ordered pair
I. parallel lines
J. substitution

9.5 Exercises

Follow your instructor's guidelines for showing your work.

For exercises 1–6, use the graph to estimate the solution of the system of equations.

1.

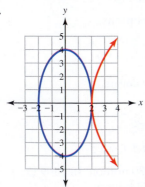

2.

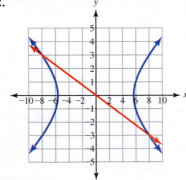

3.

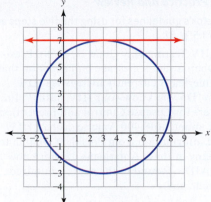

4.

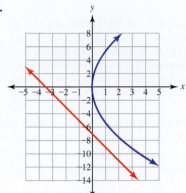

5.

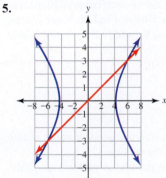

6.

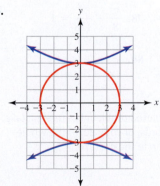

For exercises 7–14, you do not need to give the equations of the conic sections.

7. Draw a hyperbola and a circle that intersect in only one point.

8. Draw two circles that intersect in only one point.

9. Draw a parabola and a circle that intersect in two points.

10. Draw an ellipse and a circle that intersect in four points.

11. Draw a hyperbola and a line that intersect in only one point.

12. Draw a parabola and a line that intersect in only one point.

13. Draw a circle and an ellipse that intersect in two points.

14. Draw a hyperbola and a parabola that intersect in two points.

For exercises 15–32,
(a) graph the system of equations.
(b) estimate the solution(s).

15. $2x + 5y = 14$
$y = 6x - 10$

16. $2x + y = 9$
$y = 3x - 6$

17. $x^2 + y^2 = 16$
$x = 4$

18. $x^2 + y^2 = 25$
$x = 5$

19. $\dfrac{x^2}{32} + \dfrac{y^2}{8} = 1$
$x^2 = 8y$

20. $\dfrac{x^2}{8} + \dfrac{y^2}{32} = 1$
$y^2 = 8x$

21. $x^2 + y^2 = 4$
$y = -x + 2$

22. $x^2 + y^2 = 9$
$y = x + 3$

23. $\dfrac{x^2}{1} + \dfrac{y^2}{4} = 1$
$2x - y = -2$

24. $\dfrac{x^2}{1} + \dfrac{y^2}{9} = 1$
$3x - y = 3$

25. $\dfrac{x^2}{16} - \dfrac{y^2}{9} = 1$
$y = 0$

26. $\dfrac{x^2}{9} - \dfrac{y^2}{16} = 1$
$y = 0$

27. $(x - 3)^2 + (y + 1)^2 = 4$
$(x - 2)^2 + (y + 2)^2 = 36$

28. $(x + 3)^2 + (y - 2)^2 = 4$
$(x + 4)^2 + (y - 3)^2 = 49$

29. $y^2 = 4x$
$y = x$

30. $y^2 = -4x$
$y = x$

31. $x^2 = 8y$
$x^2 = -8y$

32. $x^2 = 4y$
$x^2 = -4y$

For exercises 33–50, solve the system of equations by substitution.

33. $3x + 8y = -28$
$y = -3x + 7$

34. $2x + 9y = -57$
$y = -2x - 1$

35. $y^2 = 32x$
$y = x - 10$

36. $y^2 = 64x$
$y = x - 9$

37. $x^2 = -8y$
$y = 2x - 10$

38. $x^2 = -4y$
$y = 3x - 16$

39. $x^2 + y^2 = 25$
$y = 2x - 10$

40. $x^2 + y^2 = 100$
$y = 2x - 20$

41. $x^2 + y^2 = 16$
$y = x - 9$

42. $x^2 + y^2 = 25$
$y = -x - 10$

43. $\dfrac{x^2}{4} - \dfrac{y^2}{9} = 1$
$x^2 = \dfrac{8}{3}y$

44. $\dfrac{x^2}{10} - \dfrac{y^2}{9} = 1$
$x^2 = \dfrac{20}{3}y$

45. $\dfrac{x^2}{32} + \dfrac{y^2}{8} = 1$
$x^2 = 8y$

46. $\dfrac{x^2}{32} + \dfrac{y^2}{128} = 1$
$x^2 = 2y$

47. $\dfrac{x^2}{18} + \dfrac{y^2}{50} = 1$
$x^2 = \dfrac{9}{5}y$

48. $\dfrac{x^2}{36} + \dfrac{y^2}{48} = 1$
$x^2 = \dfrac{3}{2}y$

49. $\dfrac{x^2}{18} - \dfrac{y^2}{9} = 1$
$y = \dfrac{1}{2}x$

50. $\dfrac{x^2}{32} - \dfrac{y^2}{9} = 1$
$y = -\dfrac{3}{8}x$

For exercises 51–68, solve the system of equations by elimination.

51. $2x + 3y = 8$
$5x + 4y = 27$

52. $3x + 2y = 17$
$7x + 5y = 38$

53. $6x^2 + y^2 = 150$
$x^2 + y^2 = 25$

54. $x^2 + 6y^2 = 150$
$x^2 + y^2 = 25$

55. $x^2 + y^2 = 20$
$x^2 = 8y$

56. $x^2 + y^2 = 80$
$x^2 = 16y$

57. $x^2 + y^2 = 20$
$x^2 = -8y$

58. $x^2 + y^2 = 80$
$x^2 = -16y$

59. $x^2 - 2y^2 = 162$
$x^2 = 36y$

60. $x^2 - 2y^2 = 162$
$x^2 = -36y$

61. $\dfrac{x^2}{32} + \dfrac{y^2}{8} = 1$
$x^2 = 8y$

62. $\dfrac{x^2}{8} + \dfrac{y^2}{128} = 1$
$y^2 = 32x$

63. $x^2 + 2y^2 = 8$
$x^2 + y^2 = 35$

64. $x^2 + 3y^2 = 30$
$x^2 + y^2 = 130$

65. $x^2 - 2y^2 = 8$
$x^2 + y^2 = 35$

66. $x^2 - 3y^2 = 30$
$x^2 + y^2 = 130$

67. $x^2 - y^2 = 25$
$(x - 8)^2 + y^2 = 9$

68. $x^2 + y^2 = 25$
$(x - 8)^2 + y^2 = 9$

69. The system $8x^2 + 4y^2 = 32$ and $\dfrac{x^2}{4} + \dfrac{y^2}{8} = 1$ has infinitely many solutions. Explain why.

70. The system $x^2 + y^2 = 9$ and $x^2 + y^2 = 36$ has no solutions. Explain why.

71. The function $y = 5x + 200$ represents the relationship of the number of years since 1995, x, and the total number of employees of a company, y. The function $y = 5x^2 - 55x + 300$ represents the relationship of the number of years since 1995, x, and the total number of employees of a different company, y. Find the time(s) when the two companies have the same number of employees and the number of employees.

72. The function $y = 10x + 300$ represents the relationship of the number of years since 2000, x, and the number of trout per year Mike caught fly-fishing, y. The function $y = 20x^2 - 110x + 300$ represents the relationship of the number of years since 2000, x, and the number of bull-heads per year that Richard caught fishing with corn, y. Find the time(s) when Mike and Richard caught the same number of fish and the number of fish caught.

Problem Solving: Practice and Review

Follow your instructor's guidelines for using the five steps as outlined in Section 1.5, p. 51.

73. On National HIV Testing Day (June 27, 2009), President Barack Obama sent a message on Facebook that said, "1 in 5 Americans currently living with HIV doesn't know it. Do you know your HIV status?" In 2007, the Centers for Disease Control and Prevention estimated that about 1,018,400 Americans had HIV/AIDS. Use the information from President Obama to predict how many of these Americans did not know they had HIV/AIDS. Round to the nearest hundred. (*Source:* www.cdc.gov)

74. Find the number of 25-cent tolls needed to equal $35 million.

The Triborough Bridge opened on July 11, 1936 at a cost of $60.3 million. The new Triborough Bridge Authority, which had its administrative offices at the Randall's Island toll plaza, financed $35 million of the construction costs. The bonds were backed by 25-cent tolls. Federal, state and city outlays financed the remainder of the costs. (*Source:* www.nycroads.com)

75. Find the number of students in the Chicago Public Schools (CPS) in 2010.

In 2010, approximately 51,000 CPS students, or 12%, were English Language Learners. (*Source:* www.cps.edu, Final Budget, 2010–2011)

76. The East River Suspension Bridge in New York (I-278) has two cables. The diameter of each cable is 20 in., and the length of each cable is 3104 ft. Find the total volume of both cables in cubic feet. Round to the nearest whole number (volume of a cylinder $= \pi r^2 h$, $\pi \approx 3.14$). (*Source:* www.nycroads.com)

Technology

For exercises 77–80,
(a) solve each equation for y.
(b) graph each equation. Use a standard window, zoom square. Sketch the graph, describe the window.
(c) to solve the system, find the point(s) of intersection. State the solution(s); if necessary, round to the hundredths place.

77. $x^2 - 2y^2 = 8$
$x^2 + y^2 = 35$

78. $\dfrac{x^2}{8} + \dfrac{y^2}{128} = 1$
$y^2 = 32x$

79. $(x - 6)^2 + y^2 = 9$
$x^2 + y^2 = 25$

80. $x^2 + 2y^2 = 8$
$x^2 + y^2 = 35$

Find the Mistake

For exercises 81–84, the completed problem has one mistake.
(a) Describe the mistake in words, or copy down the whole problem and highlight or circle the mistake.
(b) Do the problem correctly.

81. Problem: Solve $\dfrac{x^2}{2} + \dfrac{y^2}{8} = 1$
$y = 2x$ by substitution.

Incorrect Answer: $\dfrac{x^2}{2} + \dfrac{y^2}{8} = 1$

$$8\left(\dfrac{x^2}{2} + \dfrac{y^2}{8}\right) = 1 \cdot 8$$

$$4x^2 + y^2 = 8$$

$$4x^2 + (2x)^2 = 8$$

$$4x^2 + 4x^2 = 8$$

$$8x^2 = 8$$

$$\dfrac{8x^2}{8} = \dfrac{8}{8}$$

$$x^2 = 1$$

$$\sqrt{x^2} = \sqrt{1}$$

$$x = 1$$

$$y = 2x$$

$$y = 2(1)$$

$$y = 2$$

The solution is $(1, 2)$.

82. Problem: Solve $\begin{array}{l} x^2 + y^2 = 4 \\ x^2 - y^2 = 2 \end{array}$ by elimination.

Incorrect Answer: $x^2 + y^2 = 4$

$$\underline{+\ x^2 - y^2 = 2}$$

$$2x^2 = 6$$

$$\dfrac{2x^2}{2} = \dfrac{6}{2}$$

$$x^2 = 3$$

$$\sqrt{x^2} = \sqrt{3}$$

$$x = \sqrt{3} \quad \text{or} \quad x = -\sqrt{3}$$

83. Problem: Solve $\begin{array}{l} x^2 + y^2 = 4 \\ x^2 - y^2 = 16 \end{array}$ by elimination.

Incorrect Answer: $x^2 + y^2 = 4$

$$\underline{+\ x^2 - y^2 = 16}$$

$$2x^2 = 20$$

$$\dfrac{2x^2}{2} = \dfrac{20}{2}$$

$$x^2 = 10$$

$$x = \sqrt{10} \quad \text{or} \quad x = -\sqrt{10}$$

$x^2 + y^2 = 4$	$x^2 + y^2 = 4$
$(\sqrt{10})^2 + y^2 = 4$	$(-\sqrt{10})^2 + y^2 = 4$
$10 + y^2 = 4$	$10 + y^2 = 4$
$y^2 = -6$	$y^2 = -6$
$y = \sqrt{-6}$ or $y = -\sqrt{-6}$	$y = \sqrt{-6}$ or $y = -\sqrt{-6}$

The solutions are $(\sqrt{10}, \sqrt{-6})$, $(\sqrt{10}, -\sqrt{-6})$, $(-\sqrt{10}, \sqrt{-6})$, and $(-\sqrt{10}, -\sqrt{-6})$.

84. Problem: Solve the system $x^2 = -4y$ and $y = 2x + 5$ by substitution.

Incorrect Answer: $x^2 = -4y$

$$x^2 = -4(2x + 5)$$

$$x^2 = -8x - 20$$

$$x^2 + 20 = -8x - 20 + 20$$

$$x^2 + 8x + 20 = 0$$

$$(x - 2)(x + 10) = 0$$

$$x - 2 = 0 \quad \text{or} \quad x + 10 = 0$$

$$x = 2 \quad \text{or} \quad x = -10$$

$y = 2x + 5$	$y = 2x + 5$
$y = 2(2) + 5$	$y = 2(-10) + 5$
$y = 9$ or	$y = -15$

The solutions are $(2, 9)$ and $(-10, -15)$.

Review

85. Evaluate $f(5)$ when $f(x) = \$2000\left(1 + \dfrac{0.04}{12}\right)^{12x}$. Round to the nearest whole number.

86. Evaluate $f(3)$ when $f(x) = (-1)^x x$.

87. Evaluate $h(10)$ when $h(x) = 8x - 12$.

88. Evaluate $h(4)$ when $h(x) = \dfrac{x^2}{x}$.

SUCCESS IN COLLEGE MATHEMATICS

89. Some people think that studying mathematics helps students develop their ability to represent abstract relationships using a visual image like a graph. In Example 9 of this section, the graph represents the net sales of a company and the net sales of a competitor. The graphs represent two equations: $y = 115x + 600$ and $y = 57.5x^2 - 460x + 1520$. If you had to explain this situation to someone, would you prefer to show them the graph or show and solve the system of equations? Explain why.

90. Are you interested in *academics*? Explain.

Study Plan for Review of Chapter 9

Ask Yourself	Test Yourself	Help Yourself
Can I...	Do 9.1 Review Exercises	See these Examples and Practice Problems
write the distance formula?	1	Ex. 1, PP 1–3
use the distance formula to find the distance between two points?		
write the midpoint formula?	2	Ex. 2, PP 4–6
use the midpoint formula to find the midpoint of a line segment?		
write the standard form of a circle with center (h, k)?	3, 4	Ex. 3, 4, PP 7
write the equation of a circle in standard form, given the radius and the center (h, k)?		
given the center and a point on a circle, write its equation in standard form?	5	Ex. 5, PP 8
rewrite the equation of a circle in standard form and identify its center, radius, and sketch its graph?	7–10	Ex. 5, 6, 7, PP 9
solve the equation of a circle for y?	11	Ex. 8, PP 10, 11
use the equation of a circle or semicircle to model or solve an application?	11	Ex. 8, PP 10, 11

9.1 Review Exercises

1. Use the distance formula to find the exact distance between $(5, -8)$ and $(-3, 10)$. Do not round.

2. The endpoints of a line segment are $(6, -2)$ and $(15, 12)$. Find the midpoint of this line segment.

3. The radius of a circle is 12, and its center is $(0, 0)$. Write its equation in standard form.

4. The radius of a circle is 5, and its center is $(3, -8)$. Write its equation in standard form.

5. The center of a circle is $(5, 1)$. A point on the circle is $(10, 2)$. Write the equation of this circle in standard form.

6. Add a constant to the expression $x^2 + 10x$ to complete the square.

7. Explain why $x^2 + y^2 = -100$ is not the equation of a circle.

8. The equation of a circle is $x^2 + y^2 - 6x - 10y + 18 = 0$.
   a. Rewrite this equation in standard form.
   b. Identify the center.
   c. Identify the radius.

9. Sketch the graph of $x^2 + y^2 = 36$.

10. Sketch the graph of $(x - 4)^2 + (y + 1)^2 = 4$.

11. The diameter of a circular porthole window is 15 in. Write an equation that is a model of this window. Assume that the origin of the xy-coordinate system is on a point on the left edge of the window.

© Lagui/Shutterstock

Ask Yourself	Test Yourself	Help Yourself
Can I...	Do 9.2 Review Exercises	See these Examples and Practice Problems
given the center of an ellipse and the distance between vertices on its major and minor axes, write its equation in standard form?	12	Ex. 1, PP 1, 2
given the vertices on the major and minor axes of an ellipse, write its equation in standard form?	13	Ex 2, PP 3
sketch the graph of an ellipse?	12–14	Ex. 2, PP 3
rewrite the equation of an ellipse in standard form?	14	Ex. 3, PP 4, 5

Ask Yourself	Test Yourself	Help Yourself
Can I . . .	Do 9.2 Review Exercises	See these Examples and Practice Problems
given the equation of an ellipse, identify its center, the equations of the major and minor axes, and its vertices?	14–16	Ex. 3, PP 4, 5
use an ellipse to model or solve an application problem?	17	Ex. 4, PP 6

9.2 Review Exercises

12. The center of an ellipse is $(4, 5)$. The distance between the vertices on its horizontal major axis is 12. The distance between the vertices on its vertical minor axis is 8.
  a. Identify the vertices of this ellipse.
  b. Sketch the graph.
  c. Write the equation of this ellipse in standard form.

13. The center of an ellipse is $(-2, 1)$. The vertices on its vertical major axis are $(-2, 9)$ and $(-2, -7)$. The vertices on its horizontal minor axis are $(-8, 1)$ and $(4, 1)$.
  a. Write the equation of this ellipse in standard form.
  b. Sketch the graph.

14. The equation of an ellipse is
  $25x^2 + 16y^2 - 100x - 96y - 156 = 0$.
  a. Write this equation in standard form.
  b. Identify the center of the ellipse.
  c. Write the equation of the major axis.
  d. Write the equation of the minor axis.
  e. Sketch a graph of this ellipse.

15. The equation of an ellipse is $\dfrac{(x + 3)^2}{36} + \dfrac{(y - 4)^2}{4} = 1$.
  a. Identify the center of this ellipse.
  b. Write the equation of the major axis.
  c. Write the equation of the minor axis.
  d. Identify the vertices of this ellipse.

16. Identify the equation(s) of the ellipse with a major axis that is horizontal.
  a. $\dfrac{x^2}{36} + \dfrac{y^2}{49} = 1$
  b. $\dfrac{x^2}{49} + \dfrac{y^2}{36} = 1$
  c. $\dfrac{(x - 3)^2}{16} + \dfrac{(y - 4)^2}{1} = 1$
  d. $2x^2 + y^2 = 32$

17. An above-ground swimming pool is an ellipse. The length between its vertices on its horizontal major axis is 28 ft. The length between its vertices on its minor vertical axis is 12 ft. Assume that the origin is at the center of the pool. Write the equation of this ellipse.

SECTION 9.3   Parabolas

Ask Yourself	Test Yourself	Help Yourself
Can I . . .	Do 9.3 Review Exercises	See these Examples and Practice Problems
given the equation of a parabola, identify the direction in which a parabola opens, its vertex, two other points on its graph, the focus, the equation of the directrix, and the equation of the axis of symmetry, sketch its graph, and determine whether it is a function?	20, 21	Ex. 1, 2, 4, PP 1, 2, 5
given the vertex of a parabola, the equation of the axis of symmetry, and another point on the graph, identify the direction in which the graph opens and the form of its equation and write its equation?	19	Ex. 3, PP 3, 4
write the equation of a parabola by completing the square?	22	Ex. 4, PP 5
given a quadratic equation in vertex form, identify the vertex, the direction in which its graph opens, the equation of the axis of symmetry, the domain, and the range?	23	Ex. 5, PP 6, 7
use a parabola to model or solve an application problem?	24	Ex. 6, 7, PP 8, 9

9.3 Review Exercises

18. The vertex of a parabola is $(0, 0)$. The axis of symmetry is the y-axis. Its graph includes the point $(6, 12)$.
  a. Sketch the graph; label the vertex, focus, axis of symmetry, and directrix.
  b. Write its equation.
  c. Does this equation represent a function?

19. The vertex of a parabola is $(5, 6)$. The axis of symmetry is $x = 5$. Its graph includes the point $(13, -2)$.
  a. Sketch the graph; label the vertex, focus, axis of symmetry, and directrix.
  b. Write its equation.
  c. Does this equation represent a function?

For exercises 20–21,
(a) identify the vertex.
(b) identify p and the focus.
(c) identify two other points on the graph.
(d) write the equation of the axis of symmetry.
(e) write the equation of the directrix.
(f) sketch the graph; label the vertex, focus, axis of symmetry, and directrix.

20. $x^2 = -12y$

21. $(y - 3)^2 = 8(x - 1)$

22. Rewrite $y^2 - 8x - 6y + 17 = 0$ in the form $(y - k)^2 = 4p(x - h)$.

23. A quadratic function is $y = 3(x + 6)^2 - 11$.
   a. Identify the vertex.
   b. Write the equation of the axis of symmetry.
   c. Identify the direction in which the graph opens.
   d. Use interval notation to represent the domain.
   e. Use interval notation to represent the range.

24. A portable battery-powered parabolic microphone can pick up audio that is 900 ft away. The parabolic dish has a diameter of 24 in. The depth of the dish at the center is 6 in. Find the distance of the focus from the vertex.

SECTION 9.4 **Hyperbolas**

Ask Yourself	Test Yourself	Help Yourself
Can I...	Do 9.4 Review Exercises	See these Examples and Practice Problems
given the center, a focus, a vertex, and the equation of the transverse axis, write the equation of a hyperbola?	25	Ex. 1, PP 1
given the foci and one vertex or a focus and the vertices, write the equation of a hyperbola?	26	Ex. 2, PP 2
given the equation of a hyperbola, identify its form, the transverse axis, p, a, the foci, the vertices, write the equations of the asymptotes, and sketch the graph of the asymptotes and hyperbola?	27, 32	Ex. 3, 4 PP 3–6
use a hyperbola to solve an application problem?	33	Ex. 5, PP 7
identify an equation that represents a circle (Section 9.1), ellipse (Section 9.2), parabola (Section 9.3), or hyperbola (Section 9.4)?	28–31	Examples in Sections 9.1–9.4

9.4 Review Exercises

For exercises 25–26, write the equation of the hyperbola.

25. Center $(0, 0)$; transverse axis $y = 0$; focus $(5, 0)$; vertex $(3, 0)$

26. Focus $(0, 10)$; vertices $(0, 6)$, $(0, -6)$

27. The equation of a hyperbola is $\dfrac{x^2}{36} - \dfrac{y^2}{49} = 1$.

   a. Write the equation of the transverse axis.
   b. Identify the foci.
   c. Identify the vertices.

28. Identify the equation that represents a hyperbola.

   a. $(x - 3)^2 + (y - 4)^2 = 16$
   b. $\dfrac{x^2}{49} + \dfrac{y^2}{36} = 1$
   c. $(y - 3)^2 = 32x$
   d. $\dfrac{x^2}{49} - \dfrac{y^2}{36} = 1$

29. Identify the equation that represents a parabola.

   a. $(x - 3)^2 + (y - 4)^2 = 16$
   b. $\dfrac{x^2}{49} + \dfrac{y^2}{36} = 1$
   c. $(y - 3)^2 = 32x$
   d. $\dfrac{x^2}{49} - \dfrac{y^2}{36} = 1$

30. Identify the equation that represents a circle.

   a. $(x - 3)^2 + (y - 4)^2 = 16$
   b. $\dfrac{x^2}{49} + \dfrac{y^2}{36} = 1$
   c. $(y - 3)^2 = 32x$
   d. $\dfrac{x^2}{49} - \dfrac{y^2}{36} = 1$

31. Identify the equation that represents an ellipse.

   a. $(x - 3)^2 + (y - 4)^2 = 16$
   b. $\dfrac{x^2}{49} + \dfrac{y^2}{36} = 1$
   c. $(y - 3)^2 = 32x$
   d. $\dfrac{x^2}{49} - \dfrac{y^2}{36} = 1$

32. The equation of a hyperbola is $\dfrac{x^2}{16} - \dfrac{y^2}{25} = 1$.

   a. Identify a and b.
   b. Write the equations of the asymptotes of the graph.
   c. Sketch the graph of the asymptotes and the hyperbola.

33. The "eye" of a robot is an example of an omnidirectional vision system. Light traveling on a path to the focus of a mirror reflects to the other focus. A video camera captures the image. Software converts the round images taken by the camera into panoramic images. The equation of the hyperbola that is rotated to create the mirror is $\dfrac{x^2}{(28.095 \text{ mm})^2} - \dfrac{y^2}{(23.4125 \text{ mm})^2} = 1$. Find the distance between the vertex of the mirror and its focus. Round to the nearest thousandth. (*Source:* vision.ai.uiuc.edu, 2006)

SECTION 9.5 **Systems of Nonlinear Equations in Two Variables**

Ask Yourself	Test Yourself	Help Yourself
Can I...	Do 9.5 Review Exercises	See these Examples and Practice Problems
identify a nonlinear equation?	34–38	Ex. 1, PP 1, 2
use graphing to estimate the solution of a system of nonlinear equations?		
use substitution to solve a system of nonlinear equations?	39, 41	Ex. 2–5, PP 3–5
use elimination to solve a system of nonlinear equations?	40, 41	Ex. 6–8, PP 6–8
use a system of nonlinear equations to solve an application problem?	42	Ex. 9, PP 9

9.5 Review Exercises

34. Identify the nonlinear equations.

   a. $5x + 6y = 9$ **b.** $x^2 + y^2 = 30$

   c. $y = \dfrac{3}{4}x + 1$ **d.** $(y - 3)^2 = -8x$

   e. $\dfrac{x^2}{12} + \dfrac{y^2}{36} = 1$

35. Use graphing to solve $\begin{aligned} x^2 &= 16y \\ x^2 + y^2 &= 80 \end{aligned}$.

36. a. Graph the system $\begin{aligned} x^2 + y^2 &= 100 \\ x^2 + y^2 &= 36 \end{aligned}$.

   b. Explain why this system has no solution.

37. Identify the number of solutions of the system represented by the graph.

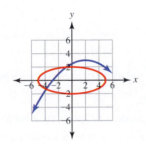

38. Identify the number of solutions for the nonlinear system represented by the graph.

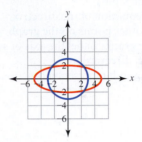

39. Solve $\begin{aligned} x^2 &= 20y \\ x^2 + y^2 &= 44 \end{aligned}$ by substitution.

40. Solve $\begin{aligned} \dfrac{x^2}{5} - \dfrac{y^2}{20} &= 1 \\ \dfrac{x^2}{17} + \dfrac{y^2}{34} &= 1 \end{aligned}$ by elimination.

41. a. Would you use substitution or elimination to solve the system $\begin{aligned} x^2 &= -4y \\ y &= 3x - 16 \end{aligned}$?

   b. Explain why.

42. The function $y = 310x + 3550$ represents the relationship of the number of years since 1996, x, and the revenues of a company, y, in thousands of dollars. The function $y = 25x^2 - 90x + 4525$ represents the relationship of the number of years since 1996, x, and the revenues of a different company, y, in thousands of dollars. Find the time(s) when the companies had the same revenues.

Chapter 9 Test

1. Use the distance formula to find the exact distance between $(7, 15)$ and $(-2, -6)$.

2. The center of a circle is $(-8, 2)$. A point on the circle is $(-15, 6)$. Write the equation of this circle in standard form.

3. Sketch the graph of $(x + 2)^2 + (y - 5)^2 = 9$.

4. The equation of an ellipse is $\dfrac{(x - 4)^2}{25} + \dfrac{(y + 2)^2}{81} = 1$.

   a. Identify the center.

   b. Write the equations of the major axis and minor axis.

   c. Identify the vertices.

   d. Sketch the graph.

5. In the autumn of 2001 in Canada, Bill Loney and Steve Irvine designed and constructed an analemmatic sundial. The sundial is elliptical. The distance between vertices on the horizontal major axis is 600 cm. The distance between vertices on the vertical minor axis is 423 cm. Write an equation that is a model of this ellipse. (*Source:* www.steveirvine.com/sundial)

6. The vertex of a parabola is $(-2, 2)$. Its axis of symmetry is $y = 2$. Its graph includes the point $(7, 14)$. Write its equation.

7. The equation of a parabola is $x^2 - 4x - 8y - 60 = 0$.
   a. Rewrite this equation in the form $(x - h)^2 = 4p(y - k)$.
   b. Identify the direction in which the graph opens.
   c. Identify the vertex.
   d. Write the equation of the axis of symmetry.
   e. Identify the focus.
   f. Write the equation of the directrix.
   g. Find two other points on the graph.
   h. Sketch the graph. Label the axis of symmetry, focus, vertex, and directrix.
   i. Does this graph represent a function?

8. Write the equation of a hyperbola with center $(0, 0)$, vertex $(3, 0)$, and focus $(8, 0)$.

9. The equation of a hyperbola is $\dfrac{y^2}{9} - \dfrac{x^2}{25} = 1$.
   a. Write the equation of the transverse axis.
   b. Identify the vertices.
   c. Identify the foci.

10. a. Write the equations of the asymptotes of $\dfrac{y^2}{25} - \dfrac{x^2}{144} = 1$.
    b. Graph the asymptotes, and sketch the graph of the hyperbola.

11. Solve $\begin{array}{c} \dfrac{x^2}{180} - \dfrac{y^2}{256} = 1 \\ x^2 + y^2 = 289 \end{array}$ by elimination.

12. Solve $\begin{array}{c} x^2 + y^2 = 68 \\ y^2 = -32x \end{array}$ by substitution.

13. A quadratic function is $y = -3(x - 4)^2 + 8$.
    a. Identify the vertex.
    b. Identify the direction in which the graph opens.
    c. Use interval notation to represent the domain.
    d. Use interval notation to represent the range.

Cumulative Review Chapters 7–9

Follow your instructor's guidelines for showing your work.

1. Copy the table below. If a number is an element of the set, mark the box with an **X**.

For Exercises 2–5, simplify. All answers must be in $a + bi$ form.

2. $(7 - 2i)(3 + 5i)$

3. $\dfrac{7}{3 - 6i}$

4. $\sqrt{-288x^2y^3}$

5. $\sqrt{-15ab}\,\sqrt{-35a}$

6. Evaluate: $\log(1000)$

7. Evaluate: $\ln(e)$

8. Evaluate: $\log(10^3)$

9. Use a calculator to evaluate $\log_3(15)$. Round to the nearest ten-thousandth.

10. Rewrite $\log(x) - \log(y) + 3\log(z)$ as a single logarithm.

11. Rewrite $\ln\left(\dfrac{a^2}{bc^3}\right)$ as an expression with more than one logarithm.

12. Use the zero product property to solve $6x^2 - x - 40 = 0$.

13. Solve $x^2 - 8x + 17 = 0$ by completing the square.

14. Write the quadratic formula.

Table for exercise 1

		Complex numbers	Imaginary numbers	Real numbers	Rational numbers	Irrational numbers	Integers	Whole numbers
a.	$4 + 3i$							
b.	9							
c.	-14							
d.	$-\dfrac{7}{8}$							
e.	$7i$							
f.	$\sqrt{7}$							
g.	$\sqrt{-16}$							

15. Use the quadratic formula to solve $5x^2 - 7x + 2 = 0$.

16. Use the quadratic formula to solve $x^2 + 3x + 5 = 0$.

17. Use any method to solve $5x^2 = 105$.

18. Use any method to solve $6x^2 - 3x = 0$.

For exercises 19–28, solve the equation. Round any irrational solutions to the ten-thousandths place.

19. $3^x = 21$

20. $\log_4(x) = 3.5$

21. $e^{2x} = 20$

22. $\ln(5x) = 6$

23. $\sqrt{4c - 12} = 6$

24. $-7b - 9 = 40$

25. $\dfrac{k}{3} - \dfrac{1}{3} = \dfrac{20}{3k}$

26. $-4 = |2x + 8| - 32$

27. $-6x^2 + 19x + 20 = 0$

28. $\dfrac{5}{6}w + \dfrac{2}{3} = \dfrac{4}{5}w - 1$

For exercises 29–32,
(a) solve the inequality.
(b) use interval notation to represent the solution.

29. $-9p + 54 < 117$

30. $x^2 + 8x - 20 \geq 0$

31. $\dfrac{8}{a + 4} > 12$

32. $20 \geq |2n + 16| - 8$

33. a. Use a table of ordered pairs to graph $y = x^2 - 4$.
   b. Use interval notation to represent the domain of this function.
   c. Use interval notation to represent the range of this function.

34. Copy and complete the table below. Use interval notation to represent the domain and range.

35. a. Rewrite $y = x^2 - 8x + 9$ in vertex form.
   b. Identify the vertex.
   c. Use interval notation to represent the domain.
   d. Use interval notation to represent the range.

36. a. Rewrite $y = -3x^2 - 6x - 20$ in vertex form.
   b. Write the equation of the axis of symmetry.
   c. Use interval notation to represent the domain.
   d. Use interval notation to represent the range.

37. Change the function $y = (x - 2)^2 + 9$ so that its graph is shifted horizontally to the left 3 and shifted vertically up 2.

38. A function is $y = \log(x)$.
   a. Use interval notation to represent the domain.
   b. Use interval notation to represent the range.
   c. Identify the x-intercept.
   d. Is this function increasing or decreasing?

39. A function is $y = 5^x$.
   a. Use interval notation to represent the domain.
   b. Use interval notation to represent the range.
   c. Identify the y-intercept.
   d. Is this function increasing or decreasing?

40. A function is $y = 0.3^x$.
   a. Use interval notation to represent the domain.
   b. Use interval notation to represent the range.
   c. Identify the y-intercept.
   d. Is this function increasing or decreasing?

41. A function is $f(x) = \log_4(x)$. Use $f^{-1}(x)$ notation to represent the inverse of this function.

42. A function is $f(x) = e^x$. Use $f^{-1}(x)$ notation to represent the inverse of this function.

43. The length of a rectangle is 6 in. less than three times its width. Its area is 45 in.^{2}. Use a quadratic equation in one variable to find its length and width.

44. The function $y = 0.0995x^2 - 1.1631x + 22.078$ represents the relationship of the number of years since 1990, x, and the rate of cesarean deliveries per 100 live births in the United States for women ages 18–29 years. Find the rate of cesarean deliveries per 100 live births for this age group in 2012. Round to the nearest tenth. (*Source:* www.cdc.gov, 2008)

45. The hypotenuse of a right triangle is 30 in. One leg is 6 in. longer than the other leg. Use a quadratic equation to find the length of each leg.

46. A utility pole is 40 ft long. When a utility pole is put up, 6 ft of the pole is buried underground. A squirrel drops a walnut from the top of the pole. The initial velocity of the nut is $0 \dfrac{\text{ft}}{\text{s}}$. The acceleration on the walnut caused by gravity is $32 \dfrac{\text{ft}}{\text{s}^2}$. Use $d = v_i t + \dfrac{1}{2}at^2$ to find the time it takes the walnut to hit the ground, ignoring air resistance. Round to the nearest tenth.

47. The concentration of hydronium ions, H_3O^+, in a solution is $7.0 \times 10^{-9} \dfrac{\text{mole}}{\text{liter}}$. Find the pH of this solution. Round to the nearest tenth.

48. The pH of a solution is 2.4. Find the hydronium ion concentration of this solution in moles per liter. Use scientific notation; round the mantissa to the tenths place.

49. Find the future value of $7000 that is invested at an interest rate of 2.25% for 5 years, compounded monthly. Round to the nearest hundredth.

Table for exercise 34

Function	Vertex	Domain	Range	Axis of symmetry	Opens up or down?	Number of real zeros
$y = -3(x - 9)^2 + 4$						
$y = 2(x + 3)^2 - 6$						

50. The function $y = 85{,}500(1.06)^x$ represents the relationship of the number of years since 2010, x, and the population of a city, y. Find the population of the city in 2017. Round to the nearest hundred.

51. Use the distance formula to find the distance between $(-6, -9)$ and $(12, 3)$. Do not round.

52. The endpoints of a line segment are $(-6, -9)$ and $(12, 3)$. Find the midpoint of this line segment.

53. The center of a circle is $(4, 2)$. A point on the circle is $(8, 10)$.
  a. Write the equation of this circle in standard form.
  b. Sketch the graph of the circle.

54. The vertices on the horizontal major axis of an ellipse are $(5, 4)$ and $(15, 4)$. The vertices on the vertical minor axis are $(10, 1)$ and $(10, 7)$.
  a. Write the equation of the major axis.
  b. Write the equation of the minor axis.
  c. Identify the center.
  d. Write the equation of the ellipse in standard form.
  e. Sketch the graph of the ellipse.

55. The vertex of a parabola is $(1, 3)$. The axis of symmetry is $x = 1$. Its graph includes the point $(9, 11)$.
  a. Write the equation of the parabola.
  b. Write the equation of the directrix.
  c. Identify the focus.
  d. Sketch the graph, including the focus, directrix, and axis of symmetry.

56. The equation of a hyperbola is $\dfrac{y^2}{4} - \dfrac{x^2}{9} = 1$.
  a. Identify the foci.
  b. Identify the vertices.
  c. Identify the transverse axis.

57. The vertices of a hyperbola are $(-8, 0)$ and $(8, 0)$. One focus is $(-12, 0)$.
  a. Write the equation of the hyperbola.
  b. Write the equations of the asymptotes.
  c. Graph the asymptotes; sketch the graph of the hyperbola.

58. Identify the equation(s) that represent(s) a circle.
  a. $x^2 + y^2 = 25$
  b. $\dfrac{x^2}{4} + \dfrac{y^2}{9} = 1$
  c. $\dfrac{x^2}{16} - \dfrac{y^2}{49} = 1$
  d. $(x - 3)^2 + (y - 4)^2 = 1$

59. In 1962, the *Mariner 2* spacecraft was sent to Venus to conduct scientific experiments designed to learn more about the atmosphere of Venus. Find the distance of the focus from the vertex in the parabolic antenna used on the spacecraft. Round to the nearest hundredth.

As the Mariner spacecraft flies past Venus, the microwave radiometer will scan its surface to detect electromagnetic radiation at two wave lengths, 13.5 and 19 millimeters. In the electromagnetic spectrum, 13.5 mm is the location of a microwave water absorption band. If there is water vapor above certain minimal concentration in the atmosphere, it will be possible to detect it. . . . The microwave radiometer is mounted on the hexagonal base of the Mariner. Both wave lengths are detected by a parabolic antenna that is 20 inches in diameter and three inches deep. (*Source:* www.jpl.nasa.gov/releases/60s, May 19, 1962)

Sequences, Series, and the Binomial Theorem

SUCCESS IN COLLEGE MATHEMATICS

Look Back

At the beginning of this course, you may have set goals about attendance, planning, studying, and seeking extra help. Now, at the end, it is important to look back and reflect on your goals, your actions, and the results. Sometimes changes in your life situation make initial goals unrealistic. Or you might not have included enough time in your schedule to achieve them. If you look back at your experiences in this class, you can use what you discover to set goals and plan for classes in the future.

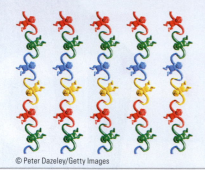

© Peter Dazeley/Getty Images

If we search the words "What comes next" on the Internet, the results will include many pattern recognition games. Pattern recognition is an important part of developing logical reasoning. In this section, we will make logical predictions about the outputs of special functions called *sequences*.

Sequences and Series

After reading the text, working the practice problems, and completing assigned exercises, you should be able to:

1. Given the domain of a sequence, use roster notation to represent the range.
2. Graph a sequence.
3. Find a formula for the nth term of a sequence.
4. Use sigma notation to write a sum.
5. Evaluate a partial sum of a sequence.
6. Use a sequence to solve an application problem.

Sequences

The domain of many of the functions we have studied is the set of real numbers. For example, the domain of $f(x) = 3x + 5$ is $(-\infty, \infty)$. This domain is **continuous**. If we identify any two points on the real number line, there is always another point between them. A **sequence** is a function in which the domain is a set of integers greater than 0. This domain is not continuous; it is **discrete**. The domain may have infinitely many elements or a finite number of elements. We can identify two inputs in a discrete domain in which there is no input between them.

The outputs of a sequence are called **terms**. The range of a sequence is the set of all its terms. To find the range, we evaluate the function for each value of the domain. We can represent a sequence with a graph.

We often use different notation to describe a function that is a sequence. Instead of using $f(x)$ notation, we use a_n. In the next example, the sequence is $a_n = n + 2$.

EXAMPLE 1 A finite sequence is $a_n = n + 2$. The domain is $\{1, 2, 3, 4, 5\}$.

(a) Use roster notation to represent the range.

SOLUTION ▶

a_1	a_2	a_3	a_4	a_5
$= 1 + 2$	$= 2 + 2$	$= 3 + 2$	$= 4 + 2$	$= 5 + 2$
$= 3$	$= 4$	$= 5$	$= 6$	$= 7$

Range: $\{3, 4, 5, 6, 7\}$

(b) Graph this sequence.

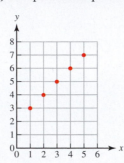

When graphing a sequence, do not draw a line or curve through the ordered pairs of the function. The domain is discrete, not continuous.

In the next example, an ellipsis shows that the range is an infinite set. Since the input variable is k, the output variable is a_k.

EXAMPLE 2 | An infinite sequence is $a_k = \dfrac{k}{2^k}$. The domain is the set of integers greater than 0.

(a) Use roster notation to represent the range. Include the first five output values.

SOLUTION ▶

a_1	a_2	a_3	a_4	a_5
$= \dfrac{1}{2^1}$	$= \dfrac{2}{2^2}$	$= \dfrac{3}{2^3}$	$= \dfrac{4}{2^4}$	$= \dfrac{5}{2^5}$
$= \dfrac{1}{2}$	$= \dfrac{2}{4}$	$= \dfrac{3}{8}$	$= \dfrac{4}{16}$	$= \dfrac{5}{32}$
	$= \dfrac{1}{2}$		$= \dfrac{1}{4}$	

Range: $\left\{ \dfrac{1}{2}, \dfrac{1}{2}, \dfrac{3}{8}, \dfrac{1}{4}, \dfrac{5}{32}, \ldots \right\}$

(b) Graph the first five terms of this sequence.

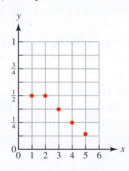

The signs of terms in some sequences alternate. In the next example, when the input value is even, the output value is positive. When the input value is odd, the output value is negative. Since the input variable is n, the output variable is a_n.

EXAMPLE 3 | A sequence is $a_n = (-1)^n n$. The domain is the set of integers greater than 0.

(a) Use roster notation to represent the range. Include the first five output values.

SOLUTION ▶

a_1	a_2	a_3	a_4	a_5
$= (-1)^1 \cdot 1$	$= (-1)^2 \cdot 2$	$= (-1)^3 \cdot 3$	$= (-1)^4 \cdot 4$	$= (-1)^5 \cdot 5$
$= -1$	$= 2$	$= -3$	$= 4$	$= -5$

Range: $\{-1, 2, -3, 4, -5, \ldots\}$

(b) Graph the first five terms of this sequence.

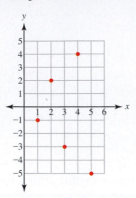

Practice Problems

1. The domain of the sequence $a_n = 2^n$ is $\{1, 2, 3, 4, 5\}$.
   a. Use roster notation to represent the range.
   b. Graph this sequence.

2. The domain of the sequence $a_k = \dfrac{k - 1}{k + 1}$ is the set of integers greater than 0.

   a. Use roster notation with at least five terms to represent the range.
   b. Graph the first five terms of this sequence.

3. The domain of the sequence $a_n = (-2)^n + n$ is the set of integers greater than 0.
   a. Use roster notation with at least five terms to represent the range.
   b. Graph the first five terms of this sequence.

Formula for the nth Term of a Sequence

The **function rule** of $f(x) = 3x + 2$ is $3x + 2$. The **formula for the nth term of the sequence** $a_n = n - 3$ is $n - 3$. Given the range of a sequence, we can use logical thinking combined with a process of guess and check to find its formula.

EXAMPLE 4   Find the formula for the nth term of the sequence with the given range.

(a) $\{4, 5, 6, 7, \ldots\}$

SOLUTION ▶   Since the domain of a sequence is $\{1, 2, 3, 4, \ldots\}$ and the range of this sequence is $\{4, 5, 6, 7, \ldots\}$, when the input value is 1, the output value is 4. Since $4(1) = 4$, a possible formula for the nth term of this sequence is $a_n = 4n$. To check whether this formula is correct, find a_n when $n = 2$. If the formula is correct, when $n = 2$, $a_n = 5$.

> **Guess:** The formula is $a_n = 4n$.
> **Check:**
> $a_2 = 4(2)$ Replace n with 2.
> $a_2 = 8$ Simplify; since the output does not equal 5, the formula is not correct.

The formula for the nth term of this sequence is not $a_n = 4n$. Try a different guess. Since $1 + 3 = 4$, a possible formula is $a_n = n + 3$.

> **Guess:** The formula is $a_n = n + 3$.
> **Check:**
> $a_2 = 2 + 3$ Replace n with 2.
> $a_2 = 5$ Simplify; since the output equals 5, the formula may be correct.

Since this formula also works for the other terms in the sequence, the formula for the nth term is $a_n = n + 3$.

(b) $\{3, 9, 27, 81, 243, \ldots\}$

▶ Since the domain of a sequence is $\{1, 2, 3, 4, \ldots\}$ and the range of this sequence is $\{3, 9, 27, 81, 243, \ldots\}$, when the input value is 1, the output value is 3. The output values are increasing very rapidly, so the formula may include an exponent. Since $3^1 = 3$, a possible formula for the nth term of this sequence is $a_n = 3^n$. To check whether this formula is correct, find a_n when $n = 2$. If the formula is correct, when $n = 2$, $a_n = 9$.

Guess: The formula is $a_n = 3^n$.
Check:

$a_2 = 3^2$ Replace n with 2.

$a_2 = \mathbf{9}$ Simplify; since the output equals 9, the formula may be correct.

Since this formula also works for the other terms in the sequence, the formula for the nth term is $a_n = 3^n$.

(c) $\{-10, 100, -1000, 10{,}000, \ldots\}$

▶ Since the domain of a sequence is $\{1, 2, 3, 4, \ldots\}$ and the range of this sequence is $\{-10, 100, -1000, 10{,}000, \ldots\}$, when the input value is 1, the output value is -10. Since the signs in the range are alternating, the formula may include a negative number raised to the nth power. Since $(-10)^1 = -10$, a possible formula for the nth term of this sequence is $a_n = (-10)^n$. To check whether this formula is correct, find a_n when $n = 2$. If the formula is correct, when $n = 2$, $a_n = 100$.

Guess: The formula is $a_n = (-10)^n$.
Check:

$a_2 = (-10)^2$ Replace n with 2.

$a_2 = \mathbf{100}$ Simplify; since the output equals 100, the formula may be correct.

Since this formula also works for the other terms in the sequence, the formula for the nth term is $a_n = (-10)^n$.

We will explore other methods for finding the formula of a sequence in the next two sections.

Practice Problems

For problems 4–6, find the formula for the nth term of the sequence with the given range.

4. $\{4, 8, 12, 16, 20, \ldots\}$ **5.** $\left\{1, \dfrac{1}{4}, \dfrac{1}{9}, \dfrac{1}{16}, \dfrac{1}{25}, \ldots\right\}$ **6.** $\{1, 4, 9, 16, \ldots\}$

Partial Sums and Series

In common use, a **series** is a number of objects or events that come one after another. The World Series in baseball is a succession of up to seven games. A television series such as *Grey's Anatomy* is a succession of episodes. However, **series** has a special meaning in mathematics. The sum of all the terms of a sequence is an **infinite series** or **series**. So a series is a sum. The result of adding only some of the terms in a sequence is a **partial sum**.

EXAMPLE 5 | Evaluate the partial sum of the first six terms of the sequence $a_n = 5n$.

SOLUTION ▶ The terms in this sequence are $\{5, 10, 15, 20, 25, \ldots\}$. The sum of the first six terms is $5 + 10 + 15 + 20 + 25 + 30 = 105$. The partial sum of the first six terms of this sequence is 105.

For convenience and clarity, we use **sigma notation** to represent a partial sum. The Greek letter *sigma* is Σ. If an infinite sequence is $\{a_1, a_2, a_3, a_4, a_5, \ldots\}$, the sum of the first four terms of this sequence is $a_1 + a_2 + a_3 + a_4$. In sigma notation, this partial sum is represented as $\displaystyle\sum_{i=1}^{4} a_i$. The **index** of this sum is i. The variable in the sequence rule must be the same as the variable used for the index. Since for $\displaystyle\sum_{i=1}^{4} a_i$, $i = 1$, the sum begins with the first term of the sequence, a_1. The number on the top of sigma tells us the final term in the sequence to add. The terms in this sum begin with a_1 and end with a_4. *Notice that i is a variable, not a number. It is not equivalent to $\sqrt{-1}$.*

EXAMPLE 6 | **(a)** Use sigma notation to represent the partial sum of the first six terms of the sequence $a_i = 5i$.

SOLUTION ▶

$$\sum_{i=1}^{n} a_i \qquad \text{Sigma notation.}$$

$$= \sum_{i=1}^{6} 5i \qquad n = 6; a_i = 5i$$

(b) Evaluate this partial sum.

▶
$$\sum_{i=1}^{6} 5i$$

$$= 5(1) + 5(2) + 5(3) + 5(4) + 5(5) + 5(6) \qquad \text{The sum of the first six terms.}$$

$$= 5 + 10 + 15 + 20 + 25 + 30 \qquad \text{Simplify.}$$

$$= 105 \qquad \text{Simplify.}$$

The partial sum of the first six terms is 105.

In the next example, we will find the formula of the nth term of a sequence and then write and evaluate a partial sum.

EXAMPLE 7 | **(a)** Use sigma notation to represent the partial sum of the first seven terms of the sequence $\{3, 9, 27, 81, 243, \ldots\}$.

SOLUTION ▶ In Example 4, we found the formula of the nth term of this sequence: $a_n = 3^n$. We need to change the variable in this formula to match the index: $a_i = 3^i$. We want the sum of seven terms, beginning with the first term.

$$\sum_{i=1}^{n} a_i$$

$$= \sum_{i=1}^{7} 3^i \qquad a_i = 3^i; n = 7$$

(b) Evaluate this partial sum.

$$\sum_{i=1}^{7} 3^i$$

$$= 3^1 + 3^2 + 3^3 + 3^4 + 3^5 + 3^6 + 3^7 \qquad \text{Partial sum of seven terms.}$$

$$= 3 + 9 + 27 + 81 + 243 + 729 + 2187 \qquad \text{Evaluate.}$$

$$= 3279 \qquad \text{Evaluate.}$$

The partial sum of the first seven terms is 3279.

We can also evaluate a partial sum when the first term is not a_1. In the next example, the partial sum begins with the fifth term.

EXAMPLE 8   Evaluate: $\displaystyle\sum_{i=5}^{7}(3i + 4)$

SOLUTION ▶

$$\sum_{i=5}^{7}(3i + 4) \qquad a_i = 3i + 4$$

$$= (3(\mathbf{5}) + 4) + (3(\mathbf{6}) + 4) + (3(\mathbf{7}) + 4) \qquad \text{Partial sum of three terms.}$$

$$= (\mathbf{15} + 4) + (\mathbf{18} + 4) + (\mathbf{21} + 4) \qquad \text{Simplify.}$$

$$= 66 \qquad \text{Simplify.}$$

Practice Problems

For problems 7–9, evaluate the partial sum.

7. $\displaystyle\sum_{i=1}^{4}(-2i + 9)$ **8.** $\displaystyle\sum_{i=3}^{6}(-2i + 9)$ **9.** $\displaystyle\sum_{i=1}^{5}\frac{(-1)^i}{2}$

Applications

We have solved many application problems by evaluating a function written in function notation. We can also solve problems by evaluating a function written in sequence notation.

EXAMPLE 9   The future value, a_n, of an investment of \$5000 that is earning 6% compounded monthly is $a_n = 5000\left(1 + \dfrac{0.06}{12}\right)^{12n}$, where n is the number of years that the money is invested. Find the future value of this investment in 3 years. Round to the nearest whole number.

SOLUTION ▶

$$a_n = 5000\left(1 + \frac{0.06}{12}\right)^{12n} \qquad \text{Original sequence.}$$

$$a_3 = \$5000\left(1 + \frac{0.06}{12}\right)^{36} \qquad \text{Replace } n \text{ with 3.}$$

$$a_3 \approx \mathbf{\$5983} \qquad \text{Round to the nearest whole number.}$$

Practice Problems

For problems 10–11, a model of the population of the United States from 1960 to 2005 is $a_n = 181,493,000(1.01)^n$, where $n = 1$ is 1960 and a_n is the number of people. Assume that this model continues to predict the population correctly.

10. Find a_{50}. Round to the nearest million.

11. What term in this sequence represents the population in 2015?

Using Technology: Graphing a Sequence

In Sequence graphing mode, we can graph up to three sequences. They are named $u(n)$, $v(n)$, and $w(n)$. This notation represents the sequences u_n, v_n, and w_n.

To change the graphing mode from Function to Sequence, press MODE. Move the cursor down to the fourth line; move right and select SEQ. Press Y=.

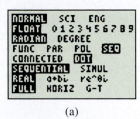
(a)

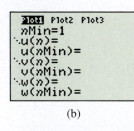
(b)

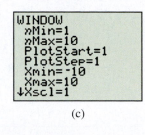
(c)

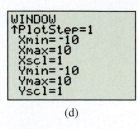
(d)

The WINDOW screen in Sequence graphing mode shows the first term to be graphed (nMin), the last term to be graphed (nMax), and the interval between terms (PlotStep). On the graph itself, n is represented by x. The value of each term in the sequence is represented by y.

EXAMPLE 10 Graph ten terms of the sequence $a_n = 3n$ beginning with the first term. Describe the window.

To enter $3n$, press 3 and then press the X,T,Θ,n key. Leave the individual sequence index values such as $u(n$Min$)$ blank. Press GRAPH. Since only three terms in the sequence are visible, we need to change the window.

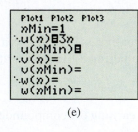
(e)

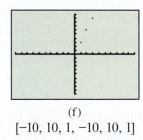
(f)

[−10, 10, 1, −10, 10, 1]

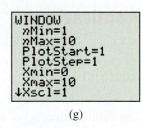
(g)

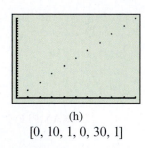
(h)

[0, 10, 1, 0, 30, 1]

The smallest value of n is 1, so let $X_{\min} = 0$. The smallest value of a_n is 3, so let $Y_{\min} = 0$. When $n = 10$, $a_n = 30$, so let $Y_{\max} = 30$. To find $Y_{\min}$ and $Y_{\max}$ on the WINDOW screen, use the arrow key to scroll down.

Practice Problems For problems 12–14, graph ten terms of each sequence. Sketch the graph; describe the window.

12. $a_n = n + 3$　　　**13.** $a_n = 4n$　　　**14.** $a_n = (-1)^n \cdot n$

For problem 15, graph five terms of the sequence. Sketch the graph; describe the window.

15. $a_n = 2^n$

10.1 VOCABULARY PRACTICE

Match the term with its description.

1. A set of ordered pairs in which each input value corresponds to exactly one output value
2. The set of the inputs of a function
3. The set of the outputs of a function
4. $\{\ldots, -3, -2, -1, 0, 1, 2, 3, \ldots\}$
5. A function in which the domain is a set of integers greater than 0
6. The output values of a sequence
7. The sum of some of the terms in an infinite sequence
8. The sum of all of the terms in an infinite sequence
9. $\sum$
10. In sigma notation, this is located below the sigma.

A. domain
B. function
C. index
D. partial sum of a sequence
E. range
F. sequence
G. series
H. set of integers
I. sigma
J. terms

10.1 Exercises

Follow your instructor's guidelines for showing your work.

For exercises 1–2, use roster notation to represent the range.

1. A finite sequence is $a_n = n - 2$. The domain of this function is $\{1, 2, 3, 4, 5\}$.

2. A finite sequence is $a_n = n - 3$. The domain of this function is $\{1, 2, 3, 4, 5\}$.

For exercises 3–8, the domain is the set of positive integers greater than 0.
(a) Use roster notation to represent the range of the sequence.
(b) Graph the first five terms of the sequence.

3. $a_n = n + 5$

4. $a_n = n + 4$

5. $a_n = (-1)^n(2n)$

6. $a_n = (-1)^n(3n)$

7. $a_n = \dfrac{n^2}{n^2 + 1}$

8. $a_n = n^2 - n$

For exercises 9–12, the domain is the set of positive integers greater than 0. Use roster notation to represent the range of the sequence.

9. $a_n = \dfrac{1}{n^2 + 1}$

10. $a_n = n^{-2}$

11. $a_n = \dfrac{2}{n^2 + 1}$

12. $a_n = n^{-2} + 1$

For exercises 13–32, find the formula for the nth term of each sequence.

13. $\{-7, -6, -5, -4, -3, \ldots\}$

14. $\{-10, -9, -8, -7, -6, \ldots\}$

15. $\{-7, 6, -5, 4, -3, \ldots\}$

16. $\{-10, 9, -8, 7, -6, \ldots\}$

17. $\{-2, -4, -6, -8, -10, \ldots\}$

18. $\{-3, -6, -9, -12, -15, \ldots\}$

19. $\left\{-1, -\dfrac{1}{4}, -\dfrac{1}{9}, -\dfrac{1}{16}, -\dfrac{1}{25}, \ldots\right\}$

20. $\left\{-1, -\dfrac{1}{8}, -\dfrac{1}{27}, -\dfrac{1}{64}, -\dfrac{1}{125}, \ldots\right\}$

21. $\{2, 5, 10, 17, 26, \ldots\}$

22. $\{4, 7, 12, 19, 28, \ldots\}$

23. $\{1, 4, 27, 256, 3125, \ldots\}$

24. $\{-1, -4, -27, -256, -3125, \ldots\}$

25. $\{3, 5, 7, 9, 11, \ldots\}$

26. $\{4, 6, 8, 10, 12 \ldots\}$

27. $\left\{\dfrac{1}{2}, \dfrac{1}{3}, \dfrac{1}{4}, \dfrac{1}{5}, \dfrac{1}{6}, \ldots\right\}$

28. $\left\{\dfrac{1}{3}, \dfrac{1}{4}, \dfrac{1}{5}, \dfrac{1}{6}, \dfrac{1}{7}, \ldots\right\}$

29. $\left\{\dfrac{2}{3}, \dfrac{3}{4}, \dfrac{4}{5}, \dfrac{5}{6}, \dfrac{6}{7}, \ldots\right\}$

30. $\left\{\dfrac{4}{3}, \dfrac{5}{4}, \dfrac{6}{5}, \dfrac{7}{6}, \dfrac{8}{7}, \ldots\right\}$

31. $\{1, 2, 4, 8, 16, \ldots\}$

32. $\{1, 3, 9, 27, 81, \ldots\}$

33. Create a formula for the nth term of a sequence. The third term must be greater than 8. Find the first five terms of this sequence.

34. Create a formula for the nth term of a sequence. The third term must be less than 8. Find the first five terms of this sequence.

35. Explain the difference between a *partial sum* and a *series*.

36. Sigma notation is also called *summation notation*. Explain why.

37. Evaluate the partial sum of the first four terms of the sequence $a_n = 2n + 3$.

38. Evaluate the partial sum of the first four terms of the sequence $a_n = 4n - 1$.

39. Evaluate the partial sum of the first five terms of the sequence $a_n = n^3$.

40. Evaluate the partial sum of the first five terms of the sequence $a_n = n^{n-1}$.

41. Use sigma notation to represent the partial sum of the first four terms of the sequence $a_i = 2i + 3$.

42. Use sigma notation to represent the partial sum of the first four terms of the sequence $a_i = 4i - 1$.

43. Use sigma notation to represent the partial sum of the first ten terms of the sequence $a_i = 2i + 3$.

44. Use sigma notation to represent the partial sum of the first ten terms of the sequence $a_i = 4i - 1$.

45. Use sigma notation to represent the partial sum of the first five terms of the sequence $\{3, 6, 9, 12, 15, \ldots\}$.

46. Use sigma notation to represent the partial sum of the first five terms of the sequence $\{5, 10, 15, 20, 25, \ldots\}$.

47. Use sigma notation to represent the partial sum of the first five terms of the sequence $\{0, -1, -2, -3, -4, -5, \ldots\}$.

48. Use sigma notation to represent the partial sum of the first five terms of the sequence $\{-2, -3, -4, -5, -6, \ldots\}$.

49. Use sigma notation to represent the partial sum of the first six terms of the sequence $\left\{3, \dfrac{3}{2}, 1, \dfrac{3}{4}, \dfrac{3}{5}, \ldots\right\}$.

50. Use sigma notation to represent the partial sum of the first six terms of the sequence $\left\{2, 1, \dfrac{2}{3}, \dfrac{1}{2}, \dfrac{2}{5}, \ldots\right\}$.

For exercises 51–62, evaluate the partial sum.

51. $\displaystyle\sum_{i=1}^{4} 6i$

52. $\displaystyle\sum_{i=1}^{4} 7i$

53. $\displaystyle\sum_{i=2}^{6} 6i$

54. $\displaystyle\sum_{i=2}^{6} 7i$

55. $\displaystyle\sum_{i=1}^{3} 5^i$

56. $\displaystyle\sum_{i=1}^{3} 4^i$

57. $\displaystyle\sum_{i=1}^{4} \dfrac{4}{i}$

58. $\displaystyle\sum_{i=1}^{4} \dfrac{5}{i}$

59. $\displaystyle\sum_{i=50}^{53} (i + 8)$

60. $\displaystyle\sum_{i=60}^{63} (i + 7)$

61. $\displaystyle\sum_{i=1}^{6} (-1)^i$

62. $\displaystyle\sum_{i=1}^{7} (-1)^i$

63. The future value in dollars, a_n, of an investment of $4000 that is earning 5.5% compounded monthly is $a_n = 4000\left(1 + \dfrac{0.055}{12}\right)^{12n}$, where n is the number of years that the money is invested. Find the future value of this investment after 4 years. Round to the nearest whole number.

64. The future value in dollars, a_n, of an investment of $3500 that is earning 7% compounded quarterly is $a_n = 3500\left(1 + \dfrac{0.07}{4}\right)^{4n}$, where n is the number of years that the money is invested. Find the future value of this investment after 3 years. Round to the nearest whole number.

65. Use roster notation to write the first five terms of the sequence in exercise 63. Round each term to the nearest whole number.

66. Use roster notation to write the first five terms of the sequence in exercise 64. Round each term to the nearest whole number.

Problem Solving: Practice and Review

Follow your instructor's guidelines for using the five steps as outlined in Section 1.5, p. 51.

67. Find the average gambling debt two years ago. Round to the nearest whole number.

Callers to the state's hotline for problem gamblers report gambling debts, on average, of $62,495. That's a jump of 54 percent from two years ago. (*Source:* www.miamiherald .com, June 27, 2009)

68. A rectangular lock on the St. Lawrence Seaway is 766 ft long, 80 ft wide, and 30 ft deep. It fills with water in 7 to 10 min. If the lock fills in 7 min, find the average rate that it is filling in gallons per second (1 ft³ ≈ 7.48 gal). Round to the nearest whole number. (*Source:* www .greatlakes-seaway.com)

69. An Idaho dairy farmer received $8.80 for 100 lb of milk. At the store, a gallon of milk cost $2.75. Find the difference in the amount that the farmer received for the milk and the amount that customers paid for it (100 lb of milk ≈ 11.5 gal). Round to the nearest hundredth. (*Source:* www.lmtribune.com, April 19, 2010)

70. a. Write a linear model including units that describes the relationship of the number of years since 2002, t, and the enrollment in the Chicago public schools.
b. Use this model to predict the enrollment in 2012.

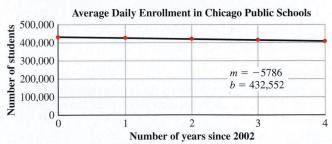

Average Daily Enrollment in Chicago Public Schools

$m = -5786$
$b = 432,552$

Source: www.cps.edu, Fiscal Year 2009

Technology

For exercises 71–74, use a graphing calculator to graph the first ten terms of the sequence. Sketch the graph; describe the window.

71. $a_n = 4n - 6$

72. $a_n = n(n + 1)$

73. $a_n = 2^n$

74. $a_n = (-1)^n(n - 1)$

Find the Mistake

For exercises 75–78, the completed problem has one mistake.
(a) Describe the mistake in words, or copy down the whole problem and highlight or circle the mistake.
(b) Do the problem correctly.

75. Problem: An infinite sequence is $a_n = n + 3$. Use roster notation to represent the domain of this sequence.

Incorrect Answer: The domain is $\{4, 5, 6, 7, 8, \ldots\}$.

76. Problem: Find the formula for the nth term of this sequence: $\{2, 4, 6, 8, 10, \ldots\}$.

Incorrect Answer: $a_n = 2^n$

77. Problem: Use sigma notation to represent the partial sum of the first five terms of the sequence $a_i = i^2$.

Incorrect Answer: Since $6 - 1 = 5$, $\sum\limits_{i=1}^{6} i^2$.

78. Problem: Evaluate: $\sum\limits_{i=8}^{10} (i + 1)$

Incorrect Answer:

$$\sum_{i=8}^{10} (i + 1)$$
$$= 2 + 3 + 4 + 5 + 6 + 7 + 8 + 9 + 10 + 11$$
$$= 65$$

Review

79. Factor: $6x^2 - 13x - 28$

80. a. Solve: $\sqrt{2x + 3} = 21$
   b. Check.

81. Solve: $3x^2 - 5x - 9 = 0$

82. Solve $\dfrac{5}{x} + \dfrac{3}{y} = \dfrac{11}{z}$ for y.

SUCCESS IN COLLEGE MATHEMATICS

83. On a final exam, students were asked, "If you could start this class over again from the beginning, what would you do differently?" How would you answer this question?

© Maxx-Studio/Shutterstock

In *sport stacking*, competitors stack special cups in pyramids. For example, a 6-stack includes six cups. The number of cups in each row from top to bottom is an example of an *arithmetic sequence*. In this section, we will learn about arithmetic sequences.

SECTION 10.2

Arithmetic Sequences

After reading the text, working the practice problems, and completing assigned exercises, you should be able to:

1. Determine whether a sequence is arithmetic.

2. Find the formula for the nth term of an arithmetic sequence.

3. Evaluate the partial sum of an arithmetic sequence.

4. Use an arithmetic sequence to solve an application problem.

Arithmetic Sequences

A sequence is a function. The domain of each sequence in this section is the set of integers greater than 0. The range of a sequence is a set of terms, the outputs of the function. The terms a_1, a_2, and a_3 are **consecutive terms** in the range of a sequence. When the difference of consecutive terms is a constant, the sequence is an **arithmetic sequence**. This constant is the **common difference**, d.

A sequence is often described by just listing the terms in the range. In the next example, the sequence is $\{2, 8, 14, 20, \ldots\}$. This is the range of the sequence.

EXAMPLE 1 The sequence $\{2, 8, 14, 20, \ldots\}$ is an arithmetic sequence. Find the common difference.

SOLUTION ▶

$a_4 - a_3$	$a_3 - a_2$	$a_2 - a_1$
$= 20 - 14$	$= 14 - 8$	$= 8 - 2$
$= 6$	$= 6$	$= 6$

The common difference, d, is 6.

The formula for the nth term of the sequence $\{2, 8, 14, 20, \ldots\}$ is $a_n = 6n - 4$. Finding this formula using guess and check is time-consuming. We can develop a general formula for the nth term of any arithmetic sequence.

a_1	a_2	a_3	a_4	a_5
	$= a_1 + d$	$= a_2 + d$	$= a_3 + d$	$= a_4 + d$
		$= (a_1 + d) + d$	$= (a_1 + 2d) + d$	$= (a_1 + 3d) + d$
		$= a_1 + 2d$	$= a_1 + 3d$	$= a_1 + 4d$

This relationship leads us to a formula for the nth term of an arithmetic sequence.

Formula for the nth Term of an Arithmetic Sequence

$$a_n = a_1 + (n - 1)d$$

where d is the common difference between consecutive terms of the sequence and a_1 is the first term.

EXAMPLE 2 $\{-4, -7, -10, -13, -16, \ldots\}$ is an arithmetic sequence.

(a) Find the common difference.

SOLUTION ▶

$a_5 - a_4$	$a_4 - a_3$	$a_3 - a_2$
$= -16 - (-13)$	$= -13 - (-10)$	$= -10 - (-7)$
$= -3$	$= -3$	$= -3$

The common difference, d, is -3.

(b) Find the formula for the nth term of this sequence.

▶

$a_n = a_1 + (n - 1)d$	Formula for the nth term of an arithmetic sequence.
$a_n = -4 + (n - 1)(-3)$	Replace a_1 with -4 and d with -3.
$a_n = -4 - 3n + 3$	Distributive property.
$a_n = -3n - 1$	Simplify; formula for the nth term of this sequence.

Finding the Formula for the nth Term of an Arithmetic Sequence, Given the Sequence

1. Choose two terms in the sequence; find the common difference.
2. Use $a_n = a_1 + (n - 1)d$; replace a_1 with the first term of the given sequence and replace d with the common difference.
3. Simplify.

A formula for the nth term of a sequence is an **explicit formula**. We can use it to find any term in the sequence. The only information we need is the first term and the common difference. Other formulas allow us to find a term if we know *any* term of the sequence and the common difference. These are **recursive formulas**. For an arithmetic sequence, the value of a term is described by the recursive formula $a_{n+1} = a_n + d$.

Finding a Term a_{n+1} in an Arithmetic Sequence, Given a Term a_n and the Common Difference, d

1. Use $a_{n+1} = a_n + d$; replace a_n with the given term and d with the common difference.
2. Simplify.

EXAMPLE 3 | The 15th term in an arithmetic sequence is 39. The common difference is 2.

(a) Find the 16th term in the sequence.

SOLUTION ▶ Since $d = 2$, we know that $a_{16} = a_{15} + 2$.

$a_{16} = \mathbf{39} + 2$

$a_{16} = \mathbf{41}$

(b) Find a_1.

$a_n = a_1 + (n - 1)d$	Formula for the nth term of an arithmetic sequence.
$39 = a_1 + (\mathbf{15} - 1)(\mathbf{2})$	Replace a_n with 39, n with 15, and d with 2.
$39 = a_1 + \mathbf{28}$	Simplify.
$\mathbf{11} = a_1$	The first term in the sequence.

(c) Find the formula for the nth term of this sequence.

$a_n = \mathbf{11} + (n - 1)(\mathbf{2})$	Replace a_1 with 11 and d with 2.
$a_n = 11 + \mathbf{2n} - \mathbf{2}$	Distribute.
$a_n = 2n + \mathbf{9}$	The formula for the nth term of this sequence.

Finding the Formula for the nth Term of an Arithmetic Sequence, Given a Term in the Sequence a_n and the Common Difference, d

1. Find a_1. Use $a_n = a_1 + (n - 1)d$; replace a_n and d. Simplify.
2. Use $a_n = a_1 + (n - 1)d$ to find the formula for the nth term; replace a_1 and d. Simplify.

Practice Problems

For problems 1–3,
(a) find the common difference of the arithmetic sequence.
(b) find the formula for the nth term of the sequence.

1. $\{5, 9, 13, 17, 21, \ldots\}$ **2.** $\{6, 5, 4, 3, 2, 1, \ldots\}$ **3.** $\{0, 6, 12, 18, 24, \ldots\}$
4. The fourth term in an arithmetic sequence is 31. The common difference is 2.
   a. Find the fifth term in the sequence.
   b. Find the formula for the nth term.

The Partial Sum of an Arithmetic Sequence

In Section 10.1, we found the partial sums of sequences by identifying the terms and adding them together. For arithmetic sequences, we can instead develop a formula to find the nth partial sum represented by the notation $S_n = \sum_{i=1}^{n} a_i$. This is the sum of n terms of the sequence, beginning with the first term, a_1. To find the formula, first write an equation for the sum of an arithmetic sequence from a_1 to a_n. The common difference, d, is added.

$$S_n = a_1 + (a_1 + d) + (a_1 + 2d) + (a_1 + 3d) + \cdots + (a_1 + (n-1)d)$$

Now write an equation for the sum of an arithmetic sequence from a_n to a_1. The common difference is subtracted.

$$S_n = a_n + (a_n - d) + (a_n - 2d) + (a_n - 3d) + \cdots + (a_n - (n-1)d)$$

Finally, add these equations together.

$$S_n = a_1 + (a_1 + d) + (a_1 + 2d) + \cdots + (a_1 + (n-1)d)$$
$$+ \; S_n = a_n + (a_n - d) + (a_n - 2d) + \cdots + (a_n - (n-1)d)$$

$$2S_n = (a_1 + a_n) + (a_1 + d + a_n - d) + (a_1 + 2d + a_n - 2d) + \cdots + (a_1 + (n-1)d + a_n - (n-1)d)$$

$2S_n = (a_1 + a_n) + (a_1 + a_n) + (a_1 + a_n) + \cdots + (a_1 + a_n)$ — Simplify; the sequence has n terms.

$2S_n = n(a_1 + a_n)$ — Rewrite addition of n terms as multiplication.

$\dfrac{2S_n}{2} = \dfrac{n(a_1 + a_n)}{2}$ — Division property of equality.

$S_n = \dfrac{n}{2}(a_1 + a_n)$ — Simplify; rewrite the fraction.

$\sum_{i=1}^{n} a_i = \dfrac{n}{2}(a_1 + a_n)$ — $S_n = \sum_{i=1}^{n} a_i$

Formula for the Partial Sum of n Terms of an Arithmetic Sequence

$$\sum_{i=1}^{n} a_i = \frac{n}{2}(a_1 + a_n)$$

EXAMPLE 4 An arithmetic sequence is $\{1, 5, 9, 13, 17, \ldots\}$.

(a) Find the formula for the nth term of this sequence.

SOLUTION ▶

$a_n = a_1 + (n-1)d$ — Formula for the nth term of an arithmetic sequence.

$a_n = \mathbf{1} + (n-1)(\mathbf{4})$ — Replace a_1 with 1 and d with 4.

$a_n = 1 + 4n - 4$ — Distributive property.

$a_n = 4n - 3$ — Simplify; this is the formula for the nth term.

(b) Evaluate the sum of the first 30 terms of the sequence.

▶

$\sum_{i=1}^{n} a_i = \dfrac{n}{2}(a_1 + a_n)$ — Formula for the partial sum of n terms.

$\sum_{i=1}^{n} a_i = \dfrac{n}{2}(a_1 + 4n - 3)$ — Replace a_n with $4n - 3$.

$\sum_{i=1}^{30} a_i = \dfrac{30}{2}(1 + (4(30)) - 3))$ — Replace n with 30 and a_1 with 1.

$$\sum_{i=1}^{30} a_i = \mathbf{15(118)} \qquad \text{Simplify.}$$

$$\sum_{i=1}^{30} a_i = \mathbf{1770} \qquad \text{Simplify.}$$

The sum of the first 30 terms is 1770.

Evaluating a Partial Sum of *n* Terms of an Arithmetic Sequence, Given the Sequence

1. Find the formula for the *n*th term of the sequence.

2. Use $\sum_{i=1}^{n} a_i = \dfrac{n}{2}(a_1 + a_n)$; replace a_n with the formula for the *n*th term of the sequence.

3. Replace *n* and a_1.

4. Simplify.

EXAMPLE 5   Evaluate the sum of the first 12 integers greater than 0.

(a) Identify the common difference and the first term.

SOLUTION ▶   The sequence is $\{1, 2, 3, 4, \dots\}$; the common difference, *d*, is 1; the first term, a_1, is 1.

(b) Find the formula for the *n*th term of this sequence.

$a_n = a_1 + (n - 1)d$	Formula for the *n*th term of an arithmetic sequence.
$a_n = \mathbf{1} + (n - 1)\mathbf{1}$	Replace a_1 and *d*.
$a_n = 1 + \mathbf{1}\mathbf{\mathit{n}} - 1$	Distributive property.
$a_n = \mathbf{\mathit{n}}$	Formula for the *n*th term of this sequence.

(c) Find the partial sum.

$\sum_{i=1}^{n} a_i = \dfrac{n}{2}(a_1 + a_n)$	Formula for the partial sum of *n* terms.
$\sum_{i=1}^{n} a_i = \dfrac{n}{2}(a_1 + \mathbf{\mathit{n}})$	Replace a_n with the formula for the *n*th term.
$\sum_{i=1}^{12} a_i = \dfrac{\mathbf{12}}{2}(\mathbf{1} + \mathbf{12})$	Replace *n* with 12 and a_1 with 1.
$\sum_{i=1}^{12} a_i = \mathbf{6(13)}$	Simplify.
$\sum_{i=1}^{12} a_i = \mathbf{78}$	Simplify.

The sum of the first 12 integers is 78.

Practice Problems

5. Evaluate the partial sum of the first 50 terms of the sequence $\{5, 8, 11, 14, 17, \dots\}$.

6. Evaluate the sum of the first 20 even integers greater than 0.

Applications

Consecutive terms in an arithmetic sequence have a constant difference. In the next example, we look at a physical representation of an arithmetic sequence.

EXAMPLE 6 | The number of seats in the first seven rows of a theater is an arithmetic sequence.

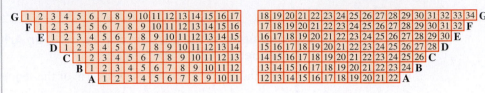

(a) Find the formula for the nth term of this sequence.

SOLUTION ▶ Each term in the sequence is the number of seats in the row. The number of seats in Row A is a_1, 22. There are two more seats in Row B than in Row A, a pattern that continues for each row. So d is 2. The sequence is $\{22, 24, 26, 28, \ldots\}$.

$$a_n = a_1 + (n - 1)d \qquad \text{Formula for the } n\text{th term of an arithmetic sequence.}$$
$$a_n = \mathbf{22} + (n - 1)(\mathbf{2}) \qquad \text{Replace } a_1 \text{ and } d.$$
$$a_n = 22 + \mathbf{2}n - \mathbf{2} \qquad \text{Distributive property.}$$
$$a_n = 2n + \mathbf{20} \qquad \text{Simplify.}$$

The formula for the nth term of this sequence is $a_n = 2n + 20$.

(b) Find the total number of seats in the first seven rows. This is the partial sum of seven terms of the sequence.

$$\sum_{i=1}^{n} a_i = \frac{n}{2}(a_1 + a_n) \qquad \text{Partial sum formula.}$$

$$\sum_{i=1}^{n} a_i = \frac{n}{2}(a_1 + \mathbf{2}n + \mathbf{20}) \qquad a_n = 2n + 20$$

$$\sum_{i=1}^{7} a_i = \frac{7}{2}(\mathbf{22} + 2(\mathbf{7}) + 20) \qquad n = 7; a_1 = 22$$

$$\sum_{i=1}^{7} a_i = \frac{7}{2}(\mathbf{56}) \qquad \text{Simplify.}$$

$$\sum_{i=1}^{7} a_i = \mathbf{196} \qquad \text{Simplify.}$$

There are 196 seats in these seven rows.

Practice Problems

7. Layers of Krispy Kreme Original Glazed® doughnuts can be stacked in an arithmetic sequence to create an unusual and high-calorie wedding cake. The first circular row contains 20 doughnuts. Each row above it contains one less doughnut. There are 17 rows.
a. Find the formula for the nth term of this sequence.
b. Find the total number of doughnuts in the cake.

Using Technology: Listing the Terms of a Sequence

To list the terms of a sequence, press MODE ; choose SEQ. To access the LIST screen, press 2nd STAT . Choose OPS; choose 5: seq(. The calculator pastes seq(on the home screen.

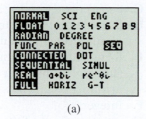

(a)

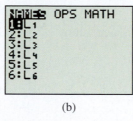

(b)

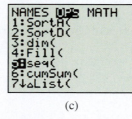

(c)

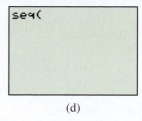

(d)

After seq(, enter information about the sequence in this order: formula of the nth term, variable in the formula, index of the beginning term, index of the end term, interval between the indices of the terms to be displayed, parenthesis.

EXAMPLE 7 List the first six terms of the sequence $a_n = 2n + 1$.

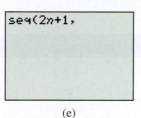

(e)

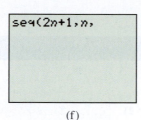

(f)

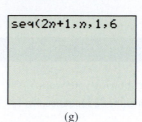

(g)

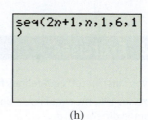

(h)

Type the formula, using the X,T,θ,n key for the variable.

Type the variable in the formula, n.

The beginning term is the first term, 1. The last term is 6.

Since we want to see each term and the interval between the indices of the terms is 1, type 1. Close parentheses.

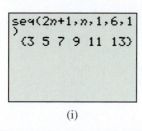

(i)

The six terms of the sequence appear. If there are too many terms to fit on one line, there will be an ellipsis after the last term. Use the cursor to move right and see the other terms.

Practice Problems For problems 8–10, find the first seven terms of each sequence. Use roster notation to list the terms.

8. $a_n = n + 3$ **9.** $a_n = 4n$ **10.** $a_n = 2^n$

10.2 VOCABULARY PRACTICE

Match the term with its description.

1. Σ

2. The set of the terms of a sequence
3. $\{\ldots, -3, -2, -1, 0, 1, 2, 3, \ldots\}$
4. A sequence in which the difference of consecutive terms is a constant
5. d
6. a_1
7. a_n
8. $\displaystyle\sum_{i=1}^{n} a_i = \frac{n}{2}(a_1 + a_n)$
9. $a_n = a_1 + (n - 1)d$
10. a_7, a_8

A. arithmetic sequence
B. common difference
C. consecutive terms
D. formula for the nth term of an arithmetic sequence
E. formula for the partial sum of n terms of an arithmetic sequence
F. integers
G. notation for the first term of a sequence
H. notation for the nth term of a sequence
I. range of a sequence
J. sigma

10.2 Exercises

Follow your instructor's guidelines for showing your work.

For exercises 1–8, the sequence is an arithmetic sequence. Find the common difference, d.

1. $\{33, 37, 41, 45, \ldots\}$
2. $\{63, 68, 73, 78, \ldots\}$
3. $\{-8, -11, -14, -17, \ldots\}$
4. $\{-9, -19, -29, -39, \ldots\}$
5. $\{-8, -5, -2, 1, \ldots\}$
6. $\{-9, 1, 11, 21, \ldots\}$
7. $\left\{\dfrac{1}{3}, \dfrac{7}{12}, \dfrac{5}{6}, \dfrac{13}{12}, \ldots\right\}$
8. $\left\{\dfrac{1}{4}, \dfrac{7}{12}, \dfrac{11}{12}, \dfrac{5}{4}, \ldots\right\}$

For exercises 9–26, the sequence is an arithmetic sequence.
(a) Find the common difference, d.
(b) Find the formula for the nth term of this sequence.

9. $\{3, 14, 25, 36, \ldots\}$
10. $\{6, 15, 24, 33, \ldots\}$
11. $\{8, -1, -10, -19, \ldots\}$
12. $\{9, -3, -15, -27, \ldots\}$
13. $\{100, 125, 150, 175, \ldots\}$
14. $\{100, 135, 170, 205, \ldots\}$
15. $\{200, 225, 250, 275, \ldots\}$
16. $\{200, 235, 270, 305, \ldots\}$
17. $\{57, 58, 59, 60, \ldots\}$
18. $\{71, 72, 73, 74, \ldots\}$
19. $\{12 \text{ seats}, 16 \text{ seats}, 20 \text{ seats}, 24 \text{ seats}, \ldots\}$
20. $\{7 \text{ days}, 14 \text{ days}, 21 \text{ days}, 28 \text{ days}, \ldots\}$

21. $\left\{\dfrac{1}{2}, 1, \dfrac{3}{2}, 2, \ldots\right\}$
22. $\left\{\dfrac{1}{3}, 1, \dfrac{5}{3}, \dfrac{7}{3}, \ldots\right\}$
23. $\left\{\dfrac{1}{7}, \dfrac{11}{28}, \dfrac{9}{14}, \dfrac{25}{28}, \ldots\right\}$
24. $\left\{\dfrac{1}{5}, \dfrac{8}{15}, \dfrac{13}{15}, \dfrac{6}{5}, \ldots\right\}$
25. $\{\$9000, \$8965, \$8930, \$8895, \ldots\}$
26. $\{\$550, \$533, \$516, \$499, \ldots\}$

For exercise 27–34,
(a) is the sequence an arithmetic sequence?
(b) explain.

27. $\{8, 20, 32, 40, \ldots\}$
28. $\{9, 15, 24, 29, \ldots\}$
29. $\{1, 2, -3, -4, 5, 6, \ldots\}$
30. $\{1, -2, 3, -4, 5, \ldots\}$
31. $\{8, 1, -6, -13, \ldots\}$
32. $\{6, -2, -10, -18, \ldots\}$
33. $\{2, 4, 8, 16, 32, \ldots\}$
34. $\{3, 9, 27, 81, 243, \ldots\}$

35. The tenth term in an arithmetic sequence is 47. The common difference is 3. Find the 11th term in this sequence.

36. The 12th term in an arithmetic sequence is 72. The common difference is 5. Find the 13th term in this sequence.

37. Without using the nth term formula, find the 14th term in the sequence in exercise 35.

38. Without using the nth term formula, find the 16th term in the sequence in exercise 36.

For exercises 39–50, each sequence is arithmetic. Find the formula for the *n*th term.

39. The 18th term is 100. The common difference is 3.

40. The 15th term is 75. The common difference is 4.

41. The 20th term is 3000. The common difference is 10.

42. The 30th term is 500. The common difference is 8.

43. The tenth term is 15. The 11th term is 17.

44. The eighth term is 21. The ninth term is 28.

45. The tenth term is 20. The 12th term is 28.

46. The 14th term is 50. The 16th term is 60.

47. The fourth term is $\dfrac{21}{2}$. The common difference is $\dfrac{1}{2}$.

48. The ninth term is $\dfrac{33}{4}$. The common difference is $\dfrac{1}{4}$.

49. The sixth term is -18. The seventh term is -20.

50. The fifth term is -40. The sixth term is -48.

For exercises 51–66, evaluate the partial sum of the given number of terms of the arithmetic sequence.

51. $a_n = 2n$; first 12 terms

52. $a_n = 3n$; first 14 terms

53. $a_n = 2n - 9$; first 12 terms

54. $a_n = 3n - 8$; first 14 terms

55. $a_n = 2n + 9$; first 12 terms

56. $a_n = 3n + 8$; first 14 terms

57. $a_n = \dfrac{1}{2}n + 20$; first 100 terms

58. $a_n = \dfrac{1}{2}n + 40$; first 100 terms

59. $\{10, 14, 18, 22, \dots\}$; first 50 terms

60. $\{20, 25, 30, 35, \dots\}$; first 50 terms

61. $\{14, 18, 22, 26, \dots\}$; first 49 terms

62. $\{25, 30, 35, 40, \dots\}$; first 49 terms

63. $\displaystyle\sum_{i=1}^{40}(6i + 8)$

64. $\displaystyle\sum_{i=1}^{80}(2i + 5)$

65. $\displaystyle\sum_{i=1}^{100} 5i$

66. $\displaystyle\sum_{i=1}^{30} 4i$

67. Evaluate the sum of the first 25 integers greater than 0.

68. Evaluate the sum of the first 15 integers greater than 0.

69. Evaluate the sum of the first 25 even integers greater than 0.

70. Evaluate the sum of the first 25 odd integers greater than 0.

71. One way to describe the formula for the partial sum of *n* terms of an arithmetic sequence is "the product of the number of terms and the average of the smallest and largest terms." Explain why this means the same thing as
$$\sum_{i=1}^{n} a_i = \frac{n}{2}(a_1 + a_n).$$

72. The article describes the method used by the famous mathematician Carl Gauss to evaluate the sum of the first 100 integers. Instead of using Gauss's method, use the formula for the partial sum of *n* terms of an arithmetic sequence to find this sum. Does your answer match Gauss's answer?

The story goes that when Carl Friedrich Gauss (1777–1855) was 10 years old, his teacher gave the pupils in an arithmetic class the problem of summing the integers from 1 to 100. Gauss came up with the answer almost immediately: 5050. He had found it by noting that the sum consists of 50 pairs of numbers, where each pair sums to 101. (*Source:* Ivars Peterson, www.sciencenews.org, 2004)

73. The first row in an auditorium has 20 seats. Each successive row has two more seats. Use a partial sum to find the total seats in the first 15 rows.

74. The top row in the pyramid quilt design has one cube. Each successive row has one more cube. Use a partial sum to find the total cubes in all seven rows.

© igl247/Shutterstock

75. Circular hay bales are stacked in an arithmetic sequence from top to bottom. The top row has three bales. Each row beneath has two more bales. Use the partial sum formula and multiplication to find how many bales can be stored in a stack that is four rows high and six rows from front to back.

76. The first row in an auditorium has 15 seats. Each successive row has three more seats. Use a partial sum to find the total seats in the first 20 rows.

Problem Solving: Practice and Review

Follow your instructor's guidelines for using the five steps as outlined in Section 1.5, p. 51.

77. A person in Lewiston, Idaho, wants to purchase a sofa from IKEA. The price of the sofa is $499. There are no IKEA stores in Idaho. The closest store is in Renton, Washington. The distance between Lewiston and Renton is 298 miles. If the sofa is ordered on-line, no sales tax will be charged and the cost of delivery is $289. To purchase the sofa in Renton requires a drive to Renton in a truck that gets 24 miles per gallon and payment of sales tax of 9.5%. Assume that the cost of gas is $3.50 a gallon. Find the difference in total cost between buying the sofa on-line and the total cost of driving to Renton to buy it. Round to the nearest whole number.

78. In June 2012, the fare for a driver and vehicle to travel by ferry from Seattle to Bremerton will be about $16.05. Find the fare on September 30, 2011, before the increases.

General Fare Increase: A 2.5 percent general fare increase is proposed to take effect Oct. 1, 2011 and an additional 3 percent general fare increase is proposed to take effect May 1, 2012. (*Source:* www.wsdot.wa.gov)

79. Find the number of people in this age group. Round to the nearest thousand.

About 215,000 people younger than 20 years have diabetes—type 1 or type 2. This represents 0.26 percent of all people in this age group. (*Source:* diabetes.niddk.nih.gov, Feb. 2011)

80. On Christmas Day 2007, a 243-pound Siberian tiger escaped from its enclosure at the San Francisco Zoo, killed one man, and mauled two others. Find the percent decrease in attendance as a result of the bad publicity from the attack. Round to the nearest percent.

The bad publicity resulted in $430,000 in canceled memberships and a loss of $1.2 million in corporate and individual donations. Attendance dropped by 20,831 to 904,000 in the 2008–2009 fiscal year. (*Source:* www.sfgate.com, June 28, 2009)

Technology

For exercises 81–84, use a graphing calculator to find the first seven terms of each sequence. Use roster notation to list the terms.

81. $a_n = 3n + 8$

82. $a_n = -12n + 15$

83. $a_n = 8n + 1$

84. $a_n = 350n$

Find the Mistake

For exercises 85–88, the completed problem has one mistake.
(a) Describe the mistake in words, or copy down the whole problem and highlight or circle the mistake.
(b) Do the problem correctly.

85. Problem: Determine whether $\{5, 8, 11, 14, 19, \ldots\}$ is an arithmetic sequence.

Incorrect Answer: Yes, this is an arithmetic sequence. Since $8 - 5 = 3$ and $11 - 8 = 3$, the common difference of consecutive terms is 3.

86. Problem: The 15th term in an arithmetic sequence is -41. The common difference is -3. Find the 16th term in this sequence.

Incorrect Answer: The 16th term is $(-41) - (-3)$, which equals -38.

87. Problem: Evaluate the partial sum of the first 20 terms of the sequence $\{1, 4, 7, 10, 13, \ldots\}$.

Incorrect Answer: $a_n = n + 3$

$$\sum_{i=1}^{20} a_i = \frac{20}{2}(1 + (20 + 3))$$

$$\sum_{i=1}^{20} a_i = 10(24)$$

$$\sum_{i=1}^{20} a_i = 240$$

88. Problem: Evaluate the sum of the odd integers less than 20.

Incorrect Answer: $a_n = 2n - 1$

$$\sum_{i=1}^{20} a_n = \frac{20}{2}(1 + 2(20) - 1)$$

$$\sum_{i=1}^{20} a_n = 10(40)$$

$$\sum_{i=1}^{20} a_n = 400$$

Review

89. Find any real zeros of $f(x) = 7x + 21$.

90. A quadratic function is $y = 2(x - 3)^2 + 1$.
   a. Use interval notation to represent the domain.
   b. Use interval notation to represent the range.

91. a. Describe the purpose of the vertical line test.
   b. Describe the purpose of the horizontal line test.

92. Use interval notation to represent the domain of
$$f(x) = \frac{x + 1}{x^2 + 5x + 4}.$$

SUCCESS IN COLLEGE MATHEMATICS

93. Did you plan and follow a schedule for studying for this and/or your other classes?

94. Do you plan to make and follow a schedule next term? Explain.

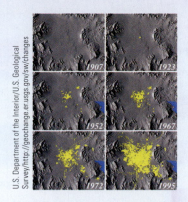

GIS (Geographic Information System) images show that the population of Las Vegas and surrounding Clark County grew exponentially from 1907 to 1995. We can use an exponential function to describe this growth. Or we can describe it using a geometric sequence. In this section, we will study this and other geometric sequences.

| SECTION 10.3 |

Geometric Sequences

After reading the text, working the practice problems, and completing assigned exercises, you should be able to:

1. Determine whether a sequence is geometric.
2. Find the formula for the nth term of a geometric sequence.
3. Find the partial sum of a geometric sequence.
4. Evaluate the sum of a geometric series.
5. Use a geometric sequence to solve an application problem.

Geometric Sequences

A geometric sequence is a function. The domain of a geometric sequence is a set of integers greater than 0. The range of a geometric sequence is the set of its terms.

In an arithmetic sequence, the *difference* of consecutive terms is a constant. In a **geometric sequence**, the *ratio* of the consecutive terms is a constant. This ratio is the **common ratio**, r.

EXAMPLE 1 Find the common ratio, r, of the geometric sequence $\{4, 8, 16, 32, \ldots\}$.

SOLUTION ▶

$$\frac{a_4}{a_3} \qquad \frac{a_3}{a_2} \qquad \frac{a_2}{a_1}$$

$$= \frac{32}{16} \qquad = \frac{16}{8} \qquad = \frac{8}{4}$$

$$= 2 \qquad = 2 \qquad = 2$$

The common ratio, r, is 2.

The terms a_n and a_{n+1} are **consecutive terms**. For a geometric sequence, their ratio is a constant: $\dfrac{a_{n+1}}{a_n} = r$. In the next example, we use this relationship to show that a sequence is geometric.

EXAMPLE 2 | Show that the sequence $a_n = 2(2)^{n-1}$ is geometric.

(a) Find a_{n+1}.

SOLUTION ▶

$$a_n = 2(2)^{n-1}$$ The formula for the nth term of the sequence.

$$a_{n+1} = 2(2)^{(n+1)-1}$$ Rewrite the formula for the term $n+1$.

$$a_{n+1} = 2(2^n)$$ Simplify.

(b) Find the ratio of consecutive terms, $\dfrac{a_{n+1}}{a_n}$.

▶

$$\dfrac{a_{n+1}}{a_n}$$ Find the ratio of consecutive terms.

$$= \dfrac{\cancel{2}(2^n)}{\cancel{2}(2)^{n-1}}$$ Replace a_{n+1} and a_n.

$$= 2^{(n)-(n-1)}$$ Simplify; quotient rule of exponents.

$$= 2^1$$ Simplify.

$$= 2$$ Simplify.

Since 2 is a constant, this is a geometric sequence.

Showing That a Sequence Is Geometric, Given the Formula for the nth Term

1. Use the formula for a_n to find a formula for a_{n+1}.

2. Find the ratio of consecutive terms $\dfrac{a_{n+1}}{a_n}$.

3. If the ratio of consecutive terms is a constant, this is a geometric sequence.

If a sequence is not geometric, the ratio of consecutive terms includes a variable, n.

EXAMPLE 3 | Show that the sequence $a_n = n^2 + 3$ is not geometric.

SOLUTION ▶

$$\dfrac{a_{n+1}}{a_n}$$ Find the ratio of consecutive terms.

$$= \dfrac{(n+1)^2 + 3}{n^2 + 3}$$ Replace a_{n+1} and a_n.

$$= \dfrac{n^2 + 2n + 1 + 3}{n^2 + 3}$$ $(n+1)^2 = n^2 + 2n + 1$

$$= \dfrac{n^2 + 2n + 4}{n^2 + 3}$$ Combine like terms.

Since this simplified expression includes at least one variable, it is not a constant. The sequence is not geometric.

As with arithmetic sequences, we can develop a formula for the nth term of a geometric sequence. Since the ratio of consecutive terms is a constant, we can build a pattern that leads to the formula.

$$r = \dfrac{a_2}{a_1}$$ The ratio of consecutive terms; solve for a_2.

$$a_1 r = a_1 \cdot \dfrac{a_2}{a_1}$$ Multiplication property of equality.

$$a_1 r = a_2$$ Simplify; a_2 is isolated.

Repeat this process with the next consecutive terms, a_2 and a_3.

$$r = \frac{a_3}{a_2} \qquad \text{The ratio of consecutive terms; solve for } a_3.$$

$$a_2 r = a_2 \cdot \frac{a_3}{a_2} \qquad \text{Multiplication property of equality.}$$

$$a_2 r = \boldsymbol{a_3} \qquad a_3 \text{ is isolated.}$$

$$\boldsymbol{a_1 r \cdot r} = a_3 \qquad \text{Replace } a_2 \text{ with } a_1 r.$$

$$\boldsymbol{a_1 r^2} = a_3 \qquad \text{Simplify.}$$

Repeat this process with the next consecutive terms, a_3 and a_4.

$$r = \frac{a_4}{a_3} \qquad \text{The ratio of consecutive terms; solve for } a_4.$$

$$a_3 r = a_3 \cdot \frac{a_4}{a_3} \qquad \text{Multiplication property of equality.}$$

$$a_3 r = \boldsymbol{a_4} \qquad \text{Isolate } a_4.$$

$$\boldsymbol{a_1 r^2 \cdot r} = a_4 \qquad \text{Replace } a_3 \text{ with } a_1 r^2.$$

$$\boldsymbol{a_1 r^3} = a_4 \qquad \text{Simplify.}$$

A pattern appears that continues to the consecutive terms a_{n-1} and a_n.

$$r = \frac{a_n}{a_{n-1}} \qquad \text{The ratio of consecutive terms; solve for } a_n.$$

$$a_{n-1} r = a_{n-1} \cdot \frac{a_n}{a_{n-1}} \qquad \text{Multiplication property of equality.}$$

$$a_{n-1} r = \boldsymbol{a_n} \qquad \text{Isolate } a_n.$$

$$\boldsymbol{a_1 r^{(n-1)-1} \cdot r} = a_n \qquad \text{Replace } a_{n-1} \text{ with } a_1 r^{(n-1)-1}.$$

$$a_1 r^{n-2} \cdot r^1 = a_n \qquad \text{Simplify.}$$

$$\boldsymbol{a_1 r^{n-1}} = a_n \qquad \text{Product rule of exponents; } a_n \text{ is isolated.}$$

This is the formula for the nth term of a geometric sequence, a_n.

Formula for the nth Term of a Geometric Sequence $a_n = a_1 r^{n-1}$

r is the common ratio between consecutive terms of the sequence, $r \neq 0$.
a_1 is the first term.

EXAMPLE 4 A geometric sequence is $\{15, 45, 135, 405, \ldots\}$.

(a) Find r.

SOLUTION ▶
$$r \qquad \text{Find the common ratio, } r.$$

$$= \frac{a_4}{a_3} \qquad a_3 \text{ and } a_4 \text{ are consecutive terms.}$$

$$= \frac{405}{135} \qquad \text{Replace terms.}$$

$$= 3 \qquad \text{Simplify.}$$

(b) Find the formula for the nth term.

▶
$$a_n = a_1 r^{n-1} \qquad \text{Formula for the } n\text{th term of a geometric sequence.}$$

$$a_n = \boldsymbol{15(3)^{n-1}} \qquad \text{Replace } a_1 \text{ and } r.$$

The formula for the nth term of this sequence is $a_n = 15(3)^{n-1}$.

(c) Use this formula to find the eighth term of the sequence.

$$a_8 = 15(3)^{8-1} \qquad \text{Replace } n \text{ with 8.}$$
$$a_8 = \mathbf{32{,}805} \qquad \text{Simplify.}$$

The eighth term in the sequence is 32,805.

> **Finding the Formula for the *n*th Term of a Geometric Sequence, Given the Sequence**
>
> 1. Find *r* by writing the ratio of two consecutive terms.
> 2. Replace a_1 and *r* in the formula for the *n*th term of a geometric sequence, $a_n = a_1 r^{n-1}$.

When the terms of a geometric sequence have alternating signs, *r* is less than 0.

EXAMPLE 5 | A geometric sequence is $\{2, -4, 8, -16, \ldots\}$.

(a) Find *r*.

SOLUTION ▶

$$r \qquad \qquad \text{Find the common ratio, } r.$$
$$= \frac{a_3}{a_2} \qquad a_2 \text{ and } a_3 \text{ are consecutive terms.}$$
$$= \frac{8}{-4} \qquad \text{Replace terms.}$$
$$= -2 \qquad \text{Simplify.}$$

(b) Find the formula for the *n*th term of this geometric sequence.

$$a_n = a_1 r^{n-1} \qquad \text{Formula for the } n\text{th term of a geometric sequence.}$$
$$a_n = \mathbf{2(-2)^{n-1}} \qquad \text{Replace } a_1 \text{ and } r.$$

The formula for the *n*th term of this sequence is $a_n = 2(-2)^{n-1}$.

(c) Use this formula to find the 20th term.

$$a_n = 2(-2)^{n-1} \qquad \text{Formula for the } n\text{th term of the sequence.}$$
$$a_{20} = 2(-2)^{20-1} \qquad \text{Replace } n \text{ with 20.}$$
$$a_{20} = \mathbf{-1{,}048{,}576} \qquad \text{Simplify.}$$

The 20th term in the sequence is $-1{,}048{,}576$.

In the next example, the common ratio is between -1 and 1: $-1 < r < 1$.

EXAMPLE 6 | A geometric sequence is $\left\{ \dfrac{1}{2}, \dfrac{1}{4}, \dfrac{1}{16}, \dfrac{1}{32}, \ldots \right\}$

(a) Find *r*.

SOLUTION ▶

$$r \qquad \qquad \text{Find the common ratio, } r.$$
$$= \frac{a_2}{a_1} \qquad a_1 \text{ and } a_2 \text{ are consecutive terms.}$$
$$= \frac{\frac{1}{4}}{\frac{1}{2}} \qquad \text{Replace terms.}$$

$$= \frac{1}{4} \div \frac{1}{2} \qquad \text{Rewrite the fraction as division.}$$

$$= \frac{1}{4} \cdot \frac{2}{1} \qquad \text{Multiply by the reciprocal.}$$

$$= \frac{1}{2} \qquad \text{Simplify.}$$

(b) Find the formula for the nth term of the geometric sequence.

▶ $a_n = a_1 r^{n-1} \qquad$ Formula for the nth term of a geometric sequence.

$$a_n = \frac{1}{2}\left(\frac{1}{2}\right)^{n-1} \qquad \text{Replace } a_1 \text{ and } r.$$

The formula for the nth term of this sequence is $a_n = \frac{1}{2}\left(\frac{1}{2}\right)^{n-1}$.

(c) Use this formula to find the eighth term.

▶ $a_n = \frac{1}{2}\left(\frac{1}{2}\right)^{n-1} \qquad$ Formula for the nth term.

$$a_8 = \frac{1}{2}\left(\frac{1}{2}\right)^{8-1} \qquad \text{Replace } n \text{ with 8.}$$

$$a_8 = \frac{1}{256} \qquad \text{Simplify.}$$

The eighth term in the sequence is $\frac{1}{256}$.

Practice Problems

For problems 1–4,
(a) find the common ratio, r, of the geometric sequence.
(b) find the formula for the nth term of this sequence.
(c) use the formula to find the tenth term.

1. $\{8, 32, 128, 512, \ldots\}$ 2. $\{-9, -27, -81, -243, \ldots\}$

3. $\{9, -27, 81, -243, \ldots\}$ 4. $\left\{\frac{1}{3}, \frac{1}{9}, \frac{1}{27}, \frac{1}{81}, \ldots\right\}$

5. The fifth term in a geometric sequence is 16. The common ratio is 2.
   a. Find the sixth term in the sequence.
   b. Find the formula for the nth term.
6. Show that the sequence $a_n = n^2 - 4$ is not geometric.

Partial Sum of a Geometric Sequence

The sum of a geometric sequence S_n from a_1 to a_n is $S_n = a_1 + (a_1 r) + (a_1 r^2) + (a_1 r^3) + \cdots + (a_1 r^{n-1})$. To develop a formula for the partial sum of n terms of a geometric sequence, we use the multiplication property of equality and multiply both sides of this equation by r.

$S_n = a_1 + (a_1 r) + (a_1 r^2) + (a_1 r^3) + \cdots + (a_1 r^{n-1})$ The sum of the sequence; $r \neq 0, r \neq 1$; Equation 1.

$rS_n = r(a_1 + (a_1 r) + (a_1 r^2) + (a_1 r^3) + \cdots + (a_1 r^{n-1}))$ Multiplication property of equality.

$rS_n = a_1 r + (a_1 r^2) + (a_1 r^3) + (a_1 r^4) + \cdots + (a_1 r^n)$ Distributive property

$rS_n = 0 + a_1 r + (a_1 r^2) + (a_1 r^3) + (a_1 r^4) + \cdots + (a_1 r^n)$ Write 0 as the first term of the sum; Equation 2.

When solving a system of equations by elimination, we use the addition property of equality and add the equations together. To develop a formula for the partial sum of n terms of a geometric sequence, we use the subtraction property of equality and subtract two equations. We then solve for S_n.

$$S_n = a_1 + (a_1 r) + (a_1 r^2) + (a_1 r^3) + \cdots + (a_1 r^{n-1})$$ Equation 1.

$$- rS_n = 0 + a_1 r + (a_1 r^2) + (a_1 r^3) + (a_1 r^4) + \cdots + (a_1 r^n)$$ Equation 2.

$$S_n - rS_n = a_1 + 0 + 0 + 0 + 0 + \cdots - a_1 r^n$$ Simplify.

$$S_n - rS_n = a_1 - a_1 r^n$$ Simplify.

$$S_n(1 - r) = a_1(1 - r^n)$$ Factor.

$$\frac{S_n(1 - r)}{(1 - r)} = \frac{a_1(1 - r^n)}{(1 - r)}$$ Division property of equality.

$$S_n = a_1\left(\frac{1 - r^n}{1 - r}\right)$$ Simplify; S_n is isolated.

$$\sum_{i=1}^{n} a_i = a_1\left(\frac{1 - r^n}{1 - r}\right)$$ Rewrite in sigma notation.

Using this formula, $\sum_{i=1}^{n} a_i = a_1\left(\frac{1 - r^n}{1 - r}\right)$, we can find the partial sum of a geometric sequence.

Formula for the Partial Sum of n Terms of a Geometric Sequence

$$\sum_{i=1}^{n} a_i = a_1\left(\frac{1 - r^n}{1 - r}\right); \qquad r \neq 0, \quad r \neq 1$$

In the next example, to evaluate the sum of n terms of a sequence, we first find the common ratio, r.

EXAMPLE 7   A geometric sequence is $\{14, 28, 56, 112, \dots\}$

(a) Find the common ratio, r.

SOLUTION ▶ r Find the common ratio, r, of two consecutive terms.

$$= \frac{a_2}{a_1}$$ a_1 and a_2 are consecutive terms.

$$= \frac{28}{14}$$ Replace a_1 and a_2.

$$= 2$$ Simplify. This is the common ratio, r.

(b) Evaluate the sum of the first nine terms.

$$\sum_{i=1}^{n} a_i = a_1\left(\frac{1 - r^n}{1 - r}\right)$$ Formula for the partial sum of n terms.

$$\sum_{i=1}^{9} a_i = (\mathbf{14})\left(\frac{1 - \mathbf{2^9}}{1 - \mathbf{2}}\right)$$ Replace n with 9, a_1 with 14, and r with 2.

$$\sum_{i=1}^{9} a_i = (14)(\mathbf{511})$$ Simplify.

$$\sum_{i=1}^{9} a_i = \mathbf{7154}$$ Simplify.

The sum of the first nine terms is 7154.

> **Evaluating the Partial Sum of *n* Terms of a Geometric Sequence, Given the Sequence**
>
> 1. Find r.
> 2. Use the formula $\sum\limits_{i=1}^{n} a_i = a_1\left(\dfrac{1 - r^n}{1 - r}\right)$; replace a_1, r, and n.
> 3. Simplify.

> **Practice Problems**
>
> 7. Evaluate the partial sum of the first ten terms of the geometric sequence $\{12, 36, 108, 972, \ldots\}$.
> 8. Evaluate the partial sum of the first twelve terms of the geometric sequence $\{-5, -25, -125, -625, \ldots\}$.

Sum of an Infinite Geometric Sequence

If $-1 < r < 1$ and $r \neq 0$, we can evaluate the sum of all of the terms in an infinite geometric sequence. The sum of all of the terms in an infinite geometric sequence is a **geometric series**.

EXAMPLE 8 A geometric sequence is $\{1, 0.5, 0.25, 0.125, \ldots\}$.

(a) Find the common ratio, r.

SOLUTION ▶ r Find the common ratio, r, of two consecutive terms.

$$= \frac{a_3}{a_2} \qquad a_2 \text{ and } a_3 \text{ are consecutive terms.}$$

$$= \frac{0.25}{0.5} \qquad \text{Replace } a_2 \text{ and } a_3.$$

$$= 0.5 \qquad \text{Simplify; this is the common ratio, } r.$$

(b) Evaluate the sum of two terms, four terms, six terms, and eight terms.

▶ $\sum\limits_{i=1}^{n} a_i = a_1\left(\dfrac{1 - r^n}{1 - r}\right)$ Formula for the partial sum of a geometric sequence.

Sum of two terms:	Sum of four terms:
$\sum\limits_{i=1}^{2} a_i = (1)\left(\dfrac{1 - (0.5)^2}{1 - 0.5}\right)$	$\sum\limits_{i=1}^{4} a_i = (1)\left(\dfrac{1 - (0.5)^4}{1 - 0.5}\right)$
$\sum\limits_{i=1}^{2} a_i = \left(\dfrac{0.75}{0.5}\right)$	$\sum\limits_{i=1}^{4} a_i = \left(\dfrac{0.9375}{0.5}\right)$
$\sum\limits_{i=1}^{2} a_i = 1.5$	$\sum\limits_{i=1}^{4} a_i = 1.875$
Sum of six terms:	**Sum of eight terms:**
$\sum\limits_{i=1}^{6} a_i = (1)\left(\dfrac{1 - (0.5)^6}{1 - 0.5}\right)$	$\sum\limits_{i=1}^{8} a_i = (1)\left(\dfrac{1 - (0.5)^8}{1 - 0.5}\right)$
$\sum\limits_{i=1}^{6} a_i = \left(\dfrac{0.984375}{0.5}\right)$	$\sum\limits_{i=1}^{8} a_i = \left(\dfrac{0.99609375}{0.5}\right)$
$\sum\limits_{i=1}^{6} a_i = 1.96875$	$\sum\limits_{i=1}^{8} a_i = 1.9921875$

As the number of terms in the partial sum increases, the value of the numerator gets closer and closer to 1 and the value of the denominator is constant, 0.5. So as the number of terms increases, the sum gets closer and closer to the quotient of $1 \div 0.5$, which equals 2.

Imagine adding an infinite number of terms of this sequence. The sum is equal to $\dfrac{a_1}{1-r}$. In this example, the sum is $\dfrac{1}{1-0.5}$, which equals 2.

Formula for the Sum of the Terms of an Infinite Geometric Sequence (a Geometric Series)

$$\sum_{i=1}^{\infty} a_1(r)^{i-1} = \frac{a_1}{1-r}, \qquad -1 < r < 1, \quad r \neq 0$$

EXAMPLE 9 Evaluate: $\displaystyle\sum_{i=1}^{\infty} (1)(0.5)^{i-1}$

SOLUTION ▶

$$\sum_{i=1}^{\infty} a_1(r)^{i-1} = \frac{a_1}{1-r} \qquad \text{Formula for the sum; } -1 < r < 1.$$

$$\sum_{i=1}^{\infty} \mathbf{(1)(0.5)}^{i-1} = \frac{1}{1-\mathbf{0.5}} \qquad \text{Replace } a_1 \text{ with 1 and } r \text{ with 0.5.}$$

$$\sum_{i=1}^{\infty} (1)(0.5)^{i-1} = \frac{1}{\mathbf{0.5}} \qquad \text{Simplify.}$$

$$\sum_{i=1}^{\infty} (1)(0.5)^{i-1} = \mathbf{2} \qquad \text{Simplify.}$$

The sum is 2.

We can also evaluate the sum when the terms in the geometric sequence are fractions.

EXAMPLE 10 A geometric sequence is $\left\{ 2, \dfrac{3}{5}, \dfrac{9}{50}, \dfrac{27}{500}, \ldots \right\}$. Find the sum of the geometric series: $2 + \dfrac{3}{5} + \dfrac{9}{50} + \dfrac{27}{500} + \cdots$

(a) Find the common ratio, r.

SOLUTION ▶

$$r = \frac{a_2}{a_1} \qquad \text{Find the common ratio, } r, \text{ of two consecutive terms; } a_1 \text{ and } a_2 \text{ are consecutive terms.}$$

$$= \frac{\frac{3}{5}}{2} \qquad \text{Replace } a_1 \text{ and } a_2.$$

$$= \frac{3}{5} \div 2 \qquad \text{Rewrite the fraction as division.}$$

$$= \frac{3}{5} \cdot \frac{\mathbf{1}}{\mathbf{2}} \qquad \text{Division is multiplication by the reciprocal.}$$

$$= \frac{3}{10} \qquad \text{Simplify.}$$

(b) Find the formula for the *n*th term.

$$a_n = a_1 r^{n-1} \qquad \text{Formula for the nth term of a geometric sequence.}$$

$$a_n = (2)\left(\frac{3}{10}\right)^{n-1} \qquad \text{Replace } a_1 \text{ and } r.$$

(c) Find the sum.

$$\sum_{i=1}^{\infty} a_1 (r)^{i-1} = \frac{a_1}{1-r} \qquad \text{Sum of a geometric series; change from n to i.}$$

$$\sum_{i=1}^{\infty} (2)\left(\frac{3}{10}\right)^{i-1} = \frac{2}{1-\dfrac{3}{10}} \qquad \text{Replace } a_1 \text{ and } r.$$

$$\sum_{i=1}^{\infty} (2)\left(\frac{3}{10}\right)^{i-1} = \frac{2}{\dfrac{7}{10}} \qquad \text{Simplify the numerator.}$$

$$\sum_{i=1}^{\infty} (2)\left(\frac{3}{10}\right)^{i-1} = 2 \div \frac{7}{10} \qquad \text{Rewrite as division.}$$

$$\sum_{i=1}^{\infty} (2)\left(\frac{3}{10}\right)^{i-1} = 2 \cdot \frac{10}{7} \qquad \text{Multiply by the reciprocal.}$$

$$\sum_{i=1}^{\infty} (2)\left(\frac{3}{10}\right)^{i-1} = \frac{20}{7} \qquad \text{Simplify.}$$

The sum of the geometric series is $\dfrac{20}{7}$.

Practice Problems

For problems 9–11,
(a) identify r.
(b) identify a_1.
(c) evaluate the sum.

9. $\displaystyle\sum_{i=1}^{\infty} (3)(0.4)^{i-1}$ **10.** $\displaystyle\sum_{i=1}^{\infty} (4)(-0.1)^{i-1}$ **11.** $\displaystyle\sum_{i=1}^{\infty} (2)\left(\frac{1}{4}\right)^{i-1}$

12. A geometric sequence is $\left\{3, 1, \dfrac{1}{3}, \dfrac{1}{9}, \ldots\right\}$. Evaluate the sum of the geometric series: $3 + 1 + \dfrac{1}{3} + \dfrac{1}{9} + \cdots$

Applications

In Chapter 8 we used exponential functions to solve application problems involving the future value of an investment, population growth, and radioactive decay. A geometric sequence is also an exponential function. However, its domain is restricted to a set of positive integers. When we compare the graphs of an exponential function and a geometric sequence, we see that the exponential function is continuous. The geometric sequence is not continuous.

EXAMPLE 11 | **(a)** Graph: $y = 3^x$ **(b)** Graph: $a_n = 3(3)^{n-1}$

SOLUTION ▶

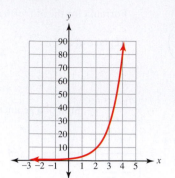

Graph of $y = 3^x$
Domain: $(-\infty, \infty)$
The graph is continuous.

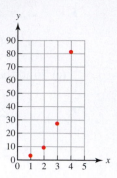

Graph of $a_n = 3(3)^{n-1}$
Domain: $\{1, 2, 3, \ldots\}$
The graph is not continuous.

We can use a geometric sequence to model a situation with a domain that is restricted to a set of positive integers. The next example is an application from Section 8.3. If the output values are rounded to a given place value, the common ratios of consecutive terms might not be exactly equal. So we also round the common ratio.

EXAMPLE 12 | The function $y = \$4500(1.08)^t$ describes the relationship of the number of years since 2010, t, and the annual tuition at a college. This college increases tuition once a year, in the fall.

(a) Find three ordered pairs for $y = \$4500(1.08)^t$, rounding the output values to the nearest whole number.

SOLUTION ▶ $\{(1, \$4860), (2, \$5249), (3, \$5669), \ldots\}$

(b) Rewrite this function as a geometric sequence with a domain of $\{1, 2, 3, \ldots\}$.

▶ $\{\$4860, \$5249, \$5669, \ldots\}$ The range of the sequence.

(c) Find the common ratio, r, rounding to the nearest hundredth.

▶ $r = \dfrac{a_2}{a_1}$ a_1 and a_2 are consecutive terms.

   $r = \dfrac{\$5249}{\$4860}$ Replace terms.

   $r \approx 1.08$ Round to the hundredths place.

(d) Find the formula for the nth term.

▶ $a_n = a_1 r^{n-1}$ nth term of a geometric sequence.
   $a_n = \$4860(1.08)^{n-1}$ $a_1 \approx \$4860$; $r \approx 1.08$

(e) Find the tuition in six years. Round to the nearest whole number.

▶ $a_n = \$4860(1.08)^{n-1}$ Formula for the nth term of a geometric sequence.
   $a_6 = \$4860(1.08)^{6-1}$ Replace n with 6.
   $a_6 = \$4860(1.08)^5$ Simplify.
   $a_6 \approx \$7141$ Simplify.

The tuition in six years will be about $7141.

Practice Problems

13. The function $D(t) = 15,000(1.19)^t$ describes the relationship of the number of years since 2007, t, and the number of products sold per year, $D(t)$.
   a. Rewrite this function as a geometric sequence with a domain of $\{1, 2, 3, \ldots\}$. Round to the nearest whole number.
   b. Find the common ratio, r, rounding to the hundredths place.
   c. Find the formula for the nth term of this sequence.
   d. Find the number of units sold in 2016. Round to the nearest hundred.

Using Technology: Finding the Partial Sum of a Sequence

To evaluate the partial sum of n terms of a sequence, the sequence is first named and stored in the memory of the calculator. Two special keys are used, STO▸ and ALPHA.

EXAMPLE 13 Name and store the first ten terms of the sequence $a_n = 2(3)^{n-1}$. Evaluate the partial sum of the first five terms.

As shown in Section 10.2, press MODE and highlight SEQ. Press 2nd STAT. Press the right arrow key to highlight OPS. Press 5. Type the sequence, the variable, the number of the beginning term, the number of the last term, and a closing parenthesis. Press ENTER. Press STO▸. Press ALPHA S 1. Press ENTER. The ALPHA key works just like the 2nd key. It activates the letters or menus printed in green. This sequence is now named S1.

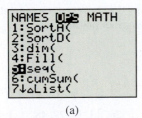
(a)

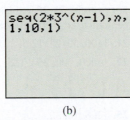
(b)

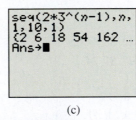
(c)

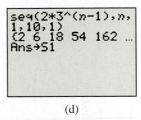
(d)

To find a partial sum of the terms in the sequence, press 2nd STAT, press the right arrow key to highlight OPS, press 6. On the screen, cumSum(appears. Press 2nd STAT. Press the down arrow to highlight S1. Type a closing parenthesis. Press ENTER.

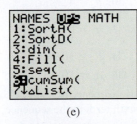
(e)

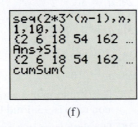
(f)

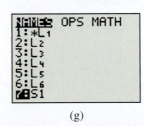
(g)

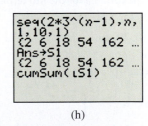
(h)

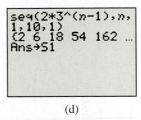
(i)

The first term listed below cumSum, 2, is the same as the first term in the sequence. The second term, 8, is the sum of the first and second terms: $2 + 6 = 8$. The third term, 26, is the sum of the first three terms: $2 + 6 + 18 = 26$. Press the right arrow key to see the sums to the right of the ellipsis. The sum of the first five terms of the sequence is 242.

Practice Problems For problems 14–16, name the sequence and store the first five terms. Find the sum of the first four terms.

14. $a_n = 3(4)^{n-1}$ 15. $a_n = n + 4$ 16. $a_n = \dfrac{n}{n+1}$

10.3 VOCABULARY PRACTICE

Match the term with its description.

1. $\sum_{i=1}^{\infty} a_1 r^{i-1} = \dfrac{a_1}{1-r}$, $-1 < r < 1, r \neq 0$

2. $\sum_{i=1}^{n} a_i = (a_1)\left(\dfrac{1-r^n}{1-r}\right)$, $r \neq 0, r \neq 1$

3. $a_n = a_1 r^{n-1}$

4. A sequence in which the difference of successive terms is a constant

5. A sequence in which the ratio of successive terms is a constant

6. $(-\infty, \infty)$

7. a_3, a_4

8. In a geometric sequence, $\dfrac{a_2}{a_1}$

9. In an arithmetic sequence, $a_2 - a_1$

10. $\sum$

A. arithmetic sequence
B. common difference
C. common ratio
D. consecutive terms
E. formula for nth term of a geometric sequence
F. formula for the partial sum of n terms of a geometric sequence
G. formula for the sum of a geometric series
H. geometric sequence
I. set of real numbers
J. sigma

10.3 Exercises

Follow your instructor's guidelines for showing your work.

For exercises 1–8, the sequence is a geometric sequence. Find the common ratio, r.

1. $\{2, 6, 18, 54, \ldots\}$

2. $\{4, 12, 36, 108, \ldots\}$

3. $\{6, 18, 54, 162, \ldots\}$

4. $\{9, 27, 81, 243, \ldots\}$

5. $\left\{-\dfrac{1}{2}, -1, -2, -4, \ldots\right\}$

6. $\left\{-\dfrac{1}{8}, -\dfrac{1}{2}, -2, -8, \ldots\right\}$

7. $\left\{\dfrac{1}{2}, \dfrac{1}{4}, \dfrac{1}{8}, \dfrac{1}{16}, \ldots\right\}$

8. $\left\{\dfrac{1}{3}, \dfrac{1}{9}, \dfrac{1}{27}, \dfrac{1}{81}, \ldots\right\}$

For exercises 9–20,
(a) find the common ratio, r, for the geometric sequence.
(b) find the formula for the nth term of the sequence.

9. $\{3, 6, 12, 24, \ldots\}$

10. $\{5, 10, 20, 40, \ldots\}$

11. $\{-3, -6, -12, -24, \ldots\}$

12. $\{-5, -10, -20, -40, \ldots\}$

13. $\left\{4, 2, 1, \dfrac{1}{2}, \ldots\right\}$

14. $\left\{6, 3, \dfrac{3}{2}, \dfrac{3}{4}, \ldots\right\}$

15. $\left\{4, -2, 1, -\dfrac{1}{2}, \ldots\right\}$

16. $\left\{6, -3, \dfrac{3}{2}, -\dfrac{3}{4}, \ldots\right\}$

17. $\left\{\dfrac{1}{3}, \dfrac{1}{6}, \dfrac{1}{12}, \dfrac{1}{24}, \ldots\right\}$

18. $\left\{\dfrac{1}{5}, \dfrac{1}{10}, \dfrac{1}{20}, \dfrac{1}{40}, \ldots\right\}$

19. $\{1, 8, 64, 512, \ldots\}$

20. $\{1, 7, 49, 343, \ldots\}$

For exercise 21–28, explain why the sequence is a geometric sequence, an arithmetic sequence, or neither.

21. $\{100, 50, 25, 12.5, \ldots\}$

22. $\{1000, 500, 250, 125, \ldots\}$

23. $\left\{3, \dfrac{5}{2}, \dfrac{7}{3}, \dfrac{9}{4}, \ldots\right\}$

24. $\left\{4, \dfrac{7}{2}, \dfrac{10}{3}, \dfrac{13}{4}, \ldots\right\}$

25. $\{2, 8, 18, 32, \ldots\}$

26. $\{3, 12, 27, 48, \ldots\}$

27. $\{-4, -1, 2, 5, \ldots\}$

28. $\{-3, -1, 1, 3, \ldots\}$

For exercises 29–36, show that the sequence is geometric. Use the method shown in Example 2.

29. $a_n = 3(2)^{n-1}$

30. $a_n = 4(3)^{n-1}$

31. $a_n = \left(\dfrac{1}{3}\right)(6)^{n-1}$

32. $a_n = \left(\dfrac{1}{2}\right)(8)^{n-1}$

33. $a_n = 2(-3)^{n-1}$

34. $a_n = 3(-4)^{n-1}$

35. $a_n = 7(0.4)^{n-1}$

36. $a_n = 5(0.6)^{n-1}$

37. The fifth term in a geometric sequence is 324. The common ratio is -3. Find the sixth term in the sequence.

38. The sixth term in a geometric sequence is -256. The common ratio is -4. Find the seventh term in the sequence.

39. Find the formula for the nth term in the sequence in exercise 37.

40. Find the formula for the nth term in the sequence in exercise 38.

41. The eighth term in a geometric sequence is 390,625. The common ratio is 5. Find the ninth term in the sequence.

42. The tenth term in a geometric sequence is 60,466,176. The common ratio is 6. Find the 11th term in the sequence.

43. Find the formula for the nth term in the sequence in exercise 41.

44. Find the formula for the nth term in the sequence in exercise 42.

For exercises 45–48, evaluate the partial sum of the given number of terms of the geometric sequence.

45. $\{5, 15, 45, 135, \ldots\}$; ten terms

46. $\{8, 16, 32, 64, \ldots\}$; eight terms

47. $\left\{12, 6, 3, \dfrac{3}{2}, \ldots\right\}$; seven terms

48. $\left\{14, 7, \dfrac{7}{2}, \dfrac{7}{4}, \ldots\right\}$; five terms

For exercises 49–54, evaluate the partial sum.

49. $\displaystyle\sum_{i=1}^{8} (-3)(2)^{i-1}$

50. $\displaystyle\sum_{i=1}^{7} (-2)(3)^{i-1}$

51. $\displaystyle\sum_{i=1}^{5} (4)(8)^{i-1}$

52. $\displaystyle\sum_{i=1}^{7} (6)(3)^{i-1}$

53. $\displaystyle\sum_{i=1}^{6} (4)(-3)^{i-1}$

54. $\displaystyle\sum_{i=1}^{11} (5)(-2)^{i-1}$

For exercises 55–60, evaluate the sum.

55. $\displaystyle\sum_{i=1}^{\infty} (-3)(0.2)^{i-1}$

56. $\displaystyle\sum_{i=1}^{\infty} (-2)(0.9)^{i-1}$

57. $\displaystyle\sum_{i=1}^{\infty} (4)(0.8)^{i-1}$

58. $\displaystyle\sum_{i=1}^{\infty} (6)(0.5)^{i-1}$

59. $\displaystyle\sum_{i=1}^{\infty} (4)(-0.5)^{i-1}$

60. $\displaystyle\sum_{i=1}^{\infty} (5)(-0.2)^{i-1}$

61. A geometric sequence is $\{16, 8, 4, 2, \ldots\}$. Evaluate the sum of the geometric series: $16 + 8 + 4 + 2 + \cdots$

62. A geometric sequence is $\{24, 12, 6, 3, \ldots\}$. Evaluate the sum of the geometric series: $24 + 12 + 6 + 3 + \cdots$

63. A geometric sequence is $\left\{200, 50, \dfrac{25}{2}, \dfrac{25}{8}, \ldots\right\}$.

Evaluate the sum of the geometric series:

$$200 + 50 + \frac{25}{2} + \frac{25}{8} + \cdots$$

64. A geometric sequence is $\left\{800, 200, 50, \dfrac{25}{2}, \ldots\right\}$.

Evaluate the sum of the geometric series:

$$800 + 200 + 50 + \frac{25}{2} + \cdots$$

65. An unstable carbon-14 atom decays to a stable nitrogen atom. For an initial sample of 2 g of carbon-14, the function $N(t) = (2)e^{(-1.2097 \times 10^{-4})t}$ describes the amount of carbon-14 in grams, $N(t)$, that remains after t years.
  a. Rewrite this function as a geometric sequence with a domain of $\{1, 2, 3, \ldots\}$. Round to the nearest millionth.
  b. Find the common ratio, r. Round to the nearest hundred thousandth.
  c. Find the formula for the nth term of this sequence.
  d. Use the sequence to find the amount of carbon-14 left in the sample after 5000 years. Round to the nearest ten-thousandth.

66. An unstable cobalt-60 atom decays to a stable nickel atom. For an initial sample of 2 g of cobalt-60, the function $N(t) = (2)e^{-0.1308t}$ describes the amount of cobalt-60, $N(t)$, that remains after t years.

a. Rewrite this function as a geometric sequence with a domain of $\{1, 2, 3, \ldots\}$. Round to the nearest ten-thousandth.
b. Find the common ratio, r. Round to the nearest ten-thousandth.
c. Find the formula for the nth term of this sequence.
d. Use the sequence to find the amount of cobalt-60 left in the sample after 5 years. Round to the nearest ten-thousandth.

67. The function $D(t) = 24,000(1.07)^t$ describes the relationship of the number of years since 2008, t, and the number of products sold each year, $D(t)$.
  a. Rewrite this function as a geometric sequence with a domain of $\{1, 2, 3, \ldots\}$. Round to the nearest whole number.
  b. Find the common ratio, r. Round to the nearest hundredth.
  c. Find the formula for the nth term of this sequence.
  d. Use the sequence to find the number of products sold in 2014. Round to the nearest hundred.

68. The function $D(t) = 8000(1.06)^t$ describes the relationship of the number of years since 2008, t, and the number of products sold each year, $D(t)$.
  a. Rewrite this function as a geometric sequence with a domain of $\{1, 2, 3, \ldots\}$. Round to the nearest whole number.
  b. Find the common ratio, r. Round to the nearest hundredth.
  c. Find the formula for the nth term of this sequence.
  d. Use the sequence to find the number of products sold in 2014. Round to the nearest hundred.

69. The function $y = 710,683(1.058)^x$ describes the relationship of the number of years since 1989, x, and the population of the city of Las Vegas and surrounding Clark County, y.
  a. Rewrite this function as a geometric sequence with a domain of $\{1, 2, 3, \ldots\}$. Round to the nearest whole number.
  b. Find the common ratio, r. Round to the nearest hundredth.
  c. Find the formula for the nth term of this sequence.
  d. Use the sequence to find the population in 2015. Round to the nearest thousand.

70. The function $y = 918,004(1.029)^x$ describes the relationship of the number of years since 1989, x, and the population of the cities of Austin and Red Rock, Texas, y.
  a. Rewrite this function as a geometric sequence with a domain of $\{1, 2, 3, \ldots\}$. Round to the nearest whole number.
  b. Find the common ratio, r. Round to the nearest hundredth.
  c. Find the formula for the nth term of this sequence.
  d. Use the sequence to find the population in 2012. Round to the nearest thousand.

71. The function $y = 5000(1 + 0.06)^x$ describes the relationship of the number of years, x, and the future value, y, of an investment of $5000 dollars. The investment is earning 6% interest, compounded annually.
    a. Rewrite this function as a geometric sequence with a domain of $\{1, 2, 3, \ldots\}$. Round to the nearest hundredth.
    b. Find the common ratio, r. Round to the nearest hundredth.
    c. Find the formula for the nth term of this sequence.
    d. Use the sequence to find the future value after 10 years. Round to the nearest whole number.

72. The function $y = 9000(1 + 0.05)^x$ describes the relationship of the number of years, x, and the future value, y, of an investment of $9000. The investment is earning 5% interest, compounded annually.
    a. Rewrite this function as a geometric sequence with a domain of $\{1, 2, 3, \ldots\}$. Round to the nearest hundredth.
    b. Find the common ratio, r. Round to the nearest hundredth.
    c. Find the formula for the nth term of this sequence.
    d. Use the sequence to find the future value after 8 years. Round to the nearest whole number.

Problem Solving: Practice and Review

Follow your instructor's guidelines for using the five steps as outlined in Section 1.5, p. 51.

73. The diameter of a circular corral for training horses is 50 ft. Find the number of fence posts needed to build the corral if the posts are to be about 8 ft apart and there is a 12-ft gate ($\pi \approx 3.14$). Round up to the nearest post.

74. In the summer of 2009, Hyundai customers who bought or leased a vehicle by August 31 received a gas card that let them buy gas at $1.49 a gallon for a year. The fuel efficiency of a 2009 Elantra is 24 mi per gallon in city driving and 33 mi per gallon for highway driving. A customer drives about 8500 mi per year in the city and 12,500 mi per year on the highway. If the average price of gas during the year is $2.90 per gallon, find the savings to the customer from using the gas card. Round to the nearest whole number. (*Sources:* www.ap.org, June 30, 2009; www.hyundaiusa.com)

75. A study by researchers from Tufts University found that about one out of five items ordered from restaurants had at least 100 more calories than the calories reported for the item on the restaurant's website. The researchers tested 269 items at 42 fast-food and sit-down restaurants. Find the number of items tested that had at least 100 more calories than reported on the website. Round to the nearest whole number. (*Source:* Urban, Accuracy of Stated Energy Contents of Restaurant Foods, *JAMA*, July 2011)

76. Find the number of births in 2009. Round to the nearest thousand.

The broad-based decline in births and fertility rates from 2007 through 2009 is now well-documented. An earlier NCHS Health E-Stat showed the overall birth count and fertility rate continuing to decline through the first 6 months of 2010. The provisional count of births in the United States for 2010 (12-month period ending December 2010) was 4,007,000. This count was 3 percent less than the number of births in 2009 and 7 percent less than the all-time high of 4,316,233 births in 2007. (*Source:* www.cdc.gov, June 15, 2011)

Technology

For exercises 77–80, use a graphing calculator to evaluate the sum of the first ten terms of each sequence.

77. $a_n = 3n + 8$

78. $a_n = -12n + 15$

79. $a_n = 8n + 1$

80. $a_n = 350n$

Find the Mistake

For exercises 81–84, the completed problem has one mistake.
(a) Describe the mistake in words, or copy down the whole problem and highlight or circle the mistake.
(b) Do the problem correctly.

81. **Problem:** Determine whether $\{6, 12, 18, 24, \ldots\}$ is a geometric sequence.

    Incorrect Answer: Yes, this is a geometric sequence. The common ratio of successive terms is 6.

82. **Problem:** The fifth term in a geometric sequence is 64. The common ratio is -2. Find the sixth term in this sequence.

    Incorrect Answer: The sixth term is $\dfrac{64}{-2} = -32$.

83. **Problem:** The formula for the nth term of a geometric sequence is $a_n = 4(3)^{n-1}$. Find the eighth term.

    Incorrect Answer: $a_n = 4(3^8)$
    $$a_8 = 26,244$$

84. **Problem:** Find the formula for the nth term of the geometric sequence $\{3, -6, 12, -24, \ldots\}$

    Incorrect Answer:

    r
    $$= \frac{a_4}{a_3}$$
    $$= \frac{-24}{12}$$
    $$= -2$$
    $$a_n = r^{n-1}$$
    $$a_n = (-2)^{n-1}$$

Review

85. Simplify: $\dfrac{2x^2 + 15x - 27}{2x^2 + 19x + 9}$

86. Simplify: $\dfrac{5}{x + 2} + \dfrac{8x}{3x - 1}$

87. Solve: $\dfrac{1}{2} - \dfrac{3}{x^2} = \dfrac{5}{2x}$

88. Solve: $\dfrac{4}{9}w + \dfrac{3}{4} = \dfrac{1}{6}w$

The diagram shows a simple pattern. A number is the sum of the two numbers above it. In this section, we will use this pattern to help us find the coefficients of a polynomial.

SECTION 10.4 The Binomial Theorem

After reading the text, working the practice problems, and completing assigned exercises, you should be able to:

1. Evaluate a factorial.
2. Find the coefficient of a term in a binomial expansion.
3. Use the binomial theorem to find a binomial expansion.
4. Use Pascal's triangle to find the coefficients of the terms of a binomial expansion.

Factorial Notation

The **factorial** of a positive integer, n, is the product of all positive integers less than or equal to n. The notation for factorial is an exclamation point, !. So $5! = 5 \cdot 4 \cdot 3 \cdot 2 \cdot 1$ and $8! = 8 \cdot 7 \cdot 6 \cdot 5 \cdot 4 \cdot 3 \cdot 2 \cdot 1$. The factorial of 0 is defined to be 1; $0! = 1$.

Factorial

$0! = 1$

If $n \geq 1$ and n is an integer, $n! = n(n - 1)(n - 2) \cdots (1)$.

EXAMPLE 1 Evaluate: $\dfrac{8!}{2!\,6!}$

SOLUTION ▶

$\dfrac{8!}{2!\,6!}$

$= \dfrac{8 \cdot 7 \cdot 6 \cdot 5 \cdot 4 \cdot 3 \cdot 2 \cdot 1}{(2 \cdot 1)(6 \cdot 5 \cdot 4 \cdot 3 \cdot 2 \cdot 1)}$ Rewrite factorials as multiplication.

$= \dfrac{8 \cdot 7}{2 \cdot 1} \cdot \dfrac{6}{6} \cdot \dfrac{5}{5} \cdot \dfrac{4}{4} \cdot \dfrac{3}{3} \cdot \dfrac{2}{2} \cdot \dfrac{1}{1}$ Find common factors.

$= \dfrac{56}{2} \cdot 1 \cdot 1 \cdot 1 \cdot 1 \cdot 1 \cdot 1$ Simplify.

$= 28$ Simplify.

When simplifying expressions that include factorials, we often "cross off" common factors in the numerator and denominator. We can do this only because a fraction with the same numerator and denominator equals 1 and because the product of a number and 1 is the number.

EXAMPLE 2 Evaluate: $\dfrac{9!}{5!\,3!}$

SOLUTION ▶

$\dfrac{9!}{5!\,3!}$

$= \dfrac{9 \cdot 8 \cdot 7 \cdot 6 \cdot 5 \cdot 4 \cdot 3 \cdot 2 \cdot 1}{5 \cdot 4 \cdot 3 \cdot 2 \cdot 1 \cdot 3 \cdot 2 \cdot 1}$ Rewrite factorials as multiplication.

$= \dfrac{9 \cdot 8 \cdot 7 \cdot 6 \cdot \cancel{5} \cdot \cancel{4} \cdot \cancel{3} \cdot \cancel{2} \cdot \cancel{1}}{\cancel{5} \cdot \cancel{4} \cdot \cancel{3} \cdot \cancel{2} \cdot \cancel{1} \cdot 3 \cdot 2 \cdot 1}$ Find common factors.

$= \dfrac{9 \cdot 8 \cdot 7 \cdot 6}{3 \cdot 2 \cdot 1}$ Simplify.

$= \dfrac{3024}{6}$ Simplify.

$= 504$ Simplify.

Practice Problems

For problems 1–4, evaluate.

1. $8!$ **2.** $\dfrac{10!}{8!\,2!}$ **3.** $\dfrac{10!}{7!\,3!}$ **4.** $\dfrac{10!}{5!\,5!}$

Binomials, Binomial Expansion, and Binomial Coefficients

A **binomial** is a polynomial with two terms such as $a + b$. If we raise a binomial to a power, $(a + b)^n$, the product of the binomials is a **binomial expansion**.

$(a+b)^0$ $= 1$	$(a+b)^1$ $= a + b$	$(a+b)^2$ $= (a+b)(a+b)$ $= a^2 + 2ab + b^2$	$(a+b)^3$ $= (a+b)(a+b)(a+b)$ $= a^3 + 3a^2b + 3ab^2 + b^3$
$(a+b)^4$ $= (a+b)(a+b)(a+b)(a+b)$ $= a^4 + 4a^3b + 6a^2b^2 + 4ab^3 + b^4$		$(a+b)^5$ $= (a+b)(a+b)(a+b)(a+b)(a+b)$ $= a^5 + 5a^4b + 10a^3b^2 + 10a^2b^3 + 5ab^4 + b^5$	

If we continue to do expansions for larger values of the exponent, n, four patterns appear.

Expansion of $(a + b)^n$

Pattern 1.   The number of terms in an expansion equals $n + 1$.

Pattern 2.   The terms in the expansion can be arranged so that, moving from left to right, the power of a in each term decreases by 1 and the power of b in each term increases by 1.

Pattern 3.   The sum of the exponents in each term equals n.

Pattern 4.   The value of the coefficient for a term in a binomial expansion is equal to $\dfrac{n!}{j!\,(n - j)!}$. In the first term, $j = 0$. Moving from left to right, the value of j increases by 1 for each term.

EXAMPLE 3   The expansion of $(a + b)^5$ is $a^5 + 5a^4b + 10a^3b^2 + 10a^2b^3 + 5ab^4 + b^5$. Show that Patterns 1–4 are true for this expansion.

SOLUTION ▶ Rewrite the expansion as $1a^5b^0 + 5a^4b^1 + 10a^3b^2 + 10a^2b^3 + 5a^1b^4 + 1a^0b^5$.

Pattern 1.   **The number of terms equals $n + 1$.**
In $(a + b)^5$, $n = 5$. Since this expansion has six terms, the number of terms equals $n + 1$.

Pattern 2.   **Moving from left to right, the power of a in each term decreases by 1 and the power of b in each term increases by 1.**
Moving from left to right, the power of a is 5, 4, 3, 2, 1, 0.
Moving from left to right, the power of b is 0, 1, 2, 3, 4, 5.

Pattern 3.   **The sum of the exponents in each term equals n.**
Moving from left to right, the sum of the exponents equals n, 5.

Pattern 4.   **The value of the coefficient of each term is $\dfrac{n!}{j!\,(n - j)!}$.**

First term: $1a^5b^0$	**Second term: $5a^4b^1$**
$\dfrac{n!}{j!\,(n - j)!}$	$\dfrac{n!}{j!\,(n - j)!}$
$= \dfrac{5!}{0!\,(5 - 0)!}$ $n = 5; j = 0$	$= \dfrac{5!}{1!\,(5 - 1)!}$ $n = 5; j = 1$
$= \dfrac{5 \cdot 4 \cdot 3 \cdot 2 \cdot 1}{1(5 \cdot 4 \cdot 3 \cdot 2 \cdot 1)}$ Evaluate factorials.	$= \dfrac{5 \cdot 4 \cdot 3 \cdot 2 \cdot 1}{1(4 \cdot 3 \cdot 2 \cdot 1)}$ Evaluate factorials.
$= 1$ Same as coefficient, 1.	$= 5$ Same as coefficient, 5.
Third term: $10a^3b^2$	**Fourth term: $10a^2b^3$**
$\dfrac{n!}{j!\,(n - j)!}$	$\dfrac{n!}{j!\,(n - j)!}$
$= \dfrac{5!}{2!\,(5 - 2)!}$ $n = 5; j = 2$	$= \dfrac{5!}{3!\,(5 - 3)!}$ $n = 5; j = 3$
$= \dfrac{5 \cdot 4 \cdot 3 \cdot 2 \cdot 1}{(2 \cdot 1)(3 \cdot 2 \cdot 1)}$ Evaluate factorials.	$= \dfrac{5 \cdot 4 \cdot 3 \cdot 2 \cdot 1}{(3 \cdot 2 \cdot 1)(2 \cdot 1)}$ Evaluate factorials.
$= 10$ Same as coefficient, 10.	$= 10$ Same as coefficient, 10.

Fifth term: $5a^1b^4$		**Sixth term:** $1a^0b^5$	
$\dfrac{n!}{j!\,(n-j)!}$		$\dfrac{n!}{j!\,(n-j)!}$	
$= \dfrac{5!}{4!\,(5-4)!}$	$n = 5; j = 4$	$= \dfrac{5!}{5!\,(5-5)!}$	$n = 5; j = 5$
$= \dfrac{5 \cdot 4 \cdot 3 \cdot 2 \cdot 1}{(4 \cdot 3 \cdot 2 \cdot 1)(1)}$	Evaluate factorials.	$= \dfrac{5 \cdot 4 \cdot 3 \cdot 2 \cdot 1}{(5 \cdot 4 \cdot 3 \cdot 2 \cdot 1)(1)}$	Evaluate factorials.
$= 5$	Same as coefficient, 5.	$= 1$	Same as coefficient, 1.

From left to right, the coefficients of the expansion are 1, 5, 10, 10, 5, 1.

These are the same coefficients predicted by $\dfrac{n!}{j!\,(n-j)!}$.

The notation $\dbinom{n}{j}$ also represents the coefficient of each term of a binomial expansion.

> **Binomial Expansion Coefficient Notation**
>
> $$\binom{n}{j} = \frac{n!}{j!\,(n-j)!} \qquad n \text{ and } j \text{ are integers greater than or equal to } 0; n \geq j$$

EXAMPLE 4   Find the coefficient of the eighth term in the binomial expansion $(a + b)^{11}$.

SOLUTION   For the first term, $j = 0$. For the eighth term, $j = 7$. To find the coefficient of the eighth term, evaluate $\dbinom{n}{j}$ for $n = 11$ and $j = 7$.

$$\binom{n}{j}$$

$$= \binom{11}{7} \qquad n = 11; j = 7$$

$$= \frac{11!}{7!\,(11-7)!} \qquad \binom{n}{j} = \frac{n!}{j!\,(n-j)!}$$

$$= \frac{11 \cdot 10 \cdot 9 \cdot 8 \cdot 7 \cdot 6 \cdot 5 \cdot 4 \cdot 3 \cdot 2 \cdot 1}{(7 \cdot 6 \cdot 5 \cdot 4 \cdot 3 \cdot 2 \cdot 1)(4 \cdot 3 \cdot 2 \cdot 1)} \qquad \text{Rewrite factorials as multiplication.}$$

$$= \frac{11 \cdot 10 \cdot 9 \cdot 8 \cdot \cancel{7} \cdot \cancel{6} \cdot \cancel{5} \cdot \cancel{4} \cdot \cancel{3} \cdot \cancel{2} \cdot \cancel{1}}{(\cancel{7} \cdot \cancel{6} \cdot \cancel{5} \cdot \cancel{4} \cdot \cancel{3} \cdot \cancel{2} \cdot \cancel{1})(4 \cdot 3 \cdot 2 \cdot 1)} \qquad \text{Find common factors.}$$

$$= \frac{11 \cdot 10 \cdot 9 \cdot 8}{4 \cdot 3 \cdot 2 \cdot 1} \qquad \text{Simplify.}$$

$$= 330 \qquad \text{Simplify.}$$

The coefficient of the eighth term is 330.

> **Practice Problems**
>
> **5.** Find the coefficient of the fourth term of the expansion of $(a + b)^9$.
>
> **6.** Find the coefficient of the sixth term of the expansion of $(a + b)^7$.

The Binomial Theorem and Pascal's Triangle

The binomial theorem summarizes the patterns of a binomial expansion.

> **Binomial Theorem**
>
> $$(a + b)^n = \binom{n}{0}a^n + \binom{n}{1}a^{n-1}b^1 + \binom{n}{2}a^{n-2}b^2 + \cdots + \binom{n}{j}a^{n-j}b^j + \cdots + \binom{n}{n}b^n$$
>
> where a and b are real numbers and n is an integer greater than or equal to 0.

Instead of multiplying $(a + b)$ by itself n times to find the expansion of $(a + b)^n$, we can use the binomial theorem to find the coefficients and the variables in each term.

EXAMPLE 5   Use the binomial theorem to expand $(a + b)^6$.

SOLUTION ▶ $(a + b)^6$

$$= \binom{6}{0}a^6 + \binom{6}{1}a^5b + \binom{6}{2}a^4b^2 + \binom{6}{3}a^3b^3 + \binom{6}{4}a^2b^4 + \binom{6}{5}ab^5 + \binom{6}{6}b^6$$

$$= \left(\frac{6!}{0!\,6!}\right)a^6 + \left(\frac{6!}{1!\,5!}\right)a^5b + \left(\frac{6!}{2!\,4!}\right)a^4b^2 + \left(\frac{6!}{3!\,3!}\right)a^3b^3 + \left(\frac{6!}{4!\,2!}\right)a^2b^4 +$$

$$\left(\frac{6!}{5!\,1!}\right)ab^5 + \left(\frac{6!}{6!\,0!}\right)b^6$$

$$= a^6 + 6a^5b + 15a^4b^2 + 20a^3b^3 + 15a^2b^4 + 6ab^5 + b^6$$

If we arrange the coefficients for a binomial expansion in n rows, a pattern emerges.

Binomial Coefficients	$(a + b)^n$	
1	$(a + b)^0$	Zeroth row
1 1	$(a + b)^1$	First row
1 2 1	$(a + b)^2$	Second row
1 3 3 1	$(a + b)^3$	Third row
1 4 6 4 1	$(a + b)^4$	Fourth row
1 5 10 10 5 1	$(a + b)^5$	Fifth row
1 6 15 20 15 6 1	$(a + b)^6$	Sixth row

This pattern is **Pascal's triangle**. Notice that the coefficients of the binomial expansion for $(a + b)^6$ in the last example match the coefficients in the sixth row of this pattern. To write Pascal's triangle, begin the top row with a single 1. (The top row is called the "zeroth row" because $n = 0$.) The next row (first row) has two numbers: 1 1. To create more rows, begin and end the row with 1. Each of the other numbers in the row is the sum of the two numbers above it.

EXAMPLE 6   **(a)** Use Pascal's triangle to find the coefficient of the third term in the expansion of $(a + b)^5$.

SOLUTION ▶   Fifth row of Pascal's triangle: 1   5   **10**   10   5   1

The coefficient of the third term is 10.

(b) Check.

▶   The binomial theorem for $n = 5$ is

$$(a + b)^5 = \binom{5}{0}a^5 + \binom{5}{1}a^4b^1 + \binom{5}{2}a^3b^2 + \binom{5}{3}a^2b^3 + \binom{5}{4}ab^4 + \binom{5}{5}b^5$$

The coefficient of the third term is $\binom{5}{2}$.

$$\binom{5}{2} \qquad n = 5; j = 2$$

$$= \frac{5!}{2!\,(5-2)!} \qquad \binom{n}{j} = \frac{n!}{j!\,(n-j)!}$$

$$= \frac{5 \cdot 4 \cdot \cancel{3} \cdot \cancel{2} \cdot \cancel{1}}{(2 \cdot 1)(\cancel{3} \cdot \cancel{2} \cdot \cancel{1})} \qquad \text{Evaluate the factorials.}$$

$$= 10 \qquad \text{Simplify.}$$

Both Pascal's triangle and the binomial theorem show that the coefficient of the third term is 10.

Although it is called Pascal's triangle, Blaise Pascal (1623–1662) was not the first person to notice the relationship between this pattern and binomial coefficients. Six hundred years earlier, Chinese and Arab mathematicians were familiar with the pattern. Other European mathematicians worked with the pattern before Pascal. Although Pascal never claimed that he originated the idea, the triangle bears his name.

EXAMPLE 7 Expand: $(2x + y)^4$

(a) Use Pascal's triangle to find the coefficient in each term.

SOLUTION ▶ The coefficients of the terms in $(a + b)^4$ are found in the fourth row of Pascal's triangle: 1   4   6   4   1.

Using the coefficients from Pascal's triangle, the expansion of $(a + b)^4$ is

$$1a^4 + 4a^3b + 6a^2b^2 + 4ab^3 + 1b^4$$

(b) Use the binomial theorem to find the variables in each term.

▶ The binomial theorem for $n = 4$ is

$$(a + b)^4 = \binom{4}{0}a^4 + \binom{4}{1}a^3b^1 + \binom{4}{2}a^2b^2 + \binom{4}{3}ab^3 + \binom{4}{4}b^4$$

For $(2x + y)^4$, $a = 2x$ and $b = y$.

$$(a + b)^4$$
$$= 1a^4 + 4a^3b + 6a^2b^2 + 4ab^3 + 1b^4 \qquad \text{Coefficients from Pascal's triangle.}$$
$$= 1(2x)^4 + 4(2x)^3y + 6(2x)^2y^2 + 4(2x)y^3 + 1y^4 \qquad a = 2x; b = y$$
$$= 1(16x^4) + 4(8x^3)y + 6(4x^2)y^2 + 8(xy^3) + 1(y^4) \qquad \text{Simplify.}$$
$$= 16x^4 + 32x^3y + 24x^2y^2 + 8xy^3 + y^4 \qquad \text{Simplify.}$$

In Example 7, we expanded a sum, $(2x + y)^4$. In the next example, we expand a difference, $(2x - y)^4$, by rewriting it as a sum, $(2x + (-y))^4$.

EXAMPLE 8 Expand: $(2x - y)^4$

(a) Use Pascal's triangle to find the coefficients.

SOLUTION ▶ Using the coefficients from the fourth row of Pascal's triangle, 1   4   6   4   1, the expansion of $(a + b)^4$ is $1a^4 + 4a^3b + 6a^2b^2 + 4ab^3 + 1b^4$.

(b) Use the binomial theorem to find the variables in each term.

▶ The binomial theorem for $n = 4$ is

$$(a + b)^4 = \binom{4}{0}a^4 + \binom{4}{1}a^3b^1 + \binom{4}{2}a^2b^2 + \binom{4}{3}ab^3 + \binom{4}{4}b^4$$

For $(2x + (-y))^4$, $a = 2x$ and $b = -y$.

$(a + b)^4$

$= 1a^4 + 4a^3b + 6a^2b^2 + 4ab^3 + 1b^4$ Coefficients from Pascal's triangle.

$= 1(2x)^4 + 4(2x)^3(-y) + 6(2x)^2(-y)^2 + 4(2x)(-y)^3 + 1(-y)^4$ $a = 2x; b = -y$

$= 1(\mathbf{16x^4}) + 4(\mathbf{8x^3})(\mathbf{-1}y) + 6(\mathbf{4x^2})(\mathbf{y^2}) + 4(2x)(\mathbf{-1}y^3) + 1(\mathbf{y^4})$ Simplify.

$= \mathbf{16x^4 - 4(8x^3)}y + 6(\mathbf{4x^2})y^2 - \mathbf{8}xy^3 + y^4$ Simplify.

$= 16x^4 - \mathbf{32x^3y + 24x^2y^2} - 8xy^3 + y^4$ Simplify.

In Example 7, we see that the expansion of $(2x + y)^4$ is $16x^4 + 32x^3y + 24x^2y^2 + 8xy^3 + y^4$. In Example 8, we see that the expansion of $(2x - y)^4$ is $16x^4 - 32x^3y + 24x^2y^2 - 8xy^3 + y^4$. Notice how the signs in the expansion of the difference alternate from positive to negative.

Practice Problems

7. Use the binomial theorem to find the coefficient of the fourth term of the expansion of $(a + b)^9$.

8. Write nine rows of Pascal's triangle. Find the coefficient of the fourth term of the expansion of $(a + b)^9$.

9. Expand $(4x + 2y)^5$. Use Pascal's triangle to find the coefficients. Use the binomial theorem to find the variables in each term.

10. Expand $(4x - 2y)^5$. Use Pascal's triangle to find the coefficients. Use the binomial theorem to find the variables in each term.

Using Technology: Finding $\binom{n}{j}$ with a Graphing Calculator

A graphing calculator can evaluate a binomial coefficient, $\binom{n}{j} = \dfrac{n!}{j!\,(n - j)!}$. However, it uses a different notation: nCr.

EXAMPLE 9 Use a graphing calculator to evaluate $\binom{9}{2}$.

Since $n = 9$, press 9. Press MATH. Move the cursor right to highlight the PRB menu. Choose 3: nCr. Press ENTER. The screen shows "9 nCr." Now since $j = 2$, press 2. Press ENTER. The value of $\binom{9}{2}$ is 36.

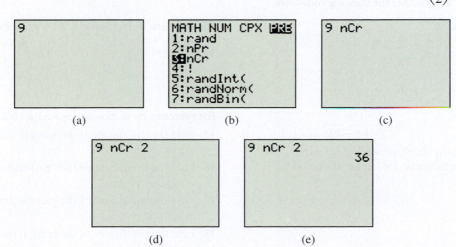

(a)   (b)   (c)

(d)   (e)

A graphing calculator can also evaluate factorials.

EXAMPLE 10 Evaluate 9!

Press 9 . Press MATH . Choose PRB. Choose 4. Press ENTER . Press ENTER . The value of 9! is 362,880.

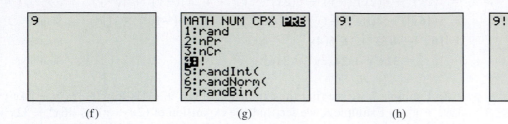

(f)   (g)   (h)   (i)

Practice Problems For problems 11–15, use a graphing calculator to evaluate.

11. $\binom{8}{3}$ 12. $\binom{6}{5}$ 13. $7!$ 14. $8!$ 15. $\dfrac{11!}{5!\,(11-5)!}$

10.4 VOCABULARY PRACTICE

Match the term with its description.

1. $(a+b)^n = \binom{n}{0}a^n + \binom{n}{1}a^{n-1}b^1 + \binom{n}{2}a^{n-2}b^2 + \cdots + \binom{n}{j}a^{n-j}b^j + \cdots + \binom{n}{n}b^n$
2. $a(b+c) = ab + ac$
3. The top row in Pascal's triangle
4. !
5. A polynomial with two terms
6. The coefficient of a term in a binomial expansion
7. Terms that have the same variables and the same exponents on each variable
8. $\{\ldots, -3, -2, -1, 0, 1, 2, 3, \ldots\}$
9. The product of a binomial multiplied by itself n times
10. This pattern is a display of the coefficients of binomial expansions.

A. binomial
B. binomial coefficient
C. binomial expansion
D. binomial theorem
E. distributive property
F. factorial notation
G. like terms
H. Pascal's triangle
I. set of integers
J. zeroth row

10.4 Exercises

Follow your instructor's guidelines for showing your work.

For exercises 1–8, evaluate.

1. $11!$
2. $12!$
3. $5!$
4. $7!$
5. $\dfrac{11!}{5!\,(11-5)!}$
6. $\dfrac{12!}{7!\,(12-7)!}$
7. $\dfrac{11!}{3!\,(11-3)!}$
8. $\dfrac{12!}{2!\,(12-2)!}$

For exercises 9–10, use the distributive property and combine like terms to expand the binomial. Do not use the binomial theorem.

9. $(a+b)^4$
10. $(a+b)^5$

For exercises 11–14, evaluate. Do not use Pascal's triangle.

11. $\binom{12}{4}$ 12. $\binom{10}{6}$

13. $\binom{4}{3}$ 14. $\binom{5}{4}$

For exercises 15–18, do not use Pascal's triangle.

15. Find the coefficient of the seventh term in the binomial expansion $(a+b)^9$.

16. Find the coefficient of the third term in the binomial expansion $(a+b)^{10}$.

17. Find the coefficient of the second term in the binomial expansion $(a+b)^{15}$.

18. Find the coefficient of the tenth term in the binomial expansion $(a+b)^{16}$.

19. Identify the number of terms in the expansion of $(a + b)^{21}$.

20. Identify the number of terms in the expansion of $(a + b)^{23}$.

21. Construct Pascal's triangle with 13 rows.

22. Construct Pascal's triangle with 14 rows.

For exercises 23–38, expand the expression. Use Pascal's triangle to find the coefficients, and use the binomial theorem to find the variables in each term.

23. $(a + b)^{12}$

24. $(a + b)^{13}$

25. $(a - b)^{12}$

26. $(a - b)^{13}$

27. $(x + y)^8$

28. $(x + y)^7$

29. $(2a + b)^3$

30. $(3a + b)^3$

31. $(c + 3d)^5$

32. $(c + 2d)^5$

33. $(5k + 3h)^6$

34. $(4d + 6f)^6$

35. $(3k + 5h)^6$

36. $(6d + 4f)^6$

37. $(3k - 5h)^6$

38. $(6d - 4f)^6$

39. The expression $(1.1)^5$ is equivalent to $(1 + 0.1)^5$.
    a. Expand this expression. Use Pascal's triangle to find the coefficients, and use the binomial theorem to find the variables in each term.
    b. Evaluate the sum of the terms. Round to the nearest ten-thousandth.

40. The expression $(1.2)^5$ is equivalent to $(1 + 0.2)^5$.
    a. Expand this expression. Use Pascal's triangle to find the coefficients, and use the binomial theorem to find the variables in each term.
    b. Evaluate the sum of the terms. Round to the nearest ten-thousandth.

41. If a number greater than 1 is a **triangular number**, then that number of balls can be arranged in a triangle. The first ten triangular numbers are 1, 3, 6, 10, 15, 21, 28, 36, 45, and 55. Draw Pascal's triangle with 12 rows. Highlight the diagonals that contains the triangular numbers.

Triangular numbers 1, 3, 6, 10

42. If a number greater than 1 is a **tetrahedral number**, then that number of balls can be arranged in a pyramid. The diagram shows that 56 is a tetrahedral number. The first nine tetrahedral numbers are 1, 4, 10, 20, 35, 56, 84, 120, and 165. Draw Pascal's triangle with 12 rows. Highlight the diagonals that contain the tetrahedral numbers.

43. Refer to exercise 41. Are all triangular numbers also binomial coefficients? Explain.

44. Refer to exercise 42. Are all tetrahedral numbers also binomial coefficients? Explain.

45. a. Draw Pascal's triangle with ten rows. Evaluate the sum of the numbers of each horizontal row.
    b. Write an exponential expression that predicts the sum of the nth row.

46. a. Draw Pascal's triangle with ten rows. Evaluate the sum of each diagonal as shown in the diagram. These sums are numbers in the **Fibonacci sequence**.

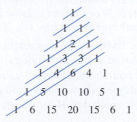

    b. The first number in the Fibonacci sequence is 0. The next numbers are the sums of the diagonals. Use roster notation to list the first 11 numbers of the Fibonacci sequence.

Problem Solving: Practice and Review

Follow your instructor's guidelines for using the five steps as outlined in Section 1.5, p. 51.

47. Find the percent decrease in beach visits. Round to the nearest percent.

    The city was a soggy mess in June, drenched with rains. . . . Central Park was pelted with 10.06 inches. . . . Beach use in June was way down, to 735,000 from the usual 2.6 million. (*Source:* www.nytimes.com, July 1, 2009)

48. A teacher in a family of four has an adjusted gross income of $38,575. Find the annual Income Based Repayment amount that the teacher must pay on her outstanding student loans.

    The annual Income Based Repayment amount is 15 percent of the difference between your Adjusted Gross Income and 150 percent of the Department of Health and Human Services Poverty Guideline for your family size and state. This amount is then divided by 12 to get the monthly IBR payment amount. (*Source:* studentaid.ed.gov)

The 2011 Poverty Guidelines for the 48 Contiguous States and the District of Columbia

Persons in family	Poverty guideline
1	$10,890
2	14,710
3	18,530
4	22,350
5	26,170
6	29,990
7	33,810
8	37,630

For families with more than 8 persons, add $3,820 for each additional person.

Source: aspe.hhs.gov

49. The Body Mass Index (BMI) is one measure of obesity. If a person is 5 ft 10 in. tall and is not overweight or obese, find his or her maximum weight. Round to the nearest whole number.

Obesity is defined as an excessively high amount of body fat or adipose tissue in relation to lean body mass. Overweight refers to increased body weight in relation to height, which is then compared to a standard of acceptable weight. BMI is a common measure expressing the relationship (or ratio) of weight-to-height. It is a mathematical formula:

$$BMI = \frac{(703)(\text{weight in pounds})}{(\text{height in inches})(\text{height in inches})}.$$ Adults with a BMI of 25 to 29.9 are considered overweight, while those with a BMI of 30 or more are considered obese. (*Source:* www.healthyamericans.org, July 2009)

50. If an adult eats 3500 calories that are not used by the body but are stored as fat, the adult gains 1 pound. Find the number of pounds an adult will gain per year (1 year = 365 days) if sleep deprivation resulted in eating extra calories each day. Round to the nearest whole number.

Participants consumed an average of 296 calories more when they were sleep-deprived compared with when they were well-rested. (*Source:* www.usatoday.com, March 23, 2011)

Technology

For exercises 51–54, use a graphing calculator to evaluate the binomial coefficient.

51. $\binom{16}{4}$

52. $\binom{16}{8}$

53. $\binom{16}{13}$

54. $\binom{11}{2}$

Find the Mistake

For exercises 55–58, the completed problem has one mistake.
(a) Describe the mistake in words, or copy down the whole problem and highlight or circle the mistake.
(b) Do the problem correctly.

55. Problem: Evaluate: $\dfrac{4!}{(4-4)!}$

Incorrect Answer: $\dfrac{4!}{(4-4)!}$

$$= \frac{4 \cdot 3 \cdot 2 \cdot 1}{0!}$$

$$= \frac{24}{0}$$

$$= \text{undefined}$$

56. Problem: Find the coefficient of the seventh term in the binomial expansion $(a+b)^{15}$. Do not use Pascal's triangle.

Incorrect Answer: $\binom{15}{7}$

$$= \frac{15!}{7!\,(15-7)!}$$

$$= 6435$$

The coefficient of the seventh term is 6435.

57. Problem: Expand the expression $(2x+y)^3$. Use Pascal's triangle to find the coefficients. Use the binomial theorem to find the variables in each term.

Incorrect Answer: Fourth row of Pascal's triangle: 1   3   3   1

$$1(2x)^3 + 3(2x)^2(y) + 3(2x)y^2 + 1y^3$$
$$= 1(2x^3) + 3(2x^2y) + 3(2xy^2) + 1y^3$$
$$= 2x^3 + 6x^2y + 6xy^2 + y^3$$

58. Problem: Expand the expression $(x+3y)^3$. Use Pascal's triangle to find the coefficients.

Incorrect Answer: Fourth row of Pascal's triangle: 1   3   3   1

$$1x^3 + 3x^2(3y) + 3x(3y)^2 + 1(3y)^3$$
$$= 1x^3 + 9x^2y + 27xy^2 + 27y^3$$
$$= x^3 + 36x^2y^2 + 27y^3$$

Review

59. Show that $f(x) = 2x + 6$ and $g(x) = \dfrac{1}{2}x - 3$ are inverse functions.

60. A function is $f(x) = \log_3(x)$.
   a. Use interval notation to represent its domain.
   b. Use interval notation to represent its range.

61. Solve $15 = 3^x$. If the solution is irrational, round to the ten-thousandths place.

62. Rewrite $3 \log_8(x) - \log_8(y)$ as a single logarithm.

SUCCESS IN COLLEGE MATHEMATICS

63. Some students read their textbook regularly, doing the examples and the practice problems. They may use the examples in the textbook to help them understand how to do an assignment. They check their answers with the answers in the back of the book. Describe how you used the book in this class.

64. Do you plan to change how you use the textbook for your next class? Explain why or why not.

Study Plan for Review of Chapter 10

SECTION 10.1　Sequences and Series

Ask Yourself	Test Yourself	Help Yourself
Can I . . .	**Do 10.1 Review Exercises**	**See these Examples and Practice Problems**
describe the difference between a discrete domain and a continuous domain?	1	Ex. 1–3, PP 1–3
use roster notation to represent the range of a sequence with a given domain?		
graph a sequence?	1	Ex. 1–3, PP 1–3
use guess and check to find the formula for the *n*th term of a sequence?	2	Ex. 4, PP 4–6
given the formula for the *n*th term, evaluate a partial sum of a sequence?	3	Ex. 5–8, PP 7–9
use a sequence to solve an application problem?	4	Ex. 9, PP 10–11

10.1 Review Exercises

1. A sequence is $a_n = 4n - 1$. Its domain is $\{1, 2, 3 \ldots\}$.
   a. Use roster notation to represent the range of the sequence.
   b. Graph the first five terms of the sequence.

2. Find the formula for the *n*th term of the sequence
   $$\left\{\frac{1}{4}, \frac{1}{5}, \frac{1}{6}, \frac{1}{7}, \ldots\right\}.$$

3. a. Use sigma notation to represent the partial sum of the first five terms of the sequence $a_n = \dfrac{2}{n}$.
   b. Evaluate this partial sum.

4. The future value, a_n, of an investment of \$10,000 that is earning 7.5% compounded quarterly is
   $$a_n = 10{,}000\left(1 + \frac{0.075}{4}\right)^{4n},$$
   where *n* is the number of years that the money is invested. Find the future value of this investment after 12 years. Round to the nearest whole number.

SECTION 10.2　Arithmetic Sequences

Ask Yourself	Test Yourself	Help Yourself
Can I . . .	**Do 10.2 Review Exercises**	**See these Examples and Practice Problems**
determine whether a sequence is an arithmetic sequence?	5, 6	Ex. 1, 2, PP 1–3
find the common difference of an arithmetic sequence?		
given an arithmetic sequence, find the formula for the *n*th term?	7, 8	Ex. 2, 4, PP 1–3
given a term and the common difference, find the formula for the *n*th term of an arithmetic sequence?	6, 9	Ex. 3, PP 4
given a term a_n and the common difference, identify a_{n+1}?	9	Ex. 3, PP 4
evaluate a partial sum of an arithmetic sequence?	10–12	Ex. 4, 5, PP 5, 6
use an arithmetic sequence to solve an application problem?	13	Ex. 6, PP 7

10.2 Review Exercises

5. Identify the arithmetic sequence(s).
   a. $\{7, 10, 13, 16, 19, \ldots\}$
   b. $\{12, 10, 7, 3, -2, \ldots\}$
   c. $\{-1, 1, 3, 5, 7, 9, \ldots\}$

6. a. Create an arithmetic sequence in which $a_1 = 3$ and the common difference is 7. Use roster notation and at least five terms to represent this sequence.
   b. Find the formula for the *n*th term of this sequence.

7. Find the formula for the *n*th term of $\{11, 15, 19, 23, \ldots\}$.

8. Find the formula for the *n*th term of $\{-1, -4, -7, -10, \ldots\}$.

9. The fifth term in an arithmetic sequence is 40. The common difference is 5.
   a. Find the sixth term in the sequence.
   b. Find the formula for the *n*th term.

10. Evaluate the partial sum of the first 30 terms of the sequence $\{10, 13, 16, 19, \ldots\}$.

11. Evaluate: $\displaystyle\sum_{i=1}^{20} (3i - 2)$

12. Evaluate the sum of the first eight integers greater than 0.

13. The bottom row in a retaining wall has 100 blocks. Each row above it has 2 fewer blocks. Use an arithmetic sequence to find the number of blocks needed to build a retaining wall that is 8 blocks high.

SECTION 10.3   Geometric Sequences

Ask Yourself	Test Yourself	Help Yourself
Can I . . .	Do 10.3 Review Exercises	See these Examples and Practice Problems
determine whether a sequence is a geometric sequence?	14	Ex. 2, 3, PP 6
given a geometric sequence, find the common ratio?	15	Ex. 1, 4–8, 10, PP 1–4
describe the difference between an arithmetic and geometric sequence?	16	PP 6
given a geometric sequence, find the formula for the nth term?	17–19	Ex. 4–6, 10, PP 1–4
given the formula for the nth term, identify a term in a geometric sequence?	19	Ex. 4–6, PP 1–4
given a term of a geometric sequence and the common ratio, identify a term and the formula for the nth term?	19	Ex 6, PP 5
use sigma notation to represent the partial sum of a geometric sequence and evaluate the partial sum?	20	Ex. 7, 8, PP 7, 8
evaluate the sum of a geometric series in which $-1 < r < 1$?	21	Ex. 9, 10, PP 9–12
use a geometric sequence to solve an application problem?	22	Ex. 12, PP 13

10.3 Review Exercises

14. Identify the geometric sequence(s).
    a. $\{10, 100, 1000, \ldots\}$ b. $\{10, 20, 30, 40, \ldots\}$
    c. $\{1, 3, 6, 10, 15, \ldots\}$ d. $\left\{\dfrac{1}{2}, \dfrac{1}{4}, \dfrac{1}{8}, \dfrac{1}{16}, \ldots\right\}$

15. A geometric sequence is $\{6, 18, 54, 162, 486, \ldots\}$. Find the common ratio.

16. Describe the difference between an arithmetic sequence and a geometric sequence.

For exercises 17–18, find the formula for the nth term of each geometric sequence.

17. $\{-2, 4, -8, 16, \ldots\}$

18. $\{4, 16, 64, 256, \ldots\}$

19. a. Create a geometric sequence in which $a_1 = 5$ and the common ratio is 4. Use roster notation and at least five terms to represent this sequence.

   b. Write the formula for the nth term of this sequence.
   c. Use the formula to identify the seventh term.

20. a. Use sigma notation to represent the sum of the first 15 terms of the geometric sequence $\{3, 9, 27, 81, \ldots\}$.
   b. Evaluate this sum.

21. Evaluate: $\displaystyle\sum_{i=1}^{\infty} (6)(0.4)^{i-1}$

22. The function $f(t) = \$95,000(1.09)^t$ describes the relationship of the number of years since 2006, t, and the assessed value in dollars, $f(t)$, of an empty building lot in a rapidly growing city. The value of the lot is assessed once a year.
   a. Rewrite this function as a geometric sequence with a domain of $\{1, 2, 3, \ldots\}$. Round to the nearest whole number.
   b. Find the value of the lot in 2015. Round to the nearest whole number.

SECTION 10.4   The Binomial Theorem

Ask Yourself	Test Yourself	Help Yourself
Can I . . .	Do 10.4 Review Exercises	See these Examples and Practice Problems
evaluate a factorial?	23	Ex. 1, 2, PP 1–4
find the coefficient of a term in a binomial expansion without using Pascal's triangle?	25	Ex. 4, PP 5, 6

Ask Yourself	Test Yourself	Help Yourself
Can I . . .	**Do 10.4 Review Exercises**	**See these Examples and Practice Problems**
use the binomial theorem to write a binomial expansion?	25	Ex. 5, PP 7
construct a given number of rows of Pascal's triangle?	27	Ex. 6, PP 8
use Pascal's triangle to find the coefficient of a term in a binomial expansion?	27	Ex. 6–8, PP 8
use the binomial theorem to find the coefficients and/or variables of a term in a binomial expansion?	26	Ex. 7, 8, PP 9, 10

10.4 Review Exercises

23. Evaluate 8!

24. a. Would you use the binomial theorem to write the expansion of $(x + 3y)^2$?
   b. Explain.

25. Find the coefficient of the sixth term of the expansion of $(a + b)^7$. Do not use Pascal's triangle.

26. Use the binomial theorem to write the binomial expansion $(a + b)^5$. Do not use Pascal's triangle to find the coefficients.

27. Write the expansion of $(c + 4d)^5$. Use Pascal's triangle to find the coefficients.

Chapter 10 Test

1. a. Determine whether the sequence $\{2, 8, 14, 20, \ldots\}$ is arithmetic, geometric, or neither.
   b. Explain.

2. a. Use sigma notation to represent the partial sum of the first four terms of the sequence $a_n = n^2 + n$.
   b. Evaluate this partial sum.

3. The eighth term in an arithmetic sequence is 36. The common difference is 4.
   a. Find the ninth term in the sequence.
   b. Find the formula for the nth term.

4. The eighth term in a geometric sequence is 384. The common ratio is 2.
   a. Find the ninth term in the sequence.
   b. Find the formula for the nth term.

5. a. Use sigma notation to represent the partial sum of the first five terms of the geometric sequence
   $$\left\{\frac{1}{6}, 1, 6, 36, \ldots\right\}.$$
   b. Evaluate this sum.

6. a. Use sigma notation to represent the partial sum of the first 30 terms of the arithmetic sequence $\{-20, -16, -12, -8, \ldots\}$.
   b. Evaluate this sum.

7. Evaluate: $\displaystyle\sum_{i=1}^{\infty} (9)(0.2)^{i-1}$

8. Evaluate: 7!

9. Evaluate $\dbinom{10}{4}$. Do not use Pascal's triangle.

10. Find the coefficient of the fifth term of the binomial expansion $(a + b)^8$. Do not use Pascal's triangle.

11. Use the binomial theorem and Pascal's triangle to write the binomial expansion of $(5x + y)^6$.

12. The function $f(t) = \$34(1.12)^t$ is a model of the average price per share of a stock in dollars, $f(t)$, as printed in the annual report. The number of years since 2005 is t. The annual report is printed once a year.
   a. Rewrite this function as a geometric sequence with a domain of $\{1, 2, 3, \ldots\}$. Round to the nearest hundredth.
   b. Assuming that the growth rate stays the same, find the price per share of stock in 2012. Round to the nearest hundredth.

13. The picture shows a desk toy made of steel marbles. The marbles can be stacked into different designs. A professor avoiding her work creates a stack with a base that is 14 marbles wide. Each row above the base has two fewer marbles. If the stack has six rows, use a partial sum to find the number of marbles in the stack.

Photo provided by Scratchgravel

Follow your instructor's guidelines for showing your work.

1. Use roster notation to represent the set of integers.

2. Simplify: $\dfrac{12x^7y}{30x^4y^9}$

3. Simplify: $(4a^2b)^3(5ab)$

4. Solve: $\dfrac{3}{4}(6x - 2) + 2 = \dfrac{1}{2}(8x + 6)$

5. Solve: $-9(2x + 1) < -10x + 63$

6. Use interval notation to represent $x < 5$.

7. Use interval notation to represent $x \geq -4$.

8. Evaluate: $(6 \times 10^{-5})(2 \times 10^{-11})$

9. Use unit analysis to change $\dfrac{45 \text{ mi}}{1 \text{ hr}}$ into feet per second.

For exercises 10–11, use the five steps.

10. In 2011, the Idaho Lottery had a network of 1150 retailers and presented a dividend of \$41.5 million to the state of Idaho. In Fiscal Year 2012, the Idaho Lottery sold \$175.8 million in products. This was a 19.5% increase over sales of products in 2011. Find the sale of products in 2011. Round to the nearest tenth of a million. (*Source:* www.idaholottery.com, July 24, 2012)

11. The tuition and fees per year for a full-time student at a certain college are \$5562. The cost of books is about \$1600. A student lives and eats most meals at his family home and estimates that he will need about \$3000 per year for transportation and other expenses. His job pays \$7.75 per hour (after taxes and other withholding). Find the number of hours he will have to work per year to pay for his tuition, fees, books, transportation and other expenses. Round to the nearest whole number.

12. Identify the slope of a horizontal line that passes through $(5, -2)$.

13. Graph: $5x + 7y = 35$

14. Find the x-intercept of the line represented by $8x - 5y = 16$.

15. Write the equation of a vertical line that passes through $(8, 5)$.

16. Write the equation in slope-intercept form of the line that passes through $(-3, -7)$ and $(-8, 3)$.

17. Graph: $y < \dfrac{3}{8}x + 3$

18. Graph: $x \leq 2$

19. Simplify: $(9x - 2)(4x^2 - 5x - 3)$

For exercises 20–25, factor completely. Identify any prime polynomials.

20. $8x^2 + 10x - 3$

21. $10ax + 15x - 8a - 12$

22. $100n^2 + 25p^2$

23. $3x^2 - 48$

24. $4u^2 + 28u + 20$

25. $8x^3 - y^3$

26. Solve: $x^3 - 2x^2 - 24x = 0$

27. The area of a rectangle is 99 cm². The length is 7 cm less than two times the width. Use a quadratic equation in one variable to find the length and the width.

28. Solve: $|3x + 9| - 12 = 15$

29. Solve: $|a + 14| = |2a - 8|$

30. Describe the difference between the conjunction of two sets and the disjunction of two sets.

31. Solve $-8 < 2x + 10 \leq 24$. Use interval notation to represent the solution.

32. Solve $|2x + 10| \geq 28$. Use interval notation to represent the solution.

33. Explain why not every relation is a function.

34. A linear function is $y = 8x - 1$. Use interval notation to represent the domain and range of this function.

35. A constant function is $y = 8$. Use interval notation to represent the domain and range of this function.

36. The vertex of the graph of a quadratic function is $(-6, 1)$. The graph opens up. Use interval notation to represent the domain and range of this function.

37. The graph represents an absolute value function. Use interval notation to represent the domain and range of this function.

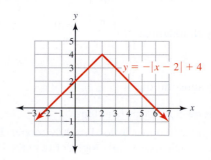

$y = -|x - 2| + 4$

38. Use an algebraic method to find any real zeros of $f(x) = x^2 + 4x - 45$.

39. Write the equation of the function represented by the graph, and identify the number of real zeros of this function.

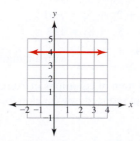

40. The graph of an absolute value function has a vertex of $(-4, 3)$ and opens down. Identify the number of real zeros of this function.

41. If $f(x) = x^2 - x + 3$, evaluate $f(-5)$.

42. A chewing gum wrapper can be folded and connected to other wrappers to make a paper chain. A person inherited a chain consisting of 20,550 wrappers and plans to add 20 wrappers to it per day. Write a linear function that represents the relationship of the total wrappers in the chain, y, and the number of days. Include the units of measurement.

43. The function $y = \left(\dfrac{75 \text{ bottles}}{1 \text{ hr}}\right)x + 825$ bottles represents the relationship of the total number of bottles in a storage area, y, after x hours of production. Find the number of bottles in the storage area after 120 hr of production.

44. A function is $y - 6 = 3(x + 4)$. Change this function to shift its graph horizontally to the right 9 units and vertically down 8 units.

45. Describe the difference between the graph of $f(x) = |x - 2| + 3$ and the graph of $g(x) = |x - 4| - 9$.

46. Use the graph to evaluate $f(2)$.

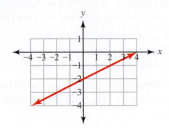

47. Solve by elimination: $\begin{array}{l} 3x + 7y = -10 \\ 5x - 6y = 54 \end{array}$

48. Solve by substitution: $\begin{array}{l} y = \dfrac{3}{4}x + 5 \\ 3x - 2y = 2 \end{array}$

For exercises 49–50, use the five steps.

49. Alloy A is 85% copper. Alloy B is 90% copper. Find the amount of each alloy to mix together to make 2500 kg of a new alloy that is 88% copper.

50. An organic farm has 540 tomato plants. There are three times as many heirloom tomato plants as modern hybrid tomato plants. Find the number of each kind of tomato plant.

51. Solve: $\begin{array}{l} 9x + 7y - 3z = 157 \\ -4x + 2y + 5z = -31 \\ 5x + 8y = 112 \end{array}$

52. Represent $\begin{array}{l} 4x + 5y = 22 \\ 2x - y = 4 \end{array}$ with an augmented matrix.

53. Solve the system of equations in exercise 52 by rewriting the matrix in reduced row echelon form. Check the solution.

54. Graph: $\begin{array}{l} y \ge \dfrac{3}{2}x - 3 \\ y \le -x + 7 \\ x \ge 0 \end{array}$

55. Use the graph in exercise 54 to estimate the vertices of the solution region.

56. Simplify: $\dfrac{x^2 + 15x + 54}{x^2 - 9} \div \dfrac{x^2 + 13x + 42}{x^2 + 10x + 21}$

57. Simplify: $\dfrac{2x - 1}{x^2 + x - 6} - \dfrac{x + 2}{x^2 + 5x + 6}$

58. Simplify: $\dfrac{\dfrac{x}{x - 2} - \dfrac{x}{x + 2}}{\dfrac{2x}{x - 2} + \dfrac{x^2}{x + 2}}$

59. Solve: $\dfrac{x}{x - 5} = \dfrac{3x}{x^2 - 7x + 10} + \dfrac{8}{x - 2}$

For exercise 60, use the five steps.

60. At a copy center, Copier A can complete a print job in 49 min. Copier B can complete the same print job in 57 min. Find the time to complete the job using both copiers. Round to the nearest whole number.

61. Use interval notation to represent the domain of $f(x) = \dfrac{x - 4}{2x^2 - 7x - 4}$.

62. Write the equation of any vertical asymptotes on the graph of $y = \dfrac{3}{x^2 - 9x - 22}$.

63. Use interval notation to represent the domain of $g(x) = \dfrac{6}{x^2 + 1}$.

64. Use an algebraic method to find any real zeros of $f(x) = \dfrac{x + 4}{2x - 5}$.

65. Explain why $y = \left(\dfrac{65 \text{ mi}}{1 \text{ hr}}\right)x$ is an example of a direct variation.

For exercise 66, use the five steps.

66. The relationship of the volume of a gold bar, x, and its mass, y, is a direct variation. A bar of gold with a volume of 5 cm³ has a mass of 96.5 g. Find the mass of a bar of gold with a volume of 13 cm³.

67. Solve $\dfrac{1}{a} + \dfrac{1}{b} = \dfrac{1}{c}$ for b.

68. Use long division to find the quotient: $\dfrac{x^2 + 7x + 12}{x + 3}$

69. Use synthetic division to find the quotient: $\dfrac{x^2 + 7x + 12}{x + 3}$

70. Solve: $\dfrac{2}{x^2 + 8x + 15} > 0$. Use interval notation to represent the solution.

For exercises 70–81, assume that factors that include variables represent nonnegative numbers.

71. Simplify: $\sqrt{24} + \sqrt{54}$

72. Simplify: $\sqrt[3]{24x^5y^6}$

73. Simplify: $\sqrt{35}\sqrt{30}$

74. Simplify: $\sqrt{\dfrac{32y}{45x}}$

75. Simplify: $\dfrac{\sqrt{5x}}{\sqrt{x}+9}$

76. Simplify: $\left(\dfrac{x^{\frac{2}{3}}}{x^{\frac{4}{5}}}\right)^{\frac{1}{2}}$

77. Solve: $a + 3 = \sqrt{15a - 5}$

78. Solve: $\sqrt{x + 8} + 9 = 2$

79. Use interval notation to represent the domain of $f(x) = \sqrt{-2x + 10}$.

80. Use an algebraic method to find any real zeros of $f(x) = \sqrt{-2x + 10}$.

81. For a cone with a volume of 32 cm³, the relationship of the height in centimeters, h, and the radius in centimeters, $f(h)$, is represented by the function $f(h) = \sqrt{\dfrac{96 \text{ cm}^3}{\pi h}}$. Using $\pi \approx 3.14$, evaluate $f(4 \text{ cm})$. Round to the nearest tenth.

82. Copy the table below. Put an X in the box if the number is an element of the set at the top of the column.

83. Solve: $x^2 + 32 = 0$

84. An Olympic swimming pool is 50 m long and 25 m wide. Find the diagonal distance between opposite corners. Round to the nearest tenth.

85. Use completing the square to solve $w^2 + 18w - 5 = 0$.

86. Use the quadratic formula to solve $3x^2 + 7x + 9 = 0$.

87. Explain how to use the discriminant to determine if the solutions of a quadratic equation in one variable are nonreal, rational, or irrational numbers.

88. Rewrite $f(x) = x^2 + 8x - 5$ in vertex form.

89. Identify the vertex of the function in exercise 88 and use interval notation to represent the domain and range of this function.

90. Use the graph of the function to estimate any real zeros of the function.

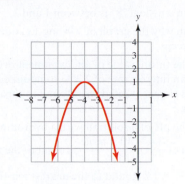

91. Solve $x^2 + 8x - 20 \le 0$. Use interval notation to represent the solution.

92. If $f(x) = \dfrac{3}{4}x - 6$, find $f^{-1}(x)$.

93. If $f(x) = -5x + 3$ and $g(x) = 9x - 2$, find $g(f(x))$.

94. If $f(x) = \log_5(x)$, identify $f^{-1}(x)$.

95. Use interval notation to represent the domain and range of $y = 3^x$.

96. Is the function $f(x) = 0.4^x$ increasing or decreasing?

97. Explain why the horizontal line test can be used to determine whether a function is one-to-one.

98. Identify the y-intercept of the function $y = 30e^x$.

99. Which graph, A or B, represents $y = 2^x$?

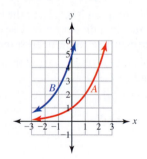

Table for exercise 82

	Complex numbers	Imaginary numbers	Real numbers	Rational numbers	Irrational numbers	Integers	Whole numbers
$\sqrt{7}$							
$\sqrt{-7}$							
$3 + 5i$							
$4i$							
-8							
$-\dfrac{8}{9}$							
0.3							
12							

100. Use interval notation to represent the domain and range of $y = \log_6(x)$.

101. Explain why $\log(78)$ is between 1 and 2.

102. Explain why the graph of $y = \log_2(x)$ does not have a y-intercept.

103. Find the future value of an investment of $2000 that earns an interest rate of 2.3%, compounded daily, for 5 years (1 year = 365 days). Round to the nearest hundredth.

104. Find the pH of a solution with a hydronium ion concentration of $\dfrac{3.4 \times 10^{-6} \text{ mole}}{1 \text{ L}}$. Round to the nearest tenth.

105. Solve $3^x = 18$. Round to the nearest ten-thousandth.

106. The function $y = 35{,}000e^{0.05x}$ represents the relationship of the number of years since 2012, x, and the population of a city. Find the population of the city in 2020. Round to the nearest hundred.

107. The radioactive decay of an isotope is represented by the function $y = Ae^{-0.025t}$, where t is the time in years and y is the amount of remaining isotope. Find the half-life of this isotope. Round to the nearest whole number.

108. Solve $\log_5(x) = 3.4$. Round to the nearest ten-thousandth.

109. Rewrite $3\log(x) - \log(y) + 7\log(z)$ as a single logarithm.

110. Rewrite $\log\left(\dfrac{a^2}{bc^3}\right)$ as a sum or difference of logarithms.

111. Solve: $\log(x) + \log(4) = 2$

112. The pH of a solution is 4.8. Find the hydronium ion concentration in moles per liter. Use scientific notation. Round the mantissa to the nearest tenth.

113. Use the distance formula to find the distance between $(8, -2)$ and $(-7, 10)$.

114. The equation of a circle is $x^2 + y^2 - 12x + 4y - 9 = 0$. Rewrite this equation in standard form.

115. Identify the center and the radius of the circle in exercise 114.

116. The equation of an ellipse is $x^2 + 16y^2 - 6x - 64y + 57 = 0$. Rewrite this equation in standard form.

117. Identify the equation of the major axis, the equation of the minor axis, and the center of the ellipse in exercise 116.

118. The vertex of a parabola is $(6, -2)$. The axis of symmetry is $y = -2$. The graph includes the point $(18, 10)$. Write the equation of this parabola.

119. The foci of a hyperbola are $(0, 8)$ and $(0, -8)$. One vertex of the hyperbola is $(0, 2)$. Write the equation of this hyperbola.

120. Solve: $\begin{array}{l} 3x - y = -2 \\ 2x^2 - y = 0 \end{array}$

121. Solve: $\begin{array}{l} x^2 + y^2 = 20 \\ x^2 = -8y \end{array}$

122. Evaluate: $\displaystyle\sum_{i=3}^{7} i^2$

123. The common difference of an arithmetic sequence is 2. The first term is 6. Write the formula for the nth term of this sequence.

124. Describe the difference between an arithmetic sequence and a geometric sequence.

125. The 10th term in a geometric sequence is 1536. The common ratio is 2. Find the 11th term in the sequence.

126. Find the coefficient of the fifth term in the binomial expansion of $(a + b)^8$.

127. Write the first four rows of Pascal's triangle.

APPENDIX 1

Reasonability and Problem Solving

To find the amount of an antibiotic solution to inject into the muscle of a patient, a nursing student wrote and solved the equation $A = \left(\dfrac{500}{1.8}\right)(250 \text{ mL})$. She looked at her answer of 69,444 mL and thought, "That can't be right!" Injecting 69,444 mL (about 35 large bottles of soda) into someone's muscle is not reasonable. *Looking back* and judging the reasonability of a proposed answer make up an essential part of solving application problems.

Reasonability

Part of looking back is deciding whether an answer is reasonable; a reasonable answer makes sense and seems possible. We can use our experience to tell us that an answer such as a car traveling a distance of 600 mi in a time of 5 min is *not* reasonable. But how do we convince someone else that our answer *is* reasonable?

In Example 1, we use *estimation* and *working backwards* to explain why the answer seems reasonable. This and the following examples do not give explanations of why the original thinking and/or the equations used to solve the problems are correct. Instead, they give explanations of why the proposed answers are reasonable solutions to the problems.

EXAMPLE 1   **Problem:** The suggested retail price of a Ford Focus sedan is $17,505. The dealer will sell the car for $16,499. Find the change in price.

Equation: $c = \$17{,}505 - \$16{,}499$

Answer: The change in price is $1006.

Explain why this answer is reasonable. To *estimate*, round the numbers in the problem so the arithmetic can be redone quickly, usually without a calculator. Rounding, $17,505 is about $17,500, and $16,499 is about $16,500. Since the difference between $17,500 and $16,500 is $1000 and $1000 is very close to the answer of $1006, the answer seems reasonable.

Working backwards from the answer to the original numbers, the sum of the change in price and the sale price should be the original price. Since $1006 + $16,499 equals $17,505, a change in price of $1006 seems reasonable.

In some situations, *estimation* is not a good method for checking. In the next example, we show that the answer seems reasonable by *working backwards* and by *doing the problem another way.*

EXAMPLE 2   **Problem:** At a warehouse store, a student can buy hamburger meat in 10-lb packages. How many 0.25-lb hamburgers can be made from this package?

Equation: $p = 10 \text{ lb} \div 0.25 \text{ lb}$

Answer: The student can make 40 hamburgers from this package.

Explain why this answer is reasonable. *Estimation* does not work well here. If we round 0.25 lb to 0.2 lb, the arithmetic is not any easier. If we round it to 0 lb, it does not work at all.

Working backwards, since $(0.25 \text{ lb})(40) = 10 \text{ lb}$, the answer of 40 hamburgers seems reasonable.

Doing the problem another way, if each hamburger is 0.25 lb (a quarter pounder), then there are four hamburgers in each pound. Since $\left(\dfrac{4 \text{ hamburgers}}{1 \text{ lb}}\right)(10 \text{ lb})$ equals 40 hamburgers, the answer seems reasonable.

In baseball, a ball in play can be anywhere *in the ballpark:* in the outfield, the infield, or in the catcher's mitt. A reasonable answer is also *in the ballpark.* We are checking whether the answer is within an acceptable or expected range of answers.

EXAMPLE 3   **Problem:** The price of a textbook is $114. The sales tax rate is 6.5%. Find the total cost to buy the textbook.

Equation: $p = \$114 + (0.065)(\$114)$

Answer: The total cost is $121.41.

Explain why this answer seems reasonable. Since the original price of the book, without tax, is $114, the bottom end of the *ballpark* is $114.

The total cost of the book is $114 plus a tax of 6.5%. Since 6.5% is less than 10%, the total cost of the book must be less than $114 plus a tax of 10%. (Why choose 10%? To find 10% of a number, just divide it by 10.) Since 10% of $114 is $11.40. and $114 + \$11.40 = \$125.40$, the top end of the *ballpark* is $125.40.

Since the answer of $121.41 is between $114 and $125.40, it is *in the ballpark*, and it seems reasonable.

For work with formulas, changing the value of just one of the variables in the formula can establish an expected range of reasonable answers.

EXAMPLE 4   **Problem:** A rectangular room is 6 ft wide and 12 ft long. Find its area.

Formula and Equation: area $= (\text{width})(\text{length})$; $A = (6 \text{ ft})(12 \text{ ft})$

Answer: The area of the room is 72 ft^2.

Explain why this answer seems reasonable. To show that the answer is *in the ballpark*, choose a value for the length that is less than 12 ft for the bottom of the ballpark, 10 ft. Choose a value for the length that is greater than 12 ft for the top of the ballpark, 15 ft. Do not change the width.

If the length is 10 ft, the area of the room is 60 ft^2. If the length is 15 ft, the area is 90 ft^2. Since 72 ft^2 is between 60 ft^2 and 90 ft^2, the answer seems reasonable.

To *do the problem another way,* divide the rectangle into twelve strips that are each 6 ft wide and 1 ft long. The area of each strip is $(6 \text{ ft})(1 \text{ ft})$, which equals 6 ft^2, and the total area is $(12)(6 \text{ ft}^2)$, which equals 72 ft^2. Since this is the same as the original answer, it seems reasonable.

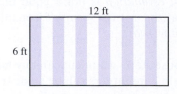

When a proportion is used to solve a problem, replace the variable with the solution and write each fraction as a decimal number. If the decimal numbers are very close to equal (any difference caused by rounding), the two ratios are the same, and the answer therefore seems reasonable.

EXAMPLE 5

Problem: In August 2009, officials expected that as many as two out of five Oregonians would have swine flu in the fall and winter flu season. If the population was about 3,800,000, find the number of Oregonians that were expected to come down with swine flu. (*Source:* lmtribune.com, Aug. 22, 2009)

Equation: $\dfrac{x}{3,800,000} = \dfrac{2}{5}$

Answer: In the fall and winter flu season, 1,520,000 Oregonians were expected to have swine flu.

Explain why this answer is reasonable. Replace x with the answer. Since $\dfrac{1,520,000}{3,800,000} = 0.4$ and $\dfrac{2}{5} = 0.4$, the answer of 1,520,000 people seems reasonable.

When *looking back* at an answer and deciding whether it is reasonable, first think about whether the answer makes sense. Use your experience and knowledge, and evaluate whether it is a possible answer to the problem situation. Then use one of the approaches in this section to explain why the solution is reasonable.

Practice Problems

For problems 1–6, explain why the answer seems reasonable.

1. **Problem:** Find the change in the population of New York.

   Since 1960, New York has lost 7.3 million residents to the rest of the country. This was partially offset by an influx of 4.8 million foreign immigrants. (*Source:* www.empirecenter.org, Aug. 2011)

   Equation: $c = 4.8 \text{ million} - 7.3 \text{ million}$

   Answer: The population of New York has decreased by 2.5 million people.

2. **Problem:** A student works 13 hours per week at a job that pays $8.25 per hour and 8 hours per week at a job that pays $10.15 an hour. Find the total amount earned at both jobs.

   Equation: $A = (13 \text{ hr})\left(\dfrac{\$8.25}{1 \text{ hr}}\right) + (8 \text{ hr})\left(\dfrac{\$10.15}{1 \text{ hr}}\right)$

   Answer: The total amount earned at both jobs per week is $188.45.

3. **Problem:** Find the number of people enrolled in community colleges who earn a degree or certificate within eight years. Round to the nearest tenth of a million.

   Nationally, 11.7 million people are enrolled in community colleges, but only about 39 percent of students earn a degree or certificate within eight years. (*Source:* www.nytimes.com, Aug. 14, 2009)

   Equation: $p = (0.39)(11.7 \text{ million people})$.

   Answer: About 4.6 million community college students earn a degree or certificate within eight years.

4. **Problem:** A room is 14 ft long, 12 ft wide, and 10 ft high. Find the volume of this room.

   Formula and equation: volume $= (\text{length})(\text{width})(\text{height})$; $V = (14 \text{ ft})(12 \text{ ft})(10 \text{ ft})$

   Answer: The volume of the room is 1680 ft^3.

5. **Problem:** Four out of five personal computers in the world run on Intel microchips. Find the number out of 16,000 PCs that run on Intel microchips. (*Source:* www.nytimes.com, Oct. 27, 2009)

   Equation: $\dfrac{4}{5} = \dfrac{x}{16,000}$

   Answer: Of the 16,000 computers, 12,800 of them run on Intel microchips.

6. **Problem:** The price of a ticket to the Free to Be Music and Arts Festival in Los Angeles is marked down 85% to $10. Find the original price of the ticket. Round to the nearest whole number. (*Source:* www.latimes.com, Aug. 2, 2011)

   Equation: $x - 0.85x = \$10$

   Answer: The original price of the ticket was about $67.

Appendix 1   Exercises

For exercises 1–24, explain why the answer seems reasonable.

1. **Problem:** At the University of North Carolina–Chapel Hill, the in-state tuition is $3865 per year. Out-of-state students pay $21,753 per year. Find the difference for 4 years of tuition for an in-state student and an out-of-state student. (*Source:* www.newsobserver.com, Aug. 25, 2009)

   Equation: $x = 4(\$21,753) - 4(\$3865)$

   Answer: The difference in tuition is $71,552.

2. **Problem:** Students at Florida International University had 557 people yo-yoing at the same time in a 2-minute period, a new world record. The previous record was 493 people. Find the difference in the number of people. (*Source:* www.miamiherald.com, Aug. 25, 2009)

   Equation: $n = 557$ people $- 493$ people

   Answer: The difference in the number of people yo-yoing is 64 people.

3. **Problem:** The scoreboard at TCF Bank Stadium at the University of Minnesota is a rectangle that is 48 ft high by 108 ft wide. Find the area of the scoreboard. (*Source:* www.twincities.com, Aug. 16, 2009)

   Equation: $A = (48\text{ ft})(108\text{ ft})$

   Answer: The area of the scoreboard is 5184 ft².

4. **Problem:** The scoreboard at Yankee Stadium is 59 ft high and 101 ft wide. Find the perimeter of the scoreboard. (*Source:* www.signindustry.com)

   Equation: $P = 2(59\text{ ft}) + 2(101\text{ ft})$

   Answer: The perimeter of the scoreboard is 320 ft.

5. **Problem:** The city of Richardson, Texas, pays monthly car allowances to employees. The cost per year for 141 employees is $842,000. Find the average monthly car allowance per employee. Round to the nearest whole number. (*Source:* www.dallasnews.com, Aug. 26, 2009)

   Equation: $A = \dfrac{\$842,000 \div 12}{141\text{ employees}}$

   Answer: The average monthly car allowance is about $498.

6. **Problem:** A light rail line in Seattle is 14 mi long, extending from Westlake Center to Tukwila. The cost of construction was $2.3 billion. Find the average cost per mile. Round to the nearest *million* dollars. (*Source:* seattletimes.nwsource.com, July 26, 2009)

   Equation: $C = \dfrac{\$2.3\text{ billion}}{14\text{ mi}}$

   Answer: The average cost per mile is about $164 million.

7. **Problem:** The regular price of a computer is $999. It is on sale at 15% off. Find the sale price.

   Equation: $P = \$999 - (0.15)(\$999)$

   Answer: The sale price is $849.15.

8. **Problem:** The number of Washington students completing the Free Application for Student Aid (FAFSA) for the 2009–2010 school year rose 20% compared to the 2008–2009 school year when 201,500 students completed the financial aid forms. Find the number of students who completed the FAFSA for 2009–2010.

   Equation: $N = 201,500$ students $+$ $(0.20)(201,500\text{ students})$

   Answer: For 2009–2010, 241,800 students completed the FAFSA.

9. **Problem:** A costumer has 25 yd of fabric. She needs $1\frac{1}{4}$ yd of this fabric for each costume. Find the number of costumes she can make from this fabric.

   Equation: $N = 25\text{ yd} \div 1\frac{1}{4}\text{ yd}$

   Answer: The costumer can make 20 costumes from this fabric.

10. **Problem:** A recipe requires $\frac{3}{4}$ cup of flour. A canister contains 9 cups of flour. Find how many recipes can be made with this flour.

    Equation: $N = 9\text{ cups} \div \frac{3}{4}\text{ cup}$

    Answer: Twelve recipes can be made with this flour.

11. **Problem:** The cost of a DVD is $35.99. Shipping is $1.99 per order plus $0.99 per item. Find the total cost (without tax) for an order of 12 DVDs.

    Equation: $C = (12 \text{ DVD})\left(\dfrac{\$35.99}{1 \text{ DVD}}\right) + \$1.99 +$
    $(12 \text{ items})\left(\dfrac{\$0.99}{1 \text{ item}}\right)$

    Answer: The total cost for an order of 12 DVDs is $445.75.

12. **Problem:** A floor seat ticket to see *Smackdown and ECW Live* at the Canton Civic Center costs $60. The convenience charge per ticket is $9.70, and the building facility charge per ticket is $1. Find the total cost for five tickets.

    Equation: $C = (5 \text{ tickets})\left(\dfrac{\$60}{1 \text{ ticket}}\right) +$
    $(5 \text{ tickets})\left(\dfrac{\$9.70}{1 \text{ ticket}}\right) + (5 \text{ tickets})\left(\dfrac{\$1}{1 \text{ ticket}}\right)$

    Answer: The total cost for five tickets is $353.50.

13. **Problem:** From 2006 to 2008, speed was a factor in 58% of crashes in Ohio caused by juvenile drivers. These drivers caused 61,784 traffic crashes. Find the number of these crashes in which speed was a factor. Round to the nearest ten. (*Source:* www.cleveland.com, Aug. 26, 2009)

    Equation: $N = (0.58)(61{,}784)$

    Answer: Speed was a factor in about 35,830 crashes caused by juvenile drivers.

14. **Problem:** Find the number of people in the U.S. population predicted for 2030. Round to the nearest tenth of a million.

    By 2030, Census Bureau data show, the 72 million people expected to be ages 65 and older will represent 19 percent of the U.S. population—up from 13 percent in 2010. (*Source:* bls.gov, Spring 2011)

    Equation: $0.19N = 72 \text{ million people}$

    Answer: The population is predicted to be about 378.9 million people.

15. **Problem:** South Salt Lake and Salt Lake City agreed to each pay $2.5 million of the $46 million cost to build two miles of streetcar line. The federal government will pay $35 million. Find the percent that the two cities are paying of the total cost of the line. Round to the nearest percent. (*Source:* www.sltrib.com, Aug. 27, 2009)

    Equation: $P = \dfrac{2(\$2.5 \text{ million})}{\$46 \text{ million}} \cdot 100\%$

    Answer: South Salt Lake and Salt Lake City will pay about 11% of the total cost of the line.

16. **Problem:** On August 2, 2011, the closing price of Harley-Davidson stock was $42.57 per share. This price was 22.8% higher than it was on January 2, 2011. Find the price on January 2, 2011. Round to the nearest hundredth. (*Source:* investor.harley-davidson.com)

    Equation: $P + 0.228P = \$42.57$

    Answer: The price on January 2, 2011 was about $34.67.

17. **Problem:** According to the Centers for Disease Control and Prevention, one out of five children in the United States is obese or overweight. Use a proportion to predict how many of 12,500 children are obese or overweight. (*Source:* www.cdc.gov)

    Equation: $\dfrac{1}{5} = \dfrac{n}{12{,}500}$

    Answer: Of 12,500 children, 2500 are overweight.

18. **Problem:** According to the National Retail Federation, four out of five Americans say that the recession is affecting their back-to-school spending and college plans. Use a proportion to predict how many of 130,000 Americans say that the recession is affecting their back-to-school spending and college plans. (*Source:* www.nrf.com, July 14, 2009)

    Equation: $\dfrac{4}{5} = \dfrac{N}{130{,}000}$

    Answer: Of 130,000 Americans, about 104,000 say that the recession is affecting their back-to-school spending and college plans.

19. **Problem:** The Spokane Valley City Council raised the motel tax by 50 cents to $2 a night. Find the percent increase in the motel tax per night. Round to the nearest percent. (*Source:* seattletimes.nwsource.com, Aug. 27, 2009)

    Equation: $P = \dfrac{\$0.50}{\$1.50} \cdot 100\%$

    Answer: The percent increase in the motel tax per night is about 33%.

20. **Problem:** The average electricity bill for a home in Austin was $88 in April 2009. During a record heat wave in July 2009, it was $235. Find the percent increase. Round to the nearest percent. (*Source:* www.statesman.com, Aug. 25, 2009)

    Equation: $P = \dfrac{\$235 - \$88}{\$88} \cdot 100\%$

    Answer: The electricity bills increased by about 167%.

21. **Problem:** As the H1N1 flu vaccine was produced in August 2009, the National Center for Immunization and Respiratory Diseases was concerned that supplies might not meet the demand. If rationing was necessary, it recommended that pregnant women, persons who live with or provide care for infants less than 6 months of age, health-care and emergency medical personnel, children between 6 months and 4 years, and children and adolescents ages 5–18 years with certain medical conditions should have priority. These groups included about 159 million people. The population of the United States in August 2009 was about 307 million people. Find the percent that the population of these groups is of the entire population of the United States Round to the nearest percent. (*Source:* www.cdc.gov, Aug. 28, 2009)

    Equation: $P = \dfrac{159 \text{ million people}}{307 \text{ million people}} \cdot 100\%$

    Answer: These groups are about 52% of the population of the United States.

22. **Problem:** On average, each year in California, there will be approximately 46,100 job openings in occupations that require a degree in science, technology, engineering, or mathematics. About 24,000 of the jobs will require at least a bachelor's degree. Find the percent of the jobs that will require at least a bachelor's degree. Round to the nearest percent. (*Source:* Institute for Higher Education Leadership & Policy, www.csus.edu/ihe, June 2009)

    Equation: $P = \dfrac{24{,}000 \text{ jobs}}{46{,}100 \text{ jobs}} \cdot 100\%$

    Answer: About 52% of the jobs will require at least a bachelor's degree.

23. **Problem:** Find the operating budget before the reductions. Round to the nearest tenth of a million.

    [The Texas Education Agency] is reducing its operating budget by $48 million, or 36 percent . . . layoffs will account for the majority of the savings. (*Source:* www.statesman.com, July 12, 2011).

    Equation: $0.36B = \$48$ million

    Answer: The operating budget before the reductions was about $133.3 million.

24. **Problem:** An Austin Energy spokesperson said that about $14.1 million in electricity bills due in July had not been paid. This was 7.8% of the total amount of bills. Find the total amount of bills. Round to the nearest tenth of a million. (*Source:* www.statesman.com, Aug. 25, 2009)

    Equation: $0.078B = \$14.1$ million

    Answer: The total amount of bills was about $180.8 million.

APPENDIX 2

Determinants and Cramer's Rule

Determinants

The **determinant** of a square matrix is a number associated with the matrix. For the 2×2 matrix $A = \begin{bmatrix} a & b \\ c & d \end{bmatrix}$, the determinant of A is $ad - bc$. There are several notations for determinants. We can write $\det(A)$ or $|A|$. In this section, we will use the notation $\det(A)$ to represent the determinant of the matrix A.

EXAMPLE 1 A matrix A is $\begin{bmatrix} 1 & 2 \\ 3 & 4 \end{bmatrix}$. Find $\det(A)$.

$\det(A)$ The determinant of A.

$= 1(4) - 2(3)$ $ad - bc$

$= 4 - 6$ Evaluate.

$= -2$ Evaluate.

To develop rules for finding determinants of larger square matrices, we use variables with subscripts. The two numbers in the subscript identify the position of the variable by row and column. For example, the variable in the first row first column of a matrix can be named a_{11}. A 2×2 matrix can be represented as $A = \begin{bmatrix} a_{11} & a_{12} \\ a_{21} & a_{22} \end{bmatrix}$.

EXAMPLE 2 A matrix A is $\begin{bmatrix} a_{11} & a_{12} \\ a_{21} & a_{22} \end{bmatrix}$. Evaluate $\det(A)$.

$\det(A)$ The determinant of A.

$= a_{11}(a_{22}) - a_{12}(a_{21})$ If $A = \begin{bmatrix} a & b \\ c & d \end{bmatrix}$, $\det(A) = ad - bc$.

We use similar notation to describe a 3×3 matrix, $B = \begin{bmatrix} b_{11} & b_{12} & b_{13} \\ b_{21} & b_{22} & b_{23} \\ b_{31} & b_{32} & b_{33} \end{bmatrix}$.

The determinant of B is

$$\det(B) = b_{11} \cdot \det\begin{bmatrix} b_{22} & b_{23} \\ b_{32} & b_{33} \end{bmatrix} - b_{12} \cdot \det\begin{bmatrix} b_{21} & b_{23} \\ b_{31} & b_{33} \end{bmatrix} + b_{13} \cdot \det\begin{bmatrix} b_{21} & b_{22} \\ b_{31} & b_{32} \end{bmatrix}$$

EXAMPLE 3   A matrix B is $\begin{bmatrix} 1 & 2 & 3 \\ -1 & 4 & 4 \\ 0 & 5 & 3 \end{bmatrix}$. Find $\det(B)$.

$$B = \begin{bmatrix} b_{11} & b_{12} & b_{13} \\ b_{21} & b_{22} & b_{23} \\ b_{31} & b_{32} & b_{33} \end{bmatrix}$$

$$\det(B) = b_{11} \cdot \det\begin{bmatrix} b_{22} & b_{23} \\ b_{32} & b_{33} \end{bmatrix} - b_{12} \cdot \det\begin{bmatrix} b_{21} & b_{23} \\ b_{31} & b_{33} \end{bmatrix} + b_{13} \cdot \det\begin{bmatrix} b_{21} & b_{22} \\ b_{31} & b_{32} \end{bmatrix}$$

$$\det(B) = 1 \cdot \det\begin{bmatrix} 4 & 4 \\ 5 & 3 \end{bmatrix} - 2 \cdot \det\begin{bmatrix} -1 & 4 \\ 0 & 3 \end{bmatrix} + 3 \cdot \det\begin{bmatrix} -1 & 4 \\ 0 & 5 \end{bmatrix}$$

$$\det(B) = 1(4 \cdot 3 - 4 \cdot 5) - 2(-1 \cdot 3 - 4 \cdot 0) + 3(-1 \cdot 5 - 4 \cdot 0)$$

$$\det(B) = 1(-8) - 2(-3) + 3(-5)$$

$$\det(B) = -8 + 6 - 15$$

$$\det(B) = -17$$

If we evaluate the determinants in

$$\det(B) = b_{11} \cdot \det\begin{bmatrix} b_{22} & b_{23} \\ b_{32} & b_{33} \end{bmatrix} - b_{12} \cdot \det\begin{bmatrix} b_{21} & b_{23} \\ b_{31} & b_{33} \end{bmatrix} + b_{13} \cdot \det\begin{bmatrix} b_{21} & b_{22} \\ b_{31} & b_{32} \end{bmatrix}$$

then $\det(B) = b_{11}b_{22}b_{33} + b_{12}b_{23}b_{31} + b_{13}b_{21}b_{32} - b_{13}b_{22}b_{31} - b_{11}b_{23}b_{32} - b_{12}b_{21}b_{33}$.

Rather than memorizing this, we can use a visual representation of the process. Write the first row of the matrix followed by the first two entries of first row. Write the second row of the matrix followed by the first two entries of the second row. Write the third row of the matrix followed by the first two entries of the third row.

$$\begin{matrix} b_{11} & b_{12} & b_{13} & b_{11} & b_{12} \\ b_{21} & b_{22} & b_{23} & b_{21} & b_{22} \\ b_{31} & b_{32} & b_{33} & b_{31} & b_{32} \end{matrix}$$

Draw in three diagonals from left to right as shown. The product of the entries in each diagonal are the first three products in $\det(B) = b_{11}b_{22}b_{33} + b_{12}b_{23}b_{31} + b_{13}b_{21}b_{32} - b_{13}b_{22}b_{31} - b_{11}b_{23}b_{32} - b_{12}b_{21}b_{33}$.

$$\begin{matrix} b_{11} & b_{12} & b_{13} & b_{11} & b_{12} \\ b_{21} & b_{22} & b_{23} & b_{21} & b_{22} \\ b_{31} & b_{32} & b_{33} & b_{31} & b_{32} \end{matrix}$$

Now draw in three diagonals from right to left as shown. The products of the entries in each diagonal are the last three products in $\det(B) = b_{11}b_{22}b_{33} + b_{12}b_{23}b_{31} + b_{13}b_{21}b_{32} - b_{13}b_{22}b_{31} - b_{11}b_{23}b_{32} - b_{12}b_{21}b_{33}$.

$$\begin{matrix} b_{11} & b_{12} & b_{13} & b_{11} & b_{12} \\ b_{21} & b_{22} & b_{23} & b_{21} & b_{22} \\ b_{31} & b_{32} & b_{33} & b_{31} & b_{32} \end{matrix}$$

EXAMPLE 4 A matrix B is $\begin{bmatrix} 1 & 2 & 3 \\ 6 & 5 & 1 \\ 1 & 8 & 4 \end{bmatrix}$. Find $\det(B)$.

$$\det(B) = b_{11}b_{22}b_{33} + b_{12}b_{23}b_{31} + b_{13}b_{21}b_{32} - b_{13}b_{22}b_{31} - b_{11}b_{23}b_{32} - b_{12}b_{21}b_{33}$$

$$
\begin{array}{ccccc}
1 & 2 & 3 & 1 & 2 \\
6 & 5 & 1 & 6 & 5 \\
1 & 8 & 4 & 1 & 8
\end{array}
\qquad
\begin{array}{ccccc}
1 & 2 & 3 & 1 & 2 \\
6 & 5 & 1 & 6 & 5 \\
1 & 8 & 4 & 1 & 8
\end{array}
$$

$$\det(B) = 1(5)(4) + (2)(1)(1) + (3)(6)(8) - (3)(5)(1) - (1)(1)(8) - (2)(6)(4)$$

$$\det(B) = 20 + 2 + 144 - 15 - 8 - 48$$

$$\det(B) = 95$$

Practice Problems

For problems 1–5, find the determinant of the matrix.

1. $A = \begin{bmatrix} 7 & 3 \\ 2 & 6 \end{bmatrix}$

2. $A = \begin{bmatrix} 8 & 3 \\ -2 & -6 \end{bmatrix}$

3. $B = \begin{bmatrix} 2 & 3 & 9 \\ 5 & 8 & 2 \\ 1 & 2 & 6 \end{bmatrix}$

4. $B = \begin{bmatrix} 4 & 1 & 2 \\ -6 & 5 & 7 \\ 1 & 0 & 1 \end{bmatrix}$

5. $B = \begin{bmatrix} 1 & 2 & 3 \\ 4 & 5 & 6 \\ 7 & 8 & 9 \end{bmatrix}$

Cramer's Rule

Cramer's rule is a method for finding the solution of a system of linear equations in which the system has the same number of equations as unknowns and it has one unique solution. It is named after Gabriel Cramer, a Swiss mathematician, who published this rule in 1750. Cramer's rule only involves one division for every variable; in a time without electronic calculators, minimizing division meant that a process took less time.

Cramer's rule includes determinants and uses similar notation. For a system of two linear equations in two variables, the variables are named x_1 and x_2; the coefficients are a_{11}, a_{12}, a_{21}, and a_{22}; and the constants are b_1 and b_2.

Cramer's Rule for a System of Two Linear Equations in Two Variables

For a system of linear equations with a single solution, $\begin{array}{l} a_{11}x_1 + a_{12}x_2 = b_1 \\ a_{21}x_1 + a_{22}x_2 = b_2 \end{array}$,

$$x_1 = \frac{\det\begin{bmatrix} b_1 & a_{12} \\ b_2 & a_{22} \end{bmatrix}}{\det\begin{bmatrix} a_{11} & a_{12} \\ a_{21} & a_{22} \end{bmatrix}} \quad \text{and} \quad x_2 = \frac{\det\begin{bmatrix} a_{11} & b_1 \\ a_{21} & b_2 \end{bmatrix}}{\det\begin{bmatrix} a_{11} & a_{12} \\ a_{21} & a_{22} \end{bmatrix}}$$

where a_{11}, a_{12}, a_{21}, a_{22}, b_1, and b_2 are real numbers.

EXAMPLE 5 Use Cramer's rule to solve $\begin{aligned} 3x_1 + 4x_2 &= 2 \\ 2x_1 - x_2 &= 5 \end{aligned}$.

(a) Find x_1.

$$x_1 = \frac{\det\begin{bmatrix} 2 & 4 \\ 5 & -1 \end{bmatrix}}{\det\begin{bmatrix} 3 & 4 \\ 2 & -1 \end{bmatrix}} \qquad x_1 = \frac{\det\begin{bmatrix} b_1 & a_{12} \\ b_2 & a_{22} \end{bmatrix}}{\det\begin{bmatrix} a_{11} & a_{12} \\ a_{21} & a_{22} \end{bmatrix}}$$

$$x_1 = \frac{2(-1) - 4(5)}{3(-1) - 4(2)} \qquad \text{If } A = \begin{bmatrix} a & b \\ c & d \end{bmatrix}, \det(A) = ad - bc.$$

$$x_1 = \frac{-22}{-11} \qquad \text{Simplify.}$$

$$x_1 = 2 \qquad \text{Simplify.}$$

(b) Find x_2.

$$x_2 = \frac{\det\begin{bmatrix} 3 & 2 \\ 2 & 5 \end{bmatrix}}{\det\begin{bmatrix} 3 & 4 \\ 2 & -1 \end{bmatrix}} \qquad x_2 = \frac{\det\begin{bmatrix} a_{11} & b_1 \\ a_{21} & b_2 \end{bmatrix}}{\det\begin{bmatrix} a_{11} & a_{12} \\ a_{21} & a_{22} \end{bmatrix}}$$

$$x_2 = \frac{3(5) - 2(2)}{3(-1) - 4(2)} \qquad \text{If } A = \begin{bmatrix} a & b \\ c & d \end{bmatrix}, \det(A) = ad - bc.$$

$$x_2 = \frac{11}{-11} \qquad \text{Simplify.}$$

$$x_2 = -1 \qquad \text{Simplify.}$$

(c) Check.

$3x_1 + 4x_2 = 2$	Use the original equation.	$2x_1 - x_2 = 5$
$3(2) + 4(-1) = 2$	Replace x_1 and x_2.	$2(2) - (-1) = 5$
$6 - 4 = 2$	Simplify.	$4 + 1 = 5$
$2 = 2$ True.	Simplify.	$5 = 5$ True.

Practice Problems

For problems 6–7,
(a) use Cramer's rule to find x_1.
(b) use Cramer's rule to find x_2.
(c) check.

6. $\begin{aligned} 7x_1 + 2x_2 &= 29 \\ 5x_1 + 3x_2 &= 27 \end{aligned}$ **7.** $\begin{aligned} 3x_1 + 4x_2 &= 18 \\ -9x_1 - x_2 &= 12 \end{aligned}$

A System of *n* Linear Equations with *n* Unknowns

We can use Cramer's rule to solve a system of three linear equations with three variables if the system has one unique solution.

Cramer's Rule for a System of Three Linear Equations in Three Variables

$$a_{11}x_1 + a_{12}x_2 + a_{13}x_3 = b_1$$

For a system of linear equations with a single solution, $a_{21}x_1 + a_{22}x_2 + a_{23}x_3 = b_2$

$$a_{31}x_1 + a_{32}x_2 + a_{33}x_3 = b_3$$

$$M_1 = \begin{bmatrix} b_1 & a_{12} & a_{13} \\ b_2 & a_{22} & a_{23} \\ b_3 & a_{32} & a_{33} \end{bmatrix}, \quad M_2 = \begin{bmatrix} a_{11} & b_1 & a_{13} \\ a_{21} & b_2 & a_{23} \\ a_{31} & b_3 & a_{33} \end{bmatrix},$$

$$M_3 = \begin{bmatrix} a_{11} & a_{12} & b_1 \\ a_{21} & a_{22} & b_2 \\ a_{31} & a_{32} & b_3 \end{bmatrix}, \quad A = \begin{bmatrix} a_{11} & a_{12} & a_{13} \\ a_{21} & a_{22} & a_{23} \\ a_{31} & a_{32} & a_{33} \end{bmatrix},$$

$$x_1 = \frac{\det(M_1)}{\det(A)}, \quad x_2 = \frac{\det(M_2)}{\det(A)}, \quad \text{and} \quad x_3 = \frac{\det(M_3)}{\det(A)}$$

$$5x_1 + 2x_2 + x_3 = 2$$

EXAMPLE 6   Use Cramer's rule to solve $4x_1 + x_2 + x_3 = -2.$

$$2x_1 + x_2 = 2$$

(a) Find x_1.

To find x_1, write the matrix M_1, evaluate its determinant, $\det(M_1)$, write the

coefficients matrix A, and evaluate its determinant, $\det(A)$. Then $x_1 = \dfrac{\det(M_1)}{\det(A)}$.

$$M_1 = \begin{bmatrix} 2 & 2 & 1 \\ -2 & 1 & 1 \\ 2 & 1 & 0 \end{bmatrix} \qquad M_1 = \begin{bmatrix} b_1 & a_{12} & a_{13} \\ b_2 & a_{22} & a_{23} \\ b_3 & a_{32} & a_{33} \end{bmatrix}$$

$\det(M_1)$

$= (2)(1)(0) + (2)(1)(2) + (1)(-2)(1)$

$\quad - (2)(-2)(0) - (2)(1)(1) - (1)(1)(2)$

$= 0 + 4 - 2 - 0 - 2 - 2$

$= -2$

$$A = \begin{bmatrix} 5 & 2 & 1 \\ 4 & 1 & 1 \\ 2 & 1 & 0 \end{bmatrix} \qquad A = \begin{bmatrix} a_{11} & a_{12} & a_{13} \\ a_{21} & a_{22} & a_{23} \\ a_{31} & a_{32} & a_{33} \end{bmatrix}$$

$\det(A)$

$= (5)(1)(0) + (2)(1)(2) + (1)(4)(1)$

$\quad - (2)(4)(0) - (5)(1)(1) - (1)(1)(2)$

$= 0 + 4 + 4 - 0 - 5 - 2$

$= 1$

$$x_1 = \frac{\det(M_1)}{\det(A)}$$

$$x_1 = \frac{-2}{1}$$

$$x_1 = -2$$

(b) Find x_2.

To find x_2, write the matrix M_2, and evaluate its determinant $\det(M_2)$. We know $\det(A)$. Then $x_2 = \dfrac{\det(M_2)}{\det(A)}$.

$$M_2 = \begin{bmatrix} 5 & 2 & 1 \\ 4 & -2 & 1 \\ 2 & 2 & 0 \end{bmatrix} \qquad M_2 = \begin{bmatrix} a_{11} & b_1 & a_{13} \\ a_{21} & b_2 & a_{23} \\ a_{31} & b_3 & a_{33} \end{bmatrix}$$

$\det(M_2)$

$= (5)(-2)(0) + (2)(1)(2) + (1)(4)(2)$

$\quad - (2)(4)(0) - (5)(1)(2) - (1)(-2)(2)$

$= 0 + 4 + 8 - 0 - 10 + 4$

$= 6$

$x_2 = \dfrac{\det(M_2)}{\det(A)}$

$x_2 = \dfrac{6}{1}$

$x_2 = 6$

(c) Find x_3.

To find x_3, write the matrix M_3 and evaluate its determinant $\det(M_3)$. We know $\det(A)$. Then $x_3 = \dfrac{\det(M_3)}{\det(A)}$.

$$M_3 = \begin{bmatrix} 5 & 2 & 2 \\ 4 & 1 & -2 \\ 2 & 1 & 2 \end{bmatrix} \qquad M_3 = \begin{bmatrix} a_{11} & a_{12} & b_1 \\ a_{21} & a_{22} & b_2 \\ a_{31} & a_{32} & b_3 \end{bmatrix}$$

$\det(M_3)$

$= (5)(1)(2) + (2)(-2)(2) + (2)(4)(1)$

$\quad - (2)(4)(2) - (5)(-2)(1) - (2)(1)(2)$

$= 10 - 8 + 8 - 16 + 10 - 4$

$= 0$

$x_3 = \dfrac{\det(M_3)}{\det(A)}$

$x_3 = \dfrac{0}{1}$

$x_3 = 0$

(d) Check: $x_1 = -2$, $x_2 = 6$, $x_3 = 0$

$5x_1 + 2x_2 + x_3 = 2$	$4x_1 + x_2 + 1x_3 = -2$	$2x_1 + x_2 = 2$
$5(-2) + 2(6) + 0 = 2$	$4(-2) + 6 + 1(0) = -2$	$2(-2) + 6 = 2$
$-10 + 12 = 2$	$-8 + 6 + 0 = -2$	$-4 + 6 = 2$
$2 = 2$ True.	$-2 = -2$ True.	$2 = 2$ True.

Cramer's rule can be extended to solve n equations with n variables. These problems are part of a linear algebra course.

Practice Problems

For problem 8,
(a) use Cramer's rule to find x_1.
(b) use Cramer's rule to find x_2.
(c) use Cramer's rule to find x_3.
(d) check.

8. $2x_1 + x_2 + x_3 = 2$
   $x_1 + 2x_2 - x_3 = -5$
   $-x_1 + x_2 + x_3 = -1$

Appendix 2 Exercises

For exercises 1–8, find the determinant of the matrix.

1. $A = \begin{bmatrix} 4 & 5 \\ 3 & 2 \end{bmatrix}$

2. $A = \begin{bmatrix} 4 & -5 \\ 3 & 2 \end{bmatrix}$

3. $A = \begin{bmatrix} 7 & 2 \\ 1 & 0 \end{bmatrix}$

4. $A = \begin{bmatrix} -4 & -2 \\ -6 & -3 \end{bmatrix}$

5. $B = \begin{bmatrix} 2 & 1 & 3 \\ 1 & 1 & 0 \\ 1 & 2 & -1 \end{bmatrix}$

6. $B = \begin{bmatrix} -2 & 1 & -1 \\ 0 & 1 & 2 \\ -3 & 1 & 1 \end{bmatrix}$

7. $B = \begin{bmatrix} 1 & 1 & 1 \\ 2 & 2 & 2 \\ -1 & -1 & -1 \end{bmatrix}$

8. $B = \begin{bmatrix} 1 & 0 & 3 \\ 2 & 1 & -1 \\ 2 & 2 & 1 \end{bmatrix}$

For exercises 9–16,
(a) use Cramer's rule to find x_1.
(b) use Cramer's rule to find x_2.
(c) check.

9. $x_1 + x_2 = 3$
   $2x_1 + x_2 = 4$

10. $x_1 + x_2 = 1$
    $2x_1 - x_2 = -4$

11. $4x_1 + x_2 = 5$
    $2x_1 - 2x_2 = -10$

12. $-1x_1 + 3x_2 = 18$
    $2x_1 - 2x_2 = -12$

13. $-3x_1 + 5x_2 = 21$
    $x_1 + 4x_2 = 10$

14. $2x_1 + 3x_2 = 9$
    $4x_1 + 5x_2 = 11$

15. $3x_1 + 7x_2 = 15$
    $2x_1 + 5x_2 = 11$

16. $3x_1 + 7x_2 = 5$
    $2x_1 + 5x_2 = 4$

For exercises 17–25,
(a) use Cramer's rule to find x_1.
(b) use Cramer's rule to find x_2.
(c) use Cramer's rule to find x_3.
(d) check.

17. $x_1 + 2x_2 + 3x_3 = 5$
    $x_1 + 3x_2 + x_3 = 7$
    $2x_1 + x_2 + 4x_3 = 4$

18. $3x_1 - x_2 + x_3 = 4$
    $x_1 + x_2 - 2x_3 = -2$
    $x_1 + 2x_2 - x_3 = 1$

19. $x_1 + 2x_2 - x_3 = 4$
    $-x_1 - x_2 - 2x_3 = -1$
    $2x_1 + x_2 + x_3 = 5$

20. $x_1 + x_2 + 4x_3 = 1$
    $3x_1 + 2x_2 + x_3 = 4$
    $-2x_1 - x_2 + x_3 = -3$

21. $-x_1 + x_2 - x_3 = -5$
    $4x_1 + x_2 + x_3 = 1$
    $x_1 + 2x_2 + 2x_3 = 2$

22. $2x_1 + x_2 - x_3 = 4$
    $x_1 + 3x_2 + 2x_3 = 2$
    $-x_1 + 2x_2 + 2x_3 = -1$

23. $x_1 + x_2 + x_3 = 6$
    $x_1 + x_2 - x_3 = 2$
    $-x_1 + x_2 + x_3 = 4$

24. $x_1 + x_2 - x_3 = 2$
    $2x_1 - x_2 + x_3 = 1$
    $x_1 - 2x_2 + 3x_3 = 2$

25. $2x_1 - x_2 + 3x_3 = -3$
    $-x_1 - x_2 + 3x_3 = -6$
    $x_1 - 2x_2 - x_3 = -2$

26. A system of equations is $\begin{array}{l} -x_1 + x_2 = 2 \\ 2x_1 - 2x_2 = -4 \end{array}$ and

$$x_1 = \frac{\det\begin{bmatrix} b_1 & a_{12} \\ b_2 & a_{22} \end{bmatrix}}{\det\begin{bmatrix} a_{11} & a_{12} \\ a_{21} & a_{22} \end{bmatrix}}.$$

a. Evaluate: $\det\begin{bmatrix} a_{11} & a_{12} \\ a_{21} & a_{22} \end{bmatrix}$

b. Explain why Cramer's rule cannot be used to solve this system of equations.

c. Use a different method to solve this system.

27. A system of equations is $\begin{array}{l} 2x_1 + x_2 - x_3 = -3 \\ x_1 + 3x_2 + x_3 = 4 \\ -x_1 + 2x_2 + 2x_3 = 7 \end{array}$ and

$$A = \begin{bmatrix} a_{11} & a_{12} & a_{13} \\ a_{21} & a_{22} & a_{23} \\ a_{31} & a_{32} & a_{33} \end{bmatrix}.$$

a. Evaluate $\det(A)$.

b. Explain why Cramer's rule cannot be used to solve this system of equations.

c. Use a different method to solve this system.

APPENDIX 3

Developing the Equations of Conic Sections

Developing the Equation of an Ellipse

EXAMPLE 1 Write the standard equation of an ellipse with center $(0, 0)$, foci $(-p, 0)$ and $(p, 0)$, major axis vertices $(-a, 0)$ and $(a, 0)$, and minor axis vertices $(0, -b)$ and $(0, b)$.
  The sum of the distances from any point on the ellipse to the foci is a constant that equals $2a$.

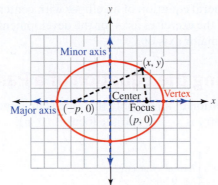

$$\text{sum of distances} = 2a$$

$$\sqrt{(x - (-p))^2 + (y - 0)^2} + \sqrt{(x - p)^2 + (y - 0)^2} = 2a$$

$$\sqrt{(x + p)^2 + y^2} - \sqrt{(x - p)^2 + y^2} + \sqrt{(x - p)^2 + y^2} = 2a - \sqrt{(x - p)^2 + y^2}$$

$$\sqrt{(x + p)^2 + y^2} = 2a - \sqrt{(x - p)^2 + y^2}$$

$$\left(\sqrt{(x + p)^2 + y^2}\right)^2 = \left(2a - \sqrt{(x - p)^2 + y^2}\right)^2$$

$$(x + p)^2 + y^2 = \left(2a - \sqrt{(x - p)^2 + y^2}\right)\left(2a - \sqrt{(x - p)^2 + y^2}\right)$$

$$(x + p)^2 + y^2 = 4a^2 - 2a\sqrt{(x - p)^2 + y^2} - 2a\sqrt{(x - p)^2 + y^2} + (x - p)^2 + y^2$$

$$(x + p)^2 + y^2 = 4a^2 - 4a\sqrt{(x - p)^2 + y^2} + (x - p)^2 + y^2$$

$$x^2 + 2px + p^2 + y^2 = 4a^2 - 4a\sqrt{(x - p)^2 + y^2} + x^2 - 2px + p^2 + y^2$$

$$x^2 - x^2 + 2px + 2px + p^2 - p^2 + y^2 - y^2 = 4a^2 - 4a\sqrt{(x - p)^2 + y^2} + x^2 - x^2 - 2px + 2px + p^2 - p^2 + y^2 - y^2$$

$$4px = 4a^2 - 4a\sqrt{(x - p)^2 + y^2}$$

$$\frac{4px}{4} = \frac{4a^2}{4} - \frac{4a\sqrt{(x - p)^2 + y^2}}{4}$$

$$px = a^2 - a\sqrt{(x - p)^2 + y^2}$$

$$px - px = a^2 - a\sqrt{(x - p)^2 + y^2} - px$$

$$0 = a^2 - a\sqrt{(x - p)^2 + y^2} - px$$

$$0 + a\sqrt{(x - p)^2 + y^2} = a^2 - a\sqrt{(x - p)^2 + y^2} + a\sqrt{(x - p)^2 + y^2} - px$$

$$a\sqrt{(x - p)^2 + y^2} = a^2 - px$$

$$\left(a\sqrt{(x - p)^2 + y^2}\right)^2 = (a^2 - px)^2$$

$$a^2((x - p)^2 + y^2) = (a^2 - px)(a^2 - px)$$

$$a^2((x - p)^2 + y^2) = a^4 - a^2px - a^2px + p^2x^2$$

$$a^2(x^2 - 2px + p^2 + y^2) = a^4 - 2a^2px + p^2x^2$$

$$a^2x^2 - 2a^2px + a^2p^2 + a^2y^2 = a^4 - 2a^2px + p^2x^2$$

$$a^2x^2 - 2a^2px + 2a^2px + a^2p^2 + a^2y^2 = a^4 - 2a^2px + 2a^2px + p^2x^2$$
$$a^2x^2 + a^2p^2 + a^2y^2 = a^4 + p^2x^2$$
$$a^2x^2 + a^2p^2 - a^2p^2 + a^2y^2 - p^2x^2 = a^4 + p^2x^2 - p^2x^2 - a^2p^2$$
$$a^2x^2 - p^2x^2 + a^2y^2 = a^4 - a^2p^2$$
$$x^2(a^2 - p^2) + a^2y^2 = a^2(a^2 - p^2)$$

From the definition of an ellipse, $b^2 = a^2 - p^2$.

$$x^2b^2 + a^2y^2 = a^2b^2$$
$$\frac{x^2b^2}{a^2b^2} + \frac{a^2y^2}{a^2b^2} = \frac{a^2b^2}{a^2b^2}$$
$$\frac{x^2}{a^2} + \frac{y^2}{b^2} = 1$$

This is the general equation of an ellipse with center at $(0, 0)$ and a major axis of $y = 0$. One of the exercises asks for the development of a similar equation for an ellipse with a major axis of $x = 0$.

Developing the Equation of a Parabola

EXAMPLE 2   Write the equation of a parabola that opens up in which the vertex is $(0, 0)$, the axis of symmetry is $x = 0$, and the focus is $(0, p)$.

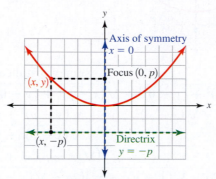

A point on the parabola is $(0, 0)$. The vertical distance from $(0, 0)$ to the focus $(0, p)$ is p. The horizontal directrix is the same distance from $(0, 0)$ in the opposite direction. The equation of the directrix is $y = -p$.

Another point on the parabola is (x, y). The point of intersection of a vertical line that passes through (x, y) and the directrix is $(x, -p)$.

distance from (x, y) to a point on the directrix = distance from (x, y) to focus

distance from (x, y) to $(x, -p)$ = distance from (x, y) to $(0, p)$

$$\sqrt{(x - x)^2 + (-p - y)^2} = \sqrt{(0 - x)^2 + (p - y)^2}$$
$$\left(\sqrt{(x - x)^2 + (-p - y)^2}\right)^2 = \left(\sqrt{(0 - x)^2 + (p - y)^2}\right)^2$$
$$(x - x)^2 + (-p - y)^2 = (0 - x)^2 + (p - y)^2$$
$$0^2 + (-p - y)(-p - y) = (-x)^2 + (p - y)(p - y)$$
$$p^2 + py + py + y^2 = x^2 + p^2 - py - py + y^2$$
$$p^2 + 2py + y^2 = x^2 + p^2 - 2py + y^2$$
$$p^2 - p^2 + 2py + 2py + y^2 - y^2 = x^2 + p^2 - p^2 - 2py + 2py + y^2 - y^2$$
$$4py = x^2$$
$$x^2 = 4py$$

Changing the direction in which the graph opens changes the equation of the parabola.

EXAMPLE 3 Write the equation of a parabola that opens to the right in which the vertex is $(0, 0)$, the axis of symmetry is $y = 0$, and the focus is $(p, 0)$.

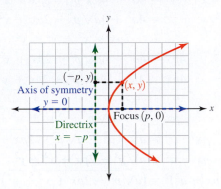

distance from (x, y) to directrix = distance from (x, y) to focus

distance from (x, y) to $(-p, y)$ = distance from (x, y) to $(p, 0)$

$$\sqrt{(-p - x)^2 + (y - y)^2} = \sqrt{(p - x)^2 + (0 - y)^2}$$

$$(\sqrt{(-p - x)^2 + (y - y)^2})^2 = (\sqrt{(p - x)^2 + (0 - y)^2})^2$$

$$(-p - x)^2 + (y - y)^2 = (p - x)^2 + (0 - y)^2$$

$$(-p - x)(-p - x) + 0^2 = (p - x)(p - x) + (-y)^2$$

$$p^2 + 2px + x^2 = p^2 - 2px + x^2 + y^2$$

$$p^2 - p^2 + 2px + 2px + x^2 - x^2 = p^2 - p^2 - 2px + 2px + x^2 - x^2 + y^2$$

$$4px = y^2$$

$$y^2 = 4px$$

Developing the Equation of a Hyperbola

EXAMPLE 4 Write the equation of a hyperbola in which the center is $(0, 0)$, the vertices are $(-a, 0)$ and $(a, 0)$, the foci are $(-p, 0)$ and $(p, 0)$, and the transverse axis is $y = 0$.

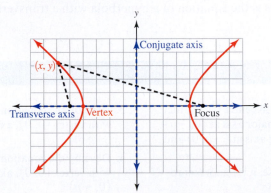

A hyperbola includes all the points (x, y) in a plane such that the absolute value of the difference of the distances from two fixed points (the foci) is a constant.

distance from (x, y) to $(-p, 0)$ − distance from (x, y) to $(p, 0)$ = $2a$

$$\sqrt{(-p - x)^2 + (0 - y)^2} - \sqrt{(p - x)^2 + (0 - y)^2} = 2a$$

$$\sqrt{p^2 + px + px + x^2 + (0 - y)^2} - \sqrt{p^2 - px - px + x^2 + (0 - y)^2} = 2a$$

$$\sqrt{p^2 + 2px + x^2 + y^2} - \sqrt{p^2 - 2px + x^2 + y^2} = 2a$$

$$\sqrt{p^2 + 2px + x^2 + y^2} - \sqrt{p^2 - 2px + x^2 + y^2} + \sqrt{p^2 - 2px + x^2 + y^2} = 2a + \sqrt{p^2 - 2px + x^2 + y^2}$$

$$\sqrt{p^2 + 2px + x^2 + y^2} = 2a + \sqrt{p^2 - 2px + x^2 + y^2}$$

$$(\sqrt{p^2 + 2px + x^2 + y^2})^2 = (2a + \sqrt{p^2 - 2px + x^2 + y^2})^2$$

$$\left(\sqrt{p^2 + 2px + x^2 + y^2}\right)^2 = \left(2a + \sqrt{p^2 - 2px + x^2 + y^2}\right)\left(2a + \sqrt{p^2 - 2px + x^2 + y^2}\right)$$

$$p^2 + 2px + x^2 + y^2 = 4a^2 + 4a\sqrt{p^2 - 2px + x^2 + y^2} + p^2 - 2px + x^2 + y^2$$

$$p^2 - p^2 + 2px + 2px + x^2 - x^2 + y^2 - y^2 = 4a^2 + 4a\sqrt{p^2 - 2px + x^2 + y^2} + p^2 - p^2 - 2px$$
$$+ 2px + x^2 - x^2 + y^2 - y^2$$

$$4px = 4a^2 + 4a\sqrt{p^2 - 2px + x^2 + y^2}$$

$$\frac{4px}{4} = \frac{4a^2}{4} + \frac{4a\sqrt{p^2 - 2px + x^2 + y^2}}{4}$$

$$px = a^2 + a\sqrt{p^2 - 2px + x^2 + y^2}$$

$$px - a^2 = a^2 - a^2 + a\sqrt{p^2 - 2px + x^2 + y^2}$$

$$px - a^2 = a\sqrt{p^2 - 2px + x^2 + y^2}$$

$$(px - a^2)^2 = \left(a\sqrt{p^2 - 2px + x^2 + y^2}\right)^2$$

$$(px - a^2)(px - a^2) = \left(a\sqrt{p^2 - 2px + x^2 + y^2}\right)\left(a\sqrt{p^2 - 2px + x^2 + y^2}\right)$$

$$p^2x^2 - a^2px - a^2px + a^4 = a^2(p^2 - 2px + x^2 + y^2)$$

$$p^2x^2 - 2a^2px + a^4 = a^2p^2 - 2a^2px + a^2x^2 + a^2y^2$$

$$p^2x^2 - a^2x^2 - 2a^2px + 2a^2px - a^2y^2 + a^4 - a^4 = a^2p^2 - 2a^2px + 2a^2px + a^2x^2 - a^2x^2 + a^2y^2 - a^2y^2 - a^4$$

$$p^2x^2 - a^2x^2 - a^2y^2 = a^2p^2 - a^4$$

$$x^2(p^2 - a^2) - a^2y^2 = a^2(p^2 - a^2)$$

$$\frac{x^2(p^2 - a^2)}{a^2(p^2 - a^2)} - \frac{a^2y^2}{a^2(p^2 - a^2)} = \frac{a^2(p^2 - a^2)}{a^2(p^2 - a^2)}$$

$$\frac{x^2}{a^2} - \frac{y^2}{p^2 - a^2} = 1$$

Now let $b^2 = p^2 - a^2$. This results in an equation that looks more like the equation of an ellipse or circle.

$$\frac{x^2}{a^2} - \frac{y^2}{b^2} = 1$$

This is the equation of a hyperbola with a transverse axis of $y = 0$.

Appendix 3 Exercises

1. Develop the equation of an ellipse with center $(0, 0)$, foci $(0, -p)$ and $(0, p)$, vertices on the major axis at $(0, -a)$ and $(0, a)$, and vertices on the minor axis at $(-b, 0)$ and $(b, 0)$.

2. In developing the equation of an ellipse, it is important to know that $b^2 = a^2 - p^2$. Explain why.

3. Develop the equation of a parabola that opens to the left with vertex $(0, 0)$, axis of symmetry $y = 0$, and focus $(-p, 0)$.

4. Develop the equation of a parabola that opens down with vertex $(0, 0)$, axis of symmetry $x = 0$, and focus $(0, -p)$.

Answers to Practice Problems

CHAPTER 1

Section 1.1

1. a.

b.

2. $\{-4, -1, 0, 2, 3\}$ **3. a.** prime   **b.** not prime
c. not prime   **d.** not prime   **e.** not prime   **f.** prime
4. a. no   **b.** yes   **5.** no   **6.** ABDE   **7.** AB   **8.** ABD
9. AB   **10.** AC   **11.** AB   **12.** AB   **13.** AB
14. a.   **b.** $(-\infty, 5]$

15. a.   **b.** $(2, \infty)$

16. a.   **b.** $[4, 10)$

17. 49   **18.** 49   **19.** 343   **20.** -343 **21.** 7   **22.** 7
23. not a real number   **24.** 3.9   **25.** -4 **26.** -63
27. -66 **28.** 343

Section 1.2

1. a. distributive property   **b.** commutative property of multiplication   **c.** distributive property   **d.** associative property of multiplication   **e.** commutative property of addition
2. $38x - 47y$ **3.** $-2x + 23y$ **4.** $-7z^2 + 30z - 15$
5. $10xy - 0.81x$ **6.** 0   **7.** 3   **8.** undefined

9. 0   **10.** -3 **11.** undefined   **12.** $\dfrac{4}{45}x + \dfrac{14}{45}$

13. $-\dfrac{47}{12}c + 5$ **14.** $-\dfrac{8}{9}h - 18$ **15.** $16x^2 + 3x$

16. $25x^{10}$ **17.** $8y^{15}$ **18.** $36a^{10}$ **19.** p^6z^4

20. $12z^3 + 4z$ **21.** $\dfrac{1}{x^5}$ **22.** $\dfrac{1}{y^{15}}$ **23.** 1   **24.** $\dfrac{z^4}{p^6}$

25. $\dfrac{1}{f^4}$ **26.** 1   **27.** 1   **28.** 42   **29.** 143   **30.** 50

Section 1.3

1. a. $x = 2$ **2. a.** $y = \dfrac{17}{12}$ **3. a.** $w = -24$

4. a. $n = 781.25$ **5. a.** $x = 70$ **6. a.** $c = 2.5$
7. a. the set of real numbers   **8. a.** no solution
9. a. no solution   **10. a.** the set of real numbers

11. a. $x = \dfrac{1}{6}$ **b.**   **c.** $\left[\dfrac{1}{6}, \dfrac{1}{6}\right]$

12. a. $x = 0$ **b.**   **c.** $[0, 0]$

13. a. the set of real numbers

b.   **c.** $(-\infty, \infty)$

14. a. $x > -2$ **c.**   **d.** $(-2, \infty)$

15. a. $x < -\dfrac{4}{3}$ **c.**   **d.** $\left(-\infty, -\dfrac{4}{3}\right)$

16. a. $x \geq \dfrac{-20}{13}$ **c.**   **d.** $\left[-\dfrac{20}{13}, \infty\right)$

17. a. $c \leq 8$ **c.**   **d.** $(-\infty, 8]$

18. a. $a > -\dfrac{27}{2}$ **c.**   **d.** $\left(-\dfrac{27}{2}, \infty\right)$

19. a. $x \leq \dfrac{33}{10}$ **c.**   **d.** $\left(-\infty, \dfrac{33}{10}\right]$

20. a. $p \geq -\dfrac{41}{14}$ **c.**   **d.** $\left[-\dfrac{41}{14}, \infty\right)$

21. a. $w < -\dfrac{1}{6}$ **c.**   **d.** $\left(-\infty, -\dfrac{1}{6}\right)$

Section 1.4

1. 5.60×10^{-4} meter   **2.** $2.99 \times 10^8 \dfrac{\text{meters}}{\text{second}}$

3. 6.2×10^3 kilograms   **4.** 4.1×10^{-2} liter

5. 2.79×10^{16} kilometer^2 **6.** $8.6 \times 10^{-6} \dfrac{\text{gram}}{\text{milliliter}}$

7. $2.45 \times 10^{-3} \dfrac{\text{kilogram} \cdot \text{meter}}{\text{second}^2}$ **8.** 5.77×10^4 liters

9. 0.007 L   **10.** 9000 g   **11.** 0.000009 m
12. 0.00008 m   **13.** 20,000 N   **14.** 300,000,000 tons
15. 1 in. = 2.54 cm   **16.** 1 L = 1000 mL

17. 1 kg = 1000 g   **18.** $22 \dfrac{\text{ft}}{\text{s}}$ **19.** $0.48 \dfrac{\text{euro}}{\text{L}}$

20. $80 \dfrac{\text{mg}}{\text{mL}}$ **21.** 832 T   **22.** $5280 \dfrac{\text{kg} \cdot \text{m}}{\text{s}}$

23. $232{,}320 \dfrac{\text{kg} \cdot \text{m}^2}{\text{s}^2}$ **24.** $0.0892 \dfrac{1}{\text{cm}^3}$ **25.** 1.28×10^2 s

26. 2.8×10^{16} km^2 **27.** $8.6 \times 10^{-6} \dfrac{\text{g}}{\text{mL}}$

28. $7.35 \times 10^{-1} \dfrac{\text{kg} \cdot \text{m}^2}{\text{s}^2}$ **29.** 5.8×10^4 L

Section 1.5

1. 755 small businesses   **2.** 90%   **3.** 22%   **4.** \$36.56
5. 11,200 people   **6.** 2.45 ERA   **7.** 945,000 tons
8. 1400 ft^2 **9.** 3,384,295 passengers

Section 1.6

1. $7x + y = 9$ **2.** $6x + y = -3$

3.

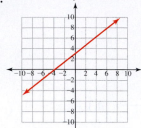

4.

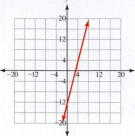

5.

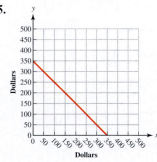

6. -2 **7.** 12 **8.** $\dfrac{475 \text{ mi}}{1 \text{ hr}}$ **9.** 0 **10.** undefined

11. $\dfrac{3}{8}$ **12.** $-\dfrac{4}{5}$ **13.** $\dfrac{1}{2}$ **14.** -3 **15.** $\dfrac{2}{5}$ **16.** -4

17. $\dfrac{6}{5}$ **18. a.** $y = -\dfrac{3}{5}x + 3$ **b.** $-\dfrac{3}{5}$ **c.** $(0, 3)$

d. Answers will vary.   **e.**

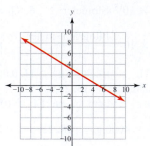

19.

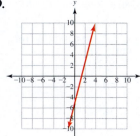

20.

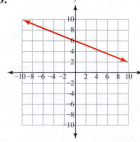

21.

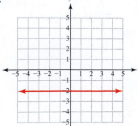

22.

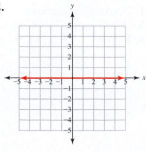

23.

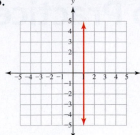

24.

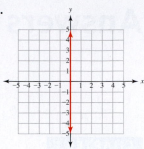

25.

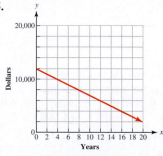

26.

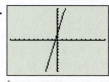

$[-10, 10, 1, -10, 10, 1]$

27.

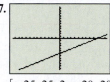

$[-25, 25, 2, -20, 20, 2]$

28.

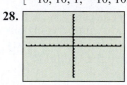

$[-10, 10, 1, -10, 10, 1]$

29.

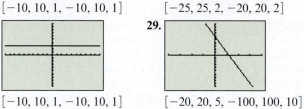

$[-20, 20, 5, -100, 100, 10]$

Section 1.7

1. $y = -8x - 1$ **2.** $y = \left(-\dfrac{\$4700}{1 \text{ year}}\right)x + \$47{,}000$

3. $y = \dfrac{3}{4}x - 1$ **4. a.** $y = -9x - 34$ **b.** $(0, -34)$

5. a. $y = \dfrac{3}{8}x - \dfrac{57}{4}$ **b.** $\left(0, -\dfrac{57}{4}\right)$ **6. a.** $y = -7x + 1$

b. $(0, 1)$ **7.** $y = 9$ **8.** $x = 2$ **9.** $y = -1$

10. $x = 7$ **11.** $y = -3x + 26$ **12.** $y = \dfrac{1}{3}x + \dfrac{23}{3}$

13. $x = 6$ **14.** $y = -4$ **15.** $y = 8$ **16.** $x = 3$

Section 1.8

1.

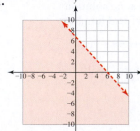

2.

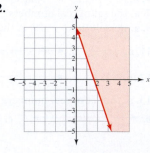

3.

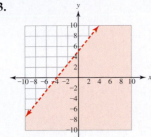

4.

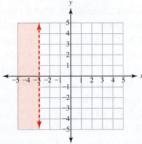

5. **6.**

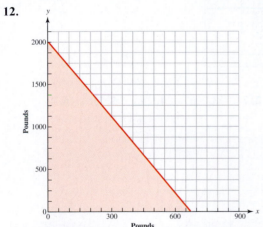

7. $y > -4x + 4$ **8.** $y \leq -\dfrac{7}{5}x - \dfrac{8}{5}$ **9.** $x < -3$

10. a. x = number of eggs, y = cups of milk
b. $6x + 8y \leq 48$
11. a. x = amount spent per month on prescription drugs
b. $x \geq \$50$

12.

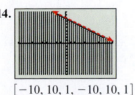

13.

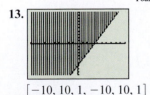

$[-10, 10, 1, -10, 10, 1]$

14.

$[-10, 10, 1, -10, 10, 1]$

CHAPTER 2

Section 2.1

1. a. polynomial   **b.** degree 3   **2. a.** not a polynomial
3. a. not a polynomial   **4. a.** polynomial   **b.** degree 9
5. a. polynomial   **b.** degree 0   **6.** $8x^2 + 10x - 12$
7. $-6z^2 + 6z - 6$ **8.** $-2x^3 - 7x^2 + 6x - 7$
9. $-3h^3 + 2h^2k - 2hk^2 - 8$ **10.** $21k^6$ **11.** $-6y^2 + 16y$
12. $-6y^2 - 5y + 56$ **13.** $5x^3 + 40x^2 - 45x$
14. $18m^3 + 60m^2 - 6m - 8$ **15. a.** $x^2 + 16x + 64$
b. $x^2 + 16x + 64$ **16. a.** $x^2 - 64$ **b.** $x^2 - 64$

17. a. $x^2 - 16x + 64$ **b.** $x^2 - 16x + 64$
18. a. $9p^2 + 30p + 25$ **b.** $9p^2 + 30p + 25$
19. a. $z^3 - 125$ **b.** $z^3 - 125$
20. a. $z^3 + 125$ **b.** $z^3 + 125$
21. a. $W + 6$ **b.** $4W + 12$ **c.**

d. $W^2 + 6W$
22. a. $2L - 25$ **b.** $6L - 50$ **c.**

d. $2L^2 - 25L$

Section 2.2

1. $14xy$ **2.** $14x^2y^2$ **3.** $14x$ **4.** 18 **5.** $3xy$ **6.** 1
7. a. $6b^3(2ab + 3)$ **8. a.** $6ac(ac^4 + 3)$
9. a. $mp^2(14mp + 3p^2 + 11m^2)$ **10. a.** prime, $20xy + 21z$
11. $(5x - 9)(3x + y)$ **12.** $(6k - 5)(2h - 1)$
13. $(a + b)(7c^2 - 2d)$ **14.** $(f + 1)(2w - 1)$
15. $(x + y)(x - y)$ **16.** $(5x + 7y)(5x - 7y)$
17. $(8x - y)^2$ **18.** $(2x - y)(4x^2 + 2xy + y^2)$
19. $(6a + 1)^2$ **20.** $6(x + 5y)^2$ **21.** prime
22. $2(x - 4y)(x^2 + 4xy + 16y^2)$ **23.** $2(x^2 + 6y + 9z)$

Section 2.3

1. $(x + 6)(2x + 5)$ **2.** $(x - 6)(2x - 5)$ **3.** prime
4. $(6x - 5)(2x - 1)$ **5.** $(x + 4)(x + 10)$
6. $(x - 7)(x - 1)$ **7.** $(x - 11)(x + 2)$
8. $(x + 1)(x + 6)$ **9. a.** 65 **b.** prime   **10. a.** -19
b. prime   **11. a.** 4 **b.** not prime   **12. a.** 216
b. prime   **13.** $(x^2 + 3)(x^2 + 5)$
14. $(x^4 - 3)(x^4 - 2)$ **15.** $(p^3 - 5)(p^3 + 9)$
16. $2(x + 7)(4x - 3)$ **17.** $3w(5w^2 + 2w - 1)$
18. $5x(x - 4)(x + 4)$ **19.** $7(x^2 + 4)$
20. $(3x + 1)(4x - 7)$ **21.** $2(a + 5)(2c + d)$
22. $4p^5(p - 6)(p + 6)$ **23.** $2(n - 3)(n + 3)(n + 5)$
24. $w^2(w^2 + 15w - 56)$ **25.** $3(x^2 - 14x - 49)$
26. $u(u + 4)^2$ **27.** $(b - 3)(b + 3)(b^2 + 7)$

Section 2.4

1. $x = -3$ or $x = -4$ **2.** $x = 0$ or $x = -3$ or $x = -4$
3. $x = 0$ or $x = 9$ **4.** $x = 0$ or $x = 5$ or $x = -5$
5. $x = -2$ or $x = 8$ **6.** $x = -4$ or $x = 6$
7. 5 ft long, 3 ft wide   **8.** 16 in. long, 7 in. wide

Section 2.5

1. 4 **2.** 3 **3. a.** $x = -13$ or $x = 13$
4. a. $x = 5$ or $x = 13$ **5. a.** $x = -7$ or $x = 3$
6. a. $a = -11$ or $a = 21$ **7. a.** $x = 7$ **8. a.** no solution
9. a. $x = -12$ or $x = -2$ **10. a.** $a = \dfrac{1}{4}$ or $a = 1$
11. a. $x = -2$ or $x = 10$

Section 2.6

1. $\{3, 5, 7\}$ **2.** $\{3, 4, 5, 6, 7, 9, 11\}$
3. a. conjunction   **b.** Answers will vary.
4. a. disjunction   **b.** Answers will vary.

5. a. 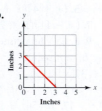 **b.** $(0, 3)$ **c.** $0 < x < 3$

6. a. ![number line] x **b.** $(-\infty, 0) \cup (3, \infty)$

7. a. ![number line] x **b.** $[-6, 2]$

8. a. ![number line] x **b.** $(0, 4)$

9. a. ![number line] x **b.** $[-8, -1]$

10. a. $1 < x \le 5$ **b.** $(1, 5]$

11. a. $-7 \le x < 7$ **b.** $[-7, 7)$

12. a. ![number line] x

b. $(-\infty, -6] \cup [1, \infty)$

13. a. ![number line] x

b. $(-\infty, -4) \cup (-1, \infty)$

14. a. ![number line] x **b.** $(-\infty, 1)$

15. a. $x \ge 0$ **b.** ![number line] x **c.** $[0, \infty)$

16. a. $x < -21$ or $x \ge 11$

b. ![number line] x

c. $(-\infty, -21) \cup [11, \infty)$ **17. a.** $x = -15$ and $x = 9$

b. $-15 < x < 9$ **c.** $(-15, 9)$ **18. a.** $x = -8$ and $x = 16$

b. $x \le -8$ or $x \ge 16$ **c.** $(-\infty, -8] \cup [16, \infty)$

19. a. There are no boundary values. **b.** the set of real numbers **c.** $(-\infty, \infty)$

CHAPTER 3

Section 3.1

1. a. $x + y = 3$ in. **b.**

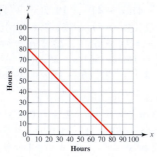

2. a. $x + y = 80$ hr **b.**

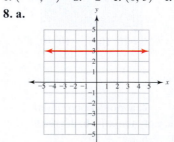

3. a.

Year	Million tons
2008	10.1
2009	9.1
2010	8.2

b.

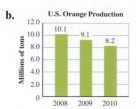

4. a. $x + y \le 200$ ft² **b.**

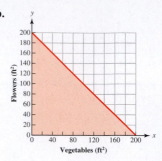

5. both a relation and a function **6.** both a relation and a function **7.** neither a relation nor a function

8. both a relation and a function **9.** a relation; a function only if none of the students have the same birth date

10. a. yes **b.** {Barb, Masoud, Galvez, Charles}

c. {May 4, May 5, May 6}

11. a. yes **b.** {Nov., Dec., Jan., Feb., Mar., Apr.}

c. {112, 94, 73, 54, 117, 89} **12.** not a function

13. function **14.** function **15.** function

16. function **17.** not a function

18.

X	Y₁
5	10
7	30
9	50
11	70
13	90
15	110
17	130

X=5

19.

X	Y₁
0	-40
20	160
40	360
60	560
80	760
100	960
120	1160

X=0

20.

X	Y₁
-20	-240
-15	-190
-10	-140
-5	-90
0	-40
5	10
10	60

X=-20

Section 3.2

1. pizza **2.** 7 **3.** -6 **4.** 23 **5.** 53 **6.** 4

7. a. **b.** $(-\infty, \infty)$

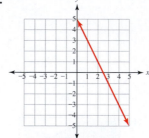

c. $(-\infty, \infty)$ **d.** -2 **e.** $(0, 5)$ **f.** 17

8. a. **b.** $(-\infty, \infty)$ **c.** $[3, 3]$

d. 0 **e.** $(0, 3)$ **f.** 3 **g.** 3 **9. a.** $x = 5$ **b.** $(-\infty, \infty)$

c. $(-\infty, \infty)$ **10.** no real zeros

11. a.

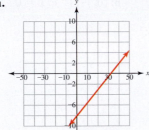

b. $x = 32$

12. a.

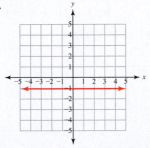

b. no real zeros

13. a. $321.30 **b.** {15 frames, 16 frames, . . . , 24 frames}
c. {$267.75, $285.60, . . . , $428.40}   **14. a.** $39.73
b. [0 HCF, 14 HCF]   **c.** [$0, $50.57]   **15.** $120
16. a. $x = 30$ meals   **b.** Answers will vary.

17. a.

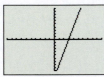

b. $x = 3$

$[-10, 10, 1, -10, 10, 1]$

18. a.

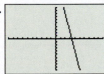

b. $x \approx 3.33$

$[-10, 10, 1, -10, 10, 1]$

19. a.

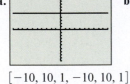

b. no zero

$[-10, 10, 1, -10, 10, 1]$

Section 3.3
1. a. 2   **b.** quadratic   **2. a.** 1   **b.** linear
3. a. 3   **b.** cubic   **4. a.** 0   **b.** constant
5. a. $a = 1, b = -8, c = 17$
b.   **c.** $(4, 1)$ **d.** up

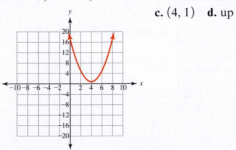

e. minimum 1   **f.** $x = 4$ **g.** $(-\infty, \infty)$ **h.** $[1, \infty)$

6. a. $a = 1, b = -2, c = -2$ **b.**

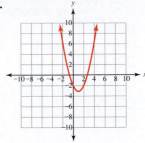

c. $(1, -3)$ **d.** up   **e.** minimum -3 **f.** $x = 1$ **g.** $(-\infty, \infty)$
h. $[-3, \infty)$ **7. a.** $a = -1, b = 2, c = 2$
b.   **c.** $(1, 3)$ **d.** down

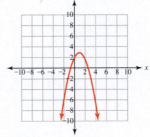

e. maximum 3   **f.** $x = 1$ **g.** $(-\infty, \infty)$ **h.** $(-\infty, 3]$
8. a. $(-\infty, \infty)$ **b.** $[4, \infty)$ **9. a.** $(-\infty, \infty)$ **b.** $(-\infty, 4]$
10. -9 **11.** 135   **12.** $428.9 billion   **13. a.** -4
b. $(-\infty, \infty)$ **c.** $(-\infty, \infty)$ **14.** $x = 3$ and $x = 9$

15. $x = \dfrac{15}{2}$ **16.** $x = 3$ **17.** $x = -1$ and $x = 1$

18. $x = 1$ **19.** $x = -2$ **20.** no real zeros   **21.** $x = 2$
22. $x = 0$

23.

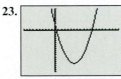

$[-10, 20, 1, -20, 10, 1]$

24.

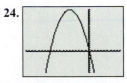

$[-20, 10, 1, -10, 20, 1]$

25.

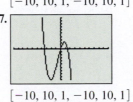

$[-10, 10, 1, -10, 10, 1]$

26.

$[-10, 10, 1, -10, 10, 1]$

27.

$[-10, 10, 1, -10, 10, 1]$

28.

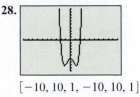

$[-10, 10, 1, -10, 10, 1]$

Section 3.4
1. a. the average rate of change of money per month in a savings account
b. the amount of money in the account at the beginning
2. $\dfrac{22 \text{ kilowatt} \cdot \text{hours}}{1 \text{ day}}$ **3. a.** the average rate of change
of money per month in the account   **b.** the amount of money in the account at the beginning

c. $y = \left(\dfrac{-\$550}{1 \text{ month}}\right)x + \$5500$ **d.** $\$2200$

e.

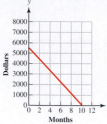

4. a. the average rate of change in height per day of the bamboo plant

b. the height of the plant at the beginning

c. $H(x) = \left(\dfrac{50 \text{ cm}}{1 \text{ day}}\right)x + 150 \text{ cm}$ **d.** 1650 cm

e.

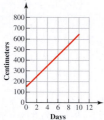

5. a. $x = 3400 \text{ hr}$ **b.** Answers will vary.

6. a. $y = \left(\dfrac{\$11.323 \text{ billion}}{1 \text{ year}}\right)x + \$98.65 \text{ billion}$

b. $\$325.1 \text{ billion}$ **7.** 19,522,000 students

8. a.

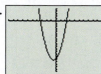
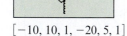

b. minimum -16 **c.** $[-16, \infty)$

$[-10, 10, 1, -20, 5, 1]$

9. a.

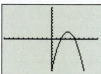

b. maximum 2.25   **c.** $(-\infty, 2.25]$

$[-10, 10, 1, -10, 10, 1]$

10. a.

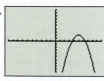

b. maximum 2   **c.** $(-\infty, 2]$

$[-10, 10, 1, -10, 10, 1]$

Section 3.5

1. a.

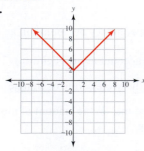

b. $(0, 2)$ **c.** $x = 0$

2. a.

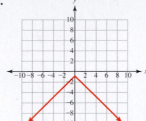

b. $(0, -1)$ **c.** $x = 0$

3. a.

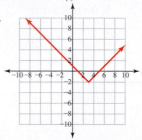

b. $(3, -2)$ **c.** $x = 3$

4. a. $(-2, 0)$ **b.** minimum 0   **c.** $(-\infty, \infty)$ **d.** $[0, \infty)$
5. a. $(-2, 3)$ **b.** maximum 3   **c.** $(-\infty, \infty)$ **d.** $(-\infty, 3]$
6. a. $(0, 4)$ **b.** maximum 4   **c.** $(-\infty, \infty)$ **d.** $(-\infty, 4]$
7. a. $(1, 1)$ **b.** minimum 1   **c.** $(-\infty, \infty)$ **d.** $[1, \infty)$
8. a. $[9, \infty)$ **b.** $(-\infty, 9]$
9. $x = -2$ and $x = 1$
10. $x = 0$ and $x = 2$
11. $x = -2$ **12.** no real zeros
13. $x = 4$ and $x = 6$
14. a. no   **b.** Answers will vary.

15. a.

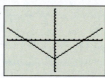

b. minimum 6   **c.** $[6, \infty)$

$[-10, 10, 1, -10, 10, 1]$

16. a.

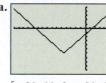

b. minimum -12 **c.** $[-12, \infty)$

$[-30, 10, 2, -20, 10, 2]$

17. a.

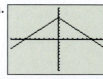

b. maximum 7   **c.** $(-\infty, 7]$

$[-10, 10, 1, -10, 10, 1]$

18. a.

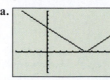

b. minimum 0   **c.** $[0, \infty)$

$[-10, 20, 2, -10, 20, 2]$

Section 3.6

1. a.

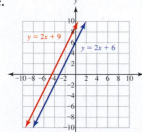

b. $f(x) = 2x + 6$

c.

2. a.

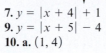

b. $f(x) = -6x + 12$

c.

3. a. $(2, 8)$

b.

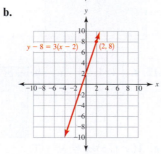

c. $y - 9 = 3(x - 4)$

d.

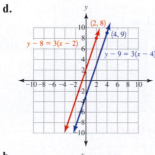

4. a. $(5, -4)$

b.

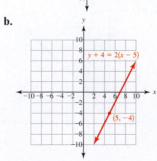

c. $y + 3 = 2(x - 7)$

d.

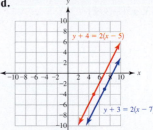

5.

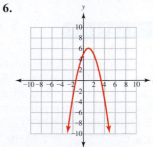

6.

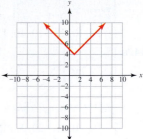

7. $y = |x + 4| + 1$ **8.** $y = |x + 2| + 4$

9. $y = |x + 5| - 4$

10. a. $(1, 4)$

b.

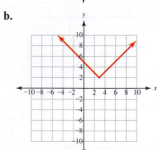

11. a. $(3, 2)$

b.

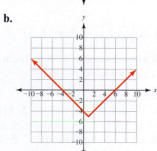

12. a. $(1, -5)$

b.

13. a. $(-3, 2)$ **b.**

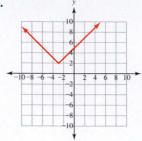

CHAPTER 4

Section 4.1

1. no solution   **2.** one solution   **3.** infinitely many solutions   **4.** one solution   **5. a.** $(2, 3)$ **6. a.** $(4, 1)$
7. a. no solution   **8. a.** infinitely many solutions   **9.** yes
10. no   **11.** (300 products, $5400)
12. (1500 products, $9000)

13. a.

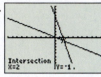

$[-10, 10, 1, -10, 10, 1]$

b. $(2, -1)$

14. a.

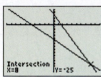

$[-10, 10, 1, -30, 10, 1]$

b. $(8, -25)$

15. a.

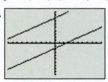

$[-10, 10, 1, -10, 10, 1]$

b. no solution

Section 4.2

1. a. $(8, -11)$ **2. a.** no solution

3. a. infinitely many solutions; $\left\{(x, y) \mid y = -\dfrac{4}{5}x + 2\right\}$

4. a. $\left(\dfrac{1}{2}, 3\right)$ **5. a.** $(8, 3)$ **6. a.** $\left(\dfrac{1}{6}, 9\right)$

7. a. no solution   **8. a.** infinitely many solutions;
$\{(x, y) \mid 2x - y = -1\}$

Section 4.3

1. 48 products; $3120   **2.** $12,000 in long-term; $24,000 in short-term   **3.** taco: $1.25; quesadilla: $3.50   **4.** 160 kg Brass A; 40 kg Brass B   **5.** 3.2 gal vinegar; 4.8 gal water
6. Add 1.92 L pure antifreeze for a total of 17.92 L.
7. The fast car will be 48 miles from Des Moines, and the slow car will be 42 miles from Des Moines.
8. The distance from the point of no return to Indianapolis is 112 miles, and the distance to Harrisburg is 380 miles.

Section 4.4

1. point   **2.** point   **3.** line   **4.** line   **5.** plane
6. point   **7.** solution   **8.** not a solution

9. a. $(-5, 4, 2)$ **10. a.** infinitely many solutions;
$\{(x, y, z) \mid x = 12z - 48, y = -5z + 21, z \text{ is a real number}\}$
11. a. no solution   **12.** 1 apple, 2 oranges, 4 bananas

Section 4.5

1. $\begin{bmatrix} 2 & 7 & | & 58 \\ 3 & -5 & | & -53 \end{bmatrix}$ **2.** $\begin{bmatrix} 2 & 7 & 3 & | & 24 \\ 1 & 8 & -2 & | & 6 \\ 0 & 1 & -1 & | & 10 \end{bmatrix}$

3. $\begin{bmatrix} 1 & 2 & | & 3 \\ 3 & 6 & | & 15 \end{bmatrix}$ **4.** $\begin{bmatrix} 1 & -2 & | & -3 \\ 3 & 6 & | & 15 \end{bmatrix}$ **5.** $\begin{bmatrix} 1 & 2 & | & 3 \\ 0 & 0 & | & 6 \end{bmatrix}$

6. $\begin{bmatrix} 1 & -2 & | & -3 \\ 0 & 12 & | & 24 \end{bmatrix}$ **7.** $\begin{bmatrix} 1 & 0 & | & 5 \\ 0 & 1 & | & 2 \end{bmatrix}$; $(5, 2)$

8. $\begin{bmatrix} 1 & 0 & 0 & | & 2 \\ 0 & 1 & 0 & | & 8 \\ 0 & 0 & 1 & | & 6 \end{bmatrix}$; $(2, 8, 6)$

9. $\begin{bmatrix} 1 & -\dfrac{1}{3} & | & 0 \\ 0 & 0 & | & 1 \end{bmatrix}$; no solution

10. a. $\begin{bmatrix} 1 & 3 & | & 5 \\ 4 & -5 & | & 3 \end{bmatrix}$ **b.** 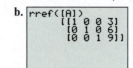   **c.** $(2, 1)$

11. a. $\begin{bmatrix} 2 & 1 & 1 & | & 21 \\ 3 & 2 & -1 & | & 12 \\ 4 & -1 & 2 & | & 24 \end{bmatrix}$

b. 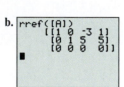   **c.** $(3, 6, 9)$

12. a. $\begin{bmatrix} 3 & 2 & 1 & | & 13 \\ 1 & 1 & 2 & | & 6 \\ 5 & 4 & 5 & | & 25 \end{bmatrix}$ **b.**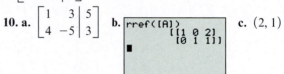

c. $\{(x, y, z) \mid x = 3z + 1, y = -5z + 5, z \text{ is a real number}\}$

13. a. $\begin{bmatrix} 2 & -1 & | & -7 \\ 4 & 2 & | & 15 \end{bmatrix}$

b. 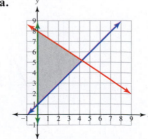   **c.** $(0.125, 7.25)$

Section 4.6

1. a.

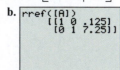

b. $(0, 1), (0, 8), (4.2, 5.2)$

2. a.

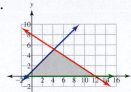

b. $(-1, 0)$, $(12, 0)$, $(4.2, 5.2)$

8. a.

b. solution   **c.** solution

3. a.

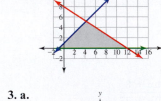

b. $(2, 2)$, $(8, 8)$, $(8, -10)$

9. a.

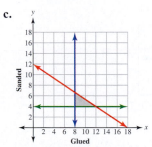

b. not a solution

4. a.

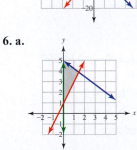

b. $(-2, -3)$, $(-2, 5)$, $(7, 5)$, $(7, -3)$

c. not a solution

10. a. $x =$ number of chairs glued;

$y =$ number of chairs sanded

b. $\left(\dfrac{10 \text{ min}}{1 \text{ chair}}\right)x + \left(\dfrac{15 \text{ min}}{1 \text{ chair}}\right)y \le 180 \text{ min}$

$x \ge 8$ chairs

$y \ge 4$ chairs

5. a.

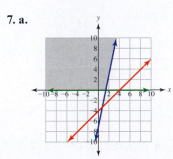

b. $(-11, 6)$, $\left(\dfrac{5}{11}, -\dfrac{60}{11}\right)$, $(10, 6)$

c.

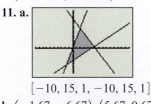

d. (8 chairs, 4 chairs), (8 chairs, 7 chairs),
(12 chairs, 4 chairs)

6. a.

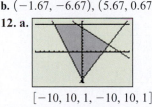

b. $(0, 1)$, $(0, 5)$, $\left(\dfrac{16}{11}, \dfrac{43}{11}\right)$

11. a.

$[-10, 15, 1, -10, 15, 1]$

b. $(-1.67, -6.67)$, $(5.67, 0.67)$, $(2, 8)$

7. a.

b. $\left(\dfrac{7}{5}, 0\right)$

12. a.

$[-10, 10, 1, -10, 10, 1]$

b. $(-1, 9)$, $(5.67, 2.33)$, $(0, -9)$, $(-6, 9)$

13. a.

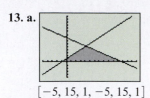

$[-5, 15, 1, -5, 15, 1]$

b. $(-1, 0), (12, 0), (4.2, 5.2)$

CHAPTER 5

Section 5.1

1. $\dfrac{x + 3}{x - 4}$ **2.** $\dfrac{-w - 3}{w - 4}$ **3.** $\dfrac{x + 1}{x - 5}$ **4.** $\dfrac{3r^8}{4p^3q}$

5. $\dfrac{x^2 - 2x + 4}{x + 3}$ **6.** $\dfrac{c^2 + 2c + 4}{c + 2}$ **7.** $\dfrac{9n^2 - 3np + p^2}{5n + 4}$

8. k **9.** $\dfrac{x - 4}{x + 4}$ **10.** $\dfrac{m}{2(m - 6)}$ **11.** $\dfrac{5}{6p^2w^4x^3y^8}$

12. a. $\dfrac{x + 8}{x - 2}$ **b.**

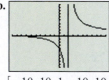

$[-10, 10, 1, -10, 10, 1]$

13. a. $\dfrac{x - 6}{x - 7}$ **b.**

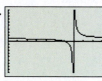

$[0, 10, 1, -10, 10, 1]$

14. a. $\dfrac{x}{x + 6}$ **b.**

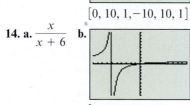

$[-10, 10, 1, -10, 10, 1]$

Section 5.2

1. a. $2^3; 2 \cdot 5$ **b.** 40 **2. a.** $2^2 \cdot 3^3 \cdot x \cdot y^2; 2^3 \cdot 3^2 \cdot x^2 \cdot y$

b. $216x^2y^2$ **3. a.** $(x + 4)(x + 5); x(x + 4)(x - 4)$

b. $x(x - 4)(x + 4)(x + 5)$

4. a. $(x - 5)(2x + 3); (x - 5)(3x + 1)$

b. $(x - 5)(2x + 3)(3x + 1)$

5. a. 12 **b.** $\dfrac{9}{12}$ pizza and $\dfrac{2}{12}$ pizza

6. a. $288xy^2$ **b.** $\dfrac{45}{288xy^2}$ and $\dfrac{28y}{288xy^2}$

7. a. $x(x - 4)(x - 2)$

b. $\dfrac{3x}{x(x - 4)(x - 2)}$ and $\dfrac{2x - 4}{x(x - 4)(x - 2)}$

8. a. $x(2x + 3)(3x - 1)$

b. $\dfrac{x}{x(2x + 3)(3x - 1)}$ and $\dfrac{6x + 9}{x(2x + 3)(3x - 1)}$

9. $\dfrac{2x + 3}{x - 9}$ **10.** $\dfrac{35h - 2k}{60h^2k^2}$

11. $\dfrac{2c^2 - c - 6}{c(c - 6)(c + 6)}$ **12.** $\dfrac{y^2 + 9y + 10}{2(y - 5)(y + 5)}$

Section 5.3

1. $\dfrac{10}{9}$ **2.** $\dfrac{2x + 2}{x - 1}$ **3.** $\dfrac{n - 3}{n + 2}$

4. $\dfrac{c}{2}$ **5.** $\dfrac{y - 2}{y + 4}$ **6.** $\dfrac{x + 4}{3(x + 2)}$

7. $\dfrac{3}{2}$ **8.** $\dfrac{6x + 8}{10x + 27}$ **9.** $\dfrac{x}{x - 3}$

10. $\dfrac{2x + 8}{3(2x + 1)}$ **11.** $\dfrac{3}{2}$ **12.** $\dfrac{6x + 8}{10x + 27}$

13. $\dfrac{x}{x - 3}$ **14.** $\dfrac{2x + 8}{3(2x + 1)}$

Section 5.4

1. a. 120 **b.** $100p - 15 = 16$ **2. a.** $8x$ **b.** $44 - 4 = x$

3. a. $(x - 4)(x - 6)$ **b.** $2x(x - 6) + 2(x - 4) = 4$

4. a. $x = 27$ **5. a.** $x = -4 \text{ or } x = 4$

6. a. $x = 0 \text{ or } x = -3$ **7. a.** $a = -4$ **8.** 181 people

9. $2{,}657{,}600$ patients   **10.** 31 min

11. Chris can empty the shelves in 6 hr.   **12.** \$6,260 million

Section 5.5

1. a.

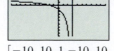

b. $(-\infty, 2) \cup (2, \infty)$

2. a.

b. $(-\infty, -3) \cup (-3, 8) \cup (8, \infty)$

3. a.

b. $(-\infty, -2) \cup (-2, 3) \cup (3, \infty)$

4. a. $(-\infty, -6) \cup (-6, 2) \cup (2, \infty)$

b. $x = -6, x = 2$ **c.** no holes

5. a. $(-\infty, -7) \cup (-7, 3) \cup (3, \infty)$

b. $x = -7$ **c.** hole at $x = 3$

6. a. $(-\infty, 0) \cup (0, \infty)$ **b.** $x = 0$ **c.** no holes

7. a. $(-\infty, -7) \cup (-7, -2) \cup (-2, \infty)$

b. $x = -7, x = -2$ **c.** no holes

8. a. $(-\infty, -8) \cup (-8, -1) \cup (-1, \infty)$

b. $x = -8$ **c.** hole at $x = -1$

9. 3 **10.** 6 **11.** undefined   **12.** 150 **13.** \$200

14. 20 mg   **15. a.** $(-\infty, 5) \cup (5, \infty)$ **b.** $x = -5$

16. a. $(-\infty, -3) \cup (-3, 8) \cup (8, \infty)$ **b.** $x = 2$

17. a. $(-\infty, -3) \cup (-3, 3) \cup (3, \infty)$ **b.** $x = 0$

18. a. $(-\infty, -2) \cup (-2, \infty)$ **b.** no real zero

19. a.

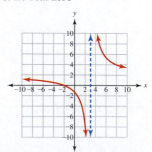

$[-10, 10, 1, -10, 10, 1]$

b. $(-\infty, 3) \cup (3, \infty)$ **c.** $x = 3$ **d.** no holes   **e.** $x = -2$

20. a.

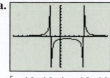

$[-10, 10, 1, -10, 10, 1]$

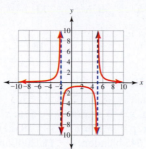

b. $(-\infty, -2) \cup (-2, 5) \cup (5, \infty)$ **c.** $x = -2, x = 5$
d. no holes   **e.** no real zeros

21. a.

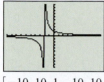

$[-10, 10, 1, -10, 10, 1]$

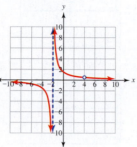

b. $(-\infty, -2) \cup (-2, 4) \cup (4, \infty)$ **c.** $x = -2$
d. hole at $x = 4$ **e.** no real zero

Section 5.6
1. a. $k = 4$ **b.** $y = 4x$ **c.** $y = 20$
d.    **e.** $y = 8$

2. a. $k = \dfrac{\$500}{1000 \text{ mi}}$ or $\dfrac{\$0.50}{1 \text{ mi}}$ **b.** $y = \left(\dfrac{\$0.50}{1 \text{ mi}}\right)x$ **c.** \$2250

d. 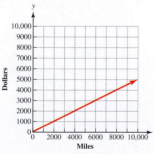   **e.** \$2000

3. a. $k = 36$ **b.** $y = \dfrac{36}{x}$ **c.** $y = 9$
d.    **e.** $y = 18$

4. a. $k = 40$ in.^{2} **b.** $y = \dfrac{40 \text{ in.}^2}{x}$ **c.** 5 in.
d. 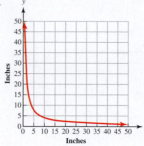   **e.** 8 in.

5. a. $k = 2$ **b.** $y = 2xz$ **c.** $y = 900$ **6. a.** $k = 9$

b. $y = \dfrac{9x}{z}$ **c.** $y = 18$ **7.** inverse   **8.** inverse

9. direct   **10.** $R = \dfrac{PV}{nT}$ **11.** $V_2 = \dfrac{V_1}{C-1}$

12. $T = 5C - 5H - 5L$

Section 5.7
1. a. divisor $7a^2bc$; dividend $21a^6b^{11}c$ **b.** $3a^4b^{10}$
2. a. divisor $3x$; dividend $15x^3 - 27x^2 + 6x - 9$

b. $5x^2 - 9x + 2 - \dfrac{3}{x}$

3. a. divisor $7a^2bc$; dividend $21a^6b^{11}c - 28a^5bc + 35$

b. $3a^4b^{10} - 4a^3 + \dfrac{5}{a^2bc}$ **4. a.** $x + 10$

5. a. $x + 5$ **6. a.** $3x + 4$ **7. a.** $x + 6 + \left(\dfrac{5}{x+2}\right)$

8. a. $5x^2 - 13x + 52 + \left(\dfrac{-205}{x+4}\right)$ **9. a.** $x + 10$

10. a. $x + 6 + \left(\dfrac{-3}{x+3}\right)$ **11.** $f(-2) = 0$; $x + 2$ is a factor

12. $f(-3) = -3$; $x + 3$ is not a factor

13. a. $x + 2 + \left(\dfrac{9}{x+2}\right)$ **b.** $f(-2) = 9$

14. a. $x - 3 + \left(\dfrac{5}{x-3}\right)$ **b.** $f(3) = 5$

15. a. $x - 9$ **b.**

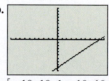

$[-10, 10, 1, -10, 10, 1]$

16. a. $x - 13 + \left(\dfrac{44}{x+2}\right)$ **b.**

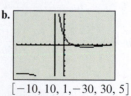

$[-10, 10, 1, -30, 30, 5]$

17. a. $x - 2$ **b.**

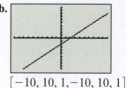

$[-10, 10, 1, -10, 10, 1]$

Section 5.8

1. a. $x = -3$ and $x = 6$ **b.** $-3 < x < 6$
c. 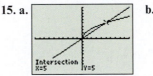 x; $(-3, 6)$

2. a. $x = -3$ and $x = 6$ **b.** $x < -3$ or $x > 6$
c. 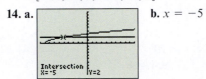 x;
$(-\infty, -3) \cup (6, \infty)$

3. a. $x \geq 3$ **b.** ⟵┼┼├─┼─┼─┼─┼⟶ x; $[3, \infty)$
  $1\ 2\ 3\ 4\ 5\ 6\ 7\ 8$

4. a. $-4 < x < 3$ or $x > 10$ **b.** $(-4, 3) \cup (10, \infty)$

5. a. $x < -4$ or $3 < x \leq 10$ **b.** $(-\infty, -4) \cup (3, 10]$

6. a. $-4 < x \leq -3$ or $x \geq 3$ **b.** $(-4, -3] \cup [3, \infty)$

CHAPTER 6

Section 6.1

1. a. index 3; radicand 27 **b.** 3
2. a. index 4; radicand 81 **b.** 3
3. a. index 2; radicand $\dfrac{49}{144}$ **b.** $\dfrac{7}{12}$
4. a. index 2; radicand -100 **b.** not a real number
5. a. index 3; radicand -125 **b.** -5
6. a. index 2; radicand 5 **b.** 2.24
7. a. index 2; radicand 169 **b.** -13 **8.** 36 **9.** 9
10. 100 **11.** 49 **12.** 8 **13.** 27 **14.** 1000
15. 64 **16.** x^4 **17.** y^5 **18.** z^6 **19.** a
20. $2x^2y^4\sqrt{23y}$ **21.** $25b\sqrt{bc}$ **22.** $3x^2\sqrt[3]{2x^2}$
23. $3a^2\sqrt[4]{5a^2b^3}$ **24.** $-2z^2\sqrt[3]{3}$ **25.** $2x + 1$

Section 6.2

1. $6\sqrt{5}$ **2.** 0 **3.** $10p\sqrt{5}$ **4.** $7\sqrt[3]{2}$ **5.** $5\sqrt[4]{x}$
6. $140\sqrt{6}$ **7.** $3x\sqrt{2} + \sqrt{15x}$ **8.** $3x\sqrt{6} + 3\sqrt{5x}$
9. $3\sqrt[3]{c^2} + 3c\sqrt[3]{2}$ **10.** $2h\sqrt{5} + \sqrt{30h} - \sqrt{210h} - 3\sqrt{35}$

Section 6.3

1. $\dfrac{3h^2\sqrt{h}}{10k}$ **2.** $\dfrac{a^2c\sqrt{5ac}}{4}$ **3.** $\dfrac{3x^3\sqrt[3]{x^2}}{4y^2}$

4. $\dfrac{5\sqrt[4]{c}}{2c^2}$ **5.** $\dfrac{x\sqrt{3} + \sqrt{15}}{3}$ **6.** $\dfrac{5\sqrt[3]{4x^2}}{2x}$

7. $\dfrac{6\sqrt[4]{y^3}}{y}$ **8.** $\dfrac{\sqrt{5ac}}{c}$ **9.** $\dfrac{\sqrt{42pw}}{10w}$

10. a. $\sqrt{y} + 5$ **b.** $\dfrac{\sqrt{xy} + 5\sqrt{x}}{y - 25}$

11. a. $\sqrt{x} + \sqrt{y}$ **b.** $\dfrac{5\sqrt{x} + 5\sqrt{y}}{x - y}$

12. a. $\sqrt{3a} - \sqrt{5b}$ **b.** $\dfrac{6a - 2\sqrt{15ab}}{3a - 5b}$

Section 6.4

1. a. $x^{\frac{4}{2}}y^{\frac{9}{2}}$ **b.** $x^2y^{\frac{9}{2}}$ **2. a.** $x^{\frac{4}{3}}y^{\frac{9}{3}}$ **b.** $x^{\frac{4}{3}}y^3$
3. a. $x^{\frac{4}{4}}y^{\frac{12}{4}}$ **b.** xy^3 **4. a.** $\sqrt{16}$ **b.** 4
5. a. $\sqrt[3]{6^2}$ **b.** $\sqrt[3]{36}$ **6.** $z^{\frac{3}{5}}$ **7.** $x^{\frac{41}{72}}$ **8.** $x^{\frac{11}{40}}y^{\frac{11}{45}}$
9. a. $h^{\frac{1}{15}}$ **b.** $\sqrt[15]{h}$ **10. a.** x **b.** x **11. a.** p **b.** p

12. a. $z^{\frac{3}{10}}$ **b.** $\sqrt[10]{z^3}$ **13.** $x^{\frac{19}{56}}$ **14.** $n^{\frac{7}{120}}$ **15.** $\dfrac{3y^{\frac{9}{44}}}{5x^{\frac{7}{72}}}$

16. $w^{\frac{15}{8}}$ **17.** $x^{\frac{13}{35}}$ **18.** $\dfrac{1}{x^{\frac{27}{40}}}$

Section 6.5

1. a. $x = 48$ **2. a.** $x = 335$ **3. a.** $x = 5$
4. a. $x = 3$ or $x = 7$ **5. a.** $x = 3$ **6. a.** $x = 23$
7. a. $p = -5$ or $p = 3$ **8. a.** $n = 29$ **9. a.** $n = 24$
10. 645 blenders **11.** 2000 in.

12. a. 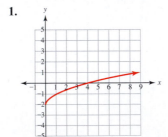 **b.** $x = 2$
$[-10, 10, 1, -10, 10, 1]$

13. a. 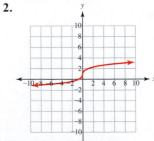 **b.** $x = -2$ or $x = 4$
$[-10, 10, 1, -10, 10, 1]$

14. a. 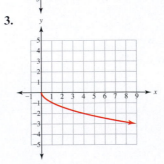 **b.** $x = -5$
$[-10, 10, 1, -10, 10, 1]$

15. a. 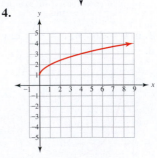 **b.** $x = 0$ or $x = 5$
$[-10, 10, 1, -10, 10, 1]$

Section 6.6

1.

2.

3.

4.

5.
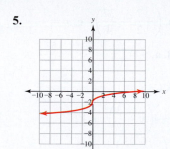

6.
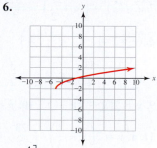

7. $[-3, \infty)$ **8.** $[2, \infty)$ **9.** $\left(-\infty, \dfrac{4}{3}\right]$

10. $[95 \text{ ft}, 10{,}050 \text{ ft}]$ **11.** $4\sqrt{15}$

12. $57 \dfrac{\text{ft}}{\text{s}}$ **13.** 31 ft **14.** $[5, \infty)$

15. $[0, \infty)$ **16.** $(-\infty, 0]$
17. a. $x = 64$ **18. a.** $x = 4$ **19.** $x = 4$
20. $x = 4$ **21.** $x = 9$
22. a.

![graph] **b.** $x \approx 3$

$[-10, 10, 1, -10, 10, 1]$
c. $y \approx 0$ **d.** $[0, \infty)$

23. a.

![graph] **b.** $x \approx 1.38$

$[-10, 10, 1, -10, 10, 1]$
c. $y \approx -4$ **d.** $[-4, \infty)$

24. a.

![graph] **b.** $x \approx 7$

$[-10, 10, 1, -10, 10, 1]$
c. $y \approx 0$ **d.** $[0, \infty)$

25. a.

![graph] **b.** $x \approx -29$

$[-40, 10, 5, -10, 10, 1]$
c. $y \approx -6$ **d.** $[-6, \infty)$

CHAPTER 7

Section 7.1
1. $3i$ **2.** $3x\sqrt{2}\,i$ **3.** $8x^4y^5\sqrt{2xy}\,i$
4. $-4x^3y^3\sqrt[3]{2y^2}$ **5.** $-2\sqrt{15}$
6. $-\sqrt{35ab}$ **7.** $-6hk\sqrt{3h}$
8. $2p^6\sqrt{77}\,i$

9.

	Complex numbers	Imaginary numbers	Real numbers	Rational numbers	Irrational numbers	Integers	Whole numbers
$6 + 2i$	X						
6	X		X	X		X	X
$\sqrt{6}$	X		X		X		
$6i$	X	X					
-6	X		X	X		X	
0	X		X	X		X	X
$\sqrt{25}$	X		X	X		X	X
$\sqrt{-25}$	X	X					

10. $10 - 13i$ **11.** $4 + 3i$ **12.** $40 + 15i$
13. $-19 - 71i$ **14.** 74 **15.** $9 - 2i$ **16.** $4 + 6i$
17. $\dfrac{4}{9} + \dfrac{1}{2}i$ **18.** $-\dfrac{7}{2}i$ **19.** $\dfrac{14}{5} - \dfrac{7}{5}i$ **20.** $\dfrac{14}{13} + \dfrac{21}{13}i$
21. i **22.** $-i$ **23.** -1 **24.** $-i$ **25.** 1 **26.** i

Section 7.2
1. a. $x = 5$ or $x = -5$ **2. a.** $b = 7i$ or $b = -7i$
3. a. $k = 2\sqrt{14}$ or $k = -2\sqrt{14}$
4. a. $m = 3\sqrt{7}\,i$ or $m = -3\sqrt{7}\,i$
5. $x = 8i$ or $x = -8i$ **6.** $x = \pm\sqrt{5}$ **7.** 394 ft
8. a. $x = 7$ or $x = 12$ **9. a.** $y = -\dfrac{4}{3}$ or $y = \dfrac{3}{2}$
10. a. $p = 0$ or $p = 56$
11. 19 cm long; 8 cm wide **12.** height 12 in.; base 8 in.

Section 7.3
1. $x^2 + 10x + 25$ **2.** $z^2 - 20z + 100$ **3.** $k^2 + 2k + 1$
4. $u^2 - 3u + \dfrac{9}{4}$ **5.** $w^2 - 9w + \dfrac{81}{4}$
6. a. $x = -7$ or $x = -3$ **7.** $z = 3$ or $z = 9$
8. $k = -1 \pm \sqrt{7}\,i$ **9.** $z = -\dfrac{3}{2} \pm \dfrac{3\sqrt{5}}{2}$
10. $p = -1 \pm \dfrac{\sqrt{11}}{2}$ **11.** $x = \dfrac{7}{4} \pm \dfrac{\sqrt{113}}{4}$
12. $x = -\dfrac{1}{6} \pm \dfrac{\sqrt{23}}{6}i$

Section 7.4
1. $x = -\dfrac{5}{4} \pm \dfrac{\sqrt{129}}{4}$ **2.** $x = -\dfrac{3}{8} \pm \dfrac{\sqrt{103}}{8}i$
3. $x = -\dfrac{7}{4}$ or $x = 1$ **4.** $x = 1$ or $x = 2$ **5.** $x = 7$
6. $x = \pm\dfrac{\sqrt{30}}{3}i$ **7. a.** 129 **b.** two **c.** irrational
8. a. -103 **b.** two **c.** nonreal **9. a.** 121 **b.** two
c. rational **10. a.** 1 **b.** two **c.** rational
11. a. 0 **b.** one **c.** rational **12. a.** -120 **b.** two
c. nonreal **13.** 59 s **14.** 11 ft long, 7 ft wide
15. 0.08 mole

Section 7.5
1. a. $(-1, -3)$ **b.** up **c.** minimum -3
d. $x = -1$ **e.** $(-\infty, \infty)$ **f.** $[-3, \infty)$ **2. a.** $(-2, 1)$
b. up **c.** minimum 1 **d.** $x = -2$ **e.** $(-\infty, \infty)$
f. $[1, \infty)$ **3. a.** $(5, -3)$ **b.** down **c.** maximum -3
d. $x = 5$ **e.** $(-\infty, \infty)$ **f.** $(-\infty, -3]$
4. $x = -2$ and $x = 0$ **5.** no real zeros

6. $x = -3$ **7. a.** 81 **b.** two **c.** rational
d. $x = \dfrac{3}{2}$ and $x = 6$ **8. a.** 133 **b.** two

c. irrational **d.** $x = -\dfrac{5}{6} \pm \dfrac{\sqrt{133}}{6}$ **9. a.** -31

b. no real zeros **10. a.** none **b.** two **c.** one **d.** two
e. none **11. a.** $(8, 1)$ **b.** up **c.** minimum 1 **d.** $x = 8$
e. $(-\infty, \infty)$ **f.** $[1, \infty)$ **g.** none **12. a.** $(-7, -6)$ **b.** down
c. maximum -6 **d.** $x = -7$ **e.** $(-\infty, \infty)$ **f.** $(-\infty, -6]$
g. none **13. a.** $y = 3(x - 6)^2 + 2$
b. $y = -2(x + 8)^2 + 4$ **c.** $y = -(x - 5)^2 - 10$
14. a. -2 **b.** 34 **15. a.** -3 **b.** -99 **16.** 127 ft
17. \$147.5 billion

Section 7.6
1. a. $y = 2x - 1$
b.

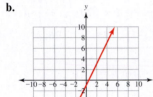

2. a. $(1, 5)$ **b.** $y - 16 = 2(x - 4)$
c.

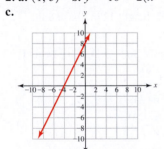

3. a. $(8, 1)$ **b.** $y = 5(x - 3)^2 - 5$
c.

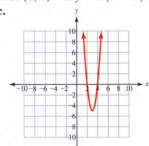

4. a. $(-7, -6)$ **b.** $y = 5(x + 4)^2 - 4$
c.

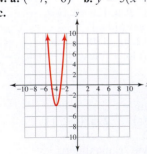

5. a. $y = (x + 4)^2 - 37$ **b.** $(-4, -37)$ **c.** $x = -4$
d. $(-\infty, \infty)$ **e.** $[-37, \infty)$ **6. a.** $f(x) = (x - 6)^2 - 29$
b. $(6, -29)$ **c.** $x = 6$ **d.** $(-\infty, \infty)$ **e.** $[-29, \infty)$

7. a. $f(x) = \left(x + \dfrac{7}{2}\right)^2 - \dfrac{37}{4}$ **b.** $\left(-\dfrac{7}{2}, -\dfrac{37}{4}\right)$ **c.** $x = -\dfrac{7}{2}$

d. $(-\infty, \infty)$ **e.** $\left[-\dfrac{37}{4}, \infty\right)$ **8. a.** $y = \left(x + \dfrac{3}{2}\right)^2 - \dfrac{53}{4}$

b. $\left(-\dfrac{3}{2}, -\dfrac{53}{4}\right)$ **c.** $x = -\dfrac{3}{2}$ **d.** $(-\infty, \infty)$ **e.** $\left[-\dfrac{53}{4}, \infty\right)$

9. $y = 2(x + 2)^2 + 1$ **10.** $y = -4\left(x - \dfrac{3}{2}\right)^2 + 12$

11. a. $y = 3\left(x + \dfrac{1}{3}\right)^2 + \dfrac{23}{3}$ **b.** $\left(-\dfrac{1}{3}, \dfrac{23}{3}\right)$

c. $x = -\dfrac{1}{3}$ **d.** up **e.** minimum $\dfrac{23}{3}$

f. $(-\infty, \infty)$ **g.** $\left[\dfrac{23}{3}, \infty\right)$ **h.** no real zeros

12. a. $y = -5\left(x - \dfrac{9}{10}\right)^2 + \dfrac{121}{20}$ **b.** $\left(\dfrac{9}{10}, \dfrac{121}{20}\right)$

c. $x = \dfrac{9}{10}$ **d.** down **e.** maximum $\dfrac{121}{20}$ **f.** $(-\infty, \infty)$

g. $\left(-\infty, \dfrac{121}{20}\right]$ **h.** two **13.** $\left(-\dfrac{3}{2}, \dfrac{21}{2}\right)$ **14.** $(-3, -16)$

15. $\dfrac{9}{20}$ m

Section 7.7
1. a. $x = 15$ **b.** $x > 15$ **c.** [number line: 13 14 15 16 17 18 19 20]
d. $(15, \infty)$
2. a. $x = -15$ **b.** $x < -15$ **c.** [number line: −19 −17 −15 −13]
d. $(-\infty, -15)$
3. a. $x = -9$; $x = 5$ **b.** $x < -9$ or $x > 5$
c. [number line: −25 −15 −5 0 5 15 20] **d.** $(-\infty, -9) \cup (5, \infty)$
4. a. $x = -7$; $x = 3$ **b.** $-7 < x < 3$
c. [number line: −9−8−7−6−5−4−3−2−1 0 1 2 3 4 5] **d.** $(-7, 3)$
5. a. $x = 0$; $x = 3$ **b.** $x < 0$ or $x > 3$
c. [number line: −5−4−3−2−1 0 1 2 3 4 5 6 7 8] **d.** $(-\infty, 0) \cup (3, \infty)$

6. a. $-\dfrac{3}{2} \le x \le 1$ **b.** $\left[-\dfrac{3}{2}, 1\right]$

7. a. $x \le -\sqrt{5}$ or $x \ge \sqrt{5}$ **b.** $\left(-\infty, -\sqrt{5}\right] \cup \left[\sqrt{5}, \infty\right)$
8. a. $x \le -5$ or $x \ge 2$ **b.** $(-\infty, -5] \cup [2, \infty)$
9. a. $-5 \le x \le 2$ **b.** $[-5, 2]$ **10. a.** the set of real
numbers **b.** $(-\infty, \infty)$ **11. a.** no real solution **b.** no real
solution **12. a.** $x = -4$ **b.** $[-4, -4]$ **13.** The width
should be between about 252 ft and about 595 ft.

CHAPTER 8
Section 8.1
1. $13x + 7$ **2.** $3x - 11$ **3.** $11 - 3x$ **4.** $-3\sqrt{x} - 2$

5. $\dfrac{2x^2 + 3x + 6}{x^2 - 1}$ **6.** $P(x) = \left(\dfrac{\$1.55}{1 \text{ bag}}\right)x$

7. a. $R(x) = \left(\dfrac{\$37}{1 \text{ product}}\right)x$ **b.** $C(x) = \left(\dfrac{\$25}{1 \text{ product}}\right)x$

c. $P(x) = \left(\dfrac{\$12}{1 \text{ product}}\right)x$ **8.** 4584 products

9. $28x^2 - 8x$ **10.** $56x^2 + 5x - 6$
11. $4x^3 + 40x^2 + 100x$ **12. a.** $40x + 70$ **b.** $40x - 1$
13. a. $\sqrt[3]{x}$ **b.** $|\sqrt[3]{x}|$ **14. a.** $81x^2 - 54x - 5$
b. $9x^2 - 72x + 19$

15. $S(R(x)) = \left(\dfrac{\$6.29}{1 \text{ lb}}\right)x$ **16.** one-to-one

17. not one-to-one   **18.** not one-to-one

19. one-to-one   **20.** not one-to-one   **21.** one-to-one

22. one-to-one

Section 8.2

1. a.

x	y
-20	2.9×10^{-10}
-10	1.7×10^{-5}
0	1
10	$59{,}049$
20	3.5×10^{9}
30	2.1×10^{14}
40	1.2×10^{19}

b.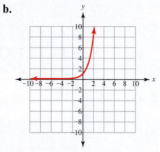

c. $(-\infty, \infty)$ **d.** $(0, \infty)$ **e.** $(0, 1)$ **f.** one-to-one

g. increasing

2. a.

x	y
-20	9.5×10^{13}
-10	$9{,}765{,}625$
0	1
10	1.0×10^{-7}
20	1.0×10^{-14}
30	1.1×10^{-21}
40	1.1×10^{-28}

b.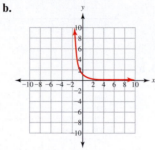

c. $(-\infty, \infty)$ **d.** $(0, \infty)$ **e.** $(0, 1)$ **f.** one-to-one

g. decreasing   **3.** $y = 15^x$ **4.** $y = 0.2^x$ **5.** 7.39

6. 0.14 **7.** 2.72 **8.** 1 **9.** 48 **10.** $48{,}000$

11. $(0, 1)$ **12.** $(0, 5)$ **13.** $624{,}000$ projects   **14.** 8

15. 20.0855 **16.** 0.01 **17.** 0.1353

18. a. 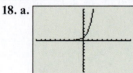   **b.** 243

$[-10, 10, 1, -10, 10, 1]$

19. a. 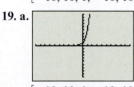   **b.** 2.56×10^{-6}

$[-10, 10, 1, -10, 10, 1]$

20. a. 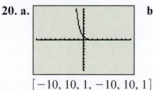   **b.** 0.0016

$[-10, 10, 1, -10, 10, 1]$

Section 8.3

1. Answers will vary.   **2.** Answers will vary.

3. Answers will vary.   **4.** $y = \dfrac{5}{3}x - \dfrac{40}{3}$

5. $y = -\dfrac{1}{9}x - \dfrac{4}{9}$ **6.** 2 **7.** -6 **8.** 3

9. 0 **10.** not a real number

11. a.   **b.** $(0, \infty)$ **c.** $(-\infty, \infty)$ **d.** $(1, 0)$

e. no y-intercept

x	y
$\dfrac{1}{9}$	-2
$\dfrac{1}{3}$	-1
1	0
3	1
9	2
27	3

12. a.   **b.** $(0, \infty)$ **c.** $(-\infty, \infty)$ **d.** $(1, 0)$

e. no y-intercept

x	y
$\dfrac{1}{16}$	-2
$\dfrac{1}{4}$	-1
1	0
4	1
16	2
64	3

Section 8.4

1. 4 **2.** 1.6990 **3.** 0 **4.** 2.3010 **5.** 1.7782

6. 4.0943 **7.** 5.9069 **8.** 3.7268 **9.** 0

10. not a real number   **11.** $1{,}084{,}050$ **12.** $3.5957; 2.4393$

13. $1{,}083{,}926.914$ **14.** x **15.** $2x$ **16.** x **17.** $4x$

18. $\log(x^2)$ **19.** $\log(y^6)$ **20.** $\ln(y^{7x})$

21. a. $\log(100 \cdot 10)$ **b.** 3 **22. a.** $\ln(45 \cdot 2)$ **b.** 4.4998

23. a. $\log\left(\dfrac{100}{10}\right)$ **b.** 1 **24. a.** $\ln\left(\dfrac{45}{9}\right)$ **b.** 1.6094

25. $\log_5(c) - \log_5(d)$ **26.** $\log_2(c) + \log_2(d)$

27. $\log\left(\dfrac{x^8 y}{z^4}\right)$ **28.** $\ln\left(\dfrac{xz^3}{y^2}\right)$

29. $5\log(x) + 3\log(y) - 2\log(z)$

30. $\ln(x) - 2\ln(y) - 4\ln(z)$

31. a. 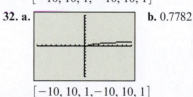   **b.** 1.6094

$[-10, 10, 1, -10, 10, 1]$

32. a.   **b.** 0.7782

$[-10, 10, 1, -10, 10, 1]$

Section 8.5

1. 324 cockroaches   **2.** \$2,846.07   **3.** \$2,847.07
4. 4.6   **5.** 115 decibels   **6.** 0.76 volt
7. a. 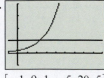   **b.** $x \approx 2.32$

$$[-1, 9, 1, -5, 20, 5]$$

8. a. 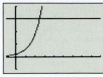   **b.** $x \approx 2.46$

$$[-1, 9, 1, -5, 20, 5]$$

9. a. 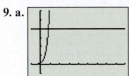   **b.** $x \approx 0.84$

$$[-1, 9, 1, -10, 30, 5]$$

10. a.   **b.** $x = 2.00$

$$[-1, 9, 1, -5, 20, 5]$$

Section 8.6

1. a. $x = 3$ **2. a.** $x \approx 2.1610$ **3. a.** $x \approx 1.2851$
4. a. $x \approx 1.3863$ **5.** 7 years 4 months
6. 22 years 0 months   **7.** 61 years   **8.** 53 days

Section 8.7

1. a. $x = 1024$ **2. a.** $x \approx 20.0855$ **3. a.** $x \approx -41.9642$
4. a. $x \approx 3162.2777$ **5. a.** $x = 432$ **6. a.** $x = 500$
7. a. $x = 10,000$ **8. a.** $x \approx 1.6601$ **9.** $\dfrac{6.3 \times 10^{-5} \text{ mole}}{1 \text{ L}}$

10. $\dfrac{3.2 \times 10^{-4} \text{ watt}}{1 \text{ m}^2}$ **11.** $\dfrac{7.06 \times 10^{-7} \text{ mole}}{1 \text{ L}}$

12. a.   **b.** $x \approx 39.8$

$$[0, 100, 10, -1, 3, 1]$$

13. a.   **b.** $x \approx 7943.3$

$$[0, 50000, 1000, -1, 5, 1]$$

14. a.   **b.** $x = 0.1$

$$[0, 1, 1, -2, 2, 1]$$

Section 9.1

1. a. 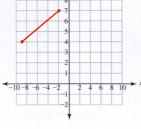   **b.** $\sqrt{58}$

2. a. 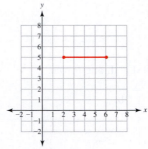   **b.** 4

3. a. 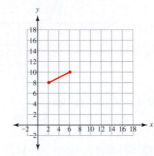   **b.** $2\sqrt{5}$

4. a. 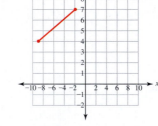   **b.** $\left(-\dfrac{11}{2}, \dfrac{11}{2}\right)$

5. a. 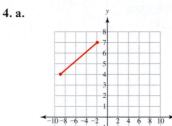   **b.** $(4, 5)$

6. a.

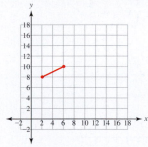

b. $(4, 9)$

7. $x^2 + y^2 = 49$ **8.** $(x + 2)^2 + (y - 3)^2 = 98$
9. a. $(x + 6)^2 + (y - 4)^2 = 64$ **b.** $(-6, 4)$ **c.** 8

10. $y = \sqrt{3364 \text{ ft}^2 - (x - 58 \text{ ft})^2}$
11. $x^2 + y^2 = 22{,}686{,}169 \text{ mi}^2$
12.

$[-15.2, 15.2, 1, -10, 10, 1]$

13.

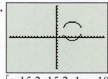

$[-15.2, 15.2, 1, -10, 10, 1]$

Section 9.2
1. a. $(5, 0); (-5, 0)$ **b.** $(0, 3); (0, -3)$ **c.** $\dfrac{x^2}{25} + \dfrac{y^2}{9} = 1$

d.

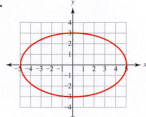

2. a. $(7, 3); (-3, 3)$ **b.** $(2, 6); (2, 0)$
c. $\dfrac{(x - 2)^2}{25} + \dfrac{(y - 3)^2}{9} = 1$

d.

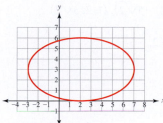

3. a. $y = 3$ **b.** $x = 6$ **c.** $(6, 3)$ **d.** $\dfrac{(x - 6)^2}{36} + \dfrac{(y - 3)^2}{4} = 1$

e.

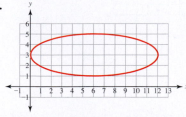

4. a. $\dfrac{(x + 3)^2}{1} + \dfrac{(y - 2)^2}{4} = 1$ **b.** $(-3, 2)$ **c.** $x = -3$
d. $y = 2$ **e.** $(-3, 0), (-3, 4)$ **f.** $(-2, 2), (-4, 2)$
g.

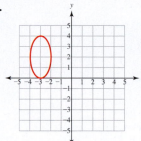

5. a. $\dfrac{(x + 3)^2}{4} + \dfrac{(y - 2)^2}{1} = 1$ **b.** $(-3, 2)$ **c.** $y = 2$
d. $x = -3$ **e.** $(-5, 2), (-1, 2)$ **f.** $(-3, 1), (-3, 3)$
g.

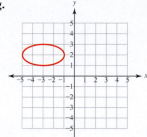

6. $\dfrac{x^2}{(1.52 \text{ AU})^2} + \dfrac{y^2}{(1.51 \text{ AU})^2} = 1$ **7.** $\dfrac{x^2}{144 \text{ ft}^2} + \dfrac{y^2}{100 \text{ ft}^2} = 1$

Section 9.3
1. a. $x^2 = 4py$ **b.** $(0, 0)$ **c.** up; $x = 0$ **d.** Answers will
vary.   **e.** $p = 4$; focus: $(0, 4)$; directrix: $y = -4$
f.   **g.** yes

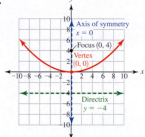

2. a. $y^2 = 4px$ **b.** $(0, 0)$ **c.** right; $y = 0$ **d.** Answers will
vary.   **e.** $p = 6$; focus: $(6, 0)$; directrix: $x = -6$
f.   **g.** no

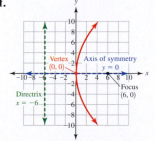

3. a. right   **b.** $y^2 = 4px$
c. $p = 5$; focus: $(5, 0)$; directrix: $x = -5$ **d.** $y^2 = 20x$

e.

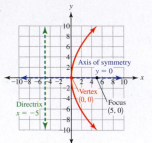

f. no

4. a. right   **b.** $(y - k)^2 = 4p(x - h)$
c. $p = 2$; focus: $(3, 4)$; directrix: $x = -1$
d. $(y - 4)^2 = 8(x - 1)$
5. a. $(y - 2)^2 = -4(3)(x - 5)$ **b.** left   **c.** $(5, 2)$ **d.** $y = 2$
e. $p = 3$; $(2, 2)$ **f.** $x = 8$ **g.** Answers will vary.
h.

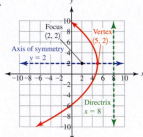

i. no

6. a. $(1, -3)$ **b.** $x = 1$ **c.** up   **d.** $(-\infty, \infty)$ **e.** $[-3, \infty)$
7. a. $(-4, 1)$ **b.** $x = -4$ **c.** down   **d.** $(-\infty, \infty)$ **e.** $(-\infty, 1]$
8. 8 in.   **9.** 8 ft 5.8 in.

Section 9.4

1. a. $\dfrac{y^2}{a^2} - \dfrac{x^2}{b^2} = 1$ **b.** $a = 3$; $p = 7$; $b = 2\sqrt{10}$

c. $\dfrac{y^2}{9} - \dfrac{x^2}{40} = 1$ **2. a.** $\dfrac{x^2}{a^2} - \dfrac{y^2}{b^2} = 1$

b. $a = 12$; $p = 14$; $b = 2\sqrt{13}$ **c.** $\dfrac{x^2}{144} - \dfrac{y^2}{52} = 1$

3. a. $\dfrac{y^2}{a^2} - \dfrac{x^2}{b^2} = 1$: transverse axis: $x = 0$
b. $a = 4$; $b = 10$; $p = 2\sqrt{29}$ **c.** $(0, -2\sqrt{29}), (0, 2\sqrt{29})$

d. $(0, -4), (0, 4)$ **4. a.** $a = 3$; $b = 4$ **b.** $y = -\dfrac{4}{3}x, y = \dfrac{4}{3}x$

c.

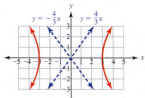

5. a. $a = 3$; $b = 4$ **b.** $y = -\dfrac{3}{4}x, y = \dfrac{3}{4}x$

c.

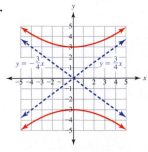

6. a. $a = 5$; $b = 3$ **b.** $y = -\dfrac{5}{3}x, y = \dfrac{5}{3}x$

c.

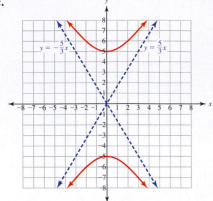

7. 8.5 mm

Section 9.5
1. $(2, 1), (6, 9)$

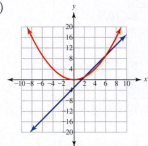

2. $(4, 1)$

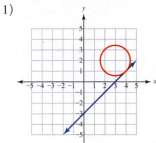

3. $(2, 1), (6, 9)$ **4.** $(0, 4), \left(\dfrac{12}{5}, -\dfrac{16}{5}\right)$ **5.** $(-4, 4), (4, 4)$

6. $(2, -5\sqrt{2}), (2, 5\sqrt{2}), (-2, -5\sqrt{2}), (-2, 5\sqrt{2})$
7. $(-2\sqrt{2}, 2), (2\sqrt{2}, 2)$
8. $(2, 6), (-2, 6), (2, -6), (-2, -6)$
9. 1999 and 2012

CHAPTER 10

Section 10.1
1. a. $\{2, 4, 8, 16, 32\}$ **b.**

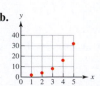

2. a. $\left\{0, \dfrac{1}{3}, \dfrac{1}{2}, \dfrac{3}{5}, \dfrac{2}{3}, \ldots\right\}$ **b.**

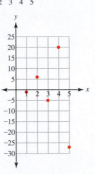

3. a. $\{-1, 6, -5, 20, -27, \ldots\}$ **b.**

4. $a_n = 4n$ **5.** $a_n = \dfrac{1}{n^2}$ **6.** $a_n = n^2$ **7.** 16   **8.** 0

9. $-\dfrac{1}{2}$ **10.** 298,000,000 people   **11.** 56th term

12.

$[0, 10, 1, 0, 13, 1]$

13.

$[0, 10, 1, 0, 45, 1]$

14.

$[0, 10, 1, -10, 10, 1]$

15.

$[0, 5, 1, 0, 40, 1]$

Section 10.2

1. a. $d = 4$ **b.** $a_n = 4n + 1$ **2. a.** $d = -1$ **b.** $a_n = -n + 7$
3. a. $d = 6$ **b.** $a_n = 6n - 6$ **4. a.** $a_5 = 33$ **b.** $a_n = 2n + 23$
5. 3925   **6.** 420   **7. a.** $a_n = -n + 21$ **b.** 204 doughnuts
8. $\{4, 5, 6, 7, 8, 9, 10\}$ **9.** $\{4, 8, 12, 16, 20, 24, 28\}$

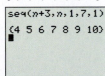

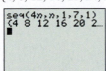

10. $\{2, 4, 8, 16, 32, 64, 128\}$

```
seq(2^n,n,1,7,1)
{2 4 8 16 32 64…
```

Section 10.3

1. a. $r = 4$ **b.** $a_n = 8(4)^{n-1}$ **c.** 2,097,152
2. a. $r = 3$ **b.** $a_n = (-9)(3)^{n-1}$ **c.** $-177,147$
3. a. $r = -3$ **b.** $a_n = 9(-3)^{n-1}$ **c.** $-177,147$
4. a. $r = \dfrac{1}{3}$ **b.** $a_n = \dfrac{1}{3}\left(\dfrac{1}{3}\right)^{n-1}$ **c.** $\dfrac{1}{59,049}$

5. a. 32   **b.** $a_n = 1(2^{n-1})$ **6.** Answers will vary.
7. 354,288   **8.** $-305,175,780$ **9. a.** $r = 0.4$ **b.** $a_1 = 3$
c. 5   **10. a.** $r = -0.1$ **b.** $a_1 = 4$ **c.** $3.\overline{63}$

11. a. $r = \dfrac{1}{4}$ **b.** $a_1 = 2$ **c.** $\dfrac{8}{3}$ **12.** $\dfrac{9}{2}$

13. a. $\{17,850 \text{ units}, 21,242 \text{ units}, 25,277 \text{ units}, \ldots\}$
b. $r \approx 1.19$ **c.** $a_n = 17,850(1.19)^{n-1}$ **d.** 71,800 units
14. 255   **15.** 26

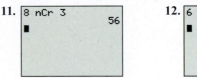

16. $2.71\overline{6}$

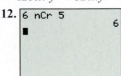

Section 10.4

1. 40,320   **2.** 45   **3.** 120   **4.** 252   **5.** 84
6. 21   **7.** 84
8. The coefficient of the fourth term is 84.

```
                        1
                    1       1
                1       2       1
            1       3       3       1
        1       4       6       4       1
      1     5      10      10       5      1
    1     6     15      20      15       6     1
   1    7     21     35      35      21      7     1
  1   8    28     56     70      56     28      8    1
 1   9    36    84    126    126     84     36     9    1
1                                                          1
```

9. $1024x^5 + 2560x^4y + 2560x^3y^2 + 1280x^2y^3 + 320xy^4 + 32y^5$
10. $1024x^5 - 2560x^4y + 2560x^3y^2 - 1280x^2y^3 + 320xy^4 - 32y^5$

11.
```
8 nCr 3
                56
```
12.
```
6 nCr 5
                 6
```

13.
```
7!
              5040
```
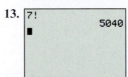

14.
```
8!
            40320
```
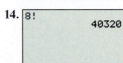

15.
```
11!/(5!*(11-5)!)
              462
```

APPENDIX 1

1–6. Answers will vary.

APPENDIX 2

1. 36   **2.** -42 **3.** 22   **4.** 23   **5.** 0
6. a. $x_1 = 3$ **b.** $x_2 = 4$ **7. a.** $x_1 = -2$ **b.** $x_2 = 6$
8. a. $x_1 = 1$ **b.** $x_2 = -2$ **c.** $x_3 = 2$

Answers to Selected Exercises

CHAPTER 1

Section 1.1 Vocabulary Practice
1. B **2.** G **3.** H **4.** A **5.** J **6.** C **7.** D **8.** F
9. I **10.** E

Section 1.1 Exercises
1.
3. 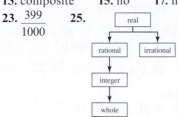 **5.** $\{-4, -1, 2, 3\}$
7. $\{-250, -200, -150, -100, -50, 0, 50, 100, 150, 200\}$
9. $\{\ldots, -3, -2, -1, 0, 1, 2, 3, \ldots\}$ **11.** prime
13. composite **15.** no **17.** no **19.** yes **21.** yes
23. $\dfrac{399}{1000}$ **25.**

```
          real
         /    \
    rational  irrational
       |
    integer
       |
     whole
```

27. ABD **29.** AB **31.** AC **33.** ABDEF **35.** AB
37. H **39.** AC **41.** see table below **43.** $(4, \infty)$
45. $(-\infty, 4)$ **47.** $[4, \infty)$ **49.** $(-\infty, 4]$
51. a. ─── **b.** $(-\infty, 7]$
53. a. ─── **b.** $[-4, \infty)$
55. a. ─── **b.** $(2, 10)$
57. a. ─── **b.** $(-\infty, -15]$
59. a. ─── **b.** $[25, 100)$ **61.** 81 **63.** 81
65. 8 **67.** -8 **69.** 9 **71.** 7 **73.** 1 **75.** 6
77. not a real number **79.** 1 **81.** -5 **85.** 2, 3, 5, 7,
11, 13, 17, 19, 23, 29, 31, 37, 41, 43, 47 **87.** -14 **89.** 8
91. -62 **93.** -32 **95.** 26 **97.** 16.03 **99.** -441
101. b. $[-5, \infty)$ **103. b.** not a real number

Section 1.2 Vocabulary Practice
1. H **2.** G **3.** F **4.** C **5.** D **6.** B **7.** A **8.** J
9. E **10.** I

Section 1.2 Exercises
1. $21w - 31y$ **3.** $-8x + 3y$ **5.** $-x - \dfrac{35}{6}$
7. $16x^2 + 15x - 4$ **9.** $-a + \dfrac{3}{2}$ **11.** $21x - 28$
13. $21x^2 - 28x$ **15.** $21x^3 - 28x^2$ **17.** $2p - 2$
19. $54f$ **21.** $-2n + 12$ **23.** $-6n - 12$ **25.** $-4g - \dfrac{31}{45}$
27. $0.7b + 1$ **29.** $-7n + 10.5p - 70z$ **33.** F **35.** A
37. D **39.** J **41.** I **43.** yes **49.** $5h^3 + 13h^2$
51. $x^{11}y^{10}$ **53.** u^4x^3 **55.** x^{12} **57.** $a^{24}b^6$ **59.** $8x^{15}y^{21}$
61. $45j^4k^2$ **63.** $\dfrac{x^3}{8}$ **65.** $\dfrac{1}{h^5}$ **67.** 3 **69.** 1 **71.** 1
73. $-69x^2 - 9$ **75.** w^6z^7 **77.** $-\dfrac{44}{15}a$ **79.** $\dfrac{17}{10}k^3$
81. $\dfrac{3b^6}{a^4}$ **83.** $\dfrac{12y^{15}}{x^{13}}$ **85.** $\dfrac{1}{v^8}$ **87.** $\dfrac{1}{v^{32}}$ **89.** $\dfrac{36x^7y^8}{175}$
91. a. z^5 **b.** z^5 **93.** -32 **95.** -264 **97. b.** x^{21}
99. b. $2h - 4$

Section 1.3 Vocabulary Practice
1. J **2.** I **3.** D **4.** F **5.** G **6.** A **7.** H **8.** B
9. C **10.** E

Section 1.3 Exercises
1. a. ─── **b.** $[5, 5]$
3. a. ─── **b.** $(5, \infty)$
5. a. ─── **b.** $(-\infty, 5)$
7. a. ─── **b.** $[5, \infty)$
9. a. ─── **b.** $(-\infty, 5]$
11. a. ─── **b.** $\left[\dfrac{3}{7}, \dfrac{3}{7}\right]$
13. a. ─── **b.** $(100, \infty)$
15. a. ─── **b.** $(-\infty, 5)$ **17. a.** $p = 29$
19. a. $n = -71$ **21. a.** $b = -2$ **23. a.** $x = \dfrac{17}{2}$
25. a. $z = -6$ **27. a.** $x = 22$ **29. a.** $c = 5$
31. a. $h = 20$ **33. a.** $w = -1.1$ **35. a.** $a = 4.5$

Table for Section 1.1 exercise 41

	Real numbers	Rational numbers	Irrational numbers	Integers	Whole numbers
$\dfrac{3}{5}$	X	X			
π	X		X		
-8	X	X		X	
$\sqrt{13}$	X		X		
$\sqrt{100}$	X	X		X	X
$-0.\overline{43}$	X	X			

37. a. $x = \dfrac{4}{3}$ **39. a.** $k = 120$ **41. a.** $a = \dfrac{5}{3}$

43. a. $k = 84$ **45. a.** $c = -\dfrac{21}{2}$ **47. a.** $q = 450$

49. a. no solution   **51. a.** the set of real numbers   **53. a.** no solution   **55. a.** the set of real numbers   **57. a.** $z = 0$

59. a. $p = \dfrac{22}{9}$ **61. a.** $h = 108$ **63. a.** $M = 40$

65. a. $n = \dfrac{32}{3}$ **67. a.** the set of real numbers

69. a. $a = -\dfrac{5}{18}$ **71. a.** $v = -16$ **73. a.** $x = 10$

75. a. $z = \dfrac{1}{11}$ **77. a.** $x = 0$ **89. a.** $x \le 8$

c. $(-\infty, 8]$ **91. a.** $x > -8$ **c.** $(-8, \infty)$
93. a. $x < 2$ **c.** $(-\infty, 2)$ **95. a.** $x \ge -2$ **c.** $[-2, \infty)$
97. a. $d < -9$ **c.** $(-\infty, -9)$ **99. a.** $n < -109$
c. $(-\infty, -109)$ **101. a.** $x > -84$ **c.** $(-84, \infty)$
103. a. $x < -72$ **c.** $(\infty, -72)$ **105. a.** $y \le 0$ **c.** $(-\infty, 0]$
107. a. $x \le -\dfrac{10}{9}$ **c.** $\left(-\infty, -\dfrac{10}{9}\right]$ **109. a.** $f < -3$

c. $(-\infty, -3)$ **111. a.** $k < 0$ **c.** $(-\infty, 0)$
113. a. $x \ge \dfrac{3}{2}$ **c.** $\left[\dfrac{3}{2}, \infty\right)$ **115. a.** $x > 0$ **c.** $(0, \infty)$

117. a. $x \le -7$ **c.** $(-\infty, -7]$ **119. a.** $x \ge 119$ **c.** $[119, \infty)$

121. a. $x > 315$ **c.** $(315, \infty)$ **123. a.** $x > \dfrac{17}{3}$ **c.** $\left(\dfrac{17}{3}, \infty\right)$

125. a. $p < 0$ **c.** $(-\infty, 0)$ **127. a.** $p > 0$ **c.** $(0, \infty)$
129. a. $p < 0$ **c.** $(-\infty, 0)$ **131. a.** $p > 0$ **c.** $(0, \infty)$

133. a. $m > \dfrac{10}{9}$ **c.** $\left(\dfrac{10}{9}, \infty\right)$ **135. a.** $k \ge -\dfrac{75}{2}$ **c.** $\left[-\dfrac{75}{2}, \infty\right)$

137. a. $w < \dfrac{41}{6}$ **c.** $\left(-\infty, \dfrac{41}{6}\right)$ **139. a.** $h > \dfrac{1}{28}$ **c.** $\left(\dfrac{1}{28}, \infty\right)$

141. a. $u > 16$ **c.** $(16, \infty)$ **143. a.** $x \ge \dfrac{39}{5}$ **c.** $\left[\dfrac{39}{5}, \infty\right)$

145. b. $x > -4$ **147. b.** $(-\infty, 5)$

Section 1.4 Vocabulary Practice
1. I   **2.** E   **3.** G   **4.** B   **5.** H   **6.** J   **7.** F   **8.** C
9. A   **10.** D

Section 1.4 Exercises
1. 7×10^{-6} meter   **3.** 2.7×10^{12} metric tons
5. 1.04×10^4 light-years   **7.** 1×10^{-4} micrograms
9. 2.5×10^1 kilonewtons   **11.** 2.79×10^{14}
13. 2.79×10^{-2} **15.** 2.79×10^4 **17.** 3.6×10^{11}
19. 3.6×10^{-9} **21.** 7.5×10^{11} **23.** 6.4×10^{-11}
25. 8×10^{21} **27.** 5×10^2 **29.** 5×10^6 **31.** 5×10^{-8}
33. 1.26×10^5 **35.** 1.26×10^{-3} **37.** 9.72×10^5
39. 3.294×10^6 **41.** 0.03 gram   **43.** 6000 meters
45. 0.8 bel   **47.** 2,000,000 tons   **49.** 0.000007 gram
51. 1 km = 1000 m   **53.** 2 c = 1 pt   **55.** 1 min = 60 s
57. 1.06 qt ≈ 1 L   **59.** 1 dollar ≈ 10.687 Mexican pesos
61. $3080 \dfrac{\text{ft}}{\text{min}}$ **63.** $40 \dfrac{\text{km}}{\text{hr}}$ **65.** $55 \dfrac{\text{ft}}{\text{s}}$ **67.** $2.5 \dfrac{\text{kg}}{\text{L}}$

69. $2.5 \dfrac{\text{kg}}{\text{L}}$ **71.** $2.0 \times 10^{-3} \dfrac{\text{AU}}{\text{s}}$ **73.** 3 capsules

75. $2.60 \dfrac{\text{U.S. dollars}}{\text{gallon}}$ **77.** 4.07×10^{13} gal   **79.** 5×10^2 s

81. 175 g oxygen   **83.** 2×10^0 s   **85.** 10^{12}

89. 1.312×10^{15} **91.** 1.312×10^{-13} **93. b.** $51 \dfrac{\text{ft}}{\text{s}}$
95. b. 6.2×10^4

Section 1.5 Vocabulary Practice
1. B   **2.** B   **3.** A   **4.** B   **5.** A   **6.** A   **7.** B   **8.** B
9. B   **10.** B

Section 1.5 Exercises
1. $600   **3.** $73.50   **5.** $3,240,000   **7.** $9900   **9.** 17 hr
11. 100 text messages   **13.** 195,233 applications
15. 125,000 gallons   **17.** $27   **19.** 1031 dentists
21. 24,300 students   **23.** 20%   **25.** The traditional plan costs $135.50 more than the PPO plan.   **27.** 193,000 people
29. 20%   **31.** Using the Discover card costs $0.14 more per year than using the AT&T Universal card.   **33.** 2.8 yd
35. 72.3 million children   **37.** 83% score   **39.** 39 gal
41. Machine 2 produces 23,128,000 ft² more newsprint than machine 1.   **43.** $34,728   **45.** 26%   **47.** $319
49. $130   **51.** $117,000   **53.** 6,218,000 positions

55. 69%   **57.** $41,347.50   **59.** $1.02 \times 10^3 \dfrac{\text{m}}{\text{s}}$

61. 19 drops per minute   **63.** $47.2 billion   **65.** 77.2 million hogs and pigs   **67.** $19,500   **69.** 342,000 ft³

71. a. 800%   **73. b.** $125   **75. b.** $4.32 \times 10^{-12} \dfrac{\text{kg} \cdot \text{m}^2}{\text{s}^2}$

Section 1.6 Vocabulary Practice
1. G   **2.** J   **3.** C   **4.** E   **5.** B   **6.** H   **7.** I   **8.** D
9. F   **10.** A

Section 1.6 Exercises
1. $12x + y = 3$ **3.** $4x + y = 39$ **5.** $15x + y = -46$
7. a.

x	y
3	0
0	6

b.

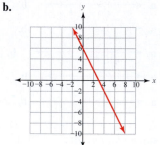

9. a.

x	y
8	0
0	−6

b.

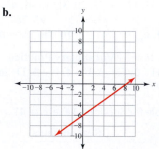

11. a.

x	y
15	0
0	−3

b.

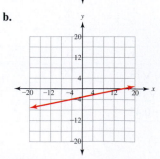

13. a.

x	y
5	0
0	−6

b.

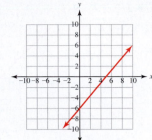

15. a.

x	y
6	0
0	$\dfrac{12}{5}$

b.

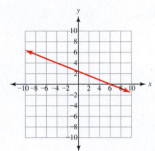

17. a.

x	y
$-\dfrac{3}{2}$	0
0	3

b.

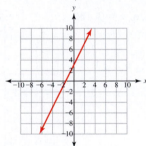

19. a.

x	y
$\dfrac{28}{3}$	0
0	−7

b.

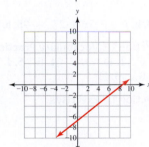

21. a.

x	y
8	0
0	8

b.

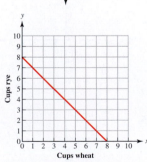

23. 12   **25.** 1   **27.** $-\dfrac{6}{5}$ **29.** $-\dfrac{3}{28}$ **31.** undefined

33. 0   **35.** $\dfrac{4}{3}$ **37.** $-\dfrac{1}{2}$ **39.** $-\dfrac{5}{48}$ **41.** $\dfrac{4}{5}$ **43.** $\dfrac{\$0}{1\ \text{year}}$

45. $\dfrac{50\ \text{m}}{1\ \text{s}}$ **47. b.** $-\dfrac{4}{9}$ **49. b.** $\dfrac{1}{3}$ **51. b.** 5   **53. b.** $-\dfrac{2}{5}$

55. b. 1   **57. b.** −1   **59. b.** 1   **61. b.** −2

63. b. undefined   **65. b.** 0   **67. b.** 1   **69.** $\dfrac{1}{3}$

71. undefined   **73.** $-\dfrac{1}{13}$ **75.** $\dfrac{8}{7}$ **77.** undefined

79. a. $y = \dfrac{2}{3}x - 3$ **b.** $\dfrac{2}{3}$ **c.** $(0, -3)$ **81. a.** $y = 3x - 1$
b. 3   **c.** $(0, -1)$

83. a. $\dfrac{5}{8}$ **b.** $(0, -6)$ **c.**

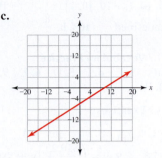

85. a. −2   **b.** $(0, 5)$ **c.**

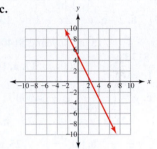

87. a. $-\dfrac{1}{2}$ **b.** $(0, 0)$ **c.**

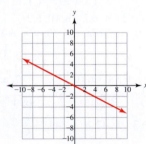

89. a. $-\dfrac{\$4000}{1\ \text{year}}$ **b.** $(0\ \text{years},\ \$16,000)$

c.

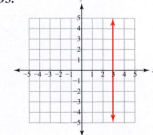

91.   **93.**

97.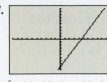

99.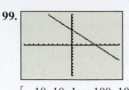

$[-10, 10, 1, -10, 10, 1]$ $[-10, 10, 1, -100, 100, 10]$

101. b. undefined   **103. b.** $\dfrac{8}{3}$

Section 1.7 Vocabulary Practice

1. G **2.** C **3.** F **4.** I **5.** D **6.** E **7.** J **8.** H
9. A **10.** B

Section 1.7 Exercises

1. $y = \dfrac{5}{8}x + 3$ **3.** $y = -6x - 4$ **5.** $y = 15$

7. $y = -7x$ **9.** $y = \left(\dfrac{\$10.25}{1\ hr}\right)x + \$80$

11. a. $(0, 3)$ **b.** 2 **c.** $y = 2x + 3$ **13. a.** $(0, 1)$ **b.** -3

c. $y = -3x + 1$ **15. a.** $(0\ years, \$1400)$ **b.** $-\dfrac{\$175}{1\ year}$

c. $y = \left(-\dfrac{\$175}{1\ year}\right)x + \$1400$ **17. a.** $(0\ years, \$4000)$

b. $\dfrac{\$8000}{1\ year}$ **c.** $y = \left(\dfrac{\$8000}{1\ year}\right)x + \$4000$ **19. a.** $y = 8x - 25$

b. $(0, -25)$ **21. a.** $y = -8x + 25$ **b.** $(0, 25)$

23. a. $y = \dfrac{2}{3}x + \dfrac{14}{3}$ **b.** $\left(0, \dfrac{14}{3}\right)$ **25. a.** $y = -3$ **b.** $(0, -3)$

27. a. $y = 0.2x + 6.2$ **b.** $(0, 6.2)$ **29. a.** 5 **b.** $y = 5x - 29$

c. $(0, -29)$ **31. a.** -4 **b.** $y = -4x + 48$ **c.** $(0, 48)$

33. a. -5 **b.** $y = -5x + 2$ **c.** $(0, 2)$ **35. a.** $\dfrac{3}{4}$

b. $y = \dfrac{3}{4}x + 18$ **c.** $(0, 18)$ **37. a.** 0 **b.** $y = -9$

c. $(0, -9)$ **39.** $x = -2$ **41.** $y = 5$ **43.** $y = 6$

45. $x = -3$ **47.** 8   **49.** $-\dfrac{1}{8}$ **51.** -1

53. $y = 9x - 58$ **55.** $y = \dfrac{3}{4}x + \dfrac{1}{2}$ **57.** $y = -\dfrac{8}{5}x - \dfrac{31}{5}$

59. $y = \dfrac{3}{2}x + 12$ **61.** $y = -\dfrac{1}{5}x - \dfrac{26}{5}$

63. $y = 5x - 26$ **65.** $y = -2x - 1$ **67.** $y = \dfrac{2}{3}x + \dfrac{23}{3}$

69. $x = 10$ **71.** $y = 11$ **73.** $y = 1$ **75.** $x = 6$

77. a. $y = 3x - 4$ **b.** $y = 3x + 12$ **79. a.** $y = -x + 4$

b. $y = -x + 5$ **81. a.** $y = -3$ **b.** $y = 1$

83. a. $x = -3$ **b.** $x = 1$

85. a. $y = \dfrac{2}{3}x + 2$ **b.**

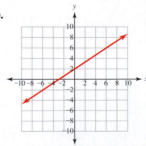

87. a. $y = -x$ **b.**

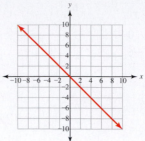

89. a. $y = -3x + 5$ **b.**

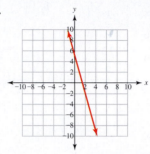

91. a. $y = 3$ **b.**

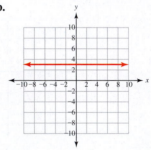

93. b. $y = -5x + 27$ **95. b.** $y = -1$

Section 1.8 Vocabulary Practice

1. F **2.** D **3.** G **4.** E **5.** J **6.** A **7.** I **8.** C
9. B **10.** H

Section 1.8 Exercises

1.

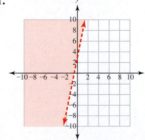

3.

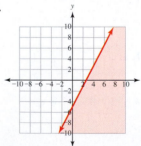

5.

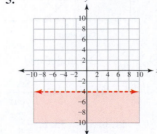

7.

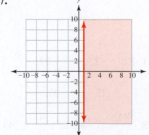

9.
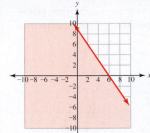

11.
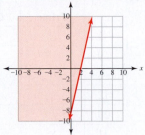

33.
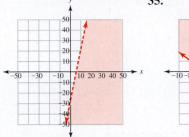

35.
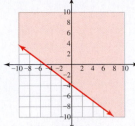

13.
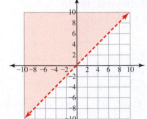

15.
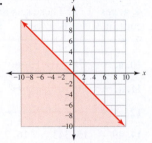

37.
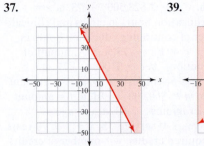

39.
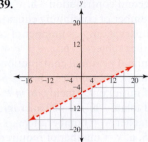

17.
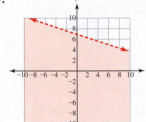

19.
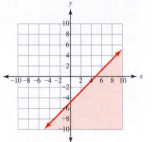

41.
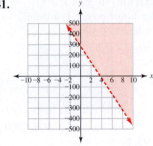

43.
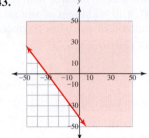

21.
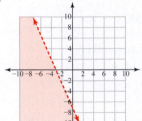

23.
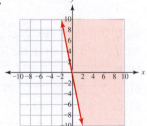

45.
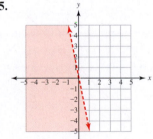

47.
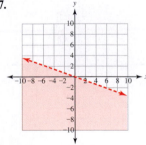

25.
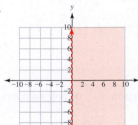

27.
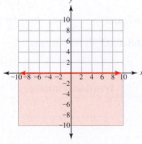

49.
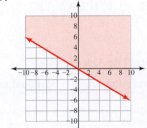

51. $y \le -2x - 1$

53. $y < \dfrac{2}{5}x - \dfrac{1}{5}$

55. $y < x$

57. $x < 1$

59. $y > -x$

61. $y \ge 2$

29.
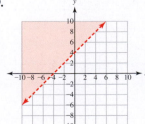

31.
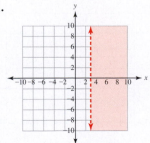

67. a. $x + y \le \$44{,}000$ **b.**

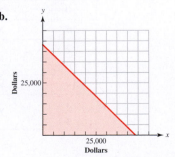

69. a. $\$0.99x + \$4.99y \le \$75$ **b.**

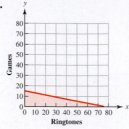

71. a. x = amount of first contribution, y = amount of second contribution **b.** $x + y \le \$28{,}500$ **73. a.** x = number of continuing education hours in ethics and clinical supervision, y = number of continuing education hours in other subjects **b.** $x + y \ge 40$ hr **75. a.** c = amount of annual personal contribution **b.** $c \ge \$1000$ **77. a.** n = number of vehicles **b.** $n \le 6$ vehicles **79. a.** m = amount of initial investment **b.** $m \ge \$50{,}000$ **81. a.** f = amount of folate **b.** $f \ge 400$ micrograms **83. a.** x = age of teacher, y = number of years of service **b.** $x + y \ge 70$ years **85. a.** x = number of required credits, y = number of credits that can be paid for with financial aid **b.** $y \le 1.25x$ **87. a.** x = amount of annual take-home pay, y = annual amount of mortgage payments **b.** $y \le 0.40x$

89.

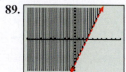

$[-10, 10, 1, -10, 10, 1]$

91.

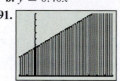

$[-2, 10, 1, 0, 20000, 2000]$

93. b.

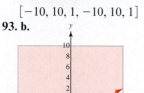

95. b.

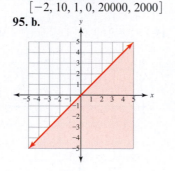

Chapter 1 Review Exercises

1. $\{-4, 0, 2\}$ **2.** $\{\ldots, -3, -2, -1, 0, 1, 2, 3, \ldots\}$

3.

	Prime	Composite	Neither prime nor composite
5	X		
0			X
-3			X
12		X	
20		X	
11	X		

4. a. yes **5.** These sets are not disjoint. Both contain 9. **6. a.** yes **8.** see table below

9.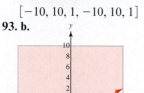

10. a. **b.** $(8, \infty)$

11. a. **b.** $(-\infty, 2]$

12. a. **b.** $(-\infty, \infty)$ **13.** 49

14. 49 **15.** 343 **16.** 7 **17.** 3 **18.** -4 **19.** 2.2 **20.** 3.5 **21.** -17 **22.** -33 **23.** -12 **24.** -16 **25.** square 4 **26.** divide 32 by 4 **27.** C **28.** A **29.** F **30.** E **31.** D **32.** B **33.** $\dfrac{2x^7}{5}$

34. $15x^{11}$ **35.** $-12x^2$ **36.** 1 **37.** $\dfrac{18p^7}{m^{16}}$ **38.** $\dfrac{9}{h^{10}}$

39. 5 **40.** $24x^3 - 12x^2$ **41.** $30x^5 + 20x^4$

42. $\dfrac{4h^6}{25k^4}$ **43.** $-14x - 58$ **44.** $10p - 3$ **45.** $\dfrac{12}{7}a + \dfrac{17}{40}$

46. $\dfrac{1}{4}w^5$ **48.** d: $\dfrac{7x^4}{y^5}$ **49.** undefined **50.** 0 **51.** 0

52. -7 **53.** 0 **54.** c, f **55.** $x = 16$

56. a. $x = 24$ **57. a.** no solution

58. a. $x = 0$ **59. a.** the set of real numbers

60. a. **b.** $[4, 4]$

61. a. **b.** $(-\infty, \infty)$

62. c **63.** b **64.** a

65. a. $x \le -10$ **c.** **d.** $(-\infty, -10]$

66. a. $x > 10$ **c.** **d.** $(10, \infty)$

67. a. $x > -\dfrac{5}{9}$ **c.** **d.** $\left(-\dfrac{5}{9}, \infty\right)$

68. 6.02×10^{23} atoms

Table for Chapter 1 Review exercise 8

	Real numbers	Rational numbers	Irrational numbers	Integers	Whole numbers
5	X	X		X	X
-1.1	X	X			
0	X	X		X	X
π	X		X		
$-\dfrac{2}{9}$	X	X			
-14	X	X		X	
$\dfrac{3}{4}$	X	X			

69. $6.673 \times 10^{-11} \dfrac{m^3}{kg \cdot s^2}$ **70. b.** 3.5×10^9

71. b. 3.5×10^{-7} **72.** 3.6×10^{-13} **73.** 9.1×10^{12}

74. 5×10^{10} **75.** 6.74×10^3 **76.** 8000 meters

77. 0.000004 liter   **78.** 1 mi = 5280 ft

79. 1000 mL = 1 L   **80.** $\dfrac{1.6 \text{ km}}{1 \text{ mi}}$ **81.** $\dfrac{1 \text{ min}}{60 \text{ s}}$ **82.** $270 \dfrac{L}{hr}$

83. 384 T   **84.** 4.24×10^0 hr   **85.** 1.0 million victims

86. \$750   **87.** 22%   **88.** 63,750 barrels   **89.** 7.0 million students   **90.** 8 gal   **91.** $3x - y = -11$

92. a.

x	y
8	0
0	−3

b.

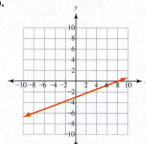

93. a.

x	y
$\dfrac{5}{3}$	0
0	5

b.

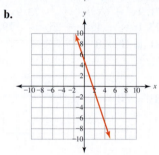

94. undefined   **95.** $-\dfrac{7}{9}$ **96.** 0   **97.** $\dfrac{2}{9}$ **98.** $-\dfrac{1}{3}$

99. a. $\dfrac{1}{8}$ **b.** −8   **100.** c   **101. a.** $y = \dfrac{6}{5}x - 4$ **b.** $\dfrac{6}{5}$

c. $(0, -4)$ **d.**

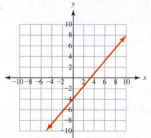

102. a. $-\dfrac{3}{5}$ **b.** $(0, 7)$ **c.**

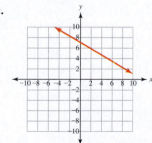

103.

Equation	Slope of the line	y-intercept of the line
$y = -3x + 1$	−3	$(0, 1)$
$x = 2$	undefined	none
$y = 9x$	9	$(0, 0)$
$y = 8$	0	$(0, 8)$
$3x - 5y = 24$	$\dfrac{3}{5}$	$\left(0, -\dfrac{24}{5}\right)$

104.

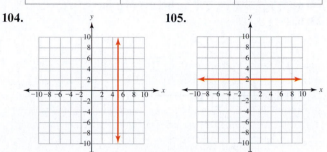

105.

106. $y = -11x + 7$ **107.** $y = \dfrac{1}{2}x + 8$

108. $y = -x + 14$ **109.** $x = 9$ **110.** $y = 6$

111. $y = 3x + 18$ **112.** $y = -\dfrac{3}{2}x + \dfrac{17}{2}$ **113.** 9

114.

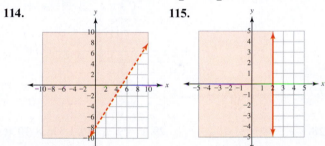

115.

116.

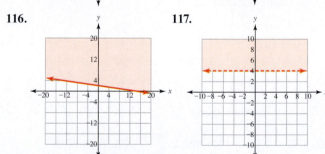

117.

118. a. $\geq$ **119.** $y \geq -2x + 7$

120. a. $x + y \geq 75$ **b.**

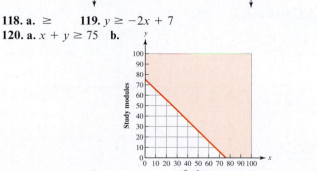

121. a. x = deaths from cardiovascular disease

b. $x \geq 800,000$ deaths

Chapter 1 Test

1. -28 **2.** 8 **3.** 64 **4.** not a real number

5. undefined **6.** 0 **7.** 1 **8.** $\dfrac{1}{36}$ **9.** -16 **10.** 16

11. 0 **12.** $147x^{13}y$ **13.** $\dfrac{9y^4}{20x^2}$ **14.** $125k^9$

15. the commutative property **16. a.** $z = 6$ **17. a.** no solution **18. a.** $p = 6$ **19. a.**

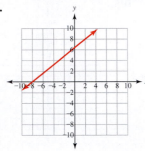

b. $(-\infty, \infty)$ **20. a.** $x > -9$ **21. a.** $x < 9$

22. a. ![number line] **b.** $(-\infty, 4]$

23. b.

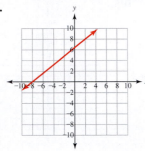

24. $\dfrac{2}{9}$ **26.** $y = -11x + 20$

27. a. $-\dfrac{5}{6}$ **b.** $(0, 7)$ **c.**

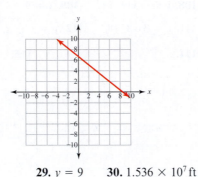

28.

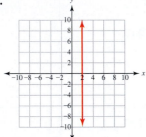

29. $y = 9$ **30.** 1.536×10^7 ft

31. 6.45×10^{-3} **32.** $3067.2 \dfrac{\text{km}}{\text{hr}}$ **33.** 24 showers

34. \$70 per credit **35.**

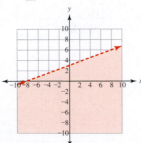

36. $y = -\dfrac{2}{3}x + 1$

CHAPTER 2

Section 2.1 Vocabulary Practice

1. E **2.** D **3.** I **4.** H **5.** G **6.** A **7.** C **8.** J
9. F **10.** B

Section 2.1 Exercises

1. a. 1 **b.** binomial **3. a.** 3 **b.** trinomial **5. a.** 1
b. monomial **7. a.** 0 **b.** monomial **13. a.** polynomial
15. a. not a polynomial **b.** A variable is in the denominator.
17. a. polynomial **19. a.** not a polynomial
b. An exponent is not an integer. **21.** $12x + 13$
23. $-2x + 17$ **25.** $35x^2 + 115x + 30$
27. $35x^2 + 95x - 30$ **29.** $q^2 + 2q + 13$
31. $q^2 - 12q + 17$ **33.** $-38h^2 + 11h - 29$
35. $3h^3 + 41h^2 + h + 11$ **37.** $11x^2 - 2xy + 8y^2$
39. $-13a^2 + 13ab - 15b^2$ **41.** $-2ab + 2b^2$
43. $x^3 - x^2 - 3x + 27$ **45.** $x^3 + x^2 - 3x - 27$
47. $x^3 + 7x^2 + 21x + 27$ **49.** $c^2 + 18c + 81$
51. $c^2 - 18c + 81$ **53.** $c^2 - 18c + 81$ **55.** $c^2 - 81$
57. $25y^2 - 9$ **59.** $-x^2 + y^2$ **61.** $2y$ **63.** $2x$
65. $a^2 + 2ab + b^2$ **67.** $a^2 - b^2$ **69.** $a^3 + b^3$
71. $9x^2 + 6xy + y^2$ **73.** $9x^2 - 6xy + y^2$ **75.** $x^3 - 216y^3$
77. $16x^2 - 40xy + 25y^2$ **79.** $8h^3 + 27k^3$

81. a. $W + 8$ **b.** **c.** $4W + 16$ **d.** $W^2 + 8W$

W

$W + 8$

83. a. $2W + 7$ **b.** **c.** $6W + 14$ **d.** $2W^2 + 7W$

W

$2W + 7$

85. a. $\dfrac{5}{4}W$ **b.** **c.** $\dfrac{9}{2}W$ **d.** $\dfrac{5}{4}W^2$

W

$\dfrac{5}{4}W$

87. \$4.9 billion **89.** $339{,}000$ ft^3 **91. b.** $-20x^4 + 8x^3$
93. b. $40x^2$ **95.** $\{0, 1, 2, 3, \ldots\}$

Section 2.2 Vocabulary Practice

1. G **2.** F **3.** D **4.** J **5.** H **6.** B **7.** E **8.** I
9. C **10.** A

Section 2.2 Exercises

1. $2 \cdot 3^2 \cdot x^2 \cdot y$ **3.** $2 \cdot 3 \cdot 7^2 \cdot a \cdot b^3$ **5.** $60x^2$ **7.** $14a$
9. $7ac$ **11.** $10xyz$ **13.** k^2z **15.** 72
17. $3(x^2 + 5x - 9)$ **19.** $3x(x^2 + 5x - 9)$
21. $5a^2(2b^2 - 9c)$ **23.** $4fg(3 + h)$
25. $x(7x^3 - 21x + 14x^4 - 1)$
27. $x(-7x^3 + 21x - 14x^4 + 1)$ **29.** $(3a + 5)(2x + y)$
31. $(2x + y)(3a + 5)$ **33.** $(3a - 5)(2x + y)$
35. $(3x + 5)(9y - 1)$ **37.** $(10p - 9)(11q - 2)$
39. $(x + z)(3x + y)$ **41.** $(x + z)(3x - y)$
43. $(-3a + q)(7a + p)$ **45.** $(b + c)(b - c)$
47. $(5y + 9)(5y - 9)$ **49.** $(5y + 9z^6)(5y - 9z^6)$
51. $(z + w)^2$ **53.** $(3z + 5)^2$
55. $(5x + y)(25x^2 - 5xy + y^2)$ **57.** $(a - 3)(a^2 + 3a + 9)$
59. $(2a - 3)(4a^2 + 6a + 9)$ **61.** $(2a + 3)(4a^2 - 6a + 9)$
63. $(4p + w)(16p^2 - 4pw + w^2)$
65. $(6b - 5c)(36b^2 + 30bc + 25c^2)$
67. $3(x + 2y)(x - 2y)$ **69.** $3(w + 2h)(3x - 5)$

71. $5(x + y)^2$ **73.** $10(x - y)(x^2 + xy + y^2)$
75. $4(a^2 + 8b^2)$ **77.** $3(xy + 4bx - fx + 3)$
79. $7(2x + y)^2$ **81.** $4(3a + 1)(3a - 1)$
83. a. $(x + 3)(x^2 + 2)$ **c.** Yes   **85. a.** πn^2 **b.** πN^2
c. $\pi N^2 - \pi n^2$ **d.** $\pi(N - n)(N + n)$ **87.** \$330,000
per mile   **89.** 17%   **91. b.** $9(x - 3y)(x + 3y)$
93. b. $(x + 1)(x + 4)$ **95.** 3, 7   **97.** 5, 5

Section 2.3 Vocabulary Practice

1. I   **2.** H   **3.** E   **4.** J   **5.** F   **6.** G   **7.** D   **8.** B
9. A   **10.** C

Section 2.3 Exercises

1. a. 81   **b.** not prime   **3. a.** -47 **b.** prime   **5. a.** 1
b. not prime   **7. a.** -7 **b.** prime   **11. a.** $(x + 5)(2x + 1)$
13. a. $(x - 5)(2x + 1)$ **15. a.** $(x + 5)(2x - 1)$
17. a. $2x^2 + 5x + 6$ **b.** $-23 < 0$ **19. a.** $(x - 5)(2x - 1)$
21. a. $(x - 5)(2x + 3)$ **23. a.** $(x + 2)(3x + 4)$
25. a. $(3x + 1)(4x + 5)$ **27. a.** $(x - 2)(3x - 2)$
29. a. $(z + 5)^2$ **31. a.** $(w - 5)^2$ **33. a.** $(a + 3)(a + 2)$
35. a. $(f + 2)(f + 9)$ **37. a.** $(b + 11)(b + 1)$
39. a. $(y - 12)(y + 3)$ **41. a.** $(c - 3)^2$
43. a. $(p - 7)(p - 8)$ **45. a.** $y^2 - 8y - 15$ **b.** 124 is not a
perfect square.   **47. a.** $(q - 6)(q - 3)$
49. a. $(x - 9)(x + 11)$ **51.** $(x^2 + 5)(x^2 + 9)$
53. $(x^2 + 7)(4x^2 - 3)$ **55.** $(3n^2 - 2)(3n^2 + 1)$
57. $(u^5 - 8)(2u^5 + 5)$ **59.** $(3a^2 - 4)^2$
61. $(5w^4 + 1)(7w^4 - 1)$ **63.** $4(h - 7)(2h + 3)$
65. $(x - 2)(x + 2)(x + 3)$ **67.** $7x(x - y)(x^2 + xy + y^2)$
69. $c(2c + 5)(3c - 7)$ **71.** $2x(x^2 + 7x + 15)$
73. $7(2p + 3)^2$ **75.** prime   **77.** $3(p - 3)(7p - 4)$
79. $-4x^5(1 + 4x^2)$ **81.** $(y - 3)(y + 3)(y - 2)$
83. $4(5ab - 10ac + b - c)$ **85.** $3w(w + 5)^2$
87. $5(x^2 + 13x - 36)$ **89.** $(c + 3)(c^2 + 16)$
91. $a(5a - 1)^2$ **93.** $(x - 3)(x + 3)$ **95. a.** 0   **b.** -36;
prime   **97.** 357% increase   **99.** \$859.27
101. b. $(x + 6)(x - 1)$ **103. b.** prime   **105.** $x = \dfrac{2}{3}$
107. zero

Section 2.4 Vocabulary Practice

1. E   **2.** I   **3.** A   **4.** D   **5.** G   **6.** H   **7.** F   **8.** B
9. C   **10.** J

Section 2.4 Exercises

1. $z = -8$ or $z = 2$ **3.** $x = 0$ or $x = -1$ or $x = \dfrac{3}{2}$
5. $d = 0$ or $d = \dfrac{7}{2}$ or $d = -\dfrac{1}{5}$ **7.** $x = 0$ or $x = -8$
9. $c = -3$ **11.** $p = -3$ or $p = 3$ **15. a.** $x = -5$ or $x = 9$
17. a. $j = 3$ or $j = -3$ **19. a.** $d = 0$ or $d = 1$ or $d = 4$
21. a. $x = -\dfrac{5}{2}$ **23. a.** $a = -\dfrac{1}{4}$ or $a = 3$
25. a. $x = -\dfrac{5}{2}$ or $x = \dfrac{5}{2}$ **27. a.** $p = -\dfrac{7}{3}$ or $p = -\dfrac{9}{2}$
29. a. $x = -\dfrac{2}{5}$ or $x = \dfrac{1}{3}$ **31. a.** $c = 0$ or $c = 7$
33. a. $c = 0$ or $c = -1$ **35. a.** $w = -\dfrac{5}{6}$ or $w = 3$
37. a. $w = -\dfrac{5}{6}$ or $w = 3$ **39. a.** $x = 8$
41. a. $m = -4$ or $m = 4$ **43. a.** $x = -\dfrac{7}{3}$ or $x = \dfrac{1}{2}$
47. a. $x = -7$ or $x = -6$ **49. a.** $f = 3$ or $f = -2$
51. a. $p = 0$ or $p = 3$ **53. a.** $p = 0$ or $p = -3$

55. a. $p = 0$ or $p = -3$ **57. a.** $v = -4$ or $v = 4$
59. a. $v = -\dfrac{4}{3}$ or $v = \dfrac{4}{3}$ **61. a.** $x = -\dfrac{2}{3}$ or $x = -\dfrac{1}{4}$
63. a. $x = -\dfrac{5}{3}$ or $x = \dfrac{7}{2}$ **65. a.** $f = 9$ or $f = -2$
67. a. $x = -1$ or $x = 0$ or $x = 4$ **69. a.** $x = -6$ or $x = 8$
71. a. $a = -4$ or $a = -2$ **73.** 9 ft long, 6 ft wide
75. 10 m long, 8 m wide   **77.** 15 ft long, 7 ft wide
79. 21 in. long, 14 in. wide   **81.** 22 cm long, 8 cm wide
83. 44 m long, 22 m wide   **85.** height = 12 ft, base = 10 ft
87. height = 6 in., base = 18 in.   **89.** 184 million Internet
users   **91.** 3.25 mi   **93. b.** $x = -3$ or $x = -2$
95. b. $x = \dfrac{3}{2}$ or $x = -7$ **97. a.**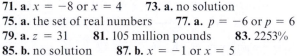
b. 6   **99.** $x = 11$

Section 2.5 Vocabulary Practice

1. A   **2.** D   **3.** D   **4.** I   **5.** C   **6.** F   **7.** E   **8.** H
9. B   **10.** G

Section 2.5 Exercises

1. 3   **3.** 6   **5.** -1 **7.** 3.5938   **9.** 31%
11. a. $p = -2$ or $p = 2$ **13. a.** $x = -6$ or $x = 6$
15. a. $x = -24$ or $x = 24$ **17. a.** $x = -11$ or $x = -7$
19. a. no solution   **21. a.** $w = -5$ or $w = 11$
23. a. $z = -1$ or $z = 7$ **25. a.** no solution
27. a. $x = 0$ **29. a.** $x = -8$ or $x = 4$
31. a. $z = -20$ or $z = 6$ **33. a.** $x = -20$ or $x = 28$
35. a. $x = -8$ or $x = 14$ **37. a.** $w = -4$ or $w = 12$
39. a. $w = -12$ or $w = 0$ **41.** no solution
43. a. $x = -\dfrac{26}{3}$ or $x = 4$ **45. a.** $x = -\dfrac{33}{5}$ or $x = \dfrac{21}{5}$
47. a. $x = -\dfrac{1}{2}$ **49. a.** $x = -\dfrac{1}{2}$ **55. a.** $c = -3$ or $c = 21$
57. a. $a = -\dfrac{10}{3}$ or $a = -2$ **59. a.** $z = 1$
61. a. no solution   **63. a.** $x = 14$ **65. a.** $z = -18$ or $z = 13$
67. a. $a = -7$ **69. a.** $x = -\dfrac{8}{3}$ or $x = -\dfrac{5}{2}$
71. a. $x = -8$ or $x = 4$ **73. a.** no solution
75. a. the set of real numbers   **77. a.** $p = -6$ or $p = 6$
79. a. $z = 31$ **81.** 105 million pounds   **83.** 2253%
85. b. no solution   **87. b.** $x = -1$ or $x = 5$
89.   **91.**

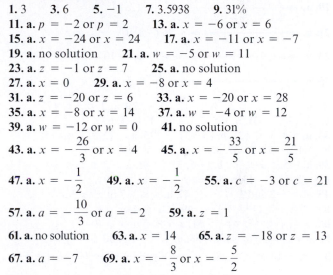

Section 2.6 Vocabulary Practice

1. C   **2.** B   **3.** G   **4.** J   **5.** A   **6.** D   **7.** F   **8.** I
9. E   **10.** H

Section 2.6 Exercises

1. a. {black, yellow, purple}
b. {red, black, blue, yellow, purple, green}

3. a. the set of real numbers greater than 4 and less than 9, $(4, 9)$
b. the set of real numbers, $(-\infty, \infty)$ **7.** disjunction
9. a. 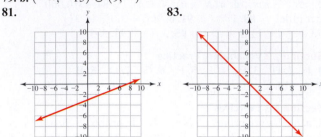 **b.**

c. **d.** $-2 \le x \le 3$

11. a. **b.**

c. **d.** $-6 \le x \le -2$

13. a.

b.

c.

15. a. **b.**

c.

17. a. **b.**

c.

19. a. **b.** $(2, 8)$

21. a. **b.** $(-4, 0]$

23. a. **b.** $[-10, -3]$

25. a. **b.** $[0, 5)$

27. a. $-12 < x \le 3$ **b.** $(-12, 3]$ **29. a.** $2 < x \le 17$
b. $(2, 17]$ **31. a.** $-5 < x < 6$ **b.** $(-5, 6)$
33. a. $5 < x < 12$ **b.** $(5, 12)$ **35. a.** $-12 \le x \le -5$
b. $[-12, -5]$ **37. a.** $-3 \le x < 5$ **b.** $[-3, 5)$
39. a. $-5 < x \le 3$ **b.** $(-5, 3]$ **41. a.** $-2 \le x < 5$
b. $[-2, 5)$ **43. a.** $-5 \le x \le 9$ **b.** $[-5, 9]$
45. $x > -2$ and $x < 9$ **47. a.** $x \le -7$ or $x \ge 3$
b. $(-\infty, -7] \cup [3, \infty)$ **49. a.** $x \le 4$ or $x \ge 7$
b. $(-\infty, 4] \cup [7, \infty)$ **51. a.** $x < 3$ or $x > 8$
b. $(-\infty, 3) \cup (8, \infty)$ **53. a.** $x = -10$ and $x = -6$
b. $x < -10$ or $x > -6$ **c.**

d. $(-\infty, -10) \cup (-6, \infty)$ **55. a.** $x = -10$ and $x = -6$
b. $-10 < x < -6$ **c.** **d.** $(-10, -6)$

57. a. $x = -6$ and $x = 12$ **b.** $-6 < x < 12$

c. **d.** $(-6, 12)$

59. a. $a = -8$ and $a = 4$ **b.** $a \le -8$ or $a \ge 4$

c. **d.** $(-\infty, -8] \cup [4, \infty)$

61. a. $x = -3$ and $x = 17$ **b.** $-3 \le x \le 17$

c. **d.** $[-3, 17]$

63. a. There are no boundary values **b.** the set of real
numbers **c.** **d.** $(-\infty, \infty)$

65. a. $w = -8$ and $w = 8$ **b.** $w < -8$ or $w > 8$

c.

d. $(-\infty, -8) \cup (8, \infty)$ **67. a.** $x = -7$ and $x = 13$

b. $x \le -7$ or $x \ge 13$ **c.**

d. $(-\infty, -7] \cup [13, \infty)$ **69. a.** $x = -3$ and $x = 3$

b. $-3 < x < 3$ **c.** **d.** $(-3, 3)$

71. a. $x = -3$ and $x = 3$ **b.** $-3 < x < 3$
c. **d.** $(-3, 3)$

73. 25,583,900 people **75.** \$6.0 billion **77. b.** $-2 < x \le 9$
79. b. $(-\infty, -15) \cup (9, \infty)$
81. **83.**

Chapter 2 Review Exercises
2. not a polynomial **3.** polynomial **4.** polynomial
6. 7 **7.** 3 **8.** 1 **10.** binomial **11.** monomial
12. trinomial **14.** $7x^2 + 7x - 19$ **15.** $-5x^2 + 3x + 3$
16. $a(b + c) = ab + ac$ **17.** $10x^3 - 31x^2 - 69x + 27$
18. $42x^5 + 54x^4 - 60x^3 + 150x^2$
19. $(a + b)(a - b) = a^2 - b^2$ **20.** $9x^2 - 25$
21. $x^2 - 30x + 225$ **22. a.** $A = LW$ **b.** $P = 2W + 2L$
23. $3L - 6$ **24.** $5W + 3$ **25.** $12W + 6$
26. $5W^2 + 3W$ **28.** $3x^2$ **29.** $3df^2$ **32.** $9ab(a^2b^4 + 3)$
33. prime, $75p^2x^2 + 49q^3y^2$ **34.** $(7k - 3)(x - y)$
35. $4(3x - 4)(3x + 4)$ **36.** $(2x + 5)(4y - k)$
37. $10(x - 3)(x + 3)$ **38.** $(z + 2)(z^2 - 2z + 4)$
39. $(3z + 5)(9z^2 - 15z + 25)$ **40.** $(x + 4)(x + 5)$
41. $(9p + 2)^2$ **42.** $6(2u + 3)^2$ **43.** $(2w - 7)(3w + 1)$
44. $(4x - 3)(x + 2)$ **45.** $(2x - 9)(x + 10)$
46. $(h - 10)(h + 3)$ **47.** $(k - 6)(k - 7)$
48. $10xy(x - 70y^2 + 7)$ **49.** $-5(h - k)(2m - 3)$
50. $6(2x^2 + 3y^2)$ **51.** $3(2a + b)(4c - 1)$
52. $2(2a - 3b)(4a^2 + 6ab + 9b^2)$
53. $2(4p - 5w)(4p + 5w)$ **54.** $(a^2 + 7)(a^2 + 8)$
55. $(x - 7)(2x + 1)$ **56.** $2(w^3 - 9)(w^3 + 2)$
58. a. 4225 **b.** not prime **59. a.** -35 **b.** prime
61. a. $x = -12$ or $x = 2$ **62. a.** $z = -\dfrac{1}{2}$ or $z = 0$ or $z = \dfrac{3}{2}$
63. a. $x = 0$ or $x = 9$ **64. a.** $p = -\dfrac{1}{4}$ or $p = \dfrac{1}{4}$
65. $3W - 5$ **66.** 11 in. long, 4 in. wide **67.** 16 cm long,
6 cm wide **68.** The price elasticity of demand is 0.5.
71. a. $x = -5$ or $x = 11$ **72. a.** no solution
73. a. $x = -13$ or $x = 2$ **74. a.** $x = -\dfrac{9}{2}$ or $x = \dfrac{7}{4}$
75. a. $\{3, 5, 11\}$ **b.** $\{1, 3, 5, 7, 8, 9, 11, 14\}$ **76.** disjunction
77. a. **b.** $(-3, 10]$

78. a. **b.** $(-\infty, 0] \cup (4, \infty)$

79. a. $-16 \le x < 8$ **b.** $[-16, 8)$ **80. a.** $x > 3$ or $x < -3$
b. $(-\infty, -3) \cup (3, \infty)$ **81. a.** $x = -11$ and $x = 19$
b. $-11 < x < 19$ **c.** $(-11, 19)$ **82.** no boundary values
b. no solution **83. a.** $x = -24$ and $x = 16$
b. $x \le -24$ or $x \ge 16$ **c.** $(-\infty, -24] \cup [16, \infty)$

Chapter 2 Test
1. $10x^3 - 53x^2 + 43x + 36$ **2.** $-3x^3 - 5x^2 + 5x + 6$
5. $3(4k + h)(2x - y)$ **6.** $k(2k + 3)(3k - 5)$

7. $(8b - 5c)(8b + 5c)$ **8.** $2(x^2 + x + 12)$
9. $(p - 3)(p^2 + 3p + 9)$ **10.** prime, $a^2 + 9a + 12$
11. $(h - 7)^2$ **12.** $(2z + 9)(3z - 7)$ **13.** prime, $x^2 + 16$

14. $7(3x - 2)(3x + 2)$ **16. a.** $x = -\dfrac{9}{2}$ or $x = 7$

17. a. $w = -10$ or $w = 0$ or $w = 10$
18. The rectangle is 13 in. long and 7 in. wide.
19. a. no solution   **20. a.** $x = -11$ or $x = -3$
21. a. $x = -12$ or $x = 11$ **22. a.** $-11 < x < 14$
b.    **c.** $(-11, 14)$

23. a. $-68 \le x \le 50$ **b.** (number line)   **c.** $[-68, 50]$

CHAPTER 3

Section 3.1 Vocabulary Practice
1. F   **2.** C   **3.** E   **4.** I   **5.** B   **6.** J   **7.** A   **8.** H
9. D   **10.** G

Section 3.1 Exercises
1. {(Time of the Preacher, 2:26), (I Couldn't Believe It Was True, 1:32), (Time of the Preacher Theme, 1:13), (Medley: Blue Rock Montana/Red . . . , 1:36), (Blue Eyes Crying in the Rain, 2:21), (Red Headed Stranger, 4:00), (Time of the Preacher Theme, 0:27), (Just As I Am, 1:48), (Denver, 0:53)}

3.

Ranking	Name of movie
1	Crash
2	The Departed
3	Mr. and Mrs. Smith
4	Walk the Line

5.

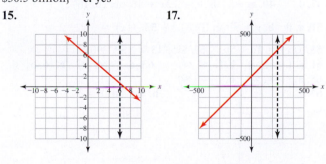

Minimum Wage

7. {(Johnson, 6622), (Hamlin, 6583), (Harvick, 6581), (Edwards, 6393), (Kenseth, 6294)}   **9.** no   **11.** {0, 1, 2, 3}
13. a. $N = $ {(2008, $45.2 billion), (2009, $19.3 billion), (2010, $30.5 billion)}   **b.** {$45.2 billion, $19.3 billion, $30.5 billion}   **c.** yes

15.

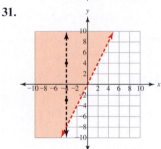

17.

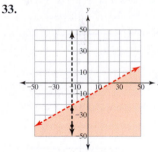

19.

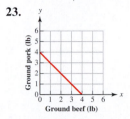

21.

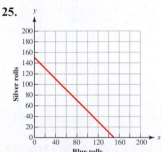

23.

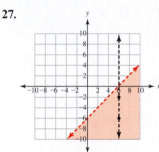

25.

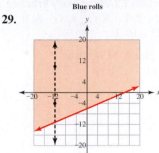

27.

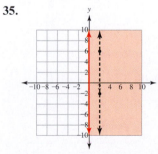

29.

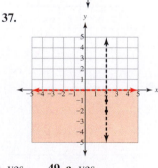

31.

33.

35.

37.

43. a. yes   **45. a.** no   **47. a.** yes   **49. a.** yes
53. a. {small, medium, large}   **b.** {$1.29, $1.69, $2.29}
55. yes   **57.** yes   **59.** no   **61.** no   **63.** yes
65. no   **67.** no   **73.** 32%   **75.** $21.5 trillion
77.

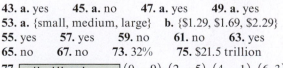

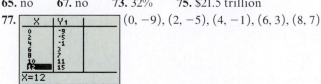

$(0, -9), (2, -5), (4, -1), (6, 3), (8, 7)$

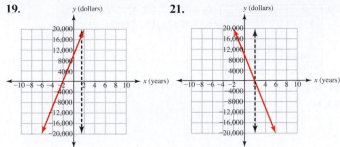

79. 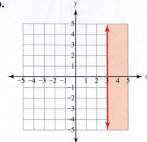 $(0, 4.4444)$, $(2, 4)$, $(4, 3.5556)$,
$(6, 3.1111)$, $(8, 2.6667)$

81. b. $\{1, 2, 6\}$

83. b.

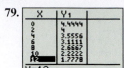

85. $x = \dfrac{10}{9}$ **87.** $y - y_1 = m(x - x_1)$

Section 3.2 Vocabulary Practice

1. J   **2.** F   **3.** C   **4.** G   **5.** B   **6.** E   **7.** D   **8.** A
9. H   **10.** I

Section 3.2 Exercises

1. no   **3.** yes   **5.** no

7. a.   **b.** $x = 0$

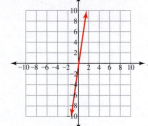

9. a.   **b.** yes

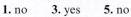

11. a.   **b.** no

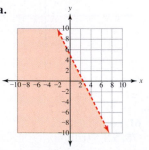

13. a.   **b.** yes

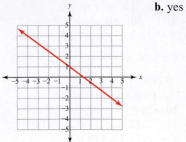

15. a.   **b.** yes

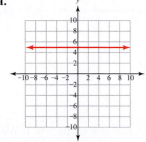

17. a.   **b.** yes

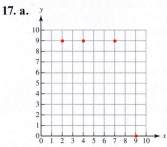

19. a.   **b.** yes

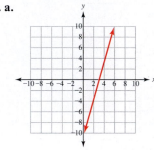

21. 5   **23.** $-\dfrac{1}{10}$ **25.** 1   **27.** 9.1   **29.** 5   **31.** 7

33. 4   **35.** -17 **37.** infinitely many ordered pairs
39. $1.42   **41. a.** 2   **b.** -6 **43. a.** -1 **b.** 2   **45.** -3
47. 0   **49. a.** -3 **b.** $y = -3x + 16$ **c.** $(0, 16)$

51. a. 0   **b.** $y = 3$ **c.** $(0, 3)$ **53.** $f(x) = \dfrac{1}{2}x - 2$

55. $f(x) = 2x - 5$ **57.** $153.13   **59.** 290 cal
61. a. $(-\infty, \infty)$ **b.** $(-\infty, \infty)$ **63. a.** $(-\infty, \infty)$ **b.** $[6, 6]$
65. a. $(-\infty, \infty)$ **b.** $(-\infty, \infty)$
67. a.   **b.** yes   **d.** $x = 5$

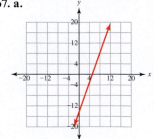

69. a.

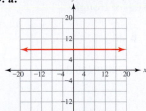

b. no

71. a.

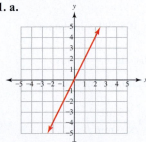

b. yes   **d.** $x = 0$

73. a. \$225   **b.** {20 books, 21 books, . . . , 49 books}
c. {\$180, \$189, . . . , \$441}   **75. a.** {1 roll, 2 rolls, . . . , 50 rolls}
b. {\$12.93, \$25.86, . . . , \$646.50}
77. a. {0 shirts, 1 shirt, . . . , 60 shirts}
b. {0 yd, 1.25 yd, . . . , 75 yd}   **79. a.** $x = 17$ months
81. a. $x = 50$ customers   **83.** \$4840   **85.** 5387 adults
87. a.

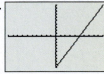

b. $x = 5$

$[-10, 10, 1, -10, 10, 1]$

89. a.

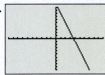

b. $x \approx 3.67$

$[-10, 10, 1, -10, 10, 1]$

91. b. $x = -5$ **93. b.** -18
95. $x = -2$ or $x = 2$ **97.** $[4, \infty)$

Section 3.3 Vocabulary Practice
1. J   **2.** F   **3.** B   **4.** G   **5.** C   **6.** H   **7.** A   **8.** D
9. E   **10.** I

Section 3.3 Exercises
1. a. 1   **b.** linear   **3. a.** 2   **b.** quadratic
5. a. 3   **b.** cubic   **11.** $(-\infty, \infty)$
13. a. $a = 1, b = 0, c = 1$
b.

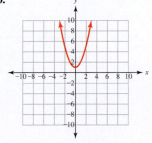

c. $(0, 1)$ **d.** up

e. minimum 1   **f.** $x = 0$ **g.** $(-\infty, \infty)$ **h.** $[1, \infty)$

15. a. $a = -1, b = 0, c = 1$
b.

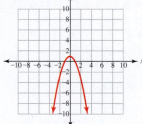

c. $(0, 1)$ **d.** down

e. maximum 1   **f.** $x = 0$ **g.** $(-\infty, \infty)$ **h.** $(-\infty, 1]$
17. a. $a = 1, b = 0, c = -6$
b.

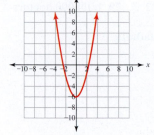

c. $(0, -6)$ **d.** up

e. minimum -6 **f.** $x = 0$ **g.** $(-\infty, \infty)$ **h.** $[-6, \infty)$
19. a. $a = 1, b = -2, c = 1$
b.

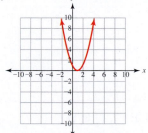

c. $(1, 0)$ **d.** up

e. minimum 0   **f.** $x = 1$ **g.** $(-\infty, \infty)$ **h.** $[0, \infty)$
21. a. $a = 1, b = 4, c = 1$
b.

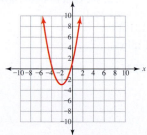

c. $(-2, -3)$ **d.** up

e. minimum -3 **f.** $x = -2$ **g.** $(-\infty, \infty)$ **h.** $[-3, \infty)$
23. a. $a = -1, b = 8, c = -13$
b.

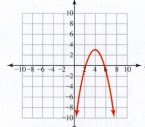

c. $(4, 3)$ **d.** down

e. maximum 3   **f.** $x = 4$ **g.** $(-\infty, \infty)$ **h.** $(-\infty, 3]$

25. a. up **27. a.** down **29. a.** $(-\infty, \infty)$ **b.** $(-\infty, 15]$
31. a. $(-\infty, \infty)$ **b.** $[-2, \infty)$ **33. a.** $(-\infty, \infty)$
b. $(-\infty, -11]$ **35. a.** $(-\infty, \infty)$ **b.** $[-9, \infty)$ **37. a.** $(3, 5)$
b. $(-\infty, \infty)$ **c.** $[5, \infty)$ **39. a.** $(-2, 3)$ **b.** $(-\infty, \infty)$
c. $(-\infty, 3]$ **41. a.** $(-5, -7)$ **b.** $(-\infty, \infty)$ **c.** $[-7, \infty)$
43. a. $(-\infty, \infty)$ **b.** $(-\infty, \infty)$ **45. a.** $(-\infty, \infty)$
b. $(-\infty, \infty)$ **47.** $x = 5$ **49.** $x = -8$ and $x = 3$

51. $x = -5$ and $x = \dfrac{3}{2}$ **53.** $x = 0$ and $x = 7$

55. $x = 0$ and $x = -3$ and $x = 9$ **57.** $x = 0$
59. no real zero **61.** $x = -4$ and $x = 7$ **63.** $x = 1$
65. $x = 2$ **67. a.** $f(-2) = -7$ **b.** $x = -4$ or $x = 4$

69. -29 **71.** 1190 **73.** -13 **75.** $\dfrac{68}{45}$

77. $a^2 + 2ah + h^2 + 3a + 3h + 2$ **79.** 1,014,000 doctors
81. 68% increase **83.** $87,000
85. a. **b.** quadratic

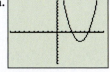

$[-10, 10, 1, -10, 10, 1]$

87. a. **b.** cubic

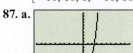

$[-10, 10, 1, -10, 10, 1]$

89. b. $(-\infty, \infty)$ **91. b.** -87 **93.** $y = 4x + 6$

95. $y = \left(\dfrac{60 \text{ mi}}{1 \text{ hr}}\right)x + 200 \text{ mi}$

Section 3.4 Vocabulary Practice
1. J **2.** G **3.** F **4.** B **5.** H **6.** E **7.** I **8.** D
9. C **10.** A

Section 3.4 Exercises
1. a. the average number of pages the student reads per day
b. the beginning number of pages **3. a.** the cost of materials
to make each product **b.** the overhead cost

5. $\dfrac{18 \text{ ft}}{1 \text{ day}}$ **7.** $\dfrac{0.7 \text{ CCF}}{1 \text{ day}}$ **9.** $\dfrac{-181 \text{ ft}^2}{1 \text{ min}}$

11. a. $y = \left(\dfrac{\$3.75}{1 \text{ day}}\right)x + \$150$ **b.** \$225

c.

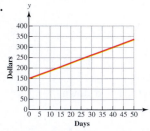

13. a. $y = \left(\dfrac{-4 \text{ oz}}{1 \text{ glass}}\right)x + 128 \text{ oz}$ **b.** $x = 32$ glasses
c. The container of milk is empty.

15. a. $L(x) = \left(\dfrac{0.02 \text{ in.}}{1 \text{ day}}\right)x + 2 \text{ in.}$ **b.** 200 days

17. a. $N(x) = \left(\dfrac{345,189 \text{ families}}{1 \text{ year}}\right)x + 2,001,000$ families
b. 16,499,000 families

19. a. $y = \left(\dfrac{2.5 \text{ cm}}{1 \text{ bird}}\right)x$ **b.** 47,000 cm of trough

21. a. $y = \left(\dfrac{0.63\%}{1 \text{ year}}\right)x + 17.72\%$ **b.** 46.7%

23. a. $y = \left(\dfrac{2,697,002 \text{ drivers}}{1 \text{ year}}\right)x + 57,869,446$ drivers
b. 230,500,000 licensed drivers

25. a. $y = \left(\dfrac{-920,105 \text{ days}}{1 \text{ year}}\right)x + 46,945,050$ days
b. $x \approx 51$ years

27. $y = \left(\dfrac{12,071 \text{ complaints}}{1 \text{ year}}\right)x + 195,790$ complaints

29. a. $y = \left(\dfrac{-2 \text{ pills}}{1 \text{ day}}\right)x + 180 \text{ pills}$ **b.** $x = 90$ days

31. 40.2% **33.** 53.4% **35.** 2200 centinewtons
37. \$2335 billion

39. a. $\left(300 \text{ m}, \dfrac{66 \text{ mL}}{\text{kg} \cdot \text{min}}\right), \left(2800 \text{ m} \dfrac{55 \text{ mL}}{\text{kg} \cdot \text{min}}\right)$

b. $\dfrac{-0.0044 \dfrac{\text{mL}}{\text{kg} \cdot \text{min}}}{1 \text{ m}} = \dfrac{-0.0044 \text{ mL}}{\text{kg} \cdot \text{min} \cdot \text{m}}$

c. $y = \left(\dfrac{-0.0044 \text{ mL}}{\text{kg} \cdot \text{min} \cdot \text{m}}\right)x + \dfrac{67.32 \text{ mL}}{\text{kg} \cdot \text{min}}$ **d.** $\dfrac{61 \text{ mL}}{\text{kg} \cdot \text{min}}$

41. \$800,000 **43.** 18 miles per hour

45. a. **b.** minimum 4 **c.** $[4, \infty)$

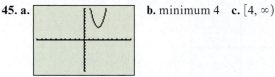

$[-10, 10, 1, -10, 10, 1]$

47. a. **b.** maximum 7.25 **c.** $(-\infty, 7.25]$

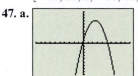

$[-10, 10, 1, -10, 10, 1]$

49. b. $y = \left(\dfrac{-\$4.60}{1 \text{ ad}}\right)x + \$500$ **51.** 213 million drivers
53. $9x^2 - 36x + 36$ **55.** $9x^2 - 30x + 31$

Section 3.5 Vocabulary Practice
1. C **2.** H **3.** B **4.** E **5.** G **6.** D **7.** A **8.** J
9. I **10.** F

Section 3.5 Exercises
1. **3.**

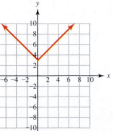

5.

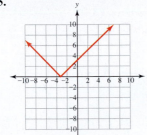

7.

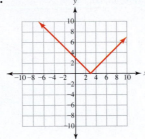

9.

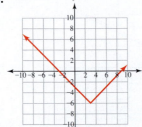

11.

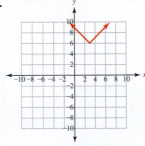

13.

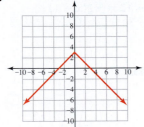

15.

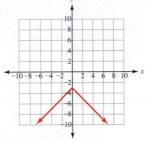

17.

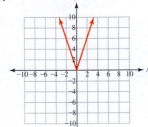

19.

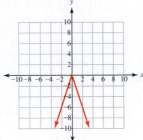

21.

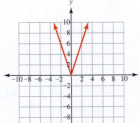

23.

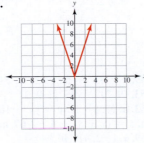

27. a. vertex $(-3, -7)$ **b.** $[-7, \infty)$ **29. a.** vertex $(0, 5)$
b. $(-\infty, 5]$ **31. a.** vertex $(0, -4)$ **b.** $[-4, \infty)$ **33.** $[2, \infty)$
35. $(-\infty, 2]$ **37.** $(-\infty, 2]$ **39.** $(-\infty, -3]$ **41.** $[0, \infty)$
43. $x = -7$ and $x = 1$ **45.** $x = -6$ and $x = -4$
47. $x = -8$ and $x = 8$ **49.** $x = -2$ and $x = 2$
51. $x = -12$ and $x = -8$ **53.** $x = -10$
55. no real zeros   **57.** $x = -7$ and $x = 1$
59. $x = -1$ and $x = 7$ **61. a.** yes   **63. a.** no

65. a. $(-\infty, \infty)$ **b.** $[-2, \infty)$ **67. a.** $(-\infty, \infty)$
b. $[-2, \infty)$ **69.** $x = -8$ **71.** $x = -\dfrac{1}{2}$ and $x = 7$
73. $x = -4$ and $x = 0$ and $x = 4$ **75.** $x = -12$ and $x = 6$
77. a. $(-\infty, \infty)$ **b.** $(-\infty, \infty)$ **79. a.** $(-\infty, \infty)$ **b.** $[4, \infty)$
81. a. $(-\infty, \infty)$ **b.** $(-\infty, -4]$ **83.** 1100 gal
85. \$3.7 million

87. a.

$[-10, 10, 1, -10, 10, 1]$

b. $(-\infty, \infty)$ **c.** minimum -4
d. $[-4, \infty)$

89. a.

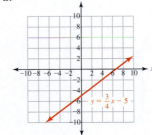

$[-20, 20, 2, -10, 20, 2]$

b. $(-\infty, \infty)$ **c.** maximum 15
d. $(-\infty, 15]$

91. b. $(-\infty, \infty)$ **93. b.** $x = -8$ and $x = 2$
95. $y = mx + b$ **97.** $y = ax^2 + bx + c$

Section 3.6 Vocabulary Practice
1. F   **2.** H   **3.** C   **4.** B   **5.** A   **6.** I   **7.** G   **8.** E
9. J   **10.** D

Section 3.6 Exercises
1. a. $\dfrac{3}{4}$ **b.** $(0, -5)$ **3. a.** -6 **b.** $(0, 1)$ **5.** 3

7. a.

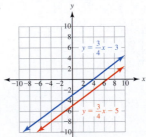

b. $f(x) = \dfrac{3}{4}x - 3$

c.

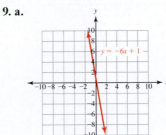

9. a.

b. $f(x) = -6x - 8$

c.

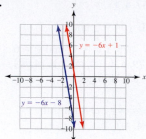

11. a.

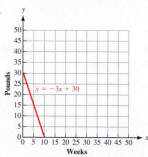

b. $y = \left(\dfrac{-3 \text{ lb}}{1 \text{ week}}\right)x + 15 \text{ lb}$

c.

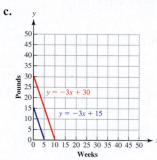

13. a. 3   **b.** $(4, 7)$ **15. a.** 3   **b.** $(4, -7)$
17. a. 3   **b.** $(-4, 7)$ **19.** 7   **21.** 4
23. a. $(12, 7)$ **b.**

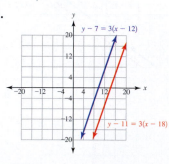

c. $y - 11 = 3(x - 18)$ **d.**

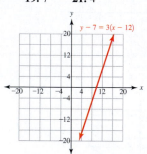

25. a. $(2, 8)$ **b.**

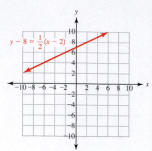

c. $y - 3 = \dfrac{1}{2}(x + 7)$ **d.**

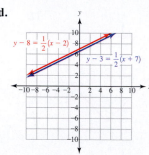

27. $y - 5 = 3(x - 4)$ **29.** $y - 9 = -3(x - 1)$
31. $y + 2 = 5(x + 4)$ **33.** $y - 12 = -2(x - 1)$
35. a. $y = -x - 1$ **b.** $y = -x - 3$

37. b. $\dfrac{2}{3}$ **39.** 2

41.

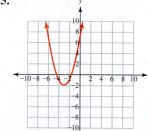

43.

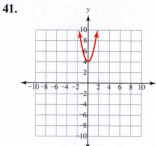

45.

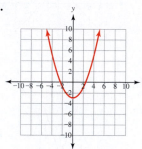

47.

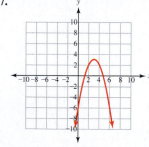

49. $y = 3x^2 + 5x - 2$ **51.** $y = 3x^2 + 5x - 14$
53. $(9, 2)$ **55.** $(-9, 2)$ **57.** $(-9, -2)$ **59.** 3   **61.** 8
63. a. $(7, 8)$ **b.** $y = |x - 10| + 10$ **65. a.** $(-6, 1)$
b. $y = |x + 8| - 4$ **67. a.** $(-10, -2)$ **b.** $y = |x + 14| + 4$
69. a. $(-2, 4)$ **b.**

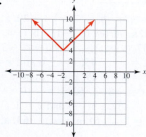

71. a. $(-2, -5)$ **b.**

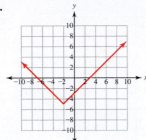

73. a. $(-3, -1)$ **b.**

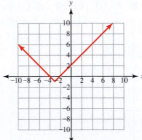

75. a. $(2, -1)$ **b.**

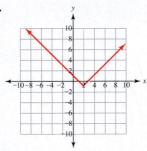

81. 84 serious adverse events
83. 260 construction workers
85. a.

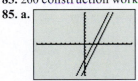

$[-10, 10, 1, -10, 10, 1]$

b. no horizontal shift
c. shift down 3

87. a.

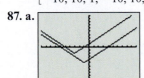

$[-10, 10, 1, -10, 10, 1]$

b. shift left 2
c. shift up 3

89. b.

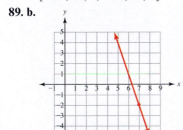

91. b. The graph will be shifted 5 units to the left.

93. $12x^5$ **95.** $\dfrac{6x^4}{25y^6}$

Chapter 3 Review Exercises
1. {(2004, 204 deaths), (2005, 180 deaths), (2006, 222 deaths) (2007, 223 deaths)} **2. a.** {2004, 2005, 2006, 2007}
b. {204 deaths, 180 deaths, 222 deaths, 223 deaths}
4. a.

Semester	Sections
Spring 2009	5 sections
Fall 2009	7 sections
Spring 2010	6 sections
Fall 2010	10 sections

b.

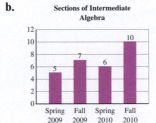

5.

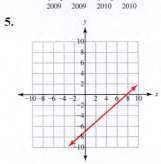

6.

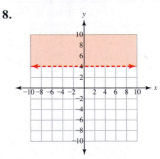

7.

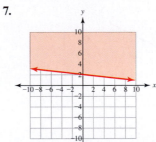

8.

9. yes **10.** {Washington, Adams, Reagan, Obama}
11.

President	Order
Washington	1st
Adams	2nd
Reagan	40th
Obama	44th

12. no **16.** -15 **17.** \$487.50 **18.** $\dfrac{2}{3}$ **19.** 4
20. a. $(-\infty, \infty)$ **b.** $(-\infty, \infty)$ **c.** $[4, 4]$ **21.** $x = \dfrac{10}{3}$

22. a.

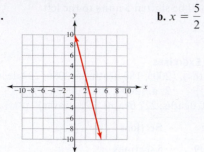

b. $x = \dfrac{5}{2}$

42.

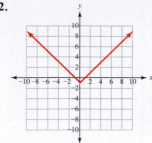

23. Lansing **24. a.** {0 boxes, 1 box, 2 boxes, . . . , 8 boxes}
b. {$0, $1.99, $3.98, . . . , $15.92} **25. a.** $x = 250$ loads
27. a. $(4, -2)$ **b.** down **c.** maximum -2 **d.** $x = 4$ **e.** $(-\infty, \infty)$
f. $(-\infty, -2]$ **28. a.** $(-4, -3)$ **b.** up **c.** minimum -3
d. $x = -4$ **e.** $(-\infty, \infty)$ **f.** $[-3, \infty)$

29.

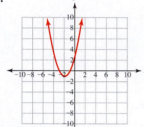

30.

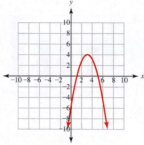

31. a. $(-\infty, \infty)$ **b.** $(-\infty, \infty)$ **c.** -58 **d.** 3
32. a. $x = -2$ and $x = 4$ **b.** $x = -2$ and $x = 4$

33. $x = 0$ and $x = 6$ and $x = 9$ **34.** no **35.** $\dfrac{12 \text{ mi}}{1 \text{ hr}}$

36. a. $y = \left(\dfrac{12 \text{ credits}}{1 \text{ semester}}\right)x + 24$ credits **b.** 60 credits

37. c. $875 **d.**

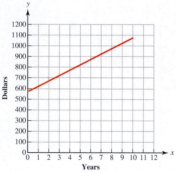

38. a. $y = \left(\dfrac{\$4}{1 \text{ map}}\right)x$ **c.** (0 maps, $0)

39. a. $y = \left(\dfrac{-44 \text{ gal}}{1 \text{ day}}\right)x + 500$ gal **d.** 11 days

40. a. $y = \left(\dfrac{\$42.724 \text{ billion}}{1 \text{ year}}\right)x + \$697.2$ billion
b. $1338 billion **41.** $362.2 billion

43. a. $(0, -4)$ **b.** $x = 0$ **c.** -4 **d.** $(-\infty, \infty)$ **e.** $[-4, \infty)$
f. $x = -4$ and $x = 4$ **44.** $x = -6$ and $x = 14$

45. a.

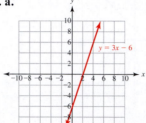

b. $h(x) = 3x + 2$

c.

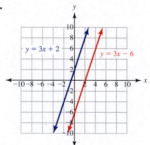

46. a.

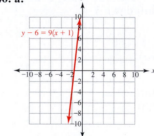

b. $y - 9 = 9(x - 6)$

c.

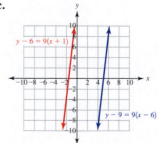

49. $y = x^2 + 5x + 7$ **50. a.** $(3, 1)$ **b.** $(-4, -2)$
51. $f(x) = |x + 1| - 4$

52. a.

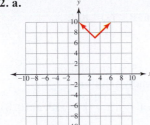

b.

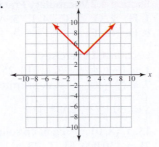

Chapter 3 Test

2. a. $\{5, 7, -2, 0\}$ **b.** $\{2, 4, -5\}$ **c.** 4

3. $\{3 \text{ packages}, 4 \text{ packages}, \ldots, 10 \text{ packages}\}$

4.

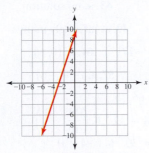

5. $x = -3$ **6.** domain $(-\infty, \infty)$, range $(-\infty, \infty)$

7. $f(-8) = -15$ **8.** 1   **9.** $f(x) = 3x + 17$

10. a. $y = \left(\dfrac{6.4 \text{ acres}}{1 \text{ hr}}\right)x + 15 \text{ acres}$ **b.** 91.8 acres

11.

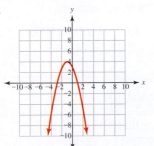

12. $(-\infty, \infty)$ **13.** $(-1, 4)$ **14.** $x = -1$

15. maximum 4   **16.** $(-\infty, 4]$ **17.** 2   **18.** -77

19. $x = -7$ and $x = 9$ **20.** \$4615 billion   **21.** $(-\infty, \infty)$

22. $(5, -2)$ **23.** $[-2, \infty)$ **24.** $x = 3$ and $x = 7$

25. $x = -8$ and $x = 4$ **27. a.** $\{(\text{Family}, \$19.99), (\text{Bronze},$

$\$9.99), (\text{Silver}, \$22.99), (\text{Gold}, \$32.99), (\text{Everything}, \$72.99)\}$

b. $\{\$9.99, \$19.99, \$22.99, \$32.99, \$72.99\}$ **c.** yes   **d.** \$22.99

Chapters 1–3 Cumulative Review

1. -200 **2.** $\dfrac{3}{5}$ **3.** 0   **4.** undefined   **5.** 1   **6.** -3

7. 1.8×10^8 **8.** 4×10^{-4} **9.** -10 **10.** $\dfrac{15c^2}{8d^2f^6}$

11. $112n^{23}$ **12.** $12c^3 - 14c^2 - 69c + 56$ **13.** $16w^2 - 9$

14. $10x^3 - 6x^2 - x - 5$ **15.** $5ax^2(2x + a - 3)$

16. $(7p + 6w)(7p - 6w)$ **17.** $(x - 4)(3x + 5)$

18. $2(x - 9)(x - 4)$ **19.** $(x + 4)(x + 5)$

20. $(2p - w)(3x + 7)$ **21.** $(2x - 3)(4x^2 + 6x + 9)$

22. a. $x = \dfrac{7}{3}$ **23. a.** $p = \dfrac{1}{12}$ **24. a.** $x = 1.27$

25. a. $k = -11$ **26. a.** no solution   **27. a.** $x = -3$ or $x = 9$

28. a. $u = -\dfrac{5}{2}$ or $u = \dfrac{1}{3}$ **29. a.** $x = -27$ or $x = 9$

30. a. no solution   **31. a.** $a > -3$ **c.** $(-3, \infty)$

32. a. $x \le -\dfrac{3}{2}$ **c.** $\left(-\infty, -\dfrac{3}{2}\right]$ **33. a.** $-10 < x < 17$

b. $(-10, 17)$ **34.** $y = -\dfrac{10}{9}x - \dfrac{11}{9}$ **35.** $x = 4$

36. $y = -\dfrac{2}{7}x + \dfrac{23}{7}$

38.

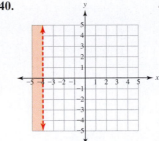

39.

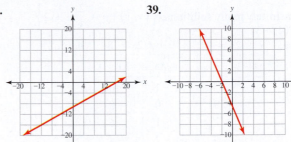

40.

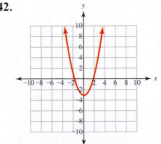

41.

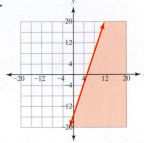

42.

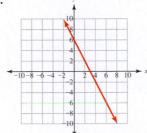

43. a. x **b.** $(-\infty, -2]$ **44.** $(-\infty, \infty)$

45. a. x **b.** $[-8, -2)$

47. a. $\{\text{gelato, ice cream, sorbet}\}$ **b.** $\{\text{chocolate, strawberry,}$
$\text{lemon}\}$ **48.** 79.8 years   **49.** 42

50.

51. $x = 3$ **52.** $x = 3$ **53. a.** $(-\infty, \infty)$ **b.** $(-\infty, \infty)$

54. 1   **55.** $f(x) = -2x - 4$ **56.** maximum 4

57. $(-\infty, \infty)$ **58.** $(-\infty, 4]$ **59.** $x = -2$ and $x = 6$

60. a. $f(x) = \left(\dfrac{\$150}{1 \text{ month}}\right)x + \$1200$ **b.** \$2550

61. $\dfrac{100 \text{ students}}{1 \text{ year}}$ **62.** length 11 in., width 4 in.

63. 20,500 tons   **64.** 128 superintendents   **65.** 40 gal

66. \$1260

CHAPTER 4

Section 4.1 Vocabulary Practice
1. H   2. G   3. F   4. D   5. B   6. F   7. B   8. A
9. C   10. E

Section 4.1 Exercises
1. $(0, 2)$ 3. $(1, 3)$ 5. no solution   7. $(3, 4)$
9. infinitely many solutions; $\left\{(x, y) \mid y = \dfrac{2}{3}x + 2\right\}$
11. $(4, -3)$
13. a. $x + 2y = 9$ $y = 2x - 8$ b. $(5, 2)$

x	y
−1	5
0	$\dfrac{9}{2}$
1	4
2	$\dfrac{7}{2}$
3	3
4	$\dfrac{5}{2}$
5	2
6	$\dfrac{3}{2}$

x	y
−1	−10
0	−8
1	−6
2	−4
3	−2
4	0
5	2
6	4

17. $(0, 3)$ 19. $(5, 4)$ 21. $(-3, -6)$
23. no solution   25. $(2, -3)$
27. infinitely many solutions; $\left\{(x, y) \mid y = \dfrac{2}{3}x + 8\right\}$
29. $\left(7, \dfrac{3}{2}\right)$ 31. $(-4, 2)$ 33. $(-4, 2)$ 35. no solution
37. $(4, 1)$ 39. $(-6, 3)$ 41. infinitely many solutions;
$\{(x, y) \mid y = 3\}$ 43. $(0, 0)$ 45. $(4, 0)$ 47. yes
49. no   51. $(150$ products, $\$8250)$ 53. cost to make
each product   55. $(3200$ products, $\$400,000)$
57. fixed cost   59. $\$24,000$ short-term; $\$12,000$ long-term
61. chicken wraps cost $\$3$; ham sandwiches cost $\$5$
63. 100 additional text messages   65. 2400 hotdogs;
600 Polish sausages   67. 12,454,000 math errors
69. $\dfrac{8042 \text{ L}}{1 \text{ hr}}$
71. a.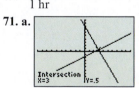
$[-10, 10, 1, -10, 10, 1]$ b. $\left(3, \dfrac{1}{2}\right)$
73. a.   b. $(36, 21)$
$[-40, 40, 4, -40, 40, 4]$
75. b. This system has infinitely many solutions;
$\left\{(x, y) \mid y = -\dfrac{3}{2}x + 2\right\}$ 77. b. The solution is $(4, 20)$.
79. a. $x = 1$ 81. a. $x = 4$

Section 4.2 Vocabulary Practice
1. H   2. G   3. C   4. I   5. A   6. B   7. J   8. E
9. D   10. F

Section 4.2 Exercises
1. a. $(-1, 4)$ 3. a. $(10, 25)$ 5. a. no solution
7. a. $(9, -6)$ 9. a. infinitely many solutions;
$\left\{(x, y) \mid y = \dfrac{6}{7}x + 11\right\}$ 11. a. $\left(\dfrac{3}{5}, -\dfrac{1}{2}\right)$ 13. a. $(0, 8)$
15. a. $(8, 0)$ 17. a. no solution   19. a. $(-6, 1)$
21. a. $(10,000, 5)$ 23. a. $(50, 18)$ 25. b. $(-8, 15)$
27. b. $(-2, -7)$ 29. b. $(2, 0)$
31. b. infinitely many solutions; $\{(x, y) \mid 5x + 6y = -12\}$
33. b. no solution   35. b. no solution   37. a. $(6, 5)$
39. a. $(-2, 4)$ 41. a. $(-3, -7)$ 43. a. no solution
45. a. $\left(\dfrac{1}{2}, \dfrac{1}{2}\right)$ 47. a. $\left(\dfrac{1}{4}, \dfrac{1}{4}\right)$ 49. a. infinitely many
solutions; $\{(x, y) \mid -6x + 5y = -17\}$ 51. a. $(0, 6)$
53. a. $(0, 0)$ 55. a. $\left(-7, -\dfrac{1}{5}\right)$ 57. a. $(-8, -6)$
59. a. $-3x + y = 1, 5x + 8y = 66$ b. $(2, 7)$
61. a. $30x + 16y = 35, 10x + 8y = 11$ b. $\left(\dfrac{13}{10}, -\dfrac{1}{4}\right)$
63. a. $3x + 4y = 40, -x + 2y = 30$ b. $(-4, 13)$
65. a. $x + y = 180, 8x + 14y = 2160$ b. $(60, 120)$
67. a. $(4, 3)$ b. $(4, 3)$ 69. a. $(1, 5)$ b. $(1, 5)$
71. a. $(0, 3)$ b. $(0, 3)$
73. a. infinitely many solutions; $\left\{(x, y) \mid y = \dfrac{3}{4}x + 2\right\}$
b. $\left\{(x, y) \mid y = \dfrac{3}{4}x + 2\right\}$ 75. a. no solution
b. no solution   81. 14.4%   83. 1125 trillion BTU
85. a. $(4, -6)$ b.   c. $(4, -6)$

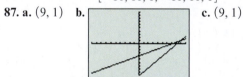

$[-10, 10, 1, -10, 10, 1]$
87. a. $(9, 1)$ b.   c. $(9, 1)$

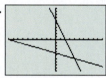

$[-10, 10, 1, -10, 10, 1]$
89. b. $(3, 5)$ 91. b. $(1, 2)$ 93. 0.058   95. $d = rt$

Section 4.3 Vocabulary Practice
1. I   2. H   3. G   4. C   5. B   6. F   7. D   8. J
9. A   10. E

Section 4.3 Exercises
1. 500 flies per week; $\$625$ 3. 546 customers; $\$2919$
5. $\$2625$ in stocks and $\$875$ in CDs   7. 1200 calories of
carbohydrates and 300 calories of protein and fat
9. hamburger $\$1.65$; order of fries $\$1.25$ 11. 833 L of Drink
A; 1167 L of Drink B   13. 4.5 L of Solution A; 0.5 L of
pure water   15. 56.5 g of rose-gold alloy; 23.5 g of red-gold
alloy   17. faster train 2.7 mi; slower train 2.3 mi
19. 2.5 hr; $\$28.13$ 21. The distance from the point of no
return to Los Angeles is 1263 mi, and the distance to Honolulu
is 993 mi.   23. 300 pounds Feed A; 450 pounds Feed B

25. 100 entrees; 50 appetizers **27.** 60 yellow daylilies; 20 red daylilies **29.** 833 lb of the 34% protein meal; 1667 lb of the 28% protein meal **31.** 1500 gal of Blend A; 1500 gal Blend B **33.** 955 lb water **35.** 18.7 lb cashews; 11.3 lb peanuts **37.** 0.07 qt stop bath concentrate; 0.93 qt water **39.** 9500 gal No. 2 Diesel **41.** 2050 min **43.** 20.1 oz 20-karat gold alloy **45.** 6.5 L of 35% antifreeze solution; 0.5 L pure antifreeze **47.** 20 oz concentrate; 108 oz water **49.** large truck 10.5 yd³; small truck 8 yd³ **51.** 300 adult tickets; 200 child tickets **53.** 15 min **57.** $23.60 per square inch **59.** $68.16 **61. b.** $x + y = 12; 0.80x + 0.925y = 0.88(12)$ **63. b.** $x + y = 40; 54.50x + 158.40y = (40)(105.60)$

Section 4.4 Vocabulary Practice

1. E **2.** I **3.** J **4.** C **5.** D **6.** G **7.** F **8.** B **9.** A **10.** H

Section 4.4 Exercises

1. infinitely many solutions **3.** one solution **5.** no solution **7.** yes **9.** no **11. a.** $(2, 5, 1)$ **13. a.** $(4, -1, -5)$ **15. a.** no solution **17. a.** infinitely many solutions **19. a.** $(3, -5, 2)$ **21. a.** infinitely many solutions **23. a.** $(2, 9, 0)$ **25. a.** no solution **27.** $5x + 8y + 0z = 20$ **29. a.** $(16, 4, 5)$ **31. a.** $(2, 0, 5)$

33. a. $\left(5, \dfrac{1}{2}, -3\right)$

37. $\left\{(x, y, z) \mid x = z + 2, y = -\dfrac{3}{2}z + 2, z \text{ is a real number}\right\}$

47. 1 cup almonds, 2 cups candies, 3 cups raisins **49.** 20 patients with shoulder injuries, 10 patients with back injuries, 12 patients with knee injuries **51.** 20 small packs, 40 medium packs, 60 large packs **53.** $10,000 in bonds, $30,000 in a certificate of deposit, $20,000 in the mutual stock fund **55.** 12.9 billion 2008 dollars **57.** 11.5% **59. b.** $(2, -4, -7)$ is a solution of this system.

$$2x + 3y + 3.5z = 142$$

61. b. $15x + 20y + 25z = 110$
$$20x + 30y + 40z = 271$$
63. $\dfrac{1}{8}$ **65.** 1

Section 4.5 Vocabulary Practice

1. F **2.** B **3.** C **4.** J **5.** H **6.** A **7.** I **8.** E **9.** G **10.** D

Section 4.5 Exercises

1. $\begin{bmatrix} 3 & 8 & | & 19 \\ 9 & 2 & | & 13 \end{bmatrix}$ **3.** $\begin{bmatrix} 2 & 5 & 1 & | & 18 \\ -6 & 0 & 5 & | & 26 \\ 2 & 7 & 3 & | & 26 \end{bmatrix}$ **5.** $\begin{bmatrix} 1 & -1 & -1 & | & 7 \\ 1 & 1 & -1 & | & 11 \\ 1 & 1 & 1 & | & 15 \end{bmatrix}$

7. $\begin{bmatrix} 1 & 2 & | & 4 \\ 9 & 3 & | & 17 \end{bmatrix}$ **9.** $\begin{bmatrix} 1 & \dfrac{-8}{3} & -5 & | & \dfrac{-17}{3} \\ -1 & 6 & 7 & | & 8 \\ 5 & 20 & 2 & | & 11 \end{bmatrix}$

11. $\begin{bmatrix} 1 & -5 & | & 13 \\ 4 & 1 & | & 11 \end{bmatrix}$ **13.** $\begin{bmatrix} 1 & 2 & | & 10 \\ 0 & -1 & | & -7 \end{bmatrix}$

15. $\begin{bmatrix} 1 & -7 & 1 & | & -6 \\ 0 & -17 & 12 & | & -59 \\ 5 & 2 & 1 & | & -3 \end{bmatrix}$ **17.** $\begin{bmatrix} 1 & 4 & | & 3 \\ 0 & -2 & | & -4 \end{bmatrix}$

19. $\begin{bmatrix} 1 & 3 & | & 5 \\ 0 & 0 & | & 21 \end{bmatrix}$ **21.** $\begin{bmatrix} 1 & 3 & | & 5 \\ 0 & 0 & | & 0 \end{bmatrix}$

25. $(-6, 5)$ **27.** $(-8, 4, 9)$ **29.** no solution **31.** infinitely many solutions **33.** no solution **35.** infinitely many solutions **37. a.** $\begin{bmatrix} 3 & 5 & | & 1 \\ 3 & 1 & | & 5 \end{bmatrix}$ **b.** $(2, -1)$

39. a. $\begin{bmatrix} 2 & -3 & | & -13 \\ 1 & 3 & | & 7 \end{bmatrix}$ **b.** $(-2, 3)$ **41. a.** $\begin{bmatrix} -2 & 3 & | & 12 \\ 2 & 1 & | & 4 \end{bmatrix}$

b. $(0, 4)$ **43. a.** $\begin{bmatrix} -2 & 3 & | & 12 \\ 2 & 1 & | & 20 \end{bmatrix}$ **b.** $(6, 8)$

45. a. $\begin{bmatrix} 4 & 3 & | & 25 \\ 2 & -1 & | & -5 \end{bmatrix}$ **b.** $(1, 7)$ **47. a.** $\begin{bmatrix} 4 & -5 & | & 4 \\ 4 & -5 & | & 3 \end{bmatrix}$

b. no solution **49. a.** $\begin{bmatrix} 3 & 5 & | & -14 \\ 9 & 2 & | & 10 \end{bmatrix}$ **b.** $(2, -4)$

51. a. $\begin{bmatrix} 2 & -1 & | & 6 \\ -4 & 2 & | & -12 \end{bmatrix}$ **b.** infinitely many solutions

53. intersecting lines **55.** parallel lines **57.** coinciding lines **59.** $\left\{(x, y) \mid x = \dfrac{1}{2}y + 3, y \text{ is a real number}\right\}$

61. a. $\begin{bmatrix} 1 & 1 & 1 & | & 6 \\ 2 & 1 & 2 & | & 9 \\ -2 & -1 & 1 & | & -3 \end{bmatrix}$ **b.** $(1, 3, 2)$

63. a. $\begin{bmatrix} 3 & 1 & 1 & | & 9 \\ -3 & 2 & 3 & | & -8 \\ 3 & 4 & 5 & | & 30 \end{bmatrix}$ **b.** no solution

65. a. $\begin{bmatrix} 1 & 1 & 1 & | & 11 \\ 2 & -1 & 1 & | & 5 \\ 4 & 2 & 4 & | & 38 \end{bmatrix}$ **b.** infinitely many solutions

67. a. $\begin{bmatrix} 2 & 1 & 1 & | & 6 \\ 2 & -1 & 2 & | & 4 \\ 6 & 2 & 1 & | & 22 \end{bmatrix}$ **b.** $(4, 0, -2)$

69. a. $\begin{bmatrix} 1 & 1 & -2 & | & 6 \\ 3 & -1 & 1 & | & 16 \\ 5 & 1 & -3 & | & 28 \end{bmatrix}$ **b.** infinitely many solutions

71. a. $\begin{bmatrix} 2 & 1 & 1 & | & 14 \\ -2 & 4 & 1 & | & -15 \\ 0 & 2 & 1 & | & -1 \end{bmatrix}$ **b.** $(8, 1, -3)$

73. $\left\{(x, y, z) \mid x = \dfrac{1}{4}z + \dfrac{11}{2}, y = \dfrac{7}{4}z + \dfrac{1}{2}, z \text{ is a real number}\right\}$

75. a. $\begin{bmatrix} 1 & 1 & 1 & | & 5 \\ 4 & -1 & 2 & | & 3 \\ 2 & 1 & 1 & | & 6 \end{bmatrix}$ **b.** $(1, 3, 1)$

77. a. $x =$ amount of 30% ammonia; $y =$ amount of water **b.** $\begin{array}{l} x + y = 5 \\ 0.30x + 0y = 0.12(5) \end{array}$ **c.** $\begin{bmatrix} 1 & 1 & | & 5 \\ 0.30 & 0 & | & 0.60 \end{bmatrix}$ **d.** $\begin{bmatrix} 1 & 0 & | & 2 \\ 0 & 1 & | & 3 \end{bmatrix}$ **e.** 2 gal commercial cleaning solution; 3 gal water **79. a.** $x =$ number of products; $y =$ dollars;

b. $\begin{array}{l} y = \left(\dfrac{\$20}{1 \text{ product}}\right)x + \$1500 \\ y = \left(\dfrac{\$40}{1 \text{ product}}\right)x \end{array}$ **c.** $\begin{bmatrix} -20 & 1 & | & 1500 \\ -40 & 1 & | & 0 \end{bmatrix}$

d. $\begin{bmatrix} 1 & 0 & | & 75 \\ 0 & 1 & | & 3000 \end{bmatrix}$ **e.** 75 products; revenue $3000

81. a. x = cost of a burrito; y = cost of a taco; z = cost of a soft drink

$$x + y + z = 5.10$$
b. $2x + y + 2z = 8.90$ **c.**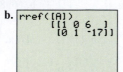
$$x + 2y + 3z = 8.60$$

d. 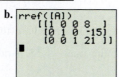   **e.** burrito \$2.70, taco \$1.30, soft drink \$1.10

83. 34,080,000 ft³   **85.** 45 gal

87. a. $\begin{bmatrix} 3 & 4 & | & -50 \\ 5 & -2 & | & 64 \end{bmatrix}$ **b.** 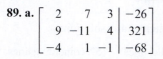   **c.** $(6, -17)$

89. a. $\begin{bmatrix} 2 & 7 & 3 & | & -26 \\ 9 & -11 & 4 & | & 321 \\ -4 & 1 & -1 & | & -68 \end{bmatrix}$ **b.**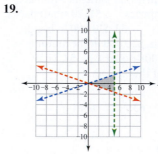

c. $(8, -15, 21)$ **91. b.** infinitely many solutions

93. b. $\begin{bmatrix} 1 & -\dfrac{5}{6} & | & -\dfrac{2}{3} \\ 2 & 1 & | & 6 \end{bmatrix}$ **95.**

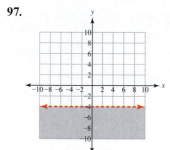

97.

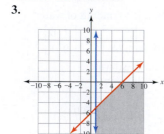

Section 4.6 Vocabulary Practice

1. F   **2.** D   **3.** G   **4.** E   **5.** J   **6.** A   **7.** I   **8.** C
9. B   **10.** H

Section 4.6 Exercises

3.

5.

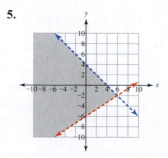

7.

9.

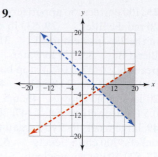

11.

13.

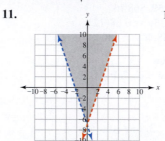

15.

17.

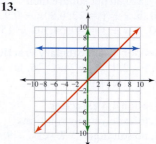

19.

21.

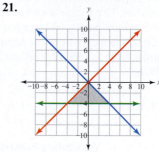

23. $(-1, -1), (-1, 5), (1, 1)$ **25.** $(-2, -2), (-2, 1),$
$(-1, 2), (2, 0)$ **27.** $(2, 10), (2, 65), (12, 10)$

29. a.   **b.** $(0, 0), (0, 2), (8, 0)$

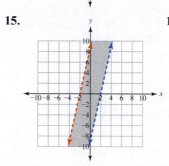

31. a.   **b.** $(2, 1), (2, 5), (4, 1)$

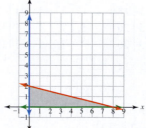

33. a. 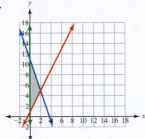　**b.** $(0, 1), (0, 11), (2, 5)$

35. a. 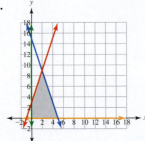　**b.** $(0, 0), (0, 3), (2, 9), (5, 0)$

37. a. 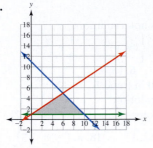　**b.** $(0, 1), (6, 5), (10, 1)$

39. a. 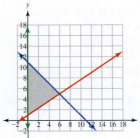　**b.** $(0, 1), (0, 11), (6, 5)$

41. a. 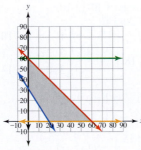　**b.** $(0, 32), (0, 60),$ $(60, 0), (20, 0)$

43. a. 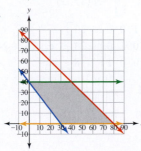　**b.** $(0, 40), (40, 40),$ $(80, 0), (30, 0)$

45. a. 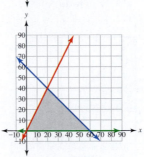　**b.** $(0, 0), (0, 60), (20, 40)$

47. a. 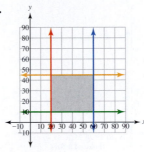　**b.** $(0, 0), (20, 40), (60, 0)$

49. a. 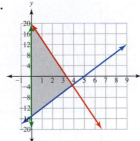　**b.** $(20, 10), (20, 45),$ $(60, 45), (60, 10)$

51. a. 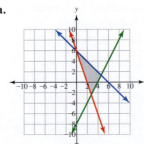　**b.** $(0, -15), (0, 20),$ $\left(\dfrac{35}{9}, -\dfrac{10}{3}\right)$

53. a. 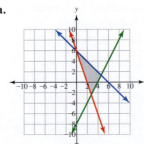　**b.** $(0, 6), \left(\dfrac{14}{3}, \dfrac{4}{3}\right),$ $\left(\dfrac{14}{5}, -\dfrac{12}{5}\right)$

59. a. $x + y \le \$7500$
$y \le \$3500$
$x \ge \$0$
$y \ge \$0$

c. $(\$0, \$0), (\$0, \$3500),$ $(\$4000, \$3500),$ $(\$7500, \$0)$

b.

61. a. $x \geq 45\%$
$x \leq 65\%$
$y \geq 35\%$
$y \leq 55\%$
$x + y \leq 100\%$

b.
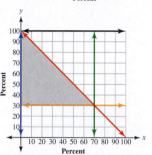

c. $(45\%, 35\%), (45\%, 55\%),$
$(65\%, 35\%)$

63. a. $x \geq 0\%$
$x \leq 70\%$
$y \geq 30\%$
$y \leq 100\%$
$x + y \leq 100\%$

b.
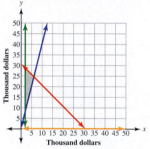

c. $(0\%, 30\%),$
$(0\%, 100\%), (70\%, 30\%)$

65. a. $x + y \leq \$30,500$
$y \geq 4x$
$x \geq \$2000$
$y \geq 0$

b.
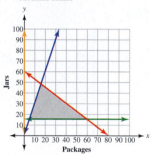

c. $(\$2000, \$8000),$
$(\$2000, \$28,500),$
$(\$6100, \$24,400)$

67. a. $\$1.50x + \$2.00y \leq \$120$
$y \leq 3x$
$y \geq 15$ jars
$x \geq 0$ packages

b.
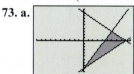

c. $(5$ packages, 15 jars$),$
$(16$ packages, 48 jars$),$
$(60$ packages, 15 jars$)$

69. 0.6 L 80% thioglycolic acid solution; 2.4 L water
71. 1.6 billion (1600 million) of today's dollars

73. a.
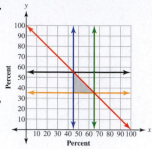

b. $(0, -9), (6, 3), (9, 0)$

$[-10, 10, 1, -10, 10, 1]$

75. a.

b. $(-6, 3), (0, 9), (-1.5, 3)$

$[-10, 10, 1, -10, 10, 1]$

77. b.
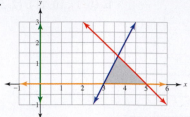

79. b.
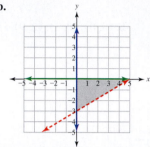

81. 8   **83.** not a real number

Chapter 4 Review Exercises
1. It is a solution.   **2.** It is not a solution.
3. infinitely many solutions; $\left\{(x, y) \mid y = \dfrac{1}{2}x - 6\right\}$
4. no solution   **5.** $(8, 5)$ **6.** $(5, 7)$
9. \$16,000 in Fund A; \$4,000 in Fund B   **12.** x
13. a. $(7, -1)$ **14. a.** no solution
15. a. infinitely many solutions; $\{(x, y) \mid y = -5x + 2\}$
16. a. $\left(\dfrac{3}{7}, \dfrac{51}{7}\right)$ **17. a.** $(-12, 1)$ **18. a.** $(-18, -3)$
19. a. $(8, 6)$ **20. a.** infinitely many solutions;
$\{(x, y) \mid x - y = 15\}$ **21. a.** no solution
22. 51.2 oz pure antifreeze; 12.8 oz water
23. 20 prints; \$2200   **24.** \$3333 on fiction;
\$1667 on nonfiction   **25.** 12 oz of the 48% olive oil
dressing; 20 oz of the 35% olive oil dressing
26. wrap \$4.25; taco \$2.25   **27.** 101 checked one piece;
24 checked two pieces   **28.** 15 min
29. a. plane   **b.** point   **c.** line   **31. a.** $(-5, 4, 8)$
32. a. infinitely many solutions   **33. a.** no solution
34. a. $(4, 3, -1)$ **35.** 22.5 bags of 10-10-10;
37.5 bags of 18-4-12; 0 bags of 20-3-6

36. $(-3, 8)$ **39.** $\begin{bmatrix} 2 & 5 & | & 1 \\ -3 & -1 & | & 18 \end{bmatrix}$ **40.** $\begin{bmatrix} 1 & 1 & 1 & | & 10 \\ 3 & 2 & -1 & | & 1 \\ 3 & 0 & 1 & | & 13 \end{bmatrix}$

41. $\begin{bmatrix} 1 & -\dfrac{5}{8} & | & -\dfrac{17}{8} \\ 9 & 1 & | & 39 \end{bmatrix}$ **42.** $\begin{bmatrix} 1 & 2 & | & -9 \\ 0 & -5 & | & 25 \end{bmatrix}$

43. a. $\begin{bmatrix} 2 & 1 & | & 33 \\ 4 & -1 & | & 57 \end{bmatrix}$ **b.** $\begin{bmatrix} 1 & 0 & | & 15 \\ 0 & 1 & | & 3 \end{bmatrix}$ **c.** $(15, 3)$

44. a. $\begin{bmatrix} 1 & 1 & 1 & | & 10 \\ 2 & 1 & -1 & | & 20 \\ 1 & 2 & 1 & | & -14 \end{bmatrix}$ **b.** $\begin{bmatrix} 1 & 0 & 0 & | & 26 \\ 0 & 1 & 0 & | & -24 \\ 0 & 0 & 1 & | & 8 \end{bmatrix}$ **c.** $(26, -24, 8)$

45. a. $\begin{bmatrix} 3 & -4 & | & 8 \\ -6 & 8 & | & -16 \end{bmatrix}$ **b.** $\begin{bmatrix} 1 & -\dfrac{4}{3} & | & \dfrac{8}{3} \\ 0 & 0 & | & 0 \end{bmatrix}$

c. $\left\{(x, y) \mid x = \dfrac{4}{3}y + \dfrac{8}{3}\right\}$ **46. a.** $\begin{bmatrix} 2 & 1 & 1 & | & 18 \\ 1 & 1 & -1 & | & 5 \\ 5 & 3 & 1 & | & 40 \end{bmatrix}$

b. $\begin{bmatrix} 1 & 0 & 2 & | & 0 \\ 0 & 1 & -3 & | & 0 \\ 0 & 0 & 0 & | & 1 \end{bmatrix}$ **c.** no solution

47. a. x = milliliters of stock solution; y = milliliters of water

b. $\begin{aligned} x + y &= 100 \\ 0.04x + 0y &= 0.01(100) \end{aligned}$ **c.** $\begin{bmatrix} 1 & 1 & | & 100 \\ 0.04 & 0 & | & 1 \end{bmatrix}$ **d.** $\begin{bmatrix} 1 & 0 & | & 25 \\ 0 & 1 & | & 75 \end{bmatrix}$

e. 25 mL 4% stock solution; 75 mL water

48. a. x = number of products; y = dollars

b. $y = \left(\dfrac{\$8}{1 \text{ product}}\right)x + \$3200$

$y = \left(\dfrac{\$24}{1 \text{ product}}\right)x$

c. $\begin{bmatrix} -8 & 1 & | & 3200 \\ -24 & 1 & | & 0 \end{bmatrix}$ **d.** $\begin{bmatrix} 1 & 0 & | & 200 \\ 0 & 1 & | & 4800 \end{bmatrix}$

e. 200 products; $4800 revenue

49. a. $\left(\dfrac{3}{2}, \dfrac{5}{2}\right)$, $(3, 4)$, $(0, 4)$

50. a.

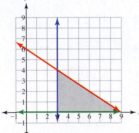

b. $(3, 0)$, $(3, 4)$, $(9, 0)$

51. a.

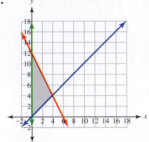

b. $(0, 0)$, $(0, 12)$, $(4, 4)$

52. a.

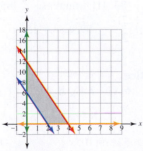

b. $(0, 6)$, $(0, 12)$, $(4, 0)$, $(2, 0)$

53. a. $\left(\dfrac{3 \text{ g}}{1 \text{ serving}}\right)x + \left(\dfrac{6 \text{ g}}{1 \text{ serving}}\right)y \le 21 \text{ g}$

$\left(\dfrac{20 \text{ g}}{1 \text{ serving}}\right)x + \left(\dfrac{5 \text{ g}}{1 \text{ serving}}\right)y \ge 35 \text{ g}$

$x \ge 0$ serving

$y \ge 0$ serving

b.

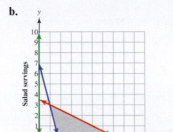

c. (1 serving stir-fry, 3 serving salad), (1.75 serving stir-fry, 0 serving salad), (7 serving stir-fry, 0 serving salad)

Chapter 4 Test

1. $(-4, -2)$ **2.** no solution **3. a.** $(2, 7)$

4. a. infinitely many solutions; $\{(x, y) \mid y = 9x - 3\}$

5. a. $(-5, 4)$ **6. a.** $\left(\dfrac{1}{2}, \dfrac{3}{4}\right)$

7. 11 oz of creamy pasta sauce; 21 oz of the other sauce

8. 6.8 acres open space; 1.7 acres planted space

9. a. $(3, 9, -2)$ **10.** 2 cups broccoli, 1 cup radishes, 1 cup cauliflower **11. a.** $(5, 1)$

12. a.

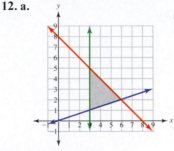

b. $(3, 1)$, $(3, 5)$, $(6, 2)$

CHAPTER 5

Section 5.1 Vocabulary Practice

1. F **2.** H **3.** C **4.** E **5.** D **6.** G **7.** B **8.** A **9.** I **10.** J

Section 5.1 Exercises

1. $\dfrac{y^6}{x^5 z^2}$ **3.** $\dfrac{11y^6}{15x^5 z^2}$ **5.** $-\dfrac{1}{2h^3 k^4}$ **7.** $\dfrac{x - 5}{x + 3}$ **9.** $\dfrac{2y + 1}{y + 6}$

11. $\dfrac{x - 4}{(x + 2)(x + 3)}$ **13.** $\dfrac{a}{2a + 7}$ **15.** $\dfrac{3w + 2}{3w^3(2w - 1)}$

17. $\dfrac{x + 5}{x - 3}$ **19.** $\dfrac{-x - 5}{x - 3}$ **21.** $\dfrac{-b - 3}{b + 10}$ **23.** $x - 3$

25. $-x + 3$ **27.** $\dfrac{4y}{5x}$ **29.** $\dfrac{x + 5}{x + 2}$ **33.** $\dfrac{x^2 + xy + y^2}{x + y}$

35. $\dfrac{x^2 - xy + y^2}{x - y}$ **37.** $\dfrac{x^2 - 2x + 4}{x + 5}$

39. $\dfrac{9x^2 + 12xy^2 + 16y^4}{3x + 4y^2}$ **41.** $\dfrac{c + 3}{c - 4}$ **43.** $\dfrac{x + 5}{x + 6}$

45. 1 **47.** $\dfrac{x + 6}{x + 9}$ **49.** $\dfrac{z - 10}{z + 3}$ **51.** $\dfrac{-z + 10}{z + 3}$

53. $\dfrac{2x + 1}{x + 5}$ **55.** $\dfrac{-2x - 1}{x + 5}$ **57.** $\dfrac{2y^6}{5x^2}$ **59.** $\dfrac{x + 1}{x + 6}$

61. $\dfrac{w + 4}{w + 3}$ **63.** $\dfrac{2}{45d^4 f^7}$ **65.** $\dfrac{x + 6}{x + 2}$ **67.** $\dfrac{-x - 6}{x + 2}$

69. $\dfrac{24c^6}{7a^{11}b}$ **71.** $\dfrac{x+2}{2x+3}$ **73.** 1 **75.** abc

77. $\dfrac{2z^2+15z+25}{(z+3)(2z+1)}$ **79.** $\dfrac{x-6}{x-7}$ **81.** $\dfrac{4u^2+18u+8}{u^2+u+3}$

83. 396 students **85.** 37%

87. a. $\dfrac{x+6}{x-8}$ **b.**

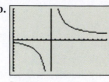

$[0, 20, 1, -10, 10, 1]$

89. a. $\dfrac{1}{3x-2}$ **b.**

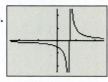

$[-2.5, 2.5, 1, -2.5, 2.5, 1]$

91. b. $\dfrac{x-2}{x-6}$ **93. b.** $\dfrac{y^2-8y+16}{(y+3)(y+1)}$ **95.** $2^2 \cdot 3^2$

97. $2^4 \cdot 3^2 = 144$

Section 5.2 Vocabulary Practice
1. G **2.** J **3.** C **4.** D **5.** E **6.** F **7.** B **8.** H
9. I **10.** A

Section 5.2 Exercises
1. $2^4 \cdot 3 \cdot x^2 \cdot y$ **3.** $3 \cdot 7 \cdot 11 \cdot x \cdot (x-1)$
5. $2 \cdot c \cdot (2c-3) \cdot (2c+3)$ **7.** $6(f-5)(f-8)$
9. $2(3y+8)(2y+5)$ **13.** $(x-9)(x-1)(x+2)$
15. $3d(d-4)$ **17.** $700a^2b^3c$ **19.** $-x(x+3)(x-3)$
21. $-(k+10)(k-3)(k+3)$ **23.** $8y^2$
25. $(x-2)(x+2)(x-4)$ **29.** $\dfrac{8}{x-6}$

31. $\dfrac{24a-35b}{90a^2b^2}$ **33.** $\dfrac{27x-10y+48xy}{48xy}$ **35.** $\dfrac{x-9}{x+9}$

37. $\dfrac{d-9}{d-3}$ **39.** $\dfrac{x+2}{x-5}$ **41.** $\dfrac{4}{2n-9}$

43. $\dfrac{3k+7}{2k-1}$ **45.** $\dfrac{3x-5}{(x+3)(x-4)}$

47. $\dfrac{-7}{(x+3)(x-4)}$ **49.** $\dfrac{8}{(x-5)(x+3)}$

51. $\dfrac{8x^2}{(x-5)(x+3)}$ **53.** $\dfrac{x+6}{x}$ **55.** $-\dfrac{15}{4p-1}$

57. $-\dfrac{3}{5x-3}$ **59.** $\dfrac{9}{3y+1}$ **61.** $\dfrac{2x+8}{2x+5}$

63. $\dfrac{7x^2+29x+14}{(x+6)(x-5)(x+2)}$ **65.** $\dfrac{c-9}{(c-5)(c+3)}$

67. $\dfrac{-2x+1}{(x-5)(x-2)(x+2)}$ **69.** $\dfrac{x}{(x+4)(x+5)}$

71. $\dfrac{x-3}{x+4}$ **73.** $\dfrac{-3x+1}{x-4}$ **75.** $\dfrac{x^2-3x-4}{(x-3)(x+3)}$

77. $\dfrac{-x^2+2x+8}{x(x+3)}$ **79.** $\dfrac{3x+1}{(x-2)(2x-3)(2x+1)}$

81. 3.8% **83.** $\dfrac{649.84 \text{ thousand owners}}{1 \text{ year}}$

85. a. $\dfrac{2}{(x-1)(x+3)}$ **b.**

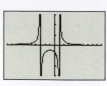

$[-10, 10, 1, -3, 3, 1]$

87. a. $\dfrac{5}{2x+7}$ **b.**

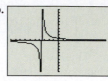

$[-10, 10, 1, -10, 10, 1]$

89. b. $\dfrac{x-3}{x-1}$ **91. b.** $\dfrac{18x-y}{30x^2y^2}$ **93.** $\dfrac{4}{27}$ **95.** -1

Section 5.3 Vocabulary Practice
1. C **2.** D **3.** F **4.** E **5.** I **6.** J **7.** B **8.** A
9. H **10.** G

Section 5.3 Exercises
1. $\dfrac{1}{4}$ **3.** $\dfrac{8}{9}$ **5.** $\dfrac{42}{5}$ **7.** 1 **9.** $\dfrac{8}{3}$ **11.** 4 **13.** $\dfrac{2}{3q}$

15. $\dfrac{4m}{5}$ **17.** $\dfrac{x^2+3x}{(x-2)(x+4)}$ **19.** $\dfrac{x^2-1}{(x-3)(x+3)}$

21. $\dfrac{5}{2}$ **23.** $\dfrac{5y-15}{24(y+3)}$ **25.** $\dfrac{3a^3+9a^2}{a-8}$ **27.** $\dfrac{a-3}{3a^2(a+2)}$

29. $\dfrac{b-1}{b-4}$ **31.** $\dfrac{21x^6-105x^5}{10(x-7)}$ **33.** $\dfrac{h+7}{h+6}$

35. $(x+1)(x+1)$ **37.** $\dfrac{35}{27}$ **39.** $\dfrac{14}{27}$ **41.** $\dfrac{1}{3}$ **43.** 2

45. $\dfrac{4x^2+x-14}{(x-2)(x-1)}$ **47.** $\dfrac{6x+3}{8x+13}$ **49.** $\dfrac{k}{k-4}$

51. $\dfrac{x+2}{x+3}$ **53.** $\dfrac{x+2}{2(x+1)}$ **55.** $\dfrac{x^2-4}{2x^2}$ **57.** $\dfrac{2x+2}{x}$

59. $-\dfrac{4}{x}$ **61.** $\dfrac{3x+1}{2x+1}$ **63.** $\dfrac{5x}{3x+4}$ **65.** $\dfrac{2x+10}{11x-8}$

67. $\dfrac{3x+22}{8x+8}$ **69.** xy **71.** $\dfrac{1}{xy}$ **73.** $\dfrac{14c+21d}{3(7c+2d)}$

75. $\dfrac{2c+3d}{5c+2d}$ **77.** 4 ohms **79.** 5 ohms

81. 77 years, 10 months old **83.** 111% increase **85. b.** $\dfrac{1}{8}$

87. b. $\dfrac{(n-2)(n+2)+7(n+2)}{(n-2)(n+2)+3(n-2)}$ **89. a.** $x=15$

91. a. $x=6$

Section 5.4 Vocabulary Practice
1. J **2.** I **3.** B **4.** D **5.** H **6.** G **7.** F **8.** A
9. E **10.** C

Section 5.4 Exercises
1. $56x+32=105$ **3.** $56+32x=105x$
5. $56x+32x^2=105$ **7.** $56x-168=3x^2-9x-30$
9. $-24x+232=3x^2-9x-30$ **11.** $29-3x=3$
15. a. $x=-\dfrac{245}{2}$ **17. a.** $p=\dfrac{10}{3}$ **19. a.** $x=42$
21. a. $k=\dfrac{20}{3}$ **23. a.** $c=-6$ **25. a.** $z=-36$

27. a. $x = \dfrac{1}{60}$ **29. a.** $w = -56$ or $w = 7$

31. a. $p = 8$ or $p = -1$ **33. a.** $a = -16$
35. a. $x = 8$ or $x = -2$ **37. a.** $a = 3$
39. a. $x = 9$ or $x = 1$ **41. a.** $m = 4$
43. a. $w = -5$ or $w = -8$ **45. a.** $c = 4$ or $w = -\dfrac{5}{3}$

47. a. no solution   **49. a.** $p = 6$ **51. a.** $x = 60$

53. a. no solution   **55. a.** $x = 3$ **57. a.** $x = \dfrac{1}{3}$

59. a. $x = -1$ or $x = 4$ **61. a.** $x = -2$ or $x = 4$
63. a. no   **65.** 10.8 cups of bleach   **67.** 718,000 gal
69. 52 million individuals   **71.** 20,080 students
75. 21 min   **77.** 45 min   **79.** 29 min   **81.** 2 hr 21 min
83. 3 hr 16 min   **87.** 30 ohms   **89.** 41.2 cm
91. 1963 hits   **93.** 2635 three-pointers   **95.** 2.1%
decrease
97. 33,200 deaths   **99. b.** $x = -2$ or $x = 8$
101. b. It takes 23 min for the two to muck out the barn.
103. $(-\infty, \infty)$ **105.** 27

Section 5.5 Vocabulary Practice
1. I   **2.** A   **3.** H   **4.** G   **5.** B   **6.** C   **7.** F   **8.** E
9. D   **10.** J

Section 5.5 Exercises

1. a. 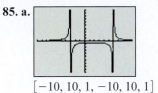   **b.** $(-\infty, 0) \cup (0, \infty)$

3. a.   **b.** $(-\infty, -1) \cup (-1, \infty)$

5. a.   **b.** $(-\infty, -1) \cup (-1, \infty)$

7. a.
b. $(-\infty, -6) \cup (-6, -4) \cup (-4, \infty)$
9. a.
b. $(-\infty, 4) \cup (4, 6) \cup (6, \infty)$
11. a. $(-\infty, 0) \cup (0, 3) \cup (3, \infty)$ **b.** $x = 0, x = 3$ **c.** no holes
13. a. $\left(-\infty, -\dfrac{3}{2}\right) \cup \left(-\dfrac{3}{2}, \dfrac{4}{5}\right) \cup \left(\dfrac{4}{5}, \infty\right)$ **b.** $x = -\dfrac{3}{2}$

c. hole at $x = \dfrac{4}{5}$ **15. a.** $(-\infty, 0) \cup (0, 3) \cup (3, \infty)$ **b.** $x = 3$

c. hole at $x = 0$ **17. a.** $(-\infty, -5) \cup (-5, 6) \cup (6, \infty)$
b. $x = -5, x = 6$ **c.** no holes
19. a. $(-\infty, -4) \cup (-4, 0) \cup (0, 7) \cup (7, \infty)$ **b.** $x = -4$,
$x = 7$ **c.** hole at $x = 0$ **21. a.** $(-\infty, 2) \cup (2, \infty)$
b. $x = 2$ **c.** no holes   **23. a.** $(-\infty, -3) \cup (-3, 3) \cup (3, \infty)$
b. $x = -3, x = 3$ **c.** no holes
25. a. $(-\infty, -2) \cup (-2, 3) \cup (3, \infty)$ **b.** $x = -2$
c. hole at $x = 3$ **27. a.** $(-\infty, \infty)$
b. no vertical asymptotes   **c.** no holes
29. a. $(-\infty, \infty)$ **b.** no vertical asymptotes   **c.** no holes
31. a. $(-\infty, \infty)$ **b.** no vertical asymptotes   **c.** no holes
33. a. $(-\infty, \infty)$ **b.** no vertical asymptotes   **c.** no holes
35. a. $(-\infty, -3) \cup (-3, \infty)$ **b.** $x = -3$ **c.** no holes
37. a. $(-\infty, -3) \cup (-3, \infty)$ **b.** no vertical asymptotes
c. hole at $x = -3$
39. a. $(-\infty, -6) \cup (-6, -2) \cup (-2, 0) \cup (0, \infty)$ **b.** $x = -6$,
$x = -2, x = 0$ **c.** no holes   **41. a.** asymptote

43. a. hole   **45. a.** asymptote   **47.** $-\dfrac{1}{2}$ **49.** -50

51. undefined   **53.** $\dfrac{1}{2}$ **55.** 5   **57.** undefined

59. 1.3 in.   **61.** about 3 in.   **63.** 2.8 hr
65. about 8 hr   **67.** $182.69   **69.** $337.50   **73.** $30
75. a. $x = 3$ **77. a.** no real zeros   **79. a.** $x = -2, x = 3$

81. 1,602,000 people   **83.** $\dfrac{\text{lb}}{\text{in.}^3}$

85. a.

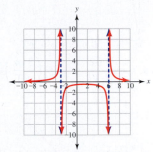

$[-10, 10, 1, -10, 10, 1]$

b. $(-\infty, -3) \cup (-3, 6) \cup (6, \infty)$ **c.** $x = 6, x = -3$
d. no holes   **e.** no real zeros
87. a.

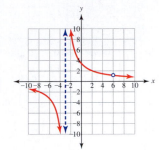

$[-10, 10, 1, -10, 10, 1]$

b. $(-\infty, -3) \cup (-3, 6) \cup (6, \infty)$ **c.** $x = -3$
d. hole at $x = 6$ **e.** no real zeros
89. b. $(-\infty, -4) \cup (-4, -3) \cup (-3, \infty)$
91. b. no real zeros   **93.** increase
95. a.

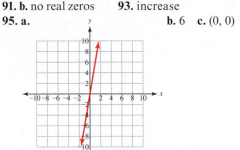

b. 6   **c.** $(0, 0)$

Section 5.6 Vocabulary Practice
1. B   **2.** A   **3.** A   **4.** B   **5.** A   **6.** A   **7.** A   **8.** B
9. A   **10.** B

Section 5.6 Exercises
1. a. $k = 3$ **b.** $y = 3x$ **c.** $y = 12$
d.   **e.** $y = 15$

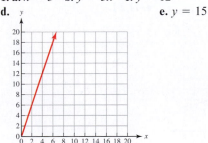

3. a. $k = 6$ **b.** $y = 6x$ **c.** $y = 24$
d.   **e.** $y = 12$

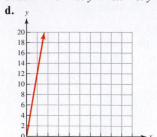

5. a. $k = \dfrac{3\ \text{oz}}{1\ \text{T}}$ **b.** $y = \left(\dfrac{3\ \text{oz}}{1\ \text{T}}\right)x$ **c.** 144 oz
d.   **e.** 96 oz

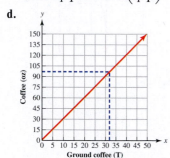

7. a. $k = 12$ **b.** $y = \dfrac{12}{x}$ **c.** $y = 3$
d.   **e.** $y = 4$

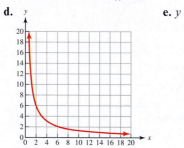

9. a. $k = 20$ **b.** $y = \dfrac{20}{x}$ **c.** $y = 4$
d.   **e.** $y = 5$

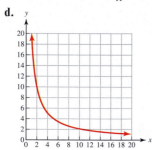

11. a. $k = 22.44\ \text{L}\cdot\text{atm}$ **b.** $y = \dfrac{22.44\ \text{L}\cdot\text{atm}}{x}$ **c.** 16.0 L
d.   **e.** 11.2 L

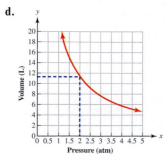

13. a. $k = 6.28$ **b.** $y = 6.28x$ **c.** 125.6 cm
d.   **e.** 94.2 cm

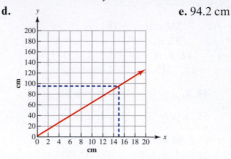

15. a. $k = \$44.431$ **b.** $y = \dfrac{\$44.431}{x}$ **c.** $3.17 per share

17. a. $k = 1.6956 \times 10^{-8}\ \text{ohm}\cdot\text{m}^2$

  b. $y = \dfrac{1.6956 \times 10^{-8}\ \text{ohm}\cdot\text{m}^2}{x}$ **c.** 8.3×10^{-3} ohm

19. a. $k = 360\ \text{min}\cdot\text{mi}$ **b.** $y = \dfrac{360\ \text{min}\cdot\text{mi}}{x}$

21. a. $k = \dfrac{12.5\ \text{g}}{1\ \text{cm}}$ **b.** $y = \left(\dfrac{12.5\ \text{g}}{1\ \text{cm}}\right)x$ **c.** 6.8 cm

23. a. $k = \dfrac{\$45}{1\ \text{ticket}}$ **b.** $y = \left(\dfrac{\$45}{1\ \text{ticket}}\right)x$ **c.** $340,875

  d. price of 1 ticket   **25. a.** $k = \dfrac{\$3144.48}{1\ \text{hr}}$

b. $y = \left(\dfrac{\$3144.48}{1\ \text{hr}}\right)x$ **c.** $37,734

d. the cost to charter the jet for 1 hr   **27.**   **a.** $k = 17.5$ mi

b. $y = \dfrac{17.5\ \text{mi}}{x}$ **c.** 21 min   **29.** $y = \dfrac{2400\ \text{micrometers}}{x}$

31. $k = 1.5$ **33.** $k = 16$ **35. a.** combined variation
b. directly   **c.** inversely   **37. a.** joint variation
39. inverse variation   **41.** direct variation

43. inverse variation   **45.** $F = \dfrac{P_m P_i}{CT}$

47. inverse variation   **49.** $T = \dfrac{PV}{nR}$

51. inverse variation   **53.** $H = \dfrac{165 - L - 2W}{2}$

55. $L = \dfrac{A - 2WH - GW}{2H + 2R}$ **57.** $E = \dfrac{A + P - AF - PF}{F}$

59. $A = \dfrac{P - EF - FP}{F - 1}$ **61.** $W = \dfrac{H + P - AT - PT}{T - 1}$

63. $d_o = \dfrac{d_i \cdot f}{d_i - f}$ **65.** $A = \dfrac{N^2 L}{478G - 2N^2 L}$

67. $E = \dfrac{SI}{D - S}$ **69.** $A = k\dfrac{L}{R}$

71. $T = \dfrac{S + 0.63162H - 6.332}{0.01089H + 0.891}$ **73.** $P = \dfrac{hH}{3H + 1}$

75. $y = \left(\dfrac{45,841\ \text{teachers}}{1\ \text{year}}\right)x + 1,976,398\ \text{teachers}$

77. 33,300 people

79. a.   **b.** direct variation

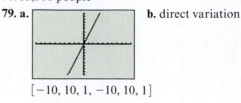

$[-10, 10, 1, -10, 10, 1]$

81. a. 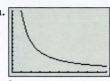   **b.** inverse variation

$[0, 1, 1, 0, 150000, 10000]$

83. b. $x_2 = \dfrac{x_1}{y - 1}$ **85. b.** $y = \dfrac{xz}{x - z}$ **87.** 107   **89.** 1

Section 5.7 Vocabulary Practice

1. J   **2.** D   **3.** I   **4.** A   **5.** B   **6.** H   **7.** C   **8.** G
9. E   **10.** F

Section 5.7 Exercises

1. a. $x^2 + 6x$ **b.** x **3. a.** $108x^6y^3z$ **b.** $48xy^9$ **5.** $\dfrac{17a^7c}{9b^5}$

7. $-\dfrac{9b^5}{17a^7c}$ **9.** $4p^5 + 6p^3 - 7 + \dfrac{5}{p}$ **11.** $\dfrac{20xy^3}{9} - 2y + 5$

13. $20xy^3z^3 - 18yz + 45$ **15. a.** $y + 3$ **17. a.** $y + 6$
19. a. $y - 3$ **21. a.** $c - 3$ **23. a.** $c + 7$

25. a. $c + 1 + \left(\dfrac{-24}{c + 3}\right)$ **27. a.** $2x + 3$

29. a. $2x - 25 + \dfrac{154}{x + 7}$ **31. a.** $x - 7$

33. a. $x - 4 + \left(\dfrac{-33}{2x - 3}\right)$ **35. a.** $5b + 2$ **37. a.** $b + 9$

39. a. $2z - 3$ **41. a.** $2w - 1$ **43. a.** $5x + 2$
45. a. $x^2 + 5x + 7$ **47. a.** $x + 3$ **49. a.** $x^2 - 8$
51. a. $a^2 - 6a + 1$ **53. a.** $b^5 + 2$ **55. a.** $x + 3$
b. $f(-6) = 0$ **57. a.** $x + 6$ **b.** $f(-3) = 0$

59. a. $y - 3$ **b.** $f(-7) = 0$ **61. a.** $c + 1 + \left(\dfrac{-24}{c + 3}\right)$

b. $f(-3) = -24$ **63. a.** $2p - 33 + \left(\dfrac{270}{p + 9}\right)$

b. $f(-9) = 270$ **65. a.** $2p + 3$ **b.** $f(9) = 0$

67. a. $4x + 31 + \left(\dfrac{114}{x - 3}\right)$ **b.** $f(3) = 114$

69. a. $4x + 7$ **b.** $f(-3) = 0$
71. a. $6x^2 + 5x + 2$ **b.** $f(4) = 0$

73. a. $6x^2 - 43x + 154 + \left(\dfrac{-624}{x + 4}\right)$ **b.** $f(-4) = -624$

75. $f(-9) = 0$; $x + 9$ is a factor
77. $f(9) = 270$; $x - 9$ is not a factor
79. $f(8) = 0$; $x - 8$ is a factor
81. $f(3) = 0$; $x - 3$ is a factor

83. $x + 7 + \left(\dfrac{5}{x + 2}\right)$; $f(-2) = 5$

85. $x + 11 + \left(\dfrac{41}{x - 2}\right)$; $f(2) = 41$

87. $x^2 + 9x + 35 + \left(\dfrac{141}{x - 4}\right)$; $f(4) = 141$

89. 5,854,000 LCD TVs   **91.** 57.5%

93. a. $x - 3$ **b.**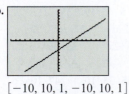

$[-10, 10, 1, -10, 10, 1]$

95. a. $x + 13 + \dfrac{80}{x - 8}$ **b.**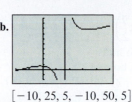

$[-10, 25, 5, -10, 50, 5]$

97. b. $5x^2 + x - \dfrac{1}{2}$ **99. b.** $2x + 3$

101.    **103. a.** $x = -\dfrac{8}{5}$ or $x = 1$

Section 5.8 Vocabulary Practice

1. I   **2.** A   **3.** F   **4.** H   **5.** G   **6.** E   **7.** C   **8.** D
9. B   **10.** J

Section 5.8 Exercises

1. $x = 0$ and $x = \dfrac{1}{4}$ **3.** $x = -7$ and $x = -4$

5. $x = -5$ and $x = -2$ **7.** $x = -2$ and $x = 0$
9. $[-6, 0]$ **11.** $(-8, -3)$ **13.** $(11, 12) \cup (12, 15)$
15. $(-\infty, -2] \cup (-1, \infty)$ **17.** $(0, 4]$
19. a. $x = -4$ and $x = 1$ **b.** $x < -4$ or $x > 1$
c. $(-\infty, -4) \cup (1, \infty)$ **21. a.** $x = -4$ and $x = 1$
b. $-4 \le x < 1$ **c.** $[-4, 1)$ **23. a.** $x = -3$ and $x = 3$
b. $x < -3$ or $x > 3$ **c.** $(-\infty, -3) \cup (3, \infty)$
25. a. $x = -29$ and $x = -7$ and $x = 4$
b. $-29 \le x < -7$ or $x > 4$ **c.** $[-29, -7) \cup (4, \infty)$
27. a. $x = 9$ and $x = 10$ **b.** $x < 9$ or $x > 10$
c. $(-\infty, 9) \cup (10, \infty)$ **29. a.** $x = -5$ and $x = -2$
b. $x < -5$ or $x > -2$ **c.** $(-\infty, -5) \cup (-2, \infty)$
31. 920% increase   **33.** $449 billion   **35. b.** $x = -9$
37. b. $(\infty, -2] \cup (1, \infty)$ **39.** 64   **41.** 1

Chapter 5 Review Exercises

1. $-\dfrac{7c^7}{17a^5b^3}$ **2.** $\dfrac{2x - 7}{x + 6}$ **3.** $\dfrac{x + 3}{x + 1}$ **4.** $\dfrac{x - 2}{(x - 1)(x + 1)}$

5. $\dfrac{7}{h^3k^3w^8}$ **6.** $\dfrac{x + 2}{3(x - 7)}$ **10.** $2^4 \cdot 3 \cdot a^2bc^3$

11. $x(x - 5)(x + 9)$ **12.** $x(x - 3)(x + 3)(2x - 5)$

13. $72h^2km$ **14.** $\dfrac{14xy^3}{105x^3y^4}$ **15.** $\dfrac{5x - 5}{(3x - 2)(x + 8)(x - 1)}$

16. $\dfrac{z + 5}{z + 4}$ **17.** $\dfrac{-h^2 + 21}{14h}$ **18.** $\dfrac{p + 1}{p - 1}$

19. $\dfrac{3x - 3}{(x - 5)(x - 2)}$ **20.** $\dfrac{-4c}{(c - 2)(c + 2)}$ **21.** $\dfrac{x + 7}{x - 5}$

22. $\dfrac{x - 6}{x + 5}$ **23.** $\dfrac{d}{3d + 1}$ **24. a.** $k = -3$ **25. a.** $x = 9$

26. a. $x = -1$ or $x = 6$ **30.** 2600 people   **31.** 50 min
32. 3.3%   **33. a.** $(-\infty, 0) \cup (0, \infty)$ **b.** $x = 0$ **c.** no holes
34. a. $(-\infty, -24) \cup (-24, \infty)$ **b.** $x = -24$ **c.** no holes

35. $\left(-\infty, -\frac{9}{2}\right) \cup \left(-\frac{9}{2}, 5\right) \cup (5, \infty)$ **b.** $x = -\frac{9}{2}$ and
$x = 5$ **c.** no holes   **36. a.** $(-\infty, -3) \cup (-3, 3) \cup (3, \infty)$
b. $x = 3$ **c.** hole at $x = -3$
37. a. $(-\infty, -2) \cup (-2, 9) \cup (9, \infty)$ **b.** $x = -2$ and $x = 9$
c. no holes   **38. a.** $(-\infty, -4) \cup (-4, 4) \cup (4, \infty)$
b. $x = -4$ **c.** hole at $x = 4$ **39. a.** $(-\infty, 5) \cup (5, \infty)$
b. $x = 5$ **c.** no holes   **40. a.** $(-\infty, -10) \cup (-10, 3) \cup (3, \infty)$
b. $x = -10$ **c.** hole at $x = 3$
41. a. $(-\infty, -5) \cup (-5, 0) \cup (0, \infty)$ **b.** $x = -5$ and
$x = 0$ **c.** no holes   **42.** 0.125   **43.** undefined
44. $x = -\frac{2}{7}$ **51. a.** inverse variation **b.** 1.2 qt

c.   **d.** about 0.80 qt

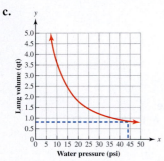

52. a. direct variation   **b.** 17,500,000 gal
c.   **d.** 14,000,000 gal

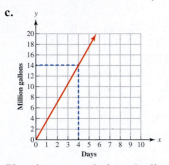

53. a. inverse variation   **b.** direct variation
c. $P = \dfrac{CV - CRT}{V}$ **d.** $C = \dfrac{PV}{V - RT}$
54. a. $k = \dfrac{0.012 \text{ in.}}{1 \text{ page}}$ **b.** $y = \left(\dfrac{0.012 \text{ in.}}{1 \text{ page}}\right)x$
55. a. $k = 40 \text{ cm}^2$ **b.** $y = \dfrac{40 \text{ cm}^2}{x}$
56. a. $k = 0.25$ **b.** direct **c.** direct **d.** $y = 80$
57. a. $k = 3$ **b.** direct **c.** inverse **d.** $y = 40$
58. direct   **59.** inverse   **60.** $\dfrac{18h^3}{25k^7}$
61. $6x + 15 - \dfrac{8}{x} + \dfrac{4}{x^2}$ **62. a.** $3x + 5$
63. a. $3x + 41 + \dfrac{276}{x - 6}$ **64.** $3x + 5$
65. a. $3x + 41 + \dfrac{276}{x - 6}$ **b.** $f(6) = 276$
66. $x - 6$ is a factor   **67.** $f(6) = 0; x - 6$ is a factor
68. $2x + 3 + \left(\dfrac{-9}{x + 6}\right); f(-6) = -9$
69. a. $x = -6$ and $x = -5$ **b.** $-6 < x < -5$
c. [number line] **d.** $(-6, -5)$
70. a. $x < -9$ or $x \geq 11$ **b.** [number line]
c. $(-\infty, -9) \cup [11, \infty)$

Chapter 5 Test
1. $\dfrac{1}{x + 2}$ **2.** $\dfrac{x^2 - 4x - 12}{4}$ **3.** $\dfrac{1}{x - 9}$
4. $\dfrac{y^2 - 4}{(y - 6)(y - 4)}$ **5.** $\dfrac{x - 3}{(x + 1)(x + 3)}$
6. $\dfrac{x^2 + 12x + 30}{(x + 3)(x + 6)(x + 7)}$ **7.** $\dfrac{-2}{6x - 1}$ **8. a.** $x = -2$
9. a. $x = 2$ or $x = 15$ **10.** $P = \dfrac{4Q + 8AQ}{7 - 3R}$ **11.** $x = -3$
13. a. $x = -2$ **b.** hole at $x = -5$ **14. a.** inversely
related **b.** decrease **c.** increase   **15. a.** Z increases
b. Z decreases **c.** inversely related **d.** inversely related
e. $H = \dfrac{KLZ}{4W}$ **16.** 19 min   **17.** 3400 people
18. $x^3 + \dfrac{3x^2}{4} - \dfrac{x}{12} - \dfrac{1}{3} + \dfrac{2}{3x}$ **19.** $2x + 7$ **20.** $5x + 8$
21. $x + 10 + \left(\dfrac{-10}{x + 3}\right); f(-3) = -10$
22. $(-\infty, -16] \cup (-8, \infty)$

CHAPTER 6

Section 6.1 Vocabulary Practice
1. E **2.** A **3.** F **4.** C **5.** J **6.** B **7.** D **8.** G
9. H **10.** I

Section 6.1 Exercises
1. a. 2 **b.** 5 **3. a.** 3 **b.** x **5.** 9 **7.** -8 **9.** 6
11. 3.3 **13.** -7.7 **15.** not a real number **17.** $\dfrac{10}{7}$
19. $\dfrac{1}{8}$ **21.** 0.1 **23.** greater than **25.** 27 ft **27.** 11.1
29. $10x^3$ **31.** $10x^3\sqrt{2}$ **33.** $10x^3\sqrt{2x}$ **35.** $4h^2k\sqrt{3h}$
37. $8c\sqrt{2ab}$ **39.** $12x^6\sqrt{2x}$ **41.** $4\sqrt{5xy}$ **43.** x
45. x **47.** x^2 **49.** a^3 **51.** $2a^2b^2\sqrt[3]{5b}$ **53.** $3\sqrt[3]{3g^2h}$
55. $-4x^4y^5\sqrt[3]{2y}$ **57.** $-8x^7y$ **59.** $2ab^2\sqrt[4]{2a}$
61. $4r^3w^3\sqrt[4]{r^3w}$ **63.** $3\sqrt[4]{3h^3k^2}$ **65.** $2x^3y^2\sqrt[4]{8y}$
67. $2x\sqrt[8]{x^7}$ **69.** $x^3\sqrt[9]{x^4}$ **71.** $a^2\sqrt[100]{a}$ **73.** $x + 5$
75. $a - 4$ **77.** $z - 1$ **79.** $2x + 3$ **81.** $2h - 9$
83. $3x - 7$ **85.** $x - 5$ **87. a.** $y = \left(\dfrac{\$99}{1 \text{ month}}\right)x + \$99$
b. \$6039 **89.** \$13,000 **91. b.** $4x^{12}\sqrt{2x}$ **93. b.** $x + 5$
95. $-2x + 8y$ **97.** $-x - 31$

Section 6.2 Vocabulary Practice
1. I **2.** A **3.** H **4.** G **5.** D **6.** J **7.** B **8.** F
9. E **10.** C

Section 6.2 Exercises
1. $11\sqrt{3}$ **3.** $-4\sqrt{5}$ **5.** $-14\sqrt{21}$ **7.** $13\sqrt[3]{x^2}$
9. $11w\sqrt{5}$ **11.** already simplified **13.** $3\sqrt{5}$
15. $3\sqrt{5x}$ **17.** $3x\sqrt{5}$ **19.** $13\sqrt{3}$ **21.** $5\sqrt[3]{2}$
23. $10\sqrt[3]{3x}$ **25.** $-\sqrt[4]{a^3}$ **27.** $-5\sqrt[3]{2k^2}$ **29.** $-\sqrt{3}$
31. $5x\sqrt[5]{2x^2}$ **33.** $x\sqrt[5]{2x^2}$ **35.** $6p\sqrt[4]{5}$
37. $2p\sqrt[4]{5} + 4p^3\sqrt[4]{5p}$ **41. a.** $y\sqrt{x}$ **b.** $x\sqrt{y}$ **c.** no
43. a. no **b.** yes **45.** $3\sqrt{10}$ **47.** $10\sqrt{6}$ **49.** $14x\sqrt{3}$
51. $p\sqrt{210}$ **53.** $2\sqrt[3]{3}$ **55.** $3\sqrt[3]{3}$ **57.** x
59. $n\sqrt[5]{n}$ **61.** $w\sqrt[101]{w^{19}}$ **63.** $\sqrt[17]{a^{16}}$ **65.** $\sqrt{6} + 12$

67. $x\sqrt{2} - x\sqrt{10}$ **69.** $2x\sqrt{3} - 2x\sqrt{5}$
71. $x^2 + 2x\sqrt{2} + 2$ **73.** $x^2 - 2$ **75.** $x^2 - 2x\sqrt{2} + 2$
77. $x - 4\sqrt{x} + 4$ **79.** $3h\sqrt{2} + \sqrt{30hk} - 2\sqrt{3hk} - 2k\sqrt{5}$
81. $5a\sqrt{3} - a\sqrt{70} - 5a\sqrt{6} + 2a\sqrt{35}$ **83.** $9z - 25$
85. $72x + 12\sqrt{2x} + 1$ **87.** $4c$ **89.** Four out of five dream of a new career while on vacation. **91.** 260%

93. b. $16x^2\sqrt{2x}$ **95. b.** $\sqrt{6x} - 8\sqrt{3}$ **97.** $\dfrac{x^m}{x^n} = x^{m-n}$

99. $\dfrac{3y^5}{10x^6}$

Section 6.3 Vocabulary Practice
1. H **2.** E **3.** I **4.** C **5.** A **6.** G **7.** A **8.** D
9. F **10.** B

Section 6.3 Exercises
1. $\dfrac{6w^4}{7p^3}$ **3.** $\dfrac{10\sqrt{m}}{9p}$ **5.** $\dfrac{3k^2}{2h^4}$ **7.** $\dfrac{b\sqrt[3]{17b}}{3m^6}$ **9.** $\dfrac{6a^3\sqrt{2ab}}{5cd^2}$

11. $\dfrac{6x^2y\sqrt{3xy}}{7z}$ **13.** $\dfrac{x^2\sqrt{3y}}{2}$ **15.** $\dfrac{a^2b\sqrt{3b}}{5}$ **17.** $\dfrac{\sqrt{xy}}{10}$

19. $\dfrac{h\sqrt[3]{5h}}{2}$ **21.** $\dfrac{x\sqrt[4]{5}}{2}$ **23.** $\dfrac{\sqrt[5]{8b^4}}{2}$ **25.** $\dfrac{5\sqrt{x}}{x}$

27. $\dfrac{5\sqrt{x}}{3x}$ **29.** $\dfrac{15 - 5\sqrt{x}}{9 - x}$ **31.** $\dfrac{15 + 5\sqrt{x}}{9 - x}$

33. $\dfrac{15\sqrt{x} + 5x}{9 - x}$ **35.** $\dfrac{7\sqrt[3]{4}}{2}$ **37.** $\dfrac{9\sqrt[4]{a^3}}{a}$ **39.** $\dfrac{3\sqrt{k} - k}{9 - k}$

41. $\dfrac{\sqrt{5x} + 9\sqrt{5}}{5}$ **43.** $\dfrac{5x + 15\sqrt{y}}{x^2 - 9y}$ **45.** $\dfrac{3x\sqrt{2} - 27\sqrt{x}}{2x - 81}$

47. $\dfrac{x + 2\sqrt{xy} + y}{x - y}$ **49.** $\dfrac{x\sqrt{3} + 3}{x^2 - 3}$ **51.** $\dfrac{\sqrt{3x} + 3}{x - 3}$

53. $\dfrac{\sqrt[4]{27}}{3}$ **55.** $\dfrac{6\sqrt{a} + 15\sqrt{b}}{4a - 25b}$ **57.** $\dfrac{6a + 15\sqrt{ab}}{4a - 25b}$

59. $\dfrac{2a\sqrt{3} + 5\sqrt{3ab}}{4a - 25b}$ **61.** $\dfrac{3a\sqrt{10} - \sqrt{42a} + 5\sqrt{15a} - 5\sqrt{7}}{6a - 25}$

63. $\dfrac{\sqrt{10x}}{5x}$ **65.** $\dfrac{3\sqrt{6pq}}{8q^2}$ **67.** $\dfrac{3\sqrt[3]{4x}}{2}$ **69.** $\dfrac{2c\sqrt[3]{25}}{5}$

71. $\dfrac{\sqrt[4]{12a^2}}{2}$ **73.** $\dfrac{\sqrt{3xyz}}{3yz}$ **75.** $\dfrac{7\sqrt[9]{x^8}}{x}$ **77.** $\dfrac{\sqrt[5]{144xy^4}}{2y}$

79. $x - y$ **85. a.** $y = \left(\dfrac{\$18.50}{1\ \text{hr}}\right)(4\ \text{hr})\left(\dfrac{x}{25\ \text{guests}}\right)$

b. $888 **87.** 7200 degrees **89. b.** $\dfrac{\sqrt{x}}{2x}$ **91. b.** $\dfrac{\sqrt[4]{40x^3}}{2x}$
93. x^6 **95.** x^2

Section 6.4 Vocabulary Practice
1. F **2.** H **3.** E **4.** B **5.** I **6.** D **7.** C **8.** G
9. A **10.** J

Section 6.4 Exercises
1. $x^{\frac{1}{2}}$ **3.** $x^{\frac{3}{2}}$ **5.** $x^{\frac{2}{3}}y^{\frac{1}{3}}$ **7.** $x^{\frac{3}{4}}$ **9.** $x^{\frac{1}{2}}$ **11.** $x^{\frac{1}{5}}$

13. $x^{\frac{1}{99}}$ **15.** $a^{\frac{1}{5}}$ **17.** $d^{\frac{1}{8}}$ **19.** $h^{\frac{4}{3}}$ **21.** a^2b^2

23. $x^{\frac{4}{5}}y^{\frac{1}{60}}$ **25.** $\sqrt[4]{m^3}$ **27.** $\sqrt[3]{bc^2}$ **29.** $\sqrt[4]{ab^5}$

31. a. $x^{\frac{4}{12}}y^{\frac{3}{12}}$ **b.** $\sqrt[12]{x^4y^3}$ **33.** $Q = \dfrac{1.486A}{n}\sqrt[3]{R^2}\sqrt{S}$

35. x^8 **37.** $a^{\frac{2}{3}}$ **39.** $x^{\frac{1}{3}}$ **41.** $x^{\frac{19}{45}}$ **43.** x^{12} **45.** $x^{\frac{1}{8}}$

47. $x^{\frac{1}{4}}$ **49.** 1 **51.** $x^{\frac{2}{7}}$ **53.** $x^{\frac{3}{4}}$ **55.** $x^{\frac{89}{90}}$

57. a. $x^{\frac{1}{9}}$ **b.** $\sqrt[9]{x}$ **59. a.** $x^{\frac{1}{15}}$ **b.** $\sqrt[15]{x}$ **61.** $\dfrac{3p^5}{4}$

63. $\dfrac{4}{3a^2}$ **65.** $a^{\frac{1}{4}}$ **67.** $\dfrac{1}{a^{\frac{1}{4}}}$ **69.** $\dfrac{1}{a^{\frac{8}{35}}}$ **71.** $x^{\frac{11}{35}}y^{\frac{23}{35}}$

73. $\dfrac{5}{21m^{\frac{17}{30}}}$ **75.** $\dfrac{1}{5x^{\frac{9}{40}}}$ **77. a.** $\dfrac{x^{\frac{3}{5}}y^{\frac{1}{5}}}{x^{\frac{1}{2}}y^{\frac{1}{2}}}$ **b.** $\dfrac{x^{\frac{1}{10}}}{y^{\frac{3}{10}}}$

79. a. $\dfrac{a^{\frac{1}{3}}b^{\frac{1}{3}}}{a^{\frac{2}{5}}b^{\frac{3}{5}}}$ **b.** $\dfrac{1}{a^{\frac{1}{15}}b^{\frac{4}{15}}}$ **81. a.** $\dfrac{12p^{\frac{1}{2}}w^{\frac{1}{2}}}{5p^{\frac{1}{3}}w^{\frac{1}{3}}}$ **b.** $\dfrac{12p^{\frac{1}{6}}w^{\frac{1}{6}}}{5}$

83. $x^{\frac{5}{24}}$ **85.** $x^{\frac{83}{72}}$ **87.** $x^{\frac{29}{8}}$ **89.** 23 children

91. 316% increase **93. b.** $x^{\frac{3}{14}}$ **95. b.** $x^{\frac{5}{7}}$ **97. a.** $x = 23$

99. a. $x = -\dfrac{3}{2}$ or $x = 5$

Section 6.5 Vocabulary Practice
1. H **2.** E **3.** G **4.** J **5.** I **6.** B **7.** F **8.** D
9. C **10.** A

Section 6.5 Exercises
1. a. $x = 64$ **3. a.** no solution **5. a.** $x = 16$
7. a. $m = 125$ **9. a.** $d = -729$ **11. a.** $x = 81$
13. a. $h = 81$ **15. a.** no solution **17. a.** $y = 5$
19. a. $y = 49$ **21. a.** $x = 212$ **23. a.** $x = 1$
25. a. $x = 15$ **27. a.** $x = 37$ **29. a.** $p = 6$
31. a. $k = -16$ **33. a.** $x = -76$ **35. a.** $x = 9$
37. a. $x = -2$ or $x = -1$ **39. a.** $y = 4$ or $y = 6$
41. a. $w = 0$ or $w = 5$ **43. a.** $w = 0$ **45. a.** $p = 2$
47. a. $x = 4$ **49. a.** $p = 4$ **51. a.** no solution
53. a. $a = 30$ **55. a.** $p = 25$ **57. a.** $y = 13$

59. a. no solution **61. a.** $x = \dfrac{4}{5}$ **63. a.** $n = 4$

65. a. $x = 1$ **67. a.** $x = 3$ **69.** 109 ft **71.** 6.7 miles per hour **73.** 35.6 ft **75.** 1.98 m² **77.** 72 ft
79. 51 ft 11 in. **81.** 1.98 s **83. a.** 81 **b.** 746 **c.** 9.1

85. $144.72 **87.** 653 gadgets **89. a.** $y = \left(\dfrac{0.123\ \text{in.}}{1\ \text{year}}\right)x$
b. 8.1 years **91.** $193.48

93. a. **b.** $x = -1$

$[-10, 10, 1, -10, 10, 1]$

95. a. **b.** $x = 6$

$[-10, 10, 1, -10, 10, 1]$

97. b. $x = 510$ **99. b.** $x = 2$ or $x = 5$ **101.** $(-\infty, \infty)$
103. $x = -1$ and $x = 6$

Section 6.6 Vocabulary Practice
1. I **2.** C **3.** G **4.** J **5.** D **6.** A **7.** H **8.** E
9. B **10.** F

Section 6.6 Exercises
1.
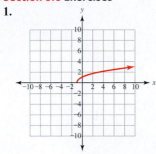

3.
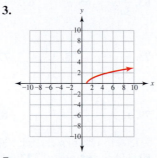

5.
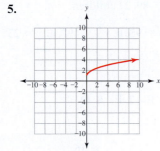

7.
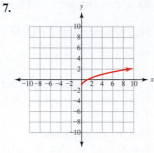

9.
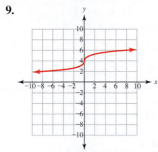

11.
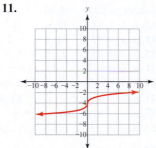

13.
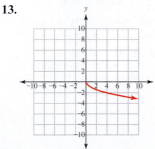

15.
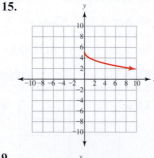

17.
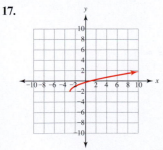

19.
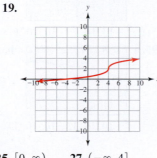

21. $[-1, \infty)$ **23.** $[1, \infty)$ **25.** $[0, \infty)$ **27.** $(-\infty, 4]$
29. $[0, \infty)$ **37.** $6\sqrt{2}$ **39.** 10 **41.** 5.4 in.

43. 531 hertz **45.** $1.12 \times 10^9 \frac{m}{s}$ **47.** grade level 11.4
49. a. $[0, \infty)$ **b.** $[6, \infty)$ **51. a.** $[0, \infty)$ **b.** $[-7, \infty)$
53. a. $[-7, \infty)$ **b.** $[0, \infty)$ **55. a.** $[-4, \infty)$ **b.** $[-7, \infty)$
57. a. $[0, \infty)$ **b.** $(-\infty, -4]$ **59.** $x = -1$ **61.** $x = 1$
63. $x = 36$ **65.** $x = 4$ **67.** $x = -8$ **69.** $x = 8$
71. $x = 3$ **73.** $(-\infty, \infty)$
75. $(-\infty, -4) \cup (-4, 4) \cup (4, \infty)$ **77.** $(-\infty, \infty)$
79. $(-\infty, \infty)$ **81.** $[7, \infty)$ **83.** $(-\infty, \infty)$
85. $(-\infty, 2) \cup (2, \infty)$ **87.** $(-\infty, \infty)$ **89.** $(-\infty, 9]$
91. 109,900 lb **93.** \$3839.80

95. a. 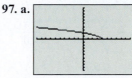 **b.** $[0, \infty)$ **c.** $[-6, \infty)$ **d.** $x = 4$

$$[-10, 10, 1, -10, 10, 1]$$

97. a. **b.** $(-\infty, 4]$ **c.** $[0, \infty)$ **d.** $x = 4$

$$[-10, 10, 1, -10, 10, 1]$$

99. b. $[8, \infty)$ **101. b.** $[0, \infty)$ **103.** $y = \frac{1}{4}x - 9$
105. $x(2x + 3)(3x - 5)$

Chapter 6 Review Exercises
1. -10 and 10 **5.** index 2; radicand $x - 4$
6. index 3; radicand $x + 2$ **7.** 8 **8.** -3 **9.** 5.92
10. not a real number **11.** x^2 **12.** $y^8\sqrt[4]{y^3}$
13. $5a^3b^3\sqrt{2bc}$ **14.** $-3x^2y^3\sqrt[3]{2x^2z^2}$ **15.** $3p + 1$
16. x **17.** $2n^2p^3\sqrt[4]{3n}$ **18.** $u^2\sqrt{21uz}$ **23.** $-7\sqrt{u}$
24. $7\sqrt{3}$ **25.** $13\sqrt{6k}$ **26.** $39x\sqrt{2x}$ **27.** $5\sqrt[3]{5xy}$
28. $ab\sqrt[4]{4b}$ **29.** $5x\sqrt{6y}$ **30.** $21ab\sqrt{2}$
31. $2p\sqrt[3]{p}$ **32.** $6c + 45c\sqrt{c}$ **33.** $h - 5$
34. $2a\sqrt{3} - 3\sqrt{6a} - 5\sqrt{2a} + 15$ **35.** $\frac{x^2\sqrt{x}}{2z^3}$ **36.** $\frac{a^2}{b^6}$
37. $\frac{7w^2\sqrt{2w}}{9p}$ **38.** d **39.** $\sqrt{x} + \sqrt{5}$
40. $\frac{p\sqrt{10} + 3\sqrt{5p}}{5p}$ **41.** $\frac{p\sqrt{10} - 3\sqrt{5p}}{2p - 9}$ **42.** $\frac{x\sqrt{10xy}}{2}$
43. $\frac{5\sqrt[3]{36x^2}}{6}$ **44.** $\frac{5\sqrt[4]{216x^3}}{6}$ **45.** $d^{\frac{5}{11}}$ **46.** $\sqrt[24]{a^{23}}$
47. $x^{\frac{5}{3}}y^4$ **51.** addition and subtraction **52.** $a^{\frac{7}{8}}$
53. $x^{\frac{13}{20}}$ **54.** $a^{\frac{2}{5}}$ **55.** $c^{\frac{19}{36}}$ **56.** $\frac{1}{9x^{\frac{1}{8}}}$ **57.** $\frac{3}{20x^{\frac{1}{2}}z^{\frac{1}{35}}}$
58. $\frac{b^{\frac{3}{8}}}{a^{\frac{1}{6}}}$ **59. a.** $\left(\frac{x^{14}}{z^{11}}\right)^{\frac{1}{2}}$ **b.** $\frac{x^7}{z^{\frac{11}{2}}}$ **60. a.** $\frac{p^{\frac{2}{4}}w^{\frac{1}{4}}}{p^{\frac{1}{2}}w^{\frac{1}{2}}}$ **b.** $\frac{1}{w^{\frac{1}{4}}}$
61. a. $\left((x^3y^4)^{\frac{1}{2}}\right)^{\frac{1}{3}}$ **b.** $x^{\frac{1}{2}}y^{\frac{2}{3}}$ **64. a.** no **66. a.** $x = \frac{55}{2}$
67. a. no solution **68. a.** $x = 57$ **69. a.** $x = -14$
70. a. $x = 20$ **71. a.** $k = -1$ or $k = 3$
72. a. $y = 2$ or $y = 6$ **73. a.** $n = 25$ **74. a.** no solution
75. 409 in.^{3}

76.

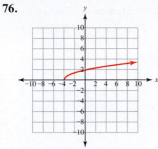

77.

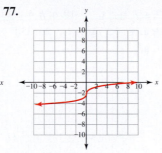

78.

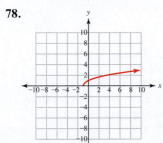

79. a. $x = 4$ **b.**

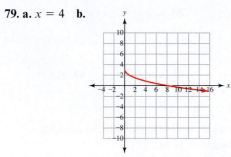

80. b. $[0, \infty)$ **c.** $[1, \infty)$ **82.** 128 ft   **83.** $x = 1$
84. $[-2, \infty)$ **85.** $[-6, \infty)$ **86.** $[7, \infty)$ **87.** $(-\infty, 4]$
88. a. no

Chapter 6 Test

1. $9a^3b^2\sqrt{2bc}$ **2.** $2x^3y^3\sqrt[3]{9xz^2}$ **3.** $\dfrac{5y^2\sqrt{3xy}}{7x^3}$

4. $16\sqrt{11}$ **5.** $14\sqrt{5}$ **6.** $21d\sqrt{2c}$ **7.** $3w\sqrt{2} - \sqrt{15w}$
8. $5\sqrt{15} + 5\sqrt{21} - 3\sqrt{10} - 3\sqrt{14}$ **9.** $2xy\sqrt[5]{8x^4}$

10. x **11.** $x^{\frac{2}{3}}$ **12.** $x^{\frac{38}{45}}$ **13.** $a^{\frac{1}{12}}$ **14.** $x^{\frac{5}{12}}y^{\frac{22}{45}}$

15. $\dfrac{2\sqrt{x} - 10}{x - 25}$ **16.** $\dfrac{4a^2\sqrt{6a}}{3}$ **17.** $x^{\frac{2}{9}}$ **18.** $\sqrt[7]{k^3}$

19. a. $\dfrac{5\sqrt{az} - 2\sqrt{z}}{4z}$ **b.** $\dfrac{3p + 3\sqrt{pw}}{p - w}$ **20. a.** $[-7, \infty)$

b. $(-\infty, -5]$ **21. a.** $x = 3$ or $x = 7$
22. a. $x = 0$ **b.** $[-4, \infty)$ **c.** $[-2, \infty)$
d.

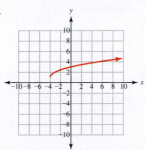

23. $x = 34$ **24.** 658 hertz

Chapters 4-6 Cumulative Review

1. $(3, -4)$ **2. a.** $\left(\dfrac{1}{2}, -5\right)$ **3.** $(-8, 11)$
4. parallel lines   **5.** coinciding lines
6. a. $(-8, 4, 18)$ **7.** $(4, 6)$
8. a.   **b.** $(4, 2), (0, 6), (0, -6)$

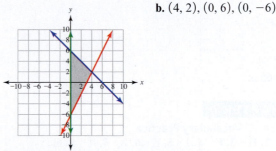

9. 1500 lb 12% nickel steel; 1000 lb 17% nickel steel
10. 37.5 lb compost; 12.5 lb sand
11. $\dfrac{x^2 + 4x}{2x + 3}$ **12.** $\dfrac{-a - 2}{a - 3}$ **13.** $\dfrac{a - 6}{a - 3}$

14. $\dfrac{2x + 1}{(x - 1)(x + 1)}$ **15.** $\dfrac{n - 9}{(n - 5)(n + 3)}$ **16.** $4ac\sqrt{15ab}$

17. $16\sqrt{2a}$ **18.** $\dfrac{8\sqrt{2xy}}{5y}$ **19.** $16x - 5$ **20.** $\dfrac{3x\sqrt{6xy}}{8y^2}$

21. $\dfrac{2\sqrt{2n} + 4\sqrt{3}}{n - 6}$ **22.** $3xy^2\sqrt[3]{2x^2}$ **23.** $2hk\sqrt[4]{3h}$

24. $\dfrac{2w^{\frac{17}{45}}}{9}$ **25.** $81xy^{\frac{3}{2}}$ **26.** $x^{\frac{10}{11}}$ **27.** $\sqrt[5]{x^4}$

28. $\sqrt{x} - \sqrt{3}$ **30.** $x - 2$ **31.** $x + 4$
32. a. $x = 5$ or $x = 20$ **33. a.** no solution
34. a. $x = 30$ **35. a.** $x = 5$ **36. a.** $x = 10$
37. a. $-4 < x < -2$ **b.**
$(-4, -2)$

38. $y = \dfrac{2xz}{5x - 3z}$ **39. a.** inverse variation

40. a. $k = 0.065$ **b.** $y = 0.065x$ **c.** $\$55.25$ **d.** the sales
tax rate   **41.** 5   **42.** 937,500 Americans   **43.** 16 min
44. a. $(-\infty, 9) \cup (9, \infty)$ **b.** $x = 9$ **c.** no holes
45. a. $(-\infty, -3) \cup (-3, 3) \cup (3, \infty)$ **b.** $x = -3$
c. hole at $x = 3$ **46. a.** $(-\infty, \infty)$ **b.** no vertical asymptotes
c. no holes   **47.** $-\dfrac{1}{4}$ **48.** $x = -5$ and $x = 4$

49. a.   **b.** $[0, \infty)$ **c.** $[2, \infty)$
d. no real zeros

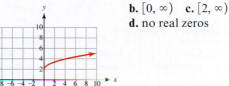

50. a.

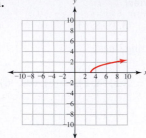

b. $[3, \infty)$ **c.** $[0, \infty)$
d. $x = 3$

51. 0.65 in.

CHAPTER 7

Section 7.1 Vocabulary Practice
1. I **2.** H **3.** F **4.** J **5.** E **6.** B **7.** A **8.** G
9. C **10.** D

Section 7.1 Exercises
1. $5i$ **3.** $10\sqrt{x}i$ **5.** $a\sqrt{6}i$ **7.** $3a\sqrt{2b}i$ **9.** $5y\sqrt{2xy}$
11. $5y\sqrt{2xy}i$ **13.** $7\sqrt{6}i$ **15.** $-7\sqrt{6}$ **17.** $3w\sqrt{2}i$
19. $-3w\sqrt{2}$ **21.** abi **23.** $-6ab\sqrt{35}$ **25.** $18a\sqrt{5bc}i$
27. $-2c\sqrt[3]{4cd}$ **29.** $4c^2\sqrt[3]{12d^2}$ **31.** $\sqrt{55}i$ **33.** $\sqrt{73}$
37. see table below **39.** $11 + 12i$ **41.** $11 - 12i$
43. $-1 + 6i$ **45.** $-1 - 12i$ **47.** $18 - 54i$
49. $-6 + 63i$ **51.** $54 - 33i$ **53.** $-6 - 63i$
55. $\frac{2}{9} + \frac{1}{6}i$ **57.** $\frac{1}{9} - \frac{1}{5}i$ **59.** $-5i$ **61.** $-\frac{7}{3}i$
63. $\frac{5}{17} - \frac{3}{17}i$ **65.** $\frac{18}{25} + \frac{24}{25}i$ **67.** $-\frac{24}{25} + \frac{18}{25}i$
69. $\frac{4}{29} + \frac{10}{29}i$ **71.** $\frac{7}{41} + \frac{22}{41}i$ **73.** $\frac{1}{2}$ **75.** $-i$ **77.** 1
79. -1 **81.** i **83.** 120 volts **85.** $10 + 7.5398i$ ohms
87. \$22,050 **89.** \$30.80 per person **91.** -32 **93.** $74i$
95. b. $\frac{8}{17} - \frac{32}{17}i$ **97. b.** $-5xy\sqrt{6}$

101. $x = -\frac{5}{2}$ or $x = \frac{5}{2}$

Section 7.2 Vocabulary Practice
1. I **2.** C **3.** G **4.** J **5.** A **6.** F **7.** H **8.** D
9. B **10.** E

Section 7.2 Exercises
1. $ax^2 + bx + c = 0$, where $a \neq 0$ **9. a.** $x = \pm 5$
11. a. $x = \pm 5i$ **13. a.** $x = \pm 3\sqrt{6}$ **15. a.** $d = \pm 2\sqrt{39}i$

17. a. $h = \pm 6$ **19. a.** $h = \pm 6i$ **21. a.** $p = \pm 9\sqrt{3}$
23. a. $x = \pm 7\sqrt{15}i$ **25.** 33 ft **27.** 9 ft 8 in. **29.** 8 ft 1 in.
31. a.

b. 9 **c.** 5 **d.** 10.3

33. $\sqrt{(x_2 - x_1)^2 + (y_2 - y_1)^2}$ **35. a.** $a = 1, b = -6,$
$c = -7$ **b.** discriminant 64; not prime
37. a. $a = 10, b = 33, c = 27$ **b.** discriminant 9; not prime
39. a. $a = 9, b = -35, c = 20$ **b.** discriminant 505; prime
41. a. $a = 7, b = 3, c = 1$ **b.** discriminant -19; prime
43. a. $x = -2$ or $x = -11$ **45. a.** $w = 6$ or $w = 2$
47. a. $z = -3$ or $z = 18$ **49. a.** $z = -18$ or $z = 3$
51. a. $x = -3$ or $x = \frac{5}{2}$ **53. a.** $p = -\frac{5}{2}$ or $p = \frac{1}{3}$
55. a. $x = \frac{1}{7}$ or $x = \frac{1}{3}$ **57. a.** $m = -\frac{5}{3}$ or $m = \frac{1}{2}$
59. a. $z = -\frac{2}{7}$ or $z = 3$ **61. a.** $x = -\frac{3}{2}$ **63. a.** $v = \frac{1}{3}$
65. a. $x = -\frac{1}{2}$ or $x = -\frac{2}{5}$ **67. a.** $x = 4$ or $x = 9$
69. 20 ft long; 18 ft wide **71.** 32 cm long; 14 cm wide
73. base = 20 in.; height = 16 in.
75. $b_1 = 10$ cm; $b_2 = 20$ cm; $h = 6$ cm
77. a. $R(x) = \left(\frac{\$7.50}{1\text{ lb}}\right)x$ **b.** $C(x) \approx \left(\frac{\$4.54}{1\text{ lb}}\right)x$
c. $P(x) \approx \left(\frac{\$2.96}{1\text{ lb}}\right)x$ **d.** \$414,400 **79.** $\frac{1\text{ death}}{17\text{ min}}$
81. a.

b. $x = -5$ or $x = 1$

$[-10, 10, 1, -10, 10, 1]$

Table for Section 7.1 exercise 37

	Complex numbers	Imaginary numbers	Real numbers	Rational numbers	Irrational numbers	Integers	Whole numbers
$\sqrt{5}$	X		X		X		
$\sqrt{-5}$	X	X					
$5i$	X	X					
$3 - 7i$	X						
9	X		X	X		X	X
0.4	X		X	X			

83. a.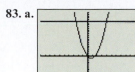

$[-10, 10, 1, -10, 25, 1]$

b. $x = -2.5$ or $x = 4$

85. b. $x = \pm 2\sqrt{5}i$ **87. b.** $x = \pm 3\sqrt{2}i$ **89.** 16   **91.** $\dfrac{9}{4}$

Section 7.3 Vocabulary Practice
1. I   **2.** E   **3.** H   **4.** B   **5.** F   **6.** G   **7.** J   **8.** D
9. C   **10.** A

Section 7.3 Exercises
1. 2   **3.** $x^2 + 12x + 36$ **5.** $u^2 - 6u + 9$
7. $z^2 + 3z + \dfrac{9}{4}$ **9.** $x^2 - 11x + \dfrac{121}{4}$
11. a. $x = -5$ or $x = -1$ **13. a.** $n = 2$ or $n = 6$
15. a. $j = -4 \pm \sqrt{33}$ **17. a.** $j = -4 \pm i$
19. a. $x = 1 \pm \sqrt{6}i$ **21. a.** $x = -1$ or $x = 3$
23. a. $a = -2$ or $a = -1$ **25. a.** $x = -10$ or $x = 1$
27. a. $x = \dfrac{11}{2} \pm \dfrac{\sqrt{109}}{2}$ **29. a.** $z = \dfrac{1}{2} \pm \dfrac{\sqrt{15}}{2}i$
31. a. $d = 3$ **33. a.** $u = -5$ **35.** $x = -\dfrac{3}{2}$ or $x = \dfrac{5}{2}$
37. $x = -\dfrac{1}{2}$ **39.** $x = -\dfrac{1}{2}$ or $x = 3$ **41.** $x = 1$ or $x = \dfrac{3}{2}$
43. $x = -\dfrac{5}{2} \pm \dfrac{\sqrt{15}}{2}$ **45.** $w = -\dfrac{3}{2} \pm \dfrac{\sqrt{177}}{6}$
47. $x = -\dfrac{5}{4} \pm \dfrac{\sqrt{113}}{4}$ **49.** $x = -\dfrac{5}{4} \pm \dfrac{3\sqrt{7}}{4}i$
51. $d = -8$ or $d = -\dfrac{9}{2}$ **53.** $x = -\dfrac{1}{3} \pm \dfrac{\sqrt{2}}{3}i$
55. $x = -1$ or $x = \dfrac{1}{3}$ **57.** $x = \dfrac{3}{2} \pm \dfrac{\sqrt{87}}{2}$
59. $x = 2 \pm \dfrac{\sqrt{14}}{2}i$ **61. a.** $x = \pm 2\sqrt{6}i$
63. a. $p = -9$ or $p = 3$ **65. a.** $x = 3 \pm \sqrt{6}$
67. a. $c = 2 \pm \sqrt{7}i$ **69. a.** $q = \pm\sqrt{89}$
71. a. $x = -\dfrac{3}{2} \pm \dfrac{\sqrt{35}}{2}$ **73. a.** $z = 0$ or $z = 64$
75. a. $z = -8$ or $z = 8$ **77. a.** $k = -1 \pm \dfrac{\sqrt{3}}{3}i$
79. a. $p = \pm 4\sqrt{2}i$ **81.** 49%   **83.** $32.10 per gram
85. a.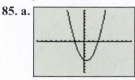

$[-10, 10, 1, -10, 10, 1]$

b. $x = -2$ or $x = 3$

87. a.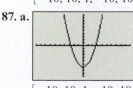

$[-10, 10, 1, -10, 10, 1]$

b. $x \approx -2.646$ or $x \approx 2.646$

89. b. $x = -3 \pm \sqrt{19}$ **91. b.** $x = -1 \pm \sqrt{7}i$
93. 1   **95.** $2\sqrt{11}i$

Section 7.4 Vocabulary Practice
1. J   **2.** D   **3.** A   **4.** G   **5.** I   **6.** E   **7.** B   **8.** C
9. H   **10.** F

Section 7.4 Exercises
1. a. 289   **b.** two   **c.** rational   **3. a.** 60   **b.** two
c. irrational   **5. a.** -63 **b.** two   **c.** nonreal   **7. a.** 0
b. one   **c.** rational   **9. a.** 1   **b.** two   **c.** rational
11. a. -196 **b.** two   **c.** nonreal   **17.** $x = -\dfrac{9}{14} \pm \dfrac{\sqrt{137}}{14}$
19. $x = -\dfrac{2}{7}$ or $x = -1$ **21.** $x = -\dfrac{9}{14} \pm \dfrac{\sqrt{479}}{14}i$
23. $w = -\dfrac{3}{4} \pm \dfrac{3\sqrt{5}}{4}$ **25.** $x = -\dfrac{3}{2} \pm \dfrac{\sqrt{5}}{2}$
27. $x = -1$ or $x = -2$ **29.** $m = \pm\dfrac{4\sqrt{3}}{3}$
31. $r = -\dfrac{1}{12} \pm \dfrac{\sqrt{191}}{12}i$ **33.** $x = -\dfrac{2}{3}$
35. $x = 0$ or $x = -8$ **37.** $x = 0$ or $x = 8$
39. $x = \pm 2\sqrt{2}i$ **41.** $x = \pm 2\sqrt{2}$ **43. a.** $x = 9$ or $x = 13$
b. $x = 9$ or $x = 13$ **c.** $x = 9$ or $x = 13$
45. a. $x = -\dfrac{5}{2}$ or $x = \dfrac{4}{3}$ **b.** $x = -\dfrac{5}{2}$ or $x = \dfrac{4}{3}$
c. $x = -\dfrac{5}{2}$ or $x = \dfrac{4}{3}$ **47. a.** $x = \pm 13$ **b.** $x = \pm 13$
c. $x = \pm 13$ **49.** 12.3 s   **51.** width = 36 in.; length = 45 in.
53. 0.033 mole   **55.** 5 ohms and 15 ohms   **57.** 2 hr 25 min
59. 4.1 s   **61.** $\dfrac{2.8 \text{ mi}}{1 \text{ hr}}$ **63.** 0.07 s; 0.29 s   **65.** 8 sides
67. 9 in. and 12 in.   **69.** 188.9 ft   **71.** $\dfrac{88 \text{ ft}}{1 \text{ s}}$
73. 3 products or 222 products   **75.** 47%   **77.** 6900 cm^3
79. b. 24; two irrational solutions   **81. b.** $x = -\dfrac{3}{2} \pm \dfrac{\sqrt{35}}{2}i$
83. $x = -4$

Section 7.5 Vocabulary Practice
1. G   **2.** I   **3.** A   **4.** D   **5.** C   **6.** F   **7.** B   **8.** E   **9.** H   **10.** J

Section 7.5 Exercises
3.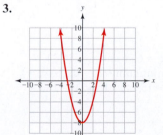

5.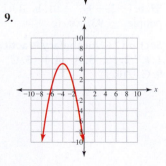

7.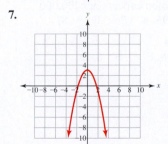

9.

11. a. $(4, 1)$ **b.** up **c.** minimum 1 **d.** $x = 4$
e. $(-\infty, \infty)$ **f.** $[1, \infty)$ **13. a.** $(4, 1)$ **b.** down
c. maximum 1 **d.** $x = 4$ **e.** $(-\infty, \infty)$ **f.** $(-\infty, 1]$
15. a. $(-2, 0)$ **b.** up **c.** minimum 0 **d.** $x = -2$
e. $(-\infty, \infty)$ **f.** $[0, \infty)$ **17. a.** $(-3, -1)$ **b.** down
c. maximum -1 **d.** $x = -3$ **e.** $(-\infty, \infty)$ **f.** $(-\infty, -1]$
19. a. $(0, 1)$ **b.** down **c.** maximum 1 **d.** $x = 0$
e. $(-\infty, \infty)$ **f.** $(-\infty, 1]$ **21. a.** $(5, -60)$ **b.** up
c. minimum -60 **d.** $x = 5$ **e.** $(-\infty, \infty)$ **f.** $[-60, \infty)$
23. $x = -2$ and $x = 8$ **25.** $x = -10$ and $x = 2$
27. $x = -4$ **29. a.** 100 **b.** two **c.** rational
d. $x = -3$ and $x = 7$ **31. a.** -111 **b.** no real zeros
33. a. 57 **b.** two **c.** irrational

d. $x = \dfrac{3}{2} + \dfrac{\sqrt{57}}{6}$ and $x = \dfrac{3}{2} - \dfrac{\sqrt{57}}{6}$

35. a. -7 **b.** no real zeros **37. a.** 0 **b.** one
c. rational **d.** $x = -4$ **39. a.** -48 **b.** no real zeros
41. a. 48 **b.** two **c.** irrational **d.** $x = 2\sqrt{3}$ and $x = -2\sqrt{3}$
43. a. two **45. a.** no real zeros **47. a.** one
49. a. no real zeros **53. a.** $(2, 5)$ **b.** down **c.** maximum 5
d. $x = 2$ **e.** $(-\infty, \infty)$ **f.** $(-\infty, 5]$
55. a. $(2, -5)$ **b.** up **c.** minimum -5 **d.** $x = 2$
e. $(-\infty, \infty)$ **f.** $[-5, \infty)$
57. a. $(-2, -5)$ **b.** up **c.** minimum -5 **d.** $x = -2$
e. $(-\infty, \infty)$ **f.** $[-5, \infty)$
59. a. $(5, 0)$ **b.** up **c.** minimum 0 **d.** $x = 5$ **e.** $(-\infty, \infty)$
f. $[0, \infty)$ **61. a.** $(0, 0)$ **b.** down **c.** maximum 0
d. $x = 0$ **e.** $(-\infty, \infty)$ **f.** $(-\infty, 0]$
63. a. $(-2, 0)$ **b.** up **c.** minimum 0 **d.** $x = -2$
e. $(-\infty, \infty)$ **f.** $[0, \infty)$ **65. a.** $(8, -1)$ **b.** down
c. maximum -1 **d.** $x = 8$ **e.** $(-\infty, \infty)$ **f.** $(-\infty, -1]$
67. a. $(-3, 0)$ **b.** down **c.** maximum 0 **d.** $x = -3$
e. $(-\infty, \infty)$ **f.** $(-\infty, 0]$ **69.** $y = 2(x - 5)^2 - 9$
71. $y = 2(x + 5)^2 + 9$ **73.** $y = 2x^2$ **75.** 25,100 stores
77. 23.9 million people **81.** 38,972,000 votes
83. 24,000 people

85. a. 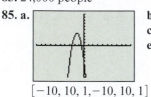 **b.** $x = -2.5$ and $x = -0.5$
c. maximum 4 **d.** $(-\infty, \infty)$
e. $(-\infty, 4]$ **f.** $f(2) = -45$

$[-10, 10, 1, -10, 10, 1]$

87. a. 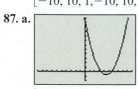 **b.** $x = 6.76$ and $x = 11.24$
c. minimum -15 **d.** $(-\infty, \infty)$
e. $[-15, \infty)$ **f.** $h(2) = 132$

$[-20, 20, 2, -40, 240, 20]$

89. b. $(-2, -9)$ **91. b.** 31.0 colon cancers per 100,000
people **93.** $y = 6x - 4$ **95.** $y - 3 = 3(x - 8)$

Section 7.6 Vocabulary Practice
1. E **2.** H **3.** A **4.** I **5.** C **6.** F **7.** B **8.** D
9. G **10.** J

Section 7.6 Exercises
1. a. $y = x - 3$ **b.**

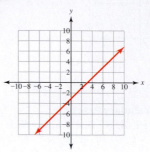

3. a. $(4, 1)$ **b.** $y + 4 = 2(x - 1)$

c.

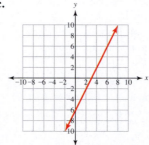

5. a. $(1, 3)$ **b.** $y = 2(x - 5)^2 - 2$

c.

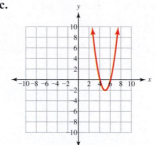

7. a. $(-2, -4)$ **b.** $y = (x + 5)^2 - 3$

c.

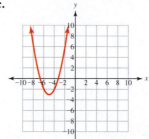

9. $(5, 7)$ **11.** $(-6, -7)$ **13.** $y = 3(x - 11)^2 + 14$
15. $y = 3(x + 3)^2 + 4$ **17.** $y = 2(x - 2)^2 - 16$
19. $y = 2(x + 14)^2 + 2$ **21. a.** $y = (x - 9)^2 + 1$
b. $y = (x - 7)^2 + 16$ **23. a.** $y = (x + 6)^2 + 3$
b. $y = (x + 8)^2 + 18$ **25. a.** $y = (x + 12)^2 - 8$
b. $y = (x + 14)^2 + 7$ **27. a.** $y = -2(x + 2)^2 - 1$
b. $y = -2(x - 3)^2 - 8$ **29. a.** $y = -2x^2$
b. $y = -2(x - 5)^2 - 7$ **31. a.** $y = -2(x + 5)^2$
b. $y = -2x^2 - 7$ **33.** $y = (x + 3)^2 + 6$
35. $y = (x - 5)^2 - 10$ **37.** $w(x) = (x + 10)^2 - 165$

39. $y = \left(x + \dfrac{5}{2}\right)^2 - \dfrac{69}{4}$ **41.** $d(t) = \left(t + \dfrac{9}{2}\right)^2 - \dfrac{49}{4}$

43. $f(t) = \left(t - \dfrac{15}{2}\right)^2 - \dfrac{345}{4}$ **45.** $y = \left(x - \dfrac{1}{2}\right)^2 + \dfrac{15}{4}$

47. $f(x) = (x + 6)^2 - 36$ **49.** $h(x) = (x - 5)^2 - 25$

51. $y = \left(x + \dfrac{1}{2}\right)^2 - \dfrac{1}{4}$ **53.** $y = (x + 1)^2 - 12$

55. $y = \left(x + \dfrac{7}{2}\right)^2 - \dfrac{49}{4}$ **57. a.** $y = (x - 10)^2 - 112$

b. $(10, -112)$ **c.** $x = 10$ **d.** up   **e.** minimum -112
f. $(-\infty, \infty)$ **g.** $[-112, \infty)$ **h.** two irrational zeros

59. a. $y = \left(x + \dfrac{5}{2}\right)^2 + \dfrac{7}{4}$ **b.** $\left(-\dfrac{5}{2}, \dfrac{7}{4}\right)$ **c.** $x = -\dfrac{5}{2}$ **d.** up

e. minimum $\dfrac{7}{4}$ **f.** $(-\infty, \infty)$ **g.** $\left[\dfrac{7}{4}, \infty\right)$ **h.** no real zeros

61. a. $f(x) = (x - 3)^2 - 9$ **b.** $(3, -9)$ **c.** $x = 3$ **d.** up
e. minimum -9 **f.** $(-\infty, \infty)$ **g.** $[-9, \infty)$ **h.** two rational
zeros   **63.** $y = 2(x + 3)^2 + 33$ **65.** $y = 2(x - 4)^2 - 2$

67. $f(x) = 2\left(x - \dfrac{9}{4}\right)^2 - \dfrac{57}{8}$ **69.** $y = 3\left(x - \dfrac{7}{2}\right)^2 - \dfrac{139}{4}$

71. $y = -(x - 10)^2 + 142$ **73. a.** $y = -(x + 10)^2 + 142$
b. $(-10, 142)$ **c.** $x = -10$ **d.** down   **e.** maximum 142
f. $(-\infty, \infty)$ **g.** $(-\infty, 142]$ **h.** two

75. a. $y = 4\left(x + \dfrac{3}{2}\right)^2 - 34$ **b.** $\left(-\dfrac{3}{2}, -34\right)$ **c.** $x = -\dfrac{3}{2}$

d. up   **e.** minimum -34 **f.** $(-\infty, \infty)$ **g.** $[-34, \infty)$ **h.** two

77. $\left(-\dfrac{5}{4}, -\dfrac{45}{4}\right)$ **79.** $(-3, -35)$ **81.** $(3, 55)$ **83.** $\dfrac{49}{5}$ m

85. \$290   **87.** $1091 \dfrac{\text{mi}}{\text{hr}}$

89. a.
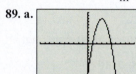
$[-10, 10, 1, -10, 10, 1]$

b. $x = 1, x = 5$ **c.** maximum 8
d. $(-\infty, \infty)$ **e.** $(-\infty, 8]$ **f.** 6

91. a.
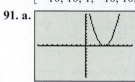
$[-10, 10, 1, -10, 10, 1]$

b. $x = 4$ **c.** minimum 0
d. $(-\infty, \infty)$ **e.** $[0, \infty)$ **f.** 4

93. b. $y = (x + 3)^2 + 2$ **95. b.** $y = (x - 9)^2 - 9$

97. a. $x > 11$ **b.** [number line: 9 10 11 12 13 14 15 16]

99. a. $x \le -4$ **b.** [number line: -9 -8 -7 -6 -5 -4 -3 -2]

Section 7.7 Vocabulary Practice
1. H   **2.** J   **3.** D   **4.** I   **5.** A   **6.** B   **7.** E   **8.** C
9. F   **10.** G

Section 7.7 Exercises
1. a. $x = 7$ **b.** $x > 7$ **c.** [number line: 5 6 7 8 9 10 11 12]

3. a. $x = 2$ **b.** $x \le 2$ **c.** [number line: -3 -2 -1 0 1 2 3 4]

5. $x = -2, x = 2$ **7.** $x = -8, x = -2$

9. $x = -9, x = 12$ **11.** $x = -\dfrac{5}{2}, x = \dfrac{4}{3}$

13. $x = -\dfrac{1}{6} + \dfrac{\sqrt{61}}{6}, x = -\dfrac{1}{6} - \dfrac{\sqrt{61}}{6}$

15. $x = \sqrt{13}, x = -\sqrt{13}$ **17.** no real boundary values
19. $x = -5$ **23.** $(-\infty, 3) \cup (5, \infty)$ **25.** $[1, 6)$
27. $(-\infty, 750] \cup (1250, \infty)$ **29. a.** $x = -5, x = 8$

b. $x < -5$ or $x > 8$ **c.** [number line: -25 -15 -5 0 5 15 25]

d. $(-\infty, -5) \cup (8, \infty)$ **31. a.** $x = -5, x = 8$

b. $-5 < x < 8$ **c.** [number line: -15 -5 0 5 15]   **d.** $(-5, 8)$

33. a. $x = -\dfrac{1}{3}, x = \dfrac{5}{3}$ **b.** $-\dfrac{1}{3} \le x \le \dfrac{5}{3}$ **c.** [number line: -2 -1 0 1 2 3]

d. $\left[-\dfrac{1}{3}, \dfrac{5}{3}\right]$ **35. a.** $x = -\dfrac{1}{3}, x = \dfrac{5}{3}$ **b.** $x \le -\dfrac{1}{3}$ or $x \ge \dfrac{5}{3}$

c. [number line: -5 -4 -3 -2 -1 0 1 2 3 4 5 6]   **d.** $\left(-\infty, -\dfrac{1}{3}\right] \cup \left[\dfrac{5}{3}, \infty\right)$

37. a. $x = -8, x = 0$ **b.** $-8 < x < 0$

c. [number line: -10 -9 -8 -7 -6 -5 -4 -3 -2 -1 0 1 2]   **d.** $(-8, 0)$

39. a. $x = -8, x = 0$ **b.** $x < -8$ or $x > 0$

c. [number line: -16 -12 -8 -4 0 4 8]   **d.** $(-\infty, -8) \cup (0, \infty)$

41. a. $x = -9$ **b.** the set of real numbers

c. [number line: -4 -3 -2 -1 0 1 2 3 4]   **d.** $(-\infty, \infty)$

43. a. $x = -9$ **b.** $x < -9$ or $x > -9$

c. [number line: -13 -11 -9 -7 -5]   **d.** $(-\infty, -9) \cup (-9, \infty)$

45. a. $x = -9$ **b.** no real solution   **c.** no real solution
d. no real solution   **47. a.** $x = -9$ **b.** $x = -9$

c. [number line: -13 -11 -9 -7 -5]   **d.** $[-9, -9]$

49. a. no boundary values   **b.** the set of real numbers

c. [number line: -4 -3 -2 -1 0 1 2 3 4]   **d.** $(-\infty, \infty)$ **51. a.** no boundary

values   **b.** no real solution   **c.** no real solution   **d.** no real
solution   **57. a.** $-8 \le x \le 2$ **b.** $[-8, 2]$

59. a. $x \le -\dfrac{1}{5}$ or $x \ge \dfrac{1}{8}$ **b.** $\left(-\infty, -\dfrac{1}{5}\right] \cup \left[\dfrac{1}{8}, \infty\right)$

61. a. $0 < x < \dfrac{9}{2}$ **b.** $\left(0, \dfrac{9}{2}\right)$ **63. a.** $x < -\dfrac{5}{2}$ or $x > -\dfrac{5}{2}$

b. $\left(-\infty, -\dfrac{5}{2}\right) \cup \left(-\dfrac{5}{2}, \infty\right)$ **65. a.** the set of real numbers

b. $(-\infty, \infty)$ **67. a.** $x < -6 - 6\sqrt{2}$ or $x > -6 + 6\sqrt{2}$
b. $(-\infty, -6 - 6\sqrt{2}) \cup (-6 + 6\sqrt{2}, \infty)$ **69. a.** $x = 6$
b. $[6, 6]$ **71.** The width is between 97 ft and 903 ft.
73. The sheet of plywood must be less than or equal to 65 in.
wide.   **75.** It is profitable to make from 6 products to 104
products.   **77.** 25%   **79.** 34,300 lb

81. a.
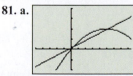
$[-100, 170, 20, -10{,}000, 20{,}000, 5000]$

b. It is profitable to make from
6 products to 104 products.

83. b. $(-\infty, \infty)$ **85. b.** $(-\infty, 3] \cup (5, \infty)$

Chapter 7 Review Exercises

1. see table below **2.** $5yz^2\sqrt{3xy}\,i$ **3.** $-4pq^2\sqrt[3]{2pq}$

4. $6\sqrt{2}\,i$ **5.** $-\dfrac{3}{2}i$ **6.** a, c, f **7. a.** yes **b.** yes

c. no **8. a.** $-10a^2b\sqrt{3b}$ **b.** $7x\sqrt{2y}\,i$ **9.** $5 + 5i$

10. $5 - 22i$ **11.** $55 + 51i$ **12.** $-6 - 30i$ **13.** $\dfrac{2}{5}i$

14. $-\dfrac{9}{13} - \dfrac{32}{13}i$ **15.** $3 + \dfrac{1}{2}i$ **16.** $-1 - 9i$ **17.** $-i$

18. i **19.** -1 **20. a.** $1 + i$ **b.** $9 - 6i$ **21.** $3 - i$

22. a. $h = \pm 3\sqrt{5}\,i$ **23. a.** $p = \pm 7\sqrt{2}$ **24. a.** $x = \pm 3\sqrt{5}\,i$

26. 13 ft 11 in. **27.** $x = -\dfrac{5}{2}$ or $x = \dfrac{1}{3}$

28. $x = 0$ or $x = 5$ **29.** $x = -2$ or $x = 0$
32. length: 12 ft, width: 9 ft **33.** 169 **34.** $\dfrac{81}{4}$

35. $(x - 9)^2$ **36.** $(x + 10)^2$ **37.** $\left(x + \dfrac{7}{2}\right)^2$

38. $x = -7 - \sqrt{5}\,i$ or $x = -7 + \sqrt{5}\,i$ **39.** $x = 3 \pm \sqrt{2}$

40. $x = \dfrac{3 - \sqrt{5}}{2}$ or $x = \dfrac{3 + \sqrt{5}}{2}$ **43.** $h = -7$ or $h = -2$

44. $x = 3 \pm \sqrt{2}\,i$ **45.** $z = \dfrac{5}{2} \pm \dfrac{\sqrt{17}}{2}$

46. $x = -\dfrac{5}{2}$ or $x = -\dfrac{3}{2}$ **47.** $x = \dfrac{3}{2} \pm \dfrac{\sqrt{13}}{2}i$

48. $x = -7$ or $x = \dfrac{1}{2}$ **49. a.** 1 **b.** two rational solutions

c. $x = 1$ or $x = \dfrac{4}{3}$ **50. a.** -119 **b.** two nonreal solutions

c. $x = \dfrac{9}{10} + \dfrac{\sqrt{119}}{10}i$ or $x = \dfrac{9}{10} - \dfrac{\sqrt{119}}{10}i$

51. a. 481 **b.** two irrational solutions

c. $x = -\dfrac{17}{16} + \dfrac{\sqrt{481}}{16}$ or $x = -\dfrac{17}{16} - \dfrac{\sqrt{481}}{16}$

52. a. 49 **b.** two rational solutions **c.** $x = 0$ or $x = -\dfrac{7}{5}$

55. $x = \dfrac{-b \pm \sqrt{b^2 - 4ac}}{2a}$, if $a \neq 0$ **56.** 4 s

57. 13 ft long, 8 ft wide **58.** $(2, -2)$ **59.** $x = 2$

60. $x = 0$ and $x = 4$ **61.** minimum -2 **62.** $(-\infty, \infty)$

63. $[-2, \infty)$ **64. a.** 169 **b.** two rational zeros

c. $x = -9$ or $x = 4$ **65. a.** -44 **b.** no real zeros

c. no real zeros **66. a.** 0 **b.** one real zero **c.** $x = 4$

67–70. see table below **71. a.** $y = (x + 2)^2 + 7$ **b.** up

c. no real zeros **72. a.** $y = -3(x + 2)^2 + 7$ **b.** down

c. two **73.** 335 million prescriptions **74.** 1

75. a. $y = (x - 4)^2 - 5$ **b.** $(4, -5)$

76. a. $f(x) = \left(x + \dfrac{5}{2}\right)^2 + \dfrac{27}{4}$ **b.** $\left(-\dfrac{5}{2}, \dfrac{27}{4}\right)$

77. a. $y = 2(x - 5)^2 - 6$ **b.** $(5, -6)$

78. a. $f(x) = 3(x + 1)^2 + 8$ **b.** $(-1, 8)$ **80.** $5\left(2x^2 - \dfrac{3}{5}x\right)$

81. a. $(5, 8)$ **b.** $y = 4(x - 12)^2 + 10$ **c.** $y = 4(x + 2)^2 + 6$

82. a. $(-3, -6)$ **b.** $g(x) = (x - 4)^2 - 4$

c. $g(x) = (x + 10)^2 - 8$ **83.** $(-3, -8)$ **84.** $(4, 5)$

85. a. $(7, 1)$ **b.** $x = 7$ **c.** minimum 1 **d.** $(-\infty, \infty)$ **e.** $[1, \infty)$

86. a. $x = -5, x = -3$ **b.** $x \leq -5$ or $x \geq -3$

c. $(-\infty, -5] \cup [-3, \infty)$ **87. a.** $x = 2 - \sqrt{11}, x = 2 + \sqrt{11}$

b. $2 - \sqrt{11} < x < 2 + \sqrt{11}$ **c.** $(2 - \sqrt{11}, 2 + \sqrt{11})$

88. a. $x = 7$ **b.** $x = 7$ **c.** $[7, 7]$ **89. a.** $x = 9$

b. $x < 9$ or $x > 9$ **c.** $(-\infty, 9) \cup (9, \infty)$ **90. a.** no

boundary values **b.** the set of real numbers **c.** $(-\infty, \infty)$

91. a. no boundary values **b.** no solution **c.** no solution

92. The width of the rectangle can be between 12 in. and 138 in.

Chapter 7 Test

1. see table on next page **2.** $\dfrac{12}{85} + \dfrac{14}{85}i$ **3.** $\dfrac{9}{17} + \dfrac{5}{17}i$

4. $\dfrac{5}{3} - \dfrac{4}{3}i$ **5.** $6\sqrt{3}\,i$ **6.** $-6x\sqrt{35xy}$ **7.** $2 - 15i$

8. $62 + 24i$ **9. a.** $x = -2$ or $x = 20$ **b.** $x = -2$ or $x = 20$

c. $x = -2$ or $x = 20$ **10. a.** $x = \pm 3\sqrt{7}\,i$

b. $x = \dfrac{5}{6} \pm \dfrac{\sqrt{71}}{6}i$ **c.** $x = 0$ or $x = 16$ **11.** 16.5 m long;

12.4 m wide **12. a.** 109 **b.** two **c.** irrational

13. see table on next page **14.** $y = 4(x - 1)^2 - 1$

15. $y = (x + 8)^2 - 95$ **16.** $y = 2\left(x + \dfrac{5}{2}\right)^2 - \dfrac{3}{2}$

17. $51.4 billion **18.** $(-\infty, -2] \cup [11, \infty)$

Table for Chapter 7 Review exercise 1

	Complex numbers	Imaginary numbers	Real numbers	Rational numbers	Irrational numbers	Integers	Whole numbers
$3i$	X	X					
$7 - 15i$	X						
$\dfrac{3}{5}$	X		X	X			
6	X		X	X		X	X
$\sqrt{11}$	X		X		X		
-41	X		X	X		X	

Table for Chapter 7 Review exercises 67–70

	Vertex	Opens up or down?	Domain	Range	Axis of symmetry	Number of real zeros
67.	$(4, 9)$	up	$(-\infty, \infty)$	$[9, \infty)$	$x = 4$	0
68.	$(-6, -21)$	up	$(-\infty, \infty)$	$[-21, \infty)$	$x = -6$	2
69.	$(4, 0)$	down	$(-\infty, \infty)$	$(-\infty, 0]$	$x = 4$	1
70.	$(-1, 50)$	down	$(-\infty, \infty)$	$(-\infty, 50]$	$x = -1$	2

Table for Chapter 7 Test exercise 1

		Complex numbers	Imaginary numbers	Real numbers	Rational numbers	Irrational numbers	Integers	Whole numbers
a.	-17	X		X	X		X	
b.	$2 - 8i$	X						
c.	$\sqrt{14}$	X		X		X		
d.	$-6i$	X	X					
e.	$\dfrac{2}{5}$	X		X	X			
f.	8	X		X	X		X	X
g.	$\sqrt{-5}$	X	X					

Table for Chapter 7 Test exercise 13

Function	Vertex	Domain	Range	Axis of symmetry	Opens up or down?	Number of real zeros
$y = 4(x - 1)^2 - 6$	$(1, -6)$	$(-\infty, \infty)$	$[-6, \infty)$	$x = 1$	up	2
$y = -2(x + 5)^2 - 3$	$(-5, -3)$	$(-\infty, \infty)$	$(-\infty, -3]$	$x = -5$	down	0

CHAPTER 8

Section 8.1 Vocabulary Practice

1. D **2.** J **3.** C **4.** A **5.** F **6.** E **7.** H **8.** G
9. B **10.** I

Section 8.1 Exercises

1. $5x + 2$ **3.** $3x - 2$ **5.** $-3x + 2$ **7.** $4x^2 + 8x$
9. $4x + 2$ **11.** $4x + 8$ **13.** $-9x - 16$ **15.** $-3x - 8$
17. $3x + 8$ **19.** $18x^2 + 60x + 48$ **21.** $18x + 32$
23. $18x + 12$ **25.** $x^2 + 13x - 2$ **27.** $8x^2 + 40x - 30$
29. $9x^2 + 42x + 39$ **31.** $63x^2 - 84x + 28$
33. $10x^2 - 3$ **35.** $-3x^2 - 2$ **37.** $-\dfrac{16}{5}x^2 + 56x - 245$
39. $\sqrt{x}$ **41.** $\sqrt{x}$ **43.** $\sqrt[9]{x^5}$ **45.** $\dfrac{x^2 + 64}{x^2 - 8x}$
47. $\dfrac{x + 8}{x}$ **51. a.** $48x - 9$ **b.** $(-\infty, \infty)$ **c.** $(-\infty, \infty)$
53. a. $x^2 + 8x + 21$ **b.** $(-\infty, \infty)$ **c.** $[5, \infty)$ **55. a.** $x^2 - 2$
b. $(-\infty, \infty)$ **c.** $[-2, \infty)$ **57. a.** $13\sqrt{3x + 12}$ **b.** $[-4, \infty)$
59. a. $-8\sqrt{-3x + 15}$ **b.** $(-\infty, 5]$ **61. a.** $2\sqrt{-2x + 6}$
b. $(-\infty, 3]$ **63. a.** $\dfrac{x + 6}{x}$ **b.** $(-\infty, 0) \cup (0, 6) \cup (6, \infty)$
65. 5209.2 million **67. a.** $R(x) = \left(\dfrac{\$15.99}{1\text{ product}}\right)x$
b. $C(x) = \left(\dfrac{\$2.30}{1\text{ product}}\right)x$ **c.** $P(x) = \left(\dfrac{\$13.69}{1\text{ product}}\right)x$
d. $\$547,600$ **69. a.** $R(x) = \left(\dfrac{\$1760}{1\text{ acre}}\right)x$
b. $C(x) = \left(\dfrac{\$814}{1\text{ acre}}\right)x$ **c.** $P(x) = \left(\dfrac{\$946}{1\text{ acre}}\right)x$ **d.** $\$4730$
71. $S(R(x)) = \left(\dfrac{\$3.99}{1\text{ lb}}\right)x$ **73.** $S(R(x)) = \left(\dfrac{\$600}{1\text{ sale}}\right)x$
75. $\$139.60$ **83. b.** no **85. a.** yes **b.** no **c.** yes
87. 205 million drivers **89.** $\$31,900$
91. b. $f(g(x)) = -35x + 65$
93. b. $f(x) - g(x) = -12x + 11$ **95.** $x = -8$ or $x = 8$
97. $x = -4$ or $x = 10$

Section 8.2 Vocabulary Practice

1. D **2.** E **3.** A **4.** J **5.** F **6.** B **7.** I **8.** G
9. H **10.** C

Section 8.2 Exercises

1. a. 4 **b.**

x	y
-20	9.1×10^{-13}
-10	9.5×10^{-7}
0	1
10	$1{,}048{,}576$
20	1.1×10^{12}
30	1.2×10^{18}
40	1.2×10^{24}

c.

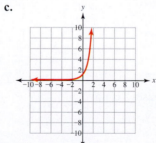

d. $(-\infty, \infty)$ **e.** $(0, \infty)$
f. $(0, 1)$ **g.** increasing

3. a. 0.4 **b.**

x	y
-20	9.1×10^{7}
-10	9536.7432
0	1
10	1.0×10^{-4}
20	1.1×10^{-8}
30	1.2×10^{-12}
40	1.2×10^{-16}

c.

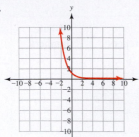

d. $(-\infty, \infty)$ **e.** $(0, \infty)$
f. $(0, 1)$ **g.** decreasing

5. a. e **b.**

x	y
-20	2.1×10^{-9}
-10	4.5×10^{-5}
0	1
10	$22{,}026.466$
20	4.9×10^{8}
30	1.1×10^{13}
40	2.4×10^{17}

c.

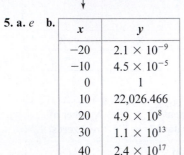

d. $(-\infty, \infty)$ **e.** $(0, \infty)$
f. $(0, 1)$ **g.** increasing

9. $y = 6^x$ **11.** $y = 0.3^x$ **13.** decreasing
15. increasing   **17.** increasing   **19.** increasing
21. a.

x	y
-2	1
-1	1
0	1
1	1
2	1
3	1
4	1

b. no   **c.** no

23. a.

x	y
0	1
-1	$\dfrac{1}{2}$
-2	$\dfrac{1}{4}$
-3	$\dfrac{1}{8}$
-4	$\dfrac{1}{16}$
-5	$\dfrac{1}{32}$
-6	$\dfrac{1}{64}$

b. y decreases   **c.** no

25. a. A **27. a.** B **29. a.** A **31. a.** A **33. a.** B
35. a. A **37. a.** A **41.** $(0, 12)$ **43.** 1331 **45.** 2662
47. 0.729 **49.** 2.916 **51.** 1102.5 **53.** 2931.35
55. 20.086 **57.** $25{,}816$ million lines of code
59. $g(x) = 2^x$ **61. a.** no   **63.** $(-\infty, \infty)$ **65.** $(0, 3)$
67. $(0, \infty)$ **69.** no x-intercept   **71.** $(-\infty, \infty)$
73. $(0, 6)$ **75.** $(0, \infty)$ **77.** no x-intercept
79. \$11.56 trillion   **81.** $10{,}100$ people

83. a. $y = \left(\dfrac{\$3}{\text{bundle}}\right)(x - 1) + \$10.35$ **b.** \$52.35

85. a.

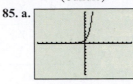

$[-10, 10, 1, -10, 10, 1]$

b. $(-\infty, \infty)$ **c.** $(0, \infty)$ **d.** 16

87. a.

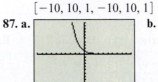

$[-10, 10, 1, -10, 10, 1]$

b. $(-\infty, \infty)$ **c.** $(0, \infty)$ **d.** 0.16

89. b. domain: $(-\infty, \infty)$; range: $(0, \infty)$ **91. b.** $\dfrac{1}{36}$
93. x **95.**

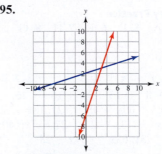

Section 8.3 Vocabulary Practice
1. F   **2.** B   **3.** E   **4.** H   **5.** J   **6.** C   **7.** I   **8.** G
9. D   **10.** A

Section 8.3 Exercises
1.

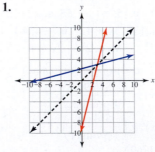

3. yes   **7.** domain $f(x)$: $(-\infty, \infty)$; range $f(x)$: $(-\infty, \infty)$;
domain $g(x)$: $(-\infty, \infty)$; range $g(x)$: $(-\infty, \infty)$

9. yes   **11.** yes   **13.** $y = \dfrac{1}{6}x + \dfrac{5}{6}$

15. $f^{-1}(x) = -\dfrac{5}{4}x + \dfrac{25}{2}$

17. a.

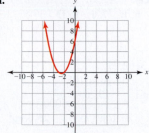

b. no

19. $(-\infty, \infty)$ **21.** $(0, \infty)$ **23.** $(0, 1)$ **25.** no x-intercept

27. $f^{-1}(x) = \log_2(x)$ **29.** $(0, \infty)$ **31.** $(-\infty, \infty)$

33. $(0, \infty)$ **35.** $(-\infty, \infty)$ **37.** no y-intercept

39. $(1, 0)$ **41.** $f^{-1}(x) = 7^x$ **43.** $(-\infty, \infty)$ **45.** $(0, \infty)$

47. 2   **49.** 7   **51.** 0   **53.** not a real number   **55.** 4

57. 1   **59.** 2   **61.** 1   **63.** -1 **65.** -1 **67.** 6

69. undefined   **71.** -2 **73.** not a real number

75. $f^{-1}(x) = \log_3(x)$

77. $f(x) = 3^x$ $f^{-1}(x) = \log_3(x)$

x	y
-2	$\dfrac{1}{9}$
-1	$\dfrac{1}{3}$
0	1
1	3
2	9
3	27

x	y
$\dfrac{1}{9}$	-2
$\dfrac{1}{3}$	-1
1	0
3	1
9	2
27	3

79.

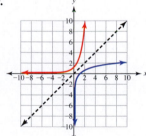

81. $f^{-1}(x) = 7^x$

83. $f(x)$ f^{-1}

x	y
$\dfrac{1}{49}$	-2
$\dfrac{1}{7}$	-1
1	0
7	1
49	2
343	3

x	y
-2	$\dfrac{1}{49}$
-1	$\dfrac{1}{7}$
0	1
1	7
2	49
3	343

85.

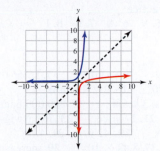

91. 3,072,000 people   **93.** 66,100 lb

95. b. $f^{-1}(x) = \log_5(x)$ **97. b.** $\log(-100)$ is not a real

number.   **99.** $y = \left(\dfrac{\$76}{1 \text{ ticket}}\right)x + \$11.50$

101. the service fee

Section 8.4 Vocabulary Practice

1. D   **2.** C   **3.** I   **4.** A   **5.** J   **6.** H   **7.** G   **8.** F

9. E   **10.** B

Section 8.4 Exercises

1. 1.6532   **3.** 3.8067   **5.** -7.5229 **7.** 2.7782

9. -0.5229 **11.** 1.9459   **13.** 5.3219   **15.** 1.6555

17. 2.8614   **19.** 2.5702   **21.** 2.7712   **23.** 1.7925

31. 56   **33.** 2226   **35.** $2x$ **37.** x **39.** $4 \log(x)$

41. $2 \log(y)$ **43.** $\log(x^8)$ **45.** $\ln(y^{2x})$ **49.** 1.5441

51. 3.5835   **53.** 0.3010   **55.** 1.6094   **57.** $\log_5(xy)$

59. $\log_5\left(\dfrac{c}{d}\right)$ **61.** $\log(xyz)$ **63.** $\log\left(\dfrac{xz}{y}\right)$

67. $\log_5(x^4 y)$ **69.** $\log\left(\dfrac{x}{yz}\right)$ **71.** $\log_5\left(\dfrac{x^4}{y}\right)$

73. $\log_5\left(\dfrac{x^3 y}{z}\right)$ **75.** $\log\left(\dfrac{6xz}{y^2}\right)$ **77.** $\ln(x^3)$

79. $\log(x) + \log(y) - \log(z)$

81. $2 \log(x) + 5 \log(y) - 8 \log(z)$

83. $\ln(x) - 8 \ln(y) - 2 \ln(z)$ **85.** no   **87.** 9.1%

89. a.

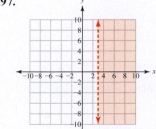

$[-10, 10, 1, -10, 10, 1]$

b. 1.0792

91. a.

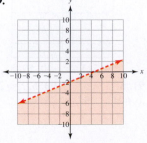

$[-10, 10, 1, -10, 10, 1]$

b. 2.7081

93. b. 3.2925   **95. b.** $\log\left(\dfrac{x^3}{y^5 z^7}\right)$

97.

99.

Section 8.5 Vocabulary Practice
1. C   **2.** J   **3.** H   **4.** G   **5.** F   **6.** A   **7.** B   **8.** I
9. D   **10.** E

Section 8.5 Exercises
1. $5955.08   **3.** $5986.09   **5.** $6351.19   **7.** $6356.08
9. $1214.38   **11.** $1208.70   **13.** 21,242 units
15. {0 years, 1 year, 2 years, 3 years}   **17.** 800,000 bacteria
19. 6,400,000 bacteria   **21.** 27 mice   **23.** 76 earwigs
25. 1.0 g   **27.** 5730 years   **29.** 42.6   **31.** 4.1%
33. 0.17%   **35.** 2.07   **37.** 5.42×10^2 **39.** increases
41. $\dfrac{133.5 \text{ mg}}{1 \text{ mL}}$ **43.** $\dfrac{0.89 \text{ ng}}{1 \text{ mL}}$ **45.** 8.4   **47.** 2.0
49. 0.4 volt   **51.** 1.3 volts   **53.** 170 decibels
55. 25 decibels   **57.** 7877 s   **59.** $k = 1.5$
61. 15.5 Swiss francs   **63.** 14,000 trips   **65.** 72 payments
67. 2085   **69.** 306 women   **71.** 25,187,000 people
73. a.

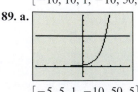

b. 122.10

[0, 10, 1, 0, 150, 15]

77. b. $11,877.82   **79. b.** 3.5   **81.** $(x^m)^n = x^{mn}$ **83.** $x = 2$

Section 8.6 Vocabulary Practice
1. H   **2.** E   **3.** G   **4.** I   **5.** B   **6.** A   **7.** B   **8.** F
9. C   **10.** D

Section 8.6 Exercises
1. $x \log(4)$ **3.** $8x \log(4)$ **7. a.** $x = 2$ **9. a.** $x = 5$
11. a. $x \approx 2.6610$ **13. a.** $x \approx 2.7081$ **17.** $t = 2$
19. $t \approx 1.1131$ **21.** $x \approx 1.3711$ **23.** $x \approx 1.0955$
27. $x \approx 0.4307$ **29.** $x \approx 0.6610$ **31.** $t \approx 10.5220$
33. $t \approx 30.0861$ **35.** $t \approx 29.9373$ **37.** $t \approx 29.8651$
39. $t \approx 13.8629$ **41.** $t \approx 13.8629$ **43.** 13 years 7 months
45. 12 years 8 months   **47.** 2012
49. 59,000,000 transistors   **51.** no   **53.** 72 m
55. 6.4 m   **57.** 3.1 m   **59.** 8 years; 2010
61. $846.984 billion   **63.** 3 hr   **65.** $\dfrac{96 \text{ mg}}{1 \text{ mL}}$ **67.** 1 year
69. 5.5 weeks   **71.** 8 years   **73.** 7 hr 18 min
75. 2 years 10 months   **77.** 9 years   **79.** 9 years 0 months
81. 73 days 18 hr   **83.** $177,800
85. 11,100 lb fertilizer A, 13,900 lb fertilizer B
87. a.   **b.** $x = 5$

[−10, 10, 1, −10, 50, 5]

89. a.   **b.** $x \approx 2.4534$

[−5, 5, 1, −10, 50, 5]

91. b. $x = 5$ **93. b.** $t \approx 2.6$ years   **95.** $\dfrac{x^m}{x^n} = x^{m-n}$;
$x \neq 0$ **97.** $x^m \cdot x^n = x^{m+n}$

Section 8.7 Vocabulary Practice
1. A   **2.** C   **3.** I   **4.** E   **5.** G   **6.** B   **7.** J   **8.** D
9. F   **10.** H

Section 8.7 Exercises
1. 10   **3.** $x = 25$ **5.** $x \approx 3162.2777$
7. $x \approx 33.1155$ **9.** $x = 128$ **11.** $x = 108$
13. $x = -108$ **15.** $x \approx 0.9061$ **17.** $x = \dfrac{1}{5000}$
19. $x = \dfrac{1}{1000}$ **21.** $x = \dfrac{1}{81}$ **23.** $\log_5(2x)$ **25.** $\log_5\left(\dfrac{x}{2}\right)$
27. $\log_5\left(\dfrac{2x}{9}\right)$ **29.** $\log_5(2x^4)$ **31.** $\log_5\left(\dfrac{x^4}{2}\right)$
33. $\log\left(\dfrac{18}{x}\right)$ **37. a.** $x = 648$ **39. a.** $x = 2592$
41. a. $x = \dfrac{1}{6}$ **43. a.** $x = 25$ **45. a.** $x = 1200$
47. a. $x \approx 4.0171$ **49. a.** $x = 100$ **51. a.** $x = 512$
53. a. $x = 20$ **55. a.** $x = -\dfrac{1}{2}$ **57. a.** $x = 1$
59. $\log(4) + \log(x)$ **61.** $2\log(4) + 2\log(x)$
63. $\log(x) - \log(4)$ **65.** $2\log(x) - \log(4)$
67. $\log(I) - \log(1 \times 10^{-12})$ **69.** 5.0×10^{-9}
71. 6.3×10^{-5} **73.** $\dfrac{1.3 \times 10^{-9} \text{ mole}}{1 \text{ L}}$
75. $\dfrac{1.3 \times 10^{-4} \text{ watt}}{1 \text{ m}^2}$ **77.** $\dfrac{5.0 \times 10^{-2} \text{ watt}}{1 \text{ m}^2}$
79. $\dfrac{0.03 \text{ mole}}{1 \text{ L}}$ **81.** $\dfrac{0.25 \text{ mole}}{1 \text{ L}}$ **83.** 464%
85. 10,583,000 people
87. a.   **b.** $x \approx 1.5811$

[0, 5, 1, 0, 2, 1]

89. a.   **b.** $x \approx 27.2991$

[0, 100, 10, −1, 5, 1]

91. b. $x = 64$ **93. b.** $\log\left(\dfrac{x}{2}\right)$ **95.** $-\dfrac{1}{3}$

97.

Chapter 8 Review Exercises
1. $10x + 1$ **2.** $4x + 11$ **3.** $21x^2 - 17x - 30$
4. $21x - 29$ **5.** $\dfrac{x + 4}{x + 6}$ **6. a.** $h(k(x)) = x^2 + 8x + 2$
b. $h(k(x)) = (x + 4)^2 - 14$ **c.** $(-\infty, \infty)$ **d.** $[-14, \infty)$

7. $\sqrt[10]{x^3}$ **11. a.** $P(x) = -\$1900x - \$150,000$ **b.** no
12. a. $(-\infty, \infty)$ **b.** $(0, \infty)$ **c.** $(0, 1)$ **d.** yes   **e.** 343
f. increasing   **g.** 1   **13. a.** $(-\infty, \infty)$ **b.** $(0, \infty)$ **c.** $(0, 1)$
d. yes   **e.** 0.343   **f.** decreasing   **g.** 1   **15.** 54.5982
16. e is less than π.   **18.** yes   **19.** a, c, d   **22.** no
23. $f^{-1}(x) = -\dfrac{1}{3}x + \dfrac{8}{3}$ **24. a.** $f^{-1}(x) = \log_{20}x$
b. $f^{-1}(x) = 17^x$ **25.** 3   **26. a.** 11   **b.** 10   **c.** e
d. e **e.** 10   **27. a.** 4   **b.** 0   **c.** 2   **d.** undefined   **e.** 2
28. a. $y = 2^x$ **b.** $y = \log_2 x$

x	y		x	y
-2	$\dfrac{1}{4}$		$\dfrac{1}{4}$	-2
-1	$\dfrac{1}{2}$		$\dfrac{1}{2}$	-1
0	1		1	0
1	2		2	1
2	4		4	2
3	8		8	3

c.

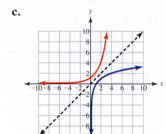

d. $(-\infty, \infty)$ **e.** $(-\infty, \infty)$
f. $(0, \infty)$ **g.** $(0, \infty)$

29. a. $(0, \infty)$ **b.** $(-\infty, \infty)$ **c.** $(1, 0)$ **d.** yes
30. a. 1.6094   **b.** 1.6128   **c.** not a real number   **d.** 2.6309
31. a. 32,940   **b.** 32,940   **32.** a, c, d, e   **33.** $\log_2(9x)$
34. $\log_2\left(\dfrac{x}{9}\right)$ **35.** $\log\left(\dfrac{x^5 z^6}{y}\right)$ **36.** $\log_8\left(\dfrac{x^3}{2}\right)$ **37.** 9%
38. 24   **39.** 9.4   **40.** \$9094.77   **41.** \$9051.72
42. 93 decibels   **43.** 9.0   **44. a.** $x \approx 1.3010$
45. a. $x \approx 2.9957$ **46. a.** $x \approx 1.5440$ **47. a.** $x \approx 2.0959$
48. 8 years 11 months   **49.** 67 days   **50.** 347 min
51. 13 years 10 months   **52.** 13.2335 hr   **53.** $x = 10^{1.2}$
54. $x = e^{1.2}$ **55.** $x = 5^{1.2}$ **56.** argument = base^{logarithm}
57. a. $x \approx 59.5259$ **58. a.** $x \approx 54.5982$
59. a. $x \approx 3162.2777$ **60. a.** $x = \dfrac{3}{2}$ **61. a.** $x \approx 7.3891$
62. $\dfrac{7.9 \times 10^{-10} \text{ mole}}{1 \text{ L}}$ **63.** $\dfrac{3.2 \times 10^{-6} \text{ watt}}{1 \text{ m}^2}$
64. 3.8×10^{19} newton · meters

Chapter 8 Test
3. $(-\infty, \infty)$ **4.** $(0, \infty)$ **5.** yes   **6.** increasing   **9.** $\dfrac{1}{49}$
10. $f^{-1}(x) = \log_7(x)$ **11.** $(0, 1)$ **12.** no x-intercept
13. 5.1240   **14.** 3.8074   **15.** not a real number
16. 20.0855   **17.** 1.2304   **18.** $(0, \infty)$ **19.** $(-\infty, \infty)$
20. yes   **21.** $(1, 0)$ **22.** no y-intercept   **23.** $x \approx 2.6397$
24. $x = 25,000$ **25.** \$7,892.43   **26.** 41.82 mm of mercury
27. decreasing   **28.** $\dfrac{6.3 \times 10^{-5} \text{ mole}}{1 \text{ L}}$ **29.** 5 years

CHAPTER 9

Section 9.1 Vocabulary Practice
1. F   **2.** J   **3.** C   **4.** E   **5.** H   **6.** I   **7.** G   **8.** A
9. B   **10.** D

Section 9.1 Exercises
1. $3\sqrt{13}$ in.   **3.** 5   **5.** 20   **7.** $3\sqrt{34}$ **9.** $2\sqrt{34}$
11. $|x_2 - x_1|$ **13. a.** $(6, 8), (6, 3)$ **b.** $5\sqrt{2}$
15. a. $(1, -3)$ **b.** $\sqrt{97}$ **17.** $3\sqrt{41}$ **19.** $\left(3, \dfrac{13}{2}\right)$
21. $(-14, 9)$ **23.** $(2, 9)$ **25.** $x^2 + y^2 = 64$
27. $(x - 3)^2 + (y - 1)^2 = 81$
29. $(x + 8)^2 + (y - 2)^2 = 400$
31.   **33.**

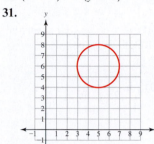   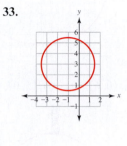

35. $(x + 4)^2 + (y - 5)^2 = 58$
37. $(x + 2)^2 + (y + 3)^2 = 72$ **39.** $x^2 + y^2 = 40$
41. a. $(x - 2)^2 + (y + 6)^2 = 49$ **b.** $(2, -6)$ **c.** 7
43. a. $(x + 5)^2 + (y - 3)^2 = 36$ **b.** $(-5, 3)$ **c.** 6
45. a. $(x + 4)^2 + (y - 1)^2 = 27$ **b.** $(-4, 1)$ **c.** $3\sqrt{3}$
47. a. $x^2 + (y - 4)^2 = 25$ **b.** $(0, 4)$ **c.** 5
49. a. $(x - 9)^2 + y^2 = 85$ **b.** $(9, 0)$ **c.** $\sqrt{85}$
51. a. $(x - 3)^2 + (y - 4)^2 = 12$ **b.** $(3, 4)$ **c.** $2\sqrt{3}$
53. a. $(x + 8)^2 + (y + 2)^2 = 12$ **b.** $(-8, -2)$ **c.** $2\sqrt{3}$
55. $y = \pm\sqrt{15 - (x - 3)^2}$ **57.** $y = \pm\sqrt{49 - (x + 6)^2}$

59. $x^2 + y^2 = 686,649,616$ mi^2 **61. a.** $r = \sqrt[3]{\dfrac{\mu P^2}{4\pi^2}}$
b. 26,256 mi   **c.** 22,293 mi
63. $y = \sqrt{210.25 \text{ in.}^2 - (x - 14.5 \text{ in.})^2}$ **65.** 13.9 in.
67. $(x - 0.9 \text{ m})^2 + y^2 = 0.81 \text{ m}^2$ **69.** 1,131,000,000 people
71. 0.2 cases per 100,000 people
73. a. $y = \sqrt{9 - (x - 2)^2} + 6, y = -\sqrt{9 - (x - 2)^2} + 6$
b.

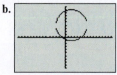

$[-15.2, 15.2, 1, -10, 10, 1]$
75. a. $y = \sqrt{25 - (x - 2)^2} + 4, y = -\sqrt{25 - (x - 2)^2} + 4$
b.

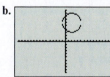

$[-15.2, 15.2, 1, -10, 10, 1]$
77. b. $(-4, 7)$ **79. b.** 8   **81. a.** $y = (x + 6)^2 - 45$
b. $(-6, -45)$ **83.** $x = -12 \pm 2\sqrt{34}$

Section 9.2 Vocabulary Practice

1. I **2.** D **3.** E **4.** F **5.** G **6.** C **7.** H **8.** J
9. A **10.** B

Section 9.2 Exercises

1. a. $(6, 0)$, $(-6, 0)$, $(0, 5)$, $(0, -5)$ **b.** $\dfrac{x^2}{36} + \dfrac{y^2}{25} = 1$
c.

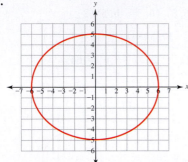

3. a. $(0, 6)$, $(0, -6)$, $(5, 0)$, $(-5, 0)$ **b.** $\dfrac{x^2}{25} + \dfrac{y^2}{36} = 1$
c.

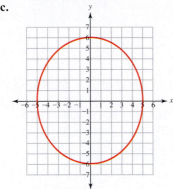

5. a. $(2, 4)$, $(2, -6)$, $(6, -1)$, $(-2, -1)$
b. $\dfrac{(x - 2)^2}{16} + \dfrac{(y + 1)^2}{25} = 1$ **c.**

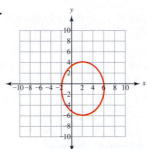

7. a. $(3, 3)$, $(-11, 3)$, $(-4, 7)$, $(-4, -1)$
b. $\dfrac{(x + 4)^2}{49} + \dfrac{(y - 3)^2}{16} = 1$ **c.**

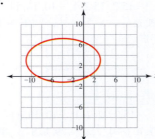

9.

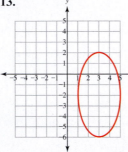

11.

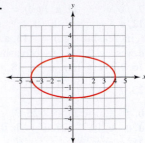

13.

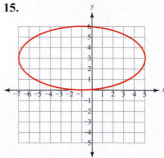

15.

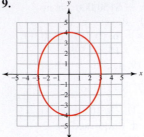

17. a. $y = 4$ **b.** $x = 11$ **c.** $(11, 4)$
d. $\dfrac{(x - 11)^2}{5^2} + \dfrac{(y - 4)^2}{1^2} = 1$ **19. a.** $y = -1$ **b.** $x = -7$

c. $(-7, -1)$ **d.** $\dfrac{(x + 7)^2}{7^2} + \dfrac{(y + 1)^2}{6^2} = 1$ **21. a.** $x = 3$

b. $y = 5$ **c.** $(3, 5)$ **d.** $\dfrac{(x - 3)^2}{2^2} + \dfrac{(y - 5)^2}{10^2} = 1$

25. $\dfrac{(x - 2)^2}{3^2} + \dfrac{(y - 3)^2}{1^2} = 1$ **27.** $\dfrac{(x + 1)^2}{4^2} + \dfrac{(y - 2)^2}{2^2} = 1$

29. $\dfrac{(x + 3)^2}{3^2} + \dfrac{(y + 1)^2}{2^2} = 1$ **31.** $\dfrac{x^2}{5^2} + \dfrac{(y + 3)^2}{10^2} = 1$

33. a. $(21, -13)$ **b.** vertical
35. a. $(26, -8)$, $(-14, -8)$ **b.** $(6, 6)$, $(6, -22)$
37. a. $(1, 3)$, $(-15, 3)$ **b.** $(-7, -1)$, $(-7, 7)$

39. a. $x = -4$ **b.** $y = 1$ **41.** $\dfrac{x^2}{(20 \text{ ft})^2} + \dfrac{y^2}{(14.5 \text{ ft})^2} = 1$

43. $\dfrac{x^2}{(24 \text{ in.})^2} + \dfrac{y^2}{(16 \text{ in.})^2} = 1$

45. $\dfrac{x^2}{(0.387 \text{ AU})^2} + \dfrac{y^2}{(0.3785 \text{ AU})^2} = 1$ **47.** 365 days

49. a. $\dfrac{x^2}{(33 \text{ in.})^2} + \dfrac{y^2}{(16 \text{ in.})^2} = 1$

b. $y = \pm \dfrac{\sqrt{278,784 \text{ in.}^2 - (256 \text{ in.}^2)x^2}}{33 \text{ in.}}$

c. $y = \dfrac{\sqrt{278,784 \text{ in.}^2 - (256 \text{ in.}^2)x^2}}{33 \text{ in.}}$

51. 1278 million tons **53.** 678 million kWh

55. a. $y = 5\sqrt{1 - \dfrac{x^2}{16}}$, $y = -5\sqrt{1 - \dfrac{x^2}{16}}$

b.

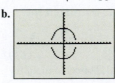

$[-15.2, 15.2, 1, -10, 10, 1]$

57. a. $y = 4\sqrt{1 - \dfrac{x^2}{25}}, \; y = -4\sqrt{1 - \dfrac{x^2}{25}}$

b.

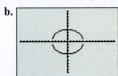

$[-15.2, 15.2, 1, -10, 10, 1]$

59. b. $(3, -4)$ **61. b.** 6   **63.** $x = -2 \pm \sqrt{5}i$

65. $x = \dfrac{5}{4} \pm \dfrac{\sqrt{95}}{4}i$

Section 9.3 Vocabulary Practice

1. E   **2.** H   **3.** D   **4.** B   **5.** C   **6.** J   **7.** A   **8.** I
9. F   **10.** G

Section 9.3 Exercises

1. a. $(4, -5)$ **b.** $x = 4$ **c.** up   **3. a.** $(-4, -5)$
b. $x = -4$ **c.** down   **5. a.** $x^2 = 4(p)y$ **b.** vertex: $(0, 0)$
c. up; $x = 0$ **e.** $p = 8$; focus: $(0, 8)$; directrix: $y = -8$
f.   **g.** yes

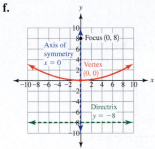

7. a. $y^2 = 4(p)x$ **b.** vertex: $(0, 0)$ **c.** right; $y = 0$
e. $p = 1$; focus: $(1, 0)$; directrix: $x = -1$
f.   **g.** no

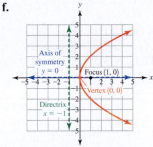

9. a. $x^2 = -4py$ **b.** vertex: $(0, 0)$ **c.** down; $x = 0$
e. $p = 10$; focus: $(0, -10)$; directrix: $y = 10$
f.   **g.** yes

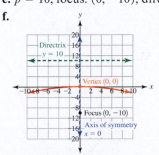

11. a. $y^2 = -4(p)x$ **b.** vertex: $(0, 0)$ **c.** left; $y = 0$
e. $p = 5$; focus: $(-5, 0)$; directrix: $x = 5$

f.   **g.** no

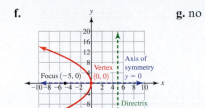

13. a. $y^2 = 4(p)x$ **b.** vertex: $(0, 0)$ **c.** right; $y = 0$
e. $p = 6$; focus: $(6, 0)$; directrix: $x = -6$
f.   **g.** no

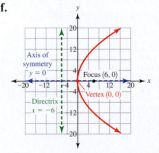

15. a. $x^2 = -4(p)y$ **b.** vertex: $(0, 0)$ **c.** down; $x = 0$
e. $p = 7$; focus: $(0, -7)$; directrix: $y = 7$
f.   **g.** yes

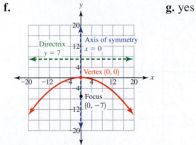

17. a. $y^2 = -4px$ **b.** vertex: $(0, 0)$ **c.** left; $y = 0$
e. $p = 8$; focus: $(-8, 0)$; directrix: $x = 8$
f.   **g.** no

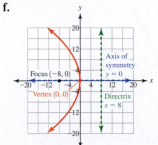

19. a. $x^2 = 4py$ **b.** vertex: $(0, 0)$ **c.** up; $x = 0$
e. $p = 2$; focus: $(0, 2)$; directrix: $y = -2$
f.   **g.** yes

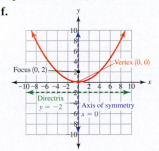

21. a. right; $y^2 = 4px$ **b.** focus: $(6, 0)$; directrix: $x = -6$
c. $y^2 = 24x$ **d.**   **e.** no

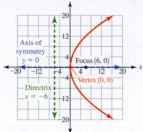

23. a. down; $x^2 = -4py$ **b.** focus: $(0, -7)$; directrix: $y = 7$
c. $x^2 = -28y$ **d.**   **e.** yes

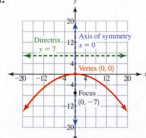

25. a. left; $y^2 = -4px$ **b.** focus: $(-8, 0)$; directrix: $x = 8$
c. $y^2 = -32x$ **d.**   **e.** no

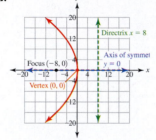

27. a. up; $x^2 = 4py$ **b.** focus: $(0, 2)$; directrix: $y = -2$
c. $x^2 = 8y$ **d.**   **e.** yes

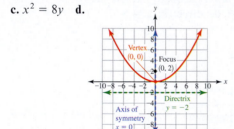

29. a. up; $x^2 = 4py$ **b.** focus: $(0, 8)$; directrix: $y = -8$
c. $x^2 = 32y$ **d.**   **e.** yes

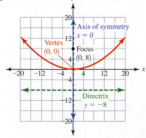

31. a. right; $y^2 = 4px$ **b.** focus: $(1, 0)$; directrix: $x = -1$
c. $y^2 = 4x$ **d.**   **e.** no

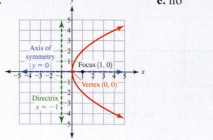

33. a. down; $x^2 = -4py$ **b.** focus: $(0, -10)$; directrix: $y = 10$
c. $x^2 = -40y$ **d.**   **e.** yes

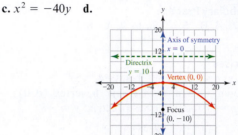

35. a. left; $y^2 = -4px$ **b.** focus: $(-5, 0)$; directrix: $x = 5$
c. $y^2 = -20x$ **d.**   **e.** no

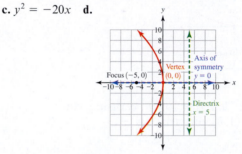

37. $(x - 1)^2 = -4(y - 1)$ **39.** $(x - 5)^2 = 36(y - 8)$
41. $(y - 8)^2 = 48(x - 5)$ **43.** $(y - 5)^2 = -80(x + 2)$
45. $(x + 4)^2 = -36(y + 2)$ **47.** $(y - 3)^2 = 9(x - 5)$

49.   **51.**

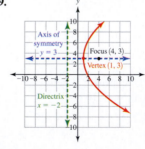

53.   **55.**

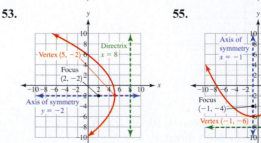

57. a. $(x - 6)^2 = -8(y - 1)$ **b.** down   **c.** $(6, 1)$ **d.** $x = 6$
e. $(6, -1)$ **f.** $y = 3$ **h.**   **i.** yes

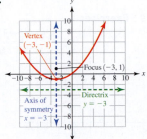

59. a. $(x + 3)^2 = 8(y + 1)$ **b.** up   **c.** $(-3, -1)$ **d.** $x = -3$
e. $(-3, 1)$ **f.** $y = -3$ **h.**   **i.** yes

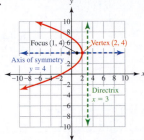

61. a. $(y - 4)^2 = -4(x - 2)$ **b.** left   **c.** $(2, 4)$ **d.** $y = 4$
e. $(1, 4)$ **f.** $x = 3$ **h.**   **i.** no

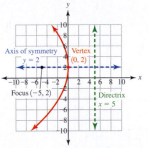

63. a. $(y - 2)^2 = -20x$ **b.** left   **c.** $(0, 2)$ **d.** $y = 2$
e. $(-5, 2)$ **f.** $x = 5$ **h.**   **i.** no

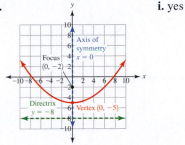

65. a. $x^2 = 12(y + 5)$ **b.** up   **c.** $(0, -5)$ **d.** $x = 0$
e. $(0, -2)$ **f.** $y = -8$ **h.**   **i.** yes

67. a. $(3, -6)$ **b.** $x = 3$ **c.** up   **d.** $(-\infty, \infty)$ **e.** $[-6, \infty)$

69. a. $(-1, -9)$ **b.** $x = -1$ **c.** down   **d.** $(-\infty, \infty)$
e. $(-\infty, -9]$ **71. a.** $(-8, 2)$ **b.** $x = -8$ **c.** up
d. $(-\infty, \infty)$ **e.** $[2, \infty)$ **73. a.** $(0, -4)$ **b.** $x = 0$
c. up   **d.** $(-\infty, \infty)$ **e.** $[-4, \infty)$ **75.** 10.125 in.
77. 6.9 in.   **79.** 1,368,000 repossessions

81. a. $y = \left(\dfrac{125 \text{ students}}{1 \text{ year}}\right)x + 3578$ students   **b.** 4203 students

83. a. $y = \dfrac{x^2 - 6x + 21}{12}$ **b.**

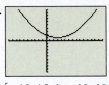

$[-10, 15, 1, -10, 10, 1]$

85. a. $y = \dfrac{x^2 + 8x - 8}{-2}$ **b.**

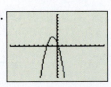

$[-40, 40, 4, -40, 40, 4]$

87. b. $(1, -3)$ **89. b.** $y = -1$ **91.** $y = mx + b$
93. slope $= \dfrac{3}{4}$; y-intercept $(0, 0)$

Section 9.4 Vocabulary Practice

1. I   **2.** B   **3.** C   **4.** A   **5.** F   **6.** E   **7.** D   **8.** G   **9.** J   **10.** H

Section 9.4 Exercises

1. a. $a = 3$; $p = 8$; $b = \sqrt{55}$ **b.** $\dfrac{x^2}{9} - \dfrac{y^2}{55} = 1$ **3. a.** $a = 3$;

$p = 8$; $b = \sqrt{55}$ **b.** $\dfrac{y^2}{9} - \dfrac{x^2}{55} = 1$ **5. a.** $a = 6$; $p = 11$;

$b = \sqrt{85}$ **b.** $\dfrac{x^2}{36} - \dfrac{y^2}{85} = 1$ **7. a.** $a = 4$; $p = 7$; $b = \sqrt{33}$

b. $\dfrac{y^2}{16} - \dfrac{x^2}{33} = 1$ **9. a.** $a = 3$; $p = 5$; $b = 4$ **b.** $\dfrac{x^2}{9} - \dfrac{y^2}{16} = 1$

11. a. $a = 3$; $p = 5$; $b = 4$ **b.** $\dfrac{y^2}{9} - \dfrac{x^2}{16} = 1$

13. a. $a = 8$; $p = 10$; $b = 6$ **b.** $\dfrac{y^2}{64} - \dfrac{x^2}{36} = 1$

15. a. $a = 24$; $p = 26$; $b = 10$ **b.** $\dfrac{x^2}{576} - \dfrac{y^2}{100} = 1$

17. A hyperbola is the intersection of a vertical plane with two cones.   **19.** A parabola is the intersection of a tilted (not vertical, not horizontal) plane with a cone.

21. a. $x = 0$ **b.** $\left(0, -\sqrt{89}\right), \left(0, \sqrt{89}\right)$ **c.** $(0, -5), (0, 5)$
23. a. $y = 0$ **b.** $\left(-\sqrt{89}, 0\right), \left(\sqrt{89}, 0\right)$ **c.** $(-5, 0), (5, 0)$
25. a. $y = 0$ **b.** $\left(-\sqrt{190}, 0\right), \left(\sqrt{190}, 0\right)$ **c.** $(-13, 0), (13, 0)$
27. a. $x = 0$ **b.** $\left(0, -2\sqrt{6}\right), \left(0, 2\sqrt{6}\right)$ **c.** $\left(0, -\sqrt{13}\right), \left(0, \sqrt{13}\right)$
31. a. $a = 36$ **b.** $p = 45$ **c.** $b = 27$
33. a. $a = 100$ **b.** $p = 125$ **c.** $b = 75$
35. a. $a = 5$ **b.** $p = 8$ **c.** $b = \sqrt{39}$

37. $y = -\dfrac{11}{13}x$, $y = \dfrac{11}{13}x$ **39.** $y = -\dfrac{1}{2}x$, $y = \dfrac{1}{2}x$

41. $y = -\dfrac{\sqrt{21}}{13}x$, $y = \dfrac{\sqrt{21}}{13}x$

43. $y = -\dfrac{\sqrt{13}}{3}x, \; y = \dfrac{\sqrt{13}}{3}x$

45. a. $y = -\dfrac{5}{6}x, \; y = \dfrac{5}{6}x$ **b.**

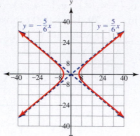

47. a. $y = -\dfrac{6}{5}x, \; y = \dfrac{6}{5}x$ **b.**

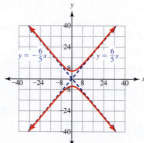

49. a. $y = -\dfrac{5}{8}x, \; y = \dfrac{5}{8}x$ **b.**

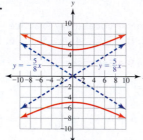

51. a. $y = -\dfrac{7}{4}x, \; y = \dfrac{7}{4}x$ **b.**

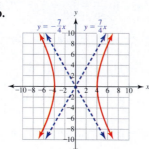

53. $\dfrac{x^2}{330{,}625 \text{ mm}^2} - \dfrac{y^2}{60{,}000 \text{ mm}^2} = 1$ **55.** \$172 million

57. 192,000 cases

59. a. $y = 5\sqrt{\dfrac{x^2}{16} - 1}, \; y = -5\sqrt{\dfrac{x^2}{16} - 1}$

b.

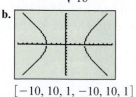

$[-10, 10, 1, -10, 10, 1]$

61. a. $y = 4\sqrt{\dfrac{x^2}{25} + 1}, \; y = -4\sqrt{\dfrac{x^2}{25} + 1}$

b.

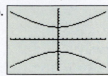

$[-10, 10, 1, -10, 10, 1]$

63. b. $(-4, 0), \; (4, 0)$ **65. b.** $x = 0$

Section 9.5 Vocabulary Practice

1. E   **2.** H   **3.** I   **4.** A   **5.** C   **6.** D   **7.** F   **8.** B
9. J   **10.** G

Section 9.5 Exercises

1. $(2, 0)$ **3.** $(3, 7)$ **5.** $(-6, -3), \; (6, 3)$

15. a.   **b.** $(2, 2)$

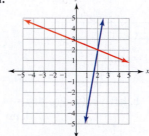

17. a.   **b.** $(4, 0)$

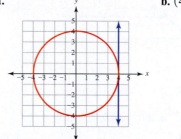

19. a.   **b.** $(-4, 2), \; (4, 2)$

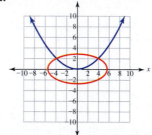

21. a.   **b.** $(2, 0), \; (0, 2)$

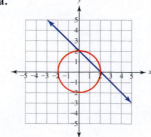

23. a.

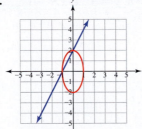

b. $(-1, 0)$, $(0, 2)$

25. a.

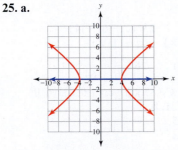

b. $(4, 0)$, $(-4, 0)$

27. a.

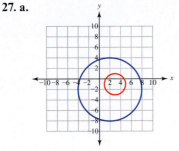

b. no real solution

29. a.

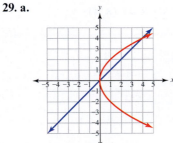

b. $(0, 0)$, $(4, 4)$

31. a.

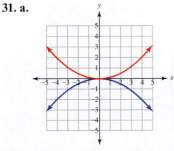

b. $(0, 0)$

33. $(4, -5)$ **35.** $(2, -8)$, $(50, 40)$

37. $(4, -2)$, $(-20, -50)$ **39.** $(5, 0)$, $(3, -4)$

41. no real solution   **43.** $(2\sqrt{2}, 3)$, $(-2\sqrt{2}, 3)$

45. $(4, 2)$, $(-4, 2)$ **47.** $(3, 5)$, $(-3, 5)$

49. $(-6, -3)$, $(6, 3)$ **51.** $(7, -2)$ **53.** $(5, 0)$, $(-5, 0)$

55. $(4, 2)$, $(-4, 2)$ **57.** $(4, -2)$, $(-4, -2)$

59. $(18, 9)$, $(-18, 9)$ **61.** $(4, 2)$, $(-4, 2)$

63. no real solution   **65.** $(-\sqrt{26}, -3)$, $(-\sqrt{26}, 3)$,

$(\sqrt{26}, -3)$, $(\sqrt{26}, 3)$ **67.** $(5, 0)$

71. 1997: 210 employees; 2005: 250 employees

73. 203,700 Americans

75. 425,000 students   **77. a.** $y = \sqrt{\dfrac{x^2 - 8}{2}}$,

$y = -\sqrt{\dfrac{x^2 - 8}{2}}$, $y = \sqrt{35 - x^2}$, $y = -\sqrt{35 - x^2}$

b. 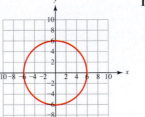   **c.** $(5.10, 3)$, $(5.10, -3)$, $(-5.10, 3)$,

$(-5.10, -3)$

$[-15.2, 15.2, 1, -10, 10, 1]$

79. a. $y = \sqrt{9 - (x - 6)^2}$, $y = -\sqrt{9 - (x - 6)^2}$,

$y = \sqrt{25 - x^2}$, $y = -\sqrt{25 - x^2}$

b. 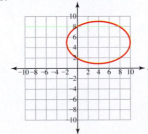   **c.** $(4.33, 2.49)$, $(4.33, -2.49)$

$[-15.2, 15.2, 1, -10, 10, 1]$

81. b. $(-1, -2)$, $(1, 2)$ **83. b.** no real solution

85. $2442   **87.** 68

Chapter 9 Review Exercises

1. $2\sqrt{97}$ **2.** $\left(\dfrac{21}{2}, 5\right)$ **3.** $x^2 + y^2 = 144$

4. $(x - 3)^2 + (y + 8)^2 = 25$ **5.** $(x - 5)^2 + (y - 1)^2 = 26$

6. $x^2 + 10x + 25$

8. a. $(x - 3)^2 + (y - 5)^2 = 16$ **b.** $(3, 5)$ **c.** 4

9.

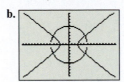

10.

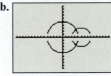

11. $(x - 7.5 \text{ in.})^2 + y^2 = 56.25 \text{ in.}^2$

12. a. major axis vertices: $(-2, 5)$, $(10, 5)$;

minor axis vertices: $(4, 1)$, $(4, 9)$

b.

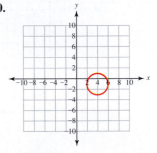

c. $\dfrac{(x - 4)^2}{36} + \dfrac{(y - 5)^2}{16} = 1$

13. a. $\dfrac{(x + 2)^2}{36} + \dfrac{(y - 1)^2}{64} = 1$

b.

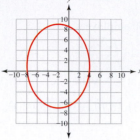

14. a. $\dfrac{(x - 2)^2}{16} + \dfrac{(y - 3)^2}{25} = 1$ **b.** $(2, 3)$ **c.** $x = 2$

d. $y = 3$ **e.**

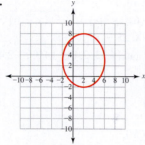

15. a. $(-3, 4)$ **b.** $y = 4$ **c.** $x = -3$ **d.** $(-9, 4), (3, 4),$

$(-3, 2), (-3, 6)$ **16.** b, c   **17.** $\dfrac{x^2}{196 \text{ ft}^2} + \dfrac{y^2}{36 \text{ ft}^2} = 1$

18. a.   **b.** $x^2 = 3y$ **c.** yes

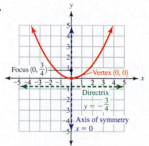

19. a.   **b.** $(x - 5)^2 = -8(y - 6)$
                                        c. yes

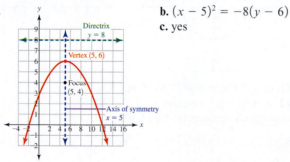

20. a. $(0, 0)$ **b.** $p = 3; (0, -3)$ **d.** $x = 0$ **e.** $y = 3$

f.

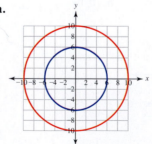

21. a. $(1, 3)$ **b.** $p = 2; (3, 3)$ **d.** $y = 3$ **e.** $x = -1$

f.

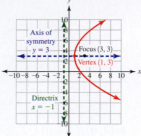

22. $(y - 3)^2 = 4(2)(x - 1)$ **23. a.** $(-6, -11)$

b. $x = -6$ **c.** up   **d.** $(-\infty, \infty)$ **e.** $[-11, \infty)$

24. 6 in.   **25.** $\dfrac{x^2}{9} - \dfrac{y^2}{16} = 1$ **26.** $\dfrac{y^2}{36} - \dfrac{x^2}{64} = 1$

27. a. $y = 0$ **b.** $(-\sqrt{85}, 0), (\sqrt{85}, 0)$ **c.** $(-6, 0), (6, 0)$

28. d   **29.** c   **30.** a   **31.** b

32. a. $a = 4; b = 5$ **b.** $y = \dfrac{5}{4}x, y = -\dfrac{5}{4}x$

c.

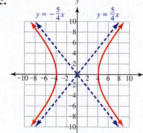

33. 8.476 mm   **34.** b, d, e

35. $(-8, 4), (8, 4)$

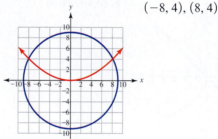

36. a.

37. 2   **38.** 4   **39.** $(-2\sqrt{10}, 2), (2\sqrt{10}, 2)$

40. $(-3, -4), (-3, 4), (3, -4), (3, 4)$ **42.** 1999; 2009

Chapter 9 Test

1. $3\sqrt{58}$ **2.** $(x + 8)^2 + (y - 2)^2 = 65$

3.

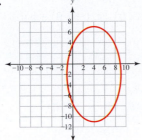

4. a. $(4, -2)$ **b.** major axis: $x = 4$, minor axis: $y = -2$
c. $(4, 7), (4, -11), (9, -2), (-1, -2)$

d.

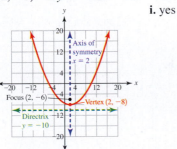

5. $\dfrac{x^2}{90{,}000 \text{ cm}^2} + \dfrac{y^2}{44{,}732.25 \text{ cm}^2} = 1$

6. $(y - 2)^2 = 16(x + 2)$

7. a. $(x - 2)^2 = 4(2)(y + 8)$ **b.** up   **c.** $(2, -8)$ **d.** $x = 2$
e. $(2, -6)$ **f.** $y = -10$

h.   **i.** yes

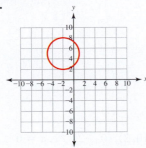

8. $\dfrac{x^2}{9} - \dfrac{y^2}{55} = 1$ **9. a.** $x = 0$ **b.** $(0, -3), (0, 3)$
c. $(0, -\sqrt{34}), (0, \sqrt{34})$

10. a. $y = -\dfrac{5}{12}x, \; y = \dfrac{5}{12}x$

b.

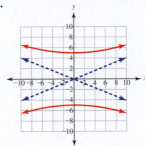

11. $(-15, -8), (-15, 8), (15, -8), (15, 8)$
12. $(-2, -8), (-2, 8)$
13. a. $(4, 8)$ **b.** down   **c.** $(-\infty, \infty)$ **d.** $(-\infty, 8]$

Chapters 7–9 Cumulative Review

1. see table below   **2.** $31 + 29i$ **3.** $\dfrac{7}{15} + \dfrac{14}{15}i$

4. $12xy\sqrt{2y}\,i$ **5.** $-5a\sqrt{21b}$ **6.** 3 **7.** 1 **8.** 3

9. 2.4650 **10.** $\log\left(\dfrac{xz^3}{y}\right)$ **11.** $2\ln(a) - \ln(b) - 3\ln(c)$

12. $x = -\dfrac{5}{2}$ or $x = \dfrac{8}{3}$ **13.** $x = 4 + i$ or $x = 4 - i$

14. $x = \dfrac{-b \pm \sqrt{b^2 - 4ac}}{2a}, \; a \neq 0$ **15.** $x = \dfrac{2}{5}$ or $x = 1$

16. $x = -\dfrac{3}{2} + \dfrac{\sqrt{11}}{2}i$ or $x = -\dfrac{3}{2} - \dfrac{\sqrt{11}}{2}i$

17. $x = -\sqrt{21}$ or $x = \sqrt{21}$ **18.** $x = 0$ or $x = \dfrac{1}{2}$

19. $x \approx 2.7712$ **20.** $x = 128$ **21.** $x \approx 1.4979$
22. $x \approx 80.6858$ **23.** $c = 12$ **24.** $b = -7$
25. $k = -4$ or $k = 5$ **26.** $x = -18$ or $x = 10$

27. $x = -\dfrac{5}{6}$ or $x = 4$ **28.** $w = -50$

29. a. $p > -7$ **b.** $(-7, \infty)$
30. a. $x \leq -10$ or $x \geq 2$ **b.** $(-\infty, -10] \cup [2, \infty)$

31. a. $-4 < a < -\dfrac{10}{3}$ **b.** $\left(-4, -\dfrac{10}{3}\right)$

32. a. $-22 \leq n \leq 6$ **b.** $[-22, 6]$

Table for Chapters 7–9 Cumulative Review exercise 1

		Complex numbers	Imaginary numbers	Real numbers	Rational numbers	Irrational numbers	Integers	Whole numbers
a.	$4 + 3i$	X						
b.	9	X		X	X		X	X
c.	-14	X		X	X		X	
d.	$-\dfrac{7}{8}$	X		X	X			
e.	$7i$	X	X					
f.	$\sqrt{7}$	X		X		X		
g.	$\sqrt{-16}$	X	X					

33. a.

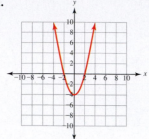

b. $(-\infty, \infty)$ **c.** $[-4, \infty)$

55. a. $(x - 1)^2 = 4(2)(y - 3)$ **b.** $y = 1$ **c.** $(1, 5)$

d.

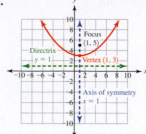

34. see table below **35. a.** $y = (x - 4)^2 - 7$ **b.** $(4, -7)$
c. $(-\infty, \infty)$ **d.** $[-7, \infty)$ **36. a.** $y = -3(x + 1)^2 - 17$
b. $x = -1$ **c.** $(-\infty, \infty)$ **d.** $(-\infty, -17]$
37. $y = (x + 1)^2 + 11$ **38. a.** $(0, \infty)$ **b.** $(-\infty, \infty)$
c. $(1, 0)$ **d.** increasing **39. a.** $(-\infty, \infty)$ **b.** $(0, \infty)$
c. $(0, 1)$ **d.** increasing **40. a.** $(-\infty, \infty)$ **b.** $(0, \infty)$
c. $(0, 1)$ **d.** decreasing **41.** $f^{-1}(x) = 4^x$
42. $f^{-1}(x) = \ln(x)$ **43.** length: 9 in.; width: 5 in.
44. 44.6 cesarean deliveries **45.** 18 in. and 24 in.

46. 1.5 s **47.** 8.2 **48.** $4.0 \times 10^{-3} \dfrac{\text{mole}}{\text{L}}$ **49.** \$7832.68

50. 128,600 people **51.** $6\sqrt{13}$ **52.** $(3, -3)$
53. a. $(x - 4)^2 + (y - 2)^2 = 80$
b.

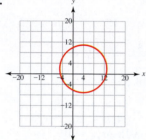

56. a. $(0, \sqrt{13}), (0, -\sqrt{13})$ **b.** $(0, 2), (0, -2)$ **c.** $x = 0$

57. a. $\dfrac{x^2}{64} - \dfrac{y^2}{80} = 1$ **b.** $y = \dfrac{\sqrt{5}}{2}x, y = -\dfrac{\sqrt{5}}{2}x$

c.

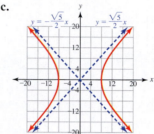

58. a, d **59.** 8.33 in.

CHAPTER 10

Section 10.1 Vocabulary Practice
1. B **2.** A **3.** E **4.** H **5.** F **6.** J **7.** D **8.** G
9. I **10.** C

Section 10.1 Exercises
1. $\{-1, 0, 1, 2, 3\}$
3. a. $\{6, 7, 8, 9, 10, \ldots\}$ **b.**

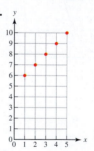

54. a. $y = 4$ **b.** $x = 10$ **c.** $(10, 4)$
d. $\dfrac{(x - 10)^2}{25} + \dfrac{(y - 4)^2}{9} = 1$

e.

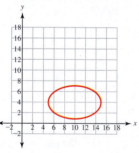

Table for Chapters 7–9 Cumulative Review exercise 34

Function	Vertex	Domain	Range	Axis of symmetry	Opens up or down?	Number of real zeros
$y = -3(x - 9)^2 + 4$	$(9, 4)$	$(-\infty, \infty)$	$(-\infty, 4]$	$x = 9$	down	2
$y = 2(x + 3)^2 - 6$	$(-3, -6)$	$(-\infty, \infty)$	$[-6, \infty)$	$x = -3$	up	2

5. a. $\{-2, 4, -6, 8, -10, \ldots\}$ **b.**

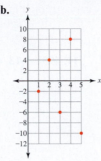

7. a. $\left\{\dfrac{1}{2}, \dfrac{4}{5}, \dfrac{9}{10}, \dfrac{16}{17}, \dfrac{25}{26}, \ldots\right\}$ **b.**

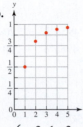

9. $\left\{\dfrac{1}{2}, \dfrac{1}{5}, \dfrac{1}{10}, \dfrac{1}{17}, \dfrac{1}{26}, \ldots\right\}$ **11.** $\left\{1, \dfrac{2}{5}, \dfrac{1}{5}, \dfrac{2}{17}, \dfrac{1}{13}, \ldots\right\}$

13. $a_n = n - 8$ **15.** $a_n = (-1)^{n-1}(n-8)$ **17.** $a_n = -2n$

19. $a_n = -\dfrac{1}{n^2}$ **21.** $a_n = n^2 + 1$ **23.** $a_n = n^n$

25. $a_n = 2n + 1$ **27.** $a_n = \dfrac{1}{n+1}$ **29.** $a_n = \dfrac{n+1}{n+2}$

31. $a_n = 2^{n-1}$ **37.** 32 **39.** 225 **41.** $\displaystyle\sum_{i=1}^{4}(2i+3)$

43. $\displaystyle\sum_{i=1}^{10}(2i+3)$ **45.** $\displaystyle\sum_{i=1}^{5}3i$ **47.** $\displaystyle\sum_{i=1}^{5}(-i+1)$ **49.** $\displaystyle\sum_{i=1}^{6}\dfrac{3}{i}$

51. 60 **53.** 120 **55.** 155 **57.** $\dfrac{25}{3}$ **59.** 238

61. 0 **63.** \$4982 **65.** {\$4226, \$4464, \$4716, \$4982, \$5263}
67. \$40,581 **69.** \$22.83

71. 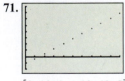 **73.**

$[0, 10, 1, -10, 40, 5]$ $\quad[0, 10, 1, 0, 1100, 100]$

75. b. $\{1, 2, 3, 4, \ldots\}$ **77. b.** $\displaystyle\sum_{i=1}^{5}i^2$ **79.** $(2x-7)(3x+4)$

81. $x = \dfrac{5 \pm \sqrt{133}}{6}$

Section 10.2 Vocabulary Practice

1. J **2.** I **3.** F **4.** A **5.** B **6.** G **7.** H **8.** E
9. D **10.** C

Section 10.2 Exercises

1. 4 **3.** −3 **5.** 3 **7.** $\dfrac{1}{4}$ **9. a.** $d = 11$ **b.** $a_n = 11n - 8$

11. a. $d = -9$ **b.** $a_n = -9n + 17$ **13. a.** $d = 25$
b. $a_n = 25n + 75$ **15. a.** $d = 25$ **b.** $a_n = 25n + 175$
17. a. $d = 1$ **b.** $a_n = n + 56$ **19. a.** $d = 4$ seats
b. $a_n = (4n)$ seats $+ 8$ seats **21. a.** $d = \dfrac{1}{2}$ **b.** $a_n = \dfrac{1}{2}n$

23. a. $d = \dfrac{1}{4}$ **b.** $a_n = \dfrac{1}{4}n - \dfrac{3}{28}$ **25. a.** $d = -\$35$

b. $a_n = -\$35n + \$9035$ **27. a.** no **29. a.** no
31. a. yes **33. a.** no **35.** 50 **37.** 59
39. $a_n = 3n + 46$ **41.** $a_n = 10n + 2800$ **43.** $a_n = 2n - 5$

45. $a_n = 4n - 20$ **47.** $a_n = \dfrac{1}{2}n + \dfrac{17}{2}$ **49.** $a_n = -2n - 6$

51. 156 **53.** 48 **55.** 264 **57.** 4525 **59.** 5400
61. 5390 **63.** 5240 **65.** 25,250 **67.** 325 **69.** 650
73. 510 seats **75.** 144 bales **77.** It costs \$155 less to
drive to Renton than to buy the sofa on-line.
79. 82,692,000 people

81. 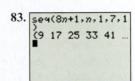 **83.**

$\{11, 14, 17, 20, 23, 26, 29\}$ $\quad\{9, 17, 25, 33, 41, 49, 57\}$
85. b. no **87. b.** 590 **89.** $x = -3$

Section 10.3 Vocabulary Practice

1. G **2.** F **3.** E **4.** A **5.** H **6.** I **7.** D **8.** C
9. B **10.** J

Section 10.3 Exercises

1. 3 **3.** 3 **5.** 2 **7.** $\dfrac{1}{2}$ **9. a.** $r = 2$ **b.** $a_n = 3(2)^{n-1}$

11. a. $r = 2$ **b.** $a_n = -3(2)^{n-1}$ **13. a.** $r = \dfrac{1}{2}$

b. $a_n = 4\left(\dfrac{1}{2}\right)^{n-1}$ **15. a.** $r = -\dfrac{1}{2}$ **b.** $a_n = 4\left(-\dfrac{1}{2}\right)^{n-1}$

17. a. $r = \dfrac{1}{2}$ **b.** $a_n = \dfrac{1}{3}\left(\dfrac{1}{2}\right)^{n-1}$ **19. a.** $r = 8$ **b.** $a_n = 1(8)^{n-1}$

21. geometric **23.** neither **25.** neither
27. arithmetic **37.** −972 **39.** $a_n = 4(-3)^{n-1}$
41. 1,953,125 **43.** $a_n = 5(5)^{n-1}$ **45.** 147,620

47. $\dfrac{381}{16}$ **49.** −765 **51.** 18,724 **53.** −728

55. −3.75 **57.** 20 **59.** $2.\overline{6}$ **61.** 32 **63.** $\dfrac{800}{3}$

65. a. $\{1.999758\text{ g}, 1.999516\text{ g}, 1.999274\text{ g}, \ldots\}$
b. $r \approx 0.99988$ **c.** $a_n = 1.999758\text{ g}(0.99988)^{n-1}$ **d.** 1.0976 g
67. a. $\{25,680\text{ units}, 27,478\text{ units}, 29,401\text{ units}, \ldots\}$
b. $r \approx 1.07$ **c.** $a_n = 25,680\text{ units}(1.07)^{n-1}$ **d.** 36,000 units
69. a. $\{751,903\text{ people}, 795,513\text{ people}, 841,653\text{ people}, \ldots\}$
b. $r \approx 1.06$ **c.** $a_n = 751,903\text{ people}(1.06)^{n-1}$
d. 3,227,000 people **71. a.** $\{\$5300, \$5618, \$5955.08, \ldots\}$
b. $r = 1.06$ **c.** $a_n = \$5300(1.06)^{n-1}$ **d.** \$8954
73. 20 posts **75.** 54 items

77. 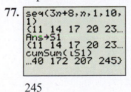 **79.**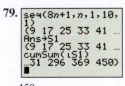

245 $\qquad$ 450

81. b. no **83. b.** 8748 **85.** $\dfrac{2x-3}{2x+1}$

87. $x = -1$ or $x = 6$

Section 10.4 Vocabulary Practice

1. D **2.** E **3.** J **4.** F **5.** A **6.** B **7.** G **8.** I **9.** C **10.** H

Section 10.4 Exercises

1. 39,916,800 **3.** 120 **5.** 462 **7.** 165 **9.** $a^4 + 4a^3b + 6a^2b^2 + 4ab^3 + b^4$ **11.** 495 **13.** 4 **15.** 84 **17.** 15 **19.** 22

21.

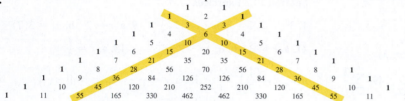

23. $a^{12} + 12a^{11}b + 66a^{10}b^2 + 220a^9b^3 + 495a^8b^4 + 792a^7b^5 + 924a^6b^6 + 792a^5b^7 + 495a^4b^8 + 220a^3b^9 + 66a^2b^{10} + 12ab^{11} + b^{12}$

25. $a^{12} - 12a^{11}b + 66a^{10}b^2 - 220a^9b^3 + 495a^8b^4 - 792a^7b^5 + 924a^6b^6 - 792a^5b^7 + 495a^4b^8 - 220a^3b^9 + 66a^2b^{10} - 12ab^{11} + b^{12}$

27. $x^8 + 8x^7y + 28x^6y^2 + 56x^5y^3 + 70x^4y^4 + 56x^3y^5 + 28x^2y^6 + 8xy^7 + y^8$

29. $8a^3 + 12a^2b + 6ab^2 + b^3$

31. $c^5 + 15c^4d + 90c^3d^2 + 270c^2d^3 + 405cd^4 + 243d^5$

33. $729h^6 + 7290h^5k + 30{,}375h^4k^2 + 67{,}500h^3k^3 + 84{,}375h^2k^4 + 56{,}250hk^5 + 15{,}625k^6$

35. $15{,}625h^6 + 56{,}250h^5k + 84{,}375h^4k^2 + 67{,}500h^3k^3 + 30{,}375h^2k^4 + 7290hk^5 + 729k^6$

37. $15{,}625h^6 - 56{,}250h^5k + 84{,}375h^4k^2 - 67{,}500h^3k^3 + 30{,}375h^2k^4 - 7290hk^5 + 729k^6$

39. a. $(1 + 0.1)^5 = 1(1^5) + 5(1^4)(0.1) + 10(1^3)(0.1^2) + 10(1^2)(0.1^3) + 5(1)(0.1^4) + 1(0.1^5)$ **b.** 1.6105

41.

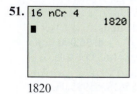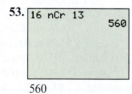

43. yes

45. a.

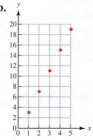

b. 2^n

47. 72% decrease **49.** 174 lb

51.

```
16 nCr 4
              1820
■
```
1820

53.

```
16 nCr 13
              560
```
560

55. b. 24 **57. b.** $8x^3 + 12x^2y + 6xy^2 + y^3$ **61.** 2.4650

Chapter 10 Review Exercises

1. a. $\{3, 7, 11, 15, 19, \ldots\}$ **b.**

2. $a_n = \dfrac{1}{n+3}$ **3. a.** $\displaystyle\sum_{i=1}^{5} \dfrac{2}{i}$ **b.** $\dfrac{137}{30}$ **4.** $24,392

5. a, c   **6. a.** $\{3, 10, 17, 24, 31, \ldots\}$ **b.** $a_n = 7n - 4$

7. $a_n = 4n + 7$ **8.** $a_n = -3n + 2$ **9. a.** 45

b. $a_n = 5n + 15$ **10.** 1605   **11.** 590   **12.** 36

13. 744   **14.** a, d   **15.** $r = 3$

17. $a_n = -2(-2)^{n-1}$ **18.** $a_n = 4(4)^{n-1}$

19. a. $\{5, 20, 80, 320, 1280, \ldots\}$ **b.** $a_n = 5(4)^{n-1}$ **c.** 20,480

20. a. $\displaystyle\sum_{i=1}^{15} 3(3)^{i-1}$ **b.** 21,523,359   **21.** 10

22. a. $\{103,550, 112,870, 123,028, \ldots\}$ **b.** $206,330

23. 40,320   **25.** 21

26. $a^5 + 5a^4b + 10a^3b^2 + 10a^2b^3 + 5ab^4 + b^5$

27. $c^5 + 20c^4d + 160c^3d^2 + 640c^2d^3 + 1280cd^4 + 1024d^5$

Chapter 10 Test

1. a. arithmetic   **2. a.** $\displaystyle\sum_{i=1}^{4}(i^2 + i)$ **b.** 40

3. a. 40   **b.** $a_n = 4n + 4$ **4. a.** 768   **b.** $a_n = 3(2)^{n-1}$

5. a. $\displaystyle\sum_{i=6}^{5} \dfrac{1}{6}(6)^{i-1}$ **b.** $\dfrac{1555}{6}$ **6. a.** $\displaystyle\sum_{i=1}^{30}(4i - 24)$ **b.** 1140

7. 11.25   **8.** 5040   **9.** 210   **10.** 70

11. $15,625x^6 + 18,750x^5y + 9375x^4y^2 + 2500x^3y^3 + 375x^2y^4 + 30xy^5 + y^6$

12. a. $\{38.08, 42.65, 47.77, \ldots\}$ **b.** $75.16

13. 54 marbles

Chapters 1–10 Cumulative Review

1. $\{\ldots, -3, -2, -1, 0, 1, 2, \ldots\}$ **2.** $\dfrac{2x^3}{5y^8}$

3. $320a^7b^4$ **4.** $x = 5$ **5.** $x > -9$ **6.** $(-\infty, 5)$

7. $[-4, \infty)$ **8.** 1.2×10^{-15} **9.** $\dfrac{66\text{ ft}}{1\text{ s}}$

10. $147.1 million   **11.** 1311 hr   **12.** 0

13.

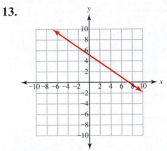

14. $(2, 0)$ **15.** $x = 8$ **16.** $y = -2x - 13$

17.

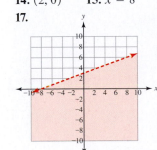

18.

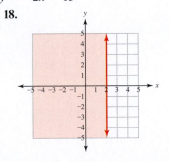

19. $36x^3 - 53x^2 - 17x + 6$ **20.** $(2x + 3)(4x - 1)$

21. $(2a + 3)(5x - 4)$ **22.** $25(4n^2 + p^2)$

23. $3(x - 4)(x + 4)$ **24.** $4(u^2 + 7u + 5)$

25. $(2x - y)(4x^2 + 2xy + y^2)$

26. $x = -4$ or $x = 0$ or $x = 6$

27. length: 11 cm; width: 9 cm   **28.** $x = -12$ or $x = 6$

29. $a = -2$ or $a = 22$ **31.** $(-9, 7]$

32. $(-\infty, -19] \cup [9, \infty)$

34. domain: $(-\infty, \infty)$; range: $(-\infty, \infty)$

35. domain: $(-\infty, \infty)$; range: $[8, 8]$

36. domain: $(-\infty, \infty)$; range: $[1, \infty)$

37. domain: $(-\infty, \infty)$; range: $(-\infty, 4]$

38. $x = -9$ and $x = 5$ **39.** $y = 4$; no real zeros

40. two real zeros   **41.** 33

42. $y = \left(\dfrac{20\text{ wrappers}}{1\text{ day}}\right)x + 20,550$ wrappers

43. 9825 bottles   **44.** $y + 2 = 3(x - 5)$

46. -1 **47.** $(6, -4)$ **48.** $(8, 11)$

49. 1000 kg Alloy A; 1500 kg Alloy B

50. 405 heirloom tomato plants; 135 modern hybrid tomato plants   **51.** $(16, 4, 5)$

52. $\begin{bmatrix} 4 & 5 & | & 22 \\ 2 & -1 & | & 4 \end{bmatrix}$ **53.** $(3, 2)$

54.

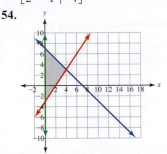

55. $(0, -3), (0, 7), (4, 3)$ **56.** $\dfrac{x+9}{x-3}$

57. $\dfrac{x+1}{(x-2)(x+3)}$ **58.** $\dfrac{4}{x^2+4}$

59. $x = 8$ **60.** 26 min

61. $\left(-\infty, -\dfrac{1}{2}\right) \cup \left(-\dfrac{1}{2}, 4\right) \cup (4, \infty)$

62. $x = 11, x = -2$ **63.** $(-\infty, \infty)$ **64.** $x = -4$

66. 250.9 g   **67.** $b = \dfrac{ac}{a-c}$ **68.** $x + 4$ **69.** $x + 4$

70. $(-\infty, -5) \cup (-3, \infty)$ **71.** $5\sqrt{6}$

72. $2xy^2\sqrt[3]{3x^2}$ **73.** $5\sqrt{42}$

74. $\dfrac{4\sqrt{10xy}}{15x}$ **75.** $\dfrac{x\sqrt{5} - 9\sqrt{5x}}{x - 81}$

76. $\dfrac{1}{x^{\frac{1}{15}}}$ **77.** $a = 2$ or $a = 7$ **78.** no solution.

79. domain: $(-\infty, 5]$ **80.** $x = 5$ **81.** 2.8 cm

82. see table below   **83.** $x = \pm 4\sqrt{2}\, i$ **84.** 55.9 m

85. $w = -9 \pm \sqrt{86}$ **86.** $x = -\dfrac{7}{6} \pm \dfrac{\sqrt{59}}{6}\, i$

88. $f(x) = (x + 4)^2 - 21$

89. vertex: $(-4, -21)$; domain: $(-\infty, \infty)$; range: $[-21, \infty)$

90. $x = -5$, $x = -3$ **91.** $[-10, 2]$

92. $f^{-1}(x) = \dfrac{4}{3}x + 8$ **93.** $g(f(x)) = -45x + 25$

94. $f^{-1}(x) = 5^x$ **95.** domain: $(-\infty, \infty)$; range: $(0, \infty)$

96. decreasing   **98.** $(0, 30)$ **99.** A

100. domain: $(0, \infty)$; range: $(-\infty, \infty)$

103. \$2243.74   **104.** 5.5   **105.** $x \approx 2.6309$

106. 52,200 people   **107.** 28 years   **108.** 237.9567

109. $\log\left(\dfrac{x^3 z^7}{y}\right)$ **110.** $2\log(a) - \log(b) - 3\log(c)$

111. $x = 25$ **112.** $\dfrac{1.6 \times 10^{-5}\ \text{mole}}{1\ \text{L}}$ **113.** $3\sqrt{41}$

114. $(x - 6)^2 + (y + 2)^2 = 49$ **115.** center: $(6, -2)$;

radius: 7   **116.** $\dfrac{(x - 3)^2}{16} + \dfrac{(y - 2)^2}{1} = 1$

117. major axis: $y = 2$; minor axis: $x = 3$; center: $(3, 2)$

118. $(y + 2)^2 = 12(x - 6)$ **119.** $\dfrac{y^2}{4} - \dfrac{x^2}{60} = 1$

120. $\left(-\dfrac{1}{2}, \dfrac{1}{2}\right)$, $(2, 8)$ **121.** $(4, -2)$, $(-4, -2)$

122. 135   **123.** $a_n = 2n + 4$ **125.** 3072   **126.** 70

127.
$$
\begin{array}{ccccccc}
 & & & 1 & & & \\
 & & 1 & & 1 & & \\
 & 1 & & 2 & & 1 & \\
1 & & 3 & & 3 & & 1
\end{array}
$$

APPENDIX 2

1. -7 **3.** -2 **5.** 2   **7.** 0   **9. a.** $x_1 = 1$ **b.** $x_2 = 2$

11. a. $x_1 = 0$ **b.** $x_2 = 5$ **13. a.** $x_1 = -2$ **b.** $x_2 = 3$

15. a. $x_1 = -2$ **b.** $x_2 = 3$ **17. a.** $x_1 = 1$ **b.** $x_2 = 2$

c. $x_3 = 0$ **19. a.** $x_1 = 3$ **b.** $x_2 = 0$ **c.** $x_3 = -1$

21. a. $x_1 = 0$ **b.** $x_2 = -2$ **c.** $x_3 = 3$ **23. a.** $x_1 = 1$

b. $x_2 = 3$ **c.** $x_3 = 2$ **25. a.** $x_1 = 1$ **b.** $x_2 = 2$ **c.** $x_3 = -1$

27. a. 0   **c.** $\left\{(x_1, x_2, x_3)\,|\,x_1 = \dfrac{4}{5}x_3 - \dfrac{13}{5}, x_2 = -\dfrac{3}{5}x_3 + \dfrac{11}{5}, \right.$

$\left. x_3 \text{ is a real number} \right\}$

Table for Chapters 1–10 Cumulative Review exercise 82

	Complex numbers	Imaginary numbers	Real numbers	Rational numbers	Irrational numbers	Integers	Whole numbers
$\sqrt{7}$	X		X		X		
$\sqrt{-7}$	X	X					
$3 + 5i$	X						
$4i$	X	X					
-8	X		X	X		X	
$-\dfrac{8}{9}$	X		X	X			
0.3	X		X	X			
12	X		X	X		X	X

Index

Page numbers in bold indicate location of terms when appearing in Definition, Summary, or Procedure boxes.

Calculator Index

Menus and Keys

Instruction and Features

Page numbers in black indicate where the given entries are used in the text or Using Technology boxes.
Page numbers in red indicate where the given entries are used in end-of-section exercises.